Integrated Population Models

Theory and Ecological Applications with R and JAGS

Integrated Population Models

Theory and Ecological Applications with R and JAGS

Michael Schaub
Swiss Ornithological Institute,
Sempach, Switzerland

Marc Kéry
Swiss Ornithological Institute,
Sempach, Switzerland

ELSEVIER

ACADEMIC PRESS
An imprint of Elsevier

Academic Press is an imprint of Elsevier
125 London Wall, London EC2Y 5AS, United Kingdom
525 B Street, Suite 1650, San Diego, CA 92101, United States
50 Hampshire Street, 5th Floor, Cambridge, MA 02139, United States
The Boulevard, Langford Lane, Kidlington, Oxford OX5 1GB, United Kingdom

Library of Congress Cataloging-in-Publication Data
A catalog record for this book is available from the Library of Congress

British Library Cataloguing-in-Publication Data
A catalogue record for this book is available from the British Library

ISBN: 978-0-323-90810-8

For information on all Academic Press publications visit our website at
https://www.elsevier.com/books-and-journals

Publisher: Mica Haley
Acquisitions Editor: Anna Valutkevich
Editorial Project Manager: Rachel Pomery
Production Project Manager: Prem Kumar Kaliamoorthi
Cover Designer: Mark Rogers

Typeset by TNQ Technologies

Contents

Foreword

Population biology is a substantial discipline composed of two important subdisciplines, population genetics and population ecology. Population genetics and closely related fields ultimately provide a bridge to molecular and organismal processes by revealing the mechanisms responsible for genetic differences among individuals within and between populations. Population ecology, on the other hand, is the study of the mechanisms responsible for changes in the distribution and abundance of individuals over space and time. These evolutionary and ecological topics are in fact interrelated. To paraphrase an eloquent statement by G. Evelyn Hutchinson, the individuals of a population acting within an ecological theatre can affect the forces of natural selection for individuals and the evolution of the population (the company of actors). These eco-evolutionary dynamics affect lower organizational levels of biology and are essential ingredients that influence higher levels of organization such as biological communities and ecosystems.

At first glance, the primary parameters affecting the distribution and abundance of individuals over space and time are so simple that a troglodyte could be a population ecologist. In fact, when I attempt to explain my work as a wildlife population ecologist to family and friends, their responses are usually something along the lines of "Oh, you count animals for a living." Sigh—a stark reminder to work on my elevator talk! I say this because when expressed as rates, the abundance of individuals at a particular location x and point in time t ($N_{x,t}$) is simply a function of the preceding local ***b***irth rate, the rate at which individuals ***i***mmigrate into the location from elsewhere, the local ***d***eath rate, and the rate at which individuals ***e***migrate out of the location to other locales, which collectively act on local abundance at the previous point in time $t-1$: $N_{x,t} = N_{x,t-1} + (\boldsymbol{b}_{x,t-1} + \boldsymbol{i}_{x,t-1} - \boldsymbol{d}_{x,t-1} - \boldsymbol{e}_{x,t-1})N_{x,t-1}$. When this simple discrete-time equation is rearranged, the "balance" of these ***bide*** demographic parameters defines the geometric rate of population growth from one time step to the next: $N_{x,t}/N_{x,t-1} = \lambda_{x,t-1} = 1 + \boldsymbol{b}_{x,t-1} + \boldsymbol{i}_{x,t-1} - \boldsymbol{d}_{x,t-1} - \boldsymbol{e}_{x,t-1}$. However, the dynamics of some (e.g., birth-flow) populations are better described in continuous time, and the demographic parameters and abundance state variable might importantly be structured by age, developmental stage, size, phenotype, or even genotype. These states of individuals might be measured continuously or with discrete categories, and in the real world, all demographic parameters are influenced by a large suite of biotic, abiotic, and anthropogenic forces. Too many, in fact, to estimate everything, and we must always deal with multiple uncertainties when making inferences about population dynamics. Alas, the job of a population ecologist is extremely challenging!

Population ecology is a mature subdiscipline of biology with a rich theory set and empirical findings to back it up. Nevertheless, there are many unknowns in population ecology and perhaps an equally great number of misunderstandings. Given that community and even ecosystem-level processes hinge on population-level processes, this should be concerning. Can we pretend to seek truths about these higher levels of biological organization by throwing demographic mechanisms under the rug and instead chase patterns in the outcomes at these higher levels? I think not. Can we work toward decreasing uncertainty and enhancing our understanding of demography to achieve a more mechanistic understanding of how multiple levels of biological organization are connected? I think so.

If we desire less uncertainty and greater understanding, population ecology must spread its wings as a quantitative force of nature while not forgetting its roots in natural history. How can we achieve this in such challenging times when all we want is to daydream of a better tomorrow? Well, I invite you to daydream of a magical place near the foot of the Swiss Alps—a small village along a glistening lake—and if you wish, grab a virtual bottle of local Riesling and a brick of virtual cheese to enjoy while viewing wintering red-crested pochards or after a long summer hike in meadows of wildflowers. Here you will find two gentlemen who work at the Schweizerische Vogelwarte, seemingly ordinary in every way, except they are most generous with their time and possess a genius for teaching the masses how to become better quantitative population ecologists. Following on the heels of their highly acclaimed book, *Bayesian Population Analysis Using WinBUGS: A Hierarchical Perspective*, Michael Schaub and Marc Kéry bring to us this blockbuster, *Integrated Population Models: Theory and Ecological Applications Using R and JAGS*.

Integrated population models (IPMs) are statistical models of population dynamics as opposed to mathematical models. Rather than estimating each ***bide*** parameter separately and then plugging them into my simple equation above to project abundance from a starting value, IPMs integrate available time-series data for demographic parameters *and* abundances. That is to say, IPMs combine data for both sides of the equation. This approach is much more efficient and seamless for estimation and modeling, with many advantages over traditional piecemeal approaches to population modeling. In practice, IPMs are often applied to structured populations, such as those structured by discrete ages or developmental stages, and are intricately connected to matrix population models. By providing estimates of the population structure and how it changes over time (and space), IPMs are ideal for empirically examining transient and stochastic dynamics.

Moreover, IPMs are used to conduct prospective and retrospective perturbation analyses, scenario projections, and population forecasting; identify metapopulation sources and sinks, density dependence, natural selection, and eco-evolutionary dynamics; and inform harvest management, habitat management, and spatial conservation prioritization. They've even been extended to include demographic interactions among multiple species and to model community dynamics. The other "IPM" commonly used to model continuously structured populations is the integral projection model and can be implemented as an integrated population model, thus forming an IIPM (or IPM^2). The possibilities of IPMs are limited only by our creativity and computing skills.

Michael Schaub and Marc Kéry's new book will help you expand your population modeling creativity and computing skills to transform this creativity into science. Building on *Bayesian Population Analysis*, which provided in-depth coverage of Bayesian demographic models that can be used as IPM submodels (e.g., mixed, state-space, capture-mark-recapture, and occupancy models), *Integrated Population Models* provides broad coverage of the underlying theory behind IPMs, including the connection to matrix population models, and a range of formal analyses that can be performed with IPMs. The book then covers a wide array of IPMs applied to real-world data, such that we can get a feel for the art of modifying an IPM to address particular objectives, available types of data, and any shortcomings of the data (e.g., missing data). With complete sets of how-to computer code provided from start to finish, the reader is never left hanging. The work of Michael and Marc, and the generosity of their time, helped reinvigorate my career, for which I will be forever thankful. I hope you will find this new book equally exciting and helpful in your career.

David N. Koons
Colorado State University
Fort Collins, Colorado
January 2021

Preface

Demographic models to understand population changes and the mechanisms underlying them and to predict these changes into the future represent a key analytical technology in an extensive range of biological sciences, including ecology, conservation biology, wildlife biology, and wildlife management. Often, multiple sources of information are available about the demography of a study species and the changes in the population size over time. A smart approach then is to exploit all the available information in a single statistical model. The joint analysis of different data sets that are informative about the demography and population dynamics of a species is the topic of *integrated population models* (IPMs) and of this book.

IPMs were first developed in fisheries science about 40 years ago (Fournier and Archibald, 1982) and were then introduced in terrestrial ecology in a seminal *Biometrics* paper by Besbeas et al. (2002). Since then, IPMs have become a very active area of research in statistical ecology and have been employed in an ever-increasing number of application papers at the interface of ecology/wildlife management and statistical science (Zipkin and Saunders, 2018). However, from the beginning, the uptake of IPMs by practitioners has been slow, despite a seminal paper in 2004 by Brooks et al. (2004) that demonstrated the first Bayesian implementation of an IPM in the widely used BUGS software. This opened up the possibility for ecologists to fit their own IPMs, something also documented by the first paper on the topic by one of us (Schaub et al., 2007). Over the last 10 years, there has been vigorous growth in IPM applications, most of them implemented in some variant of BUGS software (for an earlier review, see Schaub and Abadi, 2011; and more recently, Zipkin and Saunders, 2018). Nevertheless, it must have proven difficult or even impossible for most ecologists to translate the often highly technical descriptions of an IPM in the primary literature, such as in *Biometrics*, *Ecology*, or *Methods in Ecology and Evolution*, into an actual analysis in R or BUGS for *their own* data. We believe this is so even nowadays when code for analyses is increasingly published in electronic appendices to published papers. Our book provides you with that translation for a large number of case studies, and it attempts a comprehensive synthesis of the current state of research on IPMs. We honestly hope it will empower you to apply IPMs that interest you to data sets from your study.

In addition, owing to the vigorous growth of the IPM field, it has become difficult to obtain an overview of the burgeoning number of different classes, methods, and goals of IPM-based analyses. Despite several published reviews (Schaub and Abadi, 2011; Maunder and Punt, 2013; Zipkin and Saunders, 2018), to date, a comprehensive synthesis of IPMs does not exist, and our book aims to fill that gap.

This book gives broad coverage of the theory of this powerful class of models and therefore should appeal to statisticians and quantitative ecologists who wish to obtain an overview of this field. But at the same time, it is an applied book that aims to empower ecologists, wildlife managers, and similar professionals to fit their own IPMs with confidence. Therefore, an important part of the book consists of almost a dozen fully worked and complete example analyses presented in a richly commented tutorial style. Thus, we emphasize learning by doing and we invite you to work through fully documented example analyses. From teaching our workshops and our own progress in understanding statistical models, including IPMs, it is our experience that this is by far the best way for ecologists to learn and understand a new technique or statistical model. Our collection of fully worked examples will first help you to understand a model and then serve as a template for your own analyses, which you can start from our code and then modify to your needs. This, by the way, is exactly how most statistical analyses are conducted in practice anyway. We very rarely write code for a model from scratch, but rather, we typically start from a template for a similar model that we may have written earlier, was given to us by a colleague, or we took from the appendix of a paper, and we then adapt the code for our case. Our book also shows how IPMs can be used to understand past and predict future population dynamics, thus synthesizing various important topics in population ecology. More specifically, our book has the following emphases:

- It covers the theory of IPMs and shows how data sampled at the population level (e.g., population count data) and those from the individual level (e.g., capture-recapture data) can be jointly analyzed in an IPM to make better inferences about population processes and more accurately predict the future trajectory of a population.
- Matrix population models are used often in population ecological studies to compute life history summaries, carry out retrospective population analyses, or analyze population viability. The classic manner of conducting such analyses is by a two-step procedure: first, estimates of various demographic parameters such as survival are obtained from different data sets in separate analyses, and second, these parameter estimates are fed into a matrix population model. This two-step procedure is clumsy and can be statistically inefficient. This book explains the basics of classic matrix population models and then shows how the matrix population model in an IPM is seamlessly embedded into an estimation framework. That is, in an IPM, the population analysis and parameter estimation steps of traditional applications of matrix population models are combined in a single analysis. We demonstrate the multiple benefits that result.

- The book serves as a manual for practicing population analysts and a graduate-level text for students in ecology, conservation biology, wildlife management, and related fields. Specifically, the book makes the fitting of IPMs easily accessible to practitioners and graduate students with an accessible and not very math-heavy style of presentation and many tutorial-style illustrations of the analyses possible with an IPM in a wide range of case studies. These fully worked examples serve as a template to start your own analyses and can be modified for your own needs. Thus, our book builds a bridge between the typically far more technical descriptions of IPMs in the primary literature and the practitioners who want to use them to answer questions of scientific interest or management importance. It contains fully documented, "cookbook-style" analyses using R and the Bayesian modeling software JAGS and thus is a textbook on IPMs fit with R and JAGS software. So far, no canned software exists to fit IPMs (and quite likely there never will be), and over the last 15 years, software that adopts the BUGS language such as WinBUGS, JAGS and NIMBLE has become the standard manner in which IPMs are fit.

Michael has gained much experience with IPMs over the last 15 years in countless IPM applications and simulation studies, resulting in more than 20 papers on the topic thus far. Marc has a great deal of experience with Bayesian statistical modeling. Together, we wrote a successful earlier book, *Bayesian Population Analysis* (Kéry and Schaub, 2012), that is similar in style and concept. Therefore, we have strived to replicate that book's successful recipe in our new book on the topic of IPMs. In preparing to write this book, we taught more than 10 workshops on IPMs with a total audience of about 400 people. The experience we gained during those workshops, and the questions and comments we received, were essential contributions to this book. Although we hope this book will also be of interest to statisticians who want to learn about IPMs and population modeling in general, it was primarily written by ecologists for ecologists. Perhaps this is a natural result, but we have written the book we would like to have had when we started with the topic of integrated population modeling.

WHO SHOULD READ THIS BOOK?

We hope that any quantitative population ecologist who conducts empirical studies on population dynamics can greatly benefit from reading this book because it will foster a better understanding of a whole range of analytical methods for estimating demographic rates and population sizes. Even if you do not actually carry out an IPM, we think there is much to be learned from our book that will benefit your demographic analyses of a natural population.

As a professor in ecological research or related disciplines, including natural resource management, you will have to choose a book on which to base your class on modern population modeling using matrix population models and related approaches. You may have heard that in the new wave of these models, the matrix modeling part and the parameter estimation part can be unified into a single framework and that the result is called an IPM. You will have questions such as *"What is a comprehensive yet easy to understand text that presents both the theory and the practical application of these powerful models, ideally with open-source software such as R, in a way that my students, often lacking adequate mathematical background and with limited software experience, can still understand?"* We believe that our book fully addresses your concerns.

Finally, as a wildlife or fisheries manager, you may be working at a national park, government agency, or a similar place where you are in charge of monitoring and modeling natural resources. Perhaps you need to know, "*How many individuals from my population can be harvested in a sustainable way?*", "*Which is the best strategy to reverse the negative population trend of an endangered species into a positive one?*", "*What is the best eradication strategy for an invasive weed?*", "*How long is a small population likely to survive?*" or "*What is the main demographic driver for the decline of a given species?*" In all cases, our book will allow you to confidently apply an IPM to your data to answer these and related questions.

In particular, we hope that our book and the techniques we cover will receive a much wider uptake than hitherto in the management of hunted species, i.e., game management. Whenever we are in the United States to teach, we are impressed by the quantitative rigor of wildlife management on both the monitoring and the management side. For instance, it appears to us that in the U.S. population assessments without rigorous incorporation of false-negative errors (i.e., those arising from imperfect detection) are rare and that decision analyses in the management of hunted species are typically based on a formal population viability analysis. Also, adaptive harvest management is much more widely conducted in the U.S. We have no doubt that our book will be widely useful in such an environment where formal, quantitative population assessments are *de rigueur*, but we sincerely hope the ideas about quantitative population analysis that our book espouses will become much more widely adopted in similar circles, especially in Europe, including Switzerland.

CONVENTIONS IN THIS BOOK

The book is composed of two parts, which start after the introductory Chapter 1. Part 1 covers the theory of IPMs, including required knowledge about Bayesian inference, Bayesian computing, matrix population models, and many examples of what we call the "ingredient models" of an IPM. We introduce IPMs in two conceptually slightly different ways in the hope that you find at least one of them appealing and insightful. We highlight the benefits of IPMs but also stress their limitations and discuss the possible consequences when model assumptions are not sufficiently well met. Most of the time, we develop and use an IPM for our model species—the beautiful woodchat shrike, which also adorns the cover of our book. We think it is helpful to use the same study system consistently when explaining different topics because attention on a specific topic becomes more focused. We show how using an IPM can result in a better understanding of population dynamics and can be invaluable for more accurate population forecasting. Part 2 of the book is a rich collection of case studies, most of which have already been published by our colleagues or by us. Finally, we have a concluding chapter where we attempt a brief synthesis and offer our view about the likely future developments of the IPM enterprise.

In its style and concept, this book is similar to other books written by Marc (Kéry, 2010; Kéry and Royle, 2016, 2021) and especially our *Bayesian Population Analysis* book (Kéry and Schaub, 2012). The key style concepts are the following:

- We provide code for all analyses and figures on our book web page (see below). However, all the essential code is also printed in the book.
- We have a clear and consistent layout for all analyses, and we use fixed-width font for all code (i.e., `this is how we write R and BUGS code`). In addition, all complete BUGS code appears in a light-grey box (see below).
- We use the free software R and JAGS for all data preparation, model fitting, and output processing.
- All statistical models are specified using the powerful BUGS language, used also by other Bayesian software including NIMBLE, OpenBUGS, and WinBUGS. Hence, alternative software could be used for model fitting with hardly any modifications of the code.
- We often show how to simulate data to foster a clearer understanding of the analyzing model. Indeed, we claim that if you can simulate data under a certain model, you likely understand that model.
- We aim for light and engaging language.

We use "models" a great deal in this book, and there are different kinds. First, we have mathematical models such as matrix population models (Chapter 3). They do not attempt to estimate quantities but simply produce output based on some input parameters and calculations, often in a "what-if" spirit. Many statistical models are available for estimating parameters based on your data. Such models are, for example, a Poisson regression model (Chapter 2), a state-space, a Cormack-Jolly-Seber, occupancy, or an *N*-mixture model (Chapter 4). An IPM also belongs in that category, but it is composed of multiple, connected statistical models. We often refer to the latter as "submodels" or "component models" of an IPM, and related to that, we may write of "single-data likelihood" or "component data likelihood." Finally, we also use "model" to describe the spatial or temporal patterns in a parameter. For example, for survival in a capture-recapture model, we may adopt a model with temporal random effects or a linear trend. Hence, the meaning of the word "model" can be manifold, but we think it should become clear from the context.

COMPUTING

Throughout this book, we use R (R Core Team, 2018) and JAGS (Plummer, 2003) for data management, model fitting, and output processing, including the drawing of figures. JAGS uses the simple but very powerful BUGS language, which is also implemented in the programs WinBUGS (Lunn et al., 2013), OpenBUGS (Thomas et al., 2006), and NIMBLE (de Valpine et al., 2017; de Valpine et al., 2020a; 2020b), among others. Thus, when we refer to the BUGS *language*, we do not denote any particular software, as it could be any of those mentioned here.

To run the code in our book, you need to have a recent version of R and JAGS installed on your computer (we have used R 4.0 and JAGS 4.3). Both programs can be downloaded for free: JAGS, including its manual, from https://sourceforge.net/projects/mcmc-jags/, and R from https://www.r-project.org/. In R, you also need to install several packages. Apart from the R package that accompanies this book (`IPMbook`; see below), the most important is the interface between R and JAGS, and in this book, we use the package `jagsUI` (Kellner, 2019). Other interface packages, such as `runjags`, `R2jags`, `rjags`, or `dclone`, could also be used. These packages may differ slightly in how MCMC options are defined and in the format of the output they produce. Therefore, if you want to use one of these other interface packages for R and JAGS, you obviously would have to modify some of our code in the book, although any code that defines IPMs for JAGS will remain exactly the same.

All the code we provide must be executed in R. Some of that code will write a plain text file containing a description of the IPM or another statistical model defined in the BUGS language. JAGS uses this file to fit the model. You have to be very clear about which part of the code belongs to R and which is for JAGS. Fortunately, the syntax of both programs is very similar, and at first sight, they may even look identical. To make it absolutely clear which is which, we have added a light grey layer underneath all code belonging to the model definition file for JAGS. All the rest, in Courier New font, is R code (though note that in very rare cases we show a small number of lines of BUGS code to demonstrate something specific, and they do not have the grey layer). The text file for JAGS is created with the R function `cat`. So whenever there is JAGS code for a complete model, you will see the following construction:

```
cat(file="model1.txt", "
model {
… # Here is some code in the BUGS language
}
")
```

In this example, R writes the code within the grey layer into a text file with the name `model1.txt`. It is saved into the R working directory, which is the folder in which R is active and that you can identify using `getwd()`. You can change the R working directory with the `setwd` function in R. Especially when you start with JAGS, it can be insightful to search for the text file produced and have a look at it. In addition, when JAGS reports an error, it will indicate a line number in the model. You can open the text file to see which part of the model the error message refers to.

These model files always have the name `modelX.txt`, where X is an increasing integer that starts anew at 1 in every chapter. The output from JAGS is always stored in R objects with a name such as `outX`, where X again starts anew at 1 in each chapter. Sometimes we fit the same model to different data sets, in which case the numbers between model and output files may no longer correspond in later parts of a chapter. Moreover, we sometimes fit additional models using code you will find on the book web page, but that is not printed in the book. These model and output files are included in the numbering. Therefore, it is for instance possible to generate an output file `out9` from the model `model7.txt`.

Throughout the book, we use several other R packages for data manipulation, to produce plots, or for further analyses. Here is the complete list of packages (in alphabetic order): `AHMbook`, `denstrip`, `grid`, `ggplot2`, `IPMbook`, `jagsUI`, `lme4`, `maptool`, `MCMCglmm`, `plotrix`, `plyr`, `raster`, `RColorBrewer`, `rgdal`, `rgeos`, `scales`, `sp`, `wiqid`.

THE IPMBOOK PACKAGE

We are indebted to Mike Meredith for creating and maintaining the R package `IPMbook` that accompanies this book and that you can download from CRAN. It contains various useful R functions that we have developed to simulate data, format data in various ways, and provide initial values for model fitting (initial values can be a critical part of modeling with JAGS). It also contains all the data sets used in the book, including those for the case studies in Part 2. To get an overview of the functions and data sets provided in the package, you can use this:

```
?IPMbook
```

BOOK WEB PAGE

Additional material for the book is provided on the book web page at www.vogelwarte.ch/ipm. Besides a description of the book, you will find a link to a GitHub page maintained by Mike Meredith that provides the full code. This is handy because you can download the code in your favorite R interface and because the code is complete, while in the book only parts of code (the important ones of course) are printed. Hence, you should be able to reproduce every analysis presented and draw any graph shown in the book. The solutions to the exercises can also be found on the web page.

An important landing page for this and several related books by some colleagues and us is www.hierarchicalmodels.com. It contains additional resources and information on hierarchical modeling in ecology, discussion groups and help forums, workshops, and some particularly relevant R packages on CRAN and GitHub repositories. In particular, we would like to draw your attention to the Google group email list, where questions about topics covered in our book can be asked: https://groups.google.com/g/hmecology.

We truly hope you will find our book useful for the fitting of IPMs and to enhance your understanding of these complex hierarchical models. We hope that applying IPMs will lead to better inferences that are useful in management and conservation. Indeed, we consider one of our major contributions to the mitigation of the biodiversity crisis to be the provision of analytical knowledge and help, such that an increasing number of ecologists and wildlife managers can apply complex statistical models to better analyze their population data and make better decisions about managed populations. For us, putting all this material together has been a great learning experience, but it is clear that our understanding of IPMs is far from complete. We therefore also hope that our book will inspire much future research on the fascinating topic of integrated population models.

Michael and Marc
December 2020

Acknowledgments

We thank Panagiotis ("Takis") Besbeas, who, along with colleagues, wrote the foundational paper on integrated population models (IPMs) back in 2002 and who has continued to conduct important research on this class of models in the frequentist mode of inference ever since. We also thank Steve Brooks and his colleagues for the first translation of an IPM into the BUGS language in 2004. That translation made these models practically accessible to a much larger audience, including ourselves.

Our colleague Mike Meredith read and checked the code throughout the entire book and suggested corrections and ways to make things better, more efficient, less clumsy, or correct. He checked and improved all R functions and the structure of the data sets, maintains a GitHub site hosting the IPM book code and most of all, he created the R package that accompanies this book, called `IPMbook`. We are indebted to him for his huge contribution to this project and for his friendship, his patience, and the incredible swiftness with which he always dealt with all our queries.

We warmly thank the following people who read and commented on parts of the book: Ismael Verrastro Brack, Thomas Bregnballe, Richard Chandler, Dilsad Dagtekin, Dan Gibson, J. Nicholas Hendershot, Joseph Hightower, Cat Horswill, Pepe Jimenez, Dave Koons, Omar Lenzi, Wendy Leuenberger, Dan Linden, Mark Maunder, Rachel McCrea, Mike Meredith, Helen Moor, Josh Nightingale, Joseph Hightower, Floriane Plard, Rob Robinson, Andy Royle, Sarah Saunders, Eric Stolen, Cat Sun, Chris Sutherland, Gesa von Hirschheydt, Greg Wann, Amanda Warlick, Matt Weldy, Kristopher Winiarski and Qing Zhao. In particular, Thomas Riecke read and commented on most chapters, and his input was vital. We are grateful to Dave Koons for writing his inspiring foreword.

We also thank the persons who have made data sets available that we could use as well as offer to the world for free use: Farshid Ahrestani, Raphaël Arlettaz, Gaby Banderet (†), Radames Bionda, Thomas Bregnballe, Bob Clark, Martin Grüebler, Hans Jakober (†), Ulrich Köppen, René-Jean Monnerat, Luca Rotelli, Wolfgang Stauber (†), and Bruno Ullrich. Next, we are very grateful to the photographers for their great pictures, and in particular, that amazing young artist, Helene Rimbach, who created the fantastic drawing of the woodchat shrike for the book cover.

We are also grateful to the many people who have participated in our workshops. Our interactions with them helped improve the material compiled in this book, and in many cases, helped us come up with some content in the first place.

Our employer, the Swiss Ornithological Institute (www.vogelwarte.ch), deserves our gratitude for giving us the freedom to write this book and for the liberty in our research to pursue what we thought was important, fascinating, and beautiful.

Finally, we thank our families, particularly our wives Christine and Susana and our children Lilly, Lukas, and Gabriel, for their support, love, and patience while we wrote the book.

SPECIAL THANKS BY MICHAEL

I thank my good old friend Marc for suggesting to write this book during our unforgettable BPA workshop in Esporles in September 2012, and for going along this way together. It has been a real pleasure. I'm also very grateful to Olivier Gimenez, who introduced me to the wonderful world of IPMs 15 years ago. Finally, I would like to thank Fitsum Abadi, Todd Arnold, Rémi Fay, David Fletcher, Dave Koons, Thomas Riecke, Simone Tenan, Mitch Weegman and Floriane Plard for various inspiring collaborations on different aspects of IPMs.

SPECIAL THANKS BY MARC

I would like to thank Miguel for his friendship and help over the years and for inviting me to become a coauthor of this book, which was a wonderful learning experience for me. In addition, I feel forever indebted to my friend and colleague Andy Royle, to my mentor Jim Nichols, and to Jim Hines for so much inspiration and support over so many years. Without them and the fantastic time I spent at the legendary Patuxent Wildlife Research Center in Laurel, Maryland, I would never have become the statistical ecologist I am now. Finally, it is my great pleasure to acknowledge Andrés Calamaro for so much joy and inspiration over the years, right from the start 20 years ago, when all this hierarchical modeling began for me.

LITERATURE CITED

Besbeas, P., Freeman, S.N., Morgan, B.J.T., Catchpole, E.A., 2002. Integrating mark-recapture-recovery and census data to estimate animal abundance and demographic parameters. Biometrics 58, 540–547.

Brooks, S.P., King, R., Morgan, B.J.T., 2004. A Bayesian approach to combining animal abundance and demographic data. Anim. Biodiv. Cons. 27.1, 515–529.

de Valpine, P., Paciorek, C.J., Turek, D., Michaud, N., Anderson-Bergman, C., Obermeyer, F., Wehrhahn Cortes, C., Rodriguez, A., Temple Lang, D., Paganin, S., 2020a. NIMBLE User Manual. R package manual version 0.10.1.

de Valpine, P., Paciorek, C.J., Turek, D., Michaud, N., Anderson-Bergman, C., Obermeyer, F., Wehrhahn Cortes, C., Rodriguez, A., Temple Lang, D., Paganin, S., 2020b. NIMBLE: MCMC, Particle Filtering, and Programmable Hierarchical Modeling. R package version 0.10.1.

de Valpine, P., Turek, D., Paciorek, C.J., Anderson-Bergman, C., Lang, D.T., Bodik, R., 2017. Programming with models: writing statistical algorithms for general model structures with NIMBLE. J. Comput. Graph. Stat. 26, 403–413.

Fournier, D., Archibald, C.P., 1982. A general theory for analyzing catch at age data. Can. J. Fish. Aquat. Sci. 39, 1195–1207.

Kellner, K.F., 2019. Package 'jagsUI': A Wrapper Around 'rjags' to Streamline 'JAGS' Analyses. https://cran.r-project.org/web/packages/jagsUI/index.html.

Kéry, M., 2010. Introduction to WinBUGS for Ecologists – A Bayesian Approach to Regression, ANOVA, Mixed Models and Related Analyses. Academic Press, Burlington.

Kéry, M., Royle, J.A., 2016. Applied Hierarchical Modeling in Ecology – Analysis of Distribution, Abundance and Species Richness in R and BUGS. Volume 1 Prelude and Static Models. Elsevier/Academic Press, London.

Kéry, M., Royle, J.A., 2021. Applied Hierarchical Modeling in Ecology – Analysis of Distribution, Abundance and Species Richness in R and BUGS. Volume 2 Dynamic and Advanced Models. Elsevier/Academic Press, London.

Kéry, M., Schaub, M., 2012. Bayesian Population Analysis Using WinBUGS – A Hierarchical Perspective. Academic Press, Boston.

Lunn, D., Jackson, C., Best, N., Thomas, A., Spiegelhalter, D., 2013. The BUGS Book. A practical introduction to Bayesian Analysis. CRC Press, Boca Raton.

Maunder, M.N., Punt, A.E., 2013. A review of integrated analysis in fisheries stock assessment. Fisheries Research 142, 61–74.

Plummer, M., 2003. JAGS: A Program for Analysis of Bayesian Graphical Models Using Gibbs Sampling. Technische Universität, Wien, Vienna, Austria.

R Core Team, 2018. R: A Language and Environment for Statistical Computing. R Foundation for Statistical Computing, Vienna.

Schaub, M., Abadi, F., 2011. Integrated population models: a novel analysis framework for deeper insights into population dynamics. J. Ornithol. 152, 227–237.

Schaub, M., Gimenez, O., Sierro, A., Arlettaz, R., 2007. Use of integrated modeling to enhance estimates of population dynamics obtained from limited data. Cons. Biol. 21, 945–955.

Thomas, A., O'Hara, R.B., Ligges, U., Sturtz, S., 2006. Making BUGS open. R news 6, 12–17.

Zipkin, E.F., Saunders, S.P., 2018. Synthesizing multiple data types for biological conservation using integrated population models. Biol. Cons. 217, 240–250.

CHAPTER

INTRODUCTION

1

CHAPTER OUTLINE

1.1 POPULATION MODELING IN POPULATION ECOLOGY AND MANAGEMENT

The number of individuals in study species changes in both time and space, and these changes are broadly known as *population dynamics*. The description of these changes and the demographic and ecological mechanisms that cause them are important for many scientific questions in population ecology, conservation biology, evolutionary ecology, and disease ecology as well as in applied fields such as fisheries, pest control, and general conservation practice (Caughley, 1994; Norris, 2004; Conroy and Carroll, 2009; Mills, 2013).

In an ideal world, when attacking almost any scientific question or decision problem in population dynamics, we would have perfect knowledge about the state and location of all the individuals of a study species at a certain point in time and then be able to follow them through time. We can thus define the state s of an individual at time t and location i as the fundamental information unit in demography (see also Section 4.2), where the individual state denotes features such as sex, age, size, or a measure of quality. Individuals may survive, may disperse to another place, may reproduce, and will eventually die. Survival, movement/dispersal, and reproduction are the individual demographic processes, and they may vary by state and location and also over time. Over time, demographic processes result in changes in the spatial distribution of individuals, their state frequencies, and temporal changes in their numbers. We could depict these individuals on a map and draw such a map at different points in time. We then would see points appearing, moving, and disappearing again (Fig. 1.1). In other words, when we follow the population dynamics processes over time, we will see points that blink on and off, move, and spawn new points. Mathematically then, this becomes all about describing spatio-temporal point patterns (Illian et al., 2008; Baddeley et al., 2016; Wikle et al., 2019).

In such a simple model of a dynamic population of points, the goal of a population dynamics study can be defined as the description of the blinking, movement, and spawning behavior of these points and the attempt to understand the processes that govern these behaviors. Which demographic process is responsible for points occurring more densely in some areas than in others; or becoming scarcer over time in some areas while becoming more locally abundant in others? Perhaps we want to predict (i.e., forecast) how that cloud of blinking points looks in the future. Will the density of these points increase or decrease? What happens if we eliminate some points—will this affect the long-term dynamics of the remaining points? These questions could be translated into the real world: What are the demographic drivers of spatial and spatio-temporal variations in abundance? What are the demographic causes of an observed population change, such as a decline, over time? How large is the extinction probability of a study population? How many individuals can be removed (i.e., killed, harvested) from the population without putting its long-term viability at risk?

In practice, addressing these and many related questions requires simplifications. That is, it requires abstract models, because not having them results in "no model, no understanding," as Keyfitz and Caswell (2005) succinctly and

Integrated Population Models. https://doi.org/10.1016/B978-0-12-820564-8.00013-0

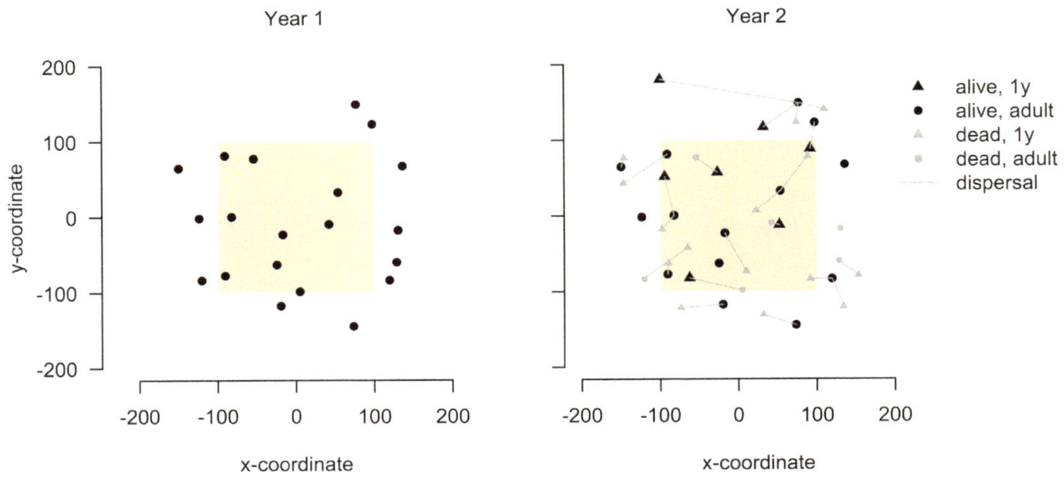

FIGURE 1.1

Conceptual outline of a population dynamics study over a 2-year period in a perfect world depicted as part of a continuous spatio-temporal point pattern (perfect world refers to perfect observability of the process). Growth of the population is governed by the demographic processes of survival, reproduction, and natal dispersal. In year 1, mature adults are shown as black dots (left). In year 2, surviving individuals are shown as black symbols, while individuals that have died since year 1 are denoted by gray symbols (right). Circles are individuals that were already mature in year 1, and triangles are individuals born in year 1. Gray lines show natal dispersal from the birth location to the breeding location. The beige quadrat defines the area of the study population, the boundaries of which are arbitrary to some degree. Dispersal results in immigration and emigration with respect to the defined population.

delightfully put it. The first and typically necessary simplification is the definition of a certain space wherein we study the individuals; we define an area and then define the individuals within its boundaries as our study population (Fig. 1.1). These individuals are the target of our inference. Rarely are the boundaries of a study area set such that all individuals of a species are included, i.e., the world population of the study species. Nearly always, a study area includes only a small fraction of the world population of a species (for a rare exception, see Sillett et al., 2012). Boundaries may be defined based on biological considerations, e.g., when the spatial distribution of the individuals is patchy, boundaries may be chosen such that one or several distinct patches are included. Boundaries can also be defined based on practical reasoning or by what we want to know, e.g., whether we are interested in population dynamics on small or large spatial scales. It is also possible to define several such areas, resulting in multiple studied populations, or perhaps colonies for colonial species such as many seabirds. Again, these areas may be defined based on biological reasoning such that, for instance, all occupied patches are defined. Alternatively, they may be defined more arbitrarily or as part of a spatially replicated study design.

If we are truly honest, we must admit that in practice, population definitions are often extremely arbitrary in terms of their location, and especially, their extent. We most often study a species where it is numerous (because we need a large sample size in our statistical analyses) and where it is easily accessible. The extent of what we call a "population" is most typically defined by the area wherein all the individuals live that our PhD students can get their hands on during one field season. Hence, the size of a population is most often a function of the logistical resources available to a researcher.

Whichever way we define a spatial area to contain the individuals that we study, it will affect our view of the fundamental point pattern and will help us understand it. By referring to a spatial area, we add one level to the basic level of the individual point in the pattern—the population level, i.e., the whole set of points as defined by our choice of study area. Recognizing the population level allows us to summarize the set of individuals; for example, we can now count the number of individuals and characterize this population by its size (N). The change in population size over time, then, is the most basic, and simplest, description of population dynamics. Properties of individuals can also be summarized by quantities such as the sex ratio, the age distribution of the individuals, or the distribution of any other classification chosen for them.

The outcomes of the individual demographic processes (survival, dispersal, reproduction) are never perfectly predictable and in this sense are stochastic. That means that we need to describe them as realizations from a chance process, which in practice is a chosen statistical distribution (see Chapter 2). For example, the survival of an individual is akin to the flipping of a loaded coin, and the load of the coin (affecting its propensity, or probability, to produce a head or a tail) will depend on characteristics of the individual as well as on time and space. That is, the outcome of the coin flip is a chance process, and we want to estimate the probability governing it from some observations of that chance process. Both dispersal and reproduction can similarly be viewed as the outcomes of chance processes, and the parameters that govern their average behavior may similarly depend on features of the individual as well as on time and location. Estimation of these demographic parameters may require scaling-up to the population level. Scaling-up from the individual to population level has especially strong effects on the way we perceive dispersal, i.e., the movement into or out of a chosen study area. Individual dispersal movements are characterized at the population level by their outcomes with respect to the defined area—a movement starts from inside or outside the area and ends inside or outside. A movement from inside to outside is called an emigration event, whereas a movement from outside to inside is called an immigration event.

A further possible simplification of the general spatio-temporal point pattern is in reference to time. In practice, the numbers and locations of individuals changes continuously over time, but in practice we are often interested in a characterization of the population at specific points in time, e.g., during the reproductive season, when individuals are more tied to certain places and may be easier to study. Therefore, we often discretize time, which also renders simpler models. To understand population dynamics, it is often sufficient to describe a population only at specific points in time.

Finally, we may make simplifications regarding demographic processes that in reality will hardly ever be identical for two individuals and are unlikely to be identical over time and in space. In our simplifications of the demographic processes, i.e., in our models, we typically make a number of constancy assumptions such that demographic processes are assumed to be identical for all individuals in the same state, to not vary in space (or in only a restricted way, e.g., as "explained" by a linear model of some covariate), to be constant over time, or to vary only in some restricted way (e.g., as a time trend).

Thus, at a fundamental level we have a population concept as a spatio-temporal point pattern that we could describe at the individual level where the demographic processes actually operate. The results of these processes are the appearance, movement, and disappearance of individuals over time and space and their branching into new individuals (i.e., reproduction). In addition, there is the population level, which is artificial to some degree but of paramount importance for our understanding and for a parsimonious description of the population. The population level is typically our starting point for thinking about models to describe and understand population dynamics.

The most fundamental models of population dynamics, however, are individual-based (Huston et al., 1988; Fahse et al., 1998), which describe the demographic performance of every individual. They do not attempt to describe effects at the population level; rather, the dynamics at the population level are an emergent property that results from processes acting at the individual level. That is, the number of individuals or their age distribution in a certain area is obtained by simple summarization, and we can define space any which way we like. These models closely resemble the fundamental point-process-based view of a population described earlier. A drawback of these models is their great computational burden, as modeling each individual separately is quite expensive as well as a conceptual burden because their descriptions are mathematically complex.

A second class of population dynamics models explicitly includes both individual- and population-level data. A spatial unit is defined, and the number of individuals and its change over time is modeled within this spatial unit that defines the population. Individuals are characterized by age or stage, and the age- or stage-specific numbers of individuals are important quantities in these models. Information about the location of individuals is reduced to whether or not they are inside or outside the limits of the population. These models are the celebrated stage-structured or matrix population models (Caswell, 2001). A further development of this class of models recognizes that individual heterogeneity in demographic performance is often substantial even among individuals within an age group or stage in a matrix population model. In these models, heterogeneity is defined in a continuous rather than categorical manner, and demographic rates may be specified that depend, in parametric form, on continuous individual traits such as size. Models that include these features are called integral projection models (Easterling et al., 2000; Coulson, 2012).

A third class of models considers only the population level and therefore does not include any demographic parameters that describe individual processes. A change in the number of individuals in a population is modeled as a function of the population growth rate only. These models are much more phenomenological and are known as time-series analyses in population dynamics (Dennis et al., 1991; Dennis and Taper, 1994; Dennis et al., 2006; Knape et al., 2013).

The use of these models may answer different and multiple questions relevant to a fundamental understanding of population dynamics, conservation, and management. Depending on the class of model used, we can address various questions. With individual-based and matrix population models, we arguably come closer to a mechanistic understanding of a population, i.e., at the level of the actual demographic processes that always operate at the individual level. But the costs of these models include increased complexity and greater demands on the data. In contrast, time-series models of population abundance represent a more phenomenological description of population dynamics but are less complex, and the data they require are usually "cheaper" in the sense that they are more widely available or logistically easier to obtain.

1.2 THE TWO-STEP APPROACH TO POPULATION MODELING

Population models contain parameters such as survival probability, productivity, or the population growth rate. In practice, these parameters are virtually never known. Sometimes we may have information from previous studies or the experience of experts, but in most cases when we study population dynamics, these parameters must be estimated from data (which must be collected in the first place). Estimation means the analysis of data from (ideally) a random sample of individuals in the population by means of a statistical model, which is an idealized description of the processes that are thought to have produced the data. Often these models must account for imperfect detection, i.e., for the fact that an individual can be alive and inside a study area but not be detected. Failure to accommodate this type of measurement error in models of animal populations can have severe consequences when inferences are made based on those models. This recognition has given rise to the development of a large branch of ecological statistics that we collectively call "capture-recapture models" (Seber, 1982; Borchers et al., 2002; Williams et al., 2002; Royle and Dorazio 2008; King et al. 2010; Royle et al., 2014; Buckland et al., 2015; McCrea and Morgan, 2015; Seber and Schofield, 2019). Such models allow us to correct for this particular measurement error in our inferences about temporal processes such as survival or dispersal that require us to follow recognizable individuals over time; they also correct for the measurement error that arises when we estimate the locations of individuals. Of course, formal estimation also means that we obtain a measure of uncertainty associated with a parameter estimate, which depends on sample size, noise in the system, and model complexity, among other factors.

The estimated parameters can then be plugged into a population model, most commonly a matrix population model, which allows the calculation of quantities such as the population growth rate or growth rate sensitivities (Caswell, 2001). Hence, inferences about population dynamics are obtained by a two-step approach in which the first step is parameter estimation, and the second step is the use of the parameter estimates obtained in step 1 as inputs in a population model.

Data sampling in population studies usually includes both individual- and population-level data. That is, typically longitudinal data on productivity and marked or otherwise recognizable individuals are sampled in addition to counts of the population, where no individuality is usually distinguished. While parameters estimated from the individual-level data are used to parameterize a population model, the data from the population level can be used to evaluate the population model developed, i.e., to gauge whether the predicted and observed population dynamics agree.

Such piecemeal analyses are not wrong, but usually they ignore the sampling (co-)variation of the estimators of the demographic rates. They are also statistically inefficient because information about individual-level processes contained in the population-level data is not exploited. In addition, the estimation uncertainty coming from the first step of the procedure must be propagated into the inferences about population dynamics produced by the second step, but proper error propagation is challenging in the two-step approach. Thus, it would be desirable to have an analytical framework that allows us to jointly analyze all available individual- and population-level data sets and avoid a two-step approach. Integrated population models do exactly that.

1.3 INTEGRATED POPULATION MODELS

An integrated population model (IPM) is a special case of an integrated model that is generally defined by a joint likelihood that shares at least one parameter across separate likelihoods (see Section 2.8). A defining feature of an IPM is a joint likelihood that is usually the product of the likelihoods for both individual- and population-level data. Shared parameters are typically demographic rates (see Fig. 5.1 and Schaub and Abadi, 2011). The idea of the IPM is usually attributed to Fournier and Archibald (1982), who developed a fishery model to analyze catch-at-age data that also allows the incorporation of other data types. Indeed, the development of integrated analyses in fisheries (particularly for what is known in that field as "stock assessment") progressed rapidly, and Maunder and Punt (2013) reviewed this active development. Interestingly, outside the fisheries world, it took about 20 years before IPMs were reinvented in a seminal paper by Besbeas et al. (2002). Since then, we have seen more or less parallel development with little exchange between so-called *integrated stock assessment models* in fisheries and *IPMs* outside the water. We think this is due to the widely different data sets, different terminology, and physical and personal separation of the research in terrestrial and aquatic (usually marine) populations. It would certainly be inspiring if there were more exchange about the IPMs in these two worlds, and we do not doubt that each side could learn a great deal from the other.

A critical contribution to the development of "terrestrial" IPMs was the first paper that adopted a Bayesian approach to model fitting with software of the BUGS family (Brooks et al., 2004). An important further step in the rise of the IPM was the study of their performance by simulation (e.g., Besbeas et al., 2009; Abadi et al., 2010; McCrea et al., 2010; Schaub and Fletcher, 2015; Riecke et al., 2019; Weegman et al., 2021), the inclusion of density dependence (Abadi et al., 2012; Besbeas and Morgan, 2012a), and the development of goodness-of-fit tests (Besbeas and Morgan, 2014; Newman et al., 2014), alternative fitting procedures (Besbeas and Morgan, 2019), and an individual-based version of these models (Chandler and Clark, 2014). Early on, Buckland et al. (2004 and 2007) highlighted the prominent role of stage-structured population models for IPMs and characterized IPMs as mathematical-statistical models that embed population modeling into inference, where the latter of course means parameter estimation from data. We think that this matrix population model-centered view of an IPM is important, but have a feeling that it became somewhat lost during the last decade.

The last 20 years (i.e., since Besbeas et al., 2002) have seen a marked increase in the number of applications of IPMs. While writing this book, we carried out extensive research on the literature about IPMs. The sources included our own database, which we had been compiling in the past, and several searches on the Web of Knowledge using the keyword phrase "integrated population model." After we had read a paper, we decided whether it dealt with an IPM in the sense defined in this book (Fig. 5.1). In addition, we restricted our selection to papers that dealt with bird, mammal, or amphibian data and excluded papers on fishes and invertebrates. We made this choice because we feared that our search for IPMs in fisheries and for invertebrates might be incomplete because of the (for us) uncommon data types and different language. By June 2020 we had compiled a list of 118 papers that met our criteria. Ten years earlier, our first review on IPMs by Schaub and Abadi (2011) had been based on a mere 25 papers, clearly showing that IPMs are increasingly being used. At various places in the book, we use our updated database comprising these 118 papers to gain insights into the current and likely future trends in the field of IPMs.

Integrated population models are complex statistical models and are usually hierarchical, i.e., models with multiple linked sources of variability and uncertainty (Berliner, 1996; Royle and Dorazio, 2008; Cressie et al., 2009; Newman et al., 2014; Hobbs and Hooten, 2015; Kéry and Royle, 2016, 2021). In an IPM, we jointly analyze multiple individual- and population-level data sets, and therefore an IPM is composed of multiple submodels. These submodels are the analytical models for each individual data set (or what we call the single-data or component-data likelihoods). In addition, all non-individual-based IPMs (i.e., unlike the new model by Chandler and Clark, 2014) contain a mathematical population model that provides a link between demographic rates and population sizes. The development of an IPM, therefore, requires knowledge about all these different models.

1.4 DEVELOPING INTEGRATED POPULATION MODELS WITH THE BUGS LANGUAGE

Early implementations of IPMs used the likelihood framework and Kalman filter (Besbeas et al., 2002; Besbeas and Morgan, 2012b; Newman et al., 2014), and this technology is inaccessible to most ecologists. In contrast, we believe that the first Bayesian implementation of an IPM (Brooks et al., 2004) does not look overly intimidating to many non-statisticians. Indeed, the implementation of even very complex models becomes conceptually fairly trivial in the BUGS language because all you need to do is stack several models on top of each other and have a few parameters that are shared between two or more of these submodels. The definition of complex joint likelihoods in a hierarchical model such as an IPM therefore comes within reach of almost any quantitative ecologist.

The BUGS language is currently implemented in WinBUGS (Lunn et al., 2000), OpenBUGS (Thomas et al., 2006), JAGS (Plummer, 2003), NIMBLE (NIMBLE Development Team, 2018), and multiBUGS (Goudie et al., 2020). For us, the BUGS language is a key invention of the BUGS Project (Lunn et al., 2000, 2013). It has proven to be supremely accessible to ecologists in understanding hierarchical models and allows them to implement existing hierarchical models as well as develop their own new models in a flexible and confident way (Link et al., 2002; McCarthy, 2007; Gimenez et al., 2009; Kéry, 2010). Hence, the Bayesian application of IPMs using BUGS software is the way to go for most ecologists who want full flexibility of model fitting without having to bring a trained statistician into their project.

Moreover, the Bayesian simulation-based approach has many additional advantages for population modeling because simulation is also the natural approach for dealing with matters of variability and uncertainty when we make inferences about their combined effects in model output, particularly including their combined effects in forecasts. The Bayesian approach also facilitates the inclusion of knowledge from elsewhere in case our own data lack some information about a process or parameter.

1.5 THIS BOOK

1.5.1 WHY THIS BOOK?

So far, no applied book about IPMs exists, nor indeed any textbook dedicated entirely to IPMs, so here we fill this gap. Our main aims when writing this book were threefold. First, we emphatically advertise IPMs as a tremendously sensible and powerful approach to population modeling. Second, we want to make IPMs as widely understood as possible among users of population modeling. And third, we aim to offer a comprehensive synthesis of Bayesian IPMs for terrestrial organisms.

1.5.2 STRUCTURE AND OVERVIEW OF THIS BOOK

The book has two main parts. In the first part, we provide the necessary theory on IPMs, including what you need to know about Bayesian statistics and computation, matrix population models, and other components of an IPM. The second part of the book is a collection of case studies.

Part 1 covers the following topics:

- In Chapter 2, we provide a brief review of the Bayesian way of learning from data about unknown quantities, about Bayesian computation, and the use of JAGS software for this computation. On the statistical side, we give a brief and applied introduction to generalized linear models, generalized linear mixed models, and integrated models (Pacifici et al., 2017; Miller et al., 2019; Chapter 10 in Kéry and Royle, 2021).
- Chapter 3 introduces stage-structured (or matrix) population models (Caswell, 2001) and their analysis in R and JAGS software. Nearly all IPMs are currently based on stage-structured models, and therefore a sound knowledge of them is important for you when you want to construct an IPM. We emphasize the definition of the life-cycle graph for your study species as a crucial part in the development of a stage-structured population model and hence also for construction of an IPM. We feel that the close connection between stage-structured population models and IPMs has perhaps not been stressed enough in the literature (despite Buckland et al., 2004, 2007). Virtually all IPMs hitherto published are simply stage-structured population models embedded in a joint likelihood for the estimation of their parameters. Importantly, all that you can do with a stage-structured population model (e.g., perturbation analysis, computation of life-history summaries), you can also do inside of an IPM; you can easily accommodate various sources of uncertainty and obtain comprehensive error propagation essentially for free. Hence, we see IPMs as a natural generalization of stage-structured population models.
- Chapter 4 is a long yet concise description of what we call the ingredients of an IPM, i.e., the specific models that we adopt for particular data sets to which we want to fit IPMs. In this chapter, we briefly describe some primary models for time series of population counts, survival, and productivity data as well as for some other data types that are sometimes utilized to form an IPM. This is a truly enormous topic, and we previously wrote an entire book with exactly this scope (Kéry and Schaub, 2012). So while we wanted to provide a fairly comprehensive catalog of the component models that most often appear inside an IPM, we really show only the simplest incarnations of these models, typically in a "null form," i.e., without covariates and without any refinements such as extra sets of random effects. For such additional complexity in these models, you can consult other books (e.g., Royle and Dorazio, 2008; King et al., 2010; Link and Barker, 2010; Kéry and Schaub, 2012; Kéry and Royle, 2016, 2021). In addition, in the case studies in Part 2 of our book (see below), you will reencounter many of these models, and often this time with some additional complexity, such as additional random effects or correlations between sets of random effects.
- In Chapter 5, we show how an IPM is constructed and fit to data using JAGS software. We show two different paths toward the construction of an IPM to explain and motivate the concept. The first path starts from a stage-structured population model in which an increasing number of hitherto fixed demographic rates are replaced by parameter estimates from models fit to actual data sets. The second path is perhaps more typical, where we start from the likelihoods of the component models and then build the joint likelihood of the IPM under an independence assumption as a simple product. In this chapter, we also introduce a favorite model species, the handsome woodchat shrike (*Lanius senator*), which also features prominently on the cover of this book. This species shares its life history with that of many other short-lived birds, so we consider it a good model species for much of avian ecology in practice. In our toy data set, we assume to have population count data, capture-recapture data, and productivity data all collected during a succession of breeding periods. Many IPMs are fit to such relatively simple data, and therefore we use this combination of data sets in our first practical introduction to an IPM.

- In Chapter 6, we discuss the benefits of IPMs in more depth. We show that parameter estimates usually become more precise, that demographic parameters for which no explicit data are available (what we call "hidden parameters"), as well as stage-structured population sizes (which may be completely latent), can be estimated in an IPM, and we explain why this is possible. We also introduce functions of our `IPMbook` package that can be used to simulate data for IPMs. Simulations then allow rigorous assessment of benefits in terms of the precision and estimability of the parameters. We also illustrate the flexibility that IPMs offer for jointly analyzing data that have been sampled in a less than ideal way—or some might say, in the real world.
- Chapter 7 is a long but important chapter that investigates the limitations of IPMs. We start by describing the typical assumptions on which IPMs normally rely and use simulation to study the consequences incurred by parameter estimates when these assumptions are violated. We also show how goodness-of-fit can be assessed for an IPM. There has been little research on model criticism for IPMs so far, which is why you will see quite a bit of novel material in this section. This also emphasizes the need for more research in this field.
- In Chapter 8, we cover IPMs that include density dependence. This is undoubtedly *the* major process in the regulation of natural populations. In an IPM, density dependence is specified for demographic parameters—density dependence at the population level (i.e., in the population growth rate) is then simply an emerging property of the underlying individual-level density-dependent demographic processes. Hence, using an IPM, we can investigate whether a population is regulated by density dependence, and in addition, we can identify the demographic mechanism through which density dependence operates.
- In Chapter 9, we show how to conduct retrospective population analyses from the output of an IPM. We illustrate a number of approaches ranging from a simple correlation analysis to the novel *transient life-table response experiments* (Koons et al., 2016). While the benefit of IPMs for simple correlation or asymptotic life-table response experiments is the ability to obtain measures of uncertainty, transient life-table response experiments require estimates of stage-specific population sizes, and these are hard to obtain without an IPM.
- Chapter 10 is the last chapter of Part 1 of the book and illustrates the use of an IPM to forecast population size while accounting for estimation uncertainty and stochasticity of the demographic processes. Such forecasts can be used for classic population viability analysis to evaluate extinction probabilities and compare the outcomes of different potential management options in "what-if" scenarios. We stress that both parameter estimation based on the data and forecasting can be achieved in a single model, with the major benefit that all uncertainties are propagated correctly and automatically.

Part 2 of the book contains a fairly large collection of case studies. We have chosen them to illustrate a diversity of data types, study systems, and taxa among birds and mammals. Most of these case studies feature classic IPMs with a stage-structured population model at their core, but we also have an example of the new type of individual-based spatial IPMs pioneered by Chandler and Clark (2014). For all case studies, we provide the data and complete code for the analysis, mostly within the book, and you can also find additional code on the book website. The emphasis in most case studies is on estimation of the parameters. Usually, we do not conduct any subsequent analyses such as retrospective or prospective analyses, as these are treated in detail in Chapters 9 and 10.

1.5.3 THE IMPORTANCE OF SIMULATION

Simulation is tremendously important in much of statistical analysis. By data simulation, we mean the generation of random numbers from a stochastic process described by a statistical model, and of course, the development of code for this complex random number generator. Data simulation is useful for multiple reasons (see Chapter 4 in Kéry and Royle, 2016 for a more detailed discussion): (1) Truth is known, so we can study whether a model provides unbiased parameter estimates, validate our code for fitting the model, or investigate the effect of certain assumption violations on parameter estimates. (2) Power analyses can be performed in a straightforward manner for any type of model or design. (3) The identifiability and estimability of parameters can be assessed in a very straightforward way. (4) Finally, and perhaps most importantly, data simulation provides proof that you have understood your model. If you know how to simulate (or assemble) data under a model, then it is highly likely that you also understand the opposite process, namely the breaking down (or disassembly) of the data using the analysis procedure. *Indeed, we use data simulation to explain a statistical model!* For instance, in Chapter 4 we will introduce most models by simulating a simple data set under that model. We believe that for nonstatisticians, data simulation is one of the most effective ways of explaining a statistical model.

We often use a simulated data set in the first part of the book. This allows us to clearly demonstrate certain features of an IPM, as truth is known, and the required assumptions are all met. Moreover, we illustrate the accuracy of parameter estimates based on simulations in which data simulation and subsequent model fitting are done repeatedly. This allows for the visualization of effects when model assumptions are violated or assessment of the precision of parameter estimates. To facilitate data simulation under an IPM, we have developed a number of functions that can be found in the `IPMbook` package.

We also use simulation to obtain parameter estimates in the Bayesian analysis of a model—Bayesian computation in BUGS software produces random numbers from the posterior distributions of the model parameters. This kind of simulation belongs to the class of Markov chain Monte Carlo (MCMC) algorithms. Statistical inference is then based on these random numbers, which can be summarized (e.g., as posterior means or posterior standard deviations) and used to make probability statements about the parameters (e.g., probability that a parameter such as a population growth rate is positive).

Finally, the prediction for unsampled spatial or temporal "areas" (the latter is called forecasting) is typically achieved based on simulation as well. In the context of forecasts, we simply let the population process in a model evolve forward in time, i.e., produce a sample of simulation trajectories beyond the temporal range of the available data, and then summarize this statistical "population" of forecast population trajectories in a suitable manner.

1.5.4 USE OF THIS BOOK IN COURSES AND FOR TEACHING

We have developed nearly all the material in this book in a series of 5-day workshops on IPMs, of which we have run 12 events for a current total audience of about 400 people over the last 6 years. This audience mostly comprised quantitative ecologists, wildlife ecologists, and statisticians. Clearly, the audience of our workshops also defines the intended readership of this book. We think this book can be useful as the basis of a course on IPMs in the curricula of such branches. Depending on what is needed, we would see Chapters 2–7 as providing the basis of a course, and then you could pick some of the case study chapters based on your particular interests. For audiences with a good grasp of statistics in general and Bayesian statistics and computation in particular, Chapter 2 can be skipped, as can Chapter 3 for audiences who already have a good knowledge of matrix population models, as well as Chapter 4 for people who are well-versed in the typical models used in population estimation, i.e., capture-recapture in the broad sense.

In addition, we are convinced that our book is excellent for self-teaching any material it contains. We have strived to write it in the form of a gentle introduction, and you can access all the code and try out and replicate anything in the book until the penny drops.

PART

THEORY OF INTEGRATED POPULATION MODELS

1

CHAPTER 2

BAYESIAN STATISTICAL MODELING USING JAGS

CHAPTER OUTLINE

2.1 INTRODUCTION

Integrated population models (IPMs) are fairly complex parametric statistical models. We usually present them as hierarchical models, i.e., as a combination of simpler statistical models that are linked through conditional probability relations (Royle and Dorazio, 2008; Cressie et al., 2009; Hobbs and Hooten, 2015; Kéry and Royle, 2016, 2021). Currently, the only way most nonstatisticians can realistically hope to fit a customized IPM to their data is by using BUGS software (Gilks et al., 1994; Lunn et al., 2000; Lunn et al., 2013; Plummer, 2003; de Valpine et al., 2017; Goudie et al., 2020). We call BUGS software any generic model-fitting software which adopts the extremely simple but powerful BUGS model-definition language. JAGS is one of several BUGS software variants (see Section 2.6) and the software of choice in our book. All BUGS software provides Bayesian inference for a statistical model except for NIMBLE (de Valpine et al., 2017), which also has functionality for maximum likelihood estimation. Although simple and fairly easy to learn, the BUGS language requires you to specify a statistical model in a much more fundamental way than you do when using canned functions such as `lm`, `glm`, or `glmer`, model-fitting functions in `unmarked` in R, or any model-fitting functionality in software such as MARK (White and Burnham, 1999), E-SURGE (Choquet et al., 2009), PRESENCE (Hines, 2006), or DISTANCE (Buckland et al., 2015).

Integrated Population Models. https://doi.org/10.1016/B978-0-12-820564-8.00014-2

Hence, to use JAGS with confidence when fitting an IPM to your data, you need a certain understanding of the things that go on under the hood in terms of both statistics and computation. This includes the following topics:

- Building parametric statistical models for simple random variables
- Using hierarchical models for random variables in sequence
- Analyzing statistical models by applying the Bayesian view of statistical inference, as well as how the major alternative method works (maximum likelihood)
- Discussing the basics of Bayesian computation
- Describing the BUGS model-definition language and its implementation in different pieces of software
- Defining generalized linear models (GLMs) and mixed models and how they are described in BUGS
- Integrating data within a statistical model fit in BUGS

This list describes what we set out to explain in this chapter. In the first part of the chapter, we provide a quick and nonmathematical overview of parametric statistical inference including the principles both of maximum likelihood and of Bayesian posterior inference. We then describe the Markov chain Monte Carlo (MCMC) family of simulation algorithms that has been instrumental in the rise of Bayesian statistics since the early 1990s. After that, we give a brief overview of widespread software that implements MCMC and related methods with which you can fit complex hierarchical statistical models such as IPMs. And finally, we focus on our current software of choice, JAGS (Plummer, 2003), which we use throughout this book.

In the second part of the chapter, we use JAGS to fit several examples of GLMs (Nelder and Wedderburn, 1972; McCullagh and Nelder, 1989). This part of the chapter is as much statistical as it is a practical introduction to using JAGS for fitting these models and to our consistent workflow when working with JAGS. In addition, we will work predominantly with simulated data in this chapter. We have repeatedly stressed how important and extremely rich the simulation of data sets can be for your understanding of anything quantitative (Section 1.5.3; Kéry, 2010, Chapter 4 in Kéry and Royle, 2016). We will not dwell on data simulation in any more detail here, but our use of simulated data throughout this chapter hopefully serves as a powerful reminder for the importance of data simulation in applied statistical modeling.

2.2 PARAMETRIC STATISTICAL MODELING

Throughout this book, we build and analyze parametric statistical models to understand the processes underlying our data sets and use them to predict events in the future or in places other than where we collected the data. In these models, we explicitly describe chance processes in terms of probability. By chance process, we mean any process that is not entirely deterministic, has unknown components, and is therefore not perfectly predictable. In this section, we first briefly review how chance processes are described in the mathematical field of probability. We then show how to build probabilistic, i.e., statistical, models to understand these processes once we have made some observations about them, i.e., collected data that we claim were produced by these chance processes. This section is very nonmathematical. If you want more mathematical details or find this material fascinating and simply want to learn more, you should definitely read up on it in a proper introduction to probability and statistics (e.g., Casella and Berger, 2002; Hobbs and Hooten, 2015).

2.2.1 DESCRIPTION OF CHANCE PROCESSES IN PROBABILITY

Variability has been called the law of life (Sir William Osler, cited in Wonnacott and Wonnacott, 1990). The fact that no two individual plants or animals are exactly identical creates a sort of randomness or unpredictability in nearly *anything* we can observe or even think about. Hence, chance is involved in virtually all things in life, including how long our commute home will take us today, the weather tomorrow, the fate of our grant proposal next year, or the possibility of health problems during the upcoming decade. Likewise, whether we see a European dipper on our bird-watching trip tomorrow is to some degree a chance event, and the number of detected territories of the bird species surveyed in our Swiss breeding bird quadrat next spring is similarly uncertain. Chance does not mean something we observe has no *cause* behind it—there is always a cause; we just do not know what it is or do not fully understand the mechanisms that produce a particular event or observation. In fact, most of the time there will be not just a single cause but rather a whole mesh of interacting causes of an observed event, and most of them are unknown to us.

But even if we do not understand all the mechanisms behind the events we are interested in, we can still describe their results. To better understand, predict, and make decisions in an uncertain world and to learn about that world in a formal way in fields such as ecology and natural resource management, it is useful to formally describe and understand uncertainty and thus to learn from our observations despite all the noise they contain. All of this is done in the field of probability, which has been called the extension of logic to all events, certain and uncertain (Lindley, 2006). Statistical inference is based on probability and can be understood as the formal method of learning from data in the face of uncertainty.

To formally describe uncertain events, probability uses the concept of a *random experiment*, which is a conceptual model for any process where some chance is involved and with a defined set of possible outcomes. (We note as an aside that "experiment" here has nothing to do with the same term in experimental design, where it usually denotes the application of some treatment and existence of controls and replication.) One example of a random experiment might be us submitting a manuscript to a scientific journal.

To formally describe a specific, measurable aspect of a random experiment, we use another concept in probability called a *random variable*, which for our example could be the editor's decision. The set of all possible outcomes of this random variable might be "reject without review," "major revision," "minor revision," and "immediate acceptance," and we could label them with the codes 1 to 4. Thus, to describe the random experiment "Paper submission," we could denote by D the discrete random variable "Editor's decision," which has four possible outcomes, $d = \{1, 2, 3, 4\}$. We note that it is customary to write the random variable in uppercase (D) and each possible realization in lowercase (d). In addition, and a little confusingly, we denote "random variable" as both the formal description of some aspect of a random experiment and any realization that comes from it. Hence, a data set with, say, a sample size of 100 will be treated as a collection of 100 random variables when forming a likelihood function for estimating parameters from it (see Section 2.8).

To describe the results of chance, i.e., the variability in D, we attach a probability to every possible outcome of the random variable. This set of probabilities will depend on many things including the journal, the quality of our research and writing, our name, the topicality of the subject, the charisma of the study species, our institution and home country, perhaps the time of year, and possibly many other factors. Let us assume that under one scenario, we can describe the variability in the random variable "Editor's decision" by the vector of probabilities $\{0.29, 0.4, 0.3, 0.01\}$. Thus, we cannot predict the realization of random variable D exactly, i.e., what happens when we submit our paper or what value d will be. However, we can say that the chance of outright rejection ($P(D = 1) = 0.29$) is much less than that of getting a second chance ($P(D = 2 \cup 3) = 0.7$), but is much greater than immediate acceptance ($P(D = 4) = 0.01$). This set of probabilities sums to 1 and is called a probability distribution, or more specifically in our example, a *probability mass function* (pmf), because it describes all possible outcomes of a discrete random variable. That is, the pmf yields $f(d) = P(D = d)$: the probability that a discrete random variable D takes on value d in a particular realization. The pmf is a fundamental description of a discrete random variable.

Discrete random variables have outcomes that may be mere unordered labels (such as the sex or hair color of an animal at birth), ordered labels (such as our "Editor's decision" random variable), or counts, which are nonnegative integers such as the number of territories detected in a breeding bird survey quadrat. Typically, some pmfs are particularly adequate for particular types of data collected in a certain way, e.g., a Bernoulli distribution for the sex of a newborn animal, a categorical distribution for the "Editor's decision" random variable, or a binomial, Poisson, or negative binomial distribution for the number of territories in the breeding bird survey. You can read much more about these and many more statistical distributions in countless statistics books, including the excellent text by Bolker (2008). All of these probability distributions are useful probability models because they help us understand the role of chance for widely different quantities of interest (i.e., random variables) and for very different settings. Virtually always, these probability models have constants that govern their specific form, and when they are unknown and estimated in a model (see below), they represent the parameters of that model.

The other major class includes continuous random variables. A continuous variable does not have a finite set of possible outcomes, and its values are measurements made on a continuous scale, such as body mass or a measure of length in space or time. To describe them, probability uses the concept of a *probability density function* (pdf), which gives the density for each possible value c of a random variable C. You can think of the density as the probability that a realization of C falls inside the interval $\{c - \Delta, c + \Delta\}$ as Δ approaches zero. A pdf has the peculiarity that the probability of any one outcome c is zero, because with infinite measurement accuracy, C has an infinite number of possible outcomes. However, you can still think of a pdf as giving the relative probability of a value somewhere in the close vicinity of c. To obtain an actual probability from a pdf, we must specify a range of events, say, m_{lower} and m_{upper} around the event c, and then integrate the pdf over this range.

Common continuous random variables are described by uniform, normal (Gaussian), or exponential densities. Exactly analogous to the pmf of a discrete random variable, pdf $f(c)$ is useful to describe how likely different values of c are for a continuous random variable C. And just as with the pmf of a discrete random variable, the pdf of a continuous random variable is usually governed by one or more constants that in an estimation setting become the parameters of a model. We also note that the terms "density" or "pdf" are often used generically for either the pdf of a continuous random variable or the pmf of a discrete random variable. This use of "pdf" can also be seen in all statistical distribution functions in R, which always begin with the letter "d" for "density."

2.2.2 PARAMETRIC STATISTICAL MODELS FOR INFERENCE ABOUT CHANCE PROCESSES

Statistical modeling or inference is based on probability, but it goes "in the other direction" (Link and Barker, 2010)—rather than describing how likely the outcomes from a certain random process are, it views some variable of interest (Y) as the outcome of a specified random process and then tries to infer features of that process from a sample of the values produced by it, i.e., from a data set at hand (Fig. 2.1). If this inference is based on a parametric statistical model, we then make the weighty claim that this random process is reasonably well approximated by a normal or any other named distribution, though we typically do not know its parameters, which we here denote by θ. Thus, a parametric statistical model is nothing more than a pmf or pdf for a random variable that we claim has produced the data set we have collected.

In addition, in most practical applications of statistical modeling, a system we are interested in cannot be described well with just a single random variable. Instead, we need two or more random variables that are linked such that the output from one affects the other(s) according to the laws of conditional probability. Statistical models for multiple random variables in sequence are called hierarchical models (Berliner, 1996; Royle and Dorazio, 2008; Cressie et al., 2009; Hobbs and Hooten, 2015; and see Royle et al., 2014, for a particularly lucid and accessible introduction for nonstatisticians). Nearly all models in this book belong to the class of hierarchical models, but in this chapter, we deal mostly with simple statistical models that describe only a single random process.

Estimation of the parameter(s) θ in a statistical model is always an essential part of statistical inference. But this is not the only analytical aspect of a statistical model. For instance, if we have an estimate of the value of θ and have observed some output Y but failed to record the associated input value X, we can use the model to estimate the values of such a missing covariate X. An example of this would be a Cormack-Jolly-Seber (CJS) model (see Section 4.5.1) in which we specify an effect on survival of the individual body mass as a covariate X. The value of X is unknown for any occasion when an individual is not caught. So in a CJS model with an effect of body mass X on survival Y, we might wish to estimate the missing values in X.

Similarly, if we have an estimate of the parameters θ in a system and have measured some covariates X but have failed to measure the associated response Y, we can use our knowledge of X and θ to obtain a guess about the value of Y we could expect. That is, we can estimate the missing response. This is exactly the typical setting in a species distribution model (SDM), where we measure the occurrence or abundance of an organism at a sample of sites and relate it to environmental covariates X at these sites. This yields an estimate of parameter θ that represents a prescription of the form "*if the habitat is such and such, then we expect this or that occurrence or abundance.*" Knowing the values of the (environmental) input variables for a similar sample of sites, we can use this "rule" to also predict the expected response Y for them and thus interpolate or extrapolate abundance or occurrence for a wider region and produce a species distribution map.

Estimation of the unknown quantities, along with assessment of the estimation uncertainty, is a major part of what we call statistical inference. Over the course of centuries, statisticians have developed formal methods for the estimation of parameters and other unknown quantities (estimands), including the method of moments, variants of least-squares and maximum likelihood, and Bayesian inference methods. In this chapter, we describe the Bayesian method of inference and contrast it first with the major non-Bayesian method of statistical inference, maximum likelihood (Edwards, 1992).

We will first illustrate maximum likelihood and Bayesian inference for parameter estimation in the simplest possible statistical model, which has exactly one unknown. Let us assume that we are frog modelers and have counted tadpoles (Fig. 2.2) in a number of natural ponds by walking along their shore at a slow pace and counting every tadpole we can see. Now we would like to estimate the true number (N) of tadpoles along the shores of each pond by correcting our counts y by dividing them by the detection probability θ. But unfortunately, θ is unknown. Hence, we decide to estimate θ in an experiment—we prepare a model pond in the form of a 2 m × 2 m kiddies' paddling pool filled with water, wherein we place some rocks, dead leaves, and a couple water plants to make the pool resemble the shores of the natural ponds in our study. We then release a known number of $N = 50$ tadpoles. We wait awhile so the tadpoles can disperse and start to behave naturally, and then we walk around our model pond and try to count as many tadpoles as possible, emulating our search protocol in the field as well as we can, e.g., spending the same average time per shore length in both. Let us assume that we arrive at a count of 20 tadpoles. The task now is to estimate the detection probability θ.

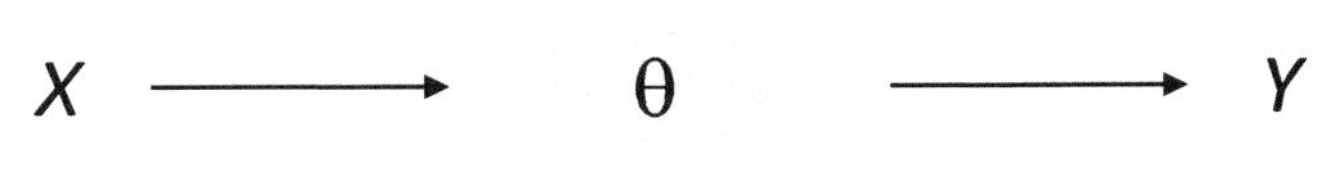

FIGURE 2.1

Sketch of a statistical model (gray box) with input X, parameter(s) θ, and output Y (inspired by Breiman, 2001); X can be thought of as covariate(s) and Y as the response. There is nothing Bayesian in sight about this model.

FIGURE 2.2

A tadpole or pollywog (from https://www.britannica.com/science/tadpole).

Of course, we all believe that we know from this information that the detection probability is 0.4, i.e., 20/50. Interestingly, this is just an intuition, albeit one based on the same implicit model as the one we use when we formally estimate the quantity with the method of maximum likelihood. To understand the principles of parameter estimation by maximum likelihood and using Bayesian posterior inference, we will now apply both methods to formally estimate the parameter θ from our data.

```
# Read the data for the tadpole experiment into R
N <- 50             # Number of tadpoles released initially
y <- 20             # Number of tadpoles detected later
```

2.3 MAXIMUM LIKELIHOOD ESTIMATION IN A NUTSHELL

Using maximum likelihood requires specification of a parametric statistical model, which is a formal probabilistic description of the random process that we claim is an adequate approximation of the random process that has produced our data. For a formal description, we may choose any of the fairly large number of probability distributions you can find in any statistics book or online resources. Without going into specific details, we simply note here that the nature of a data set and the manner in which it was collected typically suggest suitable candidates for such a probability model for our data. For the tadpoles, we have a bounded count, i.e., a nonnegative integer that cannot be higher than N unless we count the same tadpole multiple times, which we will assume does not occur. Under the assumption of no mortality and that all 50 tadpoles have identical detection probability and are detected independently, the classic probability model for data such as ours is that of a binomial distribution, the pmf of which is

$$p(y|N,\theta) = \frac{N!}{y!(N-y)!}\theta^y(1-\theta)^{N-y}.$$

This expression allows us to describe the probability of any possible outcome y in our random experiment (i.e., a count of tadpoles in a bucket where we know that $N = 50$ and where detection is imperfect) as a function of parameter θ, which is the detection probability. As a shorthand for this probability model, we will often write $y \sim Binomial(\theta, N)$, which you can read as "*y is distributed as a binomial random variable with a success probability* θ *and trial size N.*"

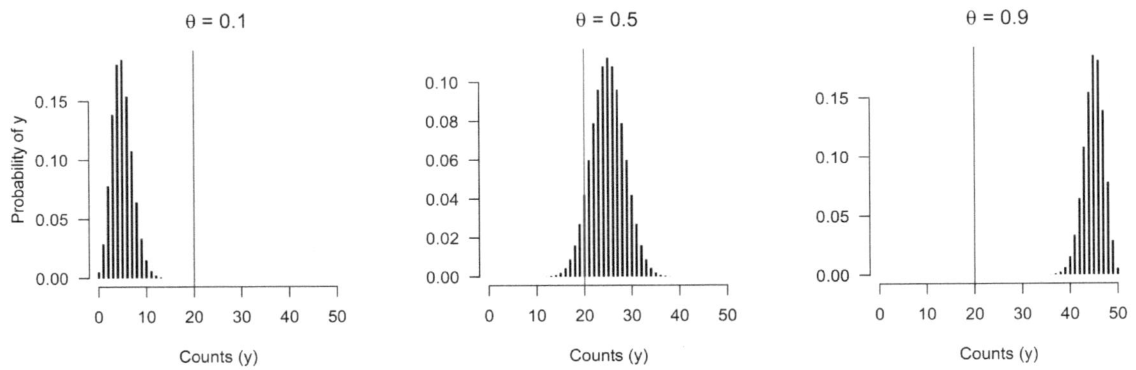

FIGURE 2.3

Binomial probability mass functions for $N = 50$ and for $\theta = 0.1$ (left), $\theta = 0.5$ (middle) and $\theta = 0.9$ (right). The observed count of 20 tadpoles is shown as a blue line in each.

In R, the function `dbinom` gives the binomial density (or more accurately, the pmf) for any possible binomial count; use `?dbinom` to read up on `dbinom` in R. The syntax is

```
dbinom(x, size, prob, log=FALSE)
```

Under the assumptions of our parametric statistical model for the chance process that we assume has produced our tadpole count in the experimental setting, we can use this function to find how likely any possible count (this is the argument called x) is for given choices of N (called "size") and θ (called "prob"). As an aside, we note that by setting the final argument `log = TRUE`, you will obtain the natural logarithm of the probability, which will become important when we use the density function for statistical estimation (see below). For instance, Fig. 2.3 shows the entire binomial probability distribution for $N = 50$ when detection probability (θ) is small (0.1), moderate (0.5), or high (0.9). We see that with $\theta = 0.1$ or $\theta = 0.9$, we would be very unlikely to observe 20 tadpoles out of 50, but this realization of a binomial random variable will not be so unlikely if we assume a detection probability of 0.5.

```
# Data for Fig. 2.3
all.possible <- 0:50                                         # All possible values
pmf1 <- dbinom(all.possible, size=50, prob=0.1)              # pmf 1
pmf2 <- dbinom(all.possible, size=50, prob=0.5)              # pmf 2
pmf3 <- dbinom(all.possible, size=50, prob=0.9)              # pmf 3
```

When we use a binomial pmf in this way, we use probability as a description of the variability of our "tadpole count" random variable. Probability can then be thought of as the relative frequency of each possible count if we were to repeat the binomial experiment a very large number of times, say 10 million times, under identical conditions. Thanks to your computer and R, it is actually very easy to perform this task using the random number generator (RNG) for a binomial density (`rbinom`).

```
# Use RNG to obtain binomial density for pmf3
hist(rbinom(10^6, size=50, prob=0.9), freq=FALSE, xlim=c(0, 50)) # Not shown
```

Statistical methods that use only this concept of probability are often called "frequentist" because of the exclusive meaning of probability as an expected frequency of the possible outcomes of a random variable in a large number of hypothetical repetitions of a study. In this use of a pmf (or likewise for the pdf of a continuous random variable), we express the probability of every possible outcome (y) of a binomial random variable as a function of the binomial sample size N and detection probability θ, i.e., the following:

$$p(y|N,\theta) = \frac{N!}{y!(N-y)!}\,\theta^y(1-\theta)^{N-y}.$$

And here comes perhaps *the* key point for your understanding of maximum likelihood—when we define a likelihood function to estimate tadpole detection probability using the method of maximum likelihood, we are using the exact same pmf! We just use it in a different manner, namely, we regard the pmf as a function of variable θ for a given N and data y! In a sense then, we use the same density but simply upside down or in reverse direction, so it is customary to call the function L and invert the order of arguments relative to that in the pmf above:

$$L(\theta|N,y) = \frac{N!}{y!(N-y)!}\,\theta^y(1-\theta)^{N-y}.$$

Hence, in the tadpole example, we use the binomial pmf as follows when we apply the maximum likelihood method to estimate θ by picking that value which results in the highest function value:

$$L(\theta|N=50, y=20) = \frac{50!}{20!(50-20)!}\theta^{20}(1-\theta)^{50-20}.$$

In summary, when we apply the method of maximum likelihood to estimate tadpole detection probability, we do the following:

1. We claim that we know exactly what chance process produced our data, i.e., we assume that y is a binomial random variable with known size N.
2. We use the binomial pmf as a likelihood function L and vary θ until we find the value of θ that leads to the highest function value when L is evaluated for the actual data (here, 20 and 50).
3. We call the value of θ that maximizes the likelihood function $L(\theta|N, y)$ for the observed data the maximum likelihood estimate (MLE). It makes the observed data most likely, i.e., produces them with the highest probability. That's all!

How should we find the MLE in practice? The simplest method is by brute force, where we literally try out a large number of possible values for θ and then take as our MLE the one that leads to the highest likelihood function value. This is what we show later in this section. In practice, however, we usually define the likelihood function algebraically and then use an optimization algorithm such as `optim` or `nlm` in R to obtain the MLEs; see Chapter 2 in each of the works of Royle and Dorazio (2008), Royle et al. (2014), and Kéry and Royle (2016) for more information, as well as Section 2.8 for several examples.

Here we want to show only the basic principle of maximum likelihood estimation, and that can be neatly illustrated by the brute-force approach. In addition, we note that there are standard methods to produce uncertainty estimates (e.g., standard errors, confidence intervals) around the MLEs. We refer you again to Chapter 2 in Royle and Dorazio (2008) or a multitude of statistics books and online resources on maximum likelihood, all of which explain this in detail. See also Sections 2.7.3 and 2.8 for a quick illustration of the derivation of the variance-covariance matrix for the MLEs from the results produced with `optim`.

In practice, statisticians and statistical software usually work with the *negative* of the *log-transform* of the likelihood function rather than with the likelihood function itself. You may wonder why. The reason for using the log-transform of the likelihood is numerical, because likelihoods can be extremely small. Computers cannot deal with values smaller than 10^{-308} that may occur for a large data set and treat them as zero values. The problem does not arise for the logarithm of the likelihood ($\log(10^{-308}) \approx -709$). The reason for working with the negative log-likelihood is that in computer programs, optimization means function minimization rather than maximization. However, working with either approach will produce the same value for the MLE, as we will now see.

For the tadpole example, we will now conduct a brute-force search for the MLE by maximizing both the likelihood and the log-likelihood and minimizing the negative log-likelihood (NLL). For all three, we first plot a profile of their values over a large range of possible values for θ and then pick as the MLE that value which leads to the highest function value of the likelihood or log-likelihood or the lowest function value for the NLL. We see that they all produce an estimate for θ of 0.4 (Fig. 2.4).

```
# Brute-force search for MLE of theta for the tadpole data set (Fig. 2.4)
try.theta <- seq(0, 1, by=0.01)                    # Values to try out
like <- dbinom(20, 50, try.theta, log=FALSE)       # Likelihood
loglike <- dbinom(20, 50, try.theta, log=TRUE)     # Log-Likelihood
negloglike <- -dbinom(20, 50, try.theta, log=TRUE) # Negative log-likelihood (NLL)

par(mfrow=c(1, 3), mar=c(5, 5, 4, 1), cex.lab=1.5, cex.axis=1.5)
plot(x=try.theta, y=like, xlab=expression(paste("Detection probability (", theta, ")")),
  ylab="Likelihood", frame=FALSE, type="p", pch=16, col="black")
abline(v=try.theta[which(like == max(like))], col="red", lwd=3)
plot(x=try.theta, y=loglike, xlab=expression(paste("Detection probability (", theta, ")")),
  ylab="Log-Likelihood", frame=FALSE, type="p", pch=16, col="black")
abline(v=try.theta[which(loglike == max(loglike))], col="red", lwd=3)
plot(x=try.theta, y=negloglike, xlab=expression(paste("Detection probability (", theta, ")")),
  ylab="Negative log-Likelihood", frame=FALSE, type="p", pch=16, col="black")
abline(v=try.theta[which(negloglike == min(negloglike))], col="red", lwd=3)
```

This is maximum likelihood estimation in a nutshell, illustrated with a trivially simple parametric model—you pick a suitable probability density for your random variable and then "turn it around" and use the identical probability function as a likelihood function by regarding it as a function of the parameters for a given data set. That is, you use this likelihood

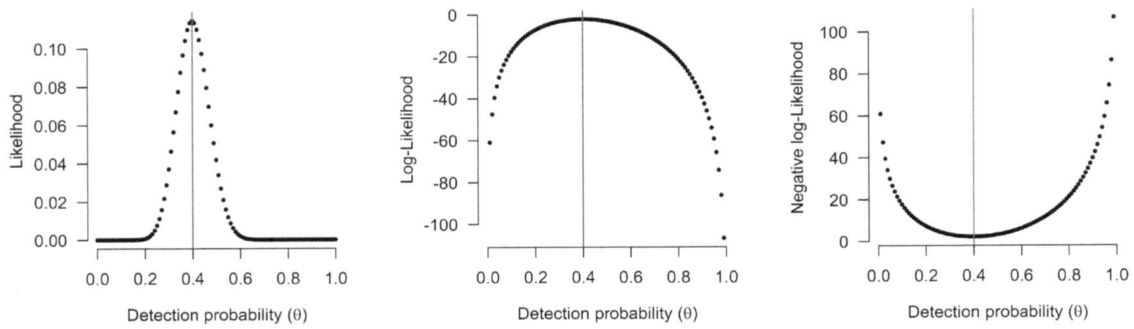

FIGURE 2.4

Profiles of the binomial likelihood function (left), the log-likelihood function (middle), and the negative log-likelihood function (right) for the tadpole example, where 20 individuals were counted of a known total population of 50, and the target of inference is the detection probability (θ) of a single tadpole. The red lines show the maximum likelihood estimate at $\widehat{\theta} = 0.4$.

function with the observed data to find the value for the unknown parameter that leads to the highest possible function value. This is the MLE. The same principles apply regardless of whether you have 1 or 100 parameters in a model.

In addition, and of importance for your understanding of the concept, note that in the tadpole example, we had only one datum (the number 20 along with the known $N = 50$), but you will very rarely fit a statistical model to a sample size of 1. So to estimate maximum likelihood for a sample of size n, you formulate the joint likelihood for the entire data set. Assuming independence, this is the product over n random variables, or alternatively, you work with the sum of the log-likelihood over n data points. See Sections 2.7.3 and 2.8 for illustrations within the contexts of several regression models.

Despite the compelling conceptual simplicity of maximum likelihood estimation, there can be several challenges when implementing it in practice. The likelihood becomes increasingly complex as the number of parameters increases, and for hierarchical models (which, broadly, you can imagine as being those with random effects), one must carry out integration to eliminate the random effects to define a marginal, or integrated, likelihood (see Chapters 2 in Royle and Dorazio, 2008; Royle et al., 2014; and Kéry and Royle, 2016). The numeric search algorithms with which MLEs are sought in practice may fail to converge, or the MLEs may come to lie at the boundary of the space within which a parameter is defined, such as 0 and 1 for a survival or detection probability. Nevertheless, maximum likelihood estimation is extremely general and very powerful for many statistical estimation problems, so much so that for most of the 20th century, it was the gold standard in statistical estimation.

2.4 BAYESIAN INFERENCE

Conceptually, the start of any Bayesian analysis is exactly the same as that of a frequentist analysis when using maximum likelihood. We view our data set as the outcome of a process influenced at least in part by chance, and more specifically as the realization of one or more random variables for which we pick one suitable pdf or more. In each pdf, we have input and output variables, and the pdf usually contains parameters that are unknown, so they must be estimated. Finally, we sometimes want to estimate missing values in the input or output, as discussed previously and depicted in Fig. 2.1.

Hence, in a Bayesian statistical analysis we start with a parametric statistical model exactly as for inference by maximum likelihood—there is nothing Bayesian or non-Bayesian about this statistical model at first. But from then on, we continue differently. To a Bayesian, it appears odd to base estimation of an unknown quantity on a probability expression about the data, which assumes that a parameter is known, i.e., in the tadpole example, on the direct use of the pmf $p(y|\theta)$. Rather, to a Bayesian, it seems more natural to base estimation on a probability expression for what is not known given what is known. That is, a Bayesian thinks it is more sensible to go after $p(\theta|y)$, which is the conditional probability of a parameter given the information in the data. Lindley (1983) sums up the recipe for every Bayesian analysis in the following three steps, and we annotate them in the context of the tadpole example:

1. What is known?	The data: observed $y = 20$ out of $N = 50$
2. What is unknown?	Probability of detection (θ)
3. What should be done?	Calculate $p(\theta|y)$

Thus, even though the data are considered the outcome of a chance process, once they are collected, they are perfectly known. In this sense, the data are fixed, and there is no longer any uncertainty about them. So to a Bayesian, it makes sense to base estimation of θ on a probability expression for something unknown based on what is known, and this is $p(\theta|y)$.

Note that above we have defined probability as a relative frequency of an event for a discrete random variable (and in a similar manner for a continuous random variable as well) as the sample size goes to infinity. But this is not the same use of probability that you now see in $p(\theta|y)$. Rather, here we encounter the unique hallmark of Bayesian inference, which is a more general notion of probability as a measure of uncertainty. Bayesians do use probability as a measure of the variability of random variables, as do frequentists. But in addition, they use probability as a measure of their level of knowledge about uncertain things such as parameter values, missing covariates, unobserved outcomes from random processes, or the truth of some hypothesis.

Thus, although the value of a parameter or any other unknown quantity will usually be a constant, the parameter is treated *as if* it were a random variable because of the dual use in Bayesian inference of probability as a measure of variation and for the quantification of imperfect knowledge. You can often read that a Bayesian views parameters as random variables. We have always found this confusing, as it may give the impression that for a Bayesian, a parameter may in some sense "bounce up and down." But in reality, it simply means that Bayesians view probability not only as a quantification of variability but also as a measure of uncertainty. This leads to treating a parameter *as if* it were a random variable. Link and Barker (2010, p. 14) give the neat example of the one-millionth digit of the number π, which is clearly fixed, but in a Bayesian attempt at guessing it would also be treated as a random variable.

So how should the probability of a parameter given the data, $p(\theta|y)$, be calculated? In principle this is easy, because from basic conditional probability, we know that the following relations hold for two random variables A and B:

$$p(A|B) = \frac{p(B|A)p(A)}{p(B)} = \frac{p(A,B)}{p(B)}.$$

This is called Bayes' rule and can be read as follows: "the conditional probability of A given B is equal to the conditional probability of B given A times the marginal probability of A, divided by the marginal probability of B." The numerator is equivalent to the joint probability of A and B, but in the following, we will use only the first form of this relationship, i.e., with $p(B|A)p(A)$ in the numerator.

Bayes' rule is the basis for Bayesian inference, but nothing about it is intrinsically Bayesian. It is an undisputed mathematical fact about the relationships among conditional, marginal, and joint probabilities (Royle et al., 2014), and it is frequently used for non-Bayesian probability calculations of observable quantities, such as in clinical testing. If you go to a hospital and are tested for some disease, this relationship may be used to calculate how likely you are to *really* have the disease if you receive a positive test result. The same relationship can also be used to estimate the probability that a site is occupied by some study species even though the species has never been detected during one or several surveys (Wintle et al., 2012).

However, in Bayesian inference, we apply Bayes' rule to unobservable quantities such as parameters. When we substitute A and B with our notations for parameter and data, respectively, we obtain the format of Bayes' rule as used for parameter estimation in Bayesian inference. It is named after Thomas Bayes (1702–61), an English clergyman and mathematician, who first formulated this relationship for statistical inference.

$$p(\theta|y) = \frac{p(y|\theta)p(\theta)}{p(y)} = \frac{p(\theta, y)}{p(y)}$$

The four quantities in Bayes' rule are as follows:

- $p(\theta|y)$ is the target of Bayesian inference. It is called the posterior distribution and is the probability of the parameter given the information in the data.
- The second quantity is the probability of the data given the parameter $p(y|\theta)$, which is the likelihood (but in Bayes' rule is always written as $p(y|\theta)$ rather than as $L(\theta|y)$.
- The third quantity is the probability of the parameter $p(\theta)$, which is called the prior distribution, or "prior" for short.
- Finally, the last quantity is the probability of the data $p(y)$. This is a constant ensuring that Bayes' rule integrates to 1 as required for a probability function. The denominator $p(y)$ is the integral of the numerator over the parameter(s) θ, $\int_\theta p(y|\theta)p(\theta)d\theta$. That is, we integrate out the parameter(s) by averaging over all possible values of θ.

Bayes' rule can be expressed alternatively as the ratio of the joint probability of all random variables (i.e., data and parameters), $p(\theta,y)$, and the probability of the data (this last element can also be written as $p(y) = \int_\theta p(\theta, y)d\theta$).

This is the foundation of the Bayesian method of making formal probabilistic statements about uncertain quantities based on the information in some data set that is analyzed. Whenever you carry out a statistical analysis using Bayesian software such as JAGS, this is what you are doing implicitly. We notice again that we are using probability to express knowledge about an uncertain quantity at two places in Bayes' rule. First in the prior $p(\theta)$, where we make an unconditional statement about what we believe are likely values of the parameter or what we want to assume about them (see below). And second, in the posterior $p(\theta|y)$, where we make a statement about the likely values of the parameter based on the information contained in the data. Both the prior and the posterior are probability distributions. This means that they integrate to 1, so we can interpret areas under their curves as a probability.

To recap, these are the formal steps in a Bayesian analysis:

1. We use probability in a broader way than we do in a frequentist analysis, i.e., we use probability as a measure of both variability *and* imperfect knowledge about uncertain quantities.
2. We treat all statistical inference (e.g., parameter estimation, testing, and estimation of missing values) as a mere probability calculation by application of Bayes' rule.
3. We do this by expressing our knowledge about a parameter before having seen the data in one probability distribution. This is the prior $p(\theta)$.
4. We then use Bayes' rule to update that knowledge with the information in the data as embodied in the likelihood function $p(y|\theta)$. The result is another probability distribution, the posterior distribution $p(\theta|y)$.

The terms "prior" and "posterior" imply a chronological sequence, but in practice, this is not really so relevant. It may be more useful to read them as "ignoring the information in our current data set" and "including the information in our current data set," respectively. This is the most basic distinction between them, and it is in fact very important to clearly distinguish prior and data. Hence, the information in the data must never be used to formulate an informative prior (see below) because this would mean using the data twice, which is forbidden.

We believe that Bayes' rule has much heuristic appeal as a model for inference or generally as a model for learning from data in an empirical study. First, it employs what could be called a "human" concept of probability. We humans love to use probability as a measure of uncertainty, e.g., when we say things such as "I'm almost 100% certain that it's not going to rain today" or "I am 99% certain that my sports team is going to win next week." When we carry out a Bayesian analysis, we are using probability in exactly the same way as a measure of uncertainty, and we make calculations about it. Next, the formalism of Bayes' rule can be described as "the posterior is proportional to the likelihood times the prior." This means that what we learn or conclude after seeing some new data, i.e., the posterior distribution, is always the result of both the data itself and prior knowledge of the unknowns. This exactly matches the way in which we humans deal with new information—we *always* consider new information in the context of what we already know. Context-free learning does not exist.

Take bird-watching, for instance, where identification exactly follows Bayesian logic. If your friend goes to the island of Mauritius and comes back to tell you he saw a live dodo bird (this is the new information you get), you will probably think he is joking, a liar, or simply drunk. Because you *know* (this is the prior, and an extremely strong prior at that!) that dodos were exterminated by humans a couple centuries ago. So the contextual information trumps the new information at hand. Another example: my friend goes bird-watching in the Jura mountains and sees a large soaring bird of prey that looks like a condor (this is the new information). However, at the same time, we know that condors occur naturally only in the Americas (this is the prior). My prior knowledge about the world distribution of condors and therefore of the a priori likelihood of a condor appearing in the Jura will dominate my conclusion based on what my friend claimed she saw. Hence, I may assume she misidentified the species. If, on the other hand, I know that she is a world-renowned expert in field identification of raptors, this might represent a quite different prior and would probably let me conclude that she has indeed seen a condor in the Jura ...albeit presumably an escapee from captivity.

What is further remarkable about Bayes' rule is that every prior can eventually be overwhelmed by empirical evidence unless we use a 0/1 prior stating that something is either totally certain or impossible. The fact that empirical evidence can trump prior beliefs is a critical feature of the scientific method, where any theory or prior belief ought to be falsifiable in principle. That is a fundamental distinction from nonscientific theories about the world, such as religion, where no amount of evidence can ever overthrow The One and Only Prior Belief. In science, we must avoid 0/1 priors, because they would be unscientific and moreover represent the end of learning. Beliefs represented by a prior that reflects impossibility or certainty can no longer be proven wrong by any evidence. This would make application of the scientific method impossible, which is perhaps *the* defining feature of science.

Interestingly, using prior information in estimation is not unique to Bayesian inference. For instance, penalizing the likelihood is one way in which we can incorporate prior knowledge in a frequentist analysis (e.g., Moreno and Lele, 2010). But the prior is particularly striking in the Bayesian method for learning from data about uncertain things—it is similar to a plug that says, "If you want to add external information into the estimation procedure, then put it in here." We will see that we do not necessarily need to input external information, but if we want to do so, Bayes' rule provides a clear and formal way of doing so.

In this sense, the prior distribution is an advantage of Bayesian inference—we can add external information if we want but can largely avoid doing so if we do not. Often, much can be gained from exploiting information about parameters that comes from outside the data set at hand. First, external information can be seen as equivalent to additional data (see the use of informative beta priors that follows below for a particularly clear illustration). By using such additional data, we base our inference on a larger data set and obtain more precise estimates, e.g., smaller standard errors. Second, we can sometimes estimate additional parameters that may not be estimable from the data alone without some external information. And third, we believe that in principle it just makes tremendous sense not to feign being stupid. An example: imagine we are carrying out a survival analysis of some large bird such as an albatross. Based on life history, we know that adult survival for such a large species must be pretty high, quite certainly considerably higher than 0.80 (McCarthy and Masters, 2005). So why should we not use this undisputed knowledge when we estimate survival from some particular data set? In contrast, MLEs do not incorporate any external information, and in this sense it could be said that when we do maximum likelihood estimation, we feign being stupid.

On the other hand, perhaps priors also have disadvantages. First, the results in a Bayesian analysis always depend on the priors; there is no escaping that. It is true that often the choice of prior has only a very small effect numerically, but the effect is still there. When the information in the data is weak and/or a parameter is hard to estimate, taking one prior over another can make a major difference in the resulting estimates. This is not wrong, but it does create ambiguity, because what are you now going to report in the Results section of your paper? So if you have a method such as maximum likelihood that does not require you to specify priors, this may sometimes be an advantage because you will have a less ambiguous story to tell in your paper.

Second, some might object to being required to *choose* priors, and this choice is to some degree always subjective. They may feel that since we have learned that science must be objective, Bayesian statistics is therefore unscientific. However, this would be complete bogus. True, you must choose your priors, but this is just one of a truly vast number of choices you must make in any scientific investigation, starting with what species to study for which ecological question, where and when to study it, what treatment levels to choose in an experiment, whether to discard some data points as "outliers," whether to log-transform your response or your covariates, which covariates to add into the analysis and in what form (e.g., up to which polynomial), what type of statistical model to adopt for inference, and on and on. It is not the choice that is unscientific, but to make an *undocumented* choice, or much, much worse still, making a choice specifically to help us get the result we want. Objectivity means that in our choices, we must be indifferent with respect to the possible outcome of our study. In addition, science has the tenet that a study must be repeatable in principle, and hence, all choices ought to be documented. This is exactly what we do with Bayesian priors—they are an intrinsic part of our analysis and therefore must be specified in the methods section or else perhaps in a table or in the computer code given in the appendix of a paper.

At any rate, for better or worse, most people who carry out Bayesian analyses behave as "cowards" and do not exploit the potential benefits of including external information and formalizing this in an informative prior. Instead, they specify so-called vague (or diffuse or weakly informative) priors in which they try to minimize the effect of the prior on the posterior. In this way, they are enjoying all the other benefits of a Bayesian analysis in practice, e.g., better ability to fit complex models (see concluding section of this chapter). But in a sense, they are closet frequentists because they obtain the same old MLEs, more or less, as people who use the method of maximum likelihood. We will see a couple times later in this chapter that in practice, Bayesian estimates are often numerically indistinguishable from MLEs when vague priors are specified. This holds for both point and interval estimates.

There has been some discussion about the lack of so-called transformation-invariance for priors in a Bayesian analysis. This means that a prior that is vague on one scale may be quite informative on another transformed scale. For instance, in binomial models, we typically model parameters on the logit scale and may then give a uniform prior for the intercept in a regression model. But at the probability scale, this same prior says that extreme values close to 0 and 1 are much more likely than intermediate values for the intercept (see e.g., Northrup and Gerber, 2018). In many cases, this will not have much effect on the posterior, but to be cautious, we like the advice of Royle and Dorazio (2008) to specify vague priors on the scale at which we like to interpret the estimates. For the example above, this would be the probability rather than the logit scale.

So let us now proceed to a Bayesian analysis of the tadpole data. Actually, we will conduct four analyses to explore the impact of the prior on the posterior. Analysis 1 will use a vague prior for the estimated parameter. For vagueness, we will use a uniform prior that covers the whole range where a probability lives, i.e., from 0 to 1. Analyses 2 to 4 will employ several kinds of informative priors, where we assume that we specified informative priors either based on expert advice from our good old friend, the frog modeler Beni, or based perhaps on the results from another similar data set, e.g., from another year, a different site, or even from a different species, all of which are deemed comparable to our system used in collecting our current data. Specifically, in Analysis 2, Beni might have said the following: "*I think that detection probability θ must be about 0.4, and I am quite certain that we can rule out values more extreme than about 0.1 and 0.8.*" In Analysis 3, we assume that Beni said the same, except that now he was confident that values outside the approximate interval (0.3, 0.5) were essentially impossible. For Analysis 4, we assume that Beni said he was quite certain that detection probability was about 0.6 and that values outside the interval (0.5, 0.7) could be more or less excluded. We note that the prior in Analysis 4 has exactly the same shape as that in Analysis 3 but mirrored at the value of 0.5. We summarize the vague and three informative priors in the first four columns of Table 2.1.

Table 2.1 Summary of four priors for the detection probability parameter in the tadpole analysis. Analyses 2 – 4 use expert knowledge about the likely value and the approximate range of the likely values. This information about the mean and approximate range then needed to be translated into the parameters of a beta distribution (which is in the last column). Note that a Beta(1, 1) is identical to a Uniform(0, 1).

Analysis	Prior type	Mean of Prior	Approximate Range of Prior	Prior Distribution
1	Vague	0.5	0–1[a]	Beta(1, 1)
2	Informative	0.4	0.1–0.8	Beta(4, 6)
3	Informative	0.4	0.3–0.5	Beta(40, 60)
4	Informative	0.6	0.5–0.7	Beta(60, 40)

[a] *This range is exact of course.*

For Analyses 2–4, we must translate these statements into a named probability distribution with specific values of the parameters, because otherwise we cannot use Beni's wisdom in a prior in our Bayesian analysis. Both the elicitation of prior information from experts and the translation of that expert knowledge into a probability distribution are important topics (O'Hagan et al., 2006), but we do not go into them here (for some ecological examples, see Martin et al., 2005; Low Choy et al., 2009; Kuhnert et al., 2010). We just note the very basics here—to serve as an informative prior for a probability parameter, we need a probability distribution for a continuous random variable on the interval 0 to 1 that may be peaked around some value to give that value more weight in the analysis. The beta distribution is such a distribution. It has two shape parameters, a and b, that have the neat interpretation of the number of successes and failures in a hypothetical additional sample from a binomial random variable and both determine the mean and the variance of a beta-distributed random variable. The mean of the beta distribution is equal to $a/(a+b)$, and the variance is $(a \times b)/((a+b)^2(a+b+1))$.

To obtain the posterior distributions for detection probability in the four analyses, we now use a trick that was fairly widespread in Bayesian analyses before the rediscovery of MCMC algorithms around about 1990 (see next section) using a so-called conjugate prior. This is a prior with the same "form" as the distribution used as a likelihood function in the Bayesian analysis. In this case, the posterior has a known distribution, and hence, analytical Bayesian solutions become possible. The combination of a binomial likelihood function with y successes and $N-y$ failures and a beta prior distribution with parameters a and b results in a beta posterior distribution with parameters $y+a$ and $N-y+b$ (Lunn et al., 2013). Hence, for Analysis 1, the posterior is a beta distribution with parameters $(20+1)$ and $(30+1)$, where you recognize the first numbers in the two parentheses as the observed data and the second numbers as the two parameters of the Beta(1, 1) prior. In the following, we compute the four posteriors analytically, and in Fig. 2.5, we plot the likelihood function, the prior distribution, and the resulting posterior distribution for all four Bayesian analyses of the tadpole example. To make the shape and location of all three more easily comparable, we scale the likelihood function such that it also integrates to 1.0. The likelihood function in all four panels is of course the same as that in Fig. 2.4.

```
theta.vals <- seq(0, 1, 0.001)

# Define likelihood function (same for all four analyses)
like <- dbinom(20, 50, theta.vals)

# Define four prior distributions
prior0 <- dbeta(theta.vals, 1, 1)
prior1 <- dbeta(theta.vals, 4, 6)
prior2 <- dbeta(theta.vals, 40, 60)
prior3 <- dbeta(theta.vals, 60, 40)

# Derive four posterior distributions
post0 <- dbeta(theta.vals, 20 + 1, 30 + 1)
post1 <- dbeta(theta.vals, 20 + 4, 30 + 6)
post2 <- dbeta(theta.vals, 20 + 40, 30 + 60)
post3 <- dbeta(theta.vals, 20 + 60, 30 + 40)
```

When using a vague prior (top left in Fig. 2.5), we see that the posterior distribution is entirely determined by the likelihood function, and this is why typically we see such a good numerical agreement between MLEs and Bayesian estimates. Indeed, the scaled posterior is so close to the scaled likelihood that it covers the latter completely, and we have to

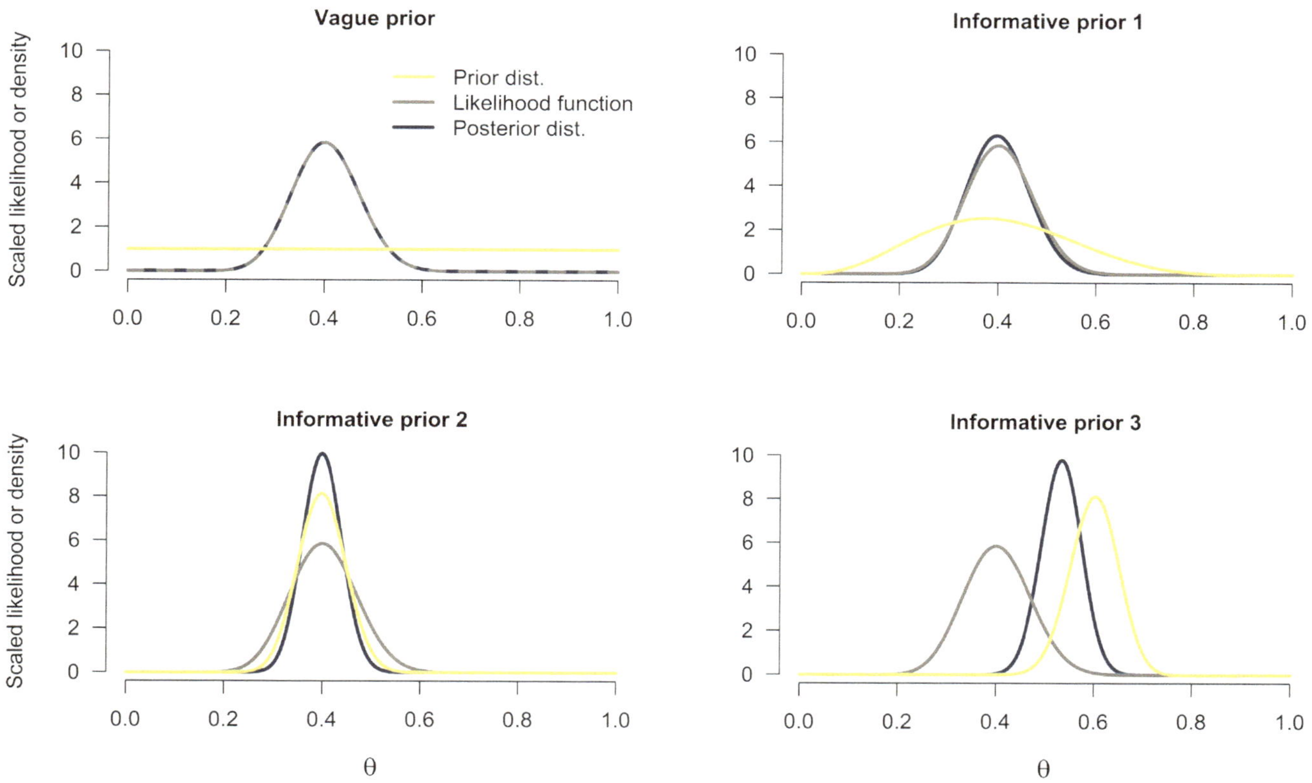

FIGURE 2.5

Four Bayesian analyses of the tadpole example where we counted 20 of 50 released individuals. The likelihood function is scaled such that the area under the curve is equal to 1 for greater ease of comparison with the prior and posterior probability densities.

show it as a dashed line to even recognize it. Using informative prior 1 is interesting, since a priori we might perhaps have expected that the information in that prior was stronger. After all, it represents quite a strong preference for some values of detection over others and essentially excludes some values almost completely. Nevertheless, the posterior is still nearly identical to the likelihood, which shows that even in a fairly small data set, the data may overwhelm a prior, i.e., have much more weight in determining location and shape of the posterior distribution. In contrast, using informative prior 2 makes a more substantial difference and shows clearly how an informative prior can act like an increased sample size by making our estimate more precise, which means the resulting posterior is more concentrated. Using this more strongly informative prior has reduced our uncertainty to a greater degree than is the case using the less strongly informative prior 1. Finally, informative prior 3 makes it particularly clear how in a Bayesian analysis, the "result," i.e., the posterior, is always a compromise between what the data say and what the prior says the parameter estimate should be. Here, the posterior lies between the likelihood and the prior, and because this prior is strongly informative, the posterior has "moved" more than halfway toward the prior.

Which of these analyses is the correct one? They are all correct, and whether you prefer one over the other depends on your objectives and on whether you believe one or the other prior is more relevant for these objectives. Nevertheless, we reiterate that by far the most typical setting of a Bayesian analysis in science and management is that in the top left of Fig. 2.5, i.e., where vague priors are adopted, and hence Bayesian estimates closely agree numerically with MLEs.

We conclude this section with a remark about how probability is used as a formal measure of uncertainty when we use Bayesian inference. In the analysis with vague priors, we started from total uncertainty about the magnitude of detection probability θ, which was represented by the flat prior distribution. This state of knowledge was then updated with the information in the data expressed by the likelihood function, resulting in a much more concentrated posterior distribution. Hence, a more concentrated distribution reflects an increased state of knowledge. The shrinking of the spread of a probability distribution that represents our knowledge about an uncertain quantity therefore reflects learning in Bayesian statistics (Fig. 2.6). Thus, there are two extremes. On the one hand, we have a uniform distribution that reflects total ignorance about the magnitude of a parameter or the desire to let the data alone speak. On the other hand, we have a spike with point mass at some value. This reflects the opposite, namely complete certainty about the value of a parameter. This is equivalent to the fixing of the value of a parameter in an analysis; there is then no longer anything to estimate about it. Any

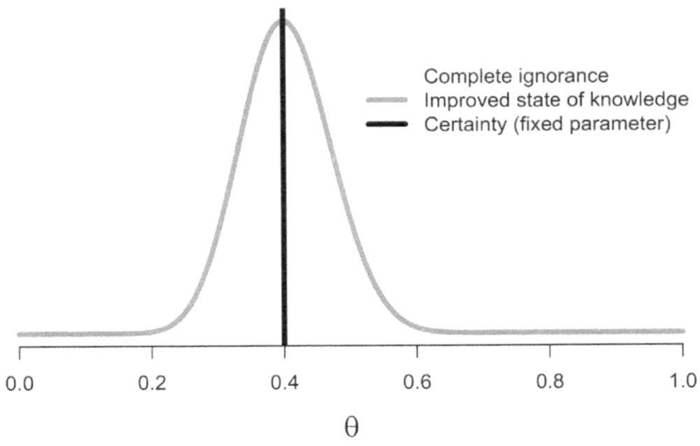

FIGURE 2.6

Summary figure that shows how probability is used as a measure of uncertainty in Bayesian inference, with three possible states of knowledge illustrated: complete ignorance (light gray), complete certainty (black), and one of an infinite number of possible intermediate states of knowledge about the magnitude of parameter θ (gray). Bayesian learning represents reduction in uncertainty by moving from the light gray to dark gray state, or generally from a state of knowledge represented by a less concentrated distribution to one represented by one that is more concentrated.

degree of the state of knowledge between ignorance and complete certainty can be expressed by a more or less concentrated probability distribution, and this is what the prior and posterior distributions in a Bayesian analysis do.

2.5 BAYESIAN COMPUTATION

So if Bayesian inference is so great, why hasn't everyone been a Bayesian since its invention in the 18th century (Efron, 1986)? Perhaps the main reason is that Bayes' rule was impossible to apply in practice for most interesting estimation problems because the required integrals could not be solved. Recall the denominator of the rule, $p(y) = \int_\theta p(y|\theta)p(\theta)d\theta$, which is simply the numerator but where we integrate over all parameters (θ) in the model. Thus, for a model with n parameters, an n-dimensional integral must be solved. This cannot usually be done analytically.

This did not change until the late 1980s and early 1990s, when statisticians discovered work that had been done by physicists some 40 (!) years earlier. This work showed that it is possible to construct simulation algorithms that produce dependent random numbers from an unknown distribution. Importantly, in these algorithms, the denominator in Bayes' rule, which contains the integral over θ, does not need to be evaluated! One may then summarize the random numbers produced by the algorithm to approximate features of the unknown distributions, such as the mean, variance, or percentiles (Metropolis and Ulam, 1949; Metropolis et al., 1953). An arbitrary degree of accuracy can be achieved by simply drawing a large enough sample. These algorithms are known as *Markov chain Monte Carlo* algorithms, and their name says what they do: they simulate (which is what statisticians call "Monte Carlo") a Markov chain, i.e., a series of autocorrelated random numbers. The rediscovery of MCMC algorithms around 1990 (Gelfand and Smith, 1990; Gelman and Rubin, 1992; Tierney, 1994) was responsible for the ensuing explosion of Bayesian inference in the statistics community.

There are numerous flavors of MCMC algorithms, but perhaps the most common classes of algorithms are Metropolis-Hastings (MH) algorithms (Metropolis and Ulam, 1949; Metropolis et al., 1953; Hastings, 1970) and Gibbs sampling (Geman and Geman, 1984); the latter is a more efficient special case of the former. The two can be combined in a single analysis; for some parameters, MH is used, while for others, Gibbs sampling is employed.

Here is a general outline of a simple MH algorithm, specifically, a random-walk MH (RW-MH) algorithm. Remember that our goal is to produce a chain of random numbers from the posterior distribution of a parameter θ, i.e., from $p(\theta|y) = p(y|\theta)p(\theta)\,/p(y)$. The way we construct the algorithm induces a one-step dependency in the series of random numbers produced by the algorithm, i.e., the output is a Markov chain of order 1.

To obtain a sample from the posterior distribution of θ, we repeat the following steps a large number T of times:

- We start with an arbitrary initial value for parameter θ, θ^0. This initial value must be possible given the prior adopted for that parameter (e.g., we cannot supply an initial value of 2 for a probability).
- We randomly perturb θ^0 and propose a new candidate value for the parameter that may potentially be added to our posterior sample. We call this candidate θ^*. Importantly, we need to use a stochastic rule for proposing θ^*, and in the random-walk MH algorithm, this is based on the current value θ^t (which at the start of the chain, of course, is θ^0). In

practice, and somewhat confusingly perhaps, a normal distribution is often used for that stochastic rule, or a candidate-generating density, i.e., we draw θ^* from a normal distribution with mean equal to θ^t and a standard deviation that we call $\sigma_{proposal}$ such that our proposal is $\theta^* \sim Normal(\theta^t, \sigma_{proposal})$. Note that this normal distribution is *not* part of the model, nor is it the prior distribution! We do not estimate anything from it but simply use a normal RNG as a useful device to serve as our candidate-value generator for serially autocorrelated random numbers. The standard deviation $\sigma_{proposal}$ has the role of a tuning parameter in the algorithm. As we will see in the following, among other things, a higher value of $\sigma_{proposal}$ will tend to induce a smaller autocorrelation than does a smaller value of $\sigma_{proposal}$ and will let the algorithm explore the parameter space more rapidly.

- At this stage, we have now two values for θ: the current value in the chain (θ^t) and the proposed new value under consideration for inclusion in our posterior sample (θ^*). In the case of our simple RW-MH algorithm with symmetric proposal density, we can compare them by computing the ratio of their posterior densities, R (while in the general case of an MH algorithm other than one with a random-walk proposal, both the numerator and the denominator of R will also contain a contribution from the proposal density):

$$R = \frac{p(\theta^*|y)}{p(\theta^t|y)} = \frac{p(y|\theta^*)p(\theta^*)/p(y)}{p(y|\theta^t)p(\theta^t)/p(y)}.$$

The difficult term in the evaluation of Bayes' rule in R is the denominator in each posterior, i.e., $p(y) = \int_\theta p(y|\theta)p(\theta)d\theta$.

But note that $p(y)$ is the same in both the numerator and the denominator of R; hence, it simply cancels out and therefore vanishes! This is a key to the magic of an MCMC algorithm.

- In the final step of our RW-MH algorithm, we use R to decide whether we add θ^* to our chain (i.e., the sample from the posterior), i.e., whether we accept or reject θ^*. Our decision rule for accepting θ^* into our posterior sample is probabilistic:

$$P(accept\ \theta^*) = \min(1, R).$$

Hence, if $R > 1$, the posterior density for θ^* is greater than that for the current value θ^t, and we immediately accept θ^* into our MCMC sample, i.e., we set $\theta^{t+1} = \theta^*$. On the other hand, if $R < 1$, the posterior density for θ^* is less than that of θ^t, but this does *not* necessarily mean we reject θ^*. Rather, we may still accept θ^* into our MCMC sample with probability R, whereas with probability 1-R, we reject θ^* and instead duplicate the current value θ^t to become our new value θ^{t+1}, i.e., we then set $\theta^{t+1} = \theta^t$. We see that in this way a value θ^t may appear multiple times in the posterior sample that we accumulate.

The magic of this recipe is that if we accept/reject with a probability according to the ratio of the posterior densities of two compared values (R) in an RW-MH algorithm, we end up sampling values in proportion to their posterior density. Thus, we will accumulate a sample of size T from the posterior distribution of our parameter θ. Therefore, we can think of an MCMC algorithm as a superfancy RNG for posterior distributions whose precise parametric form is generally unknown. In other words, *we use MCMC as an RNG for the posterior distributions of all the unknowns in the Bayesian analysis of a statistical model.*

We now illustrate this principle with the first steps in an RW-MH MCMC algorithm applied to the data in the tadpole example. Our likelihood is binomial with $y = 20$ and $N = 50$, and we assume a Beta(1, 1) = Uniform(0, 1) prior on the detection probability parameter. We conduct the computations in the logit space and choose a tuning parameter $\sigma_{proposal} = 1$ in our Gaussian candidate-generating, or proposal, density and initialize a random walk through parameter space at logit(θ) = 1. We work on the logit scale to ensure that θ is always in the range between 0 and 1, as a decent probability should be. We prepare an empty array called `ltheta`, into which we will save the MCMC samples produced.

```
# Choose initial value for logit(theta) and tuning parameters
ltheta1 <- 1          # Initial value for tadpole detection prob.
sigma_prop <- 1       # SD of Gaussian proposal distribution

# Array to hold the MCMC samples
ltheta <- numeric()

# Initial value becomes first (and 'current') value in the chain
ltheta[1] <- ltheta1
ltheta                # Our posterior sample up to now (not shown)

# Randomly perturb the current value
set.seed(1)           # To initalize your RNGs identically to ours
( ltheta_star <- rnorm(1, ltheta[1], sigma_prop) )
 [1] 0.3735462
```

We now have two values for logit(θ) to compare, and we compute the ratio of their posterior densities, ignoring the $p(y)$ term in Bayes' rule.

```
# Compute likelihood times prior evaluated for the proposed new value of ltheta
( pd_star <- dbinom(20, 50, plogis(ltheta_star)) * dbeta(plogis(ltheta_star), 1, 1) )
[1] 0.002716919

# Compute likelihood times prior evaluated for the current value of ltheta
( pd_1 <- dbinom(20, 50, plogis(ltheta[1])) * dbeta(plogis(ltheta[1]), 1, 1) )
[1] 6.951277e-07

# Compute posterior density ratio R
( R <- pd_star / pd_1 )
 [1] 3908.518
```

The proposed new value for logit(θ) yields a much larger value of the product of likelihood times prior ($R > 1$), so we immediately accept it into our MCMC sample.

```
# Add theta_star into MCMC sample
ltheta[2] <- ltheta_star
ltheta                          # Our posterior sample up to now (not shown)
```

Our posterior sample of logit(θ) now has two values, 1 and 0.3735. We repeat these steps by perturbing ltheta[2] and drawing a new proposal for logit(θ).

```
( ltheta_star <- rnorm(1, ltheta[2], sigma_prop) )
[1] 0.5571895
```

We compare the posterior densities for ltheta[2], which is the current value, and the new proposal.

```
pd_star <- dbinom(20, 50, plogis(ltheta_star)) * dbeta(plogis(ltheta_star), 1, 1)
pd_t <- dbinom(20, 50, plogis(ltheta[2])) * dbeta(plogis(ltheta[2]), 1, 1)
( R <- pd_star / pd_t )
 [1] 0.1398872
```

Now $R < 1$, so the proposed new logit(θ^*) has a much lower posterior density than the current value of logit(θ). Therefore, we accept only the proposed new value with a probability of about 0.14, and with probability of 0.86 we duplicate the current value and add it to the chain. We flip a coin to decide whether we add the new θ^* to our sample or push forward the current value of θ, θ^t to become the new value in the MCMC sample for θ.

```
( keep.ltheta_star <- rbinom(1, 1, R) )
[1] 0
```

Bad luck for this θ^*: it is rejected, and hence, the "old" value of θ in the chain is plugged in as the new, current value in the chain instead.

```
ltheta[3] <- ltheta[2]
ltheta                          # Our posterior sample up to now (not shown)
```

Now we continue by perturbing theta[3] according to the normal proposal distribution and so forth. Fig. 2.7 (left) shows the nascent Markov chain that gives our posterior samples for detection probability (on the logit scale) in the tadpole example. The following loop allows us to extend this code of our RW-MH algorithm for an arbitrary number of steps (here, we choose 60,000).

```
# Iteration 4 to T of RW-MH algorithm
T <- 60000                      # Choose chain length
for (t in 4:T){                 # Continue where we left off
  ltheta_star <- rnorm(1, ltheta[t-1], sigma_prop)
  pd_star <- dbinom(20, 50, plogis(ltheta_star)) * dbeta(plogis(ltheta_star), 1, 1)
  pd_t <- dbinom(20, 50, plogis(ltheta[t-1])) * dbeta(plogis(ltheta[t-1]), 1, 1)
  R <- min(1, pd_star / pd_t)  # Note more general solution here
  keep.ltheta_star <- rbinom(1, 1, R)
  ltheta[t] <- ifelse(keep.ltheta_star == 1, ltheta_star, ltheta[t-1])
}
```

This takes almost no time at all! We plot the results for our Bayesian estimate of tadpole detection probability (Fig. 2.7).

Because of the way this fancy RNG algorithm is constructed, samples from it form a Markov chain. If all goes well, and after some transient initial phase, the chain will eventually settle down at its stationary distribution, which is the desired

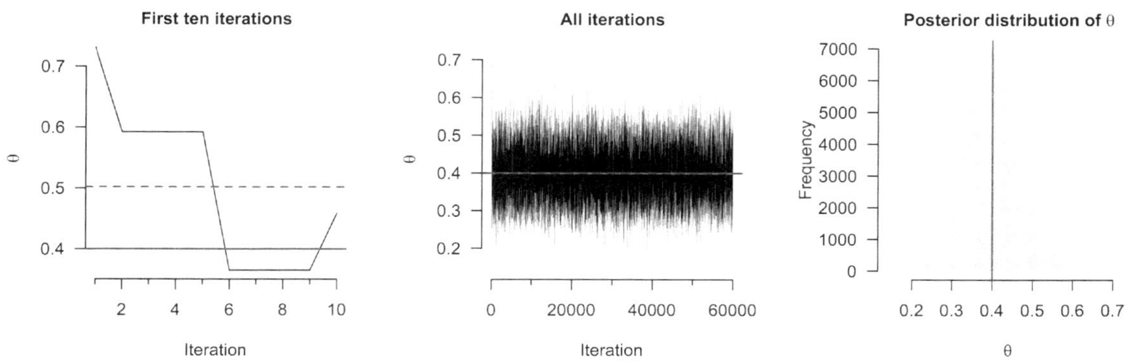

FIGURE 2.7

(Left) Trace-plot showing initial 10 iterations of a random-walk Metropolis-Hastings Markov chain Monte Carlo algorithm producing a sequence of dependent random numbers from the posterior distribution of tadpole detection probability (θ). (Middle) Trace-plot of the full sample of 60,000 posterior draws produced. (Right) Histogram of the 60,000 random numbers from the posterior distribution of tadpole detection probability (θ). Red lines show the value of the maximum likelihood estimate of θ, while blue is the Bayesian estimate based on the sample mean of the produced draws, i.e., the posterior mean. The latter is based on just 10 values in the plot on the left but based on the full posterior sample in the other plots.

posterior distribution. Each value of θ is then produced in a frequency proportional to its posterior density. The number of iterations required for the chain to reach a stationary distribution depends on the type of MCMC algorithm, the tuning parameter $\sigma_{proposal}$, the complexity of the model, and the size of the data set. Convergence is something that must always be checked when basing statistical inferences on MCMC methods. Typically, the samples in the initial transient phase of the chain are discarded as a "burn-in" and are not representative of the posterior distribution. How long the "MCMC RNG" must be "burned-in" can be judged visually by looking at a trace-plot of the draws produced as well as by calculating several numerical criteria (Brooks and Gelman, 1998).

In the `IPMbook` package, we have a function called `demoMCMC`, which is basically a wrapper around code given by Ntzoufras (2009) that implements a random-walk MCMC algorithm for a binomial proportion similar to what we have just sketched out. We use this function here to illustrate some important features of MCMC RNGs. The function with all its arguments made explicit and set to their defaults is called as follows.

```
library(IPMbook)
out <- demoMCMC(y=20, N=50, niter=25000, mu.ltheta=0, sd.ltheta=100, prop.sd=1, init=0)
```

At default, it fits a model to a binomial proportion exactly as in our tadpole problem and runs the MCMC algorithm for 25,000 iterations. In contrast to our previous sketch, it uses a normal prior for the logit-scale probability with mean = `mu.ltheta` and standard deviation = `sd.ltheta`. This is a broad normal prior distribution, which is one common choice for a vague prior. The tuning parameter $\sigma_{proposal}$ is called `prop.sd`, and `init` is the value of θ at which the algorithm is initialized.

The function produces graphical output (Fig. 2.8) that shows a time series or trace-plot of the sample of the logit of θ (top left) and of θ (top right). The blue line shows the mean of these 25,000 posterior samples. The red line shows the observed value of the proportion, 0.4, which is also the MLE for the parameter. From either plot, there is no sign of nonconvergence of the chain in its initial steps. The samples from the posterior distribution are summarized in a histogram (with overlaid density plot in blue) in the bottom left of the figure. An MCMC RNG does not produce independent draws from a distribution as does, for instance, function `rnorm` in R, but rather, each draw depends on the previous value. To emphasize this, there is a plot of the autocorrelation function (ACF) in the bottom right. It shows that in our example, the autocorrelation dies down after about 10 steps.

We can use this function to illustrate a few general features of MCMC algorithms related to chain length and choice of the tuning parameter of the candidate-generating Gaussian density. First, we must ensure that the algorithm is run for long enough that convergence is reached, i.e., that the stationary distribution is arrived at.

```
# Show convergence
tmp <- demoMCMC(y=20, N=50, niter=2500, mu.ltheta=0, sd.ltheta=100, prop.sd=0.1, init=10)
```

On execution you will see that the Markov chain has an initial transient phase of about 300 iterations before convergence is reached. Before you summarize the sample for inference about θ, you must discard these initial draws as a

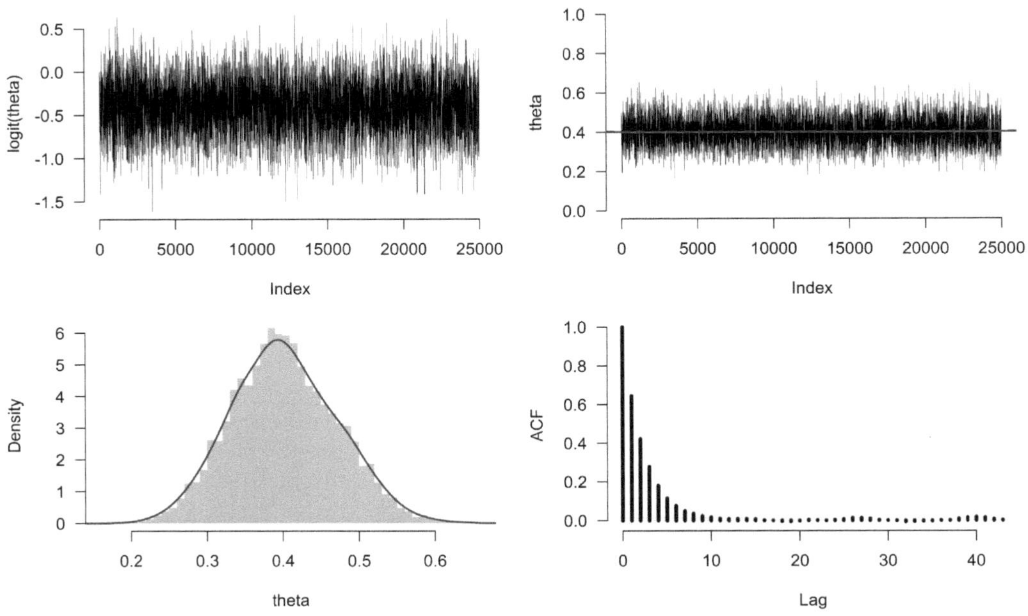

FIGURE 2.8

Graphical output from the random-walk Metropolis-Hastings Markov chain Monte Carlo algorithm in the function `demoMCMC` (main MCMC code in the function is from Ntzoufras, 2009).

burn-in. It is even possible that you do not reach convergence at all. In that case, you can run the algorithm for more steps, or alternatively increase the value of the tuning parameter $\sigma_{proposal}$; see below.

```
# No convergence within 2500 iterations
tmp <- demoMCMC(y=20, N=50, niter=2500, mu.ltheta=0, sd.ltheta=100, prop.sd=0.1, init=100)

# But convergence is reached after about 3k iterations
tmp <- demoMCMC(y=20, N=50, niter=25000, mu.ltheta=0, sd.ltheta=100, prop.sd=0.1, init=100)

# ... and you get convergence within 2500 iters with longer step length
tmp <- demoMCMC(y=20, N=50, niter=2500, mu.ltheta=0, sd.ltheta=100, prop.sd=1, init=100)
```

It is interesting to see the effects of the tuning parameter $\sigma_{proposal}$ on the resulting random numbers. You can think of this variance as governing the step length of the algorithm, which determines how quickly a chain explores the parameter space. Try running `demoMCMC` with values of `prop.sd = 0.01, 0.1, 10, 100, 1000`, and look at the trace-plots and acceptance rate shown in the console.

If you choose a large value for the tuning parameter, the parameter space is explored more rapidly and the area with high posterior density found more quickly. On the other hand, a large step size will tend to propose values with a very low posterior density, and hence most proposed values will be rejected. In this case, the current value may be replicated, possibly very many times, in a chain. This reduces the efficiency of the algorithm. There is some intermediate step length at which the algorithm turns out to be most efficient. There are technical arguments for an acceptance probability between 0.2 and 0.4 that leads to maximum efficiency of the algorithm (Carlin and Louis, 2009; Ntzoufras, 2009; Gelman et al., 2013). Tuning some "knobs" of an MCMC algorithm is something that MCMC software such as JAGS does automatically at the beginning of every analysis in the so-called adaptive phase.

The function `demoMCMC` yields the values of all input arguments and the sample of the desired size from the posterior distribution of θ, plus the acceptance probability of the algorithm, which is the proportion of proposed θs accepted.

```
str(out)

List of 7
 $ y         : num 20
 $ N         : num 50
 $ mu.ltheta: num 0
 $ sd.ltheta: num 100
 $ prop.sd   : num 1
 $ ltheta    : num [1:25000] 0 0 -0.0519 -0.0519 -0.7637 ...
 $ acc.prob : num 0.34
```

Because you can think of an MCMC algorithm as just a sort of fancy RNG, the basic result from running it is simply a stream of random numbers. You can inspect them by typing `out$ltheta`. We use them to characterize features of the posterior distribution of tadpole detection probability based on the corresponding sample statistics. This is called Monte Carlo integration. For instance, we can use any measure of location as a point estimate of detection probability, e.g., the mean, median, or mode. The mean is most commonly seen in Bayesian applications, whereas the posterior mode strictly corresponds to the MLE when a vague prior is used. We can use the sample standard deviation as an uncertainty measure of the estimate that is exactly analogous to the standard error (SE) of an MLE, and sample percentiles to give Bayesian credible intervals (CRIs), analogous to frequentist confidence intervals.

```
# Measures of central tendency to serve as point estimates
library(MCMCglmm)                                   # For function for mode
mean(plogis(out$ltheta))                            # Posterior mean
[1] 0.4007489                                       # Your result will differ slightly

median(plogis(out$ltheta))                          # Posterior median
[1] 0.4001582                                       # Your result will differ

posterior.mode(mcmc(plogis(out$ltheta)))            # Posterior mode (ditto)

     var1
0.4152145

# Measures of spread:
# - Bayesian 'variant' of standard error (= posterior SD)
# - two Bayesian credible intervals (CRI and HPDI)
sd(plogis(out$ltheta))                              # Posterior SD
[1] 0.06950494                                      # Your result will differ

quantile(plogis(out$ltheta), prob=c(0.025, 0.975)) # Symmetrical Bayesian CRI (your result will differ)
     2.5%     97.5%
0.2705029 0.5383629

HPDinterval(mcmc(plogis(out$ltheta)))               # Highest posterior density credible
                                                    # interval(HPDI); your result will differ
         lower     upper
var1 0.2672082 0.5349586
attr(,"Probability")
[1] 0.95
```

We can also compute the probability that the detection probability lies within a certain range of values. We illustrate computation of the probability that the tadpole detection probability is greater than 0.5 and find it to be about 0.07–0.08 (depending on your MCMC sample).

```
# Compute p(theta > 0.5)
mean(plogis(out$ltheta) > 0.5)
[1] 0.07584
```

This concludes our crash course on Bayesian computation by MCMC methods. Obviously, this is a very superficial treatment of this immense topic at the interface between statistics and computation. We hope it gives you a taste for what goes on under the hood of JAGS when you fit IPMs or indeed any other statistical model. For a recent textbook on using custom-built MCMC algorithms for model-fitting in ecology, see Hooten and Hefley (2019).

2.6 BUGS SOFTWARE: WinBUGS, OpenBUGS, JAGS, AND NIMBLE

It was the rediscovery of MCMC algorithms that gave Bayesian inference its substantial boost starting in the 1990s and caused a revolution in statistics (Brooks, 2003). But this revolution at first did not reach biology. The reason for this is that to code MCMC algorithms, you need to know quite a bit about statistics and computation, way more than the average ecologist or wildlife manager will understand. So those potential users could not benefit at first from the fantastic new modeling freedom that MCMC algorithms gave to statisticians. It is no exaggeration to say that the change in this was mainly due to the BUGS project. BUGS is an acronym for *Bayesian inference using Gibbs sampling* (Gilks et al., 1994; Lunn et al., 2000, 2009, 2013). The BUGS project was started by epidemiologists and statisticians in Cambridge/UK and was represented by software of the same name. At around the turn of the century, a Windows GUI named WinBUGS became available and around 2005 was rewritten in what was meant to be an open-source language, the result of which was the program OpenBUGS (Thomas et al., 2006). A little earlier, another program with similar capabilities named JAGS ("Just another Gibbs sampler") was first published (Plummer, 2003). Development for WinBUGS ceased with the appearance of OpenBUGS, and any recent development of OpenBUGS is sadly rather microscopic. Nevertheless, WinBUGS, OpenBUGS, and JAGS all still run on some of the latest machines at the time of writing. There is now also MultiBUGS as the latest development (Goudie et al., 2020). Importantly, the syntax of all four is very similar, so code written for one often runs in another with almost no modification. From a user perspective, the most important feature of these BUGS variants is the BUGS language with which the statistical models are specified. Hence, we use the term BUGS somewhat interchangeably for WinBUGS, OpenBUGS, and JAGS, although in this book, we usually mean JAGS. All three programs can be run from R using a variety of packages that serve as an interface, including `R2WinBUGS` and `R2OpenBUGS` (Sturtz et al., 2005), `R2jags` (Su and Yajima, 2015), `rjags` (Plummer, 2019), `runjags` (Denwood, 2019), `dclone` (Solymos, 2019), and `jagsUI` (Kellner, 2019).

NIMBLE is another software that uses the BUGS modeling language (de Valpine et al., 2017; Ponisio et al., 2020; de Valpine et al., 2020a, 2020b) and runs as an R package. The idea of NIMBLE is to separate the choice of model-definition language from the fitting algorithm. With the previous BUGS software (i.e., WinBUGS, OpenBUGS or JAGS), you have to use Bayesian inference. Yet, there may be good reasons for a likelihood analysis of a model, and this is exactly where NIMBLE comes in—it lets you specify a model in the BUGS language but then provides for different fitting algorithms, some of which lead to Bayesian and others to likelihood inference. In addition, NIMBLE is much more extendable; for instance, it is fairly easy to add new functions based on native R code. Finally, there are many ways in which a knowledgeable user can tune the algorithms and thereby obtain substantial gains in their speed and efficiency. This project is in full development, and we hope to see great things done to and with NIMBLE in the near future.

So what is BUGS software? BUGS is a flexible, generic Bayesian model-fitting software that is first a simple and intuitive yet extremely powerful programming language to specify a model; this is the BUGS language. Second, BUGS has an automated way of choosing among several MCMC algorithms to generate draws from the posterior distribution, and third, it lets you produce as many MCMC samples as you wish or have time to wait for. The BUGS language is perhaps the most remarkable feature. It provides a manner of specifying the likelihood of even very complex statistical models by formulating it as a series of simpler, "local" relationships between what BUGS calls nodes, which include data, parameters, latent variables, and missing values. We then define relationships between nodes, and there are exactly two such types of relationships: deterministic and stochastic. For the former, you can essentially imagine an equal sign, whereas you use the latter to define one quantity as a draw from some statistical distribution. There is a moderate array of mathematical and statistical functions, including many distribution functions. We find that the BUGS language is one of the most transparent ways in which a parametric statistical model can be defined. It has the added side effect for many BUGS users that it will assist in understanding essential statistical models such as GLMs or random-effects models. In the rest of this chapter, we give a brief introduction to statistical model-fitting using JAGS for these classes of models.

Four other relatively recent software programs that are very powerful for fitting some hierarchical models are Stan (Carpenter et al., 2017; Stan Development Team, 2020), INLA (Rue et al., 2009), ADMB (Fournier et al., 2012) with its successor TMB (Kristensen et al., 2016), and greta (Golding, 2019). Stan uses a specialized variant of MCMC to do Bayesian inference, whereas ADMB and TMB do primarily likelihood inference. INLA gives Bayesian inference using an approximation to the marginal posteriors of the parameters of a model.

We have only limited experience with either Stan or ADMB/TMB (and none with greta), but in our view, the challenge with both is that you must be able to write the likelihood of most models in their integrated or marginal form, where discrete random effects are integrated over or summed out (Chapter 2 in Kéry and Royle, 2016; Yackulic et al., 2020). All the IPMs in this book abound with such discrete random effects. Integrating them out may not be such a big deal for statisticians but would be a total showstopper for most ecologists. Therefore, we believe that Stan and ADMB/TMB are great pieces of software in the hands of a statistician, but they do not provide the great modeling freedom that BUGS or NIMBLE confer to many ecologists. Finally, INLA is primarily for complex spatial models, but it again struggles with discrete random effects and hence is not suited for IPMs.

2.7 USING JAGS TO FIT SIMPLE STATISTICAL MODELS FROM R: GENERALIZED LINEAR AND GENERALIZED LINEAR MIXED MODELS

In this section, we show how to use JAGS from R using the package `jagsUI` (Kellner, 2019) to fit the main building blocks of hierarchical models and IPMs: GLMs and generalized linear mixed models (GLMMs). We will fit six models: five GLMs and a simple GLMM. We compare Bayesian and likelihood inferences and will see that with vague priors, we typically obtain Bayesian estimates that are numerically indistinguishable from frequentist estimates for all practical purposes. In this section, we also establish the consistent layout and our usual workflow of an analysis in JAGS to which we adhere throughout the book.

This is only a crash course for using JAGS for these models. Hence, if you are new to Bayesian modeling, you should also consult other sources, e.g., Carlin and Louis (2009), King et al. (2010), Link and Barker (2010), Gelman et al. (2013), or Hobbs and Hooten (2015), or the introductions by McCarthy (2007), Kéry (2010) or Korner-Nievergelt et al. (2015). If you are new to GLMs and GLMMs, you should consider the study of a book such as that by McCullagh and Nelder (1989), Dobson and Barnett (2008), Ntzoufras (2009), or Kéry (2010), to name but a few. Interesting papers on using BUGS in ecology include those by Link et al. (2002) and Gimenez et al. (2009). Of course, at least a superficial glance at the JAGS manual, which you can download from the Internet, will be required from time to time too.

A GLM is in a sense the quintessential parametric statistical model, where the variation in a response variable is partitioned into a deterministic and a stochastic part, and where the two are linked at the scale of a transformation of the mean of the response (Nelder and Wedderburn, 1972). Importantly, effects of both continuous and categorical explanatory variables in the mean can be modeled in the fashion of a linear model (McCullagh and Nelder, 1989). Natural extensions of the GLM lead to generalized additive models (GAMs), where the linear model is replaced by splines to account for more "wiggly" relationships (Hastie and Tibshirani, 1990; Wood, 2017), or to GLMMs where one or more set(s) of parameters in the model are assumed to be the outcome of a normal distribution (GLMMs; McCulloch et al., 2008; Bolker, 2008; Bolker et al., 2009). Arguably, the GLM is the most essential model in applied statistics. So what is it?

A GLM can be defined in three steps that describe the stochastic part of a response, its deterministic part, and a transformation function that "links" the two (McCullagh and Nelder, 1989). For the stochastic part of the response, i.e., its noise, we can adopt any of the so-called *exponential family of distributions*, which include, for instance normal, Poisson, and binomial. The deterministic part of the response is called the *linear predictor* and contains the linear model for the explanatory variables, i.e., a weighted sum of the columns of a design matrix that collects the values of all covariates. These weights represent the estimated parameters, i.e., intercepts, means, and slopes. Finally, the linear model is usually not specified for the mean (or expected) response ($\mu = E(y)$) directly as in a normal linear regression but for a transformation g of the expected response that is called the *link function*. That is, the linear model is specified at the scale of $g(\mu) = g(E(y))$.

For illustration, let us assume that we have two continuous explanatory variables x_1 and x_2 that we collect in a design matrix $\boldsymbol{X}$ with three columns, where the first is a vector of 1's for the intercept of the model. If we want to use a GLM to relate the observed response of $\boldsymbol{y}$ to this set of explanatory variables, we must make choices about the following three parts of the model:

1. Stochastic part of the response—choose a suitable "data distribution" f for y: $y \sim f(\mu, \theta)$, with a mean response μ and possibly additional parameter(s) θ (e.g., for the variance of the response)
2. Link function—a transformation g of the mean (μ) of the response, i.e., $\eta = g(\mu)$, where typical choices for g are the "identity link" for a normal response (this in fact denotes the absence of a transformation of the mean), a log link for a Poisson response, and logit, cloglog, or probit links for a Bernoulli or binomial response
3. Systematic part of the response—the so-called linear predictor for the link-transformed expected response contains the linear model, e.g., $\eta = g(\mu) = \alpha + \beta_1 \times X_1 + \beta_2 \times X_2$

For the most part, throughout this section we will just fit one continuous covariate for illustration, and in Section 2.7.6, we show how to fit a categorical explanatory variable or factor.

2.7.1 POISSON GENERALIZED LINEAR MODELS

In this section, we imagine that we had surveyed 10 sites in the Jura mountains for asp vipers (Fig. 2.9) once each and counted 6, 10, 2, 7, 4, 1, 1, 2, 0, and 0 snakes. We want to take these counts as indices of relative abundance and are interested in a possible relationship between our abundance index and site elevation. The nonnegative and discrete nature of our data suggests a Poisson distribution as a natural probability model. That is, we will claim that the following pmf is a suitable description of the underlying chance process that produced these data:

$$p(Y = y|\lambda) = \frac{\lambda^y}{y!}\exp(-\lambda).$$

Here, possible values for the response y are nonnegative integers, and the single parameter λ is both the mean and the variance of the response. We will fit a Poisson GLM using JAGS assuming vague priors. For comparison, we will also fit the same model using the R function `glm`, which uses a method known as iterative reweighted least-squares (IRLS) to produce estimates identical to the MLEs for this class of model.

FIGURE 2.9

A gorgeous female asp viper (*Vipera aspis*), Départment du Doubs, French Jura mountains, August 21, 2017 (photograph by Marc Kéry).

```
# Define imaginary counts and elevation data for 10 sites
original.elev <- c(500, 400, 700, 500, 600, 800, 1000, 900, 1000, 900)
C <- c(6, 10, 2, 7, 4, 1, 1, 2, 0, 0)    # Counts

# Look at data and produce summaries of counts (not shown)
cbind(original.elev, C)
nind <- sum(C)
mean(C); sd(C); var(C); table(C)
```

We fit the following Poisson regression model for the snake count (C_i) at site i and the elevation of a site.

$$C_i \sim Poisson(\lambda_i)$$
$$\log(\lambda_i) = \alpha + \beta \times elev_i$$

These counts are considered a realization of a Poisson random variable with expected value λ, which is also indexed by i and hence is allowed to vary by site. This is the first of the three elements of the GLM definition—the stochastic part of the response C. The second and third elements are both found on line 2, where we apply a log link transformation to the expected response λ of i and then specify a linear model that corresponds to a simple linear regression of the log mean response on the elevation of each site.

We scale our covariate (`original.elev`) by centering, i.e., subtracting the mean, and by dividing the result by 1000. This makes the intercept (α) more meaningful: it becomes the expected number of snakes, on the log link scale, at the average elevation of all study sites, rather than at sea level when we do not center. In addition, the slope (β) then represents the expected change in snake counts as we move up or down by 1000 m, because this is now our unit in which a change in elevation is expressed.

```
elev <- (original.elev - mean(original.elev)) / 1000   # Center and scale
```

The JAGS software is a standalone program, not an R package (Plummer, 2003). It can be run from R via an interface package, and we use `jagsUI` (Kellner, 2019) for that purpose. The main function in `jagsUI` is called `jags` and has the following usage (from the help file).

```
jags(data, inits, parameters.to.save, model.file, n.chains, n.adapt=NULL, n.iter, n.burnin=0,
  n.thin=1, modules=c('glm'), factories=NULL, parallel=FALSE, n.cores=NULL, DIC=TRUE,
  store.data=FALSE, codaOnly=FALSE, seed=NULL, bugs.format=FALSE, verbose=TRUE)
```

The bold arguments are those that we explicitly set in nearly every analysis and include all the R objects we need to prepare before we can launch JAGS from R in this manner. In order, we perform the following:

- prepare a data set for the analysis as a list,
- prepare a text file containing the model description in the BUGS language,
- prepare a function to produce initial values,
- prepare a list with the names of the parameters that we want to estimate, and
- define the MCMC settings: the number and length of the Markov chains, burn-in and adaptation length, and subsampling or thinning rate.

When we have worked through all these preparatory R objects and steps, we use the R function `jags` to fit the model to the data using the JAGS program. Remember that you can think of JAGS as a powerful and very general RNG for the posterior distributions that arise in the Bayesian analysis of your models. Hence, the basic output from JAGS is just a stream of numbers for all the estimated quantities specified in the `parameters.to.save` list. Once JAGS has produced the random draws from each as requested in our MCMC settings, these draws are imported back into R, and `jagsUI` then organizes and summarizes them in various ways in the output object produced; see below. Sometimes, summarizing the streams of random numbers produced may require substantial computation time, for instance, when you estimate many high-dimensional quantities such as latent states or perhaps spatial predictions, and when you choose long chains. It may then be useful to use the simpler `jags.basic` function that just returns to R the posterior random numbers produced by JAGS. In addition, `jagsUI` has a useful `autojags` function that automatically runs an analysis until convergence is reached.

We start by loading the `jagsUI` package and bundle in a list the data for the analysis.

```
library(jagsUI)

# Bundle data
jags.data <- list(C=C, elev=elev, n=length(C))
str(jags.data)

List of 3
 $ C   : num [1:10] 6 10 2 7 4 1 1 2 0 0
 $ elev: num [1:10] -0.23 -0.33 -0.03 -0.23 -0.13 0.07 0.27 0.17 0.27 0.17
 $ n   : int 10
```

We love the R function `str`, because it tells us the dimension of each element in the data set. Here, we see a scalar and two vectors. Later you will also see many higher-dimensional arrays. Understanding the dimension of every object used in a model is one of greatest challenges when you want to become a BUGS modeler.

Next, we use the R function `cat` to write into the R working directory (do `getwd()` to see which one it is) a text file named "model1.txt" that contains the model description in the BUGS language. The latter is highlighted by the light-gray box. In the model file, we strive for a clear layout and therefore add section headings (with the hash sign commenting out those lines). Comments are optional but extremely useful in code. The model file contains a definition of the likelihood of the model and of the priors for all its parameters. We must define the likelihood for each individual datum separately and use a `for` loop for repeated structures. The description of the likelihood closely follows the algebraic symbolic description of the model above. Thus, when you know how to write a model in algebra including all the required indices, you will probably also be able to write the same model in the BUGS language.

We have two parameters in the model, the intercept and slope in the log-linear regression model, and we place vague priors on both. Typical vague priors are "flat" normal (i.e., with a large variance) or suitably wide uniform distributions. The normal distribution in the BUGS language is parameterized in terms of the precision, which is the inverse of the variance. Hence, our priors that have a precision of 10^{-6} (see below) correspond to a normal distribution with SD = 1000.

One should also note that for most parameters, these priors will be sufficiently wide to be called "vague," but depending on the scaling of a covariate, they need not always be so. Whether a prior is vague or informative must be ascertained anew in every analysis. This is usually done by trial and error, i.e., by refitting a model with a wider or narrower prior and checking whether the results, i.e., the main features of the posterior, change.

```
# Write JAGS model file
cat(file="model1.txt", "
model {
# Priors and linear models
# One set of vague priors: 'flat' normal
alpha ~ dnorm(0, 1.0E-06)                   # log-linear intercept
beta ~ dnorm(0, 1.0E-06)                    # log-linear slope

# Another possible set of vague priors: suitably wide uniform
# alpha ~ dunif(-100, 100)                  # log-linear intercept
# beta ~ dunif(-100, 100)                   # log-linear slope

# Likelihood for the Poisson GLM
for (i in 1:n){
  C[i] ~ dpois(lambda[i])                   # Stochastic part
  log(lambda[i]) <- alpha + beta * elev[i]  # Link function and linear pred.
# lambda[i] <- exp(alpha + beta * elev[i])  # Same written differently
}

# Derived quantities
mean.exp.count <- exp(alpha)                # Backtransformed intercept
}
")
```

For such a simple model, it is not essential to provide initial values for all parameters. In JAGS, any parameter for which no initial values are provided will be initialized by a random draw from its prior as defined in the model. However, in more complex models, it is often absolutely critical to provide initial values that are somewhat "close" to the solution or at least that do not contradict the model and the data. Hence, it is good practice to keep an eye on initial values even for simple models. We typically define a function that generates random draws from some distribution as initial values. When running JAGS with multiple Markov chains as we always do (see below), this function will then be executed once for each chain, ensuring a different start for each chain. If the chains then move to the same stationary distribution, we have increased confidence in their convergence (remember that this requires that you do not initialize all chains from the same starting values).

```
# Initial values
inits <- function(){list(alpha=rnorm(1, 0, 1), beta=rnorm(1, 0, 1))}
```

We also give a list with the names of the parameters for which we want to generate estimates or which we "monitor," i.e., for which we want the posterior draws produced to be exported from JAGS into R after the model run is completed.

```
# Parameters monitored
parameters <- c("alpha", "beta", "mean.exp.count", "lambda")
```

In the MCMC settings, we choose the number (`nc`), total length (`ni`), and burn-in length (`nb`) of chains as well as the thinning rate (`nt`) and the length of the pre-burn-in adaptive phase of the algorithm (`na`). It typically is a matter of trial and error to ensure sufficient sample size and convergence. Usually when building and debugging a model, we start with only very short settings and then run much longer and perhaps more chains for a production run of the model. Remember that we always want more than a single chain that can assess convergence using overlaid trace-plots and the Brooks-Gelman-Rubin convergence criterion.

```
# MCMC settings
ni <- 50000; nb <- 10000; nc <- 3; nt <- 10; na <- 1000
```

Once we have prepared all these "ingredients" for an analysis in JAGS, we are ready for the launch. The `parallel` argument permits running multiple chains in parallel; by default, it is set to FALSE, and we can then see the mesmerizing progress bar of JAGS.

```
# Call JAGS from R (ART <1 min)
out1 <- jags(jags.data, inits, parameters, "model1.txt", n.iter=ni, n.burnin=nb, n.chains=nc,
  n.thin=nt, n.adapt=na, parallel=FALSE)
```

For any model that takes longer than just a couple of minutes, it is highly recommended to set `parallel=TRUE`. Unfortunately, you will then be unable to marvel at the progress of your analysis—JAGS will send each chain to a separate core of your computer. For our usual three chains, this will speed up run times by about 2.5x; not quite 3x, presumably because there is some overhead at the start and the end of a run.

```
# Call JAGS from R (ART <1 min) and do parallel computation
out1 <- jags(jags.data, inits, parameters, "model1.txt", n.iter=ni, n.burnin=nb, n.chains=nc,
  n.thin=nt, n.adapt=na, parallel=TRUE)
```

This produces the R object "out1", which is an enormous list of length 24, and here, we only show some of the most important elements.

```
# Produce an overview of the R object created by jagsUI
str(out1)           # Only a small portion of total output is shown here!

List of 24
 $ sims.list    :List of 5
  ..$ alpha          : num [1:12000] -0.0859 -0.2759 0.5362 0.5113 ...
  ..$ beta           : num [1:12000] -8.13 -7.02 -6.57 -5.51 -5.05 ...
  ..$ mean.exp.count: num [1:12000] 0.918 0.759 1.709 1.667 2.062 ...
  ..$ lambda         : num [1:12000, 1:10] 5.96 3.82 7.75 5.93 6.58 ...
  ..$ deviance       : num [1:12000] 35.2 39.4 31.5 29 28.7 ...

 $ mean         :List of 5
  ..$ alpha          : num 0.623
  ..$ beta           : num -5.16
  ..$ mean.exp.count: num 1.94
  ..$ lambda         : num [1:10(1d)] 6.21 10.52 2.25 6.21 3.71 ...
  ..$ deviance       : num 30.7

[... output truncated ...]

$ summary     : num [1:14, 1:11] 0.623 -5.157 1.936 6.209 10.522 ...
  ..- attr(*, "dimnames")=List of 2
  .. ..$ : chr [1:14] "alpha" "beta" "mean.exp.count" "lambda[1]" ...
  .. ..$ : chr [1:11] "mean" "sd" "2.5%" "25%" ...

 $ samples     :List of 3
  ..$ : 'mcmc' num [1:4000, 1:14] -0.0859 -0.2759 0.5362 0.5113 0.7234 ...
  .. ..- attr(*, "dimnames")=List of 2
  .. .. ..$ : NULL
  .. .. ..$ : chr [1:14] "alpha" "beta" "mean.exp.count" "lambda[1]" ...
  .. ..- attr(*, "mcpar")= num [1:3] 10010 50000 10
  ..$ : 'mcmc' num [1:4000, 1:14] 0.9 1.091 0.935 0.579 0.518 ...
  .. ..- attr(*, "dimnames")=List of 2
  .. .. ..$ : NULL
  .. .. ..$ : chr [1:14] "alpha" "beta" "mean.exp.count" "lambda[1]" ...
  .. ..- attr(*, "mcpar")= num [1:3] 10010 50000 10
  ..$ : 'mcmc' num [1:4000, 1:14] 0.294 0.559 0.226 0.646 0.545 ...
  .. ..- attr(*, "dimnames")=List of 2
  .. .. ..$ : NULL
  .. .. ..$ : chr [1:14] "alpha" "beta" "mean.exp.count" "lambda[1]" ...
  .. ..- attr(*, "mcpar")= num [1:3] 10010 50000 10
  ..- attr(*, "class")= chr "mcmc.list"

[... output truncated ...]
```

The key elements in the output are the following: list element `samples` contains the actual posterior draws produced in each chain, i.e., the main output from JAGS, and these are then merged in `sims.list`. Hence, we will normally use the latter in any analysis wherein we need the posterior samples directly. Element `summary` contains the main part of the summary output that we also obtain when printing the object "out1"; see below. All the parts of that summary are also available in a series of list elements starting with element `mean`, which contains the posterior means for all monitored quantities. This series includes the 2.5% and 97.5% percentiles too, which are useful for 95% CRIs. Finally, you may sometimes look for the time it took to run a model; this is in `mcmc.info$elapsed.mins`.

After fitting the model, the very first step in our workflow is always to check convergence visually using the trace-plots (Fig. 2.10)—we always do this *before* looking at any results. Trace-plots are time series plots of all the draws for each parameter, with plots for multiple chains superimposed and printed in different colors. If there is no long-term trend and these plots oscillate around an imaginary horizontal line, things look good. In addition, we always look at the value of the Brooks-Gelman-Rubin convergence statistic, which is called Rhat here. This is similar to the F-test in a one-way analysis of variance testing for a chain effect. At convergence, there should no longer be any chain effect, and all the variances

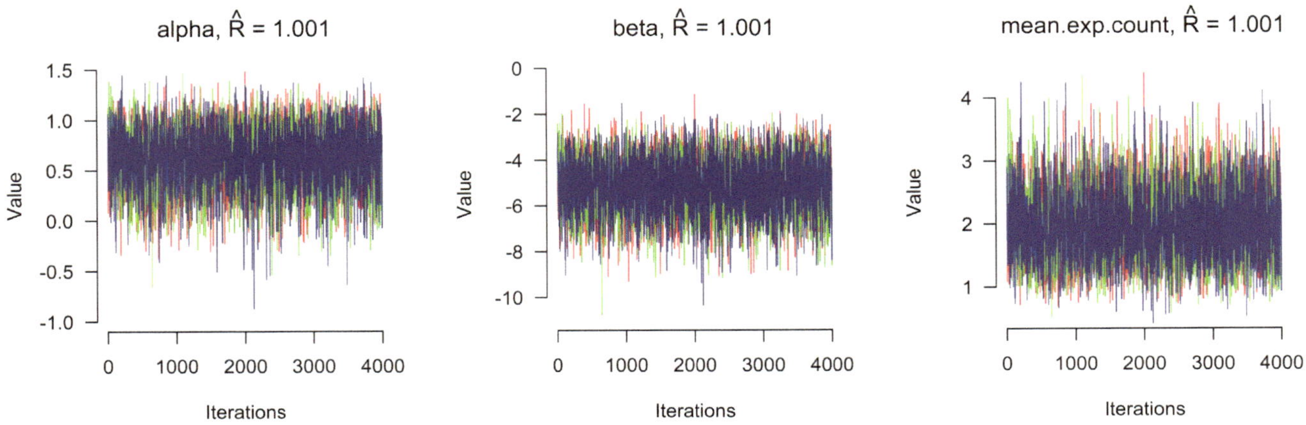

FIGURE 2.10

Trace-plots for three of the estimated quantities in the Poisson generalized linear model fit to the snake counts. The legend of each panel contains the value of the Brooks-Gelman-Rubin test for convergence (Rhat).

should be within rather than between the chains. Hence, an Rhat value of one indicates convergence, and a common rule of thumb is to take 1.1 as a critical threshold for assuming convergence. In the plots produced by the `traceplot` function in `jagsUI`, the value of Rhat is given in the title of each panel, but in addition, we also see it in the posterior summary table (see below). We also note that if you have loaded library `coda` in your current R session, you will have to write `jagsUI::traceplot(out1)`, because `coda` has its own `traceplot` function that requires a different input format.

```
jagsUI::traceplot(out1, layout=c(2, 2)) # May need jagsUI::traceplot!
```

Convergence assessment by visual inspection of trace-plots and based on the Rhat values shows no sign of non-convergence, so we proceed to producing a summary of the marginal posterior distributions of all the quantities that we put into the `parameters.to.save` list above.

```
print(out1, 3)

JAGS output for model 'model1.txt', generated by jagsUI.
Estimates based on 3 chains of 50000 iterations, adaptation = 1000 iterations (sufficient),
  burn-in = 10000 iterations and thin rate = 10, yielding 12000 total samples from the
  joint posterior.
MCMC ran for 0.081 minutes at time 2020-11-19 18:09:13.

                 mean    sd   2.5%    50%  97.5% overlap0     f  Rhat n.eff
alpha           0.619 0.280  0.027  0.637  1.114    FALSE 0.979 1.000 12000
beta           -5.178 1.140 -7.531 -5.122 -3.087    FALSE 1.000 1.000 12000
mean.exp.count  1.929 0.520  1.028  1.892  3.048    FALSE 1.000 1.000 12000
lambda[1]       6.210 1.115  4.233  6.132  8.558    FALSE 1.000 1.000 12000
lambda[2]      10.541 2.479  6.354 10.327 16.022    FALSE 1.000 1.000 10153
lambda[3]       2.238 0.550  1.271  2.206  3.414    FALSE 1.000 1.000 12000
lambda[4]       6.210 1.115  4.233  6.132  8.558    FALSE 1.000 1.000 12000
lambda[5]       3.705 0.689  2.472  3.657  5.171    FALSE 1.000 1.000 12000
lambda[6]       1.368 0.453  0.618  1.325  2.365    FALSE 1.000 1.000 12000
lambda[7]       0.529 0.283  0.141  0.478  1.229    FALSE 1.000 1.000 12000
lambda[8]       0.846 0.363  0.296  0.795  1.692    FALSE 1.000 1.000 12000
lambda[9]       0.529 0.283  0.141  0.478  1.229    FALSE 1.000 1.000 12000
lambda[10]      0.846 0.363  0.296  0.795  1.692    FALSE 1.000 1.000 12000
deviance       30.633 2.044 28.661 30.021 36.144    FALSE 1.000 1.001  3355

Successful convergence based on Rhat values (all < 1.1).
Rhat is the potential scale reduction factor (at convergence, Rhat=1).
For each parameter, n.eff is a crude measure of effective sample size.

overlap0 checks if 0 falls in the parameter's 95% credible interval.
f is the proportion of the posterior with the same sign as the mean;
i.e., our confidence that the parameter is positive or negative.

DIC info: (pD = var(deviance)/2)
pD = 2.1 and DIC = 32.721
DIC is an estimate of expected predictive error (lower is better).
```

Columns 1–5 of the actual table give the sample statistics that appear in the column titles for each parameter including the model deviance. We usually take the posterior mean as a point estimate, except when a posterior distribution is noticeably skewed, in which case the median or mode is more adequate. The meaning of the columns named "overlap0" and "f" are explained in the footer of the table, and we have described the Rhat column above. The final column in the table gives an estimate of the effective sample size for each marginal posterior. This is an estimate of the size of an equivalent independent sample that contains the same amount of information about that posterior as does our nonindependent sample of random draws. Broadly, the more serial correlation in a chain, the smaller the effective sample size will be relative to the nominal sample size.

In the header of the table, we are reminded of the MCMC settings and see the run time for the analysis, among other things. In the footer of the table, we provide some explanations and an estimate of the effective number of estimated parameters (pD) and of the deviance information criterion (Spiegelhalter et al., 2002). Both are fine for use in model selection with simple GLMs such as this one but should not be relied upon in most hierarchical models (Celeux et al., 2006; Hooten and Hobbs, 2015).

This posterior summary table presents marginal summaries of the joint posterior, where the posterior samples of each parameter are averaged over the posteriors of all other parameters. However, it is sometimes interesting to look at the joint posterior when projected in two dimensions rather than one. As an example, Fig. 2.11 shows a picture of the joint posterior of alpha and beta, which reveals a positive correlation between their estimates.

We can use either marginal or multivariate posteriors such as in this plot to play "probability games" with these posterior samples, in which we evaluate the probability that some condition is met for one or more parameters. For instance, let us assume that you had read a paper in which a theory was developed that claimed that the intercept of this relationship must be less than 0.5 and that simultaneously the slope must be more extreme, i.e., less, than - 6. We can easily test the evidence for this theory based on our analysis by computing the probability that this condition is true from our bivariate sample of the marginal posteriors of alpha and beta. This probability is estimated by the proportion of samples in the bottom left-hand quadrant defined by the red cross laid across the cloud; these posterior draws are colored in red in Fig. 2.11. This probability is estimated at about 0.18, hence, not so much, and perhaps the theory would need some revising then.

```
condition.true <- out1$sims.list$alpha < 0.5 & out1$sims.list$beta < -6
mean(condition.true)
[1] 0.1786667    # Your result will be somewhat different
```

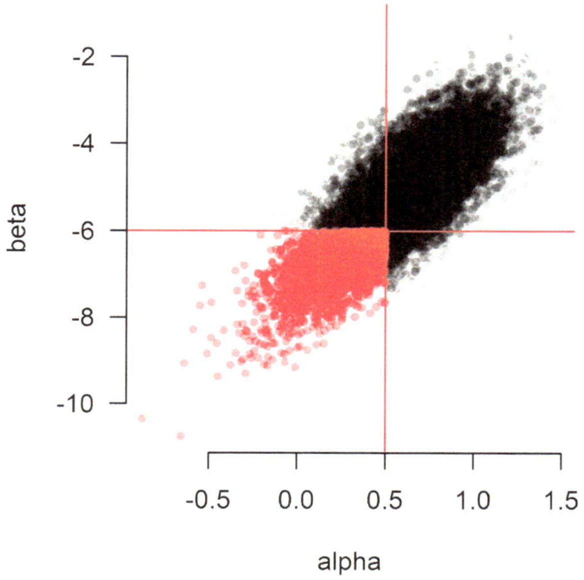

FIGURE 2.11

Joint posterior distribution of the intercept (alpha) and slope (beta) of the log-linear regression in our Poisson GLM fit to snake counts at sites with different elevation. The proportion of samples colored red among the total sample size drawn of the posterior gives the probability that alpha < 0.5 and simultaneously beta < −6.

For comparison, we now fit exactly the same model also using the IRLS method that for GLMs produces estimates that correspond to the MLEs.

```
# Get frequentist MLEs for comparison and predict intercept on natural scale
summary(fm <- glm(C ~ elev, family='poisson'))
predict(fm, type='response', newdata=data.frame(elev=0), se=TRUE)

[... output truncated ...]
Coefficients:
            Estimate Std. Error z value Pr(>|z|)
(Intercept)   0.6787     0.2705   2.509   0.0121 *
elev         -5.0198     1.1035  -4.549  5.4e-06 ***
[... output truncated ...]

$fit
       1
1.971256

$se.fit
        1
0.5331537
```

We find differences of only about 9% and 3% in the point estimates of the intercept and slope, respectively. Thus, despite the tiny sample size of 10, we obtain estimates that are numerically very similar. This is virtually always the case in Bayesian analyses with vague priors (Fig. 2.5), unless the sample size is very small relative to the complexity of a model or when a parameter is very hard to estimate (such as when it is nearly nonidentifiable). With smaller sample sizes posterior distributions tend to become more skewed and the posterior mean will then increasingly differ from the posterior mode. Note however that it is the mode that corresponds to the MLEs when vague priors are used. With increasing sample size, posteriors become more symmetric, and the estimates will then become more similar. This is also the idea behind a computational trick in frequentist statistics called "data cloning" (Lele et al., 2007): you take a data set and replicate it, say, 100 times so that the posterior distribution becomes Gaussian, and hence, the posterior mean becomes equal to the MLE.

We can check the improved agreement between MLEs and posterior means by increasing the sample size 100-fold and refitting the model in JAGS and using `glm`.

```
# New data bundle with variant of data with bigger sample size
times.bigger <- 100        # Results in sample size of 1000
newC <- rep(C, times.bigger)
newElev <- rep(elev, times.bigger)
jags.data <- list(C=newC, elev=newElev, n=length(newC))
str(jags.data)             # This is now the big data set

# Parameters monitored
parameters <- c("alpha", "beta", "mean.exp.count")

# Call JAGS from R (ART 2 min), check convergence and summarize posteriors
out2 <- jags(jags.data, inits, parameters, "model1.txt", n.iter=ni, n.burnin=nb, n.chains=nc,
  n.thin=nt, n.adapt=na, parallel=TRUE)
print(out2, 3)

                 mean    sd    2.5%     50%   97.5% overlap0 f  Rhat n.eff
alpha           0.678 0.027   0.625   0.678   0.731    FALSE 1 1.000  9482
beta           -5.021 0.111  -5.242  -5.020  -4.807    FALSE 1 1.000  7736
mean.exp.count  1.971 0.053   1.868   1.970   2.076    FALSE 1 1.000  9912

# Get frequentist MLEs for comparison (for the 100x larger sample size)
summary(fm <- glm(newC ~ newElev, family='poisson'))

Coefficients:
            Estimate Std. Error z value Pr(>|z|)
(Intercept)  0.67867    0.02705   25.09   <2e-16 ***
newElev     -5.01976    0.11035  -45.49   <2e-16 ***
```

Now we can hardly see any difference between the estimates in the analysis with JAGS and that with `glm` in R. But note the substantial difference in your waiting time for obtaining these results—you must wait for about 2 min for the Bayesian results but obtain the MLEs in no time at all. It is almost always the case that inferences based on MCMC will take much longer than that based on maximum likelihood estimation.

We often plot predicted values from a fitted model to better understand what the model is telling us about the modeled relationships or for a figure in a paper. Fig. 2.12 shows the observed data along with their expectation (λ in the model).

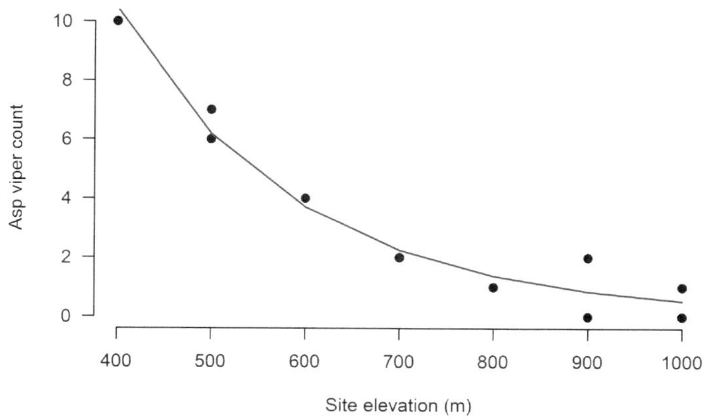

FIGURE 2.12

Observed counts of asp vipers at 10 sites in the Jura mountains (filled circles, simulated data) and prediction of the relationship between the expected count (λ) and elevation based on a log-linear model (blue line).

2.7.2 BERNOULLI GENERALIZED LINEAR MODELS

To illustrate a Bernoulli GLM, let us assume we want to estimate the apparent occupancy probability (θ), i.e., the observed proportion of occupied sites in our snake data set, ignoring imperfect detection. This has also been called a measure of the *apparent species distribution* (Kéry, 2011). We want to fit the following Bernoulli regression model for y_i, the indicator for detection or nondetection of at least one snake at site i, and the elevation of a site, which we scale again as before. We continue working with the larger variant of our data.

$$y_i \sim Bernoulli(\theta_i)$$

$$\text{logit}(\theta_i) = \log\left(\frac{\theta_i}{1-\theta_i}\right) = \alpha + \beta \times elev_i$$

In this GLM, the randomness in the binary response variable $\boldsymbol{y}$ is described by a Bernoulli probability model with a parameter called the success probability. This is also the expected response and here can be interpreted as the expected, apparent occupancy probability. The transformation typically applied to the expected response is the log of the odds (i.e., $\theta/(1-\theta)$), also known as the logit link. This is the typical link function when modeling systematic patterns in a Bernoulli or binomial random variable and leads to a model known as a logistic regression. Other link functions for Bernoulli and binomial random variables include the complementary log-log (cloglog; see Section 2.8; Royle and Dorazio, 2008; Kéry and Royle, 2016; Scharf et al., 2021), probit (Tobler et al., 2019) or log-log links (Ergon et al., 2018). The linear model could comprise really anything of your choice, but for illustration, we here only show the fitting of the effect of a single continuous covariate (elevation). For the fitting in JAGS of categorical covariates or factors, see Section 2.7.6.

When fitting the model in JAGS, we now (and in all later examples in this book) follow the consistent workflow established in the previous section. That is, we first prepare the "ingredients" required for fitting the model using the `jags` function in `jagsUI`, then we fit the model, assess chain convergence, look at the estimates, and then possibly produce a plot or form predictions. In addition, we also conduct a non-Bayesian analysis that produces MLEs that we can compare with the Bayesian estimates to observe their good agreement when vague priors are used.

```
# Turn counts into detection/nondetection data
y <- as.numeric(newC > 0); table(y)          # Not shown

# Bundle data
jags.data <- list(y=y, elev=newElev, n=length(y))
str(jags.data)

List of 3
 $ y   : num [1:1000] 1 1 1 1 1 1 1 1 0 0 ...
 $ elev: num [1:1000] -0.23 -0.33 -0.03 -0.23 -0.13 0.07 0.27 0.17...
 $ n   : int 1000
```

```
# Write JAGS model file
cat(file="model2.txt", "
model {
# Priors and linear models
alpha <- logit(mean.theta)                    # Intercept on logit link scale
mean.theta ~ dunif(0, 1)                      # Intercept on prob. scale
beta ~ dnorm(0, 1.0E-06)                      # Slope on logit link scale

# Likelihood of the Bernoulli GLM
for (i in 1:n){
  y[i] ~ dbern(theta[i])                      # Stochastic part of response
  logit(theta[i]) <- alpha + beta * elev[i]   # Link function and lin. pred.
  # theta[i] <- ilogit(alpha + beta * elev[i]) # Same written differently
}
}
")

# Initial values
inits <- function() list(mean.theta=runif(1), beta=rnorm(1))

# Parameters monitored
parameters <- c("alpha", "beta", "mean.theta")

# MCMC settings
ni <- 50000; nb <- 10000; nc <- 3; nt <- 10; na <- 1000

# Call JAGS from R (ART 1 min), check convergence and summarize posteriors
out3 <- jags(jags.data, inits, parameters, "model2.txt", n.iter=ni, n.burnin=nb, n.chains=nc,
  n.thin=nt, n.adapt=na, parallel=TRUE)
traceplot(out3)                               # May need jagsUI::traceplot
print(out3, 3)

                mean    sd    2.5%     50%   97.5% overlap0 f Rhat n.eff
alpha          2.972 0.231   2.536   2.967   3.441    FALSE 1    1 12000
beta         -12.717 1.081 -14.890 -12.699 -10.686    FALSE 1    1  8781
mean.theta     0.950 0.011   0.927   0.951   0.969    FALSE 1    1 12000
deviance     632.603 1.997 630.651 631.980 638.029    FALSE 1    1  6411
```

And here we obtain the MLEs:

```
# Get frequentist MLEs for comparison
summary(glm(y ~ newElev, family='binomial'))

Coefficients:
            Estimate Std. Error z value Pr(>|z|)
(Intercept)   2.9922     0.2363   12.66   <2e-16 ***
newElev     -12.8055     1.1039  -11.60   <2e-16 ***
```

In numerical terms, we find no difference that would be of any practical importance between the Bayesian and non-Bayesian inferences. But note that we are working with a large sample with 1000 sites. If you fit the same model to our smaller data set with only 10 sites, you will see more substantial differences. We do not show it here, but doing this will illustrate that priors in the Bayesian analysis *do* have an effect, always. With vague priors this effect is normally very small or even virtually undetectable, because the posterior distribution is dominated by the information in the data. However, when the data are less information-rich and/or a parameter is not well informed by the data in principle (e.g., when there are identifiability problems), the influence of the priors may be more readily apparent. In our case, binary data contain much less information than for instance Poisson counts, and hence, the influence of the prior will become more visible as sample size becomes small. In addition and independently from any influence of the priors, posterior distributions tend to become more skewed in smaller samples and thus the posterior mode (which corresponds to the MLE) and posterior mean will increasingly differ.

For illustrative purposes, we now form predictions. We can either calculate them inside JAGS by defining them in the model as derived quantities, or if we have all the necessary "ingredients," in R based on the posterior samples of these ingredients that were produced by JAGS. The latter here are the two parameters in the model. We have 12,000 draws from the posterior distribution of each, as we can see:

```
# Inspect the posterior draws available after our run of JAGS
str(out3$sims.list)

List of 4
 $ alpha      : num [1:12000] 2.83 2.54 2.95 2.79 2.82 ...
 $ beta       : num [1:12000] -11.5 -11.9 -12 -11.9 -11.7 ...
 $ mean.theta: num [1:12000] 0.944 0.927 0.95 0.942 0.944 ...
 $ deviance  : num [1:12000] 633 640 632 631 632 ...
```

We use the estimates of `alpha` and `beta` to obtain a prediction, or estimate, of the expected response in the model across a range of values of the covariate `elev`. This will show us the exact parametric form of the relationship between apparent occupancy probability and elevation implied by our parametric model and the data to which we fit that model. That is, we compute $\widehat{\theta} = \text{ilogit}\left(\widehat{\alpha} + \widehat{\beta} \times elev\right)$, where by the carets we denote an estimate, and ilogit is the inverse logit function. Hence, we simply need to plug our parameter estimates, `alpha` and `beta`, into the linear predictor of our model, apply the inverse link function, and repeat this calculation for a range of values of the covariate `elev`.

We could form predictions based on the posterior means of the estimated parameters $\widehat{\alpha}$ and $\widehat{\beta}$, but this would assume that their posteriors are symmetrical and moreover does not propagate the estimation uncertainty in $\widehat{\alpha}$ and $\widehat{\beta}$ into the predictions. A safer approach that also achieves that error propagation is to base the computation of a prediction or of any other derived quantity on the full posterior sample of all estimated parameters that are involved (or possibly just a random sample of sufficient size). So we show this here.

We start by preparing a variant of the covariate that contains a large number of equidistant values over the desired range of the covariate for which we want to make a prediction. This range will often be the range of the observed covariate. Importantly, this prediction covariate must be expressed on a scale *identical* to that of the actual covariate that we used in the model-fitting, i.e., we must apply exactly the same transformation to the prediction covariate as we did for the covariate used in fitting the model, i.e., for `original.elev` back in Section 2.7.1. This is what we do first.

```
# Prepare covariate for prediction
# Apply identical scaling as for the covariate used in the model fitting
mean(original.elev) # Remind ourselves of the mean in the original cov.
original.elev.pred <- seq(400, 1000, length.out=1000)   # For prediction
elev.pred <- (original.elev.pred - mean(original.elev)) / 1000
```

To obtain these predictions we carry along the entire estimation uncertainty by using all the draws from the posterior distributions of the parameters and then summarize the resulting posterior predictive distribution of the apparent occupancy probability (θ) by taking the mean, SD, and CRIs for instance.

```
# Prepare an array for holding the predictions of theta
pred1 <- array(NA, dim=c(1000, 12000))
# Fill the array: get posterior distribution of predictions
for (i in 1:12000){
  pred1[,i] <- plogis(out3$sims.list$alpha[i] + out3$sims.list$beta[i] * elev.pred)
}
# Summarize posterior distribution of theta_hat
pm <- apply(pred1, 1, mean)                              # Posterior mean
psd <- apply(pred1, 1, sd)                               # Posterior SD (like prediction SE)
CRI <- apply(pred1, 1, quantile, probs=c(0.025, 0.975)) # 95% CRI
```

We plot the posterior means along with the 95% CRI obtained from the full simulated posterior distributions of these predictions (Fig. 2.13).

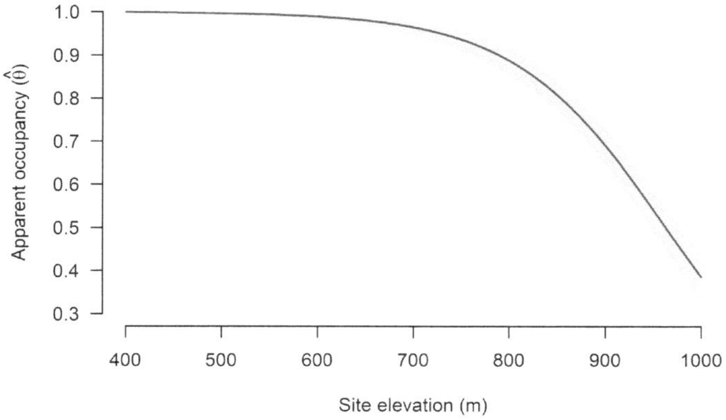

FIGURE 2.13

Predictions of the apparent occupancy probability ($\widehat{\theta}$) in relation to the covariate "Site elevation" computed from the full posterior distributions of alpha and beta. The blue line shows the posterior means, the gray layer the 95% prediction credible interval.

We now use this example to illustrate how trivially easy it is to compute even very complicated derived quantities in an MCMC-based analysis. Imagine that we wanted to compute the elevation at which the probability that apparent occupancy θ becomes less than 0.01 is greater than 0.95. In other words, what is the elevation at which you can be almost certain (here defined by a probability of 0.95) to find virtually no snakes (here defined to be 0.01 apparent occupancy) during a visit to a site?

This probability can be estimated from the posterior predictive distribution of $\hat{\theta}$, either inside a model fit in JAGS or outside it, if we have saved the posterior draws for all the necessary "ingredients." Fig. 2.13 suggests that the desired elevation is considerably greater than 1000 m, hence, we repeat our prediction of θ for larger elevations and at each value of the elevation prediction covariate evaluate the proportion of posterior draws for which $\hat{\theta} < 0.01$. We then plot both relationships and determine the value of elevation at which the probability of $\hat{\theta} < 0.01$ exceeds 0.95 for the first time. Based on our MCMC samples of $\hat{\theta}$ we can easily determine that the answer is 1391 m (Fig. 2.14). This is really easy in an MCMC-based analysis, but in a non-Bayesian analysis would require some complex bootstrapping and depend on more distributional assumptions such as the normality of the sampling distribution of the parameter estimates. In contrast, up to MCMC simulation error such predictions are exact in a Bayesian analysis.

If you imagine that time instead of elevation ran along the x-axis, so that your detection-nondetection data appeared in chronological order, you could view this analysis as a very simple population viability analysis (PVA; see Chapter 10) based on a scalar, binary population model and where imperfect detection is ignored. In such a PVA, you would estimate the first year when you are almost certain ($P > 0.95$) that the population became extinct (which you would define by a very small $\hat{\theta}$).

```
# Prepare another covariate with different range
mean(original.elev)            # Remind ourselves of the mean in the original cov.
alt.elev.pred <- 601:1600      # For prediction
elev.pred <- (alt.elev.pred - mean(original.elev)) / 1000

# Get posterior distribution of predictions of theta
pred2 <- array(NA, dim=c(length(elev.pred), 12000))
for(i in 1:12000){
  pred2[,i] <- plogis(out3$sims.list$alpha[i] + out3$sims.list$beta[i] * elev.pred)
}

# Get posterior mean of theta (theta_hat)
pm <- apply(pred2, 1, mean)  # Posterior mean
# Compute probability that theta_hat < 0.01
prob <- apply(pred2, 1, function(x) mean((x - 0.01) < 0))
# Compute minimum elevation at which prob > 0.95
min.elev <- min(alt.elev.pred[prob > 0.95])
print(min.elev)
[1] 1391
```

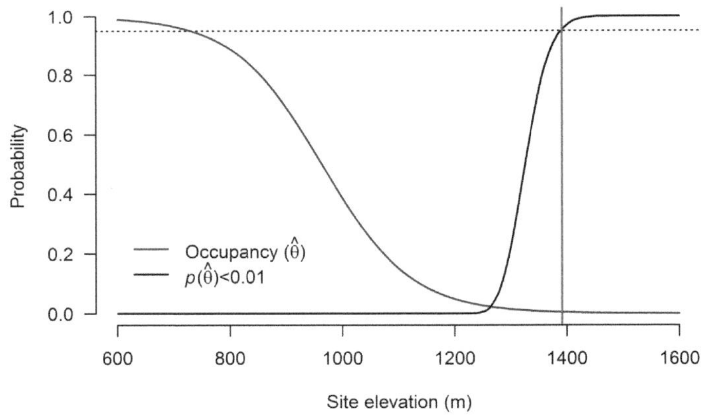

FIGURE 2.14

Predictions of apparent occupancy probability ($\hat{\theta}$, blue) and of the probability that the apparent occupancy is less than 0.01 (black) in relation to "Site elevation." The red line indicates the desired minimum elevation at which that probability first exceeds 0.95 (dotted line), which is at an elevation of 1391 m. (We note that this analysis is subject to the usual caveats about extrapolation beyond the observed range of a covariate.)

2.7.3 BINOMIAL GENERALIZED LINEAR MODELS

A binomial distribution is used for counts that can be conceptualized as a sum of N independent and identical Bernoulli trials. Thus, there are hardly any new considerations in the modeling of binomial over that of Bernoulli counts. Nevertheless, we cover this case in detail because binomial distributions appear extremely commonly in the modeling of population data. An outstanding use of the binomial distribution is as a probabilistic model for detection error in distribution and abundance data that is caused by imperfect detection (Williams et al., 2002; Royle and Dorazio, 2008; MacKenzie et al., 2018). To serve as an illustration for a binomial GLM, we now simulate a data set where we assume that we surveyed a large number of asp viper sites once each at a different time during the snake's activity period (March to October). We further assume that we miraculously knew the true population size N at each site. We simulate counts subject to detection error, which we assume varies over the season. We let detection probability (p) be lower in the summer than in spring or autumn, because higher temperatures allow the animals to remain hidden and safe from predators more often than during spring and autumn.

Specifically, we assume 1000 sites with a known population size N, which we simulate as a Poisson random variable. We then assume that each site is visited once, with visits spread out between the start of March and the end of October, i.e., between ordinal days 60–300 of the year. We choose the parameters of a second-order polynomial to describe the relationship between p and survey date. As a result, we obtain the desired U-shaped seasonal profile of p, and can then draw binomial counts with a trial size of N and the simulated detection probability p. Note that the setting of this example is simply a 1000-fold replication of the tadpole example earlier in this chapter where, moreover, instead of just an intercept, we now model a linear and quadratic effect of a covariate in our linear model (Fig. 2.15).

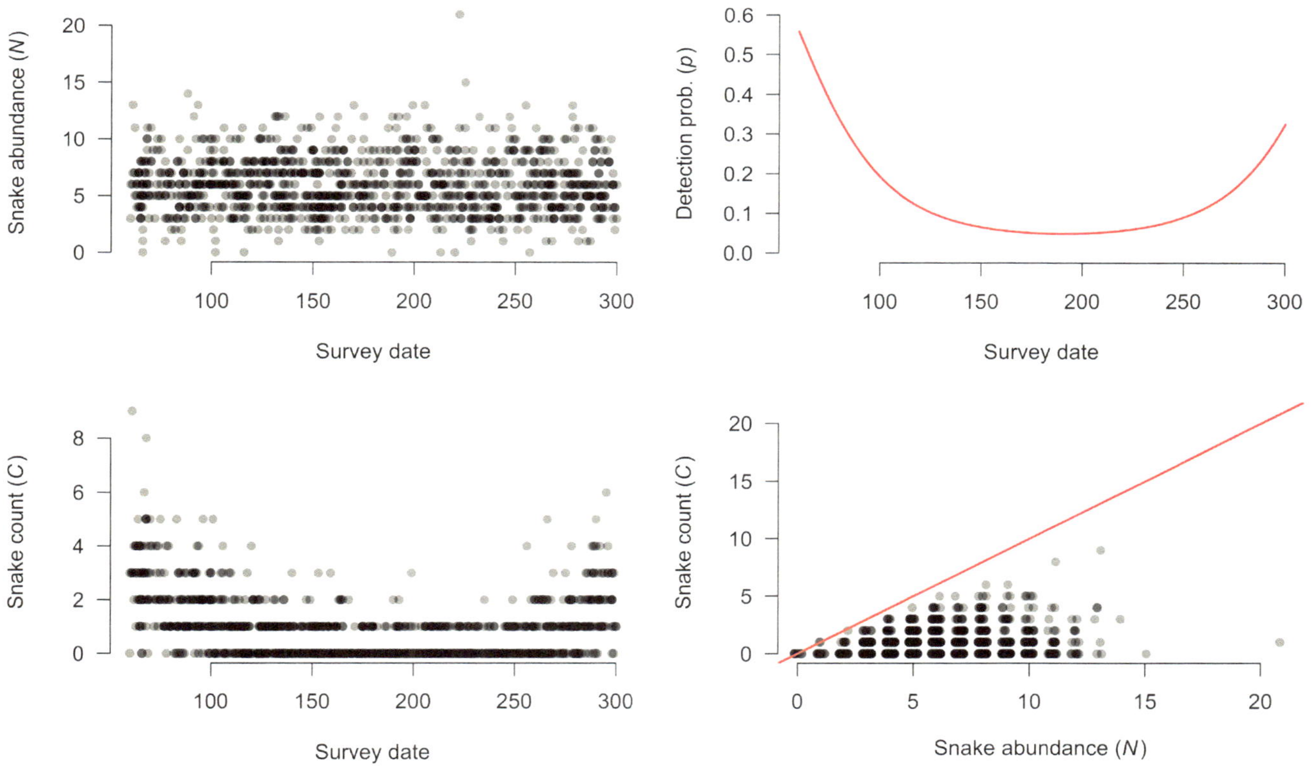

FIGURE 2.15

Relationships between the true snake abundance (N) and survey date for 1000 simulated asp viper sites (top left), between detection probability (p) and survey date (top right), between observed snake counts (C) and survey date (bottom left) and (bottom right) between observed snake counts and true population size (jittered), where the red line is the 1:1 line (if detection was perfect, all 1000 dots would sit on this line). Survey date is expressed as the ordinal day of the year.

```
# Simulate a data set of binomial counts
set.seed(99)
nsites <- 1000                                  # Number of sites

# Simulate true states
mean.abundance <- 6                             # Mean abundance across sites
N <- rpois(nsites, mean.abundance)              # Realized abundance N is Poisson
table(N)                                        # Note 7 unoccupied sites

N
  0   1   2   3   4   5   6   7   8   9  10  11  12  13  14  15  21
  7  13  38 109 141 157 144 126 101  69  47  24  14   7   1   1   1

# Simulate one count per site between March and October
# Simulate survey dates
survey.date <- sort(round(runif(nsites, 60, 300))) # Survey dates
sdate <- as.numeric(scale(survey.date))         # Scaled survey date (linear)
sdate2 <- as.numeric(sdate^2)                   # Scaled survey date (squared)

# Simulate seasonal profile of p
mean.p <- 0.05                                  # Detection probability in the middle of the season
beta1.lp <- -0.3                                # Linear effect of scaled survey date
beta2.lp <- 0.9                                 # Quadratic effect of scaled survey date
p <- plogis(qlogis(mean.p) + beta1.lp * sdate + beta2.lp * sdate2)

# Simulate one binomial count at each site
C <- rbinom(nsites, N, p)
```

We now want to fit the data-generating model to those snake counts (***C***), where ***N*** is known as the binomial index or total, and ***p*** is the detection probability of a single snake at site *i* and is the principal quantity that we estimate from the data.

$$C_i \sim Binomial(p_i, N_i)$$

$$\text{logit}(p_i) = \alpha + \beta_1 \times sdate_i + \beta_2 \times sdate_i^2$$

Here C_i is the binomial count at site *i*, where p_i and N_i, respectively, are detection probability and the true abundance at a site. The expected response of a binomial is $N_i \times p_i$, but N_i is just a known constant, and at the scale of the basic Bernoulli trial underlying any binomial count, the expected response is equal to p_i. Hence, we apply a link transformation to *p* and specify the chosen linear model at the logit scale, which again makes this a logistic regression. We adhere to our usual workflow and layout of an analysis with JAGS established earlier.

```
# Bundle data
jags.data <- list(C=C, N=N, sdate=sdate, sdate2=sdate2, n=length(C))
str(jags.data)

List of 5
 $ C     : int [1:1000] 3 0 3 9 3 2 4 3 3 1 ...
 $ N     : int [1:1000] 6 3 7 13 6 11 7 5 5 4 ...
 $ sdate : num [1:1000] -1.72 -1.72 -1.7 -1.7 -1.7 ...
 $ sdate2: num [1:1000] 2.96 2.96 2.91 2.91 2.91 ...
 $ n     : int 1000


# Write JAGS model file
cat(file="model3.txt", "
model {
# Priors and linear models
alpha <- logit(mean.p)
mean.p ~ dunif(0, 1)                            # logit-linear intercept in prob. scale
beta[1] ~ dunif(-10, 10)                        # logit-linear effect of date (linear)
beta[2] ~ dunif(-10, 10)                        # logit-linear effect of date (squared)

# Likelihood of the binomial GLM
for (i in 1:n){
  C[i] ~ dbinom(p[i], N[i])
  logit(p[i]) <- alpha + beta[1] * sdate[i] + beta[2] * sdate2[i]
  # p[i] <- ilogit(alpha + beta[1] * sdate[i] + beta[2] * sdate2[i]) # same
}
}
")
```

```
# Initial values
inits <- function(){list(mean.p=runif(1, 0, 0.3), beta=runif(2, 0, 2))}

# Parameters monitored
parameters <- c("alpha", "beta", "mean.p", "p")

# MCMC settings
ni <- 20000; nb <- 10000; nc <- 3; nt <- 2; na <- 1000

# Call JAGS from R (ART 1 min), check convergence and summarize posteriors
out4 <- jags(jags.data, inits, parameters, "model3.txt", n.iter=ni, n.burnin=nb, n.chains=nc,
  n.thin=nt, n.adapt=na, parallel=TRUE)
traceplot(out4)                        # Not shown (and may have to do jagsUI::traceplot)
print(out4, 3)

             mean    sd    2.5%     50%    97.5% overlap0 f Rhat n.eff
alpha      -2.978 0.073  -3.123  -2.977  -2.833    FALSE 1    1 13943
beta[1]    -0.304 0.032  -0.367  -0.304  -0.240    FALSE 1    1 15000
beta[2]     0.950 0.042   0.868   0.950   1.032    FALSE 1    1 11003
mean.p      0.049 0.003   0.042   0.048   0.056    FALSE 1    1 14542
p[1]        0.587 0.021   0.544   0.588   0.629    FALSE 1    1  9071
p[2]        0.587 0.021   0.544   0.588   0.629    FALSE 1    1  9071
p[3]        0.575 0.021   0.532   0.575   0.616    FALSE 1    1  9133
[... output truncated ...]
```

We have mentioned that this estimation problem is just a slightly more complex variation on the tadpole example with which we introduced maximum likelihood estimation earlier in this chapter. There, we had a sample size of $n = 1$, whereas now we have $n = 1000$. Here, we want to illustrate maximum likelihood estimation in this more typical case, and moreover, show the usual manner of obtaining the estimates, which is by specification of the negative log-likelihood (NLL) function and its minimization for the observed data using a function minimizer such as `optim` or `nlm` in R.

The likelihood to be maximized is the joint probability of all the $n = 1000$ data, regarded as a function of the parameters. Under independence of the 1000 observations, this is the product of the likelihood contribution from each data point, and hence, we can simply sum the 1000 log-likelihood contributions here and then change the sign of that sum. Next, we define the NLL of a logistic regression with two covariates and then use `optim` in R for obtaining the MLEs. In addition, we see how we can easily obtain the SEs from the so-called Hessian matrix (see Chapter 2 in Royle and Dorazio (2008), and Kéry and Royle (2016)). The Hessian is the matrix containing the second partial derivatives of the log-likelihood function with respect to the estimated parameters when evaluated at the MLEs. When the NLL is minimized as we do here, we actually obtain the *negative* of the Hessian (though this is not visible in the output from `optim`). The Hessian carries the information about the local curvature of the likelihood function around the MLE of each parameter and this is the basis for obtaining the asymptotic variance-covariance matrix (VC) of the estimates. Taking the square root of the diagonal of VC yields the asymptotic standard errors of the estimated parameters.

```
# Definition of NLL for a logistic regression with two covariates
NLL <- function(param, y, N, x1, x2){
  alpha <- param[1]
  beta1 <- param[2]
  beta2 <- param[3]
  psi <- plogis(alpha + beta1 * x1 + beta2 * x2)
  L <- dbinom(y, N, psi)               # Likelihood contribution for 1 datum
  NLL <- -sum(log(L))                  # NLL for all data points (here, 1000)
  return(NLL)
}

# Minimize NLL to obtain the MLEs and their asymptotic SEs
inits <- c(alpha=0, beta1=0, beta2=0)
sol <- optim(inits, NLL, y=C, N=N, x1=sdate, x2=sdate2, hessian=TRUE)
mle <- sol$par                         # Grab MLEs
VC <- solve(sol$hessian)               # Get variance-covariance matrix
ASE <- sqrt(diag(VC))                  # Extract asymptotic SEs
print(cbind(mle, ASE), 3)              # Print MLEs and SEs

         mle    ASE
alpha -2.979 0.0745
beta1 -0.304 0.0323
beta2  0.951 0.0423
```

We are excited that we now master maximum likelihood sufficiently well that we can fit such a useful model and obtain do-it-yourself MLEs, and moreover that our estimates are so similar numerically to those from the Bayesian analysis of the model. In addition, we can also obtain the MLEs using the IRLS method with the R function `glm`, and not surprisingly, obtain very similar estimates again.

```
# Get frequentist MLEs for comparison
summary(glm(cbind(C, N-C) ~ sdate + sdate2, family='binomial'))

Coefficients:
            Estimate Std. Error z value Pr(>|z|)
(Intercept) -2.97874    0.07448 -39.991   <2e-16 ***
sdate       -0.30354    0.03227  -9.407   <2e-16 ***
sdate2       0.95098    0.04229  22.487   <2e-16 ***
```

Finally, we note that we had added "p" into the list of monitored parameters in the analysis with JAGS. Therefore, if we want to produce predictions, we can simply use the posterior draws for p in the output object `out4` and do not need to compute the posterior draws outside JAGS in R as before.

```
# Plot seasonal detection profile: truth (red) and estimate (blue) (not shown)
plot(survey.date, p, type="l", lty=1, lwd=5, col="red", xlab="Survey date (day of year)",
  ylab="Detection probabilty", frame=FALSE, ylim=c(0, 0.6))
polygon(c(survey.date, rev(survey.date)), c(out4$q2.5$p, rev(out4$q97.5$p)), col="grey", border=NA)
lines(survey.date, p, type="l", lty=1, lwd=5, col="red")
lines(survey.date, out4$mean$p, type="l", lty=2, lwd=5, col="blue")
```

2.7.4 MULTINOMIAL GENERALIZED LINEAR MODELS

The multinomial distribution is enormously important for the statistical analysis of populations, yet it is certainly one of the least well-understood distributions among nonstatisticians. We will encounter it many times in later parts of the book as part of more complex models, such as CJS and multistate models. But interestingly, we can present the multinomial in a much simpler setting to estimate the relative proportions of two or more groups in a classification such as by habitat type. The estimation task then is to infer the cell probabilities of a multinomial, i.e., the proportions of units such as individuals or sites that fall into each group of that classification. For a known total of units, such a model can be formulated as a GLM. We will show two examples here: in the first, we consider "unstructured" cell probabilities, whereas in the second example, we illustrate the fitting of a CJS model (see Section 4.5.2) as a multinomial GLM with "structured" sets of cell probabilities. This structure is given by the algebraic expressions for the cells in an m-array. In this section, we get a little bit ahead of ourselves, but we believe it may be really insightful to make this clear connection between the CJS fit to capture-recapture data in the m-array format and the general fitting of a multinomial GLM.

Finally, we note that the multinomial is related to another distribution called a categorical distribution in a similar manner as a binomial is related to a Bernoulli distribution (Fig. 2.16). For the latter two, we first presented the simpler Bernoulli (Section 2.7.2) before the more complex binomial (Section 2.7.3), so it would be logical to also present the categorical before the multinomial. However, if you are new to these distributions, we believe it may be easier to understand if we present the multinomial here before the categorical (in Section 2.7.5.), which we surmise may be more challenging to understand even though it is conceptually simpler.

For illustration of a multinomial response model with unstructured cell probabilities, let us assume that we had 1000 snake sites that fell into four different habitat categories: cliff top (9%), forest edge (32%), rocky pasture (17%), and scree (42%).

```
# Simulate a habitat data set by drawing multinomial response with stated cell probs pi
set.seed(48)
hab.types <- c("cliff top", "forest edge", "rocky pasture", "scree")
pi <- c(0.09, 0.32, 0.17, 0.42)      # Relative probabilities, must sum to 1
habitat <- rmultinom(1, nsites, pi)
dimnames(habitat) <- list(hab.types, "Number of sites")
habitat

              Number of sites
cliff top                  79
forest edge               313
rocky pasture             181
scree                     427
```

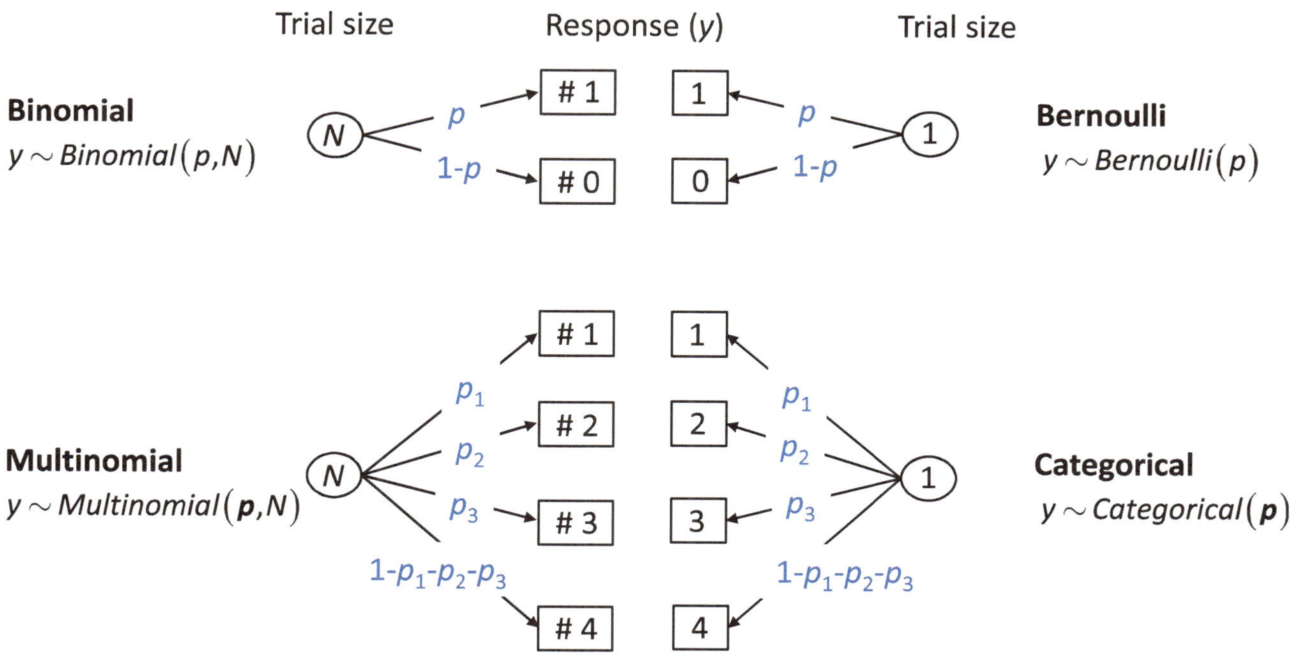

FIGURE 2.16

Relationships between four important distributions for count data that differ in trial sizes (1 or N) and their responses. Note that # indicates a frequency, i.e., # 2 shows how many times a '2' appears in y. The number of responses for the multinomial and the categorical is restricted to four in the graph, but in fact in can be any number larger than 2.

This is one realization of a multinomial "experiment": the number of sites among our 1000 snake sites that falls in the four habitat categories distinguished. The multinomial is a multivariate distribution producing a set of numbers in a single realization of the stochastic process. Our inference task now is to estimate the probabilities from these data and for this we fit a multinomial distribution to this set of count data. We can write the model in algebra as follows:

$$\boldsymbol{C} \sim Multinomial(\boldsymbol{\pi}, N)$$

$$\boldsymbol{\pi} = \left[\pi_1, \pi_2, \pi_3, \ldots, \pi_{k-1}, 1 - \sum_{i=1}^{k-1} \pi_i\right]$$

In the context of our example, $\boldsymbol{C}$ is a vector with $k = 4$ elements comprising the counts of sites in each habitat class, N is the multinomial sample size of 1000 sites, and $\boldsymbol{\pi}$ is a vector of length 4 containing the multinomial cell probabilities. In many applications of multinomial distributions as well as in ours, N is known, and only k-1 cell probabilities must be estimated. Here, we expressed the last cell probability as 1 minus the sum of the others, but we could do this with any of the other 3 cells instead.

We do not have any covariates here and fit what is known as a saturated model—one that has as many parameters as could possibly be estimated from this data set. For the cell probabilities $\boldsymbol{\pi}$ we need a prior that respects the usual constraints for this set of probabilities: their sum must be 1, and each individual probability must lie between 0 and 1. There are several possible solutions to this; see Sections 9.6 and 9.7 in Kéry and Schaub (2012). Here, we adopt a Dirichlet distribution, which we construct from a normalized set of gamma variables (Royle and Dorazio, 2008).

```
# Bundle data
jags.data <- list(C=as.numeric(habitat), k=length(habitat), N=sum(habitat))
str(jags.data)

List of 3
 $ C: num [1:4] 79 313 181 427
 $ k: int 4
 $ N: int 1000
```

Here is the model written in the BUGS language.

```
# Write JAGS model file
cat(file="model4.txt", "
model {
# Priors and linear models
# Cell probabilities for the multinomial
for (i in 1:k){                          # Loop over all cells in the multinomial
  d[i] ~ dgamma(1, 1)
  pi[i] <- d[i] / sum(d[])
}

# Likelihood for set of multinomial counts
C ~ dmulti(pi, N)
}
")

# Initial values
inits <- function(){list(d=rgamma(4, 1, 1))}

# Parameters monitored
parameters <- c("pi")

# MCMC settings
ni <- 110000; nb <- 10000; nc <- 3; nt <- 10; na <- 1000

# Call JAGS from R (ART <1 min), check convergence and summarize posteriors
out5 <- jags(jags.data, inits, parameters, "model4.txt", n.iter=ni, n.burnin=nb, n.chains=nc,
  n.thin=nt, n.adapt=na, parallel=TRUE)
traceplot(out5)                          # Not shown
print(out5, 3)

          mean    sd  2.5%   50% 97.5% overlap0 f Rhat n.eff
pi[1]    0.080 0.009 0.064 0.079 0.097    FALSE 1    1 30000
pi[2]    0.313 0.015 0.284 0.313 0.342    FALSE 1    1 12397
pi[3]    0.181 0.012 0.158 0.181 0.206    FALSE 1    1 16620
pi[4]    0.426 0.016 0.396 0.426 0.457    FALSE 1    1 30000
```

Because we have quite a large sample size and moreover fit a saturated model, our parameter estimates are fairly close to the known truth in our data simulation.

Next, we show a really cool variant of a multinomial GLM, which will be exactly the CJS model fit to the m-array format of capture-recapture data in Section 4.5.1.2. For illustration, we now simulate a radically different type of data—individual encounter histories. We assume that we conduct a capture-mark-recapture study (see Section 4.5.1) in a large asp viper population where over 4 years we capture all the snakes we can, individually mark the new ones and record the ID of those that are already marked, such that each year, we can add 30 new animals to the marked sample in the population. The target parameter is apparent survival probability. We use function `simCJS` in the `AHMbook` package to simulate such a data set under the assumptions of a CJS model. See Section 4.5.1 for more on this cornerstone model in ecological statistics.

```
# Simulate survival data: individual encounter histories
library(AHMbook)
set.seed(39)
dat <- simCJS(n.occ=4, n.marked=30, phi=0.7, p=0.4, show.plot=TRUE)
dat$ch                                   # Look at these encounter history data (not shown)

# Produce the m-array from encounter histories
Cmat <- marray(dat$ch)
Cmat
         recaptured
released  Y2    Y3    Y4    never
      Y1   7     7     1       15
      Y2   0    12     5       20
      Y3   0     0    13       36
```

Hence, of the 30 marked snakes in year 1, 7 were first recaptured in year 2, another 7 in year 3, a single one in year 4, and 15 were never recaptured. In year 2, 37 snakes were released in total (the sum of row 2); some are recaptures of snakes marked in year 1, and some are new captures. Twelve were first recaptured in year 3, 5 in year 4, and 20 were never recaptured. In year 3, 49 snakes were released, of which 13 were recaptured in year 4, and 36 were not. Please refer to Section 4.5.1.2 for more explanation as well as to Chapter 7 in Kéry and Schaub (2012).

The key idea for fitting a CJS model to these data is to treat the m-array counts for each release cohort (i.e., in each row of the m-array) as a multinomial random variable. But instead of estimating a set of unstructured cell probabilities π, we now express the cell probabilities as functions of the more fundamental probabilities of apparent survival (ϕ) and recapture (p). These functions are the expected values for each cell in the m-array. For instance, the multinomial distribution adopted for release cohort 1 has the following cell probabilities (assuming constant ϕ and p):

$$\left[\phi p,\ \phi(1-p)\phi p,\ \phi(1-p)\phi(1-p)\phi p,\ 1-\sum_{1}^{3}\ldots\right].$$

The first cell contains the probability to first be recaptured in year 2. For this, the animal must survive with probability ϕ and then be recaptured with probability p, thus this cell contains ϕp. To first be recaptured in year 3 only, the animal must survive with probability ϕ, fail to be recaptured with probability $(1-p)$ in year 2, survive again with ϕ, and then be recaptured with probability p in year 3, which yields $\phi(1-p)\phi p$. The same logic produces the probability for cell 3, which is $\phi(1-p)\phi(1-p)\phi p$. Finally, every animal in release cohort 1 must either be recaptured sometime in the years 2, 3, or 4; or else it is not recaptured. Hence, the probability for the last cell ("never recaptured") is 1 minus the sum of the first 3 cell probabilities ($\sum_{1}^{3}\ldots$). The cell probabilities for the other release cohorts can be written based on the same logic.

Let us see how this goes. We fit the simplest possible CJS model, which has both ϕ and p constant. This is also the assumption that we made in the data simulation.

```
# Compute size of release cohorts
R <- rowSums(Cmat)

# Bundle data
jags.data <- list(marray=Cmat, R=R)
str(jags.data)

List of 2
 $ marray: num [1:3, 1:4] 7 0 0 7 12 0 1 5 13 15 ...
 $ R     : num [1:3] 30 37 49


# Write JAGS model file
cat(file="model5.txt", "
model {
# Priors and linear models
phi ~ dunif(0, 1)                    # Apparent survival probability
p ~ dunif(0, 1)                      # Recapture probability

# Vectors of cell probabilities in the three multinomials
# Release cohort 1
pi1[1] <- phi * p                    # First recaptured in recap year 2
pi1[2] <- phi^2 * (1-p) * p          # First recaptured in recap year 3
pi1[3] <- phi^3 * (1-p)^2 * p        # First recaptured in recap year 4
pi1[4] <- 1 - sum(pi1[1:3])          # Never recaptured

# Release cohort 2
pi2[1] <- 0                          # Takes account of structural zero counts
pi2[2] <- phi * p
pi2[3] <- phi^2 * (1-p) * p
pi2[4] <- 1 - sum(pi2[1:3])

# Release cohort 3
pi3[1] <- 0                          # Accounts for structural zero counts
pi3[2] <- 0                          # Accounts for structural zero counts
pi3[3] <- phi * p
pi3[4] <- 1 - sum(pi3[1:3])

# Likelihood: one multinomial for each row in the m-array
marray[1,] ~ dmulti(pi1, R[1])       # MN for release cohort 1
marray[2,] ~ dmulti(pi2, R[2])       # MN for release cohort 2
marray[3,] ~ dmulti(pi3, R[3])       # MN for release cohort 3
}
")
```

```
# Initial values
inits <- function(){list(phi=runif(1, 0, 1), p=runif(1, 0, 1))}

# Parameters monitored
parameters <- c("phi", "p", "pi1", "pi2", "pi3")

# MCMC settings
ni <- 10000; nb <- 5000; nc <- 3; nt <- 5; na <- 1000

# Call JAGS from R (ART <1 min), check convergence and summarize posteriors
out6 <- jags(jags.data, inits, parameters, "model5.txt", n.iter=ni, n.burnin=nb, n.chains=nc,
  n.thin=nt, n.adapt=na, parallel=TRUE)
traceplot(out6)                    # Not shown
print(out6, 3)

           mean    sd   2.5%    50%  97.5% overlap0 f  Rhat n.eff
phi       0.786 0.095  0.603  0.783  0.967    FALSE 1 1.001  2273
p         0.376 0.079  0.241  0.369  0.540    FALSE 1 1.000  3000
pi1[1]    0.290 0.041  0.212  0.288  0.373    FALSE 1 1.000  3000
pi1[2]    0.141 0.026  0.089  0.141  0.191    FALSE 1 1.000  2950
pi1[3]    0.072 0.027  0.027  0.070  0.129    FALSE 1 1.001  2223
pi1[4]    0.497 0.061  0.375  0.497  0.617    FALSE 1 1.001  3000
[... output truncated ...]
```

We recognize the input in the simulated data, where we chose phi = 0.7 and p = 0.4. As always, with larger sample sizes, we would get closer to these values than in our fairly small sample. Remember that our model consists of three multinomials and we estimate three sets of cell probabilities; these are the basic parameters in the model. Here are our estimates (posterior means) for all three of them:

```
# Print the pi matrix: cell probabilities for all 3 release cohorts
print(rbind(out6$mean$pi1, out6$mean$pi2, out6$mean$pi3), 3)

     [,1]  [,2]   [,3]  [,4]
[1,] 0.29 0.141 0.0723 0.497
[2,] 0.00 0.290 0.1410 0.569
[3,] 0.00 0.000 0.2896 0.710
```

The cell probabilities are deterministic functions of the two fundamental parameters in the model, ϕ and p. Hence, it is for these latter parameters that we must adopt priors in our Bayesian analysis. You can check whether the first nonzero cell probability in each row of matrix π indeed equals the product of the two .... and it does (up to MCMC accuracy).

```
print(mean(out6$sims.list$phi * out6$sims.list$p), 3)
[1] 0.29
```

This concludes our introduction to the fitting of a multinomial model to sets of counts. If you find the multinomial distribution challenging, perhaps it helps to remember that in most applications in population ecology, we specify the parameters for this distribution in two steps: first, as always, we have the sets of cell probabilities (π), which give the expected proportion for each group in a classification of counts for which we adopt the multinomial distribution as a probabilistic model. In our first example, with the habitat types of asp viper sites, these cell probabilities were directly our target of inference. But much more commonly we have a second set of probability parameters that are the fundamental targets of inference. In our second example of a multinomial, where we fit a CJS model in what in essence is a multinomial GLM, we had as a response three multinomials, one for each release cohort. You saw how we expressed all sets of cell probabilities as a function of the two fundamental parameters, ϕ and p. It is in this form that you see the multinomial distribution most commonly in population ecology.

In this section, we have surreptitiously introduced the CJS model as a simple GLM because we believe it is valuable for your conceptual understanding of that model and GLMs so you can make the connection between the two. In practice, you will likely not use the code in this section but the more general code in Section 4.5.1. However, we believe that once you understand the code in this section, you will be able to better grasp the structure of the general CJS model code.

As a historical note we emphasize that the famous program SURVIV (White, 1983; Hines, 1994), i.e., the powerful precursor of program MARK (White and Burnham, 1999), was nothing but a very general and relatively user-friendly numerical solver for the multinomial distributions that arise when modeling the rows in the m-arrays in CJS, ring-recovery, multistate, and related capture-recapture models. The main challenge in SURVIV was then simply to write down the correct set of cell probabilities for each multinomial. As Jim Nichols (pers. comm.) put it: "*The ability to obtain numerical solutions for big multinomial models was a big step contributing to the explosion of capture-recapture and related modeling that took place 1983–2010. SURVIV greatly increased the pool of folks who could now play the modeling/estimation game.*" So in a sense, when you understand this section about the fitting of multinomials, you also understand the big idea of the historically important program SURVIV.

2.7.5 CATEGORICAL GENERALIZED LINEAR MODELS

In the same way that we can treat a binomial count with trial size N as an aggregation of N Bernoulli trials, a multinomial set of counts with trial size N is an aggregation of the N elemental trials of an underlying distribution with trial size 1—the categorical distribution (Fig. 2.16). In this section, we demonstrate the disaggregated form of the analysis from the start of the last section using a categorical distribution. This is an important distribution in multistate and related models later in the book.

We start by reformatting the habitat data simulated in the previous section such that for each site we obtain a numerical indicator of its habitat type.

```
# 'Stretch out' the habitat data from Section 2.7.4.
habitat                                # Check out habitat again (not shown)
habitat2 <- rep(1:4, habitat)          # Disaggregated version of the habitat data
habitat2; table(habitat2)              # Look at data (not shown)
```

The vector `habitat2` now contains a univariate version of the multivariate data in the summary in vector `habitat`. In fact, the former is the categorical covariate at the level of a site. When fitting a categorical GLM to response `habitat2` (with length 1000), we conduct what in essence is exactly the same analysis as by fitting a multinomial GLM to the aggregated response `habitat` (with length 4). So the estimated parameter is still a vector of cell probabilities π and we can adopt the same priors for them as before.

```
# Bundle data
jags.data <- list(C=habitat2, k=length(unique(habitat2)), n=length(habitat2))
str(jags.data)

List of 3
 $ C: int [1:1000] 1 1 1 1 1 1 1 1 1 1 ...
 $ k: int 4
 $ n: int 1000
```

Here's the model for the categorical response.

```
# Write JAGS model file
cat(file="model6.txt", "
model {
# Priors and linear models
# Cell probabilities
for (i in 1:k){                        # Loop over all cells
  d[i] ~ dgamma(1, 1)
  pi[i] <- d[i] / sum(d[])
}

# Likelihood
for (i in 1:n){
  C[i] ~ dcat(pi)
}
}
")

# Initial values
inits <- function(){list(d=rgamma(4, 1, 1))}

# Parameters monitored
parameters <- c("pi")

# MCMC settings
ni <- 110000; nb <- 10000; nc <- 3; nt <- 10; na <- 1000
```

```
# Call JAGS from R (ART 5 min), check convergence and summarize posteriors
out7 <- jags(jags.data, inits, parameters, "model6.txt", n.iter=ni, n.burnin=nb, n.chains=nc,
  n.thin=nt, n.adapt=na, parallel=TRUE)
traceplot(out7)                                  # Not shown
print(out7, 3)

           mean    sd   2.5%    50%  97.5% overlap0 f Rhat n.eff
pi[1]     0.080 0.009  0.064  0.079  0.097    FALSE 1    1 22081
pi[2]     0.313 0.015  0.284  0.313  0.342    FALSE 1    1 30000
pi[3]     0.181 0.012  0.158  0.181  0.206    FALSE 1    1 30000
pi[4]     0.426 0.016  0.396  0.426  0.457    FALSE 1    1 14368
```

Comparing this posterior table with that for out5 (above) you will see that indeed we obtain identical parameter estimates (up to simulation variability, or MC error, which we minimized by running very long chains). This must be so because this is really the same analysis as with multinomial model 4 at the start of the previous section. However, because now we have a sample size of 1000, whereas before we had only one of 4, it takes 138 times longer to run the model with identical MCMC settings.

```
# Compare run times of multinomial and categorical versions of analysis
out7$mcmc.info$elapsed.mins / out5$mcmc.info$elapsed.mins  # [1] 138.2222
```

Thus, in our case a multinomial and a categorical distribution are simply alternative ways of specifying the very same analysis, just that in the former we fit the model to a tabular summary of the data, whereas in the latter we fit the model to the disaggregated data, i.e., where each individual site appears as an observation. In many other applications of these statistical models, the unit of the analysis may not be a site but an individual, especially in the context of this book. Not surprisingly, the analysis for the aggregated data is far more efficient, and hence, when you have a choice you should select the multivariate variant of analysis. However, due to the aggregation over units you then lose the ability to model each individual unit independently. Hence, when you need to keep track of units individually, you must use a categorical distribution for the disaggregated data rather than a multinomial for an aggregated version of the data.

2.7.6 NORMAL LINEAR REGRESSION OR GAUSSIAN GENERALIZED LINEAR MODELS

Last in this section on GLMs, we cover a variant with normal or Gaussian errors, which is more commonly also called the normal linear model. We illustrate it with simulated data of the body mass of asp vipers and show how in a linear model we can fit the effects of a categorical explanatory variable, i.e., a factor, where we estimate one parameter for each factor level, i.e., for each different value of a classification. The modeling of the effects of factors can often be confusing, because these effects can be specified in a number of different ways, e.g., in the means, contrast treatment, or the sum-to-zero parameterizations (see Ntzoufras, 2009; Kéry, 2010; Kéry and Schaub, 2012; Kéry and Royle, 2016). One commonly used manner to specify a factor in a linear model in the BUGS language is called *nested indexing*, and we illustrate that here.

We assume that we had captured and weighed asp vipers in three populations and wanted to know whether the average mass differs among populations. For convenience we round mass data to integers. You can imagine that they are obtained with a scale with 1 g accuracy.

```
# Choose constants and create population indicator
set.seed(88)
npop <- 3                                        # Number of populations
nsnakes <- rpois(npop, 20)                       # Number of snakes caught per population
sigma <- 50                                      # Snake-to-snake variability in mass
Pop <- as.factor(rep(c("A", "B", "C"), nsnakes)) # Create population indicator (factor)

# Generate mass data
mean.mass <- round(runif(npop, 100, 300))        # Pop. average mass (g)
massA <- rnorm(nsnakes[1], mean.mass[1], sigma) # Mass data in population A
massB <- rnorm(nsnakes[2], mean.mass[2], sigma) # Mass data in population B
massC <- rnorm(nsnakes[3], mean.mass[3], sigma) # Mass data in population C
mass <- round(c(massA, massB, massC))           # Round and stack mass data

table(Pop)                                       # Look at sample size per pop. (not shown)
tapply(mass, Pop, mean)                          # Look at sample means per population
mean.mass                                        # Compare with known true means (not shown)

       A        B        C
131.6111 241.5000 246.8333
```

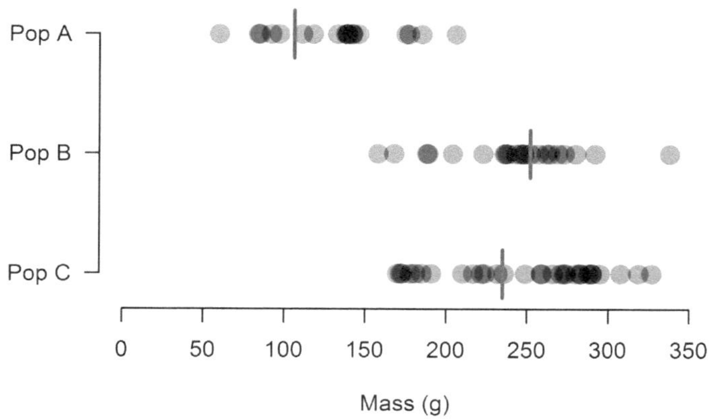

FIGURE 2.17

Body mass measurements (gray) and latent true mean mass (red lines) in three study populations of asp vipers (simulated data).

Thus, we simulate body mass data for a total of 70 snakes, with sample sizes of 18, 22, and 30 in populations A, B, and C, respectively. The true population means are 106, 252 and 235 g, respectively. However, there is individual heterogeneity in body mass of a magnitude 50 g, which is the value of the standard deviation of the normal distribution from which we draw the individual snake masses around the mean mass in their population. Hence, as a result of sampling variability the observed means do not match the true population means. Our rounding of the mass data to the nearest gram will introduce some additional variability into the data, but the magnitude of this is likely to be small, so we will ignore this variance component in the analysis. The data set is pictured in Fig. 2.17.

We want to fit a GLM with normal or Gaussian errors where we specify a categorical effect of population identity on snake mass. One simple way to write such a model is as follows.

$$y_i \sim Normal(\mu_i, \sigma^2)$$
$$\mu_i = \alpha_{pop(i)}$$

Here, y_i is body mass of snake *i*, *pop(i)* is a population indicator, or factor, with values (or levels) 1 to 3 for populations A to C. Thus, when snake *i* stems from population A, its mass measurement will be specified as a draw from a normal distribution with mean α_1, when snake *i* lives in population B it will obtain mean α_2 and when it is from population C, it obtains a mean of α_3. To fit this model in JAGS, we must first convert the factor labels into the integers from 1 to 3; this is the **only** way in which factor levels may be coded in JAGS! In the model, we will use nested indexing to specify the appropriate population-specific mean for the normal distributions of each snake *i* as a function of its population membership. Nested indexing is very similar to the way in which we just wrote the model in algebra. We note that this is the linear model underlying a one-way analysis of variance (ANOVA; Kéry, 2010). In the data simulation and the analysis, we make the standard assumption of a constant residual variance. However, we could easily account for variance heterogeneity (heteroscedasticity) among populations by double-indexing by `pop` also the parameters `tau` and `sigma`; see hashed-out parts in the code below.

```
# Bundle data
pop <- as.numeric(Pop)                 # Code factor levels as 1:3
jags.data <- list(y=mass, pop=pop, n=length(mass))
str(jags.data)

List of 3
 $ mass: num [1:70] 206 139 60 84 118 142 138 175 133 146 ...
 $ pop : num [1:70] 1 1 1 1 1 1 1 1 1 1 ...
 $ n   : int 70
```

```
# Write JAGS model file
cat(file="model7.txt", "
model {
# Priors and linear models
for (k in 1:3){
  alpha[k] ~ dnorm(0, 1.0E-06)        # Population means
# tau[k] <- pow(sigma[k], -2)         # For heteroscedasticity
# sigma[k] ~ dunif(0, 1000)
}
# Homogeneous residual variance (homoscedasticity)
tau <- pow(sigma, -2)
sigma ~ dunif(0, 1000)

# Likelihood for normal linear model
for (i in 1:n){
  y[i] ~ dnorm(mu[i], tau)            # Homoscedasticity
# y[i] ~ dnorm(mu[i], tau[pop[i]])    # Heteroscedasticity
  mu[i] <- alpha[pop[i]]
}

# Derived quantities: contrasts between population means
diff.12 <- alpha[2]-alpha[1]
diff.13 <- alpha[3]-alpha[1]
diff.23 <- alpha[3]-alpha[2]
}
")

# Initial values
inits <- function(){list(alpha=rnorm(3, 100, 300))}

# Parameters monitored
parameters <- c("alpha", "sigma", "diff.12", "diff.13", "diff.23")

# MCMC settings
ni <- 60000; nb <- 10000; nc <- 3; nt <- 5; na <- 1000

# Call JAGS from R (ART <1 min), check convergence and summarize posteriors
out8 <- jags(jags.data, inits,parameters, "model7.txt", n.iter=ni, n.burnin=nb, n.chains=nc,
  n.thin=nt, n.adapt=na, parallel=TRUE)
traceplot(out8, layout=c(2, 2)) # Not shown
print(out8, 2)

            mean    sd   2.5%    50%  97.5% overlap0    f Rhat n.eff
alpha[1] 131.60 10.53 111.10 131.48 152.59    FALSE 1.00    1 11010
alpha[2] 241.58  9.46 222.92 241.56 260.20    FALSE 1.00    1 30000
alpha[3] 246.80  8.16 230.86 246.79 262.89    FALSE 1.00    1 12920
sigma     44.60  3.92  37.69  44.32  53.01    FALSE 1.00    1 30000
diff.12  109.98 14.22  81.80 110.09 137.43    FALSE 1.00    1 13747
diff.13  115.20 13.27  88.95 115.28 141.07    FALSE 1.00    1  7177
diff.23    5.22 12.54 -19.51   5.32  30.04     TRUE 0.66    1 30000

# Compare the population mean estimates with the known truth
mean.mass                             # For population 1, 2 and 3 (or A, B and C)
[1] 106 252 235
```

Hence, all the 95% CRIs except that for the mean mass in population A contain the known truth in this example. In addition, two of the three pairwise differences between the population means do not contain zero and therefore could be considered “significant.”

For comparison, we fit the same model also using the method of least-squares, which for GLMs with normal errors produces estimates that correspond to the MLEs. Note that what is called “Residual standard error” in the output below is our sigma. We find the usual good numerical agreement between the two sets of estimates.

```
# Get frequentist MLEs for comparison
# summary(lm(mass ~ Pop))             # Treatment contrast parameterization
summary(lm(mass ~ Pop-1))             # Means parameterization of factor levels

Coefficients:
     Estimate Std. Error t value Pr(>|t|)
popA  131.611     10.315   12.76   <2e-16 ***
popB  241.500      9.331   25.88   <2e-16 ***
popC  246.833      7.990   30.89   <2e-16 ***
```

```
Residual standard error: 43.76 on 67 degrees of freedom
Multiple R-squared:  0.9639,    Adjusted R-squared:  0.9622
F-statistic: 595.7 on 3 and 67 DF,  p-value: < 2.2e-16
```

The population mean parameters in this model are estimated as fixed effects, i.e., as three unrelated constants. We could instead estimate them as random effects, in which case we would place a prior distribution on them, typically another normal, and then estimate as additional parameters the (hyper-)mean and (hyper-)variance in that prior distribution. In the last section of this chapter, we do just that, which lets us transition from a model with fixed population effects to one with population parameters treated as random effects.

2.7.7 GENERALIZED LINEAR MODELS WITH GAUSSIAN RANDOM EFFECTS

Here we quickly review specification of random effects in BUGS. Random effects are sets of parameters for which we define another, higher-level distribution, *with hyperparameters that are estimated.* This is a difference from treating them as fixed effects, as we did in the previous section, where we placed the following priors on the population means `alpha[k] ~ dnorm(0, 1.0E-06)`; there, no hyperparameters are estimated, rather, they are specified as fixed constants. There are numerous motivations for modeling a set of parameters as a realization from another stochastic process, i.e., to treat them as random effects (Bolker et al., 2009, Chapter 5 in Kéry and Royle, 2016), and we will not go into them here. Instead, we simply show the mechanics of doing it in JAGS.

For illustration, we imagine a variant of the normal linear regression from the last section with more populations and with site elevation as another explanatory variable on snake mass. When modeling the mass-elevation relationship we should accommodate the possible dependency of the mass of snakes living in the same population, by fitting a separate intercept for each population and treating it as a random effect. Hence, we specify the site-specific intercepts as draws from a normal distribution with shared and estimated hyperparameters. It is this additional distributional assumption that converts the Gaussian GLM into a Gaussian GLMM, linear mixed model, Gaussian multilevel model, or also generally a hierarchical model. Here is this GLMM written in algebra.

$$y_i \sim Normal(\mu_i, \sigma^2)$$

$$\mu_i = \alpha_{pop(i)} + \beta \times elev_i$$

$$\alpha_{pop(i)} \sim Normal(\mu_\alpha, \sigma_\alpha^2)$$

We now simulate data under the model (Fig. 2.18) and then fit the data-generating model in JAGS to illustrate the specification of random effects in the BUGS language.

```
# Set constants in simulation
npop <- 10                                        # Number of populations
grand.mean <- 200                                 # Grand mean of mass
sigma.pop <- 10                                   # SD of population variability in snake mass
beta <- -30                                       # Slope of mass on elevation
sigma.snake <- 50                                 # SD of individual variability of snake mass

# Simulate new snake mass data
set.seed(39)

# Simulate population-level data
mean.mass <- round(rnorm(npop, grand.mean, sigma.pop)) # Average mass (g) in each population at
                                                  # average elevation
elevPop <- sort(runif(npop, 200, 2000))           # 200 m to 2000 m elevation
elevPopSc <- as.numeric(scale(elevPop))
mean.pop.mass <- mean.mass + beta * elevPopSc
plot(mean.pop.mass, elevPop, pch=16)              # Not shown: population-level plot

# Simulate snake-level data
nsnakes <- rpois(npop, 20)                        # Number of snakes caught per population
mass <- numeric()
for (j in 1:npop){
  massPop <- rnorm(nsnakes[j], mean.pop.mass[j], sigma.snake)
  mass <- c(mass, massPop)
}

# Create individual-level elevation covariate and population factor
elev <- rep(elevPopSc, nsnakes)
pop <- rep(1:npop, nsnakes)
head(cbind(pop, elev, mass))
```

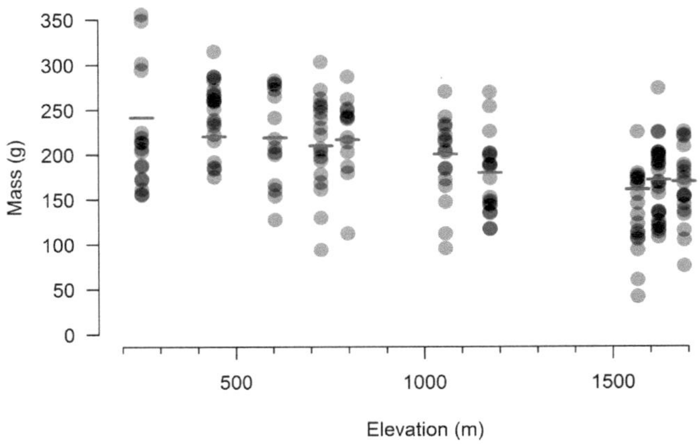

FIGURE 2.18

Simulated data of the body mass of 178 asp vipers in 10 populations along an elevational gradient. Black symbols show the mass measurements of each snake. Red symbols show the latent population means, which are linearly related to elevation but also have some population-level noise in them.

In the analysis, we use centered elevation and must supply the information about population membership of each individual snake. In the model code, for clarity of layout, we continue with a section heading entitled "Likelihood." However, we note that in a mixed model this is technically no longer entirely correct, since the likelihood maximized when fitting such a model with maximum likelihood is the so-called marginal likelihood, where the random effects are integrated out. See Chapters 2 in Royle and Dorazio (2008), Royle et al. (2014), and Kéry and Royle (2016) for more explanation on this. Nevertheless, mathematically all is of course still fine because in JAGS everything with a twiddle ($\sim$) contributes to the product of likelihood times prior.

```
# Bundle data
jags.data <- list(y=mass, pop=pop, elev=elev, npop=npop, n=length(mass))
str(jags.data)

List of 5
 $ y    : num [1:178] 214 155 203 214 348 ...
 $ pop  : int [1:178] 1 1 1 1 1 1 1 1 1 1 ...
 $ elev: num [1:178] -1.45 -1.45 -1.45 -1.45 -1.45 ...
 $ npop: num 10
 $ n    : int 178


# Write JAGS model file
cat(file="model8.txt", "
model {
# Priors and linear models
for (j in 1:npop){
  alpha[j] ~ dnorm(mu.alpha, tau.alpha) # Random effects
}
# Priors for hyper-parameters
mu.alpha ~ dnorm(0, 1.0E-06)          # Hyperprior for mean hyperparam
tau.alpha <- pow(sd.alpha, -2)
sd.alpha ~ dunif(0, 100)              # Hyperprior for sd hyperparam

# Other priors
beta ~ dnorm(0, 1.0E-06)              # Slope of mass on elevation
tau <- pow(sd, -2)
sd ~ dunif(0, 1000)                   # 1/residual variance

# 'Likelihood'
for (i in 1:n){
  y[i] ~ dnorm(mu[i], tau)
  mu[i] <- alpha[pop[i]] + beta * elev[i]
}
}
")
```

```
# Initial values
inits <- function() list(mu.alpha=rnorm(1), sd.alpha=runif(1))

# Parameters monitored
parameters <- c("mu.alpha", "sd.alpha", "alpha", "beta", "sd")

# MCMC settings
ni <- 110000; nb <- 10000; nc <- 3; nt <- 2; na <- 1000

# Call JAGS from R (ART <1 min), check convergence and summarize posteriors
out9 <- jags(jags.data, inits, parameters, "model8.txt", n.iter=ni, n.burnin=nb, n.chains=nc,
  n.thin=nt, n.adapt=na, parallel=TRUE)
traceplot(out9)                                                    # Not shown
print(out9, 2)

              mean    sd    2.5%     50%   97.5% overlap0 f Rhat  n.eff
mu.alpha    195.69  5.64  184.40  195.73  206.87    FALSE 1    1 150000
sd.alpha     11.66  7.12    0.89   10.80   28.42    FALSE 1    1   6323
alpha[1]    187.51 10.80  163.20  188.89  205.19    FALSE 1    1  23383
alpha[2]    203.56  9.62  187.42  202.46  224.92    FALSE 1    1  78358
alpha[3]    195.27  9.04  176.62  195.43  213.52    FALSE 1    1 145667
alpha[4]    195.69  8.26  179.01  195.72  212.43    FALSE 1    1 150000
alpha[5]    201.34  9.49  184.55  200.36  222.35    FALSE 1    1  93368
alpha[6]    196.33  8.31  179.67  196.26  213.38    FALSE 1    1  90852
alpha[7]    191.89  8.45  173.73  192.50  207.45    FALSE 1    1 150000
alpha[8]    183.99 11.19  159.23  185.23  201.78    FALSE 1    1  42527
alpha[9]    201.49  9.24  185.15  200.59  221.76    FALSE 1    1 130557
alpha[10]   199.74  9.79  181.62  198.99  221.20    FALSE 1    1 150000
beta        -30.76  5.83  -42.36  -30.75  -19.18    FALSE 1    1 150000
sd           47.80  2.62   43.03   47.68   53.28    FALSE 1    1 102355

# Compare the population mean estimates with the known truth
mean.mass
[1] 198 206 223 182 205 201 201 180 187 207
```

For comparison, we fit the same model also using the method of restricted maximum likelihood (REML). This is a variant of ML that is recommended for mixed models.

```
# Get frequentist restricted MLEs for comparison
library(lme4)
fm <- lmer(mass ~ elev + (1 | as.factor(pop)), REML=TRUE) # Fit model
summary(fm)                                                 # Show summary
ranef(fm)   # Give estimates of population mean deviations (not shown)

Random effects:
 Groups          Name        Variance Std.Dev.
 as.factor(Pop) (Intercept)   99.15    9.958
 Residual                   2242.97   47.360
Number of obs: 178, groups:  as.factor(Pop), 10

Fixed effects:
            Estimate Std. Error t value
(Intercept)  195.664      4.767  41.048
elev         -30.777      4.933  -6.239
```

The frequentist and Bayesian estimates we obtain are fairly similar. We must keep in mind that we would not expect the population variance estimates to be very good with data from only 10 populations.

2.8 FITTING GENERAL INTEGRATED MODELS IN JAGS

The topic of this book is IPMs, in which we link the statistical models for two or more demographic data sets through one or more shared parameters. Data integration, also called data assimilation or data fusion, is an immense topic in contemporary statistical ecology and its relevance extends far beyond its application in demographic population models (Hanks et al., 2011; Dorazio et al., 2017; Pacifici et al., 2017; Fletcher et al., 2019; Miller et al., 2019; Pacifici et al., 2019, Chapter 10 in Kéry and Royle, 2021). Statistically, an integrated model (IM) amounts to the fitting of a joint likelihood for two or more data sets, usually, but not always, under an assumption of independence. In that case, the joint likelihood of the IM is simply the product of the likelihoods of the individual data sets, which is conceptually fairly trivial. However, defining the likelihood of even simple parametric statistical models can be challenging for most ecologists, and hence, such joint likelihoods for multiple data sets will be virtually out of reach for them.

In contrast, the BUGS language makes it straightforward to specify a joint likelihood: we simply describe two or more statistical models within a single BUGS model statement and link them by some shared parameter(s). Though the big ugly joint likelihood is nowhere in sight, this amounts to the specification of such a product of the component-data likelihoods. To emphasize the generality of the approach of integrated modeling, we here provide one example that falls outside the scope of an IPM. See also Chapter 20 for another example of the specification of a general IM but this time as a part of an IPM.

BUGS is fantastic for fitting IMs in practice, but the nature of the joint likelihood is a little lost in the code. To improve our understanding of the construction of a joint likelihood as the product of the component-data likelihoods, it is very helpful to fit the same model with maximum likelihood, where we can actually see the likelihood constructed as a product, or the log-likelihood as a sum of the log-likelihoods, of the different data sets involved in the integration (Besbeas et al., 2002). Therefore, after the Bayesian solution in this section, and in keeping with most models fit in this chapter, we also use maximum likelihood to fit the IM. As a side effect, and along with Section 2.7.3, this then serves as another illustration of the principle of maximum likelihood for parameter estimation in models that are more complex than the trivial one-parameter model in the tadpole example in Section 2.3.

We consider an IM in the field of species distribution modeling (SDM) and assume we had two data sets that contain information about spatial patterns in the distribution or abundance of a species. We assume that both data sets were collected in the same broad region and that one data set contains counts, but the other consists only of detection-nondetection data, which are also known as "presence-absence" data (which of course is misleading whenever there is detection error). Here, we will assume that the two sets of sites are distinct, but the analysis could be adapted if there was some or even complete overlap among them. A natural framework for parametric statistical inference for these two data sets would be a Poisson GLM for the former and a Bernoulli GLM with logit link (i.e., a logistic regression) for the latter; see Sections 2.7.1 and 2.7.2. Thus, for the count C_i at site i we could write:

$$C_i \sim Poisson(\lambda_i)$$
$$\log(\lambda_i) = \alpha^C + \beta^C \times X_i$$

We illustrate the linear modeling of one site-specific covariate $\boldsymbol{X}$ at the log link scale for the expected count λ, and where the two parameters α^C and β^C are superscripted "C", to denote them as parameters in the model for the counts $\boldsymbol{C}$.

Similarly, for the binary detection-nondetection data y_j at site j we might write:

$$y_j \sim Bernoulli\left(\psi_j\right)$$
$$\text{logit}\left(\psi_j\right) = \alpha^y + \beta^y \times X_j$$

We again illustrate the linear modeling of one site-specific covariate $\boldsymbol{X}$ at the logit-link scale for the expected response ψ, and where the two parameters are now superscripted "y" to denote them as parameters in the model for response $\boldsymbol{y}$. In addition, note that we use different site indices i and j to indicate that we here assume we deal with different sets of sites.

Both models offer a valid characterization of the spatial variation of the distribution of the study species: one in terms of relative abundance here denoted λ (Johnson, 2008) and the other in terms of relative species occurrence probability or the apparent species distribution (Kéry, 2011), here denoted ψ. Note that the terms "relative" or "apparent" mean that true abundance or occurrence is confounded with imperfect detection. Intuitively, it seems obvious that patterns of occurrence also contain a signal of the underlying patterns of abundance, but would it be possible to formally combine the two models and thereby make use of all information about abundance that our two data sets contain?

It turns out that we can indeed do this by developing an IM that describes both data sets as different types of observations made on the same underlying process. Key to data integration here is the recognition that the detection-nondetection data are an information-poor variant of counts: they distinguish only between a count of 0 and a count of 1 or greater. Thus, the detection-nondetection data can be expressed as a summary of an underlying abundance process. There are multiple ways in which such a link between abundance and the binary detection-nondetection data can be formulated (Royle and Dorazio, 2008). Here, we specify the linkage via that vastly underused link function for Bernoulli or binomial data—the complementary log-log (cloglog) link (Section 3.3.6 in Kéry and Royle, 2016; Scharf et al., 2021). As we will see, by simply specifying that link function for the apparent occurrence probability ψ in the model for the binary detection-nondetection data, we describe the binary data in terms of an underlying Poisson process for abundance! Thus, the parameters in the linear model for ψ on the cloglog scale will automatically describe the log-linear relationship between the expectation of an assumed Poisson random variable that underlies the binary data, and the covariates in the model.

It is useful for our understanding of the cloglog link to look at it with algebra. When we model the binary data as a summary of an underlying abundance or count, the Bernoulli parameter is the probability to obtain a count greater than 0 under a Poisson abundance model. From the Poisson pmf (see Section 2.7.1), we can express this as

$$\psi = P(y=1) = P(N>0) = 1 - P(N=0) = 1 - \exp(-\lambda).$$

We can specify this relationship in BUGS directly (and below we will), but let us look now at how we can derive the cloglog link from this and how this provides an equivalent solution. Specification of a linear model at the scale of a cloglog link transformation for a Bernoulli parameter, i.e., $\text{cloglog}(\psi) = \log(-\log(1-\psi))$, is equivalent to the modeling of linear effects in $\log(\lambda)$, because we can make the following re-arrangements:

$$\begin{aligned}
&\psi = 1 - \exp(-\lambda), \text{ and therefore} \\
&\exp(-\lambda) = 1 - \psi, \text{ therefore} \\
&\log(\exp(-\lambda)) = \log(1-\psi), \text{ therefore} \\
&-\lambda = \log(1-\psi), \text{ therefore} \\
&\lambda = -\log(1-\psi), \text{ and thus, we finally obtain the following:} \\
&\log(\lambda) = \log(-\log(1-\psi)).
\end{aligned}$$

Now, what we have on the right-hand side is the cloglog link for ψ, the success probability in our Bernoulli model.

We here illustrate that "magic trick" for binary data by first simulating two count data sets and then "degrading" one of them to become binary detection-nondetection data. We then first fit a Poisson GLM to the former and a Bernoulli GLM with cloglog link to the latter individually and see how well we can recover the parameters of the original Poisson random variables that we simulated. In a second step, we develop an IM and fit it in JAGS, where we will see how trivially easy the fitting of an IM can be when using BUGS software; we simply fit two submodels within the same BUGS model statement, and one or more shared parameters are among the two submodels, where sharing means that a parameter bears the same name in both (and of course has the same meaning).

At the end of the section, we then write our own likelihood functions for all three models, the Poisson GLM, the Bernoulli GLM with cloglog link and the IM, and maximize them with `optim` in R. This will serve as another illustration of the principle of maximum likelihood for statistical estimation, and once more show the numerical match between the estimates from a Bayesian analysis with vague priors and the MLEs. But most of all, this will make much more explicit that the likelihood of the IM is the product of the likelihoods of the individual data models and that their log-likelihoods combine as a sum. In contrast, this important conceptual detail is virtually invisible in the BUGS code.

We start by simulating two count data sets with identical parameters for a log-linear regression on some environmental covariate X that here we call "elevation." We then degrade the information in Data Set 2 by turning the counts into binary detection-nondetection data. These latter are logistically "cheaper" to obtain in the field, so in our simulation, we assume that we have counts from 200 sites but detection-nondetection data from 500. We assume that detection is perfect: every animal at the site is counted and the species is detected at every occupied site. See Section 10.4 in Kéry and Royle (2021) for a variant of the IM in this section that also takes into account imperfect detection in both data sets. There, we combine a binomial N-mixture model (Royle, 2004) and a site-occupancy model (MacKenzie et al., 2002; Tyre et al., 2003) in one single integrated species distribution model (Fig. 2.19).

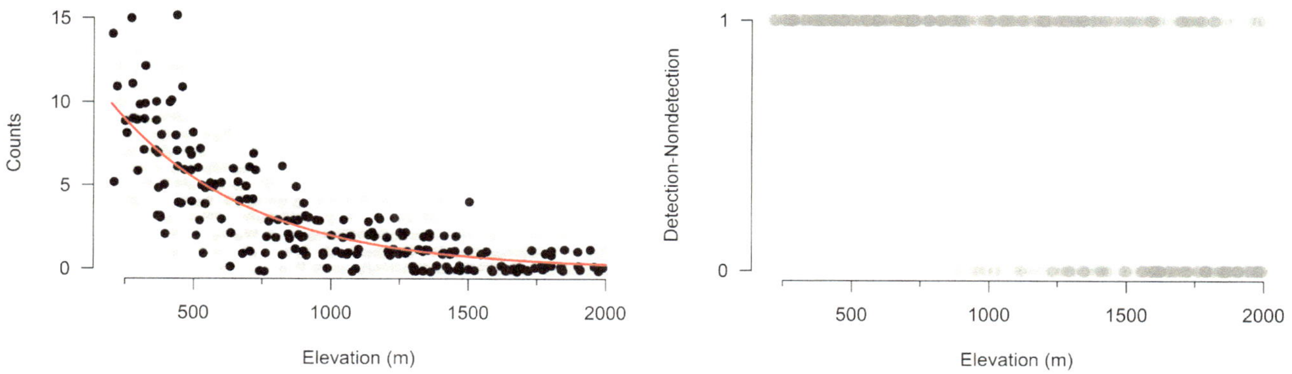

FIGTURE 2.19

Simulated count and detection-nondetection data as a function of a covariate "Elevation." Original data are Poisson counts (left) with 200 sites in Data Set 1 (black) and 500 sites in Data Set 2 (gray) and where the common expectation as a function of elevation is shown by the red curve. The counts in Data Set 2 are then "degraded" to mere detection-nondetection data (right), where a 1 denotes an observed presence (or detection) and 0 an observed absence (or nondetection). In the integrated model, we combine the counts in Data Set 1 on the left with the detection-nondetections in Data Set 2 on the right.

```
# Simulate the two data sets and plot them
# Choose sample size and parameter values for both data sets
nsites1 <- 200                          # Sample size for count data
nsites2 <- 500                          # Sample size for detection/nondetection data
mean.lam <- 2                           # Average expected abundance (lambda) per site
beta <- -2                              # Coefficient of elevation covariate on lambda

# Simulate elevation covariate for both and standardize to mean of 1000 and
# standard deviation also of 1000 m
set.seed(2016)
elev1 <- sort(runif(nsites1, 200, 2000)) # Imagine 200-2000 m a.s.l.
elev2 <- sort(runif(nsites2, 200, 2000))
selev1 <- (elev1 - 1000) / 1000
selev2 <- (elev2 - 1000) / 1000

# Create two data sets with replicated counts
C1 <- rpois(nsites1, exp(log(mean.lam) + beta * selev1))
C2 <- rpois(nsites2, exp(log(mean.lam) + beta * selev2))

# Turn count data set 2 (C2) into detection/nondetection data (y)
y <- C2                                 # Make a copy
y[y > 1] <- 1                           # Squash to binary
```

We start by fitting a Poisson GLM to Data Set 1 and a Bernoulli GLM with cloglog link to Data Set 2 to see how well we can recover the two parameters of the abundance model in each, i.e., 2 and −2 for the intercept (on the natural scale) and the slope (on the link scale), respectively. We do this with the IRLS method in R, which yields the MLEs for these models. We find estimates that are quite close to the known truth.

```
# Get MLEs for individual data sets
# Data Set 1: Poisson GLM with log link for counts
summary(fm1 <- glm(C1 ~ selev1, family=poisson(link="log")))
exp(coef(fm1)[1])                       # Estimate of lambda on natural scale from counts

Coefficients:
            Estimate Std. Error z value Pr(>|z|)
(Intercept)  0.62035    0.06213   9.985   <2e-16 ***
selev1      -2.19121    0.11691 -18.742   <2e-16 ***
[  .... truncated  .....]


(Intercept)
    1.85958

# Data Set 2: Bernoulli GLM with cloglog link for detection-nondetection
summary(fm2 <- glm(y ~ selev2, family=binomial(link="cloglog")))
exp(coef(fm2)[1])                       # Estimate of lambda on natural scale from binary data

Coefficients:
            Estimate Std. Error z value Pr(>|z|)
(Intercept)  0.59805    0.07968   7.506  6.1e-14 ***
selev2      -1.95305    0.17628 -11.079  < 2e-16 ***
[  .... truncated  .....]

(Intercept)
   1.818574
```

We now go on to fit the IM first in JAGS, where we simply stack the two GLMs inside the same BUGS model, and in both parts of the IM, choose the same names for the parameters alpha and beta. Note that this is what defines the joint likelihood as a product of the single-data likelihoods. Hence, even though you cannot see it, the likelihood of this model is thereby defined as $L_{joint} = L_1 \times L_2$, where L_1 and L_2 are, respectively, the likelihood of the Poisson GLM for the count Data Set 1 and the Bernoulli GLM for the detection-nondetection Data Set 2.

```
# Bundle data
jags.data <- list(C=C1, y=y, nsites1=nsites1, nsites2=nsites2, selev1=selev1, selev2=selev2)
str(jags.data)

List of 6
 $ C       : int [1:200] 14 5 11 9 8 15 11 9 9 6 ...
 $ y       : num [1:500] 1 1 1 1 1 1 1 1 1 1 ...
 $ nsites1: num 200
 $ nsites2: num 500
 $ selev1 : num [1:200] -0.795 -0.789 -0.78 -0.75 -0.744 ...
 $ selev2 : num [1:500] -0.793 -0.791 -0.782 -0.78 -0.774 ...


# Write JAGS model file
cat(file="model9.txt", "
model {
# Priors and linear models: shared for models of both data sets
alpha ~ dunif(-10, 10)              # Abundance intercept
mean.lam <- exp(alpha)
beta ~ dnorm(0, 0.01)

# Likelihoods for Data Sets 1 and 2
# Note identical alpha and beta for both data sets
for (i in 1:nsites1){               # Data Set 1
  C[i] ~ dpois(lambda1[i])
  log(lambda1[i]) <- alpha + beta * selev1[i]
}
for (j in 1:nsites2){               # Data Set 2
  y[j] ~ dbern(psi[j])
  cloglog(psi[j]) <- alpha + beta * selev2[j]

  # Alternative implementation of same model for Data Set 2
  # y[j] ~ dbern(psi[j])
  # psi[j] <- 1 - exp(-lambda2[j])
  # log(lambda2[j]) <- alpha + beta * selev2[j]
}
}
")

# Initial values
inits <- function(){list(alpha=runif(1), beta=rnorm(1))}

# Parameters monitored
parameters <- c("mean.lam", "alpha", "beta")

# MCMC settings
ni <- 12000; nb <- 2000; nc <- 3; nt <- 10; na <- 1000

# Call JAGS from R (ART <1 min), check convergence and summarize posteriors
out10 <- jags(jags.data, inits, parameters, "model9.txt", n.iter=ni, n.burnin=nb, n.chains=nc,
  n.thin=nt, n.adapt=na, parallel=TRUE)
traceplot(out10)                    # Not shown
print(out10, 3)

           mean    sd    2.5%     50%    97.5% overlap0 f  Rhat n.eff
mean.lam  1.902 0.079   1.747   1.901    2.061    FALSE 1 1.000  3000
alpha     0.642 0.041   0.558   0.643    0.723    FALSE 1 1.000  3000
beta     -2.127 0.079  -2.281  -2.127   -1.971    FALSE 1 1.001  2429
```

Compared with the separate analyses of each data set (i.e., the MLEs above), we seem to be doing better in the IM. In particular, we note how the posterior SDs (to be compared with the SEs above) are considerably smaller now. This means that we obtain more precise estimates in the integrated analysis because we are able to exploit the information about abundance contained in both data sets.

Next, we also try to fit the IM using maximum likelihood. Since R has no canned function for doing this, we must write our own code. We start with the likelihood code in Section 2.7.3 and adapt it first for each submodel separately, i.e., for the Poisson GLM and for the Bernoulli GLM with cloglog link. Remember from Section 2.7.3 that we must first define the negative log-likelihood (NLL) function for our model and then use `optim` to minimize this function with respect to the parameters for the observed data. Under the assumption of independence of the observations, the likelihood of each data set, L_1 and L_2, is the product over the probability of each individual observation. The joint likelihood to be maximized for the IM is then again $L_{joint} = L_1 \times L_2$.

Thus, here is code to use maximum likelihood to fit the Poisson GLM to Data Set 1, i.e., the counts alone. We also compute the SEs of the estimates from the Hessian matrix as described in Section 2.7.3 (see also Bolker, 2008, and Chapter 2 in Kéry and Royle, 2016).

```
# Definition of negative log-likelihood (NLL) for Poisson GLM
NLL1 <- function(param, y, x){
  alpha <- param[1]                # Organize parameters in a vector
  beta <- param[2]
  lambda <- exp(alpha + beta * x)# Linear predictor
  L <- dpois(y, lambda)            # Likelihood contribution for each datum
  return(-sum(log(L)))             # NLL for all observations in data set
}

# Minimize NLL
inits <- c(alpha=0, beta=0)        # Need to provide initial values
sol1 <- optim(inits, NLL1, y=C1, x=selev1, hessian=TRUE)
mle1 <- sol1$par                   # Grab the MLEs
VC1 <- solve(sol1$hessian)         # Get variance-covariance matrix
ASE1 <- sqrt(diag(VC1))            # Extract asymptotic SEs
print(cbind(mle1, ASE1), 3)        # Print MLEs and asymptotic SEs

       mle1   ASE1
alpha  0.62 0.0621
beta  -2.19 0.1169
```

Next, is the code for using maximum likelihood to fit the Bernoulli GLM with cloglog link to Data Set 2, i.e., the detection-nondetection data alone:

```
# Definition of negative log-likelihood for Bernoulli GLM with cloglog link
NLL2 <- function(param, y, x){
  alpha <- param[1]
  beta <- param[2]
  lambda <- exp(alpha + beta * x) # Linear predictor
  psi <- 1-exp(-lambda)           # cloglog link
  L <- dbinom(y, 1, psi)          # Likelihood contribution for each datum
  return(-sum(log(L)))            # NLL for all observations in data set
}

# Minimize NLL
inits <- c(alpha=0, beta=0)
sol2 <- optim(inits, NLL2, y=y, x=selev2, hessian=TRUE)
mle2 <- sol2$par                  # Grab MLEs
VC2 <- solve(sol2$hessian)        # Get variance-covariance matrix
ASE2 <- sqrt(diag(VC2))           # Extract asymptotic SEs
print(cbind(mle2, ASE2), 3)       # Print MLEs and asymptotic SEs

        mle2   ASE2
alpha  0.598 0.0794
beta  -1.953 0.1765
```

Finally, we put the two likelihoods together in a single joint likelihood for the IM for both data sets simultaneously. As shown above and under the usual independence assumption the joint likelihood for the two submodels is simply the product of the two single-data likelihoods, $L_{joint} = L_1 \times L_2$. However, we have seen in Sections 2.4 and 2.7.3 that we usually work with the log of the likelihood or the NLL, and for these, we have a sum rather than a product. This is hardly different in fact from the joint likelihood for each submodel, which is also a product over the individual likelihood contributions from each single observation in the data set or a sum over their log-likelihoods. Hence, the objective function to minimize for our IM is the following:

$$NLL_{joint} = -(\log(L_1) + \log(L_2)).$$

Let us try this out and fit our first IM using maximum likelihood.

```
# Definition of the joint NLL for the integrated model
NLLjoint <- function(param, y1, x1, y2, x2){
  # Definition of elements in param vector (shared between data sets)
    alpha <- param[1]
    beta <- param[2]
  # Likelihood for the Poisson GLM for data set 1 (y1, x1)
    lambda1 <- exp(alpha + beta * x1)
    L1 <- dpois(y1, lambda1)              # Likelihood contribution each datum
  # Likelihood for the cloglog Bernoulli GLM for data set 2 (y2, x2)
    lambda2 <- exp(alpha + beta * x2)
    psi <- 1-exp(-lambda2)
    L2 <- dbinom(y2, 1, psi)              # Likelihood contribution each datum
  # Joint log-likelihood and joint NLL: here you can see that sum!
    JointLL <- sum(log(L1)) + sum(log(L2)) # Joint LL
    return(-JointLL)                       # Return joint NLL
}

# Minimize NLLjoint
inits <- c(alpha=0, beta=0)
solJoint <- optim(inits, NLLjoint, y1=C1, x1=selev1, y2=y, x2=selev2, hessian=TRUE)

# Get MLE and asymptotic SE and print them
mleJoint <- solJoint$par                 # Grab MLEs
VC <- solve(solJoint$hessian)            # Get variance-covariance matrix
ASE <- sqrt(diag(VC))                    # Extract asymptotic SEs
print(cbind(mleJoint, ASE), 3)           # Print MLEs and asymptotic SEs

      mleJoint    ASE
alpha    0.642 0.0408
beta    -2.130 0.0804
```

Once again, both sets of estimates numerically resemble their Bayesian counterparts from above very much.

So now you understand the principle of data integration by working with a joint likelihood. To see that this latter is a product of the single-data likelihoods is easiest when working with maximum likelihood, because then the joint likelihood is clearly apparent. In contrast, in the BUGS code the joint likelihood is really hidden. However, in practice for most nonstatisticians it is much easier to specify that same joint likelihood for an IM in BUGS than when using maximum likelihood. Especially when we fit more complex IMs such as IPMs, for which maximum likelihood estimation requires use of Kalman filter techniques (Besbeas et al., 2002, 2003) or fitting algorithms for hidden Markov models (Besbeas and Morgan, 2019).

Opportunities for adopting such IMs abound in ecology and wildlife management and we expect to see increasingly more models to which to apply the principles of integrated modeling, where one or more parameter is shared between the likelihoods for two or more data sets. The power and flexibility of the BUGS language makes it almost trivially easy to define the joint likelihood that is maximized in such an IM.

2.9 WHY WE HAVE BECOME BAYESIANS...

...and why you might want to become one, too (this great quote is from Bill Link). To us, there are three big but completely separate advantages of a Bayesian analysis with BUGS software. They have to do with the Bayesian paradigm, with Bayesian computation using MCMC, and with BUGS software.

First of all, we like the Bayesian paradigm because of its "natural" concept of probability as a measure of uncertainty. It is exactly analogous to the way in which humans use probability when we say things like "*I am 99% certain that* ...." (Wade, 2000). Furthermore, if we want to introduce external information into an analysis, it can be achieved in a very straightforward way in a Bayesian analysis by specification of an informative prior.

Second, we like Bayesian computation because it makes the fitting of hierarchical models (with random effects) more straightforward, and especially, it becomes trivial to compute functions of parameters with full error propagation. This is a huge practical advantage and extends to even very complicated, derived quantities, as we have seen at the end of Section 2.7.2.

Third, and for us this is really the most important advantage of the three, is the power and flexibility of BUGS software. BUGS enables us to implement complex, completely customized statistical models, including IPMs. The hierarchical construction of the likelihood for even very complex statistical models in BUGS has given us a modeling freedom that we would never even have dreamt of before we had learned BUGS. This practical reason is the most important one for why we appear to have become Bayesians: BUGS has freed the modeler in us.

On the other hand, some might argue that we are not *real* Bayesians, because we rarely use informative priors. We might even be accused of being closet frequentists, who merely want to benefit from the power of Bayesian computation (i.e., MCMC) and its user-friendly implementation in BUGS to obtain what numerically are simply the same old MLEs as do the non-Bayesians. In addition, there are certainly also some inconveniences of Bayesian analyses with BUGS: they usually take a long time to run, model selection is perhaps even more challenging than when using maximum likelihood, the sensitivity of results to prior choice adds a kind of ambiguity not there when using maximum likelihood, and lack of identifiability is perhaps an even greater challenge in the Bayesian analysis of a complex model than in a likelihood analysis (Ogle and Barber, 2020). Therefore, when MLE is an option, we are happy to apply this (or other) non-Bayesian methods as well.

On balance, we would suggest being eclectic and choosing the most useful inference method for you and for the analytical task at hand. Usually, we may not use BUGS for a trivial problem, but BUGS is fantastic for more complex models, except when you have very large data sets; then BUGS may simply be too slow. In addition, using BUGS always has as a great side effect that you will be much more likely to actually understand the model you are fitting than when using many canned routines in R or other software. And finally, we suggest you stay tuned: in the future there will hopefully (!) be much more powerful MCMC and even likelihood-maximizing software developed that will expand beyond what even JAGS is capable of now. One of our hopes is that package developers will even more widely recognize the power of the BUGS language as a *lingua franca* for the description of hierarchical models. It is the BUGS language that is primarily responsible for the extremely wide reception of BUGS software especially among nonstatisticians. We are really pleased to see that the NIMBLE developers have recognized this and adopted BUGS as the model-definition language in their great new project (de Valpine et al., 2017).

2.10 SUMMARY AND OUTLOOK

We have discussed the formal description of "variable things" in statistics as random variables by way of a probability density or mass function (here collectively abbreviated as a pdf). A pdf is a function that assigns a probability of observing any of an exhaustive set of observations as a function of one or more fixed constants. When we build a statistical model for a data set, we treat such "variable things" as a random variable. We also call them a "response variable" for which we must pick a suitable pdf, where now the constants will become the unknown parameters that we want to estimate from our data set. Depending on the type of data set and the sampling method employed to collecting those data, different types of pdfs will be adequate. We can account for additional variability and explain variability by making these parameters functions of covariates and random effects, as in linear models, GLMs, and GLMMs. Finally, we can put the description of multiple random variables in a statistical model in a sequence organized by the conditional dependence structure of the processes involved. The result is then a hierarchical statistical model (HM) for systems with multiple stochastic components, which have both observed and unobserved random variables. IPMs are generally examples of such HMs.

When we confront an observed data set with such models, we can estimate parameters and other quantities, such as missing covariates, missing responses, and in general predictions. With a parametric statistical model, estimation can be based directly on the likelihood function, and MLE may be obtained for a classic (or frequentist) solution. Alternatively, the likelihood function can be combined with priors for all the unknown parameters in it, and the posterior distribution can be obtained for a Bayesian solution. MLEs generally require numerical optimization, whereas MCMC sampling-based algorithms are typically used for Bayesian solutions. In this chapter, we have illustrated in detail both estimation methods for one very simple model, before introducing the family of BUGS software and more specifically the program JAGS (Plummer, 2003). We then illustrated the use of JAGS by fitting each of the major building blocks for most IPMs: GLMs with a Poisson or Bernoulli, or binomial, multinomial, or categorical and with a normal distribution assumed for the observed response. We also used JAGS to fit one of the simplest HMs of the GLMM type. Finally, we looked at the general principle of building IMs for multiple data sets, by specification of a joint likelihood, in a final example of an integrated species distribution model.

This chapter was meant to be a crash course in parametric statistical modeling, Bayesian inference, and MCMC-based analyses using program JAGS. These are the statistical and computational tools we use throughout this book to fit IPMs. Coverage and depth of all these topics has necessarily been only minimal. We expect this material for you to be a refresher rather than a real first-time introduction. If any of this is totally new for you, we would suggest to you to read up on it in other, more in-depth sources. We have given pertinent citations throughout the chapter.

At the end of the chapter, we have reviewed some advantages of a Bayesian versus a maximum likelihood analysis of a statistical model. We have argued that both have their place in modern applied statistics. Nevertheless, one of the greatest assets about a Bayesian analysis in BUGS software is the BUGS language. It really drives home a major advantage of hierarchical modeling—you can and should build a statistical model in exactly the manner in which you envision the processes that underlie an observed data set. BUGS software makes hierarchical modeling completely natural, and this invites a principled and mechanistic approach to statistical modeling leading to what have been called science-based statistical models.

2.11 EXERCISES

1. Convince yourself using simulation that a uniform prior for the intercept of a logistic regression can be very informative about likely values of the intercept on the probability scale.
2. In a tadpole experiment, none of the 20 released individuals was seen. Analyze this data set with both Bayesian and maximum likelihood methods to estimate detection probability. Are the estimates and the SD or SE identical? If not, why are they different?
3. Repeat some of the analyses described with a variety of different prior distributions to examine the impacts of highly informative priors on posterior distributions.

CHAPTER

INTRODUCTION TO STAGE-STRUCTURED POPULATION MODELS

3

CHAPTER OUTLINE

3.1 INTRODUCTION

A central goal of population ecology is to achieve a detailed quantitative understanding of a population. We may look into the past to understand why a population has experienced particular dynamics (in a *retrospective* analysis), or we may look into the future to predict its likely future dynamics (in a *prospective* analysis). Some population studies focus exclusively on the *number* of individuals (population size) and its change over time. The primary interest of such studies is the population growth rate, which is the main parameter of many corresponding models. The population trend, i.e., the long-term sustained rate of population change, may be estimated as well (Dennis et al., 1991; Morris and Doak, 2002; Tuljapurkar et al., 2003). Extensions of such models may include density dependence, environmental or demographic variability ("stochasticity") (Dennis et al., 1991; Dennis and Taper, 1994; Lande et al., 2003), and observation errors (de Valpine and Hastings, 2002; Dennis et al., 2006; Pasinelli et al., 2011). Other studies aim to obtain a more mechanistic understanding of population dynamics, which requires inclusion of demographic rates in the models. Such models are extensions of more phenomenological models adjusted for population size. The typical research aims of such mechanistic models include assessment of the strength of the contribution of each demographic rate to population growth and how the latter is expected to change in the future with changes in the mean or temporal variance of a demographic rate. Clearly,

Integrated Population Models. https://doi.org/10.1016/B978-0-12-820564-8.00019-1

such population models must include an explicit link between demographic rates and population size—the population growth rate appears as a derived quantity. Often these models classify individuals into stages, typically age classes, such that individuals within a stage can be assumed to have identical demographic rates. A number of excellent textbooks about demography and population dynamics cover this kind of population model (Newton, 1998; Caswell, 2001; Morris and Doak, 2002; Skalski et al., 2005; Conroy and Carroll, 2009; Mills, 2013), the most detailed of which is the celebrated treatise by Caswell (2001). Several names exist for these models, including *matrix population model, matrix projection model, matrix-based projection model, stage-structured population model, matrix model*, and *Leslie matrix model*. In this book we will typically use "matrix population model" or simply "population model," because these are general and concise names. Integrated population models (IPMs) also make an explicit link between demography and population size and structure, and hence, matrix population models often lie at the core of an IPM, and you need sound knowledge of them to develop an IPM. In this chapter, therefore, we provide a concise introduction to matrix population models (though we make no claims of a fully comprehensive treatment of this immense topic).

The size of a population (N_t) at time t can be expressed by the following sum:

$$N_t = S_t + R_t + I_t - E_t$$

Here, S_t is the number of survivors from year $t-1$, i.e., individuals already present in the population in year $t-1$ that survive to year t. R_t is the number of local recruits, i.e., individuals born in the population in year $t-1$ that survive until year t. I_t is the number of immigrants in year t, i.e., individuals born outside the study population that reproduce within the study area for the first time. Finally, E_t is the number of individuals among the survivors from the previous year (i.e., R_t plus S_t) that emigrate to another population in year t. Hence, the population size in year t is given by the population size the year before, plus the number of new individuals resulting from local recruitment (i.e., born on site) and immigration (i.e., born elsewhere), minus the number of individuals lost as a result of death and emigration.

The population growth rate (λ_t) is a key parameter for describing population dynamics. It is defined as the rate of change in the size of a population between time $t-1$ and t:

$$\lambda_t = \frac{N_t}{N_{t-1}} = \frac{S_t}{N_{t-1}} + \frac{R_t}{N_{t-1}} + \frac{I_t}{N_{t-1}} - \frac{E_t}{N_{t-1}}.$$

The ratios on the right-hand side of the equation are per-capita numbers and represent the demographic rates. Written in that way, we obtain

$$\lambda_t = \frac{N_t}{N_{t-1}} = s_{t-1} + r_{t-1} + i_{t-1} - e_{t-1}.$$

The population growth rate is the sum of survival rate s_{t-1} (i.e., survival probability from time $t-1$ to t; here, the leading time step provides the time index), recruitment rate r_{t-1}, and immigration rate i_{t-1}, minus emigration rate e_{t-1}.

We make a number of important observations here. First, it is obvious that the rate of population change, i.e., the population growth rate, is a function of exactly four demographic rates: recruitment, immigration, survival, and emigration. Hence, whenever the population growth rate changes, it *must* be due to a change in at least one of the four demographic rates. Thus, this equation provides a formal link between demography and population growth that allows us to study demographic causes of population change or stability. However, not every change in a demographic rate need necessarily result in a change in the population growth rate. For instance, if one rate increases while another decreases by a corresponding amount, we will observe no change in the population growth rate even though the demography of the population has changed. Thus, changes in demographic rates are a necessary but not sufficient condition for a change in the population growth rate. Second, population *size* can change over time even if the population growth *rate* (or demographic rates) does not. To see why, consider a population with population growth rate λ that is constant over time and calculate population size in each year. Starting with a population size of N_0, the population size is $N_1 = \lambda N_0$ in year 1, $N_2 = \lambda N_1 = \lambda^2 N_0$ in year 2, and $N_t = \lambda N_{t-1} = \lambda^t N_0$ in year t. Only if $\lambda = 1$ does the population size not change; if $\lambda > 1$, the population grows exponentially, and if $\lambda < 1$, it declines asymptotically to zero (Fig. 3.1). Finally, we note a "notational conflict" around the population growth rate, which is traditionally expressed as λ, which is also used to represent the parameter of a Poisson distribution. Later in the book, we will see models with both a growth rate and Poisson expectation and will then use λ for the former.

To this point, we have assumed that demographic rates are constants and do not vary over time or along other dimensions. However, temporal variation in demographic rates is common. For example, annual survival of barn owls (*Tyto alba*) is much lower in years with harsh winters than in years with mild winters (Altwegg et al., 2006), and reproduction of fat dormice (*Glis glis*) occurs only in years with beech mast seeding (Pilastro et al., 2003), resulting in temporal variation in annual survival and reproductive output, respectively. Although the resulting population size fluctuates, it still grows exponentially. However, real-world populations cannot grow forever, but instead require a feedback between their size and

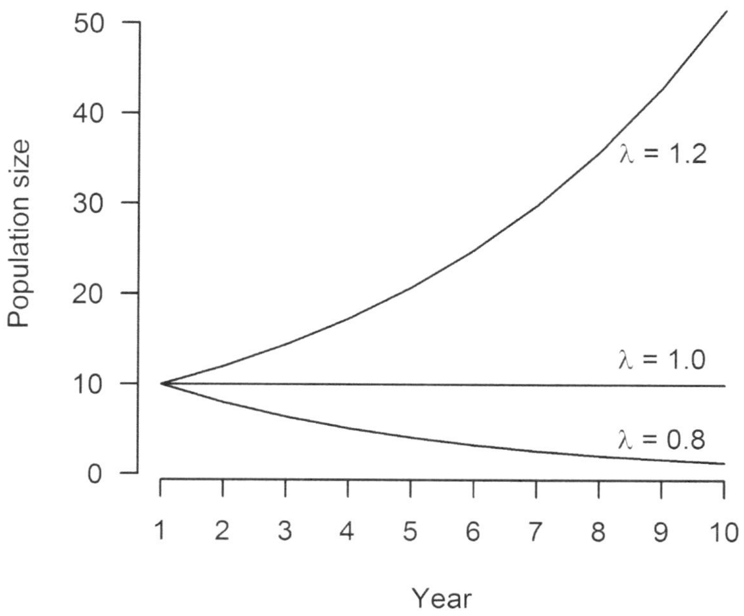

FIGURE 3.1

Expected trajectories of an exponentially growing population with three different values of the population growth rate (λ) that are constant over time. Only when the population growth rate is exactly 1 does the population size not change. Otherwise, the population either increases exponentially ("explodes") or declines asymptotically to zero (goes extinct).

demographic rates to keep them within bounds and avoid population explosion or extinction. This feedback represents the regulation of a population. A fundamental aim in population ecology is to understand how populations are regulated, i.e., to identify the limiting factors of population growth (Newton, 1998). One fundamental mechanism providing feedback is *density dependence*. For our population model, this means that for at least one demographic rate, there must be a negative relationship with population size (or a positive relationship with the emigration rate). Several mechanisms may be responsible for such feedback (Fretwell and Lucas, 1969). One important example is competition, where some individuals of a territorial species are forced to occupy territories of lower and lower quality as the population grows, thereby decreasing the average territory quality. This will result in a decline in average productivity at the population level (Rodenhouse et al., 1997). Competition may also result in a smaller territory size at a higher population density, and as a consequence, available resources are reduced for each individual on average, which may again lead to reduced productivity (Rodenhouse et al., 2003; Sillett et al., 2004; Nevoux et al., 2011). In addition to productivity, survival may change as a function of density, e.g., if competition forces wintering individuals to stay at lower-quality sites where they experience increased mortality (Gill et al., 2001; Newton, 2004). In these cases, we speak of density-dependent productivity or survival, but in reality there may be a density-dependent feedback loop with any of the demographic rates mentioned earlier. We note that a positive relationship between demographic rates and population size is also possible and is called the "Allee effect" (Courchamp et al., 1999), but Allee effects are typically restricted to small populations and species with certain social systems (Kramer et al., 2009).

3.2 AGE- AND STAGE-STRUCTURED POPULATION MODELS

The population model developed in the previous section is one of the simplest possible. It distinguishes between different types of individuals solely in terms of "where they come from." This model may be useful in some circumstances, but it makes the restrictive assumption that all individuals within a population are identical demographically and thus have identical survival probabilities, produce an identical number of offspring, and have identical dispersal. Clearly, such assumptions of homogeneous demographic rates for all individuals can never be strictly true. For example, survival typically changes with the age of an individual—it increases from birth, reaches a maximum, and declines thereafter (i.e., there is senescence; Gaillard et al., 1994; Loison et al., 1999; Tavecchia et al., 2001). Reproductive rates often also change with age. Many species only reach sexual maturity after two or more years (Harvey and Zammuto, 1985; Hadley et al., 2006), and after the start of reproduction, reproductive performance may increase with age and then decline at older ages (Forslund and Pärt, 1995; Aubry et al., 2009). However, age is not the only factor with a large impact on demographic rates. For example, many adult albatrosses can only reproduce every second year (Converse et al., 2009; Gauthier et al., 2012), and thus the reproductive output of an individual in year t depends on whether it reproduced in year $t-1$, or to put it in other words, what state (breeder, skipped breeder) the bird was in during the previous year. Other examples refer to the size of organisms with indeterminate growth (fish, reptiles, plants), in which demographic performance is not only a function of age but more generally a function of a "stage" defined by size. In most species, demography changes with age or stage, and realistic population models ought to account for these patterns.

The population model developed above also requires time to be discrete; that is, we describe the population at specific time points. The interval between these time points is called the projection interval. It is often 1 year in vertebrate species and temperate latitudes, but you may define other intervals as well. Continuous-time population models are possible, but they are mathematically much more complex and perhaps less adequate for seasonally reproducing vertebrate species (Hastings, 1997). Therefore, in this book we only cover discrete-time models.

Age-structured population models have a long history going back to Leslie (Leslie, 1945), who formalized their mathematics. Stage-structured population models are a generalization of age-structured population models and received their first mathematical treatment by Lefkovitch (1965); see Caswell (2001) for more on their history. Strictly speaking, a Leslie matrix model is an age-structured population model, whereas a stage-structured model is a Lefkovitch matrix model, though the term Leslie matrix model is often used for both.

In the following, we illustrate the construction of both age-structured and stage-structured population models.

3.2.1 FROM A LIFE-CYCLE GRAPH TO POPULATION EQUATIONS

A life-cycle graph is a graphical depiction of the life cycle of a species and can be constructed for either age-structured or stage-structured models. It is composed of nodes that represent the age/stage classification of individuals and arrows connecting these nodes that represent all possible demographic transitions within the chosen projection interval. The arrows are labeled with the transition probabilities given by demographic processes such as survival, reproduction, growth, or a combination thereof. The process of reproduction is usually expressed by productivity or fecundity, which we define by the annual number of independent offspring per female with a given stage class (Etterson et al., 2011). In our experience, it is advisable to start the development of any age/stage-structured population model, and thus of any IPM, with the construction of a life-cycle graph—it enforces more transparency with respect to the assumptions and simplifications we make about the demography of our study species. Furthermore, the formulae that describe age/stage-specific population growth can be derived easily from such a graph. If you got your graph right, making errors when deriving these equations will be difficult.

We next illustrate the development of life-cycle graphs and derivation of age- and stage-specific population sizes using the charismatic woodchat shrike (*Lanius senator*, Fig. 3.2), a species that will feature prominently at various places in this book. Woodchat shrikes breed mainly in the Mediterranean region in Western Europe and during the winter migrate to sub-Saharan Africa (Lefranc and Worfalk, 1997). This splendid bird mainly feeds on invertebrates such as beetles or grasshoppers, and unfortunately, its populations are declining in most parts of its distributional range (Sanderson et al., 2006). The woodchat shrike has a life cycle typical of many passerine bird species (Schaub and Ullrich, 2021). It starts to reproduce as a 1-year-old. Productivity is lower for a 1-year-old than for older females, and we assume that either all females reproduce in every year or that productivity includes a component for any females that skip reproduction in a given year. Juvenile survival (i.e., survival from fledging to age 1) is lower than adult survival (i.e., from age 1 onwards). But before we can draw a life-cycle graph based on this life-history information for the shrike, we must make a number of decisions.

First, remember that the size of any population may change from one day to the next. In birth-pulse populations, where reproduction of all or most individuals occurs at a particular time of the year, the pattern of change in the total population follows a typical pattern (Fig. 3.3; note that by "total" we mean all individuals, not just those holding a territory). Just after reproduction, the size of the population is largest; it then declines with deaths that occur until the start of the next reproductive period. Population size then increases again fairly sharply because of reproduction. Thus, population size follows a sawtooth pattern when plotted continuously over time. Yet for the models in this book, we discretize time and hence need to decide at which points in time we want to observe the population and how long the projection interval is. In birth-pulse populations, where reproduction occurs in one particular period of the year (and not continuously during the entire year as in birth-flow populations), the projection interval is typically 1 year, and there are two basic options for the

FIGURE 3.2

A beautiful male woodchat shrike (*Lanius senator*; photograph by Ruedi Aeschlimann).

observations. First, we can observe the population just before the reproductive period when its size is lowest, in which case we develop a pre-birth-pulse model. This is often called a model for a pre-birth-pulse census or a prebreeding census, but we note that the commonly used term "census" is a misnomer in natural populations where complete enumerations are essentially never possible and where, in reality, "survey" is the correct term. Alternatively, we can observe the population just after reproduction is completed when the total size of the population is largest and build a post-birth-pulse model, also known as a model for a post-birth-pulse "census" or postbreeding "census". As we will see later, inferences about population dynamics are not affected by this decision. What is different between these two types of models, however, is the identity of specific age/stage classes—some occur only in one type of model and not the other. Therefore, your choice of model type depends on your research interests. For IPMs, pre-birth-pulse models are often more practical because they result in more intuitive indexing of time and age (see below). In any case, it is important to stick to one of the two possibilities and not confuse them. In our experience, students with limited previous exposure to these models often have difficulty differentiating them. Therefore, we spend some time here to describe them.

Depending on our research interest, we can also discretize time in different ways. For example, we might want to understand population changes within a year and therefore construct a seasonal model (Hostetler et al., 2015; Rushing et al. 2017). Because the woodchat shrike is migratory, we thus might want to study the dynamics of population size not only on the breeding but also on the wintering grounds. Such models require more detailed information about demographic rates, such as seasonal survival probabilities.

Second, we must decide whether to model only one sex or to construct a two-sex model. Often, population models are female-based, thus describing only females and assuming unlimited availability of males for females, or expressed differently, that males do not affect the long-term dynamics of the population. Of course, this assumption is never strictly

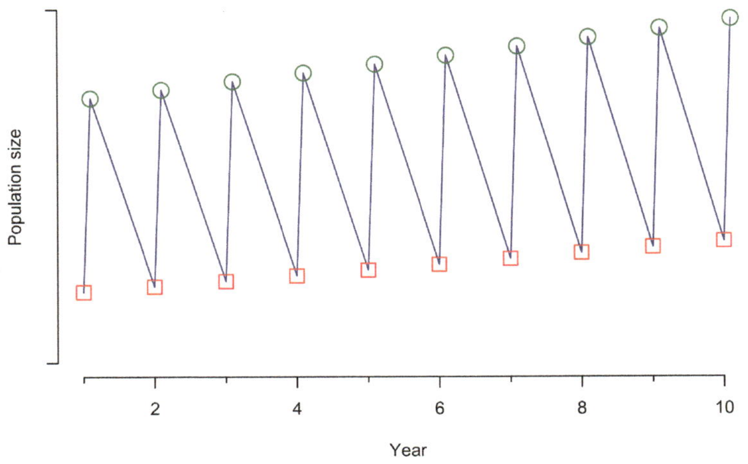

FIGURE 3.3

Schematic showing the development of the total size of a birth-pulse population. The blue line shows the total population size in continuous time, red squares indicate the lowest population size in each year just before the birth-pulse (which is the abundance measured by a prebreeding "census"), and green dots indicate the highest population size in each year just after the birth-pulse (which is the abundance recorded by a postbreeding "census"). See main text for why "census" is a misnomer for most natural populations.

true, but for monogamous species (and sometimes for those with a more complex mating system), it is often a useful approximation. We will use mostly female-based models but also have examples for two-sex IPMs in Chapters 13 and 15. Two-sex models usually require the formulation of a mating function and therefore become more complex (Caswell, 2001; Tenan et al., 2016).

Third, we need to decide whether to include dispersal, and if so, whether we have multiple populations among which we want to model exchanges of individuals or rather a single population and model immigration and emigration. Models including explicit descriptions of dispersal can easily become fairly complicated. Many models in this book do not explicitly include dispersal and thus as demographic processes consider only recruitment and survival (but see Chapters 11, 14, 16, and 18). We therefore typically assume that individuals are completely philopatric, i.e., remain within a specified study area. Because dispersal is affected by spatial scale, we could also say that we model the world population of our target species—dispersal beyond this world population is impossible, of course. Alternatively, we may say that we assume zero net immigration, i.e., that immigration and emigration cancel out exactly. Therefore, both are ignored in our life-cycle graphs here.

Equipped with this knowledge, which includes a full understanding of these assumptions, we are now ready to draw our first life-cycle graph. For illustrative purposes, here we develop life-cycle graphs for woodchat shrikes for all four possible types of models: age-structured pre-birth-pulse; stage-structured pre-birth-pulse; age-structured post-birth-pulse, and stage-structured post-birth-pulse. We develop formulae to describe age/stage-specific population dynamics for all of them. In all cases, we consider a female-based model with a projection interval of 1 year, assume geographical closure (the absence of immigration and emigration), and we have the following demographic rates:

- juvenile survival from t to $t+1$ ($s_{j,t}$),
- adult survival from t to $t+1$ ($s_{a,t}$),
- productivity of 1-year-old females in year t ($f_{1,t}$), and
- productivity of older ("adult") females in year t ($f_{a,t}$).

Because our model is female-based, productivity is defined by the number of female fledglings produced per reproducing female of a given age class. This means that the total number of fledglings is twice that number if the fledgling sex ratio is even.

3.2.2 AGE-STRUCTURED PRE-BIRTH-PULSE MODEL

We start with the construction of the life cycle for an age-structured pre-birth-pulse model for the woodchat shrike. To ensure that we do not confuse pre- and post-birth-pulse models, we find it useful to always start with the definition of the youngest age class present at the time of our surveys. In the case of a pre-birth-pulse model, we look at our population *just before* reproduction starts, and thus the youngest individuals are approximately 1 year old (not exactly, but very close). We label the node of this age class *1y*. Clearly, we need more nodes in our graph, but how many? For a strictly age-structured model, we have to define the oldest possible age class, i.e., we must assume that no individual can become older than this age class. This may seem like a difficult decision, but in practice, we can simply add more than enough age classes; having too many does no harm. For the population model that we develop here, we assume that woodchat shrikes can reach a maximal age of 4 years. This may be a bit low, but we make this choice partly for illustrative purposes to prevent the graphs and formulae from becoming too unwieldy. But in principle, it would be easy to extend the maximum age to, say, 10 years, which for this species may be more adequate. Once we have defined the youngest and the oldest age classes, we find that our graph has four nodes and label them *1y*, *2y*, *3y*, and *4y* (with obvious meanings). We here show two variants of all four types of life-cycle graphs. In the first variant (Fig. 3.4A), which we call the *full life-cycle graph*, we draw nodes for every year.

Next, we connect the nodes with arrows. We start with the survival part, since this is easier. From year t to $t+1$, 1-year-old individuals can only make one transition—to become 2-year-old individuals (i.e., they cannot remain in the 1-year age class, nor can they transition to become a 3-year-old at once). The probability associated with this transition is the adult survival probability ($s_{a,t}$), and we label the arrow accordingly. Next, we draw the other survival transitions from nodes *2y* and *3y* in year t to nodes *3y* and *4y*, respectively, in year $t+1$, and label them $s_{a,t}$. Since we have decided that our shrikes can at most become 4 years old, there is no arrow leaving node *4y*.

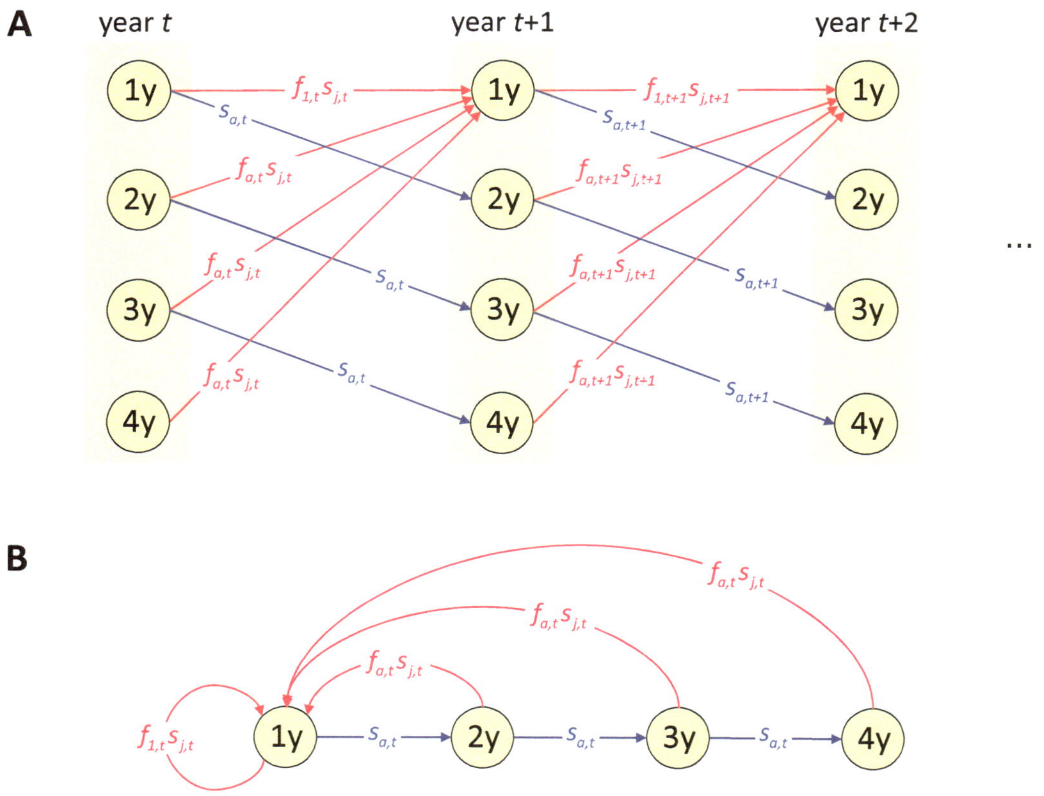

FIGURE 3.4

Two variants of a life-cycle graph of an age-structured pre-birth-pulse model for the woodchat shrike: (A) full life-cycle graph (ellipsis on the right indicates that more years would be possible), (B) simplified (or collapsed) life-cycle graph. The maximum age of a shrike is here assumed to be 4 years. Recruitment processes contain a contribution of productivity and survival and are depicted in red, while pure survival processes are shown in blue.

After drawing the survival arrows, we go on to draw the arrows representing recruitment. This can be a little trickier because recruitment is composed of two underlying demographic processes: productivity and survival of the fledged offspring. Because *1y* individuals already reproduce, there is an arrow leading from node *1y* in year t to node *1y* in year $t+1$, but how do we label it? We find it useful to start with the demographic processes in chronological order. In a pre-birth-pulse model, the first process is reproduction, and thus we have $f_{1,t}$. Be careful with the index for time and age—in this graph, reproduction happens in year t and refers to 1-year-old individuals. The next process is the survival of the fledglings produced, which is $s_{j,t}$. The label of the arrow is the product of the two—$f_{1,t}\,s_{j,t}$. Two-year-old individuals also contribute to recruitment; hence, there is an arrow from node *2y* in year t back to node *1y* in year $t+1$ with a label $f_{a,t}$. Fledglings must survive until year $t+1$ to enter the class of *1y* birds, and thus the complete transition is labeled $f_{a,t}\,s_{j,t}$. The same recruitment also applies for analogous individuals over 2 years old because they all have the same productivity. Now we have described all possible transitions from year t to year $t+1$. The transitions from year $t+1$ to year $t+2$ are analogous; only the year index of the demographic rates is different ($t+1$ instead of t, Fig. 3.4A).

Usually we will draw a simplified form of the life-cycle graph (Fig. 3.4B). The key difference from the full graph is that redundant nodes in different years are collapsed. Thus, a transition from node A to node B always means the transition from node A in year t to node B in year $t+1$ in the simplified form the of the life-cycle graph. Because of this, self-loops (say, from A to A) can occur in the simplified life-cycle graph. In our example, a self-loop is present for node *1y* because 1-year-old individuals in year t can produce 1-year-old individuals in year $t+1$.

The development of the formulae for the age-specific population sizes in year $t+1$ given the population sizes in year t and the demographic rates is now straightforward—to obtain the number of individuals in a particular node, we sum over all the arrows that enter it, and each term in that sum is the product of the node from where the arrow originates and the transition probability, which itself will consist of one demographic rate or the product of many demographic rates. This applies regardless of whether we use the full or simplified life-cycle graph. Denoting the age-specific population sizes with N, the number of 1-year-old individuals (i.e., in node *1y*) in year t is $N_{1y,t}$. The age-specific population sizes in year $t+1$ are then

$$
\begin{aligned}
N_{1y,t+1} &= N_{1y,t} f_{1,t} s_{j,t} + N_{2y,t} f_{a,t} s_{j,t} + N_{3y,t} f_{a,t} s_{j,t} + N_{4y,t} f_{a,t} s_{j,t} \\
N_{2y,t+1} &= N_{1y,t} s_{a,t} \\
N_{3y,t+1} &= N_{2y,t} s_{a,t} \\
N_{4y,t+1} &= N_{3y,t} s_{a,t}
\end{aligned}
$$

We can also write exactly the same model using linear algebra in vector and matrix form. Population size is now a vector whose elements are the age-specific population sizes. We need to find a matrix in such a way that the multiplication of this matrix with the population size vector of year t yields the population size vector in year $t+1$ (as above). We then obtain

$$
\begin{bmatrix} N_{1y,t+1} \\ N_{2y,t+1} \\ N_{3y,t+1} \\ N_{4y,t+1} \end{bmatrix} = \begin{bmatrix} f_{1,t}s_{j,t} & f_{a,t}s_{j,t} & f_{a,t}s_{j,t} & f_{a,t}s_{j,t} \\ s_{a,t} & 0 & 0 & 0 \\ 0 & s_{a,t} & 0 & 0 \\ 0 & 0 & s_{a,t} & 0 \end{bmatrix} \begin{bmatrix} N_{1y,t} \\ N_{2y,t} \\ N_{3y,t} \\ N_{4y,t} \end{bmatrix}.
$$

The transition matrix is the famous Leslie matrix; it is composed of demographic rates. The population model can also be written as $\mathbf{N}_{t+1} = \mathbf{A}_t \times \mathbf{N}_t$, where $\mathbf{N}_t$ is the population vector in year t, $\mathbf{A}_t$ is the Leslie matrix, and $\times$ denotes matrix multiplication. These are all simply different ways to describe the same process, and thus they are entirely equivalent. However, when fitting IPMs in BUGS, we typically use the former formulation, where we have one equation for each defined age class.

3.2.3 STAGE-STRUCTURED PRE-BIRTH-PULSE MODEL

The age-structured population model has two disadvantages: the maximum age must be defined, and all age classes until that age must appear explicitly. Depending on the species, this may result in very large life-cycle graphs, a large number of equations, or a high-dimensional Leslie matrix. In the life-cycle graphs (Fig. 3.4), we see that individuals in nodes *2y* and *3y* are assumed to be demographically identical; all the arrows leaving and entering these nodes are identical. Therefore we can combine these nodes and label the resulting new node *ad*, since it refers to "adult" birds in the sense that they are older than 1 year. If we relax the assumption of a maximum age of 4 and instead assume that survival beyond age 4 is $s_{a,t}$ and reproduction remains at $f_{a,t}$, we can then subsume node *4y* into the new node *ad* as well. A new transition from node *ad* in year *t* to node *ad* in year *t*+1 now becomes possible, meaning that there is no longer any predefined maximum age. The resulting full and simplified life-cycle graphs are shown in Fig. 3.5. These are now *stage-structured* life-cycle graphs because we no longer classify individuals simply by chronological age but rather by the two stages *1y* and *ad*. Strictly speaking, the model contains a mixture of a pure age class (*1y*) and a stage that contains multiple age classes (*ad*). In the simplified life-cycle graph, there is now an additional self-loop on the node *ad*.

Exactly as with the preceding age-structured model, we can easily read the graph and develop the formulae for the stage-specific population sizes:

$$N_{1y,t+1} = N_{1y,t} f_{1,t} s_{j,t} + N_{ad,t} f_{a,t} s_{j,t}$$
$$N_{ad,t+1} = N_{1y,t} s_{a,t} + N_{ad,t} s_{a,t}$$

Likewise, the linear algebra version of the model is fully analogous as before:

$$\begin{bmatrix} N_{1y,t+1} \\ N_{ad,t+1} \end{bmatrix} = \begin{bmatrix} f_{1,t} s_{j,t} & f_{a,t} s_{j,t} \\ s_{a,t} & s_{a,t} \end{bmatrix} \begin{bmatrix} N_{1y,t} \\ N_{ad,t} \end{bmatrix}$$

The stage-structured population model looks simpler than the fully age-structured model. The transition matrix of stage-structured models is often referred to as a Lefkovitch matrix (Lefkovitch, 1965).

3.2.4 AGE-STRUCTURED POST-BIRTH-PULSE MODEL

Next, we move to a post-birth-pulse model (Fig. 3.6A), where we observe the population immediately after reproduction. The youngest individuals at this point are juveniles (fledglings), which we denote as *juv*. In the age-structured model, we again make the assumption that the maximum possible age of the shrikes is 4 years, i.e., the last possible reproduction occurs at age 4. In the post-birth-pulse model, the oldest age class is therefore node *3y*. This is confusing, isn't it? But the reason for this is that individuals of this age class must first survive 1 more year to become 4 years old, and then they reproduce one last time. Thus, although we only have a node *3y* in our model, individuals are 4 years old when they reproduce last, and this is consistent with our assumption about the maximum age. We therefore define the four nodes *juv*, *1y*, *2y*, and *3y*.

We next draw the arrows and start with those representing survival. From node *juv* to *1y*, there is an arrow labeled $s_{j,t}$ because it refers to the survival of juvenile individuals. From nodes *1y* and *2y* in year *t* to nodes *2y* and *3y* in year *t*+1, we have adult survival $s_{a,t}$. For recruitment, we have an arrow from node *juv* in year *t* to node *juv* in year *t*+1. To correctly

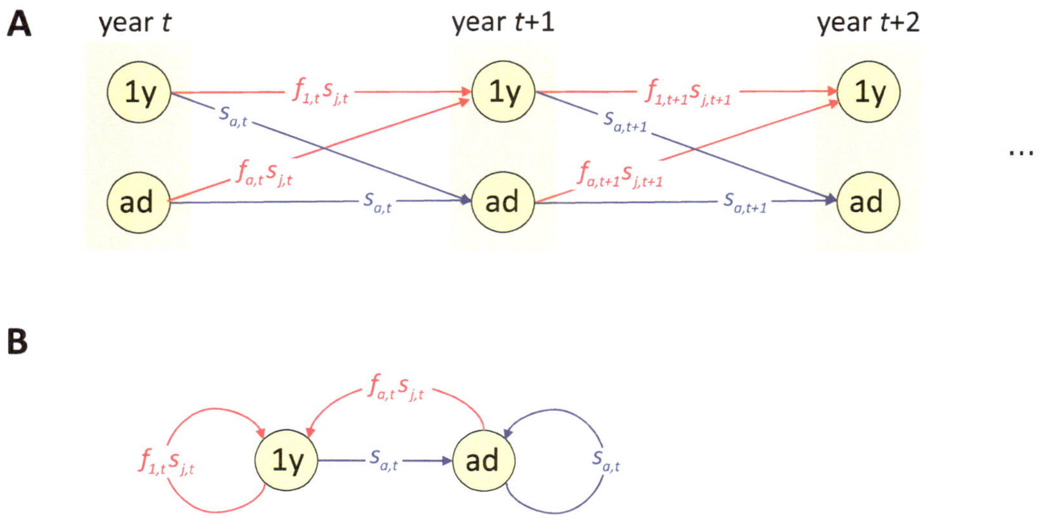

FIGURE 3.5

Life-cycle graphs of a stage-structured pre-birth-pulse model for the woodchat shrike: (A) full life-cycle graph (ellipsis on the right indicates that more years would be possible), (B) simplified life-cycle graph. Recruitment processes are depicted in red, pure survival processes in blue.

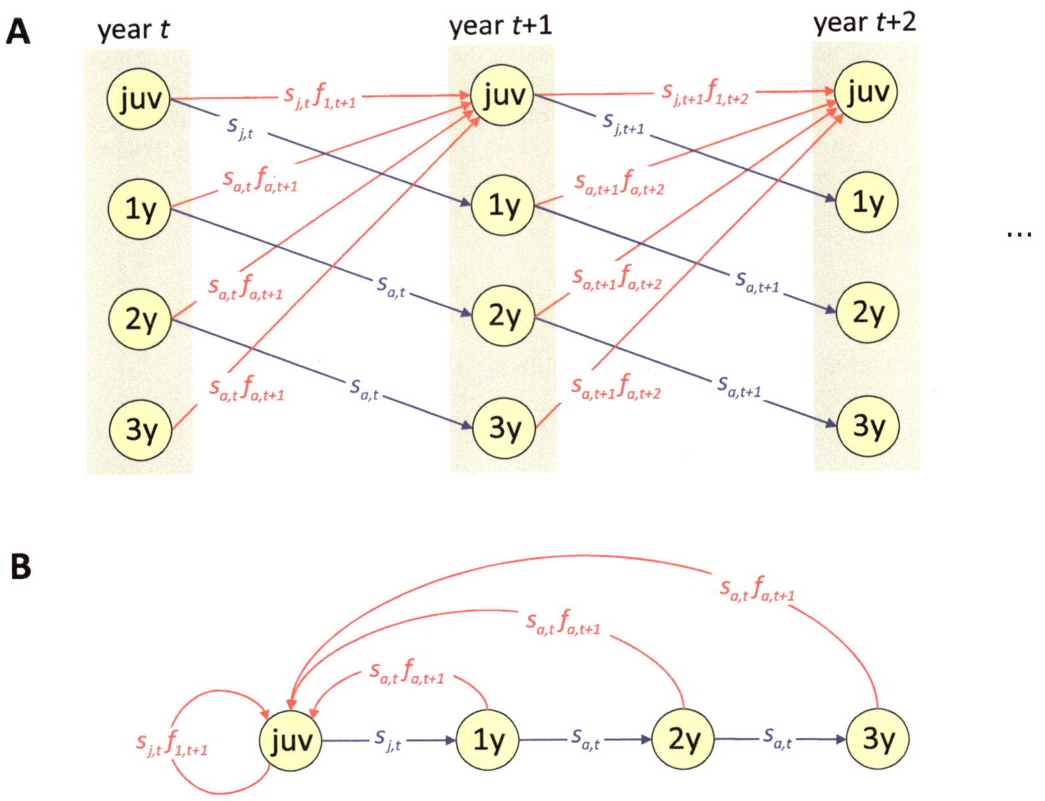

FIGURE 3.6

Life-cycle graphs of an age-structured post-birth-pulse model for the woodchat shrike: (A) Full life-cycle graph (ellipsis on the right indicates that more years would be possible), (B) simplified life-cycle graph. The maximum age is assumed to be 4 years. Recruitment processes are depicted in red, pure survival processes in blue.

determine the labels of that arrow, we again follow the involved demographic processes chronologically. We first have a survival process: a *juv* in year t must survive until $t+1$ and does so with probability $s_{j,t}$. Next, we have reproduction. A surviving individual is now 1 year old, and reproduction occurs in year $t+1$. Therefore, the correct reproduction parameter is $f_{1,t+1}$, and the complete label of the arrow is $s_{j,t}f_{1,t+1}$ (note the different time indices for s and f!). The other recruitment arrows are similar, the only differences being that survival refers to individuals at least 1 year old, and therefore $s_{a,t}$ applies, and that productivity refers to females older than 1 year. You can see that construction of a post-birth-pulse model is trickier because the indexing of age and time is more involved. On the other side, the postbreeding model may be more natural because juveniles appear explicitly as a node, whereas in the prebreeding model, juveniles "disappear," i.e., do not appear as a node in the graph.

Based on the full life-cycle graph, the construction of the simple life-cycle graph is not very challenging (Fig. 3.6B). The post- and pre-birth-pulse life-cycle graphs look the same in terms of the number of nodes and type of arrows in our example, the only difference being in the labels of the nodes and arrows. However, this is not always the case, and you may find examples where the number of nodes differs between post- and pre-birth-pulse models for the same data set.

Derivation of age-structured population sizes can be accomplished easily from the life-cycle graph:

$$N_{juv,t+1} = N_{juv,t}s_{j,t}f_{1,t+1} + N_{1y,t}s_{a,t}f_{a,t+1} + N_{2y,t}s_{a,t}f_{a,t+1} + N_{3y,t}s_{a,t}f_{a,t+1}$$
$$N_{1y,t+1} = N_{juv,t}s_{j,t}$$
$$N_{2y,t+1} = N_{1y,t}s_{a,t}$$
$$N_{3y,t+1} = N_{2y,t}s_{a,t}$$

Written in matrix algebra with population vectors and the Leslie matrix, we obtain

$$\begin{bmatrix} N_{juv,t+1} \\ N_{1y,t+1} \\ N_{2y,t+1} \\ N_{3y,t+1} \end{bmatrix} = \begin{bmatrix} s_{j,t}f_{1,t+1} & s_{a,t}f_{a,t+1} & s_{a,t}f_{a,t+1} & s_{a,t}f_{a,t+1} \\ s_{j,t} & 0 & 0 & 0 \\ 0 & s_{a,t} & 0 & 0 \\ 0 & 0 & s_{a,t} & 0 \end{bmatrix} \begin{bmatrix} N_{juv,t} \\ N_{1y,t} \\ N_{2y,t} \\ N_{3y,t} \end{bmatrix}.$$

3.2.5 STAGE-STRUCTURED POST-BIRTH-PULSE MODEL

Finally, we develop a stage-structured post-birth-pulse model. In the full life-cycle graph, we must first add an arrow from node *3y* in year t to node *3y* in $t+1$, and in the simple life-cycle graph, we add a self-loop for node *3y*. This relaxes the assumption of a maximum age of 4. Next, we recognize that all arrows leaving and entering nodes *1y*, *2y*, and *3y* are identical. Therefore, these nodes can be combined, and we label the resulting new node *ad*. The corresponding full and simplified life-cycle graphs are shown in Fig. 3.7.

For stage-specific population sizes, we obtain

$$N_{juv,t+1} = N_{juv,t}s_{j,t}f_{1,t+1} + N_{ad,t}s_{a,t}f_{a,t+1}$$
$$N_{ad,t+1} = N_{juv,t}s_{j,t} + N_{ad,t}s_{a,t}$$

or using linear algebra,

$$\begin{bmatrix} N_{juv,t+1} \\ N_{ad,t+1} \end{bmatrix} = \begin{bmatrix} s_{j,t}f_{1,t+1} & s_{a,t}f_{a,t+1} \\ s_{j,t} & s_{a,t} \end{bmatrix} \begin{bmatrix} N_{juv,t} \\ N_{ad,t} \end{bmatrix}.$$

All equations derived thus far allow the calculation of the number of individuals in each age/stage-specific class of the shrike population at any point in time, which is the time when the birds start egg laying in each year (for a prebreeding "census") or when they fledge young (for a postbreeding "census"). We need to know the demographic rates and age/stage-structured population sizes associated with these discrete points in time. The total population size at time t is simply the sum over the age/stage-specific population sizes at t. Provided that the demographic rates do not change as a function of population size (i.e., there is no density dependence), both the total population size and the age/stage-specific population sizes are growing asymptotically at a constant rate—the population growth rate—and the population shows either an

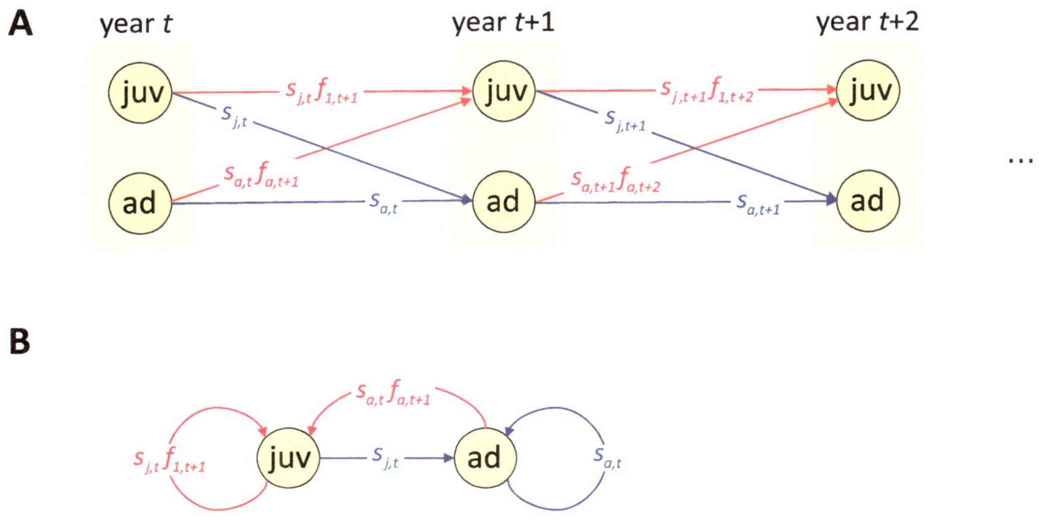

FIGURE 3.7

Life-cycle graphs of a stage-structured post-birth-pulse model for the woodchat shrike: (A) full life-cycle graph (ellipsis on the right indicates that more years would be possible), (B) simplified life-cycle graph. Recruitment processes are depicted in red, pure survival processes in blue.

exponential increase or a decline (Fig. 3.1). With age/stage-structured population models, or more generally with any matrix population model, we define an explicit link between population size and demography, and these models are the base of nearly all population modeling studies for vertebrates with a birth-pulse. In contrast, population models for species with a birth-flow (i.e., reproduction may occur more or less continuously during a year) require continuous formulations of time. They are more complex and not treated here. However, we note that as an approximation with discrete time, birth-pulse models with a projection interval shorter than 1 year may be used. Birth-flow populations are generally fairly rare for most animals or plants outside of the tropics.

3.3 CLASSICAL ANALYSIS OF A MATRIX POPULATION MODEL

Typical objectives in the analysis of age- and stage-structured population models are the calculation of quantities that can be used to describe the population dynamics (life-history summaries), to understand the demographic causes of past dynamics, or to predict (forecast) future population trajectories. Typical quantities of interest are the asymptotic population growth rate, stable stage distribution, stage-specific reproductive values, generation time (*GT*), net reproductive rate (R_0), and growth rate sensitivities and elasticities. Provided that information about all the demographic rates in the Leslie/Lefkovitch matrix is available, all of these can be calculated from the matrix using well-established methods (Caswell, 2001). If all demographic rates are constant and known without error, most of these quantities can be, and typically are, calculated from the transition matrix using matrix algebra (Caswell, 2001). However, as soon as the demographic rates become unknown and must be estimated, or if environmental stochasticity, demographic stochasticity, or density dependence is included in the model, simulations are usually required to estimate these quantities. Simulation also allows for straightforward assessment of the precision of these quantities.

In this section, we illustrate how the population growth rate and a few other important life-history summaries are calculated. We start with the simplifying assumptions that the demographic rates are constant over time (i.e., that we have neither environmental stochasticity nor density dependence), that the demographic rates are known perfectly (i.e., there is no estimation error), and that we can ignore the discrete nature of individuals (i.e., there is no demographic stochasticity). Later, we will relax each of these assumptions and show how to include environmental stochasticity, demographic stochasticity, estimation errors, and density dependence into our inferences from the transition matrix. We illustrate the computations with a female-based, stage-structured, pre-birth-pulse population model for the woodchat shrike as developed in Section 3.2.3 (see Fig. 3.5B). We assume that juvenile survival (s_j, survival from fledging until the age of 1 year) is 0.3, adult survival (s_a, annual survival from the age of 1 year onwards) is 0.55, productivity of 1-year-old females (f_1, number of female fledglings per 1-year-old female and year) is 1.3, and productivity of older females is $f_a = 1.8$. In Section 3.4, we illustrate how the same analyses can be carried out using Markov chain simulation—i.e., Markov chain Monte Carlo (MCMC)—methods. Methods for assessing demographic causes of population change and making population projections (i.e., forecasting) are treated in more detail in Chapters 8 and 9, respectively.

3.3.1 ANALYSIS OF A MATRIX POPULATION MODEL WITHOUT STOCHASTICITY AND PARAMETER UNCERTAINTY

In the following, we show the traditional computation of the asymptotic population growth rate, stable stage distribution, and sensitivities of the population growth rate to changes in the demographic rates. We could use for this R packages such as `popbio` (Stubben and Milligan, 2007) or `demogR` (Jones, 2007), but we think it is insightful to see how these life-history summaries can be obtained using standard R functions. We start by defining the demographic rates and transition matrix **A**.

```
# Define the demographic rates
sj <- 0.3          # Juvenile survival
sa <- 0.55         # Adult survival
f1 <- 1.3          # Number of female fledglings per 1-year old female
fa <- 1.8          # Number of female fledglings per adult female

# Define the transition matrix A (inspired by woodchat shrike demography)
A <- matrix(c(f1 * sj, fa * sj, sa, sa), ncol=2, byrow=TRUE)
dimnames(A) <- list(c("1y", "ad"), c("1y", "ad"))
A                  # print the matrix
      1y   ad
1y  0.39 0.54
ad  0.55 0.55
```

Asymptotic population growth rate: The asymptotic population growth rate (usually denoted λ) is the rate at which the population described by matrix **A** grows when the stage distribution has become stable, i.e., after an initial *transient phase*. To experience what asymptotic and transient behavior mean, we start with an arbitrary initial population vector, project the population for a number of years, and compute the annual growth rate of each stage class and the population as a whole as well as the actual stage distribution. We first define the initial population vector.

```
N1 <- c(10, 1)
```

Thus, we assume an extremely skewed stage distribution for illustrative purposes. We then project the population for 7 years, each time computing the realized growth rates of each stage class and the entire population and the realized stage distribution. We then plot these quantities.

```
T <- 7
N <- matrix(NA, nrow=2, ncol=T)            # Population sizes
gr <- matrix(NA, nrow=3, ncol=T-1)         # Growth rates
N[,1] <- N1
for (t in 2:T){
  N[,t] <- A %*% N[,t-1]
  gr[1,t-1] <- N[1,t] / N[1,t-1]
  gr[2,t-1] <- N[2,t] / N[2,t-1]
  gr[3,t-1] <- (N[1,t] + N[2,t]) / (N[1,t-1] + N[2,t-1])
}
sr <- N[1, ] / colSums(N)                  # State distribution (proportion of 1y)
```

From Fig. 3.8A, we see the stage-specific population sizes grow at identical rates from about year 4 onwards. The realized stage-specific population growth rates are identical, and they are identical also to the growth rate of the total population size (Fig. 3.8B). The realized proportion of individuals in the two stages changes from year 1 to 3 but thereafter remains stable (Fig. 3.8C). Hence, in this population, the transient phase of population growth is exceptionally short, and the population very soon reaches the equilibrium dynamics given by the asymptotic values of both the growth rate and the stable stage distribution. Many population statistics describe the *asymptotic properties* of population dynamics. As we will see later, natural populations hardly ever reach the asymptotic phase (i.e., they are never in the stable stage distribution) because they are continuously "perturbed" away from asymptopia by nonconstant demographic rates and demographic stochasticity. However, asymptotic population statistics are often very useful and good approximations, in particular for short-lived life cycles that reach a stable stage distribution very quickly, as is the case for our woodchat shrike example. It is important, though, that you fully understand the distinction between the asymptotic and transient behavior of a population.

The length of the transient phase, i.e., the time until the population starts growing at the asymptotic rate, is related to the *damping ratio*, which is the ratio of the largest (dominant) to second-largest eigenvalue of transition matrix **A** (Caswell, 2001, p. 95). The damping ratio is proportional to *GT*; with a shorter *GT*, convergence is attained more quickly (Ezard et al., 2010). As an illustration of a longer transient phase, we look at a population of the greater horseshoe bat (*Rhinolophus*

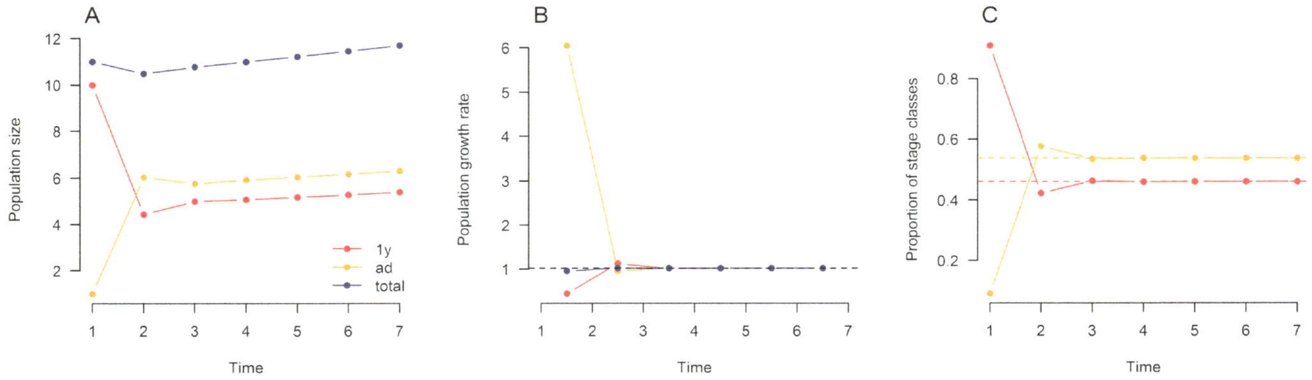

FIGURE 3.8

Trajectories of the stage-specific and total population sizes (A), stage-specific and total population realized growth rates (B), and realized stage distributions (C) computed from the matrix population model for woodchat shrikes under constant demographic rates and absent demographic stochasticity. The transient phase takes just 3 years or so; afterward, the asymptotic phase starts, which is characterized by a stable stage distribution and constant and identical (across stage classes) population growth rates. In (B and C), horizontal dashed lines indicate the asymptotic values given by the eigenanalysis (see text).

ferrumequinum), a long-lived bat that starts to reproduce at 2 years of age. Survival increases until age 3 ($s_1 = 0.5$, $s_2 = 0.8$, $s_3 = 0.9$), and productivity is lower in 2-year-olds ($\rho_1 = 0.5$) than in older females ($\rho_2 = 0.9$). The life-cycle graph contains three stage classes (1-year-old females, 2-year-old females, more than 2-year-old females), and we work again with a prebreeding "census" model. The life-cycle graph of a two-sex model for the greater horseshoe bat is shown in Chapter 13 (Fig. 13.2), but here we use a simplified, single-sex version only.

As expected, the transient phase is longer in the greater horseshoe bat than in the woodchat shrike, namely about 6 years (Fig. 3.9). And it would take even longer for a rhino, an albatross, or a human population as well as for other very long-lived species with a high age at maturity.

Arguably, the most important asymptotic property of a population is the asymptotic population growth rate, which is the rate at which it grows once the stable stage distribution is reached. To obtain it, we can start from an arbitrary population vector and project the population for a large number of years (e.g., J) and calculate the asymptotic population growth rate as $\mathbf{N}_{J+1}/\mathbf{N}_J$. Alternatively, linear algebra yields the asymptotic population growth rate as the dominant (i.e., maximum) eigenvalue of transition matrix $\mathbf{A}$ (Caswell, 2001, p. 72). Calculation of the dominant eigenvalue is based on the solution of the so-called characteristic equation.

$$\det(\mathbf{A} - \lambda \mathbf{I}) = 0$$

Here, $\mathbf{A}$ is the transition matrix, $\mathbf{I}$ the identity matrix (i.e., a square matrix with ones on the diagonal and zeros elsewhere), and det the determinant. After insertion of transition matrix $\mathbf{A}$, we obtain

$$\det\left(\begin{bmatrix} f_1 s_j & f_a s_j \\ s_a & s_a \end{bmatrix} - \begin{bmatrix} \lambda & 0 \\ 0 & \lambda \end{bmatrix}\right) = 0,$$

and after some rearrangements, we obtain

$$\lambda^2 - \lambda(s_a + s_j f_1) + s_j s_a (f_1 - f_a) = 0.$$

Plugging in the demographic rates of the woodchat shrikes as defined above, we obtain the following quadratic equation:

$$\lambda^2 - 0.94\lambda - 0.0825 = 0.$$

Because $\mathbf{A}$ is a 2×2 matrix, the characteristic equation is quadratic in λ, and there are therefore two solutions, $\lambda_1 = -0.08081757$ and $\lambda_2 = 1.02081757$. The dominant eigenvalue λ_2 yields the asymptotic population growth rate and tells us that our shrike population will eventually grow by about 2.1% annually. Note that in simpler cases (e.g., when $f_1 = f_a$), it is possible to obtain algebraic solutions for the characteristic equation.

We can also use R functions for eigenanalyses.

```
# Compute asymptotic population growth rate (lambda)
lambda <- max(Re(eigen(A)$values))
[1] 1.020818
```

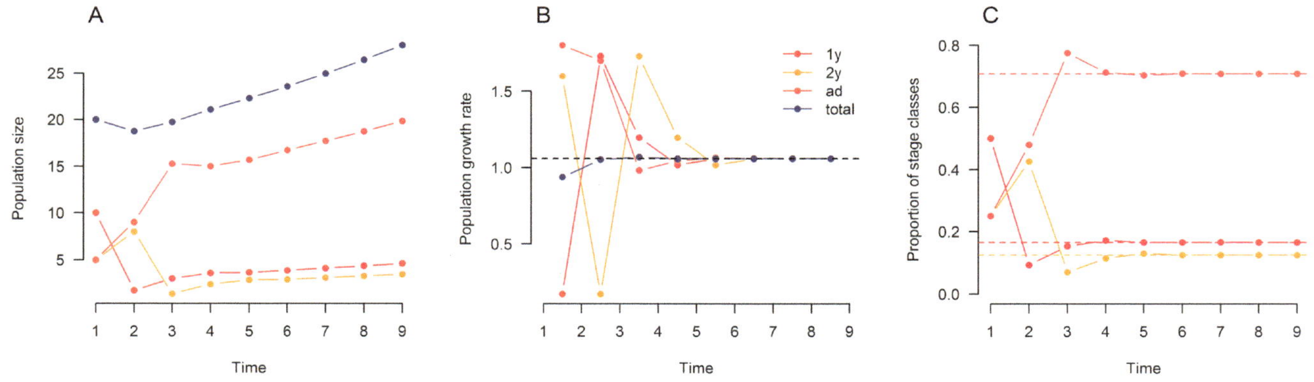

FIGURE 3.9

Development of stage-specific and total population sizes (A), stage-specific and total population realized growth rates (B), and realized stage distributions (C) calculated from the matrix population model for greater horseshoe bats assuming constant demographic rates and the absence of demographic stochasticity. The horizontal dashed lines show the asymptotic population growth rate and stable stage distribution from the dominant eigenvalue and right eigenvector, respectively, of the transition matrix **A**.

For a better understanding, it may be helpful for you to execute this from the inside out (starting at `eigen(A)`) and inspect all the intermediate results leading up to the value of `lambda`.

Stable stage distribution: This is the distribution of individuals by stage once the population grows at the asymptotic rate. When all demographic rates are constant, most stage-structured populations will eventually converge to a stable stage distribution and grow at the asymptotic population growth rate (Cohen, 1979), and we have seen this behavior in our two examples (see Figs. 3.8 and 3.9). The *strong ergodic theorem* states that the long-term dynamics of a population is determined by the population growth rate, coverges to a stable stage structure, and becomes independent of the initial stage structure if **A** is *primitive* (Caswell, 2001, p. 79). Most population matrices are primitive, but in some cases they are not. If **A** is *imprimitive*, a population will cycle, and if **A** is *reducible*, the long-term dynamics will depend on the initial conditions, and thus is not *ergodic*—see Caswell (2001) for examples.

For populations with primitive **A**, we can calculate the stable stage distribution based on the population size vector after it is projected for a sufficient number of time steps, i.e., until convergence. Alternatively, the stable stage distribution is given by the right eigenvector corresponding to the dominant eigenvalue of **A**. For this, we first determine which eigenvalue is dominant and then take the corresponding eigenvector.

```
u <- which.max(Re(eigen(A)$values))
[1] 1
revec <- Re(eigen(A)$vectors[,u])
[1] -0.6503047 -0.7596734
```

We then standardize the eigenvector to obtain the stable stage distribution.

```
revec/sum(revec)
[1] 0.4612162 0.5387838
```

Thus, once the population has reached asymptotic growth, about 46% of the population will be 1 year old, and about 54% will be 2 years old or older. Remember that this stage structure refers to the time just before reproduction because we have defined our population model for a prebreeding "census." Unlike the asymptotic population growth rate, the stable stage distribution depends on to the choice of the "census" period (see Exercise 3.1).

Stage-specific reproductive values: This measure expresses the relative contributions of each stage to long-term population growth and thus in a sense is a measure of how "important," or valuable, a stage is for population growth. Reproductive values can be decomposed into a component of current reproduction and another for future reproduction (residual reproductive value); this is useful for the study of life-history trade-offs (Stearns, 1992). They can be calculated as the left eigenvector corresponding to the dominant eigenvalue of transition matrix **A**. Often they are scaled to sum to 1 for the proportional reproductive values or such that the reproductive value of the youngest age class is 1, so the reproductive values of the older age classes can then be directly compared with that of the youngest age class, which serves as the reference.

```
u <- which.max(Re(eigen(A)$values))
levec <- Re((eigen(A)$vectors)[u,])
levec / sum(levec)
[1] 0.4631655 0.5368345
```

In our two-stage-class model, the reproductive value of the younger age class is only slightly less than that of the older age class—in asymptopia, 1-year-old woodchat shrikes contribute less to population growth than do adults, which are about (0.54/0.46 =) 1.16 times more valuable for population growth than 1-year-old individuals. Note that the reproductive values in this example are numerically almost identical to the stable stage distribution. This is a peculiarity of our example and not generally the case.

Net reproductive rate: The R_0 is the expected, or mean, number of offspring by which a newborn shrike will be replaced when it dies. This measure expresses population growth per *generation* rather than per projection interval as does λ. One way to calculate R_0 is based on the Lotka equation (Lotka, 1907; Caswell, 2001, p. 126).

$$R_0 = \sum_{i=1}^{\infty} l_i f_i$$

Here, l_i is the probability of survival from birth to stage class i, and f_i is the productivity of stage class i. In our shrike example, the net reproductive value is $R_0 = s_j f_1 + s_j s_a f_a + s_j s_a^2 f_a + s_j s_a^3 f_a + \ldots + s_j s_a^{\infty} f_a = s_j \left(f_1 + f_a \sum_{i=1}^{\infty} s_a^i \right)$. In R, we approximate this sum by taking 100 to be "infinity"; this is enough to obtain an accurate value.

```
i <- 1:100                  # 100 as our approximation to infinity
R0 <- sj * f1 + sj * fa * sum(sa^i)
[1] 1.05
```

Thus, when it dies, each shrike is expected to be replaced by 1.05 new shrikes.

Generation time is often used to characterize the life history of a species by the slow—fast gradient (Oli and Dobson, 2003; Gaillard et al., 2005). Several definitions of *GT* exist (Caswell, 2001, p. 128). The most common is arguably the mean age of mothers of offspring produced by a population in the stable stage distribution. It is calculated as the sum over all stage classes as

$$G = \sum_{i=1}^{\infty} i l_i f_i \lambda^{-i},$$

where l_i and f_i are defined as above, and the sum goes to infinity, but evaluation to some upper bound (e.g., 100 as above) will suffice in practice. For our shrikes, we obtain

$$G = s_j f_1 \lambda^{-1} + 2 s_j s_a f_a \lambda^{-2} + 3 s_j s_a^2 f_a \lambda^{-3} + \ldots = s_j \left(f_1 \lambda^{-1} + f_a \sum_{i=2}^{\infty} i s_a^{i-1} \lambda^{-i} \right),$$

which is calculated in R as follows:

```
Q <- 100                    # Our approximation to infinity
i <- 2:Q
G <- sj * f1 / lambda + sj * fa * sum(i * sa^(i-1) * lambda^(-i))
[1] 2.339834
```

Thus, the *GT* of a shrike is a little over 2 years. Another definition of *GT* is the time needed for a population to increase by the R_0. Hence,

$$GT = \frac{\log(R_0)}{\log(\lambda)}.$$

```
GT <- log(R0) / log(lambda)
[1] 2.368012
```

Typically, these two widespread measures of *GT* will produce very similar results (see Caswell, 2001, p. 129 for calculations from the transition matrix).

Sensitivity and elasticity (perturbation analysis): The population growth rate is a function of multiple demographic rates. As a measure of the effects of each, we can compute the sensitivity and elasticity, which express the expected change in the growth rate when the value of a demographic rate is varied by a small amount. Such analyses are often referred to as *perturbation analyses* (Caswell, 2000, 2001). A typical application for a perturbation analysis is to predict population dynamics when future changes in demographic rates occur; this is a *prospective perturbation analysis*. For a species of conservational concern, we may want to use appropriate management to alter the demographic rate that affects population growth particularly strongly (though in addition, we may weight this by the cost of such management). Or for getting rid of an invasive species, we need to know the most efficient strategy: is population growth reduced most strongly when survival is decreased through culling, or can we reduce it more efficiently by lowering reproductive success, e.g., by destroying nesting sites? Related to such perturbation analyses is population viability analysis, which is covered in Chapter 10. Another application of perturbation analyses is the assessment of how strongly each demographic rate has contributed to population dynamics observed in the past. Such *retrospective perturbation analyses* are treated in more detail in Chapter 9.

For all perturbation analyses, sensitivities play a central role. The two commonly used measures are proper sensitivity (S) and elasticity (E; and note that E here means something entirely different from Chapter 2, where it denoted an expectation). Sensitivity measures the effect of *absolute* changes in demographic rates (e.g., by 0.01), while elasticity measures the effect of *relative* changes (e.g., by 1%). Specifically, the sensitivity of the asymptotic population growth rate (λ) to changes in element $a_{i,j}$ of matrix **A** is the first partial derivative of λ with respect to $a_{i,j}$.

$$S(a_{i,j}) = \frac{\partial \lambda}{\partial a_{i,j}}$$

The elasticity is a scaled version of sensitivity.

$$E(a_{i,j}) = \frac{\partial \lambda}{\partial a_{i,j}} \frac{a_{i,j}}{\lambda}$$

In both ratios, you can recognize the limit of a small change in the growth rate λ per small change in element $a_{i,j}$. In E, the additional term serves to scale the sensitivity such that we deal with relative changes in $a_{i,j}$.

Empirical computation of S and E is based on the two eigenvectors—the sensitivity of the population growth rate to changes in element $a_{i,j}$ is the product of ith element of the stable stage distribution and the jth element of the reproductive value vector (Caswell, 2001, p. 208). In R, the sensitivity matrix is computed as

```
senmat <- levec %*% t(revec)

          [,1]      [,2]
[1,] 0.4228963 0.4940192
[2,] 0.4901602 0.5725957
```

and the elasticity matrix as

```
elasmat <- senmat * A / lambda

          1y        ad
1y 0.1615661 0.2613301
ad 0.2640904 0.3085053
```

Typically, we do not want to know the sensitivity or elasticity of the population growth rate to changes in an element of the transition matrix directly (e.g., to element $a_{1,1}$ which is $f_1 s_j$) but rather to changes in the actual demographic rates of which the element is composed; these are called "lower-level sensitivities/elasticities" (Caswell, 2001, p. 232). To compute them, we must sum all elements of the sensitivity matrix and weigh them with the first partial derivative of the transition matrix with respect to the focal demographic rate. Thus, the sensitivity of the population growth rate to a small change in demographic rate x is

$$S(x) = \sum_{i,j} \frac{\partial \lambda}{\partial a_{i,j}} \frac{\partial a_{i,j}}{\partial x},$$

and the elasticity is

$$E(x) = \frac{x}{\lambda} \sum_{i,j} \frac{\partial \lambda}{\partial a_{i,j}} \frac{\partial a_{i,j}}{\partial x}.$$

Let us use R to calculate growth rate sensitivity and elasticity to changes in juvenile survival in the shrikes, ($S(s_j)$ and $E(s_j)$). We first have to calculate the first partial derivatives of the transition matrix with respect to juvenile survival.

$$\frac{\partial \begin{bmatrix} f_1 s_j & f_a s_j \\ s_a & s_a \end{bmatrix}}{\partial s_j} = \begin{bmatrix} f_1 & f_a \\ 0 & 0 \end{bmatrix}$$

Next, the matrix of derivatives is multiplied with the sensitivity matrix, and the elements of the resulting matrix are summed.

```
derivmat <- matrix(c(f1, fa, 0, 0), ncol=2, byrow=TRUE)
ssj <- sum(senmat * derivmat)       # Sensitivity
esj <- sj / lambda * ssj            # Elasticity
ssj; esj
[1] 1.439
[1] 0.4228963
```

Finally, we can represent sensitivities and elasticities graphically: we numerically change one demographic rate by a small positive or negative value (Δs_j) and recalculate the asymptotic population growth rate. A plot of the resulting population growth rates against the absolute change in the demographic rate shows the sensitivity—it is the slope of this relationship (Fig. 3.10A). This slope, and hence the sensitivity, can be calculated as $\Delta\lambda/\Delta s_j$ (see Section 3.3.3). Similarly, the slope of the population growth rate against a relative change in a demographic rate shows the elasticity (Fig. 3.10B). Because the relationship between the asymptotic population growth rate and a variable demographic rate is not necessarily linear (Caswell, 2001, p. 208), Δs_j must be kept small.

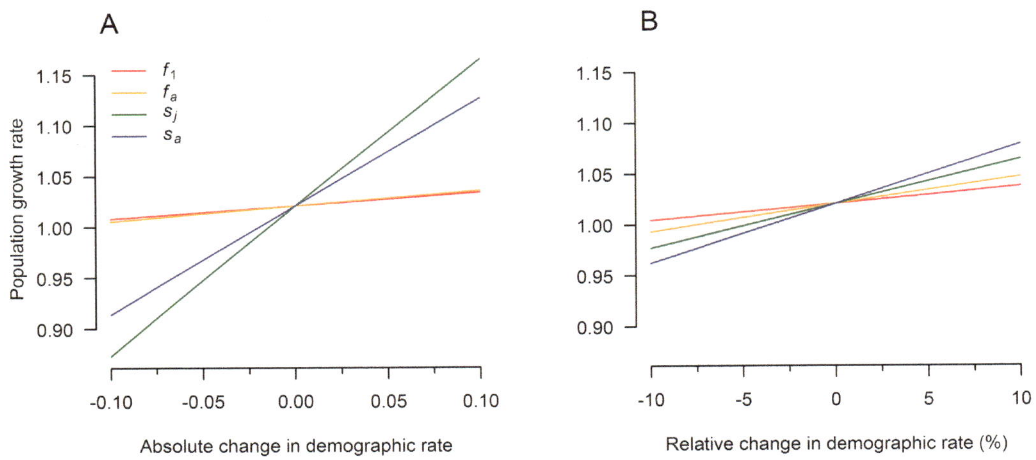

FIGURE 3.10

Visualizations of sensitivity (A) and elasticity (B)—Relationship between the population growth rate and absolute and relative changes in demographic rates, respectively.

3.3.2 ANALYSIS OF A MATRIX POPULATION MODEL WITH PARAMETER UNCERTAINTY

Demographic rates are usually not known perfectly and must be estimated. For instance, an estimate of survival probabilities may have been obtained from fitting a Cormack-Jolly-Seber model to capture-recapture data (see Section 4.5). The uncertainty associated with the estimate (e.g., as represented by the SE) then ought to be propagated into any function of the demographic rates, such as all the population summaries we computed in the preceding section. Several approaches are possible for this uncertainty or error propagation, including approximations such as the Taylor series expansion, also called the delta rule (Caswell, 2001, p. 300; Powell, 2007), and various simulation methods. The latter are certainly the most powerful methods and include bootstrapping (Brault and Caswell, 1993; Caswell, 2001, p. 304) and Monte Carlo simulations (Caswell et al., 1998). Here, we focus on Monte Carlo simulations because they are easy to perform in general and share similarities with the Bayesian MCMC methods we use for parameter estimation throughout this book.

The principle of error propagation by Monte Carlo simulation is fairly easy. The only difficulty is that the uncertainty associated with a demographic parameter must be described with a statistical distribution. The Monte Carlo simulation works as follows:

1. Draw a value for each demographic rate from its appropriate statistical distribution using point and uncertainty estimates. The latter are given by the standard error of the estimates.
2. Use those draws to calculate the asymptotic population growth rate and/or any life-history summary of interest.
3. Repeat these steps many times (e.g., 1000s), and save the results each time.
4. Calculate the mean and the standard deviation of the quantities of interest as a point estimate and its standard error.

Here, let's assume that we have estimated age-specific survival probabilities of the woodchat shrikes from a capture-recapture data set. We assume the same means (or point estimates) as before, $s_j = 0.3$ and $s_a = 0.55$, and standard errors (or in a Bayesian analysis, posterior standard deviations) of $\mathrm{SD}(s_j) = 0.05$ for juvenile and $\mathrm{SD}(s_a) = 0.03$ for adult survival. When we simulate survival probabilities from a distribution, we must respect the constraint that the simulated values must fall within the permissible range of 0 to 1. If the point estimates are far from these bounds and the SDs are small, we could sample a normal distribution with these values for the mean and SD directly. In general, however, a safer option is to sample a normal distribution on a transformed scale, e.g., the logit scale. This requires that the mean and standard deviation are transformed to this scale, or that they are obtained on this scale in the first place. In the case of the logit scale, the mean is $\mathrm{logit}(x) = \log\left(\frac{x}{1-x}\right)$, while use of the delta rule tells us that the standard deviation is $SD(\mathrm{logit}(x)) \cong \frac{SD(x)}{x(1-x)}$. A third possibility that does not require any transformation is the use of the beta distribution. The function `rbeta2` from our `IPMbook` package produces random draws from a beta distribution based on a given mean and standard deviation.

For productivity, we also assume that we have estimates; the means (i.e., point estimates) are as before, $f_1 = 1.3$ and $f_a = 1.8$, with a standard error of $\mathrm{SE}(f_1) = 0.3$ and $\mathrm{SE}(f_a) = 0.1$, respectively. Productivity has no upper bound but must be positive. A possible distribution to generate values, therefore, is the log-normal distribution. Again, if the SEs are small enough relative to the point estimate, we could sample a normal distribution directly. Since the SEs in our example are indeed small, there is an extremely low risk that a negative value is generated; hence, we proceed without a log transformation.

For illustration, we now compute several life-history summaries such as the asymptotic population growth rate, the stable stage distribution, lower-level sensitivities, net reproductive rate, and generation time in the presence of parameter uncertainty. The following code performs Monte Carlo simulations to calculate the precision or uncertainty of the asymptotic population growth rate and other life-history summaries. In addition to the Monte Carlo SEs, we use percentiles of the simulated sampling distributions to obtain a 95% CI.

```
# Define mean and SD of the demographic parameters
mean.sj <- 0.3       # Point estimate of juv. survival
se.sj.e <- 0.03      # Uncertainty of juv. survival as SE on natural scale
mean.sa <- 0.55      # Point estimate of ad. survival
se.sa.e <- 0.015     # Uncertainty of ad. survival as SE on natural scale
mean.f1 <- 1.3       # Point estimate of productivity of 1y females
se.f1.e <- 0.3       # Uncertainty of 1y productivity as SE on natural scale
mean.fa <- 1.8       # Point estimate of productivity of ad. females
se.fa.e <- 0.1       # Uncertainty of ad. productivity as SE on natural scale

# Define number of simulations, vectors and matrices to store results
nsim <- 100000
lambda <- R0 <- GT <- numeric(nsim)
stable.stage <- matrix(NA, ncol=2, nrow=nsim)
sensitivity <- matrix(NA, ncol=4, nrow=nsim)

# Generate demographic values from beta and normal distributions
library(IPMbook)
sj.sim <- rbeta2(nsim, mean.sj, se.sj.e)
sa.sim <- rbeta2(nsim, mean.sa, se.sa.e)
f1.sim <- rnorm(nsim, mean.f1, se.f1.e)
fa.sim <- rnorm(nsim, mean.fa, se.fa.e)

# Perform Monte Carlo simulations
for (s in 1:nsim){
  if(s %% 1000 == 0) {cat(paste("*** Simrep", s, "***\n"))}  # Counter

  # Transition matrix
  A <- matrix(c(sj.sim[s] * f1.sim[s], sj.sim[s] * fa.sim[s], sa.sim[s], sa.sim[s]),
    ncol=2, byrow=TRUE)
  eigenA <- eigen(A)

  # Asymptotic population growth rate
  lambda[s] <- max(Re(eigenA$values))

  # Stable stage distribution
  u <- which.max(Re(eigenA$values))
  revec <- Re(eigenA$vectors[,u])
  stable.stage[s,] <- revec / sum(revec)

  # Lower level sensitivities
  levec <- Re(solve(eigenA$vectors)[u,])
  senmat <- levec %*% t(revec)
  derivmat <- matrix(c(f1.sim[s], fa.sim[s], 0, 0), ncol=2, byrow=TRUE)
  sensitivity[s,1] <- sum(senmat * derivmat)
  derivmat <- matrix(c(sj.sim[s], 0, 0, 0), ncol=2, byrow=TRUE)
  sensitivity[s,2] <- sum(senmat * derivmat)
  derivmat <- matrix(c(0, sj.sim[s], 0, 0), ncol=2, byrow=TRUE)
  sensitivity[s,3] <- sum(senmat * derivmat)
  derivmat <- matrix(c(0, 0, 1, 1), ncol=2, byrow=TRUE)
  sensitivity[s,4] <- sum(senmat * derivmat)

  # Net reproductive rate
  i <- 1:100
  R0[s] <- sj.sim[s] * f1.sim[s] + sj.sim[s] * fa.sim[s] * sum(sa.sim[s]^i)

  # Generation time
  GT[s] <- log(R0[s]) / log(lambda[s])
}
```

Table 3.1 Mean, Monte Carlo SE, and 95% CI of population growth rate and some life-history summaries.

Parameter	Mean	Monte Carlo SE	95% CI
Population growth rate	1.022	0.062	0.910–1.154
Stable stage distribution (1y)	0.460	0.032	0.398–0.524
Stable stage distribution (adults)	0.540	0.032	0.476–0.602
Growth rate sensitivity of s_j	1.465	0.194	1.113–1.871
Growth rate sensitivity of f_l	0.129	0.023	0.089–0.177
Growth rate sensitivity of f_a	0.149	0.010	0.130–0.168
Growth rate sensitivity of s_a	1.060	0.037	0.984–1.129
Net reproductive rate	1.052	0.148	0.778–1.353
Generation time	2.385	0.191	2.045–2.795

We calculate mean and SE of the life-history summaries (Table 3.1) and prepare a plot showing frequency distributions of some quantities (Fig. 3.11).

You may note some sampling error associated with these simulation-based solutions. That is, unless we initialize the random number generators (RNGs) (using `set.seed(x)` with some common value for x), you will not obtain exactly the same estimates as we do. Nevertheless, with 100,000 draws, the Monte Carlo error will be very small, and estimates of the mean and SE of the life-history summaries should be pretty accurate. However, if you want greater accuracy, you can simply draw more simulation replicates. In Fig. 3.11, we see that the frequency distribution of these quantities is fairly symmetric, and thus the SE is an appropriate metric to describe estimation uncertainty.

3.3.3 ANALYSIS OF A MATRIX POPULATION MODEL WITH ENVIRONMENTAL STOCHASTICITY

In real life, demographic rates are rarely constant but rather vary over time, which is called "environmental stochasticity" in the matrix population model literature. The magnitude of temporal variation in demographic rates can be substantial (Gaillard et al., 2000; Altwegg et al., 2007), and calculation of the population growth rate and life-history summaries needs to account for this variation (Boyce et al., 2006). All calculations so far have assumed constant demographic rates that will produce biased values in the presence of temporal variation in demographic rates. Projection of the population is arguably the easiest way to calculate population growth rate and life-history summaries when environmental stochasticity occurs (Caswell, 2001; Morris and Doak, 2002), and we demonstrate this here.

We start with the calculation of the so-called stochastic population growth rate. The principle is fairly easy—population size is projected for a large number of years, and the annual population growth rates are calculated. In a homogeneous stochastic environment where the patterns of variation and covariation are stationary, the logarithm of population size becomes asymptotically normal, and the average growth rate converges to a fixed value (Tuljapurkar and Orzack, 1980), the stochastic population growth rate, which is calculated as the geometric mean of the annual growth rates. We must project the population for a very large number of years; the precision of the stochastic growth rate increases with the number of years.

There are two challenges for this simulation. First, how can we obtain demographic rates for many years? A demographic study never run over, say, 1000 years. Yet we may have estimates of the mean and of the temporal process variability of demographic rates that we have obtained by models with temporal random effects. This permits us to draw as many "annual" values of demographic rates as we like. Importantly, when temporal random effects are estimated, sampling variation is eliminated or "controlled for" (Gould and Nichols, 1998). For the calculation of stochastic population growth rates, we must include only process variance (i.e., temporal variability) that is not contaminated with sampling variability. If the demographic rates are temporally correlated (e.g., because in a favorable year both juvenile and adult survival is increased), the temporal correlation should be estimated (Link and Barker, 2005; Schaub et al., 2013; Riecke et al., 2019). It should then be included in the projection because such correlations can have a strong impact on population growth (Fieberg and Ellner, 2001). Alternatively, we may have annual estimates of demographic rates from a decent number of years. To obtain values for more years, we may resample them with replacement. Sampling can be set up such that

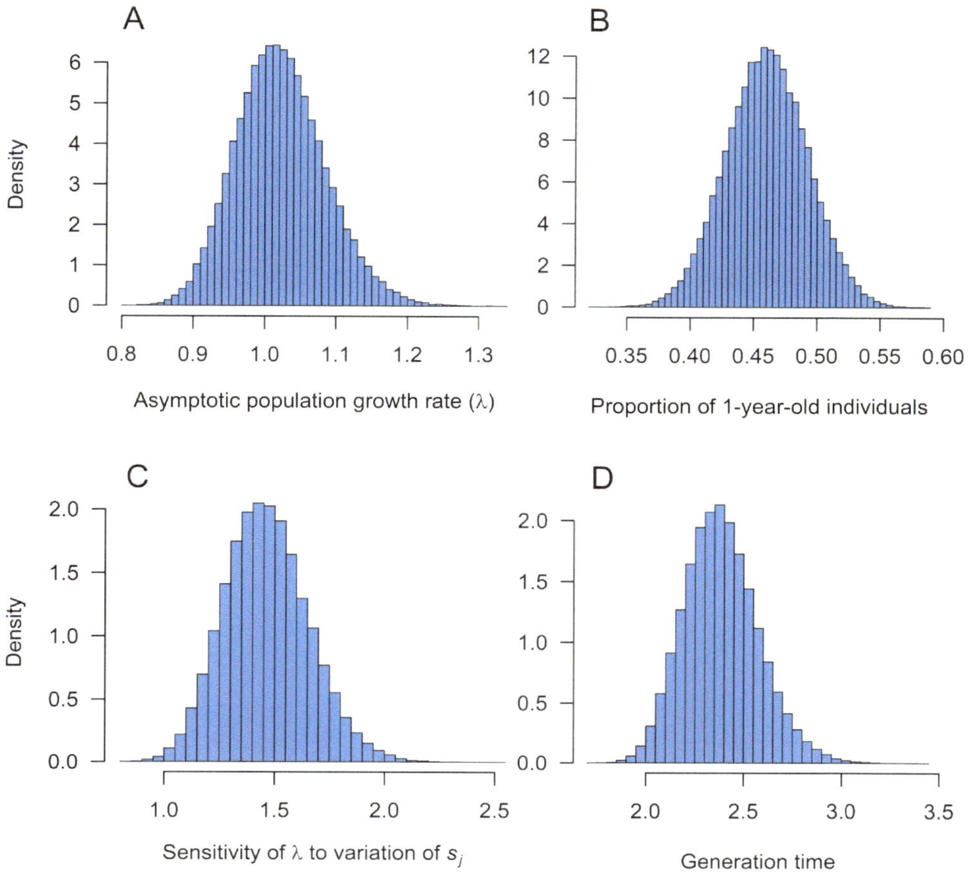

FIGURE 3.11

Frequency distribution of the asymptotic population growth rate (A), of the proportion of 1-year-old individuals (B), of growth rate sensitivity to variation in juvenile survival (C), and of generation time (D) as obtained from 100,000 Monte Carlo draws.

correlations among demographic rates are accounted for (if we sample years). In both approaches, we assume that the conditions we have observed during our study are also typical for the future, i.e., that we project a process that is stationary in terms of both mean and covariance.

To project a population over time, we must initialize it at some values, and hence, a second challenge is that stage-specific population sizes are typically unknown. Yet we can define any arbitrary stage-specific population vector provided that its sum is larger than zero. As in any Markov chain, the total size of the population will be influenced by the choice of the initial stage-specific population sizes. However, population trajectory and growth rate will eventually become independent from these initial values (Caswell, 2001, p. 283). Indeed, if the environmental stochastic process is stationary (has constant mean and variability), the stage structure of the population converges to a fixed stationary distribution. To remove the potential impact of the choice of the population vector in the first year, the first growth rates can be discarded for the calculation of the mean as a sort of "burn-in" of this Markov chain.

Assume for now that we have produced a large number (T) of samples of demographic rates ($s_{j,t}$, $s_{a,t}$, $f_{1,t}$, $f_{a,t}$) from which we construct year-specific population matrices $\mathbf{A}_t$, and that the stage-specific population vector in year 1 is $\mathbf{N}_1$. We also assume that the different demographic rates vary over time independently, i.e., that temporal covariances are zero. The stage-specific population sizes in year t ($t = 1 \ldots T$) are then calculated as $\mathbf{N}_{t+1} = \mathbf{A}_t\mathbf{N}_t$.

Annual population growth rates are calculated from the population vectors. The annual growth rates in our shrike example with two age classes are calculated as

$$\lambda_t = \frac{N_{1,t+1} + N_{2,t+1}}{N_{1,t} + N_{2,t}}.$$

The growth rate is often expressed on the log scale ($r_t = \log(\lambda_t)$). In our example, this yields

$$r_t = \log\left(N_{1,t+1} + N_{2,t+1}\right) - \log\left(N_{1,t} + N_{2,t}\right).$$

The stochastic population growth rate (either as λ_s or as r_s) is then calculated either as the geometric mean of the annual growth rates (λ_t) or as the arithmetic mean of the annual growth rates on the log scale (r_t). Typically, the first u draws of the chain of annual growth rates are discarded as a burn-in to prevent the estimates from being influenced by the arbitrary choice of the initial conditions, i.e. $\mathbf{N}_1$:

$$\lambda_s = \left(\prod_{t=u}^{T} \lambda_t\right)^{\frac{1}{T-u}} \quad \text{or } r_s = \frac{1}{T-u}\sum_{t=u}^{T} r_t, \text{ while } r_s = \log(\lambda_s).$$

The following R code calculates the stochastic population growth rate on the log scale (r_s).

```
# Define mean and temporal variability (SD) of the demographic parameters
mean.sj <- 0.3          # Mean juvenile survival (probability scale)
sd.sj.t <- 0.25         # Temporal variability on the logit scale
mean.sa <- 0.55         # Mean adult survival (probability scale)
sd.sa.t <- 0.07         # Temporal variability on the logit scale
mean.f1 <- 1.3          # Mean productivity of 1y old females
sd.f1.t <- 0.3          # Temporal variability on the natural scale
mean.fa <- 1.8          # Mean productivity of adult females
sd.fa.t <- 0.3          # Temporal variability on the natural scale

# Define the number of years with predictions and burn-in length
T <- 100000             # Length of Markov chain
u <- 1000               # Length of burn-in period
```

Models with temporal random effects often use a normal distribution on a transformed scale, such as the logit for survival or the log for productivity (Section 4.4). Therefore, any resulting estimate of the magnitude of temporal variability refers to the chosen scale, and to draw values, it is advisable to use the normal distribution on that scale as well. Because all demographic rates are drawn independently, we assume that they vary independently from one another over time, or more technically, that the process covariances are zero. It is possible to relax this assumption when estimates of the temporal covariances are available (see Section 6.4). The following code performs the desired simulations.

```
# Generate demographic values from normal distributions
sj <- plogis(rnorm(T, qlogis(mean.sj), sd.sj.t))
sa <- plogis(rnorm(T, qlogis(mean.sa), sd.sa.t))
f1 <- rnorm(T, mean.f1, sd.f1.t)
fa <- rnorm(T, mean.fa, sd.fa.t)

# Define population matrix and initial stage-specific population sizes
N <- matrix(NA, nrow=2, ncol=T+1)
N[,1] <- c(1, 1)

# Project population forwards
r <- numeric(T)
for (t in 1:T){
  if(t %% 1000 == 0) {cat(paste("*** Simrep", t, "***\n")) }  # Counter
  A <- matrix(c(sj[t] * f1[t], sj[t] * fa[t], sa[t], sa[t]), ncol=2, byrow=TRUE)
  N[,t+1] <- A %*% N[,t]
  r[t] <- log(sum(N[,t+1])) - log(sum(N[,t]))               # Annual population growth rate
  N[,t+1] <- N[,t+1] / sum(N[,t+1])                           # Scale N to avoid numerical overflow
}
```

The post-burn-in mean of the annual growth rates is an estimate of the stochastic population growth rate.

```
mean(r[u:T])
[1] 0.01944997
```

As always when using simulation, remember that each of you will obtain slightly different solutions depending on the initial value of your RNG; hence, your solution will differ slightly from ours because of Monte Carlo error. We could eliminate this effect by using either chains of "infinite" length or a seed.

As seen above, we can also express the stochastic population growth rate on the natural scale for lambda.

```
exp(mean(r[u:T]))
[1] 1.01964
```

The stochastic population growth rate (λ_s or r_s) is always smaller than the asymptotic population growth rate (λ or r, respectively), although the means of the demographic rates are exactly the same. The only difference is that the demographic rates are temporally variable. This result is general and not just specific to our example (for proofs, see Tuljapurkar and Orzack (1980)).

The precision of the stochastic population growth rate could again be evaluated using Monte Carlo simulation. It increases with the number of projected years (T). But since we can make the precision as small as we like, it is not generally a very meaningful quantity because it does not tell us anything about uncertainty in the demographic estimates obtained from a study population.

The sensitivity and elasticity of the stochastic population growth rate to changes in the demographic parameters are also computed by simulation (Caswell, 2001, p. 401; Morris and Doak, 2002). The population is projected multiple times for a large number of years, first using the actual estimates of the demographic rates, and second, where one demographic rate is changed by a small amount. Each time the stochastic growth rate is calculated, and the difference between the two growth rates tells us how sensitive it is to a small change in that demographic rate. Specifically, we change the demographic parameter (a_j) whose growth rate sensitivity we want to calculate by a small amount ($a_j^* = a_j - \Delta$, where Δ is e.g., 0.001) and calculate the stochastic population growth rates (r_s and r_s^*), first using a_j and then using a_j^*. Sensitivity is the ratio of the difference of the two stochastic population growth rates (on the natural scale) to the change in the demographic rate ($S(a_j) = (\exp(r_s^*) - \exp(r_s))/\Delta$). Elasticity is the scaled sensitivity: $E(a_j) = S(a_j) \times a_j/\exp(r_s)$. The following code provides an example of calculating the sensitivity and elasticity of stochastic population growth rate to changes in juvenile survival.

```
# Generate demographic values from normal distributions
sj <- plogis(rnorm(T, qlogis(mean.sj), sd.sj.t))
sa <- plogis(rnorm(T, qlogis(mean.sa), sd.sa.t))
f1 <- rnorm(T, mean.f1, sd.f1.t)
fa <- rnorm(T, mean.fa, sd.fa.t)

# Define population matrix and initial stage-specific population sizes
N <- N.star <- matrix(NA, nrow=2, ncol=T+1)
N[,1] <- N.star[,1] <- c(1, 1)

# Project population and calculate stochastic population growth rate
delta <- 0.001                                    # Magnitude of perturbation (= "small change")
r <- r.star <- numeric(T)
for (t in 1:T){
  if(t %% 1000 == 0) {cat(paste("*** Simrep", t, "***\n")) }    # Counter
  # Projection using sj
  A <- matrix(c(sj[t] * f1[t], sj[t] * fa[t], sa[t], sa[t]), ncol=2, byrow=TRUE)
  N[,t+1] <- A %*% N[,t]
  r[t] <- log(sum(N[,t+1])) - log(sum(N[,t]))                  # Annual population growth rate
  N[,t+1] <- N[,t+1] / sum(N[,t+1])                            # Scale N to avoid numerical overflow
  # Projection using sj.star
  A.star <- matrix(c((sj[t] + delta) * f1[t], (sj[t] + delta) * fa[t], sa[t], sa[t]), ncol=2,
    byrow=TRUE)
  N.star[,t+1] <- A.star %*% N.star[,t]
  r.star[t] <- log(sum(N.star[,t+1])) - log(sum(N.star[,t]))  # Annual population growth rate
  N.star[,t+1] <- N.star[,t+1] / sum(N.star[,t+1])        # Scale N.star to avoid numerical overflow
}

# Compute stochastic sensitivity (juvenile survival)
(exp(mean(r.star[u:T])) - exp(mean(r[u:T]))) / delta
[1] 1.450347

# Compute stochastic elasticity (juvenile survival)
(exp(mean(r.star[u:T])) - exp(mean(r[u:T]))) / delta * mean.sj / exp(mean(r[u:T]))
[1] 0.4263838
```

Sensitivity and elasticity of the asymptotic and stochastic growth rates to changes in juvenile survival are very similar. This is typically the case unless the temporal variability becomes very large (Altwegg et al., 2007).

3.3.4 ANALYSIS OF A MATRIX POPULATION MODEL WITH DEMOGRAPHIC STOCHASTICITY

Demographic stochasticity arises from random variation in the realized number of births and deaths in a population and is ultimately due to the discrete nature of individuals—e.g., a jaguar can have 2 or 3 cubs but not 2.5. We introduce demographic stochasticity by drawing these numbers from discrete-valued statistical distributions with expectations given by the chosen survival or recruitment. That is, by treating the numbers of survivors, recruits, and so on as discrete random variables, there will be variability around their expectation in the realized values. This random variability is demographic stochasticity. Demographic stochasticity can be an important driver of population growth (Sæther et al., 2007). To illustrate how it works, an example may be helpful. We can ask how many of 10 woodchat shrikes survive until the next year given a survival probability of $s_a = 0.55$. To address this question, we can conduct 10 coins flips with a coin that produces heads 55% of times, and count how many heads (= survival events) we obtain. Because such a loaded coin is difficult to obtain, use of the binomial RNG in R is more convenient.

```
rbinom(n=1, size=10, prob=0.55)
[1] 4
```

Four individuals have survived in this particular realization of the chance experiment. When we repeat this 20 times, we obtain quite variable numbers of survivors (and you obtain a different set of numbers, remember).

```
rbinom(n=20, size=10, prob=0.55)
[1] 4 6 7 6 7 4 4 7 9 3 5 5 6 6 1 6 2 6 1 6
```

The number of survivors now varies between 1 and 9, and the most frequent value is 6. The expected value of the experiment is 5.5 (= 10 × 0.55), but we never actually obtain this number of survivors because it is impossible for a shrike to be half alive. Yet if we ignore demographic stochasticity, we conduct calculations in a population model with the expected value of this random variable, i.e., with 5.5 individuals. We also note that the number of surviving individuals varies between trials, although the survival probability was the same in the 20 trials. Put differently, binomial counts with sample size 20 and success probability 0.55 naturally vary from trial to trial.

Demographic stochasticity (i.e., the difference between the realized and the expected value of a discrete-valued random variable in population dynamics) is present in all populations, but it is particularly relevant for population modeling when the population size is small (Lande, 1993). We can demonstrate this again using the example from before. We now repeat the generation of surviving individuals 1000 times and each time calculate the *realized* proportion of surviving individuals, i.e., the ratio of the number of surviving shrikes and the initial number of shrikes (10). We then repeat this experiment with progressively larger population sizes. Finally, we plot frequency distributions of the realized survival probabilities (Fig. 3.12).

```
size <- c(10, 25, 50, 75, 100, 250, 500, 750, 1000, 5000, 10000)
s <- matrix(NA, nrow=1000, ncol=length(size))
for (i in 1:length(size)){
  s[,i] <- rbinom(n=1000, size=size[i], prob=0.55) / size[i]
}
boxplot(s, ylab="Realized proportion of surviving birds",
  xlab="Population size (number of females)", names=size, outline=FALSE,
  col="dodgerblue", las=1, frame=FALSE)
abline(h=0.55, col="red", lwd=1)
```

Realized survival is much more variable when the population consists of a small rather than large number of shrikes, although the expected survival (i.e., survival probability) is exactly the same in all cases. Smaller populations will therefore experience stronger temporal variability in the realized proportion of survivors than larger populations even if the underlying parameter governing the survival process (= survival probability) is identical at each time step. This holds not only for survival but also for all demographic processes that can be modeled by discrete-valued random variables, and as a result, also to the population growth rate—smaller populations vary relatively more in size than larger populations.

In general, to include demographic stochasticity in a population model, we must simulate the demographic performance of each individual. This requires adoption of adequate statistical distributions for discrete random variables. We have already seen that the binomial distribution is an appropriate choice for the survival process: of N individuals that each survive independently with probability s, we can never observe more survivors than N and on average will observe Ns survivors. For productivity, the Poisson distribution is a natural first choice because it has no upper bound and a minimum value of 0, although it has sometimes been found to be too variable (Ridout and Besbeas, 2004; Lynch et al., 2014). Another requirement for the analysis of a matrix population model with demographic stochasticity is the definition of the stage-specific population vector in the first year. Because the "magnitude" of demographic stochasticity depends on the population size, the size of this vector now matters for the trajectory of a population affected by demographic stochasticity. Therefore we chose realistic values corresponding to our population of interest.

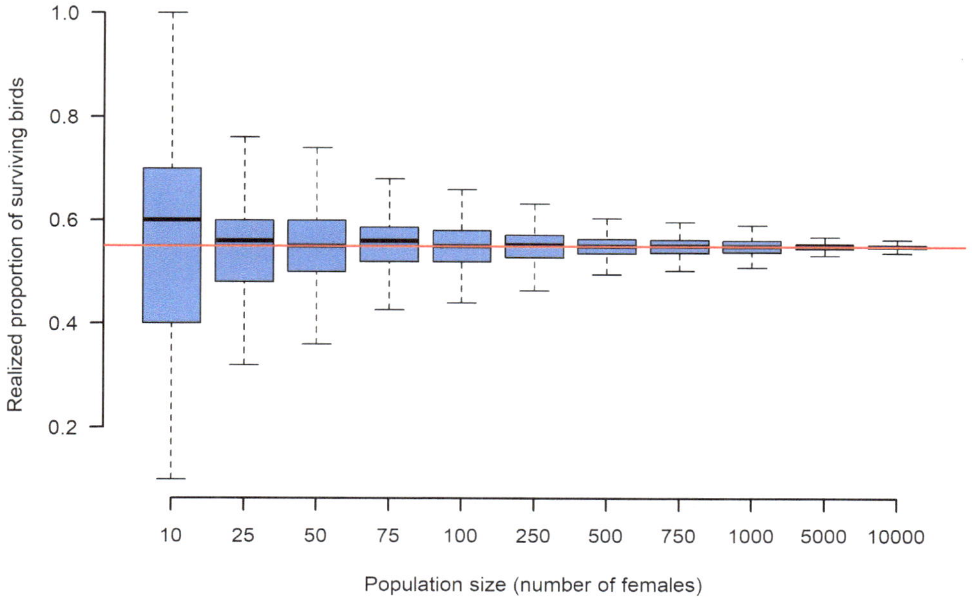

FIGURE 3.12

An illustration of demographic stochasticity: boxplot of realized proportions of surviving birds generated from 1000 binomial trials when the survival probability (i.e., the binomial success probability) is constant at 0.55 and population size (i.e., the binomial total) increases from 10 to 10,000. We see that smaller populations are much more affected by demographic stochasticity, i.e., their realized survival rate is much more variable.

A practical issue for the calculations is the possible extinction of a simulated population. When a population is small and we account for the discrete nature of individuals, it can happen at some time step that no individual survives, and no individual is recruited. The population size then drops to zero, and the population goes extinct. In contrast, in the absence of demographic stochasticity, a population can never go extinct. Population size may become very, very small in the long term, but never *exactly* zero. Mathematically, this means that we do not need to bother with extinction in a population model without demographic stochasticity because the population model still works, and the population growth rate is always defined. By contrast, as soon as the population size becomes zero, the population will remain at this size of zero provided there is no immigration. Consequently, the population growth rate is no longer defined because of division by zero or an attempt to take the log of zero. When we model a population including demographic stochasticity, we must therefore account for the possibility that the population goes extinct and restrict the calculation of the population growth rate and other life-history summaries to that part of the population trajectory before extinction. The study of extinction is an important field in conservation biology (Beissinger, 2002), and we will see more about this topic in Chapter 10, where we use an IPM to estimate extinction probability and related parameters. But for now, it is important only that we consider the possibility of population extinction in the calculation of the population growth rate.

We now know all we need to know to simulate a population with demographic stochasticity and can therefore start with an illustration of how such a population evolves over time. The algorithm is the following:

1. Define the stage-specific number of individuals in the first year.
2. Loop over time: determine the number of individuals in each year using appropriate probability distributions for discrete random variables that reflect the demographic processes in each year.
3. Calculate the population growth rate in each year.

The following R code performs these calculations. We assume that a population consists of 20 females in year 1, of which 10 are 1 year old and 10 are 2 years old or older. Moreover, we use a single probability distribution (the Poisson) for the recruitment process. The recruitment process really consists of two distinct demographic processes: the production of fledglings (for which a Poisson looks natural) and the survival of these fledglings until the next year (for which we naturally would adopt a binomial under the usual independence assumption). Hence, it would also be possible to separate these two processes. However, here we opt for the simpler solution, which is to model their product as a Poisson random variable. Nevertheless, we will show an example for the former later in the book (Chapter 13).

```
# Define mean of the demographic parameters
mean.sj <- 0.3
mean.sa <- 0.55
mean.f1 <- 1.3
mean.fa <- 1.8

# Define the number of years over which we let the population evolve
T <- 200

# Define population matrix and initial stage-specific population sizes
N <- matrix(NA, nrow=2, ncol=T+1)
N[,1] <- c(10, 10)

# Project population
r <- numeric(T)
for (t in 1:T){
  N[1,t+1] <- rpois(1, mean.sj * mean.f1 * N[1,t] + mean.sj * mean.fa * N[2,t])
  N[2,t+1] <- rbinom(1, sum(N[,t]), mean.sa)
  if (sum(N[,t+1]) == 0) break      # Stop calculation if pop. extinct
  r[t] <- log(sum(N[,t+1])) - log(sum(N[,t]))
}
mean(r)                              # Compute stochastic growth rate
[1] 0.02552

# Plot graphs with estimated population growth rates and size
layout(matrix(1:2, 1, 2, byrow=TRUE), widths=c(1, 1), heights=1, TRUE)
plot(r, type="l", lwd=1.5, ylab="Annual population growth rate", xlab="Time", axes=FALSE)
axis(1)
axis(2, las=1)
abline(h=mean(r))
mtext("A", at=1, cex=1.5)
plot(N[1,], type="l", lwd=1.5, ylab="Population size", xlab="Time", axes=FALSE,
  ylim=range(N, na.rm=TRUE))
axis(1)
axis(2, las=1)
mtext("B", at=3, cex=1.5)
lines(N[2,], lwd=1.5, col="red")
legend("topleft", lwd=c(1.5, 1.5), col=c("black", "red"), legend=c("1-year-old", ">1-year-old"), bty="n")
```

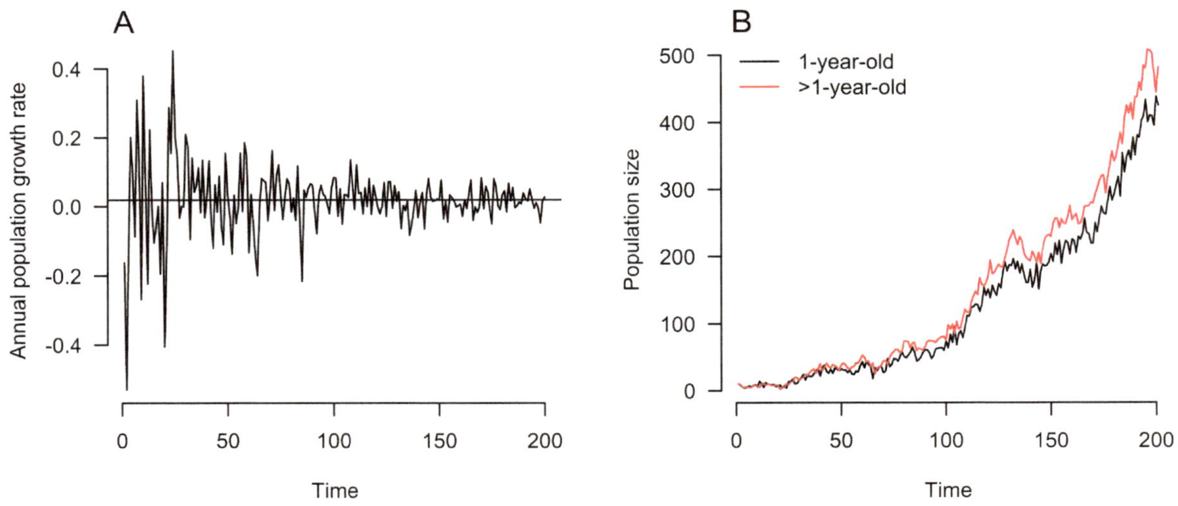

FIGURE 3.13

Trajectory of estimated annual population growth rate (A, with average superimposed as a horizontal line) and the estimated, stage-specific population sizes (B) in a woodchat shrike population subject to demographic stochasticity. The demographic parameters are constant over time and not subject to estimation error. The starting population size is 20 females, with half each in the two age classes.

In Fig. 3.13A, we notice that the growth rate is much more variable in the earlier rather than later years of the simulation. This is exactly due to demographic stochasticity—as the population increases in size, its growth rate becomes less variable, since the effects of demographic stochasticity diminish.

Although we project the population for 200 years, the current result (for any given year) is a single realization of the stochastic process represented by the model. If we need a measure of the mean and the precision of the population growth rate, we need to repeat this projection many times. When we do that, we can also evaluate how many times the population goes extinct within a specified time frame. The frequency of such extinction events can be interpreted as the extinction probability. Here, we repeat the projection 100,000 times.

```
# Define mean of the demographic parameters
mean.sj <- 0.3
mean.sa <- 0.55
mean.f1 <- 1.3
mean.fa <- 1.8

# Define the number of years with predictions and the Monte Carlo setting
T <- 200
nsim <- 100000

# Define population matrix and initial stage-specific population sizes
N <- array(NA, dim=c(2, T+1, nsim))
N[,1,] <- c(10, 10)
r <- matrix(NA, nrow=T, ncol=nsim)
alive <- matrix(NA, nrow=T, ncol=nsim)
mean.r <- numeric(nsim)

# Project population
for (s in 1:nsim){
  if(s %% 1000 == 0) {cat(paste("*** Simrep", s, "***\n")) }  # Counter
  for (t in 1:T){
    N[1,t+1,s] <- rpois(1, mean.sj * mean.f1 * N[1,t,s] + mean.sj * mean.fa * N[2,t,s])
    N[2,t+1,s] <- rbinom(1, sum(N[,t,s]), mean.sa)
    if (sum(N[,t+1,s]) == 0) break
    r[t,s] <- log(sum(N[,t+1,s])) - log(sum(N[,t,s]))
    alive[t,s] <- t
  } #t
  mean.r[s] <- mean(r[min(alive[,s], na.rm=TRUE):max(alive[,s], na.rm=TRUE),s])
} #s
```

The annual population growth rates of all simulated populations are shown in Fig. 3.14A. Variability of the growth rates is greater at the beginning than later. This pattern occurs because most populations are increasing, and the impact of demographic stochasticity on these population declines, which is a typical feature of demographic stochasticity.

The mean and standard deviation of the stochastic population growth are computed as follows.

```
# Mean and SD of the population growth rate
mean(mean.r)
[1] -0.009097617
sd(mean.r)
[1] 0.05689681
```

We notice that the mean across all 100,000 simulated populations is markedly different from the mean of a single realization of the same process that we had before. The reason for this discrepancy is that the former includes the mean growth rates of populations that went extinct. The frequency distribution of the growth rates is strongly skewed and even bimodal (Fig. 3.14B), and the arithmetic mean is not a good description of this distribution. Often we want to obtain a population growth rate for only those populations that have survived over the defined time threshold, which is computed next.

```
# Mean and SD of the population growth rate for populations that did not go extinct
not.extinct <- which(!is.na(alive[T,]))
mean(mean.r[not.extinct])
[1] 0.01966304
sd(mean.r[not.extinct])
[1] 0.006095664
```

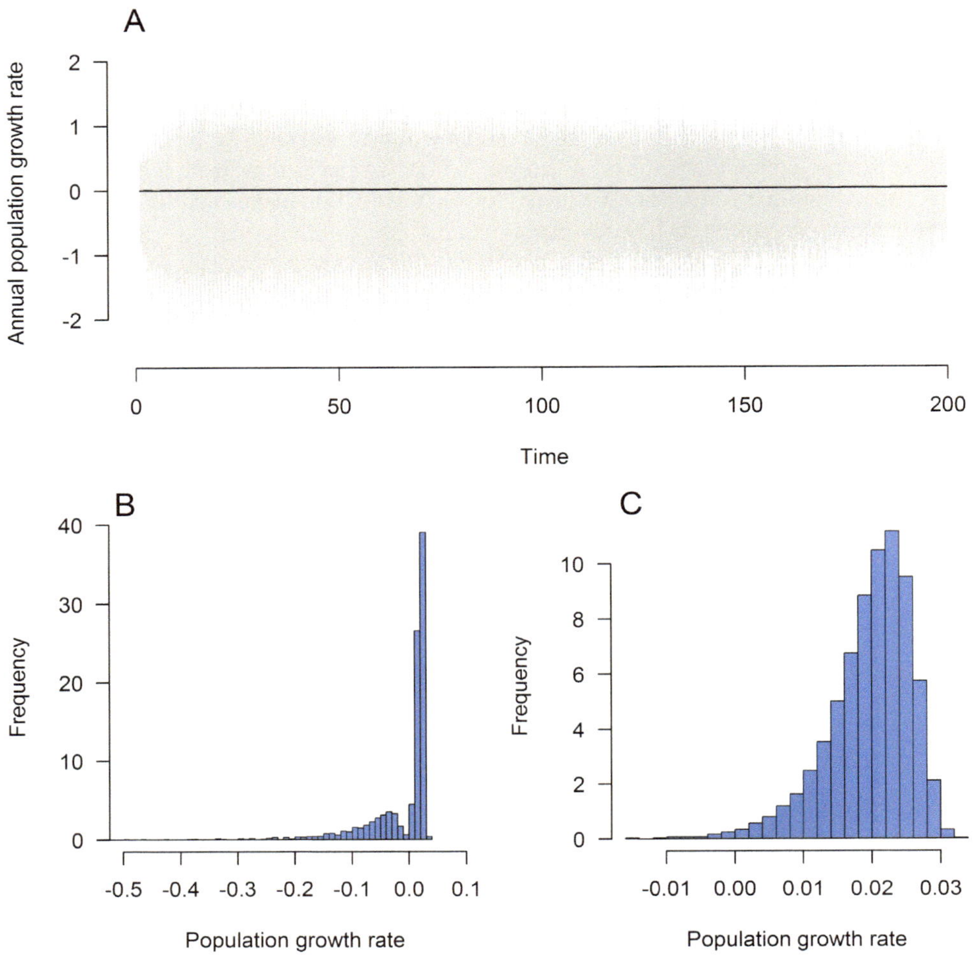

FIGURE 3.14

Population growth rates of a shrike population simulated with demographic stochasticity. (A) Realized annual population growth rates of all 100,000 populations (gray) along with the mean (horizontal black line) across all populations. (B) Frequency distribution of the mean growth rates of all populations. (C) Frequency distribution of the mean growth rates of the nonextinct populations only. Units of the *y*-axes in (B and C) are in multiples of 1000.

The mean population growth rate of the nonextinct populations is now closer to that of the single realization and to the asymptotic population growth rate (recall, it is $\log(1.0208) = 0.0206$). In addition, the nonextinct populations have a much smaller variability than those that did go extinct. The frequency distribution of the population growth rates of the nonextinct populations is unimodal but slightly skewed (Fig. 3.14C).

The simulations also allow us to evaluate the extinction probability.

```
# Extinction probability (after T years)
sum(is.na(alive[T,])) / nsim
[1]  0.28917
```

We could compute other statistics related to the extinction probability, such as the expected time to extinction or the sensitivity of the extinction probability to changes in demographic rates or initial population size. Here we define extinction as the event that a population drops exactly to zero, but in practice we often choose an extinction threshold greater than zero. Quasi-extinction is then defined as the event that a population size drops below that threshold. We will discuss these issues more in Chapter 10.

```
# Plot graph with population growth rates (Fig. 3.14 and 3.15)
layout(matrix(c(1, 1, 2, 3), 2, 2, byrow=TRUE), widths=c(1, 1), heights=c(1, 1), TRUE)
par(mar=c(4, 4, 2, 1), las=1, cex=1.1)
plot(r[,1], type="l", lwd=0.5, ylab="Annual population growth rate", xlab="Time",
  ylim=range(r[which(!is.na(alive))]), col="lightgrey", axes=FALSE)
axis(1); axis(2)
for (s in 2:nsim){
  lines(r[!is.na(alive[,s]),s], lwd=0.5, col="lightgrey")
}
lines(apply(r, 1, mean, na.rm=TRUE), lwd=1.5)
mtext("A", at=0, cex=1.5)
a <- hist(mean.r, nclass=100, col="dodgerblue", main="", xlab="Population growth rate",
  xlim=c(-0.5, 0.1), axes=FALSE)
axis(1)
axis(2, at=c(0, 10000, 20000, 30000, 40000), labels=c(0, 10, 20, 30, 40))
mtext("B", at=a$mids[26], cex=1.5)
a <- hist(mean.r[not.extinct], nclass=25, col="dodgerblue", main="",
  xlab="Population growth rate", axes=FALSE)
axis(1)
axis(2, at=c(0, 2000, 4000, 6000, 8000, 10000), labels=c(0, 2, 4, 6, 8, 10))
mtext("C", at=a$mids[1], cex=1.5)
```

Other life-history summaries such as the stage distribution or growth rate sensitivities can also be computed by adapting the code in a way analogous to what we did in Section 3.3.3.

3.3.5 ANALYSIS OF A MATRIX POPULATION MODEL WITH MULTIPLE SOURCES OF STOCHASTICITY AND PARAMETER UNCERTAINTY

A further development of the population model, of course, is to include both types of stochasticity (environmental and demographic) as well as parameter uncertainty. Conceptually this is not a big deal, but it does require more bookkeeping for all modeled quantities. To illustrate such a model, we expand on the previous examples by assuming that the means of survival and productivity are subject to estimation error and that in addition we have an estimate of the temporal variability of the true values of these rates (i.e., ignoring the estimation error for the sake of simplification). We again initialize the population at 20 females (see Exercise 3.3 for accommodating estimation error in the initial population sizes as well). To estimate the population growth rate and life-history summaries, we again need to project many populations over a large number of years and then summarize their mean behavior. Here is pseudocode for the algorithm:

1. Draw a value of each demographic rate from a specified statistical distribution using estimated means and sampling variance (i.e., estimation uncertainty, not process variance).
2. Draw annual demographic rates from appropriate distributions reflecting temporal variability using the means obtained in step 1 and the temporal variability (which is assumed to be known without error).
3. Let the population evolve over time: determine the number of individuals in each year using appropriate statistical distributions that reflect the demographic processes in each year, thereby accounting for demographic stochasticity. Use the values of the annual rates obtained in step 2.
4. Repeat steps 1–3 a large number of times, i.e., for many populations.

We define the parameter values that we will use.

```
# Define mean, measurement error and temporal variability of the demographic parameters
mean.sj <- 0.3       # Mean value of juv. survival
sd.sj.e <- 0.005     # Uncertainty of mean juv. survival as SD on natural scale
sd.sj.t <- 0.25      # Temporal variability of juv. survival as SD on logit scale
mean.sa <- 0.55      # Mean value of ad. survival
sd.sa.e <- 0.005     # Uncertainty of mean ad. survival as SD on natural scale
sd.sa.t <- 0.07      # Temporal variability of ad. survival as SD on logit scale
mean.f1 <- 1.3       # Mean value of productivity of 1y females
sd.f1.e <- 0.05      # Uncertainty of mean productivity as SD on natural scale
sd.f1.t <- 0.3       # Temporal variability of productivity as SD on natural scale
mean.fa <- 1.8       # Mean value of productivity of adult females
sd.fa.e <- 0.03      # Uncertainty of mean productivity as SD on natural scale
sd.fa.t <- 0.3       # Temporal variability of productivity as SD on natural scale
```

```
# Define the number of years with predictions and the Monte Carlo setting
T <- 200              # Number of years (projection time frame)
nsim <- 100000        # Number of replicate populations simulated

# Define population matrix and initial stage-specific population sizes
N <- array(NA, dim=c(2, T+1, nsim))
N[,1,] <- c(10, 10)
r <- matrix(NA, nrow=T, ncol=nsim)
alive <- matrix(NA, nrow=T, ncol=nsim)
mean.r <- numeric(nsim)
# Project population
for (s in 1:nsim){                                          # Loop over replicate populations
  if(s %% 100 == 0) {cat(paste("*** Simrep", s, "***\n")) }   # Counter
  # Generate a mean of the demographic rates (subject to measurement error)
  msj <- rbeta2(1, mean.sj, sd.sj.e)
  msa <- rbeta2(1, mean.sa, sd.sa.e)
  mf1 <- rnorm(1, mean.f1, sd.f1.e)
  mfa <- rnorm(1, mean.fa, sd.fa.e)

  # Generate annual demographic rates (subject to temporal variability)
  sj <- plogis(rnorm(T, qlogis(msj), sd.sj.t))
  sa <- plogis(rnorm(T, qlogis(msa), sd.sa.t))
  f1 <- pmax(0, rnorm(T, mf1, sd.f1.t))                     # Avoids negative values
  fa <- pmax(0, rnorm(T, mfa, sd.fa.t))                     # Dito

  # Project population (include demographic stochasticity)
  for (t in 1:T){                                           # Loop over years
    N[1,t+1,s] <- rpois(1, sj[t] * (f1[t] * N[1,t,s] + fa[t] * N[2,t,s]))
    N[2,t+1,s] <- rbinom(1, sum(N[,t,s]), sa[t])
    if (sum(N[,t+1,s]) == 0) break
    r[t,s] <- log(sum(N[,t+1,s])) - log(sum(N[,t,s]))
    alive[t,s] <- t
  } #t
  mean.r[s] <- mean(r[min(alive[,s], na.rm=TRUE):max(alive[,s], na.rm=TRUE),s])
} #s
```

Sometimes, this may throw an error, since the function `rbinom` in R crashes when the `size` parameter becomes larger than about 2^{31}. A workaround would be to exclude demographic stochasticity after the population has reached a very large size because then it simply does not matter anymore. However, for simplicity, we suggest you merely execute the complete code again until it completes without problems. The mean population growth rate is calculated next.

```
mean(mean.r)
[1] -0.01562592
sd(mean.r)
[1] 0.06659048
```

The distribution of the mean stochastic population growth rates is strongly skewed (Fig. 3.15B), and the mean may not be a good descriptor of this distribution. Therefore, we may want to compute the mean population growth rate of non-extinct populations whose frequency distributions are unimodal and nearly symmetrical (Fig. 3.15C).

```
not.extinct <- which(!is.na(alive[T,]))
mean(mean.r[not.extinct])
[1] 0.02353411
sd(mean.r[not.extinct])
[1] 0.01242122
```

The extinction probability becomes important for the computation of the mean stochastic population growth rate. For nonextinct populations, it is now larger than when stochasticity is ignored. This may not agree with our intuition but happens when we ignore the growth rates of the extinct populations. We can modify the computation of the mean stochastic population growth, as we think it is useful for our purposes (e.g., including the annual growth rates of the extinct populations until their extinction).

```
# Extinction probability (after T years)
sum(is.na(alive[T,])) / nsim
[1] 0.38023
```

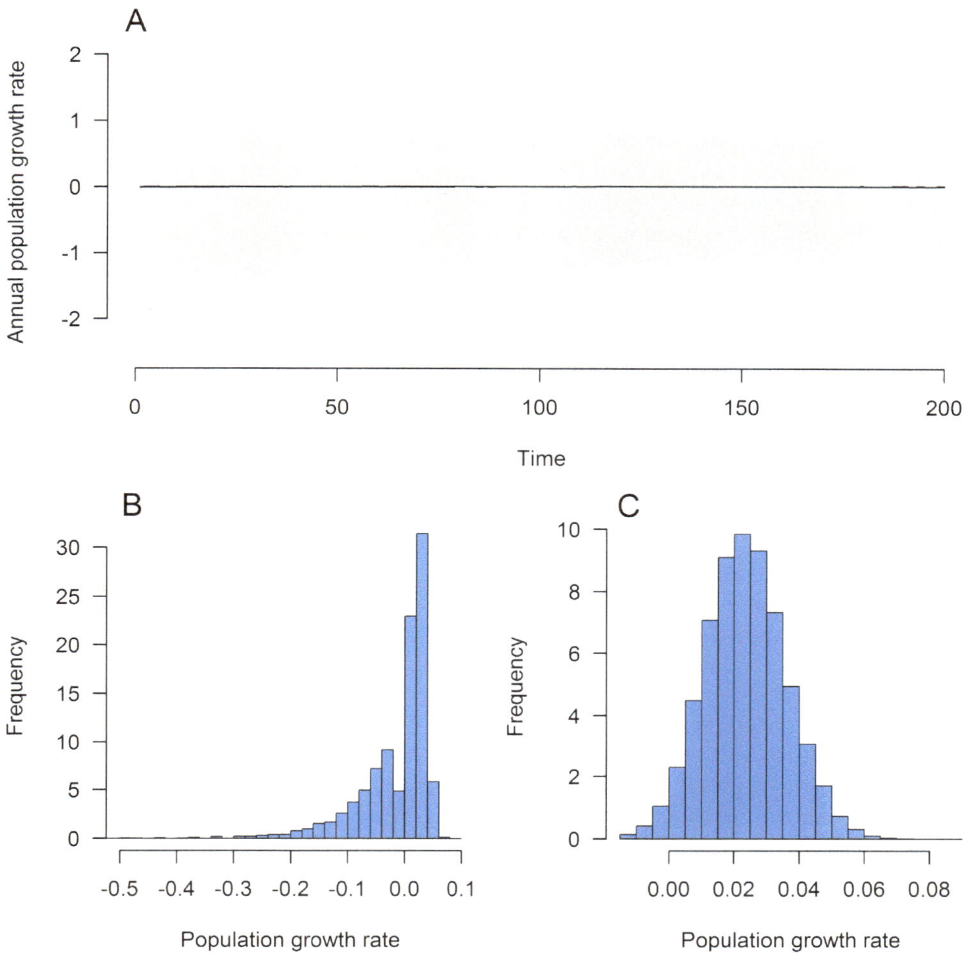

FIGURE 3.15

Population growth rates in a model that includes environmental and demographic stochasticity as well as estimation error (i.e., parameter uncertainty). (A) Realized annual population growth rates of all 100,000 populations (gray) along with the mean (black) across the populations. (B) Frequency distribution of the mean growth rates of all populations. (C) Frequency distribution of the mean growth rates of nonextinct populations. Units of the *y*-axes in (B and C) are in multiples of 1000.

3.3.6 MATRIX POPULATION MODELS WITH DENSITY DEPENDENCE AND DEMOGRAPHIC STOCHASTICITY

All the population models presented in this chapter so far have assumed that population growth is independent of the size of the population, i.e., that population growth is exponential. However, natural populations can never grow without limit over a longer time frame because resources such as food, nest sites, and mates always become limited at some point. Therefore, natural populations will always be subject to density dependence, i.e., as population size increases, the population growth rate must decrease (Newton, 1998; Hixon et al., 2002; Bonenfant et al., 2009; Knape and de Valpine, 2012). If a population is regulated by density, the population growth rate λ will fluctuate around 1 (or 0 if expressed as r).

Many different mechanisms may result in density-dependent population growth, but basically at least one demographic rate must decline with increasing population size. In other words, there must be a negative feedback between population size and at least one demographic rate. The population model may then be written in matrix notation as follows: $\mathbf{N}_{t+1} = \mathbf{A}_{\mathbf{N}_t} \times \mathbf{N}_t$, where by $\mathbf{A}_{\mathbf{N}_t}$ we mean that the transition matrix $\mathbf{A}$ depends on $\mathbf{N}_t$. Thus, the elements of the transition matrix and the demographic rates now depend on the population size $\mathbf{N}$ at time t. The resulting model is nonlinear, and population growth rate and other life-history summaries can no longer be computed from eigenvalues and eigenvectors. The easiest way to study the behavior of a population model that includes density dependence is again with simulation.

There are various different mathematical forms with which the relationship between population size and a demographic rate can be described (e.g., Ricker, Gompertz, Beverton-Holt, theta-logistic (Gilpin and Ayala, 1973; Lande et al., 2003)). This is not surprising, since there is literally an infinite number of possible monotonic and negative curves relating a demographic rate and population size. For simplicity, a linear relationship is assumed here (see also Chapter 8).

The typical behavior of most populations with density dependence is that their sizes reach an equilibrium density (typically referred to as the *carrying capacity*) and then remain there (provided, of course, that they are not perturbed any further). Some populations under density dependence may also show cycles or even chaotic behavior (Caswell, 2001, p. 504). Of course, at equilibrium, the population growth rate is $\lambda = 1$ (or $r = 0$), and thus there is not much interest in estimating this quantity when there is density dependence. However, there might be an interest in estimating carrying capacity or the time needed until the population reaches the carrying capacity. Because population size obviously matters when density dependence is modeled, we typically include demographic stochasticity. Therefore, we need simulations to assess the behavior of the population.

We here assume that the available territories in our woodchat shrike population differ in quality such that reproductive success also differs between individuals. High-quality territories produce a larger number of fledglings and are occupied in most years regardless of the actual population size. In contrast, low-quality territories produce fewer fledglings and are occupied only when population size is high and all higher-quality territories are already occupied. The resulting (population-level) annual fledging productivity will then decline with increasing population size. This mechanism of population regulation has been called *site-dependent population regulation* or *buffer effect* and has been found repeatedly in territorial species (Rodenhouse et al., 1997; Gill et al., 2001). In our simulation, we will not describe each individual territory but rather add a functional relationship between population size and fledging production, since this is the net result of heterogeneity in territory quality and fledgling output. We also assume that 1-year-old females are weaker competitors than older females and thus specify stronger density dependence for f_1 than for f_a. We assume that survival probabilities are unaffected by population size and are constant over time. Thus, in this simulation we assume that density dependence acts only on productivity, not on survival. We initialize our population with 20 females, let the population evolve over 200 years, and repeat this for 1000 replicate populations. Understanding the R code for these simulations should pose no major difficulty. Perhaps we should note that productivity in each year can be calculated only when the population size is known; thus, its computation needs to be included in the loop over time.

```
# Define means of the demographic rates and strength of density dependence
mean.sj <- 0.3
mean.sa <- 0.55
f1.int <- 2.3           # Productivity of 1y females when population size is 0
f1.beta <- -0.02        # Strength of density dependence on productivity of 1y
                        # This is simply the slope of the regression of f1 on N
fa.int <- 2.3           # Productivity of adult females when population size is 0
fa.beta <- -0.01        # Strength of density dependence on productivity of adults


# Define the number of years with predictions
T <- 200
nsim <- 1000

# Define population matrix and initial stage-specific population sizes
N <- array(NA, dim=c(2, T+1, nsim))
N[,1,] <- c(10, 10)
r <- f1 <- fa <- matrix(NA, nrow=T, ncol=nsim)

# Project population
for (s in 1:nsim){
  if(s %% 100 == 0) {cat(paste("*** Simrep", s, "***\n")) }  # Counter
  for (t in 1:T){
    f1[t,s] <- f1.int + f1.beta * sum(N[,t,s])
    fa[t,s] <- fa.int + fa.beta * sum(N[,t,s])
    N[1,t+1,s] <- rpois(1, mean.sj * (f1[t,s] * N[1,t,s] + fa[t,s] * N[2,t,s]))
    N[2,t+1,s] <- rbinom(1, sum(N[,t,s]), mean.sa)
  } #t
} #s
```

From Fig. 3.16, we see the expected patterns and variabilities of population size and productivity: population size approaches the carrying capacity and then remains more or less stable, and fluctuations are due to demographic stochasticity in this moderately small population. In contrast, productivity declines with increasing population size and then stabilizes as the carrying capacity is attained. As an estimate of carrying capacity, we can use the mean of the population sizes after 200 time intervals.

```
mean(N[1,T+1,] + N[2,T+1,])
[1] 52.687
```

The average fecundities at carrying capacity are

```
mean(f1[T,], na.rm=TRUE)
[1] 1.237938
mean(fa[T,], na.rm=TRUE)
[1] 1.768969
```

In addition, we also note that one of the simulated populations went extinct (Fig. 3.16A). The extinction probability (evaluated from the 1000 replicate populations) is small, though.

```
mean(N[1,T+1,] + N[2,T+1,] == 0)
[1] 0.001
```

It would be straightforward to also include environmental stochasticity, parameter uncertainty, density dependence in other demographic rates, or different functional forms of density dependence. The sensitivity of the carrying capacity to changes in the demographic rates can also be calculated using Monte Carlo simulations in an analogous way as shown before.

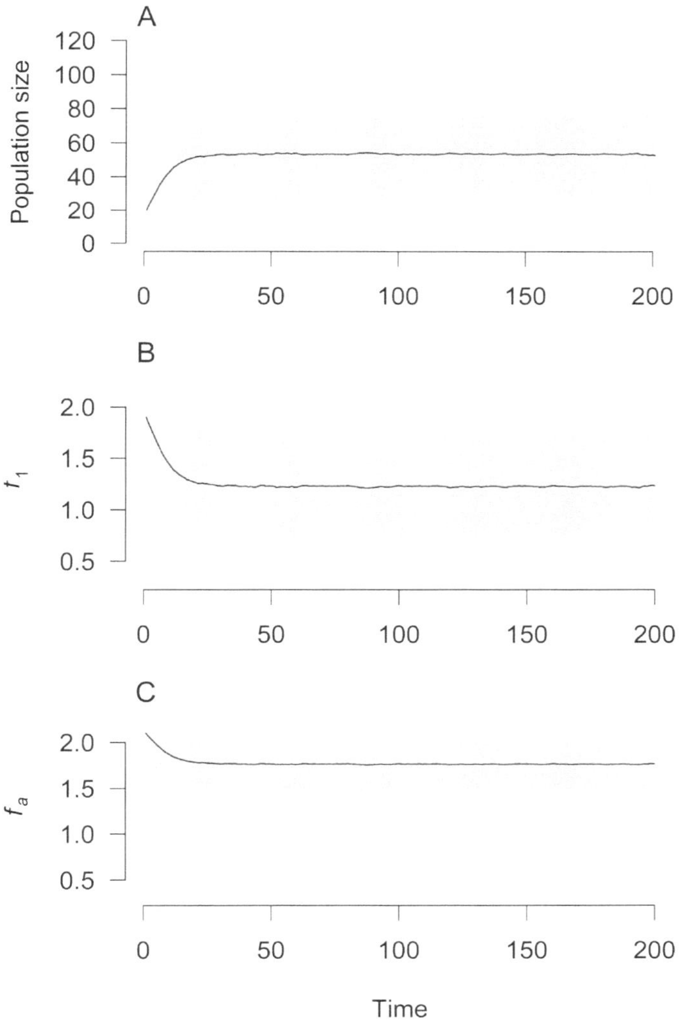

FIGURE 3.16

Trajectories of (A) population size and productivity of 1-year-old (B) and adult (C) females for our experimental population of woodchat shrikes subject to density-dependent productivity and demographic stochasticity (survival is density-independent). Plot shows the trajectories for 1000 simulated populations (gray lines) and their means (black line).

3.4 ANALYSIS OF MATRIX POPULATION MODELS WITH MARKOV CHAIN MONTE CARLO SOFTWARE

We have seen how fundamentally important simulation is for the analyses of all but the simplest population models. Interestingly, simulation techniques are also widely used in modern statistical modeling for parameter estimation, especially the powerful MCMC technique (Brooks et al., 2011; Lunn et al., 2013). Here, we make the transition to using such MCMC software for doing all of the classical analyses of matrix population models described in the previous section. We will see that this has some advantages in itself because of course any analysis done with MCMC is inherently simulation-based. But much more importantly, in this book we use MCMC for Bayesian parameter estimation in statistical models; for many examples of the power of Bayesian population analyses, see for instance Kéry and Schaub (2012) and Kéry and Royle (2016, 2021). So the main reason for why we now transition to doing simple analyses of matrix population models with MCMC software (specifically, with the JAGS variant of BUGS software) is that this will later allow us the seamless integration of matrix population modeling and parameter estimation, and this integration is one of the key defining features of an IPM.

To illustrate the use of MCMC methods implemented in JAGS, we start as before with a model that includes neither stochasticity nor any uncertainty associated with the demographic rates, and later will relax these assumptions. The goal is again the computation of the population growth rate and of various life-history summaries. The principle applied when matrix population models are analyzed with JAGS is that the population is projected over a couple of years, as is necessary in the classical approach when environmental or demographic stochasticity is included. From the projected population trajectories, the stage-specific population sizes and many quantities of interest can be derived. A key advantage of using the MCMC machinery for computing these quantities is that the uncertainty of the population growth rate or of life-history summaries is automatically and properly accommodated when the demographic parameters contain estimation errors and/or are subject to stochasticity. Admittedly, this approach is a bit clumsy for simple cases, but the nice feature is that this general approach is valid for virtually any kind of complexity that we may want to add to our model. Moreover, the models that we develop here can be extended to IPMs in a very natural way and we hope that this approach makes the link between an IPM and matrix population models particularly transparent.

3.4.1 ANALYSIS OF A MATRIX POPULATION MODEL WITHOUT STOCHASTICITY AND PARAMETER UNCERTAINTY

First, we again assume that the demographic rates of our woodchat shrikes are constant over time and perfectly known, i.e., there is no parameter uncertainty. Unfortunately, some matrix operations such as eigenvalues are not available in BUGS or JAGS (though these and many more are so in NIMBLE; NIMBLE Development Team, 2018). Hence, we will not use matrix multiplication for projecting the population, but the stage-specific population equations instead. As we have seen already in Section 3.2.3, the number of 1-year-old individuals in year $t+1$, $N_{1,t+1}$, is a function of the modeled quantities in year t (note that we use 1 and 2 as indices for the stage structured population sizes, while they were 1y and ad in Section 3.2.3.):

$$N_{1,t+1} = N_{1,t} f_1 s_j + N_{2,t} f_a s_j.$$

Likewise, the number of individuals at least 2 years old in year $t+1$ ($N_{2,t+1}$) is

$$N_{2,t+1} = N_{1,t} s_a + N_{2,t} s_a = \left(N_{1,t} + N_{2,t}\right) s_a.$$

To project the population size, we must know or make a decision about the stage-specific population size at time $t = 1$. As this is generally unknown, we start with arbitrary values. From theory we know (and we have demonstrated in Section 3.3.1) that if the demographic rates are constant, the population will eventually grow at a constant rate once it has reached stationarity, i.e., the stable stage distribution, and that under very mild assumptions, convergence to stationarity is virtually certain (Caswell, 2001, p. 79). That is, if the stable stage distribution is known, we need only project the population for a single time step to obtain an estimate of the asymptotic growth rate. But since we do not know the stable stage distribution, we must start with an arbitrary population vector and project the population for a number of time steps until the Markov chain reaches stationarity. We have seen before that this can happen very swiftly in simple models. If the stable stage distribution is obtained after T projections, an estimate of the asymptotic population growth rate is

$$\lambda = \frac{N_{1,T+1} + N_{2,T+1}}{N_{1,T} + N_{2,T}} = \frac{N_{1,T+1}}{N_{1,T}} = \frac{N_{2,T+1}}{N_{2,T}}.$$

In practice, we do not know T, but we can simply project the population long enough and see when the growth rate estimates stabilize.

In the following BUGS model file, the population is projected over T years, and the asymptotic population growth rate is calculated. The data consist of the demographic parameters and the desired number of years of population projection (T). Here, T can be small ($T = 5$), but when we model species with longer generation times, or if our model contains stochasticity, longer (or much longer) runs are needed to obtain estimates with reasonable precision.

```
# Bundle data
sj <- 0.3
sa <- 0.55
f1 <- 1.3
fa <- 1.8
jags.data <- list(sj=sj, sa=sa, f1=f1, fa=fa, T=5)

# Write JAGS model file
cat(file="model1.txt", "
model {
# Model for initial state
N[1,1] <- 1
N[2,1] <- 1

# Loop over time
for (t in 1:T){
  # Population projection
  N[1,t+1] <- sj * (f1 * N[1,t] + fa * N[2,t])
  N[2,t+1] <- sa * (N[1,t] + N[2,t])

  # Calculation of population quantities
  # Annual (realized) population growth rate
  ann.growth.rate[t] <- (N[1,t+1] + N[2,t+1]) / (N[1,t] + N[2,t])
}
lambda <- ann.growth.rate[T] # gr in final interval is our estimate of lambda
}
")


# Parameters monitored
parameters <- c("ann.growth.rate", "lambda", "N")

# MCMC settings
ni <- 1; nt <- 1; nb <- 0; nc <- 1; na <- 0

# Call JAGS (ART <1 min) and summarize results
out1 <- jags(jags.data, NULL, parameters, "model1.txt", n.adapt=na, n.chains=nc, n.thin=nt,
  n.iter=ni, n.burnin=nb, DIC=FALSE)
print(out1, 3)

                    mean sd  2.5%   50% 97.5% overlap0 f
ann.growth.rate[1] 1.015 NA 1.015 1.015 1.015    FALSE 1
ann.growth.rate[2] 1.021 NA 1.021 1.021 1.021    FALSE 1
ann.growth.rate[3] 1.021 NA 1.021 1.021 1.021    FALSE 1
ann.growth.rate[4] 1.021 NA 1.021 1.021 1.021    FALSE 1
ann.growth.rate[5] 1.021 NA 1.021 1.021 1.021    FALSE 1
lambda             1.021 NA 1.021 1.021 1.021    FALSE 1
N[1,1]             1.000 NA 1.000 1.000 1.000    FALSE 1
N[2,1]             1.000 NA 1.000 1.000 1.000    FALSE 1
N[1,2]             0.930 NA 0.930 0.930 0.930    FALSE 1
N[2,2]             1.100 NA 1.100 1.100 1.100    FALSE 1
[ ...output truncated... ]
N[1,6]             1.017 NA 1.017 1.017 1.017    FALSE 1
N[2,6]             1.188 NA 1.188 1.188 1.188    FALSE 1
```

The estimates of the last four annual population growth rates are identical; hence, the Markov chain reaches its stationary distribution almost immediately. This agrees with our earlier findings (see Fig. 3.8). We use the values from the last time step as our estimates of the asymptotic population growth rate (`lambda`). This value is exactly identical to what we obtained from the dominant eigenvalue of the transition matrix (Section 3.3.1). The standard deviations (the Bayesian analogs to standard errors) are not computed, and missing values (NAs) are reported instead. This is a consequence of our choice of a single MCMC draw only; it is not possible to calculate a standard deviation from a single number. It makes sense that there is no standard deviation for the computed quantities because the model is deterministic and includes neither stochasticity nor parameter uncertainty.

We shall make a few remarks about the code, the function `jags` in the R package `jagsUI` (Kellner, 2019), and our MCMC settings. First, we turn off calculation of the deviance information criterion when `jags` is called because the deviance cannot be calculated in this model (there is no unexplained part in the response of a deterministc model). Second, our results are for a model without parameter uncertainty and demographic stochasticity; thus, we do not need replicated calculations of the population growth rate. Yet the MCMC machinery makes replicates (simulations). We choose MCMC settings such that as few simulations as necessary are performed. Therefore, we don't do a burn-in, we ask for one chain only, and we iterate the MCMC algorithm for a single step only. Convergence assessment would be meaningless in this particular case. Third, it is not necessary to provide initial values explicitly. By calling `inits=NULL` in the `jags` function, we tell JAGS to generate initial values on its own. Finally, we note that this is not yet a Bayesian analysis of the model in the usual sense of a Bayesian analysis where we estimate something from data. There is no uncertainty about anything in this deterministic model without parameter uncertainty, and we simply use the MCMC machinery as a useful device for calculation (Lunn et al., 2013).

We can now extend this code to calculate other life-history summaries from the same data. Applying the same logic as for the asymptotic population growth rate, we use the stage distribution in the final year T as an estimate of the stable stage distribution. The sensitivity of the population growth rate to changes in juvenile survival may be calculated as stochastic sensitivities (Section 3.3), although there is no stochasticity here. However, stochastic sensitivities are always correct even in the absence of environmental stochasticity. Finally, we compute the net reproductive rate (R_0) and the generation time (GT) based on the formulae in Section 3.3.1.

```
# Write JAGS model file
cat(file="model2.txt", "
model {
# Model for initial state
N[1,1] <- 1
N[2,1] <- 1

# Loop over time
for (t in 1:T){
  # Population projection
  N[1,t+1] <- sj * (f1 * N[1,t] + fa * N[2,t])
  N[2,t+1] <- sa * (N[1,t] + N[2,t])

  # Calculation of population quantities
  # Annual (realized) population growth rate
  ann.growth.rate[t] <- (N[1,t+1] + N[2,t+1]) / (N[1,t] + N[2,t])

  # Scaled annual stage distributions
  stage.distr[1,t] <- N[1,t+1] / (N[1,t+1] + N[2,t+1])
  stage.distr[2,t] <- N[2,t+1] / (N[1,t+1] + N[2,t+1])
}
lambda <- ann.growth.rate[T]
stable.stage.distr <- stage.distr[,T]

# Sensitivity and elasticity of lambda to changes in sj
delta <- 0.001                    # Size of perturbation
N.star[1,1] <- 1
N.star[2,1] <- 1
for (t in 1:T){
  N.star[1,t+1] <- (sj + delta) * (f1 * N.star[1,t] + fa * N.star[2,t])
  N.star[2,t+1] <- sa * (N.star[1,t] + N.star[2,t])
  ann.growth.rate.star[t] <- (N.star[1,t+1] + N.star[2,t+1]) / (N.star[1,t] + N.star[2,t])
}
s.sj <- (ann.growth.rate.star[T] - ann.growth.rate[T]) / delta
e.sj <- s.sj * sj / lambda

# Calculation of net reproductive rate (R0)
for (i in 1:100){
  u[i] <- pow(sa, i)
}
R0 <- sj * f1 + sj * fa * sum(u[])

# Calculation of generation time (GT)
GT <- log(R0) / log(lambda)
}
")
```

```
# Parameters monitored
parameters <- c("lambda", "stable.stage.distr", "s.sj", "e.sj", "R0", "GT")

# MCMC settings
ni <- 1; nt <- 1; nb <- 0; nc <- 1; na <- 0

# Call JAGS (ART <1 min) and summarize results
out2 <- jags(jags.data, NULL, parameters, "model2.txt", n.adapt=na, n.chains=nc, n.thin=nt,
  n.iter=ni, n.burnin=nb, DIC=FALSE)
print(out2, 3)

                       mean sd  2.5%   50% 97.5% overlap0 f
lambda                1.021 NA 1.021 1.021 1.021    FALSE 1
stable.stage.distr[1] 0.461 NA 0.461 0.461 0.461    FALSE 1
stable.stage.distr[2] 0.539 NA 0.539 0.539 0.539    FALSE 1
s.sj                  1.454 NA 1.454 1.454 1.454    FALSE 1
e.sj                  0.427 NA 0.427 0.427 0.427    FALSE 1
R0                    1.050 NA 1.050 1.050 1.050    FALSE 1
GT                    2.368 NA 2.368 2.368 2.368    FALSE 1
```

All quantities computed are identical to those that we obtained using the standard matrix population methods (Section 3.3.1). So remember, for all the calculations performed so far, there is no actual benefit to using MCMC methods—but there will be one as soon as we include parameter uncertainty in our model, and this comes next.

3.4.2 ANALYSIS OF A MATRIX POPULATION MODEL WITH PARAMETER UNCERTAINTY

Demographic rates are generally not known perfectly and must be estimated. We therefore typically have a point estimate, e.g., the maximum likelihood estimate (MLE) or posterior mean, and an associated measure of uncertainty such as the standard error for MLEs or the posterior standard deviation for a Bayesian point estimate. Now we want to estimate the population growth rate and life-history summaries while accounting for that estimation uncertainty. In other words, we want to propagate the estimation uncertainty associated with the parameters in our model into derived quantities, such as the population growth rate.

The traditional approach to this is to use the delta rule or simulations (Section 3.3.2). Using the MCMC machinery in BUGS software, we can do such simulations in a straightforward way, and we don't have to specifically program any simulation algorithm ourselves. We only need to change our model such that the demographic rates are no longer fixed values but rather characterized by a distribution that itself represents the uncertainty about the parameter. This distribution is characterized by a mean and a measure of spread that reflects the parameter uncertainty. Using Bayesian estimation software such as BUGS, we can express the parameter estimation error (i.e., the uncertainty around the value of a parameter) by a statistical distribution for the demographic parameter. Technically, in the BUGS model this will look exactly like a prior distribution, but it does not act as a prior because there is still no estimation involved in this model. What looks like a prior is in fact just a call to an RNG for a specified statistical distribution (Lunn et al., 2013). We must choose these RNGs (and their arguments) such that no impossible values for the demographic rates can be generated. For survival probabilities, this means that the values drawn must be in the interval between 0 and 1, and for productivity, only positive values must be produced. Here, we use a beta distribution for the survival probabilities and a normal for productivity. Note that the specification of these RNGs is equivalent to the specification of the distributions for the demographic rates from which we generated values in the classical Monte Carlo approach in Section 3.3.2. We start by specification of the means and associated measures of uncertainty for the four demographic rates in our shrike example.

```
# Define mean and SD of the demographic parameters
mean.sj <- 0.3       # Point estimate of juv. survival
se.sj.e <- 0.03      # Uncertainty of juv. survival as SE on natural scale
mean.sa <- 0.55      # Point estimate of ad. survival
se.sa.e <- 0.015     # Uncertainty of ad. survival as SE on natural scale
mean.f1 <- 1.3       # Point estimate of productivity of 1y females
se.f1.e <- 0.3       # Uncertainty of productivity as SE on natural scale
mean.fa <- 1.8       # Point estimate of productivity of adult females
se.fa.e <- 0.1       # Uncertainty of productivity as SE on natural scale
```

The data bundle includes the shape parameters for the beta distributions that we calculate using function `getBeta2Par` from the `IPMbook` package and the standard deviations of productivity. As the normal distribution in JAGS is parameterized in terms of the precision, we calculate it for the two productivity measures.

```
# Bundle data
jags.data <- list(alpha.sj=getBeta2Par(mean.sj, se.sj.e)[1], beta.sj=getBeta2Par(mean.sj,
  se.sj.e)[2], alpha.sa=getBeta2Par(mean.sa, se.sa.e)[1], beta.sa=getBeta2Par(mean.sa, se.sa.e)[2],
  mean.f1=mean.f1, tau.f1=1/se.f1.e^2, mean.fa=mean.fa, tau.fa=1/se.fa.e^2, T=50)
```

The code to analyze a matrix population model with uncertainty in the demographic parameters looks almost exactly as before. The only difference is that we specify distributions for the demographic parameters from which one value is randomly drawn at each iteration of the MCMC algorithm.

```
# Write JAGS model file
cat(file="model3.txt", "
model {
# Random number generators (RNGs)
sj ~ dbeta(alpha.sj, beta.sj)       # These only *look* like priors
sa ~ dbeta(alpha.sa, beta.sa)       # ... but they are not
f1 ~ dnorm(mean.f1, tau.f1)         # ... as there is no estimation in this model
fa ~ dnorm(mean.fa, tau.fa)

# Initialize the population size nodes
N[1,1] <- 1
N[2,1] <- 1

# Loop over time
for (t in 1:T){
  # Population model
  N[1,t+1] <- sj * (f1 * N[1,t] + fa * N[2,t])
  N[2,t+1] <- sa * (N[1,t] + N[2,t])

  # Annual (realized) growth rate
  ann.growth.rate[t] <- (N[1,t+1] + N[2,t+1]) / (N[1,t] + N[2,t])

  # Scaled stage distributions
  stage.distr[1,t] <- N[1,t+1] / (N[1,t+1] + N[2,t+1])
  stage.distr[2,t] <- N[2,t+1] / (N[1,t+1] + N[2,t+1])
}
lambda <- ann.growth.rate[T]
stable.stage.distr <- stage.distr[,T]

# Sensitivity and elasticity of lambda to changes in sj
delta <- 0.001                      # size of perturbation
N.star[1,1] <- 1
N.star[2,1] <- 1
for (t in 1:T){
  N.star[1,t+1] <- (sj + delta) * (f1 * N.star[1,t] + fa * N.star[2,t])
  N.star[2,t+1] <- sa * (N.star[1,t] + N.star[2,t])
  ann.growth.rate.star[t] <- (N.star[1,t+1] + N.star[2,t+1]) / (N.star[1,t] + N.star[2,t])
}
s.sj <- (ann.growth.rate.star[T] - ann.growth.rate[T]) / delta
e.sj <- s.sj * sj / lambda

# Calculation of net reproductive rate (R0)
for (i in 1:100){
  u[i] <- pow(sa, i)
}
R0 <- sj * f1 + sj * fa * sum(u[])

# Calculation of generation time (GT)
GT <- log(R0) / log(lambda)
}
")

# Parameters monitored
parameters <- c("lambda", "stable.stage.distr", "s.sj", "e.sj", "R0", "GT")
# MCMC settings
ni <- 100000; nt <- 1; nb <- 0; nc <- 1; na <- 0

# Call JAGS (ART <1 min) and summarize results
out3 <- jags(jags.data, NULL, parameters, "model3.txt", n.adapt=na, n.chains=nc, n.thin=nt,
  n.iter=ni, n.burnin=nb, DIC=FALSE)
print(out3, 4)

                         mean     sd   2.5%    50%  97.5% overlap0 f
lambda                 1.0223 0.0627 0.9102 1.0188 1.1553    FALSE 1
stable.stage.distr[1] 0.4603 0.0323 0.3979 0.4600 0.5241    FALSE 1
stable.stage.distr[2] 0.5397 0.0323 0.4759 0.5400 0.6021    FALSE 1
s.sj                   1.4658 0.1935 1.1135 1.4567 1.8693    FALSE 1
e.sj                   0.4273 0.0452 0.3438 0.4258 0.5203    FALSE 1
R0                     1.0516 0.1490 0.7798 1.0453 1.3619    FALSE 1
GT                     2.3842 0.1910 2.0426 2.3727 2.7911    FALSE 1
```

Now our model is no longer strictly deterministic, since there is the stochastic element involved by drawing random values from the specified distributions to account for parameter uncertainty. Hence, in the posterior summary the column containing the posterior SD no longer has zero values as in the previous example but rather the correct errors representing the uncertainty about the estimated quantities. To visualize the results, we plot the simulated distributions of some of the computed quantities (Fig. 3.17).

Because we need to perform simulations here, we increase the number of iterations in the MCMC settings to obtain 100,000 draws from the simulated distributions of the quantities of interest. It is not yet necessary to include a burn-in or have multiple chains, because convergence is not an issue in this analysis. After all, it is not yet a formal Bayesian analysis (because we still don't formally estimate anything), but we merely use the MCMC machinery available in JAGS to perform stochastic simulations.

The simulated distributions of the target quantities (Fig. 3.17) and the frequency distribution of the target quantities from the simulation that we obtained in Section 3.3.2 are almost completely identical, as shown by density plots in Fig. 3.18. Slight differences are due to sampling variability, or Monte Carlo error; if the number of draws is increased, both distributions will eventually converge to become identical. This means that with either method, the classical Monte Carlo simulation and the Bayesian MCMC software, we obtain identical results. This is not a surprise, of course, because mathematically we are doing the same thing, but for us it is important to emphasize this agreement. An advantage of the MCMC algorithm is that the simulation part is already implemented in JAGS; thus, we obtain the measures of uncertainty without a need to code our own simulation.

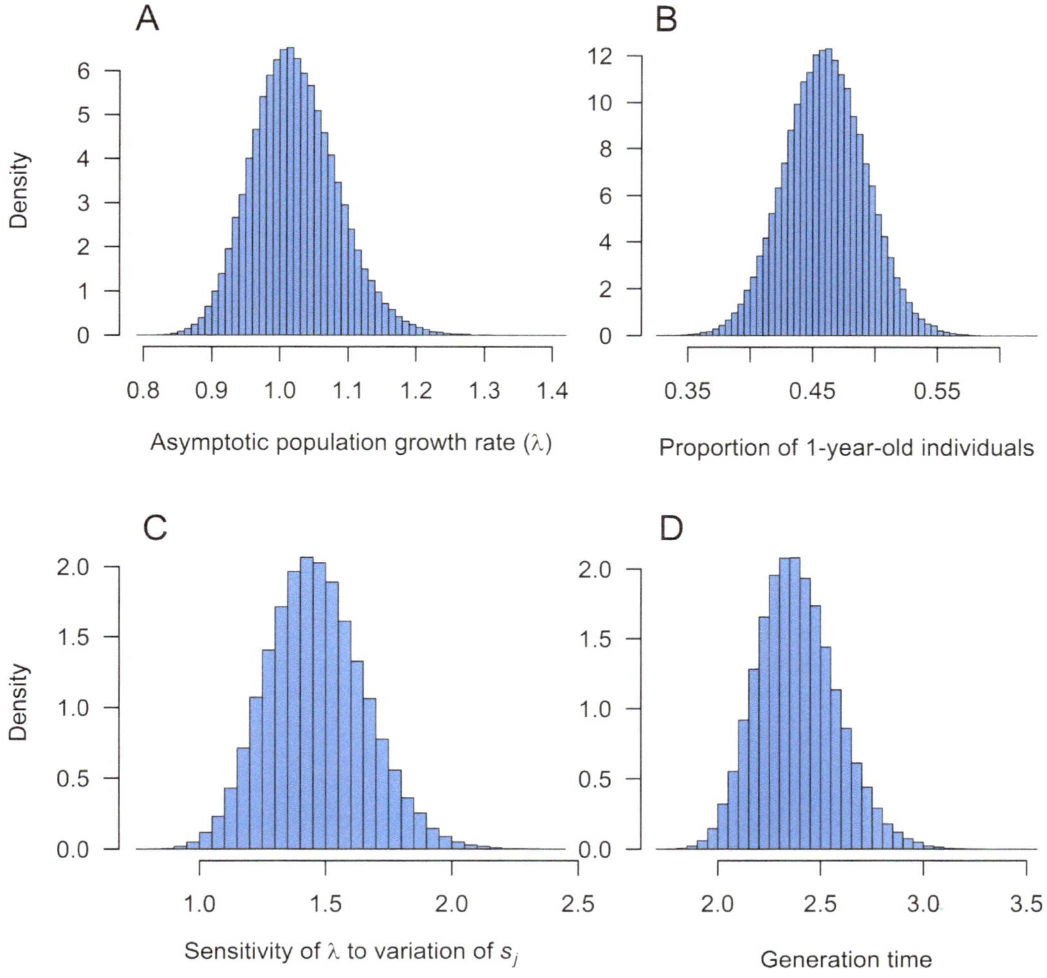

FIGURE 3.17

Simulated distributions of the asymptotic population growth rate (A), of the proportion of 1-year-old individuals (B), of growth rate sensitivity to variation in juvenile survival (C) and of generation time (D) obtained using MCMC software with 100,000 draws.

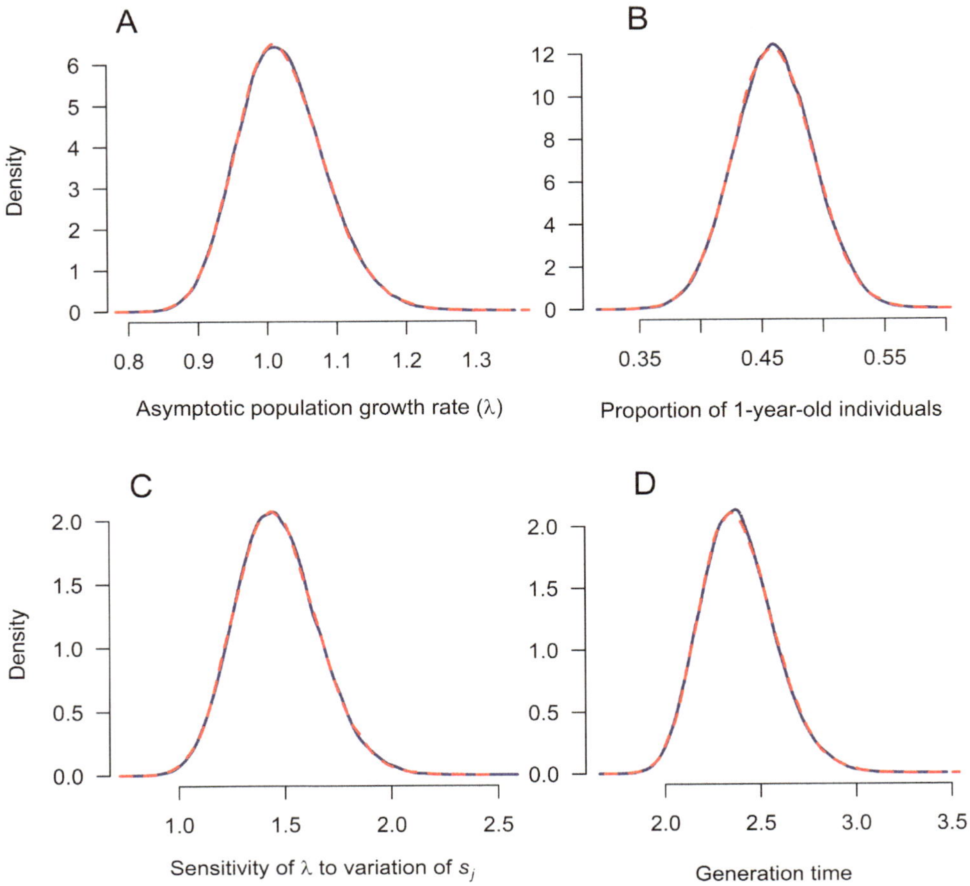

FIGURE 3.18

Density plots (smoothed frequency distributions) of some life-history summaries computed using classical Monte Carlo simulation (blue lines) and Markov chain Monte Carlo simulation in JAGS (dashed red lines): (A) asymptotic population growth rate, (B) proportion of 1-year-old individuals, (C) growth rate sensitivity of juvenile survival, (D) generation time.

3.4.3 ANALYSIS OF A MATRIX POPULATION MODEL WITH ENVIRONMENTAL STOCHASTICITY

We next illustrate a case where the demographic rates are subject to environmental stochasticity, i.e., when there is temporal variation, the magnitude of which is assumed to be known perfectly. Computation of the stochastic population growth rate and of other life-history summaries is done in the same way as we have already seen for the classical analyses (Section 3.3.3): we need to project the population for a long time, and from the stage-specific population trajectories, we can calculate the quantities of interest.

First we define the mean demographic rates (using the same values as before) and their temporal variability expressed as standard deviations. We assume that there are no estimation errors in the parameters (i.e., no parameter uncertainty).

```
# Define mean and temporal variability (SD) of the demographic parameters
mean.sj <- 0.3          # Mean juv. survival on probability scale
sd.sj.t <- 0.25         # Temporal variability of sj on the logit scale
mean.sa <- 0.55         # Mean ad. survival on probability scale
sd.sa.t <- 0.07         # Temporal variability of sa on the logit scale
mean.f1 <- 1.3          # Mean productivity of 1y females on natural scale
sd.f1.t <- 0.3          # Temporal variability of f1 on the natural scale
mean.fa <- 1.8          # Mean productivity of adult females on natural scale
sd.fa.t <- 0.3          # Temporal variability of fa on the natural scale
```

We will project the population for $T = 21{,}000$ years and discard the first $u = 1000$ years as a burn-in of the Markov chain.

```
# Bundle data
jags.data <- list(mean.sj=mean.sj, mean.sa=mean.sa, mean.f1=mean.f1, mean.fa=mean.fa,
  sd.sj.t=sd.sj.t, sd.sa.t=sd.sa.t, sd.f1.t=sd.f1.t, sd.fa.t=sd.fa.t, T=21000, u=1000)
```

The BUGS code differs from the previous examples in important ways. We now need to index the demographic rates by time. To simulate the annual estimates of demographic rates we specify appropriate distributions. That is, we specify an annual value of a demographic rate as a draw from a normal distribution with the known mean and temporal variability. Because we need to impose the usual range constraints for a probability, we draw the annual values of survival on the logit scale and then back-transform, whereas for productivity, we work directly on the natural scale; both of these we do as previously discussed in the classical analysis. In addition, as in the classical analysis, we compute the stochastic population growth rate as well as growth rate sensitivity and elasticity to variation in juvenile survival.

```
# Write JAGS model file
cat(file="model4.txt", "
model {
# Calculate precision for temporal variability of demographic rates
tau.logit.sj <- pow(sd.sj.t, -2)
tau.logit.sa <- pow(sd.sa.t, -2)
tau.f1 <- pow(sd.f1.t, -2)
tau.fa <- pow(sd.fa.t, -2)

# Use of RNG to accomodate temporal variability of demographic rates (process variability)
for (t in 1:T){
  sj[t] <- ilogit(logit.sj[t])      # Backt. from logit to natural scale
  logit.sj[t] ~ dnorm(logit(mean.sj), tau.logit.sj)
  sa[t] <- ilogit(logit.sa[t])      # Backt. from logit to natural scale
  logit.sa[t] ~ dnorm(logit(mean.sa), tau.logit.sa)
  f1[t] ~ dnorm(mean.f1, tau.f1)
  fa[t] ~ dnorm(mean.fa, tau.fa)
}

# Model for initial state
N[1,1] <- 1
N[2,1] <- 1

# Loop over time
for (t in 1:T){
  # Population model
  N[1,t+1] <- sj[t] * (f1[t] * N[1,t] + fa[t] * N[2,t])
  N[2,t+1] <- sa[t] * (N[1,t] + N[2,t])

  # Annual growth rate on log scale
  r.annual[t] <- log(N[1,t+1] + N[2,t+1]) - log(N[1,t] + N[2,t])
}
r <- mean(r.annual[u:T])
lambda <- exp(r)

# Sensitivity and elasticity of lambda to changes in sj
delta <- 0.001                      # Size of perturbation
N.star[1,1] <- 1
N.star[2,1] <- 1
for (t in 1:T){
  N.star[1,t+1] <- (sj[t] + delta) * (f1[t] * N.star[1,t] + fa[t] * N.star[2,t])
  N.star[2,t+1] <- sa[t] * (N.star[1,t] + N.star[2,t])
  r.annual.star[t] <- log(N.star[1,t+1] + N.star[2,t+1]) - log(N.star[1,t] + N.star[2,t])
}
r.star <- mean(r.annual.star[u:T])
s.sj <- (exp(r.star) - lambda) / delta
e.sj <- s.sj * mean.sj / lambda
}
")
# Parameters monitored
parameters <- c("r", "lambda", "s.sj", "e.sj")

# MCMC settings
ni <- 1; nt <- 1; nb <- 0; nc <- 1; na <- 0

# Call JAGS (ART <1 min) and summarize results
out4 <- jags(jags.data, NULL, parameters, "model4.txt", n.adapt=na, n.chains=nc, n.thin=nt, n.iter=ni,
  n.burnin=nb, DIC=FALSE)
print(out4, 4)

         mean sd   2.5%    50%  97.5% overlap0 f
r      0.0191 NA 0.0191 0.0191 0.0191    FALSE 1
lambda 1.0193 NA 1.0193 1.0193 1.0193    FALSE 1
s.sj   1.4486 NA 1.4486 1.4486 1.4486    FALSE 1
e.sj   0.4263 NA 0.4263 0.4263 0.4263    FALSE 1
```

We see again that the simulated means of the monitored parameters agree well with the estimates obtained from the classical Monte Carlo analysis.

Sometimes the temporal variability of the demographic rates is unknown, but we have point estimates from a number of years. In this case, we can incorporate into the model the information about this temporal variation in the rates and incorporate the resulting uncertainty into derived quantities such as the stochastic population growth rate as follows: we randomly pick the value of 1 year, project the population, and repeat this a large number of times. For illustration, here is a list with the demographic rates that are assumed known from 10 years.

```
sj.t <- c(0.303, 0.300, 0.288, 0.310, 0.306, 0.291, 0.303, 0.281, 0.314, 0.303)
sa.t <- c(0.546, 0.545, 0.547, 0.545, 0.547, 0.556, 0.540, 0.551, 0.548, 0.550)
f1.t <- c(1.56, 1.15, 1.38, 1.89, 1.06, 0.69, 1.16, 1.10, 1.48, 1.28)
fa.t <- c(2.03, 1.69, 1.95, 1.98, 1.63, 1.43, 1.63, 1.45, 1.79, 1.91)
```

We create vectors of these demographic rates such that the number of years corresponds to *T*. We pick years at random first, and then construct vectors with the demographic rates.

```
T <- 21000                 # Number of predicted years
t <- 1:10                  # Number of years with actual data
year <- sample(t, T, replace=TRUE)
sj <- sj.t[year]
sa <- sa.t[year]
f1 <- f1.t[year]
fa <- fa.t[year]
```

Since we randomly draw the same years for all parameters, any temporal correlations between demographic parameters will be preserved. This is a desired feature of this simulation method because any positive correlation between the demographic rates will result in smaller population growth rates, while negative correlations tend to dampen fluctuations of the annual growth rates (Tuljapurkar and Orzack, 1980; Fieberg and Ellner, 2001).

```
# Bundle data
jags.data <- list(sj=sj, sa=sa, f1=f1, fa=fa, T=21000, u=1000)
```

The BUGS code needs a few changes. Because the demographic rates are now known for all *T* years, we do not need the section with their simulation.

```
# Write JAGS model file
cat(file="model5.txt", "
model {
# Model for initial state
N[1,1] <- 1
N[2,1] <- 1

# Loop over time
for (t in 1:T){
  # Population model
  N[1,t+1] <- sj[t] * (f1[t] * N[1,t] + fa[t] * N[2,t])
  N[2,t+1] <- sa[t] * (N[1,t] + N[2,t])

  # Annual (realized) growth rate on log scale
  r.annual[t] <- log(N[1,t+1] + N[2,t+1]) - log(N[1,t] + N[2,t])
}
r <- mean(r.annual[u:T])
lambda <- exp(r)

# Sensitivity and elasticity of lambda to changes in sj
delta <- 0.001             # size of perturbation
N.star[1,1] <- 1
N.star[2,1] <- 1
for (t in 1:T){
  N.star[1,t+1] <- (sj[t] + delta) * (f1[t] * N.star[1,t] + fa[t] * N.star[2,t])
  N.star[2,t+1] <- sa[t] * (N.star[1,t] + N.star[2,t])
  r.annual.star[t] <- log(N.star[1,t+1] + N.star[2,t+1]) - log(N.star[1,t] + N.star[2,t])
}
r.star <- mean(r.annual.star[u:T])
s.sj <- (exp(r.star) - lambda) / delta
e.sj <- s.sj * mean(sj) / lambda
}
")
```

```
# Parameters monitored
parameters <- c("r", "lambda", "s.sj", "e.sj")

# MCMC settings
ni <- 1; nt <- 1; nb <- 0; nc <- 1; na <- 0

# Call JAGS (ART <1 min) and summarize results
out5 <- jags(jags.data, NULL, parameters, "model5.txt", n.adapt=na, n.chains=nc, n.thin=nt,
  n.iter=ni, n.burnin=nb, DIC=FALSE)
print(out5, 4)

         mean sd   2.5%     50%  97.5% overlap0 f
r      0.0061 NA 0.0061 0.0061 0.0061    FALSE 1
lambda 1.0061 NA 1.0061 1.0061 1.0061    FALSE 1
s.sj   1.4112 NA 1.4112 1.4112 1.4112    FALSE 1
e.sj   0.4207 NA 0.4207 0.4207 0.4207    FALSE 1
```

All point estimates are similar to before.

3.4.4 ANALYSIS OF A MATRIX POPULATION MODEL WITH DEMOGRAPHIC STOCHASTICITY

The next type of stochasticity that we intend to include in the population model analyzed with BUGS is demographic stochasticity. Recall that demographic stochasticity stems from the stochastic nature of demographic processes operating on individuals, and it requires the specification of discrete-valued distributions that mimic these processes. The resulting population sizes will be discrete and the initial population size then matters. As in the classical analyses, we need to project the population subject to demographic stochasticity for a number of years. From the simulated population trajectories, we can then obtain life-history summaries as well as our estimates of the population growth rate and extinction probability.

A potential difficulty when fitting models including demographic stochasticity is that population size can drop to zero, i.e., a population may go extinct. Using the `equals` function in JAGS, we can monitor extinction events: in each year and for each simulation replicate (i.e., simulated population), we can check whether the population size is equal to zero. Once a population is extinct, we can no longer estimate the population growth rate or sensitivities because this would involve a division by zero. In these cases, JAGS will crash. One way to avoid this problem is to use JAGS only for the population projection and the extinction probability and to use R rather than JAGS to calculate the population growth rate and all life-history summaries. This gives us the possibility of using the proper *if* statement to avoid any division by zero.

```
# Define mean of the demographic parameters
mean.sj <- 0.3
mean.sa <- 0.55
mean.f1 <- 1.3
mean.fa <- 1.8

# Define the number of years with predictions and the Monte Carlo setting
T <- 200

# Define initial stage-specific population sizes
N1 <- 10
N2 <- 10

# Bundle data
jags.data <- list(sj=mean.sj, sa=mean.sa, f1=mean.f1, fa=mean.fa, T=T, N1=N1, N2=N2)

# Write JAGS model file
cat(file="model6.txt", "
model {
# Model for initial state
N[1,1] <- N1
N[2,1] <- N2

# Loop over time
for (t in 1:T){
  # Population model
  N[1,t+1] ~ dpois(sj * (f1 * N[1,t] + fa * N[2,t]))
  N[2,t+1] ~ dbin(sa, (N[1,t] + N[2,t]))
  extinct[t] <- equals(N[1,t+1] + N[2,t+1], 0)   # Determines whether
    # population is still thriving (extinct = 0) or went extinct (extinct = 1)
}
}
")
```

An initial value can be supplied for at least one stochastic node, although for this example it is not necessary. However, more complicated models may not run if we let JAGS generate initial values. Population size after the initial year is stochastic (due to demographic stochasticity); hence, we can give initial values for those nodes. In our model, the population size variable is defined as a matrix, and thus we need to define a matrix with the same dimension for the initial values. But since population size in the first year is deterministic, no initials must then be given, and we therefore set NAs in the first year.

```
# Initial values
Ninit <- matrix(10, nrow=2, ncol=T+1)
Ninit[,1] <- NA
inits <- function(){list(N=Ninit)}

# Parameters monitored
parameters <- c("N", "extinct")

# MCMC settings
ni <- 50000; nt <- 1; nb <- 0; nc <- 1; na <- 0
```

We want to replicate the projection of the population 50,000 times, corresponding here to that number of MCMC iterations. No burn-in is necessary, nor are multiple chains, since convergence is again not an issue. This is still not a Bayesian analysis; we are just using the MCMC algorithm for stochastic simulations, not for parameter estimation.

```
# Call JAGS (ART <1 min) and summarize results
out6 <- jags(jags.data, inits, parameters, "model6.txt", n.chains=nc, n.thin=nt, n.iter=ni,
  n.burnin=nb, DIC=FALSE)
print(out6, 4)

                  mean       sd 2.5%   50%    97.5% overlap0 f
N[1,1]         10.0000   0.0000   10  10.0   10.000    FALSE 1
N[2,1]         10.0000   0.0000   10  10.0   10.000    FALSE 1
N[1,2]          9.3110   3.0491    4   9.0   16.000    FALSE 1
N[2,2]         10.9922   2.2313    7  11.0   15.000    FALSE 1
[ ...output truncated... ]
extinct[1]      0.0000   0.0000    0   0.0    0.000    FALSE 1
extinct[2]      0.0000   0.0000    0   0.0    0.000    FALSE 1
extinct[3]      0.0000   0.0000    0   0.0    0.000    FALSE 1
[output truncated... ]
extinct[198]    0.2876   0.4526    0   0.0    1.000     TRUE 1
extinct[199]    0.2876   0.4527    0   0.0    1.000     TRUE 1
extinct[200]    0.2878   0.4527    0   0.0    1.000     TRUE 1
```

We are particularly interested in the estimates of the stochastic population growth rate. This associated calculation is carried out outside of JAGS in R to avoid the abovementioned numerical problems. This code is very similar to that for the classical analysis (Section 3.3.4).

```
# Calculation of the stochastic population growth rate and of the extinction probability
# outside JAGS
dimensions <- dim(out6$sims.list$extinct)
r.annual <- matrix(NA, nrow=dimensions[2], ncol=dimensions[1])
r <- lambda <- numeric(dimensions[1])
for (s in 1:dimensions[1]){
  for (t in 1:dimensions[2]){
    # Calculate annual growth rate on log scale
    if (out6$sims.list$extinct[s,t] == 1) break
    r.annual[t,s] <- log(out6$sims.list$N[s,1,t+1] + out6$sims.list$N[s,2,t+1]) -
      log(out6$sims.list$N[s,1,t] + out6$sims.list$N[s,2,t])
  } #t
r[s] <- mean(r.annual[which(out6$sims.list$extinct[s,] == 0),s])
lambda[s] <- exp(r[s])
} #s
```

Now we can easily calculate the mean population growth rate and its precision as follows:

```
mean(r)
 [1] -0.008533999
sd(r)
[1] 0.05592566
```

We could produce the same plot as Fig. 3.14 (not shown here) and would then notice the same thing: the frequency distribution of the population growth rates is strongly skewed and bimodal because the extinct populations have different growth rates on average from those of the populations that survive for 200 years. We may want to calculate the mean population growth rates of only the surviving populations.

```
mean(r[out6$sims.list$extinct[,T]==0])
[1] 0.01973898
sd(r[out6$sims.list$extinct[,T]==0])
[1] 0.006009628
```

3.4.5 ANALYSIS OF A MATRIX POPULATION MODEL WITH MULTIPLE SOURCES OF STOCHASTICITY AND PARAMETER UNCERTAINTY

Finally, we illustrate the combination of multiple types of stochasticity (environmental, demographic) and parameter uncertainty using MCMC simulations. As in Section 3.3.5, we assume that we have estimates of the mean and the temporal variance of the demographic parameters and that their means are measured with known errors. Our goal is estimation of the population growth rate and extinction probability while fully accommodating all sources of uncertainty.

```
# Define mean, measurement error and temporal variability of the demographic parameters
mean.sj <- 0.3                              # Mean juv. survival on probability scale
se.sj.e <- 0.005                            # Uncertainty of mean sj as SE on natural scale
sd.sj.t <- 0.25                             # Temporal variability of sj as SD on logit scale
mean.sa <- 0.55                             # Mean ad. survival on probability scale
se.sa.e <- 0.005                            # Uncertainty of mean saas SE on natural scale
sd.sa.t <- 0.07                             # Temporal variability of sa as SD on logit scale
mean.f1 <- 1.3                              # Mean productivity of 1y females on natural scale
se.f1.e <- 0.05                             # Uncertainty of f1 as SE on natural scale
sd.f1.t <- 0.3                              # Temporal variability of f1 as SD on natural scale
mean.fa <- 1.8                              # Mean productivity of adult females on natural scale
se.fa.e <- 0.03                             # Uncertainty of fa as SE on natural scale
sd.fa.t <- 0.3                              # Temporal variability of fa as SD on natural scale

# Define the number of years with predictions and the Monte Carlo setting
T <- 200

# Bundle data
N1 <- 10
N2 <- 10
jags.data <- list(alpha.sj=getBeta2Par(mean.sj, se.sj.e)[1], beta.sj=getBeta2Par(mean.sj,
  se.sj.e)[2], alpha.sa=getBeta2Par(mean.sa, se.sa.e)[1], beta.sa=getBeta2Par(mean.sa,
  se.sa.e)[2], mean.f1=mean.f1, mean.fa=mean.fa, se.f1.e=se.f1.e, se.fa.e=se.fa.e,
  sd.sj.t=sd.sj.t, sd.sa.t=sd.sa.t, sd.f1.t=sd.f1.t, sd.fa.t=sd.fa.t, T=T, N1=N1, N2=N2)

# Write JAGS model file
cat(file="model7.txt", "
model {
# Use of RNG to account for measurement error (uncertainty)
# in the means of the demographic rates on the natural scale
mean.sj.e ~ dbeta(alpha.sj, beta.sj)
mean.sa.e ~ dbeta(alpha.sa, beta.sa)
mean.f1.e ~ dnorm(mean.f1, tau.f1.e)
mean.fa.e ~ dnorm(mean.fa, tau.fa.e)

# Calculate precision of the measurement error of productivity
tau.f1.e <- pow(se.f1.e, -2)
tau.fa.e <- pow(se.fa.e, -2)

# Calculate precision for the temporal variability of demographic rates
tau.logit.sj.t <- pow(sd.sj.t, -2)
tau.logit.sa.t <- pow(sd.sa.t, -2)
tau.f1.t <- pow(sd.f1.t, -2)
tau.fa.t <- pow(sd.fa.t, -2)

# Use of RNG to accomodate temporal variability of demographic rates
# (process variability)
for (t in 1:T){
  sj[t] <- ilogit(logit.sj[t])              # Backtransformation from logit scale
  logit.sj[t] ~ dnorm(logit(mean.sj.e), tau.logit.sj.t)
  sa[t] <- ilogit(logit.sa[t])              # Backtransformation from logit scale
  logit.sa[t] ~ dnorm(logit(mean.sa.e), tau.logit.sa.t)
  f1[t] ~ dnorm(mean.f1.e, tau.f1.t)T(0,)   # Truncation to positive values
  fa[t] ~ dnorm(mean.fa.e, tau.fa.t)T(0,)   # Truncation to positive values
}
```

```
# Model for initial state
N[1,1] <- N1
N[2,1] <- N2

# Loop over time
for (t in 1:T){
  # Population model
  N[1,t+1] ~ dpois(sj[t] * (f1[t] * N[1,t] + fa[t] * N[2,t]))
  N[2,t+1] ~ dbin(sa[t], (N[1,t] + N[2,t]))
  extinct[t] <- equals(N[1,t+1] + N[2,t+1], 0)   # Determines whether
    # population is still thriving (extinct = 0) or went extinct (extinct = 1)
}
}
")

# Parameters monitored
parameters <- c("N", "extinct")
```

Since we include parameter uncertainty and demographic stochasticity, we need repeated simulations of the population and therefore choose 50,000 iterations. We do not need burn-in or multiple chains because we are not estimating anything—we are still using only the MCMC in JAGS for error propagation and conducting stochastic simulations.

```
# MCMC settings
ni <- 50000; nt <- 1; nb <- 0; nc <- 1; na <- 0

# Call JAGS (ART <1 min) and summarize results
out7 <- jags(jags.data, NULL, parameters, "model7.txt", n.adapt=na, n.chains=nc, n.thin=nt,
  n.iter=ni, n.burnin=nb, DIC=FALSE)
print(out7, 4)

                  mean          sd 2.5%   50%      97.5% overlap0 f
N[1,1]         10.0000      0.0000   10  10.0     10.000    FALSE 1
N[2,1]         10.0000      0.0000   10  10.0     10.000    FALSE 1
N[1,2]          9.3800      3.7034    3   9.0     17.000    FALSE 1
N[2,2]         10.9935      2.2543    7  11.0     15.000    FALSE 1
[ ... output truncated ... ]
extinct[1]      0.0000      0.0000    0   0.0      0.000    FALSE 1
extinct[2]      0.0000      0.0000    0   0.0      0.000    FALSE 1
extinct[3]      0.0000      0.0000    0   0.0      0.000    FALSE 1
[ ... output truncated ... ]
extinct[198]    0.3838      0.4863    0   0.0      1.000     TRUE 1
extinct[199]    0.3842      0.4864    0   0.0      1.000     TRUE 1
extinct[200]    0.3846      0.4865    0   0.0      1.000     TRUE 1
```

We again compute the stochastic population growth rates outside of JAGS in R. We do this only for the average and the surviving populations.

```
# Calculation of the stochastic population growth rate outside JAGS
dimensions <- dim(out7$sims.list$extinct)
r.annual <- matrix(NA, nrow=dimensions[2], ncol=dimensions[1])
r <- lambda <- numeric(dimensions[1])
for (s in 1:dimensions[1]){
  for (t in 1:dimensions[2]){
    # Calculate annual growth rate on log scale
    if (out7$sims.list$extinct[s,t] == 1) break
    r.annual[t,s] <- log(out7$sims.list$N[s,1,t+1] + out7$sims.list$N[s,2,t+1]) -
      log(out7$sims.list$N[s,1,t] + out7$sims.list$N[s,2,t])
  } #t
r[s] <- mean(r.annual[which(out7$sims.list$extinct[s,] == 0),s])
lambda[s] <- exp(r[s])
} #s

mean(r)
[1] -0.01570117
sd(r)
[1] 0.06555304
```

```
# Population growth rate of populations that did not go extinct
mean(r[out7$sims.list$extinct[,T]==0])
[1] 0.02353183
sd(r[out7$sims.list$extinct[,T]==0])
[1] 0.01234406
```

As expected, the mean and standard deviation of the population growth rate of the surviving populations are very similar to the values that we obtained from our simulations in R (Section 3.3.5).

3.4.6 MATRIX POPULATION MODELS WITH DENSITY DEPENDENCE AND DEMOGRAPHIC STOCHASTICITY

Coding a model with density dependence and demographic stochasticity should not be particularly challenging at this point—we simply combine what we learned in the previous sections. We forecast the population for 200 years, monitor its size, extinction events, and productivity rates, and repeat the projection 1000 times, corresponding to the number of replicate populations.

```
# Define means of the demographic rates and strength of density dependence
mean.sj <- 0.3
mean.sa <- 0.55
f1.int <- 2.3          # Productivity of 1y females when population size is 0
f1.beta <- -0.02       # Strength of density dependence on 1y productivity
fa.int <- 2.3          # Productivity of adult females when population size is 0
fa.beta <- -0.01       # Strength of density dependence on ad productivity

# Define the number of years with predictions
T <- 200

# Bundle data
jags.data <- list(sj=mean.sj, sa=mean.sa, f1.int=f1.int, f1.beta=f1.beta, fa.int=fa.int,
  fa.beta=fa.beta, T=T)

# Write JAGS model file
cat(file="model8.txt", "
model {
# Model for initial state
N[1,1] <- 10
N[2,1] <- 10

# Loop over time
for (t in 1:T){
  # Calculate actual productivity
  f1[t] <- f1.int + f1.beta * (N[1,t] + N[2,t])
  fa[t] <- fa.int + fa.beta * (N[1,t] + N[2,t])

  # Population model
  N[1,t+1] ~ dpois(sj * (f1[t] * N[1,t] + fa[t] * N[2,t]))
  N[2,t+1] ~ dbin(sa, (N[1,t] + N[2,t]))
  extinct[t] <- equals(N[1,t+1] + N[2,t+1], 0)   # Determines whether
    # population is still thriving (extinct = 0) or went extinct (extinct = 1)
}
}
")

# Initial values
N <- matrix(NA, nrow=2, ncol=T+1)
N[,2:(T+1)] <- 10
inits <- function(){list(N=N)}

# Parameters monitored
parameters <- c("N", "f1", "fa", "extinct")

# MCMC settings
ni <- 1000; nt <- 1; nb <- 0; nc <- 1; na <- 0

# Call JAGS (ART <1 min) and summarize results
out8 <- jags(jags.data, inits, parameters, "model8.txt", n.adapt=na, n.chains=nc, n.thin=nt,
  n.iter=ni, n.burnin=nb, DIC=FALSE)
print(out8, 4)
```

```
                mean     sd    2.5%     50%   97.5% overlap0 f
N[1,1]       10.0000 0.0000 10.0000 10.000 10.0000    FALSE 1
N[2,1]       10.0000 0.0000 10.0000 10.000 10.0000    FALSE 1
N[1,2]       11.9210 3.3838  6.0000 12.000 19.0000    FALSE 1
N[2,2]       11.0280 2.2804  7.0000 11.000 15.0000    FALSE 1
N[1,3]       13.4220 4.2102  6.0000 13.000 22.0000    FALSE 1
N[2,3]       12.6680 3.3665  6.0000 12.000 19.0250    FALSE 1
[ ... output truncated ... ]
f1[1]         1.9000 0.0000  1.9000  1.900  1.9000    FALSE 1
f1[2]         1.8410 0.0814  1.6795  1.840  2.0000    FALSE 1
f1[3]         1.7782 0.1257  1.5195  1.780  2.0200    FALSE 1
f1[4]         1.7197 0.1594  1.4000  1.720  2.0200    FALSE 1
[ ... output truncated ... ]
fa[1]         2.1000 0.0000  2.1000  2.100  2.1000    FALSE 1
fa[2]         2.0705 0.0407  1.9897  2.070  2.1500    FALSE 1
fa[3]         2.0391 0.0628  1.9097  2.040  2.1600    FALSE 1
fa[4]         2.0099 0.0797  1.8500  2.010  2.1600    FALSE 1
[ ... output truncated ... ]
```

We could draw a graph similar to Fig. 3.16 but instead compute the carrying capacity and age-specific fecundities at carrying capacity.

```
# Carrying capacity
mean(rowSums(out8$sims.list$N[,,T+1]))
[1] 53.382

# Productivity at carrying capacity
mean(out8$sims.list$f1[,200])
[1] 1.22968
mean(out8$sims.list$fa[,200])
[1] 1.76484
```

As before, we find these estimates to be very similar to those obtained with traditional simulation methods (Section 3.3.6). Other life-history summaries could be calculated as we did in previous sections. Obviously, and finally, we could also code an analysis that accommodates all parameter uncertainty, environmental and demographic stochasticity, and density dependence.

3.5 SUMMARY AND OUTLOOK

The core of most IPMs is a stage-structured (or matrix) population model that provides the linkage between the demography of individuals and the trajectory of a population. Hence, it is crucial that you have a really good working knowledge of matrix population models to be able to fully harness the power of IPMs. Indeed, IPMs can be seen as one important direction in the extension of matrix population models over the past 20 years. In this chapter, we have given a concise summary of the traditional analysis of matrix population models and emphasized the connection between matrix population models and IPMs. We have started from the knowledge about the demography of a study species and explained how this leads to the life-cycle graph and finally to the population equations. In this mathematical formalization of population dynamics, several key decisions must be made. Arguably, the most prominent are the choice of the "census" (or actually, survey) period and how the population is structured into age or stage classes. These decisions must be made primarily in the light of the available data, how (at which time with respect to a birth-pulse) data are collected, and of course, your research questions or management problems. In principle, it is straightforward to develop stage-structured population models, and plenty of excellent literature exists (e.g., Caswell, 2001; Morris and Doak, 2002), but nevertheless, mistakes are often found (Kendall et al., 2019).

A key assumption of stage-structured population models is that all individuals within a class are identical in terms of their demography. This assumption is bound to be violated to some degree for every population, as individuals differ demographically beyond what can be captured by the chosen stage classification. One possible avenue to relax this assumption is to introduce more stages, but typically there are restrictions due to limited available data or knowledge. Another possibility is to relate *individual* demographic performance to a measurable *continuous, individual trait*, such as body size, and to formulate an *integral projection model* (Easterling et al., 2000; Rees et al., 2014), or what to us is "the other IPM."

Stage-structured population models require information about demographic rates. The primary source of course is the data that have been sampled, from which we may estimate survival, productivity, or dispersal probabilities. We emphasize that you can view an IPM as a stage-structured population model whose demographic rates are directly estimated from several data sets in a single step. In other words, an IPM is typically a matrix population model embedded into an estimation framework to obtain values for demographic rates from different types of data. In Chapter 4, we provide a catalog of various such models with which demographic rates can be estimated from data. In Chapter 5, we jointly analyze multiple demographic models, where the link is provided by a common process model, that is, a matrix population model. The result of this will be an IPM.

3.6 EXERCISES

1. Compute the stable stage distribution of a woodchat shrike population when a postbreeding "census" is assumed. The demographic rates are the following: juvenile survival (0.30), adult survival (0.55), productivity of 1-year-old females (1.3), productivity of adult females (1.8). Assume that the demographic rates are known without error and that they are constant over time. Ignore demographic stochasticity. Use JAGS for computation.
2. Compute the stochastic population growth rate of a common swift (*Apus apus*) population and its sensitivity to changes in all underlying demographic rates. We know the mean and the temporal variation (SD on natural scale) of the following demographic rates: juvenile survival: 0.4 (SD: 0.05); survival in the second year: 0.7 (SD: 0.02); adult (i.e., year 2+) survival 0.8 (SD: 0.01); probability to start reproduction at the age of 2 years: 0.5 (SD: 0.1); productivity per reproducing female: 1.7 (0.4). Use JAGS for this exercise.
3. Calculate the extinction probability of a lapwing (*Vanellus vanellus*) population and assess how much uncertainty about the initial population size affects the outcome. We assume that lapwings start to reproduce when 2 years old, survival in the first year is 0.5, survival after their first birthday is 0.8, and productivity per female is 1.7. The size of the population is between 8 and 12 breeding pairs. Extinction probability shall be calculated with respect to 25 years, and the demographic rates are assumed to be constant. Use JAGS for the calculations.

CHAPTER

COMPONENTS OF INTEGRATED POPULATION MODELS

4

CHAPTER OUTLINE

Integrated Population Models. https://doi.org/10.1016/B978-0-12-820564-8.00020-8

4.1 INTRODUCTION

In this chapter, we give an overview of the *components*, or what we like to think of as the *ingredients*, of an integrated population model (IPM). In an IPM, we recognize different ecological and demographic processes such as survival, reproduction, immigration, and emigration. These are usually observed only incompletely through our observations or measurements made on a population, so they will be at least partially latent, i.e., observed imperfectly. Moreover, both these ecological processes and the measurements we take of them contain chance elements. Therefore, we like to describe both the ecological process and the observation process as two stochastic processes in sequence that together produce our observed data sets. The result is a hierarchical model (Royle and Dorazio, 2008; Cressie et al., 2009; Kéry and Schaub, 2012; Hobbs and Hooten, 2015; Kéry and Royle, 2016, 2021). In the time-series literature this class of models is also known as a state-space model (SSM) (Buckland et al., 2004, 2007; Newman et al., 2014), or if all system states and measurements are discrete, a hidden Markov model (HMM; King, 2014; McClintock et al., 2020).

Hierarchical models are a vast class of models and include traditional mixed models and generalized linear mixed models (GLMMs) as their best-known example and special case. These hierarchical models use random effects as a phenomenological description of some sort of "overdispersion," but these random effects hardly ever have a well-defined biological meaning. Models of this type are covered in many sources, including Gelman and Hill (2007), Bolker et al. (2009), Cressie et al. (2009) and Barker et al. (2018). They have also been called *implicit* hierarchical models (Royle and Dorazio, 2008; Kéry and Schaub, 2012) because their random effects have only an implicit biological meaning.

Another class of hierarchical models is that which Royle and Dorazio (2008) have called *explicit* hierarchical models. Here, the random effects do have a well-defined ecological interpretation and represent such quantities as population abundance, the alive/dead state of an individual or a nest, or the size of a brood, all of which may be observed with some associated error in our data sets. To us, such explicit, or science-based, hierarchical models have a distinctly more mechanistic feel, and therefore also have a more scientific flavor, in a sense, than do the more phenomenological, implicit hierarchical models. If we have a choice and there is no strong reason against their adoption, our preference is always for an explicit over an implicit hierarchical model.

In an explicit hierarchical model, we gain the clear interpretation of the random effects by including in the model an explicit, or mechanistic, description of the observation process that underlies our observed data set(s). The observation process links the true underlying process of interest, such as the size of a population (abundance) or the alive/dead state of an individual, which is formulated in terms of biologically meaningful quantities, with our observations or measurements. The observed data can therefore be thought of as a noisy or "dirty" picture of the underlying true states. The task of the observation or *measurement error* model is to statistically "clean" our observations such that proper estimates of the true underlying states of interest result from it.

Thus, we often use specialized statistical models with parameters that have a clear scientific interpretation (Royle and Dorazio, 2008), and it is these explicit hierarchical models that we cover as much as possible in this chapter on IPM ingredients. There is a massive body of literature on the estimation and modeling of population demography such as population size and survival with mechanistic measurement error description in terms of false negatives and false positives. This includes many thousands of pages in major treatises such as those by Seber (1982), Buckland et al. (2001, 2004, 2015), Borchers et al. (2002), Williams et al. (2002), Royle and Dorazio (2008), King et al. (2010), Link and Barker (2010), Kéry and Schaub (2012), Royle et al. (2014), Hobbs and Hooten (2015), McCrea and Morgan (2015), MacKenzie et al. (2018), Kéry and Royle (2016, 2021) and Seber and Schofield (2019), to name but a few. These textbooks cover estimation techniques such as capture-recapture, distance sampling, occupancy modeling, spatial capture-recapture (SCR), and *N*-mixture modeling, among many others. To give even a brief overview of all these topics would be far beyond the scope of a single book, let alone a chapter within it.

What we do attempt here is to give a conceptual overview and review of some key models that have been, or could or should be, used in an IPM. Thus, we provide a catalog of models from which you can pick ingredients when building an IPM for your particular data sets. Because BUGS is the language in which we build our IPMs in this book, we give BUGS code for the implementation of at least simple versions of all these models. For more detailed coverage, you should consult more specialized literature such as the works just cited. For model implementation in the BUGS language, i.e., in Win-BUGS or JAGS, please check out Royle and Dorazio (2008), King et al. (2010), Link and Barker (2010), Kéry and Schaub (2012), Royle et al. (2014), and Kéry and Royle (2016, 2021).

We will explain most models in words and simple algebra. In addition, we have emphasized many times (Kéry, 2010; Kéry and Schaub, 2012; Kéry and Royle, 2016, 2021) that we are convinced that explaining a statistical model to nonstatisticians may be best done via the simulation of a data set under that model. Thus, throughout this chapter we also use data simulation to explain a model. Finally, for all but a few models, we also give BUGS code and present all the code for a complete analysis of a simple data set. We hope that the material in this chapter can serve as a "store" from which you can pick the base ingredients including BUGS code for building your IPMs. In addition, for many models, we provide further illustrations within the case studies in Part 2 of this book; otherwise, you can find them in other books, especially in our *Bayesian Population Analysis* book (Kéry and Schaub, 2012). In a sense, the "BPA," as well as its planned Second Edition, is a very much enlarged version of this chapter.

In this chapter, you will find introductions to the following classes of models. To model population count data, we present:

- Basic Gaussian SSMs for population counts of unmarked animals (de Valpine and Hastings, 2002; de Valpine, 2003; Lindley, 2003; Buckland et al., 2004; Dennis et al., 2006; Knape, 2008; Humbert et al., 2009; Newman et al., 2014; Auger-Méthé et al., 2016)
- "Demographic" SSMs of the type developed by Dail and Madsen (2011) as a more mechanistically realistic alternative to Gaussian SSMs
- Dynamic occupancy models (MacKenzie et al., 2003; Royle and Kéry, 2007) and binomial N-mixture models (Royle, 2004a, c) to obtain population count data for an IPM which are corrected for coverage and detection bias

To model productivity survey data, we present:

- Poisson generalized linear models (GLMs) and GLMMs for brood size data
- Nest survival models (Mayfield, 1975; Dinsmore et al., 2002), which statistically are closely related to survival models such as the Cormack-Jolly-Seber (CJS) model. However, topically they belong in this section because they are often needed to properly estimate a component of productivity

To model survival survey data, we present:

- CJS models (Cormack, 1964; Jolly, 1965; Seber, 1965; Lebreton et al., 1992; Nichols, 1992) for survival estimation from the capture-recapture data of marked animals
- Ring-recovery models (Brownie et al., 1985) for survival estimation from recoveries of dead marked animals
- Multistate models (Arnason, 1972, 1973; Hestbeck et al., 1991; Brownie et al., 1993; Schwarz et al., 1993; Lebreton et al., 2009) for estimation of survival and state-transition probabilities from capture-recapture data of marked animals with discrete, time-varying categorical covariates
- The combination of capture-recapture and dead-recovery data (Burnham, 1993), and very briefly, the powerful extension of multistate models—multievent models (Pradel, 2005)

Finally, we offer a brief introduction to SCR models (Efford, 2004; Borchers and Efford, 2008; Royle and Young, 2008; Royle et al., 2014) that we believe will be increasingly adopted in IPMs (Chandler and Clark, 2014). Our specific example is the simplest possible spatial closed-population model for estimation of abundance.

In addition, we discuss modeling solutions for the case where instead of actual data you have estimates from a separate analysis of some data (i.e., outside the IPM) and want to use those, along with their associated measures of uncertainty such as the standard errors (SEs) or posterior standard deviation (SD), instead of the actual data in an IPM or indeed any other statistical model. This can be a useful "trick," especially when you are using JAGS and want to use some external data in a model such as, say, survival estimates with their SEs taken from a published article. We discuss this technique in the context of the population count part of an IPM, where it becomes important as a way to indirectly specify a more mechanistic observation model such as capture-recapture or distance sampling. It can thus serve as a workaround for specification of an observation model for population count data that is more robust than the simple normal distribution adopted in Gaussian SSMs (see Sections 4.3.2 and 4.3.3).

Furthermore, we present models for special topics that are especially relevant for productivity data: truncation of the response ("the unknown unknowns," or when some values are not observed at all) and censoring ("the known unknowns," or when all values are observed, but some or all have a meaning like "at least as high" or "between this and that value"). We also present zero inflation of a response here. As we often stress, GLMs are fundamental to all these models. In Chapter 2, we have covered them in more detail, specifically for Poisson, Bernoulli, binomial, normal, and multinomial responses. So if you find anything related to them challenging in this chapter, please revisit the relevant section(s) of that chapter.

4.2 OVERVIEW OF THE KEY TYPES OF DATA AND ASSOCIATED MODELS THAT GO INTO AN IPM

Rather than throwing a long list of seemingly unrelated models at you, we would like to first make some connections between them, especially in the context of how they are typically used in an IPM. Thus, in this section we give a broad overview of the types of data that typically go into an IPM and some features of the associated models.

In terms of the demographic modeling of unitary organisms such as animals, the fundamental unit of information is the state of individual i at site or location s during time t, that is, *the state of an individual at a certain point in space and time*. The most fundamental individual state is certainly the alive/dead distinction. The next most important states are arguably sex and age, but a myriad of other states may be important for research questions or management objectives, including body size and condition, reproductive state, and disease state. However, the data rarely come to us in the highest possible resolution. Instead, we typically obtain them, or choose to model them, in one of several possible schemes of aggregation.

In addition, virtually all demographic data come with an associated *observation* or *measurement error*, and our models may accommodate the underlying observation (sub)model to different degrees of realism about the mechanisms that generated the data. Here, we start by first exploring both topics (aggregation and measurement error) before giving a broad overview of the three main types of data and some associated models used in IPMs: population survey data (i.e., population counts), productivity data, and survival data.

4.2.1 LEVELS OF INFORMATION AND DATA AGGREGATION

Among the three fundamental dimensions of demographic data, we typically see some degree of aggregation along some or all of the individual, space, and time dimensions. For instance, in one of the simplest possible demographic models, we may simply distinguish a certain number of individuals that are counted in an area over each of several breeding seasons. Such a data set is the result of strong aggregation over all three dimensions. For instance, we no longer distinguish the individuals in any way other than simply noting that they are alive and not dead; we ignore whether they are male or female, juvenile or adult, fat or lean, and so on. Nor do we keep track of the actual locations of these individuals other than whether they are inside our study area. And finally, these data may also aggregate the results of multiple individual surveys (for instance, at a daily scale) over a longer period that we call a "breeding season."

However, in most demographic models, we will not work with such scalars that ignore any state differences or individual locational information and that lump all data over a possibly fairly long period. Instead, we will work with models that are state-structured in some way deemed useful for the modeling objective, e.g., by age or sex. This is the typical setting of a matrix population model (Chapter 3) and what we typically see in most current IPMs, including those in this book. Much finer resolution of the individual states—in the extreme, up to the case where each individual is in its own class as in integral projection models (Easterling et al., 2000) or individual-based models (Fahse et al., 1998)—would also be possible. But this is currently still rare, especially in the context of IPMs, although Plard et al. (2019) developed an IPM that is also an integral projection model. Another type of IPM is adopted in an individual-based spatial IPM (Chandler and Clark, 2014; Chandler et al., 2018); see Chapter 21. These IPMs retain the states of the individuals in their full information resolution, i.e., do not aggregate along the individual dimension of the demographic data.

Along the spatial dimension of our data, we may have counts that are also structured by location, for instance, by subarea or geographical stratum, or by subpopulation or colony; note that the former are more arbitrary subdivisions, whereas the latter are based more on biology. In the extreme, we may again keep track of the location of every single individual as we do in SCR models (Efford, 2004; Borchers and Efford, 2008; Royle and Young, 2008; Royle et al., 2014).

Finally, different degrees of aggregation are possible along the third dimension of demographic information: time. For instance, rather than lumping all observations within a breeding season or year, we may distinguish the early from the late breeding season or a breeding from a nonbreeding period. Sometimes, we may even work with a much finer temporal resolution such as a week, day, or still shorter time interval if this is needed for our modeling objectives and if the available data allow us to define such a fine temporal scale.

Any degree of aggregation always involves information loss; hence, aggregation may be inherently undesirable in some sense. On the other hand, dealing with demographic information at a highly disaggregated scale is not unlike the adoption of a very highly parameterized model—it will not always be sensible for the chosen modeling objectives. In addition, disaggregated data mean larger data sets, and this in turn will increase the time required to fit models to them. As a result, we often want to aggregate the information in some degree. In practice, we can imagine there may be an optimal degree of information aggregation, which among other things will depend on the size of a data set, the modeling objectives, the available computational resources, and the statistical and computational know-how of the analyst. Finally, we may sometimes be the ones to choose the level of aggregation for our modeled data, but in many other cases, this choice may already have been imposed in some way, e.g., by the computer person who set up the database that stores the demographic data in a long-term demographic study.

One important direction in the development of demographic models in general, and of IPMs in particular, is the increasing use of more disaggregated data. For instance, an IPM that no longer aggregates all individuals but rather models individual differences along a continuous scale as in integral projection models has been called IPM^2 by Plard et al.(2019). Another example is the increasing incorporation of space, either by stratification of a study area (e.g., McCrea et al., 2010; Ahrestani et al., 2017; see also Chapter 19) or by describing the exact locations of all individuals by point process models as we do in SCR models (Efford, 2004; Royle et al., 2014); see the seminal paper by Chandler and Clark (2014) and Chapter 21. Finally, along the time dimension, we increasingly see models that distinguish seasonal (i.e., subannual and year-round) processes rather than lumping all demographic processes into a single annual time resolution (e.g., Rushing et al., 2017; see also Chapter 15).

We believe it is conceptually valuable for you to distinguish different levels of aggregation of demographic information because it is something you may sometimes be able to choose and therefore optimize with respect to your modeling objectives. In addition, this scheme may be helpful for you to recognize the relationships among different variants of IPMs that use various aggregation schemes for their data.

4.2.2 OBSERVATION OR MEASUREMENT ERRORS IN OUR DATA SETS

Typically, we cannot observe the demographically interesting data perfectly in all three of the dimensions just discussed. Rather, our data will almost always be afflicted by some kind of observation or measurement error. For instance, and perhaps most commonly, we may fail to observe an individual at all, resulting in a *false-negative* observation. Such errors always occur with some frequency whenever the probability of detection does not equal 1. The ubiquity of this type of measurement error in any kind of demographic field data has given rise to an enormous number of statistical models in capture-recapture, distance sampling, occupancy modeling, and related fields (e.g., Williams et al., 2002; Royle et al., 2014; Buckland et al., 2015; MacKenzie et al., 2018). The converse are *false-positive* errors, which are most often due to species misidentification or multiple counts of the same individual. So far, these errors have received less attention, but recent work suggests that they may be much more common in ecological field data than what is believed or hoped for (Miller et al., 2011). In addition, contrary to false-negative errors, the biasing effects of false positives on parameter estimates in demographic models do not go away when we expend more effort in field or lab work. Quite to the contrary, whenever the false-positive probability does not equal 0, things actually become worse as we look harder, i.e., more rather than fewer errors accumulate (this is the *Third Law of Capture-Recapture*; Kéry and Royle, 2021). Clearly, both types of error mechanisms will add uncertainty into the state assessment of an individual as well as into the counts of any aggregation adopted for the individual information. In addition, there is a myriad of other possible ways in which the state of an individual can be recorded with an error, even if an individual is detected. For instance, a male can be taken for a female or we can simply be unsure about its sex.

Although observation errors are perhaps most striking along the individual dimension of the demographic information, they can also occur along the space and time dimensions. For instance, there will always be some degree of error associated with the measured location of an individual. Often, this locational error may be small and irrelevant, but when an individual is at the edge of a study area, even a small location error may be relevant because it may determine whether the individual adds to the count of your population or should be ignored. Along the temporal dimension, the exact time when some state change happens is typically unknown, e.g., when an individual dies, changes its location, or starts to reproduce. This may add uncertainty when such data are used in a demographic model and may have to be accounted for with additional complexity in the model.

One typical kind of temporal aggregation is when multiple surveys take place within a breeding season and thus in principle, the typical repeated measurements for application of capture-recapture or occupancy modeling would be available. But then only the aggregated result is recorded in a database, e.g., whether an individual was detected at least once during the breeding season or whether a site was recorded as occupied during at least one visit (Monneret et al., 2018; Kéry et al., 2021). Unfortunately, this type of temporal aggregation discards valuable information on the observation process underlying your data set that would be required in formally modeling false-negative or false-positive detection errors to correct for them. Hence, it should be avoided whenever possible.

In our models, we may choose to ignore such errors or accommodate them explicitly. If we add an explicit measurement error component to our demographic model, different degrees of mechanistic detail are possible and represent different hypotheses about the measurement error processes that produced our demographic data. SSMs and hierarchical models in general make an explicit distinction between the latent, true state of a system and the observed state.

Hence, they are ideally suited to represent a probabilistic (sub)model for the true states and a linked or conditional (sub) model for the observations. The state model is where the scientific or management interest typically lies, and the observation model is where we describe the measurement error process. This second part of the model is more of a "nuisance" and represents unwanted additional complexity that is required to "get right" those things in the model that we are most interested in. We can think of the observation model as a statistical method of "cleaning" our data for noise or bias.

Interestingly, not all observation errors have the same effect on our modeling when we use aggregated information. For instance, in a scalar population model, any uncertainty about the state of an observed individual other than the alive/dead distinction is irrelevant. Likewise, typical nonspatial demographic models implicitly may only distinguish two spatial strata (inside and outside the study area), and for them, most location errors are again irrelevant. Thus, aggregation of demographic information has one additional benefit—to some degree, it makes observation errors go away. In contrast, the modeling of more highly resolved demographic information, e.g., in the context of individual-based or SCR models, makes our models not only computationally more expensive but also more vulnerable to the effects of observation errors along the individual, spatial, or temporal dimensions of the modeled data.

4.2.3 LEVELS OF AGGREGATION AND MEASUREMENT ERROR IN POPULATION SIZE SURVEY DATA

As we will see, in most current IPMs, we have maximum aggregation of the population survey data with only a single count per chosen time step, such as a breeding season. Sometimes we do have state-classified counts, e.g., for juveniles and adults or for males and females (e.g., Tavecchia et al., 2009; Lee et al., 2015; Wilson et al., 2016). Spatial stratification of counts is also possible, e.g., for different subregions in a wider area of interest (Ahrestani et al., 2017). Alternatively, individual-based spatial IPMs may be viewed as representing the highest possible degree of disaggregation for the population count submodel in an IPM. In these models, we actually do away with modeling population counts and instead model each individual essentially as a separate class (Chandler and Clark, 2014; Sun et al., 2019; see Chapter 21).

In terms of the observation process, right from the beginning of the "terrestrial era" of IPMs with Besbeas et al. (2002), a distinction was made between the true state of a population and the observed state, i.e., the population counts, by adoption of a Gaussian SSM or sometimes with a Poisson or log-normal distribution adopted for the counts (see Section 4.3.1). Twenty years later, this is still done in the vast majority of IPMs, as well as in most examples in this book. However, as we will see, modeling the observation process by a simple symmetric noise process such as represented by a normal distribution is fairly phenomenological and cannot correct the state estimates for systematic patterns in time in the rates of observation error (see Sections 4.3.2 and 7.4). More explicit representations of the observation process for the count data would thus be desirable, but unfortunately, this may be challenging because extra information—for instance, in the form of repeated measurements or distance measurements—are required to estimate detection probability (Williams et al., 2002), and even more information is needed when we also must accommodate false-positive errors. The additional information required is frequently lacking.

But even if we do have the required extra information (e.g., from capture-recapture), then owing to the way in which models for population size estimation are typically implemented in a Bayesian framework by data augmentation (DA; Royle et al., 2007), it is not straightforward to obtain estimates of the time series of population size ($\widehat{N}$) and then use them directly in a population model as part of one and the same IPM. With data augmentation, the $\widehat{N}$'s are derived quantities, i.e., deterministic functions of other parameters. JAGS views them as fixed and does not allow us to declare them as a random variable in another part of the same model.

There are several possible ways around this. The first is that it should be possible to write the likelihood of an IPM with a capture-recapture submodel for population counts directly and without the need for data augmentation. Then, we could do proper estimation of the $\widehat{N}$'s and use them in the population model of the IPM all at once. The second solution is to conduct two analyses: in a first step, we estimate $\widehat{N}$ in a capture-recapture, distance sampling, *N*-mixture, or related type of model. In a second step, we incorporate these estimates into the IPM in the usual place of the population counts, and importantly, carry along their estimation uncertainty. This represents a two-step variant of an IPM with a highly mechanistic observation model for the population count portion. A third possible solution is to replace the Gaussian SSM with a variant of a Dail-Madsen (2011) model, i.e., instead of a Gaussian SSM, we specify what we call a "demographic SSM" for counts (see also Section 1.7 in Kéry and Royle, 2021). In this population model, we use the usual extra information to estimate detection probability (Dail and Madsen, 2011; Zipkin et al., 2014a; Zipkin et al., 2014b; Sollmann et al., 2015) and thus estimate abundance $\widehat{N}$ directly in the context of a classical model of population growth that distinguishes some or all of the four components in the birth–immigration–death–emigration scheme (Hostetler and Chandler, 2015). That is, this is a model for estimating $\widehat{N}$ that has parameters for survival and recruitment with which we can then link other data sets in the usual fashion of an IPM. As we will see, the price to pay is that we need multivariate counts, i.e., counts from more than one and typically at least 5–10 sites.

In this chapter, we first present the traditional Gaussian SSM. Then, we illustrate the two-step variant of fitting an IPM (Besbeas et al., 2003; Dennis et al., 2021) by incorporating into one model (here the Gaussian SSM) the estimates along with their variances (squared SEs or posterior SDs) or the variance-covariance matrix from another model (here, an *N*-mixture model). Finally, we explore a transition from the Gaussian SSM to an SSM with discrete-valued distributions and explicitly specified demographic mechanism. There are multiple ways that we can build such "demographic SSMs" (e.g., Buckland et al., 2004, 2007; Thomas et al., 2005). One useful variant is the demographic SSM of Dail and Madsen (2011; Hostetler and Chandler, 2015), which we cover last.

4.2.4 LEVELS OF AGGREGATION AND MEASUREMENT ERROR IN PRODUCTIVITY SURVEY DATA

Productivity is a complex biological process that contains elements such as the probability that a female reproduces at a given time and location, the number of offspring produced per reproductive event, and the number of reproductive events that a female performs during a breeding season; see Section 4.4. Productivity data, i.e., data that measure these different processes, are usually modeled at a very high level of aggregation in most current IPMs. For example, most often as the number of young per female or even as an annual total of young produced by a known number of females in the entire study area. Sometimes we have stratification into a few classes by age or other stages that classify mothers, and additional spatial or temporal stratification is also possible. The estimation of breeding probabilities typically requires data on marked individuals, which are often analyzed with capture-recapture or related models, and hence, a mechanistic observation model is implemented (Bonnet and Naulleau, 1996; Pradel et al., 1997b; Souchay et al., 2014). The number of breeding attempts per season is one of the most difficult productivity parameters to estimate (Etterson et al., 2011). When individuals are marked, applications of multistate capture-recapture models allow the estimation of probabilities of multiple brooding while accounting for imperfect detection (Kendall et al., 2019). However, this approach has rarely been adopted, presumably because the required data are for the most part lacking. Instead, it is mostly assumed that there is no measurement error in this quantity (Husby et al., 2009; Hoffmann et al., 2015; Schaub et al., 2015; see also Chapter 16). Another approach to estimate the number of breeding attempts is based on the modeling of the frequency of laying dates of observed broods without a correction for imperfect detection (Cornulier et al., 2009).

In terms of the observation error model for the actual number of offspring per reproduction event, it is here where we tend to see the lowest level of mechanistic detail in nearly all current IPMs. Many IPMs specify a Poisson or normal GLM that fails to make any distinction between the observed and true state of the number of young in a nest. In many situations this may be justified because these data are often obtained with very little measurement error. For instance, think about counting the number of young when checking nest boxes of cavity-nesting birds. Often, nestlings are also ringed at that time, so there should be no error in recording brood size. But there may be other situations where a more elaborate observation model for the productivity data could be usefully adopted. For instance, productivity for peregrine falcons (*Falco peregrinus,* see Chapter 13) is typically assessed by counting young in cliff nests by visual observation through telescopes at distances up to several kilometers. False negatives and even false positives appear as a very real possibility under that protocol, and hence, some kind of capture-recapture model to estimate the number of young might be a good idea. This might require the collection of extra information, such as repeated counts or simultaneous counts by two independent observers as for a binomial *N*-mixture model (Royle, 2004c).

Two interesting features in the observation process, for counts of young in particular, are truncation and censoring. For instance, in large-scale population surveys of peregrine falcons, people typically conduct one to two surveys during the nesting period. If the nestlings are hidden or have fledged already, they may go undetected. Alternatively, failure to detect any nestlings may be the result of brood failure. Faced with this uncertainty when failing to detect a nestling, observers may be reluctant to record a zero for productivity and thus implicitly record a missing value instead. This may then lead to a systematic undercounting of failed broods relative to successful broods and result in a positive bias in the observed productivity caused by partial zero truncation of the productivity data. Such nonignorable missingness must then be accommodated in a model for productivity.

A fairly common feature in the observation process for brood size in species similar to raptors, such as the peregrine, is that we may only be able to record a brood size such as "at least two young." For instance, whenever the number of young produced by a pair is assessed after fledging, it is quite unlikely that all young are seen simultaneously and that the number of young can be counted reliably. Then, *observing*, say, two fledged young may well mean that three or even four have fledged. Indeed, we may simply be able to record a successful nesting by the observation of a single fledged young, which again might mean that in reality, as many as four have fledged. Such counts of fledged young may be best treated as right-censored data, i.e., as giving only the minimum number of fledged young. Taking such censored counts at face value would result in bias toward low estimates of productivity. Hence, censoring must be properly accommodated in your model.

A final approach to get an integrative measure of productivity is the modeling of age ratios of randomly shot or captured individuals after the young have become independent from their parents (Arnold, 2018; Specht and Arnold, 2018). Imperfect detection matters if its magnitude differs among age classes. Similar approaches have been used in IPMs, where a correction due to differential detection of age classes can be applied (Péron and Koons, 2012; Koons et al., 2017; Zhao et al., 2019).

4.2.5 LEVELS OF AGGREGATION AND MEASUREMENT ERROR IN SURVIVAL SURVEY DATA

In typical IPMs, we work with individual-level survival data that are completely aggregated along the spatial dimension. This allows very fine modeling of patterns among individuals (e.g., by sex, age, or body condition) but averages over any spatial patterns and over any fine-scale temporal patterns that may exist in the survival process. For instance, we are then unable to distinguish between differences in monthly survival. In survival models in general (i.e., not necessarily as part of an IPM), increasing stratification is possible for time (Tavecchia et al., 2002; Gauthier et al., 2001; Giavi et al., 2014; Rushing, 2019) and for space at a large scale (Royle and Dubovsky, 2001; Saracco et al., 2010, Chapter 3 in Kéry and Royle, 2021) as well as in a spatially explicit, individual manner (Gardner et al., 2010; Schaub and Royle, 2014; Ergon and Gardner, 2014). Increased temporal and spatial stratification is more often observed in contemporary IPMs (e.g., Chandler and Clark, 2014; Rushing et al., 2017; Sun et al., 2019)—see also Chapters 15 and 21.

In terms of the observation model, most existing IPMs have used what could be called a mechanistically fully explicit SSM for survival data, such as a CJS or a ring-recovery model. Both represent a detailed description of the false-negative measurement error process that produces the observed survival data. This is in sharp difference to either the population survey or the productivity survey data in an IPM, where such mechanistically explicit observation models are rare or absent in most contemporary IPMs.

4.3 MODELS FOR POPULATION SIZE SURVEYS

4.3.1 GAUSSIAN STATE-SPACE MODELS

In a Gaussian SSM, we usually deal with a single time series of counts y_t recorded at times t with regular intervals (de Valpine and Hastings, 2002; Buckland et al., 2004; Newman et al., 2014; Auger-Méthé et al., 2021). SSMs are hierarchical models that decompose the variance in the observed counts into one contribution from the underlying true dynamics of the state N_t, called the *state process* or *process model* (or here, the *population model*), and another contribution from an *observation process*, which describes how an observation at time t is linked with the true state at that time. As in all typical SSMs, the population dynamics is described in terms of the initial state of the system and by a probabilistic "rule" for the state transitions from one time step to the next. Both the state and the observation processes are stochastic, and hence the model is also a hierarchical model, where the latent states $\mathbf{N}$, or in our formulation of the model below, the interannual population growth rates $\boldsymbol{\lambda}$, represent the random effects (the residuals are also random but are not usually termed random effects).

The main choices in the fitting of an SSM are related to how we describe the state and the observation processes, including the statistical distributions adopted for the two random variables. Gaussian SSMs adopt two normal distributions, but we will see later that with Bayesian methods we can transition to other distributions fairly easily, e.g., to a log-normal or a Poisson distribution for the observation model (see Chapter 2). See also the derivation of a "demographic SSM" starting from a Gaussian SSM in Section 1.7 in Kéry and Royle (2021) and the Dail-Madsen (DM) state-space model for multivariate counts in Section 4.3.6.

The simplest population dynamics model is arguably one that assumes exponential population growth (but see Jamieson and Brooks, 2004; Knape, 2008; Hostetler and Chandler, 2015, for models with density dependence). The SSM is defined by the following three equations, where N_t and y_t are the true and observed population sizes at time t, respectively, λ_t is the population growth rate in the time interval from t to $t+1$, ε_t is the residual at time t, and $f(\theta)$ is some distribution with parameter(s) θ:

$$N_1 \sim f(\theta) \qquad \text{Initial state: population size at } t=1$$

$$N_{t+1} = \lambda_t N_t \qquad \text{State process: transition rule, population model}$$

$$y_t = N_t + \varepsilon_t \qquad \text{Observation process: observation/residual error}$$

We model the state at the first modeled time point (the "initial state") as a random variable but cannot model it in the Markovian manner as we do in later periods, thus resulting in slightly different treatment with an expression for a general probability density function $f(\theta)$. Stochasticity is included in the model by making two distributional assumptions about the population growth rate $\boldsymbol{\lambda}$ (not to be confused with the expectation of a Poisson random variable; see later) and the observation errors ε_t:

$$\lambda_t \sim Normal(\mu_\lambda, \sigma^2_\lambda) \quad \text{Stochastic population growth rate}$$

$$\varepsilon_t \sim Normal\left(0, \sigma^2_y\right) \quad \text{'Residual': stochastic errors associated with the observed data}$$

Here, σ^2_λ is the process variance and σ^2_y the observation variance, or the observation error. Interestingly, the process variance represents genuine *variability*, while the observation variance will represent at least in part mere *uncertainty*, and we use a variance to describe both. The latter is the residual of the SSM and thus includes true observation errors as well as lack of fit in the process model.

To include covariates in this model, we could, for instance, model $\log(\lambda_t)$ as a function of, say, winter hardness for a covariate in the state process (Pasinelli et al., 2011; Koons et al., 2015; Furrer et al., 2016). Alternatively, we could write the observation process as $y_t = N_t + \beta X_t + \varepsilon_t$ for a (typically centered) covariate X_t assumed to affect the magnitude of observation error, such as wind speed during a survey for a bird species detected mostly by ear.

Next, we replicate this algebraic description of the model using data simulation code in R, which is taken mostly from Chapter 5 in the BPA book (Kéry and Schaub, 2012) and results in the data depicted in Fig. 4.1. Choose the same seed for the random number generator (RNG) if you want to produce identical data as we do.

```
# Choose constants
nyears <- 25                                             # Number of years
N1 <- 30                                                 # Abundance at t = 1
mu.lam <- 1.02                                           # Mean of the distribution of lambda
sig2.lam <- 0.02                                         # Variance of the distribution of lambda
sig2.y <- 400                                            # Variance of observation error

# Simulate true system state
N <- numeric(nyears)
N[1] <- N1                                               # Set initial abundance
set.seed(1)                                              # Initialize the RNGs
lambda <- rnorm(nyears-1, mu.lam, sqrt(sig2.lam)) # Draw random lambdas
for (t in 1:(nyears-1)){
  N[t+1] <- lambda[t] * N[t]                             # Propagate population size forwards
}

# Simulate observations
eps <- rnorm(nyears, 0, sqrt(sig2.y))                    # Draw random residuals
y <- N + eps                                             # Add residual (error) to value of true state
```

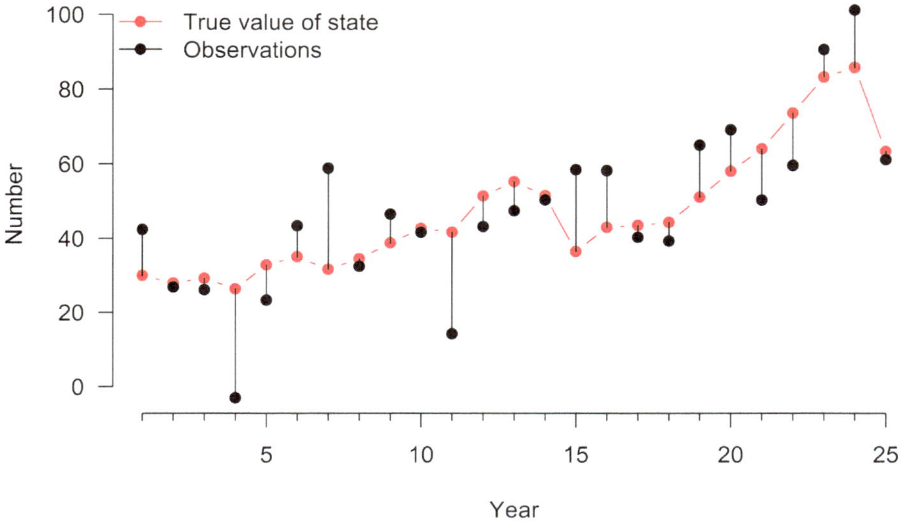

FIGURE 4.1

True system state (population size ***N***) and observed system state (measurements, observations, or counts ***y***) with residuals ("observation error") shown as black lines.

We now fit the data-generating model in JAGS.

```
library(IPMbook); library(jagsUI)

# Data bundle
jags.data <- list(y=y, T=length(y))
str(jags.data)

List of 2
 $ y: num [1:25] 42.4 26.82 26.11 -3.06 23.27 ...
 $ T: int 25

# Write JAGS model file
cat(file="model1.txt", "
model {
# Priors and linear models
mu.lam ~ dunif(0, 10)                 # Prior for mean growth rate
sig.lam ~ dunif(0, 1)                 # Prior for sd of growth rate
sig2.lam <- pow(sig.lam, 2)
tau.lam <- pow(sig.lam, -2)
sig.y ~ dunif(0.1, 100)               # Prior for sd of observation process
sig2.y <- pow(sig.y, 2)
tau.y <- pow(sig.y, -2)

# Likelihood
# Model for the initial population size: uniform priors
N[1] ~ dunif(0, 500)

# Process model over time: our model of population dynamics
for (t in 1:(T-1)){
  lambda[t] ~ dnorm(mu.lam, tau.lam)
  N[t+1] <- N[t] * lambda[t]
}

# Observation process
for (t in 1:T){
  y[t] ~ dnorm(N[t], tau.y)
}
}
")

# Initial values
inits <- function(){list(sig.lam=runif(1, 0, 1))}

# Parameters monitored
parameters <- c("lambda", "mu.lam", "sig2.y", "sig2.lam", "sig.y", "sig.lam", "N")

# MCMC settings
ni <- 200000; nb <- 10000; nc <- 3; nt <- 100; na <- 5000

# Call JAGS from R (ART <1 min), check convergence and summarize posteriors
out1 <- jags(jags.data, inits, parameters, "model1.txt", n.iter=ni, n.burnin=nb, n.chains=nc,
  n.thin=nt, n.adapt=na, parallel=TRUE)
traceplot(out1)                       # Not shown
print(out1, 3)

               mean     sd    2.5%     50%   97.5% overlap0 f  Rhat n.eff
lambda[1]     1.002  0.156   0.591   1.033   1.264    FALSE 1 1.001  2659
lambda[2]     1.012  0.145   0.656   1.035   1.280    FALSE 1 1.002  3990
[... output truncated ...]
lambda[23]    1.071  0.108   0.871   1.057   1.331    FALSE 1 1.000  3275
lambda[24]    0.952  0.151   0.577   1.004   1.149    FALSE 1 1.000  5700
mu.lam        1.053  0.036   0.989   1.050   1.140    FALSE 1 1.001  5700
sig2.y      246.009 96.637  98.641 229.619 477.841    FALSE 1 1.002  1145
sig2.lam      0.024  0.038   0.000   0.009   0.132    FALSE 1 1.012  1073
sig.y        15.403  2.958   9.932  15.153  21.860    FALSE 1 1.002  1157
sig.lam       0.119  0.098   0.005   0.095   0.363    FALSE 1 1.003  1611
N[1]         28.222  8.115  15.371  27.046  47.294    FALSE 1 1.000  5700
N[2]         27.535  6.468  15.820  27.159  41.492    FALSE 1 1.001  5700
[... output truncated ...]
```

We see in Fig. 4.2 that in our case, the estimates do a pretty good job recovering the input values, i.e., the true states. Note that the credible interval refers to the true state and *not* to the observations. Hence, we don't need to worry about a number of counts that lie outside the blue band.

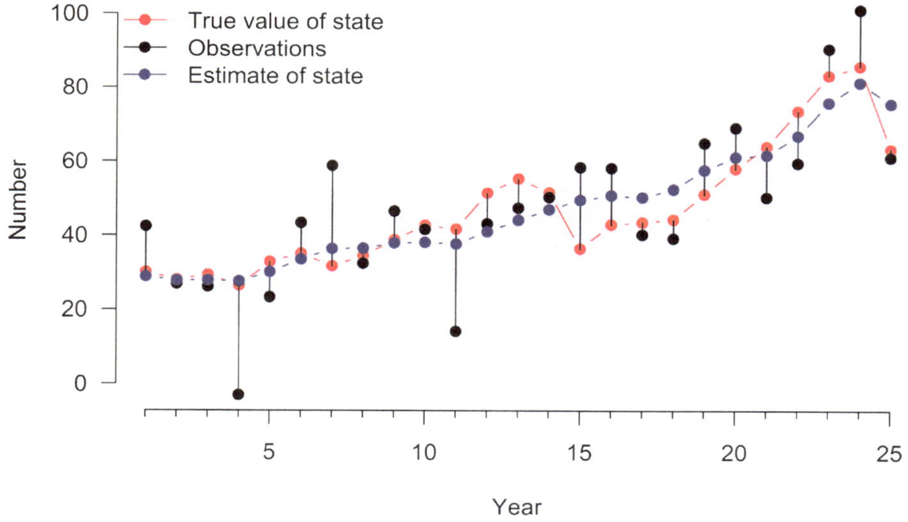

FIGURE 4.2

True system state (population size ***N***), observed system state (population counts ***y***) with residuals (observation error + lack of fit of state model), and the state estimates along with 95% credible bands (blue).

Although the model did well in our example, separating process variability and observation/residual error can be challenging in other situations. One example is when the residual error becomes too large relative to process variability or when time series are short (Auger-Méthé et al., 2016). Even with our data simulation, you may get less satisfactory model performance for other realizations of the stochastic process that is simulated, i.e., for different arguments in the call to `set.seed`. We have found that it may often help to force the posterior of the observation error variance just a very slight bit away from zero by setting the lower bound of its uniform prior at some small value greater than zero and not choosing priors too vague for the process variance. For instance, when choosing a U(0,10) for `sig.lam` above, mixing will usually become terrible. But as always, having a data simulator in R lets you conduct your own simulation studies to investigate the circumstances under which the performance of a model breaks down.

This model is a variant of the one currently adopted for the population count data in virtually all IPMs, except that most times, the process model in an IPM is more complex because it is a matrix population model with two or more states and with a transition matrix that is a function of demographic rates such as survival probability and productivity. These latter then serve as the "plugs" for the other data sets in the IPM that directly inform about these demographic rates. But the observation model is exactly the same—a normal distribution that assumes symmetric observation errors on either side of the true population size, or somewhat analogously, one that is log-normal or Poisson (see Section 6.6.5).

The simple observation model in the Gaussian SSM (as well as similar choices such as log-normal or Poisson) assumes symmetric noise in the observation process but posits that the population count data are unbiased with respect to true population abundance. This observation model will often work well in that it tries to separate the true state from observation error and results in some smoothing of the estimated states with respect to the more jagged time series of the actual observations. However, this observation model has some clear limitations, and there are several ways in which it may break down; see also Section 7.4. One major case for the performance of this model becoming unsatisfactory is when the measurement errors are not well represented by a simple, unbiased noise process such as represented by a normal, log-normal, or Poisson distribution. This is what we demonstrate next.

4.3.2 EFFECTS OF "EVIL" PATTERNS IN THE MEASUREMENT ERROR OF A GAUSSIAN STATE-SPACE MODEL

We now want to illustrate two situations in which the Gaussian observation process may be inadequate. For this, we simulate a data set with a more mechanistically realistic observation process by assuming a population that is counted subject to false-negative detection error. In the first case, we assume a constant undercounting of 20% on average, whereas in the second case, we assume that the undercounting increases from 20% to 90%, on average, over the course of the study. That is, we will take the simulated states N from above and produce two further data sets by adopting a binomial observation model with $p = 0.8$ in the first case and with a linear decrease in p from 0.8 to 0.1 in the second. Then we fit the same Gaussian SSM from above to both data sets and compare truth and estimates. The simulated states (N) are not integers, so we round them before we use them as binomial totals in our new observation process (and ignore the resulting, minor additional variance in the data). This results in two new data sets, where one time series is simply smaller than the true values *N* by a constant fraction, while the other suggests a population decline rather than the increase that we know to be the truth (Fig. 4.3).

```
# Simulate observations in Problem Case 1: p is constant at 0.8
set.seed(24)
p1 <- 0.8
y1 <- rbinom(length(N), round(N), p1)

# Simulate observations in Problem Case 2: p declines from 0.8 to 0.1
p2 <- seq(0.8, 0.1, length.out=length(N))
y2 <- rbinom(length(N), round(N), p2)

# Bundle the two new data sets
jags.data1 <- list(y=y1, T=length(y1))
jags.data2 <- list(y=y2, T=length(y2))

# MCMC settings
ni <- 100000; nb <- 50000; nc <- 3; nt <- 50; na <- 5000

# Fit the same Gaussian SSM to both new data sets
# Call JAGS from R (ART <1 min), check convergence and summarize posteriors
out2 <- jags(jags.data1, inits, parameters, "model1.txt", n.iter=ni, n.burnin=nb, n.chains=nc,
  n.thin=nt, n.adapt=na, parallel=TRUE)
out3 <- jags(jags.data2, inits, parameters, "model1.txt", n.iter=ni, n.burnin=nb, n.chains=nc,
  n.thin=nt, n.adapt=na, parallel=TRUE)

traceplot(out2)                        # Not shown
traceplot(out3)                        # Not shown
print(out2, 3)                         # Not shown
print(out3, 3)                         # Not shown
```

We plot the two series of estimated states under both models (Fig. 4.4). Not entirely surprisingly perhaps, we clearly see that the Gaussian SSM is unable to correct the state estimates for either a systematic, proportionally constant under-counting (Case 1, with detection probability $p = 0.8$) and even less for a temporally changing degree of undercounting (Case 2, where p declines from 0.8 to 0.1). With constant detection probability <1, we underestimate the population size by a constant proportion of $1-p$. In addition, because we simulated the data with a constant *proportion* of animals missed, the observation error in absolute terms (i.e., the difference between the red trajectory and the blue polygon) increases for larger population sizes. In Case 2, we also underestimate the population size, and this bias increases so much with the decline in detection probability over time that we ultimately diagnose a population decline.

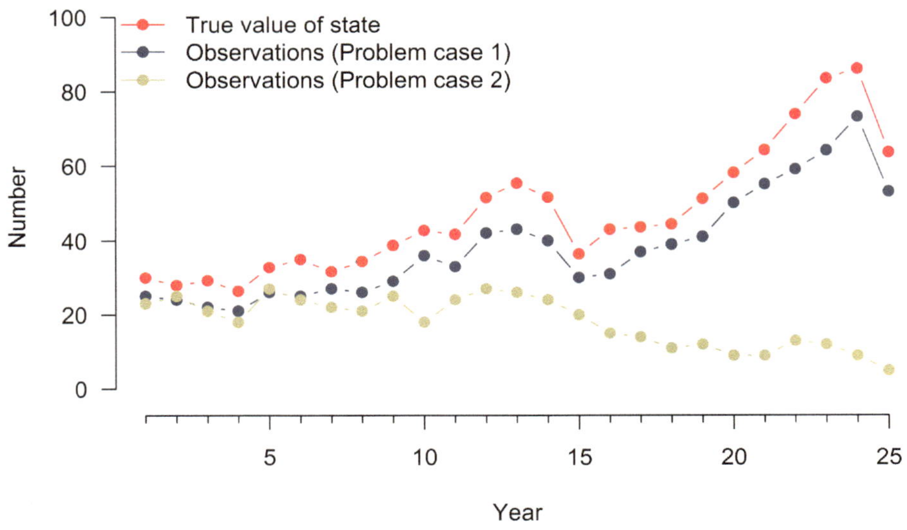

FIGURE 4.3

True system state (population size ***N***, the same as before) and observed system states (population counts ***y***) under two problematic cases simulated with a binomial observation process: in the first, we assume a constant detection probability of $p = 0.8$, while in the second, we assume a decline in p from 0.8 to 0.1.

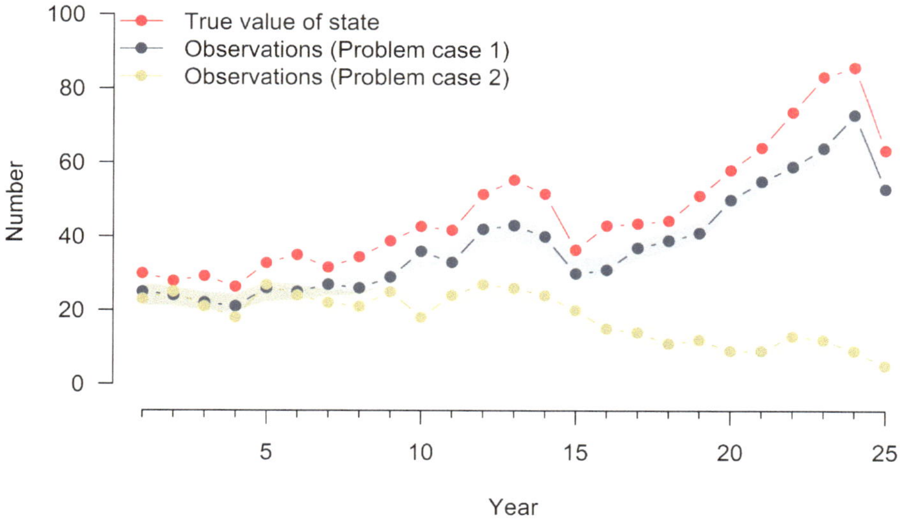

FIGURE 4.4

True system states (population size ***N***) and observed system states (population counts ***y***) under two problematic cases simulated with a binomial observation process. The gray and ochre polygons show the credible bands for the estimated states under the Gaussian SSM (model 1) fit to these data sets.

None of the alternative observation models commonly adopted for population counts in an IPM (i.e., log-normal or Poisson; see Section 6.6.5) would be immune to the effects of such "evil patterns" in the observation process. However, for the problem of Case 1, when detection probability remains constant over time, the temporal pattern of population sizes is retrieved correctly; hence, the population growth rates or demographic rates when this model is specified as part of an IPM remain unbiased. This is true even when the detection probability varies randomly over time, as we show in Section 7.4. Nevertheless, this section is meant to serve as a cautionary note that, yes, we do distinguish between a state and an observation process in an SSM, but, no, not every SSM always provides a suitable description of the observation errors encountered in population ecology and wildlife management.

This recognition serves as a motivation to explore alternative formulations of the observation model in an SSM for population counts, and this is what we do next. We will present two methods to achieve this. In the first method, we conduct two analyses, where in the first step we use a capture-recapture, distance sampling, *N*-mixture, or similar model to properly estimate a time series of abundance while correcting for imperfect detection and possibly also other components of the observation error. Then, in the second step, we use these estimates along with their uncertainty in a Gaussian or other SSM, either alone or when this SSM is part of an IPM. This is what we cover in Section 4.3.3. We note, however, that this approach is fully general. Hence, our code can serve as a template for the incorporation in any statistical model of estimates from another estimation procedure. In addition, by uncertainty we mean either the SEs or the variance-covariance matrix from likelihood inference or the analogous posterior SDs and the variance-covariance matrix in Bayesian inference. Thus, you may easily incorporate estimates from an analysis in MARK (White and Burnham, 1999) into a statistical model fit in JAGS.

The second method to gain insurance against "evil patterns" in the observation process underlying our population count data is to obtain spatial replication of the population counts and adopt one particular variant of what we call a "demographic" SSM. This is the dynamic *N*-mixture model of Dail and Madsen (2011), which we cover in Section 4.3.6, but first examine in Section 4.3.5 by making a transition from the Gaussian SSM and replacing the normal distributions with discrete-valued distributions that are arguably more adequate for abundance and count data.

In between, in Section 4.3.4, we have a deeper look at the processes that often produce population count data especially for territorial species. We showcase the power of dynamic occupancy models (MacKenzie et al., 2003) in conjunction with Bayesian missing-value imputation to correct population counts for the biasing effects of incomplete coverage of a study area and of imperfect detection.

4.3.3 USE OF ESTIMATES FROM ANOTHER ANALYSIS IN A GAUSSIAN STATE-SPACE MODEL OR AN IPM

In this section we show what you can do when you want, or must, use in your IPM the results from some other analysis that provides you with estimates of population size along with some associated measure of uncertainty, such as the variance (i.e., the squared SEs or posterior SDs) or possibly the whole variance-covariance (VC) matrix of these estimates (see also

Section 6.6.4). These might be unbiased estimates of population size from a capture-recapture, distance sampling or binomial *N*-mixture model, or they could even be estimates of some population index that is not necessarily unbiased with respect to true population size, but is assumed to have a constant proportionality to it, and has a measure of uncertainty associated with it. We will illustrate with our Gaussian SSM and assume that you are doing this modeling as part of an IPM. However, in the same way you could also incorporate into any other statistical model the estimates from an external analysis with proper accommodation of the additional uncertainty. This is really an exercise in proper error propagation when using some uncertain quantities (here, parameter estimates with their uncertainty) in any kind of parametric statistical model. Alternatively, we can also view this section as providing an example of how to use a data-based informative prior in a Bayesian analysis. In the context of an IPM, what we cover here is what is in the paper by Besbeas et al. (2003).

When fitting an IPM with BUGS software, there are at least three motivations for wanting to master the material in this section. First, instead of the actual data on population size, you may simply have some estimates with their SEs or posterior SDs and you wonder how to incorporate these into your IPM in place of the population count data. Second, you may have nice data on population size that allow you to actually *estimate* detection probability and therefore obtain unbiased estimates of population size. However, it is not clear to us at present if or how this could be done inside our IPM specified in JAGS. The reason is that when estimating population size using data augmentation, we treat it as a derived quantity that is a deterministic function of estimated parameters. However, in the IPM, we want to treat population size as a stochastic quantity and JAGS will not allow both in the same model. Hence, you must fit your other model first and obtain the population size estimates along with a measure of uncertainty, and then you can incorporate them into your Gaussian SSM in place of the population counts. Now, you no longer need to be worried about any evil patterns in detection probability (p) that could derail your inferences about the true population trajectory because the "population count data" that went into it are corrected for p and any patterns in it that you modeled in the first analysis.

Finally, entirely apart from the specific scenario of putting estimates of population size into your IPM, you may simply have some existing demographic estimates that you deem relevant for your IPM and that you want to incorporate in your model in a formal way (McCarthy and Masters, 2005). For instance, you might have a table of annual survival estimates along with their SEs published in a journal article and you want to incorporate these into your population model. For the purpose of this section, it doesn't matter whether these are the result of Bayesian or frequentist model fitting. For instance, they could have been produced by program MARK (White and Burnham, 1999) or package `unmarked` (Fiske and Chandler, 2011). You simply need a set of estimates along with the SEs, posterior SDs, or VC matrix. Here we show how to use them as part of another parametric statistical model. To illustrate the use of external parameter estimates along with their uncertainty, we first simulate a data set where abundance can be estimated while correcting for imperfect detection. Then, we use these estimates in a Gaussian SSM along with just the SEs (or more precisely, the posterior SDs) or the full VC matrix.

We assume that we have data from the Swiss breeding bird survey, or MHB (Monitoring Häufige Brutvögel) (see descriptions all through Kéry and Royle, 2016, 2021), with two to three surveys per breeding season in each of 267 1 km^2 quadrats. We simulate a look-alike MHB data set over 25 years for an imaginary species that we will call a jay, assuming a slight increase in its populations on average, with site-level heterogeneity in both the expected abundance and the rate of that increase and with imperfect detection. Then we fit to these data a simple *N*-mixture model with annual variation in both expected abundance and detection probability, obtain an estimate of the posterior variance of the estimates of the total annual abundance for all 267 quadrats combined, as well as an estimate of the full VC matrix. Then, we use these estimates in a Gaussian SSM and thus carry forwards either the posterior variances alone or the full VC matrix. For data simulation we use our `simMHB` function. We play around with changed values to function arguments to see what it does and then generate one data set to which we fit an *N*-mixture model. We note a temporary change in our notation: here we use lambda for the expected value of a Poisson random variable, where usually in this book we use it to denote the population growth rate.

```
library(IPMbook); library(jagsUI)
?simMHB                                 # Check out help function

# Some examples of how the simMHB function works
# Explicit default values for all function arguments
str(dat <- simMHB(nsites=267, nsurveys=3, nyears=25, mean.lam=1, mean.beta=0.03,
  sd.lam=c(0.5, 0.05), mean.p=0.6, beta.p=0.1, show.plot=TRUE))
str(dat <- simMHB())                    # Same, implicit
str(dat <- simMHB(nsites=1000))         # More sites
str(dat <- simMHB(nsurveys=10))         # More surveys
str(dat <- simMHB(nyears=50))           # More years
str(dat <- simMHB(mean.lam=5))          # Higher mean abundance
str(dat <- simMHB(mean.beta=-0.03))     # Population declines
str(dat <- simMHB(sd.lam=c(0, 0)))      # No site variability in lambda
str(dat <- simMHB(mean.p=1))            # Perfect detection
str(dat <- simMHB(mean.p=0.6, beta.p=0))    # Constant p = 0.6
str(dat <- simMHB(mean.p=0.6, beta.p=-0.2)) # Declining p
str(dat <- simMHB(show.plot=FALSE))     # No plots (when used in simulations)
```

Once we are satisfied that the function does sensible things we generate one data set with constant $p = 0.6$ and then fit a multiseason, static binomial N-mixture model (Royle, 2004c) to these data (see Chapter 2 in Kéry and Royle, 2021).

```
# Create one data set for our hypothetical surveys of jays
set.seed(1982)
str(MHBdata <- simMHB(nsites=267, nsurveys=3, nyears=25, mean.lam=1, mean.beta=0.03,
  sd.lam=c(0.5, 0.05), mean.p=0.6, beta.p=0, show.plot=TRUE))


# Bundle and summarize data set
jags.data <- with(MHBdata, (list(C=C, nsites=dim(C)[1], nsurveys=dim(C)[2], nyears=dim(C)[3])))
str(jags.data)

List of 4
 $ C        : int [1:267, 1:3, 1:25] 0 0 0 2 0 0 0 1 0 3 ...
 $ nsites   : int 267
 $ nsurveys: int 3
 $ nyears   : int 25

# Write JAGS model file
cat(file = "model2.txt", "
model {
# Priors
for (t in 1:nyears){                                # Loop over years
  lambda[t] ~ dunif(0, 100)                         # Expected abundance
  p[t] ~ dunif(0, 1)                                # Detection probability
}

# Ecological model for true abundance
for (i in 1:nsites){                                # Loop over 267 sites
  for (t in 1:nyears){                              # Loop over 25 years
    N[i,t] ~ dpois(lambda[t])
    # Observation model for replicated counts
    for (j in 1:nsurveys){                          # Loop over 3 occasions
      C[i,j,t] ~ dbin(p[t], N[i,t])
    } #j
  } #t
} #i
# Total abundance across all surveyed sites as a derived quantity
for (t in 1:nyears){
  totalN[t] <- sum(N[,t])
}
}
")

# Initial values
Nst <- apply(MHBdata$C, c(1, 3), max, na.rm=TRUE) + 1
Nst[Nst == '-Inf'] <- 1
nyears <- dim(MHBdata$C)[3]
inits <- function() list(N=Nst, lambda=runif(nyears))

# Parameters monitored
parameters <- c("lambda", "p", "totalN")

# MCMC settings
ni <- 5000; nb <- 1000; nc <- 3; nt <- 4; na <- 1000

# Call JAGS from R (ART 3 min), check convergence and summarize posteriors
out4 <- jags(jags.data, inits, parameters, "model2.txt", n.iter=ni, n.burnin=nb, n.chains=nc,
  n.thin=nt, n.adapt=na, parallel=TRUE)
traceplot(out4)                                     # Not shown
print(out4, 2)

             mean      sd     2.5%      50%    97.5% overlap0 f  Rhat n.eff
totalN[1]   258.159  6.900  246.000  258.000  273.000    FALSE 1 1.000  3000
totalN[2]   231.374  8.227  217.000  231.000  250.000    FALSE 1 1.001  1133
totalN[3]   238.130  6.574  227.000  238.000  252.000    FALSE 1 1.001  2609
[... output truncated ...]
totalN[23]  418.273 11.255  399.000  417.000  443.000    FALSE 1 1.000  3000
totalN[24]  477.464 11.496  457.000  477.000  502.000    FALSE 1 1.001  2138
totalN[25]  522.101 13.020  499.000  521.000  551.000    FALSE 1 1.001  1565
```

We compute the variance-covariance matrix of `totalN` based on our draws from the posterior distribution. We find that the covariances between the estimates of `totalN` are generally small, which means that the estimates are not correlated much.

```
vcMat <- array(NA, dim=c(nyears, nyears))
for (i in 1:nyears){
  for (j in 1:nyears){
    vcMat[i,j] <- cov(out4$sims.list$totalN[,i], out4$sims.list$totalN[,j])
  } #j
} #i
print(round(vcMat,1), 1)            # Look at that beast! (not shown)
```

So now that we have 25 annual estimates of our hypothetical jays of total population size over the 267 sample quadrat combined, along with their Bayesian analogue to an SE (i.e., the posterior SD) as well as an estimate of the full VC matrix of the set of estimates; how can we use these in further modeling, e.g., in a Gaussian SSM alone or as a part of an IPM?

The important fact to realize is that these are estimates, not data. Hence, they come with an estimation uncertainty that we need to propagate properly into any statistical analysis where we want to further use them. That is, we need to account for this additional uncertainty when doing "statistics on statistics" (Link, 1999). The simplest way, perhaps, is to ignore the covariances and only carry along the variances for this. This will often be adequate and it may be easier to understand. Moreover, it is perhaps more widely useful because there may be many cases where we can only obtain estimates with associated variances, but not the full VC matrix, for instance from a table in a paper. After that, we show the more general solution that carries along the full uncertainty about a set of parameter estimates contained in the VC matrix and not just its diagonal (which is what we do when we only use the variances). In both cases, our analysis assumes that the marginal posterior distributions (or the sampling distributions when this is a frequentist analysis of the parameter estimates) are approximately normal and that their joint distribution is multivariate normal. Therefore, we look at how well a normal distribution can approximate the posterior distributions of `totalN` and (when executing the next bit of code) find it to be reasonable.

```
# Plot posterior distributions of totalN (results not shown)
par(mfrow=c(3, 2))
for (t in 1:25){
  hist(out4$sims.list$totalN[,t], freq=FALSE, col='grey', main=paste('totalN in year', t))
  lines(density(out4$sims.list$totalN[,t]), col='blue', lwd=3)
  curve(dnorm(x, mean=out4$mean$totalN[t], sd=out4$sd$totalN[t]),
    range(out4$sims.list$totalN[,t]), col='red', lwd=3, add=TRUE)
  browser()
}
```

To understand how we model these estimates along with their uncertainty, which is a sort of additional measurement error, we start by writing in a slightly different way the observation model that we assumed in the basic Gaussian SSM in Section 4.3.1:

$$y_t \sim Normal\left(N_t, \sigma_y^2\right),$$

where y_t and N_t are respectively the measurements and the true states at t and σ_y^2 is the observation variance, which quantifies the magnitude of the variance of the model residuals and accounts both for lack of fit in the process model (which we do not repeat here) and for measurement error.

When instead of actual observations $\mathbf{y}$ we use estimates $\widehat{\mathbf{N}}$ of population size, we must account for an additional variance component in $\widehat{\mathbf{N}}$ due to estimation error in the analysis that produced the $\widehat{\mathbf{N}}$. The magnitude of this estimation variance in $\widehat{\mathbf{N}}$ is given by the VC matrix or the variances if we choose to ignore the covariances. Hence, we can write

$$\widehat{N}_t \sim MVN\left(y_t^*, \boldsymbol{\Sigma}\right) \text{ and}$$

$$y_t^* \sim Normal\left(N_t, \sigma_{y^*}^2\right),$$

i.e., the $\widehat{N}_t$ are assumed to be draws from a multivariate normal (MVN) distribution with the estimate of the VC matrix $\boldsymbol{\Sigma}$ obtained in the analysis that produced the population size estimates $\widehat{N}_t$. These estimates of $\widehat{N}_t$ are assumed to be centered on the mean vector y_t^*, for which we assume another normal distribution such that it varies symmetrically around the true system state N_t with variance $\sigma_{y^*}^2$. Note the asterisk superscript that we employ to emphasize that y_t^* is not the same as the observations y_t in Section 4.3.1 above. Rather, the y_t^* account for the lack of fit in the state model for N_t. We can also interpret the first equation above as implying a precision-weighting for the population size estimates $\widehat{N}_t$ such that more precisely estimated values get more weight when we use them in the Gaussian SSM.

We note that there is one additional component of uncertainty that we ignore here: that associated with the *estimate* of the VC matrix, or the variances, of the population size estimates $\widehat{N}_t$. This could also be accounted for in the analysis; see Sauer and Link (2002).

Thus, the first equation accommodates the first-stage analysis estimation uncertainty (measurement/estimation error) while the second accounts for the remaining residual error (lack of fit) and excludes the first-stage estimation error. Make sure that you understand the meaning of the different random variables related to population size: $\widehat{N}_t$ is the estimated population size originating from the first-stage analysis (i.e., here from the N-mixture model); these are treated as the "data" in the SSM. The N_t are the estimated population sizes in the state-space model as before and are one of the main target outputs from our analysis of the Gaussian SSM. Finally, y_t^* is the expected population size after correction for first-stage estimation error, where the latter is quantified by $\mathbf{\Sigma}$.

As discussed above, we will often only have access (or only want to account) for the diagonal of the VC matrix, i.e., for the variances, and ignore any correlations among the estimates. In that case, we can write the first equation as

$$\widehat{N}_t \sim Normal\left(y_t^*, SE\left(\widehat{N}_t\right)^2\right),$$

and each annual estimate in $\widehat{N}_t$ is then assumed to be a draw from an independent normal distribution centered on y_t^* and with the variance given by the square of the SE of the estimate, if it comes from a frequentist analysis, or of the posterior SD if it comes from a Bayesian analysis.

Now, we first fit the Gaussian SSM to these estimates $\widehat{N}_t$ while accounting for the associated estimation uncertainty as represented by the estimation variance only and hence ignoring the covariances.

```
# Data bundle
jags.data <- list(Nhat=out4$mean$totalN, var.Nhat=out4$sd$totalN^2, T=length(out4$mean$totalN))
str(jags.data)

List of 3
 $ Nhat     : num [1:25(1d)] 258 231 238 264 237 ...
 $ var.Nhat: num [1:25(1d)] 47.6 67.7 43.2 79.3 38 ...
 $ T        : int 25

# Write JAGS model file
cat(file="model3.txt", "
model {
# Priors and linear models
mu.lam ~ dunif(0, 10)                   # Prior for mean growth rate
sig.lam ~ dunif(0, 10)                  # Prior for sd of growth rate
sig2.lam <- pow(sig.lam, 2)
tau.lam <- pow(sig.lam, -2)

sig.ystar ~ dunif(0, 10000)             # Prior for sd of observation process
sig2.ystar <- pow(sig.ystar, 2)
tau.ystar <- pow(sig.ystar, -2)

# Likelihood
# Model for the initial population size: uniform priors
N[1] ~ dunif(0, 500)

# Process model over time: our model of population dynamics
for (t in 1:(T-1)){
  lambda[t] ~ dnorm(mu.lam, tau.lam)
  N[t+1] <- N[t] * lambda[t]
}

# Observation process for the abundance estimates with their SEs
for (t in 1:T){
  Nhat[t] ~ dnorm(ystar[t], tau.se[t])
  tau.se[t] <- 1/var.Nhat[t]            # Assumed known and given by variance of Nhat
  ystar[t] ~ dnorm(N[t], tau.ystar)
}
}
")
```

```
# Initial values
inits <- function(){list(sig.lam = runif(1, 0, 1))}

# Parameters monitored
parameters <- c("lambda", "mu.lam", "sig2.ystar", "sig2.lam", "sig.ystar", "sig.lam", "N", "ystar")

# MCMC settings
ni <- 50000; nb <- 25000; nc <- 3; nt <- 25; na <- 10000

# Call JAGS from R (ART <1 min), check convergence and summarize posteriors
out5 <- jags(jags.data, inits, parameters, "model3.txt", n.iter=ni, n.burnin=nb, n.chains=nc,
  n.thin=nt, n.adapt=na, parallel=TRUE)
traceplot(out5)                     # Not shown
print(out5, 3)                      # Not shown
```

This model converges very swiftly, and we compare the true states with the estimates provided by the *N*-mixture model and those from the Gaussian SSM when fed with the latter (Fig. 4.5).

Finally, here is the variant of this analysis with first-stage error propagation that carries along the entire VC matrix, not just its diagonal. For this, we will directly specify a multivariate normal (MVN) distribution for the set of 25 estimates along with their 25×25 VC matrix. In JAGS, we use a parameterization of the MVN distribution directly in terms of the VC matric. We found this model somewhat hard to fit and had to do some trial and error with the initial values. Importantly, we initialized the estimates for the population growth rates lambda near the estimates from the previous model.

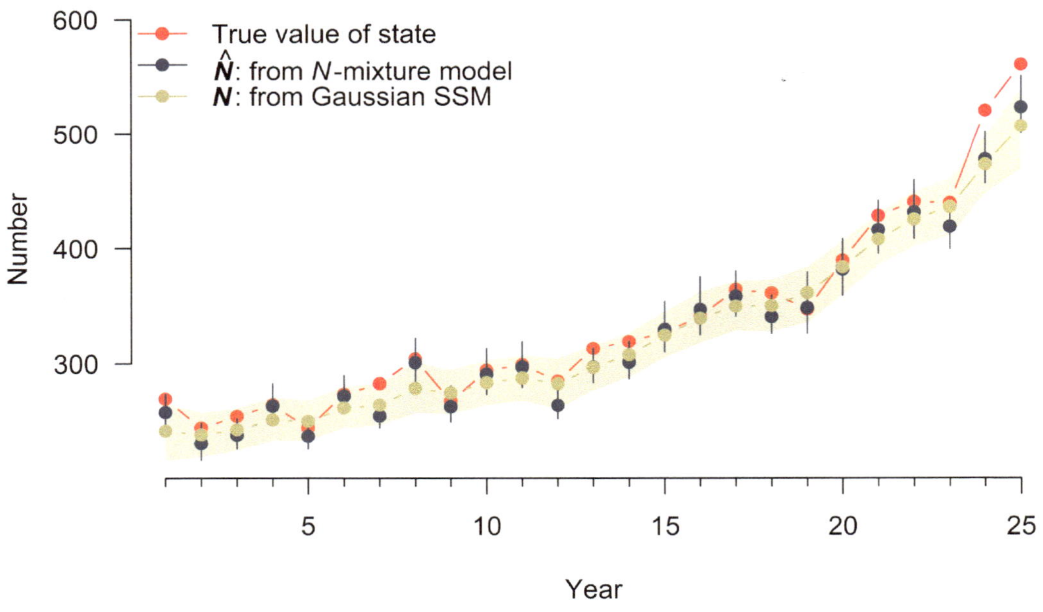

FIGURE 4.5

True system state (red; total jay population in 267 Swiss sample quadrats), estimates of that total population size under a binomial *N*-mixture model (gray, the 95% CRI are shown by vertical lines) and estimates of total population size when the latter are incorporated into our Gaussian SSM along with their variances (i.e., the squared posterior SDs) to accommodate estimation uncertainty from the *N*-mixture model (yellow dots show posterior means, the yellow band the 95% CRI).

```
# Data bundle
jags.data <- list(Nhat=out4$mean$totalN, vcMat.Nhat=vcMat, T=length(out4$mean$totalN))
str(jags.data)

List of 3
 $ Nhat      : num [1:25(1d)] 258 231 238 264 237 ...
 $ vcMat.Nhat: num [1:25, 1:25] 47.606 0.943 0.71 -0.144 -0.673 ...
 $ T         : int 25


# Write JAGS model file
cat(file="model4.txt", "
model {
# Priors and linear models
mu.lam ~ dunif(0, 10)                 # Prior for mean growth rate
sig.lam ~ dunif(0, 10)                # Prior for sd of growth rate
sig2.lam <- pow(sig.lam, 2)
tau.lam <- pow(sig.lam, -2)

sig.ystar ~ dunif(0, 10000)           # Prior for sd of observation process
sig2.ystar <- pow(sig.ystar, 2)
tau.ystar <- pow(sig.ystar, -2)

# Likelihood
# Model for the initial population size: uniform priors
N[1] ~ dunif(0, 500)

# Process model over time: our model of population dynamics
for (t in 1:(T-1)){
  lambda[t] ~ dnorm(mu.lam, tau.lam)
  N[t+1] <- N[t] * lambda[t]
}

# Observation process for the abundance estimates with their VC matrix
Nhat[1:T] ~ dmnorm.vcov(ystar[1:T], vcMat.Nhat[1:T, 1:T])
for (t in 1:T){
  ystar[t] ~ dnorm(N[t], tau.ystar)
}
}
")

# Initial values
Nst <- c(out5$mean$N[1], rep(NA, 24))
lamst <- rnorm(24, out2$mean$lambda, 0.1)
inits <- function(){list(N=Nst, lambda=lamst, sig.lam=runif(1, 0, 1))} # Works fine

# Parameters monitored
parameters <- c("lambda", "mu.lam", "sig2.ystar", "sig2.lam", "sig.ystar", "sig.lam", "N", "ystar")

# MCMC settings
ni <- 50000; nb <- 20000; nc <- 3; nt <- 10; na <- 5000

# Call JAGS from R (ART 11 min), check convergence and summarize posteriors
out6 <- jags(jags.data, inits, parameters, "model4.txt", n.iter=ni, n.burnin=nb, n.chains=nc,
  n.thin=nt, n.adapt=na, parallel=TRUE)
traceplot(out6)                      # Not shown
print(out6, 3)                       # Not shown

# Compare posterior means and SDs under the two models
print(cbind('pm (model 2)'=out5$mean$N, 'psd (model 2)'=out5$sd$N, 'pm (model 3)'=out6$mean$N,
  'psd (model 3)'=out6$sd$N), 3)

      pm (model 2) psd (model 2) pm (model 3) psd (model 3)
 [1,]          242         13.01          244         12.82
 [2,]          239          9.80          240          9.61
 [... output truncated ...]
 [24,]         472         13.06          473         12.64
 [25,]         506         17.85          507         17.67
```

We compare the estimates (posterior means and posterior SDs) of N_t obtained from both models in a table. The estimates are quite similar in terms of both the posterior means and the SDs. However, this may not always be so when comparing the two types of error propagation because in our case the covariances between the estimates were only small. Greater differences, especially in the posterior SDs, could presumably be expected when the covariances are larger.

This concludes our coverage of the proper manner of "doing statistics on statistics" (Link, 1999)—the use in a statistical analysis of the estimates from another statistical analysis, where carrying along the uncertainty associated with the estimates of the first analysis into the second analysis is required for proper error propagation. The resulting models may not be the easiest to grasp at first sight. However, being able to properly do error propagation in a statistical model where you use estimates and their variances from another analysis is very useful. Hence, we think it is good if you spend the effort to really understand it.

In the particular case of when you have estimates of population size corrected for imperfect detection and possibly also for false-positive errors, this method of error propagation will enable you to fit, in two steps, what in effect is a more realistic observation model in your IPM than the simple Gaussian SSM fit to the raw counts.

4.3.4 CORRECTION OF POPULATION COUNT DATA FOR COVERAGE BIAS AND DETECTION BIAS

The typical format of the population count data used in an IPM is highly aggregated: usually, we have a single count per year or other chosen point in time. This count is seldom questioned much and in part of the literature is even called a "census" (e.g., Besbeas et al., 2002); although a census really is a perfect enumeration of a population and hence something virtually impossible to achieve in nature. In reality, this bare number may hide a great amount of detail, and its use at face value will sweep under the rug the existence of multiple nuisance processes that may add into these counts noise or bias with respect to the true size of the population of interest. For instance, was the study area the same size over all years? In the common case of a territorial species, were all territories surveyed in every year? Or if not all could be visited in every year, was at least the same proportion visited in every year? And was that a random selection of all known territories or were "better" territories (e.g., those known to be occupied in previous years) targeted primarily? And in the case of a visited territory, what is the chance that territory occupancy was correctly assessed, i.e., what is detection probability (p)? If $p \neq 1$, can we expect the absence of any systematic temporal patterns so that the trends in our time series of population counts will not be biased with respect to the true abundance trajectory of the study population? And what about the possibility of false positives, e.g., when a transient individual is seen in an unoccupied territory and this is then erroneously added to the population count?

We believe that time spent to find answers to some or all of these questions underlying your population count data in an IPM is time spent well. Hence, we next look into two key processes that often underlie your population count data and that may lead to coverage bias and (non-)detection bias, and we showcase occupancy models as a powerful analytical framework to prevent such bias. We here ignore the additional issue of a specific kind of nonrandom coverage bias known as *preferential sampling* (Conn et al., 2017; Monneret et al., 2018; Kéry et al., 2021, Section 4.11 and 6.4 in Kéry and Royle, 2021) as well as false-positive errors due to species misidentification and other processes (Royle and Link, 2006; Miller et al., 2011, Chapter 7 in Kéry and Royle, 2021). Basically, this section is a call for the wider adoption of formal population estimation methods to produce the numbers that we feed into an IPM as "the population count data." We present one formal estimation method, dynamic occupancy models (MacKenzie et al., 2003; Royle and Kéry, 2007), to deal with coverage and nondetection bias in population counts of a territorial species (though other estimation methods may be useful for nonterritorial species). They yield bias-corrected estimates of true annual population size that may then be input into your IPM using the methods just presented.

In site-occupancy models (MacKenzie et al., 2002; Tyre et al., 2003) we describe the presence or absence of a species at a sample of "sites." The usual open version of the model (MacKenzie et al., 2003) describes transitions between these states among successive "seasons," where "season" in temperate latitudes typically refers to a year or breeding season. Occupancy models are perhaps *the* most natural estimation framework for surveys of any kind of territorial animals where we consider a territory to be a site that can be in one of the two states "occupied" and "not occupied." The multistate version of the model (Royle and Link, 2005; Nichols et al., 2007; MacKenzie et al., 2009) generalizes this framework to more than two states, e.g., empty territories and those occupied by single birds and by pairs, or territories occupied by pairs with and without reproduction.

Occupancy models have been used increasingly to understand the occupancy dynamics of sets of territories of large, cliff-nesting raptors and owls, where the natural definition of a site is a cliff complex that can at most have one pair, e.g., in golden eagles (*Aquila chrysaetos*) (Martin et al., 2009a, b; Mizel et al., 2018), peregrine and gyr falcons (*Falco rusticolus*) (Bruggeman et al., 2015, 2016; Monneret et al., 2018; Anderson et al., 2019; Kéry et al., 2021), or European eagle owls (*Bubo bubo*) (León-Ortega et al., 2017, Section 6.4 in Kéry and Royle, 2021). Occupancy models have also been used for territory occupancy dynamics of tree-nesting Northern spotted (*Strix occidentalis*) and Barred owls (*S. varia*) (MacKenzie et al., 2003) and for individual nests in seabird colonies (Bled et al., 2011). Territories in bird populations can be remarkably stable over the years (e.g., sparrowhawk (*Accipiter nisus*) in Newton, 2010; nightingale (*Luscinia megarhynchos*) in Roth and Amrhein, 2009), such that one can reasonably define some patch of land as a (potential) territory that is more or less stable over the duration of the studied populations (e.g., 10–20 years).

In summary, it appears that territories with fairly stable locations are the rule rather than the exception in plenty of species and that therefore occupancy models become a natural statistical inference framework for them. Here, we describe one particular variant of an occupancy model that is especially relevant for the goals of population size estimation with proper accommodation of coverage and detection bias. However, we emphasize that this is a very large class of models about which entire monographs have been written (MacKenzie et al., 2018; Kéry and Royle, 2021).

The dynamic occupancy model of MacKenzie et al. (2003) is a classical state-space model or HMM that distinguishes an ecological process, i.e., the dynamics of a latent discrete territory occupancy state, and an observation process, which describes how the true state is related to our observations in a territory. The state model is conceptually identical to that of the Gaussian SSM in Section 4.3.1: we describe the initial condition of the territory system and then the state-transition rules among the chosen time intervals. We can describe transitions as stochastic processes governed by two parameters: the probabilities of colonization and persistence (or alternatively, extinction). The full model can be written as a hierarchical model with four linked probability statements, where $z_{i,t}$ defines the occupied/unoccupied state of territory i in year t, and ψ, ϕ, γ and p are the probabilities of initial occupancy, persistence, colonization and detection, respectively (and note that γ in this model is a different parameter from the γ in the dynamic N-mixture model of Dail and Madsen (2011) in Section 4.3.6.).

$$z_{i,1} \sim Bernoulli(\psi) \qquad \text{Initial state of territory } i \text{ in year } 1$$

$$z_{i,t+1} \sim Bernoulli\left(z_{i,t}\phi + (1 - z_{i,t})\gamma\right) \qquad \text{State process: occupancy dynamics}$$

$$y_{i,t} \sim Bernoulli\left(z_{i,t}p\right) \qquad \text{Observation process: observation error}$$

Thus, if territory i is occupied at time t (i.e., $z_{i,t} = 1$), it will be occupied again at time $t+1$ with probability ϕ, whereas if it is unoccupied at time t (i.e., $z_{i,t} = 0$) it will become occupied at time $t+1$ with probability γ.

To be able to estimate all four types of parameters, we need some extra information that usually comes in the form of an additional level of replication according to a robust design (Pollock, 1982), such that some territories in some years are repeatedly visited within a short period over which we assume the state cannot change, i.e., we require the "closure assumption" to hold (Chapter 4 in Kéry and Royle, 2021). Ideally all territories have some replication in all years because the parameter estimates become more precise, but this is not necessary, provided that there is no systematic pattern in the number of visits per territory (e.g., we must not visit "better" territories more often or at all). Indexing these short-term replicate measurements of presence/absence as j, the observation model becomes

$$y_{i,t,j} \sim Bernoulli\left(z_{i,t}p\right) \qquad \text{Observation process: observation error}$$

We can add covariates into all four parameter types as usual via a link transformation such as the logit.

In our application here, the key inferential objective is population size. This is given by the sum over z, i.e., the number of occupied territories, for which the model provides a natural estimator corrected for nondetection error. However, in probably all field studies there is a second process that adds error into the sum-of-z estimator of population size: that not all territories can be visited in every year due to logistical limitations. This will add some coverage bias into our measure of population size unless this process is properly accounted for. Moreover, the proportion of territories visited at least once in a year will usually vary over time and not infrequently will change over time. For instance, the proportion of sites visited per year may increase because more volunteers enroll in a program over time. If the resulting trend in the coverage bias is not accommodated in your model for population size, your population trends will be biased, and this bias will filter through into some or all parameter estimates in an IPM.

Technically, what we have in most population studies will be missing values for the occupancy states of some territories in some years, and they need to be "imputed" in any analysis that aims at an unbiased estimator of population size. That is, we need to estimate the presence/absence state of the territories not visited in a year and then account for the resulting additional uncertainty in the population size estimator. Estimation of missing values in a statistical analysis is a very large field (Rubin, 1976; Gelman et al., 2013) and we can barely scratch the surface here. In short, we need a model for the basic process of interest and then sometimes also a second model for the missing-value-generating process, unless this latter can be ignored. Here, the basic process model for territories that lack visits is the same as that for territories that were visited: it is simply the colonization/persistence dynamics just described. For the missing-value-generating model, we here assume that sites are visited purely at random and not based on any site feature related in any way to the presence or absence of a pair at a territory. In the former case the missing-value-generating process is said to be *ignorable* and no additional modeling is required to account for the process that generates NAs.

The estimated occupancy state of a territory not visited in a year is then a simple function of the three parameters in the state model: initial occupancy, colonization, and persistence. Its uncertainty can be assessed trivially easily in a Bayesian analysis using MCMC and this then gets propagated to further functions of z such as the total population size in a year. Thus, random missing-value imputation to correct population size estimates for coverage bias is really easy in an MCMC-based analysis.

Nevertheless, we emphasize that in practice, the assumption of ignorable missingness when imputing the occupancy state of a territory without visits is a big one. Indeed, it may frequently be violated in field studies, where "good territories" typically tend to be visited first and "bad territories" last, such that territories with missing values will on average be less likely to be occupied than those without missing values, i.e., which were visited at least once. Ignoring the effects of such preferential, or adaptive, sampling will bias the results from a simple missing-value imputation model that assumes territory occupancy observations are missing at random (Conn et al., 2017; Monneret et al., 2018; Kéry et al., 2021, Sections 4.11 and 6.4 in Kéry and Royle, 2021; Kéry et al., 2021). In this section we ignore preferential sampling, but in practice you will need to check for it and correct for it when you find it to obtain unbiased estimates of population size.

We continue by simulating territory visitation data under the simplest possible dynamic occupancy model, where all four parameters are constants. We could easily simulate a much more complex data set with all sorts of structure in these parameters, e.g., by adding effects of covariates and hidden structure (random effects); see Chapter 4 in Kéry and Royle (2021) and `AHMbook` functions `simDynocc` and `simDynoccSpatial`. But remember that we use data simulation also as a concise way of explaining the basics of a model or a stochastic process in general. Hence, here we avoid any added complexity and just show the naked model structure when we simulate some data.

We assume a total of 100 territories of the European jay that are visited twice a year for a total of 12 years. The probability of initial occupancy (ψ), persistence (ϕ), colonization (γ) and detection (p) are 0.66, 0.9, 0.2 and 0.4, respectively. The first three values are chosen such that a dynamic equilibrium results according to $\psi^{eq} = \gamma/(\gamma + (1-\phi))$. Further, with two independent surveys the aggregated detection probability, i.e., the probability to detect a jay at least once in an occupied territory per year, is $1-(1-p)^2 = 0.64$.

To mimic a situation where survey effort increases over time, we assume that initially only 30% of all territories were surveyed, but in the final year 90% are, with a linear increase in between, and that in each year a random sample of territories was visited twice (remember that we ignore preferential sampling here). We will generate this feature of the detection/nondetection data by turning the detection/nondetection data into NAs for an appropriate proportion of territories chosen randomly in each year.

```
# Choose constants in simulation
nterritory <- 100                          # Number of territories (= sites)
nvisit <- 2                                # Number of short-term replicate visits
nyear <- 12                                # Number of years
psi <- 0.66                                # Initial occupancy probability
phi <- 0.9                                 # Persistence probability
gamma <- 0.2                               # Colonization probability
p <- 0.4                                   # Detection probability (per visit)

# State or ecological process
z <- array(NA, dim=c(nterritory, nyear)) # Empty presence/absence matrix

# Set initial presence/absence (in year 1)
set.seed(24)
z[,1] <- rbinom(nterritory, 1, psi)

# Propagate presence/absence forwards via transition rule
for (i in 1:nterritory){
  for (t in 1:(nyear-1)){
    z[i,t+1] <- rbinom(1, 1, z[i,t] * phi + (1-z[i,t]) * gamma)
  } #t
} #i

# Plot the true presence/absence pattern (Fig. 4.6)
mapPalette <- colorRampPalette(c("white", "#C7B76BFF"))
image(x = 1:nyear, y=1:nterritory, z=t(z), col=mapPalette(10), axes=TRUE, xlab="Year",
  ylab="Territory")
```

Fig. 4.6 shows the true state of the system, which we would like to observe perfectly, but in field studies essentially never get to see completely, owing to various features of the logistics of field work, which then determine part of the observation process. We now add into our data simulation two components of the observation process, which when unaccounted for in the models later will lead to nondetection bias and coverage bias. We simulate them in this order here but could also do the reverse.

We start by simulating a first data set with the effects of imperfect detection in the two hypothetical visits to each territory, where detection probability is 0.4 in each. Up to here, we still assume each territory gets two visits per year.

```
# Observation process (1): Imperfect detection
y1 <- array(NA, dim=c(nterritory, nvisit, nyear))
for (i in 1:nterritory){
  for (t in 1:nyear){
    y1[i,,t] <- rbinom(nvisit, 1, z[i,t] * p)
  } #t
} #i
```

Second, we simulate a second data set where we build in the effects of incomplete site visitation, in addition to the effects of imperfect detection in data set 1. We do this by turning some of these observed detection/nondetection data into missing values (NAs). We assume that effort greatly increases over the years from 30% surveyed sites in the first year to 90% in the last. We simulate this by defining a probability for two visits taking place at a site per year, and based on that, drawing a Bernoulli random number called "visit." Then, we divide the observed detection/nondetection data (i.e., `y`) by the value of "visit." If the value of "visit" is 1, two visits take place, and the division is by 1 and nothing changes in `yobs` (into which we save a copy of `y`). However, if the value of "visit" is 0, no visit takes place, we have a division by 0 and get "NaN" or "Inf," depending on whether the value of `y` is 0 or 1, which we both turn into NAs.

```
# Observation process (2): Incomplete coverage of territory
pvisit <- seq(0.3, 0.9, length.out=nyear) # Visitation probability
y2 <- y1                                  # Make a copy
for (i in 1:nterritory){
  for (t in 1:nyear){
    visit <- rbinom(1, 1, pvisit[t])      # 1 if visit takes place
    y2[i,,t] <- y1[i,,t] / visit
  } #t
} #i
y2[!is.finite(y2)] <- NA
y2[1:11,,1:3]                             # Look at data from first 11 sites and 3 years
```

As a result, we have two variants of observed data. In `y1` we have the detection/nondetection data when all territories are visited twice a year and there is only detection bias, whereas `y2` contains the "final" detection/nondetection data when a variable fraction of sites is visited each year, with NAs coding for surveys that failed to take place. Hence, these data are affected by both detection bias and coverage bias.

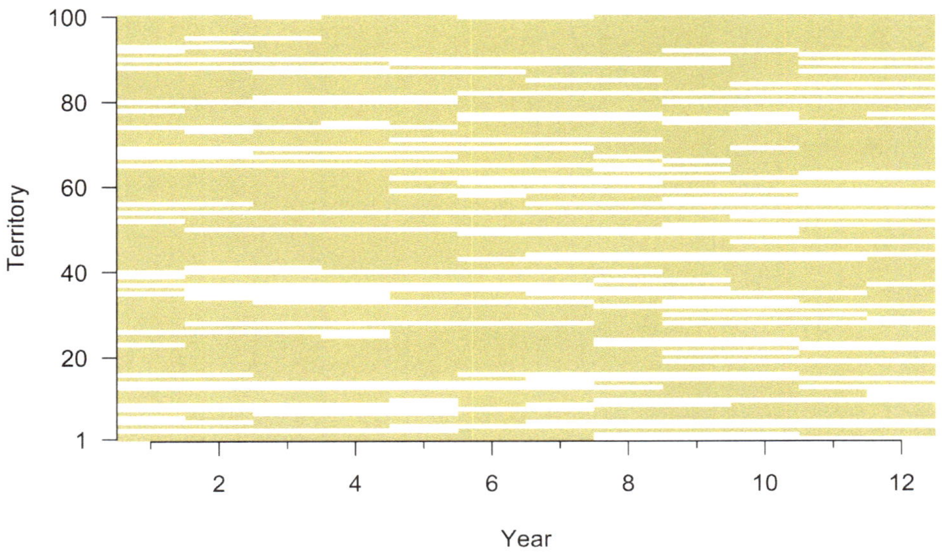

FIGURE 4.6

Simulated true system state (the z matrix) in an imaginary population study of the European jay with a total of 100 territories over 12 years: occupied territories are shown in greenish gray and unoccupied territories are shown in white.

We compare the true population size (i.e., number of occupied territories) with the counts under the two sampling regimes (Fig. 4.7 top). We see that both imperfect detection and additional incomplete coverage will bias low the raw, observed population counts relative to the true population size trajectory. In our particularly "evil" pattern for the coverage bias, the observed population size will be increasing, despite the population stability that we know is the truth.

```
# True population size (= number of occupied territories)
trueN <- apply(z, 2, sum)
# Observed population size if you had visited each territory twice a year
obsN1 <- apply(apply(y1, c(1, 3), max), 2, sum)
# Observed population size if you not had visited each territory twice a year
tmp <- apply(y2, c(1, 3), max, na.rm=TRUE)
tmp[tmp == -Inf] <- NA
obsN2 <- apply(tmp, 2, sum, na.rm=TRUE)
```

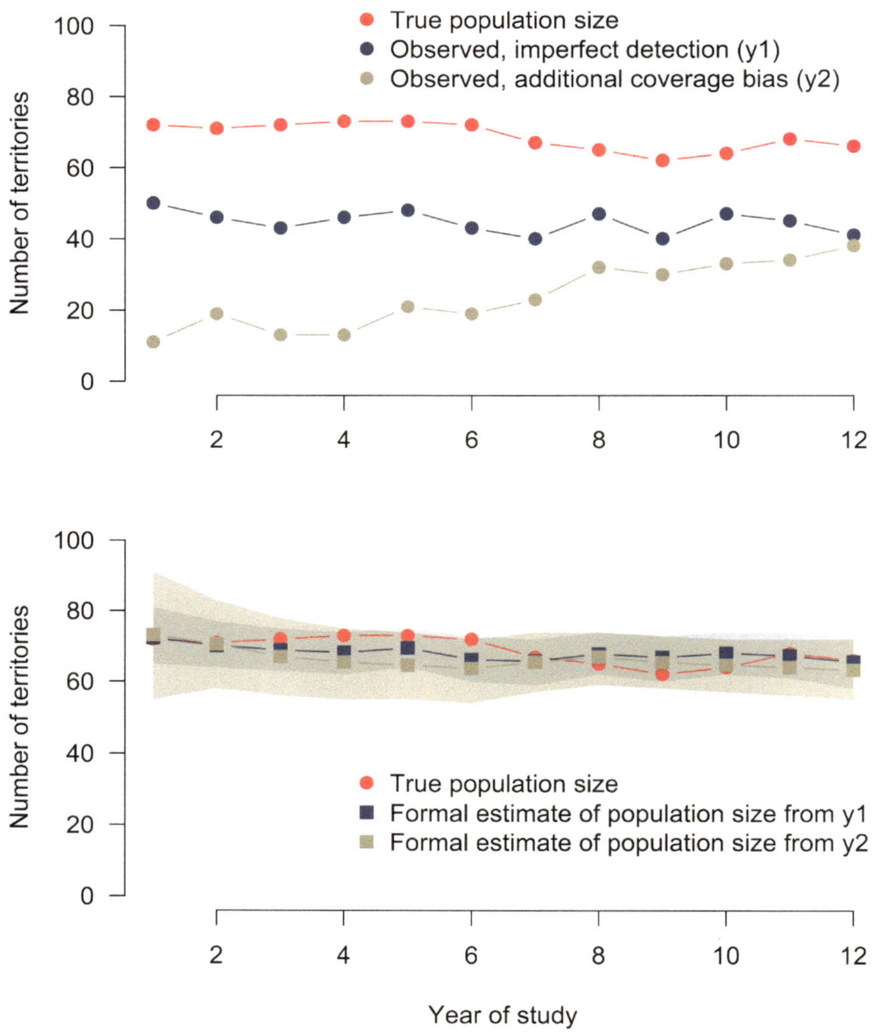

FIGURE 4.7

(top) Comparison of truth and the observations: True size (i.e., number of pairs or number of occupied territories) of a simulated population size of European jays, compared with what you would observe under the impact of nondetection bias only (i.e., without any coverage bias) and with what you will actually observe under the impacts of both nondetection bias and coverage bias. (bottom) Comparison of truth and estimates of population size under a dynamic occupancy model. The latter are corrected for bias due to both imperfect detection and incomplete coverage. Squares are posterior means, the bands show the 95% CRI.

We next fit the dynamic occupancy model to both variants of the observed data and compare the resulting estimates of population size that are corrected for imperfect detection and for incomplete coverage.

```
# Data bundle
jags.data <- list(y=y1, nterritory=nterritory, nvisit=nvisit, nyear=nyear)
str(jags.data)

List of 4
 $ y         : num [1:100, 1:2, 1:12] 0 0 0 0 0 0 0 0 0 0 ...
 $ nterritory: num 100
 $ nvisit    : num 2
 $ nyear     : num 12

# Write JAGS model file
cat(file = "model5.txt", "
model {
# Priors and linear models
psi ~ dunif(0, 1)
phi ~ dunif(0, 1)
gamma ~ dunif(0, 1)
p ~ dunif(0, 1)

# Likelihood
for (i in 1:nterritory){
  # Model for initial occupancy state
  z[i,1] ~ dbern(psi)
  # Model for occupancy state dynamics
  for (t in 1:(nyear-1)){
    z[i, t+1] ~ dbern(z[i, t] * phi + (1 - z[i, t]) * gamma)
  } #t
} #i

# Observation process
for (i in 1:nterritory){
  for (j in 1:nvisit) {
    for (t in 1:nyear){
      y[i,j,t] ~ dbern(z[i, t] * p)
    } #t
  } #j
} #i

# Derived quantities: population size
# correcting for both coverage bias and nondetection bias
for (t in 1:nyear){
  ntot[t] <- sum(z[,t])                # Number of occupied territories = population size
}
}
")

# Initial values
zst <- array(1, dim=c(nterritory, nyear))
inits <- function(){list(z=zst)}

# Parameters monitored
parameters <- c("psi", "phi", "gamma", "p", "ntot", "z")

# MCMC settings
ni <- 6000; nb <- 1000; nc <- 3; nt <- 5; na <- 1000

# Call JAGS from R (ART <1 min), check convergence and summarize posteriors
out7 <- jags(jags.data, inits, parameters, "model5.txt", n.iter=ni, n.burnin=nb, n.chains=nc,
  n.thin=nt, n.adapt=na, parallel=TRUE)
traceplot(out7)                        # Not shown
print(out7, 3)
```

Next, we fit the same model to our second data set. Only the data are different compared with our previous analysis; all the rest can be recycled from before.

```
# Data bundle
jags.data <- list(y=y2, nterritory=nterritory, nvisit=nvisit, nyear=nyear)
str(jags.data)                          # Summary of data set (not shown)

# Call JAGS from R (ART <1 min), check convergence and summarize posteriors
out8 <- jags(jags.data, inits, parameters, "model5.txt", n.iter=ni, n.burnin=nb, n.chains=nc,
  n.thin=nt, n.adapt=na, parallel=TRUE)
traceplot(out8)                         # Not shown
print(out8, 3)                          # Not shown
```

In Fig. 4.7 (bottom) we see that our dynamic occupancy model did an excellent job at estimating the true population size in our jay study by correcting for both the bias induced in the observations by imperfect detection (in both data sets) and by incomplete coverage (in data set 2). We also see that incomplete coverage leads to increased uncertainty in the population size estimates and that this uncertainty is much greater at the start of the study when fewer territories were assumed to be surveyed.

Of course, both of these are best-case scenarios, e.g., because the data-generating model matched the data analysis model and p was not too small. In real life, you cannot expect to always do just as well. However, one main point of this section is that what we call "the population count data" in an IPM are typically the outcome of several observation processes and that these data may very well be noisy and biased. Provided that you can distinguish territories that are more or less stable over the years, you can use dynamic occupancy models to correct your population size estimates for both imperfect detection and incomplete coverage as just shown. If you have preferential sampling (not random as we have simulated), you must accommodate that in your model as well. So in practice, there may be additional complexities to be addressed. On the other hand, if you have no repeated visits within a year but feel confident that detection probability is close to 1, you can ignore the observation model and still use this model to correct for coverage bias, with or without correction for preferential sampling (Monneret et al., 2018; Kéry et al., 2021, Section 4.11 in Kéry and Royle, 2021; Kéry et al., 2021). The other main point of this section is that occupancy models can be fantastic for obtaining improved estimates of population size, which you can then use in an IPM with proper propagation of the estimation errors as shown in Section 4.3.3.

4.3.5 TRANSITIONING FROM GAUSSIAN TO DISCRETE-VALUED STATE-SPACE MODELS FOR POPULATION COUNTS

At the start of Section 4.3. we have seen some limitations of the Gaussian SSM as a model of the observation process for the count data in an IPM. We pointed out the desirability of a more "mechanistic" observation model, e.g., one that can accommodate imperfect detection as it is now so commonly done in capture-recapture, distance sampling or in binomial N-mixture models (Williams et al., 2002; Royle and Dorazio, 2008; Royle et al., 2014; Buckland et al., 2015). Doing "statistics on statistics" provides one solution as we have seen in Section 4.3.3. However, a different and perhaps better, or at least more elegant, approach might avoid the Gaussian assumptions altogether and replace them with more sensible discrete-valued distributions. This might then perhaps enable us to use a binomial distribution in the observation model, as is done in almost all of population size estimation including capture-recapture (Williams et al., 2002; Royle and Dorazio, 2008; McCrea and Morgan, 2015). In this section we develop such a discrete-valued variant of an SSM to emphasize the conceptual similarity between the traditional Gaussian SSMs and one particular, existing discrete-valued SSM due to Dail and Madsen (2011), which we cover in the following section.

Historically, the Gaussian SSMs were the first choice for population survey data (Besbeas et al., 2002; de Valpine and Hastings, 2002) mostly for their mathematical and computational tractability. However, we could conceivably also adopt a model that explicitly accommodates the discreteness of population size and counts, e.g., by using two Poisson distributions for the stochastic components in the SSM.

$$
\begin{aligned}
&N_1 \sim f(\theta) && \text{Initial state: population size at } t = 1 \\
&N_{t+1} \sim Poisson(\lambda_t N_t) && \text{State process: transition rule, population model} \\
&y_t \sim Poisson(N_t) && \text{Observation process: observation/residual error}
\end{aligned}
$$

This model has one fewer parameter because there is no free parameter for the magnitude of the observation errors. In contrast, the process variability is now composed of two parts: the variance of the Gaussian population growth rate (λ), which we still assume to be drawn from a normal distribution (not shown here), and the Poisson variability. The former variance corresponds to environmental stochasticity (see Sections 3.3.3 and 3.4.3), the latter to demographic stochasticity (see Sections 3.3.4 and 3.4.4). As a quick check on identifiability, we try out this model with one simulated data set to which we fit the data-generating model.

```
# Choose constants
nyear <- 25                                  # Number of years
N1 <- 30                                     # Initial abundance
mu.lam <- 1.02                               # Mean of the distribution of lambda
sig2.lam <- 0.02                             # Variance of the distribution of lambda

# Simulate true system state
N <- numeric(nyear)
N[1] <- N1                                   # Set initial abundance
set.seed(1)                                  # Initialize the RNGs
lambda <- rnorm(nyear-1, mu.lam, sqrt(sig2.lam)) # Draw random lambda
for (t in 1:(nyear-1)){
  N[t+1] <- rpois(1, lambda[t] * N[t])       # Propagate population size forwards
}

# Simulate observations
y <- rpois(nyear, N)                         # Observation error is now Poisson noise

# Data bundle
jags.data <- list(y=y, T=length(y))
str(jags.data)

List of 2
 $ y: int [1:25] 26 26 40 38 24 21 16 33 25 31 ...
 $ T: int 25

# Write JAGS model file
cat(file = "model6.txt", "
model {
# Priors and linear models
mu.lam ~ dunif(0, 10)                        # Prior for mean growth rate
sig.lam ~ dunif(0, 2)                        # Prior for sd of growth rate
sig2.lam <- pow(sig.lam, 2)
tau.lam <- pow(sig.lam, -2)

# Likelihood
# Model for the initial population size: uniform priors
N[1] ~ dunif(0, 500)
# Process model over time: our model of population dynamics
for (t in 1:(T-1)){
  lambda[t] ~ dnorm(mu.lam, tau.lam)
  N[t+1] ~ dpois(N[t] * lambda[t])
}

# Observation process
for (t in 1:T){
  y[t] ~ dpois(N[t])
}
}
")

# Initial values
inits <- function(){list(sig.lam=runif(1, 0, 1), mu.lam=runif(1, 0.1, 2),
  N=round(runif(nyear, 20, 40)))}

# Parameters monitored
parameters <- c("mu.lam", "sig2.lam", "sig.lam", "N")

# MCMC settings
ni <- 20000; nb <- 10000; nc <- 3; nt <- 10; na <- 1000

# Call JAGS from R (ART <1 min), check convergence and summarize posteriors
out9 <- jags(jags.data, inits, parameters, "model6.txt", n.iter=ni, n.burnin=nb, n.chains=nc,
  n.thin=nt, n.adapt=na, parallel=TRUE)
traceplot(out9)                              # Not shown
print(out9, 3)                               # Not shown
```

This model converges fine, and we see in Fig. 4.8 that it appears to do a decent job at recovering the underlying true states $\boldsymbol{N}$. However, we are not entirely sure about its usefulness in practice. For instance, it has just one parameter to quantify the size of the process variability, and it lacks a free parameter for the magnitude of the observation errors. On the other hand, it does preserve the discreteness of the system. Perhaps it can serve as a useful steppingstone to a different discrete-valued variant of an SSM?

For instance, we may think of the stochastic population growth rates λ_t broadly as akin to the simplest possible transition rule for population size from t to $t+1$. Thus, we might perhaps replace λ_t by more complex transition models that better reflect population dynamics, as for instance the Leslie matrix in matrix population model (see Chapter 3). Another possibility to extend the model to become more realistic would be to decompose the population growth rate into a contribution from a survival and another from a recruitment process (Nichols et al., 2000), e.g., as in a model that we will see in Section 6.3 or as in a dynamic N-mixture model (Dail and Madsen, 2011; see next section) and to account for imperfect detection in the observation process (and where E denotes an expectation).

$N_1 \sim f(\theta)$	Initial state: abundance at $t = 1$
$E(N_{t+1}) = \phi_t N_t + \rho_t N_t$	State process: transition rule, population model
$E(y_t) = N_t p_t$	Observation process: observation/residual error

We could allow for stochasticity by using statistical distributions for all of initial abundance, survival, recruitment and the observation processes.

$N_1 \sim Poisson(\gamma)$	Stochastic initial abundance
$S_{t+1} \sim Binomial(\phi_t, N_t)$	Stochastic survival process
$R_{t+1} \sim Poisson(N_t \rho_t)$	Stochastic recruitment process
$N_{t+1} = S_{t+1} + R_{t+1}$	Total population size
$y_t \sim Binomial(p_t, N_t)$	Stochastic observation process

Here, S_{t+1} and R_{t+1} are, respectively, the number of survivors and recruits from time t to $t+1$, γ is the expected initial abundance, ϕ_t and ρ_t are survival probability and per-capita recruitment rate at time t, and p_t is detection probability at that time. We see that now we have partitioned the population growth rate λ_t into the sum $\phi_t + \rho_t$.

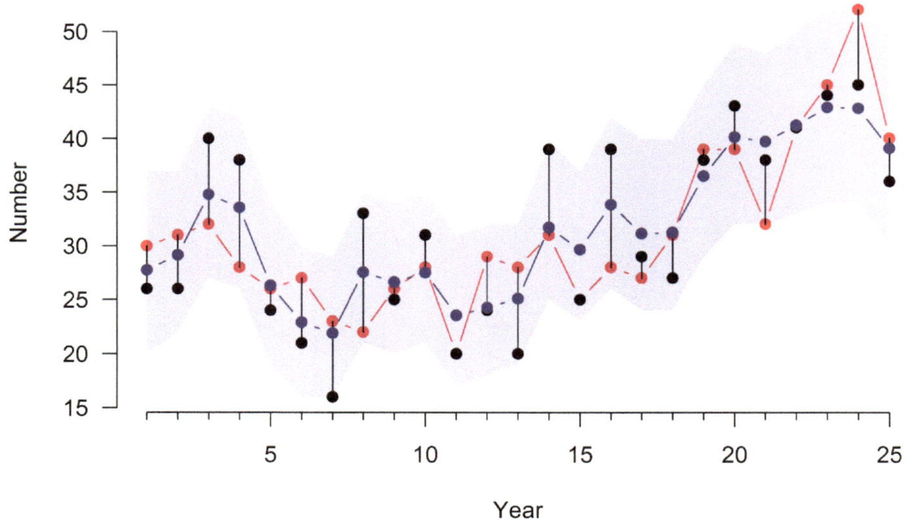

FIGURE 4.8

Comparison of the true population size (red), the observed counts (black) and in blue the estimated population size with 95% CRI under our new discrete-valued state-space model where the two main Gaussian distributions were replaced with Poisson distributions.

Clearly, this model will not be identifiable from a single time series of counts; see Section 6.3. But we show this extension here mainly for conceptual reasons: it may clarify ways in which population count data could be linked with a model for demographic data (e.g., recruitment or survival surveys), which are described in terms of the demographic rates (such as ϕ_t or ρ_t) while adopting a more mechanistic observation model than that in the traditional Gaussian SSM. The dynamic N-mixture model of Dail and Madsen (2011) in the following section is essentially this very same model, but fit to multivariate time series of counts.

4.3.6 THE "DEMOGRAPHIC" STATE-SPACE MODEL OF DAIL AND MADSEN

N-mixture models (Royle, 2004b, c) are hierarchical models that conceptually are very similar to site-occupancy models in estimating the states of "sites," correcting for imperfect detection, based on data from spatially stratified populations. However, instead of characterizing the state of a site by one of just two (as in conventional occupancy models) or of a small number of states (as in multistate occupancy models), we now characterize the state of a site by local abundance of some species.

There are several types of N-mixture models, e.g., binomial and multinomial, corresponding to the distribution chosen for the observation part of the model. Binomial N-mixture models (Royle, 2004c) are for replicated counts of unmarked individuals, whereas multinomial N-mixture models (Wyatt, 2002; Royle, 2004a, b; Dorazio et al., 2005; Royle et al., 2007) are for spatially replicated capture-recapture, removal sampling or distance sampling data; see Chapters 6 and 7 in Kéry and Royle (2016) for comprehensive monographs of the binomial and the multinomial N-mixture models, respectively. Here we highlight the binomial N-mixture model because the data to fit it, counts of unmarked individuals, are usually much easier to obtain than those under a protocol leading to an analysis by a multinomial N-mixture model. However, we emphasize that use in an IPM of a multinomial N-mixture model for, say, spatially stratified capture-recapture or distance sampling data would proceed in an exactly analogous way in just a slight variant of the binomial N-mixture model presented here.

Basically, with the binomial N-mixture model we estimate abundance (N) from counts (C) correcting for detection probability (p). With data from multiple years, as will be the case in IPM applications, we can either link the estimates across the years or keep them independent. Royle and Dorazio (2008) and Kéry et al. (2009) show examples of multiyear analyses that do not link the years in a dynamic fashion. In contrast, linking the years essentially means specifying a population dynamics model for the abundance at a site in consecutive years. Dail and Madsen (2011) developed such a dynamic N-mixture model, where abundances are linked via a basic population dynamics model with parameters for apparent survival and recruitment, exactly as we had in our wish-list model in Section 4.3.5.

To illustrate the dynamic N-mixture model for potential use as part of an IPM, we simulate an example data set under the simplest possible model and ignore any potential extensions such as different distributions for abundance, the addition of covariate effects and random effects; see Chapter 2 in Kéry and Royle (2021) for a recent overview of DM models (we use the terms, dynamic N-mixture and DM model, as synonymous).

In the classic Dail-Madsen (2011) model, we model and estimate abundance at a collection of sites over multiple years using counts of unmarked individuals while correcting for imperfect detection. Typically, there is some replication of counts at a site within a year over a short period when abundance is assumed to remain constant (the closure assumption again), but parameters can also be estimated with single-visit data if the parametric assumptions of the model are met sufficiently well; see also Section 1.7 in Kéry and Royle (2021). In the population model, from year 2 onwards, abundance at a site in a year is expressed as the sum of survivors and recruits. Thus, the abundance at site i in year $t+1$ is expressed conditional on the abundance in year t and on parameters for apparent survival (ϕ) and per-capita recruitment (ρ), or $N_{i,t+1} = N_{i,t}\phi_t + N_{i,t}\rho_t$ (one may also formulate the model with absolute recruitment instead).

Similarly as the conceptually related dynamic occupancy model in Section 4.3.4, we can summarize the DM model using four probability statements, where N and C denote the true and observed local abundance, respectively, γ, ϕ, ρ, and p are, respectively, the expected abundance in year 1, apparent survival probability, per-capita recruitment rate, and detection probability, and i, j and t are the indices for site, within-year replicate and year.

$$N_{i,1} \sim Poisson(\gamma) \qquad \text{Initial abundance at site } i \text{ in year 1}$$

$$S_{i,t+1} \sim Binomial(\phi, N_{i,t}) \qquad \text{Survival process}$$

$$R_{i,t+1} \sim Poisson(N_{i,t}\rho) \qquad \text{Recruitment process}$$

$$N_{i,t+1} = S_{i,t+1} + R_{i,t+1} \qquad \text{Abundance} = \text{survivors} + \text{recruits}$$

$$C_{i,j,t} \sim Binomial(p, N_{i,t}) \qquad \text{Observation process}$$

In a reduced version of the model, the two demographic processes are collapsed into a single parameter, the population growth rate $\lambda = \phi + \rho$ (Hostetler and Chandler, 2015). We can add covariates via a link transformation as in any GLM, e.g., the log-link for rate parameters γ and ρ, and the logit-link for probability parameters ϕ and p.

We now simulate data under the simplest possible such model, where all four parameters are constants, and as for the dynamic occupancy model above, assume a total of 100 sites visited twice each over a total of 12 years. We assume the following values for the four model parameters: $\gamma = 5$, $\phi = 0.82$, $\rho = 0.22$ and $p = 0.4$, representing a growing population because $\lambda = \phi + \rho = 1.04$. Note that we think of the number of pairs or of females as exchangeable in this simulation.

```
# Choose constants in simulation
nsite <- 100                              # Number of sites
nvisit <- 2                               # Number of short-term replicate visits
nyear <- 12                               # Number of years
gamma <- 5                                # Initial expected abundance
phi <- 0.82                               # Apparent survival probability
rho <- 0.22                               # Recruitment rate (per-capita)
p <- 0.4                                  # Detection probability (per individual)

# State or ecological process
# Simulate true system state
N <- array(NA, dim=c(nsite, nyear))       # Empty abundance matrix

# Set initial abundance (year 1)
set.seed(24)
N[,1] <- rpois(nsite, gamma)
head(N, 11)                               # Look at first 11 sites (not shown)

# Propagate abundance forwards via transition rule = pop.dyn.model
for (t in 1:(nyear-1)){
  N[,t+1] <- rbinom(nsite, N[,t], phi) + rpois(nsite, N[,t] * rho)
}
head(N, 11)                               # Look at first 11 sites (not shown)
```

The N matrix is a fundamental quantity comparable to the z matrix in a dynamic occupancy model. It describes the true abundance state of each site in every year; hence, we can sum it up across rows to obtain the abundance in all sites combined in every year.

As we did in the simulation of occupancy data, we would strongly argue that the data as created so far (i.e., N) will rarely be a faithful representation of counts of abundance made under field conditions. Hence, we add the effects of imperfect detection in our data simulation.

```
# Observation process: Simulate counts with imperfect detection
C <- array(NA, dim=c(nsite, nvisit, nyear))
for (i in 1:nsite){
  for (t in 1:nyear){
    C[i,,t] <- rbinom(nvisit, N[i,t], p)
  } #t
} #i

# Look at data in year 6 and compare with truth (for illustration; not shown)
cbind(Truth=N[,6], 'Count 1'=C[,1,6], 'Count 2'=C[,2,6])
```

We compare the true and observed number of pairs/females. For the observed number, we sum the maximum counts at each site.

```
trueN <- apply(N, 2, sum)                 # True number of pairs
obsN <- apply(apply(C, c(1, 3), max), 2, sum) # Observed number of pairs
```

We illustrate model fitting in JAGS where the processes in the data simulation match those in the data analysis exactly.

```
# Data bundle
jags.data <- list(C=C, nsite=nsite, nvisit=nvisit, nyear=nyear)
str(jags.data)

List of 4
 $ C      : int [1:100, 1:2, 1:12] 2 0 2 3 3 3 2 4 2 2 ...
 $ nsite : num 100
 $ nvisit: num 2
 $ nyear : num 12
```

```
# Write JAGS model file
cat(file = "model7.txt", "
model {
# Priors and linear models
gamma ~ dunif(0, 10)                    # Expected abundance in year 1
phi ~ dunif(0, 1)                       # Apparent survival
rho ~ dunif(0, 1)                       # Per-capita recruitment rate
p ~ dunif(0, 1)                         # Per-individual detection probability

# Likelihood
for (i in 1:nsite){
  # Model for the initial population size
  N[i,1] ~ dpois(gamma)
  # Process model over time: our model of population dynamics
  for (t in 2:nyear){
    S[i,t] ~ dbinom(phi, N[i,t-1])    # Survivors
    R[i,t] ~ dpois(N[i,t-1] * rho)    # Recruits
    N[i,t] <- S[i,t] + R[i,t]
  } #t
} #i

# Observation process
for (i in 1:nsite){
  for (j in 1:nvisit){
    for (t in 1:nyear){
      C[i,j,t] ~ dbinom(p, N[i,t])
    } #t
  } #j
} #i

# Derived quantities
for (t in 1:nyear){
  ntot[t] <- sum(N[,t])                # Total abundance across all sites
}
}
")
```

Finding initial values for the DM model that satisfy JAGS and do not lead to contradictions with model and observed data is a particularly dark science (J.A. Royle, pers. comm.). So be prepared for some trial and error … and sometimes even utter defeat. It is especially the $S + R = N$ part in the model, along with the binomial observation model that implies $N \geq C$, which impose constraints on the possible values that all these quantities can take and that may easily be violated for certain sets of initial values.

A workaround that may solve these problems for this model is to left-truncate the estimated population size with the maximum observed counts at i and t. Hence, in the state process model we would write

```
N[i,t] ~ dsum(S[i,t], R[i,t])T(maxC[i,t], )
```

and provide the maximum count per site and year (maxC) as data (T. Riecke, pers. comm).

```
# Initial values
# Need to get inits for N that do not contradict model and data!
Nst <- apply(C, c(1,3), max) + 5
Nst[,2:nyear] <- NA
Rst <- apply(C, c(1,3), max) + 1      # Observed max. counts + 1 as inits
Rst[,1] <- NA
inits <- function(){list(N=Nst, R=Rst)}

# Parameters monitored
parameters <- c("gamma", "phi", "rho", "p", "ntot", "N")

# MCMC settings
ni <- 200000; nb <- 100000; nc <- 3; nt <- 100; na <- 5000

# Call JAGS from R (ART 45 min), check convergence and summarize posteriors
out10 <- jags(jags.data, inits, parameters, "model7.txt", n.iter=ni, n.burnin=nb, n.chains=nc,
  n.thin=nt, n.adapt=na, parallel=TRUE)
traceplot(out10, c("gamma", "phi", "rho", "p", "ntot"))
print(out10)                          # Not shown

# Compare truth and estimates
truth <- c('gamma'=gamma, 'phi'=phi, 'rho'=rho, 'p'=p, Ntot=apply(N, 2, sum))
print(cbind(truth, out10$summary[1:16,c(1:3, 7)]), 3)

        truth    mean      sd    2.5%   97.5%
gamma    5.00   4.919  0.3520   4.240   5.646
phi      0.82   0.800  0.0264   0.744   0.847
rho      0.22   0.240  0.0267   0.193   0.298
p        0.40   0.396  0.0201   0.356   0.435
Ntot1  476.00 490.508 26.9491 441.000 548.025
Ntot2  509.00 516.968 27.4614 467.000 576.000
Ntot3  537.00 546.495 28.5580 494.000 608.000
Ntot4  562.00 559.367 29.8402 507.000 623.000
Ntot5  585.00 576.243 31.1611 520.000 643.000
Ntot6  623.00 601.033 32.5235 542.000 671.000
Ntot7  646.00 636.714 33.3183 577.975 709.025
Ntot8  655.00 659.010 34.3807 596.000 733.025
Ntot9  675.00 681.968 35.7977 617.000 758.000
Ntot10 691.00 694.363 37.7847 624.975 774.000
Ntot11 736.00 728.986 39.6545 657.000 812.000
Ntot12 773.00 758.156 42.9149 680.000 850.000
```

Thus, in our best-case scenario we see excellent estimates of the four parameters and total annual population size. So we could do either of two things now if this were part of the real analysis with an IPM. First, we could take the estimates of `Ntot` with their VC matrix and plug them into our Gaussian SSM. But perhaps the more elegant and statistically efficient solution would be to directly use the DM model as the core of our demographic model in the IPM. This model can be extended to include age- or general stage-structure (Zipkin et al., 2014a, b, Chapter 2 in Kéry and Royle, 2021), so it can directly serve as your population model in an IPM. It has all the usual demographic parameters such as apparent survival or recruitment to which you can link the other submodels in a typical IPM. See Chapters 19 and 20 for IPMs where the DM model instead of a Gaussian SSM is the main "motherboard".

Provided that you do have spatially replicated time series of counts then we think that the DM model can be a fairly natural choice for the basic structure in an IPM. However, it must be said that at least the raw DM model (that is, not augmented with other data sets as in an IPM) can be extremely sensitive to violation of its parametric assumptions, especially in terms of the resulting estimates for apparent survival and recruitment. For instance, it is very commonly found that the former are estimated too high and consequently the latter too low (Bellier et al., 2016, 2018). In addition, we have often found it very challenging or even near-impossible to find initial values that did not let JAGS choke on a *"Node inconsistent with parents"* error. Like much of the material in this section, therefore, this is somewhat experimental material, and it will be useful to explore the practical usefulness of the DM model in IPMs more in the future.

4.4 MODELS FOR PRODUCTIVITY SURVEYS

Reproduction is a key process to include in any population model. Population models typically require the *number of offspring produced per female and year* (Caswell, 2001; Streby et al., 2014) which we term "productivity." This appears to be an easy quantity to assess in a field study, but if we look at it more closely, we realize that "productivity" is actually composed of multiple stochastic processes and the estimation of each of them can be fairly challenging (Ricklefs and Bloom, 1977; Powell et al., 1999; Kendall and Wittmann, 2010; Etterson et al., 2011; Streby et al., 2014; Tenan et al., 2021). If we are interested in mean productivity only and not in the variability (i.e., ignoring demographic stochasticity), it is usually not necessary to estimate all these components for use in the population model. However, it is important to be aware of them in order to use in the population model an adequate measure of productivity and one that can be accurately estimated from your field data. There is a diversity of terms related to reproduction and we must be clear about their meaning; otherwise, they may be wrongly used in the population model. We first introduce one possible decomposition of productivity into the subprocesses involved and then sketch out methods to estimate them. This will not cover all possible facets of productivity. In addition, our illustration is inspired by avian productivity because this is what we know best, and other decomposition of "productivity" may be useful for other taxa. Nevertheless, we believe that the main principles should become clear.

We find it useful to envision three major components of the productivity of a population. The first is the proportion of females that actually reproduce in a given year. The second component is the number of offspring that a reproducing female produces per brood (or more generally per reproductive attempt). And the last component is the number of broods per female and year. The product of these three components gives the expected annual number of offspring produced per female. Let us now take a closer look at each of these components.

The proportion of reproducing females in a year typically depends on female age because younger females may not be mature yet and therefore fail to reproduce. Therefore, in population models we often include an age-specific probability of first reproduction to model the recruitment of individuals into the reproductive stages. This parameter can be estimated using multistate capture-recapture models (Section 4.5.2), when individuals marked as young (and therefore of known age) are recaptured when reproducing (Pradel et al., 1997b; Pradel and Lebreton, 1999). But even mature females may not all reproduce in every year. The probability with which a mature female reproduces is known as breeding propensity and can again be estimated with multistate or multievent capture-recapture models when sightings of reproducing and non-reproducing females are available (Converse et al., 2009; Souchay et al., 2014; Catlin et al., 2019).

The second major component of productivity is the number of offspring produced per reproducing female and brood. Importantly, it can be assessed at different stages (i.e., points of time) of the annual cycle (Fig. 4.9). For example, we can record the number of offspring at birth (e.g., as clutch or litter size), when they are ringed (e.g., as number of nestlings), when they leave the nest (e.g., as number of fledglings) or even the number of newly independent young. At the most extreme we may count the offspring at the time when they recruit into the breeding population, i.e., when they reproduce for the first time.

The number of offspring declines monotonically from birth to any of the later stages. The rate of this decline is governed by the survival probability in the processes involved, which for birds are known by period-specific names, such as hatching success, fledging success, or postfledging survival (Fig. 4.9). It is interesting to think about how the distribution of the number of offspring changes across successive stages. Clutch or litter size has no zeroes, or have you ever seen a clutch with no eggs? Then, the embryos in the eggs will start to develop, but some will die. In addition, a predator may find the nest and eat all the eggs. Both processes result in a shift of the initial distribution to the left and in an accumulation of zeroes. We have similar changes during the nestling period up to fledging; the distribution of the number of offspring is further shifted to the left and more zeroes are accumulated. This continues in an analogous manner until the young become independent or recruit into the breeding population. The resulting distributions can be rather complex (Fig. 4.9) and cannot easily be approximated by a single, named statistical distribution. At a minimum, one might need a zero-inflated discrete-valued distribution to account for the frequency of extra zeroes that have become accumulated over several stages in this compound process.

To estimate the number of offspring we must decide at which stage we want to sample the productivity data and understand the survival processes that have operated up to that point. We also must know whether the sampled data are complete with respect to the actual offspring distribution that the sequence of survival processes has created. Quite often

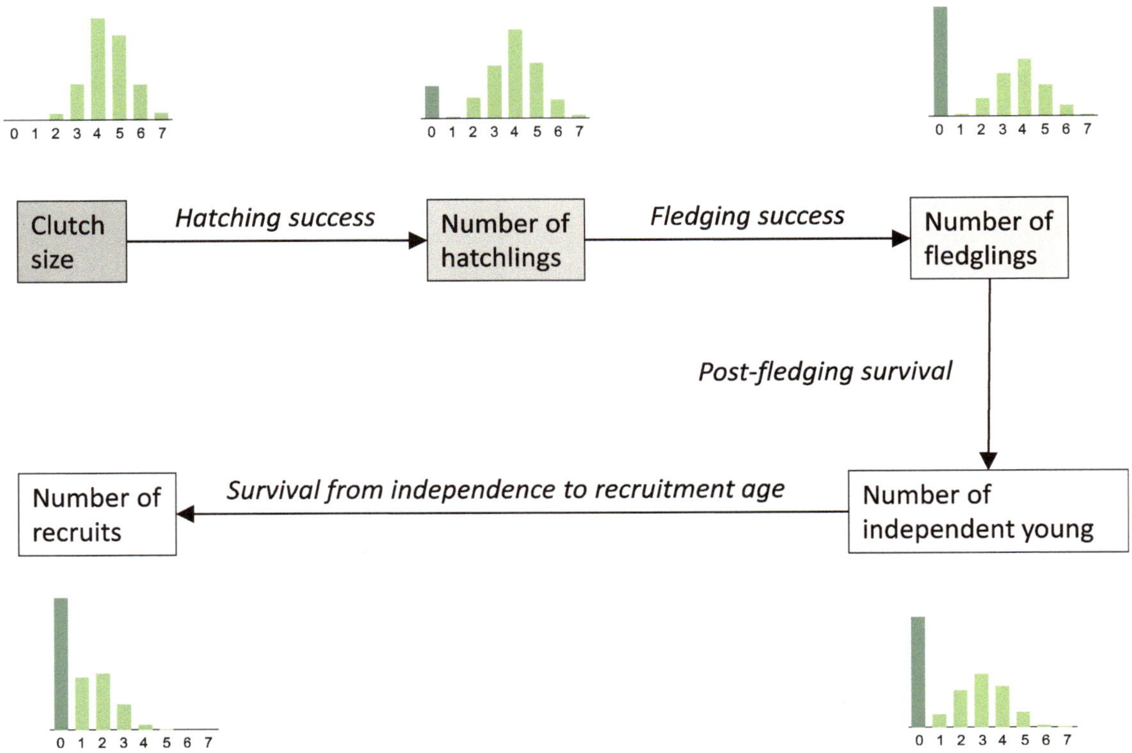

FIGURE 4.9

Changes in the frequency distributions of "productivity" when expressed at different chronological stages between laying of a clutch and recruitment of surviving offspring into the breeding population. There is a monotonic decline in the number of offspring and a continuous accumulation of zeroes over time. The boxes show the terminology at different stages, while the decay parameters (in italic) are all survival probabilities measured over periods of different length. For the hatching and fledging success, survival here includes individual survival in the nest and accounts for the effects of nest predation. Typical resulting distributions are shown for each stage, with the zeroes depicted in darker green. These zeroes are often not, or only incompletely observed, which means that the observed distributions may be zero-truncated or partially so. This schematic is drawn for birds but the same concept applies to many other taxa.

we lack some zeroes, e.g., because broods that failed early may fail to be observed. We also must decide which statistical distribution we use for the data analysis. The most accurate solution might be to model each process separately, but this would require sample data at each stage, rather than just a single one. For the number of offspring at any given point in time, there is generally no statistical distribution that always exactly matches the true distribution, and we must seek a good approximation. Potential candidates are the Poisson, with and without censoring and truncation, the normal distribution, as well as mixture distributions such as zero-inflated normal or Poisson. In Sections 4.4.1–4.4.3 we illustrate the use of Poisson distributions.

The third component of productivity is the number of reproduction attempts that a reproducing female performs in a year. Some bird species have multiple broods within a year, but the number of broods produced per female typically varies. When a brood fails, a female may produce a replacement clutch, but again, not every failure is followed by a replacement. The estimation of these probabilities usually requires that the female identity is known for a number of broods. Binomial models either with or without correction for imperfect detection can be adopted (Husby et al., 2009; Arnold et al., 2010; Hoffmann et al., 2015; Tenan et al., 2021). A different approach to estimate the annual number of broods per female is based on the frequency distribution of the laying dates and does not require female identities to be known (Cornulier et al., 2009; Robinson et al., 2014). The idea is that the observed distribution of the laying dates is a mixture of laying dates of first, second and possibly later broods. Adopting a two- or three-component mixture of normal or similar distributions can then permit estimation of the fraction of broods that are first and second broods and so on and hence derive the expected number of broods per female.

A decomposition of the overall productivity into its components is fairly commonplace in bird studies (Mattsson and Cooper, 2007; Kendall and Wittmann, 2010; Etterson et al., 2011; Oppel et al., 2013), but is sometimes done also for other vertebrates (Rivalan et al., 2005, 2006; Oosthuizen et al., 2019). Fig. 4.10 graphically summarizes a possible decomposition of productivity in birds starting from a mature female. The sequence of processes is governed by *breeding propensity* in a Bernoulli process, *nest success* (proportion of nests with ≥ 1 fledging) in another Bernoulli, the *number of fledglings* (sometimes referred to as brood size) in a Poisson or similar process (see above), renesting probability in a Bernoulli and probability to start a second brood in yet another Bernoulli process.

A different approach to estimate productivity is based on age ratios in the population after reproduction (Hanson, 1963; Flanders-Wanner et al., 2004; Zimmerman et al., 2010; Arnold, 2018; Specht and Arnold, 2018). This results in a measure of the mean productivity per female that integrates all the processes described above. Typical data sources for age ratios are mist-netting projects, and hunter surveys or counts of waterfowl at migration staging or wintering areas. To exploit age ratio data, it must be possible to age individuals at a distance or in the hand when captured or shot, and the sampling must be unbiased with respect to age. The latter can be difficult to achieve because individuals of different age classes may have different capture probabilities, may be differentially preferred by hunters, or may migrate at different times or to different places. If a biased sample is present, it may sometimes be possible to obtain a correction, e.g., by the inclusion of a recovery probability when age ratios of hunted individuals are used (Péron and Koons, 2012; Koons et al., 2017). Age ratios have been used several times in IPMs as the primary source of productivity data (e.g., Péron and Koons, 2012; Koons et al., 2017; Zhao et al., 2019).

In the subsequent sections we focus on estimation of two components of productivity. The first one is perhaps the most crucial, or the one that we always need when modeling productivity: the number of offspring produced per female and brood. The second is the estimation of the nest success, i.e., the probability that at least one fledgling is produced per initiated nest.

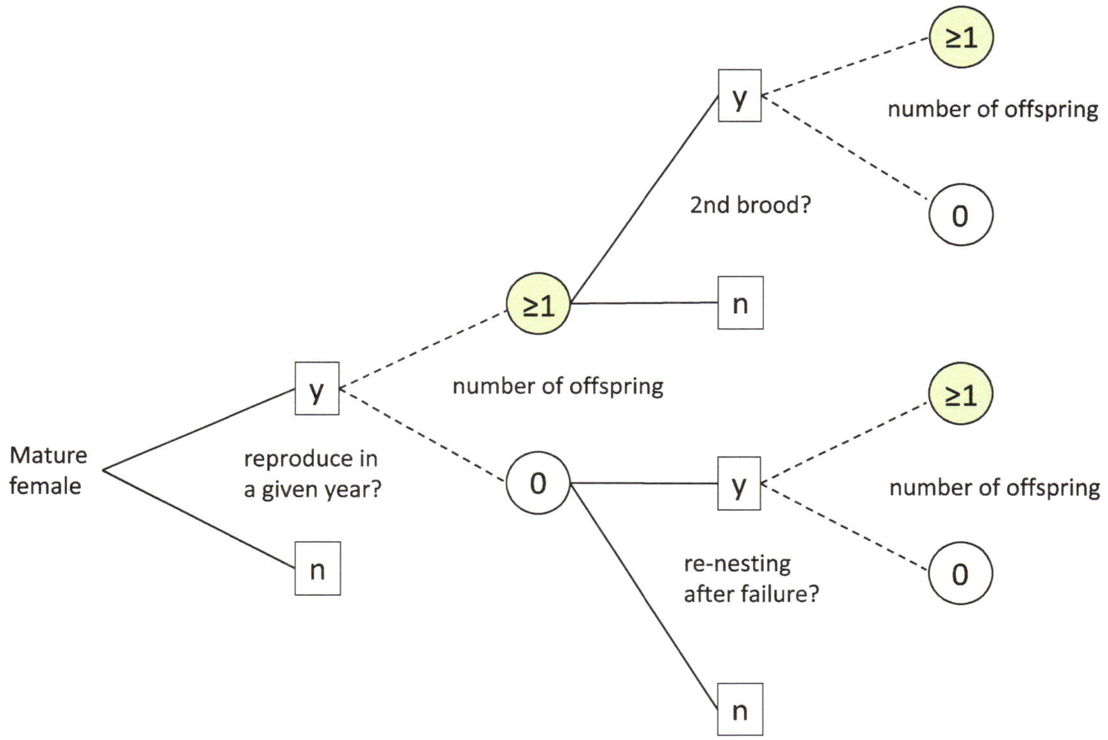

FIGURE 4.10

Possible processes for modeling the annual productivity of a mature female bird that may produce two clutches per breeding season. The solid lines can be modeled with Bernoulli or binomial models, the dashed lines with a sequence of binomial and Gaussian models or by zero-inflated Poisson models.

4.4.1 POISSON MODELS FOR BROOD SIZE DATA

The typical scenario is that we have some counts of young per brood, nest, or female, i.e., we observe data that result in a frequency distribution at some of the stages depicted in Fig. 4.9. As we have just discussed there are theoretical reasons to believe that there is probably no simple parametric statistical distribution that can exactly describe this frequency distribution and therefore our parametric choice will typically be an approximation only. The purpose of the following three sections is the illustration of possible approximations.

We start with a Poisson model for brood size. We simulate some data for illustration, but now assuming a Poisson random variable for brood counts and with some overdispersion due to brood effects, which we simulate as a Gaussian "residual" on the log-link scale.

```
# Choose constants in simulation
nbrood <- 1000            # Number of broods with young counted
brood.mean <- 1.5         # Average brood size
sd.brood <- 0.3           # log-linear brood random effect

# Draw Poisson random numbers
set.seed(24)
expNyoung <- exp(log(brood.mean) + rnorm(nbrood, 0, sd.brood))
C <- rpois(nbrood, expNyoung)
table(C)

C
  0   1   2   3   4   5   6   7   8
231 330 237 134  39  16   7   3   3
```

When faced with productivity data like these, a typical approach in an IPM is to just fit a Poisson GLM, either to this disaggregated format of the data, or to the aggregated format where we model the sum of all young and specify the known number of broods as an offset. Sometimes, we may also specify in the model some random effects resulting in a GLMM; for a zero-inflated Poisson see Section 4.4.2. Here, we quickly illustrate all three approaches and for the GLMM simply adopt the same random brood effects with which we simulated the data. Hence, the two Poisson GLMs are misspecified in the sense that they ignore the overdispersion at the brood level, while the GLMM is the data-generating model. In the mean of all three models, we simply fit an intercept, corresponding to the average number of young produced per brood. In real life you may also have covariates such as age of the mother or other individual or environmental effects, but this simply means adding a linear model in the link-scaled mean response.

We can fit completely separate models inside the same JAGS model statement. We do this here for simplicity and specify three responses:

- The disaggregated data as simulated to which we fit a Poisson GLM.
- The aggregated data to which we also fit a Poisson GLM that is the identical model and should result in the same estimates.
- A copy of the disaggregated data to which we fit a Poisson log-normal model, which is a GLMM because we add in a Gaussian brood random effect.

We call the count of young `C` and when fitting two models to the same data set in a single JAGS model statement we need to make a copy of that data set.

```
# Data bundle
jags.data <- list(C1=C, sumC1=sum(C), C1copy=C, nbrood=nbrood)
str(jags.data)

List of 4
 $ C1    : int [1:1000] 0 0 1 1 4 1 0 2 0 1 ...
 $ sumC1 : int 1529
 $ C1copy: int [1:1000] 0 0 1 1 4 1 0 2 0 1 ...
 $ nbrood: num 1000
```

```
# Write JAGS model file
cat(file="model8.txt", "
model {
# Priors and linear models
rho1 ~ dunif(0, 5)                          # Mean brood size in model 1
rho2 ~ dunif(0, 5)                          # Mean brood size in model 2
rho3 ~ dunif(0, 5)                          # Mean brood size in model 3
tau.rho3 <- pow(sd.rho3, -2)
sd.rho3 ~ dunif(0, 3)                       # Brood-level overdispersion in model 3

# Likelihoods for three separate models
# Model 1: Poisson GLM for disaggregated data
for (i in 1:nbrood){
  C1[i] ~ dpois(rho1)
}

# Model 2: Poisson GLM for aggregated data
sumC1 ~ dpois(rho2 * nbrood)

# Model 3: Poisson GLMM for aggregated data with brood-level overdispersion
for (i in 1:nbrood){
  C1copy[i] ~ dpois(pois.mean[i])
  log(pois.mean[i]) <- logmean[i]
  logmean[i] ~ dnorm(log(rho3), tau.rho3)
}
}
")

# Initial values
inits <- function(){list(rho1=runif(1, 0.5, 2.5), rho2=runif(1, 0.5, 2.5),
  rho3=runif(1, 0.5, 2.5))}

# Parameters monitored
parameters <- c("rho1", "rho2", "rho3", "sd.rho3")

# MCMC settings
ni <- 60000; nb <- 10000; nc <- 3; nt <- 10; na <- 1000

# Call JAGS from R (ART 3 min), check convergence and summarize posteriors
out11 <- jags(jags.data, inits, parameters, "model8.txt", n.iter=ni, n.burnin=nb, n.chains=nc,
  n.thin=nt, n.adapt=na, parallel=TRUE)
traceplot(out11)                            # Not shown
print(out11, 3)

           mean    sd   2.5%    50%  97.5% overlap0 f  Rhat n.eff
rho1      1.530 0.039  1.454  1.529  1.609    FALSE 1 1.000 15000
rho2      1.530 0.039  1.453  1.530  1.606    FALSE 1 1.000 15000
rho3      1.464 0.047  1.372  1.464  1.560    FALSE 1 1.002  1145
sd.rho3   0.292 0.061  0.138  0.297  0.393    FALSE 1 1.031   156
```

Up to MC error, the estimates under models 1 and 2, Poisson GLMs for the respective disaggregated and aggregated data are identical, while the point estimate under the data-generating model, the Poisson GLMM with brood-level overdispersion, is slightly different, but its estimates are similar to the data-generating values of 1.5 and 0.3.

Some variant of Poisson GLM or GLMM is by far the most common model adopted for productivity data that come in the form of a count of young. For some species with only a single young per brood such as vultures, albatrosses or whales, a Bernoulli or binomial GLM would be adopted instead, where the former is for the disaggregated and the latter for the aggregated format of the data. Multistate models can also be useful for modeling productivity especially of such species that have small brood sizes (large variability in brood size will increase the number of states that are required and may make use of a multistate model too cumbersome).

In what follows, we deal with some specialties that are particularly relevant for counts in the context of the modeling of productivity data: zero inflation, zero truncation, right-censoring, and underdispersion. In our case, truncation and censoring are both features of the observation process underlying a productivity data set. They must be accommodated in the model to prevent biased estimates as we will see. Hence, imperfect detection is not the only important feature of the observation processes that produce data in population ecology that may require proper modeling. Truncation and censoring can be confusing sometimes, but the difference between the two is, to paraphrase Donald Rumsfeld, that between an unknown and a known unknown. Under an observation process producing truncated data, some values of the response cannot be observed at all, and hence, in a sense, they are unknown unknowns because we do not know how many and what they are. In contrast, when a response is censored, we observe all the data, but some are not observed "quite right" as we will see. In this sense, they are known unknowns because we have them all but are just unsure about what they are.

4.4.2 ZERO INFLATION IN BROOD SIZE DATA

We have seen that the sequence of survival processes involved in "productivity" leads to the accumulation of zeroes in the frequency distribution of offspring number data. Thus, you may well find that the frequency of zeroes is not well described by a simple Poisson distribution, even perhaps if you account for brood-level overdispersion as in the Poisson GLMM in the previous section. Therefore, we here illustrate one possible solution to this, which is the adoption of a zero-inflated distribution that can be constructed hierarchically as the combination a Bernoulli distribution with another distribution, which we take to be a Poisson again (though it might also be a normal, see Chapter 11). The Bernoulli then describes a first failure process, and the conditional Poisson describes brood size for those broods in the "success" group from the first process. This construction allows two sorts of zeroes: those coming from the initial Bernoulli process and those coming from the Poisson process. This might be seen as a little strange and an alternative would be adoption of a hurdle distribution, where the second distribution is zero-truncated so that all brood failures can only be ascribed to the Bernoulli process (Potts and Elith, 2006). In practice adoption of a zero-inflated distribution often has no physical basis, i.e., it does not usually accurately represent the processes producing a zero-inflated response. Yet if we are sure that our data are free of observation errors and contain all occurring zeroes, the probability of generating a zero may have a biological meaning (e.g., nest failure rate). In many cases, however, adoption of a zero-inflated distribution is usually a quick and dirty solution to excess zeroes. Hence, for this descriptive goal we would not worry about the choice between a zero-inflated and the hurdle-variants of a Poisson with excess zeroes and here simply demonstrate the former. So we simulate a zero-inflated Poisson (ZIP) variant of our data and then fit the data-generating model to them. We choose the success probability to be 0.7, which means that we will generate about 300 additional zeroes when simulating 1000 broods. Other than that, we repeat everything from the previous section, i.e., we also add in some extra-Poisson variation. This means that in fact we simulate the data under the parametric assumptions of a ZIP log-normal GLMM.

```
# Choose constants in simulation
nbrood <- 1000                    # Number of broods with young counted
theta <- 0.7                      # Success probability in Bernoulli process
brood.mean <- 1.5                 # Average brood size in conditional Poisson process
sd.brood <- 0.3                   # Overdispersion in conditional Poisson process

# Simulate Bernoulli process dividing broods in failures and 'potential successes'
set.seed(46)
z <- rbinom(nbrood, 1, theta)    # z = 1 means 'potential success'

# Draw conditional Poisson random numbers with overdispersion
expNyoung <- z * exp(log(brood.mean) + rnorm(nbrood, 0, sd.brood))
Cx <- rpois(nbrood, expNyoung)
table(Cx)

Cx
  0   1   2   3   4   5   6   7   8
466 234 167  76  31  14   9   1   2
```

This is only about 230 additional zeroes compared with our Poisson before, but of course we have a proportional reduction in the numbers of zeroes coming from the Poisson now. To fit a ZIP in BUGS is really trivial once you have seen the manner in which we create these data.

```
# Data bundle
jags.data <- list(C=Cx, nbrood=nbrood)
str(jags.data)

List of 2
 $ C     : int [1:1000] 0 1 1 0 4 1 0 1 0 0 ...
 $ nbrood: num 1000
```

```
# Write JAGS model file
cat(file="model9.txt", "
model {
# Priors and linear models
theta ~ dunif(0, 1)                          # Success probabiliy in Bernoulli process
rho ~ dunif(0, 5)                            # Mean brood size in conditional Poisson
tau.rho <- pow(sd.rho, -2)
sd.rho ~ dunif(0, 3)                         # Overdispersion in conditional Poisson

# Likelihood
# Note this is a zero-inflated Poisson log-normal GLMM .... cool !
for (i in 1:nbrood){
  z[i] ~ dbern(theta)                        # Zero-inflation process
  C[i] ~ dpois(z[i] * cond.pois.mean[i])     # Conditional Poisson
  log(cond.pois.mean[i]) <- log.condmean[i]
  log.condmean[i] ~ dnorm(log(rho), tau.rho) # Overdispersion

}
}
")

# Initial values
inits <- function(){list(z=rep(1, nbrood), rho=runif(1, 0.5, 2.5))}

# Parameters monitored
parameters <- c("theta", "rho", "sd.rho")

# MCMC settings
ni <- 110000; nb <- 10000; nc <- 3; nt <- 100; na <- 1000

# Call JAGS from R (ART 6 min) and check convergence
out12 <- jags(jags.data, inits, parameters, "model9.txt", n.iter=ni, n.burnin=nb, n.chains=nc,
  n.thin=nt, n.adapt=na, parallel=TRUE)
traceplot(out12)                             # Not shown
print(out12, 3)

            mean     sd     2.5%      50%     97.5% overlap0 f  Rhat n.eff
theta      0.748  0.039    0.676    0.745     0.832    FALSE 1 1.000  3000
rho        1.297  0.111    1.077    1.296     1.515    FALSE 1 1.000  3000
sd.rho     0.443  0.081    0.282    0.443     0.604    FALSE 1 1.002  1053
```

These estimates look not entirely bad, though they do not show a very striking similarity with the truth in the data simulation (i.e., 0.7, 1.5 and 0.3). We therefore conducted a quick simulation with 100 replicated data sets and varying values of the truth and found results that suggested the estimators were unbiased. Hence, if you need to accommodate zero inflation in a distribution, you see how easy this is when constructed in a hierarchical manner.

4.4.3 ZERO TRUNCATION IN BROOD SIZE DATA

In many raptor population surveys, volunteers repeatedly visit known territories and try to count the young in the nest from a distance that may be up to several kilometers away. This can be much more challenging than counting chicks in a nest box. Different processes may add errors and noise in the resulting data that, unless accounted for in the model, may lead to biased assessments of productivity. One of the challenges is partial zero truncation of the counts of young. This may arise as follows. When a brood fails, we do not see any young (anymore). However, the same observation of zero young may also result when a nest was visited too late in the season. Then, the young have simply fledged and dispersed in the surroundings or were inactive and invisible in the vicinity of the nest during the visit. Faced with this uncertainty, a conscientious observer might therefore decide to simply record nothing, i.e., a missing value. The result of this then is that zeroes are significantly underrepresented among the observed counts of young. That is, we have partial zero truncation and an analysis ignoring this would overestimate productivity because we "lack some zeroes." When such a problem is suspected in a productivity data set, one solution is to discard *all* zero observations and simply fit a zero-truncated variant of the original model. We illustrate this here with a variant of our simulated data set and show that proper accommodation of zero truncation enables us to recover improved estimates of the mean productivity, while ignoring zero truncation leads to overestimation of the mean productivity in the population, i.e., we will then estimate the mean number of offspring per *successful* brood.

First, we make a copy of our counts of young and then assume there was uncertainty about the true fate of a brood in 60% of all brood failures. That is, we drop 60% of all zero brood sizes from our data set (and achieve this by turning some zeroes into NAs).

```
# Create a variant of the data with partial zero truncation
set.seed(68)
C1 <- C                                # Copy data set
table(C1)                              # Remind ourselves of the actual data
prop.uncertain <- 0.6                  # Proportion of zeros dropped
zeros <- which(C1==0)
toss <- zeros[rbinom(n=length(zeros), size=1, prob=prop.uncertain) == 1]
C1 <- C1[-toss]                        # Toss out some of the zeros
table(C1)                              # Frequency dist. of new brood size data

C1
  0   1   2   3   4   5   6   7   8
 89 330 237 134  39  16   7   3   3

mean(C); mean(C1)                      # Compare sample means
[1] 1.529
[1] 1.782051
```

So we went from 231 down 89 zeroes in the new data set and the sample mean in the partially zero-truncated version of the data (`C1`) is too high because brood failures are now systematically underrepresented. We will again fit two models to this data set or a variant thereof, again within a single JAGS model statement. In the first model, we take the data set `C1` with partial zero truncation and fit a Poisson distribution, while in the second, we toss out all remaining zeroes and fit a GLM with zero-truncated Poisson.

```
# Make another copy of C1 data and kick out remaining zeroes as well
C2 <- C1[C1 > 0]                       # Make a copy and toss out remaining zeroes

# Data bundle
jags.data <- list(C1=C1, C2=C2, nC1=length(C1), nC2=length(C2))
str(jags.data)

List of 4
 $ C1 : int [1:858] 0 0 1 1 4 1 2 0 1 3 ...
 $ C2 : int [1:769] 1 1 4 1 2 1 3 1 1 1 ...
 $ nC1: int 858
 $ nC2: int 769

# Write JAGS model file
cat(file="model10.txt", "
model {
# Priors and linear models
rho1 ~ dunif(0, 3)                     # Expected brood size in model 1
rho2 ~ dunif(0, 3)                     # Expected brood size in model 2

# Likelihoods
# Model 1: Standard Poisson GLM
for (i in 1:nC1){
  C1[i] ~ dpois(rho1)
}

# Model 2: Poisson GLM with response truncated at one (i.e., no zeroes)
for (i in 1:nC2){
  C2[i] ~ dpois(rho2)T(1,)
}
}
")
```

```
# Initial values
inits <- function(){list(rho1=runif(1, 0.5, 2.5), rho2=runif(1, 0.5, 2.5))}

# Parameters monitored
parameters <- c("rho1", "rho2")

# MCMC settings
ni <- 60000; nb <- 10000; nc <- 3; nt <- 10; na <- 1000

# Call JAGS from R (ART 3 min), check convergence and summarize posteriors
out13 <- jags(jags.data, inits, parameters, "model10.txt", n.iter=ni, n.burnin=nb, n.chains=nc,
  n.thin=nt, n.adapt=na, parallel=TRUE)
traceplot(out13)                    # Not shown
print(out13, 3)

              mean    sd    2.5%     50%   97.5% overlap0 f Rhat n.eff
rho1         1.784 0.046   1.694   1.784   1.874    FALSE 1    1  9064
rho2         1.580 0.053   1.481   1.579   1.685    FALSE 1    1  5698
```

We see in Fig. 4.11 that dropping all remaining zeroes and fitting a Poisson GLM with a zero-truncated response considerably reduces the error in the estimated mean productivity in the population. Under this model we can even estimate the number of zeroes expected under the standard Poisson as follows and compare it with the true number in the original data and with that in the version of the data set that is corrupted by partial zero truncation.

```
# Number of zeroes in data set corrupted by partial zero-truncation
sum(C1==0)
[1] 89

# Expected number of zeroes from model accommodating zero-truncation
round(nbrood * dpois(0, out13$mean$rho2))
[1] 206

# True number of zeroes in the original data set C
sum(C==0)
[1] 231
```

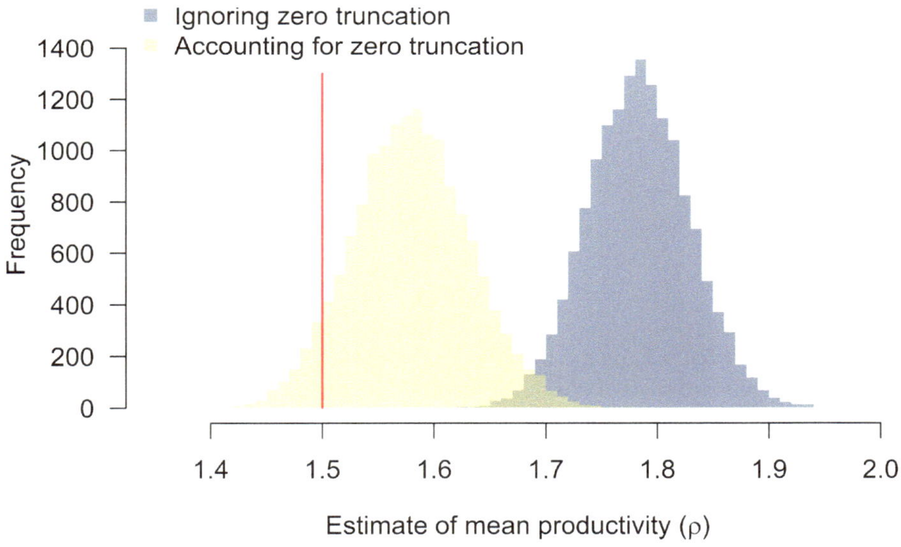

FIGURE 4.11

Estimates (posterior distributions) of mean productivity in a standard Poisson GLM (ignoring truncation) fit to a partially zero-truncated response and in a Poisson GLM with zero truncation fit to a variant of the data set where all zeroes were discarded. Vertical red line shows the truth of 1.5.

Thus, by adopting the zero-truncated Poisson likelihood we do a reasonable job at recovering the (partially) missing zeroes. How well this method performs and how much better it does than the naive approach of ignoring partial truncation will depend on the size of the data set relative to the complexity of the model, the proportion of zeroes dropped out (i.e., the degree of partial zero truncation) and on how well the data follow the assumed parametric distribution.

When zeroes are excluded from the data and analyzed with a simple Poisson or normal model, we obtain an estimate of the mean number of offspring *per successful brood*, rather than for the population as a whole. This parameter is often reported in bird studies. In an IPM it is only useful if in addition, we can estimate the probability that a nest is successful (nest success), which we show in Section 4.4.6.

4.4.4 CENSORING IN BROOD SIZE DATA

In the example of truncated brood size above, brood sizes of zero (i.e., failures) did occur, but they were never observed or only some were. In contrast, with a censored response we will obtain all the data, but for some (or even all) we will not know their precise values. Censoring is most well-known in event-history analyses such as continuous time survival models, where we distinguish between left-, right-, and interval-censoring. This denoted cases where an event such as death happens before the start or after the end of a study, or where it is only known to have occurred during some time interval, respectively (Landes et al., 2020). In the context of productivity data, we are faced with right-censoring when some brood counts X have a meaning of "X young or more."

As a motivating example we imagine again a population survey of a large raptor where young in the nest must be counted from a distance or in some cases even after fledging. It is easy then to overlook one or even several young. Hence, seeing, say, two fledged young peregrines may often mean that three or four have actually fledged. If we take such right-censored data in our productivity model at face value, we will underestimate productivity. A better approach is to accommodate such censoring in the model. Often the observer is able to judge whether her count of young is likely accurate or would be better considered as censored at the observed value. This is important information that must be recorded in the field and then entered in a database.

In this section we illustrate two examples of such censoring. In the first, we assume that some proportion of our 1000 brood sizes are known to be right-censored and then account for this in the model. In the second case we assume that for some broods we can only distinguish the failure or success of a nest. This is an extreme case of right-censoring, where all we know is that brood size was either 0, or greater than 0. We show how we can easily fit a joint model to both portions of the data set. In our first example, let us assume that 300 among the 439 "larger" broods of two or more were censored on the right, which we simulate by subtracting 1 young from the true value.

```
# Simulate right-censored Poisson observations
set.seed(1)
C3 <- C                                                 # Make another copy of the data set
sum(C3 >= 2)                                            # How many broods 2 or greater ? -- 439
large.broods <- which(C3>=2)                            # Index them
censored.broods <- sample(large.broods, 300)            # Randomly select 300
C3[censored.broods] <- C3[censored.broods] - 1          # Censor these 300
mean(C)                                                 # True mean
mean(C3)                                                # Mean of C3 biased low
[1] 1.529
[1] 1.229
```

We follow Link and Barker (2010, p. 241−243) and model the censored observations by replacing them with a missing value. At the same time, we introduce a known censoring indicator d with a value of 1 for a censored and 0 for an uncensored observation. The original and partly censored brood size is treated as the censoring threshold, i.e., it becomes a covariate needed to calculate the value of the binary censoring indicator. This calculation uses the step function that returns a value of 1 if its argument is nonnegative and 0 if it is negative. The argument of the step function is the difference between the observed or estimated brood size and the value of the censoring threshold. We need to prepare these data first.

```
C4 <- C3                                                # Make another copy
C4[censored.broods] <- NA                               # Censored are NA'd out
d <- as.numeric(is.na(C4))                              # Binary censoring indicator
```

The core of the model in JAGS (below) will be this:

```
for (i in 1:nbrood){
  C4[i] ~ dpois(rho)                                    # These are the data
  prob.censored[i] <- step(C4[i] - cens.threshold[i])   # Prob of censoring
  d[i] ~ dbern(prob.censored[i])                        # Censoring indicator is response
}
```

Hence, to accommodate right-censoring, we specify a Poisson density for all brood sizes, regardless of whether they are uncensored (and therefore are nonmissing in `C4`) or whether they are censored and represented as NAs in `C4`. Then, the ultimate response in the model is the censoring indicator `d`, which is equal to 1 when we know a response to be censored and equal to 0 when it is not. The brood size data (`C4`) and the censoring indicator (`d`) are linked by the probability of censoring (`prob.censored`), which we define to be a step function of the data `C4` (where censored values are set to NA) and the censoring threshold. We must choose the latter in such a way that we get the correct binary values of `prob.censored`: 1 and 0 for censored and uncensored values of brood size, respectively. We achieve this by choosing the threshold equal to the observed value in `C3` for the values that are censored (and NAs in `C4`), but equal to the observed value plus some small value for the noncensored values in `C3`.

```
# Compute censoring threshold
# Equal to observed value for censored
# Equal to observed value plus some small value for non-censored
cens.threshold <- C3
cens.threshold[d==0] <- C3[d==0] + 0.1
cbind('Truth'=C, 'Obs w/cens'=C3, 'Modelled data (C4)'=C4, 'Cens. indicator (d)'=d, cens.threshold)[1:6,]
                                                        # Look at first 6 rows

     Truth Obs w/cens Modelled data (C4) Cens. indicator (d) cens.threshold
[1,]     0          0                  0                   0            0.1
[2,]     0          0                  0                   0            0.1
[3,]     1          1                  1                   0            1.1
[4,]     1          1                  1                   0            1.1
[5,]     4          3                 NA                   1            3.0
[6,]     1          1                  1                   0            1.1
```

To better understand this, start by considering the first brood, which has size 0 and is not censored. With a censoring threshold set at a value just slightly greater than the observed value of 0, we get `step(0 - 0.1) = step(-0.1) = 0`, and a Bernoulli trial with this success probability produces a value of the censoring indicator (`d`) equal to 0, as required. Next, look at the first censored value, which is for brood 5, which in reality had 4 young, but in our observations (in `C3`) is censored at 3. Here, we choose a censoring threshold of 3, so we get `step(C4[5] - 3)`. Of course, the value of `C4[5]` is not known and must be estimated, but the censoring indicator is equal to 1, so the censoring probability as defined will have to be 1, and in this way, the information is passed to JAGS that the true value of `C4[5]` must be equal to or greater than 3.

```
# Data bundle
jags.data <- list(C4=C4, d=d, cens.threshold=cens.threshold, nbrood=nbrood)
str(jags.data)

List of 4
 $ C4            : num [1:1000] 0 0 1 1 NA 1 0 NA 0 1 ...
 $ d             : num [1:1000] 0 0 0 0 1 0 0 1 0 0 ...
 $ cens.threshold: num [1:1000] 0.1 0.1 1.1 1.1 3 1.1 0.1 1 0.1 1.1 ...
 $ nbrood        : num 1000


# Write JAGS model file
cat(file="model11.txt", "
model {
# Prior
rho ~ dunif(0, 5)

# Likelihood
for (i in 1:nbrood){
  C4[i] ~ dpois(rho)
  prob.censored[i] <- step(C4[i] - cens.threshold[i])
  d[i] ~ dbern(prob.censored[i])
}
}
")

# Initial values
C4st <- rep(NA, nbrood)
C4st[is.na(C4)] <- C3[is.na(C4)]
inits <- function() list(rho=runif(1), C4=C4st)
```

We need to specify initial values for the missing elements in the response variable, otherwise JAGS throws the dreaded *"Node inconsistent with parents"* error.

```
# Parameters monitored
parameters <- c("rho", "C4", "prob.censored")

# MCMC settings
ni <- 2000; nb <- 1000; nc <- 3; nt <- 2; na <- 1000

# Call JAGS from R (ART <1 min), check convergence and summarize posteriors
out14 <- jags(jags.data, inits, parameters, "model11.txt", n.iter=ni, n.burnin=nb, n.chains=nc,
  n.thin=nt, n.adapt=na, parallel=TRUE)
traceplot(out14)                          # Not shown
print(out14, 3)

                        mean     sd    2.5%     50%   97.5% overlap0 f  Rhat n.eff
rho                    1.446  0.042   1.363   1.445   1.531    FALSE 1 1.001  1500
C4[1]                  0.000  0.000   0.000   0.000   0.000    FALSE 1    NA     1
C4[2]                  0.000  0.000   0.000   0.000   0.000    FALSE 1    NA     1
C4[3]                  1.000  0.000   1.000   1.000   1.000    FALSE 1    NA     1
C4[4]                  1.000  0.000   1.000   1.000   1.000    FALSE 1    NA     1
C4[5]                  3.459  0.733   3.000   3.000   5.000    FALSE 1 1.000  1500
C4[6]                  1.000  0.000   1.000   1.000   1.000    FALSE 1    NA     1
[... output truncated ...]

# Remind ourselves of mean observed brood size ignoring censoring
mean(C3)
[1] 1.229
```

So in the observed data, where 300 observations were censored, the mean was 1.229, but in our model, we estimate mean productivity at 1.444, much closer to the known truth of 1.5, which furthermore is included in the 95% CRI. We also get estimates of the censored values. For instance, brood 5 was censored at 3 (and because we simulated these data, we know it actually had 4 young), and our estimate is 3.459. This is not perfect but an improvement over the erroneous use of 3 in a model and likely over an approach that would simply discard all censored data.

Modeling censored observations works quite well, provided that the parametric model is adequate, there are not too many censored observations, that the "stunted" part of the abundance distribution is not too large, and that we know which observation is truncated. Better results still are obtained if we have one or several covariates that affect the responses: if their value is observed for the censored response, this will improve our estimates. We believe that censoring occurs quite frequently in population studies (see Chapter 11) but is often either swept under the rug or else the censored data are dropped from an analysis. We find it remarkable how we can improve on either approach by a mere two additional lines of code in JAGS.

In the second part of this section, we turn to a special case of censoring where we assume that for some broods, we can merely distinguish a failure from a success. Provided that this is a random sample of all broods, we can easily combine these data in a joint analysis of brood size, as we show now.

```
C5 <- C                                   # Make another copy of the data
# Assume last 200 broods only known to be a success or failure
C5part1 <- C5[1:800]                      # The unaltered counts
C5part2 <- as.numeric(C5[801:1000]>0)     # Success/failure indicators
```

It would not be wrong to only use part 1 of the data in your productivity model, but this approach would be inefficient because you would discard the information about productivity contained in part 2 of the data. Intuitively, these must also contain some information about brood size, even if it comes with much reduced "resolution." The solution to using the information in both parts of the data is to specify for part 2 a binary regression with success probability equal to the probability to produce a value greater than 0 under the estimated Poisson distribution. The probability of a zero under a Poisson distribution is simply $\exp(-\rho)$, so the probability of a value greater than zero is 1 minus that. Alternatively, we could also specify a cloglog link that achieves exactly the same: it allows us to describe a binary response in terms of an underlying Poisson process that is imperfectly observed such that we are only able to distinguish values of zero from values greater than zero.

We next fit again two independent models inside one JAGS model statement. The first is the Poisson regression for part 1 of the data. The second is the integrated model, where we merge the information in parts 1 and 2 by assuming the same underlying process (i.e., the same parameter `rho`), but a different observation model (for such more general integrated models see Chapter 10 in Kéry and Royle, 2021). The integrated model uses the information about brood size contained in all 1000 observations rather than only that in the first 800, as does the first model.

```
# Data bundle
jags.data <- list(C5part1copy=C5part1, C5part1=C5part1, C5part2=C5part2)
str(jags.data)

List of 3
 $ C5part1copy: int [1:800] 0 0 1 1 4 1 0 2 0 1 ...
 $ C5part1    : int [1:800] 0 0 1 1 4 1 0 2 0 1 ...
 $ C5part2    : num [1:200] 0 1 0 1 1 1 1 1 0 1 ...

# Write JAGS model file
cat(file="model12.txt", "
model {
# Priors
rho1 ~ dunif(0, 5)                        # For model part 1
rho2 ~ dunif(0, 5)                        # For model part 2

# Likelihood
# First model: Poisson GLM for C5 (part 1)
for (i in 1:800){
  C5part1copy[i] ~ dpois(rho1)
}

# Second, integrated model: Poisson GLM for C5 (part 1) plus a Bernoulli GLM for C5 (part 2) with
#   appropriate definition of the success probability as a function of the same rho
for (i in 1:800){                         # Submodel for part 1 of the data
  C5part1[i] ~ dpois(rho2)
}
for (i in 1:200){                         # Submodel for part 2 of the data
  C5part2[i] ~ dbern(1 - exp(-rho2))
}
}
")

# Initial values
inits <- function() list(rho1=runif(1), rho2=runif(1))

# Parameters monitored
parameters <- c("rho1", "rho2")

# MCMC settings
ni <- 5000; nb <- 1000; nc <- 3; nt <- 4; na <- 1000

# Call JAGS from R (ART <1 min), check convergence and summarize posteriors
out15 <- jags(jags.data, inits, parameters, "model12.txt", n.iter=ni, n.burnin=nb, n.chains=nc,
  n.thin=nt, n.adapt=na, parallel=TRUE)
traceplot(out15)                          # Not shown
print(out15, 3)

            mean    sd   2.5%    50%  97.5% overlap0 f  Rhat n.eff
rho1       1.545 0.045  1.461  1.545  1.638    FALSE 1 1.001  1173
rho2       1.526 0.041  1.445  1.526  1.608    FALSE 1 1.001  1053
```

Thus, in the integrated model, where we use the data from all 1000 broods, our estimate of mean brood size is improved in terms of both the mean and the SD over that from the simple Poisson regression only for those 800 broods with the accurate counts. Whether these patterns are systematic or just arose by chance in this particular data set would of course again have to be studied by simulation with replicated data sets, i.e., in a proper simulation. We dare guessing that the pattern is systematic.

4.4.5 UNDERDISPERSION

At this stage, it may perhaps come as a surprise to you when we tell you that brood sizes are very commonly underdispersed relative to a Poisson distribution. That is, in real life they have fewer large values than what a Poisson with given mean lets us expect. For instance, the Poisson we have used for illustration purposes in this section had mean 1.5, which approximately corresponds to the average brood size of an eagle owl (*Bubo bubo*). In our example data set we had values up to 8, but real eagle owls never produce more than 4 young. So does this mean that all the material in this section is obsolete? Obviously, we do not think so. The main reason the Poisson is still often a good starting point for a model for many productivity data sets is the relative rarity of values that are "too big" for the chosen Poisson. For instance, in our example Poisson, fewer than 2% of the broods would be too big for an eagle owl (to see this, do `1-sum(dpois(0:4, 1.5))` in R). The most likely result of adopting a Poisson distribution that is overdispersed with respect to the real data is that you overstate the uncertainty a bit.

However, there are a number of possible alternatives also. For instance, you can use a variant of a Poisson that can accommodate underdispersion, such as a Conway-Maxwell Poisson or the generalized Poisson (Ridout and Besbeas, 2004; Lynch et al., 2014; Brooks et al., 2019). Quite commonly, one still sees a normal distribution being used for brood size, which has the variance as a parameter to control its spread and hence is more flexible than a standard Poisson. It may be slightly embarrassing to use it for a discrete response, but in practice this is unlikely to do much harm especially when there are more than about 5 observed classes. Ridout and Besbeas (2004) have described an approach specifically for under-dispersed productivity data; see also Robinson et al. (2012) for an application of their approach using JAGS. Finally, sometimes one might even use a simple multinomial distribution, which amounts to the estimation of the proportions of each brood size. However, this may become awkward or impossible if you also want to model effects of individual covariates in the mean, or need the results from the fitted model for downstream analyses such as life-table response experiments (see Chapter 9).

Overall, it is important that you are aware of this interesting fact about brood size, and it may be intriguing to hypothesize about the biological mechanisms underlying this fairly general pattern of underdispersion. However, we do not think that the main modeling objectives in an IPM project are very sensitive to your parametric choices in the productivity part of the model. The adoption of demographic stochasticity of productivity in the population model also requires a statistical distribution (see Section 3.4.4). We expect that the choice of this distribution has an effect on the estimates at low population size, but the optimal treatment of demographic stochasticity in an IPM is an as yet open question.

4.4.6 NEST SURVIVAL MODELS

The primary goal of nest survival models is the estimation of nest success, which is the probability that at least one fledgling is produced by an initiated nest. Nest failure may be due to predation of the nest or of one or both parents, parasitism by another species or starvation of the nestlings if they are not fed enough. Nest success is composed of the successes during the incubation and the nestling phases, and the processes governing the successes may be different. Therefore, the two components of overall nest success are often modeled separately.

Data about nest survival are collected from active nests that must be visited repeatedly. Time intervals between visits do not have to be identical and may be more than 1 day. The status of the nest is coded as "alive" or "dead" and is recorded at each visit. A nest that is alive means that it contains at least one egg or one nestling that is alive and that it is not abandoned by the parents.

If nest success is expressed by the proportion of successful nests in our sample, it will often be positively biased because nests that failed at an early stage are less likely to appear in our sample (i.e., to be detected at all) than nests that failed late or are successful. Hence, it is difficult to obtain a sample of nests that is random with respect to fate. This problem has been recognized for a long time and Harold Mayfield was the first to formalize a solution by developing models that can yield unbiased estimates of nest success (Mayfield, 1961, 1975). The key idea is that daily survival is estimated from the time when a nest is detected and enters the studied sample, and nest success is then derived as the product of the daily nest survival probabilities over the entire period, which may be incubation only or also include the nestling period. The original models assumed constant daily nest survival probabilities and that nest failure had occurred at the midpoint between the date when the nest was last recorded alive and the date when failure was first detected. Later, models were improved to relax these assumptions and to allow daily survival to depend on the age of a nest and on temporal covariates (Johnson, 1979; Hensler and Nichols, 1981; Aebischer, 1999; Dinsmore et al., 2002; Rotella et al., 2004). Furthermore, the inclusion of nest random effects allows to accommodate nonindependence of nest survival due to individual or site effects (He, 2003; Cao et al., 2008, 2009; Royle and Dorazio, 2008; Schmidt et al., 2010; Converse et al., 2013).

The data collected for a nest survival study include the date of a visit to a nest, the information about the status of that nest at the visit and possibly also the age of the nestlings (at least during the visit when a nest is found). Thus, data for each nest i are typically summarized as the date when the nest is first detected (f_i), the last day when nest i was encountered alive (j_i), the day when nest i was visited for the last time (k_i) and whether nest i was alive during this last visit (we write $x_i = 0$ for a failed nest and $x_i = 1$ for a successful nest). The most flexible and straightforward way to analyze these data is simply by the adoption of a CJS model (see Section 4.5.1) where we drop the submodel for recapture probability. This approach was proposed by Royle and Dorazio (2008) and Schmidt et al. (2010) and we illustrate it here.

There are three challenges in the analysis of such data. First, not all nests are detected on day 1, i.e., at the laying date of the first egg. Therefore, the estimation of daily survival starts on the day when a nest is discovered and thus the model conditions on first detection. We can also say that our data are left-censored because the event modeled (failure) may have happened before the "subject" (here, a nest) enters the study (Landes et al., 2020). Second, unless nests are visited daily, the exact date of failure is unknown, i.e., the observed failure dates are typically interval-censored. However, the date of failure can be estimated from the model based on the daily survival probabilities without the need for any assumptions about when the failure occurred as was assumed in the original nest survival models (Johnson, 1979). Finally, the eventual fate may not be known for all nests, which happens when a nest recorded as alive during the last visit is not revisited again until the fledging date of the nestlings. In this case, the data are right-censored, i.e., we know only that the nest has survived *at least* until the last date when it was observed alive. We can eliminate the resulting uncertainty by not modeling the fate of the nest after the last visit. Overall nest success is calculated as the product over the required number of days until fledging of the daily nest survival probabilities. To give a simple example: with constant daily nest survival probability s (i.e., no effects of time and nest age) total nest success is estimated as s^T for a species that requires T days from egg laying to fledging of the young.

We illustrate a nest survival model using a data set on wrynecks (*Jynx torquilla*, Fig. 4.12) that was analyzed by Geiser et al. (2008). The data contain information about 181 wryneck broods collected in the Valais (Swiss Alps) from 2002 to 2006. Dates are recorded such that day 1 is May 16. These data were collected in a nest box study, and thus theoretically all broods should have been detected from the first day already. However, nest boxes were not visited daily. Moreover, wrynecks may lay their eggs in old nests of other species that have reproduced in the box before and an active wryneck nest may therefore be overlooked. Thus, there is a risk to miss early failed broods and we therefore treat the data as if they were found on the actual day when nestlings were present. We load the data and have a look at them.

FIGURE 4.12

A wonderful wryneck (*Jynx torquilla*, photograph by Daniele Occhiato).

```
library(IPMbook); library(jagsUI)
data(wryneck)
str(wryneck)

'data.frame':   181 obs. of  5 variables:
 $ f   : int  49 15 23 23 47 11 23 15 65 2 ...
 $ j   : int  53 32 37 40 66 30 37 30 71 18 ...
 $ k   : int  53 32 37 40 66 30 37 30 72 18 ...
 $ x   : int  1 1 1 1 1 1 1 1 0 1 ...
 $ age: int  16 2 4 2 2 2 5 2 5 2 ...

tail(wryneck)

     f  j  k x age
176  5 25 25 1   1
177 53 56 60 0   1
178  9 29 29 1   1
179 26 46 46 1   1
180 55 75 75 1   1
181 20 35 35 1   6
```

As a reading example, nest 176 was first detected on day 5 when the nestling age was 1 day, and it was last checked on day 25 when the nestlings had fledged, so it was coded as a success ($x = 1$). Nest 177 was detected on day 53 when the nestlings were 1 day old, and it was checked for the last time with nestlings seen alive on day 56. Then, it was checked again on day 60 when the brood was found to have failed (hence, $x = 0$). From this information we can produce daily encounter histories, such that 1 indicates visits when the nest was alive, 0 shows visits when the nest was found to have failed and NA means missing data. The states of a nest at occasions with missing data (i.e., dates without visits) can then be estimated. The following code does this reformatting of the original data into encounter histories.

```
# Identify failed broods
fail <- which(wryneck$x==0)

# Create encounter histories
y <- matrix(NA, nrow=length(wryneck$f), ncol=max(wryneck$k))
for (i in 1:length(wryneck$f)){
  y[i,wryneck$f[i]] <- 1
  y[i,wryneck$j[i]] <- 1
}
for (i in 1:length(fail)){
  y[fail[i],wryneck$k[fail[i]]] <- 0
}
```

Let us look at the encounter histories of the two broods 176 and 177:

```
y[176,]
[1] ... NA  1 NA NA ... NA NA  1 NA NA ...

y[177,]
[37] ... NA NA  1 NA NA  1 NA NA NA  0 NA NA ...
```

In the encounter history of brood 176 there are NA before the nest was detected and after the nestlings fledged, i.e., the nest was successful. In the model (below) we will define the likelihood in a loop only between the first and last visit and thus NAs outside this loop will not be estimated, nor will they provide any information about daily survival. For nest 177 there are NAs between the last visit when the nest was alive and the visit when the failure was detected, so there is uncertainty about the exact date when the nest failed. The data required to fit the model are these encounter histories, the dates of the first and last visits to each nest, the number of nests and the number of days a brood requires from hatching to fledging (T), which is assumed to be a constant, and is needed to scale daily survival to total nest survival.

We fit a nest survival model where the daily nest survival is a linear function of nestling age. Often daily nest survival increases with nestling age, for instance because nestlings become less sensitive to food shortage, severe weather, or predators. The data bundle contains the age covariables as well as the other needed data and information.

```
# Bundle data
jags.data <- with(wryneck, list(y=y, f=f, k=k, n.nest=nrow(y), T=21, age=age))
str(jags.data)

List of 6
 $ y      : num [1:181, 1:81] NA NA NA NA NA NA NA NA NA NA ...
 $ f      : int [1:181] 49 15 23 23 47 11 23 15 65 2 ...
 $ k      : int [1:181] 53 32 37 40 66 30 37 30 72 18 ...
 $ n.nest: int 181
 $ T      : num 21
 $ age    : int [1:181] 16 2 4 2 2 2 5 2 5 2 ...
```

The likelihood of the nest survival model is similar to that of the CJS model (Section 4.5.1), except that here we assume perfect detection and hence have no observation submodel. As always, the model also has the priors and linear models section where we define our linear models for the primary parameters and the chosen priors. At the end of the code, we compute nest success as a derived parameter that is a function of daily survival and duration *T*.

```
# Write JAGS model file
cat(file="model13.txt", "
model {
# Priors and linear models
for (i in 1:n.nest){
  for (t in f[i]:(k[i]-1)){
    phi[i,t] <- phia[age[i] + t - f[i]]
  } #t
} #i

for (a in 1:T){
  phia[a] <- ilogit(alpha + beta * a)
}
alpha ~ dnorm(0, 0.001)
beta ~ dnorm(0, 0.001)

# Likelihood
for (i in 1:n.nest){
  for (t in (f[i]+1):k[i]){
    y[i,t] ~ dbern(phi[i,t-1] * y[i,t-1])
  } #t
} #i

# Derived parameter: nest success
nu <- prod(phia[1:T])
}
")

# Initial values
inits <- function(){list(alpha=runif(1, 4, 5), beta=runif(1, 0, 0.1))}

# Parameters monitored
parameters <- c("phia", "nu", "alpha", "beta")

# MCMC settings
ni <- 3000; nb <- 1000; nc <- 3; nt <- 1; na <- 1000

# Call JAGS from R (ART 1 min) and check convergence
out16 <- jags(jags.data, inits, parameters, "model13.txt", n.iter=ni, n.burnin=nb, n.chains=nc,
  n.thin=nt, n.adapt=na, parallel=TRUE)
traceplot(out16)                              # Not shown
print(out16, 3)

            mean    sd  2.5%    50%  97.5% overlap0     f  Rhat n.eff
phia[1]    0.971 0.008 0.953  0.972  0.985    FALSE 1.000 1.001  1650
phia[2]    0.973 0.007 0.958  0.974  0.985    FALSE 1.000 1.001  1633
[... output truncated ...]
phia[20]   0.993 0.003 0.987  0.993  0.997    FALSE 1.000 1.001  5757
phia[21]   0.993 0.003 0.987  0.994  0.997    FALSE 1.000 1.001  4989
nu         0.733 0.036 0.659  0.733  0.800    FALSE 1.000 1.000  3483
alpha      3.479 0.319 2.871  3.470  4.137    FALSE 1.000 1.002  1361
beta       0.078 0.032 0.016  0.078  0.141    FALSE 0.995 1.001  2152
```

As expected, there is a slight but clear increase with nestling age in the daily nest survival probabilities (Fig. 4.13). The coefficient of age (`beta`) is positive indicating that nest failure becomes less likely as nestlings get closer to fledging. Overall success during the nestling stage is estimated at 0.73, which means that about three of four nests with hatched young produce at least one fledgling.

This model offers great flexibility in modeling all sorts of covariate effects. For example, we can include year effects if the data were collected in different years, and the yearly parameters can be treated as either fixed or random. Individual random effects could also be included, e.g., if some nest boxes were repeatedly occupied and we may want to account for possible nonindependence due to location or parent ID. Daily nest survival could also be modeled as a function of temporal (e.g., daily temperatures or precipitation, Geiser et al., 2008) or spatial covariates (e.g., territory quality).

This basic model offers interesting possibilities for extensions. In particular the use of a multistate instead of the single-state model allows compelling additional insights into nest survival. For example, we often sample data not only during the nestling stage, but also during the incubation stage. Interest then lies in estimating the success of each stage (Miller et al., 2017). Another development is the inclusion of the cause of nest failure (Darrah et al., 2018), so that the proportion of causes of nest loss can be estimated and modeled. In the context of conservation this can be interesting because it may then be possible to mitigate a particular cause of nest failure through some management action and we might want to quantify its effect on population dynamics. Without going into details, we briefly sketch out these two applications.

The required data for the two-stage model including the incubation and the nestling periods are, for each stage, the day when a nest is seen alive for the first and last times as well as the day of the last visit to a nest along with the observed fate. Encounter histories are constructed from this information in a similar way as for the simple nest survival model shown above. But now we denote with a 1 the days when a nest is visited and found alive for the first and the last times during the incubation stage, with a 2 the days when the nest is visited and found alive for the first and last times during the nestling stage. In addition, a 3 indicates the day when a nest is observed to have failed during the incubation stage and a 4 is the same for the nestling stage. All other days in the encounter history are filled with NAs. These missing values mean that the alive/dead state of a nest is latent and will be estimated in the model. To model these data, we use a multistate capture-recapture model (Section 4.5.2) where we drop the observation submodel. The structure of this model is defined by the state-transition matrix, where rows denote the states on day t and columns the states on day $t+1$. We construct the likelihood by looping from the first visit to the last visit at each nest and the transition matrix is defined as follows:

$$\mathbf{\Psi} = \begin{bmatrix} s_i(1-\psi) & s_i\psi & 1-s_i & 0 \\ 0 & s_n & 0 & 1-s_n \\ 0 & 0 & 1 & 0 \\ 0 & 0 & 0 & 1 \end{bmatrix}_t .$$

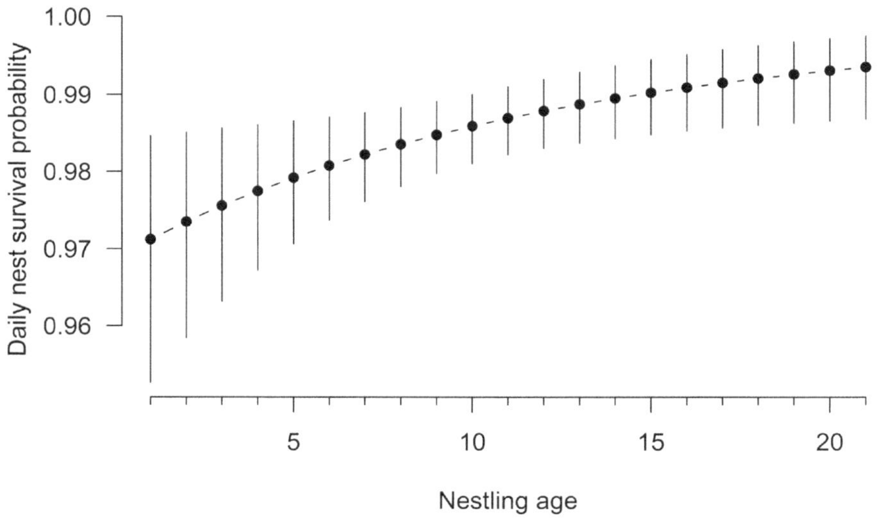

FIGURE 4.13

Daily nest survival probability of wrynecks as a function of nestling age (note that the average age at fledging, T, is 21 days).

Here, s_i is the daily nest survival probability during the incubation stage, s_n is the daily nest survival probability during the nestling stage and ψ is the daily hatching probability of the brood.

To implement the model with the cause of failure (Darrah et al., 2018), the data are again coded in a similar way as for the simple CJS model variant above. The only difference now is that the occasion when a failure is observed is not coded with a 0, but with a 2 for the first cause of failure (e.g., predation) and with a 3 for another cause of failure (e.g., abandonment), and so on if you want to distinguish more causes. We again use a multistate model with perfect observation of the nest states during the visits, and define the following transition matrix:

$$\boldsymbol{\Psi} = \begin{bmatrix} s & (1-s)\alpha & (1-s)(1-\alpha) \\ 0 & 1 & 0 \\ 0 & 0 & 1 \end{bmatrix}_t .$$

Here, now, s is the daily nest survival probability and α is the proportion of nests that failed due to the first reason (e.g., predation) among the failed nests. This model can also be extended to more than two causes of nest failure (Darrah et al., 2018).

Conceptually, it seems possible to accommodate also uncertainty regarding covariates or states. For example, when the age of the nest is unknown, it might still be possible to fit age-dependent models (Colchero and Clark, 2012). If the cause of failure is unknown for some nests, it would be straightforward to extend the multistate model from above to become a multievent model (Pradel, 2005); see Section 4.5.6. Such a model would have a further component to allow for errors in the state assignments.

4.5 MODELS FOR SURVIVAL SURVEYS

Survival estimation forms a big branch of ecological statistics, and a wide array of models have been developed for a large number of protocols corresponding to a large number of ways in which data informative about survival probability can be collected in the field (Williams et al., 2002; Sandercock, 2020). The estimation of survival usually requires knowing how many among a sample of individuals are still alive after some time interval. The ratio of the number of individuals still alive and the initial number of individuals is an estimator of survival over that time interval. This implies that animals can be distinguished individually and that the probability to detect a known individual that is alive is perfect. In the wild, the latter can typically only be guaranteed under very specific sampling protocols, such as satellite or radio tracking. However, under most sampling protocols the detection of marked or otherwise individually known individuals is not perfect (Kellner and Swihart, 2014), and we must estimate a "detection" or encounter probability to correct survival estimates for bias by nondetection.

The most frequently used sampling protocol for estimating survival in discrete time is capture-recapture: we catch a random sample of individuals, mark them individually and then release them back into the population. This procedure is repeated multiple times and each time new (i.e., unmarked) individuals get a mark and the identity of the recaptured individuals (i.e., those already marked) is recorded. The time interval between these so-called "capture occasions" corresponds to the time span over which survival is estimated. This is often a 1-year period but could also be a month if we want to estimate monthly survival. Importantly, a "recapture" does not necessarily require a physical capture. If individuals are marked with a conspicuous mark, such as a color ring, the identity of an individual may also be determined by visual observation from a distance, in which case we refer to a "resighting." Sometimes animals wear natural marks, such as spots and stripes in large cats (e.g., an ocelot Trolle and Kéry, 2003) or striking skin patterns in some amphibians and reptiles, which can serve to individually identify each animal, and that again can typically be recorded without ever capturing any animal. And more exciting yet: nowadays, genetics is increasingly used to identify animals, e.g., based on blood or tissue samples of recaptured animals but especially also with noninvasive genetic sampling, when animal parts such as shed skins for snakes or feces are analyzed. In this case, capture-recapture statistical methods are commonly applied to data sets where none of the individuals is ever even seen by an investigator! Thus, for these models there is conceptually no difference between "recapture" and "resighting," and hence, sometimes the more general term "re-encounter" is used. The data collected in this way are live recaptures and typically known as capture-recapture data, life encounter data, or more precisely as or single-state capture-recapture data. If additional categorical and time-varying information about the individuals is recorded, such as a geographical location or reproductive state, the data are referred to as multistate capture-recapture data.

A slightly different sampling protocol for survival estimation, but also in discrete time, is based on the collection of marked individuals with known time of death. Again, a random sample of individuals is marked, and we record the time of death of those individuals found dead. Obviously not all marked individuals will be found when they die, and thus the probability of detection of a dead individual must be estimated to obtain unbiased survival estimates. This type of data is known as dead-recovery, ring-recovery, band-recovery, or tag-recovery data.

Live recaptures and dead recoveries yield different estimates of survival. The sampling of live recaptures is restricted to a specific study area. Individuals that leave it and never return are said to have permanently emigrated and cannot be distinguished from dead individuals. Hence, analysis of live recaptures usually produces an estimate of the probability that an individual survives *and* remains in the study area. This is not true survival but is the so-called *apparent* or *local* survival. The inclusion of auxiliary data such as locations at which individuals are encountered within the study area (i.e., SCR data) or of sightings outside the study area permit estimation of true survival even from live recaptures (Barker, 1997; Ergon and Gardner, 2014; Schaub and Royle, 2014). In contrast to live recaptures, the sampling of dead-recovery data is usually not restricted to a specific area and therefore estimates of true survival result from their analysis.

Increasingly, we also see survival models formulated in continuous time (Barbour et al., 2013; Fouchet et al., 2016; Ergon et al., 2018; Murray and Bastille-Rousseau, 2020). They have a number of advantages, but conceptually they are probably harder to understand than their counterparts formulated for discrete time and so we do not cover them here. However, they could be incorporated into an IPM just as well as the survival models that we cover here.

To understand the models for analyzing life encounter or dead-recovery data we find it useful to distinguish between a binary time series of the alive/dead state (z) of an individual and a linked time series, also binary, of detection/nondetection (y) of the individual (for life encounters) or of the detection/nondetection of its death in a specific time interval (for dead recoveries). Conceptually, in each interval between two occasions that define our temporal discretization, we can imagine that every individual flips a loaded coin to decide whether it survives for one more occasion (i.e., $z = 1$) or dies (i.e., $z = 0$), i.e., makes the transition from the life to the death state. That coin is loaded and so its probabilities may not only be 0.5, but anything between 0 and 1, and the parameter governing this transition is the survival probability that we want to estimate from the observed data. Our observations of the marked individuals are not perfect; hence, an individual alive during an occasion (with true state $z = 1$) may fail to be recorded, leading to an *observed* state of 0 (i.e., a nondetection, $y = 0$). Analogously, for dead-recovery data an individual that dies during an occasion, i.e., makes the transition from state $z = 1$ to state $z = 0$, may go undetected (i.e., produce an observation $y = 0$). The process of recording a marked individual as either alive or recover it freshly dead can be described as another coin flip and now the parameter to be estimated quantifies the complement of a sort of measurement error: it is the recapture or recovery probability, respectively, in the analysis of capture-recapture and dead-recovery data.

In simple survival models, we only distinguish two individual states, or alternatively, a single alive state; hence, these models are also called single-state models. However, often it may be useful to distinguish more than a single state for a live individual and the associated models are called multistate models (Arnason, 1972; Hestbeck et al., 1991; Brownie et al., 1993; Lebreton et al., 2009). They stratify survival and detection by state and can estimate state-transition rates as an additional set of parameters. Multistate models can be generalized to so-called multievent models (Pradel, 2005). These allow for uncertainty or errors in state assignments of the detected individuals.

In this section on the modeling of data from survival surveys, we cover the famous CJS model for the analysis of single-state capture-recapture data, multistate models for multistate capture-recapture data, and ring-recovery models for dead-recovery data and briefly introduce multievent models. With the exception of the last one, we always present two formulations of these models. The first is the state-space or hierarchical formulation, which is conceptually easier to understand and allows maximum flexibility in the modeling of individual variation in the parameters. The second is the formulation using a multinomial likelihood for the so-called m-array aggregation of the individual capture history data. It is computationally much more efficient but does not offer exactly the same flexibility and is harder to grasp in the details. In each case we start with the simulation of the corresponding data, by which we hope to additionally clarify these models conceptually. Our introduction to these models is fairly brief and we refer you to Williams et al. (2002), King et al. (2010), Kéry and Schaub (2012), McCrea and Morgan (2015) and Seber and Schofield (2019) for a more in-depth treatment of these models.

4.5.1 CORMACK-JOLLY-SEBER MODEL FOR CAPTURE-RECAPTURE DATA

The CJS is a corner-stone model in ecological statistics and allows to estimate and model apparent survival probability (ϕ) of a set of marked individuals correcting for recapture probability (p), starting after (i.e., conditional on) first capture of an individual (Cormack, 1964; Jolly, 1965; Seber, 1965; Lebreton et al., 1992, Chapter 7 in Kéry and Schaub, 2012). There are two main ways to formulate the model. The first, and perhaps more natural, approach is specification of the model through a hierarchical construction of the likelihood (or an SSM; Gimenez et al., 2007, 2012; Royle, 2008; King, 2012), and the second, is the specification of a multinomial likelihood for an aggregation of the data known as the m-array (Burnham et al., 1987). We present both and start with the state-space formulation. We also sketch out how covariate effects can be modeled and present in particular the modeling of age-dependent survival.

4.5.1.1 State-Space Formulation

In developing the CJS model, we distinguish the true ($z_{i,t}$) from the observed state (our capture-recapture data, $y_{i,t}$) of individual i at occasion t and denote as f_i the occasion when individual i is first captured and marked. We describe the CJS model as a special case of a general SSM (Gimenez et al., 2007; Royle, 2008; King, 2012) where the initial state is known and therefore not stochastic. Then, we can write the model likelihood in the following three expressions:

$z_{i,f(i)} = 1$ Initial state (deterministic): individual known to be alive at f_i

$z_{i,t+1} \sim Bernoulli(z_{i,t}\phi_{i,t})$ State process: stochastic transitions with apparent survival probability

$y_{i,t} \sim Bernoulli(z_{i,t}p_{i,t})$ Observation process: stochastic observation with recapture probability

We now translate this model definition into R code to simulate one data set under that very model. In the simulation, we generate the fate of a set of individuals over a number of years, starting with the occasion when each is first captured and marked, after which they survive with probability ϕ and are recaptured with probability p. In our demonstration of the model, both probabilities are assumed to be identical for all individuals and occasions, but in general you can simulate and model individual and temporal variation by covariates; see below.

```
# Choose values for data simulation
nmarked <- 10                          # Number of marked individuals at each occasion
nyears <- 11                           # Number of years
phi <- 0.8                             # Constant apparent survival probability
p <- 0.4                               # Constant recapture probability
```

An individual first caught on the last year of the study has no opportunity of being recaptured during the study. Hence, such individuals do not carry any information about survival and can be discarded from the analysis. So in our simulation, we choose first-capture occasions only up to the penultimate year.

```
# Determine occasion when an individual first captured and marked
f <- rep(1:(nyears-1), each=nmarked)
nind <- length(f)                      # Total number of marked individuals

# State or ecological process
z <- array(NA, dim=c(nind, nyears)) # Empty alive/dead matrix

# Initial conditions: all individuals alive at f(i)
for (i in 1:nind){
  z[i,f[i]] <- 1
}

set.seed(1)                            # Initialize the RNGs in R
# Propagate alive/dead process forwards via transition rule:
# Alive individuals survive with probability phi
for (i in 1:nind){
  for (t in (f[i]+1):nyears){
    z[i,t] <- rbinom(1, 1, z[i,t-1] * phi)
  } #t
} #i
head(z);  tail(z)                      # Not shown: look at start and end of z
```

As in the related occupancy models (Section 4.3.4), the z matrix is a fundamental quantity also in a CJS model, although in traditional expositions it is not emphasized much: it gives the true life history of each individual in terms of the survival process, and hence, we could produce some interesting statistics from it, such as longevity.

In other fields such as medical statistics, we might hope to directly observe the z matrix, but in almost all applications of survival analysis in ecology and wildlife management, individuals are never observed perfectly, and hence, any reasonable survival model must account for imperfect detection. This is what we explain next with our data simulation: the recapture process. After the initial marking, marked individuals that are alive are recaptured with probability p.

```
# Observation process: simulate observations
y <- array(0, dim=c(nind, nyears))
for (i in 1:nind){
  y[i,f[i]] <- 1
  for(t in (f[i]+1):nyears){
    y[i,t] <- rbinom(1, 1, z[i,t] * p)
  } #t
} #i
```

Let us look at the simulated data set and make some comparisons between the true state and the observed data for a sample of individuals. We do not print the results of the next couple of lines of code, but we think that it may be insightful for you to look especially at the latter and understand the relationship between the quantities z (latent) and $\boldsymbol{y}$ (observed data).

```
y                               # Complete simulated capture-recapture data set (not shown)
for (i in 1:10){                # Look at true and observed states of first 10 individuals
  print(rbind("True state (z)" = z[i,], "Observed state (y)" = y[i,]))
  browser()
}
```

We are now ready to fit the simplest possible CJS model in JAGS for illustration. For the data bundle, we need to provide the capture-recapture data matrix, its dimensions and the vector $\boldsymbol{f}$ that indicates the marking occasion for each individual.

```
# Data bundle
jags.data <- list(y=y, f=f, nind=nind, nyears=ncol(y))
str(jags.data)

List of 4
 $ y     : num [1:100, 1:11] 1 1 1 1 1 1 1 1 1 1 ...
 $ f     : int [1:100] 1 1 1 1 1 1 1 1 1 1 ...
 $ nind  : num 100
 $ nyears: num 11
```

We develop the SSM to fit the CJS model to the data. The likelihood of the model is written in the most general way, that is, apparent survival and recapture are defined as two matrices with elements indexed by i and t. Hence, both parameters could vary among individuals and over time. To fit a model with constant rates, we specify a linear model where we assign the same two constants (`phi.const` and `p.const`) to each element in the two parameter matrices and then place priors on these parameters.

```
# Write JAGS model file
cat(file="model14.txt", "
model {
# Priors and linear models
phi.const ~ dunif(0, 1)          # Vague prior for constant phi
p.const ~ dunif(0, 1)            # Vague prior for constant p

for (i in 1:nind){               # Loop over individuals
  for (t in f[i]:(nyears-1)){ # Loop over time intervals/occasions
    phi[i,t] <- phi.const        # Here we model pattern in phi ...
    p[i,t] <- p.const            # ... and p
  } #t
} #i

# Likelihood
for (i in 1:nind){
  # Define latent state at first capture
  z[i,f[i]] <- 1
  for (t in (f[i]+1):nyears){
    # State process
    z[i,t] ~ dbern(z[i,t-1] * phi[i,t-1])
    # Observation process
    y[i,t] ~ dbern(z[i,t] * p[i,t-1])
  } #t
} #i
}
")
```

As for the occupancy and N-mixture models earlier in this chapter, we must provide initial values for the latent states z that do not contradict either the model or the data because otherwise JAGS will crash. The function `zInit` in the `IPMbook` package generates such initial values for single-state capture-recapture data.

```
# Initial values
inits <- function(){list(z=zInit(y))}
```

Finally, we define the list with the parameters we like to estimate, choose the MCMC settings and launch the model.

```
# Parameters monitored
parameters <- c("phi.const", "p.const")       # Could also add "z"

# MCMC settings
ni <- 3000; nb <- 1000; nc <- 3; nt <- 1; na <- 1000

# Call JAGS from R (ART < 1 min) and check convergence
out17 <- jags(jags.data, inits, parameters, "model14.txt", n.iter=ni, n.burnin=nb, n.chains=nc,
  n.thin=nt, n.adapt=na, parallel=TRUE)
traceplot(out17)                              # Not shown
print(out17, 3)

              mean    sd    2.5%    50%   97.5% overlap0 f  Rhat n.eff
phi.const    0.830 0.028   0.774  0.830   0.884    FALSE 1 1.002   834
p.const      0.414 0.038   0.341  0.414   0.490    FALSE 1 1.001  1567
```

This model runs quickly and converges easily. The estimates of survival and recapture probability agree well with the values chosen to simulate the data. Recall that we would need to repeat many times the data simulation and the parameter estimation to judge whether the model produces unbiased and accurate parameter estimates. We know from theory that maximum likelihood as well as Bayesian inference with vague priors for this model produces unbiased parameter estimates when the assumptions are met and for moderate to large samples, but constructing a short simulation may be a useful teaching tool for readers.

This model requires some homogeneity assumptions. For example, in our model we assume that survival and recapture is identical for all individuals and occasions. In life encounter data some behaviors of the marked individuals can lead to the violation of the homogeneity assumptions. For instance, when some individuals are not really part of the marked population, but are mere transients, i.e., appear only during one occasion and then are gone. Their apparent survival is effectively zero. Another problematic type of behavior may affect capture. Depending on the sampling protocol, some individuals may become trap-happy or trap-shy after an earlier capture, i.e., their capture probability changes after they are captured once or more times. Specific goodness-of-tests have been developed to detect these violations from statistics that can be computed from the data set (Pradel, 1993; Pradel et al., 1997a; Choquet et al., 2009a). Alternatively, we may conduct goodness-of-fit tests directly as part of the Bayesian model fitting when using posterior predictive checks (see Section 7.2 for an example). When it is found that a model does not fit, we can expand it, i.e., make it more complex, to accommodate for the cause of the assumption violation.

Survival and recapture can both be modeled as a function of covariates or with random effects by appropriate specification of linear models. Conceptually we can model effects along the temporal and along the individual dimension of the parameter matrices $\phi_{i,t}$ and $p_{i,t}$, and effects can be fixed, random or explained by covariates. Both parameters are probabilities and hence must lie between 0 and 1. Therefore, linear models are typically specified on the scale of a link function such as the logit. Here we sketch out just a few examples of possible models in algebra. You can find code for them in the BUGS language in the case studies and in Chapter 7 of Kéry and Schaub (2012).

As a first example, consider a model along the temporal dimension of the survival parameter matrix. We assume survival to be identical for all individuals, but want it to be variable over time, with time effects random. That is, we want to estimate an overall mean survival and the magnitude of temporal variation. Hence, we can write

$$\text{logit}(\phi_{i,t}) = \mu + \varepsilon_t$$
$$\varepsilon_t \sim Normal(0, \sigma^2).$$

Here, μ is mean survival on the logit-scale, ε_t is the effect of year t, and σ^2 is the temporal variance. There is a sense in which the realized survivals $\phi_{i,t}$ are now not completely independent from each other, but rather they are assumed (on the link scale) to be drawn from a normal distribution characterized by its mean and the temporal variance. To complete the model definition we provide priors for μ or $\text{logit}(\mu)^{-1}$, and for σ^2 or σ. Examples of models with temporal random effects along with BUGS code can be found in Sections 10.5.1, 11.5 and 12.5.

Often, we want to relax the assumption that survival is identical for all individuals and to model individual effects, or individual heterogeneity. Then, we specify a model along the individual axis of our survival or recapture parameter matrix. For example, imagine that we want to estimate a different survival for males and females. Then, we can define g_i as a categorical variable with two levels (1: male, 2: female) indicating the group membership of each individual. One possible linear model for survival is then

$$\phi_{i,t} = \beta_{g(i)},$$

where $\beta_1$is the survival of males and β_2 is that of females. No link function is necessary here (or in other words, the identity link is adopted) provided that we supply priors for $\boldsymbol{\beta}$ with support in the range between 0 and 1. Slightly more complex examples of models with a sex effect can be found in Sections 13.5 and 15.5.

A further common pattern in survival that we often want to model is age dependence. Over the life-time of an individual, survival in many taxa first increases with age, then reaches a maximum, and finally declines again (Loison et al., 1999; Tavecchia et al., 2001; Bouwhuis et al., 2012). Therefore, we might want to estimate different survival parameters for each age class. This can again be achieved by specification of a linear model on our primary parameter matrix $\phi_{i,t}$. The model requires a covariate that contains the information about the age class of each individual i at each time t. We can construct such a covariate provided that we know the age of each individual when first captured and based on the age structure adopted in the model. As an example, let us assume that we have capture-recapture data of woodchat shrikes collected over 20 years and that they were marked either as nestling or as adults, the latter meaning they are at least 1 year old. We load these data from the `IPMbook` package and have a look at them.

```
library(IPMbook); library(jagsUI)
data(woodchat5)
str(woodchat5)

List of 4
 $ ch   : num [1:1902, 1:20] 1 1 1 1 1 1 1 1 1 1 ...
 $ age  : num [1:1902] 2 2 2 2 2 2 2 2 2 2 ...
 $ repro: num [1:929, 1:3] 6 2 2 5 3 5 3 2 3 2 ...
 $ count: num [1:20] 91 119 131 88 139 145 148 116 112 106 ...
```

This data set also contains count and productivity data, but here we are interested only in the capture-recapture data (`ch`) and the information about the age class at marking (`age`), with 1 indicating a nestling and 2 an adult. We create a matrix $x_{i,t}$ that indicates the age class at time t for each individual i.

Function `createAge` computes such a matrix based on the information about the first-marking occasion of each individual along with its age class at that time and the numbers of capture occasions and of age classes specified in your model. The age structure is implemented in increments of 1 year up to the maximum number of age classes specified and then remains at that value until the end of the study. Thus, for example, if we distinguish three age classes and an individual is marked as a nestling (and so is in age class 1), its row in the age matrix will be `1, 2, 3, 3, 3`.... This tells us that it was in age class 1 in the first occasion, in age class 2 in the second occasion, and in age class 3 in all following occasions.

Because individuals marked in the last occasion do not provide any information about survival, we remove them from the data set. This is optional and helps to reduce the run time of a model. Then we apply the `createAge` function to generate the age matrix $\boldsymbol{x}$. We want just two age classes (juvenile survival vs. adult survival), and so we set the maximum number of age classes at 2.

```
f <- getFirst(woodchat5$ch)
last <- which(f==ncol(woodchat5$ch))
f <- f[-last]
ch <- woodchat5$ch[-last,]
age <- woodchat5$age[-last]

x <- createAge(f, age, 20, 2)
```

We look at $\boldsymbol{x}$ for individuals 50 and 51 to better understand the age matrix. These were marked as adult and nestling, respectively. The age structure specified in $\boldsymbol{x}$ makes sense (hopefully ...).

```
x[50:51,]                        # Most columns omitted
     [,1] [,2] [,3] [,4] ...
[1,]    2    2    2    2 ...
[2,]    1    2    2    2 ...
```

Now that we have the matrix $\boldsymbol{x}$ indicating the age class, we can model survival as

$$\phi_{i,t} = \beta_{x(i,t)}$$

where β_1 is survival of the first age class (juvenile survival) and β_2 is survival of the second age class (adult survival). As for the case of sex-dependent survival before, we do not need a link function in the model now. We bundle the data, write the BUGS code, and launch the model.

```
# Data bundle
jags.data <- list(y=woodchat5$ch[-last,], f=f, nind=length(f), nyears=ncol(woodchat5$ch), x=x)
str(jags.data)

List of 5
 $ y     : num [1:1799, 1:20] 1 1 1 1 1 1 1 1 1 1 ...
 $ f     : int [1:1799] 1 1 1 1 1 1 1 1 1 1 ...
 $ nind  : int 1799
 $ nyears: int 20
 $ x     : num [1:1799, 1:19] 2 2 2 2 2 2 2 2 2 2 ...


# Write JAGS model file
cat(file="model15.txt", "
model {
# Priors and linear models
for (j in 1:2){
  beta[j] ~ dunif(0, 1)                # Vague priors for age-dependent survival
}
p.const ~ dunif(0, 1)                  # Vague prior for constant p

for (i in 1:nind){
  for (t in f[i]:(nyears-1)){
    phi[i,t] <- beta[x[i,t]]           # Linear model for phi ...
    p[i,t] <- p.const                  # ... and p
  } #t
} #i

# Likelihood
for (i in 1:nind){
  # Define latent state at first capture
  z[i,f[i]] <- 1
  for (t in (f[i]+1):nyears){
    # State process
    z[i,t] ~ dbern(z[i,t-1] * phi[i,t-1])
    # Observation process
    y[i,t] ~ dbern(z[i,t] * p[i,t-1])
  } #t
} #i
}
")

# Initial values
inits <- function(){list(z=zInit(jags.data$y))}

# Parameters monitored
parameters <- c("beta", "p.const")

# MCMC settings
ni <- 3000; nb <- 1000; nc <- 3; nt <- 1; na <- 1000

# Call JAGS from R (ART 5 min) and check convergence
out18 <- jags(jags.data, inits, parameters, "model15.txt", n.iter=ni, n.burnin=nb, n.chains=nc,
  n.thin=nt, n.adapt=na, parallel=TRUE)
traceplot(out18)                       # Not shown
```

This model has a fairly substantial run time because the computation of the likelihood loops over all individuals and occasions. But convergence is good and only a moderate number of MCMC iterations is necessary.

```
print(out18)

           mean    sd   2.5%    50%  97.5% overlap0 f  Rhat n.eff
beta[1]   0.304 0.017  0.273  0.304  0.339    FALSE 1 0.998  6000
beta[2]   0.542 0.014  0.513  0.542  0.569    FALSE 1 1.002  6000
p.const   0.603 0.021  0.563  0.604  0.641    FALSE 1 1.003  3450
```

The concepts introduced here for the modeling of the matrices containing the survival and recapture probabilities allow much flexibility: any kind of effect along the individual and the time axes, as well as along the interactions among the two (such as age effects) can be modeled. We can treat effects to be fixed (such as sex) or random (such as could be time), and we can model the parameters as functions of covariates that are categorical (e.g., sex), or continuous (no example shown, but this could be the amount of rainfall along the time axis). One potential difficulty is that parameters might not always be identifiable (e.g., a model with an individual fixed effect on survival). Parameter identifiability can be studied by different methods including symbolic algebra (Gimenez et al., 2004; Cole et al., 2012) or by inspection of the overlap between prior and posterior (Gimenez et al., 2009), though in practice the simplest approach is to use simulation (see Section 6.3 and Chapter 13 for examples of a simulation-based assessment of parameter identifiability in IPMs). Second, it may take a long time to run these models to convergence, particularly if the number of individuals is large. This is a particular drawback if you fit the CJS model inside an IPM, which is often computationally expensive anyway. Luckily, there is another, more efficient way to fit a CJS model to capture-recapture data, and we show that next.

4.5.1.2 Multinomial Formulation

For a more efficient implementation of the CJS model we first aggregate the capture histories and then specify a product-multinomial likelihood for parameter estimation. Data aggregation is carried out by breaking the capture history of an individual into fragments defined by a release occasion and the following recapture occasion. Hence, the number of such fragments equals the number of captures of an individual. For example, we break the capture history `[1 0 1 0]` into the two fragments `[1 0 1 0]` and `[0 0 1 0]`. The first fragment says that the individual was released on the first occasion and then recaptured on the third. The second fragment tells us that the individual was released again on the third occasion, but then never recaptured again. We then tally up the numbers of all fragments for all individuals and summarize their frequency in what is known as an m-array (Burnham et al., 1987). This is a matrix with rows that indicate the occasion of releases and columns that refer to the occasion of first recapture. An additional column is put at the end and contains the number of individuals that are never recaptured. In addition, when fitting the model to a data set in the m-array aggregation, we also need a vector that contains the number of individuals released at each release occasion. A schematic for an m-array for single-state capture-recapture data over four occasions is shown in Table 4.1.

We now go back to our simulated data set `y` and create the m-array using the function `marray` from the `IPMbook` package and have a look at it.

```
marr <- marray(y)
print(marr)
```

```
          recaptured
released Y2 Y3 Y4 Y5 Y6 Y7 Y8 Y9 Y10 Y11 never
      Y1  3  2  1  2  0  0  0  0   0   0     2
      Y2  0  5  1  3  0  0  0  0   0   0     4
      Y3  0  0  3  4  3  1  1  0   0   0     5
      Y4  0  0  0  7  2  1  0  0   0   0     5
      Y5  0  0  0  0 12  1  1  3   0   0     9
      Y6  0  0  0  0  0 10  3  3   0   0    11
      Y7  0  0  0  0  0  0  9  5   1   1     7
      Y8  0  0  0  0  0  0  0  7   6   2     9
      Y9  0  0  0  0  0  0  0  0   5   6    17
     Y10  0  0  0  0  0  0  0  0   0   8    14
```

Table 4.1 The m-array representation for a single-state capture-recapture data set with four capture occasions. For a reading example, $m_{1,3}$ is the number of capture-history fragments `[1 0 1 0]` occurring in the data. Note that the m-array is fully defined with its "central" part (columns 3–5) and one of either column 2 or 6. The typical format of the m-array contains the "central" part plus the last column, see also the R output just above.

		Numbers at First Re-encounter Occasion			
Release Occasion	**Numbers Released**	**2**	**3**	**4**	**Numbers Never Re-encountered**
1	R_1	$m_{1,2}$	$m_{1,3}$	$m_{1,4}$	$R_1 - \sum_{j=2}^{4} m_{1,j}$
2	R_2	0	$m_{2,3}$	$m_{2,4}$	$R_2 - \sum_{j=3}^{4} m_{2,j}$
3	R_3	0	0	$m_{3,4}$	$R_3 - m_{3,4}$

In the m-array we can see, for example, that three individuals released on the first occasion were recaptured on the second occasion, whereas one individual released on the first occasion was first recaptured on occasion 4 and hence failed to be recaptured on occasions 2 and 3. Finally, two individuals released on occasion 1 were never recaptured. The total number of releases per occasion is not provided directly in the m-array but can be computed by taking the row sums of the matrix.

```
rel <- rowSums(marr)
print(rel)

Y1  Y2  Y3  Y4  Y5  Y6  Y7  Y8  Y9 Y10
10  13  17  15  26  27  23  24  28  22
```

You may wonder why on occasions 2 to 10 there are more than 10 released individuals even though we simulated the data with only 10 newly marked individuals per occasion. But of course, this is because in the m-array "released" means newly marked *or* recaptured and released. For example, we see in the m-array that there are three individuals released on occasion 1 that were recaptured on occasion 2. These three were then released again on occasion 2 along with the 10 newly marked individuals for a total of 13.

For each row in the m-array we model the frequency distributions across columns using a multinomial distribution:

$$\boldsymbol{m}_t \sim Multinomial(\pi_t, R_t),$$

where $\boldsymbol{m}_t$ is the t-th row in the m-array, with the last element being the number of individuals in release cohort t that were never recaptured, R_t is the total number of individuals released at time t, and π_t is a vector with the cell probabilities that distributes the R_t individuals into vector $\boldsymbol{m}_t$. These cell probabilities in vector π_t represent the expected frequencies of each capture-history fragment and they are expressed as functions of (i.e., replaced by) the apparent survival and recapture probabilities that we are interested in estimating. Table 4.2 shows these probabilities for a data set with four occasions. π_t contains the cell probabilities for release cohort t and is found in row t of the table.

We bundle the data and have a look at them.

```
# Bundle data
jags.data <- list(marr=marr, rel=rel, nyears=ncol(marr))
str(jags.data)

List of 3
 $ marr   : num [1:10, 1:11] 3 0 0 0 0 0 0 0 0 0 ...
 $ rel    : num [1:10] 10 13 17 15 26 27 23 24 28 22
 $ nyears: int 11
```

The BUGS code for the capture-recapture model written using the multinomial distribution consists of two parts as usual, denoted by the two subheadings in bold face. A core part of the likelihood is the definition of the multinomial cell probabilities in π as given in Table 4.2. This likelihood is written in a very general way such that any kind of temporal effects on survival and recapture can be modeled without any change needed, by simply adapting the code under the `# Prior and linear models` section.

Table 4.2 Schematic for the probabilities in the m-array for a Cormack-Jolly-Seber model for a study with four occasions. The parameters are apparent survival between occasions t and $t+1$ (ϕ_t) and recapture probability on occasion $t+1$ (p_t). Fitting the CJS model to a data set represented by this table amounts to maximizing a likelihood that is the product of three multinomials (one for each release occasion), with their cell probabilities in columns 2–5.

	Recapture Occasion			
Release Occasion	**2**	**3**	**4**	**Never Recaptured**
1	$\phi_1 p_1$	$\phi_1(1-p_1)\phi_2 p_2$	$\phi_1(1-p_1)\phi_2(1-p_2)\phi_3 p_3$	$1-\sum$ rel. occ. 1
2	0	$\phi_2 p_2$	$\phi_2(1-p_2)\phi_3 p_3$	$1-\sum$ rel. occ. 2
3	0	0	$\phi_3 p_3$	$1-\phi_3 p_3$

Note that you will sometimes see a different indexing of time for the recapture probability such that the first recapture parameter on occasion 2 is denoted p_2.

```
# Write JAGS model file
cat(file="model16.txt", "
model {
# Priors and linear models
phi.const ~ dunif(0, 1)
p.const ~ dunif(0, 1)

for (t in 1:(nyears-1)){
  phi[t] <- phi.const
  p[t] <- p.const
}

# Likelihood
# Define the multinomial likelihood
for (t in 1:(nyears-1)){
  marr[t,1:nyears] ~ dmulti(pi[t,], rel[t])
}
# Define the cell probabilities of the m-array
for (t in 1:(nyears-1)){
  # Main diagonal
  q[t] <- 1 - p[t]              # Probability of non-recapture
  pi[t,t] <- phi[t] * p[t]
  # Above main diagonal
  for (j in (t+1):(nyears-1)){
    pi[t,j] <- prod(phi[t:j]) * prod(q[t:(j-1)]) * p[j]
  } #j
  # Below main diagonal
  for (j in 1:(t-1)){
    pi[t,j] <- 0
  } #j
} #t
# Last column: probability of non-recapture
for (t in 1:(nyears-1)){
  pi[t,nyears] <- 1-sum(pi[t,1:(nyears-1)])
}
}
")

# Initial values
inits <- function(){list(phi.const=runif(1, 0, 1))}

# Parameters monitored
parameters <- c("phi.const", "p.const")

# MCMC settings
ni <- 3000; nb <- 1000; nc <- 3; nt <- 1; na <- 1000

# Call JAGS (ART <1 min), check convergence and summarize posteriors
out19 <- jags(jags.data, inits, parameters, "model16.txt", n.iter=ni, n.burnin=nb, n.chains=nc,
  n.thin=nt, n.adapt=na, parallel=TRUE)
traceplot(out19)                # Not shown
print(out19)

             mean    sd   2.5%    50%  97.5% overlap0 f  Rhat n.eff
phi.const   0.831 0.027  0.777  0.832  0.881    FALSE 1 1.001  5290
p.const     0.415 0.037  0.345  0.414  0.488    FALSE 1 1.001  3714
```

In comparison with the state-space formulation (Section 4.5.1.1), the model runs extremely fast, convergence is excellent and we obtain essentially identical estimates.

Compared with the state-space formulation of the CJS model, the multinomial formulation allows the same modeling along the time axis: that is, we can model fixed or random time effects, and include temporal covariates that are categorical, continuous or both. However, we are somewhat more restricted in the modeling of individual effects. Because we break apart individual histories into capture-history fragments, the individual identity is lost. This is why survival and recapture parameters now only have a time, but not an individual index. However, it is still possible to model some individual heterogeneity, simply by specifying groups and creating one m-array for each group. For example, survival can be modeled as a function of sex, which then requires that we create one m-array for males and another for females; see Schaub and Ullrich (2021) for an example with the iconic species in this book.

Some care is needed for modeling age effects on the parameters with the multinomial likelihood of the CJS model. When individuals are released after recapture, their age will change compared with the age at which they were marked or recaptured on previous occasions. This information must be carried on in the analysis to estimate age effects by specifying as many m-arrays as there are age classes. Consider a simple example with two age classes in survival, as we have seen in Section 4.5.1.1. In this case we need two m-arrays, one for individuals released as juveniles and another for individuals released as adults. When an individual marked as a juvenile on occasion 1 is recaptured on occasion 3, it will first contribute to the juvenile m-array (in element m_{13}). But when it is released again, it is an adult and so the next fragment in its encounter history will contribute to the adult m-array.

You can construct such m-arrays using our function `marrayAge`, which has three arguments: the capture history matrix, a vector indicating the age class of each individual at marking (i.e., when first released) and the maximum number of age classes distinguished in our model. For illustration, we again use the woodchat shrike data and compute the two m-arrays.

```
data(woodchat5)
marr <- marrayAge(ch=woodchat5$ch, age=woodchat5$age, mAge=2)
```

The object `marr` is now a 3-dimensional array with the last dimension referring to the two age classes and the first two dimensions being the usual dimensions of the m-array, with the last column containing the number of individuals in a release cohort that are never recaptured. When fitting the model, the multinomial cell probabilities will differ for the juvenile and the adult m-array (Table 4.3).

```
# Bundle data and produce data overview
jags.data <- list(marr.j=marr[,,1], marr.a=marr[,,2], nyears=dim(marr)[2], rel.j=rowSums(marr[,,1]),
  rel.a=rowSums(marr[,,2]))
str(jags.data)

List of 5
 $ marr.j: num [1:19, 1:20] 8 0 0 0 0 0 0 0 0 0 ...
 $ marr.a: num [1:19, 1:20] 16 0 0 0 0 0 0 0 0 0 ...
 $ nyears: int 20
 $ rel.j : num [1:19] 51 53 55 65 73 66 61 76 65 75 ...
 $ rel.a : num [1:19] 36 39 44 61 61 50 43 61 51 53 ...
```

The BUGS code for this age-dependent model is now slightly more complex—for each age class, a likelihood statement must be written and cell probabilities differ between age classes according to Table 4.3.

Table 4.3 Multinomial cell probabilities for the m-arrays when survival is age-dependent and we distinguish two age classes. Parameters are juvenile and adult survival between occasion t and $t+1$ ($\phi_{j,t}$ and $\phi_{a,t}$, respectively) and recapture probability on occasion $t+1$ (p_t).

	Recapture Occasion			
Release Occasion	**2**	**3**	**4**	**Never Recaptured**
A. Cell Probabilities in the m-array for Individuals Marked as Juvenile				
1	$\phi_{j,1}p_1$	$\phi_{j,1}(1-p_1)\phi_{a,2}p_2$	$\phi_{j,1}(1-p_1)\phi_{a,2}(1-p_2)\phi_{a,3}p_3$	$1-\sum$ rel. occ. 1
2	0	$\phi_{j,2}p_2$	$\phi_{j,2}(1-p_2)\phi_{a,3}p_3$	$1-\sum$ rel. occ. 2
3	0	0	$\phi_{j,3}p_3$	$1-\phi_{j,3}p_3$
B. Cell Probabilities in the m-array for Individuals Marked or Released as Adult				
1	$\phi_{a,1}p_1$	$\phi_{a,1}(1-p_1)\phi_{a,2}p_2$	$\phi_{a,1}(1-p_1)\phi_{a,2}(1-p_2)\phi_{a,3}p_3$	$1-\sum$ rel. occ. 1
2	0	$\phi_{a,2}p_2$	$\phi_{a,2}(1-p_2)\phi_{a,3}p_3$	$1-\sum$ rel. occ. 2
3	0	0	$\phi_{a,3}p_3$	$1-\phi_{a,3}p_3$

```
# Write JAGS model file
cat(file="model17.txt", "
model {
# Priors and linear models
phij.const ~ dunif(0, 1)
phia.const ~ dunif(0, 1)
p.const ~ dunif(0, 1)

for (t in 1:(nyears-1)){
  phij[t] <- phij.const
  phia[t] <- phia.const
  p[t] <- p.const
}

# Capture-recapture data (CJS model with multinomial likelihood)
# Define the multinomial likelihood
for (t in 1:(nyears-1)){
  marr.j[t,1:nyears] ~ dmulti(pi.j[t,], rel.j[t])
  marr.a[t,1:nyears] ~ dmulti(pi.a[t,], rel.a[t])
}
# Define the cell probabilities of the m-arrays
for (t in 1:(nyears-1)){
  # Main diagonal
  q[t] <- 1 - p[t]              # Probability of non-recapture
  pi.j[t,t] <- phij[t] * p[t]
  pi.a[t,t] <- phia[t] * p[t]
  # Above main diagonal
  for (j in (t+1):(nyears-1)){
    pi.j[t,j] <- phij[t] * prod(phia[(t+1):j]) * prod(q[t:(j-1)]) * p[j]
    pi.a[t,j] <- prod(phia[t:j]) * prod(q[t:(j-1)]) * p[j]
  } #j
  # Below main diagonal
  for (j in 1:(t-1)){
    pi.j[t,j] <- 0
    pi.a[t,j] <- 0
  } #j
} #t
# Last column: probability of non-recapture
for (t in 1:(nyears-1)){
  pi.j[t,nyears] <- 1-sum(pi.j[t,1:(nyears-1)])
  pi.a[t,nyears] <- 1-sum(pi.a[t,1:(nyears-1)])
}
}
")

# Initial values
inits <- function(){list(phij.const=runif(1, 0, 1))}

# Parameters monitored
parameters <- c("phij.const", "phia.const", "p.const")

# MCMC settings
ni <- 3000; nb <- 1000; nc <- 3; nt <- 1; na <- 1000

# Call JAGS (ART <1 min), check convergence and summarize posteriors
out20 <- jags(jags.data, inits, parameters, "model17.txt", n.iter=ni, n.burnin=nb, n.chains=nc,
  n.thin=nt, n.adapt=na, parallel=TRUE)
traceplot(out20)                 # Not shown
print(out20)

              mean    sd  2.5%   50% 97.5% overlap0 f  Rhat n.eff
phij.const   0.304 0.016 0.274 0.304 0.337    FALSE 1 1.001  6000
phia.const   0.542 0.014 0.515 0.542 0.570    FALSE 1 1.002  1302
p.const      0.603 0.020 0.564 0.603 0.644    FALSE 1 1.000  6000
```

Fitting the CJS model using the multinomial rather than the state-space formulation of the likelihood is much faster, and the estimates are identical up to Monte Carlo error. Because of this we often use the multinomial likelihood to model capture-recapture data inside an IPM.

This completes our quick introduction to the estimation of survival from single-state capture-recapture data. These data could also be analyzed with a Jolly-Seber (JS) model (Jolly, 1965; Seber, 1965; Williams et al., 2002; Seber and Schofield, 2019) which, in addition to survival, permits estimation of population size and recruitment. The JS model is unconditional, i.e., the complete capture history and not just the part after the initial capture is used for parameter estimation. Therefore, we estimate the probability to catch an unmarked individual in addition to that for an already marked one and because we cannot estimate them separately (that model would not be identifiable from these data alone), we must assume that they are identical. This assumption can be hard to meet in practice and moreover is difficult to test, and as a result the JS model is used much less often than the CJS model. BUGS code for the JS model can be found in Chapter 11 of Link and Barker (2010) and in Chapter 10 of Kéry and Schaub (2012) and a more advanced variant that is spatially explicit is showcased in Chapter 21.

Sometimes data are sampled under a so-called robust design, which means that repeated subsamples (secondary occasions) are taken during each (primary) occasion and the population is assumed to be closed among secondary but open among the primary occasions (Pollock, 1982; Williams et al., 2002; Seber and Schofield, 2019). These secondary occasions provide additional information about the capture and recapture probabilities and thereby allow estimation of population size, recruitment and temporary emigration (Kendall et al., 1995; Kendall and Bjorkland, 2001). BUGS code for this parameter-rich model is provided in Rankin et al. (2016) and Riecke et al. (2018).

Capture-recapture data and the associated models are common and frequently used components of an IPM. For further examples in our book see the case studies in Chapters 11, 13, 14, 16 and 19.

4.5.2 MULTISTATE CAPTURE-RECAPTURE MODELS

When we sample capture-recapture data we often collect more information from a captured individual than only recording its capture. If this additional information is an individual categorical variable that may change over time, we speak about "states" of an individual. These can be geographical locations, breeding status, or age classes, to name just a few. The resulting multistate capture-recapture data can be analyzed with multistate capture-recapture models (Arnason, 1972, 1973; Hestbeck et al., 1991; Nichols et al., 1992; Brownie et al., 1993; Schwarz et al., 1993; Lebreton et al., 2009) which are a generalization of the single-state CJS model in Section 4.5.1 (Lebreton et al., 1999). These models allow the estimation of state-specific survival and recapture probabilities as well as state-transition probabilities. Like the CJS model, also the multistate model can be analyzed using either the state-space likelihood or the multinomial likelihood. We start by demonstrating the state-space formulation because it is more intuitive to understand.

4.5.2.1 State-Space Formulation

To develop a multistate capture-recapture model we distinguish again between the latent state $z_{i,t}$ and the observed state $y_{i,t}$ for individual i at occasion t. The latent state $z_{i,t} = s$ if individual i is in state s on occasion t, where $s = \{1, 2, \ldots, S\}$ and S-1 is the total number of live states. If individual i on occasion t is dead, it is assigned to state S. The observed states typically include all latent states of the live individuals, but models can also be formulated that include unobserved states. The observations y do not, however, include the dead state, that is, in this model we assume that we do not observe any dead individuals (in contrast to dead-recovery models in Section 4.5.3). Generally, the list of observed states is $y = \{1, \ldots, O\}$, where $O-1$ is the number of observed states of live individuals, and we code by O an individual that is "not recaptured." We also assume that the state of an observed individual is recorded without error. The only measurement error accommodated in this model, therefore, is that associated with nondetection of an individual and not with the state assignment of an individual that is detected. This is the major difference between multistate and multievent models (Section 4.5.5); the latter do accommodate also state assignment uncertainty for an individual when it is detected, even during the first occasion of a study.

To develop a multistate capture-recapture model using the state-space formulation we first determine the state of an individual i when it is marked (s_i) and then probabilistically model the transitions of the latent states over time along with our observations using categorical distributions. We can write the model in the following three expressions:

$$z_{i,f(i)} = s_i \qquad \text{Initial state: individual known to be in state } s \text{ at marking}$$

$$z_{i,t+1} \sim Categorical(\mathbf{\Omega}_{z_{i,t},1\ldots S,i,t}) \qquad \text{State transition process}$$

$$y_{i,t} \sim Categorical(\mathbf{\Theta}_{z_{i,t},1\ldots O,i,t}) \qquad \text{Observation process}$$

To most ecologists, the categorical distribution will be fairly obscure, but it is simply a multinomial distribution with a trial size of one and thus is similar to what the Bernoulli is to the binomial distribution (see Fig. 2.16). To define a categorical, we need a vector with cell probabilities that sum to one, e.g., $\boldsymbol{g} = \left\{ g_1, g_2, \ldots, g_{S-1}, 1 - \sum_{i=1}^{S-1} g_i \right\}$. The outcome of a single realization of the categorical distribution is a scalar; for example, it is x with probability g_x.

In a multistate model, the arguments of the categorical distributions that define our state-transition and observation processes both become four-dimensional arrays that contain the state-transition and recapture probabilities. The two first dimensions index the states and the last two individual and occasion. The definition of the structure of these two arrays is a key step in developing a multistate capture-recapture model and we suggest to explicitly show them whenever you describe such a model, as they uniquely define your multistate capture-recapture model. We recommend to first develop a list with all latent states and another with the observed states. The state-transition matrix then defines how latent states evolve over time in a Markovian fashion and the observation matrix links the latent (i.e., true) states with the observed states.

To illustrate, we use an example where we assume to have captured individuals at two locations between which they can move. Our interest is in estimating location-specific apparent survival, movement probabilities and location-specific recapture probabilities. The list of the latent states therefore is "alive at location A," "alive at location B," and "dead." The list of observed states would typically include "recaptured at location A," "recaptured at location B," and "not recaptured." Now we can build the transition and observation matrices. The state-transition matrix describes the evolution of the latent states from occasion t to occasion $t+1$, i.e., the departure (or "from") state is in the rows and the arrival (or "to") state in the columns, and therefore, probabilities sum to 1 across columns of a row (the matrix is said to be row-stochastic):

$$\boldsymbol{\Omega}_{1\ldots3,1\ldots3,i,t} = \begin{bmatrix} \phi_{A,i,t}(1-\psi_{AB,i,t}) & \phi_{A,i,t}\psi_{AB,i,t} & 1-\phi_{A,i,t} \\ \phi_{B,i,t}\psi_{BA,i,t} & \phi_{B,i,t}(1-\psi_{BA,i,t}) & 1-\phi_{B,i,t} \\ 0 & 0 & 1 \end{bmatrix}$$

Here, $\phi_{A,i,t}$ is the probability that individual i at site A in occasion t survives until "just before" occasion $t+1$ at site A and $\phi_{B,i,t}$ is the probability that individual i at site B in occasion t survives until "just before" occasion $t+1$ at site B. There are two movement parameters: $\psi_{AB,i,t}$ is the probability that individual i at site A on occasion t that had survived until "just before" $t+1$ moves to site B where it is present at time $t+1$ and $\psi_{BA,i,t}$ is the reverse: that individual i at site B on occasion t that had survived to "just before" $t+1$ moves to site A at $t+1$. Hence, $\phi_{A,i,t}\psi_{AB,i,t}$ is the probability that individual i at site A on occasion t is at site B on occasion $t+1$. This parameterization requires that individual i first must survive for most of the interoccasion interval at site A and then moves very quickly just before the following occasion. Other parametrizations where movement occurs temporally before survival are also possible. The dead state is said to be absorbing. This means that once an individual is in this state, it cannot escape from it until the end of the study. Therefore, the transition probabilities for the state dead in row 3 are just 0 and 1.

Next, we turn our attention to the observation matrix that describes the "mapping" of the true states (in the rows) on the observed states (in the columns). Below is such an observation matrix for our hypothetical study:

$$\boldsymbol{\Theta}_{1\ldots3,1\ldots3,i,t} = \begin{bmatrix} p_{A,i,t} & 0 & 1-p_{A,i,t} \\ 0 & p_{B,i,t} & 1-p_{B,i,t} \\ 0 & 0 & 1 \end{bmatrix}$$

Here we estimate two parameters: recapture probability of individual i on occasion t at site A ($p_{A,i,t}$) and that of individual i on occasion t at site B ($p_{B,i,t}$). This matrix is again row-stochastic, but depending on your model, it need not necessarily be square, because the numbers of latent and of observed states may be different. An interesting detail for your understanding of the matrix is the element in the last row of the last column. It contains the probability that a dead individual (in the third row) is not recaptured (third column). As we have no dead recoveries in our data, this probability must be 1: no dead individual is ever observed.

Now we have defined our multistate capture-recapture model, so next we simulate a data set under this model and then develop BUGS code to analyze it. For illustration, we assume that all probabilities are identical for all individuals and constant over time and that during each of 6 years we mark a total of 50 individuals at each of the two sites. Moreover, we pick values for survival, movement, and recapture probabilities.

```
# Choose constants in simulation
nmarked <- 50                   # Number of marked individuals each year and site
nyears <- 6                     # Number of years
phiA <- 0.8                     # Apparent survival probability when at site A
phiB <- 0.7                     # Apparent survival probability when at site B
psiAB <- 0.3                    # Probability to move from A to B
psiBA <- 0.5                    # Probability to move from B to A
pA <- 0.7                       # Recapture probability when at site A
pB <- 0.4                       # Recapture probability when at site B
```

Individuals marked on the last occasion do not provide any information about the parameters, so that in our simulation we choose first-capture occasions only up to the penultimate year.

```
# Determine occasion when an individual first captured and marked
f <- rep(rep(1:(nyears-1), each=nmarked), 2)
nind <- length(f)               # Total number of marked individuals

# Construct the transition probability matrix (TPM), above called OMEGA
# Includes the dead state as state number 3
# Departure state in rows (time t), arrival state in columns (t+1)
TPM <- matrix(c(
      phiA * (1-psiAB), phiA * psiAB,     1-phiA,
      phiB * psiBA,     phiB * (1-psiBA), 1-phiB,
      0,                0,                1       ), nrow=3, byrow=TRUE)

# Construct the observation probability matrix (OPM), above called THETA
# True state is in rows, observed state is in columns
# Includes nondetection as observation event number 3
OPM <- matrix(c(
      pA, 0,  1-pA,
      0,  pB, 1-pB,
      0,  0,  1       ), nrow=3, byrow=TRUE)
```

After these preliminaries, we are ready to simulate the data under the multistate model.

```
# State or ecological process
# Simulate true system state
z <- array(NA, dim=c(nind, nyears)) # Empty alive/dead matrix

# Initial conditions: all individuals alive at f(i)
initial.state <- c(rep(1, nind/2), rep(2, nind/2))
for (i in 1:nind){
  z[i,f[i]] <- initial.state[i]
}

set.seed(2)                     # Initialize the RNGs in R
# Propagate alive/dead process forwards via transition rule (=TPM=OMEGA)
for (i in 1:nind){
  for (t in (f[i]+1):nyears){
    departure.state <- z[i,t-1]
    arrival.state <- which(rmultinom(1,1, TPM[departure.state,])==1)
    z[i,t] <- arrival.state
  } #t
} #i
z                               # Not shown, but useful if you do look at this
```

That is the true state, but this is not what we are usually able to observe. We need now to simulate the observations, accounting for false-negative detection errors.

```
# Observation process: simulate observations using observation matrix OPM (=THETA)
y <- array(3, dim=c(nind, nyears))
for (i in 1:nind){
  y[i,f[i]] <- z[i,f[i]]
  for (t in (f[i]+1):nyears){
    true.state <- z[i,t-1]
    observed.state <- which(rmultinom(1,1, OPM[true.state,])==1)
    y[i,t] <- observed.state
  } #t
} #i
```

Let's look at the simulated data set; note that label 1 is an individual observed in state A, 2 is an observation of state B and 3 denotes "not captured". In standard multistate capture-recapture data we would typically code the observed state "not recaptured" with a 0. However, in BUGS we use the codes of the observed states as indices in the model, and hence, they must be integers that go from 1 up to the number of latent states plus 1. Thus, when you deal with real data, make sure that you replace the 0 with the appropriate nonzero integer code for "not recaptured."

```
head(y)

     [,1] [,2] [,3] [,4] [,5] [,6]
[1,]    1    1    1    3    3    3
[2,]    1    1    3    3    3    3
[3,]    1    1    1    3    3    3
[4,]    1    3    1    3    3    3
[5,]    1    1    2    2    1    1
[6,]    1    3    3    3    3    3
```

We are ready now to fit the simplest possible multistate model in JAGS for illustration. We start with the data bundle and then write the model in BUGS language.

```
# Data bundle
jags.data <- list(y=y, f=f, nind=nind, nyears=ncol(y))
str(jags.data)

List of 4
 $ y     : num [1:500, 1:6] 1 1 1 1 1 1 1 1 1 1 ...
 $ f     : int [1:500] 1 1 1 1 1 1 1 1 1 1 ...
 $ nind  : int 500
 $ nyears: num 6
```

Our BUGS code of the multistate model has three major parts, "priors and linear models," "Define state-transition and observation matrices" and the "Likelihood." The likelihood part is written in a general way such that it does not need to be modified regardless of which multistate model is run. In the "Define state-transition and observation matrices" section, the structure of the model is specified that may change from model to model. Finally, as always, the "priors and linear models" section is where we define how we want to model the primary parameters of the model and their priors.

```
# Write JAGS model file
cat(file="model18.txt", "
model {
# Priors and linear models
for (t in 1:(nyears-1)){
  phiA[t] <- mean.phi[1]
  phiB[t] <- mean.phi[2]
  psiAB[t] <- mean.psi[1]
  psiBA[t] <- mean.psi[2]
  pA[t] <- mean.p[1]
  pB[t] <- mean.p[2]
}
for (u in 1:2){
  mean.phi[u] ~ dunif(0, 1)     # Priors for mean state-specific survival
  mean.psi[u] ~ dunif(0, 1)     # Priors for mean transitions
  mean.p[u] ~ dunif(0, 1)       # Priors for mean state-specific recapture
}

# Define state-transition and observation matrices
for (i in 1:nind){
  # Define probabilities of state S(t+1) given S(t)
  #   (Transition probability matrix, or OMEGA)
  for (t in f[i]:(nyears-1)){
    TPM[1,i,t,1] <- phiA[t] * (1-psiAB[t])
    TPM[1,i,t,2] <- phiA[t] * psiAB[t]
    TPM[1,i,t,3] <- 1-phiA[t]
    TPM[2,i,t,1] <- phiB[t] * psiBA[t]
    TPM[2,i,t,2] <- phiB[t] * (1-psiBA[t])
    TPM[2,i,t,3] <- 1-phiB[t]
    TPM[3,i,t,1] <- 0
    TPM[3,i,t,2] <- 0
    TPM[3,i,t,3] <- 1
```

```
    # Define probabilities of observation O(t) given S(t)
    #   (Observation probability matrix, or THETA)
    OPM[1,i,t,1] <- pA[t]
    OPM[1,i,t,2] <- 0
    OPM[1,i,t,3] <- 1-pA[t]
    OPM[2,i,t,1] <- 0
    OPM[2,i,t,2] <- pB[t]
    OPM[2,i,t,3] <- 1-pB[t]
    OPM[3,i,t,1] <- 0
    OPM[3,i,t,2] <- 0
    OPM[3,i,t,3] <- 1
  } #t
} #i

# Likelihood
for (i in 1:nind){
  # Define latent state at first capture
  z[i,f[i]] <- y[i,f[i]]
  for (t in (f[i]+1):nyears){
    # State process: draw S(t) given S(t-1)
    z[i,t] ~ dcat(TPM[z[i,t-1], i, t-1,])
    # Observation process: draw O(t) given S(t)
    y[i,t] ~ dcat(OPM[z[i,t], i, t-1,])
  } #t
} #i
}
")
```

The model requires initial values for the latent states that must not contradict either the model (e.g., no states that are impossible under the model), or the data (e.g., when an individual is seen at site 2 on some occasion, its latent state must then be initialized in state 1). In complicated models it can be quite tricky to find initial values that JAGS finds acceptable. Here we provide a function that creates suitable initial values for our specific model. Other multistate capture-recapture model may need an adaptation of this function.

```
# Function to create a matrix of initial values for latent state z
zInitMS <- function(ch, f){
  states <- max(ch, na.rm=TRUE)
  known.states <- 1:(states-1)
  v <- which(ch==states)
  ch[v] <- sample(known.states, length(v), replace=TRUE)
  for (i in 1:nrow(ch)) ch[i,1:f[i]] <- NA
  return(ch)
}

# Initial values
inits <- function(){list(z=zInitMS(y, f))}

# Parameters monitored
parameters <- c("mean.phi", "mean.psi", "mean.p")

# MCMC settings
ni <- 3000; nb <- 1000; nc <- 3; nt <- 1; na <- 1000

# Call JAGS from R (ART 3 min), check convergence and summarize posteriors
out21 <- jags(jags.data, inits, parameters, "model18.txt", n.iter=ni, n.burnin=nb, n.chains=nc,
  n.thin=nt, n.adapt=na, parallel=TRUE)
traceplot(out21)                        # Not shown
print(out21)

              mean    sd   2.5%    50%  97.5% overlap0 f  Rhat n.eff
mean.phi[1]  0.864 0.021  0.821  0.864  0.903    FALSE 1 1.002   957
mean.phi[2]  0.813 0.028  0.756  0.814  0.866    FALSE 1 1.001  6000
mean.psi[1]  0.120 0.019  0.086  0.119  0.159    FALSE 1 1.001  1503
mean.psi[2]  0.230 0.026  0.183  0.229  0.283    FALSE 1 1.001  6000
mean.p[1]    0.694 0.032  0.630  0.695  0.754    FALSE 1 1.004   885
mean.p[2]    0.463 0.042  0.383  0.462  0.549    FALSE 1 1.001  2843
```

The model runs relatively slowly, but convergence is acceptable. The estimated parameters show reasonable agreement with the values used to simulate the data.

Multistate capture-recapture models make additional assumptions compared with the single-state capture-recapture models (Pradel et al., 2003; Pradel et al., 2005; Lebreton et al., 2009). Goodness-of-fit tests for them are well developed and software to compute them is available (Pradel et al., 2003; Choquet et al., 2009a; Gimenez et al., 2018).

4.5.2.2 Multinomial Formulation

We can also analyze multistate capture-recapture data with the multinomial likelihood in a similar way as we did with the CJS model (Section 4.5.1.2) with the advantage of faster computation time and at the price of a loss of the ability to model truly individual effects. For this, the data again need to be summarized in the m-array format, i.e., the observed capture histories are broken up into fragments the frequencies of which enter into the m-array. The multistate m-array is created from the multistate capture-recapture data in an analogous way as we have seen for the single-state capture-recapture data. The difference to the single-state case is that we must keep track of the states in which individuals are released and first recaptured. Specifically, the elements of the multistate m-array are the state-specific numbers of released individuals in each occasion (R_t^s: number of released individuals at occasion t in state s) and the number of individuals *first* recaptured in state r given release at a specific occasion in state s ($m_{t,j}^{s,r}$: number of individuals first recaptured at occasion j in state r given that they have been released at occasion t in state s, Table 4.4).

The multistate m-array data are then analyzed with a series of multinomial distributions,

$$\boldsymbol{m}_{t,}^{s,} \sim Multinomial\left(\boldsymbol{\pi}_{t,}^{s,}, R_t^s\right),$$

where $\boldsymbol{m}_{t,}^{s,}$ is a vector with the number of recaptured and not-recaptured individuals released in state s at time t, R_t^s is the number of released individuals in state s at time t, and $\boldsymbol{\pi}_{t,}^{s,}$ is a vector with the cell probabilities to be estimated that distribute the R_t^s individuals into vector $\boldsymbol{m}_{t,}^{s,}$. A cell probability $\pi_{t,j}^{s,r}$ of the m-array represents the probability that an individual released in state s at time t is next recaptured in state r at time j. But as for the CJS model, this probability is again expressed as a function of state-specific survival, recapture, and movement probabilities. Expressions for the probabilities $\pi_{t,j}^{s,r}$ can become quite complex, in particular when there are multiple occasions between release and subsequent recapture.

Table 4.4 Schematic of a multistate m-array data representation for a multistate data set with two states and four capture occasions. Note that the multistate m-array aggregation of a data set will not usually contain the vector of release totals R.

			Numbers at First Re-encounter Occasion (State of Re-encounter)						
			2		3		4		
Release Occasion	**State of Release**	**Number Released**	**(1**	**2)**	**(1**	**2)**	**(1**	**2)**	**Numbers Never Re-encountered**
1	1	R_1^1	$m_{1,2}^{1,1}$	$m_{1,2}^{1,2}$	$m_{1,3}^{1,1}$	$m_{1,3}^{1,2}$	$m_{1,4}^{1,1}$	$m_{1,4}^{1,2}$	$R_1^1 - \sum_{j=2,r=1}^{4,2} m_{1,j}^{1,r}$
1	2	R_1^2	$m_{1,2}^{2,1}$	$m_{1,2}^{2,2}$	$m_{1,3}^{2,1}$	$m_{1,3}^{2,2}$	$m_{1,4}^{2,1}$	$m_{1,4}^{2,2}$	$R_1^2 - \sum_{j=2,r=1}^{4,2} m_{1,j}^{2,r}$
2	1	R_2^1	0	0	$m_{2,3}^{1,1}$	$m_{2,3}^{1,2}$	$m_{2,4}^{1,1}$	$m_{2,4}^{1,2}$	$R_2^1 - \sum_{j=3,r=1}^{4,2} m_{2,j}^{1,r}$
2	2	R_2^2	0	0	$m_{2,3}^{2,1}$	$m_{2,3}^{2,2}$	$m_{2,4}^{2,1}$	$m_{2,4}^{2,2}$	$R_2^2 - \sum_{j=3,r=1}^{4,2} m_{2,j}^{2,r}$
3	1	R_3^1	0	0	0	0	$m_{3,4}^{1,1}$	$m_{3,4}^{1,2}$	$R_3^1 - \sum_{r=1}^{2} m_{3,4}^{1,r}$
3	2	R_3^2	0	0	0	0	$m_{3,4}^{2,1}$	$m_{3,4}^{2,2}$	$R_3^2 - \sum_{r=1}^{2} m_{3,4}^{2,r}$

Adapted from Williams, B.K., Nichols, J.D., Conroy, M.J., 2002. Analysis and Management of Animal Populations. Modeling, Estimation, and Decision Making. Academic Press, San Diego.

In that case, there is uncertainty about the states of each individual during all occasions without recapture. However, use of matrix multiplication simplifies computation of these probabilities.

We define $\boldsymbol{\Psi}_t$ to be the state-transition matrix between occasion t and $t+1$, $\boldsymbol{p}_t$ is a state-specific vector of re-encounter probabilities at occasion $t+1$, and $\boldsymbol{q}_t$ is a vector with state-specific probabilities of non-encounter (i.e., $\boldsymbol{q}_t = 1-\boldsymbol{p}_t$). Further, we define $D(\boldsymbol{x})$ to be a square matrix with diagonal $\boldsymbol{x}$. For our example with two locations between which individuals may move we get the following matrices:

$$\boldsymbol{\Psi}_t = \begin{bmatrix} \phi_{A,t}(1-\psi_{AB,t}) & \phi_{A,t}\psi_{AB,t} \\ \phi_{B,t}\psi_{BA,t} & \phi_{B,t}(1-\psi_{BA,t}) \end{bmatrix},$$

$$D(\boldsymbol{p}_t) = \begin{bmatrix} p_{A,t} & 0 \\ 0 & p_{B,t} \end{bmatrix} \text{ and } D(\boldsymbol{q}_t) = \begin{bmatrix} 1-p_{A,t} & 0 \\ 0 & 1-p_{B,t} \end{bmatrix}.$$

The cell probabilities in π of the multinomial m-array are calculated by matrix multiplication based on the elements defined above as shown in the following table:

	Re-encounter Occasion			
Release Occasion	**2**	**3**	**4**	**Never Re-encountered**
1	$\boldsymbol{\Psi}_1 D(\boldsymbol{p}_1)$	$\boldsymbol{\Psi}_1 D(\boldsymbol{q}_1)\boldsymbol{\Psi}_2 D(\boldsymbol{p}_2)$	$\boldsymbol{\Psi}_1 D(\boldsymbol{q}_1)\boldsymbol{\Psi}_2 D(\boldsymbol{q}_2)\boldsymbol{\Psi}_3 D(\boldsymbol{p}_3)$	$1-\Sigma_1$
2	0	$\boldsymbol{\Psi}_2 D(\boldsymbol{p}_2)$	$\boldsymbol{\Psi}_2 D(\boldsymbol{q}_2)\boldsymbol{\Psi}_3 D(\boldsymbol{p}_3)$	$1-\Sigma_2$
3	0	0	$\boldsymbol{\Psi}_3 D(\boldsymbol{p}_3)$	$1-\Sigma_3$

The transition matrix $\boldsymbol{\Psi}_t$ corresponds to the state-transition matrix $\boldsymbol{\Omega}_{1...3,1...3,i,t}$ that we defined above for the implementation of the state-space formulation of the model (Section 4.5.2.1), but there are two main differences. First, now we do not include the dead state and as a consequence $\boldsymbol{\Psi}_t$ is no longer row-stochastic, i.e., now rows do not sum to 1. Second, full individual effects can no longer be modeled because the individual identity is "lost" as in the single-state m-array.

The matrix containing the re-encounter probabilities $D(\boldsymbol{p}_t)$corresponds to the observation matrix $\boldsymbol{\Theta}_{1...3,1...3,i,t}$ of the state-space formulation, but there are again slight differences. Now, we do not include the "not recaptured" observation type, which again leads to a matrix where rows do not sum to 1. And in this part of the model, individual identity is lost.

We now have the knowledge necessary to analyze our multistate capture-recapture data in the m-array formulation. We first construct the m-array from the individual capture-recapture data, and for that can again use function `marray`. It requires that the multistate capture-recapture data are in the usual form, i.e., non-encounters are coded with a 0, and the observed states with sequentially increasing integers, e.g., with 1, 2 and 3 when there are three observed states, while labels 1, 3, and 4 or A, B and C would not work. By default, the `marray` function assumes there are no unobservable states, but it works also when there are unobservable states, in which case we must specify their number. Unobservable states are labeled automatically. For example, if there are three observable and one unobservable state, state 4 corresponds to the unobservable state.

In our simulated data we have coded by 3 an individual that is not recaptured. We need to replace the 3 with a 0, and then look at some capture histories to improve our understanding of this formatting business.

```
y[y==3] <- 0

y[45:55,]
      [,1] [,2] [,3] [,4] [,5] [,6]
 [1,]    1    0    1    0    0    0
 [2,]    1    1    1    0    0    0
 [3,]    1    1    0    2    0    0
 [4,]    1    0    1    1    0    0
 [5,]    1    1    0    1    0    0
 [6,]    1    1    1    2    1    1
 [7,]    0    1    0    0    1    0
 [8,]    0    1    0    0    0    0
 [9,]    0    1    1    2    0    0
[10,]    0    1    1    1    0    0
[11,]    0    1    0    0    1    1
```

This looks good. We next compute the multistate m-array and again have a look at it.

```
marr <- marray(y)
marr
        recaptured
released Y2.S1 Y2.S2 Y3.S1 Y3.S2 Y4.S1 Y4.S2 Y5.S1 Y5.S2 Y6.S1 Y6.S2 never
   Y1.S1    34     0     8     1     1     1     0     0     1     0     4
   Y1.S2     0    17     8     2     3     2     1     1     0     0    16
   Y2.S1     0     0    51     3     9     2     3     1     0     1    14
   Y2.S2     0     0     5    24     8     8     1     1     1     1    18
   Y3.S1     0     0     0     0    56    10    13     5     3     1    34
   Y3.S2     0     0     0     0     5    32     6     3     4     2    28
   Y4.S1     0     0     0     0     0     0    72     3    14     3    40
   Y4.S2     0     0     0     0     0     0    14    33    15     5    38
   Y5.S1     0     0     0     0     0     0     0     0    85     9    66
   Y5.S2     0     0     0     0     0     0     0     0    10    22    65
```

The row and column lables of the produced table show to which occasion (Y for years) and state (S) an entry refers to. The m-array corresponds to our expectations. The capture histories have six occasions and two states, thus there are (6-1) * 2 = 10 possible rows (releases) and columns (recaptures). In addition, there is the last column that contains the number of released individuals that are never recaptured. The vector containing total number of released individuals is not provided but can be calculated as the row sums from the m-array.

Next, we bundle the data. Obviously, we need the m-array, the number of released individuals and the number of capture occasions. In addition, we need to provide the number of states (2 in our case) and two matrices ("zero" and "ones"): an all-zero matrix and an identity matrix with size given by the number of states. These matrices are used to perform the matrix multiplications in JAGS. If you do not understand them entirely, there is no need for you to worry about them. You need them, but our code JAGS will take care to do the right thing automatically for any multistate model you may wish to fit.

```
# Bundle data
# Calculate the number of states
ns <- length(unique(as.numeric(y))) - 1
jags.data <- list(marr=marr, nyears=ncol(y), rel=rowSums(marr), ns=ns, zero=matrix(0, ns, ns),
  ones=diag(ns))
str(jags.data)

List of 6
 $ marr : num [1:10, 1:11] 34 0 0 0 0 0 0 0 0 0 ...
 $ nyears: int 6
 $ rel  : num [1:10] 50 50 84 67 122 80 132 105 160 97
 $ ns   : num 2
 $ zero : num [1:2, 1:2] 0 0 0 0
 $ ones : num [1:2, 1:2] 1 0 0 1
```

Presumably, the BUGS code for the model will look a bit scary. However, the main part of the code is just the calculation of the cell probabilities for the multinomial distributions, and it is written in a general way such that you do not need to touch it, regardless of which multistate model you want to fit. The only parts of the code that need to be modified when you fit a different multistate model are the section "Prior and linear models" and "Define state-transition and re-encounter probabilities." The former defines the linear model for the parameters and their priors and the latter is the definition of the basic structure of the multistate model, i.e., the state-transition matrix ($\mathbf{\Psi}$) and the vector with the recapture probabilities ($\boldsymbol{p}$).

We write the model, provide the usual information about initial values, parameters to be monitored and MCMC specification, and launch the model.

```
# Write JAGS model file
cat(file="model19.txt", "
model {
# Priors and linear models
for (t in 1:(nyears-1)){
  phiA[t] <- mean.phi[1]
  phiB[t] <- mean.phi[2]
  psiAB[t] <- mean.psi[1]
  psiBA[t] <- mean.psi[2]
  pA[t] <- mean.p[1]
  pB[t] <- mean.p[2]
}

for (u in 1:2){
  mean.phi[u] ~ dunif(0, 1)     # Priors for mean state-specific survival
  mean.psi[u] ~ dunif(0, 1)     # Priors for mean transitions
  mean.p[u] ~ dunif(0, 1)       # Priors for mean state-specific recapture
}

# Define state-transition and re-encounter probabilities
for (t in 1:(nyears-1)){
  psi[1,t,1] <- phiA[t] * (1-psiAB[t])
  psi[1,t,2] <- phiA[t] * psiAB[t]
  psi[2,t,1] <- phiB[t] * psiBA[t]
  psi[2,t,2] <- phiB[t] * (1-psiBA[t])

  po[1,t] <- pA[t]
  po[2,t] <- pB[t]
}

# From here onwards, no changes needed regardless of which model is fitted
# Calculate probability of non-encounter (dq) and reshape the array for the encounter
# probabilities
for (t in 1:(nyears-1)){
  for (s in 1:ns){
    dp[s,t,s] <- po[s,t]
    dq[s,t,s] <- 1-po[s,t]
  } #s
  for (s in 1:(ns-1)){
    for (m in (s+1):ns){
      dp[s,t,m] <- 0
      dq[s,t,m] <- 0
    } #s
  } #m
  for (s in 2:ns){
    for (m in 1:(s-1)){
      dp[s,t,m] <- 0
      dq[s,t,m] <- 0
    } #s
  } #m
} #t

# Define the multinomial likelihood
for (t in 1:((nyears-1)*ns)){
  marr[t,1:(nyears *ns-(ns-1))] ~ dmulti(pi[t,], rel[t])
}

# Define cell probabilities of the multistate m-array
# Matrix U: product of probabilities of state-transition and non-encounter (needed because
# there is no product function for matrix multiplication in JAGS)
for (t in 1:(nyears-2)){
  U[(t-1)*ns+(1:ns),(t-1)*ns+(1:ns)] <- ones
  for (j in (t+1):(nyears-1)){
    U[(t-1)*ns+(1:ns),(j-1)*ns+(1:ns)] <- U[(t-1)*ns+(1:ns),(j-2)*ns+(1:ns)] %*% psi[,t,] %*%
      dq[,t,]
  } #j
} #t
U[(nyears-2)*ns+(1:ns), (nyears-2)*ns+(1:ns)] <- ones
```

```
for (t in 1:(nyears-2)){
  # Diagonal
  pi[(t-1)*ns+(1:ns),(t-1)*ns+(1:ns)] <- U[(t-1)*ns+(1:ns),(t-1)*ns+(1:ns)] %*% psi[,t,] %*%
    dp[,t,]
  # Above main diagonal
  for (j in (t+1):(nyears-1)){
    pi[(t-1)*ns+(1:ns),(j-1)*ns+(1:ns)] <- U[(t-1)*ns+(1:ns),(j-1)*ns+(1:ns)] %*% psi[,j,] %*%
      dp[,j,]
  } #j
 } #t
pi[(nyears-2)*ns+(1:ns),(nyears-2)*ns+(1:ns)] <- psi[,nyears-1,] %*% dp[,nyears-1,]

# Below main diagonal
for (t in 2:(nyears-1)){
  for (j in 1:(t-1)){
    pi[(t-1)*ns+(1:ns),(j-1)*ns+(1:ns)] <- zero
  } #j
} #t

# Last column: probability of non-recapture
for (t in 1:((nyears-1)*ns)){
  pi[t,(nyears*ns-(ns-1))] <- 1-sum(pi[t,1:((nyears-1)*ns)])
} #t
}
")

# Initial values
inits <- function(){list(mean.phi=runif(2, 0, 1))}

# Parameters monitored
parameters <- c("mean.phi", "mean.psi", "mean.p")

# MCMC settings
ni <- 3000; nb <- 1000; nc <- 3; nt <- 1; na <- 1000

# Call JAGS from R (ART <1 min), check convergence and summarize posteriors
out22 <- jags(jags.data, inits, parameters, "model19.txt", n.iter=ni, n.burnin=nb, n.chains=nc,
  n.thin=nt, n.adapt=na, parallel=TRUE)
par(mfrow=c(2, 2)); traceplot(out22)      # Not shown
print(out22)

              mean    sd   2.5%    50%   97.5% overlap0 f  Rhat n.eff
mean.phi[1]  0.863 0.021  0.822  0.863  0.902    FALSE 1 1.000  6000
mean.phi[2]  0.813 0.029  0.756  0.813  0.867    FALSE 1 1.000  6000
mean.psi[1]  0.119 0.020  0.085  0.118  0.161    FALSE 1 1.000  6000
mean.psi[2]  0.230 0.027  0.180  0.229  0.285    FALSE 1 1.002  2185
mean.p[1]    0.694 0.032  0.630  0.695  0.755    FALSE 1 1.000  6000
mean.p[2]    0.466 0.043  0.385  0.465  0.553    FALSE 1 1.000  6000
```

The chains for all parameters have converged well and up to MC error the posteriors are identical to those obtained from the state-space formulation of the multistate model. However, compared with the state-space formulation, the m-array formulation runs *much* faster. The reason for the increased speed is that the dimension of the data is now only 10×11, whereas with the state-space formulation it was 500×6. Convergence is typically obtained with a lower number of MCMC iterations with the multinomial likelihood than with the state-space likelihood, so it is much more efficient. Therefore, we generally recommend using the multinomial formulation of this model, unless the state-space formulation is required. An example for the latter may be when individual covariates such as body size are analyzed. Furthermore, if the number of states becomes very large, then the dimension of the m-array also greatly increases, and the computation time can become a burden also when using the multinomial likelihood. In that case it may be necessary to fit the marginal likelihood of the data were the latent states are eliminated (Turek et al., 2016; Yackulic et al., 2020).

With the multinomial likelihood the modeling of covariates is restricted in principle to those that vary temporally, as we have seen already for the single-state case. Differences among individuals must be coded as group effects and can be implemented by producing m-arrays for each group. This works well when the number of groups is relatively low (e.g., for sex), but becomes impractical for larger numbers of groups and will not be practically possible for truly individual effects. Then the use of the state-space likelihood is advised.

In the context of multistate models, we have not shown so far how to implement age dependence. However, we have exactly the same options as illustrated for the single-state capture-recapture models. That is, we can create an individual covariate that indicates the age of each individual at every occasion and use that as a covariate when using the state-space likelihood. Or we work with the multinomial likelihood and create an age-dependent multistate m-array. However, the most straightforward approach for implementation of age structure in a multistate model is by defining states that correspond to age classes. For example, when we want to estimate age-dependent survival with two age classes (same model as in Section 4.5.1.1), we define the latent states "juvenile," "adult" and "dead" and the observations "seen as juvenile," "seen as adult" and "not seen." The state-transition and observation matrices are then

$$\boldsymbol{\Omega}_{1\ldots3,1\ldots3,i,t} = \begin{bmatrix} 0 & \phi_{j,i,t} & 1-\phi_{j,i,t} \\ 0 & \phi_{a,i,t} & 1-\phi_{a,i,t} \\ 0 & 0 & 1 \end{bmatrix} \text{ and } \boldsymbol{\Theta}_{1\ldots3,1\ldots3,i,t} = \begin{bmatrix} 0 & 0 & 1 \\ 0 & p_{i,t} & 1-p_{i,t} \\ 0 & 0 & 1 \end{bmatrix}.$$

Aging is now implemented by the transition of the juvenile to the adult state—a juvenile at time t has a probability of zero of again being a juvenile at time $t+1$, but if it survives (governed by juvenile survival), it is an adult on the next occasion. If the multinomial likelihood is chosen to fit this model, the corresponding transition matrix and the vector with the recapture probabilities is this:

$$\boldsymbol{\Psi}_t = \begin{bmatrix} 0 & \phi_{j,t} \\ 0 & \phi_{a,t} \end{bmatrix} \text{ and } \boldsymbol{p}_t = \begin{bmatrix} 0 \\ p_t \end{bmatrix}.$$

In both cases the data are coded in such a way that a 1 indicates release as a juvenile and a 2 a release as an adult. Hence, the age information is provided directly by the data.

Multistate models are a common component model in IPMs. This is not surprising given the large flexibility that these models offer. To give a couple of examples, they were used for modeling movement between sites (McCrea et al., 2010; Péron et al., 2010b; see also Chapter 18), for the modeling of recruitment that extends over several age classes (Szostek et al., 2014; Margalida et al., 2020), and to estimate productivity (Regehr et al., 2018) or mortality causes (Tenan et al., 2012).

4.5.3 DEAD-RECOVERY DATA

There is another type of data on marked individuals that also provides information about survival: the so-called "dead recoveries." Observers retrieve and report marked individuals that were harvested (i.e., killed) or died from natural causes. As not all individuals that die will be found and reported, we must estimate the recovery probability to obtain unbiased estimates of survival probability.

The typical sampling protocol for this type of data is to mark a sample of individuals and to collect information about the time of death of those marked individuals found dead. The latter information typically comes from members of the public who find and report dead marked individuals. For harvested species hunters usually provide the bulk of data. Dead-recovery data are often collected in birds (Baillie and Schaub, 2009; Du Feu et al., 2016), but also in other taxa such as fish (Nater et al., 2020) or mammals (Servanty et al., 2010). Important primary sources for this class of models are Brownie et al. (1985) and Williams et al. (2002); for the model in general and its Bayesian implementation in particular, see Brooks et al. (2000a, b, 2002), Barry et al. (2003), Gimenez et al. (2007) and Chapter 8 in Kéry and Schaub (2012).

There is a bewildering variety of synonyms for this type of data and the associated models, including "dead-recovery," "mark-recovery," "ring-recovery," "band-recovery" and "tag-recovery." We can again implement these models in the state-space or in the multinomial formulation, and we show both for a simple data set based on simulations.

4.5.3.1 State-Space Formulation

For the development of the dead-recovery model in the state-space formulation we distinguish again between the true ($z_{i,t}$) and the observed state ($y_{i,t}$, our data) of individual i at occasion t and denote as f_i the occasion at which individual i was marked. The latent state matrix z is defined as for the single-state capture-recapture model: it is $z_{i,t} = 1$ when an individual i at time t is alive and $z_{i,t} = 0$ when it is dead. The observed state, however, has a slightly different meaning compared with the capture-recapture models: whenever an individual is marked or recovered dead, $y_{i,t} = 1$, and otherwise $y_{i,t} = 0$. Hence, $y_{i,t} = 1$ at some other than the marking occasion indicates that individual i was found and recovered as dead in the interval from $t-1$ to t, and hence, we assume that this is also when it died. This is an important assumption, and we say that the individual has "recently died." To describe the dead-recovery model as an SSM (Gimenez et al., 2007; Royle, 2008; King, 2012) we condition on the initial state (that is, we say that we know an individual is alive when marked and released and do not model its state probabilistically) and define the model as a linked sequences of Bernoulli trials for each ensuing time interval. We write the model by the following three expressions.

$z_{i,f(i)} = 1$ Initial state (deterministic): individual known to be alive at f_i

$z_{i,t+1} \sim Bernoulli(z_{i,t} s_{i,t})$ State process: stochastic transitions with survival probability

$y_{i,t} \sim Bernoulli((z_{i,t-1} - z_{i,t}) r_{i,t})$ Observation process: stochastic observation with recovery probability

The parameters for the Bernoulli trials are the true survival probability s and recovery probability r. Survival in this model is not restricted to any spatial area and thus we assume that true survival is estimated. For some applications it is interesting to decompose the recovery process into different subprocesses with their associated probabilities, such as the probability that a dead individual is actually detected, is retrieved and is reported (Williams et al., 2002). We note again the implicit assumption that recovery is immediate, i.e., in the time interval when an individual died and not later.

We now simulate a dead-recovery data set under a simple model. We simulate the fate of a number of individuals over a number of years, starting at the occasion when they are first captured and marked, after which they survive with probability s and when dead are recovered with probability r. Both probabilities are assumed to be identical for all individuals and constant over time in our example.

```
# Choose constants in simulation
nmarked <- 100                      # Number of marked individuals at each occasion
nyears <- 11                        # Number of years
s <- 0.8                            # Survival probability
r <- 0.2                            # Recovery probability
```

An individual first caught in the last year of the study has no opportunity of being recovered during the study; hence, such individuals cannot contribute any information about survival and are ignored in the analysis. Therefore, in our simulation we choose first-capture occasions only up to the penultimate year.

```
# Determine occasion when an individual first captured and marked
f <- rep(1:(nyears-1), each=nmarked)
nind <- length(f)                   # Total number of marked individuals

# State or ecological process
# Simulate true system state
z <- array(NA, dim=c(nind, nyears)) # Empty alive/dead matrix

# Initial conditions: all individuals alive at f(i)
for (i in 1:nind){
  z[i,f[i]] <- 1
}

set.seed(3)                         # Initialize the RNGs in R
# Propagate alive/dead process forwards via transition rule
# Alive individuals survive with probability s
for (i in 1:nind){
  for (t in (f[i]+1):nyears){
    z[i,t] <- rbinom(1, 1, z[i,t-1] * s)
  } #t
} #i
z                                   # Look at the true state (not shown, but insightful)

# Observation process: simulate observations
y <- array(0, dim=c(nind, nyears))
for (i in 1:nind){
  y[i,f[i]] <- 1
  for(t in (f[i]+1):nyears){
    y[i,t] <- rbinom(1, 1, (z[i,t-1]-z[i,t]) * r)
  } #t
} #i
```

Let's look at the simulated data set.

```
y                                   # Complete simulated data set(not shown)

# Compute the total number of dead recoveries
sum(rowSums(y, na.rm=TRUE)==2)

[1] 131
```

The data bundle contains the dead-recovery data, the vector with the information about the marking occasion of each individual and the dimensions of the data. We bundle the data and produce a summary.

```
# Data bundle
jags.data <- list(y=y, f=f, nind=nind, nyears=ncol(y))
str(jags.data)

List of 4
 $ y      : num [1:1000, 1:11] 1 1 1 1 1 1 1 1 1 1 ...
 $ f      : int [1:1000] 1 1 1 1 1 1 1 1 1 1 ...
 $ nind   : int 1000
 $ nyears: num 11
```

The BUGS code for this model consists of the usual sections and elements. The likelihood is again written in a very general way so that no changes are needed for fitting any kind of dead-recovery model. We define the structure of the parameter matrices by a linear model with an intercept only (i.e., that survival and recovery are constant over time and identical for all individuals) and place priors on the estimated quantities.

```
# Write JAGS model file
cat(file="model20.txt", "
model {
# Priors and linear models
s.const ~ dunif(0, 1)                    # Vague prior for constant s
r.const ~ dunif(0, 1)                    # Vague prior for constant r

for (i in 1:nind){                       # Loop over individuals
  for (t in f[i]:(nyears-1)){            # Loop over time intervals/occasions
    s[i,t] <- s.const                    # Here model pattern in s ...
    r[i,t] <- r.const                    # ... and r
  } #t
} #i

# Likelihood
for (i in 1:nind){
  # Define latent state at first capture
  z[i,f[i]] <- 1
  for (t in (f[i]+1):nyears){
    # State process
    z[i,t] ~ dbern(z[i,t-1] * s[i,t-1])
    # Observation process
    y[i,t] ~ dbern((z[i,t-1]- z[i,t]) * r[i,t-1])
  } #t
} #i
}
")
```

As is typical when we estimate latent states, we must provide initial values that do not contradict the model and the data. Our function `zInitDR` creates adequate initial values based on the dead-recovery data. Next, we define the usual settings for model fitting and launch the model.

```
# Initial values
inits <- function(){list(z=zInitDR(y))}   # Function in IPMbook package

# Parameters monitored
parameters <- c("s.const", "r.const")

# MCMC settings
ni <- 30000; nb <- 5000; nc <- 3; nt <- 10; na <- 5000

# Call JAGS from R (ART 13 min), check convergence and summarize posteriors
out23 <- jags(jags.data, inits, parameters, "model20.txt", n.iter=ni, n.burnin=nb, n.chains=nc,
  n.thin=nt, n.adapt=na, parallel=TRUE)
traceplot(out23)                          # Not shown
acf(out23$sims.list$s.const)              # Check out autocorrelation for s (not shown)
print(out23, 3)

           mean    sd  2.5%   50% 97.5% overlap0 f  Rhat n.eff
s.const   0.777 0.037 0.709 0.775 0.852    FALSE 1 1.015  1992
r.const   0.198 0.024 0.159 0.195 0.256    FALSE 1 1.015   811
```

We need long chains for the model to converge because of the strong autocorrelation. The estimates agree well with the values used to simulate the data. We note that convergence is harder to obtain when there are fewer recoveries.

4.5.3.2 Multinomial Formulation

Dead-recovery data can also be analyzed with the more efficient multinomial formulation. The data must again be formatted as an m-array; though this time this is the recovery m-array. It summarizes the number of individuals marked at a given occasion and recovered at each later occasion and how many were never recovered.

When the dead-recovery data are in the m-array format, they can be analyzed with the multinomial likelihood, and thus,

$$\boldsymbol{m}_t \sim Multinomial(\pi_t, R_t),$$

where $\boldsymbol{m}_t$ is a vector with the occasion-specific numbers of recovered and not-recovered individuals marked at time t, π_t is a vector containing the multinomial cell probabilities that themselves are expressed as functions of s and r, and R_t is the total number of marked individuals at time t, or the size of the release cohort. The cell probabilities in an m-array for a dead-recovery data with time-dependence assumed for both survival and recovery probabilities four occasions are these:

	Recovery Occasion			
Release Occasion	**2**	**3**	**4**	**Never Recovered**
1	$(1-s_1)r_1$	$s_1(1-s_2)r_2$	$s_1s_2(1-s_3)r_3$	$1-\sum$ rel. occ. 1
2	0	$(1-s_2)r_2$	$s_2(1-s_3)r_3$	$1-\sum$ rel. occ. 2
3	0	0	$(1-s_3)r_3$	$1-(1-s_3)r_3$

As an illustration, the multinomial cell probabilities of the third release cohort are $\pi_3 = [0 \quad 0 \quad (1-s_3)r_3 \quad 1-(1-s_3)r_3]$.

We adopt our usual workflow to fit this model. First, we create the dead-recovery m-array, which is computed with our function `marrayDead`.

```
marr <- marrayDead(y)
```

We then bundle the necessary data into a list and write the model in the BUGS language.

```
# Data bundle
jags.data <- list(marr=marr, rel=rowSums(marr), nyears=ncol(y))
str(jags.data)

List of 3
 $ marr  : num [1:10, 1:11] 4 0 0 0 0 0 0 0 0 0 ...
 $ rel   : num [1:10] 100 100 100 100 100 100 100 100 100 100
 $ nyears: int 11

# Write JAGS model file
cat(file="model21.txt", "
model {
# Priors and linear models
s.const ~ dunif(0, 1)            # Vague prior for constant s
r.const ~ dunif(0, 1)            # Vague prior for constant r

for (t in 1:(nyears-1)){         # Loop over time intervals/occasions
  s[t] <- s.const                # Here model pattern in s ...
  r[t] <- r.const                # ... and r
} #t

# Likelihood
for (t in 1:(nyears-1)){
  marr[t,1:nyears] ~ dmulti(pi[t,], rel[t])
}
# Define the cell probabilities of the m-array
for (t in 1:(nyears-1)){
  # Main diagonal
  pi[t,t] <- (1-s[t])*r[t]
  # Above main diagonal
  for (j in (t+1):(nyears-1)){
    pi[t,j] <- prod(s[t:(j-1)])*(1-s[j])*r[j]
  } #j
  # Below main diagonal
  for (j in 1:(t-1)){
    pi[t,j] <- 0
  } #j
} #t
```

```
# Last column: probability of non-recovery
for (t in 1:(nyears-1)){
  pi[t,nyears] <- 1-sum(pi[t,1:(nyears-1)])
} #t
}
")

# Initial values
inits <- function(){list(s.const=runif(1))}

# Parameters monitored
parameters <- c("s.const", "r.const")

# MCMC settings
ni <- 3000; nb <- 1000; nc <- 3; nt <- 1; na <- 3000

# Call JAGS from R (ART <1 min), check convergence and summarize posteriors
out24 <- jags(jags.data, inits, parameters, "model21.txt", n.iter=ni, n.burnin=nb, n.chains=nc,
  n.thin=nt, n.adapt=na, parallel=TRUE)
traceplot(out24)              # Not shown
print(out24)

          mean    sd  2.5%   50%  97.5% overlap0 f  Rhat n.eff
s.const  0.771 0.035 0.701 0.771  0.840    FALSE 1 1.001  3343
r.const  0.195 0.022 0.157 0.193  0.244    FALSE 1 1.001  4078
```

Up to MC error, the parameter estimates are identical to those under the state-space formulation. However, convergence is obtained much easier, and the run time is much shorter. Hence, once again we see the multinomial formulation to be more efficient than the state-space formulation of a survival model.

The modeling of temporal and individual effects or with covariates is done in the same way as was sketched out for the life encounter data, and the modeling of age effects follows the same principles, too. An example for a dead-recovery model with age dependence is presented in Chapter 12. One important fact about parameter identifiability must be kept in mind when modeling age effects: if only individuals in the juvenile age class are marked, age-dependent survival can only be estimated under the assumption that the ring-recovery probability does not depend on age; otherwise, only adult survival probability is estimable (Anderson et al., 1985; Cole et al., 2012). A marking strategy targeted at juveniles only is quite common in bird studies—ringing nestlings is typically much easier than ringing adults. However, the identifiability problem is solved if life resightings later in life of the individuals marked as nestlings are also available or probably also in the context of an IPM when some additional information comes from the population count data—although that latter notion should best be tested by simulation.

Although a nuisance parameter, it can be important to adequately model the recovery probability. For example, for British birds it was shown to decline over time for many species (Robinson et al., 2009). The probability to find a marked dead individual depends clearly on the mortality cause: individuals that die due to human-related causes such as hunting or collisions with human infrastructure have a (much) higher recovery probability than do individuals that die due to natural causes such as predation or starvation (Schaub and Pradel, 2004). Such heterogeneity can become especially problematic if the relative frequency of mortality causes changes over time or with age. Patterns unaccounted for in the nuisance parameter r may then result in spurious patterns in survival or their masking. If the cause of mortality of the recovered individuals is known, a multistate model can be formulated to estimate cause-specific mortality and to account for such heterogeneity in recovery probabilities (Schaub and Pradel, 2004; Schaub, 2009; Servanty et al., 2010; Nater et al., 2020).

Dead recoveries are an often-used data source in IPMs, starting from the seminal work by Besbeas et al. (2002), which analyzed data on lapwings (*Vanellus vanellus*) and gray herons (*Ardea cinerea*). Typical applications are for larger birds with a moderately high ring-recovery probability such as ducks (Péron and Koons, 2012; Koons et al., 2017; Zhao et al., 2019), but dead-recovery data were also used in IPMs for songbirds (Baillie et al., 2009; Robinson et al., 2012).

4.5.4 JOINT ANALYSIS OF CAPTURE-RECAPTURE AND DEAD-RECOVERY DATA

The joint analysis of capture-recapture and dead-recovery data permits estimation of true survival and site fidelity because of the different protocols used to collect live recapture (or resighting) and dead-recovery data. Typically, recaptures are obtained from a clearly defined, restricted study area and therefore the data collected provide information about apparent survival only. Dead recoveries, by contrast, are sampled by members of the public and can in theory originate from any area to which the marked individuals move to. Consequently, they allow the estimation of true survival, free of the confounding with fidelity. The combination of the two data types therefore provides the ability to separately estimate true survival and permanent emigration, or the complement of the latter, site fidelity. Such a model was developed by Burnham (1993) and was later expressed in the framework of a multistate model by Lebreton et al. (1999). Here we show the joint analysis of live recaptures and dead recoveries using a multistate model with the multinomial formulation.

To illustrate this model, we use data on marked white storks (*Ciconia ciconia*) collected for 16 years (1986–2001) in Germany. White storks are long-lived, migratory birds that live in open agricultural landscapes. The goal of this exercise is the estimation of true survival of adult white storks and their fidelity to the study area, which is the former German Democratic Republic (GDR). Chicks were ringed and surviving individuals resighted at the nest when they raised a brood in the study area. In addition, some ringed storks were recovered dead. To avoid the estimation of survival of younger storks and to keep the model simple, we restrict the analysis to individuals resighted in the GDR at least once as >2-year-olds. This restriction reduced the original data set to 691 adult white storks, of which 19 were recovered dead. These data were analyzed together with a similar data set of Polish storks by Schaub et al. (2005) to study the impact of environmental conditions in the nonbreeding area on age-dependent survival.

To develop our multistate model, we start by defining the states. We distinguish four states, "alive, inside the study area," "recently dead," "alive, outside the study area," and "long dead" (see below for the distinction between the two kinds of dead states) and construct the state-transition matrix.

$$\boldsymbol{\Psi}_t = \begin{bmatrix} s_t f_t & 1-s_t & s_t(1-f_t) & 0 \\ 0 & 0 & 0 & 1 \\ 0 & 1-s_t & s_t & 0 \\ 0 & 0 & 0 & 1 \end{bmatrix}$$

Here, s_t is true survival probability and f_t is the probability to remain in the study area, i.e., site fidelity probability. Note that the states are ordered in the matrix as given above. The vector with the encounter probabilities is

$$\mathbf{P}_t = \begin{bmatrix} p_t \\ r_t \\ 0 \\ 0 \end{bmatrix},$$

where p_t is the probability to resight a marked stork in the study area in year t, and r_t is the probability that a stork that died between $t-1$ and t is found and its ring is reported to the German ringing office.

The states "alive, outside" and "long dead" are both unobservable because no resighting activities were carried out outside the study area and because by assumption dead recoveries can only be obtained during the time interval in which that event occurred, but not later. You may wonder why we need the state "long dead," as we already defined one state with dead individuals, "recently dead." The reason is twofold. First, dead individuals can be recovered only once. Therefore, we have the absorbing "long dead" state that is unobservable and to which all "recent dead" individuals transition during the time interval following their death. Second, recovery is only possible for individuals that have died in the current time interval, but not for those that have been dead for a longer period already. This is ensured by the deterministic "movement" of the recently dead individuals to the "long dead" state after 1 year.

The multistate encounter histories contain a 1 when a stork was seen alive in the study area, a 2 when it was recovered dead, and a 0 otherwise. We load this data set and have a look at it.

```
library(IPMbook); library(jagsUI)
data(stork)
str(stork)

int [1:691, 1:16] 0 0 0 0 0 0 0 0 0 0 ...
 - attr(*, "dimnames")=List of 2
  ..$ : NULL
  ..$ : chr [1:16] "1986" "1987" "1988" "1989" ...
```

The next step is to construct the multistate m-array. We use function `marray` again and this time specify the presence of two unobservable states. The dimension of the resulting m-array is now already 60 × 61.

```
marr <- marray(stork, unobs=2)
```

We then write the model in the BUGS language. Things that need to be changed compared with the code in Section 4.5.2.2 are the definitions of the state-transition matrix and of the vector with the encounter probabilities, as well as the usual modifications for the linear models of the parameters and the priors. The definition of the state-transition matrix now extends over 16 code lines because there are four states and so the square matrix has 16 elements. We want to fit a model with site fidelity and recovery probabilities that are both constant over time and with time-dependent resighting and survival probabilities. For illustration, we adopt fixed time effects for resighting probability, but random time effects for survival probability. We write this model, provide the usual ingredients, and launch it.

```
# Bundle data
unobs <- 2                                            # Number of unobserved states
ns <- length(unique(as.numeric(stork))) - 1 + unobs  # Total num. states
jags.data <- list(marr=marr, nyears=ncol(stork), rel=rowSums(marr), ns=ns,
  zero=matrix(0, ns, ns), ones = diag(ns))
str(jags.data)

List of 6
 $ marr : num [1:60, 1:61] 1 0 0 0 0 0 0 0 0 0 ...
 $ nyears: int 16
 $ rel   : num [1:60] 4 0 0 0 15 0 0 0 30 0 ...
 $ ns    : num 4
 $ zero : num [1:4, 1:4] 0 0 0 0 0 0 0 0 0 0 ...
 $ ones : num [1:4, 1:4] 1 0 0 0 0 1 0 0 0 0 ...

# Write JAGS model file
cat(file="model22.txt", "
model {
# Priors and linear models
for (t in 1:(nyears-1)){
  logit.s[t] ~ dnorm(mu, tau)                         # Means parameterization of random effect
  s[t] <- ilogit(logit.s[t])
  f[t] <- mean.f
  r[t] <- mean.r
  p[t] ~ dunif(0, 1)
}

mean.s ~ dunif(0, 1)                                  # Prior for mean survival
mean.f ~ dunif(0, 1)                                  # Prior for mean site fidelity
mean.r ~ dunif(0, 1)                                  # Prior for mean recovery

mu <- logit(mean.s)
sigma ~ dunif(0, 10)                                  # Prior for temporal variability of survival
tau <- pow(sigma, -2)

# Define state-transition and observation probabilities
for (t in 1:(nyears-1)){
  psi[1,t,1] <- s[t] * f[t]                           # State-transitions
  psi[1,t,2] <- 1-s[t]
  psi[1,t,3] <- s[t] * (1-f[t])
  psi[1,t,4] <- 0
  psi[2,t,1] <- 0
  psi[2,t,2] <- 0
  psi[2,t,3] <- 0
  psi[2,t,4] <- 1
  psi[3,t,1] <- 0
  psi[3,t,2] <- 1-s[t]
  psi[3,t,3] <- s[t]
  psi[3,t,4] <- 0
  psi[4,t,1] <- 0
  psi[4,t,2] <- 0
  psi[4,t,3] <- 0
  psi[4,t,4] <- 1

  po[1,t] <- p[t]                                     # Observation probabilities
  po[2,t] <- r[t]
  po[3,t] <- 0
  po[4,t] <- 0
}

# From here onwards, no changes needed regardless of which model is fitted
# Calculate probability of non-encounter (dq) and reshape the array for the encounter
# probabilities
for (t in 1:(nyears-1)){
  for (s in 1:ns){
    dp[s,t,s] <- po[s,t]
    dq[s,t,s] <- 1-po[s,t]
  } #s
```

```
  for (s in 1:(ns-1)){
    for (m in (s+1):ns){
      dp[s,t,m] <- 0
      dq[s,t,m] <- 0
    } #s
  } #m
  for (s in 2:ns){
    for (m in 1:(s-1)){
      dp[s,t,m] <- 0
      dq[s,t,m] <- 0
    } #s
  } #m
} #t

# Define the multinomial likelihood
for (t in 1:((nyears-1)*ns)){
  marr[t,1:(nyears *ns-(ns-1))] ~ dmulti(pi[t,], rel[t])
}

# Define cell probabilities of the multistate m-array
# Matrix U: product of probabilities of state-transition and non-encounter (needed because
# there is no product function for matrix multiplication in JAGS)
for (t in 1:(nyears-2)){
  U[(t-1)*ns+(1:ns),(t-1)*ns+(1:ns)] <- ones
  for (j in (t+1):(nyears-1)){
    U[(t-1)*ns+(1:ns),(j-1)*ns+(1:ns)] <- U[(t-1)*ns+(1:ns),(j-2)*ns+(1:ns)] %*% psi[,t,] %*%
      dq[,t,]
  } #j
} #t
U[(nyears-2)*ns+(1:ns), (nyears-2)*ns+(1:ns)] <- ones

for (t in 1:(nyears-2)){
  # Diagonal
  pi[(t-1)*ns+(1:ns),(t-1)*ns+(1:ns)] <- U[(t-1)*ns+(1:ns),(t-1)*ns+(1:ns)] %*% psi[,t,] %*%
    dp[,t,]
  # Above main diagonal
  for (j in (t+1):(nyears-1)){
    pi[(t-1)*ns+(1:ns),(j-1)*ns+(1:ns)] <- U[(t-1)*ns+(1:ns),(j-1)*ns+(1:ns)] %*% psi[,j,] %*%
      dp[,j,]
  } #j
 } #t
pi[(nyears-2)*ns+(1:ns),(nyears-2)*ns+(1:ns)] <- psi[,nyears-1,] %*% dp[,nyears-1,]

# Below main diagonal
for (t in 2:(nyears-1)){
  for (j in 1:(t-1)){
    pi[(t-1)*ns+(1:ns),(j-1)*ns+(1:ns)] <- zero
  } #j
} #t

# Last column: probability of non-recapture
for (t in 1:((nyears-1)*ns)){
  pi[t,(nyears*ns-(ns-1))] <- 1-sum(pi[t,1:((nyears-1)*ns)])
} #t
}
")


# Initial values
inits <- function(){list(mean.f=runif(1, 0, 1))}

# Parameters monitored
parameters <- c("mean.s", "mean.f", "mean.r", "sigma", "s", "p")

# MCMC settings
ni <- 10000; nb <- 5000; nc <- 3; nt <- 1; na <- 5000

# Call JAGS from R (ART 1 min) and check convergence
out25 <- jags(jags.data, inits, parameters, "model22.txt", n.iter=ni, n.burnin=nb, n.chains=nc,
  n.thin=nt, n.adapt=na, parallel=TRUE)
par(mfrow=c(3, 3)); traceplot(out25)          # Not shown
print(out25)
```

```
              mean    sd    2.5%     50%   97.5% overlap0 f  Rhat n.eff
mean.s       0.793 0.047   0.717   0.788   0.890    FALSE 1 1.004   716
mean.f       0.907 0.049   0.809   0.911   0.989    FALSE 1 1.004   660
mean.r       0.043 0.011   0.025   0.041   0.068    FALSE 1 1.007  1047
sigma        0.291 0.158   0.021   0.272   0.667    FALSE 1 1.015   180
s[1]         0.792 0.066   0.660   0.791   0.916    FALSE 1 1.002  1217
[... output truncated ...]
s[15]        0.798 0.065   0.669   0.797   0.924    FALSE 1 1.001  3594
p[1]         0.488 0.230   0.112   0.461   0.945    FALSE 1 1.000 10916
[... output truncated ...]
p[15]        0.720 0.071   0.593   0.716   0.870    FALSE 1 1.002  2696
```

The model converges quite swiftly. The estimate of the mean adult survival probability of white storks is close to that in Schaub et al. (2005), although here we used only a subset of the original data. The estimate of site fidelity is very high as expected because breeding dispersal is known to be small in this species. Apparent survival can be calculated as the product of true survival and site fidelity, $\phi = sf$, which yields an average of 0.72 (SD: 0.017). The recovery probability is much lower than the resighting probability, as is often found to be the case.

Joint modeling of capture-recapture and dead-recovery data is carried out quite often (Altwegg et al., 2003; Duriez et al., 2009; Frederiksen and Bregnballe, 2000). McCrea et al. (2014) have developed a goodness-of-fit test for this model. In addition, a number of published IPMs have used both types of data to estimate survival and sometimes also site fidelity (e.g., Tavecchia et al., 2009; Borysiewicz et al., 2009; Reynolds et al., 2009; Plard et al., 2020; Margalida et al., 2020).

4.5.5 MULTIEVENT MODELS

Multievent capture-recapture models are a powerful generalization of multistate capture-recapture models that allow for state uncertainty not only in terms of the imperfect detection of marked individuals, but also in terms of state assignment of observed individuals, which is now allowed to contain errors or to be uncertain. The basic idea of this model has been formulated by Roger Pradel (Pradel, 2005, 2009), for a review see Gimenez et al. (2012). Multievent models can be useful for various sources of state uncertainty, such as unknown sex (Pradel et al., 2008) or uncertain disease status of observed, marked individuals (Conn and Cooch, 2009), pair fidelity (Culina et al., 2013) or uncertain breeding status (Desprez et al., 2011; Catlin et al., 2019; Couet et al., 2019). The model can also be useful to account for individual heterogeneity by specification of a number of latent classes of individuals. Such a model is similar to a finite mixture model (Pledger et al., 2003), but it is more general because group membership of an individual is allowed to change over time in a dynamical manner, and the parameters governed that dynamics are estimated. A model with individual heterogeneity is useful to study senescence in survival or other state-transition probabilities (Péron et al., 2010a). Finally, a "memory model" (Hestbeck et al., 1991; Brownie et al., 1993), i.e., a relative of a multistate model in which survival or a state-transition probability does not only depend on the current state at time t, but also on the state at the previous time $t-1$, can be formulated as a multievent model (Rouan et al., 2009).

Multievent models are commonly fit in the classical framework with the powerful software E-SURGE (Choquet et al., 2009b), but the use of Bayesian methods is possible as well. Published BUGS code can be found, for example, in Tenan et al. (2017) and Catlin et al. (2019). As far as we know, a multievent model cannot be written using a multinomial formulation but instead requires a state-space formulation. So far, multievent models have only rarely been used as a component of an IPM (Tenan et al., 2017; Jan et al., 2019), but we imagine that their frequency will increase in the future.

4.6 INTRODUCTION TO SPATIAL CAPTURE-RECAPTURE MODELING

As in much of ecological statistics, also for IPMs we observe an increasing trend toward more explicitly spatial model formulations; see, e.g., Chandler and Clark (2014) and Chapters 20 and 21. Therefore, here we provide a basic introduction to the principles of SCR models (Efford, 2004; Borchers and Efford, 2008; Royle and Young, 2008; Royle et al., 2014; Borchers et al., 2015; Borchers and Marques, 2017). In SCR models, we use repeated detections and nondetections of individually identifiable animals *along with the locations of their detections and typically also with the locations of a set of "detectors"* such as camera traps or hair snare-traps to describe the location, state, and fate of *all* individuals in a defined area including those that were never detected. Depending on the type of data and model, SCR permits inferences about essentially any kind of demographic quantity of interest (e.g., Williams et al., 2002; Seber and Schofield, 2019) and the estimates are frequently improved over their nonspatial capture-recapture analogues. In addition, the individual-based formulation of SCR models lends itself very naturally for them to become the "motherboard" for combining any other kind of demographic data in an IPM fashion. Hence, the basic SCR model is the spatially explicit analogue of, and replacement for, the usual matrix population models in a nonspatial and especially nonindividual-based IPM (and you may want to read this sentence multiple times to fully grasp what we believe SCR models will represent for many future IPMs).

When moving to spatially explicit formulations of a demographic model, additional complexities arise both in the state and the observation parts of a hierarchical model and these may at first be challenging to grasp. Therefore, we here explain the principles of SCR using our full complement of explanatory tools for applied statistics: words, data simulation in R, algebra, and BUGS code. We first walk you through the simulation of one of the simplest possible SCR data sets (a closed-population, spatially explicit capture-recapture model) and then through its analysis using the model fit in JAGS. Since their invention by Efford (2004), there has been an enormous development in SCR models and so we put particular emphasis on our usual disclaimer that we can only scratch the surface here. However, arguably the two most challenging, new aspects when moving from nonspatial to spatial forms of capture-recapture models are (1) the individual-based formulation of the state process in terms of a spatial or spatio-temporal point process model, and (2) typically also a spatially explicit description of the detection process in terms of a detection function that depends on unobserved distances between some measure of an animal's average location and the set of "traps". We can see both of them in our minimalistic SCR example in this section.

There are a number of different field designs that produce data amenable to SCR modeling, and the three most typical are the following:

- Distance sampling is a very widely used framework for estimation of density (Buckland et al., 2015) and of vital rates as well (Sollmann et al., 2015, Chapters 8 and 9 in Kéry and Royle, 2016). In its typical form, the distance between observer and object is recorded and the decline in the number of detected objects with increasing distance is used to estimate detection probability. We do not cover distance sampling here, but simply mention that it can be viewed as a sort of primitive SCR model (Borchers et al., 2015; Borchers and Marques, 2017, Chapter 11 in Kéry and Royle, 2021).
- Animals occur in some continuous area and are detected by an array of point-like traps or detectors. The typical example is in camera-trap studies, where individuals are recognizable by their natural variation in stripes or other marks and are detected at fixed camera locations. Other examples include hair snares that may produce genetic ID data (see Chapter 21) and sometimes also microphone (or autonomous recording unit) arrays used as a "trapping grid." Sometimes the continuous space is discretized in the analysis, for instance to easily clip out nonhabitat or to model environmental effects on density.
- Animals occur in some continuous area and can be detected potentially anywhere in this area and possibly even without any notion of a trapping grid ("search-encounter design," Royle et al., 2011, Chapter 15 in Royle et al., 2014; Schaub and Royle, 2014, Chapter 11 in Kéry and Royle, 2021). This is a useful design because it corresponds to the areal searches so common in the sampling of many taxa such as birds or reptiles, or the scats of grouse or large carnivores.

SCR models do not require temporally replicated observations to estimate detection probability separately from demographic parameters, as do the traditional capture-recapture models (Williams et al., 2002, Chapter 6 in Kéry and Schaub, 2012). Rather, "spatial recaptures" of individuals are sufficient, such as when scats of capercaillies (*Tetrao urogallus*) are genotyped and some individuals are detected at multiple locations (Mollet et al., 2015; Augustine et al., 2020a). This is a huge design advantage because it may allow greatly reduced costs of field work by only having to survey an area once.

In this section we examine the simplest SCR model known as SCR0 (Royle et al., 2014), which is analogous to the simplest possible nonspatial capture-recapture model known as M0 (Otis et al., 1978). The main goal is the estimation of population density. We simulate SCR data step by step and comment on what we do. As always, simulating data is one of the best ways to exactly understand the stochastic processes represented by a model. Hence, by simulating the data set, you will understand the fundamentals of SCR models in the simplest possible setting. After data simulation, seeing the model in the BUGS language often results in a feeling of "déjà-vu."

We start by simulating data under a design where we get data for a number of N animals in a study area and assume a simple Bernoulli observation process where an individual is either detected or not during a specific time interval. In their book and in the associated R package `scrbook`, Royle et al. (2014) have a function called `simSCR0` to simulate simple SCR data and another function called `SCR0bayes` to fit model SCR0 using JAGS. Here we dissect their function `simSCR0` and beautify it a little by adding comments and plots. What we do here is what you may often want to do when faced with a piece of computer code too complicated for you to just read and understand at once: you execute it line by line as far as possible and inspect each object created along the way. In the end you will know exactly how things are computed and what these objects mean.

```
library(scrbook) # From sites.google.com/site/spatialcapturerecapture/
?simSCR0         # Open help page
```

Don't worry if you are unable to install the `scrbook` package with your version of R; it is not necessary for what follows. In this section, we will simulate the same type of data that we also get when executing the next line of code:

```
str(dat <- simSCR0(N=100, K=20, discard0=TRUE, array3d=FALSE, rnd=2013))
```

We imagine that we are simulating the data from a population study of the Eurasian lynx (Fig. 4.14) in the wild and wonderful Jura mountains along the French-Swiss border. We choose population size, number of trap nights (occasions) and the RNG seed.

```
N <- 100          # Population size: number of individuals living in state space
K <- 20           # Number of trap nights
rnd <- 2013       # Random number seed
```

We lay out a regular grid of camera traps. The methods work equally well with irregular grids, but as usual it is easier to simulate stuff in a regular way.

```
traplocs <- cbind(sort(rep(1:5, 5)), rep(1:5, 5)) # Set up trapping grid
J <- nrow(traplocs)                               # Number of traps (ntraps)
```

We add a buffer around the trapping grid area and compute the coordinates of the upper and lower bounds of *X* and *Y*. We add this buffer because in real life, in a grid we will typically trap also some animals that reside outside the trapping grid. In the simulation, we call this combined area the state space, which is meant to include all possible locations where a lynx could live that we could possibly detect in our trapping grid. We characterize where a lynx lives by a single point: the center of its home range or what we call the *activity center* (AC).

FIGURE 4.14

A Eurasian lynx (*Lynx lynx*) mother with two kittens photographed by a camera trap in the Swiss Jura mountains (photograph by Alain Mabille).

Sometimes we see the terms "activity range" and "home range" or even "territory" being used as synonyms in the context of SCR models. They may be in some situations but certainly not always. In the context of SCR, "activity range" is most accurate because what we deal with is the center of an animal's activity that is detectable to our type of trap. That is, "activity" is defined by the detection method, and may be an animal walking along a trail, visiting a bait station, defecating, or singing. That *detectable activity* may not take place in the entirety of the individual's home range. For instance, some bird species may only be detectable when they sing, and singing only occurs around the nest and not in other parts of the home range that are used for feeding only.

```
# Set up state space
buffer <- 2                              # Buffer size = 2 units
Xl <- min(traplocs[, 1] - buffer)        # X lower
Xu <- max(traplocs[, 1] + buffer)        # X upper
Yl <- min(traplocs[, 2] - buffer)        # Y lower
Yu <- max(traplocs[, 2] + buffer)        # Y upper
```

Next, we populate this area with *N* lynx individuals by randomly placing in it the individual ACs. So it is here where we simulate that first defining feature of an SCR model, the realization from a point process, which is a stochastic process that produces random locations in space (Illian et al., 2008; Baddeley et al., 2016). For some point process models such as the commonly used Poisson point process, the number of points as well as their locations are both random quantities, but here we simulate a binomial point process where the number of points is a quantity fixed at *N*.

```
# Simulate point process by generating coordinates of ACs
set.seed(rnd)                            # Set seed
sx <- runif(N, Xl, Xu)
sy <- runif(N, Yl, Yu)
S <- cbind(sx, sy)
head(S)                                  # Look at coordinates of activity centers (ACs) - not shown
```

The second defining feature of most SCR models is that detection probability of an individual is assumed to depend on the distance between some feature of the animal's location and a trap. Of course, the precise location of an animal will change as it wanders around in its home range, but we characterize the individual's average location by the center of its home range or activity range. This is *S* in the code. We next compute the matrix that contains the distances of each animal's AC to each of the 25 traps.

```
library(AHMbook)                         # In case scrbook does not work
D <- AHMbook::e2dist(S, traplocs)        # Distances between AC and traps
str(D)                                   # Shows a 100 x 25 matrix

# Plot realization of spatial point process: ACs and trapping grid
# and variability in the distance between AC and traps (Fig. 4.15)
plot(traplocs, xlim=c(Xl, Xu), ylim=c(Yl, Yu), pch=16, axes=FALSE, xlab='x-coordinate',
  ylab='y-coordinate', cex=1)
points(S, col="red", pch=1, cex=0.8)
axis(2); axis(1)
```

Now we are done with the simulation of the biological part of the system (the spatial point pattern, i.e., the spatial configuration of lynx territories in the study area) and the physical setup of the trapping apparatus (Fig. 4.15). Next, we choose the parameters of the function that for each trap relates detection probability to its distance from every animal's AC. This is exactly like distance sampling, except that in distance sampling we deal with the *observed distances* between the *actual* animal location and an observer, whereas in SCR, it is the *estimated distance* between some *average location* of the individual (i.e., the AC) and a trap (Borchers and Marques, 2017). We assume that if a trap is placed right at the location of the AC of an individual, there is some maximum (or baseline) detection probability p_0, and that from this zero distance away, detection probability declines monotonically with increasing distance between trap and AC.

We could entertain all sorts of functional forms to describe this decline in detection, but one sensible function that has become the standard in SCR is a half-normal kernel, which you can think of as the right half of a normal distribution with a mean of zero. It is characterized by the SD or scale parameter "sigma" which governs the rate of decline in detection probability with increasing distance between AC and trap. To help your intuition with this relationship between sigma and detection probability, we plotted the detection function for the given baseline detection p_0 (about 0.08) for a range of values of sigma (Fig. 4.16 left). The detection function resulting from the value of sigma in this simulation is shown with the black line. We see that higher values of sigma mean a less steep decline in detection probability with increasing animal-trap distance, and therefore, for a given home-range size of an animal, a larger "zone of influence" of a trap.

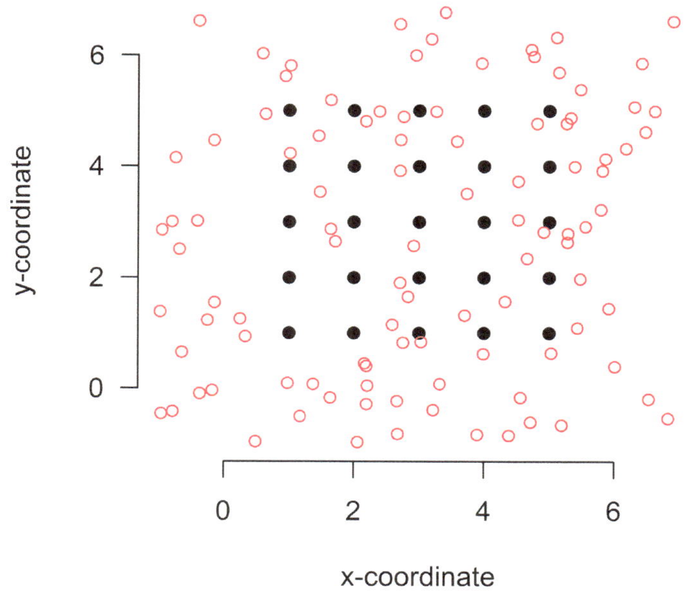

FIGURE 4.15

Point locations of 100 simulated lynx activity centers (open red circles) and trapping array of dimension 5 × 5 (solid black circles), with a buffer of 2 units. The red circles are a realization of a spatial point process, which is a hallmark of an SCR model.

In our actual simulation, we choose the parameters for `alpha0` = logit(p_O) and `sigma` and then compute the matrix of capture probabilities, `probcap`, which has dimension number of individuals x number of traps. For each of the N individuals (in the rows) it contains the detection probability for each of the J traps (in the columns of the matrix).

```
# Define parameters of detection probability
alpha0 <- -2.5               # Logit(baseline detection probability right on a trap)
plogis(alpha0)               # p0: detection probability for AC right ON a trap: 0.0758
sigma <- 0.5                 # Scale parameter of half-normal kernel
alpha1 <- 1 / (2 * sigma^2)  # Coefficient on distance

# Compute probability of detection for each combination of animal and trap
probcap <- plogis(alpha0) * exp(-alpha1 * D^2)
str(probcap)                 # This is a 100 x 25 matrix
```

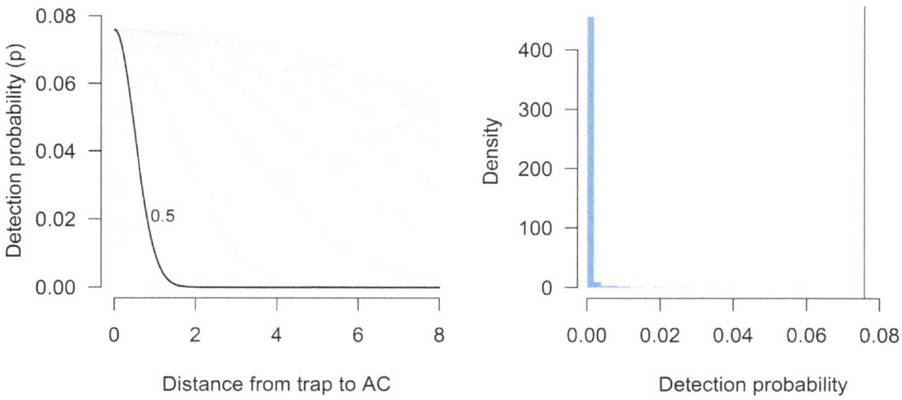

FIGURE 4.16

(left) The relationships between detection probability at a trap and its distance from the activity center (AC) of an animal for nine different values of the scale parameter sigma, with the value of 0.5 used in our simulation shown in black. (right) Frequency distribution of detection probability for all combinations of traps and individual activity centers (ACs). The black vertical line shows the value of the baseline detection probability (p_O), i.e., the maximum detection probability achieved when a trap is placed right in the middle of an activity range, i.e., when the location of a trap coincides with an AC. Note x-axis is zero-truncated to allow showing the very rare high values of p_O in the trapping array.

What is the range of detection probabilities (per occasion and for each animal-trap combination) for this realization of the home-range process and the chosen trap array? For most trap-lynx combinations, the probability of detection is essentially zero, so in Fig. 4.16 (right) we only show the histogram for a part of the range of `probcap`. The black line gives the baseline detection probability p_0 at trap-AC distance zero.

Now that we have set up the biology in our simulation (i.e., distributed the N lynx over some continuous space) and the trapping apparatus (i.e., the J traps, with their "zone of influence" defined by the detection function), we run the traps for K occasions/nights, resulting in binary detection/nondetection events for each individual, trap and night.

```
y3d <- array(NA, dim=c(N, J, K))              # Dimension is individual x trap x night
for (i in 1:nrow(y3d)){                       # Loop over individuals
  for (j in 1:J){                             # Loop over traps
    y3d[i,j,] <- rbinom(K, 1, probcap[i,j]) # Bernoulli trials
  } #j
} #i
```

When p does not change among occasions, you will get exactly the same information when aggregating these data to a 2-dimensional matrix of detection frequencies, i.e., the number of trap nights out of K that lynx i was detected in trap j. We create this format of the data as well.

```
y2d <- apply(y3d, c(1, 2), sum)               # Binomial detection frequencies
str(y2d)
```

The data set created contains the truth, which we normally won't be able to see fully because we would not know how many animals were never captured in any of the traps and therefore have an all-zero detection history for every trap. We summarize the detection frequency matrix and see that the vast majority of the individual/trap combinations are zeroes; the biggest "catch" was one individual detected in a particular trap on five nights.

```
table(y2d)
y2d
   0    1    2    3    4    5
2441   45   10    2    1    1
```

We tally up the total number of detections (over all 25 traps) for the $N = 100$ simulated lynx.

```
table(apply(y2d, 1, sum))
 0  1  2  3  4  5  6
65 10 15  4  3  2  1
```

Hence, 65 lynx were never detected during the 20 simulated trap nights, and only 35 were ever detected. Of these, 10 were detected in a single trap night, and 1 animal was detected in six trap nights (all traps combined). Let's check the number of different traps that the 35 animals detected at least once were detected in. We first create a copy of `y2d` that only distinguishes detection and nondetection.

```
yy <- y2d
yy[yy > 1] <- 1
table(apply(yy, 1, sum))
 0  1  2  3  4
65 17 13  4  1
```

That is perhaps somewhat sobering: 17 animals were detected in a single trap only, i.e., had no spatial recaptures, while 13 animals were detected in two different traps, four in three, and a single animal in four. The tabulation of spatial recaptures is crucial for SCR modeling because in a data set without spatial recaptures, you typically cannot fit such models.

We make a spider plot where we join the home-range centers of each detected animal with the traps it was detected in. First, however, we get rid of the data for those animals that were never detected. From now on, `y` will be the observed detection frequency matrix that contains the number of trap nights that each of the detected individuals (in the rows of the matrix) was detected in each of the set traps (which are in the columns).

```
totalcaps <- apply(y2d, 1, sum)
nobs <- sum(totalcaps > 0)
y <- y2d[totalcaps > 0, ]
dimnames(y) <- list(1:nrow(y), paste("trap", 1:ncol(y), sep=""))
str(y)

int [1:35, 1:25] 0 0 0 0 1 0 0 0 0 0 ...
 - attr(*, "dimnames")=List of 2
  ..$ : chr [1:35] "1" "2" "3" "4" ...
  ..$ : chr [1:25] "trap1" "trap2" "trap3" "trap4" ...
```

We plot the 35 lynx detected at least once with filled red circles and those never detected with open red circles (Fig. 4.17). For this, we make a version of the ***S*** array (which contains the AC coordinates) for the observed animals only and then build the plot one detected animal at a time; hopefully, this helps your understanding of the process.

```
Sobs <- S[totalcaps > 0,]                         # Activity center coordinates of detected animals
op <- par(ask=TRUE)
for (i in 1:nrow(y)){                             # Loop over individuals
  plot(traplocs, xlim=c(Xl, Xu), ylim=c(Yl, Yu), pch=16, xlab='x-coordinate', ylab='y-coordinate',
    frame=FALSE)
  points(S[totalcaps == 0,], col='red', pch=1)  # Never captured
  points(S[totalcaps > 0,], col='red', pch=16)  # Captured >0 times
  for (k in 1:i){
    tmp <- which(y[k,] > 0)                       # Detected by which trap
    for (j in 1:length(tmp)){
      segments(Sobs[k,1], Sobs[k,2], traplocs[tmp,1], traplocs[tmp,2], col='gray', lty=1)
    } #j
  } #k
  for (j in 1:length(tmp)){
    segments(Sobs[i,1], Sobs[i,2], traplocs[tmp,1], traplocs[tmp,2], col='red', lwd=2)
  } #j
  title(main=paste('Lynx', i, 'caught in: ', paste(names(tmp), collapse=", ")))
} #i
par(op)
```

We note a couple of interesting observations: the first is that the 35 detected lynx live on average much closer to the trapping array than do the 65 lynx that escaped detection. Of course, this is a consequence of the distance-dependent decline in detection probability. Second, despite this, some individuals were detected at a trap even though their AC was quite far away from the grid. Third, there were even some overlooked individuals that lived right in the middle of the trapping grid. Moreover, their ACs lie sometimes surprisingly close to the nearest traps.

We will now use JAGS to fit the data-generating model SCR0 to our detection frequency matrix y. We prepare the data first and this includes a data augmentation step, where we add to the observed data matrix a portion for a large number of potential but undetected individuals. There is no difference at all with data augmentation in the SCR case compared with the traditional, nonspatial closed-population models (see Chapters 6 and 10 in Kéry and Schaub, 2012): this is a handy way to induce a discrete uniform prior on population size N in an MCMC-based analysis (see also Link (2013) for a variant that

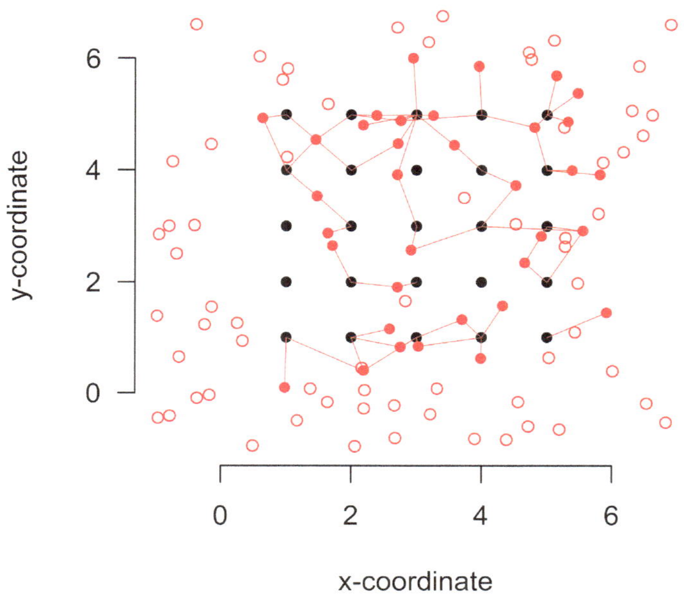

FIGURE 4.17

A spider plot showing the traps (solid black circles) that each lynx (the 35 solid red circles) was detected in by the red lines joining trap and lynx. Open red circles denote the 65 lynx that were never trapped anywhere in the trapping grid during the 20 trap nights in our simulation exercise.

sometimes has some statistical advantages but no longer has that exact interpretation). Here are the data that we need, with some redundancy for the sake of a better overview. In the analysis we define again the state space, which you can think of as the area from which all animals ever detectable inside the trapping grid must have come from. We choose to define the state space in an identical manner in the analysis as we did in the data simulation above, but this need not always be so in SCR models.

```
nind <- nrow(y)                                          # Observed number of individuals
J <- nrow(traplocs)                                      # Number of traps
xlim <- c(Xl, Xu)                                        # x limits of state-space
ylim <- c(Yl, Yu)                                        # y limits of state-space
A <- (max(xlim) - min(xlim)) * (max(ylim) - min(ylim)) # State-space area

# Data augmentation
M <- 200                                                 # Allow for up to M-35=165 undetected lynx
y <- rbind(y, matrix(0, nrow=M - nind, ncol=ncol(y)))

# Bundle data
jags.data <- list(y=y, traplocs=traplocs, K=K, M=M, J=J, xlim=xlim, ylim=ylim, A=A)
str(jags.data)

List of 8
 $ y       : num [1:200, 1:25] 0 0 0 0 1 0 0 0 0 0 ...
 $ traplocs: int [1:25, 1:2] 1 1 1 1 1 2 2 2 2 2 ...
 $ K       : num 20
 $ M       : num 200
 $ J       : int 25
 $ xlim    : num [1:2] -1 7
 $ ylim    : num [1:2] -1 7
 $ A       : num 64
```

We fit the most basic SCR model with every parameter a constant. It can be described clearly and succinctly as a hierarchical model in just four lines (Royle et al., 2014, p. 154—155). Note that below $\boldsymbol{x}_j$ denotes the trap location coordinates of trap j and is called "traplocs" in our BUGS code.

$y_{i,j}\big|\boldsymbol{s}_i \sim Binomial\left(z_i p_{i,j}, K\right)$ These are the observed detection frequency data

$p_{i,j} = p_0 \exp\left(-\alpha d_{i,j}^2\right) = p_0 \exp\left(-\left(\frac{1}{2\sigma^2}\right)\left\|\boldsymbol{x}_j - \boldsymbol{s}_i\right\|^2\right)$ Detection probability depends on distance $\left(d_{i,j}\right)$ between trap $\left(\boldsymbol{x}_j\right)$ and individual AC $\left(\boldsymbol{s}_i\right)$

$\boldsymbol{s}_i \sim Uniform(\mathbf{S})$ Point process model for location of activity centers

$z_i \sim Bernoulli(\psi)$ Model for data augmentation variables

A couple of notes about this kind of model may be interesting. At the level of the observed data, we have a binomial distribution; hence, we can describe the model as a sort of complicated logistic regression, or binomial GLM. However, we have two kinds of random effects in it: $\boldsymbol{s}_i$ and z_i. The former ($\boldsymbol{s}_i$) are continuous: they are the x and y coordinates of the AC for each lynx in the state space. The latter (z_i) are binary: they are the data augmentation variables, which are equal to 1 for a lynx that belongs to the population living in the state space and equal to 0 for a lynx that does not. Thus, our model is a peculiar kind of a random-effects logistic regression or something somewhat similar to a strange binomial GLMM; however, the random effects of our model have a uniform and a Bernoulli distribution, respectively, instead of the normal distributions in a GLMM. And perhaps most astonishing is the fact that the prior for the ACs, $\boldsymbol{s}_i$, does not even have a hyperparameter estimated!

We translate this model into BUGS, and as usual, clearly recognize the algebraic description of the model. For the detection intercept (`p0`) and scale (`sigma`), we provide vague priors on the probability scale for the former and on the scale of coefficient of distance squared for the latter.

```
# Write JAGS model file
cat(file="model23.txt", "
model {
# Priors and linear models
p0 ~ dunif(0, 1)                                    # Baseline detection probability
alpha ~ dnorm(0, 0.1)                               # 1 / (2*sigma^2) (coef of distance)
sigma <- sqrt(1 / (2 * alpha))                      # Half-normal scale
psi ~ dunif(0, 1)                                   # Data augmentation parameter

# Likelihood
for (i in 1:M){                                     # Loop over all M=200 individuals
  z[i] ~ dbern(psi)                                 # Data augmentation variable
  # State model: point process model
  s[i,1] ~ dunif(xlim[1], xlim[2])                  # x-coord of activity centers
  s[i,2] ~ dunif(ylim[1], ylim[2])                  # y coord of activity centers
  # Observation model: p ~ distance between trap and estimated AC
  for (j in 1:J){                                   # Loop over all traps
    d2[i,j] <- pow(s[i,1] - traplocs[j,1], 2) + pow(s[i,2] - traplocs[j,2], 2) # Distance squared
    p[i,j] <- z[i] * p0 * exp(-alpha * d2[i,j])     # Detection prob.
    y[i,j] ~ dbin(p[i,j],K)                         # Observed detection frequencies
  } #j
} #i

# Derived quantities
N <- sum(z[])                                       # Population size in state-space
D <- N / A                                          # Density over state-space
}
")
```

We add a couple of derived quantities so we can estimate them with full error propagation. First, population size N is a function of the data augmentation variable z, while density D is the ratio of N over the area of the state space (A, which is provided as data). Note that `sigma` is also derived quantity, but for simplicity we defined it in the prior section of the model.

As often for random-effects models in BUGS, it can be important to initialize the chains for the random effects at reasonable places, i.e., not too far away from their posterior means. Thus, we choose starting values for the two sets of random effects, $\boldsymbol{s}$ and $\boldsymbol{z}$. As always when doing data augmentation in JAGS, we give initial values for the data augmentation variables that must not contradict the observed data.

```
# Function to generate starting values
sst <- cbind(runif(M, xlim[1], xlim[2]), runif(M, ylim[1], ylim[2]))
for (i in 1:nind){
  if (sum(y[i, ]) == 0)
  next
  sst[i, 1] <- mean(traplocs[y[i, ] > 0, 1])
  sst[i, 2] <- mean(traplocs[y[i, ] > 0, 2])
}
zst <- c(rep(1, nind), rep(0, M - nind))
inits <- function() {list(p0=runif(1), alpha=runif(1, 1, 2), s=sst, z=zst)}

# Parameters to save
parameters <- c("alpha", "sigma", "N", "D", "s", "z")

# MCMC settings
ni <- 15000; nb <- 5000; nc <- 3; nt <- 5; na <- 1000

# Call JAGS (ART 7 min), check convergence and summarize posteriors
out26 <- jags(jags.data, inits, parameters, "model23.txt", n.iter=ni, n.burnin=nb, n.chains=nc,
  n.thin=nt, n.adapt=na, parallel=TRUE)
traceplot(out26)                         # Not shown
print(out26, 3)

          mean     sd   2.5%    50%   97.5% overlap0     f  Rhat n.eff
alpha    1.878  0.308  1.323  1.860   2.533    FALSE 1.000 1.002  1022
sigma    0.521  0.043  0.444  0.519   0.615    FALSE 1.000 1.002  1102
N       78.992 11.405 59.000 78.000 104.000    FALSE 1.000 1.001  1236
D        1.234  0.178  0.922  1.219   1.625    FALSE 1.000 1.001  1236
s[1,1]   3.006  0.377  2.258  3.007   3.740    FALSE 1.000 1.000  6000
s[2,1]   4.720  0.331  4.091  4.706   5.406    FALSE 1.000 1.000  4089
s[3,1]   5.226  0.408  4.416  5.242   5.972    FALSE 1.000 1.001  2284
[... output truncated ...]
```

In the posterior summary table, we see decent parameter estimates. For our particular realization of the simulated stochastic process, at about 80, the estimate of population size (N) undershoots somewhat, but it is a considerable improvement over the observed number of 35 lynx (Fig. 4.18 left). True density in our simulation is 1.56, and our estimate is at 1.24, which is again a bit low, but better than the observed value of 35 animals in a trapping grid with nominal area 16, which results in a naive estimate of density that is substantially too high at 2.19 (Fig. 4.18 right). The reason for the overestimation in the observed density is that lynx with an AC outside the trapping grid are also detected, and they should not "count" when forming the ratio of population size over trapping grid area as an estimator of density.

As a first side comment, we repeat that a single estimate that fails to be right on target does not mean that it is biased—you have to repeat the process and then look at the average of the estimates. Mike Meredith (pers. comm.) ran a simulation with 500 replicates and found that the mean number of animals captured was 42 (so we are a bit unlucky with our RNG seed). The mean posterior mean of N was 102, so any bias is small, and 98% of the 95% CRIs included the true value of 100. Mike commented that this was not too bad given the small number of traps deployed.

As a second side comment we note that ever since Dice (1938), mammologists had understood that our naive density estimator above would be an overestimate unless some buffer strip was added around the trapping grid. Consequently, much effort went into understanding how to best choose the width of that buffer (Karanth and Nichols, 1998; Parmenter et al., 2003). The usual *ad hoc* solution was to add either 0.5 or 1 times the mean maximum distance moved for animals with spatial recaptures. SCR then did away with the need for any such *ad hoc* choice by estimating the density explicitly based on the movement/encounter model. So in this sense it may be misleading if we take the naive density estimator above, i.e., without any buffer strip, as an observed benchmark because no reasonable mammologists would have done this anyway. However, this is not true for other taxa such as birds, where nonspatial density estimators adopted typically are in a pre-Dice (1938) conceptual state, i.e., they do not recognize the need for either the addition of some buffer strip or adoption of formal SCR methods.

An important part of the output from the model are the estimates of the ACs, i.e. the $\boldsymbol{s}$. We can plot their locations or summarize them to obtain estimates of the population size in different parts of the state space and produce a density map. We start by producing a map of the posterior samples of the estimated ACs of the 35 lynx detected at least once and for the 65 lynx that were never detected (Fig. 4.19). For the latter we must plot only those posterior draws of $\boldsymbol{s}_i$ for which $z_i = 1$, i.e., only plot the draws of $\boldsymbol{s}$ for "real" animals.

Apart from their considerable visual appeal worthy of the oeuvre of many a contemporary artist the plots in Fig. 4.19 can teach us several things. First of all, note that the colored clouds do not show the *extent* of the estimated home ranges, but only the uncertainty about their central *location*, i.e., the ACs. We could construct an estimator of the extent of the simple activity range implied in our movement model, i.e., by the detection function. This estimator must include both the location of the AC (i.e., $\boldsymbol{s}$) and an estimate of the extension of the home range (Royle et al., 2014), which in our model is governed by the half-normal parameter `sigma` (see next paragraph for this). In the left figure, we see that there is more uncertainty about the location of the home ranges when they are further away from the traps (which is where the information about their location comes from), i.e., at the edge rather than in the center of the trapping grid. In the right figure we see that the ACs for the animals that escaped detection are estimated to be relatively far from the trapping grid, which makes sense because the fact that they were not detected makes them more likely to have a home range far from the traps (although this will of course not hold for every single home range with a missed owner, as we saw above).

Finally, it will often be informative to produce a density map (or at least it will be a good promotion for your paper, since a common marketing wisdom can be altered slightly into "mapping sells"). A density map can be constructed in

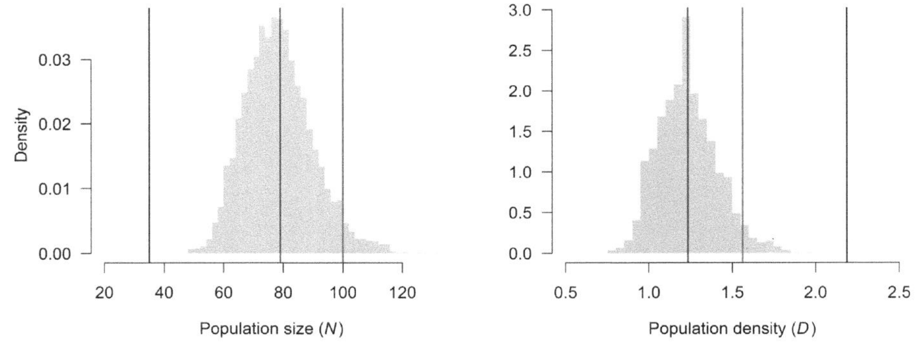

FIGURE 4.18

Posterior distributions of population size (left) and density (right). Black lines show one variant of what could be called the observed values of population size and density, blue lines give the estimated values (posterior means), and red is the truth in the data simulation.

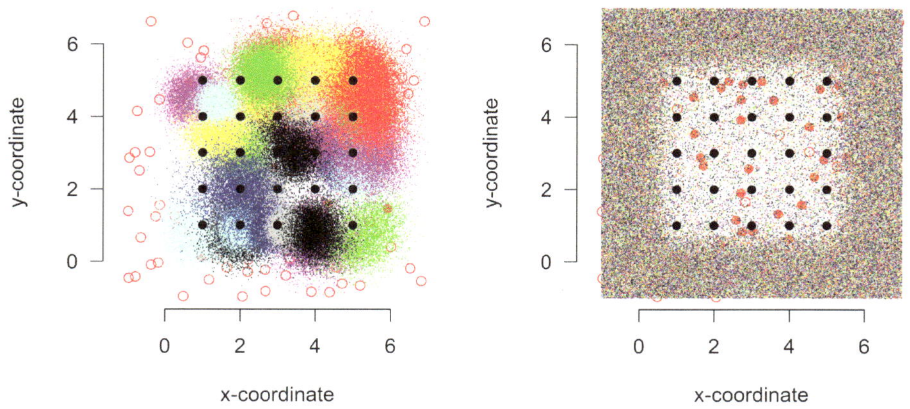

FIGURE 4.19

Maps of the posterior distributions of the AC locations for the 35 simulated lynx that were detected (left) and for those 65 animals that were never detected (right). The true locations of the activity centers (ACs) of the detected and undetected lynx are depicted with solid and open red circles, respectively, and the 16 traps with solid black circles.

multiple ways. The usual method is to summarize the posterior samples of the ACs (see Chapter 21 and Section 5.11.1 in Royle et al., 2014). However, it could be argued that a better characterization of density would be based on the actual animal locations, which are a function of both the ACs and the movement/encounter model specified in your SCR model. In practice, both solutions will probably depict similar spatial patterns, but density maps based on the ACs only will show sharper patterns than those based on the estimates of actual locations. Here, we show both and acknowledge code provided by Mike Meredith.

For both approaches, we decide on a suitable resolution for a pixel grid to summarize density variation in the state space and then tally up the number of posterior samples of $\boldsymbol{s}$ (or of simulated locations $\boldsymbol{u}$; see below) that lie within each pixel of that grid. We must again count only those samples of $\boldsymbol{s}$ or $\boldsymbol{u}$ for which the corresponding draw of z is equal to 1, i.e., representing a "real" lynx in the data augmentation scheme. In Fig. 4.20, we see considerable small-scale spatial variation in density. Both characterizations of density give broadly similar patterns, but as expected, the one based on simulated animal locations is considerably more spread out due to the additional variance coming from the movement model.

It is important to realize that in a Bayesian SCR model, a homogeneous point process density does *not* imply that the actual density, in terms of the realized locations of the ACs or the animal locations, will be uniform in space. The reason for this is that the homogenous point process intensity serves as a weak prior, which, when combined with the observed data, allows plenty of spatial variation in the posterior of the realized density.

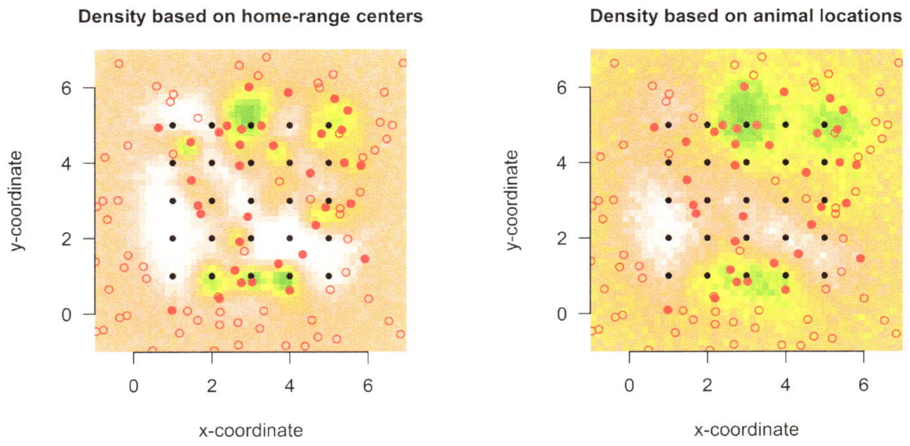

FIGURE 4.20

Two variants of a lynx density map in our simulated representation of the Jura mountain study with the trapping grid shown in solid black and activity centers of the 35 observed lynx as solid and those of the 65 undetected lynx shown as open red circles. (left) Estimated density map based on the home-range centers. (right) Estimated density map based on simulated animal locations. To help you understand patterns in this map, please also go back and study Fig. 4.17, which shows the spatial and individual information available to the model.

```
# Produce a density map based on the estimated ACs (Fig. 4.20 left)
par(mfrow=c(1, 2))
cl <- terrain.colors(20, rev=TRUE)
# Extract posterior samples for s and z
z <- out26$sims.list$z
s <- out26$sims.list$s
s[z == 0] <- NA      # 'Remove' ghost individuals
niter <- nrow(z)
# Choose a grid
xg <- seq(xlim[1], xlim[2], , 50)
yg <- seq(ylim[1], ylim[2], , 50)
# Associate each draw with its pixel
Lx1 <- cut(s[, , 1], breaks=xg, include.lowest=TRUE)
Ly1 <- cut(s[, , 2], breaks=yg, include.lowest=TRUE)
# Tally up the number of samples per pixel: a density table
D.hat1 <- table(Lx1, Ly1) / niter
sum(D.hat1)          # Check, should be the same as the posterior mean of N

# Make a plot of that density table
image(xg, yg, D.hat1, col=cl, xlab='x-coordinate', ylab='y-coordinate', asp=1,
  main='Density based on home-range centers')
points(Sobs, col='red', pch=16)
points(traplocs, pch=16)
points(S, col='red')
```

For a density map based on the estimated animal locations, $\boldsymbol{u}$, we must first use our posterior draws of the ACs, $\boldsymbol{s}$, along with those of the parameter that characterizes our bivariate normal movement model around the former, i.e., the scale parameter `sigma`, to simulate the posterior predictive distribution of the location $\boldsymbol{u}_i$ of every animal i in the population. Then, we can summarize $\boldsymbol{u}$ instead of $\boldsymbol{s}$ before.

```
# Produce a density map based on the estimated locations (Fig. 4.20 right)
# Collect up the stuff we need
AC <- out26$sims.list$s
sigma <- out26$sims.list$sigma
nind <- ncol(z)
# Do the posterior predictions of location
u <- array(NA, dim=dim(AC))
set.seed(1)
for (i in 1:niter){
  # Generate coordinates from normal distribution
  x <- rnorm(nind, AC[i,,1], sigma[i])
  y <- rnorm(nind, AC[i,,2], sigma[i])
  # Animals outside state space get back into the fold
  x <- ifelse(x > Xu, 2*Xu - x, x)
  x <- ifelse(x < Xl, 2*Xl - x, x)
  y <- ifelse(y > Xu, 2*Yu - y, y)
  y <- ifelse(y < Xl, 2*Yl - y, y)
  u[i,,] <- cbind(x, y)
}
u[z == 0] <- NA     # 'Remove' ghost individuals

# Associate each draw with its pixel
Lx2 <- cut(u[,,1], breaks=xg, include.lowest=TRUE)
Ly2 <- cut(u[,,2], breaks=yg, include.lowest=TRUE)
# Tally up the number of samples per pixel: a density table
D.hat2 <- table(Lx2, Ly2) / niter
sum(D.hat2)          # Check, should be the same as the posterior mean of N

# Make a plot of that density table
image(xg, yg, D.hat2, col=cl, xlab='x-coordinate', ylab='y-coordinate', asp=1,
  main='Density based on animal locations')
points(Sobs, col='red', pch=16)
points(traplocs, pch=16)
points(S, col='red')
```

This concludes our quick tour through the newly discovered universe of SCR models: we have looked in detail at the simplest possible model, called SCR0, which allows the estimation of population density as the main demographic parameter of interest. It has only three parameters: `psi`, which in the context of data augmentation conceptually corresponds to population size N, and the two parameters of the detection function. Clearly, we could now extend this model in multiple dimensions. First, we can implement more complicated detection models that can account for effects such as behavioral response, individual heterogeneity or time (Otis et al., 1978). All of this would simply mean to specify linear models for a link-transformed parameter in the detection model, and you may do this for the baseline detection (p_0) or for the scale parameter (σ) separately, or even considering they might be correlated (Efford and Mowat, 2014). Hence, and as for the rest of all population modeling with JAGS, at least half of the power of JAGS stems from the ease with which, within a given class of models, you can harness the powers of linear modeling to your data set and your questions.

In addition, a large number of additional extensions are possible to this most basic SCR model including:

- Open models of both the Jolly-Seber type, i.e., with both survival and recruitment (e.g., Gardner et al., 2010; Chandler and Clark, 2014; Augustine et al., 2020a), and the CJS type with survival only (e.g., Ergon and Gardner, 2014; Schaub and Royle, 2014); see also Gardner et al. (2018).
- Different formulations of the movement model across years in an open model; see Gardner et al. (2018) for a comparison between constant, independent and temporally correlated ACs over time.
- Accommodation of noncircular activity ranges (Royle et al., 2013; Sutherland et al., 2015; Efford, 2019).
- Uncertain individual identification (McClintock et al., 2013, 2015; Augustine et al., 2018, 2020b; Maronde et al., 2020).
- Inhomogeneous point processes, where the expected density of ACs is allowed to be variable, e.g., as a function of the habitat.
- Various formulations for the observation model, e.g., Poisson rather than Bernoulli when the actual number of detections per occasion can be distinguished.
- Observation models formulated in continuous time (Borchers et al., 2014; Dorazio and Karanth, 2017).
- Spatial modeling of components of the recruitment model (Chandler et al., 2018).
- An increasing number of types of integrated models, for instance, where telemetry data are jointly analyzed to obtain improved estimates of the scale parameter in the detection function (Sollmann et al., 2013; see also Chapter 10 in Kéry and Royle, 2021).

In just under 20 years of vigorous growth, SCR modeling has become a very large field. The fundamental difference between all of these models and their nonspatial counterparts is that both the state and most commonly also the observation processes in these hierarchical models are specified in a spatially explicit manner: for the state process, we adopt a spatial or spatio-temporal point process, whereas for the observation process, we typically specify a covariate effect of the latent distance between the center of an animal's activity range and each "trap." A sound understanding of these two fundamentals of SCR, which we have covered in this section, is essential for mastering these more complex SCR models. In the context of IPMs, the use of SCR as component models has only just began (Chandler and Clark, 2014; Chandler et al., 2018; Sun et al., 2019, Chapter 21), but we think that we will see this more frequently in the future.

4.7 SUMMARY AND OUTLOOK

This has been a fairly giant chapter, in which we have attempted to give a broad coverage of the individual statistical models that may be used as part of an IPM. These are like the "Lego pieces" in your box, which you can combine in a modular fashion to build a larger model. Hence, it is good if you have a reasonable overview of these models in your head. It is essential that you have a solid understanding of the individual models that you use in your IPM. The natural approach to building an IPM is to first fit the constituent submodels to each data type separately, before combining them all together in the IPM (see also Chapter 7). Therefore, we view your understanding of this broad catalog of models as essential for you so that you can make a wise pick of the type of model needed for a particular type of data set. However, this catalog is clearly not exhaustive and we refer you again to some of the usual key sources for demographic, population and species distribution models such as Williams et al. (2002), Royle and Dorazio (2008), King et al. (2010), Link and Barker (2010), Kéry and Schaub (2012), Newman et al. (2014), Royle et al. (2014), Buckland et al. (2015), McCrea and Morgan (2015), MacKenzie et al. (2018), and Seber and Schofield (2019), as well as to some key references for their implementation using BUGS, such as Royle and Dorazio (2008), Kéry and Schaub (2012), Royle et al. (2014), and Kéry and Royle (2016, 2021).

Schaub and Abadi (2011) define an IPM as an integrated statistical model that embeds a population dynamics model into statistical inference for at least two data sets, one of which must be a population survey data set. Most published IPMs so far have combined population (or abundance) survey data with either survival or with survival and productivity survey data. Therefore, we have organized our model overview in this chapter around this classification of the three types of data. Of course, there are also others, and we also have another subsection for other data types, but nevertheless it seems that most data types used in nonindividual-based IPMs do fit into these three broad classes. In contrast, for the novel individual-based, spatial IPMs of the type pioneered by Chandler and Clark (2014), this scheme no longer works so well, since the individual-based formulation, with their associated SCR data, makes the population count data somewhat obsolete.

We have also looked at the fundamental types of data sets in each data class. We have seen that the fundamental unit of information in a demographic study is the state of an individual at a given point in space and time, but we usually have to, or choose to, work with a version of this information that is aggregated along some or all of the three dimensions of the individual, space, and time. Aggregation and disaggregation of the demographic information in a model may both have advantages in a given modeling project, and hence, it is useful if you understand this concept of "aggregating information" and can use it as needed.

We emphasize the topic of space quite a bit because we think that spatial IPMs will greatly increase in the future. We have covered three broad types of models where space appears more explicitly: multistate models (Section 4.5.2), where we typically distinguish just a handful of sites, N-mixture and occupancy models (briefly, in Sections 4.3.3 and 4.3.4) that are usually fitted to data sets that distinguish an intermediate number of sites (e.g., 20 to 1000) and finally SCR models (Section 4.6), which usually model the fate of individuals in continuous space, i.e., conceptually with an infinite number of sites. SCR models can be viewed as the types of models in population ecology with the finest spatial resolution (Chandler et al., 2018).

We have also talked about the observation process that is usually in between the true states of interest and the observations that we get. That is, the truism that in nature we rarely ever can observe the things of interest perfectly. Therefore, in our models for populations we need submodels for the observation process to statistically clean the noisy and biased observations so that improved estimates of the latent states result. For various practical and historical reasons, the degree of mechanistic detail varies considerably in the observation models adopted for population count data, productivity data and survival survey data that we typically combine in an IPM. This is particularly true for a key element in the observation process: imperfect detection. It is customarily included in the survival models in an IPM, but rarely in any of the other data types (although it must be said that it may be less crucial for the modeling of productivity data than it is for population counts and survival data).

There are other features of the observation process, or observation errors, that may affect our data and that we may want to accommodate in our models, and in the context of productivity data, we have covered in detail zero inflation, zero truncation, censoring, and underdispersion. We have seen that provided that the right data are available, it is surprisingly straightforward to model these types of observation errors in BUGS.

We have seen that there are different ways in which we can specify an observation process in an SSM. For instance, in the mechanistic manner as a sort of capture-recapture model, where we usually adopt a binomial observation model, where each individual may be detected with a probability p estimated from the data (some more recent such models may also contain a parameter for the converse type of error: individual or species misclassification, or double counting). To us, this latter type of specialized SSMs represents the gold standard in quantitative population ecology: these are the "explicit" hierarchical models of Royle and Dorazio (2008). It is the explicit, mechanistic representation of the observation process that buys us the clear interpretation of the parameters of the state process model, e.g., we get true instead of a vaguely defined "relative abundance," a true species distribution instead of some proxy or mere index of the occurrence probability, and real survival probability instead of a mere return rate.

In most published IPMs, this rigorous distinction is only made in the models for the survival survey data, where a CJS or ring-recovery model is the rule. Both models describe the observation process in terms of probabilities of recapture or recovery, yielding a clear definition of the parameters in the state model as apparent or true survival probability. In contrast, in the models for population survey data, no parameter p is usually in sight, thus, for this data type we usually see only implicit hierarchical models, or traditional SSMs. Finally, as mentioned above productivity survey data are not normally modeled using an SSM approach, and hence, there cannot be a parameter p in a model for productivity survey data, either. We believe that zero truncation and censoring are present in many productivity data sets but may currently be ignored often.

Overall, we expect that over time, people will apply the SSM approach and especially also the "explicit" hierarchical model approach increasingly also for the population and productivity survey data sets in their IPMs. However, given the goals of an IPM analysis, it remains to be seen whether the accompanying increase in the complexity of the IPMs will be paid off by improvements in the inferences from these models.

We have covered two classes of models for spatially stratified data: occupancy models and N-mixture models. It is straightforward to use these models to obtain abundance estimates that fully correct for imperfect detection. Estimates of annual population size from these "explicit" hierarchical models can then be incorporated into IPMs along with their partial or full estimation uncertainty represented by the diagonals (i.e., the variance of estimates) or the full VC matrix. We are intrigued by the thought that it should be possible to avoid such a two-step approach and instead integrate an occupancy or N-mixture model directly as a module in an IPM, but so far this has proven somewhat elusive due to subtle differences in the meaning of the parameters as fixed or random quantities in the different parts of the models when implemented in JAGS. However, we have presented a "demographic" SSM due to Dail and Madsen (2011), which requires multivariate population count data and wherein estimation of abundance and demographic modeling in terms of vital rates is jointly achieved. This model appears like a prime candidate for a more mechanistic representation for the population count data in an IPM than the traditional Gaussian SSMs.

We have briefly covered SCR (Efford, 2004; Borchers and Efford, 2008; Royle and Young, 2008; Royle et al., 2014) as a further model class that may be used as an ingredient to an IPM or indeed as the "motherboard" of an individual-based spatial IPM (Chandler and Clark, 2014; Chandler et al., 2018; Sun et al., 2019; Section 10.8 in Kéry and Royle, 2021). We think that in the future we will see more applications of IPMs that include spatially explicit submodels or are entirely spatially explicit. However, we also think that there may be some limitations to universal adoption of SCR models for IPMs having to do with the greater data requirements and sometimes dramatically increased computational burden.

In this chapter, for pedagogical reasons we have only showed the bare-bones versions of each model to clarify for you as much as possible the fundamental structure of each model. Especially in the second part of this book, we will use many of the models of the catalog in this chapter to construct IPMs. There, you will see more complex versions of the models in this chapter, and we will include covariates, random effects, covariances between parameters and more than just two or three states in multistate models, to name but a few extensions to the basic variants of these models from this chapter.

4.8 EXERCISES

1. Given the mean (1.02) and temporal variability (log-scale variance = 0.02) of the population growth rate in an exponential model of growth, simulate one population trajectory of length 20 years starting from a population of size 50. Then, simulate a count data set from that population using a binomial distribution with a detection probability of 0.6, hence assuming that imperfect detection is the sole mechanism producing the observation error in the counts. Then, write and fit an SSM that corresponds exactly to the data simulation and judge whether the parameters appear to be identifiable.
2. Using the wryneck nest survival data (`data(wryneck)` in the `IPMbook` package) fit a model than includes an additive effect of the hatching date on daily nest survival in addition to the age effects (days since hatching). Do early nests have higher survival than late nests?
3. Use the woodchat shrike capture-recapture data (`data(woodchat5)` in the `IPMbook` package) and first fit a model with age-dependent survival and age-independent recapture probability. Next, fit a model in which survival is also a function of the population counts, i.e., in which you add a form of density dependence. Is there any evidence that survival changes as a function of this index of population size?
4. Use the mark-resighting-recovery data set of the storks (`data(stork)` in the `IPMbook` package). Write a new integrated model that uses as one data source the capture-resighting data, and as another source the dead-recovery data. Each of these data sets has the same number of marked individuals, but one then only includes the resightings and the other only the dead recoveries. True survival is a shared parameter between the two likelihoods in this integrated model. Compare the parameter estimates with those from the joint model developed in Section 4.5.4. Why are the posterior standard deviations different?

CHAPTER

INTRODUCTION TO INTEGRATED POPULATION MODELS

5

CHAPTER OUTLINE

5.1 INTRODUCTION

In Chapter 3, we covered how to analyze matrix population models using either conventional methods, or importantly for what follows here, Markov chain Monte Carlo (MCMC) methods in JAGS. Analysis of these models permits the estimation of a host of quantities that are interesting in population ecology and wildlife management, such as asymptotic or stochastic population growth rates, growth rate sensitivities, and generation times. In all these calculations we first assumed that the demographic rates were perfectly known, but we also saw how we can account for parameter uncertainty or measurement error associated with the parameters. Often, you will have access to the actual demographic data on which such demographic rate estimates are based, and in Chapter 4, we covered a wide array of models with which such data sets can be analyzed to obtain estimates of demographic rate parameters or population size.

Traditionally, population modeling has been carried out with a two-step approach: first, demographic data are analyzed and demographic rates estimated, and second, these estimates are fed into a matrix population model to calculate the abovementioned and other quantities of interest. However, a more efficient and elegant approach is to perform both steps simultaneously in one joint analysis, i.e., to develop an *integrated population model* (IPM). The main focus of this chapter is to introduce IPMs, which offer important advantages such as increased precision of parameter estimates, the ability to estimate additional parameters that cannot be estimated using a two-step analysis, and honest propagation of all uncertainty into every estimated quantity in an analysis. In Chapter 6, we will look at these advantages in more detail.

In this chapter, we provide two alternative introductions to IPMs. The first is in Sections 5.2 and 5.3, and the second is in Section 5.4. The first introduction to an IPM is perhaps a little harder to follow at first, but it is important for your conceptual understanding of an IPM. The following is an outline of Sections 5.2 and 5.3, where we progressively develop our first IPM as an extension of a matrix population model:

- First, in Section 5.2.1, we analyze capture-recapture data alone to estimate the mean and uncertainty of survival probabilities and then use them in a matrix population model to estimate the asymptotic population growth rate while

Integrated Population Models. https://doi.org/10.1016/B978-0-12-820564-8.00021-X

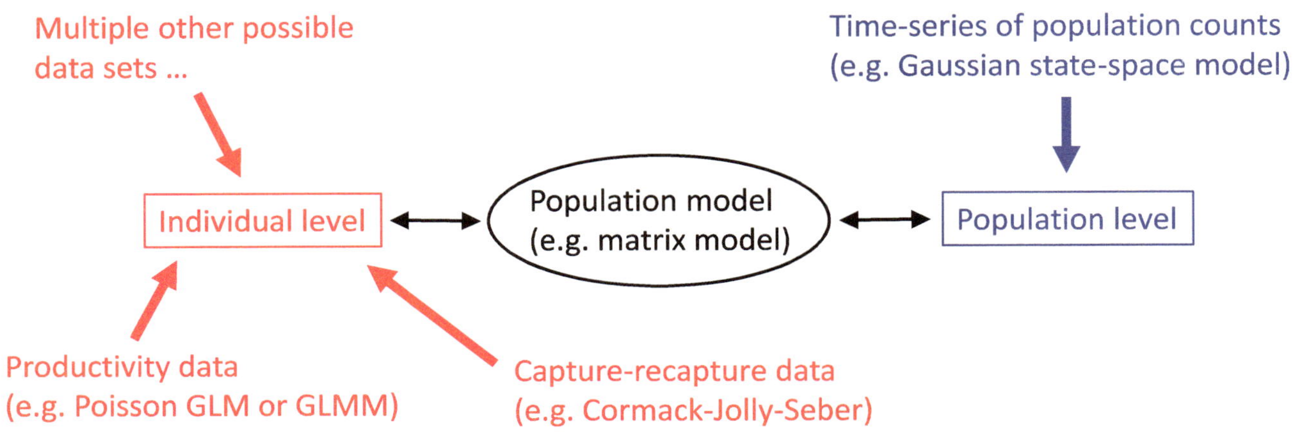

FIGURE 5.1

Graphical representation of an integrated population model (IPM). Data sets from the individual level (red) and population level (blue) are jointly analyzed, and the connection between these levels is represented by a population model (black), which typically is a matrix population model. Abbreviations: *GLM*, generalized linear model; *GLMM*, generalized linear mixed model.

accommodating this uncertainty. This will not yet be an integrated model because we analyze only a single data set—the capture-recapture data. The estimate of the population growth rate from the matrix population model will be a derived parameter.

- Second, in Section 5.2.2, we analyze productivity data in addition, and in the matrix population model, we use the resulting point estimate and its uncertainty to obtain an improved estimate of the asymptotic population growth rate and its associated uncertainty. We call this analysis a joint analysis because we analyze two data sets jointly. However, no information about any parameter is shared between the two data sets, so we do not (yet) call it an integrated (population) model. We simply achieve proper error propagation for estimation of the population growth rate when using information from both data sets.
- Third, in Section 5.3, we include population count data in the joint model, for which we specify a Gaussian state-space model *with a process model that is exactly our matrix population model.* The demographic rates provide the links of the population count data set with the other two data sets. This model is a true IPM *sensu* Schaub and Abadi (2011). Information among data sets is now shared: all the estimates (demographic rates, population sizes) are informed by more than a single data set and have the nature of weighted averages.

We like to emphasize that IPMs can be viewed as an extension of matrix population models because models for the analysis of demographic and population count data are almost always combined using matrix population models at their core (Thomas et al., 2005). Our first approach to developing an IPM highlights this fact. It will become obvious that we can derive all the same quantities (e.g., life-history summaries) and carry-out the same perturbation analyses from an IPM as we can from a matrix population model. We may therefore define an IPM as a demographic population model (e.g., a matrix population model) embedded in several submodels to simultaneously estimate parameters and analyze the population model itself. Alternatively, because all IPMs as defined in this book and by Schaub and Abadi (2011) partly contain population count data (or alternatively, estimates of population size or population indices) for which we adopt a state-space model, we can also say that IPMs are demographic state-space models coupled with separate (sub)models for the demographic data. An IPM makes use of data from both the individual and the population level, and the matrix population model connects the two levels (Fig. 5.1).

In Section 5.4, we use an alternative three-step approach to explain and develop an IPM. This is the way one usually develops an IPM in practice, but in this approach the matrix population model features much less prominently. Using this three-step approach, we develop a slightly more complex IPM that includes demographic and environmental stochasticity (i.e., variability). We also take a closer look at the joint likelihood of an IPM and discuss the key assumptions of the model. Finally, we present a number of simulations to assess the parameter accuracy of these models.

5.2 FEEDING DEMOGRAPHIC DATA INTO THE ANALYSIS OF A MATRIX POPULATION MODEL

The goal of Sections 5.2 and 5.3 is to develop an IPM with a matrix population model as its core. We start by using code for a matrix population model (from Section 3.4.1) to which we first add multiple data sets from the individual level and then from the population level. As an example, we continue using the woodchat shrike data introduced in Chapter 3, where we assumed that we had capture-recapture, productivity, and population count data from a 20-year study (these are simulated data; see Section 5.5 for more details). The underlying, constant survival probabilities are 0.3 for juveniles and 0.55 for adults, and productivity is 2.6 for first-time breeders and 3.6 for experienced breeders. Now, our productivity data are assumed to represent the total number of fledglings per female (i.e., male and female combined), whereas in Chapter 3, we assumed that we directly had the number of female fledglings per adult female. In our models here, we will assume that woodchat shrikes do not disperse outside the study area, and thus the estimated survival probabilities will be true rather than apparent survival, and we will not accommodate dispersal in the population model.

5.2.1 USING CAPTURE-RECAPTURE DATA IN A MATRIX POPULATION MODEL

We load the combined data sets but only use the capture-recapture data at first.

```
library(IPMbook); library(jagsUI)
data(woodchat5)
str(woodchat5)
```

```
 $ ch    : num [1:1902, 1:20] 1 1 1 1 1 1 1 1 1 1 ...
 $ age   : num [1:1902] 2 2 2 2 2 2 2 2 2 2 ...
 $ repro: num [1:929, 1:3] 6 2 2 5 3 5 3 2 3 2 ...
  ..- attr(*, "dimnames")=List of 2
  .. ..$ : NULL
  .. ..$ : chr [1:3] "Reproduction" "Year" "Age of mother"
 $ count: num [1:20] 91 119 131 88 139 145 148 116 112 106 ...
```

Object `ch` contains the capture histories of 1902 individuals over 20 years, and object `age` is a vector indicating whether individual i was marked as a juvenile (`age[i] = 1`) or an adult (`age[i] = 2`). We use a multinomial likelihood to analyze the capture-recapture data (Section 4.5.1.2) because of its computational efficiency. Function `marrayAge` generates an age-dependent m-array based on the individual capture histories and the vector with age information at marking.

```
marr <- marrayAge(woodchat5$ch, woodchat5$age)
```

Our main goal is the estimation of the asymptotic population growth rate λ and its precision by using point estimates and their uncertainty of survival probabilities that we obtain from this data set. We will also include age-specific productivity values in the matrix population model, which we now suppose were taken from the literature ($f_1 = 2.6, f_a = 3.6$), and we assume them to be known without error for now. Our population model is female-based; hence, we assume an even fledgling sex ratio and divide productivity in the population model by 2. The matrix population model corresponds to a prebreeding "census" female-based model (Section 3.4.1), and its life-cycle graph is shown in Fig. 3.5B. As a template, we start with BUGS code for the estimation of the asymptotic population growth rate (from Section 3.4.2), where we had the possibility to include uncertainty about the demographic parameters. There, we added information about the uncertainty in survival by using a random number generator (RNG) from a statistical distribution from which the survival values were drawn. Now, we include a model to analyze the capture-recapture data, and both survival point estimates and their uncertainty stem directly from the resulting draws of the posterior distribution. That is, we combine code that estimates age-specific survival from the m-array data (Section 4.5.1.2) with the code for the matrix population model from Section 3.4.2. We project the population for $T = 15$ years, calculate the annual population growth rates, and use the last of those as our estimate of the asymptotic population growth rate. We won't include temporal variability in our model now but instead simply assume that the survival probabilities of both age classes are constant over the 20-year period. We need to define priors for the survival and recapture parameters, as these are estimated quantities, and we use uniform distributions in the range between 0 and 1. Note that although these priors look superficially similar to the RNGs we used in Section 3.4.1, they are conceptually very different because now they are associated with an estimated parameter. Note also that it is imperative to use the same name for survival (i.e., `mean.sj`, `mean.sa`) in the different parts of the model because otherwise there would be no link between them and no information would be shared then.

Our data consist of the m-arrays along with the number of released individuals, the information about productivity, and the number of years for which we want to project the population to obtain an estimate of the population growth rate.

```
# Bundle data and produce data overview
jags.data <- list(marr.j=marr[,,1], marr.a=marr[,,2], n.occasions=dim(marr)[2],
  rel.j=rowSums(marr[,,1]), rel.a=rowSums(marr[,,2]), mean.f=c(2.6, 3.6), T=15)
str(jags.data)

List of 7
 $ marr.j      : num [1:19, 1:20] 8 0 0 0 0 0 0 0 0 0 ...
 $ marr.a      : num [1:19, 1:20] 16 0 0 0 0 0 0 0 0 0 ...
 $ n.occasions: int 20
 $ rel.j       : num [1:19] 51 53 55 65 73 66 61 76 65 75 ...
 $ rel.a       : num [1:19] 36 39 44 61 61 50 43 61 51 53 ...
 $ mean.f      : num [1:2] 2.6 3.6
 $ T           : num 15

# Write JAGS model file
cat(file="model1.txt", "
model {
# Priors and linear models
mean.sj ~ dunif(0, 1)
mean.sa ~ dunif(0, 1)
mean.p ~ dunif(0, 1)

for (t in 1:(n.occasions-1)){
  sj[t] <- mean.sj
  sa[t] <- mean.sa
  p[t] <- mean.p
}

# Capture-recapture data (CJS model with multinomial likelihood)
# Define the multinomial likelihood
for (t in 1:(n.occasions-1)){
  marr.j[t,1:n.occasions] ~ dmulti(pr.j[t,], rel.j[t])
  marr.a[t,1:n.occasions] ~ dmulti(pr.a[t,], rel.a[t])
}
# Define the cell probabilities of the m-arrays
for (t in 1:(n.occasions-1)){
  # Main diagonal
  q[t] <- 1 - p[t]            # Probability of non-recapture
  pr.j[t,t] <- sj[t] * p[t]
  pr.a[t,t] <- sa[t] * p[t]
  # Above main diagonal
  for (j in (t+1):(n.occasions-1)){
    pr.j[t,j] <- sj[t] * prod(sa[(t+1):j]) * prod(q[t:(j-1)]) * p[j]
    pr.a[t,j] <- prod(sa[t:j]) * prod(q[t:(j-1)]) * p[j]
  } #j
  # Below main diagonal
  for (j in 1:(t-1)){
    pr.j[t,j] <- 0
    pr.a[t,j] <- 0
  } #j
} #t
# Last column: probability of non-recapture
for (t in 1:(n.occasions-1)){
  pr.j[t,n.occasions] <- 1-sum(pr.j[t,1:(n.occasions-1)])
  pr.a[t,n.occasions] <- 1-sum(pr.a[t,1:(n.occasions-1)])
}

# Model for population dynamics: a simple matrix population model
# Define initial stage-specific population sizes
N[1,1] <- 1
N[2,1] <- 1

# Loop over time
for (t in 1:T){
  # Population projection
  N[1,t+1] <- N[1,t] * mean.f[1] / 2 * mean.sj + N[2,t] * mean.f[2] / 2 * mean.sj
  N[2,t+1] <- (N[1,t] + N[2,t]) * mean.sa
  # Annual growth rate
  ann.growth.rate[t] <- (N[1,t+1] + N[2,t+1]) / (N[1,t] + N[2,t])
}
lambda <- ann.growth.rate[T]
}
")
```

```
# Initial values
inits <- function(){list(mean.sj=runif(1, 0, 1), mean.sa=runif(1, 0, 1))}

# Parameters monitored
parameters <- c("mean.sj", "mean.sa", "mean.p", "lambda")
```

In the MCMC settings, we include a burn-in and run multiple chains because we estimate parameters for survival and recapture probabilities and must ensure that these estimates are based on converged chains.

```
# MCMC settings
ni <- 3000; nb <- 1000; nc <- 3; nt <- 1; na <- 1000

# Call JAGS (ART <1 min), check convergence and summarize posteriors
out1 <- jags(jags.data, inits, parameters, "model1.txt", n.iter=ni, n.burnin=nb, n.chains=nc,
  n.thin=nt, n.adapt=na)
traceplot(out1)
print(out1, 3)

          mean    sd  2.5%   50% 97.5% overlap0 f  Rhat n.eff
mean.sj  0.304 0.016 0.274 0.304 0.336    FALSE 1 1.001  6000
mean.sa  0.542 0.014 0.514 0.542 0.569    FALSE 1 1.001  2640
mean.p   0.604 0.021 0.563 0.604 0.644    FALSE 1 1.001  6000
lambda   1.018 0.027 0.966 1.018 1.073    FALSE 1 1.001  6000
```

Convergence looks good, and the estimates of the survival probabilities and asymptotic population growth rate agree well with the data-generating values (Section 5.2). We obtain an assessment of the uncertainty around the asymptotic population growth rate due to the estimation uncertainty, or measurement error, of juvenile and adult survival. The uncertainty assessment is exact because it is a function of the posterior distributions of the two survival probabilities on which it depends. Alternative methods to obtain a measure of the uncertainty of a derived quantity such as the asymptotic population growth rate (e.g., Monte Carlo simulation or delta method; see Section 3.4.2) make distributional assumptions and are therefore more approximate. Naturally, we could also have computed other life-history summaries such as growth rate sensitivities, stable age ratios, or generation time. They are all derived quantities, and their computation can be included in the JAGS code as shown in Section 3.4.

5.2.2 COMBINING CAPTURE-RECAPTURE AND PRODUCTIVITY DATA IN A MATRIX POPULATION MODEL

Next, we include productivity data in our matrix population model and adopt a simple Poisson regression model for them. We combine this new bit of code with the code just developed. The productivity data are stored in the R object `repro`, which is a matrix where each row corresponds to one brood. The first column is the number of fledglings (productivity), the second column is the year, and the last column gives the age class of the mother when she raised that brood (1 is for a 1-year-old, and 2 is for a 2-year-old or older breeder). As we want to model productivity as a function of age class, we only need the first and third columns.

Including the productivity data requires only slight adaptations to the model code—we just add a Poisson regression likelihood with vague priors for all parameters. Naturally, our data bundle now includes the productivity data as well.

```
# Bundle data and produce data overview
jags.data <- list(marr.j=marr[,,1], marr.a=marr[,,2], n.occasions=dim(marr)[2],
  rel.j=rowSums(marr[,,1]), rel.a=rowSums(marr[,,2]), J=woodchat5$repro[,1],
  age=woodchat5$repro[,3], T=15)
str(jags.data)

List of 8
 $ marr.j     : num [1:19, 1:20] 8 0 0 0 0 0 0 0 0 0 ...
 $ marr.a     : num [1:19, 1:20] 16 0 0 0 0 0 0 0 0 0 ...
 $ n.occasions: int 20
 $ rel.j      : num [1:19] 51 53 55 65 73 66 61 76 65 75 ...
 $ rel.a      : num [1:19] 36 39 44 61 61 50 43 61 51 53 ...
 $ J          : num [1:929] 6 2 2 5 3 5 3 2 3 2 ...
 $ age        : num [1:929] 1 1 1 1 1 1 1 1 1 1 ...
 $ T          : num 15
```

```
# Write JAGS model file
cat(file="model2.txt", "
model {
# Priors and linear models
mean.sj ~ dunif(0, 1)
mean.sa ~ dunif(0, 1)
mean.p ~ dunif(0, 1)
mean.f[1] ~ dunif(0, 10)
mean.f[2] ~ dunif(0, 10)

for (t in 1:(n.occasions-1)){
  sj[t] <- mean.sj
  sa[t] <- mean.sa
  p[t] <- mean.p
}

# Productivity data (Poisson regression model)
for (i in 1:length(J)){
  J[i] ~ dpois(mean.f[age[i]])
}

# Capture-recapture data (CJS model with multinomial likelihood)
# Define the multinomial likelihood
for (t in 1:(n.occasions-1)){
  marr.j[t,1:n.occasions] ~ dmulti(pr.j[t,], rel.j[t])
  marr.a[t,1:n.occasions] ~ dmulti(pr.a[t,], rel.a[t])
}
# Define the cell probabilities of the m-arrays
for (t in 1:(n.occasions-1)){
  # Main diagonal
  q[t] <- 1 - p[t]            # Probability of non-recapture
  pr.j[t,t] <- sj[t] * p[t]
  pr.a[t,t] <- sa[t] * p[t]
  # Above main diagonal
  for (j in (t+1):(n.occasions-1)){
    pr.j[t,j] <- sj[t] * prod(sa[(t+1):j]) * prod(q[t:(j-1)]) * p[j]
    pr.a[t,j] <- prod(sa[t:j]) * prod(q[t:(j-1)]) * p[j]
  } #j
  # Below main diagonal
  for (j in 1:(t-1)){
    pr.j[t,j] <- 0
    pr.a[t,j] <- 0
  } #j
} #t
# Last column: probability of non-recapture
for (t in 1:(n.occasions-1)){
  pr.j[t,n.occasions] <- 1-sum(pr.j[t,1:(n.occasions-1)])
  pr.a[t,n.occasions] <- 1-sum(pr.a[t,1:(n.occasions-1)])
}

# Model for population dynamics: a simple matrix population model
# Define initial stage-specific population sizes
N[1,1] <- 1
N[2,1] <- 1

# Loop over time
for (t in 1:T){
  # Population projection
  N[1,t+1] <- N[1,t] * mean.f[1] / 2 * mean.sj + N[2,t] * mean.f[2] / 2 * mean.sj
  N[2,t+1] <- (N[1,t] + N[2,t]) * mean.sa
  # Annual growth rate
  ann.growth.rate[t] <- (N[1,t+1] + N[2,t+1]) / (N[1,t] + N[2,t])
}
lambda <- ann.growth.rate[T]
}
")

# Initial values
inits <- function(){list(mean.sj=runif(1, 0, 1), mean.sa=runif(1, 0, 1))}

# Parameters monitored
parameters <- c("mean.sj", "mean.sa", "mean.p", "mean.f", "lambda")
```

```
# MCMC settings
ni <- 3000; nb <- 1000; nc <- 3; nt <- 1; na <- 1000

# Call JAGS (ART <1 min), check convergence and summarize posteriors
out2 <- jags(jags.data, inits, parameters, "model2.txt", n.iter=ni, n.burnin=nb, n.chains=nc,
  n.thin=nt, n.adapt=na)
traceplot(out2)
print(out2, 3)

              mean    sd    2.5%     50%   97.5% overlap0 f  Rhat n.eff
mean.sj      0.304 0.016   0.273   0.304   0.337    FALSE 1 1.000  6000
mean.sa      0.542 0.014   0.514   0.542   0.570    FALSE 1 1.002  1998
mean.p       0.604 0.021   0.562   0.604   0.643    FALSE 1 1.001  6000
mean.f[1]    2.677 0.079   2.527   2.675   2.832    FALSE 1 1.001  5055
mean.f[2]    3.685 0.089   3.517   3.685   3.866    FALSE 1 1.001  2360
lambda       1.030 0.029   0.973   1.030   1.089    FALSE 1 1.001  5372
```

In addition to survival, we also obtain parameter estimates for productivity. As expected, the uncertainty around the asymptotic population growth rate is slightly larger than under model 1, where we assumed that productivity was known perfectly. Now, the uncertainty in the asymptotic population growth rate stems from the uncertainty associated with both survival and productivity.

So far, we have jointly analyzed capture-recapture and productivity data and linked them to our matrix population model to estimate the asymptotic population growth rate. Thus, we have carried out what could be called a joint analysis but not yet what we call an *integrated* analysis. The uncertainty due to data sampling and parameter estimation (measurement errors) is correctly and automatically propagated into the uncertainty of the asymptotic population growth rate or any other derived quantity. The traditional way to estimate the population growth rate using the two data sets would be a two-step approach where first, both data sets are analyzed separately to obtain point estimates of all parameters along with their SE (or posterior SD), and second, a matrix population model would be set up with these estimates followed by Monte Carlo simulations to obtain the mean and uncertainty of the population growth rate as shown in Section 3.3.2. Using a joint analysis, the error propagation occurs "automatically," and more accurately because it directly uses the posterior draws from the posterior distributions of the demographic rates.

5.3 OUR FIRST INTEGRATED POPULATION MODEL

As a final step toward our first IPM, we now also include our data from the population level in the model, i.e., the population count data, for which we adopt a demographic state-space model (Section 4.3.1). That is, a Markovian model that distinguishes a process and an observation submodel. Population size at any time t is determined by the demographic rates and population size at $t-1$; therefore, it is evident that the population count data also contain some information about the demographic rates in addition to information about population size. Inclusion of the population count data therefore results in a fundamentally different model—previously, we simply made error propagation for the population growth rate, but now the estimate of the population growth rate is in part also influenced by this additional source of information. The information contained in multiple data sets about parameters in common is now shared in this truly integrated population model.

However, despite this crucial difference, the code is still very similar to that of the previous model. The main differences are that the matrix population model in the previous model now becomes the process model in the state-space model and that a simple Gaussian observation model for the population count data is added. Furthermore, we replace the fixed stage-specific population sizes in the first year by parameters with a prior because we estimate these quantities as well. Because the observed population count in year 1 is about 200 (this is the sum of both stage classes) but we lack more specific information, we specify uniform priors in the range 1−300 for both stage classes. Note that we must exclude 0 because a total population size of 0 is not allowed under an exponential model of population growth. Also note that we assume Gaussian errors for the population counts, as this is a customary choice in the IPM literature (e.g., Besbeas et al., 2002), but you could choose other observation models such as log-normal or Poisson (see Section 6.6.5).

In the JAGS code, we add a new section, "Derived parameters", in which we compute annual population growth rates and total population sizes as a function of other estimates in the model. The data bundle now includes the population count data. Because the temporal dimension of the population count data corresponds to that of the capture-recapture data, the variable T is no longer needed.

```
# Bundle data and produce data overview
jags.data <- list(marr.j=marr[,,1], marr.a=marr[,,2], n.occasions=dim(marr)[2],
  rel.j=rowSums(marr[,,1]), rel.a=rowSums(marr[,,2]), J=woodchat5$repro[,1],
  age=woodchat5$repro[,3], C=woodchat5$count)
str(jags.data)
```

```
List of 8
 $ marr.j      : num [1:19, 1:20] 8 0 0 0 0 0 0 0 0 0 ...
 $ marr.a      : num [1:19, 1:20] 16 0 0 0 0 0 0 0 0 0 ...
 $ n.occasions: int 20
 $ rel.j       : num [1:19] 51 53 55 65 73 66 61 76 65 75 ...
 $ rel.a       : num [1:19] 36 39 44 61 61 50 43 61 51 53 ...
 $ J           : num [1:929] 6 2 2 5 3 5 3 2 3 2 ...
 $ age         : num [1:929] 1 1 1 1 1 1 1 1 1 1 ...
 $ C           : num [1:20] 91 119 131 88 139 145 148 116 112 106 ...

# Write JAGS model file
cat(file="model3.txt", "
model {
# Priors and linear models
mean.sj ~ dunif(0, 1)
mean.sa ~ dunif(0, 1)
mean.p ~ dunif(0, 1)
mean.f[1] ~ dunif(0, 10)
mean.f[2] ~ dunif(0, 10)

for (t in 1:(n.occasions-1)){
  sj[t] <- mean.sj
  sa[t] <- mean.sa
  p[t] <- mean.p
}

sigma ~ dunif(0.5, 100)
tau <- pow(sigma, -2)

# Population count data (state-space model)
# Model for the initial stage-specific population sizes: uniform priors
N[1,1] ~ dunif(1, 300)
N[2,1] ~ dunif(1, 300)

# Process model over time: our model of population dynamics
for (t in 1:(n.occasions-1)){
  N[1,t+1] <- N[1,t] * mean.f[1] / 2 * mean.sj + N[2,t] * mean.f[2] / 2 * mean.sj
  N[2,t+1] <- (N[1,t] + N[2,t]) * mean.sa
}

# Observation model
for (t in 1:n.occasions){
  C[t] ~ dnorm(N[1,t] + N[2,t], tau)
}

# Productivity data (Poisson regression model)
for (i in 1:length(J)){
  J[i] ~ dpois(mean.f[age[i]])
}

# Capture-recapture data (CJS model with multinomial likelihood)
# Define the multinomial likelihood
for (t in 1:(n.occasions-1)){
  marr.j[t,1:n.occasions] ~ dmulti(pr.j[t,], rel.j[t])
  marr.a[t,1:n.occasions] ~ dmulti(pr.a[t,], rel.a[t])
}
# Define the cell probabilities of the m-arrays
for (t in 1:(n.occasions-1)){
  # Main diagonal
  q[t] <- 1 - p[t]    # Probability of non-recapture
  pr.j[t,t] <- sj[t] * p[t]
  pr.a[t,t] <- sa[t] * p[t]
  # Above main diagonal
  for (j in (t+1):(n.occasions-1)){
    pr.j[t,j] <- sj[t] * prod(sa[(t+1):j]) * prod(q[t:(j-1)]) * p[j]
    pr.a[t,j] <- prod(sa[t:j]) * prod(q[t:(j-1)]) * p[j]
  } #j
  # Below main diagonal
  for (j in 1:(t-1)){
    pr.j[t,j] <- 0
    pr.a[t,j] <- 0
  } #j
} #t
```

```
# Last column: probability of non-recapture
for (t in 1:(n.occasions-1)){
  pr.j[t,n.occasions] <- 1-sum(pr.j[t,1:(n.occasions-1)])
  pr.a[t,n.occasions] <- 1-sum(pr.a[t,1:(n.occasions-1)])
}

# Derived parameters
# Annual population growth rate
for (t in 1:(n.occasions-1)){
  ann.growth.rate[t] <- (N[1,t+1] + N[2,t+1]) / (N[1,t] + N[2,t])
}
# Total population size
for (t in 1:n.occasions){
  Ntot[t] <- N[1,t] + N[2,t]
}
}
")
```

Finding initial values that work for JAGS can sometimes be a serious pain with IPMs (see later), but in our simple example, this poses no problems. It is enough to provide initials for just one survival probability.

```
# Initial values
inits <- function(){list(mean.sj=runif(1, 0, 0.5))}

# Parameters monitored
parameters <- c("mean.sj", "mean.sa", "mean.p", "mean.f", "N", "sigma", "ann.growth.rate", "Ntot")

# MCMC settings
ni <- 40000; nb <- 10000; nc <- 3; nt <- 3; na <- 3000

# Call JAGS (ART 2 min), check convergence and summarize posteriors
out3 <- jags(jags.data, inits, parameters, "model3.txt", n.iter=ni, n.burnin=nb, n.chains=nc,
  n.thin=nt, n.adapt=na)
traceplot(out3)
print(out3, 3)

                       mean     sd    2.5%     50%   97.5% overlap0 f  Rhat n.eff
mean.sj               0.298  0.010   0.279   0.298   0.317    FALSE 1 1.002  1175
mean.sa               0.539  0.012   0.515   0.539   0.564    FALSE 1 1.003   857
mean.p                0.608  0.019   0.571   0.608   0.645    FALSE 1 1.000  9184
mean.f[1]             2.671  0.076   2.526   2.670   2.822    FALSE 1 1.000  7172
mean.f[2]             3.674  0.085   3.510   3.672   3.844    FALSE 1 1.001  4281
N[1,1]               47.751 32.297   2.676  43.470 110.762    FALSE 1 1.006   371
N[2,1]               62.412 29.082   6.701  65.476 106.317    FALSE 1 1.006   380
N[1,2]               53.101  4.429  44.608  53.092  61.675    FALSE 1 1.004   499
N[2,2]               59.402  4.247  51.433  59.259  68.046    FALSE 1 1.002   897
[ ...output truncated... ]
N[1,20]              70.264  4.264  61.993  70.211  78.927    FALSE 1 1.000 12067
N[2,20]              79.456  4.399  71.016  79.381  88.332    FALSE 1 1.001  2774
sigma                16.976  3.149  12.116  16.516  24.414    FALSE 1 1.000 10870
ann.growth.rate[1]    1.023  0.041   0.945   1.026   1.087    FALSE 1 1.006   363
ann.growth.rate[2]    1.016  0.007   1.003   1.015   1.029    FALSE 1 1.002   979
[ ...output truncated... ]
ann.growth.rate[18]   1.016  0.005   1.006   1.016   1.027    FALSE 1 1.000 18640
ann.growth.rate[19]   1.016  0.005   1.006   1.016   1.027    FALSE 1 1.000 18640
Ntot[1]             110.163  7.608  95.745 109.918 125.721    FALSE 1 1.001  1648
Ntot[2]             112.503  6.546  99.570 112.493 125.646    FALSE 1 1.000 11582
[ ...output truncated... ]
Ntot[19]            147.325  7.114 133.524 147.263 161.763    FALSE 1 1.000 30000
Ntot[20]            149.720  7.854 134.538 149.609 165.765    FALSE 1 1.000 30000
```

The MCMC chains converge fairly quickly, and the parameter estimates look good. We note that even though the two age classes (i.e., yearlings and adults) could not be distinguished in our population counts, we obtain estimates of age-specific population sizes. The ability of IPMs to estimate population structure is an important asset, in particular for retrospective and prospective population analyses; see Chapters 9 and 10. Comparing population counts and population estimates, we see that the latter do not vary so much over time as do the population counts (Fig. 5.2). In fact, the estimated population trajectory follows a slightly curved line corresponding to an exponentially growing population. The reason for this behavior is that our population model assumes the absence of any stochasticity, which of course is a strong simplification of the dynamics of a real population.

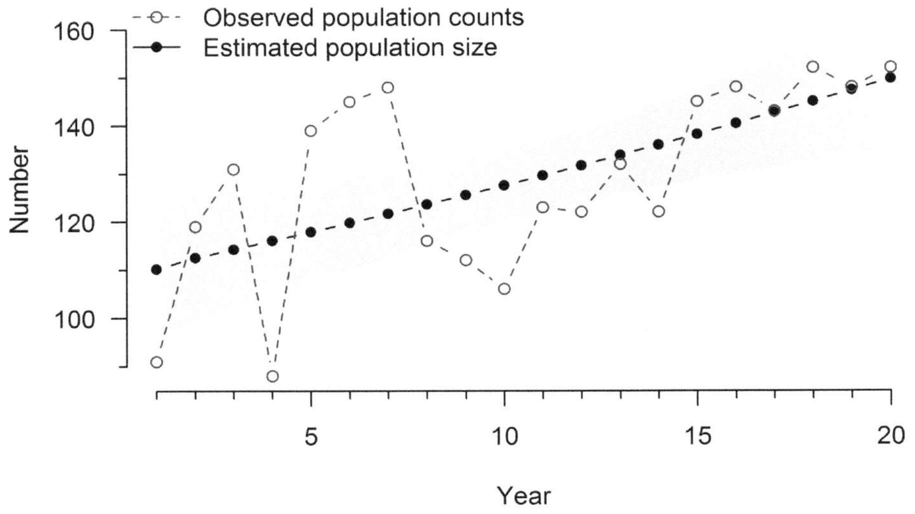

FIGURE 5.2

Estimated total population size (in black, with 95% credible intervals) from the first simple integrated population model (IPM) for woodchat shrikes along with the observed population counts (in blue).

This is now the first genuine IPM we have developed in this book! Although there are many technical details that you must master to fit your own IPMs, these models are conceptually not that complicated. They are simply a combination, with the matrix population model at their core, of several models for data at both the individual and the population level.

In Section 5.2, we combined capture-recapture and productivity data only, and we called this a joint analysis: the two data sets were jointly analyzed, but there were no shared parameters—each provides information about separate parameters, i.e., survival and productivity. In that model, the population growth rate is a derived quantity with an uncertainty due to the estimation of survival and productivity. But by adding the population count data, we now conduct an *integrated* analysis, which is fundamentally different. The population count data contain information about survival, productivity, and age-specific population sizes. Thus, estimates of some parameters are informed by more than a single data set; they are shared. In the BUGS language, by simply using identical names for some demographic parameters that appear in more than one data likelihood, we achieve the "integration" (or merging, or fusion) of these data sets into a single model. The resulting parameter estimates are informed by all the data sets—indeed they are weighted averages where the weights depend on the information content of the data sets about a parameter. See Chapters 6 and 7 for more about this topic.

5.4 THE THREE-STEP APPROACH TO INTEGRATED POPULATION MODELING

In the two previous sections, we developed our first IPM starting from a matrix population model that was supplemented firstly with demographic data and secondly with population count data. We intended for this first practical explanation of an IPM to make it very clear that an IPM is generally (though not always) a matrix population model complemented with data from both the individual and the population level. Consequently, all life-history summaries that you can calculate from a matrix population model (Caswell, 2001) can also be calculated as derived quantities from an IPM, and you obtain "automatic" uncertainty estimates along with them.

However, IPMs have often been traditionally explained or developed using a different approach in which the connection to a matrix population model may be a bit less obvious. This alternative explanation of an IPM may be more intuitive in some other ways, and especially, more practically useful. Therefore, we now present this second explanation, or derivation, of an IPM. It consists of three basic steps (Schaub and Abadi, 2011, Schaub 2020) and will likely be your preferred practical approach when you want to develop an IPM for your own data. These three steps are as follows:

1. Develop a model that links demographic rates to population size(s).
2. Formulate the likelihood for each available data set (submodel) separately.
3. Formulate the joint likelihood and analyze it for inference.

We now look at each step in detail. We will use the same woodchat shrike data sets as before but build a slightly different IPM that includes demographic and environmental stochasticity, whereas the IPM in the previous section was purely deterministic.

5.4.1 DEVELOPMENT OF A MODEL THAT LINKS DEMOGRAPHIC DATA WITH POPULATION SIZE

Our first step is the development of a population model that describes changes in the age- or stage-specific population sizes over time as a function of the demographic rates. Or in other words, a model that links demographic rates with population size in the manner of a matrix population model, as introduced in Chapter 3 and just above also in the current chapter. We must make a number of decisions, e.g., decide how many age classes or which states we want to include, whether to base the model on a pre- or postbreeding "census," whether to model females only or also include males, and whether to include demographic and environmental stochasticity (i.e., variability) or density dependence. There is much flexibility in the definition of a population model, and your decisions about these structures are crucial. Any inference from a fitted IPM is conditional on the assumption that your choices in this respect result in an adequate representation of the structure and dynamics of your population. You may check some of these assumptions against observed data—e.g., it is fairly easy to test whether environmental variability should be included in the model. But such tests are more challenging for other assumptions; for instance, it is difficult to rigorously evaluate whether your model must consider both sexes (and this of course also depends on the availability of data about both). Your choices about model structure will typically depend on your research or management questions and prior knowledge of the species, and of course, it will depend strongly on the data available. As we have stressed before, we highly recommend that you start model building by first drawing a life-cycle graph of the species from which the population equations are derived, though we do not show this step here.

For illustration, we use the same woodchat shrike data set as before. The species is monogamous, and demographic rates are similar among the sexes, so we construct a female-based model. We assume a prebreeding "census" and include two age classes, 1-year-old ("1y") and older ("ad") individuals. You have seen the life-cycle graph already as Fig. 3.5B, and thus we write down the population equations directly here:

$$N_{1y,t+1} = N_{1y,t}\frac{f_1}{2}s_j + N_{ad,t}\frac{f_a}{2}s_j$$

$$N_{ad,t+1} = \left(N_{1y,t} + N_{ad,t}\right)s_a$$

We want our IPM to include both demographic stochasticity and environmental stochasticity/variability. Demographic stochasticity is modeled by specifying appropriate statistical distributions with expected values equal to these equations. Because environmental stochasticity is just a possibly confusing name for "temporal variability," we can include it by simply letting the demographic parameters vary over time. We denote this by adding an index t for time for all parameters. Thus, we obtain:

$$N_{1y,t+1} \sim Poisson\left(N_{1y,t}\frac{f_{1,t}}{2}s_{j,t} + N_{ad,t}\frac{f_{a,t}}{2}s_{j,t}\right)$$

$$N_{ad,t+1} \sim Binomial\left(s_{a,t}, N_{1y,t} + N_{ad,t}\right)$$

There are different options for the modeling of temporal variability in the demographic rates. We can choose fixed or random time effects with or without correlation among demographic rates (e.g., across age/stage classes). Here we illustrate the specification of uncorrelated random effects. We cannot recognize the type of time effects from the population equations above, but the BUGS code that follows will clarify our choice.

5.4.2 FORMULATION OF THE LIKELIHOOD FOR EACH AVAILABLE DATA SET SEPARATELY

The second step toward our IPM consists of the formulation of the likelihoods for all available data sets. We will sometimes refer to them as "submodels," but often simply as "models," even though they are part of another larger "model." The choice of likelihood for each component data set requires a good knowledge of the range of sensible statistical models, and hence in Chapter 4, we introduce those that are perhaps most important in the context of the IPMs dealt with in this book.

IPMs necessarily include data that contain information about population size and its change over time, i.e., population counts, population indices, harvest sizes, or similar. Hence, we will always have a model that links these measures of population size to the demographic rates in our population model. State-space models are well suited for this purpose because they describe an imperfectly observed state that evolves over time as a first-order Markov process (de Valpine and Hastings, 2002; Buckland et al., 2004; Newman et al., 2014). In the context of an IPM, the state-space model consists of the state-process equations that describe the dynamics of the number of individuals at time $t+1$, stratified by age or stage class, as a function of the age-/stage-specific number of individuals at time t and the demographic rates. Generally, the state-process equations are exactly equivalent to the population model defined in the first step of this approach toward building the IPM. The state in the first year cannot be described in a recursive manner, and therefore a model that describes the initial conditions of the population in the first year is needed. In a Bayesian analysis, we estimate these quantities and therefore place priors on them. Typically, we want the priors to express a lack of information (i.e., use vague priors), so here we use a prior that gives equal support to a wide range of possible values. For the woodchat shrikes, the observed population count in year 1 was 91, and thus we choose discrete uniform priors (*DU*) between 1 and 300 for each age class (note again that 0 is not possible under our exponential population model).

$$N_{1y,1} \sim DU(1, 300)$$

$$N_{ad,1} \sim DU(1, 300)$$

We choose this prior because we intend to include demographic stochasticity in our model, and hence we need a prior that has integer support only. Other choices such as a Poisson or negative binomial distribution are also possible. If your model contains more than just two age classes, it may be useful to choose slightly more informative priors for the initial stages. For example, you can calculate the stable stage distribution for the matrix population model when parametrized with reasonable average values for the demographic rates and use this information to specify slightly informed priors for the initial numbers in each stage class. See Besbeas and Morgan (2012b) for a similar approach in a frequentist analysis. A prior sensitivity analysis then allows you to assess the degree to which your inference is affected by the choice of these priors for the initial conditions (often, we would expect very little effect when different forms of vague priors are compared).

The state-space model also contains the observation equations that link the observed population counts with the true, latent population abundance. This submodel should ideally reflect the sampling design that produced these counts. In our IPM, we assume that a single survey in each year produced our shrike counts (C_t) and that they are subject to both imperfect detection and possible double-counting, and hence we model them using a normal distribution:

$$C_t \sim Normal\left(N_{1y,t} + N_{ad,t}, \sigma^2\right),$$

where σ^2 is the constant residual error that contains both lack of fit of the process model and observation error. In our example, we assume that birds cannot be aged; hence, the sum of the numbers of the two age classes is treated as the observed random variable. Other models for the observation process are possible, as we shall see in Section 6.6.5. We note that although there may also be observation models in the submodels for other data sets in an IPM (e.g., capture-recapture data), the term "observation model" typically refers to the state-space model for the population counts.

The likelihood of the state-space model (L_{SS}) is the product of the three likelihoods for the initial state, the state process, and the observation process:

$$L_{SS}\left(\boldsymbol{N}, s_j, s_a, f_1, f_a, \sigma^2 | \boldsymbol{C}\right) = L_I(\boldsymbol{N}_1) \times L_S\left(\boldsymbol{N}_{2...T}, s_j, s_a, f_1, f_a\right) \times L_O\left(\boldsymbol{N}, \sigma^2 | \boldsymbol{C}\right),$$

where L_I is the likelihood of the population size at occasion 1, L_s is the likelihood of the state process, and L_o is the likelihood of the observation process, and bold letters denote matrices and vectors. Specifically, $\boldsymbol{N}_1$ is age-specific population size in the first year, $\boldsymbol{N}_{2...T}$ contains the age-specific population sizes from years 2 to T, and $\boldsymbol{N}$ is the age-specific population size for the complete time series, i.e., from 1 to T. The state-space model includes all the parameters to be estimated, and it is possible in principle to attempt estimation of the parameters based on this model from a single time series of population counts. However, most parameters will then not be separately estimable, and the estimates will strongly depend on the chosen priors (see Section 6.3). To render the parameters estimable, we must add further information, i.e., more data sets that are directly informative about demographic rates. And of course, this is exactly what we do in the IPM.

Our second data set is the capture-recapture data, for which we adopt a Cormack-Jolly-Seber model (Williams et al., 2002; Section 4.5.1). We model the age-specific survival probabilities with uncorrelated temporal random effects and define

$$\text{logit}(s_{j,t}) \sim Normal\left(\mu_{s_j}, \sigma^2_{s_j}\right) \text{ and}$$

$$\text{logit}(s_{a,t}) \sim Normal\left(\mu_{s_a}, \sigma^2_{s_a}\right),$$

where μ_{s_j} and μ_{s_a} are the means, and $\sigma^2_{s_j}$ and $\sigma^2_{s_a}$ the temporal variances, of juvenile and adult survival probabilities, respectively, on the logit scale. We assume recapture probability to be constant over time ($p_t = \overline{p}$), and for faster computation, we again work with the multinomial likelihood for the data summarized in the m-array (m, Section 4.5.1.2). Written symbolically, the likelihood for the capture-recapture data set is $L_{CJS}(\boldsymbol{s_j}, \boldsymbol{s_a}, \boldsymbol{p}|\boldsymbol{m})$.

Our third data set is the productivity data, for which we here adopt a Poisson regression, again with uncorrelated random time effects for the parameters of both age classes:

$$\log(f_{1,t}) \sim Normal\left(\mu_{f_1}, \sigma^2_{f_1}\right) \text{ and}$$

$$\log(f_{a,t}) \sim Normal\left(\mu_{f_a}, \sigma^2_{f_a}\right),$$

where μ_{f_1} and μ_{f_a} are the means, and $\sigma^2_{f_1}$ and $\sigma^2_{f_a}$ the temporal variances, of productivity of first-year and older females, respectively, on the log scale. The likelihood of this Poisson regression model is written symbolically as $L_P(\boldsymbol{f_1}, \boldsymbol{f_a}|\boldsymbol{J})$. Here, $\boldsymbol{J}$ are the productivity data that also include the categorical variable indicating the age class of the mother.

Now we have completed step 2 in our development of the IPM, which consists of deriving the likelihoods for each component data set that we intend to integrate into a single model. Writing the likelihood for a data set means that, as a BUGS user, you code the model in BUGS, as shown in Chapter 4 for a wide range of models.

5.4.3 FORMULATION OF THE JOINT LIKELIHOOD

The third and final step is the formulation of a joint likelihood—this forms the likelihood for the full IPM. Under an independence assumption for all the data sets, the joint likelihood for the IPM is simply the product of the likelihoods for each individual data set just discussed:

$$L_{IPM}(\boldsymbol{N}, \boldsymbol{s_j}, \boldsymbol{s_a}, \boldsymbol{p}, \boldsymbol{f_1}, \boldsymbol{f_a}, \sigma^2|\boldsymbol{m}, \boldsymbol{J}, \boldsymbol{C}) =$$

$$L_I(N_1) \times L_S(\boldsymbol{N_{2\ldots T}}, \boldsymbol{s_j}, \boldsymbol{s_a}, \boldsymbol{f_1}, \boldsymbol{f_a}) \times L_O(\boldsymbol{N}, \sigma^2|\boldsymbol{C}) \times L_{CJS}(\boldsymbol{s_j}, \boldsymbol{s_a}, \boldsymbol{p}|\boldsymbol{m}) \times L_P(\boldsymbol{f_1}, \boldsymbol{f_a}|\boldsymbol{J})$$

Inference for the IPM is based on this joint likelihood. A frequentist would maximize it directly using algorithms such as the Kalman filter (Besbeas et al., 2002, 2003, 2005; Besbeas and Freeman, 2006; Gauthier et al., 2007; Tavecchia et al., 2009; Péron et al., 2010a, b; Besbeas and Morgan, 2012a, b; de Valpine, 2012), Monte Carlo kernel likelihood, bridge sampling (de Valpine, 2012), or methods for fitting hidden Markov models (Besbeas and Morgan, 2019). A Bayesian places priors on all parameters and uses the Bayes rule and MCMC (e.g., Brooks et al., 2004) or particle filtering methods (Finke et al., 2019) to approximate the posterior distributions of all quantities of interest. De Valpine (2012) summarized several frequentist methods and compared maximum likelihood estimates with posteriors from Bayesian methods and found no differences that would lead to differing biological conclusions.

At this point, we would like to make three comments about the joint likelihood. First, we clearly see that information among different data sets is shared: the likelihood component of the state-space model contains some parameters that also appear in the likelihoods for the capture-recapture and productivity data (i.e., $\boldsymbol{s_j}$, $\boldsymbol{s_a}$, $\boldsymbol{f_1}$ and $\boldsymbol{f_a}$). The combination of the component (or single) data likelihoods to form the joint likelihood of the IPM can be conveniently illustrated using a *directed acyclic graph* (DAG, Fig. 5.3). In such a graph, all known quantities (typically data) are represented as squares, while all unknown quantities (estimated parameters, missing data, predictions, etc.) are shown as circles. Arrows link nodes and indicate dependencies among different quantities in the model. The component data models (i.e., submodels) are shown by large boxes, and the unknown quantities included in such boxes indicate that they are part of the corresponding model. Quantities that appear in multiple boxes are informed by more than one data set. For instance, in Fig. 5.3, you can see that information about juvenile survival originates from the capture-recapture data and the population count data.

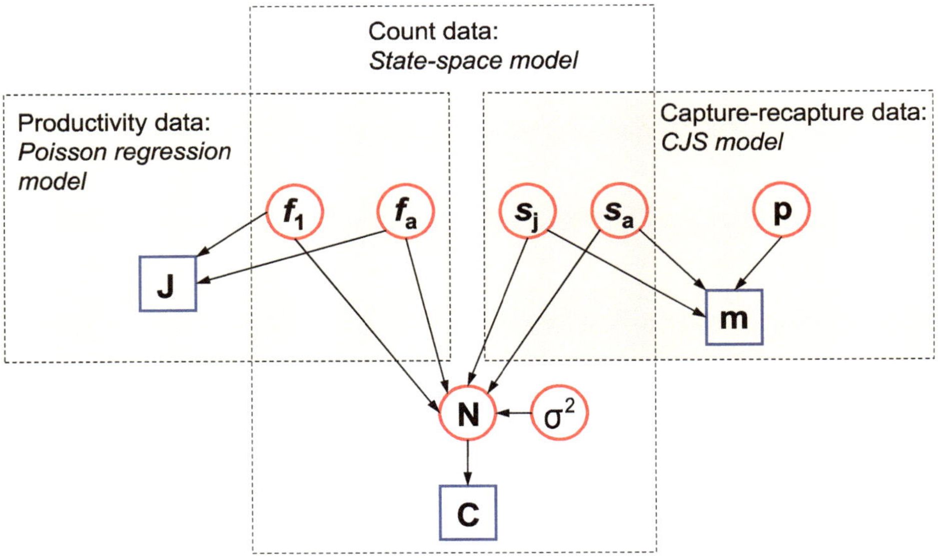

FIGURE 5.3

Simplified graphical representation of the integrated population model (IPM) for the woodchat shrike data as a *directed acyclic graph* (DAG) with nodes representing observed or unobserved quantities in the model: observed data are symbolized as blue squares, and all estimated quantities (here, parameters) are shown as red circles, while arrows show dependencies among the nodes. Large boxes show the individual submodels for the three data sets. Parameters appearing in two submodels represent those estimated from the information in the two data sets. See the text for more information on notation. Note that customarily, a DAG would also include the priors for all top-level parameters, but we have left them out to avoid clutter. Abbreviation: *CJS*, Cormack-Jolly-Seber.

Second, in the BUGS code we are about to develop, we will *not* need to write the joint likelihood explicitly as a single "ugly" expression. Rather, the BUGS language allows us to define the joint likelihood in a hierarchical manner by merely specifying "local" relationships between all the nodes in the DAG of an IPM. The integration occurs implicitly by giving the same name to quantities that appear in different parts of the separate likelihood statements for the individual data sets combined in the IPM. For example, a likelihood component of the state-space model for the observed population counts might be a Poisson model for the number of 1-year-old individuals in year $t+1$ and time constant demographic rates:

```
N[1,t+1] ~ dpois(N[1,t] * f1 / 2 * sj + N[2,t] * fa / 2 * sj)
```

Another component might be the Poisson regression model for the productivity data:

```
J[i] ~ dpois(f1)
```

In both likelihoods, the same productivity parameter (i.e., f_1) is included, so the two likelihoods become connected, and information about the common parameter f_1 is shared. That is, our estimate of f_1 is informed by the productivity data *and* by the population counts—it is an average of the information in the two data sets. As the information content in the two data sets about f_1 is usually not equal, the point estimate of f_1 from the IPM will in fact be a weighted average, and it is not *a priori* clear how much weight each data set will be given. In Section 6.2.2, we will see how to investigate this weighting issue.

In contrast, if we were to use another parameter for productivity, say f_x, in the Poisson regression model for productivity data,

```
J[i] ~ dpois(fx)
```

the two likelihoods would become separated, and information-sharing would no longer take place—f_x would then be estimated from the productivity data alone and f_1 only from the population counts via the state-space model.

The ability to define complex joint likelihoods by describing simple local relationships in the BUGS language is a key advantage of any software that uses the BUGS language for model description. In this way, it becomes relatively easy to develop even very complicated models, IPMs or otherwise, especially for nonstatisticians.

Our third comment is that the formulation of the joint likelihood as a product of the component data likelihoods requires that the data are independent (Besbeas et al., 2002; Brooks et al., 2004; Abadi et al., 2010). The independence assumption is a big omnibus assumption that precludes anything that would introduce dependence into the data of the different likelihoods. It would be hard to enumerate all of those cases. However, independence is probably reasonable when the data arise by independent sampling of distinct populations or even from independent sampling of the same population.

In practice, it may occur that the independence assumption does not fully hold. Whether a violation of the independence assumption affects the inference is context-dependent. It seems that the violation often has only negligible effects on parameter estimates (Abadi et al., 2010; Schaub and Fletcher, 2015; Weegman et al., 2021), but in other examples, some effects were found (Besbeas et al., 2009). Here, we assume that the data are independent, and we refer to Section 7.3.2 for a closer look at this assumption in most IPMs.

5.4.4 WRITING THE BUGS CODE FOR THE INTEGRATED POPULATION MODEL

The discrete uniform distribution, which we intend to use as a prior for the initial population size, is not a standard distribution in JAGS, so we implement it via the categorical distribution. The categorical distribution requires a vector of length U with the cell probabilities for each of the U possible values (which must sum to 1) as an argument. Our function `dUnif` generates the required vector of cell probabilities for specification of a discrete uniform prior between A and B.

The data we will use are almost exactly the same as those in Section 5.3 but with two slight differences. First, we use the annual productivity data, since we now want to model temporal variability also. And second, we provide as data the cell probabilities for the discrete uniform prior (which you can check will sum to 1).

```
# Bundle data and produce data overview
jags.data <- list(marr.j=marr[,,1], marr.a=marr[,,2], n.occasions=dim(marr)[2],
  rel.j=rowSums(marr[,,1]), rel.a=rowSums(marr[,,2]), J=woodchat5$repro[,1],
  year=woodchat5$repro[,2], age=woodchat5$repro[,3], C=woodchat5$count, pNinit=dUnif(1, 300))
str(jags.data)

List of 10
 $ marr.j     : num [1:19, 1:20] 8 0 0 0 0 0 0 0 0 0 ...
 $ marr.a     : num [1:19, 1:20] 16 0 0 0 0 0 0 0 0 0 ...
 $ n.occasions: int 20
 $ rel.j      : num [1:19] 51 53 55 65 73 66 61 76 65 75 ...
 $ rel.a      : num [1:19] 36 39 44 61 61 50 43 61 51 53 ...
 $ J          : num [1:929] 6 2 2 5 3 5 3 2 3 2 ...
 $ year       : num [1:929] 1 1 1 1 1 1 1 1 1 1 ...
 $ age        : num [1:929] 1 1 1 1 1 1 1 1 1 1 ...
 $ C          : num [1:20] 91 119 131 88 139 145 148 116 112 106 ...
 $ pNinit     : num [1:300] 0.00333 0.00333 0.00333 0.00333 0.00333 ...
```

What remains to be done is to write the full IPM in JAGS. Again, we use code for the individual data sets from Chapter 4 as a template and integrate these submodels by giving identical names to those parameters that occur in multiple data sets. Although we have chosen a different approach to develop the IPM than that of Section 5.3, the resulting code is identical except for the inclusion of demographic and environmental stochasticity in the current model that we omitted for ease of explanation of the example in Section 5.3.

```
# Write JAGS model file
cat(file="model4.txt", "
model {
# Priors and linear models
for (t in 1:(n.occasions-1)){
  logit.sj[t] ~ dnorm(mu.sj, tau.sj)
  sj[t] <- ilogit(logit.sj[t])       # Back-transformation from logit scale
  logit.sa[t] ~ dnorm(mu.sa, tau.sa)
  sa[t] <- ilogit(logit.sa[t])       # Back-transformation from logit scale
  p[t] <- mean.p
}
```

```
for (t in 1:n.occasions){
  log.f[1,t] ~ dnorm(mu.f[1], tau.f[1])
  f[1,t] <- exp(log.f[1,t])          # Back-transformation from log scale
  log.f[2,t] ~ dnorm(mu.f[2], tau.f[2])
  f[2,t] <- exp(log.f[2,t])          # Back-transformation from log scale
}

mean.sj ~ dunif(0, 1)
mu.sj <- logit(mean.sj)              # Logit transformation
mean.sa ~ dunif(0, 1)
mu.sa <- logit(mean.sa)              # Logit transformation
sigma.sj ~ dunif(0, 3)
tau.sj <- pow(sigma.sj, -2)
sigma.sa ~ dunif(0, 3)
tau.sa <- pow(sigma.sa, -2)

for (j in 1:2){
  mean.f[j] ~ dunif(0, 10)
  mu.f[j] <- log(mean.f[j])          # Log transformation
  sigma.f[j] ~ dunif(0, 3)
  tau.f[j] <- pow(sigma.f[j], -2)
}

mean.p ~ dunif(0, 1)

sigma ~ dunif(0.5, 100)
tau <- pow(sigma, -2)

# Population count data (state-space model)
# Model for initial stage-spec. population sizes: discrete uniform priors
N[1,1] ~ dcat(pNinit)
N[2,1] ~ dcat(pNinit)

# Process model over time: our model of population dynamics
for (t in 1:(n.occasions-1)){
  N[1,t+1] ~ dpois(N[1,t] * f[1,t] / 2 * sj[t] + N[2,t] * f[2,t] / 2 * sj[t])
  N[2,t+1] ~ dbin(sa[t], N[1,t] + N[2,t])
}

# Observation model
for (t in 1:n.occasions){
  C[t] ~ dnorm(N[1,t] + N[2,t], tau)
}

# Productivity data (Poisson regression model)
for (i in 1:length(J)){
  J[i] ~ dpois(f[age[i],year[i]])
}

# Capture-recapture data (CJS model with multinomial likelihood)
# Define the multinomial likelihood
for (t in 1:(n.occasions-1)){
  marr.j[t,1:n.occasions] ~ dmulti(pr.j[t,], rel.j[t])
  marr.a[t,1:n.occasions] ~ dmulti(pr.a[t,], rel.a[t])
}
# Define the cell probabilities of the m-arrays
for (t in 1:(n.occasions-1)){
  # Main diagonal
  q[t] <- 1 - p[t]                   # Probability of non-recapture
  pr.j[t,t] <- sj[t] * p[t]
  pr.a[t,t] <- sa[t] * p[t]
  # Above main diagonal
  for (j in (t+1):(n.occasions-1)){
    pr.j[t,j] <- sj[t] * prod(sa[(t+1):j]) * prod(q[t:(j-1)]) * p[j]
    pr.a[t,j] <- prod(sa[t:j]) * prod(q[t:(j-1)]) * p[j]
  } #j
  # Below main diagonal
  for (j in 1:(t-1)){
    pr.j[t,j] <- 0
    pr.a[t,j] <- 0
  } #j
} #t
```

```
# Last column: probability of non-recapture
for (t in 1:(n.occasions-1)){
  pr.j[t,n.occasions] <- 1-sum(pr.j[t,1:(n.occasions-1)])
  pr.a[t,n.occasions] <- 1-sum(pr.a[t,1:(n.occasions-1)])
}

# Derived parameters
# Annual population growth rate (added 0.001 to avoid possible division by 0)
for (t in 1:(n.occasions-1)){
  ann.growth.rate[t] <- (N[1,t+1] + N[2,t+1]) / (N[1,t] + N[2,t] + 0.001)
}
# Total population size
for (t in 1:n.occasions){
  Ntot[t] <- N[1,t] + N[2,t]
}
}
")
```

We provide initial values, list the parameters we want to estimate, choose the MCMC settings, and launch the model in JAGS, now choosing parallel computation to speed things up.

```
# Initial values
inits <- function(){list(mean.sj=runif(1, 0, 0.5))}

# Parameters monitored
parameters <- c("mean.sj", "mean.sa", "mean.f", "mean.p", "sigma.sj", "sigma.sa", "sigma.f",
  "sigma", "sj", "sa", "f", "N", "ann.growth.rate", "Ntot")

# MCMC settings
ni <- 12000; nb <- 2000; nc <- 3; nt <- 2; na <- 1000

# Call JAGS (ART 1 min), check convergence and summarize posteriors
out4 <- jags(jags.data, inits, parameters, "model4.txt", n.iter=ni, n.burnin=nb, n.chains=nc,
  n.thin=nt, n.adapt=na, parallel=TRUE)
traceplot(out4)
print(out4, 3)
```

```
                       mean     sd     2.5%      50%     97.5% overlap0 f  Rhat n.eff
mean.sj               0.301  0.014    0.275    0.301     0.327    FALSE 1 1.001  2027
mean.sa               0.542  0.016    0.511    0.542     0.573    FALSE 1 1.000  5642
mean.f[1]             2.673  0.083    2.502    2.673     2.829    FALSE 1 1.006  1000
mean.f[2]             3.675  0.110    3.468    3.672     3.900    FALSE 1 1.012   185
mean.p                0.605  0.019    0.567    0.605     0.643    FALSE 1 1.000 15000
sigma.sj              0.114  0.077    0.007    0.103     0.291    FALSE 1 1.011   203
sigma.sa              0.122  0.086    0.005    0.109     0.318    FALSE 1 1.001  4371
sigma.f[1]            0.041  0.033    0.000    0.034     0.120    FALSE 1 1.001  2706
sigma.f[2]            0.064  0.041    0.003    0.061     0.153    FALSE 1 1.002  1090
sigma                13.028  3.575    7.287   12.584    21.169    FALSE 1 1.001  1491
sj[1]                 0.305  0.027    0.253    0.304     0.365    FALSE 1 1.005   980
[ ... output truncated ... ]
sj[19]                0.304  0.025    0.256    0.303     0.360    FALSE 1 1.001 13466
sa[1]                 0.554  0.036    0.490    0.548     0.641    FALSE 1 1.001  9699
[ ... output truncated ... ]
sa[19]                0.538  0.032    0.468    0.539     0.604    FALSE 1 1.001  2810
f[1,1]                2.723  0.159    2.462    2.704     3.114    FALSE 1 1.002  5080
f[2,1]                4.012  0.400    3.513    3.905     5.014    FALSE 1 1.003  1260
[ ... output truncated ... ]
f[1,20]               2.691  0.140    2.435    2.683     3.010    FALSE 1 1.002  3466
f[2,20]               3.814  0.229    3.442    3.783     4.342    FALSE 1 1.004   576
N[1,1]               41.533 28.882    2.000   37.000   103.000    FALSE 1 1.002  1220
N[2,1]               60.414 26.969    6.000   64.000   103.000    FALSE 1 1.002  1080
[ ... output truncated ... ]
N[1,20]              71.405  8.953   54.000   71.000    90.000    FALSE 1 1.000 15000
N[2,20]              79.945  7.909   65.000   80.000    95.000    FALSE 1 1.001  5014
ann.growth.rate[1]    1.130  0.106    0.936    1.124     1.351    FALSE 1 1.001  3874
[ ... output truncated ... ]
ann.growth.rate[19]   1.019  0.068    0.890    1.019     1.158    FALSE 1 1.000 15000
Ntot[1]             101.947  9.925   84.000  101.000   123.000    FALSE 1 1.000 12582
[ ... output truncated ... ]
Ntot[20]            151.349 10.377  131.000  151.000   173.000    FALSE 1 1.000 15000
```

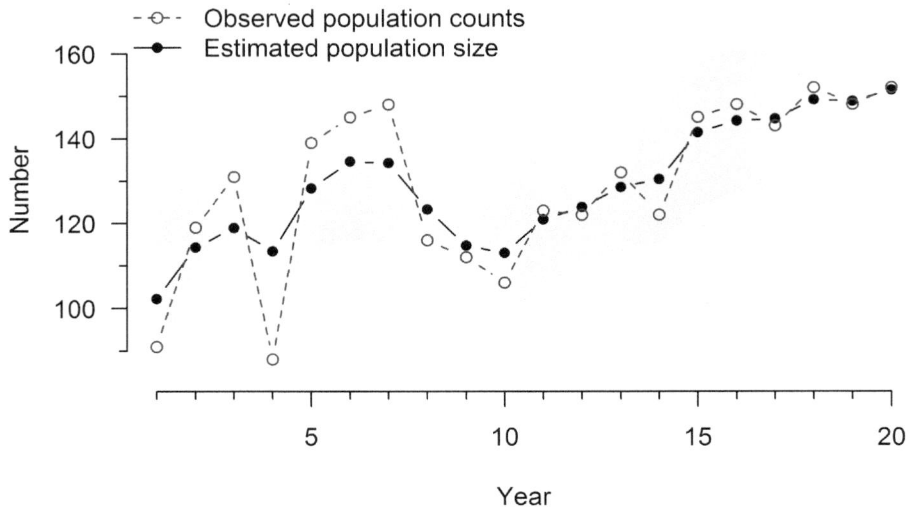

FIGURE 5.4

Estimated total population size (in black; with 95% credible intervals) from an integrated population model for woodchat shrikes including demographic and environmental stochasticity and observed population counts (in blue).

The posterior summaries of the parameters show a good resemblance with the values used for data simulation, and convergence is obtained relatively quickly. As before, we compare the estimated population sizes with the observed population counts and note much better agreement now (Fig. 5.4)—the trajectory of N is no longer simply a curved line. The reason for this is that our current model is more flexible due to the inclusion of demographic and environmental stochasticity. We also note that the magnitude of the observation error has declined somewhat, from 16.976 to 13.028. This highlights again that what is called "observation error" in the state-space model is in fact the residual error of the IPM: it measures the magnitude of the discrepancy between the population count data and the population sizes that we would expect under the chosen population model.

In summary, in Sections 5.2–5.4 we have developed our first two IPMs using two different approaches, or two different explanations, of what an IPM is and how it is constructed. In the first (in Sections 5.2 and 5.3), we have emphasized that the core of most IPMs is simply a matrix population model. However, the second approach (the three-step approach in Section 5.4) is probably the more practically intuitive approach and the one you will usually adopt when developing your IPMs. Both models produced parameter estimates that looked satisfactory; however, this is not yet a confirmation that they will also be accurate. Whether a model produces "good" (e.g., unbiased, or precise) estimates cannot be judged by the analysis of a single data set—we need simulations with many replicates. This is the topic of the next section.

5.5 SIMULATION ASSESSMENT OF A SIMPLE INTEGRATED POPULATION MODEL

In this section, we assess the quality of the estimates of the demographic rates and population trajectories under the two IPMs just developed. In the assessment, we also include a third IPM that accounts for demographic stochasticity but not for environmental stochasticity. This third IPM has demographic rates that are constant over time and has features of both IPMs developed in the previous sections. In this section, we also demonstrate how you can simulate data for an IPM. Data simulation is incredibly useful for applied statistical analysis (see Chapter 4 in Kéry and Royle, 2016), in part because it enforces a much greater understanding of your model but also for many practical reasons. In our assessment, the data simulation process agrees very closely with one of the data analysis models. The model assumptions represented by the third model are therefore met, and we will see how parameter estimation behaves under this best-case scenario. The model assumptions are violated slightly by the two other data-analyzing models, and we will see how this affects the quality of their estimates. In Exercise 3, you will encounter additional scenarios where there is a much greater mismatch between data simulation and the data analysis models. This will allow you to gauge the effects of more severe violations of the model assumptions.

5.5.1 SIMULATING DATA UNDER AN INTEGRATED POPULATION MODEL

Simulation of data for an IPM proceeds with two basic steps: first, we simulate the true dynamics of a population, and second, we simulate the sampling processes that lead to the "observed" data sets that we then use in the analysis. To simulate the population dynamics, we here use an individual-based setup in which the fate of each individual is independent from all others and governed by some "rules." These rules are the demographic processes (survival and reproduction in our case), for which we must first pick parameter values. As a result of this part of the simulation, we have complete knowledge about the life history of every individual ever alive in the population during the entire study period. That is, we know its year of birth and death and the number of offspring in every single year of its life. In the second step, we sample different data sets from the population generated. For example, we can produce capture-recapture data, for which we must first specify capture and recapture probabilities. Or we can create population count data under some specific sampling protocol, and for this, we again must specify values for the parameters that govern these sampling processes (e.g., observation variances).

Before we start sampling data from the population, we must decide whether we want to sample independent or dependent data. We want to produce independent data sets here, and one way to do this in our simulation is to ensure that no individual is present in more than one data set. Note, however, that independence is merely ensured in this way in our data simulation, but this is *not* the definition of independence among the data sets in an IPM (see Section 7.3.2). In our simulation, we can either split the population (i.e., the individual life histories) randomly into as many parts as the number of data sets we want to simulate (e.g., three, if we want to simulate population counts, capture-recapture, and productivity data). Alternatively, we can simulate multiple populations with identical parameter values. Each split or replicate population is then used to produce just one type of data. In the following, we introduce R functions from the `IPMbook` package that we use for data simulation.

The function to simulate the true dynamics of a population is called `simPop`. It is quite general, can deal with a variety of life-history features, and allows the inclusion of effects of environmental and demographic stochasticity. For the former, we must specify values for the demographic rates for each year. Demographic stochasticity is included explicitly because the individual fates and offspring production are represented by a Bernoulli and Poisson random variable, respectively. The function also allows flexibility in the number of age or stage classes and includes as demographic processes survival, productivity, and breeding probability, which can be age-, stage-, or time-dependent. The simulated population may receive immigrants, and thus we must specify as another function argument their number in every year and the age at which they enter the population. The key function arguments are a vector with the age- or stage-specific population sizes in year $t = 1$, matrices with the age-, stage-, or time-specific survival and breeding probabilities and productivity. The latter can also be a vector if survival, breeding probability, or productivity is constant over time and varies just by age, or a scalar if we want them to be constant over time and across age classes. The key function outputs are tables with the life-history information for each simulated individual.

Assumptions made in the function include the following. First, the function generates the female part of a population with the exception of the production of the offspring, which includes both males and females. As a consequence, the offspring sex ratio at birth is another function argument. Second, permanent emigration is not simulated explicitly. If you want permanent emigration to occur in your simulated population, you will need to change the survival parameters to become apparent survival (i.e., reduce them by an amount that is equal to the permanent emigration rates). Third, individual effects cannot be included, i.e., all individuals alive in a given age or stage at a given occasion are assumed to have identical demographic rates. Fourth, the function assumes the absence of any density-dependent mechanisms in the population. Clearly, if you want to simulate a population for which these assumptions don't make sense, you will need to write your own simulation function or adapt this one to include what you need.

Before we call the function, we first pick values for the function arguments, i.e., age-specific survival and breeding probabilities and values for productivity, initial population sizes, and the number of years over which we want to simulate population dynamics. Our aim is to produce data mimicking a woodchat shrike population. Hence, we assume two age classes for survival and productivity and that all individuals breed in every year. For simplicity, we assume that the demographic rates are constant over time; hence, in each year we apply the same values. Note that as with most R functions, `simPop` uses default values for arguments when no values are provided. The default value for breeding probability is 1, for number of immigrants is 0, and for the sex ratio of newborn individuals is 0.5. All of them correspond to our intended simulation setup and are therefore not specified explicitly when we call the simulation function.

```
# Pick values for the function arguments
T <- 20                # Number of years
phi <- c(0.3, 0.55) # Age specific survival probabilities (juv, adult)
f <- c(2.6, 3.6)     # Age-specific productivity (1y, older)
Ni <- c(50, 50)      # Initial pop. size for each age class (1y, older)

# Apply the function and produce data overview
set.seed(111167)     # To initialize the RNGs at the same place
pop <- simPop(Ni=Ni, phi=phi, f=f, nYears=T)
str(pop)
```

The function creates a list with 14 objects. The first 7 are summaries of the values used for the simulation. The last 7 summarize information about the simulated population. They are explained in more detail below (see also the help file at `?simPop`).

```
List of 14
 $ Ni          : num [1:2] 50 50
 $ phi         : num [1:3, 1:19] 0.3 0.55 0.55 0.3 0.55 0.55 0.3 0.55 ...
 $ f           : num [1:2, 1:20] 2.6 3.6 2.6 3.6 2.6 3.6 2.6 3.6 2.6 3.6 ...
 $ pBreed      : num [1:2, 1:20] 1 1 1 1 1 1 1 1 1 1 ...
 $ sex.ratio   : num [1:20] 0.5 0.5 0.5 0.5 0.5 0.5 0.5 0.5 0.5 0.5 ...
 $ Im          : num [1:20] 0 0 0 0 0 0 0 0 0 0 ...
 $ ageOfIm     : num [1:20] 1 1 1 1 1 1 1 1 1 1 ...
 $ state       : num [1:3492, 1:20] 1 1 1 1 1 1 1 1 1 1 ...
  ..- attr(*, "dimnames")=List of 2
  .. ..$ : NULL
  .. ..$ : chr [1:20] "Y1" "Y2" "Y3" "Y4" ...
 $ imYear      : logi [1:3492] NA NA NA NA NA NA ...
 $ reprod      : num [1:3492, 1:20, 1:3] 5 2 2 1 0 1 1 1 2 2 ...
  ..- attr(*, "dimnames")=List of 3
  .. ..$ : NULL
  .. ..$ : chr [1:20] "Y1" "Y2" "Y3" "Y4" ...
  .. ..$ : chr [1:3] "F" "M" "Age"
 $ N           : num [1:6, 1:20] 50 50 100 158 336 0 49 52 101 165 ...
  ..- attr(*, "dimnames")=List of 2
  .. ..$ : chr [1:6] "1-Year" "2-Year" "totAdults" "BornF" ...
  .. ..$ : chr [1:20] "Y1" "Y2" "Y3" "Y4" ...
 $ breeders    : num [1:3, 1:20] 50 50 100 49 52 101 46 56 102 50 ...
  ..- attr(*, "dimnames")=List of 2
  .. ..$ : chr [1:3] "1-Year" "2-Year" "totBreeders"
  .. ..$ : chr [1:20] "Y1" "Y2" "Y3" "Y4" ...
 $ totAdults   : Named num [1:20] 100 101 102 108 109 116 115 121 124 ...
  ..- attr(*, "names")= chr [1:20] "Y1" "Y2" "Y3" "Y4" ...
 $ totBreeders: Named num [1:20] 100 101 102 108 109 116 115 121 124 ...
  ..- attr(*, "names")= chr [1:20] "Y1" "Y2" "Y3" "Y4" ...
```

A first important object that summarizes the simulated population is `state`. It is a matrix with dimension equal to the number of individuals by the number of years and with entries that contain the age/stage class of each individual in every year. For an example, let's have a look at the life history of female 488 (which we truncate after year 10—because it would just continue being dead).

```
pop$state[488, 1:10]
 Y1  Y2  Y3  Y4  Y5  Y6  Y7  Y8  Y9 Y10
 NA  NA   0   1   2   2  -1  NA  NA  NA
```

This individual is born in year 3 (indicated by a "0"), she is 1 year old in year 4 (indicated by "1"), and in years 5 and 6 she is older than 1 year (indicated by "2"). In year 7, she becomes recently dead, which is indicated by "-1" and means that she died between years 6 and 7. With the exception of the nestling ("0") and the recently dead stages ("-1"), stage definitions correspond to those of a prebreeding "census" model. Therefore, the maximum number of stage classes is given by the length of the vector for the age/stage-specific initial population sizes (`Ni`). Thus, in our case, stage "1" corresponds to a 1-year-old individual and stage "2" to a >1-year-old individual.

The object `imYear` is a vector indicating the year of immigration for each female. An NA denotes a female that is not an immigrant, which is the case for all females in our example.

The object `reprod` summarizes the reproductive output (i.e., the number of nestlings) of each female in the population. It is a 3-D array with dimensions equal to the number of individuals, the number of years, and 3. For illustration, we inspect the productivity summary of female 488.

```
pop$reprod[488,1:10,]
     F  M Age
Y1  NA NA  NA
Y2  NA NA  NA
Y3  NA NA  NA
Y4   3  2   1
Y5   5  3   2
Y6   2  0   2
Y7  NA NA  NA
Y8  NA NA  NA
Y9  NA NA  NA
Y10 NA NA  NA
```

We see that this individual produced three female (column "F") and two male nestlings (column "M") in year 4 (row "Y4") when she was in age/stage class "1" (column "Age"). In the next year, she was in age/stage class "2" and produced five female and three male nestlings. In year 6, she only produced two female nestlings.

The object `N` summarizes a matrix of the annual number of individuals in each of the different age/stage classes. Let's have a look at the first 10 years of our created population.

```
pop$N[,1:10]
           Y1  Y2  Y3  Y4  Y5  Y6  Y7  Y8  Y9 Y10
1-Year     50  49  46  50  53  57  53  61  61  51
2-Year     50  52  56  58  56  59  62  60  63  62
totAdults 100 101 102 108 109 116 115 121 124 113
BornF     158 165 167 164 171 167 193 193 192 207
BornT     336 346 321 337 328 356 380 369 395 389
Im          0   0   0   0   0   0   0   0   0   0
```

For instance, we see that in year 2 (column "Y2"), 49 females are in age/stage class 1 ("1-Year"), 52 in age/stage class 2 ("2-Year"), the total adult population size is 101 ("totAdults"), a total of 346 nestlings were produced in the population ("BornT"), of which 165 were females ("BornF"), and no immigrant entered the population ("Im").

In a table, the object `breeders` summarizes the number of females reproducing annually per age/stage class. Because we assumed that all females at least 1 year old are reproducing, the numbers per age/stage classes are the same as the corresponding numbers in the object `N`. Yet if the probability to reproduce is <1, these numbers will differ. The two objects `N` and `breeders` allow flexibility in the way the count data are obtained: they could contain all females of a population, or only the reproducing ones, potentially stratified by age/stage class.

```
pop$breeders[,1:10]
             Y1  Y2  Y3  Y4  Y5  Y6  Y7  Y8  Y9 Y10
1-Year       50  49  46  50  53  57  53  61  61  51
2-Year       50  52  56  58  56  59  62  60  63  62
totBreeders 100 101 102 108 109 116 115 121 124 113
```

The last two objects, `totAdults` and `totBreeders`, are the same as those already provided in tables `N` and `breeders` and are basically included to more easily address them in simulations.

The next step is to sample data from the simulated population. Because we assume independent data sets in our IPM (see above), we first create three replicates of a population using identical demographic rates. We no longer fix a seed when the function is applied to ensure that different replicate populations are indeed produced.

```
pop1 <- simPop(Ni=Ni, phi=phi, f=f, nYears=T)
pop2 <- simPop(Ni=Ni, phi=phi, f=f, nYears=T)
pop3 <- simPop(Ni=Ni, phi=phi, f=f, nYears=T)
```

Next, we create the population count data from the first population using the function `simCountNorm`. We assume that we count the number of breeding pairs in our study area in every year and that these counts are unbiased on average with respect to the total population size. We therefore pick a value for the magnitude of the observation error, which is assumed to be identical in every year, and the function then uses a normal (or Gaussian) distribution to simulate the survey data. As arguments, the function uses a vector for the population size being counted and a vector or scalar for the observation standard deviation. As output, it provides the specified sigmas and counts rounded to the nearest integer.

```
# Pick a value of the observation error (SD) for the population survey
sigma <- 10

# Create the population survey data and produce data overview
count <- simCountNorm(N=pop1$totB, sigma=sigma)
str(count)

List of 2
 $ sigma: num [1:20] 10 10 10 10 10 10 10 10 10 10 ...
 $ count: num [1:20] 111 84 85 94 116 132 115 89 82 61 ...
```

Perhaps you are not entirely happy with the manner in which we generated the population counts because we assume that the counts are correct on average and that they contain errors due to individuals that were missed as well as from double-counting. Perhaps you feel that often only the overlooking of individuals is an issue (i.e., only false negatives occur, and no false positives) and that the binomial distribution would therefore be a more adequate model for generating these counts. Admittedly, the normal distribution may not always be reasonable as an observation model for population count data. We stick to this assumption for simplicity, however, because then we can use exactly the same state-space models as we introduced previously in Chapter 5 (and because they are adopted in most IPMs that have been published to date). Later, we will see other observation models for the population count data and present a function that creates count data using the binomial distribution.

Next we create the capture-recapture data from the second population using the function `simCapHist`. We first define the capture probability for nestlings and adults, although this is not a parameter that will later be estimated in the model. To be clear, we define capture probability as the probability of catching an individual for the first time and marking it.

Specification of an initial capture probability less than 1 is virtually imperative to avoid an implausible situation in which all individuals in a population are marked. Second, we specify recapture probability (which *is* a model parameter), which is defined as the probability of encountering an individual that was marked on an earlier occasion. Capture and recapture probabilities may differ, and sometimes different techniques are used to generate the capture and recapture data (e.g., physical capture with color banding for "capture" and resighting from a distance for "recapture"). We here assume that recapture probability is higher than the initial capture probability. Both the capture and the recapture probabilities can be supplied as matrices, with rows for age or stage classes and columns for time. If the probabilities are constant over time, they can also be supplied as vectors (for age/stage structure) or scalars (without age/stage structure). As arguments, the function `simCapHist` needs these probabilities, the maximum number of age/stage classes we want to distinguish, and of course, the life-history information of each individual.

```
# Pick values for capture and recapture probabilities
cap <- 0.4          # Initial capture probability (same for juv. and adults)
recap <- 0.6        # Recapture probability

# Create the capture histories and produce data overview
ch <- simCapHist(state=pop2$state, cap=cap, recap=recap, maxAge=2)
str(ch)

List of 5
 $ cap    : num [1:3, 1:20] 0.4 0.4 0.4 0.4 0.4 0.4 0.4 0.4 0.4 0.4 ...
 $ recap  : num [1:2, 1:19] 0.6 0.6 0.6 0.6 0.6 0.6 0.6 0.6 0.6 0.6 ...
 $ maxAge: num 2
 $ ch     : num [1:2645, 1:20] 1 1 1 1 1 1 0 1 0 1 ...
 $ age    : num [1:2645] 2 2 2 2 2 2 2 2 2 2 ...
```

This function creates a total of five objects. The first three summarize the parameters used for simulation of the capture-recapture data. The capture and recapture probabilities are organized as matrices with rows referring to the age/stage classes and columns to years. The number of age/stage classes for the capture probabilities is given by the age/stage structure of the simulated population (two in our case) plus one, the latter due to the nestlings that can be captured and marked as well. The dimensions of the recapture matrix are reduced compared with those of the capture matrix. A nestling cannot be recaptured as a nestling, and no recapture is possible in year 1. The object `maxAge` allows control of the number of age/stage classes that can be defined when individuals are captured for the first time and marked. In our example, there could be three different classes (1: nestling; 2: 1-year-old ("1-Year"); and 3: >1-year-old ("2-Year"), but we assume that the age of individuals ≥ 1 year is not determined at capture. The maximum number of age classes is therefore two. Object `ch` is a matrix with the capture histories, and object `age` is a vector with the information about each individual's age/stage class when it is marked (captured for the first time). From objects `ch` and `age`, we tally up the age-dependent m-arrays using the function `marrayAge`.

```
# Create m-arrays
marr <- marrayAge(ch$ch, ch$age)
```

Finally, we simulate productivity data using the third replicate population simulated with function `simProd`. We assume that a nest is detected with a certain probability, and that when found, the number of fledglings is recorded without error. Thus, we specify the probability that a nest is found and here assume that it is identical over the years. Again, this probability will not be estimated later as a parameter in the model but is needed simply to avoid implausible data where the entire population can be surveyed. Function `simProd` needs the inclusion probabilities and the life history of each individual as arguments.

```
# Pick probability to find a brood
pprod <- 0.3

# Create productivity data and produce data overview
pro <- simProd(reprod=pop3$reprod, pInclude=pprod)
str(pro)

List of 4
 $ pInclude     : num [1:20] 0.3 0.3 0.3 0.3 0.3 0.3 0.3 0.3 0.3 0.3 ...
 $ females.only: logi FALSE
 $ prod.ind     : num [1:705, 1:3] 2 2 0 3 5 3 5 4 5 3 ...
  ..- attr(*, "dimnames")=List of 2
  .. ..$ : NULL
  .. ..$ : chr [1:3] "Productivity" "Year" "Age of mother"
 $ prod.agg     : num [1:20, 1:2] 117 98 89 99 103 96 89 131 96 129 ...
  ..- attr(*, "dimnames")=List of 2
  .. ..$ : chr [1:20] "1" "2" "3" "4" ...
  .. ..$ : chr [1:2] "Juveniles" "Surveyed broods"
```

The function creates four objects. The first object `pInclude` repeats the chosen nest-finding probabilities for each year. Object `females.only` is a logical argument that allows you to simulate only the female nestlings when set to `TRUE`. The main function outputs are productivity data provided in two formats. Object `prod.ind` contains the full individual-level data, where the number of nestlings of each brood is provided along with the year in which it is recorded as well as the age of the mother. The object `prod.agg` contains the aggregated productivity data, which is the total numbers of nestlings and surveyed broods in each year. The former format contains all necessary information, but the aggregated data results in a simpler model and faster run times, although it allows modeling of temporal effects only and not of any effects associated with the mother. Here we use the individual-level data because we want to model productivity depending on the age of the mother.

5.5.2 SIMULATION RESULTS

Now we have all the data sets that we need and can analyze them with three IPMs, i.e., a completely deterministic IPM (as in Section 5.3, here called IPM_1), an IPM with time-constant demographic rates but demographic stochasticity (called IPM_2), and an IPM that includes environmental and demographic stochasticity (as in Section 5.4.4, called IPM_3). We will simulate 1000 data sets, analyze them with all three IPMs, and save the outputs. As an assessment of how well the parameters of the IPMs are estimated, we present boxplots of the posterior means and compare them with the data-generating values. This enables us to visually assess the bias and precision of the IPM estimators.

Under the deterministic model (IPM_1), all demographic rates and the population growth rate are for the most part estimated without any bias (Fig. 5.5). Interestingly, the residual error σ of IPM_1 is larger than the observation error that we

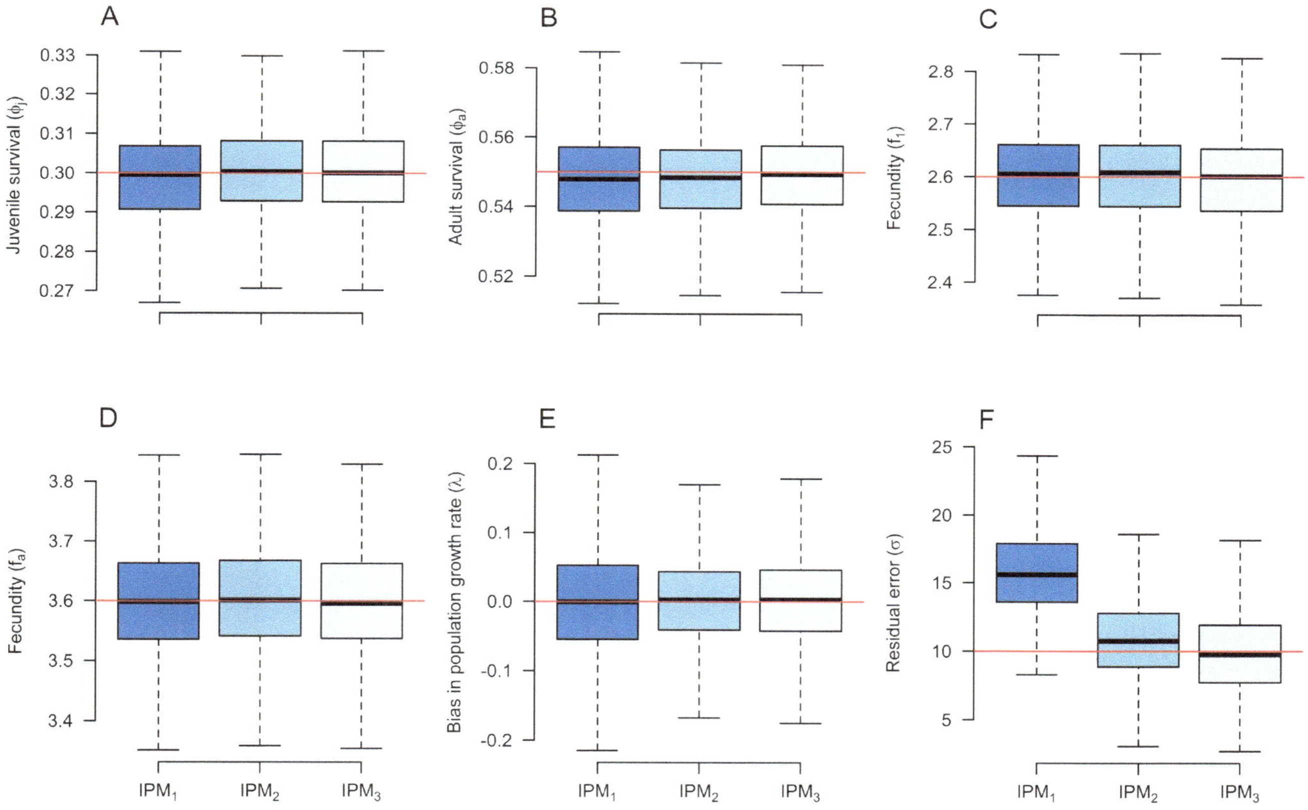

FIGURE 5.5

Boxplots of the posterior means of parameters from three different integrated population models (IPMs) (IPM_1: deterministic; IPM_2: with demographic stochasticity; IPM_3: with demographic and environmental stochasticity) fit to 1000 simulated data sets. The horizontal red lines indicate the values used in the data generation. For the population growth rate, bias is computed as the difference between the estimated and true annual population growth rates.

picked for simulating the population count data (Fig. 5.5F). We have repeatedly mentioned that the former contains not only pure observation errors associated with the counts but also any lack of fit of the model with respect to the data-generating model. IPM_1 assumes that demographic stochasticity is absent, but the data were simulated with demographic stochasticity. The resulting lack of fit is manifest in the increased residual error.

Under IPM_2, which accounts for demographic stochasticity but has constant demographic rates, the parameter estimates appear unbiased (Fig. 5.5). The variability of the demographic rates under IPM_2 is very similar to that under IPM_1. However, the variability of the population growth rates is slightly reduced under IPM_2, as it allows for more flexibility than IPM_1. A major difference compared with the deterministic IPM_1 is evident in the residual error σ, with estimates now being much closer to the value of the observation error used for simulating the population count data. IPM_2 includes demographic stochasticity and therefore matches the data-generation process perfectly. There is no lack of fit, and the residual error of the IPM contains pure observation error only.

The analyses of the same data sets with IPM_3, which adopts demographic and environmental stochasticity, show a picture very similar to the results of IPM_2: all the demographic rates appear to be estimated without any bias (Fig. 5.5). This shows that IPMs provide "good" parameter estimates when the underlying assumptions are met (Besbeas and Freeman, 2006; Abadi et al., 2010; McCrea et al., 2010; Schaub and Fletcher, 2015). For IPM_1, not all assumptions are met, but since the population size is reasonably large, the demographic stochasticity present in the data but not accounted for in the analyzing model does not have a serious impact. Moreover, the IPM has some flexibility due to the residual error in its state-space part of the model. This residual blows up under IPM_1 and accounts for some degree of the lack of fit. The increase of the residual error of an IPM that does not capture an important feature of the modeled population, is a useful property that may be helpful for diagnosing major structural problems in a model. Such a diagnosis based on the residual error is only possible when an observation model with a dispersion parameter such as the normal or log-normal distribution is chosen. When Poisson or binomial errors are chosen, a comparison of the predicted and observed population sizes can be used for the same purpose. IPM_3 allows for more flexibility than needed for the data (temporal variability of the demographic rates). It overfits the data, but this appears to be harmless in the current case (see also Section 7.3).

5.6 OUTLOOK AND SUMMARY

We have introduced the first IPM and have derived and explained it in two different ways. In the first, we started from a matrix population model and progressively added more data sets along with likelihoods for estimating certain parameters from those data sets as part of a single model fit in JAGS. We emphasized that the inclusion of the population count (or similarly, population index) data is one crucial part of an IPM, because the process model for these data allows integration (i.e., connection) with the other data sets. The process model in our case, and in most IPMs published to date, is a matrix population model containing survival probabilities and fecundities, some of which also appear as parameters in the likelihoods for the other data sets. In the integrated model, these parameters are shared between different data sets and hence are jointly informed by all of them.

The second way to think about, or introduce, an IPM is similar to the first. Here also, we first define a population model that links demographic rates to a vector of age- or stage-specific population sizes. We then proceed by defining the likelihoods for each of the available data sets, and finally by developing the joint likelihood from which inference is obtained under the IPM. In the BUGS language, we do not need to explicitly write this joint likelihood in a single "ugly" expression—this is a major practical advantage, especially for nonstatisticians, when fitting an IPM. The integration of the different data sets is straightforward; we simply give the same name to parameters that appear in different component data likelihoods, which themselves are "stacked" in a single large model described in the BUGS modeling language. Clearly however, you still need to understand the different models to analyze each individual data set, as well as have a good practical understanding of matrix population models.

In the final section of this chapter, we have demonstrated by means of simulation that IPMs provide unbiased parameter estimates when the model assumptions are sufficiently met. We have also seen how the "residual error" in the state-space model for the population counts combines uncertainty due to the observation process with any lack of fit of the process model. In the literature, we often see the slightly misleading term "observation error" for this parameter, and it is important that we fully understand its true meaning.

Next, to deepen our theoretical and practical understanding of IPMs, in Chapter 6, we deal in more detail with their benefits. Then, in Chapter 7, we investigate in more depth their key assumptions and the consequences when those assumptions are violated.

5.7 EXERCISES

1. Develop an IPM for the woodchat shrike data that does not include demographic stochasticity and that assumes environmental stochasticity for productivity only and not for survival. Use independent fixed time effects for productivity of first-time and experienced breeders, and fit a prebreeding "census" model.
2. Develop an IPM for the woodchat shrike data with the same features as in Exercise 1, but now adopt a postbreeding "census" population model.
3. Develop an IPM for the woodchat shrike data sets, but now assume erroneously that 1-year-old individuals do not yet reproduce. You may also include demographic stochasticity (but don't add temporal variability, i.e., environmental stochasticity). Compare the counts and estimated population sizes, and compare the residual error with that obtained from Exercise 1. What do you observe, and how can you interpret your observations?

CHAPTER

BENEFITS OF INTEGRATED POPULATION MODELING

6

CHAPTER OUTLINE

6.1 INTRODUCTION

In Chapter 5, we developed and fitted our first integrated population models (IPMs). That is, we built statistical models to jointly analyze multiple data sets at both the individual and the population level that contain information about the demographic parameters in the population model. Compared with a separate, piecemeal analysis of such data sets, the application of an IPM is more complex in a sense—for instance, you will see several case studies in Part 2 of the book where the BUGS code extends over several pages. We have emphasized at various places in this book that for multiple data sets that are informative about at least one shared parameter, an integrated analysis is simply a very sensible thing to do. Inferences under an integrated analysis represent an "average" based on the information from all the analyzed data sets combined, and therefore, they become more coherent in that sense. Nevertheless, you may still wonder about the specific benefits of using IPMs for your data analysis and whether the potential benefits outweigh the costs associated with greater model complexity and the effort needed to learn how to use IPMs.

In this chapter, we use examples and simple simulations to illustrate four important benefits of IPMs:

1. increased precision of parameter estimates,
2. estimation of parameters for which no explicit data are available,
3. estimation of process covariation between two or more demographic rates,
4. estimation of population structure and its change over time.

In Section 6.6, we stress the great flexibility of IPMs for demographic analyses in general: you can jointly analyze different data sets sampled at different temporal or spatial scales, be creative regarding the types of data that you combine, and owing to the Bayesian mode of analysis, easily include external information that may be an expert's guess, an estimate obtained in another analysis, or information that stems from the published literature.

Integrated Population Models. https://doi.org/10.1016/B978-0-12-820564-8.00022-1

6.2 PARAMETER ESTIMATES WITH INCREASED PRECISION

We have seen the symbolic formulation of the joint likelihood of an IPM as the product of the likelihoods of the individual data sets combined (Section 5.4.3) and its graphical representation as a directed acyclic graph (DAG; Fig. 5.3). Both reveal that some parameters appear in different components of the joint likelihood, i.e., they are informed by more than a single data set. For example, in the woodchat shrike IPM, juvenile survival occurs in the likelihoods of both the capture-recapture and state-space models, and the information about such shared parameters is pooled. We would therefore expect improved precision of the estimates of those parameters to which multiple data sets contribute information, because more information usually translates into greater precision of the inferences from a statistical model.

6.2.1 EXPERIENCING A GAIN IN PRECISION IN A SIMPLE SIMULATION

We now conduct a simple simulation exercise to show you this expected gain in precision. We again use our woodchat shrike example with input parameters identical to those in Chapter 5 but with a slightly simplified model: we assume identical productivity of $f = 3.1$ for all individuals (rather than dependent on the mother's age) and generate data for only 10 years (rather than 20). All the other features of the data simulation and model are identical. We generate three independent data sets (capture-recapture, productivity, and population counts) and analyze them in different combinations using four models: (1) an IPM for all three data sets jointly, (2) a Cormack-Jolly-Seber (CJS) model for just the capture-recapture data, (3) a Poisson regression model for just the productivity data, and (4) a state-space model for just the population counts. For simplicity, we keep all demographic parameters constant over time in all four models and do not include demographic stochasticity. Hence, our IPM here differs from the IPM in Section 5.3 only by its lack of age-dependent productivity.

You can find the simulation code on the book web page and take the code for the analysis of the capture-recapture and productivity data alone directly from the IPM code. We write the state-space model for the population counts alone such that annual population growth rates rather than demographic rates are estimated (model as in Section 4.3.1). We simulated 1000 replicates of all three data sets and then fit all four models. The average annual sample sizes across all simulations were about 60 and 20 newly marked juveniles and adults, respectively, and 55 monitored broods, while population size averaged about 110 breeding pairs. Here, we summarize the coefficient of variation (CV) of the main parameters as a measure of the precision of the estimates.

The precision of all demographic parameters was improved (i.e., the CV was reduced) for an estimate under the IPM compared with that under the corresponding single-data model (Fig. 6.1). The relative gains in precision differed among the four parameters and were greatest for juvenile survival (62 % increase in precision), less for adult survival (13%) and

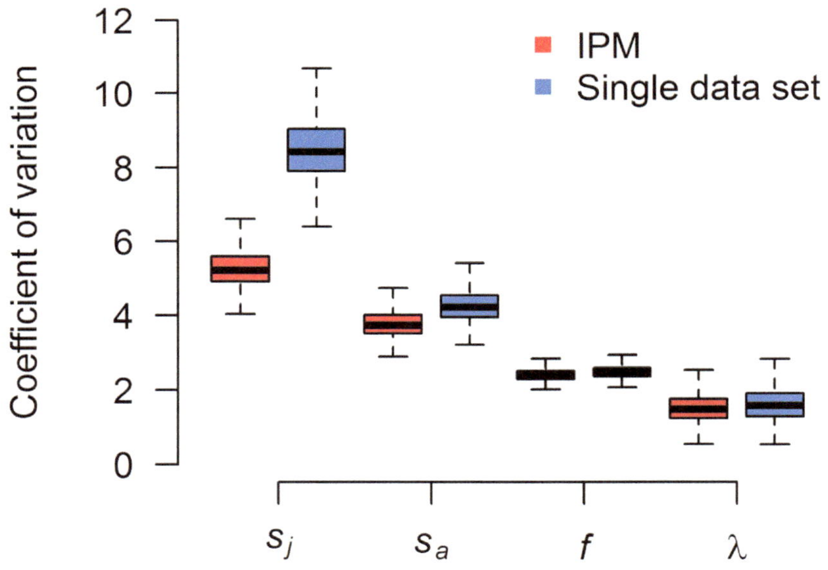

FIGURE 6.1

Comparison of the precision of the estimates under an IPM and the corresponding single-data analysis for juvenile and adult survival (s_j, s_a), productivity (f), and population growth rate (λ) for 1000 simulation replicates. The coefficient of variation is a measure of imprecision; hence, smaller is better.

population growth rate (9%), and only marginal for productivity (2%). The reason for these gains in precision under an IPM is intuitively obvious—population counts contain some information about all demographic processes in the population, and this information can be exploited with an IPM. Hence, in a sense, the sample size is increased in the IPM compared with single-data models, and increased sample size typically leads to greater precision of an estimate.

Such increases in precision in IPMs have been documented in a number of simulations and empirical studies (Besbeas et al., 2002; Brooks et al., 2004; Abadi et al., 2010a; Eacker et al., 2017). In one extensive simulation study, Abadi et al. (2010a) found that the gain in precision was greatest when sample size, n, was small. This makes intuitive sense when thinking about the definition of the standard error of a mean as $SD/\sqrt{n}$, which changes most for a small n. Besbeas et al. (2002) and Brooks et al. (2004) showed that the precision of survival estimates was only slightly improved under an IPM that combined ring-recovery and population count data compared with survival estimates under a ring-recovery model alone. The reason for this modest improvement was that the population count data contain only relatively weak information about survival, so most information in the IPM comes from the ring-recovery data anyway. Borysiewicz et al. (2009) demonstrated that gains in precision were more pronounced in multisite than single-site IPMs. These results suggest that the beneficial effects on the precision of estimates for integrated population-level data (i.e., population counts) are greater when the information about those parameters in the individual-level data is sparse. Therefore, we would expect the greatest benefits from integration in complicated models with many parameters, which we often require in ecological studies, or when sample sizes are small, as is typically the case in studies on rare species of conservational concern (Schaub et al., 2007; Oppel et al., 2014). Increased precision of parameter estimates can also be important if you are interested in temporal patterns in the parameters such as trends, because the power to detect such patterns is increased.

6.2.2 WHERE DOES THE INFORMATION COME FROM?

The increased precision of parameters that are directly informed by multiple data sets may also filter through to other parameters that are only informed by a single data set (Brooks et al., 2004). It is then natural to wonder about how the information is "flowing" from data to parameter estimates, or in other words, to ask the following question: What is the influence of each data set on the parameter estimates? In complex models such as an IPM, this is not a priori clear. Information flow can be studied by changing the information content of each data set in turn, which is closely related, but not identical, to changing the weights of the data sets in an IPM (Maunder and Punt, 2013). The amount of information about parameters contained in a data set depends in part on the sample size. Thus, by changing the sample size, we can change the amount of information in a data set. Fitting an IPM with an increased size for one data set allows us to explore its effect on the precision of each parameter. In this way, we can study the sensitivity of the estimates of one parameter to each type of data included in an IPM. Hence, we conduct another simulation study using the same demographic processes as before to demonstrate the impact of each data set and the "flow of information" in the IPM.

We again generate the three data sets for capture-recapture, productivity, and population counts. To increase the sample size of a data set we simply clone it 10 times, i.e., we create and then merge 10 copies of the same data set. Next, we fit the IPM using four different combinations of the original data and their "expanded versions": (1) no data set is cloned (denoted "original"), (2) only the capture-recapture data set is cloned, but the productivity and count data are not ("10× CR"), (3) only the productivity data are cloned ("10× Prod"), and (4) only the population count data are cloned ("10× Count").

The analyzing models need some slight adaptations compared with the original models; for the cloned data, the associated likelihood is specified in a loop with one iteration per clone. We performed 1000 simulations, and as a measure of precision, computed the CV of the estimates of the demographic parameters and of the population growth rate. Intuitively, we would expect the greatest increase in precision for that parameter for which a type of data contributes direct information, e.g., survival when capture-recapture data are cloned. Moreover, if increased precision filters through to other parameters, we expect improvements for other demographic parameters as well.

The simulation yielded the expected results: with cloned capture-recapture data, we observed substantial gains in precision in the estimates of juvenile and adult survival with a respective average reduction in their CVs of 54% and 65% compared with the original analysis (Fig. 6.2). We found analogous patterns for cloned productivity data (68% smaller CV of f) and for cloned population count data (76% smaller CV of λ).

We have mentioned several times that population count data contain information about all demographic processes. Hence, it is unsurprising that the precision of all parameters was improved when population counts were cloned 10 times (Fig. 6.2). However, the gain in precision for the two survival probabilities (11% and 3% smaller CVs for juvenile and adult survival, respectively) was not substantial, and it was marginal at best for productivity (0.5% smaller CV). This shows that in our simulated population, most information about the demographic parameters stems from the capture-recapture and productivity data rather than from the population count data.

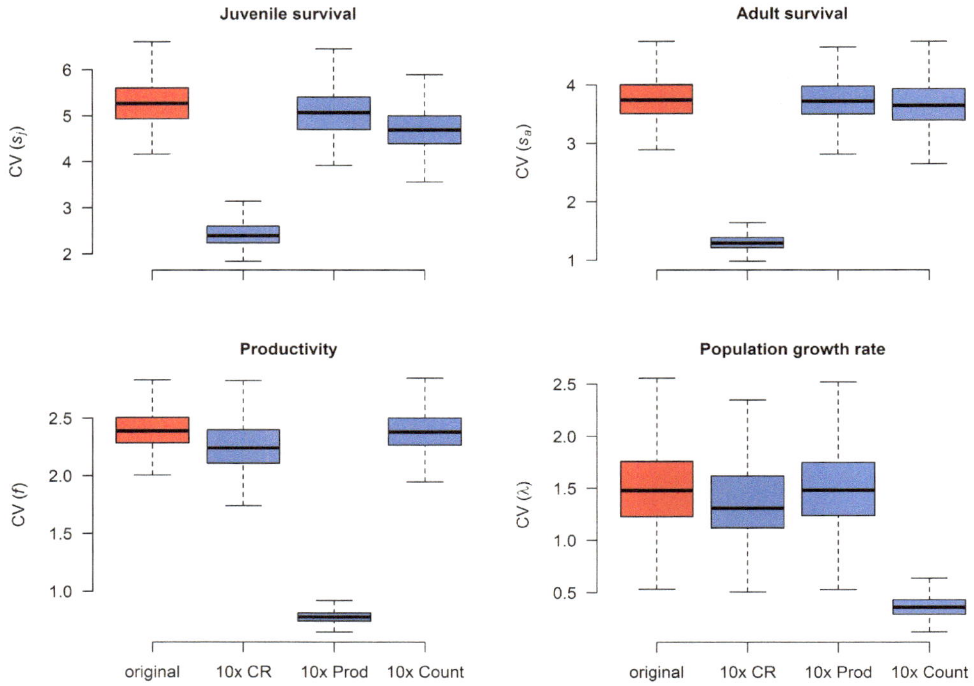

FIGURE 6.2

Sensitivity of parameter estimates for a 10-times-increased sample size of the three data types in an IPM with a classical combination of survival and productivity data with population counts: boxplots of the coefficients of variation (CVs) of juvenile survival (s_j), adult survival (s_a), productivity (f), and population growth rate (λ) for 1000 simulated combinations of the three data sets. Red denotes parameters estimated from the original analysis with no data set cloned, while blue denotes parameters estimated in scenarios where capture-recapture (10× CR), productivity (10× Prod), or population count (10× Count) data were cloned 10 times.

Interestingly, we also note some "flow of information" among the parameters. For instance, the CV of productivity became 5% smaller when the sample size of the capture-recapture data was increased (Fig. 6.2). Why would that be? After all, the capture-recapture data do not contain any direct information about productivity. The reason is that population count data contain information about productivity via the number of recruits, which is a function of productivity and juvenile survival. This information can be exploited more efficiently and productivity can be estimated more accurately when juvenile survival is also estimated more precisely.

Of course, these precise findings are specific to the simulated life history and data sets and models we have used. It is possible that in other cases, the information flow and sensitivities of parameter estimates to data types may be different. Therefore, we have shown one typical case and provided you with simple tools to investigate questions about the flow of information in an IPM. The simulations as shown here but tailored to your data and context are a powerful way to understand where information is coming from in an IPM and to see how estimator precision is affected by the sample sizes of each data type.

Knowing where the information comes from and assessing the impact of each data set on parameter estimates is important for your general understanding of complex models that jointly analyze different data sets. It is also helpful for judging the impact of a model's potential lack of fit. If one submodel does not fit the data well (e.g., there is some heterogeneity that is not adequately described such that the point estimates become biased), we may be interested in knowing how much this affects inferences under the integrated model. For instance, assume that there is no information flow among combinations of data sets and their associated "typical" parameter (e.g., survival for capture-recapture data). In such a case, the lack of fit of, say, the capture-recapture model will only be a problem for survival estimates and not necessarily for any other parameter estimates. In contrast, a typical IPM always has some flow of information "sideways" (that is, to parameters other than the "typical" parameter). Therefore, these may be affected by lack of fit as well, and the degree to which this happens will depend on the kind of flow of information and its strength.

The increase in the precision of parameter estimates just demonstrated and information flow among parameters in an IPM results from the combination of the information about the jointly analyzed parameters in the data sets. This also means that the point estimates obtained under an IPM are *weighted averages*, and the weights are proportional to the information content about a parameter in each data set. Weighting occurs automatically in all IPMs in this book, and we think that such a weighting based on information content is sensible (Maunder and Piner, 2017). If the data sets are consistent in the sense of the "common demography assumption" (see Section 7.2.3), and the observation models of each data set are adequate, the weighted averages obtained in an IPM will be unbiased, and on average, the point estimates will not change as a function of data weights (Maunder and Piner, 2017).

We could also choose custom weights for a data set. Such data weighting schemes are much discussed (and frequently adopted) in IPMs conducted for fishery stock assessments (Francis, 2011; Maunder and Punt, 2013), where some authors recommend data weighting when sample sizes differ substantially or if one has less trust in the larger data sets (Francis, 2011, 2017). It is also possible, for example, to estimate optimal weights by cross-validation (Maunder and Piner, 2015; Fletcher et al., 2019). The main goal of weighting is to avoid bias due to misleading information from a data set in which one has less trust or that "tells a different story"—for instance, when there is a spatio-temporal mismatch between different data sets (see Section 6.6.3). However, Maunder and Piner (2017) argue that the principal problem then is simply mis-specification of the integrated model. It would be better to develop a correctly specified model (i.e., one that "explains away" any such mismatch) instead of introducing weights. Hence, we generally prefer to avoid custom weighting but think more research is needed on the topic.

6.3 ESTIMATION OF DEMOGRAPHIC PARAMETERS FOR WHICH THERE ARE NO EXPLICIT DATA

Another, and in many practical applications perhaps *the* main, advantage of IPMs is their ability to estimate demographic parameters for which no specific data are available. An example is the estimation of survival without capture-recapture or ring-recovery data. This is possible because population count data contain information about all demographic processes, and this information can be extracted by the IPM. In this section, we first demonstrate how this works, and second, we study how far we can go—i.e., how many parameters without explicit data we can estimate. Finally, we present results of our review of published IPMs (Section 1.3) in which parameters without explicit data were estimated.

To illustrate how we can estimate a demographic parameter for which no explicit data are available (also called *hidden* parameter; Tavecchia et al., 2009), we again use our woodchat shrike example. To keep things simple, we use a deterministic model without demographic or environmental stochasticity and assume identical productivity for all females. We further assume that population count and capture-recapture data are available but that productivity data are not. The joint likelihood of the IPM then contains elements of the state-space and capture-recapture likelihoods but no component for productivity data because there are none. But productivity is still a parameter in the model:

$$L_{IPM}(\boldsymbol{N}, s_j, s_a, p, f, \sigma^2 | \boldsymbol{m}, \boldsymbol{y}) = L_I(N_1) \times L_O(\boldsymbol{N}, \sigma^2 | \boldsymbol{y}) \times L_S(\boldsymbol{N}_{2...T}, s_j, s_a, f) \times L_{CJS}(s_j, s_a, p | \boldsymbol{m}).$$

The code for fitting this IPM requires only a slight change to the code in the original IPM for all three data sets (Section 6.2, where we refer to the book website); we simply remove the Poisson regression model for the productivity data. However, the prior for productivity must be kept in the model because we still estimate that parameter.

```
# Write JAGS model file
cat(file="model1.txt", "
model {
# Priors and linear models
mean.sj ~ dunif(0, 1)
mean.sa ~ dunif(0, 1)
mean.p ~ dunif(0, 1)
mean.f ~ dunif(0, 10)

for (t in 1:(n.occasions-1)){
  sj[t] <- mean.sj
  sa[t] <- mean.sa
  p[t] <- mean.p
}

sigma ~ dunif(0.5, 100)
tau <- pow(sigma, -2)
```

```
# Population count data (state-space model)
# Model for the initial population size: uniform priors
N[1,1] ~ dunif(1, 300)
N[2,1] ~ dunif(1, 300)

# Process model over time: our model of population dynamics
for (t in 1:(n.occasions-1)){
  N[1,t+1] <- mean.f / 2 * mean.sj * (N[1,t] + N[2,t])
  N[2,t+1] <- mean.sa * (N[1,t] + N[2,t])
}

# Observation model
for (t in 1:n.occasions){
  C[t] ~ dnorm(N[1,t] + N[2,t], tau)
}

# Capture-recapture data (CJS model with multinomial likelihood)
# Define the multinomial likelihood
for (t in 1:(n.occasions-1)){
  marr.j[t,1:n.occasions] ~ dmulti(pr.j[t,], rel.j[t])
  marr.a[t,1:n.occasions] ~ dmulti(pr.a[t,], rel.a[t])
}
# Define the cell probabilities of the m-arrays
for (t in 1:(n.occasions-1)){
  # Main diagonal
  q[t] <- 1 - p[t]    # Probability of non-recapture
  pr.j[t,t] <- sj[t] * p[t]
  pr.a[t,t] <- sa[t] * p[t]
  # Above main diagonal
  for (j in (t+1):(n.occasions-1)){
    pr.j[t,j] <- sj[t] * prod(sa[(t+1):j]) * prod(q[t:(j-1)]) * p[j]
    pr.a[t,j] <- prod(sa[t:j]) * prod(q[t:(j-1)]) * p[j]
  } #j
  # Below main diagonal
  for (j in 1:(t-1)){
    pr.j[t,j] <- 0
    pr.a[t,j] <- 0
  } #j
} #t
# Last column: probability of non-recapture
for (t in 1:(n.occasions-1)){
  pr.j[t,n.occasions] <- 1-sum(pr.j[t,1:(n.occasions-1)])
  pr.a[t,n.occasions] <- 1-sum(pr.a[t,1:(n.occasions-1)])
}

# Derived parameters
# Annual population growth rate
for (t in 1:(n.occasions-1)){
  ann.growth.rate[t] <- (N[1,t+1] + N[2,t+1]) / (N[1,t] + N[2,t])
}
# Total population size
for (t in 1:n.occasions){
  Ntot[t] <- N[1,t] + N[2,t]
}
}
")
```

We load a new simulated woodchat shrike data set, inspect it briefly, construct the age-dependent m-array, and bundle the data (minus the productivity data!) into our list.

```
library(IPMbook); library(jagsUI)
data(woodchat6)
str(woodchat6)

List of 6
 $ ch   : num [1:947, 1:10] 1 1 1 1 1 0 1 1 1 0 ...
 $ age  : num [1:947] 2 2 2 2 2 2 2 2 2 2 ...
 $ count: num [1:10] 110 104 100 85 85 71 118 112 91 104
 $ J    : num [1:10] 147 144 132 131 178 178 235 169 177 186
 $ B    : num [1:10] 48 48 45 37 53 56 74 59 55 60
 $ f    : num [1:535] 4 7 5 5 1 5 4 1 3 0 ...

marr <- marrayAge(woodchat6$ch, woodchat6$age)
```

```
# Bundle data
jags.data <- list(marr.j=marr[,,1], marr.a=marr[,,2], n.occasions=dim(marr)[2],
  rel.j=rowSums(marr[,,1]), rel.a=rowSums(marr[,,2]), C=woodchat6$count)

# Initial values
inits <- function(){list(mean.sj=runif(1, 0, 0.5))}

# Parameters monitored
parameters <- c("mean.sj", "mean.sa", "mean.p", "mean.f", "sigma", "N", "ann.growth.rate", "Ntot")

# MCMC settings
ni <- 40000; nb <- 10000; nc <- 3; nt <- 3; na <- 2000

# Call JAGS (ART <1 min), check convergence and summarize posteriors
out1 <- jags(jags.data, inits, parameters, "model1.txt", n.iter=ni, n.burnin=nb, n.chains=nc,
  n.thin=nt, n.adapt=na, parallel=TRUE)
traceplot(out1)
print(out1, 3)

                        mean     sd    2.5%     50%   97.5% overlap0 f  Rhat n.eff
mean.sj                0.253  0.022   0.213   0.253   0.297    FALSE 1 1.000 11961
mean.sa                0.567  0.022   0.523   0.567   0.610    FALSE 1 1.000 30000
mean.p                 0.604  0.031   0.543   0.604   0.664    FALSE 1 1.000 30000
mean.f                 3.413  0.384   2.717   3.394   4.238    FALSE 1 1.000 30000
sigma                 18.669  6.076  10.877  17.415  33.982    FALSE 1 1.000 14425
N[1,1]                49.371 29.747   3.123  47.859 103.593    FALSE 1 1.003   752
N[2,1]                50.932 29.696   3.375  51.263 103.410    FALSE 1 1.003   719
[... output truncated ... ]
N[1,10]               41.636  6.521  28.889  41.594  54.558    FALSE 1 1.000 30000
N[2,10]               54.722  6.078  42.418  54.759  66.695    FALSE 1 1.000 30000
ann.growth.rate[1]     0.996  0.023   0.948   0.996   1.040    FALSE 1 1.000 30000
[... output truncated ... ]
ann.growth.rate[9]     0.996  0.023   0.948   0.996   1.040    FALSE 1 1.000 30000
Ntot[1]              100.304 11.905  78.043  99.764 125.839    FALSE 1 1.000 30000
[... output truncated ... ]
Ntot[10]              96.358 11.774  72.446  96.441 119.201    FALSE 1 1.000 30000
```

We see that the productivity estimate is very similar to that obtained from an IPM fit to all three data sets (the results are not shown, but you can do this as a bonus exercise). However, the posterior standard deviation is about four times larger now (0.294 vs. 0.065); hence, we estimate productivity with much reduced precision. This was to be expected because the information about productivity now comes exclusively from the population count data. Such reduced precision for hidden parameters in an IPM is typical (Tavecchia et al., 2009; Abadi et al., 2010b; Lahoz-Monfort et al., 2014; Schaub and Fletcher, 2015).

So when an IPM allows the estimation of hidden parameters for which no demographic data are available, how far we can go? For instance, can we still estimate demographic rates when we have only population count and productivity but no capture-recapture data? Or when we have only population count data? Can we still estimate anything meaningful in these contexts? Well, once again, simulation can provide an answer—we can simply try it out! We fit variants of the IPM to such reduced combinations of our data sets (or to population count data alone), save the estimates, and repeat this for many simulated data sets. In the model code, we must again delete some bits for the likelihoods of data that are no longer there. First we develop a model for productivity and population count data only. The joint likelihood is

$$L_{IPM}\left(\boldsymbol{N}, s_j, s_a, \boldsymbol{p}, \boldsymbol{f}, \sigma^2 | \boldsymbol{J}, \boldsymbol{y}\right) = L_I(N_1) \times L_O\left(\boldsymbol{N}, \sigma^2 | \boldsymbol{y}\right) \times L_S\left(\boldsymbol{N}_{2...T}, s_j, s_a, \boldsymbol{f}\right) \times L_P(\boldsymbol{f} | \boldsymbol{J}).$$

Writing the BUGS code should pose no difficulties. We bundle the data and launch the model in JAGS. We notice that the mixing is worse for this model, and we increase the length of the chains in our Markov chain Monte Carlo (MCMC) settings.

```
# Write JAGS model file
cat(file="model2.txt", "
model {
# Priors and linear models
mean.sj ~ dunif(0, 1)
mean.sa ~ dunif(0, 1)
mean.f ~ dunif(0, 10)

sigma ~ dunif(0.5, 100)
tau <- pow(sigma, -2)
```

```
# Population count data (state-space model)
# Model for the initial population size: uniform priors
N[1,1] ~ dunif(1, 300)
N[2,1] ~ dunif(1, 300)

# Process model over time: our model of population dynamics
for (t in 1:(n.occasions-1)){
  N[1,t+1] <- mean.f / 2 * mean.sj * (N[1,t] + N[2,t])
  N[2,t+1] <- mean.sa * (N[1,t] + N[2,t])
}

# Observation model
for (t in 1:n.occasions){
  count[t] ~ dnorm(N[1,t] + N[2,t], tau)
}

# Productivity data (Poisson regression model)
nJ ~ dpois(n.rep * mean.f)

# Derived parameters
# Annual population growth rate
for (t in 1:(n.occasions-1)){
  ann.growth.rate[t] <- (N[1,t+1] + N[2,t+1]) / (N[1,t] + N[2,t])
}
# Total population size
for (t in 1:n.occasions){
  Ntot[t] <- N[1,t] + N[2,t]
}
}
")

# Bundle data
jags.data <- list(n.occasions=length(woodchat6$count), nJ=sum(woodchat6$J),
  n.rep=sum(woodchat6$B), count=woodchat6$count)

# Initial values
inits <- function(){list(mean.sj=runif(1, 0, 0.5))}

# Parameters monitored
parameters <- c("mean.sj", "mean.sa", "mean.f", "sigma", "N", "ann.growth.rate", "Ntot")

# MCMC settings
ni <- 1000000; nb <- 50000; nc <- 3; nt <- 500; na <- 2000

# Call JAGS (ART 1 min), check convergence and summarize posteriors
out2 <- jags(jags.data, inits, parameters, "model2.txt", n.iter=ni, n.burnin=nb, n.chains=nc,
  n.thin=nt, n.adapt=na, parallel=TRUE)
traceplot(out2)
print(out2, 3)

                       mean     sd   2.5%    50%   97.5% overlap0 f  Rhat n.eff
mean.sj               0.317  0.183  0.018  0.318   0.619    FALSE 1 1.001  1201
mean.sa               0.500  0.286  0.029  0.500   0.966    FALSE 1 1.001  1205
mean.f                3.136  0.076  2.991  3.134   3.289    FALSE 1 1.000  5700
sigma                18.668  5.912 10.963 17.410  33.405    FALSE 1 1.000  4729
N[1,1]               49.994 29.343  3.387 48.571 103.446    FALSE 1 1.000  5265
N[2,1]               50.019 29.069  2.962 50.145 101.707    FALSE 1 1.000  4450
[ ... output truncated ... ]
N[1,10]              48.257 28.429  2.759 47.690  99.152    FALSE 1 1.001  1426
N[2,10]              48.477 28.350  2.683 48.059  99.484    FALSE 1 1.001  1260
ann.growth.rate[1]    0.997  0.022  0.950  0.997   1.041    FALSE 1 1.000  5700
[ ... output truncated ... ]
ann.growth.rate[9]    0.997  0.022  0.950  0.997   1.041    FALSE 1 1.000  5700
Ntot[1]             100.014 11.773 77.965 99.496 124.997    FALSE 1 1.001  5700
[ ... output truncated ... ]
Ntot[10]             96.734 11.539 73.067 96.764 119.929    FALSE 1 1.000  5700
```

We see that the estimates of productivity and population growth rates are very similar to those from the IPMs that we fit to this data set earlier. The posterior means of survival are also quite close to the true values that were used to simulate this data set (remember, they are 0.3 for juvenile and 0.55 for adult survival), but we note low precision for both.

Before we compare the estimates of the different models graphically, we go one step further and fit a model that attempts to estimate all demographic parameters from population count data alone. We would no longer call this model an IPM because it uses just a single data set. It is still a state-space model but with the process model written in terms of the demographic rates of a stage-structured (or matrix) population model. Its likelihood is

$$L_{SSM}(\boldsymbol{N}, s_j, s_a, \boldsymbol{p}, \boldsymbol{f}, \sigma^2|\boldsymbol{y}) = L_I(\boldsymbol{N}_1) \times L_O(\boldsymbol{N}, \sigma^2|\boldsymbol{y}) \times L_S(\boldsymbol{N}_{2...T}, s_j, s_a, \boldsymbol{f}).$$

```
# Write JAGS model file
cat(file="model3.txt", "
model {
# Priors and linear models
mean.sj ~ dunif(0, 1)
mean.sa ~ dunif(0, 1)
mean.f ~ dunif(0, 10)

sigma ~ dunif(0.5, 100)
tau <- pow(sigma, -2)

# Population count data (state-space model)
# Model for the initial population size: uniform priors
N[1,1] ~ dunif(1, 300)
N[2,1] ~ dunif(1, 300)

# Process model over time: our model of population dynamics
for (t in 1:(n.occasions-1)){
  N[1,t+1] <- mean.f / 2 * mean.sj * (N[1,t] + N[2,t])
  N[2,t+1] <- mean.sa * (N[1,t] + N[2,t])
}

# Observation model
for (t in 1:n.occasions){
  count[t] ~ dnorm(N[1,t] + N[2,t], tau)
}

# Derived parameters
# Annual population growth rate
for (t in 1:(n.occasions-1)){
  ann.growth.rate[t] <- (N[1,t+1] + N[2,t+1]) / (N[1,t] + N[2,t])
}
# Total population size
for (t in 1:n.occasions){
  Ntot[t] <- N[1,t] + N[2,t]
}
}
")

# Bundle data
jags.data <- list(n.occasions=length(woodchat6$count), count=woodchat6$count)

# Initial values
inits <- function(){list(mean.sj=runif(1, 0, 0.5))}

# Parameters monitored
parameters <- c("mean.sj", "mean.sa", "mean.f", "sigma", "N", "ann.growth.rate", "Ntot")

# MCMC settings
ni <- 1000000; nb <- 50000; nc <- 3; nt <- 500; na <- 2000

# Call JAGS (ART 1 min), check convergence and summarize posteriors
out3 <- jags(jags.data, inits, parameters, "model3.txt", n.iter=ni, n.burnin=nb, n.chains=nc,
  n.thin=nt, n.adapt=na, parallel=TRUE)
traceplot(out3)
print(out3, 3)
```

```
                      mean      sd   2.5%    50%   97.5% overlap0 f  Rhat n.eff
mean.sj              0.352   0.267  0.018  0.277   0.937    FALSE 1 1.000  5700
mean.sa              0.586   0.285  0.039  0.620   0.982    FALSE 1 1.002   951
mean.f               3.447   2.623  0.166  2.737   9.228    FALSE 1 1.003   729
sigma               18.860   6.173 11.089 17.473  34.453    FALSE 1 1.000  5700
N[1,1]              49.962  29.036  3.965 48.763 102.803    FALSE 1 1.000  4570
N[2,1]              50.352  29.215  3.399 50.355 102.760    FALSE 1 1.000  4231
[ ... output truncated ... ]
N[1,10]             39.787  28.355  1.320 35.676  96.392    FALSE 1 1.002   970
N[2,10]             56.734  28.456  3.851 59.533 102.348    FALSE 1 1.002  1048
ann.growth.rate[1]   0.996   0.023  0.948  0.997   1.039    FALSE 1 1.001  5700
[ ... output truncated ... ]
ann.growth.rate[9]   0.996   0.023  0.948  0.997   1.039    FALSE 1 1.001  5700
Ntot[1]            100.314  12.037 78.599 99.593 126.060    FALSE 1 1.001  5700
[ ... output truncated ... ]
Ntot[10]            96.521  11.738 73.273 96.525 119.578    FALSE 1 1.000  5700
```

Very long MCMC chains are now necessary to obtain convergence. The estimated population growth rate is very similar to what we obtained in the IPM, but the estimates of the demographic parameters do not look so good; in particular, their precision is extremely poor. To visualize the results from all models, we plot the posterior distributions of the demographic rates and stage-specific and total population sizes and compare them all.

Under the IPMs that use either all three data sets or combine capture-recapture and population count data, survival in both age classes and the population size are all well estimated with posterior distributions that show a distinct peak near the true values (Fig. 6.3). Productivity is estimated well when either all, capture-recapture plus count, or productivity plus count data are used. Yet in the latter scenario, survival and stage-specific population sizes are not estimated well.

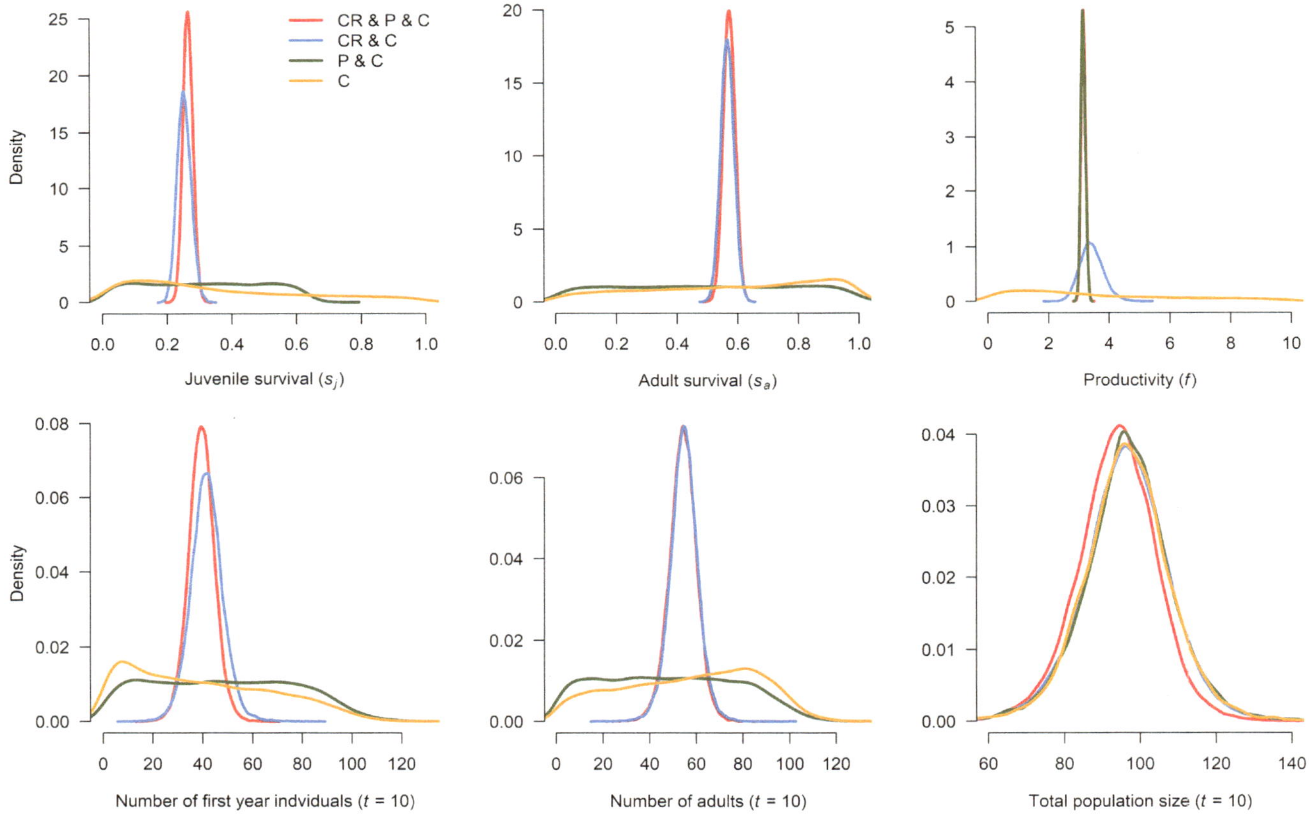

FIGURE 6.3

Posterior distributions of the estimates of juvenile and adult survival, productivity, the number of first-year and adult individuals, and total population size at the last occasion under four different models using different data sets (*CR*: capture-recapture data; *P*: productivity data; *C*: population count data). Note these are the results from fitting these models to a single data set (or a combination of data sets); no replicated data sets are involved here (as in the two previous figures).

When only population counts are available, none of the demographic rates can be properly estimated—although the posterior means happen to be not that much different from the data-generating true values (see the posterior summary above), the posterior distributions are extremely diffuse, and indeed, they hardly differ from the prior distributions (not shown). Estimates of the stage-specific population sizes are adequate when at least population count and capture-recapture data are available, whereas the estimates of the total population size are acceptable under all four scenarios. This means that if only population counts are available, we can obtain decent estimates of total population size (under the specified observation model) and hence also of the population growth rate. However, neither demographic rates nor stage-specific population sizes can be estimated. Thus, the decomposition of total population size into stage-specific population sizes or of the population growth rate into demographic contributions from survival and productivity requires additional information about demographic rates and thus more data than just population count data.

These are the results from a single set of data and may not always hold. Hence, we conducted a simulation study with 1000 replicates to gauge the generality of these findings. Fig. 6.4 confirms the main findings from Fig. 6.3: in the absence of productivity data, estimates of productivity rates under the IPM are unbiased but less precise than in the scenario where productivity data are available. Without capture-recapture data, only productivity and population growth rate can be estimated reliably in the setting of our simulation, and the estimates of the two age-specific survival parameters are biased and imprecise. Finally, the demographic state-space model fit to population count data alone produces biased, imprecise, and essentially useless estimates of all demographic rates.

In summary, it appears that with an IPM, one hidden demographic parameter can be estimated but not two or more. This is because, loosely speaking, we estimate the hidden demographic parameter "by difference" based on the population growth rate. The population growth rate in our example is a simple function of the demographic rates, namely: $\lambda = s_j \times \frac{f}{2} + s_a$. In the IPM using all three data sets, we have information about all the parameters separately from the different data sets: population counts are informative about the population growth rate, capture-recapture data about the two survival probabilities, and productivity data about productivity. Without productivity data, productivity can be estimated as $f = \frac{2(\lambda - s_a)}{s_j}$ because the three quantities of the right-hand side can be estimated based on the population count and capture—recapture data. Of course, productivity in an IPM is not really estimated in this way, but conceptually this helps to see why it is possible to obtain such an estimate when the three other components are "known" (i.e., can be estimated). Now, with only productivity and population count data but absent any capture-recapture data, no direct information about juvenile and adult survival is available, and we end up with a single equation and two unknowns. In other words, juvenile and adult survival must take on values such that the two equations $s_a = \lambda - s_j \times \frac{f}{2}$ and $s_j = \frac{2(\lambda - s_a)}{f}$ are both satisfied. Given that productivity and population growth rate are "known," there is an infinite number of possible combinations of juvenile and adult survival that satisfy the two equations. Therefore, the two cannot be estimated separately (although we may surmise that a single survival parameter in a model without age structure in survival should be estimable in that case). In terms of the likelihood surface, such a model has a flat ridge, i.e., the likelihood is the same for infinitely many combinations of the two parameters. Besbeas et al. (2002) and Cole and McCrea (2016) have formally shown that a state-space model with demographic parameters fit only to population count data is parameter-redundant, i.e., not all demographic rates are separately identifiable.

In the Bayesian mode of analysis, we always have the possibility to use informative priors to obtain nicely peaked posterior distributions. In the productivity plus population count scenario, for example, if you can use an informative prior for either of the two survivals, you can also estimate the single remaining model survival parameter in this combination of available data sets. But of course, the result will be sensitive to the chosen prior. An alternative for reducing the number of estimated parameters from two (one too many) to one (which is OK) would be to assume that juvenile survival is a constant, known fraction of adult survival; see Section 6.6.6 and Koons et al. (2017). However, it is unclear how often such a scenario would be sensible.

Sometimes you may wonder whether it is at all worthwhile within an IPM to include a smallish data set that is informative about some demographic parameter. Inclusion of such a data set will increase the complexity of the model, so the question then becomes whether this is compensated by the increased precision of the parameters. For illustration, we reanalyze the woodchat shrike data set but now vary the proportion of productivity data used from 0% to 100%, meaning that we use anything between none to all of the available productivity data.

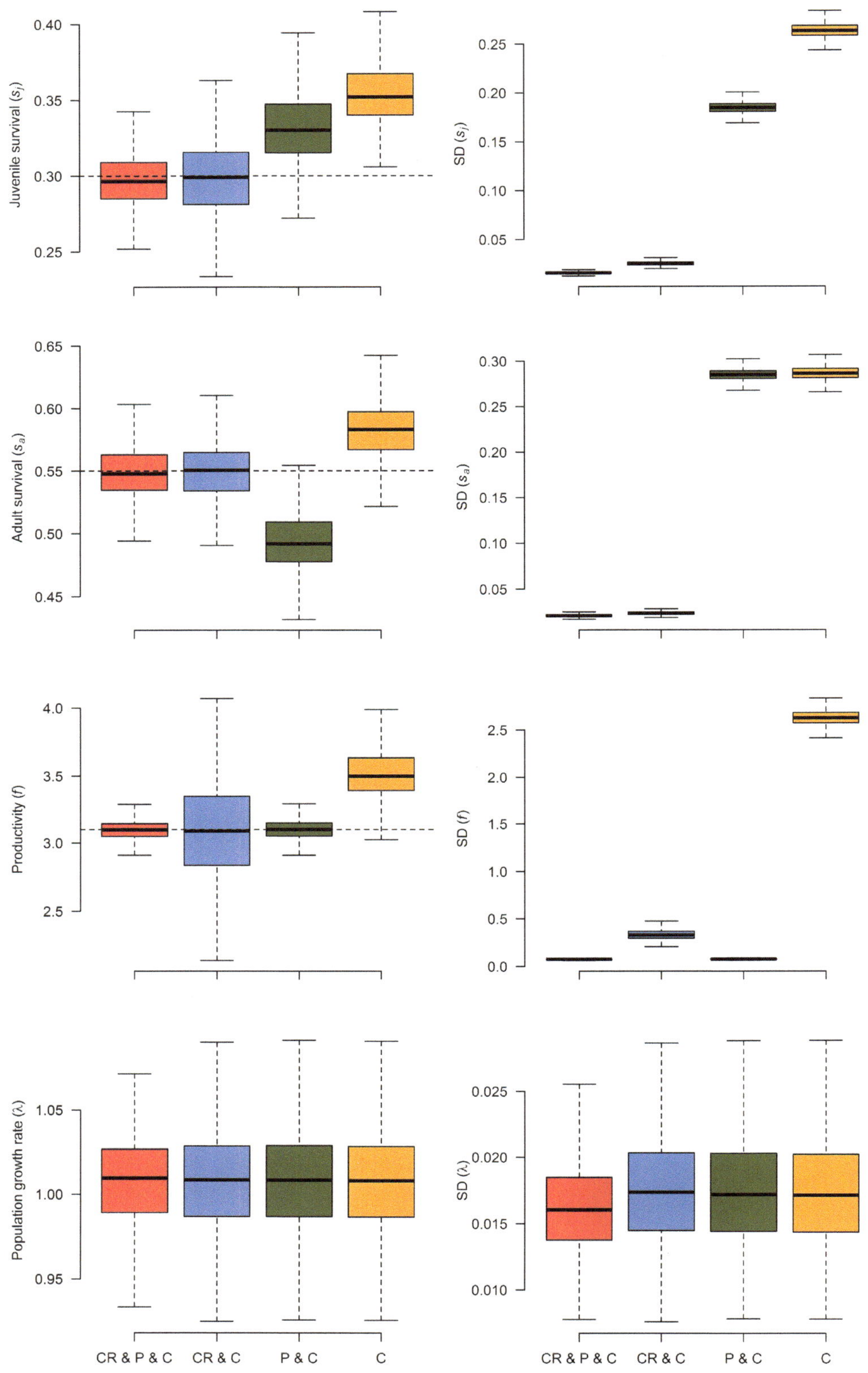

FIGURE 6.4

Boxplots of posterior means and posterior standard deviations (SDs) of juvenile (s_j) and adult survival (s_a), productivity (f), and population growth rate (λ) from four different demographic models using different combinations of data sets: (1) IPM with capture-recapture, productivity, and population count data (CR & P & C), (2) IPM with capture-recapture and population count data (CR & C), (3) IPM with productivity and population count data (P & C), and (4) demographic state-space model with population count data (C). These boxplots summarize posterior means from 1000 simulations. The horizontal dashed lines in the first three rows indicate the parameter values used for data simulation, i.e., truth.

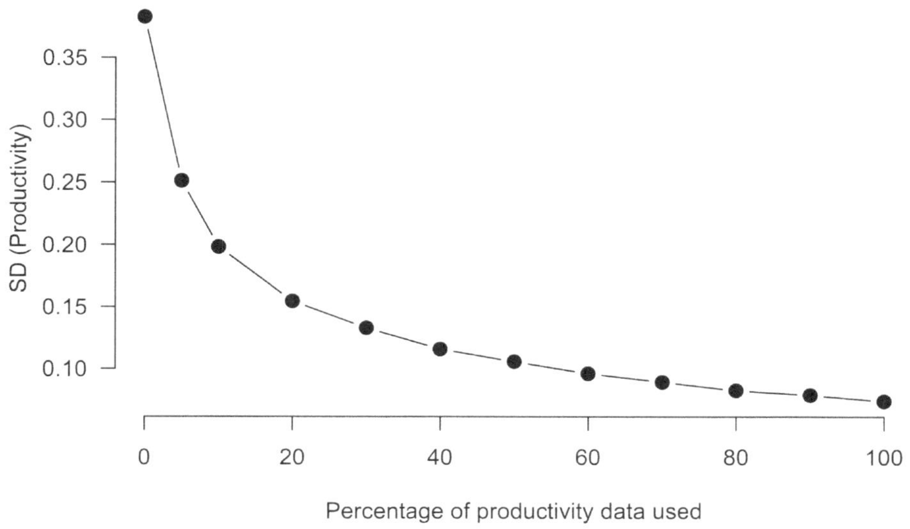

FIGURE 6.5

Precision of the estimate of productivity (posterior standard deviation) as a function of the sample size of the productivity data. On the *x*-axis, 0% denotes no inclusion of productivity data, 5% the inclusion of a sample of 21 broods, and so forth, while with 100%, the full sample of 424 broods enters into the analysis.

Not entirely surprisingly, the precision of the productivity parameter depends on the sample size of the productivity data. This can again be related to the magnitude of the SE of a mean that decreases with sample size n as $SD/\sqrt{n}$ (Fig. 6.5). Thus, if you have any data that directly informs about a demographic process, and the focus of interest is the demographic rates, you should seriously consider using them in the IPM even if the sample size is small. A similar conclusion was obtained by a thorough study using an empirical data set where the available data were thinned out (Lahoz-Monfort et al., 2014). We note, however, that a different conclusion may be arrived at when your main interest is in the population growth rate, which may be just as well estimated with the count data alone. Simulation provides the basis for making an informed decision.

The simulation results demonstrate that IPMs allow the unbiased estimation of hidden parameters for which no explicit data were collected (Fig. 6.4). It is also possible to model such a hidden parameter as a function of covariates (Besbeas et al., 2002; Abadi et al., 2010b), to use penalized splines (Besbeas and Morgan, 2012), or to specify temporal random effects for it to estimate the mean and variance of the process (Schaub et al., 2013).

Clearly, the ability of IPMs to estimate hidden parameters in a demographic model is very useful because frequently we wish to model population dynamics but lack data about one demographic process. IPMs were used in different contexts particularly for this purpose. For example, one of the first IPMs was fit to ring-recovery and population count data of British lapwings (*Vanellus vanellus*, Besbeas et al., 2002). The authors had no data about productivity, but the IPM allowed the estimation of productivity. In populations studied at relatively small spatial scales (which is most often the case...), emigration and immigration become important processes for the local dynamics of the population. Emigration is already accounted for when using capture-recapture data, but immigration is not, and it can be estimated using an IPM (Abadi et al., 2010b; Schaub and Fletcher, 2015). This is a tremendous advantage because there is mounting evidence that immigration is a key demographic process that must be included for valid inferences about the dynamics of populations at small to moderately large spatial scales, which is where nearly all populations are studied in practice (Abadi et al., 2010b; Schaub et al., 2013; Altwegg et al., 2014; Szostek et al., 2014; Tempel et al., 2014; Duarte et al., 2016; Lieury et al., 2016; Weegman et al., 2016; Brommer et al., 2017; Wilson et al., 2017; Woodworth et al., 2017; Tenan et al., 2017; Millon et al., 2019).

In our review of IPMs for terrestrial vertebrates ($n = 118$ studies), the majority of studies estimated a hidden parameter ($n = 85$, 72%), most commonly immigration, but closely followed by productivity (Table 6.1). Other demographic parameters, such as survival at younger ages, recruitment, or breeding propensity, were also estimated as hidden parameters but much less often. Sometimes, authors have estimated one extra parameter that represented a combination of several hidden parameters (i.e., demographic processes). This single hidden parameter has no clear biological meaning but is required in the IPM as a "correction factor." For instance, Robinson et al. (2014) fit IPMs to UK bird data at large spatial scales using population indices and ring-recovery, nest survival, and productivity data. Their extra parameter represented the product of breeding propensity, the number of nesting attempts, and postfledging survival. It was a requirement of the model to ensure unbiased estimates of the other parameters.

Table 6.1 Overview of the estimation of hidden parameters in contemporary integrated population models (IPMs) based on our review of 118 published IPMs on terrestrial vertebrates using real data. A hidden parameter is one that is estimated without explicit data.

Hidden Parameter	Number of Studies
Immigration	37
Productivity	31
Young age survival	6
Recruitment	4
Breeding propensity	1
Sex ratio at birth	1
Emigration	1
"Correction" factor	3
No hidden parameter estimated	33

Despite the ability of an IPM to provide estimates of a hidden parameter, we should perhaps not get too excited. For at least the following four reasons, the estimates of such parameters must be carefully evaluated before they can be believed.

First, we must check carefully whether the hidden parameters are truly informed by the available data sets. As shown earlier, we can in general estimate one hidden parameter but not several of them (Lahoz-Monfort et al., 2014; Cole and McCrea, 2016; Besbeas and Morgan, 2019). However, estimation of additional hidden parameters may be possible when structured population counts are used (Tavecchia et al., 2009). Generally, in more complex IPMs it may not be a priori clear which parameter could be estimated as hidden. Parameter estimability must then be checked carefully by either simulations or the use of formal tests for parameter identifiability (Catchpole and Morgan, 1997; Cole and McCrea, 2016). Furthermore, prior sensitivity analyses, in particular for hidden parameters, are useful to understand any "information" induced by the priors. An illustrative example for the use of prior information happened unintentionally. In Abadi et al. (2010b), we used capture-recapture and population count data and found that two hidden parameters could be estimated: immigration and productivity. However, this appeared to "work" only because the prior for productivity was in fact informative even though we had thought that a uniform distribution between 0 and 5 was vague (Besbeas and Morgan, 2019). Thus, a careful evaluation of the inferences under an IPM in the presence of hidden parameters is always required.

Second, if one can estimate a hidden parameter, the precision is often relatively low. This reduces the power to detect factors that influence the parameter (e.g., when we want to model it by covariates). We have shown that the inclusion of even small amounts of data about a parameter in question can result in strong improvements in its precision (Fig. 6.5). Hence, if there are any data about such a hidden parameter, you can try to include them in the model to partially "unhide" the parameter and improve the quality of its estimate.

Third, as a consequence of the typically low precision of the estimate of a hidden parameter, it can be difficult to obtain adequate estimates of the temporal pattern of the associated process. Lahoz-Monfort et al. (2014) provide a good illustration of this in their comparison of annual estimates of productivity of a seabird obtained from an IPM using capture-recapture, ring-recovery, productivity, and population count data with estimates from the identical IPM that had ignored the productivity data. This revealed that annual estimates of productivity as a hidden parameter were imprecise and biased, but the mean over time was an adequate estimate of the average productivity. In addition, in the latter analysis, the estimated number of breeding pairs matched the population counts almost perfectly, much better than in the analysis that used all available data sets. This suggested that the model with hidden productivity was too flexible and presumably resulted in overfitting of population size.

Finally, to obtain unbiased estimates of a hidden parameter, all the model assumptions must be met sufficiently well. If a demographic rate is not modeled adequately, for instance because survival increases over time but our model assumes constant survival over time, the estimate of the hidden parameter will be biased or may show an erroneous temporal pattern (Schaub and Fletcher, 2015; Riecke et al., 2019b). Goodness-of-fit assessments of the submodels with available data are therefore important (Section 7.2). We further discuss the consequences of model misspecification for parameter accuracy in Chapter 7.

6.4 ESTIMATION OF PROCESS CORRELATION AMONG DEMOGRAPHIC PARAMETERS

Often, demographic rates in our population model will vary over time and may not do so independently of one another. Positive correlation typically occurs if the drivers for the different rates are completely or partially the same. For example, in a year with plenty of food, survival might be enhanced in addition to productivity. We might then be interested in estimating the degree of temporal correlation among two or more demographic rates. If there is such correlation, it ought to be included in the model. This is particularly true when forecasting the future population trajectory, for instance, in a population viability analysis (Fieberg and Ellner, 2001). Another practical reason that inclusion of process correlation may be a good idea is that in this way estimates of, say, juvenile survival can borrow strength from (i.e., be improved by) the information used to estimate adult survival, much as in univariate random-effects modeling.

The estimation of process correlation can be challenging because sampling correlation may swamp process correlation, but a hierarchical modeling approach may help (Link and Barker, 2005). However, if you want to estimate process correlation between parameters informed from different data sets, adoption of a hierarchical model is only possible when the data sets are analyzed jointly, i.e., within a single integrated model. For example, we can estimate the temporal variability of survival based on capture-recapture data and temporal variability of productivity from productivity data, but their process correlation can only be properly estimated when both data sets are analyzed jointly. Hence, combining data sets in an IPM lets us explicitly model the process cross-correlations for two or more parameters while properly accommodating sampling correlations among all of them.

Estimation of process correlations usually proceeds by estimation of the variance-covariance matrix in a multivariate normal distribution specified for two or more parameters, typically on a transformed scale such as the logit or the log. We illustrate this using our woodchat shrike example for the same classical IPM scenario as before, where we have capture-recapture, productivity, and population count data. However, we have now simulated the data with positive process correlation between juvenile and adult survival and between adult survival and productivity. Accurate estimation of process correlation and temporal variability requires large sample sizes both in terms of the number of years and individuals involved (Riecke et al., 2019a), and hence, we generate 20 years of data from a fairly large population.

We want to estimate process correlation between juvenile and adult survival and between the two survival probabilities and productivity. We first collect all demographic rates (on log- and logit-transformed scales) into matrix $\boldsymbol{\varepsilon}$:

$$\begin{aligned}\text{logit}(s_{j,t}) &= \varepsilon_{1,t}\\ \text{logit}(s_{a,t}) &= \varepsilon_{2,t}\\ \log(f_t) &= \varepsilon_{3,t}\end{aligned}$$

We then place a multivariate normal (MVN) distribution on $\boldsymbol{\varepsilon}$:

$$\boldsymbol{\varepsilon} \sim \text{MVN}(\boldsymbol{\mu}, \boldsymbol{\Sigma}),$$

where $\boldsymbol{\mu}$ is the vector of means and $\boldsymbol{\Sigma}$ is the variance-covariance matrix, as follows:

$$\boldsymbol{\Sigma} = \begin{bmatrix} \sigma^2_{s_j} & \sigma_{s_j,s_a} & \sigma_{s_j,f} \\ \sigma_{s_a,s_j} & \sigma^2_{s_a} & \sigma_{s_a,f} \\ \sigma_{f,s_j} & \sigma_{f,s_a} & \sigma^2_f \end{bmatrix}.$$

Here, σ^2_x denotes the temporal variance of parameter x on the transformed scale and $\sigma_{x,y}$ is the temporal covariance between parameters x and y (and $\sigma_{x,y} = \sigma_{y,x}$). The process correlations are then given by

$$\rho_{s_j,s_a} = \frac{\sigma_{s_j,s_a}}{\sqrt{\sigma^2_{s_j}\sigma^2_{s_a}}},\quad \rho_{s_j,f} = \frac{\sigma_{s_j,f}}{\sqrt{\sigma^2_{s_j}\sigma^2_f}} \quad \text{and} \quad \rho_{s_a,f} = \frac{\sigma_{s_a,f}}{\sqrt{\sigma^2_{s_a}\sigma^2_f}}.$$

Estimation of a variance-covariance matrix can be challenging because the matrix must be positive-definite (Steward, 1976). This implies particular constraints for its elements (Barnard et al., 2000; Budden et al., 2007) that must be reflected by our choice of prior distributions in a Bayesian analysis. Several possibilities exist, and here we illustrate a solution for a reparameterization of the variance-covariance matrix in terms of temporal SDs and process correlation coefficients, as recommended in a recent simulation study (Riecke et al., 2019a). In addition, we briefly discuss some alternatives at the end of this section. JAGS permits direct specification of the variance-covariance matrix of a multivariate normal distribution using `dmnorm.vcov`. We parameterize the variance-covariance matrix in terms of the temporal SDs and correlation coefficients and then place suitable uniform priors on both. We analyze a further variant of the woodchat shrike data set that we simulated under an assumption of positive correlation between juvenile and adult survival and between adult survival and productivity. We load these data, create the m-array, bundle them into a list, and write our model.

```
library(IPMbook); library(jagsUI)
data(woodchat64)
marr <- marrayAge(woodchat64$ch, woodchat64$age)

# Bundle data
jags.data<- list(marr.j=marr[,,1], marr.a=marr[,,2], n.occasions=dim(marr)[2],
  rel.j=rowSums(marr[,,1]), rel.a=rowSums(marr[,,2]), C=woodchat64$count, J=woodchat64$J,
  B=woodchat64$B)

# Write JAGS model file
cat(file="model4.txt", "
model {
# Priors and linear models
for (t in 1:(n.occasions-1)){
  logit(sj[t]) <- eps[1,t]
  logit(sa[t]) <- eps[2,t]
  log(f[t]) <- eps[3,t]
  eps[1:3,t] ~ dmnorm.vcov(mu[], sigma2[,])
  p[t] <- mean.p
}

mu[1] <- logit(mean.sj)
mu[2] <- logit(mean.sa)
mu[3] <- log(mean.f)

mean.sj ~ dunif(0, 1)
mean.sa ~ dunif(0, 1)
mean.f ~ dunif(0, 10)
mean.p ~ dunif(0, 1)

# Reparameterize sigma2 in terms of SD and rho
for (i in 1:3){
  sigma2[i,i] <- sigma[i] * sigma[i]
}
for (i in 1:2){
  for (j in (i+1):3){
    sigma2[i,j] <- sigma[i] * sigma[j] * rho[i+j-2]
    sigma2[j,i] <- sigma2[i,j]
  } #j
} #i

# Specify priors for SD and rho
for (i in 1:3){
  sigma[i] ~ dunif(0, 5)
  rho[i] ~ dunif(-1,1)
}

sigma.res ~ dunif(0.5, 100)
tau.res <- pow(sigma.res, -2)

# Population count data (state-space model)
# Model for the initial population size: uniform priors
N[1,1] ~ dunif(1, 300)
N[2,1] ~ dunif(1, 300)

# Process model over time: our model for population dynamics
for (t in 1:(n.occasions-1)){
  N[1,t+1] <- f[t] / 2 * sj[t] * (N[1,t] + N[2,t])
  N[2,t+1] <- sa[t] * (N[1,t] + N[2,t])
}

# Observation model
for (t in 1:n.occasions){
  C[t] ~ dnorm(N[1,t] + N[2,t], tau.res)
}
```

```
# Capture-recapture data (CJS model with multinomial likelihood)
# Define the multinomial likelihood
for (t in 1:(n.occasions-1)){
  marr.j[t,1:n.occasions] ~ dmulti(pr.j[t,], rel.j[t])
  marr.a[t,1:n.occasions] ~ dmulti(pr.a[t,], rel.a[t])
}
# Define the cell probabilities of the m-arrays
for (t in 1:(n.occasions-1)){
  # Main diagonal
  q[t] <- 1 - p[t]            # Probability of non-recapture
  pr.j[t,t] <- sj[t] * p[t]
  pr.a[t,t] <- sa[t] * p[t]
  # Above main diagonal
  for (j in (t+1):(n.occasions-1)){
    pr.j[t,j] <- sj[t] * prod(sa[(t+1):j]) * prod(q[t:(j-1)]) * p[j]
    pr.a[t,j] <- prod(sa[t:j]) * prod(q[t:(j-1)]) * p[j]
  } #j
  # Below main diagonal
  for (j in 1:(t-1)){
    pr.j[t,j] <- 0
    pr.a[t,j] <- 0
  } #j
} #t
# Last column: probability of non-recapture
for (t in 1:(n.occasions-1)){
  pr.j[t,n.occasions] <- 1-sum(pr.j[t,1:(n.occasions-1)])
  pr.a[t,n.occasions] <- 1-sum(pr.a[t,1:(n.occasions-1)])
}

# Productivity data (Poisson regression model)
for (t in 1:(n.occasions-1)){
  J[t] ~ dpois(f[t] * B[t])
}
}
")

# Initial values
inits <- function(){list(mean.p=runif(1, 0, 0.5))}

# Parameters monitored
parameters <- c("mean.sj", "mean.sa", "mean.p", "mean.f", "sj", "sa", "f", "N", "sigma.res",
  "sigma2", "rho")

# MCMC settings
ni <- 60000; nb <- 10000; nc <- 3; nt <- 10; na <- 2000

# Call JAGS (ART 11 min), check convergence and summarize posteriors
out4 <- jags(jags.data, inits, parameters, "model4.txt", n.iter=ni, n.burnin=nb, n.chains=nc,
  n.thin=nt, n.adapt=na, parallel=TRUE)
traceplot(out4)
print(out4, 3)

                mean     sd    2.5%     50%   97.5% overlap0     f  Rhat n.eff
 [ ...output truncated... ]
sigma.res     11.305  4.138   3.261  11.057  20.272    FALSE 1.000 1.004   563
sigma2[1,1]    0.044  0.031   0.005   0.037   0.123    FALSE 1.000 1.003   684
sigma2[2,1]    0.019  0.017  -0.009   0.016   0.059     TRUE 0.912 1.002  1551
sigma2[3,1]    0.001  0.015  -0.030   0.001   0.030     TRUE 0.534 1.003   839
sigma2[1,2]    0.019  0.017  -0.009   0.016   0.059     TRUE 0.912 1.002  1551
sigma2[2,2]    0.045  0.036   0.004   0.037   0.139    FALSE 1.000 1.008   395
sigma2[3,2]    0.017  0.017  -0.011   0.015   0.054     TRUE 0.890 1.004   832
sigma2[1,3]    0.001  0.015  -0.030   0.001   0.030     TRUE 0.534 1.003   839
sigma2[2,3]    0.017  0.017  -0.011   0.015   0.054     TRUE 0.890 1.004   832
sigma2[3,3]    0.047  0.022   0.020   0.043   0.101    FALSE 1.000 1.001 15000
rho[1]         0.486  0.324  -0.250   0.535   0.948     TRUE 0.912 1.005   420
rho[2]         0.030  0.307  -0.548   0.029   0.637     TRUE 0.534 1.004   582
rho[3]         0.406  0.309  -0.267   0.440   0.905     TRUE 0.890 1.008   255

round(out4$mean$sigma2, 3)    # Posterior means of var-covar matrix
      [,1]  [,2]  [,3]
[1,] 0.044 0.019 0.001
[2,] 0.019 0.045 0.017
[3,] 0.001 0.017 0.047
```

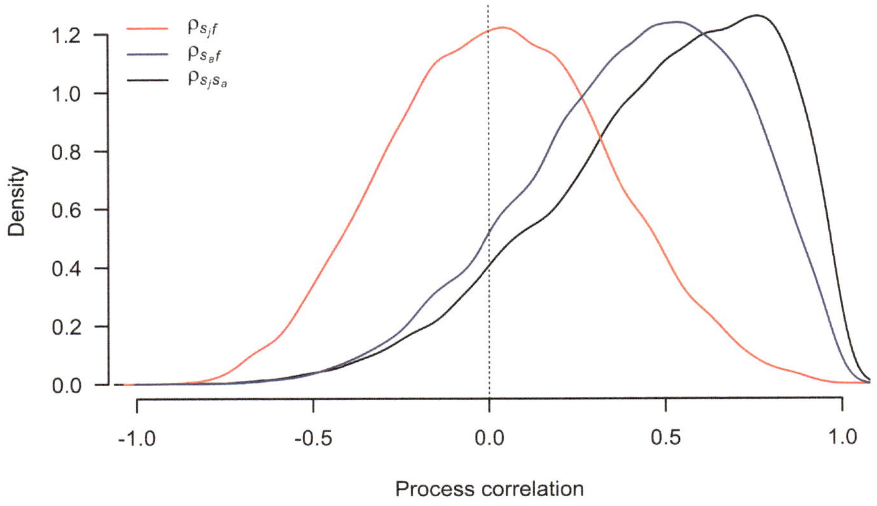

FIGURE 6.6

Posterior distributions of three process correlation coefficients in a simulated woodchat shrike data set. The cross-correlation between juvenile and adult survival is clearly positive, while the cross-correlations between adult survival and productivity and between juvenile survival and productivity are centered more closely near zero (emphasized by the vertical dotted line).

The model converges relatively quickly. The posterior distributions of the three process correlations suggest strong positive cross-correlations for both juvenile and adult survival and productivity and adult survival but indicates independence between productivity and juvenile survival (Fig. 6.6). The magnitude of the temporal process variation appears to be similar for all three parameters. However, we must keep in mind that comparison of temporal variabilities of different probability parameters is challenging because they might be constrained by the maximum possible variability, which is a function of the mean (for probabilities; Gaillard and Yoccoz, 2003), and because they may be expressed on different scales (Altwegg et al., 2007).

An appealing aspect of this approach to the modeling of a variance-covariance matrix is that we can specify priors directly for biologically meaningful and interpretable quantities: temporal process variation and cross-correlations. Hence, it would appear natural to use uniform priors on a suitable range for the temporal standard deviation and on the $(-1, 1)$ range for the process correlations. However, for the latter, there is a twist: owing to the need for a variance-covariance matrix to be positive-definite, the possible values of the correlations are constrained in a complicated way (Barnard et al., 2000; Budden et al., 2007). Hence, $U(-1, 1)$ priors are no longer possible when the dimension of the matrix exceeds 2. The MVN distribution implemented in JAGS in `dmnorm.vcov` enforces the positive-definite constraint through the likelihood so that you do not need to enforce it through the prior distribution (M. Plummer, pers. comm., via T. Riecke, also pers. comm.). Hence, when specifying a $U(-1,1)$ prior for process correlations in a MVN likelihood, we are really specifying the "most uniform prior possible given the dimensions of the covariance matrix."

There are a number of alternative ways in which we can specify priors for the variance-covariance matrix $\boldsymbol{\Sigma}$ in an MVN distribution. A typical choice has long been an inverse Wishart distribution, which is the conjugate distribution to the MVN (Link and Barker, 2005; Kéry and Schaub, 2012). However, this prior tends to be informative (Alvarez et al., 2016; Riecke et al., 2019a). An alternative and more flexible choice is the *scaled* inverse Wishart (Gelman and Hill, 2007; Alvarez et al., 2016), which is implemented in JAGS as `dscaled.wishart`. This distribution is a multivariate generalization of a scaled gamma distribution and requires a scale parameter s_i for each demographic rate i and degrees of freedom df. It enforces a half-t distribution for σ_i with scale parameter s_i and degrees of freedom df. With $df = 2$, all process correlations ρ have priors that are marginally uniform (but with unknown range), whereas for greater values of df, more prior mass is concentrated near 0 (Plummer, 2017).

Process correlations among demographic rates have repeatedly been estimated using IPMs (Schaub et al., 2013; Eacker et al., 2017; Koons et al., 2017; Woodworth et al., 2017; Tenan et al., 2017; Genovart et al., 2018), with all of them using inverse Wishart priors (for code, e.g., see the work of Schaub et al., 2013). However, based on results of simulations by Alvarez et al. (2016) and Riecke et al. (2019a), we now recommend using the approach described above.

6.5 ESTIMATION OF POPULATION STRUCTURE

A substantial practical benefit of IPMs is that they permit obtaining estimates of stage-structured population sizes. That is, they enable you to get detailed insights into the stage- or age-structure of a population and how that structure changes over time. We have seen this already in our simple woodchat shrike examples, where we estimated the annual number of 1-year-old and older females although female age in the field could not be determined, resulting in an "unstructured" population count. This stage structure could also be enlarged if we wished to do so, for example by including 1-year-old, 2-year-old, 3-year-old, and older individuals. The estimates of stage-structured population sizes at time t are based on the Markovian process model and thus are functions of the demographic rates and the stage-structured population sizes at time $t-1$. Consequently, they are conditional on the specified process model and do not account for potential uncertainty about the population structure assumed in the process model. This is an important caveat for our inferences about population structure. If you are uncertain about the structure of the process model (e.g., whether all or only a fraction of the 1-year-old females start to reproduce in our shrike example), you can perform sensitivity analyses and compare results from IPMs with different process models. In the absence of any clues about which population model is more likely, equal-weight model averaging (Dormann et al., 2018) may perhaps yield more robust estimates of population structure.

Knowledge about the population structure and its change over time is useful for different purposes (reviewed by Plard et al., 2019), and here we briefly summarize four of them. First, understanding how population structure changes with varying population size can suggest possible mechanisms of population regulation (see also Chapter 9). Schaub et al. (2013) showed that the proportion of immigrants in a population of red-backed shrikes (*Lanius collurio*) declined with increasing population size, indicating that immigrants contributed to population regulation.

Second, knowledge of population structure permits a better understanding of possible mechanisms of density dependence. Typically, not all individuals in a population contribute equally or are equally sensitive to density dependence. For example, density dependence is primarily mediated via effects in 1-year-old individuals in great tits (*Parus major*), and the recruitment rate of younger individuals occurs under stronger density dependence control than that for older individuals (Gamelon et al., 2016). Such insights are only possible if the population structure and its changes over time are known, and IPMs can offer this information. We return to IPMs with density dependence in Chapter 8.

Third, nonbreeders (often called floaters) can make up a significant part of a population and may strongly affect its dynamics and regulation (Kenward et al., 2000; Lopez-Sepulcre and Kokko, 2005; Katzner et al., 2011; Penteriani et al., 2011; Lee et al., 2017). However, estimation of the number of nonbreeders in a population is typically very challenging because they are often not present in the surveyed breeding population and therefore escape possible detection. IPMs allow estimation of the number of floaters in a population and assessment of their role for population fluctuations and regulation, even if observed population counts are restricted to reproducing, i.e., nonfloating, individuals. To account for them, the population model must explicitly include a stage of floaters, or nonbreeders. In addition, estimates of age-specific breeding probabilities are required, which typically originate from multistate capture-recapture data of individuals of known age where the observed states are "nonbreeding" and "breeding." Re-encounters can also be restricted to breeding individuals only (Pradel et al., 1997). Another possible source of information about age-specific breeding probabilities is radio tracking data (Kenward et al., 1999; Oppel and Powell, 2010). McCrea et al. (2010) developed an IPM for cormorants (*Phalacrocorax carbo*) and were able to estimate the number of nonbreeders in the population by treating them as an unobservable state (Kendall and Nichols, 2002; Lebreton et al., 2009). In a similar IPM, Pfeiffer and Schaub (in prep.) studied the impact of floaters on population dynamics of red kites (*Milvus milvus*). Because floaters were typically unobserved, the authors had to make some restrictive assumptions to estimate the parameters associated with this stage, e.g., that survival only differs by age, but not between floater and breeding birds. Floating red kites were found to be about as numerous as breeding individuals in this population, and floaters affected population dynamics most likely via competition for food during the chick-rearing period. In a similar model for the bearded vulture (*Gypaetus barbatus*) in the Pyrenees, Margalida et al. (2020) found that floaters were about equally numerous as were the territorial age classes. Finally, in the case study on the Jura mountain peregrine falcon population (*Falco peregrinus*, Chapter 12), we estimated the proportion of floaters using the same principles. Interestingly, Newton and Rothery (2001) used population count data and information about age-specific survival of sparrowhawks (*Accipiter nisus*) to estimate floater population size. They constructed a population model including floaters and known demographic rates and tuned the probability of an individual to be a floater until the predicted number of breeders matched the observed number of bre eders. In a sense, an IPM does exactly the same, but in a statistically more formal manner.

Fourth, modeling transient dynamics requires knowledge about population structure and its change over time (Koons et al., 2005; Ezard et al., 2010; Stott et al., 2011). By transient dynamics, we denote fluctuations of a population when away from its stable stage distribution (Section 3.3.1). Because populations are rarely truly in a stable stage distribution (Clutton-Brock and Coulson, 2002; Hastings, 2004), they are likely to exhibit transient dynamics most of the time. Understanding transient dynamics is therefore of paramount importance for our understanding of population dynamics in general (Hastings, 2001; Hastings, 2004; Ozgul et al., 2009). IPMs provide estimates of stage-specific population sizes and therefore offer the possibility to study transient dynamics directly. An example is the retrospective analysis of the realized population growth rate to identify the demographic drivers of observed population changes (Koons et al., 2016, 2017). This is an important topic, and we illustrate IPM-based transient life-table response experiments in Chapter 9.

6.6 FLEXIBILITY

An important benefit of IPMs is their tremendous flexibility in many respects, e.g., the types of data included, the types of population models adopted, the types of observation models, the degree to which the different data sets may overlap in time, and the possible use of prior information in Bayesian inference. Clearly, you can get a feeling for this flexibility simply by browsing through the case studies of this book, but here we explicitly highlight some features we think are of particular importance.

6.6.1 DIVERSITY OF DATA TYPES COMBINED IN AN IPM

The canonical IPM example in this book combines capture-recapture, productivity, and population count data because these data sets are often collected in population studies. However, IPMs are flexible in the types of data used. We define an IPM as a demographic model that combines data from the population and individual levels (Fig. 5.1), but we have remained silent about what kind of data these may be. Our review identified 118 IPMs on vertebrate populations (excluding fisheries, sorry) published to date with real data with a great diversity of different data types (Table 6.2). Generally, the

Table 6.2 Frequency of use in IPMs of different types of data from the population and individual levels in our review of 118 studies (see Section 1.3).

Type of Data	Number of Studies	Examples
Population level		
Unstructured population counts	77	Abadi et al. (2010b), Lieury et al. (2015)
Unstructured population index	25	Brooks et al. (2004), Péron and Koons (2012)
Structured population counts	6	Lee et al. (2015), Wilson et al. (2016)
Structured population index	1	Eacker et al. (2017)
Age-at-harvest data	2	Conn et al. (2008), Fieberg et al. (2010)
Occupancy data	2	Chandler and Clark (2014), Tempel et al. (2014)
Number of clutches	1	Duarte et al. (2017)
Detection-nondetection data	2	Sanderlin et al. (2019), Sun et al. (2019)
Distance sampling	3	Hoyle and Maunder (2004), Schmidt et al. (2020)
Individual level		
Capture-recapture data	75	Schaub et al. (2007), Paquet et al. (2019)
Ring-recovery data	40	Baillie et al. (2009), Zimmerman et al. (2017)
Mark-resight data	1	Johnson et al. (2010)
Spatially explicit capture-recapture data	3	Chandler and Clark (2014), Chandler (2018), Sun et al. (2019)
Radio tracking data	21	Coates et al. (2018), Ross et al. (2018)
Age/sex ratios	6	Zhao et al. (2019), Saracco and Rubenstein (2020)
Age/sex at death	12	Taylor and Udevitz (2015), Koons et al. (2017)
Hunting bag	2	Lee et al. (2015), Schmidt et al. (2020)
Productivity data (generally)	69	Altwegg et al. (2014), Genovart et al. (2018)
Nest cards (components of productivity)	5	Robinson et al. (2014), Tempel et al. (2014), Schaub et al. (2015)

main condition for possible inclusion of a data set in an IPM is that it must inform at least one parameter that typically belongs to the process (i.e., population) model. It is also possible to integrate data that are informative only about some parameters in the observation models. An example might be some data that help estimation of p in the CJS submodel or of some parameter in the observation model for the population counts (see discussion in Chapter 10 in Kéry and Royle, 2021). Sometimes a data set can be informative about both the individual and population levels. For example, age-at-harvest data can be used at the individual level, as these data provide information about demographic rates, and they can also be used at the population level because they provide information about population size and structure.

6.6.2 UNEQUAL TEMPORAL COVERAGE OF DATA SETS—MISSING VALUES IN CERTAIN YEARS

It is not uncommon for the data sets combined in an IPM to cover different ranges of years. IPMs can still be fit to such data (e.g., Brooks et al., 2004; Mosnier et al., 2015); indeed, the joint analysis is beneficial because under certain assumptions it allows one to estimate parameters of interest for the combined period covered by the data sets. Unequal temporal coverage of data sets results in data gaps in some data sets with respect to the combined period, so we basically have to deal with a missing value problem. Next, we conceptually describe such an analysis with differential temporal coverage of the data sets, illustrate with an example, and look into more details.

The population model must be specified for the entire time period for which we want to make inferences. Obviously, all demographic parameters need to be defined for the same time period as well and will be estimated when we fit the model. In contrast, the specification of the likelihoods that link each data set to one or more parameters needs some adaptations; basically, the time loops must exactly correspond to those time periods for which each data set is available. On fitting such an IPM, population size will be estimable for the entire time period owing to the Markovian structure of the population model and the estimated demographic parameters. Conceptually, this should work even for periods when no data are available at all. However, as we have seen in Section 6.3, demographic rates are not separately estimable in the complete absence of data at the individual level. Similar estimation problems may occur when data are missing during some periods. Therefore, additional modeling assumptions about the demographic rates for those time periods are usually required. For instance, we may have to specify these demographic rates as constant over time or that they follow a deterministic time trend, obey a regression relationship with some known covariates, or vary over time according to random effects.

We next illustrate with an example of how data with unequal temporal coverage can be combined. We assume that we have woodchat shrike data for a total of 20 years but that none of the three data types covers the entire 20-year period. Population counts are available from years 1–18 but with years 5–8 and 15–17 missing (i.e., in addition to years 19 and 20). Productivity data are available from years 7–15. Finally, capture-recapture data are available from years 10–20, but in year 15 only new individuals were ringed, while no resightings were conducted. Fig. 6.7 provides a schematic overview of data set availability. We see that over the years, a variable combination of data sets was available and that in years 5 and 6, no data at all were available.

Here is our recipe for writing code for the IPM with missing data. First, we develop an IPM assuming complete coverage of years 1–20 for all data sets and specify some model for the temporal variation of all demographic rates, e.g., constancy, random effects, or covariate effects. Here we assume temporal random effects. Second, we adapt the length of the time loops in the likelihoods of each data type, i.e., in the observation models for the population count, capture-recapture, and productivity data. There is some serious potential for confusion in this step because the lengths of the time loops and their indices will differ. In our example the length of the observation model in the population count model is

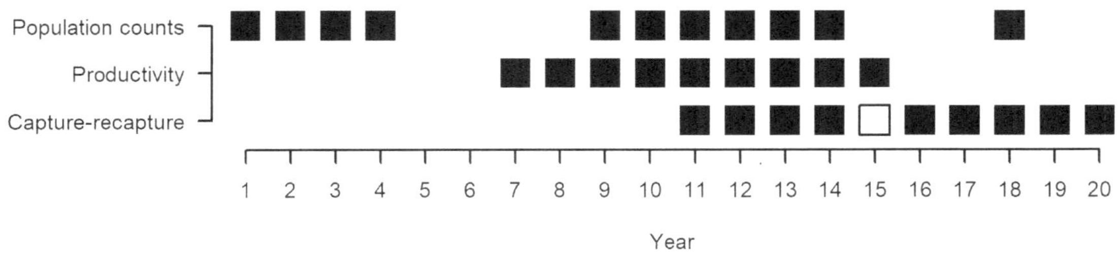

FIGURE 6.7

Fitting an integrated population model to data sets with missing years and temporal mismatch between the data sets: schematic overview of the years with population counts, productivity, and capture-recapture data available in our illustration. Filled squares denote years with data, the open square is the single year when no resighting data were collected, and the absence of any square denotes the complete absence of a data type.

18 years, the length for the capture-recapture data 10 years, and that for productivity data 9 years. Therefore, including the full year range, we need four different variables that define the lengths of periods with available data. We define them in our code as `n.occasionT` for the full study duration and `n.occasionsC`, `n.occasionsCR`, and `n.occasionsP` for the length of the population count, capture-recapture, and productivity data, respectively. A further tricky part is to correctly match the time indices of the demographic rates with the numbering of the years in these loops. In the process model, all demographic rates are defined and therefore have indices covering the full period, which here is from 1 to 19. However, in the likelihoods, they originally have different time indices because these loops are shorter. Therefore, we must modify the time indices of the demographic rates in the three likelihoods. For example, the first productivity parameter in the productivity likelihood must refer to year 7 (Fig. 6.7), and thus we add `6` to the index of productivity. Similarly, the first survival parameter in the capture-recapture likelihood refers to year 11, so we add `10` in the time index of survival. No change in the time index is necessary for the recapture probabilities. However, in year 15, recapture effort was zero, and hence we fix recapture probability at zero in that year. We use variable `recap` as an indicator for years with and without recapture effort. Recapture probability then becomes a function of this categorical variable, and we can give a prior for this parameter in the years with capture effort and fix it to zero in the other years. We load the data, construct the m-array, and define the additional vectors and time scalars in the data bundle statement.

```
library(IPMbook); library(jagsUI)
data(woodchat66)
marr <- marrayAge(woodchat66$ch, woodchat66$age)

# Bundle data
jags.data <- list(marr.j=marr[,,1], marr.a=marr[,,2], rel.j=rowSums(marr[,,1]),
  rel.a=rowSums(marr[,,2]), C=woodchat66$count, J=woodchat66$J, B=woodchat66$B, n.occasionsT=20,
  n.occasionsC=length(woodchat66$count), n.occasionsCR=dim(marr)[2],
  n.occasionsP=length(woodchat66$J), recap=c(rep(1,3),2,rep(1,5)))

# Write JAGS model file
cat(file="model5.txt", "
model {
# Priors and linear models
# Temporal random effects for demographic rates (full time)
for (t in 1:(n.occasionsT-1)){
  logit(sj[t]) <- mu[1] + eps.sj[t]
  eps.sj[t] ~ dnorm(0, tau.sj)
  logit(sa[t]) <- mu[2] + eps.sa[t]
  eps.sa[t] ~ dnorm(0, tau.sa)
  log(f[t]) <- mu[3] + eps.f[t]
  eps.f[t] ~ dnorm(0, tau.f)
}

for (t in 1:(n.occasionsCR-1)){
  p[t] <- mean.p[recap[t]]
}
mean.p[1] ~ dunif(0, 1)             # Prior for years with recapture
mean.p[2] <- 0                      # Fix to zero for year without recaptures

mu[1] <- logit(mean.sj)
mu[2] <- logit(mean.sa)
mu[3] <- log(mean.f)
mean.sj ~ dunif(0, 1)
mean.sa ~ dunif(0, 1)
mean.f ~ dunif(0, 10)

sigma.sj ~ dunif(0, 5)
tau.sj <- pow(sigma.sj, -2)
sigma.sa ~ dunif(0, 5)
tau.sa <- pow(sigma.sa, -2)
sigma.f ~ dunif(0, 5)
tau.f <- pow(sigma.f, -2)

sigma ~ dunif(0.5, 100)
tau <- pow(sigma, -2)

# Population count data (state-space model)
# Model for the initial population size: uniform priors
N[1,1] ~ dunif(1, 300)
N[2,1] ~ dunif(1, 300)
```

```
# Process model over time (full time period)
for (t in 1:(n.occasionsT-1)){
  N[1,t+1] <- f[t] / 2 * sj[t] * (N[1,t] + N[2,t])
  N[2,t+1] <- sa[t] * (N[1,t] + N[2,t])
}

# Observation model (for years with counts only)
for (t in 1:n.occasionsC){
  C[t] ~ dnorm(N[1,t] + N[2,t], tau)
}

# Capture-recapture data (CJS model with multinomial likelihood)
# For years with CR data only
for (t in 1:(n.occasionsCR-1)){
  marr.j[t,1:n.occasionsCR] ~ dmulti(pr.j[t,], rel.j[t])
  marr.a[t,1:n.occasionsCR] ~ dmulti(pr.a[t,], rel.a[t])
}
# Define the cell probabilities of the m-arrays
for (t in 1:(n.occasionsCR-1)){
  # Main diagonal
  q[t] <- 1 - p[t]                 # Probability of non-recapture
  pr.j[t,t] <- sj[t+10] * p[t]    # increase index of survival
  pr.a[t,t] <- sa[t+10] * p[t]    # to match with population model
  # Above main diagonal
  for (j in (t+1):(n.occasionsCR-1)){
    pr.j[t,j] <- sj[t+10] * prod(sa[((t+1):j)+10]) * prod(q[t:(j-1)]) * p[j]
    pr.a[t,j] <- prod(sa[(t:j)+10]) * prod(q[t:(j-1)]) * p[j]
  } #j
  # Below main diagonal
  for (j in 1:(t-1)){
    pr.j[t,j] <- 0
    pr.a[t,j] <- 0
  } #j
} #t
# Last column: probability of non-recapture
for (t in 1:(n.occasionsCR-1)){
  pr.j[t,n.occasionsCR] <- 1-sum(pr.j[t,1:(n.occasionsCR-1)])
  pr.a[t,n.occasionsCR] <- 1-sum(pr.a[t,1:(n.occasionsCR-1)])
}

# Productivity data (Poisson regression model)
# For years with productivity data only
for (t in 1:(n.occasionsP-1)){
  J[t] ~ dpois(f[t+6] * B[t])
}

# Derived quantities: total population size (full time period)
for (t in 1:n.occasionsT){
  Ntot[t] <- N[1,t] + N[2,t]
}
}
")

# Initial values
inits <- function(){list(mean.p=c(runif(1, 0, 0.5), NA))}

# Parameters monitored
parameters <- c("mean.sj", "mean.sa", "mean.p", "mean.f", "sigma.sj", "sigma.sa", "sigma.f", "sigma",
  "sj", "sa", "f", "N", "Ntot")

# MCMC settings
ni <- 40000; nb <- 10000; nc <- 3; nt <- 30; na <- 2000

# Call JAGS (ART <1 min), check convergence and produce figure
out5 <- jags(jags.data, inits, parameters, "model5.txt", n.iter=ni, n.burnin=nb, n.chains=nc,
  n.thin=nt, n.adapt=na, parallel=TRUE)
traceplot(out5)
```

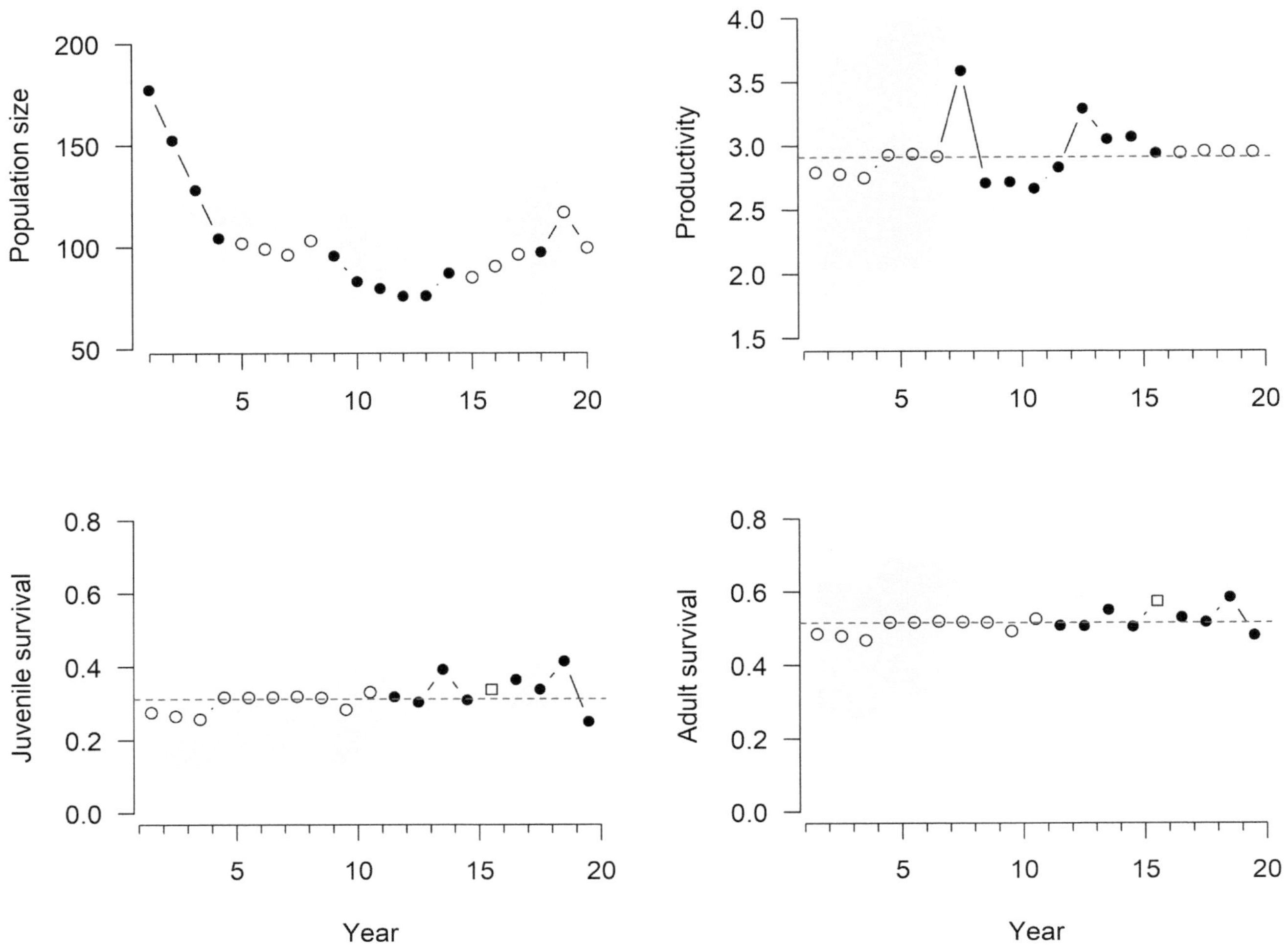

FIGURE 6.8

Estimates of population size and demographic rates in a simulated woodchat shrike population when there are some "missing years" (according to the scheme in Fig. 6.7). Filled circles show the posterior means with the associated 95% credible intervals shown as a gray layer. Dashed red lines show the estimated hypermeans of the random-effects model for the demographic rates. Open circles refer to years without data, and the open square indicates the single year without resighting data.

We do not present a posterior summary but instead produce a plot with estimates of population size and demographic rates for the full 20 years (Fig. 6.8). We see that in years when some data are missing, estimation uncertainty increases. For instance, look at survival probability at the beginning or productivity at the beginning and end of the 20-year period. Uncertainty also increases for the population size estimates in the years without any population count data. One fundamental difference between the uncertainty of population size N and that of the demographic rates is that the former increases over time, whereas the latter does not. The reason for this is the Markovian structure of the population model, which leads to an increase in the uncertainty of N over time. In contrast, our model for the demographic parameters has simple random effects and lacks any autoregressive structure (although we have the flexibility to do that—see also Chapter 12 and Johnson and Hoeting, 2003, for examples). Therefore, estimation uncertainty in one year does not carry over to the following year and so forth as it does for N. Point estimates of the demographic rates in "interior missing years" are generally similar to the hypermeans of a parameter, though they show some temporal variability because they are also informed by the population counts. In contrast, for "trailing missing years," i.e., for a prediction into the future when no information from population counts is available, these point estimates are exactly identical to the hypermeans (see Chapter 10).

Thus, dealing with unequal temporal coverage of data sets and hence with "missing years" is not particularly difficult in Bayesian IPMs and offers much flexibility, but it may require some additional assumptions. In essence, we must specify a model for the parameters associated with these years. Here we made the stationarity assumption for the demographic processes, which often seems reasonable. When population size is predicted into the future (as we will see in Chapter 10), we typically apply exactly the same principles as we have seen here, and the behavior of the estimates will be similar. The only difference from our example is that for a projection into the future, i.e., a forecast, we have no "real" data at all.

6.6.3 TIME POINTS OF DATA COLLECTION DO NOT MATCH

The core of an IPM is usually a matrix population model that we either specify for a pre- or a postbreeding "census." Typically, we collect all data such that they match the "census" period. In all our examples so far, the data were assumed to be collected during the reproductive period, so that they all agree exactly with a population model for either a pre- or a postbreeding "census." Thus, we monitored productivity and counted the number of breeding pairs in a given reproduction period (year) and collected capture-recapture data in such a way that we obtain annual estimates of survival.

However, it is not uncommon that different data sets in an IPM are *not* all collected during the same period of the year. For example, we may collect population count data during the breeding period and mark individuals, but obtain age-ratio data in autumn when hunting takes place. To use such data in an IPM, we then must account for a possibly different age-ratio during breeding and in autumn. This temporal scaling is best achieved via the estimation of seasonal survival—in the sketched example we would need to estimate survival from breeding to autumn and from there to the subsequent breeding period again. Often, seasonally different survival probabilities cannot be estimated and only an estimate of annual survival (s) is available. Under the assumption of constant survival over a year, this could then be scaled to obtain an estimate of seasonal survival as $s^{\frac{u}{12}}$, where u is the length in months of a season. See Koons et al. (2017) for an example with lesser scaup (*Aythya affinis*).

Sometimes population or capture-recapture surveys are conducted multiple times during a year. To use such data, we must then specify a seasonal matrix population model. For example, if a population is counted during reproduction in spring and then again in winter, we will specify two seasonal transition matrices that alternate within and are repeated across the years. The first transition matrix describes the dynamics between spring and winter and therefore includes productivity and survival. The second describes the dynamics between winter and the start of the breeding season in the next spring, and therefore only includes survival. So far, there are very few published examples of such seasonal IPMs: see Hostetler et al. (2015), Zimmerman et al. (2017), and Rushing et al. (2017). Our case study on a grouse population in Chapter 15 also illustrates such a seasonal IPM.

6.6.4 USING ESTIMATED INDICES INSTEAD OF COUNTS FOR POPULATION-LEVEL DATA

Sometimes a population is not counted, but rather an index of its size with an associated measure of uncertainty is available through the use of some statistical model for the original, raw data. While it is generally better to use the raw data directly if these are available (Maunder, 2004), we may sometimes still need to use indices. Perhaps they are all we have or else incorporation in the IPM of the full analysis of the raw data may be too complex or computationally expensive.

When using an index of population size in an IPM, it is obvious that we obtain an estimated population index, but not an estimate of true population size. If no measure of uncertainty associated with the population index is available, our IPM does not need any changes compared with an IPM that uses raw population counts; we just replace the counts by the indices in the data and then must acknowledge that we will somewhat understate the uncertainty in our inferences due to omission of the estimation uncertainty in the index. However, we will perhaps not want to include demographic stochasticity in our IPM because its effects truly make sense only when the units of the population counts correspond to biological individuals. In contrast, an index can be arbitrarily scaled.

If a measure of uncertainty is available for an index, both should be incorporated in the observation model for the population-level data. Let's denote the estimated index in year t with $\hat{i}_t$, with associated uncertainty as $SE\left(\hat{i}_t\right)$. We could then use a normal distribution and model the measurement error in the index as $\hat{i}_t \sim Normal\left(N_t, SE\left(\hat{i}_t\right)^2\right)$ where N_t is the "true" population size as it appears in the matrix population model. This approach has been implemented in several IPMs (Robinson et al., 2014; McCaffery and Lukacs, 2016; Koons et al., 2017; Ross et al., 2018; Zhao et al., 2019). However, because the population-level error in the state-space model is in fact the residual error of the entire IPM (Section 5.4.2), such a specification implies that the whole uncertainty in the IPM is determined by the index estimation error $SE\left(\hat{i}_t\right)$, which is unlikely to be true. A better approach therefore is to specify an observation model that allows additional noise to accommodate such lack of fit in the IPM (see Section 4.3.3). We then first estimate a "true" annual population index y_t as $\hat{i}_t \sim Normal\left(y_t, SE\left(\hat{i}_t\right)^2\right)$. Second, we describe how it relates to the "true" population size, $y_t \sim Normal(N_t, \sigma^2_{res})$, where

$SE\left(\hat{i}_t\right)^2$ is the estimation error variance of the index estimate and σ^2_{res} is the additional, residual error term of the IPM (i.e., excluding the first-stage error represented by the SEs). This approach to the incorporation of measurement error in the observation model of an IPM has been used by Fay et al. (2019) and Jacobson et al. (2020), and by McCarthy and Masters (2005) in a general context.

6.6.5 OBSERVATION MODELS FOR POPULATION-LEVEL DATA

There is a range of options for defining the observation models for the population count data. Ideally, this part of the IPM should reflect the way in which the population count data were collected. However, this observation model is usually the lowest-level model in an IPM defined through a hierarchical model (see Section 4.3). Hence, it represents the residual and therefore absorbs any lack of fit in all the other parts of the model. This has implications for the choice of the observation model.

We can distinguish at least two ways in which the data at the population level can be collected. First, it may not be possible to estimate detection probability, as when there is only a single population count per year (and site) or when an index of population size available. Alternatively, counts may be conducted in such a way that it is possible to estimate the true population size, correcting for imperfect detection (and possibly also false positives) using any of a number of specialized sampling protocols such as distance sampling (Buckland et al., 2001; Kéry and Royle, 2016; Schmidt and Robison, 2020), double observer approaches (Nichols et al., 2000), or replicated counts (Royle, 2004).

Here, we focus on the former situation when we cannot formally estimate imperfect detection. In this case, we can use different distributions to model population count or index data. The normal distribution is used very often in IPMs, but alternatives include the Poisson or log-normal distributions. Their implementation as an observation model for the population-level data in an IPM is fairly trivial and only small adaptations are necessary when you start from a template with a normal distribution. For the Poisson distribution no dispersion parameter is needed, and for the log-normal distribution we simply model the log of the population counts or indices and therefore transform these input data first. The dispersion parameter of the log-normal distribution refers to the logarithm of the population counts and therefore we should select a prior accordingly.

No doubt, at this stage you will wonder about the effects of the choice of the observation model for the population counts on the inferences under an IPM. Because the observation model links the observed population counts with an estimate of true population size we expect that if changes occur in the inferences, they will primarily be seen in the estimate of total population size. We examine these questions using simulated data and again use the woodchat shrike data set. We fit three different IPMs with temporal random effects on all demographic rates and with normal, Poisson, and log-normal observation models for the population count data.

Fig. 6.9 shows that the estimates of total population size and their associated uncertainties are remarkably similar under all three observation models, although the credible intervals tend to be narrowest with the Poisson observation model. The

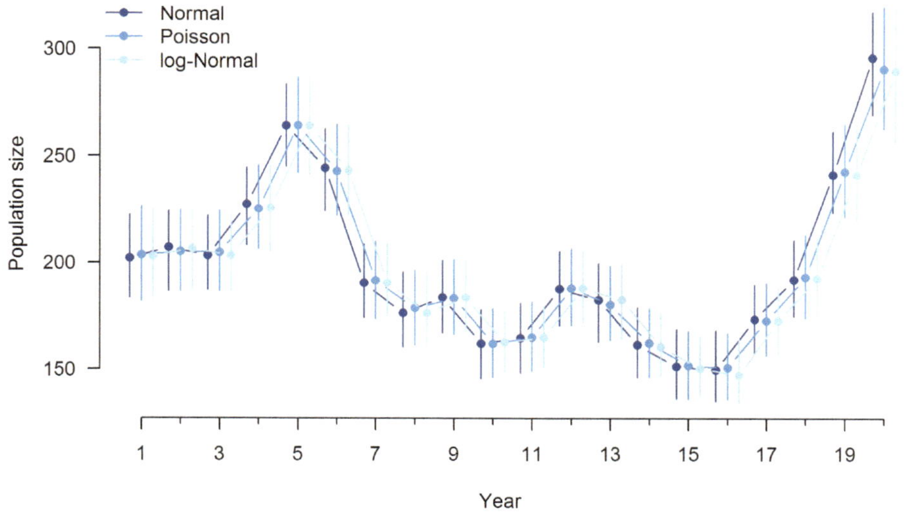

FIGURE 6.9

Estimates of total population sizes under three IPMs with different observation models in the woodchat shrike example (posterior means with 95% CRI).

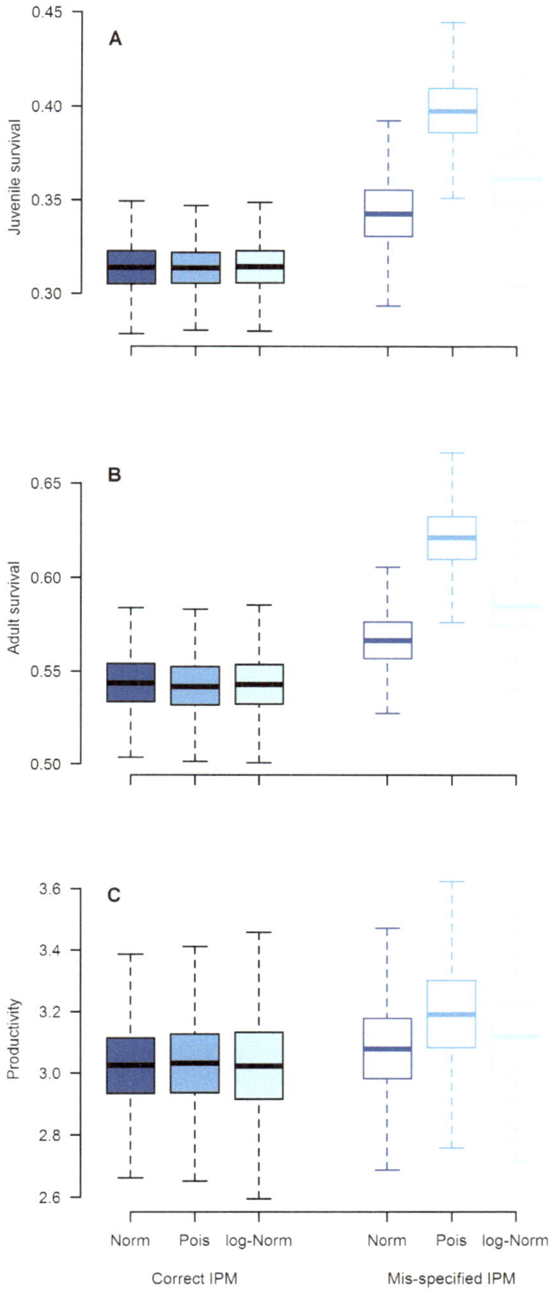

FIGURE 6.10

Boxplots of the posteriors of demographic rates ((A) juvenile survival, (B) adult survival, and (C) productivity) obtained under three different IPMs, each with correct and misspecified process models (see text for details) and normal (Norm), Poisson (Pois), or log-normal (log-Norm) observation models for the population counts.

estimates of the demographic rates are also nearly identical under all models (Fig. 6.10). These results confirm our common practical experience that these three observation models typically lead to nearly identical inferences, although sometimes convergence may be better for one than for another, and often the Poisson observation model converges best.

However, the choice of the observation model can matter when there is lack of fit in a model component. We present the effects of model misspecification in more detail in Chapter 7, but here show one example in the context of observation models.

Because the observation error for the population counts represents the residual error, we would expect it to increase when there is lack of fit. Observation models with a free dispersion parameter (e.g., normal and log-normal) are more flexible in this respect than observation models without a dispersion parameter, such as the Poisson distribution. Therefore, we expect that the former would behave in a different manner than the latter when the model does not fit in some component. Let us corroborate these expectations by fitting IPMs with different observation models and with a misspecified process model. We continue working with the same data set, and as an illustration, misspecify the process model by assuming that first-year individuals do not reproduce. In this case, the misspecified process equation for 1-year-old individuals becomes

```
N[1,t+1] <- f[t] / 2 * sj[t] * N[2,t]
```

We write the BUGS code for these models, fit them to the same data as before, and inspect the population trajectories. Estimated population size under an IPM with Poisson observation model closely follows the observed population counts and true population size (Fig. 6.11). In contrast, there is a severe mismatch in most years between estimates and counts under the two IPMs with a separate dispersion parameter in the observation model (i.e., the normal and log-normal). Fig. 6.10 shows estimates of demographic rates and compares them with those from the three IPMs with the correctly specified process model. While the means of all three demographic rates are essentially identical for the different observation models when the process model is specified correctly, we see clear differences when the process model is misspecified. Posterior means are much higher under the Poisson observation model, whereas under the normal observation model they agree more closely with the posterior means from the correctly specified IPMs. The misspecified IPM with log-normal observation errors also produces higher posterior means of demographic rates but less strongly than in the IPM with the Poisson observation model.

We believe these results can be understood by the differential weighting of the information from the data sets combined in the IPM. With a correctly specified process model, information about the demographic rates is identical in the demographic data and in the population count data, and thus the weighing has no impact on the posterior means. However, this changes when one model component in the IPM is misspecified. In our example, the process model is misspecified, but the demographic data and population counts contain information about the correct dynamics and demographic rates. The predicted dynamics are based on the correct demographic rate, but using the misspecified population model, they differ from the observed dynamics based on the population counts. Conflicting information is averaged in the IPM, and the weights implied in these averages become important. These weights change with the magnitude of the residual error in the IPM, and hence the type of observation model becomes important. In the observation models with a free dispersion

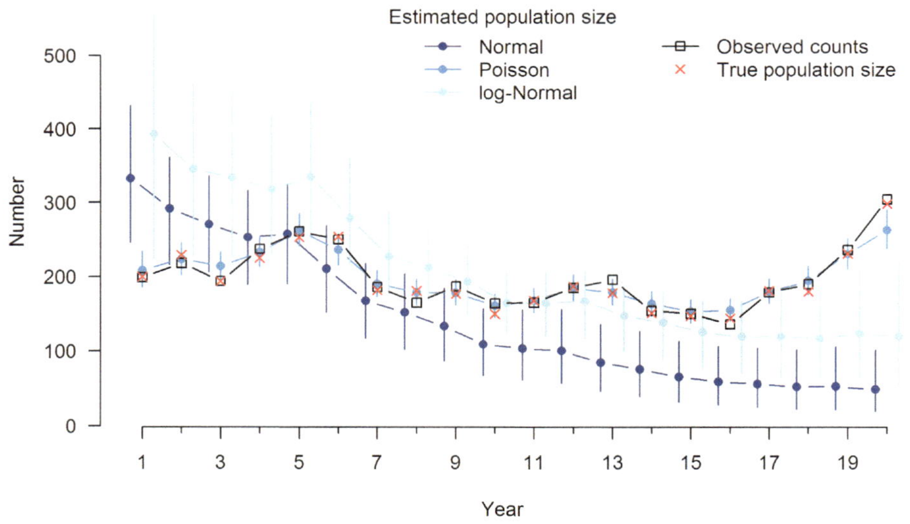

FIGURE 6.11

Estimates of total population size (posterior means with 95% CRIs) under three misspecified IPMs with different observation models in the woodchat shrike data. All three population models erroneously assume that shrikes only start to reproduce in year 2. For comparison, the observed population counts and true population size are also shown.

parameter (i.e., normal or log-normal), the weight of the count data declines in a misspecified IPM. This is achieved by a strong increase in the residual standard deviation in the observation model. In our example, in the IPM with the correctly specified process model and normal observation model, it is estimated at 17.0, but there is a fivefold increase to 92.4 in the IPM with the misspecified process model. The estimates of all demographic rates are much less sensitive to model misspecification because the information about them comes mostly from the capture-recapture and productivity data.

When the Poisson observation model is adopted, the information in the population counts is not downweighted. The information about the demographic rates extracted from the population counts is conditional on the specified process model, which when misspecified does not allow reproduction of 1-year-old individuals. Hence, estimates of the demographic rates must be increased to "get the observed population growth rate right." The result then is an average between the information contained in the demographic data (which would tend to give "correct" estimates) and the information from the population counts (which will lead to distorted estimates).

Alternatively, we can describe this behavior as follows: the demographic data (capture-recapture and productivity) provide "correct" information about the demographic rates, whereas in our case, the population count data provide biased information because the information extracted is conditional on the misspecified process model. Under an observation model with a free dispersion parameter, the residual variance will blow up. As a consequence, the information in the population counts will be downweighted, and thus the estimates of the demographic rates will contain almost entirely information stemming from the demographic data sets. The Poisson distribution, which lacks a free dispersion parameter, does not have this flexibility. Consequently, there is no downweighting of the information contained in the population counts. The resulting estimates of the demographic rates are therefore much more strongly affected by the erroneous information extracted from the population counts.

Thus, the choice of the observation model affects the weight of information about the demographic processes that comes from the data at the population level (i.e., population counts), and this can be problematic if the model is misspecified. It seems that in this sense, using the normal distribution in the observation model is a safer option than a Poisson because it allows downweighting the information from the counts. This was beneficial for the estimation of the demographic rates in our example when the process model was misspecified. However, if there is only a misspecification in the models for the demographic rates (e.g., if we fail to account for immediate trap-response in the capture-recapture data), we may *not* want to downweight the information in the population counts but rather the information in the capture-recapture data. The Poisson observation model may be a better option in that case. Self-evidently, we should always try to fit a correctly specified IPM (Maunder and Piner, 2017), as the weighting of the data sets (likelihoods) then has no impact on the posterior means. We therefore must be able to diagnose lack of fit and identify the causes for any lack of fit so that it can be remedied. Strong sensitivity of estimates to the chosen observation model for the likelihood of the population counts may give an indication for lack of fit in an IPM. We will see more on the effects of misspecified IPMs and on diagnostics in Chapter 7.

A further flexible feature in the observation model of the population-level part of an IPM is the option to include structured population counts (Tavecchia et al., 2009; Johnson et al., 2010; Lee et al., 2015; Taylor and Udevitz, 2015; Weegman et al., 2016; Wilson et al., 2016; Eacker et al., 2017; Hatter et al., 2017). A structured count is a vector of population counts for different types of individuals, typically age classes. For example, Eacker et al. (2017) distinguished yearling and adult moose (*Alces alces*) in aerial surveys. Structured population counts, especially when age-structured, contain more information about demographic rates than unstructured population counts—indeed, it is possible to estimate demographic rates from a structured population census (i.e., a count without errors) alone without any demographic data (inverse methods, Caswell, 2001, p. 142; Gross et al., 2002; Link et al., 2003; Wielgus et al., 2008; Udevitz and Gogan, 2012; Marolla et al. in prep.). Conceptually, it is easy to modify the observation model for structured population counts—basically, we write a different observation model for each observed class. We have full flexibility in defining these observation models as just discussed, e.g., the dispersion parameters might be identical or different for the different stages. We recommend utilizing structured population counts whenever available, as they may add substantially more information than unstructured counts, especially for the estimation of hidden parameters; see Exercise 6.1 for a demonstration.

6.6.6 INFORMATIVE PRIORS AND SEQUENTIAL ANALYSES

The vast majority of all published IPMs have used Bayesian inference (98 of the 118 studies in our review), presumably because of the ease of implementation of complex models, especially when using BUGS software. However, the ability to include external, quantitative information about some parameters in a Bayesian analysis is another potentially great benefit. Such extra information may stem from the literature, from another study or from experts and is added to the analysis by use of informative priors.

Although most published IPMs have used vague priors, use of all available information about some quantity or process is not only commonplace in scientific learning but just makes tremendous sense, and Bayesian inference provides a formal way to do so in a rigorous way (Low Choy et al., 2009; Lemoine, 2019). In our opinion, such statistical learning *across* a sequence of studies by the use of informative priors is greatly underused in science in general as well as in ecology and wildlife management. Not only do such priors usually lead to increased precision of parameter estimates, but they can also render estimable otherwise unidentifiable parameters. Alternatively, they may also enable you to run very complex analyses (e.g., for large data sets) sequentially by breaking the big problem into smaller pieces, thereby reducing an originally intractable computational burden.

Conceptually, it is straightforward to include informative priors in a Bayesian analysis, and hence in an IPM, but there are at least two practical challenges. First, any prior must be defined by using a named statistical distribution, and it is not always obvious how to translate some knowledge about an estimated parameter into the parameters of such a statistical distribution. Useful guidelines about the formulation of informative priors can be found in Martin et al. (2005), McCarthy and Masters (2005), Low Choy et al. (2009) and Lemoine (2019). Second, you would typically want to evaluate the impact of the priors on the inference by a sensitivity analysis. This requires refitting a model with different priors to gauge the degree to which the estimates (i.e., the posterior in a Bayesian analysis) are influenced by the data and how much by the a priori assumptions in the form of the informative priors. Prior sensitivity analyses should not only be performed when informative priors are used but also be part of a rigorous evaluation process of the results. It is not unusual to find that priors presumed to be vague sometimes turn out to be quite informative.

Only a minority of IPMs in our review (24 of the 98 Bayesian ones) have so far made explicit use of informative priors, and this was almost always done to resolve issues of parameter identifiability. For example, Koons et al. (2017) modeled population dynamics of North American lesser scaup and wanted to include the knowledge that productivity is lower in first-time than in experienced breeders. Productivity was a hidden parameter, and hence, without additional information, it would only have been possible to estimate a single average of productivity (see Section 6.3) but not of age-specific productivity. Therefore, one important and *known* aspect of life-history variation would have been neglected. Koons et al. (2017) therefore used published information about age-specific variation in lesser scaup productivity to determine parameters of a beta distribution ($\boldsymbol{u}$) that reflected the proportional reduction in productivity of first-time relative to experienced breeders, specifically $\alpha \sim Beta(u_1, u_2)$. Moreover, they compiled information about mean productivity (υ) of experienced breeders and expressed that by the parameter of a normal distribution used as an informative prior, i.e., $F_2 \sim Normal(\upsilon, \sigma_v)$. Mean productivity of first-time breeders was then expressed as $F_1 = \alpha F_2$.

In this example, just one demographic process was unknown, and estimation was facilitated using informative priors. However, there are also population models that depend much more heavily on external information expressed as informative priors or even as fixed parameters as the extreme case of an informative prior. Examples are provided by studies that basically use counts, perhaps structured by age but without demographic data. As we have seen in Section 6.3 (Fig. 6.4), demographic rates in such models cannot usually be estimated separately when vague priors are used for the types of IPMs considered in this book. Use of informative priors can render identifiable the parameters of such models (Millar and Meyer, 2000; Thomas et al., 2005; Matthiopoulos et al., 2014).

Informative priors could also be useful in a sequential analysis, where we would first estimate parameters from one or several data sets and then use the results, i.e., the resulting posteriors, to form informative priors for the population model. If the posteriors from the first analyses can be exactly translated into priors that represent a named statistical distribution, the resulting posteriors at the end of such an analytical sequence will be identical to what one would obtain if one analyzed all data at once in a single IPM. Typical reasons for performing sequential analyses instead of fitting one large model all at once include computational benefits or lack of access to the original raw data, as mentioned above.

A final reason that slightly (or weakly) informative, rather than uninformative (or vague) priors are often useful is that they typically greatly enhance the rates of convergence of MCMC algorithms (Hoyle and Maunder, 2004). For instance, restricting the range of priors for SDs (e.g., use of a uniform(0, 10) instead of a uniform(0, 100)), of regression coefficients or of random-effects hypermeans (e.g., normal(0, 0.01) instead of normal(0, 0.0001)) can produce much better mixing or sometimes downright enable the convergence of a model fit in JAGS. These slightly narrower ranges are still reasonably wide and should usually not be considered informative, but it is best to check that using a prior sensitivity analysis, by inspecting the posterior (it should not hit a hard bound when using uniform priors) or by comparison of priors and posteriors. When the reason for our choice of slightly informative priors is simply computational efficiency, we would not want these priors to affect the posteriors in any substantial way.

6.7 SUMMARY AND OUTLOOK

We have stressed several advantages of IPMs that are of great practical importance in analyses of population dynamics in ecology and management. The most relevant is probably that the information about demographic processes from the population count data is exploited, resulting in increased precision of parameters and often also in the ability to estimate hidden parameters, i.e., those about which no explicit data are available. Integrated analyses of data sets ensure that model assumptions and parameter estimates are consistent, that all uncertainty is properly propagated throughout the analysis, and that the correlations among parameters are preserved (Maunder, 2004). Therefore, it makes tremendous sense to jointly analyze different data sets in an IPM. We think that in the future, IPMs will become entirely natural in any analysis of population dynamics where multiple data sets are available, so much so that perhaps the term "integrated population model" will simply be replaced by "population model," because everybody will then be conducting IPMs when they have the right kind of data. Bayesian inference, especially when models are implemented in the current *lingua franca* of Bayesian hierarchical modeling, BUGS, also offers tremendous benefits such as (relatively) simple code and the ability to include prior information.

The price of an IPM is that it is usually more complex than single-data models. Moreover, additional assumptions must be met for sound inference. In Chapter 7, we take a closer look at the assumptions of IPMs and explore the effects of their violation on parameter estimates.

6.8 EXERCISES

1. Investigate the gain in precision when access to structured counts is available compared with when you have simple scalar counts without stage structure. Explore these questions in the context of the woodchat shrike data set, assuming you have capture-recapture data and structured counts, i.e., vectors of counts of 1-year-old and 2-year-and-older individuals. For comparison, also fit a model that uses the unstructured counts, i.e., the sum of the structured counts. Construct a model with constant demographic rates across time but accommodating demographic stochasticity. Use the data that are provided in the solutions.
2. Investigate the effect of the prior for the initial population sizes on parameter estimates, i.e., conduct a prior sensitivity analysis. Use the data set `woodchat6` for this purpose, and study (1) how sensitive parameter estimates are to alternative priors and (2) whether prior sensitivity changes when all data sets (capture-recapture, population counts, productivity) are used or when only capture-recapture data and population counts are used.

CHAPTER

ASSESSMENT OF INTEGRATED POPULATION MODELS

7

CHAPTER OUTLINE

7.1 INTRODUCTION

In Chapters 5 and 6, we extensively discussed the many benefits of integrated population models (IPMs). However, these benefits require that your model is adequate for your data, or in other words, that your IPM is not misspecified for your data sets and the intended use of the model. In this respect, IPMs are like other parametric statistical models—they are based on simplifying assumptions, and the violation of these assumptions could become a problem for your intended use of the model (e.g., prediction or explanation; Shmueli 2010; Tredennick et al., 2021). In part, this depends on which assumption is violated and how bad that violation is. In this chapter, we will take a closer look at the main assumptions made by IPMs, show how we can formally assess them, and make copious use of simulation to study the effects of violating them. For any parametric statistical model, goodness-of-fit assessment is a key for the assessment of assumption violations in a model.

So far, goodness-of-fit (GoF) testing of integrated models in general, and IPMs in particular, is an area that has received only scant research (Piner et al., 2011; Maunder and Punt, 2013; Besbeas and Morgan, 2014; Newman et al., 2014; Maunder and Piner, 2015; Carvalho et al., 2017; Plard et al., 2021). Currently, no omnibus test is available for "the GoF" of an IPM (Newman et al., 2014), and perhaps there never will be one. We emphasize that any integrated model requires assumptions beyond those of nonintegrated models, even for component data sets alone, and these assumptions are necessary for the integration of the component data sets. We also demonstrate that simulation is valuable for studying the effects of assumption violations on parameter estimates and that estimates of hidden parameters are particularly sensitive to assumption violations. Finally, we show how parameter accuracy is affected when models are misspecified in terms of the underlying population dynamics, and we study the effects of under- and overfitting and of misspecified observation models.

IPMs use state-space models to link population dynamics parameters with observed population counts. State-space models decompose the variability of the observed quantity (typically the population counts) into one part due to the change in the system state over time (i.e., the state process) and another part due to the observation error (i.e., the observation process). This variance decomposition is not always straightforward, as each component is often only weakly identifiable (Knape et al., 2011; Besbeas and Morgan, 2017). The estimation of observation variance typically works better in an IPM than in a classical state-space model for count data alone because the IPM includes additional information about

Integrated Population Models. https://doi.org/10.1016/B978-0-12-820564-8.00003-8

the state process, namely, the demographic parameters that are informed by the additional data sets. However, this advantage may be diminished if the state-process model is formulated too flexibly—e.g., by adding hidden parameters or including temporal random effects on many demographic rates, or conversely, by adopting a state model that is too rigid. In the former case, we say that the model is overfitted (contains too many parameters), and variability in the observed population counts may erroneously be attributed to variability in the state process. In the latter case, the model is said to be underfitted (contains too few parameters), and this may lead to biased estimates.

Finally, we assess the effects of a misspecified observation model for population-level data. The population counts or any other population-level data are typically not censuses—i.e., they are subject to various measurement errors. The nature of these errors differs depending on how the data were collected. These errors can be modeled explicitly if the necessary additional information (e.g., repeated counts, distance information, a subset of error-free data) is available. However, we often have only a single population count without any extra information such as distance. Use of a state-space model allows researchers to eliminate some variability due to observation error, but it is not possible to model observation error processes explicitly. This has implications for the interpretion of the estimates of population size. We also assess the effects of temporally variable observation errors on parameter estimates.

Many of the insights we present in this chapter were gained by data simulation, which (among many other things) is a powerful method to study the effects of assumption violations (see Chapter 4 in Kéry and Royle, 2016). However, our conclusions are context-dependent to some degree; hence, it may be useful to run your own simulations tailored to your own data and doubts about model assumptions.

7.2 ASSUMPTIONS OF INTEGRATED POPULATION MODELS

As with any statistical model, an IPM relies on assumptions that must be met sufficiently to produce accurate parameter estimates for sound inference. However, assumptions will never be completely met when empirical data sets are analyzed (remember the famous *"all models are wrong"* quote by George Box), and the crucial question is whether the "wrong" model can still deliver inferences that are useful for the intended purpose of the model (and we believe that it often will!). The validity of a model can be checked by comparing observed data with data predicted by the model: if these are similar, we assume that the model is valid. GoF testing is a formal way of model checking, and many tests are available for different statistical models and for testing different aspects of a model. The goal of GoF testing is to identify a model that does not have assumption violations that would result in systematically erroneous inferences (Conn et al., 2018).

IPMs rely on a range of assumptions, and we find it useful to distinguish them into two sets: the first is the set of assumptions made in each submodel (i.e., the component data likelihoods), and the second is the set required for integrating the component data likelihoods to form the joint likelihood. Formal GoF tests are available for many component data likelihoods. Importantly, if these tests are applied to each submodel inside an IPM, we test both types of assumptions, because IPM parameter estimates are weighted averages (Section 5.3). To test only the first type of assumptions, GoF must be applied to component data sets alone, i.e., outside the IPM. However, this is possible only for individual-level data and not for population-level data, because the state-space model adopted for them is specified in terms of demographic rates that can only be estimated in the IPM. A GoF test for the population-level data within the IPM is, in a sense, an overall GoF test of the IPM. In this chapter, we apply tests only to component data sets within the IPM but add some comments about the other option from time to time.

In the following, we first focus on the assumptions of the component data likelihoods in typical IPMs. We introduce posterior predictive checks as a flexible method to assess the model fit, and as an example, we show how the violation of specific model assumptions can be detected by this technique. We will see that sample size plays a crucial role for the power of any GoF test, including posterior predictive checks. Second, we discuss the two major assumptions needed for the integration of component data sets: the *independence assumption* and what we term the *common demography assumption*. We use simulation to show how a violation of these assumptions can be diagnosed and to illustrate the consequences of such violations on parameter estimates. Throughout, we use IPMs with and without a hidden parameter to emphasize the greater sensitivity to assumption violations of IPMs with hidden parameters.

7.2.1 ASSUMPTIONS MADE FOR COMPONENT DATA LIKELIHOODS

Each submodel (component data likelihood) relies on a set of assumptions. Because many different submodel types can be combined in an IPM (see Chapter 4), we will not address the assumptions for all of them but will focus on the most typical submodels: Cormack-Jolly-Seber (CJS), Poisson regression, and Gaussian state-space models. Classical (frequentist) GoF tests are available for all of them, e.g., Pollock et al. (1985) for the CJS model and Pradel et al. (2003) for multistate models. These could be used, but here we focus on Bayesian methods for model checking. The Bayesian approach offers several possibilities for model checking, such as prior predictive checks, cross-validation, or posterior predictive checks

(Conn et al., 2018). We focus on posterior predictive checks (Gelman et al., 1996; Gelman et al., 2004; Link and Barker, 2010; Gelman and Shalizi, 2013; Lunn et al., 2013; Newman et al., 2014; Conn et al., 2018) because of the (comparative) conceptual simplicity and ease of implementation of this method. We note, however, that posterior predictive checks have been criticized because of their low power to detect model violations (Conn et al., 2018); more powerful GoF tests for IPMs could certainly be formulated (see, e.g., Carvalho et al., 2017; Maunder and Piner, 2017).

7.2.1.1 Principle of Posterior Predictive Checks

The basic idea of posterior predictive checks is this: when a model fits the observed data adequately—i.e., when the underlying model assumptions are met sufficiently well—it should be possible to generate, by simulation, a new replicate data set using the model structure and estimated parameters similar to those of the observed data set. Observed and replicate data sets can be compared graphically (e.g., with histograms), but for a more formal comparison, posterior predictive checks compute one or several *discrepancy measures* or *test statistics* from observed and replicate data sets. Depending on the type of data, different choices that focus on different features of the data are possible. They may help in diagnosing a potential problem which could lead to the formulation of a better fitting model. Comparison of discrepancy measures or tests statistics is made for each Markov chain Monte Carlo (MCMC) draw such that the full uncertainty associated with the parameter estimate is propagated. In practice, a posterior predictive check works by the following steps for each MCMC iteration (i):

- Calculate a discrepancy measure $D(y^{obs}, E(y)_i)$ between the observed data, y^{obs}, and the corresponding expected (or fitted) values, $E(y)_i$. Alternatively, you can simply compute some statistic of the observed data $T(y^{obs})$.
- Generate a new data set, y_i^{rep}, under your model based on the estimated parameter values for MCMC iteration i.
- Calculate the same discrepancy measure $D(y_i^{rep}, E(y)_i)$, but this time for the replicate data, y_i^{rep}. Alternatively, calculate the same statistic for the replicate data $T(y_i^{rep})$.

Using MCMC, we will accumulate posterior samples of both $D(y^{obs}, E(y)_i)$ and $D(y_i^{rep}, E(y)_i)$, and for an adequate model, the two distributions should be similar. The Bayesian p-value (p_B) is defined as the probability that the replicate data are more extreme than the observed data (Gelman et al., 2004), and hence that the discrepancy measure of the replicate data is as large as or larger than the discrepancy measure of the observed data, i.e., $p_B = P(D(y_i^{rep}, E(y)_i) \geq D(y^{obs}, E(y)_i))$. Therefore, for a fitting model, we expect a Bayesian p-value of approximately 0.5, whereas values close to 0 or 1 indicate an ill-fitting model (either under- or overfitting, respectively). Alternatively, if we have computed a test statistic, we will end up with a single value for the observed data $T(y^{obs})$ but a sample of the posterior distribution of that test statistic computed from the replicate data $T(y_i^{rep})$. We then assess whether the test statistic for the observed data lies within the 95% credible interval of the test statistic for the replicate data.

Posterior predictive checks are attractive because they are conceptually straightforward, and moreover, quite easy to compute, as we will see. They also have clear disadvantages, however—first, they may be too conservative (i.e., not reject the hypothesis of a fitting model often enough, i.e., when they should) because they use the data twice: once to estimate the parameter values and again to simulate a new data set. Second, as with all GoF tests, their power for small sample sizes is quite low, as we will see later. Finally, they are not "calibrated" and are only "qualitative" in the sense that we cannot define critical thresholds of a Bayesian p-value computed from a posterior predictive check. That is, we cannot answer the question of how close to 0/1 is "too close" for a fitting model. See Gelman et al. (2004) or Conn et al. (2018) for more information.

We note that the Bayesian p-value is a direct generalization of the concept of a parametric bootstrap (Dixon, 2006) to the Bayesian mode of inference, where instead of having a single value that represents our parameter estimate, we have a whole "population" of such values, namely, the draws from the posterior distribution of that parameter. Seeing this analogy may help with your understanding of posterior predictive checks.

7.2.1.2 Application of Posterior Predictive Checks

For illustration, we use the woodchat shrike data set from Chapter 5. We apply posterior predictive checks for each submodel in the IPM, i.e., for CJS, Poisson regression, and state-space models. As already mentioned, it is possible to define different discrepancy measures that focus on specific assumption violations. Here, we use omnibus discrepancy measures that help to diagnose lack of fit globally but also mention a couple of targeted discrepancy functions for testing specific assumption violations.

Important assumptions of CJS models are the homogeneity of survival and recapture probabilities (i.e., all individuals in a group or age class have identical probabilities), that survival and recapture of one individual is independent from that of all other individuals, that the marked individuals are a random sample from the population of interest, and that no marks are lost or misread in the field or mistranscribed (Burnham et al., 1987; Lebreton et al., 1992; Cooch and White, 2020). Some of these assumptions must be fulfilled while collecting the data and cannot be tested—e.g., one must make sure that all marks are correctly recorded in the field. Other assumptions, such as the homogeneity of survival and recapture, can be tested by specific GoF tests (Pollock et al., 1985; Choquet et al., 2009; Gimenez et al., 2018; Cooch and White, 2020), and a different model structure must be adopted if they are violated (Pradel, 1993; Pradel et al., 1997). An omnibus discrepancy measure to assess the fit of a CJS model formulated for the m-array data is the Freeman-Tukey statistic (Brooks et al., 2000):

$$D_{FT} = \sum_{i,j} \left(\sqrt{m_{i,j}} - \sqrt{E(m_{i,j})} \right)^2,$$

where $m_{i,j}$ is the cell of row i and column j of the observed m-array, and $E(m_{i,j})$ is the cell of the corresponding m-array of the expected values. The Freeman-Tukey statistic is not a ratio of observed to expected data, and this makes it less sensitive than the Pearson χ^2-statistic to small expected values (Chapter 2 in Kéry and Royle, 2016). Because the capture-recapture data are aggregated in the m-array format, we are limited in the possible types of tests. Use of the individual capture histories in the state-space formation of the model (Section 4.5.1.1) offers more flexibility, but some aggregation is necessary, as we cannot directly test the GoF of binary data (McCullagh and Nelder, 1989). For example, specific GoF tests have been developed to detect transients or an immediate trap-response (Pradel, 1993; Pradel et al., 1997) based on χ^2-statistics applied to specific summaries of the capture histories. These discrepancy measures could be useful for posterior predictive checks. They would be cumbersome to code in JAGS but could more easily be obtained in R using posterior draws from JAGS (see Exercise 7.1). Here, we focus solely on the omnibus test.

The next submodel we evaluate in our example is the Poisson regression for the productivity data. An underlying assumption is that the response is independent and identically Poisson distributed. A commonly used fit measure for generalized linear models is the deviance (D_d), which for a Poisson regression is:

$$D_d = 2\sum_i y_i \log\left(\frac{y_i}{E(y_i)}\right) - (y_i - E(y_i)),$$

where y_i is the observed number of fledglings for each brood i, and $E(y_i)$ is the expected number of fledglings for that brood. This omnibus test is sensitive to several kinds of assumption violations. A more specific test is based on the assumption of the Poisson distribution that the mean and variance are identical. We can calculate the variance/mean ratio for the observed data and compare it with the variance/mean ratio in simulated data:

$$T_{vm} = \frac{\text{var}(\boldsymbol{y})}{\text{mean}(\boldsymbol{y})}.$$

This test statistic should be sensitive to both over- and underdispersion.

Finally, we evaluate the assumptions of the state-space models. Important assumptions in general are that the process, i.e., the matrix population model in our IPM, and observation models are correctly specified, the distributional assumptions of the state and the observation processes are met, and these processes are uncorrelated (Harvey, 2014; Newman et al., 2014). The state-space model within an IPM is formulated in terms of the underlying demographic rates, and thus we might expect that the state-space model shows lack of fit if the demographic rates are not correctly estimated. The fit of the state-space model can therefore be viewed as an omnibus test for the complete IPM. The GoF assessment of state-space models is not well developed and is challenging, because the model contains latent variables that cannot be directly compared with any observed data and because prediction errors accumulate over time (Newman et al., 2014; Auger-Méthé et al., 2016). Newman et al. (2014) suggest χ^2 or Freeman-Tukey statistics as suitable discrepancy measures, whereas Besbeas and Morgan (2014) use the mean absolute percentage error, which we also use here:

$$D_{mape} = \frac{100}{n} \sum_{t=1}^{n} \left| \frac{y_t - E(y_t)}{y_t} \right|.$$

Here, y_t is the observed and $E(y_t)$ the expected population count in year t, and the constant $100/n$ may also be omitted. As an additional targeted GoF test, we compute the number of times the observed time series switches between short-term ups and downs (Gelman and Hill, 2007). This tells us whether the model produces the same degree of jaggedness in the expected counts as what we observe in the data.

We have now defined several GoF measures and will apply them in an IPM with demographic stochasticity and constant demographic rates fit to a woodchat shrike data set. Because that data set was simulated under just those assumptions, we do not expect any lack of fit. We first load the data, compute the m-arrays, and bundle the data. We then define the model (IPM_1) in the BUGS language and include the computations for the GoF tests that we just discussed.

Key quantities that must be computed for posterior predictive checks are the expected values for each data set, simulated data sets under the fitted model, and discrepancy measures. Each is straightforward to compute, and we have added comments in the code to flag the places where they are defined. The simulation of data sets might seem complex but is actually quite easy—we just “recycle” the likelihood statement.

```
# Load data
library(IPMbook); library(jagsUI)
data(woodchat5)
marr <- marrayAge(woodchat5$ch, woodchat5$age)

# Bundle data
jags.data <- with(woodchat5, list(marr.j=marr[,,1], marr.a=marr[,,2], n.occasions=dim(marr)[2],
  rel.j=rowSums(marr[,,1]), rel.a=rowSums(marr[,,2]), J=repro[,1], year=repro[,2], age=repro[,3],
  C=count, pNinit=dUnif(1, 300)))
str(jags.data)

List of 10
 $ marr.j      : num [1:19, 1:20] 8 0 0 0 0 0 0 0 0 0 ...
 $ marr.a      : num [1:19, 1:20] 16 0 0 0 0 0 0 0 0 0 ...
 $ n.occasions: int 20
 $ rel.j       : num [1:19] 51 53 55 65 73 66 61 76 65 75 ...
 $ rel.a       : num [1:19] 36 39 44 61 61 50 43 61 51 53 ...
 $ J           : num [1:929] 6 2 2 5 3 5 3 2 3 2 ...
 $ year        : num [1:929] 1 1 1 1 1 1 1 1 1 1 ...
 $ age         : num [1:929] 1 1 1 1 1 1 1 1 1 1 ...
 $ C           : num [1:20] 91 119 131 88 139 145 148 116 112 106 ...
 $ pNinit      : num [1:300] 0.00333 0.00333 0.00333 0.00333 0.00333 ...

# Write JAGS model file
cat(file="model1.txt", "
model {
# Priors and linear models
mean.sj ~ dunif(0, 1)
mean.sa ~ dunif(0, 1)
mean.p ~ dunif(0, 1)
mean.f[1] ~ dunif(0, 10)
mean.f[2] ~ dunif(0, 10)

for (t in 1:(n.occasions-1)){
  sj[t] <- mean.sj
  sa[t] <- mean.sa
  p[t] <- mean.p
}

for (t in 1:n.occasions){
  f[1,t] <- mean.f[1]
  f[2,t] <- mean.f[2]
}

sigma ~ dunif(0.5, 1000)
tau <- pow(sigma, -2)

# Population count data (state-space model)
# Model for the initial population size: discrete uniform priors
N[1,1] ~ dcat(pNinit)
N[2,1] ~ dcat(pNinit)

# Process model over time: our model of population dynamics
for (t in 1:(n.occasions-1)){
  N[1,t+1] ~ dpois(N[1,t] * f[1,t] / 2 * sj[t] + N[2,t] * f[2,t] / 2 * sj[t])
  N[2,t+1] ~ dbin(sa[t], N[1,t] + N[2,t])
}

# Observation model
for (t in 1:n.occasions){
  C[t] ~ dnorm(N[1,t] + N[2,t], tau)
}

# Assessing the fit of the state-space model
# 1. Compute fit statistic for observed data
# 1.1. Discrepancy measure: mean absolute error
for (t in 1:n.occasions){
  C.exp[t] <- N[1,t] + N[2,t]                    # Expected counts
  Dssm.obs[t] <- abs((C[t] - C.exp[t]) / C[t])   # Discrepancy measure
}
```

```
Dmape.obs <- sum(Dssm.obs)
# 1.2. Test statistic: number of turns
for (t in 1:(n.occasions-2)){
  Tt1.obs[t] <- step(C[t+2] - C[t+1])
  Tt2.obs[t] <- step(C[t+1] - C[t])
  Tt3.obs[t] <- equals(Tt1.obs[t] + Tt2.obs[t], 1)
}
Tturn.obs <- sum(Tt3.obs)

# 2. Compute fit statistic for replicate data
# 2.1. Discrepancy measure: mean absolute error
for (t in 1:n.occasions){
  C.rep[t] ~ dnorm(N[1,t] + N[2,t], tau)               # Generate replicate data
  Dssm.rep[t] <- abs((C.rep[t] - C.exp[t]) / C.rep[t]) # Discrepancy measure
}
Dmape.rep <- sum(Dssm.rep)
# 2.2. Test statistic: number of turns
for (t in 1:(n.occasions-2)){
  Tt1.rep[t] <- step(C.rep[t+2] - C.rep[t+1])
  Tt2.rep[t] <- step(C.rep[t+1] - C.rep[t])
  Tt3.rep[t] <- equals(Tt1.rep[t] + Tt2.rep[t], 1)
}
Tturn.rep <- sum(Tt3.rep)

# Productivity data (Poisson regression model)
for (i in 1:length(J)){
  J[i] ~ dpois(f[age[i], year[i]])
}

# Assessing the fit of the Poisson regression model
# 1. Compute fit statistic for observed data
# 1.1. Discrepancy measure: deviance
for (i in 1:length(J)){
  J.exp[i] <- f[age[i],year[i]]                        # Expected data
  D.obs[i] <- J[i] * log(J[i]/J.exp[i]) - (J[i] - J.exp[i])
}
Dd.obs <- sum(D.obs)
# 1.2. Compute test statistic variance-mean ratio
Tvm.obs <- pow(sd(J),2) / mean(J)

# 2. Compute fit statistic for replicate data
# 2.1. Discrepancy measure: mean absolute error
for (i in 1:length(J)){
  J.rep[i] ~ dpois(f[age[i], year[i]])                 # Generate replicate data
  D.rep[i] <- J.rep[i] * log(J.rep[i]/J.exp[i]) - (J.rep[i] - J.exp[i])
}
Dd.rep <- sum(D.rep)
# 2.2. Compute test statistic variance-mean ratio
Tvm.rep <- pow(sd(J.rep),2) / mean(J.rep)

# Capture-recapture data (CJS model with multinomial likelihood)
# Define the multinomial likelihood
for (t in 1:(n.occasions-1)){
  marr.j[t,1:n.occasions] ~ dmulti(pr.j[t,], rel.j[t])
  marr.a[t,1:n.occasions] ~ dmulti(pr.a[t,], rel.a[t])
}
# Define the cell probabilities of the m-arrays
for (t in 1:(n.occasions-1)){
  # Main diagonal
  q[t] <- 1 - p[t]                                     # Probability of non-recapture
  pr.j[t,t] <- sj[t] * p[t]
  pr.a[t,t] <- sa[t] * p[t]
  # Above main diagonal
  for (j in (t+1):(n.occasions-1)){
    pr.j[t,j] <- sj[t] * prod(sa[(t+1):j]) * prod(q[t:(j-1)]) * p[j]
    pr.a[t,j] <- prod(sa[t:j]) * prod(q[t:(j-1)]) * p[j]
  } #j
  # Below main diagonal
  for (j in 1:(t-1)){
    pr.j[t,j] <- 0
    pr.a[t,j] <- 0
  } #j
} #t
```

```
# Last column: probability of non-recapture
for (t in 1:(n.occasions-1)){
  pr.j[t,n.occasions] <- 1-sum(pr.j[t,1:(n.occasions-1)])
  pr.a[t,n.occasions] <- 1-sum(pr.a[t,1:(n.occasions-1)])
}

# Assessing the fit of the capture-recapture model
# 1. Compute fit statistic for observed data
# 1.1. Discrepancy measure: Freeman-Tukey statistic
for (t in 1:(n.occasions-1)){
  for (j in 1:n.occasions){
    marr.j.exp[t,j] <- pr.j[t,j] * rel.j[t]              # Expected values
    marr.a.exp[t,j] <- pr.a[t,j] * rel.a[t]              # Expected values
    Dcjs.obs[t,j] <- pow(pow(marr.j[t,j], 0.5) - pow(marr.j.exp[t,j], 0.5), 2)
    Dcjs.obs[t+n.occasions-1,j] <- pow(pow(marr.a[t,j], 0.5) - pow(marr.a.exp[t,j], 0.5), 2)
  } #j
} #t
DFT.obs <- sum(Dcjs.obs)

# 2. Compute fit statistic for replicate data
# 2.1. Discrepancy measure: Freeman-Tukey statistic
for (t in 1:(n.occasions-1)){
  marr.j.rep[t,1:n.occasions] ~ dmulti(pr.j[t,], rel.j[t]) # Generate replicate data
  marr.a.rep[t,1:n.occasions] ~ dmulti(pr.a[t,], rel.a[t])
    for (j in 1:n.occasions){
      Dcjs.rep[t,j] <- pow(pow(marr.j.rep[t,j], 0.5) - pow(marr.j.exp[t,j], 0.5), 2)
      Dcjs.rep[t+n.occasions-1,j] <- pow(pow(marr.a.rep[t,j], 0.5) - pow(marr.a.exp[t,j], 0.5), 2)
  } #j
} #t
DFT.rep <- sum(Dcjs.rep)
}
")

# Initial values
inits <- function(){list(mean.p = runif(1))}

# Parameters monitored
parameters <- c("mean.sj", "mean.sa", "mean.p", "mean.f", "N", "sigma", "Dmape.obs", "Dmape.rep",
  "Tturn.obs", "Tturn.rep", "Dd.obs", "Dd.rep", "Tvm.obs", "Tvm.rep", "DFT.obs", "DFT.rep")

# MCMC settings
ni <- 12000; nb <- 2000; nc <- 3; nt <- 4; na <- 2000

# Call JAGS (ART 1 min), check convergence and summarize posteriors
out1 <- jags(jags.data, inits, parameters, "model1.txt", n.iter=ni, n.burnin=nb, n.chains=nc,
  n.thin=nt, n.adapt=na, parallel=TRUE)
traceplot(out1)
print(out1, 3)

                 mean     sd     2.5%      50%    97.5% overlap0 f  Rhat n.eff
mean.sj         0.301  0.012    0.278    0.300    0.324    FALSE 1 1.000  7500
mean.sa         0.541  0.013    0.516    0.541    0.566    FALSE 1 1.000  5682
mean.p          0.606  0.019    0.568    0.606    0.643    FALSE 1 1.000  7307
mean.f[1]       2.672  0.077    2.524    2.673    2.822    FALSE 1 1.000  7500
mean.f[2]       3.676  0.087    3.507    3.678    3.846    FALSE 1 1.000  7500
N[1,1]         44.689 30.084    2.000   40.000  106.000    FALSE 1 1.002   872
N[2,1]         60.384 27.928    6.000   63.000  105.000    FALSE 1 1.002   773
N[1,2]         54.207  7.187   40.000   54.000   69.000    FALSE 1 1.001  1946
... [output truncated] ...
N[1,20]        71.730  8.106   56.000   72.000   88.000    FALSE 1 1.001  1384
N[2,20]        80.911  7.218   67.000   81.000   95.000    FALSE 1 1.000  7500
sigma          14.149  3.413    8.660   13.713   21.872    FALSE 1 1.000  7500
Dmape.obs       1.753  0.316    1.196    1.733    2.430    FALSE 1 1.000  7500
Dmape.rep       1.836  0.610    0.958    1.733    3.291    FALSE 1 1.000  7500
Tturn.obs      12.000  0.000   12.000   12.000   12.000    FALSE 1   NaN     1
Tturn.rep      11.384  1.913    8.000   11.000   15.000    FALSE 1 1.000  7500
Dd.obs        526.625  1.018  525.655  526.301  529.451    FALSE 1 1.000  7500
Dd.rep        505.500 23.990  458.561  505.476  553.758    FALSE 1 1.000  7500
Tvm.obs         1.123  0.000    1.123    1.123    1.123    FALSE 1   NaN     1
Tvm.rep         1.078  0.054    0.976    1.077    1.187    FALSE 1 1.000  7500
DFT.obs        40.387  1.377   38.306   40.197   43.570    FALSE 1 1.001  2632
DFT.rep        44.305  4.988   35.164   44.060   54.851    FALSE 1 1.000  7500
deviance     4291.706  7.552 4276.959 4291.616 4306.746    FALSE 1 1.000  7500
```

We look at the posterior predictive checks graphically and calculate Bayesian p-values. We produce a scatterplot of the discrepancy measure of observed data against the discrepancy measure of replicate data. For a model that fits the data well, we expect an unstructured cloud of points where approximately half the points lie above and the other half below the $y = x$ line. The function `pp.check` of the `jagsUI` package produces such a figure, but here we use custom code. For the test statistic of the replicate data, we can produce a histogram and indicate the value of this statistic for the observed data with a vertical line. If this line is at the limit or even outside the support of the histogram, it suggests an ill-fitting model.

Fig. 7.1 shows the results of our posterior predictive checks and suggests that all three submodels fit the data adequately: the dots are scattered more or less equally on both sides of the $y = x$ line, and therefore the Bayesian p-values (i.e., the proportion of points above that line) are not too far from 0.5. The variability of the discrepancy measures for the replicate data is larger than for the observed data for each of the three submodels. This is a common observation and no reason for concern. The additional uncertainty in the replicate data originates from the simulation (King et al., 2010, p. 138). Moreover, there is a lower limit for the discrepancy measure of the observed data, which is well pronounced for the Poisson regression and the capture-recapture models. Such a limit is not always present and may depend, for example, on the chosen discrepancy measure or the shapes of priors and posteriors (King et al., 2010, p. 138). This is not a reason for concern either. The targeted tests look satisfactory: the number of observed switches corresponds well with that in the replicate data, and the observed variance/mean ratio is similar to that in the recplicate data. Not surprisingly, therefore, there is no lack of fit in any of the model components for this simulated data set.

7.2.1.3 Sensitivity of Posterior Predictive Checks to Diagnose Misspecified IPMs

We have just seen the outcome of posterior predictive checks applied in an IPM that fits the data. It will now be interesting to see how well these tests perform in a misspecified IPM and how the power depends on sample size. An IPM can be misspecified in many ways; we cannot be exhaustive here but instead just explore two scenarios.

In the first misspecified IPM (IPM_2), we erroneously assume that both survival and productivity have no age structure. Consequently, both the CJS and the Poisson regression models are misspecified for our data set. The second misspecified IPM (IPM_3) assumes that 1-year-old shrikes do not reproduce. Consequently, we only have productivity data for shrikes that are at least 2 years old. In this case, the CJS and Poisson regression models are correctly specified for the data, but the process model—i.e., the matrix population model—is misspecified. We fit these two models and compute the same posterior predictive checks as before but no longer show the specific test statistics of the state-space and the Poisson regression models. The code for these two models is provided on the book web page.

In addition, we repeat this exercise with a variant of the same woodchat shrike data set that is five times larger (`woodchat7`) to illustrate how the power of GoF tests depends on sample size. You can again find the code on the book web page.

We first inspect the results for the original data set (`woodchat5`). Under misspecification of the CJS and Poisson regression models (IPM_2), the tests for the Poisson regression and CJS models indicate lack of fit (Fig. 7.2E and F), but the test for the state-space model does not (Fig. 7.2D). Thus, our posterior predictive checks were successful in diagnosing lack of fit of the relevant model components. Moreover, the results show that misspecification of models at the individual-level does not lead to the rejection of the posterior predictive test for the state-space model for the population-level data. This latter model uses "biased" information from the misspecified models at the individual level, and hence we might a priori expect that the state-space model also shows lack of fit. However, we see in this example that this is not necessarily the case. If this were a real analysis, we would now have to determine why the Poisson regression and CJS models do not fit and modify them accordingly (if possible). This may be an iterative process.

Next, we inspect the results of the posterior predictive checks for the case when the population model is misspecified (IPM_3). Fig. 7.2 (bottom panels G, H, I) shows that no test indicates lack of fit. While this was expected a priori for the tests of the CJS and Poisson regression models, we had hoped that the test for the state-space model might indicate lack of fit. This shows what we have observed as almost a rule in our applied work with IPMs: the fit of the model to the population counts is almost always excellent, which suggests that fit assessments of this part of the model have extremely limited power.

The results of the posterior predictive checks differ strongly for the data set that is five times larger. For the fitting model IPM_1, the results remain the same—none of the tests show any lack of fit (Fig. 7.3A–C). However, for the misspecified IPM_2, all test components now show lack of fit (Fig. 7.3D–F), including (newly) that of the state-space model (Fig. 7.2D). For the misspecified IPM_3, the posterior predictive check of the state-space model now also shows lack of fit, but this is not so for the other two submodels (Fig. 7.3G–I). Clearly, the power to detect departures from model assumptions increases with increasing sample size.

We now try to understand why the posterior predictive checks tests are not always successful in indicating lack of fit. For this purpose, we inspect the estimated model parameters more closely. Specifically, we compare the estimated demographic rates under the three IPMs for the small and the large sample size scenarios. In addition, we include estimates of survival and productivity obtained from fitting correctly specified CJS and Poisson regression models to the corresponding

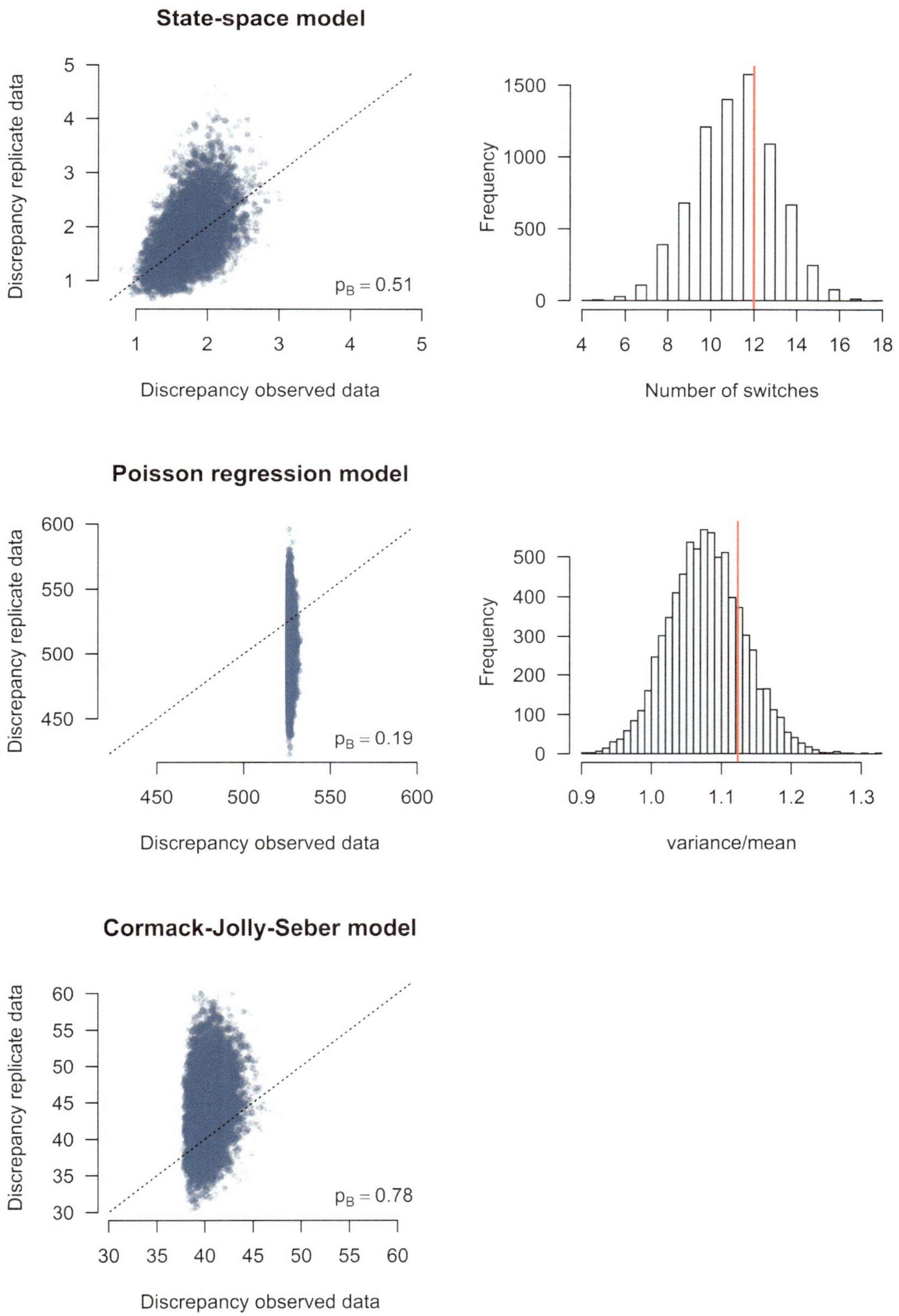

FIGURE 7.1

Posterior predictive checks using scatterplots of discrepancy measures for observed and replicate data in the different submodels as well as histograms of test statistic of replicate data for the first integrated population model, IPM_1. The discrepancy measures are the mean absolute error, the deviance, and the Freeman-Tukey statistic for the state-space, the Poisson regression, and the Cormack-Jolly-Seber model, respectively. Bayesian *p*-values (p_B) are given for each discrepancy measure. The $y = x$ line is the dotted black line. The vertical red lines show the test statistics of the observed data.

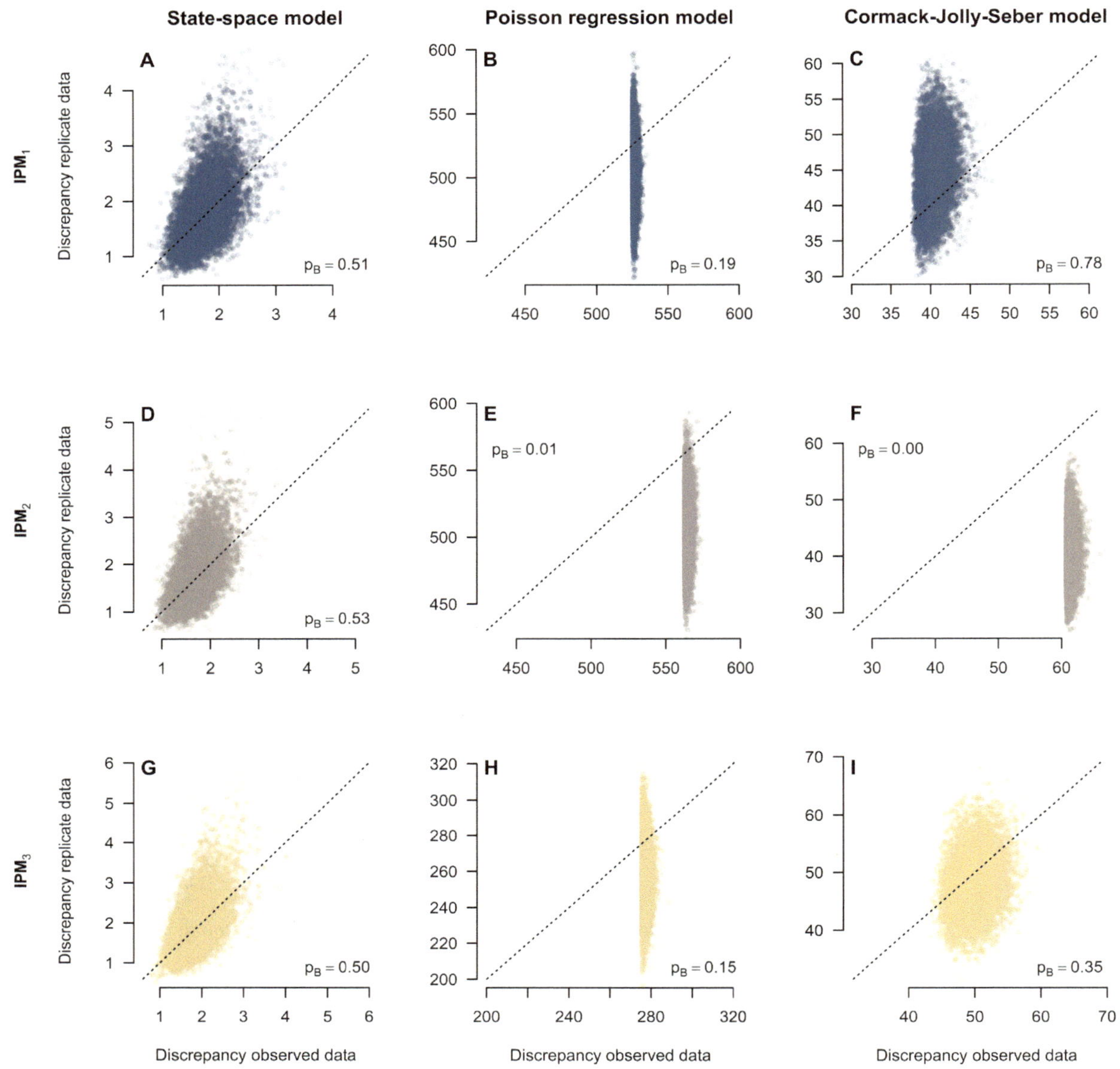

FIGURE 7.2

Scatterplots of discrepancy measures for replicate and observed data (posterior predictive checks) for the different submodels (first column: state-space model; second column: Poisson regression model; third column: Cormack-Jolly-Seber model) of different integrated population models (IPMs, differentiated by row) under the small sample size scenario. The $y = x$ line is the dotted black line. IPM_1 is correctly specified (blue), IPM_2 is misspecified in terms of the age structure of survival and productivity (gray), and IPM_3 is misspecified with respect to age at first reproduction (yellow).

component data sets alone. The latter estimates are not influenced by the population count data and thus provide information about demographic rates that only originates from the demographic data models.

Comparison of the demographic rates (Fig. 7.4) is insightful in several ways. First, the estimates from the fitting IPM_1 are quite similar to the estimates we obtained from the separate analysis of the component data regardless of sample size. This was to be expected, since the underlying demographic rates in these simulated data sets are identical.

Second, the estimates of the demographic rates from IPM_2 with the wrong age structure are quite different from those under models with the correct age structure (Fig. 7.4), and this finding does again not depend on sample size. This is not unexpected and is why the corresponding posterior predictive checks indicate lack of fit (Figs. 7.2E and F; 7.3E and F). Conversely, the estimated population size is close to the population count data when sample size is small (Fig. 7.5), but not

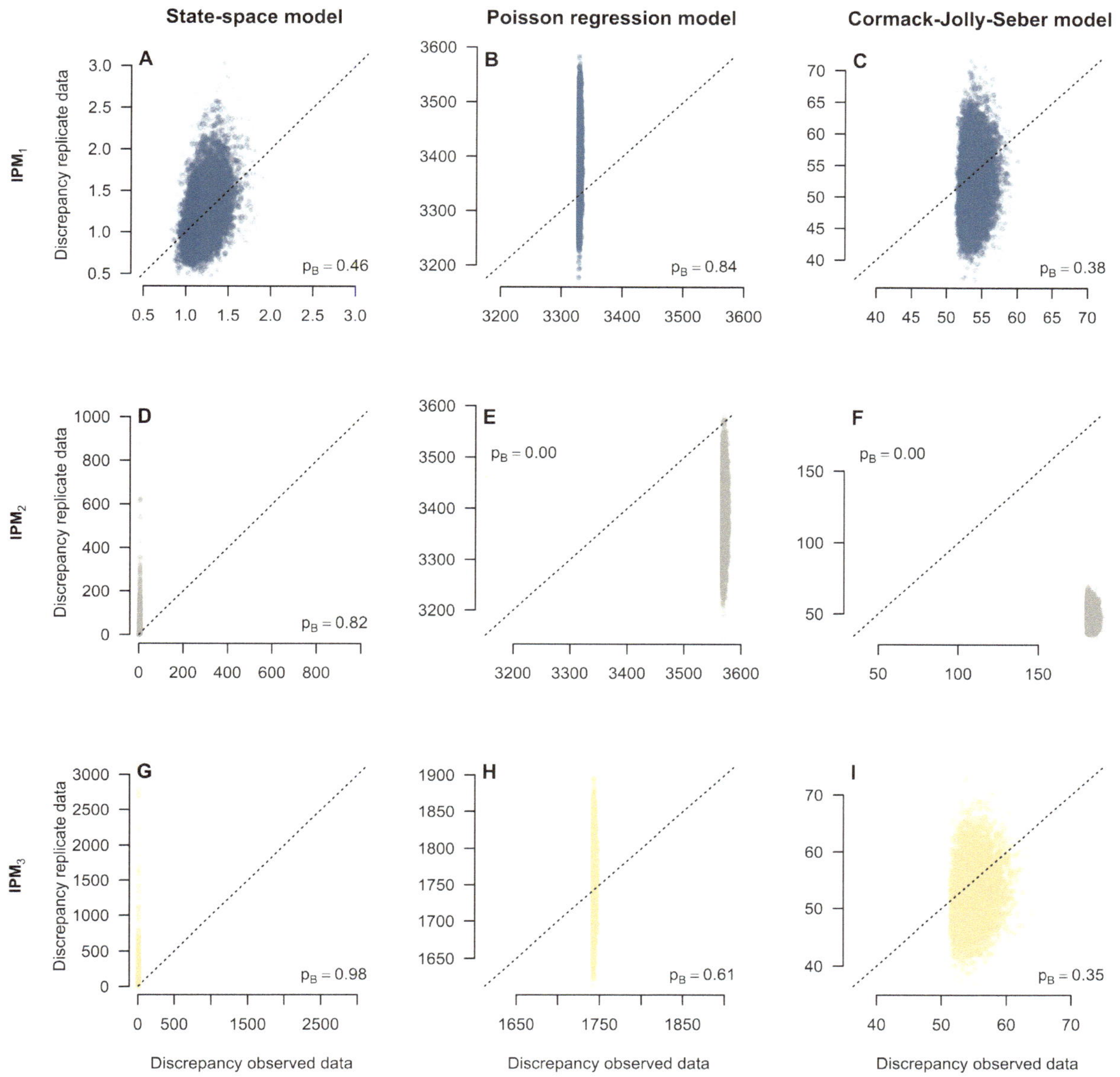

FIGURE 7.3

Scatterplots of discrepancy measures for replicate and observed data (posterior predictive checks) for the different submodels (first column: state-space model; second column: Poisson regression model; third column: Cormack-Jolly-Seber model) of different integrated population models (IPMs, differentiated by row) under the large sample size scenario. The $y = x$ line is the dotted black line. IPM_1 is correctly specified (blue), IPM_2 is misspecified in terms of the age structure of survival and productivity (gray), and IPM_3 is misspecified with respect to age at first reproduction (yellow).

when sample size is large. Consequently, the posterior predictive check indicates lack of fit for the state-space model when sample size is large but not when it is small, as we have seen before.

Third, with small sample size, all estimates of demographic rates from IPM_3 with misspecified age at first reproduction are larger than those under the component data analyses alone and the correctly specified IPM_1 (Fig. 7.4). This is because the population projection that would occur using unbiased demographic rates is quite different from the observed projection. Indeed, the asymptotic population growth rate of a population model assuming an age of first reproduction of 2 years and using unbiased demographic rates equals 0.87. This would indicate a strong population decline while the observed population is increasing (Fig. 7.5). The IPM aims to match the two, which is achieved in IPM_3 by pushing up the demographic rates. Because the estimates under an IPM are weighted averages (Section 6.2.1), the higher estimates of demographic rates under IPM_3 are due to the information about demographic rates contained in the population counts. Thus, misspecification of the matrix population model leads to bias in the demographic rates (note that we speak of bias as

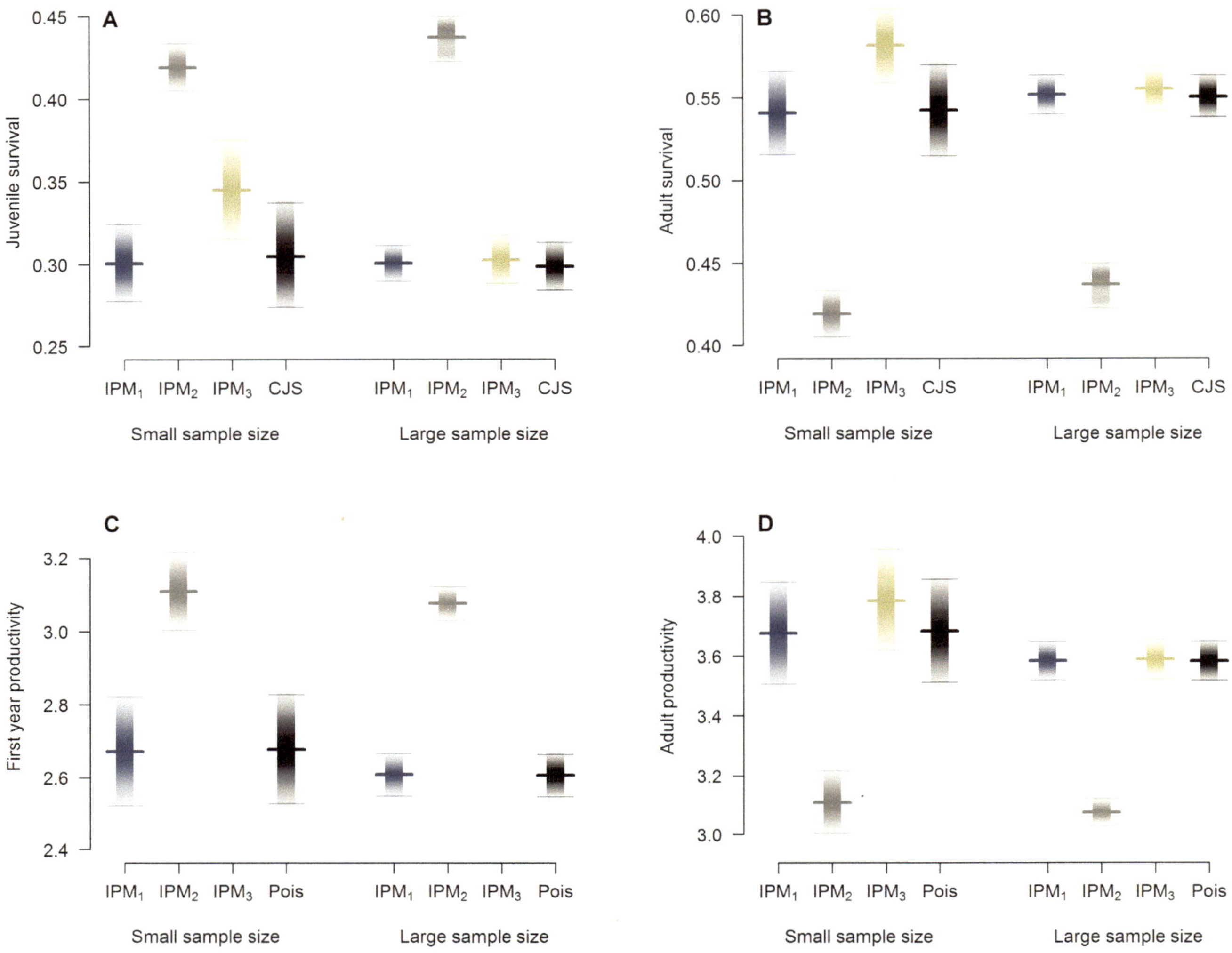

FIGURE 7.4

Estimates of demographic rates obtained from four different models with variable sample sizes. The darkness of the strips is proportional to the posterior density. The horizontal lines show the 95% credible intervals (thin lines) and posterior means (thick lines). The first integrated population model, IPM_1, is correctly specified (blue), IPM_2 is misspecified in terms of the age structure of survival and productivity (gray), and IPM_3 is misspecified in terms of age at first reproduction (yellow). The black strips (fourth from left) are estimates obtained from fitting the CJS and Poisson regression models (Pois) to the corresponding component data sets.

a systematic error in the estimation under the misspecified model, although from the analysis of a single data set, we could not technically see this). Nevertheless, none of the posterior predictive checks indicated any lack of fit (Fig. 7.2G–I) due to their generally low power. All this changes when we move to the larger sample size. The information content about the demographic rates in the individual data sets is then increased, which leads to weighted averages of the demographic rates that are close to the true values (Fig. 7.4). The resulting population projection is then decidedly different from the population count data (Fig. 7.5), and thus the posterior predictive check for the state-space model can "see" the lack of fit.

7.2.1.4 Posterior Predictive Checks for IPMs With a Hidden Parameter

The ability to estimate hidden parameters is a major benefit of IPMs (Chapter 6). However, we have already mentioned that inferences about a hidden parameter must be carefully evaluated, because the estimate of a parameter may compensate for the combined lack of fit in other model parts (Tavecchia et al., 2009; Schaub and Fletcher, 2015; Riecke et al., 2019). We now want to study the performance of posterior predictive checks to detect assumption violations in IPMs with a hidden parameter. We use the same data sets and similar models as in the previous section, but we ignore the productivity data and estimate productivity as a hidden parameter (which, of course, then precludes a posterior predictive check for

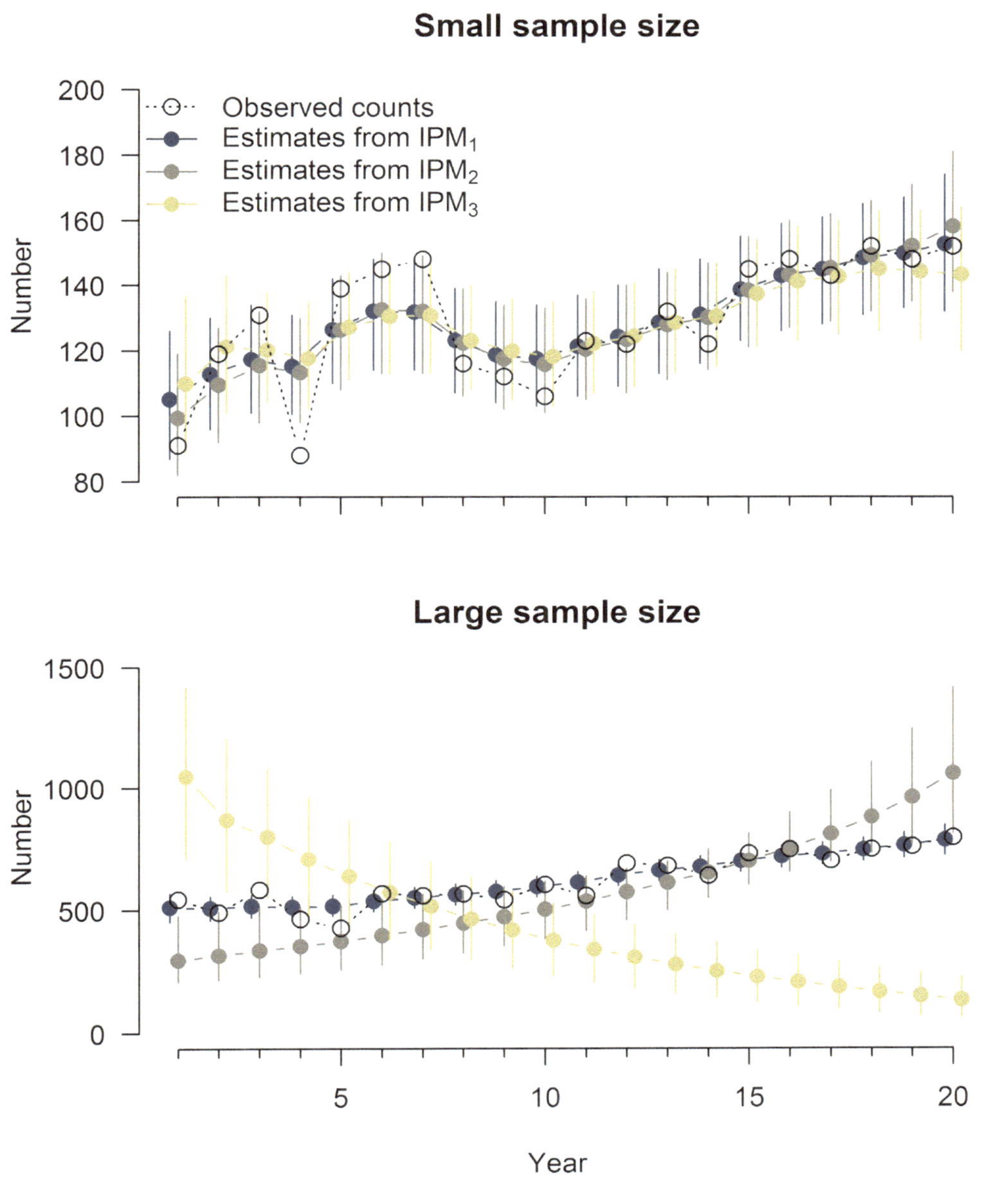

FIGURE 7.5

Comparison of the estimated trajectories of population size under the three integrated population models (IPMs) and the observed population counts for a small and large data set (with 95% credible intervals shown as vertical lines). IPM_1 is correctly specified (blue), IPM_2 is misspecified in terms of the age structure of survival and productivity (gray), and IPM_3 is misspecified in terms of the age at first reproduction (yellow).

productivity). In contrast to the previous IPMs we exclude the age effect on productivity such that only a single hidden parameter is estimated that is identifiable (see Section 6.3).

First, we analyze the smaller woodchat shrike data set (`woodchat5`) and again fit three models: IPM_4 is correctly specified, IPM_5 is misspecified in the CJS model, which lacks age-dependence, and IPM_6 is misspecified in the population process model, which assumes that shrikes start to reproduce only at age two. Second, we fit the same models to the larger data set (`woodchat7`) to study the effects of sample size on the power of posterior predictive checks. The code for these models can again be found on the book web page.

Regardless of sample size, the posterior predictive checks tell us the same story: only the test of the CJS submodel in IPM_5 indicates lack of fit (Figs. 7.6D and 7.7D), while the other tests suggest no lack of fit (Figs. 7.6 and 7.7). We had

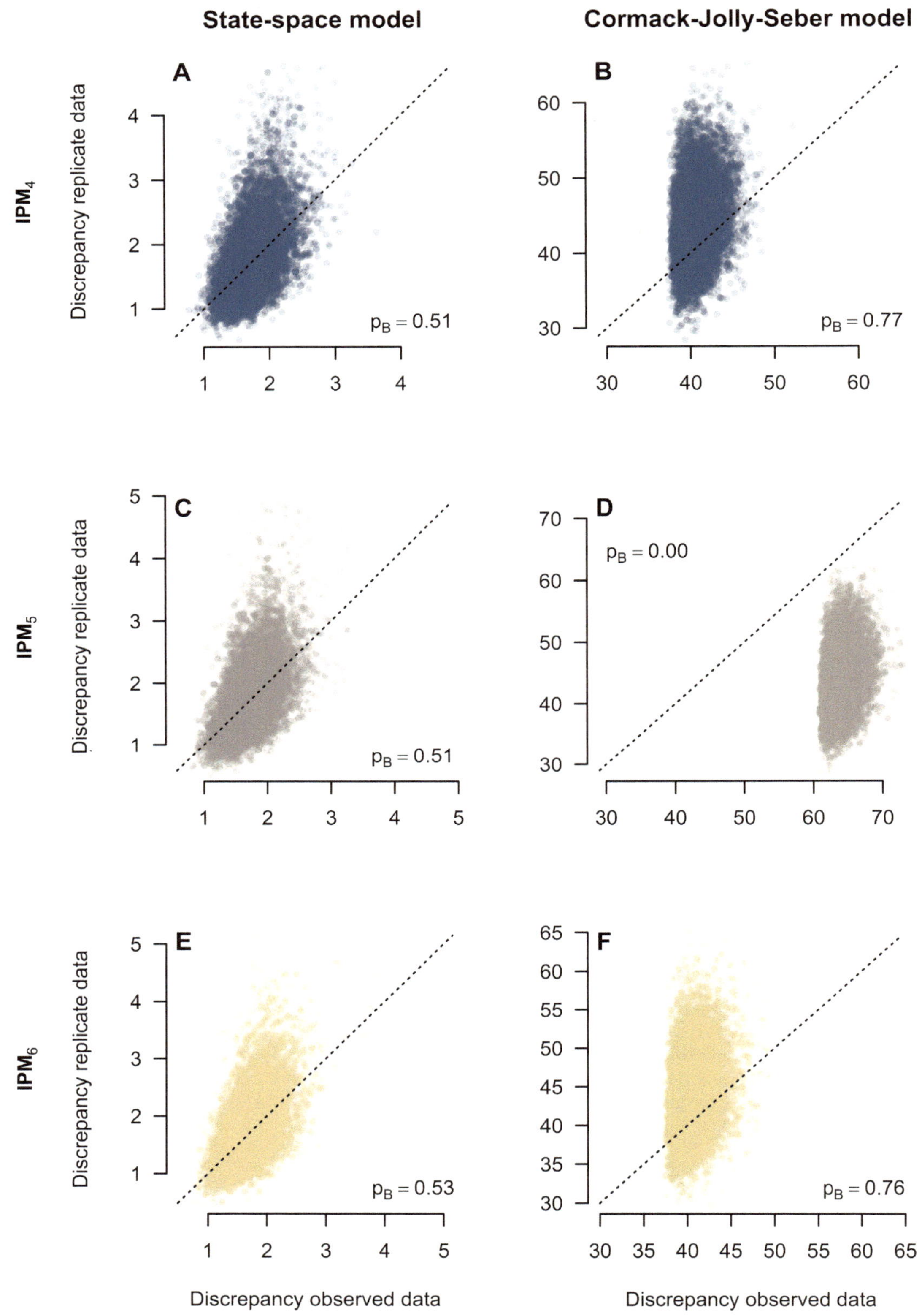

FIGURE 7.6

Scatterplots of discrepancy measures for replicate and observed data (posterior predictive checks) for the different submodels (first column: state-space model; second column: Cormack-Jolly-Seber model) of different integrated population models (IPMs, differentiated by row) without productivity data under the small sample size scenario. The $y = x$ line is the dotted black line. IPM_4 is correctly specified (blue), IPM_5 is misspecified in terms of the age structure of survival and productivity (gray), and IPM_6 is misspecified in terms of the age at first reproduction (yellow). In all three IPMs, productivity is estimated as a hidden parameter.

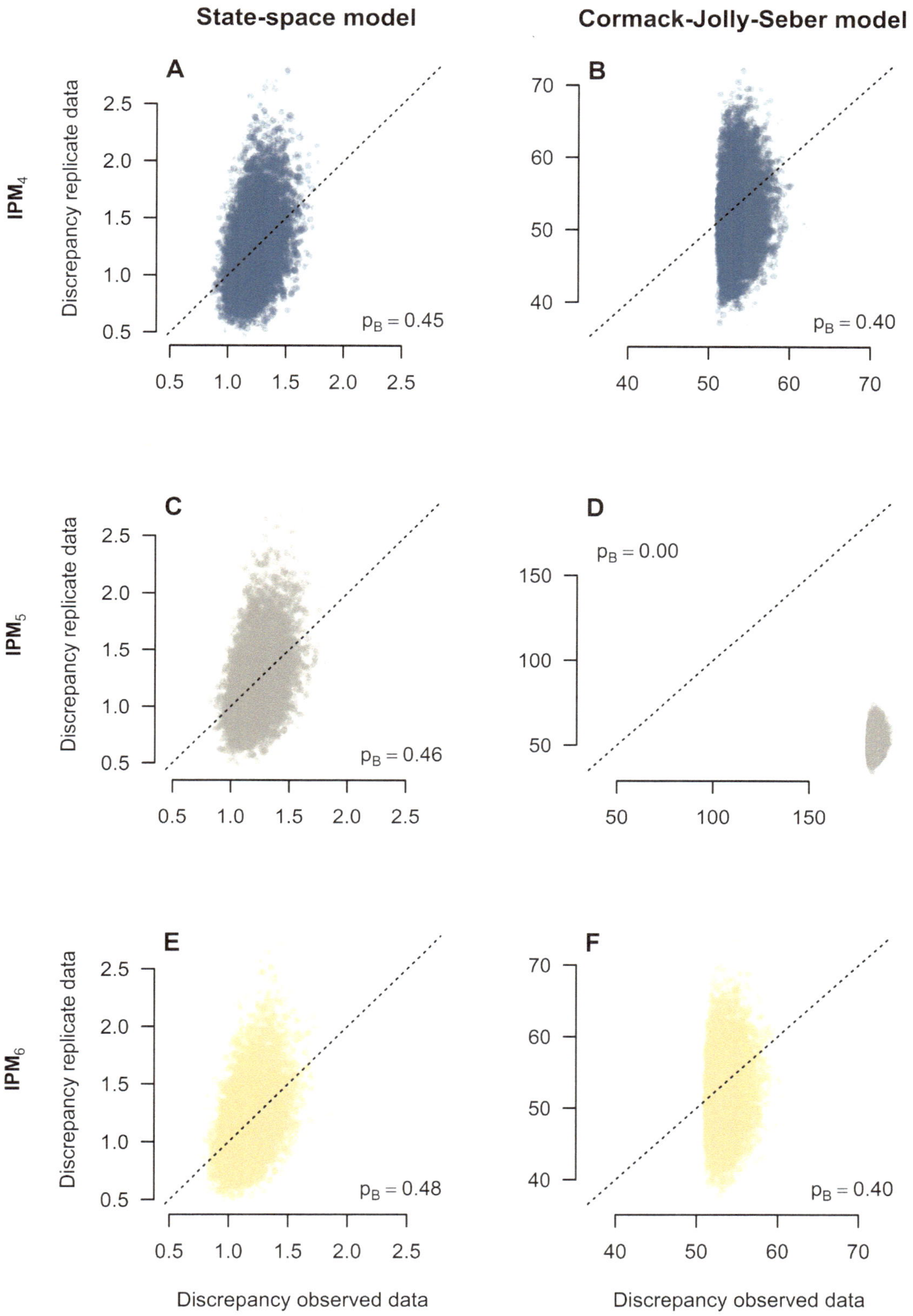

FIGURE 7.7

Scatterplots of discrepancy measures for replicate and observed data (posterior predictive checks) for the different submodels (first column: state-space model; second column: Cormack-Jolly-Seber model) of different integrated population models (IPMs, differentiated by row) without productivity data under the large sample size scenario. The $y = x$ line is the dotted black line. IPM_4 is correctly specified (blue), IPM_5 is misspecified in terms of the age structure of survival and productivity (gray), and IPM_6 is misspecified in terms of the age at first reproduction (yellow). In all three IPMs, productivity is estimated as a hidden parameter.

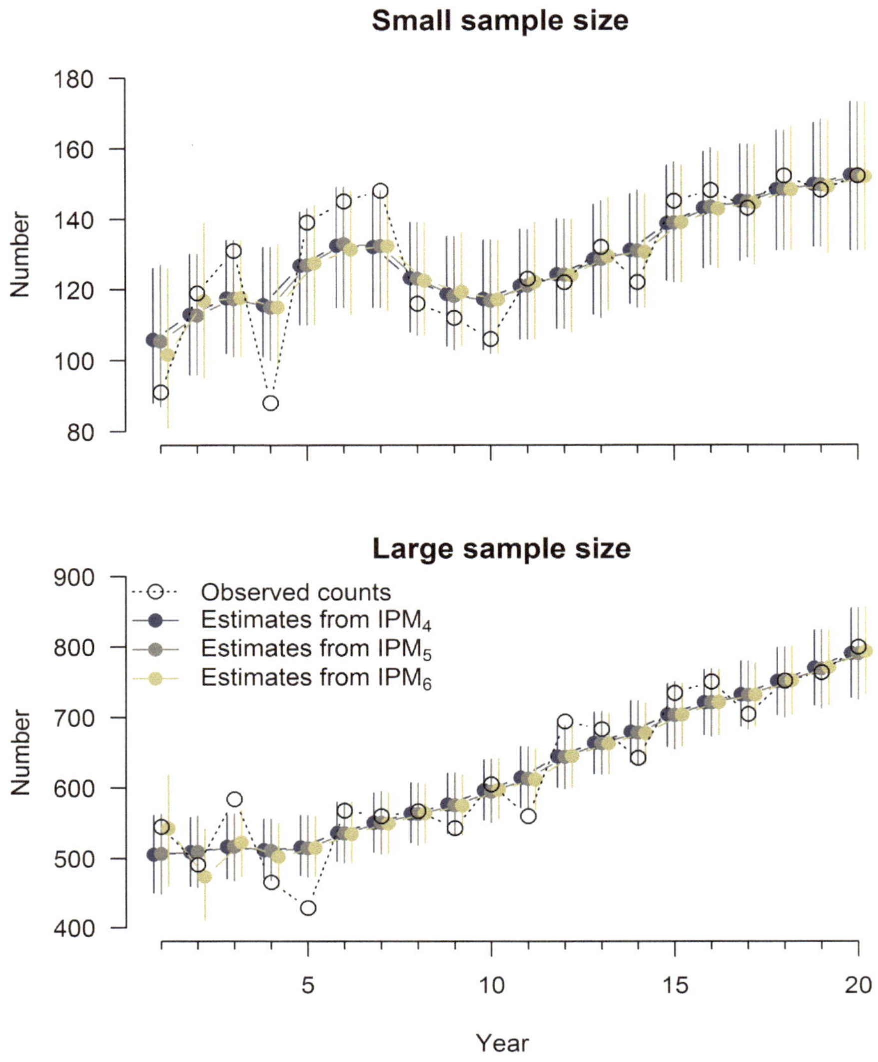

FIGURE 7.8

Comparison of the observed population counts and estimated population trajectories under the three fitted integrated population models (IPMs) with productivity estimated as a hidden parameter and fit to the small and the large data set (with 95% credible intervals). IPM_4 is correctly specified (blue), IPM_5 is misspecified in terms of the age structure of survival and productivity (gray), and IPM_6 is misspecified in terms of the age at first reproduction (yellow).

expected this result for the correctly specified IPM_4, but not necessarily for the misspecified IPMs. In contrast to the misspecified IPM without a hidden parameter (Section 7.3.1.3), we now find acceptable fits of the state-space models when the CJS or process models are misspecified (Fig. 7.6C and E), even for the large sample size (Fig. 7.7C and E). Indeed, the estimated population trajectories under all models and for both sample sizes agree well with the observed data (Fig. 7.8). We may wonder how such good agreement is possible when either survival is not correctly estimated (IPM_5, with indication of lack of fit of the CJS model) or the population model is clearly wrong (IPM_6). This is because the hidden productivity parameter is estimated in such a way that a close fit between estimated population size and observed population count is achieved. The resulting estimates of productivity differ substantially among the models (Fig. 7.9): they are on target in IPM_4 (see Sections 6.3 and 6.6.5 for simulation assessments) but off in the misspecified IPM_5 and IPM_6.

Based on the posterior predictive checks, we would conclude that the fit of IPM_4 and IPM_6 is fine, but that is not the case for IPM_5, for which we might try to identify the reason for the lack of fit of the CJS model and perhaps modify the model accordingly. More problematic is our conclusion about the absence of lack of fit in IPM_6. The only indication that something may be wrong is the estimate of productivity. If we know the species, we may find the estimate suspicious because it is much higher than expected.

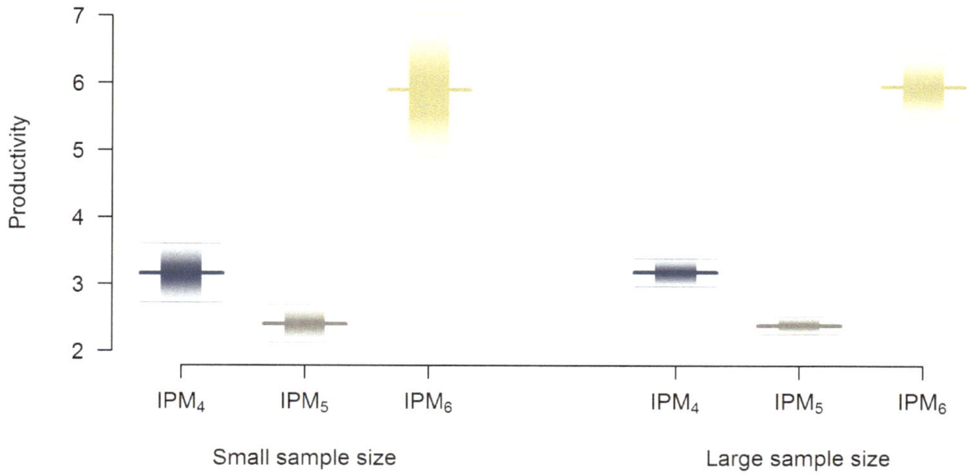

FIGURE 7.9

Estimates of posteriors of productivity under three integrated population models (IPMs) with small and large sample sizes. The darkness of the strips is proportional to the posterior densities. The horizontal lines show the 95% credible intervals (thin lines) and the posterior means (thick lines). IPM_4 is correctly specified (blue), IPM_5 is misspecified in terms of the age structure of survival and productivity (gray), and IPM_6 is misspecified in terms of the age at first reproduction (yellow).

This example emphasizes that diagnosis of lack of fit in IPMs with a hidden parameter is much more challenging, since the estimate of the hidden parameter may compensate to some degree for any lack of fit in other parts of the model (Tavecchia et al., 2009; Schaub and Fletcher, 2015; Riecke et al., 2019). GoF tests for the state-space model (at the highest, i.e., population, level) therefore become devoid of any power. Thus, the "commonsense GoF test" becomes particularly important—that is the one where we ask whether the estimate of a parameter makes any biological sense, especially that of the hidden parameter.

7.2.2 THE INDEPENDENCE ASSUMPTION

The formation of the joint likelihood of an IPM as the product of the likelihoods of the component data sets (Section 5.4) is only valid if the latter are *statistically independent* from each other. Independence in statistical terms has a clear definition. Two variables x_1 and x_2 are independent if their joint probability density $f(x_1, x_2)$ is the product of the two marginal densities:

$$f(x_1, x_2) = f(x_1)f(x_2).$$

There are several possibilities to test whether two variables are independent. Many are based on the idea of quantifying the amount of information obtained from one variable about the other. This amount of information should be zero if the two are statistically independent.

Strictly speaking, the independence assumption in an IPM precludes anything that would introduce dependence in the data of the different component (single data) likelihoods. It would be difficult to be exhaustive about the mechanisms that could introduce dependence. Dependence among data sets may arise if all, or partly, the same individuals occur in more than one data set. For example, if the same individuals are part of a capture-recapture data set and also in the count data set and in a year many individuals die, this would show up in both data sets and thus violate the independence assumption. Moreover, because the "shared" individuals partly provide the same information twice, the precision of the estimates might be inflated. A further mechanism that can create dependence is the sampling from the population. If the sampling of one data set affects who is sampled for a second data set, these data sets may become statistically dependent. If the joint likelihood in an IPM uses dependent data sets (see Chapter 21 for the sole exception) and is expressed as a simple product of the component data likelihood, as we do in virtually all of the examples in this book, the resulting parameter estimates may not be accurate, i.e., they may be biased, or their precision may be overstated.

However, it is true to say that both the mechanisms for and the effects of a violation of the independence assumption in IPMs are not yet very well understood. A handful of studies have attacked the topic with simulations in which the same individuals were included in more than one data set, i.e., where data sets were "overlapping" in a sense. In practice, some data overlap is almost the rule in an IPM and naturally happens more often in studies with a small spatial scale and with small population and sample sizes. Therefore, it is important to understand the consequences of this on the parameter estimates.

Abadi et al. (2010) used simulations to study the effect of such an "overlap" among multiple data sets. To generate independent data sets they simulated three populations with the same underlying demographic parameters and sampled the capture-recapture data from the first, productivity data from the second, and population counts from the third. To generate "overlapping," dependent data sets, they just simulated a single population and sampled from it all three data sets. Depending on the size of the simulated population and the sampling intensity, a variable number of individuals appeared in two or all three data sets. The larger the size of the simulated population and the lower the sampling intensity, the lower then is the number of individuals occurring in multiple data sets and therefore the lower was the degree of dependence assumed to be. For both the dependent and the independent data sets Abadi et al. (2010) then fitted IPMs similar to the woodchat shrike IPM that we use extensively in this book and compared their estimates. Parameter estimates were unbiased, and their precision hardly differed between dependent and independent data set analyses, even though the degree of data overlap was strong in their simulation study. These results also held for IPMs that used population count and capture-recapture data alone; thus, where productivity was estimated as a hidden parameter.

Weegman et al. (2021) carried out a similar study as Abadi et al. (2010), but included in their simulations much stronger dependence and considered temporally variable demographic rates. In the most extreme case, they had only one capture-recapture data set, from which a population index was derived as the number of captured individuals per year. These two completely overlapping data sets were then analyzed in an IPM and productivity was estimated as a hidden parameter. Astonishingly, simulation results showed that even in this extreme case of data overlap parameter estimates were unbiased and had an accuracy similar to that when the model was fit to independent data.

In a further study, Besbeas et al. (2009) simulated dead-recovery and population count data that did or did not share any individuals. They found that the mean squared error was increased when dependent as opposed to independent data sets were analyzed. This effect was more pronounced when the dead-recovery rate was larger, i.e., when the data overlap was increased.

The big question, then, is why are the conclusions from the three studies regarding the effect of overlapping data sets different? It appears that the effect of data overlap in an IPM is context specific and depends on how much information each data set provides about shared parameters. If the information from one data set dominates, as is often the case for survival when capture-recapture data are used, data overlap has hardly any effect. The capture-recapture data provide most of the information about survival, whereas the population count data contribute much less. It then does not matter so much whether this sparse information coming from the population counts originates from partially the same individuals, i.e., is not completely independent. In contrast, if both data sets provide more even amounts of information about a shared parameter, independence matters more. This was the case of the simulations in Besbeas et al. (2009). Knowing how much information a data set contributes about a shared parameter is therefore important for judging whether data overlap matters in an IPM and therefore our discussion of the "flow of information" in Section 6.2.2 is relevant.

In the abovementioned three simulation studies there was no dependency between demographic processes, i.e., survival and productivity were not correlated. However, in the real world there will often be a positive association between survival and productivity due to individual heterogeneity—good individuals have higher survival *and* higher reproductive output and vice versa (Reznick et al., 2000). This can create additional dependency among data sets that may be magnified by overlapping data. Thus, the results of the three simulation studies may not tell the whole story. Further assessments about the effects of violations of the independence assumption in IPMs that focus on different mechanisms (e.g., individual heterogeneity) are therefore welcome.

Based on our experience, we believe that violation of the independence assumption induced by "overlapping" data is rarely a serious problem. Strong overlap of data sets is only expected in small-scale studies with a small population size and a strong sampling intensity (i.e., high capture rates). Dependency in such cases can be reduced by ensuring independent sampling of the component data sets. In many studies with decent population size, however, the overlap of the data sets is expected to be low and consequently the effect on the parameter estimates is limited. Unfortunately, any violation of the independence assumption in an IPM cannot be diagnosed by GoF tests. In cases where data overlap is suspected or other mechanisms that can result in data dependency may be an issue for your own data, we recommend carrying out a simulation study tailored to your own study to gauge potential effects of the violation.

If the violation of the independence assumption is a problem, we may have to develop an IPM explicitly for dependent data sets. Such a model differs from the classical IPMs for independent data sets that we have seen so far in this book by explicitly considering the complete population, where some individuals do, and others do not, appear in each component data set. This leads to a (much) more complex model that can be computationally expensive to fit. So far, we are aware of only four studies that use such models, and the continued development and exploration of such models will be important in the future.

Maunder (2004) analyzed samples from a population composed of marked and unmarked individuals resulting in capture-recapture data and in counts of individuals. Some counted individuals were marked and some were not. The first likelihood component of the capture-recapture data was conditional on the number of individuals counted and included among other things population size as a parameter. The second likelihood component for the count of individuals also included population size as a parameter. Under the assumption that capture probabilities of marked and unmarked individuals were the same, population size was estimable, and the two likelihood components were thereby linked. Maunder's model is related to mark-resight models (McClintock and White, 2009; McClintock et al., 2019) but is an open model that includes a description of population dynamics.

Chandler and Clark (2014) published another IPM that explicitly accounted for dependence among data sets. They had an array of baited traps and collected hair samples of Louisiana black bears (*Ursus americanus*). In some years the data were genotyped, resulting in (individual) spatial capture-recapture (SCR) data, whereas in others they were not—it was simply noted whether a trap was visited by at least one bear, resulting in "cheaper" occupancy (or species-level detection/nondetection) data. These two data sets were clearly not independent, because the same individuals may have been included in both data sets. Chandler and Clark (2014) therefore developed an individual-based spatial IPM with a spatial point process as a process model at its core. This included each individual explicitly, which contrasts with the process models in the classical IPMs based on summaries of individuals (number of individuals of an age/stage class) aggregated over some spatial area. The two observation models (for the SCR and the occupancy data sets) conditioned on the same process model, and thus the dependency of the data sets was accounted for. They used data augmentation to describe all individuals in the population, regardless of whether they were ever captured (identified explicitly) in the study. This model is included as a case study in this book (Chapter 21). A similar model was adopted by Chandler et al. (2018).

Lee et al. (2015) studied a population of Svalbard reindeer (*Rangifer tarandus*) by integrating capture-recapture, harvest, and population count data. The data sets were judged to be dependent because many marked individuals were also included in the population counts. As it was noted during the population counts how many marked and unmarked individuals were seen, it was possible to account for the data overlap: in the observation model of the state-space model, only the counts of the unmarked individuals were modeled, but the state-process model included the total population, i.e., both unmarked and marked individuals. The number of marked individuals could be estimated, based on the estimated survival probabilities and the known number of marked individuals. Because the marked individuals were excluded from the population counts, they provided information about survival only via the capture-recapture model, and thus data overlap was accounted for in this neat analysis.

Finally, Béliveau (2016) developed a slightly different approach to accommodate data set dependence in an IPM. In this study, the modeled populations again consisted of marked and unmarked individuals, which allowed the specification of two observation models that condition on the same process model. Unlike the individual-based model in Chandler and Clark (2014), the process model of Béliveau is based on summary data in the form of age-specific numbers of marked and unmarked individuals. Béliveau (2016) applied her model to a greater horseshoe bat (*Rhinolophus ferrumequinum*) data set analyzed previously (Schaub et al., 2007; see also Chapter 13) in an IPM assuming independence. Her new model provided slightly higher and more precise estimates of adult survival and lower estimates of productivity compared with the original analysis.

7.2.3 THE COMMON DEMOGRAPHY ASSUMPTION

In addition to the independence assumption, an IPM also needs the assumption that the component data sets stem broadly from the *same* population, i.e., have the same underlying demographic patterns. In other words, the data sets combined must not provide "conflicting" information that could lead to conflicting inferences about the population state or process (Maunder and Piner, 2017; Saunders et al., 2019; Plard et al., 2021). This assumption, which we call the *common demography assumption*, directly follows from the construction of the joint likelihood of an IPM, where we say that one or several demographic parameters are identical in different component data likelihoods.

There are several ways in which the common demography assumption might be violated. A prominent reason is that the different data sets stem from different areas or from different time periods. The dynamics of these populations or during these periods may differ, resulting in spatio-temporal variability in the underlying demographic rates. Saunders et al. (2019) identified a spatio-temporal mismatch of component data sets in a study of the American woodcock (*Scolopax minor*) and developed an IPM that successfully overcame a violation of the common demography assumption. A further possible reason for a violation of the common demography assumption is by the following facet of a spatial mismatch, when capture-recapture data are collected in a smaller spatial area than the population counts. Capture-recapture data provide information about apparent survival, which is the probability of true survival and of not leaving the study area (i.e., not dispersing), whereas the population count data that refer to a larger spatial unit are informative about true survival. Combining apparent and true survival in an IPM therefore leads to the violation of the common demography assumption. The common demography assumption is generally more likely to be violated when disparate data sets from different location and/or time periods are jointly analyzed than when the component data all originate from the same population. Indeed, there is typically a trade-off between two major assumptions of an IPM: a larger overlap of the component data may more easily lead to a violation of the independence assumption, but at the same time the common demography assumption is then more likely to hold, and vice versa.

We next explore consequences of the violation of the common demography assumption on parameter estimates. These consequences are context specific—they depend on the magnitude of the difference in the demography in the component data sets and on the weights these have in the integrated model (see Section 6.2.2). Using simulation, we show what can happen when the common demography assumption is violated and we investigate how well we can detect this assumption violation with posterior predictive checks.

Our simulation is again inspired by the woodchat shrike life history. The idea is to simulate five different woodchat shrike populations, from which different data sets are sampled such that there is no data overlap, and the independence assumption holds perfectly. Moreover, data are simulated in such a way that all the assumptions made for the component data likelihoods are met. The first three of these populations have exactly the same demographic rates (juvenile survival (s_j) = 0.3, adult survival (s_a) = 0.55, productivity (f) = 3.1), resulting in an average population growth rate of $\lambda = 1.015$. From the first population, a capture-recapture data set is sampled, from the second, a productivity data set, and from the third, a time series of counts is obtained. We use an IPM for these data sets that will serve as a benchmark when the common demography assumption (along with all other assumptions) is fulfilled. The last two woodchat shrike populations are created using different demographic rates from those in the first three populations. In population 4, juvenile survival and productivity are still identical to those of populations 1 to 3, but adult survival is higher ($s_j = 0.3$, $s_a = 0.64, f = 3.1$), such that population 4 grows at a higher rate ($\lambda = 1.105$). In population 5, the two survival probabilities are the same as before, but productivity is now higher ($s_j = 0.3$, $s_a = 0.55, f = 3.7$), which again results in an increased population growth rate of $\lambda = 1.105$. From each of these two populations, we sample a population count data set. We then jointly analyze the capture-recapture data from population 1, the productivity data from population 2, and the population count data from population 3, 4, or 5 with an IPM. The underlying population dynamics differs among component data sets in the latter two analyses, thus violating the common demography assumption. Finally, we repeat these analyses without the productivity data to see whether the effects of the violation of the common demography assumption differ when our IPM has a hidden parameter. Finally, we conduct all this both for a small sample scenario with an initial population size of 100 individuals and for a larger sample size scenario with an initial population size of 500 to investigate whether our conclusions are sensitive to sample size.

The IPMs adopted for the analysis of these data sets are very similar to the woodchat shrike IPMs that we have seen before (Sections 6.2 and 6.3): they assume constant demographic rates, allow for demographic stochasticity, and specify Gaussian residual errors in the state-space model. To assess model fit, we compute posterior predictive checks with the mean absolute percentage error as a discrepancy measure for the state-space model, with Freeman-Tukey statistic for the CJS model, and when applicable, with the deviance for the Poisson regression model.

For each design point in the simulation, we simulate 500 replications and summarize posterior means to assess bias. Bias is the difference between the mean of the estimated posterior means and the value used to simulate the data. Because several data sets are combined in an IPM, a parameter estimate is a weighted average of the information that each data set contains about that parameter, but the weight is unknown. The same applies for the true value of a parameter. If the data simulating values are identical across data sets, weighting does not matter, and consequently, the true value is known and the bias defined. However, if the data simulation values differ across data sets, i.e., when the common demography assumption is violated, the true value is unknown because the weights are unknown. Consequently, we cannot compute bias. However, we expect to obtain parameter estimates that lie somewhere between the two different data-generating values. We now carry out these simulations and refer you to the book web page for the full code.

The results of the simulations are summarized in Fig. 7.10. There is much to observe and discuss about these results, and we review each scenario in sequence. First, we see that the parameter estimates are unbiased when the common demography assumption holds, regardless of whether productivity is estimated as a hidden parameter or not (Fig. 7.10, solid and open blue boxes). Of course, this result was expected, and we have seen this several times before (Sections 5.5.2 and 6.3).

When the common demography assumption is violated, bias occurs in one or several parameters, and there is a difference depending on whether a hidden parameter is estimated. We start by the inspection of the scenario where the population counts originate from a population with higher adult survival than the population from where the capture-recapture data are collected (Fig. 7.10, gray boxes). Without an estimated hidden parameter, juvenile survival is positively biased, adult survival is within the expected bounds provided by the two true values (0.55, 0.64), productivity is unbiased, and the population growth rate is very close to the value in the population from where the count data were collected (Fig. 7.10, solid gray boxes). An unexpected result is the positive bias in juvenile survival. We offer the following heuristic explanations. Remember that estimates in IPM are weighted averages. As the population count data originate from a population with a higher growth rate than the populations from where the demographic data were sampled, the estimated

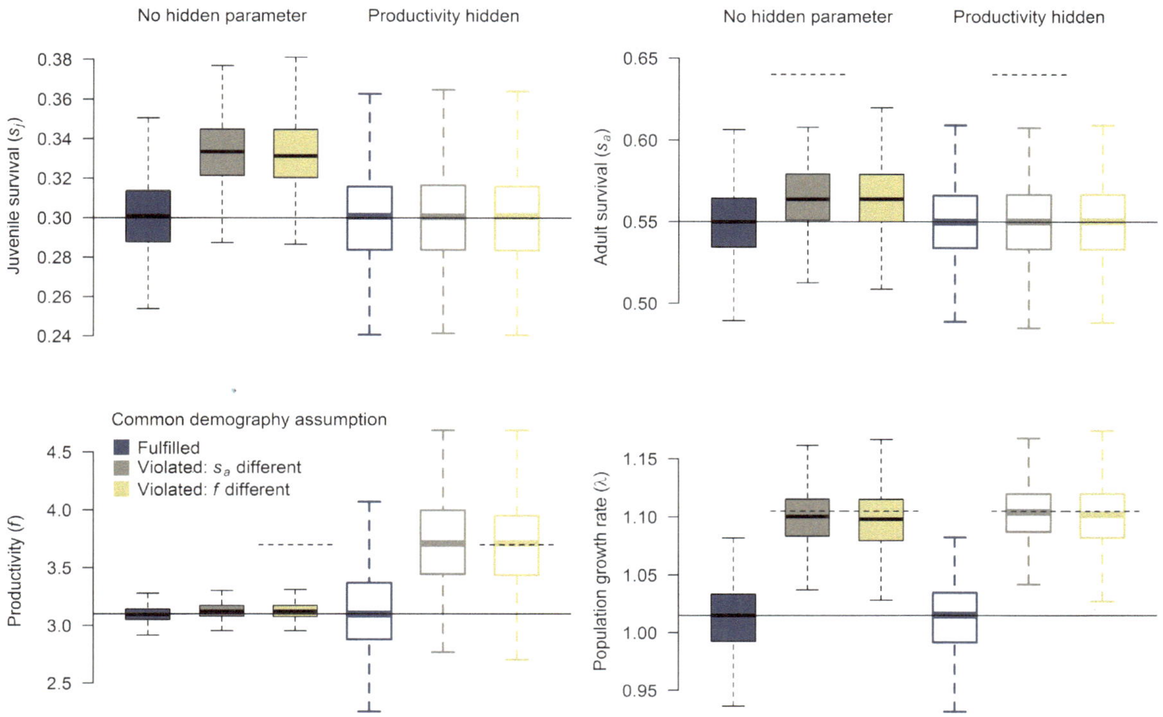

FIGURE 7.10

Boxplots of 500 posterior means of the demographic rates and population growth rate when the common demography assumption is either fulfilled or violated. The filled boxes are for integrated population models (IPMs) that use all three data sets and have no hidden parameter, while the open boxes are for IPMs with productivity as a hidden parameter. The solid horizontal lines show the values used in the generation of the demographic data. The dashed horizontal lines indicate the values of the demographic rates for the generation of the population counts when the common demography assumption is violated.

demographic rates are expected to be pulled upwards. By how much depends on the strength of the information in the demographic data. In our example it is not possible to pull productivity upwards, because the productivity data are "strong," i.e., contain much information. As expected, adult survival is pulled upwards somewhat but nowhere near the value in the population from where the count data originate ($s_a = 0.64$), because the capture-recapture data say that adult survival is $s_a = 0.55$. The slight pull upwards of adult survival is not "enough" to make the resulting, expected population trajectory match the observed one. Therefore, juvenile survival is also pulled upwards. In contrast to the productivity data, the capture-recapture data are less "strong," and therefore it is possible to shift the estimates of juvenile survival upwards. Furthermore, there is a dependence between the two survival probabilities and recapture probability. The pulling upwards of adult survival will lead to lower recapture probabilities and consequently to higher juvenile survival. Both mechanisms are likely to contribute to the observed bias in juvenile survival.

We observe completely different results in the model where productivity is a hidden parameter (Fig. 7.10, open gray boxes). Because there is no longer any independent information about productivity, this is now the "weakest" parameter in the model, and there is no resistance to shift its estimate. Indeed, productivity is estimated "by difference," meaning that the estimate is such that together with the estimated survival probabilities, a population growth rate results that corresponds to the observed one, i.e., that of the population count data. We can also see that the estimated population growth rate is slightly closer to that from where the population count data originated than that in the first analysis where we do have the productivity data. Juvenile survival is unbiased when productivity is hidden, and for adult survival, we obtain the estimate that corresponds to the value in the population from where the capture-recapture data are sampled. For the estimate of adult survival, we had expected a value somewhere between 0.55 and 0.64, but with a hidden parameter in the model, there is no longer any shift upwards. Therefore, when we have a hidden parameter in an IPM, we can say that the count data lose the power to shift unhidden parameters and affect the hidden parameter instead.

We see very similar results in the third scenario, where the population counts originate from a population with higher productivity (Fig. 7.10, yellow boxes). A slight difference is that the expected pulling upwards of productivity is not observed, since the productivity data are very "strong." To compensate, juvenile survival is again shifted upwards and therefore becomes biased as in the previous scenario, and likewise for adult survival, which is biased positively. This bias in both disappears when productivity is estimated as a hidden parameter (Fig. 7.10, open yellow boxes).

As for the GoF assessment, we find our posterior predictive checks to be insensitive to the violation of the common demography assumption (Fig. 7.11). The Bayesian p-values for the capture-recapture data in the IPMs without hidden

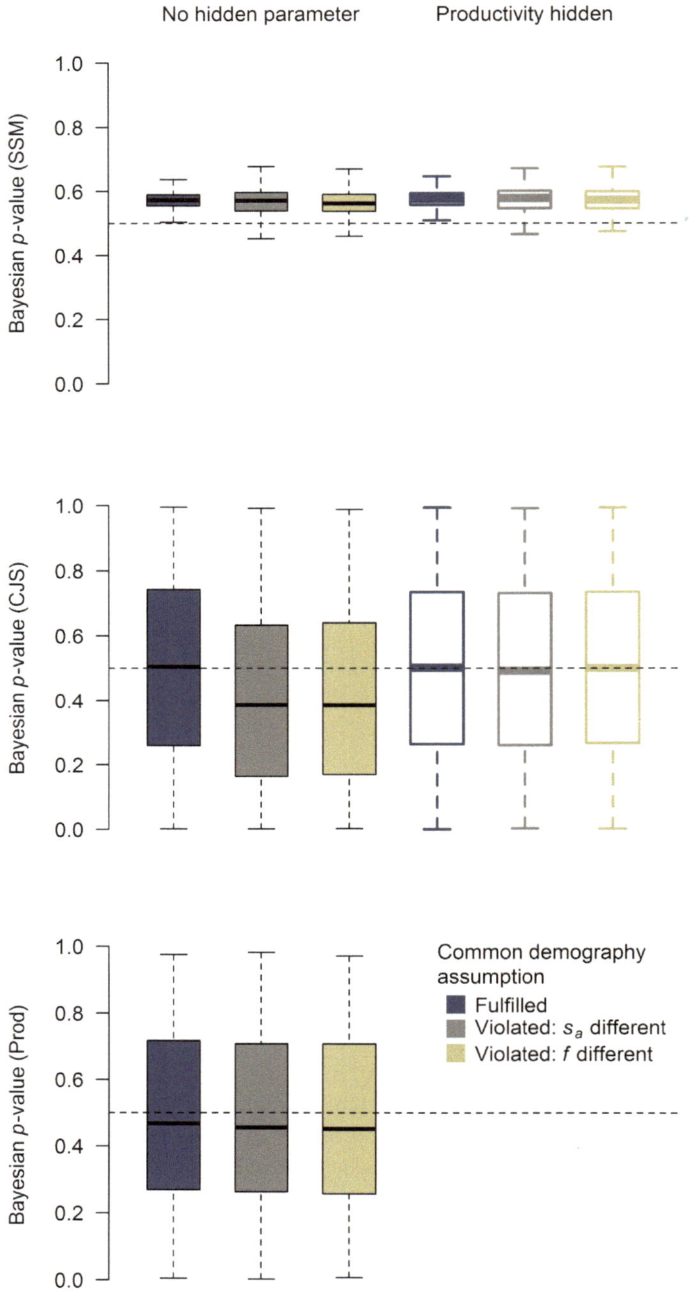

FIGURE 7.11

Boxplots of Bayesian p-values for the different posterior predictive checks in the state-space submodel (SSM, mean absolute percentage error used as a discrepancy measure), the capture-recapture submodel (CJS, Freeman-Tukey), and the Poisson regression submodel (Prod, deviance) in different integrated population models (IPMs) when the common demography assumption is fulfilled or violated. The solid boxes refer to IPMs that use all three data sets and have no hidden parameter, and open boxes refer to IPMs without productivity data.

parameter are slightly further away from 0.5 when the common demography assumption is met than when it is violated. The posterior predictive checks measure how well a simulated data set corresponds to the observed data. Data simulation using biased parameter should result in different data than those of the observed one, and therefore we can expect lack of fit for the CJS model for these IPMs. If the IPM contains a hidden parameter, the estimated survival with respect to the capture-recapture data will be unbiased. The Bayesian p-values will then be close to 0.5 and fail to indicate any lack of fit.

With a fivefold increased sample size, we find identical patterns (and show the results only on the website). The only exception is the result of the posterior predictive checks, which for the CJS model are more suggestive of a lack of fit ($p_B \approx 0.2$) for the IPM without a hidden parameter and with a violated common demography assumption. Clearly, the power to detect assumption violations increases with sample size.

Different patterns of demographic rates can be expected when the size of some but not all data sets is increased (see also Section 6.2.2). For example, if only the size of the capture-recapture data set is increased, we expect the estimated survival probabilities to be closer to the data-generating values from where the capture-recapture data were sampled and different estimates of productivity or an inflation of the residual error in the state-space model (see Section 6.6.5). The code and the results for the large sample sizes are not shown, but full code is on the book web page.

The common demography assumption is an important assumption in an IPM. When the different data sets tell a different story about a parameter, the parameter estimated is a weighted average of the components. We think this need not be a fundamental problem as long as we are aware of it. What is more surprising, and indeed more problematic, is that the violation of the common demography assumption can sometimes result in bias of the parameters informed by data sets with no conflicting information. In our simulation example, this was the case for juvenile survival. Therefore, we should generally worry about violation of the common demography assumption.

Unfortunately, a violation of the common demography assumption is not easily detected. In an IPM without any hidden parameter, posterior predictive checks for component data sets may indicate the violation of this assumption, but unfortunately, these tests often lack power. Furthermore, a comparison of specific parameter estimates obtained from the IPM and separate analysis of component data sets can be helpful to diagnose a violation of the assumption (Plard et al., 2021): if the estimates are clearly different, a common demography assumption violation might be the reason. For example, survival estimates obtained from the capture-recapture data alone would have resulted in estimates of juvenile survival of approximately 0.3 and adult survival of approximately 0.55. The estimates from the IPM were different (higher in our case) when the common demography assumption was violated. However, the process behind the violation of the assumption may remain unknown. We found the same results regarding changes of survival when either adult survival or productivity provided the conflicting information.

An alternative way to assess the common demography assumption is likelihood component profiling (Maunder and Starr, 2001; Francis, 2011; Wang et al., 2014). The idea is to compare the likelihood profiles of the component data within an IPM with respect to a common parameter. Different profiles indicate violation of the common demography assumption.

When an IPM contains a hidden parameter, it appears neither possible to detect a violation of the common demography assumption using posterior predictive checks, nor by a comparison between parameter estimates from the IPM with those of a single data analysis. Thus, GoF tests are unable to reveal any issues with model structures when we are estimating hidden parameters. We should keep this caveat in mind and list it as a potential drawback of an IPM with a hidden parameter.

If the common demography assumption is violated in an IPM there are several possible ways forwards (Saunders et al., 2019). First, we might not do anything, because we may argue that the estimates obtained are weighted averages and thus reflect a balanced knowledge about the studied system. However, as we have shown, this can lead to biased parameter estimates and is therefore not recommended. To defend such an approach, it might be useful to conduct a simulation study specifically tailored to the study system to decide whether any major bias may occur.

Second, we might reduce the amount of information that the less-trusted data set contributes to the IPM. The classical way for doing this is to weight the data (Francis, 2011; Fletcher et al., 2019). Data weighting is often done in models of integrated fisheries stock assessment (Francis, 2017). Another, less common approach is data thinning, which may be useful in particular when massive amounts of citizens science data are combined with much smaller, say, demographic, data sets (Sullivan et al., 2014; Robinson et al., 2018; Saunders et al., 2019). In a sense, this is just a different method of weighting the data sets. A last possibility is to include expert knowledge by specifying an informative prior for one or more parameters (Saunders et al., 2019). An informative prior introduces more knowledge about a parameter and therefore automatically weighs down the information contained in the data.

The third and final option is the modification of the IPM such that the common demography assumption becomes fulfilled (Maunder and Piner, 2017; Saunders et al., 2019). In our view this is often the best way to proceed, because it eliminates the reason for the assumption violation (Maunder and Piner, 2017). If we know which demographic parameter is informed by conflicting information, we can change the state-process model and introduce a new parameter. For example, if we know that there is conflicting information about adult survival from the capture-recapture and the population count data (corresponding to the solid gray boxes in Fig. 7.10), we replace parameter s_a in the process model by a new parameter, say s_a^*, where s_a is only informed by the capture-recapture data and s_a^* is estimated as a hidden parameter—it is informed by the population counts. We will then obtain two estimates of adult survival, both of which are unbiased with respect to the data-generating processes from which the capture-recapture and population count data are sampled. Furthermore, the bias originally observed for juvenile survival will then disappear. This option seems attractive at first look. However, we will end with the challenge of having to decide which value of adult survival should be used for inference. Moreover, if we correct the wrong demographic process, thinking it provides the conflicting information, our conclusions will be in error. Had we erroneously introduced a new parameter about productivity in the process model although adult survival was causing the conflict, we would have concluded that productivity in the counted population was different from that in the population from where the productivity data originate. Moreover, we would have obtained an estimate of adult survival that corresponds to the value of the population from where the capture-recapture data originate, and we would not realize that adult survival in the counted population was in fact different.

The most important step to avoid a violation of the common demography assumption must be done at the data sampling stage, or when we decide which data sets to combine in an IPM. We should combine only those data that are likely to tell a coherent story, or one that can be made coherent by 'modeling away' any differences. This evaluation is not easy, since an apparently incoherent story may simply be due to model misspecification (Maunder and Piner, 2017). If the component data sets are sampled from a small population, it is unlikely that the common demography assumption is violated—but then the independence assumption is more likely to be violated. By contrast if data from different populations or parts of a larger population are jointly analyzed (e.g., population indices from a national monitoring program and ring recoveries from more localized ringing activities or component data sets covering quite different time periods), the risk of violating the common demography assumption is increased, but in turn the independence assumption is more likely to hold. Based on our current understanding we believe a violation of the common demography assumption to be more problematic for unbiased parameter estimation in an IPM than the violation of the independence assumption.

7.2.4 CONCLUSIONS ABOUT INTEGRATED POPULATION MODEL ASSUMPTIONS

We have discussed the main assumptions needed when using an IPM, as well as GoF testing for this model class. We have illustrated assumptions and GoF tests by case studies and using simulated data. Our results are context specific to some degree, and of course we could not explore all possible types of assumption violations. For example, we have not examined the effects of several assumptions being violated simultaneously. Therefore, we could provide here only a glimpse of what can happen when assumptions are not satisfied, but there is still much to be learned about this topic for IPMs.

We covered posterior predictive checks as one method to test the violation of model assumptions. These tests are straightforward to implement and therefore perhaps the easiest way to assess the GoF of Bayesian IPMs. Posterior predictive checks include parameter uncertainty and the specified prior distribution. They are very flexible so they can be used for any kind of model and tailored to specific aspects of model fit by choosing appropriate discrepancy measures or test statistics. (Although it must be said that the choice of an appropriate discrepancy measure or test is an art in itself; A. Royle, pers. comm.). A disadvantage of the technique is their generally low power (Conn et al., 2018; Plard et al., 2021), which moreover varies with the chosen discrepancy measure. Measures based solely on the data (e.g., proportion of zeros in a data set) have more power than measures that also depend on model parameters (e.g., deviance; Lunn et al., 2013). Therefore, it may be good to use a range of different discrepancy measures (Newman et al., 2014).

In case lack of fit is diagnosed, one should try to identify the reason for it and fit an expanded model that relaxes the assumption that was violated. Models with more flexibility (which means they have more parameters) fit better, but their predictive ability will be reduced compared with simpler models (Hooten and Hobbs, 2015); this is the usual bias-variance trade-off in statistics. The goal of GoF testing is not to find a perfectly fitting model but one that does not violate assumptions and thus does not result in systematic errors (Conn et al., 2018).

GoF testing for IPMs is still in its infancy and we are aware of only two studies about this important topic. Besbeas and Morgan (2014) studied frequentist methods of GoF testing for IPMs, and Plard et al. (2021) explored posterior predictive checks and comparisons of parameter estimates from IPMs with corresponding component data models to diagnose lack of fit. A range of tests and ideas have been developed for integrated fishery stock assessment models that are very similar to IPMs (Maunder and Punt, 2013; Maunder and Piner, 2015; Carvalho et al., 2017; Maunder and Piner, 2017). Common approaches to assess model fit are residual and retrospective analyses (Carvalho et al., 2017), but the methods are similar to posterior predictive checks that have also been proposed (Piner et al., 2011). The paper by Maunder and Piner (2017) contains many useful information about causes and effects of misspecification, and methods of diagnosis, in integrated fisheries stock assessment models.

To date, most empirical studies using IPMs have not reported results of GoF tests. In our review we found 118 IPMs applied to bird, mammal, or amphibian data sets, but only 33 (28%) reported results from any GoF tests. The techniques used most were either posterior predictive checks or classical tests applied to component data using specialized software such as U-CARE (Table 7.1). Because parameter accuracy can be compromised in misspecified IPM, it is important to worry about GoF testing for IPMs. We think that the rarity of GoF testing in studies with IPMs is largely a consequence of the lack of well-developed GoF techniques for this model class.

So far, there has been very little research about the effects of assumption violations in IPMs and about ways to detect them. Based on our experience and on published information (Piner et al., 2011; Besbeas and Morgan, 2014; Carvalho et al., 2017; Maunder and Piner, 2017; Plard et al., 2021), here we give some recommendations for assessing the fit of IPMs. We hope that in the future, these can be corroborated or refuted as well as added to by more studies on this important topic:

- Start GoF assessment with the separate analysis of the component data sets when possible. The models adopted for these (e.g., CJS model and Poisson generalized linear model) are typically much better understood and various specialized diagnostics are often available. Assess the model fit and save the parameter estimates so that they can be compared with the corresponding estimates from the IPM. If you find lack of fit, modify the models accordingly.
- Fit the IPM using a structure of the component likelihood identified in the previous step and reassess the fit of each likelihood component (e.g., as we have shown in Section 7.2.1.).
- When using posterior predictive checks, try out a range of different discrepancy measures or targeted test statistics.
- If any model component shows lack of fit, try to identify the reason for it. This is often not easy because the GoF tests applied in the IPM are affected by all data/likelihoods that comprise the IPM.
- Changes in the results of the test component when applied inside or outside of the IPM and/or major changes in the parameter estimates in the component data likelihood and in the IPM may suggest the violation of the common demography assumption or a misspecified observation model (Section 7.4).
- Importantly, the fit of all model components should be assessed and not just the fit of the state-space model (Newman et al., 2014; Besbeas and Morgan, 2014; Carvalho et al., 2017).
- A good agreement between the population count data and the estimated population trajectory (i.e., in the GoF test of the state-space model) is typically not an indication that the IPM fits, but a disagreement almost certainly indicates a problem, and we would generally not trust such a model.
- The power of the GoF test components for the state-space model seems to be particularly low.
- Do not use overly flexible models for assessing the fit. Models without environmental and/or demographic stochasticity might be more useful to identify a misspecified population model.
- If the IPM contains hidden parameters, GoF tests applied to the count data set cannot detect lack of fit. It is therefore important to judge whether the estimates of hidden parameters are reasonable using the "commonsense GoF test." If they are not, then this almost certainly indicates lack of fit.
- Be aware that posterior predictive checks usually have low power (Conn et al., 2018; Plard et al., 2021). If the data sets have small size it may be almost impossible to detect lack of fit. Fit assessment of component data sets using classical

Table 7.1 Frequency of different types of goodness-of-fit (GoF) tests applied in 118 integrated population models fitted to actual bird, mammal, and amphibian data. Only 33 studies reported results for GoF testing at all.

GoF Test Applied	Frequency	Example References
Posterior predictive checks	15	Brooks et al. (2004), Rhodes et al. (2011), Schaub et al. (2015)
Frequentist test to single data likelihoods (typically for the capture-recapture data using U-CARE, Release, or MARK)	13	Péron et al. (2010), Tempel et al. (2014), Wilson et al. (2017)
QQ-plots of residuals	2	Besbeas et al. (2002), McCrea et al. (2010)
Squared loss statistic	1	Sanderlin et al. (2019)
Calibrated simulation	1	Besbeas and Morgan (2014)
Pearson residuals	1	Besbeas and Freeman (2006)
None	84	

GoF tests, which are especially well developed for capture-recapture models and have more power, are useful, although these tests do not exactly evaluate the fit of the IPM.
- Use simulations tailored to your data sets to explore the accuracy of parameters or the sensitivity of the chosen GoF tests in a misspecified model.

We have given a rather nonmathematical treatment of GoF testing for IPMs. We hope to have provided a feeling for what might be detectable with GoF tests and why such tests are often not sensitive to every assumption violation in your data set. We have emphasized that there are still many open questions, we are far from a deeper understanding of the topic in the context of IPMs, and more research on the topic will be quite welcome.

Finally, we add the following idea, which was pointed out to us by Mark Maunder. We have stressed in many places that what is often called the observation or sampling variance in a state-space model would in fact be better called the residual variance because it is composed of both an observation (or sampling) variance and of an additional variance component due to any lack of fit of the state model. In a correctly specified IPM where all assumptions are met, there is no lack of fit. Consequently, in such a model the residual variance should be identical, or very close, to the observation variance. Therefore, if the observation variance can be quantified somehow (for instance using external data in a separate analysis), its magnitude can be compared with the residual variance of the IPM, and any difference must indicate lack of fit. Note that we use this idea explicitly in model development when the population counts/index are estimated from some data and all we have is a point estimate and a standard error (Sections 4.3.3 and 6.6.4).

7.3 UNDER- AND OVERFITTING

Under a fitting IPM we can expect unbiased parameter estimates that allow valid inference. However, it is not always easy to know whether an IPM truly fits, because the power of GoF tests can be low when sample size is small or because only incomplete tests may be available to diagnose violations of model assumptions. Therefore, we must develop a gut feeling for what happens if there is unmodeled heterogeneity in the data, i.e., if there is unexplained structure in the data that is not accounted for in the model, which is then said to be *underfitted*. Alternatively, we want to know what happens when our model has too much structure, which is also known as *overfitting*. We pay particular attention to the accuracy of hidden parameters when they are estimated in an under- or overfitted model. The precision of these parameters is often low, and we may therefore expect that their estimates are particularly sensitive to under- and overfitting.

Under- and overfitting can occur for all parameter types of an IPM, and it would be impossible for us to cover all possible cases. Here, we just look at these topics in terms of the temporal structure in two parameters. As is often the case, we use a simulation approach where we create data sets that mimic a woodchat shrike population to illustrate how well parameters are estimated in the presence of under- and overfitting. We focus on the temporal structure of productivity and adult survival and produce four data sets that differ in whether productivity and/or adult survival are constant or vary over time. The four data sets are each analyzed with four IPMs that again differ in their assumptions about constancy or time-variation in productivity and adult survival. The result is 16 scenarios with underfitting models (when the data contain more structure than what our model assumes), with overfitting models (when the data contain less structure than assumed by the model), combinations between the two (because productivity might be overfitted and adult survival underfitted) and with some models that fit perfectly.

In all simulated data sets, all the other parameters are kept constant. To ensure that both the independence and the common demography assumptions hold, for each scenario we initially simulate three populations with identical demographic rates and sample one type of data from each. The initial population size is 100 females, and we project the population for 10 years with the mean demographic parameters chosen such that the population increases slightly on average: $s_j = 0.3$, $s_a = 0.55$, $f = 3.1$. In scenario 1, both demographic rates are constant over time. In scenario 2, adult survival varies over time, but productivity is constant. To produce temporal variability, for each year we pick, with replacement, a value at random from the vector $\{0.45, 0.46, 0.47, \ldots, 0.65\}$. In scenario 3, adult survival is constant but productivity is temporally variable, where for each year we pick with replacement a random value from the vector $\{2.10, 2.15, 2.20, \ldots, 4.10\}$. In scenario 4, both adult survival and productivity are temporally variable and simulated in the same manner as described.

We assume that either all three types of data or that only capture-recapture and population counts (but no productivity data) are available. Consequently, we explore two types of IPMs: IPM_1 jointly analyzes all data types and lacks a hidden parameter, whereas IPM_2 only uses capture-recapture and population count data and estimates productivity as a hidden parameter. For each type of IPM we develop four models that differ in whether adult survival and/or productivity is constant or variable over time (with fixed time effects). All the other parameters are assumed to be constant. The IPMs account for demographic stochasticity and we assume a Gaussian observation model. For each scenario, we analyze 500 data sets for a total of 8000 analyzed data sets. The simulation code can be found on the book web page as usual. We focus our attention on the estimates of productivity and adult survival and see how accurately they are estimated by inspecting relative bias and root mean squared error (RMSE).

Results from the IPMs without a hidden parameter indicate that productivity was unbiased in all scenarios (Fig. 7.12, gray boxes). The variability of the relative bias increased when a parameter was either variable over time or when a model was fit that specified time-variation in a parameter. The RMSE of productivity was larger when productivity was overfitted compared with when it was not (Fig. 7.13). Underfitting of productivity did not result in a larger RMSE compared with productivity estimates from a fitting model.

Results from the IPMs with productivity treated as a hidden parameter differed markedly. There was no bias when IPMs were used that estimated temporally constant productivity, but a slightly positive bias was apparent when productivity was estimated as temporally variable (Fig. 7.12, open boxes). The RMSE of productivity was markedly increased when estimated as a hidden parameter compared with when it was informed by explicit data (Fig. 7.13). RMSE was particularly large under a model with temporally variable productivity. The magnitude of the RMSE did not differ between under- and overfitting models (see, e.g., third row of Fig. 7.13, open boxes).

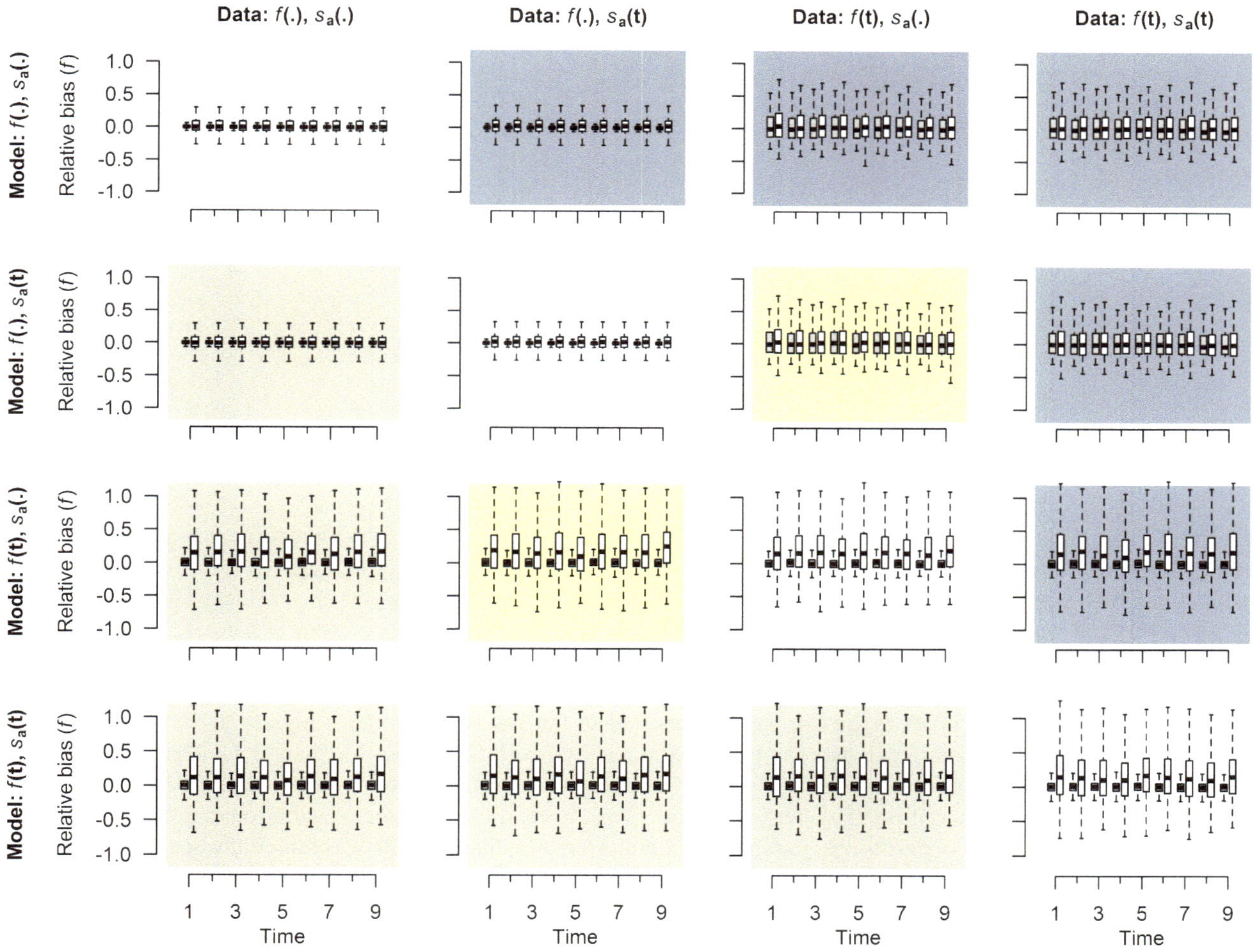

FIGURE 7.12

Boxplots of the relative bias of productivity estimates over time with data analysis model shown in the rows and data simulation model shown in the columns. Gray boxes indicate integrated population models (IPMs) using all three data sets and white boxes the IPM with productivity estimated as a hidden parameter. Gray panels indicate underfitting, blue panels overfitting models and in the yellow panels include models in which one parameter is over- and the other is underfitted. The white panels indicate scenarios where data simulation and data analysis models match, resulting in a perfectly fitting model.

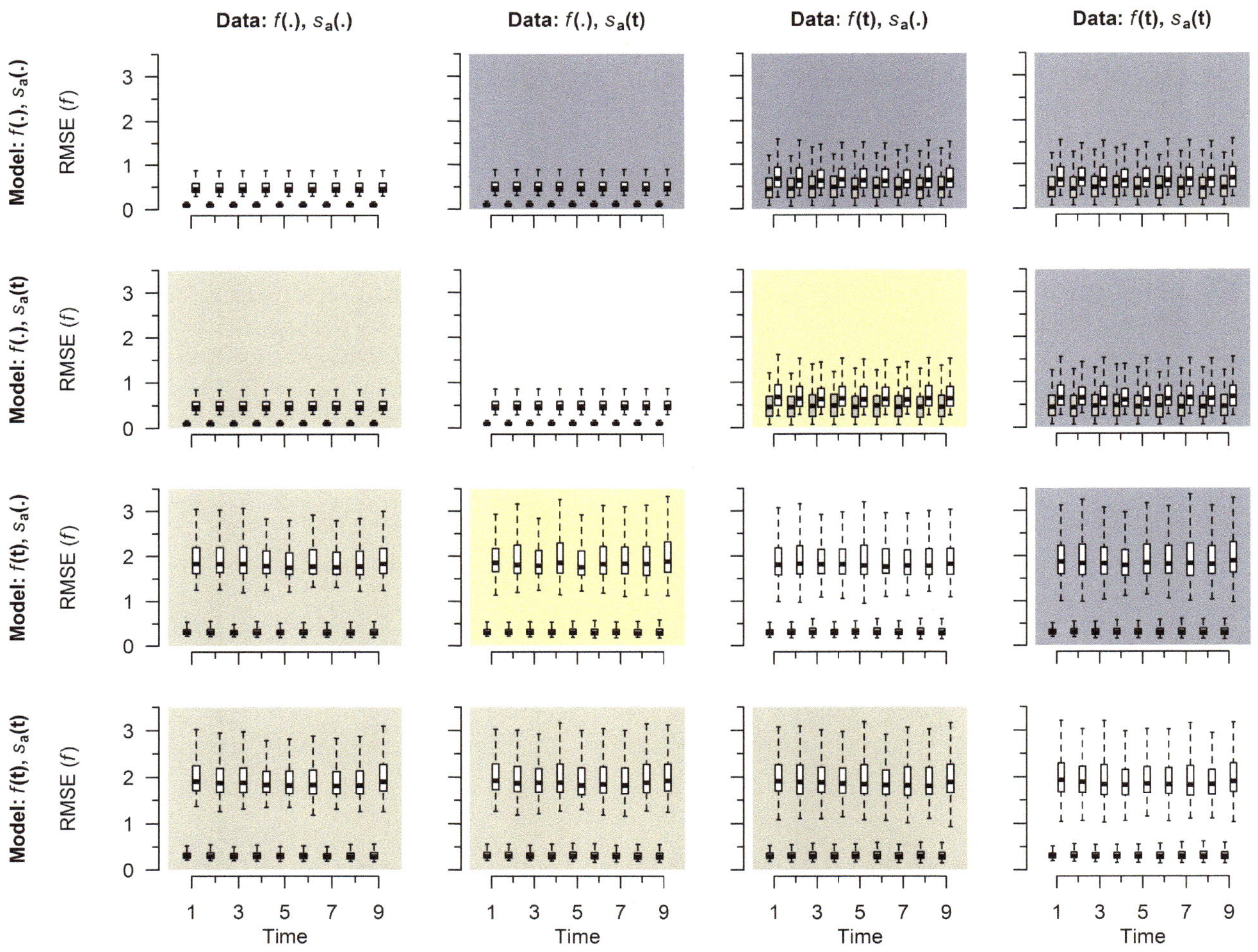

FIGURE 7.13

Boxplots of the root mean squared error (RMSE) of productivity over time. For further explanation, see legend to Fig. 7.12.

Survival was unbiased in all scenarios, regardless of whether productivity was estimated as a hidden parameter or whether the model for survival was under- or overfitting (Fig. 7.14). The variability of the relative bias in survival increased as soon as survival was either temporally variable or when a model with temporally variable survival was fitted, compared with when survival was constant over time. The RMSE of survival showed a similar pattern: it increased when either true survival or the estimated survival was temporally variable compared with when it was constant (Fig. 7.15). Under- and overfitting resulted in similar effects.

The effects of overfitting in IPMs are not well understood so far, but Besbeas and Morgan (2017) found that the estimates of the demographic parameters were not much affected by overfitting; see also Chapter 11. Generally, we expect increased uncertainty if the process model is specified assuming temporal variation when in reality it is constant (overfitting), but not bias. This was confirmed by our simulation. Conversely, if there is process variation in the population that is not adequately modeled (i.e., underfitting), parameter estimates may be biased. This was not confirmed in our simulation because the simulated process variation was unstructured (i.e., random). If there is structured process variation (e.g., a trend or correlated variation), we expect biased parameter estimates. This is similar to the way the effect of unmodeled detection probability in the count data is expressed in the parameter estimates (Section 7.4). Our simulations showed that the variability of the estimates increased when either there truly was temporal variation in the parameter or when the IPM specified such variation. We see clearly that accuracy of productivity is low when estimated as a hidden parameter, particularly when it is assumed to be temporally variable. In practice, this means that we may not always be able to identify the correct temporal pattern of a hidden parameter. This problem was already discussed by Tavecchia et al. (2009) and later confirmed by Lahoz-Monfort et al. (2014). We expect that the accuracy is improved when temporal random effects instead of fixed effects are used and when sample size is larger, but more research in this direction is needed.

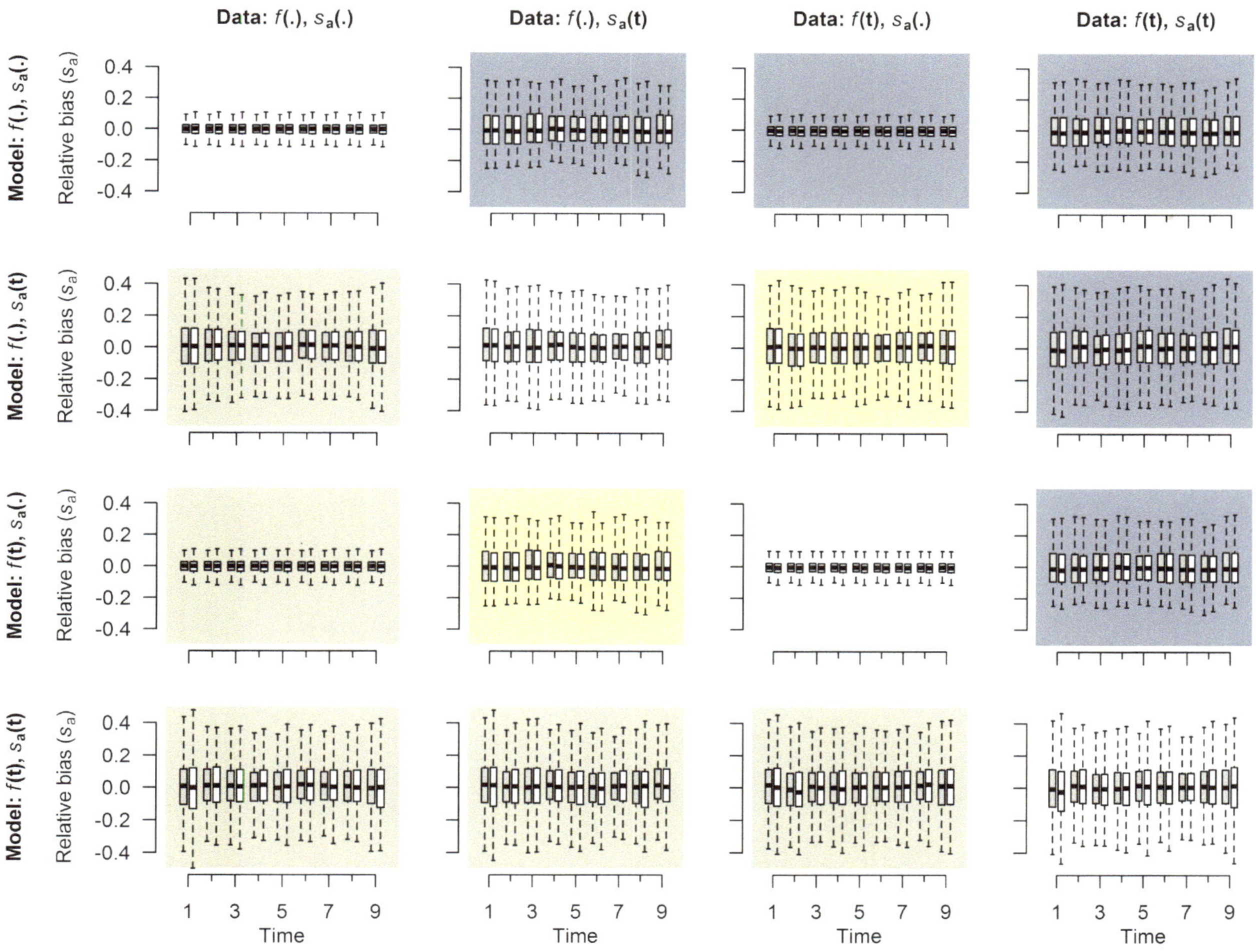

FIGURE 7.14

Boxplots of the relative bias of adult survival over time. For further explanation, see legend to Fig. 7.12.

7.4 EFFECTS OF A MISSPECIFIED OBSERVATION MODEL

Often, the goal of a population analysis with an IPM is not only the estimation of the demographic rates but also that of population size over time. The population count or population index data are typically subject to observation errors, because only in special cases is it possible to actually enumerate (i.e., to census) all individuals in a population. When a state-space model is adopted for population count data, it is sometimes claimed that one thereby "gets rid of observation errors," thus suggesting that the resulting estimates of population sizes are unbiased. However, this is generally not true (see Section 4.3.2), and it is extremely important to understand the correct interpretation of the estimates of "population size" in an IPM. We also need to gain an understanding of what happens to estimates from an IPM when the observation model for population-level data is misspecified.

All IPMs as defined in our book have at least one data set from the population level (Fig. 5.1), but there are different options for this data type. Some IPMs have used population indices derived from territory mapping data or aerial transect counts (e.g., Besbeas et al., 2002; Péron and Koons, 2012; Schaub et al., 2012; Zhao et al., 2019), detection/nondetection data of breeders (Sanderlin et al., 2019) or capture-recapture data from constant ringing effort sites (Cave et al., 2010), but most have used counts of individuals either from a single (e.g., Szostek et al., 2014; Lieury et al., 2016; Tenan et al., 2016; Woodworth et al., 2017) or from several sites (Borysiewicz et al., 2009; McCrea et al., 2010; Péron et al., 2010; Ahrestani et al., 2017). Only very rarely were stage-structured counts used (Tavecchia et al., 2009).

When an IPM is fitted to dimensionless population indices (as for instance in Besbeas et al., 2002) it is a priori clear that the estimated "population size" does not reflect true population size, but must also be interpreted as a population index. Hence, such population estimates have a relative, but not an absolute meaning. If all works well (e.g., our model is correctly specified), these indices will be tightly correlated with the true population size. However, the vast majority of

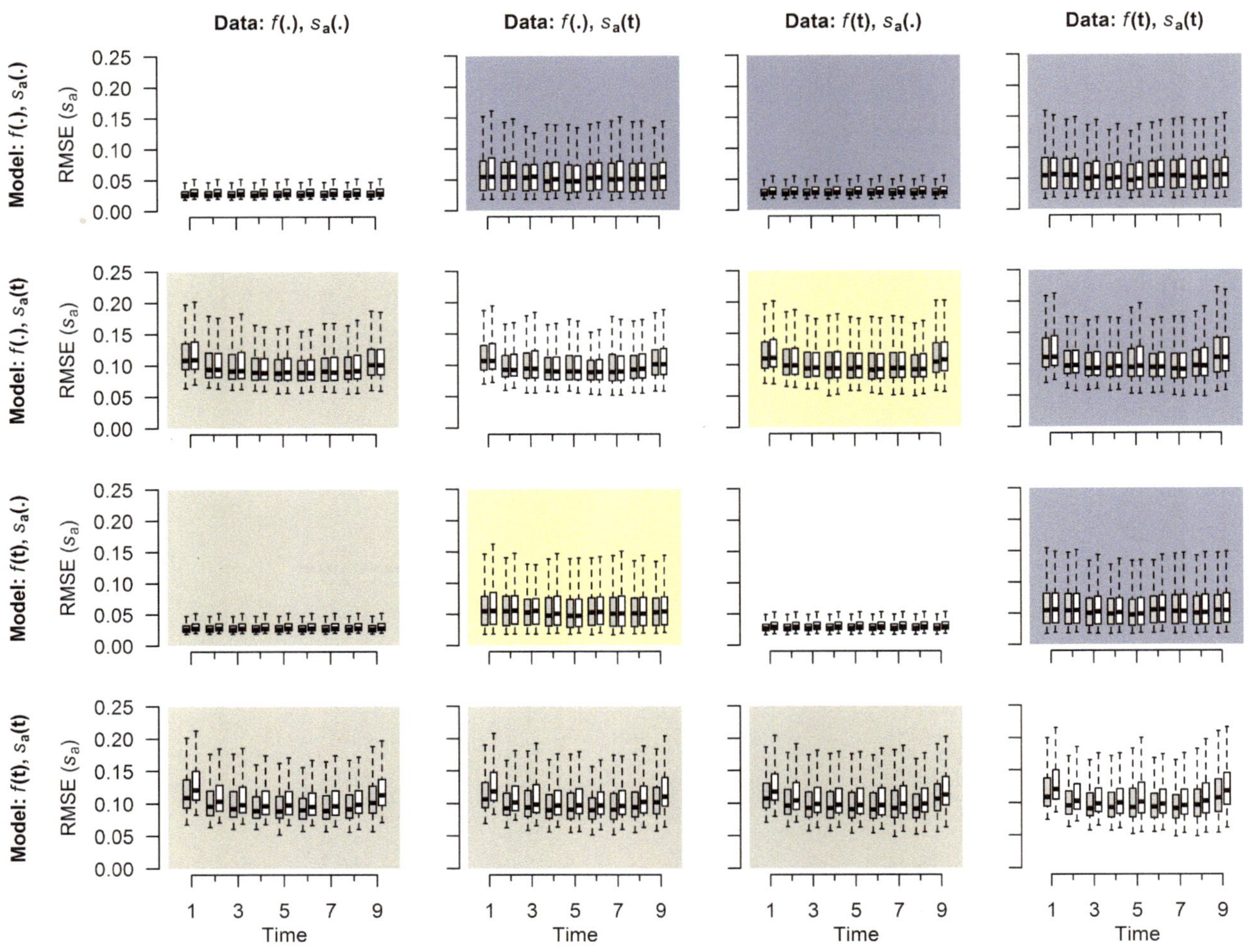

FIGURE 7.15

Boxplots of the root mean squared error (RMSE) of adult survival over time. For further explanation, see legend to Fig. 7.12.

IPMs are applied to actual population count data. These data are usually subject to various counting errors, and we may wonder how successful IPMs are in accounting for these errors. There are basically two types of counting errors (see Chapter 1 in Kéry and Royle, 2016, 2021): we may miss individuals who are "there" (i.e., imperfect detection leading to false-negative errors), or we may add in a count individuals who are "not there," which represents a false-positive error. Two major mechanisms for the latter are multiple counting of the same individual (for instance because an individual moves around during a survey), or species misidentification, when we count as our target species individuals of another species. These types of errors are not mutually exclusive and may occur simultaneously in a single count.

The first error, represented by imperfect detection, is perhaps usually the dominant error in animal and plant surveys and a host of sampling protocols and associated analytical methods have been designed to account for this type of error (e.g., capture-recapture: Williams et al., 2002; Royle et al., 2014; distance sampling: Buckland et al., 2001; double observer protocol: Nichols et al., 2000; replicated counts: Royle, 2004; depletion and catch curve analyses: Maunder and Piner, 2015). These methods can be applied to appropriately sampled population-level data and the estimated population size along with an associated measure of uncertainty can then be used as data in an IPM (see Section 6.6.4 and Fay et al., 2019). Alternatively, we may specify an IPM in which the estimation of population size is directly a part of the observation model (e.g., Sanderlin et al., 2019). Either way we obtain estimates of true population size from such an IPM. This would be the ideal situation, but the application of these approaches and hence the estimation of true population size needs more information than just a single annual population count. Often this information is not available. In addition, specification of such models may not always be easy in practice.

Estimation and therefore correction for the other type of error, leading to false-positives, is much more challenging statistically, though some progress has been made in recent years for occupancy models (Royle and Link, 2006; Miller et al., 2011; Chambert et al., 2015; MacKenzie et al., 2018; Kéry and Royle, 2021). In principle, it should be possible to specify such models also as part of an IPM, but this does not seem to have been done so far.

In this section, we explore how well IPMs estimate population size and demographic rates when population count data are subject to counting errors and we have no additional information that allows explicit modeling of these errors. Again, we use simulation to understand these scenarios. We simulate data sets that differ in the way the population count data are collected and then analyze them together with individual-level data in standard IPMs. This allows us to gauge the sensitivity of the estimates to various types of counting errors.

We consider four different scenarios to sample population count data. In the first we assume that counts are correct on average, but that the data are noisy (i.e., have errors in both "directions") because some individuals may be missed, and others may be counted multiple times. We simulate these count data using a normal distribution. In all other scenarios we assume that counting errors are just due to imperfect detection, but the probability of detection shows three different patterns over time: (1) constant, (2) randomly different in each year, and (3) with a trend over the years. We generate these count data with binomial distributions with a trial size that is the true population size and detection probability $p < 1$ using the function `simCountBin` of the `IPMbook` package. For all scenarios we also simulate independent capture-recapture and productivity data. In the first sets of simulations, we assume we have collected all three types of data (capture-recapture, productivity, counts) and we do not estimate any hidden parameter, while in the second set of simulations, we ignore the productivity data and estimate productivity as a hidden parameter. The parameter values for the simulation again mimic a slightly increasing woodchat shrike population ($s_j = 0.3$, $s_a = 0.55$, $f = 3.1$, all constant over time) starting with 100 females, and we project it over 10 years.

We then fit IPMs to these data sets assuming constant demographic rates and allowing for demographic stochasticity. We specify the observation model for the observed counts C_t as a normal distribution, i.e., $C_t \sim Normal(N_{1,t} + N_{2,t}, \sigma^2)$. In our first scenario, the observation model is therefore correctly specified, but it is misspecified in the three other scenarios. We fit IPM_1 to all three data sets and IPM_2 to the capture-recapture and population count data. These two IPMs are identical to the IPMs that we used already in Section 7.3. All code for the simulation, model fitting and figure making is provided on the book web page.

We first inspect estimates of demographic rates obtained from IPM_1, which has no hidden parameter. All demographic rates as well as the population growth rate appear unbiased when the count data are generated with a normal distribution (Fig. 7.16). This is no surprise, as we have seen repeatedly that IPMs produce unbiased parameter estimates when the model assumptions are fulfilled.

More interesting are the scenarios where the counts are subject to imperfect detection. When detection probability is either constant or varies randomly over time, all demographic rates as well as the population growth rate are again unbiased (Fig. 7.16). Their accuracy (RMSE) is also similar compared with those of the scenario when count data are generated by the normal distribution (Fig. 7.17). The only exception is the reduced accuracy of the population growth rate when detection probability varies over time.

However, we note a marked difference when detection probability has an increasing trend over time. Survival of both age classes now shows a positive bias as does the population growth rate (Fig. 7.16). The RMSE becomes inflated for these parameters, and this is most pronounced for the population growth rate. The reason for the positive bias in survival is that the counts are saying that the population increases strongly, whereas based on the demographic data alone, we would expect a moderate increase only. Because the estimates in the IPM are weighted averages, we obtain a bias in survival. This bias occurs in survival and not in productivity because the (capture-recapture) data for survival contain relatively less information. Hence, we see the same result as we have already seen in our simulations to study the effects of the violation of the common demography assumption (Section 7.2.3): the demographic data tell a different story than do the count data. The difference is that now the common demography assumption is in fact not violated, but the sampling protocol producing the population-level data induces a pattern in the counts that our state-space model was unable to correct for adequately.

Results from IPM_2, which estimates productivity as a hidden parameter, are largely similar to those from IPM_1, but differ in some important ways. First, as before the parameters are unbiased when population counts are correct on average (as when they are generated by the normal distribution) and when they are subject to imperfect detection with detection probability either constant or variable over time in a random manner (Fig. 7.16). Parameter accuracy across these different underlying observation models is similar, though with a slightly greater RMSE of the population growth rate when detection probability is variable over time. Not surprisingly, the accuracy of the estimates of productivity under IPM_2 is much reduced compared with IPM_1 when productivity data are included. However, the results differ again when detection probability has a trend over time: as before when no hidden parameter was estimated, the estimate of the population growth rate is positively biased (Fig. 7.16), but this time there is also a positive bias in productivity, while survival is unbiased in both age classes. The RMSE shows the same pattern: it is greater for productivity and population growth rate, but not for survival (Fig. 7.17). The bias is now transferred to the estimates of productivity because there is no independent information about productivity, and hence this is the "weakest" parameter. The hidden parameter compensates for any kind of inconsistency among the data sets (Section 7.2.3). Now the inconsistency is produced by the increasing trend in detection probability that could not be accounted for in the observation model for the count data.

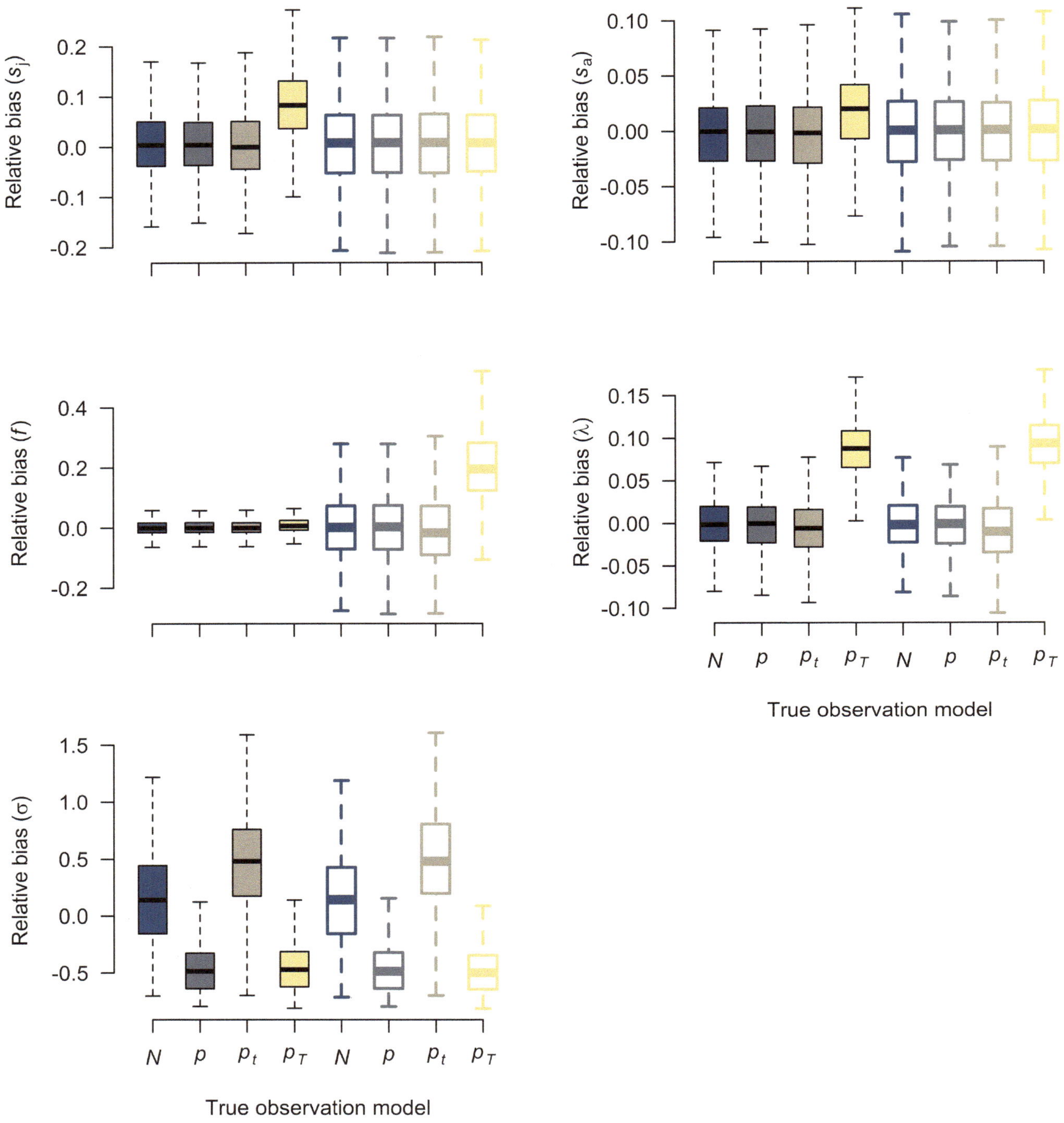

FIGURE 7.16

Boxplots of the relative bias of posterior means for juvenile survival (s_j), adult survival (s_a), productivity (f), population growth rate (λ) and the residual error (σ) obtained from 500 simulations. The results are obtained by fitting the first integrated population model (IPM), IPM_1, to all three data sets (solid boxes) and by fitting IPM_2 where productivity is estimated as a hidden parameter (open boxes). The population counts analyzed were N, generated by a normal distribution; p, generated by a binomial distribution with constant detection probability; p_t, generated by a binomial distribution with detection probabilities that vary randomly over time; and p_T, generated by a binomial distribution with linearly increasing detection probability.

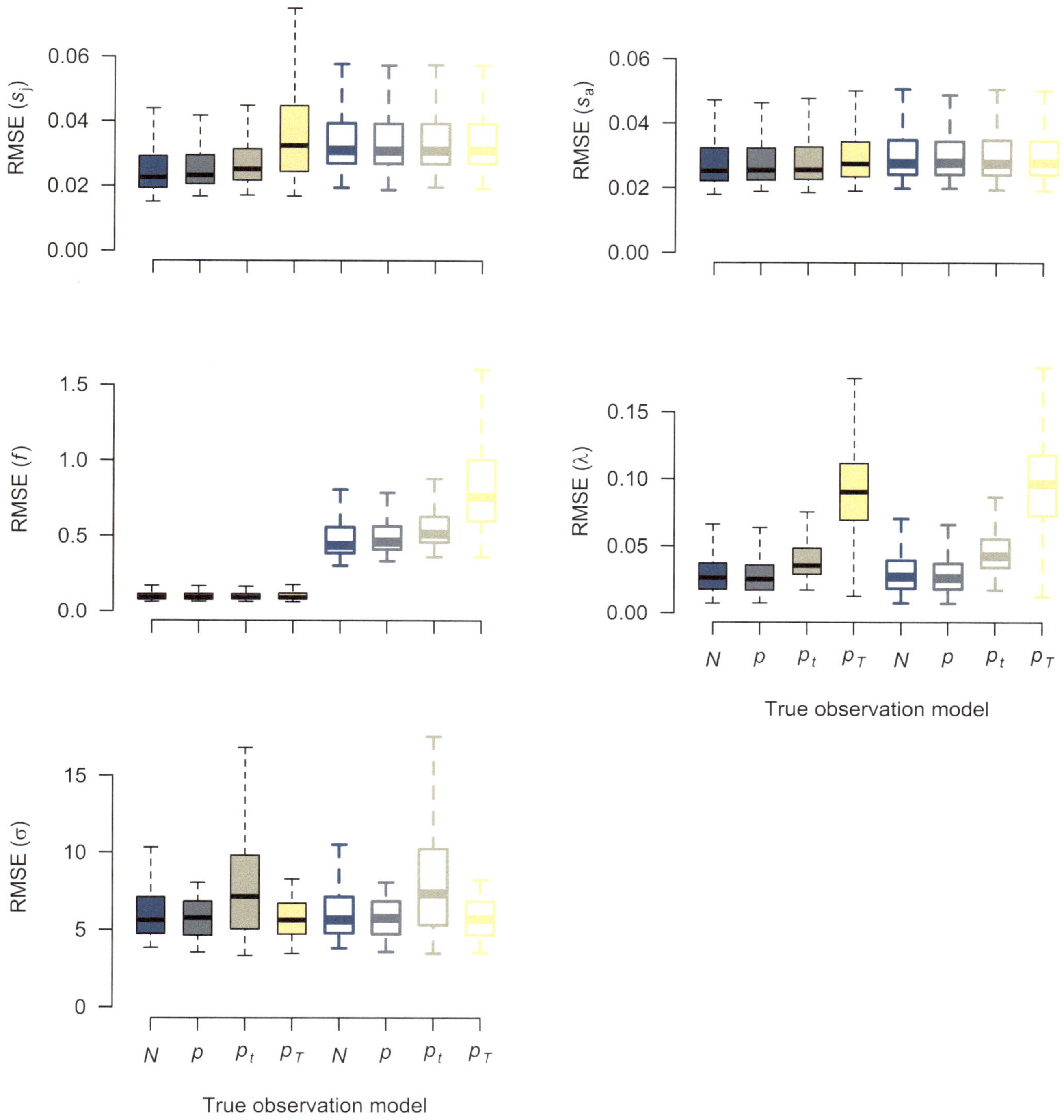

FIGURE 7.17

Boxplots of root mean squared errors of juvenile survival (s_j), adult survival (s_a), productivity (f), population growth rate (λ) and the residual error (σ) obtained from 500 simulations. The results are obtained by fitting the first integrated population model (IPM), IPM_1, to all three data sets (solid boxes) and by fitting IPM_2 where productivity is estimated as a hidden parameter (open boxes). The population counts analyzed were N, generated by a normal distribution; p, generated by a binomial distribution with constant detection probability; p_t, generated by a binomial distribution with detection probabilities that vary randomly over time; and p_T, generated by a binomial distribution with linearly increasing detection probability.

The estimated residual (or observation) errors show some interesting patterns across the studied scenarios (Fig. 7.16). Specification of the normal observation model leads to slight positive bias of the residual error variance, which we think is because there is some lack of fit, as the joint analysis combines data sets from different populations. With constant detection probability, the residual error becomes negatively biased, which can be explained by the fact that counts are smaller than the true population size and therefore their variability also declines. The coefficient of variation computed as the residual error divided by the average population size is similar in both cases, showing that the error is comparable in size. When detection probability varies randomly over time, the residual error is positively biased again. This is expected, because there is additional random noise in the counts that the observation model must account for. Interestingly, the residual error is negatively biased again when the detection probability shows a trend over time. These patterns are identical regardless of whether the IPM contains a hidden parameter or not. Moreover, the accuracy of the residual error is similar for all scenarios (Fig. 7.17), it just tends to be reduced when detection probability varies over time randomly.

Next, we inspect the estimated population sizes under the different scenarios and models. As expected, when the population counts are correct on average, the estimated population sizes are unbiased (Fig. 7.18). In contrast, when the counts are subject to imperfect detection, the estimated population sizes are negatively biased (Fig. 7.18). The magnitude of bias depends on detection probability. Indeed, the population size estimates correspond to the expected counts, i.e., the product of true population size and detection probability. Expected counts are shown by the dashed magenta lines in Fig. 7.18. When detection probability is either constant or varies randomly over time, the estimated population size and the true population size have the same temporal pattern. The resulting population growth rates are identical (Fig. 7.16). Thus, despite the negative bias in the population size, we can reliably make valid inferences about relative population changes. By contrast, if the detection probability has a deterministic trend, the temporal patterns of estimated and true population sizes differ, and we cannot make reliable inferences about population changes. These results are the same regardless of the presence or absence of a hidden parameter in the model.

The finding that imperfect detection matters for population size estimation is certainly not new (Buckland et al., 2001; Borchers et al., 2002; Williams et al., 2002; Royle and Dorazio, 2008; Kéry and Schaub, 2012; Royle et al., 2014; McCrea and Morgan, 2015; Kéry and Royle, 2016; Monroe et al., 2019; Seber and Schofield, 2019). Specification of the state-space model within an IPM is useful to account for some variation in detection (counting errors) but cannot solve the basic problem of undercounting in the presence of imperfect detection or overcounting in the presence of false-positive errors. If imperfect detection shows no deterministic pattern (i.e., is either constant or is stationary with random noise only), we can at least use the population size estimates to infer population change (see Chapter 1 in Kéry and Royle, 2021). If imperfect detection is constant, the only drawback is that population size estimates are negatively biased. The price for accounting for the random variation in imperfect detection is additional uncertainty in our model: the accuracy of the population growth rate and of the residual error is reduced. However, the state-space model cannot account for the nonrandom variability induced by imperfect detection with a deterministic pattern such as a trend. In that case, not only population sizes, but often also demographic rates become biased. The detection and correction of systematic patterns in detection is a major issue in fisheries and has been subject of much research there (Maunder and Punt, 2004).

We also performed posterior predictive checks with the discrepancy measures introduced in Section 7.2. It was sobering to see that none of the tests indicated any lack of fit (results are not shown). Initially, we might have hoped that these checks were sensitive to a misspecified observation model. However, these tests are not sensitive because the fit between the observed counts and the estimated population size is excellent regardless of the adequacy of the observation model. The estimated survival probabilities are biased under IPM_1 when the detection probability for the count data shows an increasing trend, so we can expect that the corresponding GoF test component would indicate lack of. However, it did not do so here, which is likely due to the relatively small sample size and the resulting lack of power of the test.

Hence, if we have no means to detect any misspecification of the observation model in an IPM, what should we do? We think that the best way is to adapt the sampling protocol for the population-level data in such a way that detection probability can be estimated. We can then account for imperfect detection and model, and therefore correct for, any pattern in detection probability as an integral part of the IPM (Chandler et al., 2018; Saunders et al., 2019; Schmidt and Robison, 2020). When you design a new study, we strongly suggest giving this issue some serious thought. However, in many cases, the data for an IPM have already been collected, which leaves us with the problem. Depending on how the count data were collected, we may have an idea about the likely magnitude of detection error. For example, in an intense small-scale study, most breeding individuals are likely to be detected, resulting in high detection probability and a close match of the counts and true population size. In other studies in which an index of population size is used, it is clear that an IPM does not estimate true population size. Perhaps the most serious problem is knowing whether detection has any deterministic patterns, because no reliable inference may then be possible unless this pattern is modeled adequately. We may be tempted to replace the normal observation model in the IPM by a binomial model, i.e., $C_t \sim Bin(p_t, N_{1,t} + N_{2,t})$ and to try and estimate detection probability p. Unfortunately, if we have just a single annual count *from a single population*, detection probability is not identifiable (though it may become so when replicated time series of counts are available, as in the model

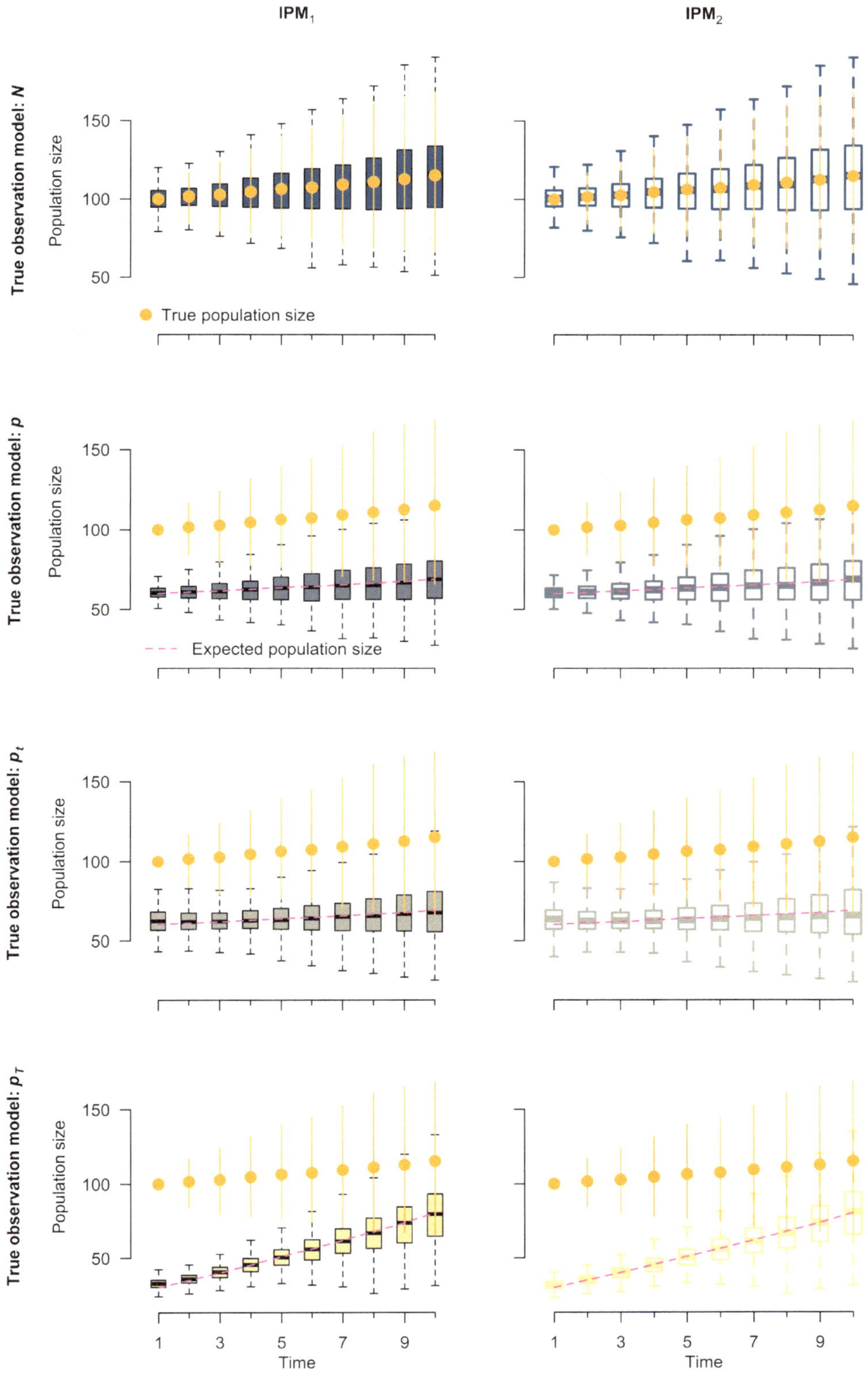

FIGURE 7.18

Boxplots of posterior means of population sizes obtained from 500 data sets. The results are obtained by fitting the first integrated population model (IPM), IPM$_1$, to all three data sets (left column) and by fitting IPM$_2$ when productivity is estimated as a hidden parameter (right column). The panel rows correspond to different ways how the count data were simulated (*N*: using the normal distribution; *p*: using the binomial distribution with constant detection probability; p_t: using the binomial distribution with detection probabilities that vary randomly over time; p_T: using the binomial distribution with linearly increasing detection probability). The orange dots and vertical lines are the mean and 95% credible interval of the true population sizes. The dashed magenta lines show the expected counts, i.e., the product of detection probability and of true population size.

of Dail and Madsen, 2011). Because the parameter estimates under an IPM with population counts subject to a deterministic detection pattern look similar as when the common demography assumption is violated, we can use the same methods for diagnosis as in Section 7.2.3, in particular any mismatch between the estimates of demographic rates when estimated in the IPM and when estimated in the component data set alone is indicative. However, this also means that we may not be able to discriminate between a violation of the common demography assumption and a violation of the assumed observation model due to systematic patterns in the detection errors.

7.5 OUTLOOK AND SUMMARY

IPMs are powerful tools for inference about population dynamics, but as any statistical model may produce inaccurate parameter estimates when their assumptions are violated to a sufficiently strong degree. Therefore, it is important to know these assumptions, whether and how they can be tested, how robust parameter estimates are likely to be when assumptions are violated, and how your IPM can be modified to rectify any lack of fit. These simulations and others also emphasize that IPMs are not a panacea for data that are too weak or too strongly discrepant.

In addition to the assumptions required for each component data set, an IPM makes additional assumptions when multiple data sets are integrated. For typical IPMs, these data sets must be statistically independent, and they must "tell the same story" about the dynamics of a population. We have illustrated the use of posterior predictive checks to detect assumption violations. The main idea is to formulate posterior predictive checks for each component likelihood and then none of them should indicate any lack of fit. The fit of the state-space model (i.e., for the population-level data) can be viewed as an overall test of the IPM, because the process model is informed by the different individual-level data. However, such a test is unfortunately quite insensitive to most assumption violations because a myriad of combinations of demographic rates can result in the same observed population trajectory. It is only in exceptional cases, and usually with large sample sizes, that we may be able to pick up lack of fit at the level of the observed population trajectory. In addition, when an IPM with a hidden parameter is adopted, the power to detect lack of fit for the state-space model goes down to virtually zero, because the hidden parameter will compensate for most lack of fit from all other model parts.

We have also seen that parameters of ill-fitting IPMs can often be biased. Because IPM estimates are weighted averages of the information from the different data sets, bias typically appears for parameters least informed by component data, or the "weakest" parameters. This is most evident when a hidden parameter is included in the model and compensates for lack of fit.

GoF tests are not well explored or developed for IPMs, and many open questions remain. Based on our experience, we have offered some advice about how to check IPMs. We have also stressed that not only should the fit of the IPM be assessed but also that of the "ingredient models" for each component data set separately (i.e., outside the IPM). Major changes in the parameter estimates or of the GoF assessments between the separate and integrated analyses may indicate a lack of fit.

In this chapter, we have made extensive use of simulations to study the consequences of ill-fitting IPMs. Simulations can be tailored to a specific study to assess, (1) whether a GoF test adopted can indeed detect lack of fit and (2) to gauge how much parameter estimates are affected when the model does not fit. Such simulation assessments are particularly useful because our knowledge about the sensitivity of GoF tests is limited and because the latter typically have low power to detect departures of the data from the assumptions embodied by the model.

In summary, this chapter represents the melding of a little bit of established knowledge, a moderate volume of new results presented in simulation studies, and a good deal of common sense gained by many years of practical fitting of IPMs to countless real and simulated data sets. We believe that the recommendations given in this chapter are useful in many ways but emphasize again that the assessment of complex statistical models such as an IPM is a particularly challenging topic and one that requires much more study.

7.6 EXERCISES

1. Established frequentist GoF tests for the CJS model (Burnham et al., 1987) are based on summaries of the capture histories. There are several such types of tests that assess whether the survival and recapture probabilities are the same in all marked individuals (e.g., tests 3.SR, 2.CT). These tests are specific to certain types of model violations, e.g., the occurrence of transients or trap-dependence and are therefore particularly useful. For each of these tests the data are tabulated in specific ways and it is tested by χ^2-tests whether the tabulated numbers deviate from expected numbers calculated under the hypothesis of homogeneous rates. We can use these tests either directly, using the R package `R2ucare` (Gimenez et al., 2018), or implement them into a Bayesian assessment. Assess the GoF of the CJS model fitted to the capture-recapture data from the `woodchat5` data set using these specific tests in the Bayesian framework, i.e., do posterior predictive checks using these tests as discrepancy measures.

CHAPTER

INTEGRATED POPULATION MODELS WITH DENSITY DEPENDENCE

8

CHAPTER OUTLINE

8.1 INTRODUCTION

In a stable environment, the population size of most species typically remains within fairly narrow bounds over long periods (Sibly et al., 2005; Brook and Bradshaw, 2006). Population size usually fluctuates because of stochastic effects but will not show any long-term trend (i.e., increase or decrease). This implies that populations are regulated in some way such that their growth rates decline when population size increases and increase when population levels decline. Such a negative feedback process in population dynamics is called density dependence. Sometimes, it is also termed negative density dependence because there is a negative relationship between growth rate and population size. Rarely, there is also positive density dependence, and this is known as the Allee effect (Courchamp et al., 1999). Here, we deal only with the classical case of negative density dependence. However, studying Allee effects is no different from studying the classical case; we simply work instead with the expectation of a positive density effect in our models.

Often, negative feedback between population growth and population size (or density) result from competition for limited resources such as food, mates, or nesting sites. Competition typically results in behavioral changes by individuals, and this in turn alters the demographic performance of some or all individuals within the population (Sutherland and Norris, 2002). The actual mechanisms leading to this pattern can vary greatly depending on the species, population, or type of limiting resource. For example, the territory size of black-throated blue warblers (*Setophaga caerulescens*) became smaller when population density was higher, and this had negative effects on reproductive output (Sillett et al., 2004). Body mass in large herbivores tends to decline because increasing population size results in food limitations that then have negative effects on survival and recruitment (Bonenfant et al., 2009). These mechanisms are known as *interference competition*—naturally, each individual on average has access to less of a limited resource when population sizes increases. Another mechanism underlying density dependence does not affect all individuals equally—stronger individuals suffer less or not at all, while the demographic performance of weaker individuals goes down as population size goes up. This type of competition can arise, for example, when territories or migration staging sites differ in quality: dominant individuals will then tend to establish themselves at the higher-quality sites and achieve better demographic performance, while subordinate individuals can only occupy lower-quality sites and will suffer from reduced demographic performance. This widespread mechanism is known as *site-dependent population regulation* or the *buffer effect* (Rodenhouse et al., 1997; Gill et al., 2001). It may affect survival (Gill et al., 2001; Pettorelli et al., 2003), clutch size (Dhondt et al., 1992), breeding success (Nilsson, 1987; Rodenhouse et al., 2003; Gunnarsson et al., 2005), or dispersal (Andreassen and Ims, 2001). It is also possible that different mechanisms acting on different demographic rates may occur simultaneously. For example, the Mauritius kestrel (*Falco punctatus*) is regulated through site-dependence affecting breeding success and through interference competition affecting juvenile mortality (Nevoux et al., 2011). Predation can also result in density dependence but requires a behavioral change

Integrated Population Models. https://doi.org/10.1016/B978-0-12-820564-8.00010-5

(or functional response) by the predator such that it kills an increasing fraction of the prey species as the number of prey individuals increases (Whitfield, 2003). Finally, a limitation on shelter space may also cause increasing predation-induced mortality with increased population size (Holbrook and Schmitt, 2002).

Understanding the mechanisms underlying density-dependent population regulation is a key question in population ecology (Lack, 1954; Hixon et al., 2002). Moreover, it is of paramount importance for the prediction of future population size and therefore highly relevant for population management. For example, the functional form of density dependence (see below) affects the rate of sustainable harvest of populations (Sæther et al., 1996; Kokko and Lindström, 1998; Lebreton, 2005). The presence of density dependence is one of the requirements for allowing hunting mortality to be compensated by other causes of mortality. Hence, we need to understand density dependence in order to assess the degree to which harvesting mortality is compensatory (Lebreton, 2005; Péron, 2013). Density dependence also plays a key role in predicting population trajectories into the future and to assess their extinction risk. The forecasts of both population size and extinction risk estimates can be very sensitive to the type and functional form of density dependence in a demographic model (Ginzburg et al., 1990; Mills et al., 1996; Sæther et al., 1998, 2004). Hence, a thorough understanding and an accurate estimation of density dependence is crucial for any kind of population management. This includes applications in conservation biology as well as in the management of harvested species.

The study of density dependence is challenging for a number of reasons (summarized by Lebreton and Gimenez, 2013). First, density dependence can be a very complex process because it may manifest itself in multiple demographic rates (e.g., recruitment, productivity, or survival), occur at different times of the year (e.g., during the breeding season or at nonbreeding sites, such as migration staging, or wintering areas), and vary in strength over time (e.g., if food availability is fluctuating, it may be limiting in one year but not in another). Finally, the functional form of density dependence may be nontrivial. For instance, it may be a straight line, a concave or convex curve, or a step function. In addition, density dependence may affect the whole population, or more commonly perhaps, only part of it. In addition, this part may be variable, for example, if only young individuals are affected and the population structure varies over time.

Second, there are statistical challenges to measuring density dependence, and that is the focus of the remainder of this section. These statistical challenges differ somewhat between the case where density dependence is studied at the level of the population (i.e., in trajectories of population size) and the case where it is studied at the level of the individual (i.e., in demographic rates). In both, evidence about density dependence can be obtained by a negative relationship between the population growth rate or demographic rates and population size using correlation or regression analysis (Dennis et al., 1991). However, population size is typically not known perfectly but rather must be estimated, or more often, only raw counts are available. Both estimates and raw counts have associated errors, and if these are not accounted for, statistical tests for density dependence in analyses of population trajectories will be too liberal, that is, density dependence will be erroneously detected too often (Shenk et al., 1998; Freckleton et al., 2006; Lebreton and Gimenez, 2013). In contrast, statistical tests for density dependence in analyses of demographic rates will be too conservative, i.e., will not detect density dependence often enough when in fact it is present (Lebreton, 2009).

Lebreton and Gimenez (2013) therefore recommend abandoning approaches that do not account for observation errors when testing for the presence of density dependence. State-space models can account for observation errors (Section 4.3.1) and they can be specified to include density dependence (Williams et al., 2003; Dennis et al., 2006). Unfortunately, the parameters of such models are typically only weakly identifiable (Polansky et al., 2009; Lebreton and Gimenez, 2013); in particular, when the observation error is substantial, estimates of density dependence are no longer reliable at all (Knape, 2008). Even if time series of population counts are available from multiple populations, estimation of the strength of density dependence remains difficult (Bellier et al., 2016). Integrated population models (IPMs) are suitable for the study of density dependence because they can overcome problems faced by many conventional analyses: population size can be estimated, and uncertainty can be fully considered. Moreover, in IPMs we can express the effects of density at the level of demographic rates, which yields fairly mechanistic insights into the demographic mechanisms causing density dependence to also be observed at the population level.

Using IPMs, density dependence can be studied in various ways. First, you can fit an IPM without density-dependent mechanisms but allowing demographic rates to be variable over time. Then, you assess density dependence *a posteriori* by regressing estimates of population size on estimates of demographic rates (e.g., Johnson et al., 2010; Schaub et al., 2013; Gamelon et al., 2016, 2017, 2019; Brommer et al., 2017; Wilson et al., 2017; Genovart et al., 2018). Second, you may formulate a link between a demographic rate and the *observed* population size (i.e., population counts) in an IPM (Besbeas et al., 2009; Tavecchia et al., 2009; Péron and Koons, 2012; Péron et al., 2012). However, by doing this, the power to detect density dependence is reduced because the observed population size is subject to observation errors. A main reason for using observed rather than estimated population sizes is that the latter can result in nonlinearity, and this is very challenging to deal with in the frequentist framework (but see Besbeas and Morgan, 2012). Finally, you can formulate an

IPM that includes an explicit relationship between one or more demographic rates and the *estimated* population size. Such models have been developed by Abadi et al. (2012) and Besbeas and Morgan (2012) and have since been applied a few times (e.g., Taylor and Udevitz, 2015; Hatter et al., 2017; Coates et al., 2018; Zhao et al., 2019; Schaub 2020). In this chapter, we illustrate a simple IPM with density dependence and focus on the questions of whether there is evidence of density dependence at the population level and about the demographic mechanisms that induce density dependence.

8.2 DENSITY DEPENDENCE IN RED-BACKED SHRIKES

We demonstrate the application of an IPM with density dependence using a real data set of red-backed shrikes (*Lanius collurio*, Fig. 8.1), a migratory passerine bird species. Our goal is to estimate the strength of density dependence in the demographic processes of apparent survival, productivity, and immigration as well as to measure the strength of density dependence at the population level. The data were collected over 36 years (1971–2006) in a study area of 18 km^2 in Southern Germany by Wolfgang Stauber and Hans Jakober, who generously made them available. A long time series is generally important for accurate inferences about density dependence (Brook and Bradshaw, 2006), which makes this data set particularly valuable.

The data collected are typical for a bird population study. They include capture-recapture data of nestlings and adults, counts of the number of breeding pairs, and productivity data from nest surveys. Immigration is an important demographic process in a relatively small-scale study such as this one. As a result, the IPM must be specified in such a way that immigration can be estimated (Abadi et al., 2010). A shorter subset of this data set was used by Abadi et al. (2012) to introduce an IPM with density dependence. The complete time series was then analyzed in Schaub et al. (2013), where we found evidence of density-dependent population regulation and identified immigration as the chief demographic process responsible for it, although productivity also contributed. The assessment of density dependence was performed *a posteriori* by correlation analysis. This approach to the study of density dependence is fine for identifying the demographic rates under density-dependent regulation, but we cannot directly estimate the strength of density dependence nor make

FIGURE 8.1

Red-backed shrike (*Lanius collurio*) male sitting on a viper's bugloss (*Echium vulgare*; photograph by Daniel Ottmann).

comparisons between studies or project the future population size. Therefore, we here illustrate another approach where we directly estimate the strength of density dependence by regressing a demographic rate parameter θ on the estimated population size:

$$g(\theta_t) = \alpha + \beta N_t + \varepsilon_t$$
$$\varepsilon_t \sim Normal(0, \sigma^2)$$

Here, g is an appropriate link function for $\boldsymbol{\theta}$, α is an intercept, β is the strength of density dependence, N_t is the estimated population size in year t, and ε_t is a normally distributed residual for year t that captures all remaining temporal variability in $\boldsymbol{\theta}$ that is not due to variation in population size (i.e., density dependence). This model of density dependence adopts a simple linear function of N and is referred to as the Ricker model (Dennis and Taper, 1994). Other possible functional forms of density dependence such as the Gompertz, theta-logistic, or threshold models could also be considered (Dennis and Taper, 1994; Rockwood, 2006; Besbeas and Morgan, 2012; Coates et al., 2018), but here we just illustrate the simplest parametric form of density dependence.

8.2.1 GENERAL POPULATION MODEL

We define a prebreeding "census", female-based population model for the shrikes and distinguish between three types of females. Surviving adults (S) are females at least 2 years of age that reproduced in at least one previous year at the study site, local recruits (R) are females 1 year of age that were born in the study area, and immigrants (I) are females at least 1 year of age that were born outside the study area and have never reproduced before in the study area. We write the following stochastic population equations:

$$R_t \sim Poisson\left(\frac{f_{t-1}}{2}\phi_{j,t-1}(R_{t-1} + S_{t-1} + I_{t-1})\right),$$
$$S_t \sim Binomial(\phi_{a,t-1}, (R_{t-1} + S_{t-1} + I_{t-1})),$$
$$I_t \sim Poisson(\Omega_t),$$

where f_t is productivity per female in year t, $\phi_{j,t}$ and $\phi_{a,t}$ are juvenile and adult apparent survival probabilities, respectively, and Ω_t is the expected number of immigrants in year t. The last item is a hidden parameter, as there is no other data set that is informative about immigration, apart from the population counts (see Sections 9.2 or 10.6, or Chapter 11, for further discussion about the estimation of immigration). The total population size (N_t) is given by the sum of the number of all three types of individuals (i.e., $N_t = R_t + S_t + I_t$). We use the log-normal distribution as an observation model to allow for a relative scaling of the error with population size, $C_t \sim logNormal(\log(N_t), \sigma_r^2)$, where σ_r^2 is the residual error, which is assumed to be constant on the log scale.

8.2.2 MODELING DENSITY DEPENDENCE IN SURVIVAL AND PRODUCTIVITY

Survival and productivity are both assumed to be linearly related to the estimated population size, and the slope of these relationships provides an estimate of the strength of density dependence for each. For the two survival probabilities, we use $\text{logit}(\phi_t) \sim Normal\left(\alpha_\phi + \beta_\phi\left(N_t - \overline{C}\right), \sigma_\phi^2\right)$, and for productivity, $\log(f_t) \sim Normal\left(\alpha_f + \beta_f\left(N_t - \overline{C}\right), \sigma_f^2\right)$. Note that we subtract the mean observed population count ($\overline{C}$) from the estimated population sizes for computational reasons—the MCMC chains in the analysis mix much better when population sizes are quasi-centered in this way. In addition, this means that the intercepts can be interpreted as the estimated value of the demographic rates on the link scale when the population size is at its observed average $\overline{C}$. The β parameters represent the strength of density dependence in the demographic rates, i.e., at the individual level. In this model, we do not specify density dependence at the population level. Instead, the resulting density-dependent patterns in the trajectory of population size constitute an emerging property that is due to the effects of density dependence at the level of the individual. Moreover, at first, we do not specify any density dependence on the number of immigrants, but rather, we will do so in the subsequent example.

We start by loading the data set and briefly inspect what's in it.

```
# Load the red-backed shrike data and produce data overview
library(IPMbook); library(jagsUI)
data(redbacked)
str(redbacked)    # Not shown
```

The productivity data are vectors containing the annual numbers of surveyed broods (`B`) and the fledglings they produced (`J`). The population count data contain the number of active broods on June 15 of each year. This may be greater than

`B` if the outcome of some broods has remained unknown, or lower than `B` if `B` contains broods that failed previously or were initiated after June 15. The capture-recapture data are formatted as annual m-arrays (with two age classes: released as nestling or as adult) and only include the females. We bundle the data into a list and formulate the IPM. We use the function `with` in the data bundle. This is handy because we can use the names of the lower hierarchy to address elements of a list.

```
# Bundle data
jags.data <- with(redbacked, list(n.occasions=ncol(marr.a), marr.j=marr.j, marr.a=marr.a,
  rel.j=rowSums(marr.j), rel.a=rowSums(marr.a), C=count, J=J, B=B, pNinit=dUnif(1, 50),
  mean.C=mean(count)))
str(jags.data)

List of 10
 $ n.occasions: int 36
 $ marr.j     : num [1:35, 1:36] 2 0 0 0 0 0 0 0 0 0 ...
 $ marr.a     : num [1:35, 1:36] 6 0 0 0 0 0 0 0 0 0 ...
 $ rel.j      : num [1:35] 61 66 45 85 56 120 90 69 54 45 ...
 $ rel.a      : num [1:35] 35 39 17 34 32 47 45 46 37 27 ...
 $ C          : num [1:36] 47 62 62 64 72 74 68 65 38 45 ...
 $ J          : num [1:36] 112 114 88 167 108 223 171 127 102 84 ...
 $ B          : num [1:36] 29 46 50 56 65 62 63 60 33 41 ...
 $ pNinit     : num [1:50] 0.02 0.02 0.02 0.02 0.02 0.02 0.02 0.02 ...
 $ mean.C     : num 53.4

# Write JAGS model file
cat(file="model1.txt", "
model {
# Priors and linear models
# Models for demographic rates
for (t in 1:(n.occasions-1)){
  logit(phij[t]) <- alpha[1] + beta[1] * (N[t] - mean.C) + err.phij[t]
  err.phij[t] ~ dnorm(0, tau.phij)
  logit(phia[t]) <- alpha[2] + beta[2] * (N[t] - mean.C) + err.phia[t]
  err.phia[t] ~ dnorm(0, tau.phia)
  logit(pj[t]) <- logit.b0.pj + err.pj[t]
  err.pj[t] ~ dnorm(0, tau.pj)
  logit(pa[t]) <- logit.b0.pa + err.pa[t]
  err.pa[t] ~ dnorm(0, tau.pa)
}
for (t in 1:n.occasions){
  log(f[t]) <- alpha[3] + beta[3] * (N[t] - mean.C) + err.f[t]
  err.f[t] ~ dnorm(0, tau.f)
  log(omega[t]) <- log.b0.om + err.om[t]
  err.om[t] ~ dnorm(0, tau.om)
}

# Priors for variance parameters (hyperparameters)
tau.phij <- pow(sigma.phij, -2)
sigma.phij ~ dunif(0, 10)
sigma2.phij <- pow(sigma.phij, 2)
tau.phia <- pow(sigma.phia, -2)
sigma.phia ~ dunif(0, 10)
sigma2.phia <- pow(sigma.phia, 2)
tau.f <- pow(sigma.f, -2)
sigma.f ~ dunif(0, 10)
sigma2.f <- pow(sigma.f, 2)
tau.om <- pow(sigma.om, -2)
sigma.om ~ dunif(0, 10)
sigma2.om <- pow(sigma.om, 2)
tau.pj <- pow(sigma.pj, -2)
sigma.pj ~ dunif(0, 10)
sigma2.pj <- pow(sigma.pj, 2)
tau.pa <- pow(sigma.pa, -2)
sigma.pa ~ dunif(0, 10)
sigma2.pa <- pow(sigma.pa, 2)
tau ~ dgamma(0.001, 0.001)
sigma2 <- 1 / tau

# Priors for the mean of immigration and resighting rates
b0.om ~ dunif(0, 75)
b0.pj ~ dunif(0, 1)
b0.pa ~ dunif(0, 1)
```

```
# Priors for regression parameters of the demographic rates
for (j in 1:3){
  alpha[j] ~ dnorm(0, 0.001)
  beta[j] ~ dunif(-5, 5)
}

# Back-transformations
log.b0.om <- log(b0.om)
logit.b0.pj <- logit(b0.pj)
logit.b0.pa <- logit(b0.pa)

# Population count data (state-space model)
# Model for initial stage-spec. population sizes: discrete uniform priors
R[1] ~ dcat(pNinit)                                    # Local recruits
S[1] ~ dcat(pNinit)                                    # Surviving adults
I[1] ~ dpois(omega[1])                                 # Immigrants

# Process model over time: our model of population dynamics
for (t in 2:n.occasions){
  R[t] ~ dpois(f[t-1] / 2 * phij[t-1] * N[t-1])        # No. local recruits
  S[t] ~ dbin(phia[t-1], N[t-1])                       # No. surviving adults
  I[t] ~ dpois(omega[t])                               # No. immigrants
}

# Observation model
for (t in 1:n.occasions){
  N[t] <- S[t] + R[t] + I[t]                           # Total number of breeding females
  logN[t] <- log(N[t])
  C[t] ~ dlnorm(logN[t], tau)
}

# Capture-recapture data (CJS model with multinomial likelihood)
# Define the multinomial likelihood
for (t in 1:(n.occasions-1)){
  marr.j[t,1:n.occasions] ~ dmulti(pr.j[t,], rel.j[t])
  marr.a[t,1:n.occasions] ~ dmulti(pr.a[t,], rel.a[t])
}
# Define the cell probabilities of the m-arrays
for (t in 1:(n.occasions-1)){
  # Main diagonal
  qj[t] <- 1-pj[t]                                     # Probability of non-recapture (juv)
  qa[t] <- 1-pa[t]                                     # Probability of non-recapture (ad)
  pr.j[t,t] <- phij[t] * pj[t]
  pr.a[t,t] <- phia[t] * pa[t]
  # Above main diagonal
  for (j in (t+1):(n.occasions-1)){
    pr.j[t,j] <- phij[t] * prod(phia[(t+1):j]) * qj[t]*prod(qa[t:(j-1)]) * pa[j] / qa[t]
    pr.a[t,j] <- prod(phia[t:j]) * prod(qa[t:(j-1)]) * pa[j]
  } #j
  # Below main diagonal
  for (j in 1:(t-1)){
    pr.j[t,j] <- 0
    pr.a[t,j] <- 0
  } #j
} #t
# Last column: probability of non-recapture
for (t in 1:(n.occasions-1)){
  pr.j[t,n.occasions] <- 1-sum(pr.j[t,1:(n.occasions-1)])
  pr.a[t,n.occasions] <- 1-sum(pr.a[t,1:(n.occasions-1)])
}

# Productivity data (Poisson regression model)
for (t in 1:n.occasions){
  J[t] ~ dpois(B[t] * f[t])
}
}
")
```

The apparent survival parameters are denoted phij for juvenile and phia for adult survival. Earlier analyses of these data sets (Schaub et al., 2013) revealed recapture probability of 1-year-old females to be different from that of older females. We implement the same structure and model the recapture probabilities of both age classes with independent temporal random effects to account for possibly varying sampling effort over the years. We specify initial values, list the monitored parameters, select the MCMC settings, and launch JAGS.

```
# Define initial values
inits <- function() {list(b0.om=runif(1,0,1), b0.pj=runif(1,0.2,0.7), b0.pa=runif(1,0.1,0.6),
  sigma.pa=runif(1,0,1), I=rep(10, 36))}

# Define parameters to be monitored
parameters <- c("phij", "phia", "f", "omega", "b0.om", "b0.pj", "b0.pa", "sigma2.phij",
  "sigma2.phia", "sigma2.f", "sigma2.om", "sigma2.pj", "sigma2.pa", "R", "S", "I", "N", "pa", "pj",
  "sigma2", "alpha", "beta")

# MCMC settings
ni <- 30000; nb <- 10000; nc <- 3; nt <- 4; na <- 1000

# Call JAGS (ART 31 min), check convergence and produce figures
out1 <- jags(jags.data, inits, parameters, "model1.txt", n.iter=ni, n.burnin=nb, n.chains=nc,
  n.thin=nt, n.adapt=na, parallel=TRUE)
traceplot(out1)

# Compute the probability that the betas are negative
mean(out1$sims.list$beta[,1] < 0)
[1] 0.2459
mean(out1$sims.list$beta[,2] < 0)
[1] 0.2258
mean(out1$sims.list$beta[,3] < 0)
[1] 0.7961

# Calculate population growth rate
lam <- out1$sims.list$N[,-1] / out1$sims.list$N[,-ncol(out1$sims.list$N)]
```

The model takes a while to run but converges quite well. We focus our attention directly on the parameters for the strength of density dependence, beta, and inspect their posterior distributions (Fig. 8.2). For productivity, the posterior mean of beta is less than zero, but the probability that this parameter is negative is relatively low at 0.80. The posterior means of beta for the two survival probabilities are both positive, but the probabilities of these parameters indeed being positive are only 0.75 and 0.77. Therefore, there is not much evidence that either productivity or survival is negatively affected by population size in this study of red-backed shrikes.

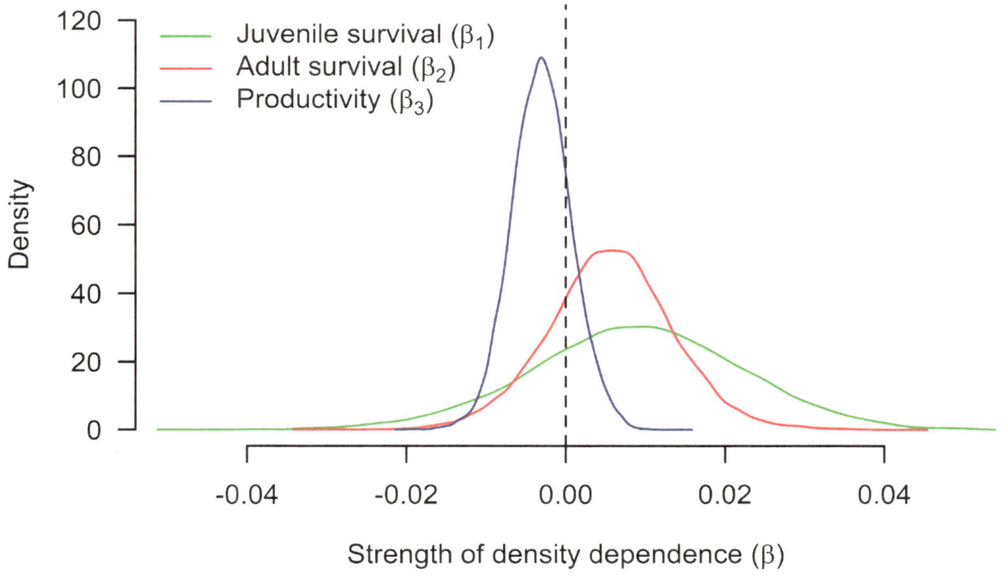

FIGURE 8.2

Posterior distributions of the parameters representing the strength of linear density dependence on three demographic parameters in a real (i.e., not simulated) population of red-backed shrikes. Note that the y-axis shows the posterior density and not the population density.

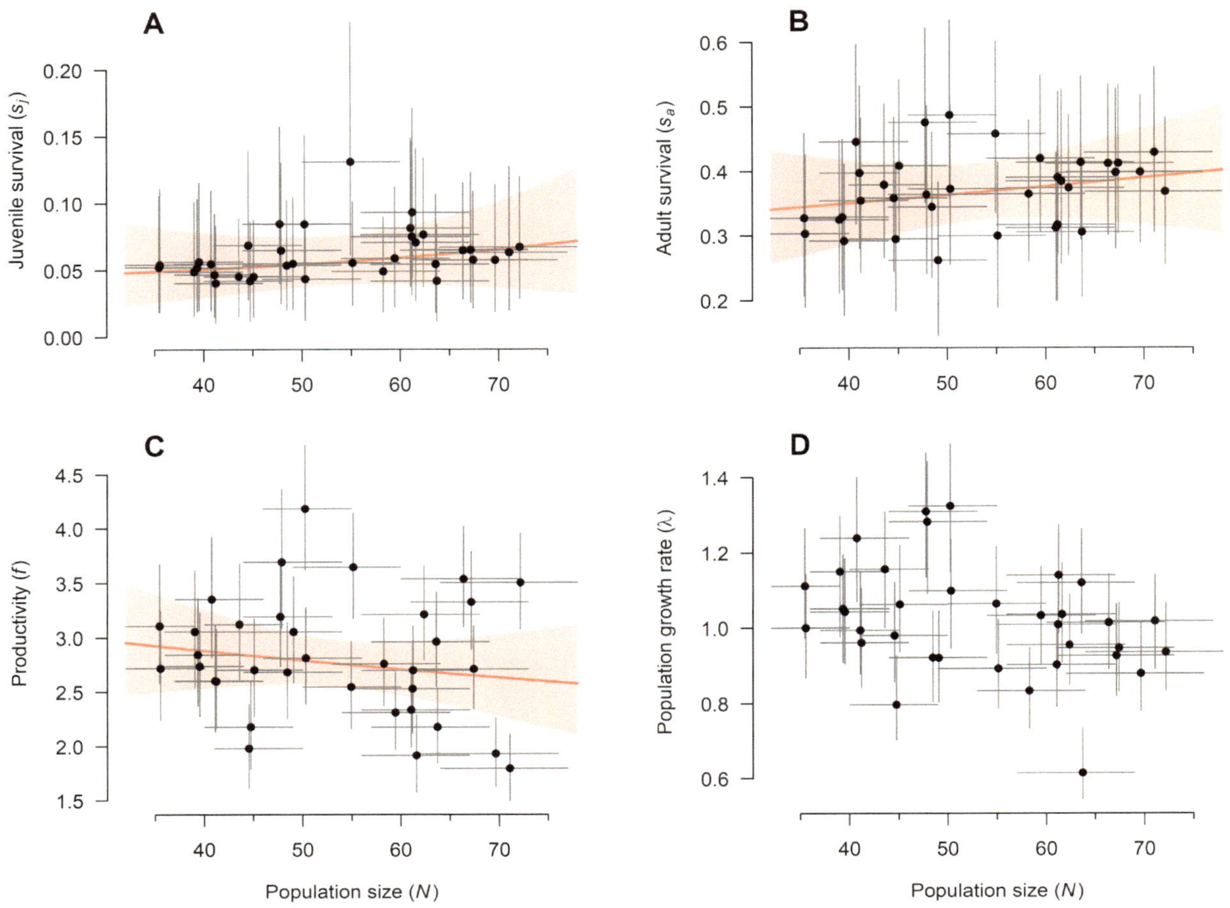

FIGURE 8.3

Annual estimates of demographic rates plotted against estimated population size in a German population of red-backed shrikes (A–C). Filled circles show posterior means, and lines show 95% credible intervals. Red regression lines show the predictions based on the estimated regression coefficients. Panel D shows the relationship between the estimates of population size and population growth rate.

We plot the estimates of the annual demographic rates against those of population size to visualize these relationships (Fig. 8.3). We notice the high annual variability of the demographic rates and consequently the weak evidence for density dependence in survival and productivity.

8.2.3 ASSESSING DENSITY DEPENDENCE AT THE POPULATION LEVEL

Fig. 8.3D shows the estimates of population growth rate plotted against those of population size. There seems to be a negative relationship suggesting density dependence at the population level. However, the two are not independent because the estimate of N_t appears on both the x-axis alone and the y-axis as a ratio with N_{t-1}. This problem is related to what is called a part-whole correlation in studies of allometry (Christians, 1999) and here will induce a negative correlation even in the absence of any density dependence (Dennis and Taper, 1994; Lebreton and Gimenez, 2013). As a result, we cannot use simple regression models or correlation to test for density dependence. A proper regression-based test of density dependence at the population level therefore requires knowledge about the distribution of the slope in the absence of any density dependence. We can obtain this by simulation. Analogously to the bootstrap suggested by Dennis and Taper (1994), we carry out the following steps for each MCMC draw i:

1. Fit a model of pure exponential growth (i.e., without density dependence) with random noise to the estimated population sizes ($\widehat{\boldsymbol{N}}$):

$$\log\left(\frac{\widehat{\boldsymbol{N}}_{i,2\ldots T}}{\widehat{\boldsymbol{N}}_{i,1\ldots(T-1)}}\right) \sim Normal\left(\widehat{r}_i, {}^{r}\widehat{\sigma}_i^2\right).$$

2. Use the parameter estimates obtained (population growth rate $\widehat{r}$, temporal variability of growth rate $\widehat{\sigma}^2$) to simulate a new population trajectory ($\boldsymbol{N}$*) starting from the estimated population size in year 1 ($\widehat{\boldsymbol{N}}_1$):

$$N^*_{i,2} = \widehat{N}_{i,1} \exp\left(\widehat{r}_i + \varepsilon_{i,1}\right).$$

From $t = 2$ onwards, use

$$N^*_{i,t+1} = N^*_{i,t} \exp\left(\widehat{r}_i + \varepsilon_{i,t}\right)$$

$$\varepsilon_{i,t} \sim Normal\left(0, {}^r\widehat{\sigma}^2_i\right)$$

3. Fit a model of density dependence to the simulated population trajectory $\boldsymbol{N}$*: this gives us the null or reference distribution of the estimates of density dependence ($\boldsymbol{\beta}$*), when in reality there is no density dependence in the abundance time series:

$$\log\left(\boldsymbol{N}^*_{i,2\ldots T}\right) \sim Normal\left(\alpha^*_i + (1 - \beta^*_i)\log\left(\boldsymbol{N}^*_{i,1:(T-1)}\right), \sigma^{2^*}_i\right).$$

4. At the same time, fit a model of density dependence to the estimated population size:

$$\log\left(\widehat{\boldsymbol{N}}_{i,2\ldots T}\right) \sim Normal\left(\alpha_i + (1 - \beta_i)\log\left(\widehat{\boldsymbol{N}}_{i,1:(T-1)}\right), \sigma^2_i\right).$$

Finally, we compare the estimates of density dependence obtained from the simulated population sizes where we know there is no density dependence ($\boldsymbol{\beta}$*) with those obtained from the actual population sizes where there may be density dependence but we do not know ($\boldsymbol{\beta}$). This yields the probability that the observed effect of density dependence is more negative (i.e., "more significant") than the simulated one under the null hypothesis where there is no density dependence. The following code performs these calculations.

```
N <- out1$sims.list$N
T <- ncol(N)
n <- nrow(N)

r <- sigma.r <- alpha.star <- beta.star <- sigma.star <- alpha <- beta <- sigma <- numeric(n)
N.star <- matrix(NA, nrow=n, ncol=T)
for (i in 1:n){     # loop over all MCMC draw
  if(round(i %% 500 ) == 0) {cat(paste('** Processing draw', i, '**\n'))}
  # 1. Estimate the growth parameters under the exponential model (i.e., without DD)
  z <- lm(log(N[i,-1] / N[i,-T]) ~ 1)
  r[i] <- coef(z)
  sigma.r[i] <- summary(z)$sigma
  # 2. Simulate population trajectories under the exponential growth model
  N.star[i,1] <- N[i,1]
  for (t in 1:(T-1)){
    N.star[i,t+1] <- N.star[i,t] * exp(r[i] + rnorm(1, 0, sigma.r[i]))
  } #t
  # 3. Fit the DD model to simulated population sizes from step 2
  z <- lm(log(N.star[i,-1]) ~ log(N.star[i,-T]))
  alpha.star[i] <- coef(z)[1]
  beta.star[i] <- coef(z)[2]-1
  sigma.star[i] <- summary(z)$sigma
  # 4. Fit DD model to estimated population sizes (from the IPM)
  z <- lm(log(N[i,-1]) ~ log(N[i,-T]))
  alpha [i] <- coef(z)[1]
  beta[i] <- coef(z)[2]-1
  sigma[i] <- summary(z)$sigma
} #i
```

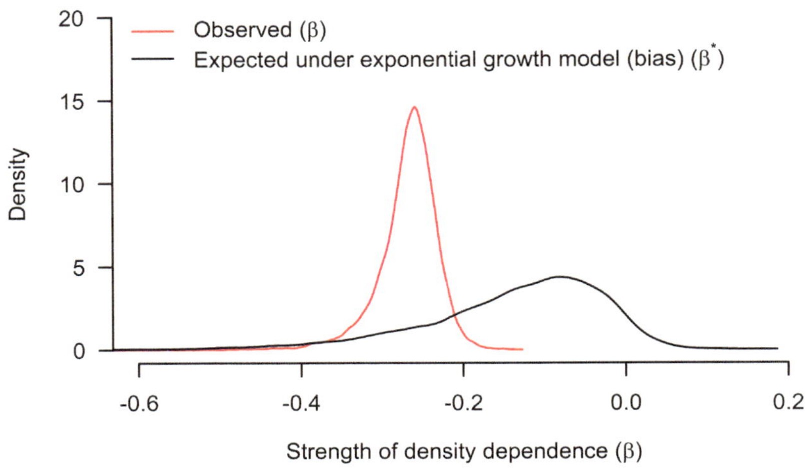

FIGURE 8.4

Testing for density dependence in a regression model while correcting for the part-whole aspect of the problem: Posterior distributions of the estimated strength of density dependence at the population level for the actual data (red), and the reference distribution (black), which shows the expected estimate in the absence of density dependence. The probability that the posterior distribution for the actual data is smaller than the reference distribution under H_0 where there is no density dependence is 0.86.

To visualize the results, we plot the posterior distribution of the observed estimate of density dependence and the reference distribution in the absence of density dependence in N (Fig. 8.4). The posterior distribution of the coefficient of density dependence for the actual N is quite different from and more negative than the distribution of the expected coefficient of density dependence under the pure exponential growth model without density dependence. A formal test confirms our impression: the probability of density dependence being present in the red-backed shrike population trajectory is relatively high.

```
mean(beta < beta.star)
[1] 0.8665333
```

8.2.4 MODELING DENSITY DEPENDENCE IN IMMIGRATION

Given that we find some evidence for density dependence at the level of the population but essentially none at the level of the individual for productivity or survival, we were wondering whether immigration might be related to population density. We have not included density dependence on immigration so far because this is more involved than the modeling of density dependence in other demographic rates. Immigration can be included in the population model as an absolute number as we have done here, or as a rate where the immigration rate is defined as the ratio of the number of immigrants and the population size in the previous year, i.e., $\omega_t = I_t/N_{t-1}$ (Schaub and Fletcher, 2015). Consequently, we may express an effect of density dependence on either one. If we focus on the immigration rate, there is the difficulty of an intrinsic dependency between the rate and population size just because of the way the immigration rate is defined (i.e., again related to the part-whole regression just discussed). Therefore, working with the rate in the context of density dependence does not seem ideal. However, if instead we focus on the number of immigrants and relate it to population size, there is a similar problem because population size also includes the immigrants. Therefore, we need a clear idea about how potential density dependence on the number of immigrants might work and what kind of hypothesis we want to test. In the case of red-backed shrikes, we hypothesize that density dependence works via a buffer effect mediated by heterogeneity in territory quality (Rodenhouse et al., 1997; Gill et al., 2001). We assume that the best territories are occupied first such that for the remaining available territories, quality declines with increasing population size. Moreover, we assume that the locals—i.e., the surviving adults and the local recruits—are stronger competitors for territories (perhaps because of their local knowledge) and that they preferentially occupy territories of higher quality. Consequently, with an increasing number of locals, the average quality and number of immigrants are expected to decline for empty territories available for immigrants. Nevertheless, we do not necessarily expect a negative relationship between the numbers of immigrants and locals. Rather, as the number of locals increases, we expect a diminishing returns of the effect of the number of immigrants. The model that we specify to express this hypothesis is

$$\log(I_t) \sim Normal(\alpha_I + \beta_I \log(S_t + R_t), \sigma_I^2).$$

We use the log of the number of locals ($S + R$) because this allows an easy test of whether the number of immigrants increased at a lower rate than the number of locals does (i.e., $\beta_I < 1$). We note that JAGS would crash if $R + S$ becomes zero, since the logarithm is then not defined. So in some cases you may need to add a small number to that sum (e.g., 1), but this turned out not to be necessary in our analysis. We write the BUGS code for this model and fit it to the shrike data.

```
# Write JAGS model file
cat(file="model2.txt", "
model {
# Priors and linear models
# Models for demographic rates
for (t in 1:(n.occasions-1)){
  logit(phij[t]) <- alpha[1] + beta[1] * (N[t] - mean.C) + err.phij[t]
  err.phij[t] ~ dnorm(0, tau.phij)
  logit(phia[t]) <- alpha[2] + beta[2] * (N[t] - mean.C) + err.phia[t]
  err.phia[t] ~ dnorm(0, tau.phia)
  logit(pj[t]) <- logit.b0.pj + err.pj[t]
  err.pj[t] ~ dnorm(0, tau.pj)
  logit(pa[t]) <- logit.b0.pa + err.pa[t]
  err.pa[t] ~ dnorm(0, tau.pa)
}
for (t in 1:n.occasions){
  log(f[t]) <- alpha[3] + beta[3] * (N[t] - mean.C) + err.f[t]
  err.f[t] ~ dnorm(0, tau.f)
  log(omega[t]) <- alpha[4] + beta[4] * log(S[t] + R[t]) + err.om[t]
  err.om[t] ~ dnorm(0, tau.om)
}

# Priors for variance parameters (hyperparameters)
tau.phij <- pow(sigma.phij, -2)
sigma.phij ~ dunif(0, 10)
sigma2.phij <- pow(sigma.phij, 2)
tau.phia <- pow(sigma.phia, -2)
sigma.phia ~ dunif(0, 10)
sigma2.phia <- pow(sigma.phia, 2)
tau.f <- pow(sigma.f, -2)
sigma.f ~ dunif(0, 10)
sigma2.f <- pow(sigma.f, 2)
tau.om <- pow(sigma.om, -2)
sigma.om ~ dunif(0, 10)
sigma2.om <- pow(sigma.om, 2)
tau.pj <- pow(sigma.pj, -2)
sigma.pj ~ dunif(0, 10)
sigma2.pj <- pow(sigma.pj, 2)
tau.pa <- pow(sigma.pa, -2)
sigma.pa ~ dunif(0, 10)
sigma2.pa <- pow(sigma.pa, 2)
tau ~ dgamma(0.001, 0.001)
sigma2 <- 1 / tau

# Priors for the mean of resighting rates
b0.pj ~ dunif(0, 1)
b0.pa ~ dunif(0, 1)

# Priors for regression parameters of the demographic rates
for (j in 1:4){
  alpha[j] ~ dnorm(0, 0.001)
  beta[j] ~ dunif(-5, 5)
}

# Back-transformations
logit.b0.pj <- logit(b0.pj)
logit.b0.pa <- logit(b0.pa)

# Population count data (state-space model)
# Model for initial stage-spec. population sizes: discrete uniform priors
R[1] ~ dcat(pNinit)                                    # Local recruits
S[1] ~ dcat(pNinit)                                    # Surviving adults
I[1] ~ dpois(omega[1])                                 # Immigrants
```

```
# Process model over time: our model of population dynamics
for (t in 2:n.occasions){
  R[t] ~ dpois(f[t-1]/2 * phij[t-1] * N[t-1])      # No. local recruits
  S[t] ~ dbin(phia[t-1], N[t-1])                    # No. surviving adults
  I[t] ~ dpois(omega[t])                            # No. immigrants
}

# Observation process
for (t in 1:n.occasions){
  N[t] <- S[t] + R[t] + I[t]                        # Total number of breeding females
  logN[t] <- log(N[t])
  C[t] ~ dlnorm(logN[t], tau)
}

# Capture-recapture data (CJS model with multinomial likelihood)
# Define the multinomial likelihood
for (t in 1:(n.occasions-1)){
  marr.j[t,1:n.occasions] ~ dmulti(pr.j[t,], rel.j[t])
  marr.a[t,1:n.occasions] ~ dmulti(pr.a[t,], rel.a[t])
}
# Define the cell probabilities of the m-arrays
for (t in 1:(n.occasions-1)){
  # Main diagonal
  qj[t] <- 1-pj[t]                                  # Probability of non-recapture (juv)
  qa[t] <- 1-pa[t]                                  # Probability of non-recapture (ad)
  pr.j[t,t] <- phij[t] * pj[t]
  pr.a[t,t] <- phia[t] * pa[t]
  # Above main diagonal
  for (j in (t+1):(n.occasions-1)){
    pr.j[t,j] <- phij[t] * prod(phia[(t+1):j]) * qj[t] * prod(qa[t:(j-1)]) * pa[j] / qa[t]
    pr.a[t,j] <- prod(phia[t:j]) * prod(qa[t:(j-1)]) * pa[j]
  } #j
  # Below main diagonal
  for (j in 1:(t-1)){
    pr.j[t,j] <- 0
    pr.a[t,j] <- 0
  } #j
} #t
# Last column: probability of non-recapture
for (t in 1:(n.occasions-1)){
  pr.j[t,n.occasions] <- 1-sum(pr.j[t,1:(n.occasions-1)])
  pr.a[t,n.occasions] <- 1-sum(pr.a[t,1:(n.occasions-1)])
}

# Productivity data (Poisson regression model)
for (t in 1:n.occasions){
  J[t] ~ dpois(B[t] * f[t])
}
}
")

# Define initial values
inits <- function() {list(b0.pj=runif(1,0.2,0.7), b0.pa=runif(1,0.1,0.6), sigma.pa=runif(1,0,1),
  I=rep(10, 36))}

# Define parameters to be monitored
parameters <- c("phij", "phia", "f", "omega", "b0.pj", "b0.pa", "sigma2.phij", "sigma2.phia", "sigma2.f",
  "sigma2.om", "sigma2.pj", "sigma2.pa", "R", "S", "I", "N", "pa", "pj", "sigma2", "alpha", "beta")

# MCMC settings
ni <- 110000; nb <- 10000; nc <- 3; nt <- 20; na <- 1000

# Call JAGS (ART 139 min), check convergence and summarize posteriors
out2 <- jags(jags.data, inits, parameters, "model2.txt", n.iter=ni, n.burnin=nb, n.chains=nc,
  n.thin=nt, n.adapt=na, parallel=TRUE)
traceplot(out2)
```

We need considerably longer chains for this model than for the previous one to reach satisfactory convergence.

```
print(out2, 3)
```

```
               mean     sd    2.5%     50%    97.5% overlap0     f  Rhat n.eff
phij[1]       0.065  0.025   0.026   0.061    0.126    FALSE 1.000 1.000 15000
phij[2]       0.072  0.027   0.031   0.068    0.137    FALSE 1.000 1.000  7918
... [output truncated] ...
phij[34]      0.062  0.021   0.028   0.060    0.110    FALSE 1.000 1.001  3062
phij[35]      0.062  0.023   0.025   0.059    0.116    FALSE 1.000 1.000  6145
phia[1]       0.369  0.057   0.259   0.368    0.488    FALSE 1.000 1.000  4193
phia[2]       0.326  0.054   0.218   0.328    0.430    FALSE 1.000 1.000  8420
... [output truncated] ...
phia[34]      0.395  0.052   0.298   0.393    0.506    FALSE 1.000 1.000 15000
phia[35]      0.384  0.055   0.280   0.381    0.499    FALSE 1.000 1.001  2657
fec[1]        3.696  0.330   3.083   3.687    4.370    FALSE 1.000 1.000 14422
fec[2]        2.523  0.217   2.121   2.517    2.962    FALSE 1.000 1.000 15000
... [output truncated] ...
fec[35]       3.321  0.231   2.886   3.314    3.787    FALSE 1.000 1.000 15000
fec[36]       3.065  0.229   2.637   3.056    3.533    FALSE 1.000 1.000 15000
omega[1]     27.684  2.988  21.668  27.717   33.504    FALSE 1.000 1.000 15000
omega[2]     30.750  3.314  25.235  30.428   38.259    FALSE 1.000 1.000 10076
... [output truncated] ...
omega[35]    32.034  3.666  25.190  31.864   39.659    FALSE 1.000 1.003   806
omega[36]    30.940  3.413  24.122  30.908   37.913    FALSE 1.000 1.004   580
b0.pj         0.442  0.069   0.316   0.438    0.588    FALSE 1.000 1.000 11575
b0.pa         0.645  0.039   0.569   0.645    0.721    FALSE 1.000 1.000  6175
sigma2.phij   0.240  0.179   0.004   0.206    0.684    FALSE 1.000 1.002  1189
sigma2.phia   0.097  0.066   0.006   0.084    0.259    FALSE 1.000 1.001  3182
sigma2.fec    0.051  0.015   0.028   0.048    0.087    FALSE 1.000 1.000 15000
sigma2.om     0.009  0.013   0.000   0.004    0.044    FALSE 1.000 1.004  5289
sigma2.pj     0.351  0.533   0.001   0.170    1.703    FALSE 1.000 1.000  6434
sigma2.pa     0.077  0.107   0.000   0.039    0.384    FALSE 1.000 1.009   466
R[1]         10.800  6.582   1.000  10.000   24.000    FALSE 1.000 1.000  6269
R[2]          6.738  3.352   1.000   6.000   14.000    FALSE 1.000 1.000 13740
... [output truncated] ...
R[35]         7.528  3.475   2.000   7.000   15.000    FALSE 1.000 1.001  2581
R[36]         6.673  3.363   1.000   6.000   14.000    FALSE 1.000 1.000  7694
S[1]         10.708  6.560   1.000  10.000   24.000    FALSE 1.000 1.000 15000
S[2]         19.683  3.902  12.000  20.000   27.000    FALSE 1.000 1.000 10316
... [output truncated] ...
S[35]        26.945  4.501  18.000  27.000   36.000    FALSE 1.000 1.000 10698
S[36]        25.381  4.504  17.000  25.000   35.000    FALSE 1.000 1.000  6205
I[1]         26.254  4.517  17.000  26.000   35.000    FALSE 1.000 1.001  5823
I[2]         35.011  4.432  26.000  35.000   44.000    FALSE 1.000 1.000 15000
... [output truncated] ...
I[35]        32.919  4.739  23.000  33.000   42.000    FALSE 1.000 1.001  2209
I[36]        30.141  4.559  21.000  30.000   39.000    FALSE 1.000 1.001  2355
N[1]         47.762  2.400  44.000  48.000   53.000    FALSE 1.000 1.000 15000
N[2]         61.431  2.750  56.000  62.000   67.000    FALSE 1.000 1.000 15000
... [output truncated] ...
N[35]        67.391  3.000  61.000  68.000   73.000    FALSE 1.000 1.000 11563
N[36]        62.195  2.836  57.000  62.000   68.000    FALSE 1.000 1.000 15000
pa[1]         0.633  0.070   0.475   0.637    0.765    FALSE 1.000 1.003   801
pa[2]         0.633  0.072   0.470   0.638    0.766    FALSE 1.000 1.002  4871
... [output truncated] ...
pa[34]        0.672  0.068   0.553   0.664    0.830    FALSE 1.000 1.006   899
pa[35]        0.651  0.065   0.522   0.649    0.788    FALSE 1.000 1.001  3339
pj[1]         0.469  0.133   0.238   0.454    0.793    FALSE 1.000 1.000 15000
pj[2]         0.437  0.119   0.205   0.433    0.703    FALSE 1.000 1.001 15000
... [output truncated] ...
pj[34]        0.441  0.117   0.217   0.436    0.696    FALSE 1.000 1.001  4079
pj[35]        0.454  0.127   0.223   0.443    0.750    FALSE 1.000 1.000 11030
sigma2.c      0.002  0.002   0.000   0.002    0.009    FALSE 1.000 1.007  6627
alpha[1]     -2.833  0.181  -3.204  -2.827   -2.494    FALSE 1.000 1.001  2407
alpha[2]     -0.540  0.091  -0.720  -0.540   -0.365    FALSE 1.000 1.000 12706
alpha[3]      1.014  0.041   0.932   1.015    1.094    FALSE 1.000 1.000  8618
alpha[4]      2.488  0.628   1.091   2.559    3.523    FALSE 1.000 1.021   137
beta[1]       0.006  0.013  -0.020   0.006    0.032     TRUE 0.667 1.001  5265
beta[2]       0.002  0.007  -0.012   0.002    0.017     TRUE 0.614 1.002   924
beta[3]      -0.003  0.004  -0.010  -0.003    0.004     TRUE 0.799 1.000  8207
beta[4]       0.275  0.197  -0.056   0.253    0.709     TRUE 0.939 1.021   136
```

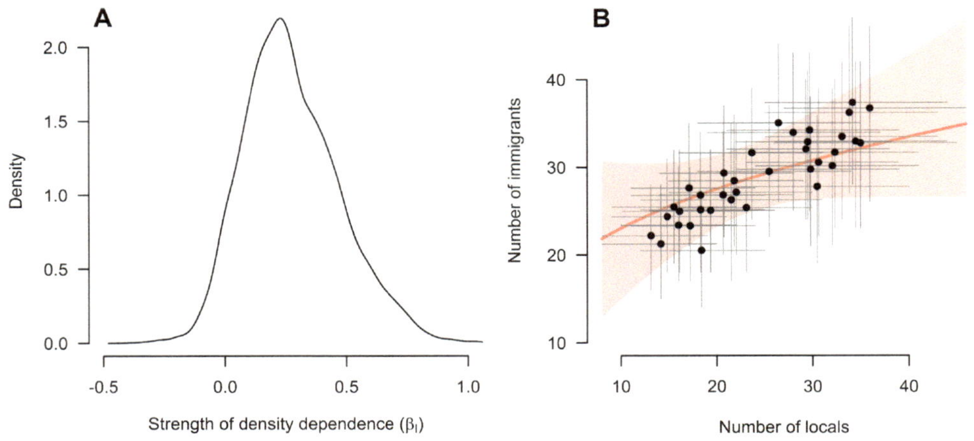

FIGURE 8.5

(A) Posterior distribution of the strength of density dependence on immigration (i.e., the slope of a regression of the number of immigrants on the number of locals) in German red-back shrikes. (B) Predicted number of immigrants as a function of the estimated number of locals (surviving adults plus local recruits). Filled circles and lines show posterior means and 95% credible intervals for the numbers, and the red line with polygon shows the posterior mean and 95% credible intervals of the predicted relationship.

The two parameters for the estimated strength of density dependence on productivity and survival are very similar to the estimates obtained under the first model above, indicating that the inclusion of density dependence on immigration did not much affect the other parts of the model. This was to be expected because in both IPMs we added temporal random effects on all demographic rates that will likely soak up all unexplained variability. The posterior distribution of the relationship between the number of immigrants and of locals, β_I, in the text and `beta[4]` in the code is shown in Fig. 8.5A. Most of the mass is positive, suggesting that the number of immigrants increased with an increasing number of locals. However, and importantly here, the 95% credible interval does not include 1. This suggests that the number of immigrants was increasing at a lower rate than the number of locals. The estimated number of immigrants plotted against the estimated number of locals shows this exact relationship (Fig. 8.5B), which supports our hypothesis that immigration contributed to the regulation of this red-backed shrike population. It is obvious, however, that this mechanism only helps to *reduce the increase* in the population and cannot fully regulate its size. If the number of locals increases, the total population continues to increase even in the absence of any immigration. Therefore, additional density-dependent mechanisms in survival, productivity, or emigration are likely to kick in when the population increases further.

8.3 ADVANTAGES OF IPMS FOR THE STUDY OF DENSITY DEPENDENCE

The inclusion of density dependence in an IPM is relatively straightforward. In its simplest form, we just need to specify a link-scaled linear model for a demographic rate and use an appropriate measure of "density" as an explanatory variable, either alone or alongside other explanatory variables. Despite its simplicity, this approach offers a number of advantages that we stress below while we also note some challenges.

First, the "density dependence variable" (typically population size) is usually unknown and must be estimated free from any observation errors. An appealing feature of the application of an IPM is that this can be estimated within the same model. Hence, its estimation uncertainty is fully accounted for when estimating the strength of density dependence.

Second, as we model density dependence in demographic rates, we develop a more mechanistic understanding of density dependence than we do with the more phenomenological, or descriptive, insights we gain when we merely model time series of population counts. Obviously, even the application of an IPM will not tell us exactly how density dependence works in the field in terms of the actual interactions between individuals, but at least we know on which demographic mechanisms we ought to focus further studies to better understand this process.

Third, exploring density dependence with an IPM offers great flexibility in the exact manner in which density dependence can be modeled. In our examples, as a model for density dependence we have simply adopted a linear model on the appropriate link scale, but we could easily specify other functional relationships, including nonlinear models such as the Gompertz model, which uses $\log(N_t)$ as a predictor, or the theta-logistic model (Dennis and Taper, 1994; Dennis et al., 2006; Sæther et al., 2008). An interesting but quite different model for density dependence was applied for the productivity of British gray herons (*Ardea cinerea*; Besbeas and Morgan, 2012). They use a threshold model with a step function and estimate different average productivities above and below some threshold population size. The threshold, however, could not be estimated but rather had to be chosen *a priori*. Besbeas and Morgan (2012) fitted the model using maximum likelihood and performed model selection to find the threshold value best supported by the data. Their model could also be extended to include multiple thresholds. In Exercise 8.1, you will be asked to formulate such a threshold model and will see that the threshold can be estimated for a simple case. It is also possible to construct models in which the strength of density dependence is not constant over time but varies in some way—for example, with the availability of a specific resource. Using such a model provided evidence that the strength of density dependence on productivity in redheads (*Aythya americana*) was modulated by the availability of breeding ponds (Péron et al., 2012). However, we think it is possible that sometimes such models with varying slopes on density may become too flexible, and one should carefully check whether all of their parameters are identifiable for a particular data set.

Fourth, there is flexibility in terms of the "density dependence" variables being used. Commonly, total population size is used because it reflects the assumption that resources are equally partitioned among all individuals in the population. However, it is likely that not all individuals in the population have the same competitive strength but rather that some individuals have greater access to resources than do others. Consequently, individuals may contribute differentially to density dependence. Assuming that the competitive ability of an individual is basically determined by its age, Charlesworth (1972) introduced the notion of the "critical age." This is the age class that contributes most to density dependence in the population as a whole. Until recently, it was very hard to study this concept empirically because it requires knowledge of the population structure and its change over time. But the application of an IPM now enables Charlesworth's (1972) hypothesis to be investigated. Gamelon et al. (2016) fitted an IPM with four female age classes to great tit (*Parus major*) data. Density dependence was tested for by using a linear model of the form $g(\theta_t) \sim Normal(\alpha + \beta_1 N_{1,t} + \beta_2 N_{2,t} + \beta_3 N_{3,t} + \beta_4 N_{4,t}, \sigma^2)$ for each demographic rate θ. The β parameters estimate the strength of density dependence that the four different age classes induced on rate θ. They found that the youngest age class imposed the strongest density dependence, suggesting that they were most competitive in this short-lived species. In other species, nonbreeders (often called floaters) constitute a significant part of the population and might contribute to density dependence (Newton, 1998; Lopez-Sepulcre and Kokko, 2005; Lee et al., 2017). However, the number of floaters can typically not be counted in the field, and thus the impact of floaters on population dynamics has so far received very little empirical study. IPMs offer the possibility to estimate the number of floaters and therefore to assess their impact on demographic rates. Thus, because in general the population structure (e.g., by age class) and its change over time is estimated explicitly in IPMs, single elements of this structure or the sum of two or more elements can be used to model density dependence. We believe that using IPMs allows us to study density dependence to a degree of detail that was hitherto simply not possible. Although it seems that this promise has yet to be widely realized, we are confident that this will change in the near future.

Fifth, in a way similar to the application of IPMs to study density dependence, we can also use them to study interspecific competition. Of course, this requires the availability of the right kind of data for at least two species. Péron and Koons (2012) developed such a multispecies IPM for two duck species, the canvasback (*Aythya valisineria*) and the related redhead (*A. americana*). The demographic rates for each were modeled as a function of the population sizes of both species, yielding effects of density dependence and intraspecific competition. Specifically, demographic rate θ of species A was modeled as $g(\theta_{A,t}) \sim Normal(\alpha + \beta N_{A,t} + \gamma N_{B,t}, \sigma^2)$, where β is the strength of density dependence, and γ indicates how strongly the demographic rate θ of species A is affected by the population size of species B. Péron and Koons (2012) found that canvasback productivity was negatively affected by redhead population size, which they explained by brood parasitism. Barraquand and Gimenez (2019) expanded on that and developed an "*integrated community model*" with which predator-prey interactions can be studied. Their model included effects of prey density on predator productivity and of predator density on prey survival. They used simulation to test how well the demographic rates and parameters of these functional relationships could be estimated. Parameter estimates were more precise and less biased when population counts, capture-recapture, and productivity data from predator and prey species were available than when only time series of population counts were analyzed. Their study also demonstrated that estimation worked well even if some data sets (e.g., capture-recapture data of a prey species) were missing. Gamelon et al. (2019) studied the intra- and interspecific density dependence (i.e., competition) of great tits (*P. major*) and blue tits (*Cyanistes caeruleus*). They included the age classes of both species in the density-dependence relationships. In that way, they could study how different age classes contributed to intra- and interspecific density dependence. They found that population regulation of great tits was mostly the result of intraspecific density dependence, while intra- and interspecific competition contributed similarly to the population regulation of the smaller blue tit. The inclusion of age-specific effects greatly improved the predictions of population size. Multispecies IPMs offer interesting possibilities but are very data-hungry and have seen only a small handful of applications so far.

Sixth, we ideally should know the level of resources an individual has at its disposal and model the relationships between demographic rates and this variable. But in most cases, the type of resource and its quantity are unknown, and hence, population size is used as a proxy. If the quantity of the resource does not vary over time, then the variability of the quantity of the resource available per individual is fully determined by the variability of population size. If we have a measure of the annual total amount of a critical resource A_t, it can be included in an IPM as $g(\theta_t) \sim Normal\left(\alpha + \beta\frac{A_t}{N_t}, \sigma^2\right)$, where β estimates how strongly demographic rate θ is affected by the quantity of the resource available per individual. To our knowledge, such a model has never been applied in the context of an IPM.

IPMs offer much flexibility in modeling intra- and interspecific density dependence, but multiple challenges lay ahead. First, when density dependence is formulated in the model explicitly as we have done for the red-backed shrikes, convergence in JAGS may be hard to achieve. As always, potential solutions include an increase in chain length and specification of slightly more informative priors. In addition, we can assess density dependence outside of the IPM by specifying demographic rates with either fixed or (preferentially) random time effects in the IPM. Then, posterior draws of the demographic rates and population size can be used to compute either correlation coefficients (Schaub et al., 2013) or linear regression coefficients (Gamelon et al., 2016) outside of BUGS in R as a means to assess density dependence.

Second, the choice of the particular model used for inference about density dependence can be a challenge. This starts with the formulation of an appropriate relationship between a demographic rate and population size. For instance, should this be linear or should we adopt a step function? If there are several reasonable formulations of the relationship, it is a matter of model selection to decide which one to use. Unfortunately, model selection is not straightforward in Bayesian IPMs (but see King et al., 2010, or Hooten and Hobbs, 2015). If our uncertainty only refers to which demographic parameter is affected by density dependence, we can proceed as we have done with the red-backed shrike example, where we included density dependence in all parameters. Inference can then be based on the estimated probability that the parameter representing the strength of density dependence is negative.

8.4 SUMMARY AND OUTLOOK

Density dependence is a crucial regulation mechanism in population dynamics, but it is challenging to study empirically. IPMs offer a number of interesting possibilities to overcome several important challenges. As IPMs include both the individual and population level, they can be used to assess density dependence at the population level and simultaneously investigate through which demographic mechanisms it is imposed at the individual level. Population size and structure are estimated, and therefore, there is no risk of flawed inference due to observation errors (Freckleton et al., 2006; Lebreton and Gimenez, 2013). IPMs are flexible in the manner in which density dependence can be formulated (e.g., via a Ricker or Gompertz model) and through which predictor it is imposed (e.g., total population size, population size per unit resource, or the size of a specific age class). Typically, density dependence is formulated only for the demographic rates in IPMs. The observed density dependence at the population level is then just the consequence of density dependence at the individual level.

Basically, there are two ways in which IPMs can be used to study density dependence. The first possibility is to include density dependence explicitly in an IPM via the specification of relationships between one or several demographic rates and a measure of density, typically population size. The alternative possibility is to fit an IPM in which demographic rates are allowed to vary over time, and then density dependence is tested for *a posteriori* by correlation or regression approaches. The former is more formal and has the advantage that the population can be forecast under density dependence. The advantage of the latter is easier model specification and usually better convergence of the MCMC chains.

Interestingly, in the same way as density dependence, IPMs can be used to study interspecific interactions, such as competition between species or predator-prey interactions. This approach is relatively new, and we foresee an increasing number of studies adopting it (see also Chapter 8 in Kéry and Royle, 2021, for more on the principles of species interaction modeling with JAGS).

So far, only a few published IPM papers have explicitly included density dependence. This may have to do with the additional complexity of the resulting model. It must also be said that even if a population is regulated by density dependence, the latter needs not always be included in a population model including an IPM. To ensure unbiased estimates of demographic rates and of population size, it is just necessary to allow the demographic rates to vary over time. However, for a more mechanistic understanding of population dynamics, for better forecasts into the future, and for applications of population modeling for harvested populations, the inclusion of density-dependent feedbacks in a model of population dynamics is imperative. IPMs are a powerful tool to study populations under density dependence.

8.5 EXERCISES

1. Using the red-backed shrike data set, fit an IPM that induces a threshold model (two levels) for density dependence in productivity. Try to specify the model in such a way that the threshold is estimated.
2. Fit an IPM to the red-backed shrike data set that uses a Gompertz model for density dependence on productivity. How do the estimates differ from those of the Ricker model?

CHAPTER

RETROSPECTIVE POPULATION ANALYSES

9

CHAPTER OUTLINE

9.1 INTRODUCTION

A fundamental goal in studies of population dynamics is to understand the demographic reasons for population changes, i.e., why population size changes over time, as well as population regulation, i.e., what keeps population size within usually fairly tight bounds. In short, we want to identify and understand the drivers of population dynamics. However, a myriad of factors can have an impact on the size, composition, and spatial configuration of a population. Hence, identifying the main drivers and quantifying their effects are far from easy tasks. Changes in individual demographic performance translate into changes in demographic rates and population structure at the population level, and this may then lead to changes in the population growth rate (Sutherland, 1996).

Factors affecting population dynamics are often classified as deterministic or stochastic. The former are generally related to population density and include interspecific competition. Once the relationship between population density and a demographic rate is known, we can determine the dynamics of a population that is only affected by its density provided that the mechanisms underlying density dependence do not change over time. In contrast, stochastic effects include factors such as food availability, parasites, predation, or weather; their effects can typically not be perfectly predicted. That is, even if we know how a stochastic factor affects a demographic rate, predictions of population dynamics will always be accompanied by an uncertainty because the future values of a stochastic effect such as the weather are unknown and at best must themselves be estimated. Populations are typically affected by both deterministic and stochastic factors, and these may also interact (Barbraud and Weimerskirch, 2003; Wang et al., 2006).

Changes in demographic rates and population structure affect population dynamics because the population growth rate is a function of the demographic rates and population structure. The strength of the impact depends on two things: first, the sensitivity of the population growth rate to changes in demographic rates or in the population structure, and second, the magnitude of these changes (Caswell, 2000, 2007). There are two broad types of analyses to assess the effects of demographic rates on the trajectory of a population as quantified by the population growth rate: *prospective* and *retrospective*.

What is typically described as a *prospective* population analysis aims to understand how a population will react in terms of its growth rate to a hypothetical change in its demography (Caswell, 2000). Therefore, a prospective analysis focuses on growth rate sensitivities and elasticities (Caswell, 2001). Both can be regarded as a "translator" of changes occurring at the level of individual demographic rates to changes at the population level, i.e., of population size. Growth rate sensitivities can be calculated analytically or by simulation, as we have seen in Chapter 3. There, we carried out the calculations in simpler demographic analyses of data sets that were not integrated in the sense of an integrated population model (IPM), but exactly the same calculations are possible as part of an IPM. Elasticity is proportional sensitivity—it is more useful in comparisons of the strength of the impact of demographic rates measured at different scales, such as survival versus productivity. An important advantage of the Bayesian computation of sensitivities and elasticities as part of an IPM is that we "automatically" obtain a measure of the uncertainty associated with them. The vast majority of published growth rate sensitivities and

Integrated Population Models. https://doi.org/10.1016/B978-0-12-820564-8.00009-9

elasticities come without any uncertainty assessment and may therefore be of limited value. Prospective analyses can be useful in the context of conservation to identify efficient management strategies (Crouse et al., 1987; Ehrlen and van Groenendael, 1998; Caswell, 2000; Heppell et al., 2000). Moreover, growth rate sensitivities play an important role in evolutionary studies because the population growth rate is a measure of fitness, and consequently, growth rate sensitivities determine the strength of selection on demographic rates and traits related to them (Lande, 1982; Stearns, 1992).

Another type of prospective analysis is specifically targeted at predicting the trajectory of a population into the future, i.e., in making forecasts, often under different management scenarios and sometimes with the goal of assessing the probability of chance extinction of the population. This type of forward-looking population analysis has become known by the term *population viability analysis*. We cover it in Chapter 10.

In contrast to the various types of prospective analyses, the goal of a *retrospective* analysis is to understand the actual numerical changes in a particular population over a specified time period in terms of the temporal patterns of the demographic rates during that same period (Caswell, 2000; Cooch et al., 2001). A major difference between a prospective and a retrospective analysis is that in the latter, the (more theoretical) sensitivity of the population growth rate to changes in demographic rates or population structure is combined with the actual magnitude of the changes that have taken place during that period. For example, the population growth rate of an albatross may be very sensitive to changes in adult survival (Rivalan et al., 2010), but if adult survival has hardly fluctuated in the past, it cannot have contributed much to observed changes in the population growth rate.

Multiple approaches are available to conduct retrospective population analyses. They include the study of correlations between demographic rates and the population growth rate (Robinson et al., 2004; Freeman et al., 2007; Schaub et al., 2012), life-stage simulation analysis (Wisdom et al., 2000; Eacker et al., 2017), path analysis (Woodworth et al., 2017), classical life-table response experiments (LTREs) (Horvitz et al., 1997; Caswell, 2001; Cooch et al., 2001), or the recently developed transient LTREs (Koons et al., 2016, 2017). Each one relies on assumptions that may differ among approaches and can be used to address slightly different questions. All retrospective analyses can be carried out based on the typical inferences under an IPM: annual estimates of the demographic rates and stage-specific population sizes.

In this chapter, we introduce retrospective analyses in the context of an IPM and focus on correlations and on the classical and novel tLTREs. Life-stage simulation analysis and retrospective analysis via path analysis based on results from an IPM have recently been described in detail by Eacker et al. (2017) and Woodworth et al. (2017), respectively, so we do not cover these here.

We conduct all computations for the retrospective analyses in this chapter in R, i.e., after fitting the IPM in JAGS, as derived quantities. In this sense, these analyses are disconnected from the fitting of an IPM—they could be applied if we knew the demographic rates and population size in a population from some source other than an IPM. However, once more, we emphasize that one key advantage of carrying out retrospective analysis based on the posterior draws obtained in a Bayesian application of an IPM is that the uncertainty of all target quantities can be assessed in a trivially easy way by simply computing them for each posterior draw. Moreover, these analyses can benefit from the two key benefits of an IPM: increased precision of parameter estimates and sometimes the ability to estimate additional parameters.

9.2 CORRELATIONS BETWEEN DEMOGRAPHIC RATES AND POPULATION GROWTH

A first approach to investigate the impact of the temporal variability of demographic rates on temporal variability in the population growth rate is the calculation of a simple correlation coefficient between annual estimates of the demographic rates and those of the population growth rate (Robinson et al., 2004; Schaub et al., 2013; Brommer et al., 2017). Obviously, the stronger that correlation, the more a demographic rate has contributed to population change. This approach is very intuitive and useful for ranking the demographic rates according to their impact on the variability of the population growth rate. A disadvantage is that the correlation coefficients have no specific meaning nor do they sum to 1, which renders comparisons across species or studies difficult.

Computation of correlation coefficients is easy, but two challenges arise when using this approach. First, we need accurate estimates of time-varying demographic rates and population growth rates. Demographic rates should not be temporally autocorrelated; otherwise, the independence assumption for the calculation of a correlation coefficient is violated. They are best estimated from models with temporal random effects that are independent from each other (Burnham and White, 2002). Thus, we will fit an IPM that specifies temporal random effects for all demographic parameters. Second, in the computation of the correlation coefficients, we must accommodate the uncertainty associated with the parameter estimates. Thus, correlation coefficients between a demographic rate and the population growth rate must be calculated for each Markov chain Monte Carlo (MCMC) draw of the posterior distributions of both to obtain the posterior distribution of the correlation coefficient (e.g., Baillie et al., 2009; Schaub et al., 2013; Tempel et al., 2014; Saunders et al., 2018).

FIGURE 9.1

A hoopoe (*Upupa epops*) with a field cricket as prey (*Gryllus campestris*; photograph by www.marcelburkhardt.ch).

To illustrate the three types of retrospective analyses highlighted in this chapter, we analyze real data on hoopoes (*Upupa epops*) collected in the famous Valais valley in the Swiss Alps during 2002–17 (Arlettaz et al., 2010; Schaub et al., 2012; Plard et al. 2020a). The hoopoe is a medium-sized, migratory, cavity-nesting bird species (Fig. 9.1). Although not a passerine and quite a bit larger, the hoopoe has a broadly similar life history to that of the wonderful woodchat shrike that features so prominently in this book, except that the hoopoe is often double brooding (Hoffmann et al., 2015), and productivity depends on female age such that first-year individuals are less likely to reproduce successfully (Plard et al., 2018). The species can be studied readily by the installation of nest boxes, which hoopoes happily accept (Arlettaz et al., 2010). Each year, nest boxes are checked several times to collect information about occupancy, clutch size, and the numbers of hatchlings and fledglings. Nestlings are ringed and adults captured and ringed when they arrive at the nest box to feed their young. Here, we use the data on females only.

As an index of population size, we cannot use the total number of broods recorded within a year because of double brooding and because not all breeding adults are individually known. Even though capture effort in the study is intense, detection is not perfect (as it hardly ever is). Therefore, for each brood we defined the starting date (when the first egg was laid) and the end date (when the nestlings fledged or the brood failed) and then recorded the maximum number of simultaneously active broods in a season. The resulting measure of population size is likely a slight underestimation of the total number of breeding pairs because there might be very early and very late broods that do not overlap temporally or pairs that fail to breed at all. However, we think that this index is an accurate approximation of the temporal pattern of the number of breeding hoopoe pairs in the Valais study area.

The annual total reproductive output of a female could be decomposed into different components such as clutch size of first and second broods, hatching and fledging success of first and second broods, and the probability of double brooding. However, to keep the model simple, we just modeled the annual total number of young produced per female. Because the reproductive output depends on female age, we classified it by female age class. A total of 1040 broods with known ID of the mother (558 first-year and 482 older females) produced a total of 5957 fledglings (3070 from first-year and 2887 from older mothers). Hoopoes can be sexed morphologically when they are at least 1 year old but not at the nestling stage. During some years, we determined the nestling sex genetically and found an even sex ratio (Schaub et al., 2012).

Therefore, in our model we generally assume an even nestling sex ratio and probabilistically assign a sex to each nestling that was never recaptured. This resulted in 3374 capture histories of females marked as nestling and 470 capture histories of females marked as ≥ 1 year old.

Our hoopoe study area extends over 62 km^2, but the population is geographically open, and thus, our population model must accommodate emigration from and immigration into the population. Permanent emigration is already accounted for in the apparent survival estimated from the capture-recapture data, but dealing with immigration requires explicit attention. We can estimate immigration either as a rate or an absolute number (of immigrants) by an extension of a classical IPM (Abadi et al., 2010; Schaub and Fletcher, 2015). Conceptually, the information to do so is contained in a comparison of the observed population count data with the expected population trajectory under the demographic rates of the local individuals. That is, immigration is estimated as a "difference" and is a hidden parameter. We define immigrants as 1-year-old females born outside the study area and breeding for the first time within the study area. If an immigrant survives for a year and remains in the study area, it transitions into the adult state. The state process model therefore includes three types of individuals (N_1: 1-year-old local recruits; N_2: $\geq$ 2-year-old individuals who have reproduced in the study area in an earlier year already (adults); N_3: immigrants).

$$N_{1,t+1} \sim Poisson\left(\frac{f_{1,t}}{2}\phi_{j,t}(N_{1,t}+N_{3,t})+\frac{f_{2,t}}{2}\phi_{j,t}N_{2,t}\right)$$

$$N_{2,t+1} \sim Binomial(\phi_{a,t}, (N_{1,t}+N_{2,t}+N_{3,t}))$$

$$N_{3,t+1} \sim Poisson((N_{1,t}+N_{2,t}+N_{3,t})\omega_t)$$

Here, $f_{1,t}$ is the productivity of locally born and immigrant individuals, $f_{2,t}$ is the productivity of adults, ϕ_j is apparent juvenile and ϕ_a apparent adult survival, and ω is immigration expressed as a per-capita rate relative to the total population size the year before. We assume that immigrants are 1 year old and consequently have the same productivity as local 1-year-olds. This seems a reasonable assumption, as the natal dispersal is stronger than breeding dispersal (Bötsch et al., 2012) and hence, immigrants should be mostly 1-year-olds.

To compute the correlation coefficients between the demographic rates and the population growth rate (λ), we endow the former with temporal random effects and express the latter as the ratios of total population sizes in subsequent years. We use a normal observation, or residual, error in the state-space model and choose 0.4 as a lower bound of the uniform prior for the standard deviation of the residual error. This helps in avoiding the posterior getting stuck at 0. Other than that, the code for the IPM is entirely analogous to the other IPMs you have seen so far in the book.

We load the hoopoe data and briefly inspect their content:

```
# Load the hoopoe data and produce data overview
library(IPMbook); library(jagsUI)
data(hoopoe)
str(hoopoe)

List of 5
 $ ch       : int [1:3844, 1:16] 0 0 0 0 0 0 0 0 0 0 ...
  ..- attr(*, "dimnames")=List of 2
  .. ..$ : NULL
  .. ..$ : chr [1:16] "2002" "2003" "2004" "2005" ...
 $ age      : int [1:3844] 1 1 1 1 1 1 1 1 1 1 ...
 $ count    : num [1:16] 34 46 68 93 88 87 85 78 82 84 ...
 $ reproAgg:List of 4
  ..$ J1: num [1:16] 73 188 243 320 261 222 206 154 278 220 ...
  ..$ J2: num [1:16] 51 83 101 182 256 226 206 278 237 244 ...
  ..$ B1: num [1:16] 15 23 36 50 44 37 39 30 48 48 ...
  ..$ B2: num [1:16] 6 13 12 23 38 38 32 41 39 41 ...
 $ reproInd:List of 3
  ..$ f   : int [1:1092] 6 11 11 3 15 6 11 12 7 6 ...
  ..$ id  : num [1:1092] 483 486 530 531 532 533 534 535 536 538 ...
  ..$ year: int [1:1092] 2002 2002 2002 2002 2002 2002 2002 2002 2002 ...
```

The data contain the capture histories for all 3844 individuals (`ch`) during 2002–17, each individual's age at marking (`age`; 1: nestling; 2: $\geq$ 1-year-old), the annual population index just described (`count`), and data on productivity in two formats (aggregated per year and at the scale of each individual). Here, we use the aggregated productivity data, namely, the annual number of fledglings produced by 1-year-old (`J1`) and older females (`J2`) and the annual number of mothers for which productivity is recorded (`B1`: 1-year-old, `B2`: >1-year-old).

We first convert the individual capture histories into age-dependent m-arrays because we do not have individual effects in survival in our analysis and want to increase the speed of model fitting. We then bundle the data, define the model, initial values, MCMC setting, and parameters we want to monitor, and launch the model in JAGS.

```
# Produce age-dependent m-arrays
marr <- marrayAge(hoopoe$ch, hoopoe$age, 2)
marr.j <- marr[,,1]
marr.a <- marr[,,2]

# Bundle data
jags.data <- with(hoopoe$reproAgg, list(marr.j=marr.j, marr.a=marr.a, n.occasions=ncol(marr.j),
  rel.j=rowSums(marr.j), rel.a=rowSums(marr.a), J1=J1, B1=B1, J2=J2, B2=B2, count=hoopoe$count,
  pNinit=dUnif(1, 50)))
str(jags.data)                          # Remind ourselves of how the data look like


# Write JAGS model file
cat(file="model1.txt", "
model {
# Priors and linear models
# For mean and variance parameters
mean.logit.phij <- logit(mean.phij)
mean.phij ~ dunif(0, 1)
mean.logit.phia <- logit(mean.phia)
mean.phia ~ dunif(0, 1)
mean.log.f1 <- log(mean.f1)
mean.f1 ~ dunif(0, 10)
mean.log.f2 <- log(mean.f2)
mean.f2 ~ dunif(0, 10)
mean.log.omega <- log(mean.omega)
mean.omega ~ dunif(0, 3)
mean.p ~ dunif(0, 1)
mean.pj ~ dunif(0, 1)

sigma.phij ~ dunif(0, 3)
tau.phij <- pow(sigma.phij, -2)
sigma.phia ~ dunif(0, 3)
tau.phia <- pow(sigma.phia, -2)
sigma.omega ~ dunif(0, 3)
tau.omega <- pow(sigma.omega, -2)
sigma.f1 ~ dunif(0, 3)
tau.f1 <- pow(sigma.f1, -2)
sigma.f2 ~ dunif(0, 3)
tau.f2 <- pow(sigma.f2, -2)

sigma ~ dunif(0.4, 20)                  # lower bound >0
tau <- pow(sigma, -2)

# Linear models for demographic parameters
for (t in 1:(n.occasions-1)){
  logit.phij[t] ~ dnorm(mean.logit.phij, tau.phij)
  phij[t] <- ilogit(logit.phij[t])
  logit.phia[t] ~ dnorm(mean.logit.phia, tau.phia)
  phia[t] <- ilogit(logit.phia[t])
  log.omega[t] ~ dnorm(mean.log.omega, tau.omega)
  omega[t] <- exp(log.omega[t])
  p[t] <- mean.p
  pj[t] <- mean.pj
}
for (t in 1:n.occasions){
  log.f1[t] ~ dnorm(mean.log.f1, tau.f1)
  f1[t] <- exp(log.f1[t])
  log.f2[t] ~ dnorm(mean.log.f2, tau.f2)
  f2[t] <- exp(log.f2[t])
}

# Population count data (state-space model)
# Model for initial stage-spec. population sizes: discrete uniform priors
N[1,1] ~ dcat(pNinit)
N[2,1] ~ dcat(pNinit)
N[3,1] ~ dcat(pNinit)
```

```
# Process model over time: our model of population dynamics
for (t in 1:(n.occasions-1)){
  N[1,t+1] ~ dpois(f1[t] / 2 * phij[t] * (N[1,t] + N[3,t]) + f2[t] / 2 * phij[t] * N[2,t])
  N[2,t+1] ~ dbin(phia[t], (N[1,t] + N[2,t] + N[3,t]))
  N[3,t+1] ~ dpois((N[1,t] + N[2,t] + N[3,t]) * omega[t])
}

# Observation model
for (t in 1:n.occasions){
  count[t] ~ dnorm(Ntot[t], tau)
  Ntot[t] <- N[1,t] + N[2,t] + N[3,t]    # total population size
}

# Productivity data (Poisson regression model)
for (t in 1:n.occasions){
  J1[t] ~ dpois(f1[t] * B1[t])
  J2[t] ~ dpois(f2[t] * B2[t])
}

# Capture-recapture data (CJS model with multinomial likelihood)
# Define the multinomial likelihood
for (t in 1:(n.occasions-1)){
  marr.j[t,1:n.occasions] ~ dmulti(pr.j[t,], rel.j[t])
  marr.a[t,1:n.occasions] ~ dmulti(pr.a[t,], rel.a[t])
}
# Define the cell probabilities of the m-arrays
for (t in 1:(n.occasions-1)){
  # Main diagonal
  qj[t] <- 1 - pj[t]
  q[t] <- 1 - p[t]
  pr.j[t,t] <- phij[t] * pj[t]
  pr.a[t,t] <- phia[t] * p[t]
  # Above main diagonal
  for (j in (t+1):(n.occasions-1)){
    pr.j[t,j] <- phij[t] * prod(phia[(t+1):j]) * qj[t] * prod(q[t:(j-1)]) * p[j] / q[t]
    pr.a[t,j] <- prod(phia[t:j]) * prod(q[t:(j-1)]) * p[j]
  } #j
  # Below main diagonal
  for (j in 1:(t-1)){
    pr.j[t,j] <- 0
    pr.a[t,j] <- 0
  } #j
} #t
# Last column: probability of non-recapture
for (t in 1:(n.occasions-1)){
  pr.j[t,n.occasions] <- 1 - sum(pr.j[t,1:(n.occasions-1)])
  pr.a[t,n.occasions] <- 1 - sum(pr.a[t,1:(n.occasions-1)])
} #t

# Derived parameters
# Annual population growth rate
for (t in 1:(n.occasions-1)){
  lambda[t] <- Ntot[t+1] / Ntot[t]
}
}
")

# Initial values
inits <- function(){list(mean.phij=runif(1, 0.05, 0.2), mean.phia=runif(1, 0.3, 0.5))}

# Parameters monitored
parameters <- c("mean.phij", "sigma.phij", "mean.phia", "sigma.phia", "mean.omega", "sigma.omega",
  "mean.f1", "sigma.f1", "mean.f2", "sigma.f2", "mean.p", "mean.pj", "sigma", "phij", "phia", "omega",
  "f1", "f2", "N", "Ntot", "lambda")

# MCMC settings
ni <- 20000; nb <- 10000; nc <- 3; nt <- 4; na <- 2000

# Call JAGS (ART 1 min), check convergence and summarize posteriors
out1 <- jags(jags.data, inits, parameters, "model1.txt", n.iter=ni, n.burnin=nb, n.chains=nc,
  n.thin=nt, n.adapt=na, parallel=TRUE)
traceplot(out1)
print(out1, 3)
```

```
                mean     sd    2.5%     50%   97.5% overlap0 f  Rhat n.eff
mean.phij      0.123  0.012   0.100   0.123   0.148    FALSE 1 1.009   249
sigma.phij     0.280  0.103   0.102   0.272   0.509    FALSE 1 1.001  3609
mean.phia      0.364  0.020   0.325   0.364   0.406    FALSE 1 1.000  7500
sigma.phia     0.175  0.107   0.012   0.163   0.419    FALSE 1 1.005   553
mean.omega     0.281  0.047   0.187   0.282   0.372    FALSE 1 1.009   244
sigma.omega    0.236  0.190   0.005   0.196   0.701    FALSE 1 1.009   440
mean.f1        5.463  0.286   4.919   5.457   6.041    FALSE 1 1.000  5840
sigma.f1       0.192  0.046   0.122   0.186   0.303    FALSE 1 1.002  4224
mean.f2        6.137  0.381   5.422   6.125   6.946    FALSE 1 1.000  7500
sigma.f2       0.226  0.054   0.144   0.218   0.354    FALSE 1 1.000  7500
mean.p         0.794  0.033   0.727   0.795   0.856    FALSE 1 1.000  7500
mean.pj        0.604  0.040   0.527   0.604   0.682    FALSE 1 1.003   629
sigma          3.871  2.636   0.477   3.375  10.200    FALSE 1 1.001  2590
phij[1]        0.124  0.025   0.079   0.122   0.177    FALSE 1 1.004   539
[ ... output truncated ... ]
phij[15]       0.108  0.023   0.066   0.108   0.154    FALSE 1 1.001  2463
phia[1]        0.362  0.042   0.277   0.361   0.449    FALSE 1 1.001  7500
[ ... output truncated ... ]
phia[15]       0.403  0.050   0.331   0.394   0.524    FALSE 1 1.001  1945
omega[1]       0.336  0.119   0.173   0.310   0.656    FALSE 1 1.001  1814
[ ... output truncated ... ]
omega[15]      0.284  0.077   0.138   0.281   0.458    FALSE 1 1.003  1776
f1[1]          5.095  0.511   4.128   5.079   6.132    FALSE 1 1.000  7500
[ ... output truncated ... ]
f1[16]         5.206  0.415   4.440   5.187   6.054    FALSE 1 1.000  7500
f2[1]          7.822  1.000   6.096   7.758   9.987    FALSE 1 1.000  4283
[ ... output truncated ... ]
f2[16]         6.231  0.434   5.404   6.215   7.116    FALSE 1 1.000  7500
N[1,1]        10.957  8.469   1.000   9.000  31.000    FALSE 1 1.001  1481
N[2,1]        15.212  9.165   1.000  14.000  33.000    FALSE 1 1.002  1205
N[3,1]        11.794  8.591   1.000  10.000  31.000    FALSE 1 1.001  2304
[ ... output truncated ... ]
N[1,16]       14.032  4.215   6.000  14.000  23.000    FALSE 1 1.000  3704
N[2,16]       20.046  3.843  13.000  20.000  28.000    FALSE 1 1.000  7500
N[3,16]       14.007  4.527   6.000  14.000  23.525    FALSE 1 1.001  2786
Ntot[1]       37.963  5.226  32.000  36.000  52.000    FALSE 1 1.000  3053
[ ... output truncated ... ]
Ntot[16]      48.085  3.864  40.000  48.000  57.000    FALSE 1 1.000  7500
lambda[1]      1.252  0.143   0.937   1.278   1.500    FALSE 1 1.000  5035
[ ... output truncated ... ]
lambda[15]     0.976  0.087   0.793   0.980   1.150    FALSE 1 1.001  7500
```

To assess the strength of the contribution of the temporal variability of the demographic rates to the temporal variability of the population growth rate, we use the accumulated posterior draws to calculate the correlations between them as a derived quantity. We could have specified these correlations directly in the BUGS code (similar to what we do in the BUGS code to calculate the population growth rate λ). However, often doing this outside of BUGS in R using the MCMC-produced posterior draws is more practical, for instance because one can use existing R functions such as `cor` here. By carrying out these calculations for each MCMC draw rather than just for the posterior means, we properly propagate all uncertainty into the computation of the correlations. That is, in this way we directly obtain the posterior distributions of the correlation coefficients.

```
n.draws <- out1$mcmc.info$n.samples          # Determine number of MCMC draws
corr <- matrix(NA, ncol=5, nrow=n.draws)     # Create object to hold results
draws <- out1$sims.list                      # Dig out MCMC samples
for (s in 1:n.draws){                        # Loop over all MCMC draws and get correlations
  corr[s,1] <- cor(draws$phij[s,], draws$lambda[s,])
  corr[s,2] <- cor(draws$phia[s,], draws$lambda[s,])
  corr[s,3] <- cor(draws$omega[s,], draws$lambda[s,])
  corr[s,4] <- cor(draws$f1[s,1:15], draws$lambda[s,])
  corr[s,5] <- cor(draws$f2[s,1:15], draws$lambda[s,])
}

# Calculate posterior summaries for the correlation coefficients
# Posterior means
apply(corr, 2, mean)
# 95% credible intervals
cri <- function(x) quantile(x, c(0.025, 0.975))
apply(corr, 2, cri)
```

We summarize the posterior distributions of the five correlation coefficients in Table 9.1. The credible intervals of all correlation coefficients are wide. We find that the population growth rate was positively correlated with all five

Table 9.1 Correlation coefficients of the population growth rate of Swiss hoopoes with demographic rates calculated including and ignoring the estimation uncertainty associated with each involved parameter. For the former, the posterior means and the 95% credible intervals are shown.

	Correlations Including Uncertainty		Correlations Excluding Uncertainty
	Mean	**95% CRI**	**Mean**
Juvenile survival (ϕ_j)	0.52	0.11; 0.81	0.70
Adult survival (ϕ_a)	0.21	−0.33; 0.66	0.38
Immigration (ω)	0.29	−0.33; 0.81	0.81
Productivity 1y (f_1)	0.56	0.29; 0.79	0.65
Productivity ad (f_2)	0.52	0.24; 0.75	0.61

demographic rates (as expected), most strongly with productivity of the younger females, but only a little less with productivity of experienced breeders and apparent juvenile survival. This suggests that the temporal variation of productivity and apparent juvenile survival, i.e., components of local recruitment, had a stronger impact on hoopoe population dynamics than did the temporal variation of adult survival and immigration.

Sometimes such correlations are calculated without taking account of the estimation uncertainty by basing them on the posterior means of the demographic rates and the population growth rate. We note that such correlations may give substantially different results compared with the more comprehensive solutions that do accommodate the uncertainty (Table 9.1). Ignoring uncertainty, the strongest correlation appears to be between immigration rate and population growth, so we would wrongly conclude that hoopoe population dynamics is mainly driven by the variation of immigration. The correlation coefficients from the two approaches may differ because of the nonzero sampling correlation between

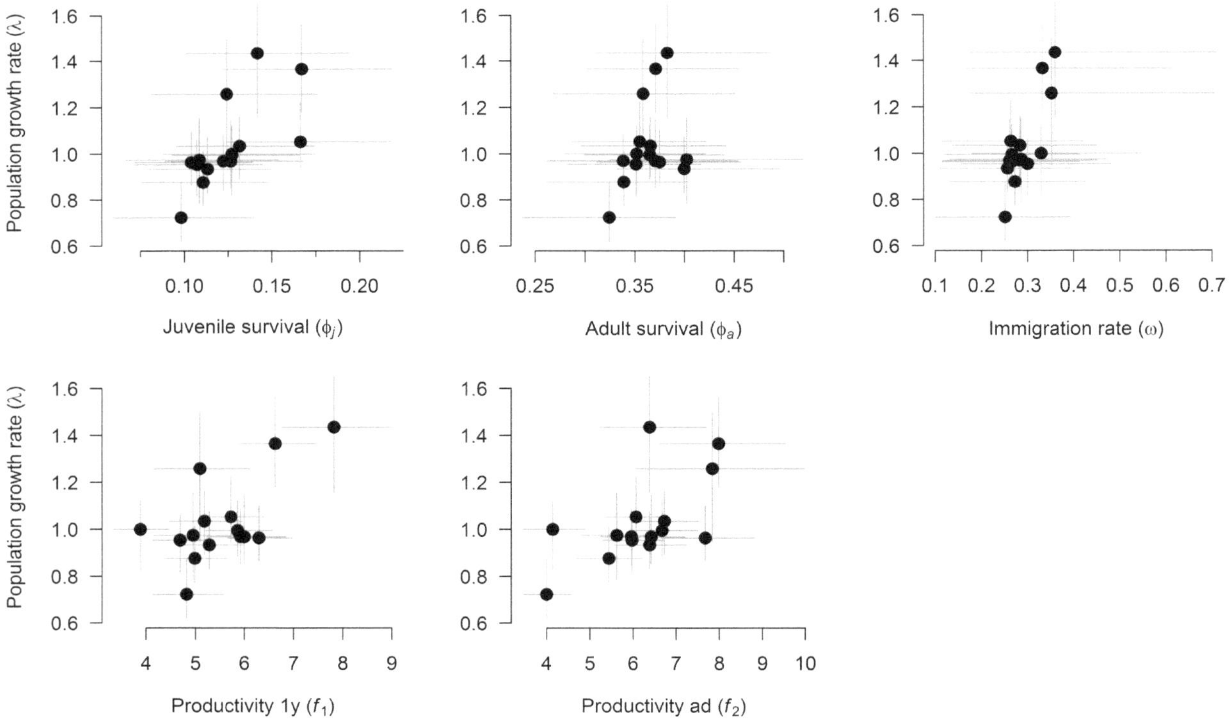

FIGURE 9.2

Scatterplot of the estimates of the population growth rate (λ) against those of each demographic rate in the integrated population model for Swiss hoopoes (posterior means and 95% credible intervals).

demographic rates and population growth rate and because of skewed posteriors. Hence, it is always safer to include parameter uncertainty in the computation of any derived quantity, not just because we want to have a measure of the uncertainty associated with the estimate of the derived quantity but also because the inferences can be very different from the "cheap" approximation based on the posterior means.

We visualize the relationships between the population growth rate (λ) and demographic rates in Fig. 9.2, and not only show the posterior means but also the 95% CRIs. The positive associations of most demographic parameters with λ seem to be mostly due to 3 years with high values of λ. These correspond to the initial 4 study years when the hoopoe population showed a very strong increase after the installation of the first nest boxes (Arlettaz et al., 2010). We therefore also repeated this analysis without these start-up years and found that the correlations remained high for apparent juvenile survival and the productivity of experienced breeders, whereas those for the other parameters become less important. Hence, the main conclusion that the dynamics of this population was driven by local recruitment appears to be robust.

9.3 LIFE-TABLE RESPONSE EXPERIMENTS

A more formal approach to understand drivers of past population dynamics is the so-called life-table response experiment (LTRE), which has a history dating back to the 1950s (Birch, 1953) and was further developed by Hal Caswell later (Caswell, 1989, 1996, 2010; Horvitz et al., 1997). The basic idea is to define a linear model with population growth rate as the response (dependent) variable and demographic rates as the explanatory (independent) variables. However, in contrast to the usual linear model, no coefficients (slopes) need to be estimated because they are given by the growth rate sensitivities. The approach was first applied in experimental settings, e.g., to study the effect of pesticides (Daniels and Allan, 1981), which is the reason it received the very specific and slightly odd name "life-table response experiment." However, LTREs can be applied much more broadly and are very well suited for observational studies.

The linear model in an LTRE can be specified in at least three different ways depending on the study design and the question asked (Caswell, 2001, p. 258). The *fixed-design* approach is suited to understanding how much the difference of each demographic rate has contributed to the difference in the population growth rate of two or more units of comparison, e.g., when comparing populations or developmental modes of individuals (Levin et al., 1987). *Random-design* approaches allow decomposition of the total temporal variability in the population growth rate into components due to the temporal variability of the demographic rates and the covariation among them. For example, the variability of adult fertility was found to contribute most to temporal variance of the growth rate in a killer whale (*Orcinus orca*) population (Brault and Caswell, 1993). Finally, if we want to know how the population growth rate has changed with respect to an environmental variable and how much each demographic rate has contributed to this change, a *regression design* can be applied (Caswell, 2001). Here, we illustrate only the random design because its purpose is the same as the correlative approach in Section 9.2. The calculations are performed with estimates of demographic rates obtained in an IPM. As always, they can therefore be carried out either inside the BUGS code during model fitting or outside in R afterward (see above).

Our goal is to decompose the temporal variance of the population growth rate λ into contributions due to the temporal variation and covariation of the demographic rates in the model. Because λ is a function of demographic rates, we can apply the Taylor series approximation (delta rule, Powell, 2007) to calculate the temporal variance of λ. Including only first-order terms often results in a good approximation and yields

$$\mathrm{var}(\lambda_{1,t}) \approx \sum_{i,j} \mathrm{cov}(\theta_i, \theta_j) \left.\frac{\partial\lambda}{\partial\theta_i}\frac{\partial\lambda}{\partial\theta_j}\right|_{\overline{\theta_{i,j}}} \tag{9.1}$$

Here, $\boldsymbol{\theta}$ is a vector with the demographic rates, $\mathrm{cov}(\theta_i, \theta_j)$ is element (i, j) in the variance-covariance matrix of the demographic rates, and $\left.\frac{\partial\lambda}{\partial\theta_i}\right|_{\overline{\theta_{i,j}}}$ are the growth rate sensitivities calculated with respect to the mean of the demographic rates. Note that $\lambda_{1,t}$ is the asymptotic population growth rate in year t given by the dominant eigenvalue (hence the first subscript) of the projection matrix parameterized with the demographic rates of year t (see Section 3.3.1). It typically differs from the realized population growth rate, which is the ratio of the total number of individuals in two consecutive years. Hence, this variance decomposition tells us how much the *observed* temporal variability (and covariation) of the demographic rates has contributed to variability of the *asymptotic* population growth rate.

For demonstration, we use the hoopoe data from the previous section and carry out all required calculations based on the fitted model object `out1` in Section 9.2. You may have to recreate it if you no longer have it in an R workspace.

To apply Eq. 9.1, we must calculate the sensitivities of the asymptotic population growth rate (λ_1) with respect to all five demographic parameters as well as the variance-covariance matrix of the demographic rates. Growth rate sensitivities can be calculated using simulation by perturbing each of the demographic rates by a small amount and evaluating the changes in the asymptotic population growth rate (see Section 3.4.1). The sensitivities are evaluated at the means of each demographic rate, which in our random-effects analysis means at the values of their hypermeans (e.g., `mean.sj`). Recall that the asymptotic λ_1 is what we get as the annual λ for demographic rates that are constant over time and for a population at its stable stage distribution, which it is when projected over a sufficient number of time steps. In our example, the stable stage distribution is obtained after about five annual time steps. The following code calculates the growth rate sensitivities. All calculations are performed for each MCMC draw to obtain the posterior distributions of all quantities.

```
# Calculation of growth rate sensitivities
delta <- 0.001                          # (Small) size of perturbation
T <- 5                                  # Number of years with projections
n.draws <- out1$mcmc.info$n.samples     # Number of MCMC draws

# Define arrays to store output
N.ref <- N.star.phij <- N.star.phia <- N.star.om <- N.star.f1 <- N.star.f2 <- array(NA,
  dim=c(n.draws, 3, T))
N.ref[,,1] <- N.star.phij[,,1] <- N.star.phia[,,1] <- N.star.om[,,1] <- N.star.f1[,,1] <-
  N.star.f2[,,1] <- 1
r.annual.ref <- r.annual.star.phij <- r.annual.star.phia <- r.annual.star.om <- r.annual.star.f1 <-
  r.annual.star.f2 <- matrix(NA, nrow=n.draws, ncol=T)
lambda <- numeric()
sens <- matrix(NA, nrow=n.draws, ncol=5)
draws <- out1$sims.list                 # store MCMC draws in a new object to simplify calculation

# Loop over all MCMC draws and project in time
for (s in 1:n.draws){                   # Loop over all MCMC draws
  for (t in 1:(T-1)){                   # Loop over all time steps
    # Calculate asymptotic population growth rate based on stage-specific population sizes
    N.ref[s,1,t+1] <- draws$mean.phij[s] * (draws$mean.f1[s] *
      (N.ref[s,1,t] + N.ref[s,3,t]) + draws$mean.f2[s] * N.ref[s,2,t]) / 2
    N.ref[s,2,t+1] <- draws$mean.phia[s] * sum(N.ref[s,,t])
    N.ref[s,3,t+1] <- draws$mean.omega[s] * sum(N.ref[s,,t])
    r.annual.ref[s,t] <- log(sum(N.ref[s,,t+1])) - log(sum(N.ref[s,,t]))

    # Sensitivity with respect to juvenile survival
    N.star.phij[s,1,t+1] <- (draws$mean.phij[s] + delta) / 2 * (draws$mean.f1[s] *
      (N.star.phij[s,1,t] + N.star.phij[s,3,t]) + draws$mean.f2[s] * N.star.phij[s,2,t])
    N.star.phij[s,2,t+1] <- draws$mean.phia[s] * sum(N.star.phij[s,,t])
    N.star.phij[s,3,t+1] <- draws$mean.omega[s] * sum(N.star.phij[s,,t])
    r.annual.star.phij[s,t] <- log(sum(N.star.phij[s,,t+1])) - log(sum(N.star.phij[s,,t]))

    # Sensitivity with respect to adult survival
    N.star.phia[s,1,t+1] <- draws$mean.phij[s] / 2 * (draws$mean.f1[s] *
      (N.star.phia[s,1,t] + N.star.phia[s,3,t]) + draws$mean.f2[s] *  N.star.phia[s,2,t])
    N.star.phia[s,2,t+1] <- (draws$mean.phia[s] + delta) * sum(N.star.phia[s,,t])
    N.star.phia[s,3,t+1] <- draws$mean.omega[s] * sum(N.star.phia[s,,t])
    r.annual.star.phia[s,t] <- log(sum(N.star.phia[s,,t+1])) - log(sum(N.star.phia[s,,t]))
```

```
    # Sensitivity with respect to immigration
    N.star.om[s,1,t+1] <- draws$mean.phij[s] / 2 * (draws$mean.f1[s] * (N.star.om[s,1,t] +
      N.star.om[s,3,t]) + draws$mean.f2[s] * N.star.om[s,2,t])
    N.star.om[s,2,t+1] <- draws$mean.phia[s] * sum(N.star.om[s,,t])
    N.star.om[s,3,t+1] <- (draws$mean.omega[s] + delta) * sum(N.star.om[s,,t])
    r.annual.star.om[s,t] <- log(sum(N.star.om[s,,t+1])) - log(sum(N.star.om[s,,t]))

    # Sensitivity with respect to productivity 1y
    N.star.f1[s,1,t+1] <- draws$mean.phij[s] / 2 * ((draws$mean.f1[s] + delta) *
      (N.star.f1[s,1,t] + N.star.f1[s,3,t]) + draws$mean.f2[s] * N.star.f1[s,2,t])
    N.star.f1[s,2,t+1] <- draws$mean.phia[s] * sum(N.star.f1[s,,t])
    N.star.f1[s,3,t+1] <- draws$mean.omega[s] * sum(N.star.f1[s,,t])
    r.annual.star.f1[s,t] <- log(sum(N.star.f1[s,,t+1])) - log(sum(N.star.f1[s,,t]))

    # Sensitivity with respect to productivity ad
    N.star.f2[s,1,t+1] <- draws$mean.phij[s] / 2 * (draws$mean.f1[s] * (N.star.f2[s,1,t] +
      N.star.f2[s,3,t]) + (draws$mean.f2[s]  + delta) * N.star.f2[s,2,t])
    N.star.f2[s,2,t+1] <- draws$mean.phia[s] * sum(N.star.f2[s,,t])
    N.star.f2[s,3,t+1] <- draws$mean.omega[s] * sum(N.star.f2[s,,t])
    r.annual.star.f2[s,t] <- log(sum(N.star.f2[s,,t+1])) - log(sum(N.star.f2[s,,t]))
  } #t

  lambda[s] <- exp(r.annual.ref[s,T-1])

  # Growth rate sensitivities
  sens[s,1] <- (exp(r.annual.star.phij[s,T-1]) - lambda[s]) / delta
  sens[s,2] <- (exp(r.annual.star.phia[s,T-1]) - lambda[s]) / delta
  sens[s,3] <- (exp(r.annual.star.om[s,T-1]) - lambda[s]) / delta
  sens[s,4] <- (exp(r.annual.star.f1[s,T-1]) - lambda[s]) / delta
  sens[s,5] <- (exp(r.annual.star.f2[s,T-1]) - lambda[s]) / delta
} #s
```

The MCMC draws of the five growth rate sensitivities are stored in the matrix `sens`. The next quantity we must calculate is the variance-covariance matrix of the demographic rates, which tells us how strongly each demographic rate has varied over time and how much it has covaried with each of the others. It is therefore important that only pure temporal variability be included and no other variability, such as sampling variability. This requirement is met when we compute it from an analysis with temporal random effects because they result in a proper separation of process (temporal) and sampling variability (Burnham and White, 2002). We can either estimate temporal variances and covariances by making use of a multivariate normal distribution (see Section 6.4) or compute the variance-covariance matrix from the annual estimates (i.e., the "shrunk" estimates). Here, we adopt the latter approach because scale transformations can be avoided. In a statistical model, variances are typically estimated on a transformed scale such as the logit or log, but in an LTRE we need them on the natural scale. Scale transformations are possible, but they are not always satisfactory, especially for the skewed posterior distributions typical for variances. The following bit of R code calculates the posterior distribution of the variance-covariance matrix based on the annual "shrunk" estimates. Note that productivity for the last year is excluded to ensure that all vectors have the same length.

```
# Calculate variance-covariance matrix of the demographic rates
vcov <- array(NA, dim=c(n.draws, 5, 5))
for (s in 1:n.draws){
  vcov[s,,] <- var(cbind(draws$phij[s,], draws$phia[s,], draws$omega[s,], draws$f1[s,1:15],
    draws$f2[s,1:15]))
}
```

The final step toward the random-effects LTRE is the calculation of the contribution of each demographic rate to λ_1, i.e., the application of Eq. 9.1. These contributions appear in the form of an array where the first dimension refers to the MCMC draws and the remaining two dimensions to the number of demographic rates (here five). A demographic rate can contribute by its pure temporal variability (on the diagonal) and by pairwise covariation with the other demographic rates (these are the off-diagonal elements).

```
# Calculate contribution matrix for every posterior draw
cont <- array(NA, dim=c(n.draws, 5, 5))
for (s in 1:n.draws){
  cont[s,,] <- vcov[s,,] * outer(sens[s,], sens[s,])
}
```

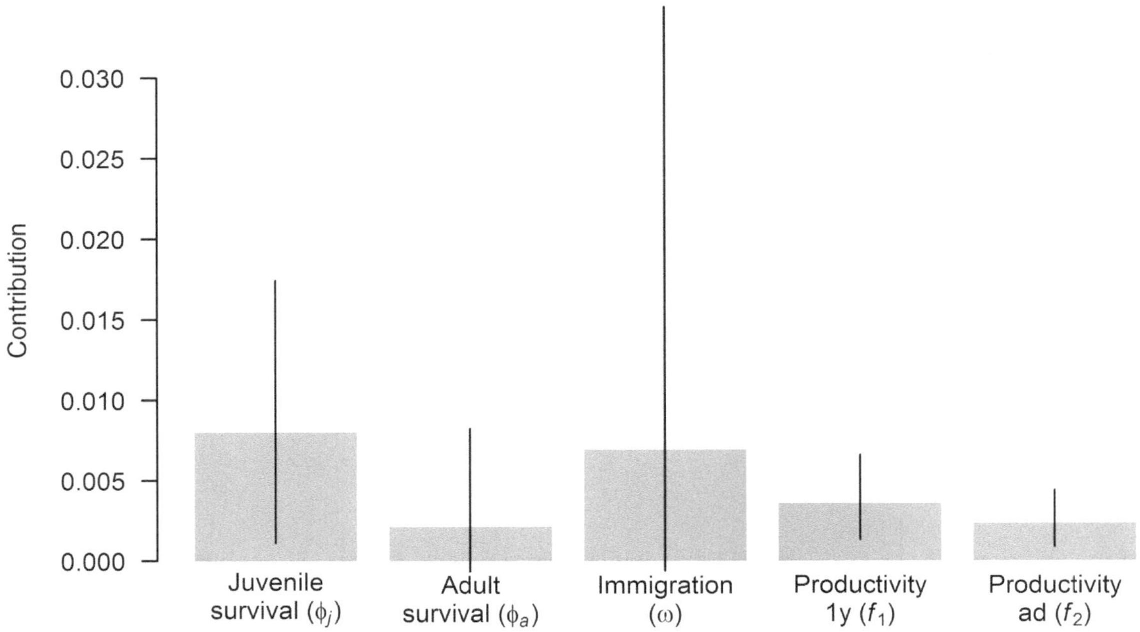

FIGURE 9.3

Estimates (posterior means with 95% credible intervals) of the contributions of the realized temporal variability in each demographic rate to the temporal variability of the asymptotic population growth rate (λ_1) under the life-table response experiment (LTRE) applied to the Swiss hoopoe integrated population model from Section 9.2.

Finally, we summarize the contributions of each demographic rate to the asymptotic growth rate λ_1. This is done by summing over all the elements in which a demographic rate is involved. For instance, the sum of the first row gives the total contribution to λ_1 of the temporal variability of apparent juvenile survival. We calculate the posterior means and the 95% credible intervals and produce Fig. 9.3, which shows us that juvenile survival has contributed more to λ_1 than the other demographic rates have. Immigration has contributed almost as much, whereas the contributions of age-dependent productivity and adult survival were smaller. However, the uncertainties around the point estimates are considerable, especially for immigration. We conclude that immigration probably had some effect on population growth but that longer time series of data would be needed to clarify its contribution to the population dynamics of Swiss hoopoes.

These results differ to some degree from those obtained by the correlative approach above. It still appears that elements of local recruitment have been important drivers of the population, but the impacts of adult survival and immigration are less clear due to their large uncertainty. The uncertainty in the LTRE approach is typically much greater than in the correlative approach. However, as the LTRE is a more formal approach, we think it perhaps may be preferred over the correlation analyses. Moreover, we also note that the two approaches do not tell us exactly the same thing. The correlative approach compares the realized temporal variation over some time period between demographic rates and the realized λ; in contrast, the LTRE approach compares the realized temporal variation between the demographic rates and the asymptotic λ.

We can also use the contribution matrix to determine how much a demographic rate has contributed through its own temporal variability and how much it has done so via pairwise temporal covariation with the other demographic rates. The calculations required for this are simple, and we show their result in Fig. 9.4. For immigration and the two survival probabilities, the "own" temporal variability was more important than was the covariation with other demographic rates. In contrast, the total contribution of age-specific productivity was more strongly driven by covariation with other parameters than by their own temporal variabilities.

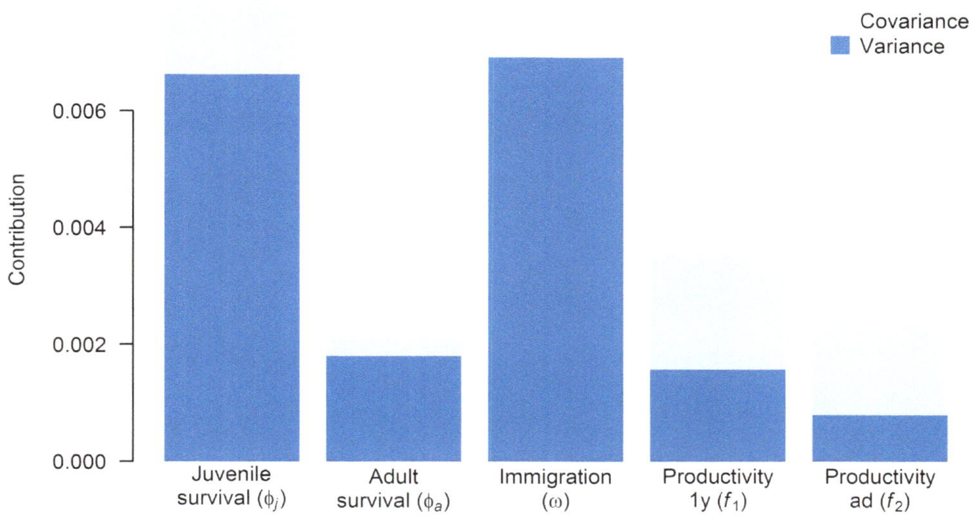

FIGURE 9.4

Contribution of the realized temporal variability and of the realized temporal covariation among demographic rates to the temporal variability of the asymptotic population growth rate in the life-table response experiment (LTRE) applied in the Swiss hoopoe integrated population model from Section 9.2. Note that 95% credible intervals are also available but were omitted to avoid clutter.

9.4 TRANSIENT LIFE-TABLE RESPONSE EXPERIMENTS

In contrast to populations in the lab, natural populations are exposed to continually variable environments. Therefore, they are unlikely to ever converge to a stable stage distribution but rather will typically remain in a permanent transient state (Hastings, 2004). Importantly, the dynamics of a population that is not in a stable stage distribution depends *not* only on demographic rates but also on the actual stage distribution (Koons et al., 2005; Ozgul et al., 2009; Ellis and Crone, 2013; Gamelon et al., 2014), as we have shown in Section 3.3.1. However, the asymptotic results from an LTRE as in Section 9.3 rely on the assumption that a population is in its stable age or stage distribution. Therefore, for a more accurate understanding of the drivers of the realized population dynamics over some specified duration, we must also account for population stage structure. For this, Koons et al. (2016) developed *transient* LTREs (tLTREs). They are suitable for identifying the demographic drivers in populations that are *not* in a stable stage distribution. The focus of tLTRE analysis is the *realized* population growth rate, i.e., the ratio of population sizes in subsequent years.

There are three different variants of tLTREs. The first tLTRE is a decomposition of the temporal variance of the realized population growth rate due to the realized temporal variation in the demographic rates and in the population structure. This corresponds to the correlation analysis in Section 9.2 and to the random-design asymptotic LTRE in Section 9.3. In the second type of tLTRE, we ask how much realized changes in the demographic rates and the population structure of successive time steps have contributed to changes in the realized population growth rate between successive time steps. Finally, the third type of tLTRE allows the assessment of demographic contributions to the difference in population growth of two time periods of equal length.

All tLTRE variants require annual estimates of demographic rates and stage-structured population sizes. It is nearly impossible to obtain the latter in any type of analysis other than an IPM. Thus, adoption of an IPM is almost a requirement for the application of tLTREs. Koons et al. (2017) apply all three types of tLTRE to better understand the demographic drivers of the population dynamics of the lesser scaup (*Aythya affinis*, Fig. 9.5) in North America. This publication contains a thorough explanation of these approaches and provides complete R code for all calculations. Here, we only demonstrate the first two types of tLTRE by applying them to the results of our hoopoe IPM.

wWe start with the first type of tLTRE, where we decompose the temporal variance of the realized population growth rate (λ_t) into components due to the realized variability and covariation of the demographic rates and of population structure. We start by defining vector $\boldsymbol{\theta}_t$, whose elements are the demographic rates and the proportional stage-structured population sizes (n_t) of year t, where proportional means that they sum to one for each year. Using the Taylor series approximation, we can calculate the variance of the realized population growth rate as a function of demographic rates and population structure as

$$\mathrm{var}(\lambda_t) \approx \sum_{i,j} \mathrm{cov}\left(\theta_{i,t}, \theta_{j,t}\right) \left.\frac{\partial\lambda_t}{\partial\theta_{i,t}} \frac{\partial\lambda_t}{\partial\theta_{j,t}}\right|_{\overline{\theta_{i,j}}} \tag{9.2}$$

FIGURE 9.5

A lesser scaup male (*Aythya affinis*; photograph by Robert Royse).

Note the great similarity between Eq. (9.2) for the tLTRE and Eq. (9.1) for the classical LTRE. Key differences are that now the realized population growth is analyzed (which we denote by λ_t) and that the population structure is included in $\boldsymbol{\theta}$. We again have two basic components: the entries of the variance-covariance matrix for all elements in $\boldsymbol{\theta}$ and the growth rate sensitivities. The latter are now the sensitivities of the realized population growth rate with respect to the demographic rates.

To derive these growth rate sensitivities, we note that the realized population growth rate can be written as

$$\lambda_t = \frac{\sum \boldsymbol{n}_{t+1}}{\sum \boldsymbol{n}_t} = \frac{\sum \mathbf{A}_t \boldsymbol{n}_t}{\sum \boldsymbol{n}_t} \tag{9.3}$$

where $\mathbf{A}_t$ is the projection matrix and $\boldsymbol{n}_t$ the vector containing the proportional population sizes in year t. We can rewrite Eq. (9.3) using the equations for the stage-specific population sizes (Section 9.2) in the hoopoe population as

$$\lambda_t = \frac{\frac{f_{1,t}}{2}\phi_{j,t}\left(n_{1,t} + n_{3,t}\right) + \frac{f_{2,t}}{2}\phi_{j,t}n_{2,t} + \left(\phi_{a,t} + \omega_t\right)\left(n_{1,t} + n_{2,t} + n_{3,t}\right)}{n_{1,t} + n_{2,t} + n_{3,t}}.$$

The sensitivities of λ_t with respect to the demographic parameters and population structure are calculated as the first derivative of λ_t with respect to all parameters. We use R function D (for "derivatives") to get these sensitivities symbolically.

```
lambda <- expression(((n1 + n3) * 0.5 * phij * f1 + n2 * 0.5 * phij * f2 + (phia + om) *
  (n1 + n2 + n3)) / (n1 + n2 + n3))                           # Define lambda

D(lambda, "phij")                                             # Get derivative for juvenile apparent survival
# ((n1 + n3) * 0.5 * f1 + n2 * 0.5 * f2)/(n1 + n2 + n3)
```

We only show the derivative of juvenile survival for illustration, but we do not need to calculate them for all parameters that are the elements of $\boldsymbol{\theta}$ explicitly. The R function `eval` performs the numerical calculation of the sensitivities automatically. Sensitivities are evaluated at the means of proportional population sizes and of all demographic rates for each MCMC draw. We first compute these values and then do the numerical calculation of the sensitivities.

```
# Calculate proportional population sizes
n.years <- length(out1$mean$phij)
n1 <- draws$N[, 1, 1:n.years] / draws$Ntot[, 1:n.years]
n2 <- draws$N[, 2, 1:n.years] / draws$Ntot[, 1:n.years]
n3 <- draws$N[, 3, 1:n.years] / draws$Ntot[, 1:n.years]

# Extract the mean demographic rates and population sizes and store them in a list
mu <- list(phij=draws$mean.phij, phia=draws$mean.phia, om=draws$mean.omega, f1=draws$mean.f1,
  f2=draws$mean.f2, n1=apply(n1, 1, mean), n2=apply(n2, 1, mean), n3=apply(n3, 1, mean))

# Calculate sensitivities
sens <- matrix(NA, n.draws, 8)
colnames(sens) <- c("phij", "phia", "om", "f1", "f2", "n1", "n2", "n3")
sens[,"phij"] <- eval(D(lambda, "phij"), envir=mu)
sens[,"phia"] <- eval(D(lambda, "phia"), envir=mu)
sens[,"om"] <- eval(D(lambda, "om"), envir=mu)
sens[,"f1"] <- eval(D(lambda, "f1"), envir=mu)
sens[,"f2"] <- eval(D(lambda, "f2"), envir=mu)
sens[,"n1"] <- eval(D(lambda, "n1"), envir=mu)
sens[,"n2"] <- eval(D(lambda, "n2"), envir=mu)
sens[,"n3"] <- eval(D(lambda, "n3"), envir=mu)
```

The final step is the calculation of the contribution of realized temporal process (co)variation of the demographic rates and of population structure to the temporal variation of realized population growth rates.

```
# Define matrix to store results
cont <- matrix(NA, nrow=n.draws, ncol=8)
colnames(cont) <- c("phij", "phia", "om", "f1", "f2", "n1", "n2", "n3")

# Calculate contributions for each demographic rate and stage-structured population size at each MCMC draw
for (s in 1:n.draws){
  dp_stoch <- cbind(draws$phij[s,], draws$phia[s,], draws$omega[s,], draws$f1[s,1:n.years],
    draws$f2[s,1:n.years], n1[s,], n2[s,], n3[s,])
  # Derive process variance and covariance among demographic parameters using shrunk estimates of
      # demographic rates and proportional pop. sizes
  dp_varcov <- var(dp_stoch)
  sensvec <- sens[s, ]
  # Calculate demographic contributions
  contmatrix <- dp_varcov * outer(sensvec, sensvec)
  cont[s, ] <- rowSums(contmatrix)
}
```

Fig. 9.6 shows the total contributions of demographic rates and population structure. We find patterns similar to those we did in the classical LTRE (look again at Fig. 9.4). Population structure had hardly any impact on the dynamics of the hoopoe population. This was to be expected because the transient phase of a species with the "fast" life history of the hoopoe is short. The realized population growth rate is therefore similar to the asymptotic growth rate calculated as the dominant eigenvalue of $\mathbf{A}_t$. In species with longer phases of transient dynamics, e.g., species with larger generation times, the impact of population structure becomes stronger, and the results of transient and classical LTRE will differ more (Koons et al., 2016).

Our next goal is to understand how much differences in the demographic rates and population structure of successive years have contributed to the difference in realized population growth rates ($\Delta\lambda_t$) of successive years. One might be particularly interested in the change in population growth between successive time steps following a management action or a specific environmental perturbation. This second type of tLTRE is similar to the fixed-design approach in a classical LTRE (Caswell, 2001). We again define a vector $\boldsymbol{\theta}_t$ that collects the demographic rates and population sizes for all stages in the population model. The difference in the realized population growth rate between years t and $t+1$ is (Koons et al., 2016)

$$\Delta\lambda_t = \sum_i (\theta_{i,t+1} - \theta_{i,t}) \frac{\partial\lambda_t}{\partial\theta_{i,t}}\bigg|_{\bar{\theta}_i}.$$

The elements of the differences in successive population growth rates are the differences of $\boldsymbol{\theta}$ between successive time steps and the growth rate sensitivities of $\boldsymbol{\theta}$, which are now evaluated at the mean values between successive time steps. For

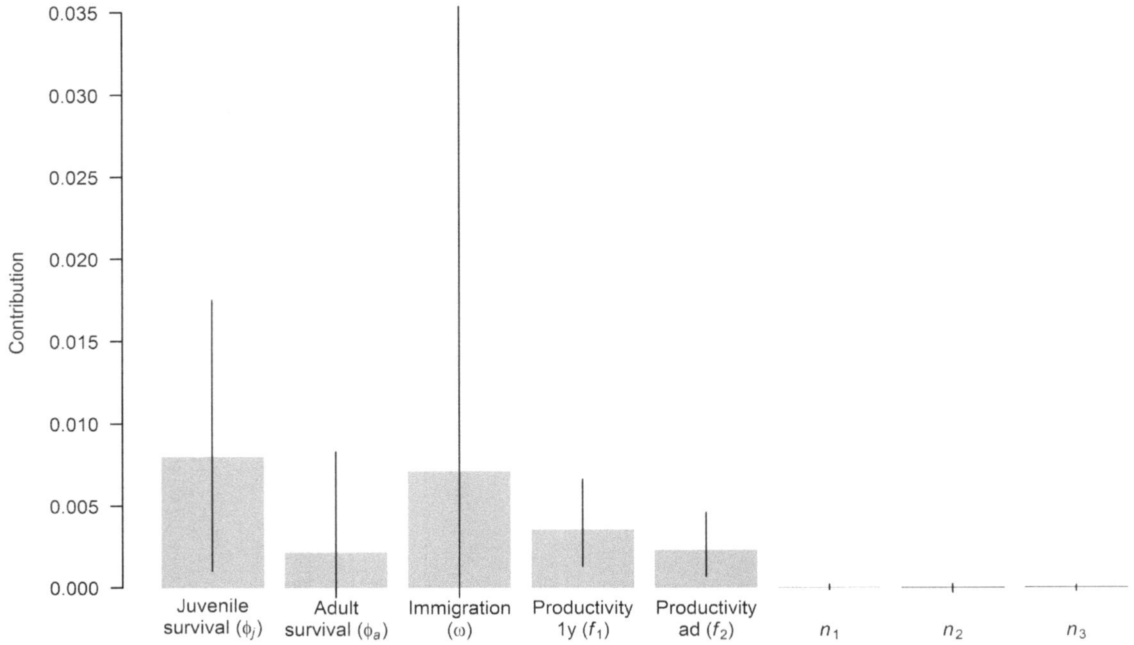

FIGURE 9.6

Example of the first type of a transient life-table response experiment (tLTRE) for the Swiss hoopoes: Contribution of the realized temporal variation of the demographic rates and of proportional population structure (n) to temporal variability of the realized population growth rate. The vertical lines show the limits of the 95% credible intervals.

the application, we first must calculate differences and means of demographic rates and population structure between successive time steps, and second, the sensitivities.

```
# Compute differences and means of demographic rates and population structure between successive years
diff <- array(NA, dim=c(n.draws, n.years-1, 8))
dimnames(diff)[[3]] <- c("phij", "phia", "om", "f1", "f2", "n1", "n2", "n3")
# Function to compute differences over successive time steps
getDiff <- function(x) x[,2:n.years] - x[,1:(n.years-1)]
diff[,,"phij"] <- getDiff(draws$phij)
diff[,,"phia"] <- getDiff(draws$phia)
diff[,,"om"] <- getDiff(draws$om)
diff[,,"f1"] <- getDiff(draws$f1)
diff[,,"f2"] <- getDiff(draws$f2)
diff[,,"n1"] <- getDiff(draws$N[,1,])
diff[,,"n2"] <- getDiff(draws$N[,2,])
diff[,,"n3"] <- getDiff(draws$N[,3,])

# Function to compute means over successive time steps, store them in a list
getMn <- function(x) (x[,2:n.years] + x[,1:(n.years-1)]) / 2
means <- list(phij=getMn(draws$phij), phia=getMn(draws$phia), om=getMn(draws$om),
  f1=getMn(draws$f1), f2=getMn(draws$f2), n1=getMn(draws$N[,1,]), n2=getMn(draws$N[,2,]),
  n3=getMn(draws$N[,3,]))

# Compute sensitivities
senss <- array(NA, dim=c(n.draws, n.years-1, 8))
dimnames(senss)[[3]] <- c("phij", "phia", "om", "f1", "f2", "n1", "n2", "n3")
senss[,,"phij"] <- eval(D(lambda, "phij"), envir=means)
senss[,,"phia"] <- eval(D(lambda, "phia"), envir=means)
senss[,,"om"] <- eval(D(lambda, "om"), envir=means)
senss[,,"f1"] <- eval(D(lambda, "f1"), envir=means)
senss[,,"f2"] <- eval(D(lambda, "f2"), envir=means)
senss[,,"n1"] <- eval(D(lambda, "n1"), envir=means)
senss[,,"n2"] <- eval(D(lambda, "n2"), envir=means)
senss[,,"n3"] <- eval(D(lambda, "n3"), envir=means)
```

The final step is the calculation of the contributions of the differences between successive years in the demographic parameters and population structures.

```
conts <- diff * senss
```

The results of this analysis are best presented graphically, where the difference in population growth between successive years is expressed as the sum of the contributions of the differences of the demographic rates and of population structure between successive years (see Fig. 9.7).

The changes in the realized population growth between successive years in Swiss hoopoes could best be explained by changes in juvenile survival and productivity, which contributed positively in some years and negatively in others (Fig. 9.7). Changes in adult survival and immigration contributed less, and changes in population structure made virtually no contribution. Thus, we see again that short-term changes in elements of local recruitment were the main demographic reason for short-term changes in the population growth of this hoopoe population.

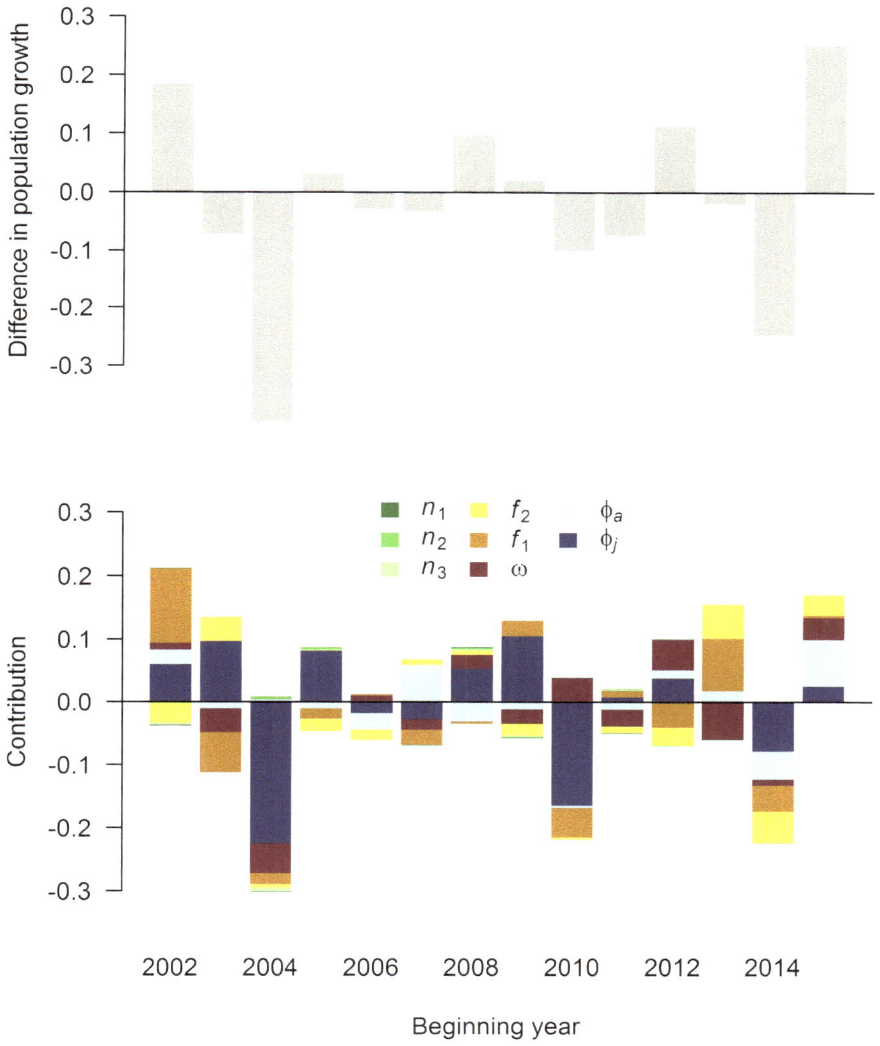

FIGURE 9.7

Example of the second type of a transient life-table response experiment (tLTRE) for the Swiss hoopoes: changes in the population growth rates between successive years (top panel) and the contributions of the changes in demographic rates and of population structure (lower panel).

9.5 SUMMARY AND OUTLOOK

The focus of the this chapter was to present a demographic population analysis that looks backward in time and tries to understand why a population has shown the observed dynamics. Such analyses are called retrospective population analyses, and they should allow a fundamental understanding of the demographic drivers of population dynamics. The key elements are growth rate sensitivities and the temporal variability of demographic rates. Improved analyses may also include the degree to which demographic rates are temporally correlated and changes in population structure. There are several approaches to retrospective population analyses. The simplest is a correlation of interannual population growth rates with demographic rates, whereas more formal approaches include various forms of so-called *life-table response experiments*. All retrospective analyses are based on estimated demographic rates and possibly population structure, and hence, these calculations can be regarded as representing derived quantities in the Bayesian analysis of an IPM with MCMC techniques. Thus, they can be computed outside of an IPM in R based on the posterior draws for all required "ingredients," and thus no changes in an IPM code are necessary to carry out these analyses. The output of an IPM is particularly useful for the novel *transient life-table response experiments* because they require estimates of population structure, and this is exactly something that IPMs provide and that is near-impossible to estimate in any other type of statistical analysis. Moreover, measures of the uncertainties of all target quantities are straightforward to compute in an MCMC-based analysis.

When measures of uncertainty for the LTRE statistics are computed, we typically find massive imprecision. This may be seen as a little frustrating because inference may be hampered, but it is clear that this is simply an honest assessment of the strength of evidence in our analysis. In the past, such quantities were often presented without any measures of uncertainty, but we now have the means to do better. We think the uncertainty mostly originates from the difficulties in estimating temporal variation and covariation in demographic rates. Thus, longer data sets will be beneficial in this respect, once more emphasizing the importance of long-term population studies in ecology (Magurran et al., 2010).

In Chapter 10, we introduce prospective population analyses. There, we ask the question about how the population size will develop in the future.

9.6 EXERCISES

1. Fit an IPM to the hoopoe data set and assess the demographic drivers using the classical LTRE. Develop an IPM in which the temporal correlation among demographic rates is explicitly estimated. Does the inference regarding demographic drivers change compared with the conclusions from Section 9.3?
2. The results of LTRE and tLTRE are typically associated with pronounced uncertainty. Sketch out a simulation study that aims to determine how strongly study duration and varying levels of temporal variability of demographic rates contributes to these uncertainties.

CHAPTER

POPULATION VIABILITY ANALYSIS

10

CHAPTER OUTLINE

10.1 INTRODUCTION

There are many cases where we want to manipulate the dynamics of a population. For instance, we may want to make more common a rare species that is of conservation concern, or the opposite; we may want to eradicate an invasive, introduced species. Yet another, and really frequent, case is that of harvested species, e.g., ungulates that are hunted for their meat or as a trophy in a terrestrial environment or fish stocks that are exploited for leisure or industry in rivers, lakes, or the sea, and where regulations must be decided upon to ensure the long-term sustainability of the harvest. Broadly, all these examples can be subsumed under the term *population management* (Williams et al., 2002; Mills, 2013; Lebreton and Gaillard, 2016).

Unfortunately, in the vast majority of cases, population management is conducted on a "feel/belief" basis and without any scientifically defensible procedure or empirical basis. That is, people manage a population based on what they *think* is right and often without any quantitative understanding of how their management actions may affect the demography of the target species (Mills, 2013). We don't think the "feel/belief" method is always detrimental in population management (or indeed in any other kind of decision making). However, if we have some demographic information about our species, we can do much better by adding the quantitative rigor of a formal population analysis into our management decisions. In this chapter, we deal with the application of population analysis in the context of the management of a population. This has become known as *population viability analysis* (PVA), but in fact is a very broad set of methods to apply demographic population analysis in conservation and in the management of invasive or harvested species.

Effective population management requires the definition of an objective for a population that is ideally formulated in quantitative terms. We also need some knowledge about how a population is likely to develop in the future. The archetypal goal in the context of conservation is to prevent a population of a desirable species from reaching size zero or to help it to reach some minimal size at which it may be deemed safe from chance extinction. We therefore need to know whether our population is in danger at all and whether any actions are necessary to increase its size. A popular measure of "danger" is the *extinction probability* of a population, which is the conceptual foundation of the modern red list criteria of the IUCN (Lande et al., 2003; Mace et al., 2008). For a population classified as endangered, we may want to know which conservation action is best to most efficiently improve its status. Such an assessment requires estimation of extinction probability and predictions of the future development of a population under different management options. The most efficient management option is identified by the greatest reduction in the extinction probability, perhaps weighted by its cost.

Integrated Population Models. https://doi.org/10.1016/B978-0-12-820564-8.00004-X

In the context of harvesting, our objective may be the achievement of a maximum sustainable number of animals that can be killed. A look into the future behavior of a population is therefore necessary to determine how many individuals can be removed from the population in a sustainable way. That is, so that the long-term survival of the population is not at risk or that its size does not drop below a predefined level. In contrast, for a pest species, the goal may be population extinction or attaining a low population density that is deemed acceptable. All these goals require a quantitative understanding of how population size changes when some action is taken.

In all these cases, we need to be able to predict how a population evolves into the future either under the current conditions or in scenarios of demographic, environmental, or other possible changes. To achieve this goal, a set of techniques broadly known as *population viability analysis*, or PVA, was developed in the 1980s in the context of conservation biology (Shaffer, 1981; Gilpin and Soulé, 1986). Interestingly, in two different fields, wildlife management and fisheries, very similar population analyses date back even earlier and were employed to predict sustainable harvests (Lebreton, 2005). In this chapter, we focus on PVA, but we emphasize the close relationship with models for sustainable harvesting and fisheries stock assessments. They all require quantitative knowledge about the demography of your population to predict its likely future development and possibly all of this as a function of different management actions.

There is no unique definition of a PVA. Morris and Doak (2002) define a PVA as a "quantitative method to predict the likely future status of a population," and Reed et al. (2002) define it as "an assessment of risk that a population reaches some threshold in some time." The former definition is broader and includes any method to assess the future status of a population, whereas the latter is more specific and focuses primarily on extinction probability and related quantities. However, predicting population trajectories into the future, i.e., making a population forecast, is a central element in both definitions.

The reasons for population extinctions can be broadly classified into two groups: first, extinction may result from the steady decline in a population due to a change in the environmental conditions that occurred in the past and is still persisting. This has been called the "declining population paradigm" in the seminal paper of Caughley (1994). For such populations, unless environmental conditions improve, extinction is certain, and the only uncertainty is when exactly the extinction is going to happen. To help such a population, we need to know the reasons for its decline and therefore have to investigate the past development of the population. Naturally, retrospective population analyses (Chapter 9) have an important role to play in this context. Predicting the future development of a population is then a mean to tell whether an improvement in the environmental conditions through some management action is sufficient to prevent extinction or to reduce the extinction risk to an acceptable level.

Second, extinction may result from chance events, e.g., due to demographic and environmental variability (called "stochasticity" in the PVA literature), environmental catastrophes or the loss of genetic diversity. Such extinctions may happen even if the demographic rates are high enough on average to ensure a population growth rate greater than 1. Naturally, chance events have greater impacts on small than on larger populations (Gilpin and Soulé, 1986), and hence, this view has been called the "small population paradigm" (Caughley, 1994). In contrast to a declining population, extinction of such a small population is not certain. Therefore, we want to estimate the likelihood of population extinction, and for this, population models again play a major role. If we find that extinction is likely, we typically want to identify the most efficient management actions to reduce that extinction risk.

PVA in a narrower sense deals with extinction probability and related quantities (Reed et al., 2002) and therefore requires population projection into the future. Ideally, projection would be based on a model that includes the major genetic and demographic features of a population. Genetic aspects can be important because small populations lose genetic heterozygosity due to inbreeding depression and genetic drift (Lande, 1988; Allendorf and Ryman, 2002), which in turn can affect the demography of individuals (Keller, 1998; Kéry et al., 2000, 2001; Luijten et al., 2002) or reduce their evolutionary ability to adapt to future environmental change. Ideally, we also have some knowledge about the demographic parameters, such as the mean and the variability of demographic rates, or the mechanisms and magnitude of density dependence, all of which are important for the construction of a population model to be used in a PVA. Most PVAs conducted in the past have excluded genetic aspects, most of all because knowledge about how the loss of genetic heterozygosity affects demographic rates is rarely available (Reed et al., 2002). To a lesser degree, this may be because genetics is likely to be less important to extinction than is demography, in particular in the short term (Lande, 1988; Boyce, 1992). Here we deal only with demographic PVA, but we do not want to claim that genetic aspects for population viability are unimportant. Another class of demographic PVA is based on population count data only (Dennis et al., 1991; Morris and Doak, 2002; Lande et al., 2003), and we don't cover such simpler, count-based PVAs, for which typically some state-space model for a time series of counts is employed. Instead, we illustrate the classical PVA, which is based on some sort of age- or stage-structured, stochastic population model, which is the focus of most integrated population models (IPMs).

In this chapter, we illustrate the use of an IPM to predict the likely future trajectory of a study population (or of multiple populations). We will see that a PVA is a very natural and relatively minor extension to any IPM. We discuss some challenges of demographic PVAs and show how Bayesian IPMs can help to cope with them. We do not provide a full review of PVA here—for that, see Boyce (1992), Beissinger and Westphal (1998), Beissinger (2002), and Reed et al. (2002).

10.2 CHALLENGES FOR DEMOGRAPHIC POPULATION VIABILITY ANALYSIS

Demographic PVAs deal with the future development of a population. One major challenge, therefore, is to predict a stochastic system based on incomplete knowledge about that system. Various forms of uncertainty exist, and they are often classified into four types (Williams, 1997; Nichols et al., 2011).

Process uncertainty refers to uncertain population development due to demographic and environmental stochasticity. If we can describe stochasticity by some rules (typically by statistical distributions), we can make statements about the expected behavior of a population, but this is unavoidably accompanied by some uncertainty. Typically, it has to be assumed that the distributions of the demographic rates are stationary, which means that they remain the same on average in the future. If the stationarity assumption is unlikely, process uncertainty greatly increases. Indeed, for a nonstationary system, prediction into the future may become impossible.

Structural uncertainty refers to our uncertainty about the structure of a system. For example, we may not know the precise functional form of density dependence, how many age classes a population model should have, or whether hunting mortality is additive to natural mortality or partially compensated. The effects of structural uncertainty are typically assessed in PVAs through sensitivity analyses that check how different key model outputs such as estimates of extinction probability become when the model structure is varied.

Parameter uncertainty denotes the fact that we never know the demographic parameters and population size with certainty—all of them must be estimated from data. Uncertainty about parameters can also result from observer bias or from differential opinions among experts about parameters for which no data are available.

Finally, *partial controllability* refers to management actions. It means first, that we fail to know exactly the demographic responses of a population to some management action and second, whether a management action is executed exactly as planned or prescribed. All four types of uncertainty will typically affect the accuracy of population predictions in the management context, such as in a PVA. It is important to include them as much as possible to adequately express our current state of knowledge about the population and get predictions with an honest assessment of the uncertainty associated with them.

Soon after their invention PVAs received severe criticism for several reasons (reviewed in Morris and Doak, 2002; Lande et al., 2003). Here we briefly summarize a few important points and explain how a PVA based on an IPM can help to overcome these problems to some degree.

A first criticism was that PVAs are sometimes performed for species for which adequate data are simply lacking. It is often true that for rare species with small populations detailed demographic information is lacking, and thus estimates of demographic rates are highly uncertain and may be biased. Ignoring parameter uncertainty is not recommended, otherwise the outcome of a PVA becomes too optimistic in the sense that we claim to be more certain than we actually are (McGowan et al., 2011). Standard practice in a PVA is the inclusion of parameter uncertainty, but predictions of future population sizes can then have such a high level of uncertainty that they may become practically meaningless (Ludwig, 1999). In IPMs, we exploit the available information more efficiently than in the piecewise analyses of the demographic data sets that are the basis of typical classical PVAs. Parameter estimates from IPMs tend to be more precise (Section 6.2). Therefore, PVAs based on IPMs may be beneficial when data are scarce.

A second criticism was that the estimates of the temporal variability of the demographic rates were often contaminated with sampling variability. This then resulted in an overestimation of the process variability and in too high extinction risks estimated in a PVA. Fortunately, models with temporal random effects (see Sections 2.7.7 and 4.5.1.1) produce corrected estimates of temporal variability that are free of the effects of sampling variability (Gould and Nichols, 1998; Link, 1999; Burnham and White, 2002). The demographic rates may also be temporally correlated with each other. Such a dependency must be included in PVAs (Fieberg and Ellner, 2001) to avoid bias in the estimated extinction probability, which may be positive or negative depending on the precise correlation structure among the demographic rates. The multivariate normal (MVN) distribution can be used to deal with temporal variability of and covariance among demographic rates (Link and Barker, 2005), and it is fairly straightforward to fit a MVN distribution in an IPM (Section 6.4). It is even possible to estimate process correlations with hidden parameters. Thus, using IPMs to conduct PVAs, we can probably alleviate this criticism somewhat.

A third criticism was that density dependence is often ignored in PVAs. This may be justified in some cases when population size is very low but surely not in general. Density dependence usually dampens population fluctuations and can greatly reduce the probability of extinction (Ginzburg et al., 1990). Even the form of density dependence can affect extinction probability (Mills et al., 1996; Middleton and Nisbet, 1997). In IPMs, inclusion of density dependence is straightforward in principle (Chapter 8), although the identification of the functional form of density dependence remains a challenge, as is the quantification of potential Allee effects (i.e., of positive density dependence; Courchamp et al., 1999; Sæther et al., 2008).

A fourth criticism revolves around spatial heterogeneity and dispersal, which can stabilize population fluctuations but also result in more complex dynamics (Beissinger and Westphal, 1998). Inclusion of the spatial structure of a population or set of populations is another important aspect of a PVA that is often neglected. Whether spatial structure can be included depends to a large extent on the quality of the available data. IPMs have been developed for spatially structured populations that allow the estimation of dispersal among populations (Chapter 18; Borysiewicz et al., 2009; Péron et al., 2010a,b; McCrea et al., 2010). IPMs also allow the estimation of immigration as a hidden parameter (Chapters 8 and 9; Schaub et al., 2015; Seward et al., 2019). Both features may be useful for an IPM-based PVA. Spatially explicit IPMs (Chapter 21; Chandler and Clark, 2014; Chandler et al., 2018a,b; Sun et al., 2019) are likely to offer interesting possibilities for PVAs, but so far, they have not been used for that purpose.

A fifth criticism pointed out that the results of PVAs are sometimes presented without declaring any measures of uncertainty. Clearly, PVAs must account for process stochasticity and parameter uncertainty—and ideally, also for structural uncertainty and partial controllability—all of which engender uncertainty about future population size. This uncertainty must be laid on the table—presenting a single value of a predicted, future population size without an associated measure of uncertainty is not scientifically defensible (Ludwig, 1996b; Beissinger and Westphal, 1998; Reed et al., 2002). On the other hand, as outlined above, inferences from PVAs that include process and parameter uncertainty are often very imprecise, and the uncertainty increases more the further we project a population into the future (Fieberg and Ellner, 2000). Therefore, we must try to learn something even from very uncertain predictions if at all possible.

A final and related criticism was that the results of a PVA that honestly include process stochasticity and parameter uncertainty may become meaningless because they are simply too uncertain to be practically useful (Ludwig, 1999). However, inferences can be made more meaningful if projections are not made too far into the future (Petchey et al., 2015). In addition, we can compare relative instead of absolute extinction risks to rank the effectiveness of different management options. Such relative risks are often more meaningful than are the single values of absolute risk under a certain management action, because process stochasticity and parameter uncertainty may cancel out to some degree when different management actions are compared in the same PVA (McCarthy et al., 2003).

We think many of these criticisms are justified to some degree. Being aware of them has resulted in guidelines about how to construct better PVAs and when to apply PVAs at all (Morris and Doak, 2002; Ralls et al., 2002). Fundamentally, model choice, in the sense of the degree of complexity that should be included in a model, ought to be determined by an interplay between the richness of the available data and the intended use of a model. A more complex model may be more realistic, but the associated gain in accuracy may be more than offset by a decrease in precision (this is the usual bias-variance trade-off in statistics). In addition, in a more complex model, we may lack data for many parameters, and this will add uncertainty to the model. Despite this, we think that PVAs should generally accommodate both process and parameter uncertainty. This can be done formally, if data are available, or via sensitivity analyses, i.e., by fitting and comparing several models that are based on different assumptions about the magnitude of parameters about which one has no data. If data and knowledge about a population are very scarce, it may not be possible to conduct a reliable PVA, and the benefit of model construction will then be restricted to a heuristic value—that it forces us to think much more clearly about the processes governing our population. Beissinger and Westphal (1998) are certainly correct in stating that "*doing a credible PVA requires good data and good ecological modeling*."

10.3 BAYESIAN POPULATION VIABILITY ANALYSIS

To date, most PVAs have been carried out using frequentist methods, but some Bayesian PVAs were developed early on (Ludwig, 1996b; Wade, 2002; Maunder, 2004) and later (McCaffrey et al., 2012; Heard et al., 2013; Hegg et al., 2013; Green and Bailey, 2015; Hobbs et al., 2015). The benefits of a Bayesian over a frequentist PVA are that (1) probability statements about the predicted/projected population are obtained directly in the Bayesian view of statistics, (2) the propagation of process and parameter uncertainty into the population projections (and any other estimated quantity) is straightforward and direct when using Markov chain Monte Carlo (MCMC) methods, and (3) the inferences can directly be used in formal decision or related analyses. Next, we briefly discuss each of these points.

The key fundamental difference between Bayesian and frequentist inference is the definition of probability. In Bayes, probability is used as a general measure of uncertainty about parameters, predictions, or any other unknown quantity. Hence, Bayesian PVAs directly allow one to make statements such as "*the probability that our population drops to below 3 individuals within 15 years is 0.34*," "*the probability that our population declines is 0.85*" or "*the probability that population size will be larger in 20 years when management option A is taken compared to when no action is taken is 0.74*." Such statements are obviously very relevant and are precisely what a manager would like to get from a PVA. By contrast, probability in frequentist inference is defined as a long-run frequency in hypothetical replicates of our study, including the study population. Any probability statement therefore refers to the data, i.e., how likely is it to observe such and such a data set given some parameters. We can never directly make probability expressions about parameters or other uncertain quantities associated with our model. In a frequentist PVA, we may also make statements like "the probability that the population will drop to below 3 individuals within 15 years is 0.34." However, this means that if we *could* somehow observe and measure, say, one million replicate populations and analyze each of them to see how likely they drop below a size of 3, about 340,000 of them would fall to below 3 within 15 years. This is not a direct statement about our particular population of interest as we can do when using Bayesian inference.

As outlined in Section 10.2, a reliable PVA should accommodate both process and parameter uncertainty. Propagating parameter uncertainty in an MCMC-based Bayesian PVA is straightforward, as it just needs the simulation of the population dynamics for many different combinations of parameter values drawn from their joint posterior distribution. This results in a posterior distribution of the desired quantity (e.g., population size) that directly includes both parameter and process uncertainty (Ludwig, 1996b; Wade, 2002). The combination of process *variability* (or stochasticity) and parameter *uncertainty* is straightforward because they are both expressed in a common currency: with probability distributions. Accounting for these forms of uncertainty in classical PVAs is less straightforward and more difficult, although it can be done (Dennis et al., 1991; Morris and Doak, 2002; McGowan et al., 2011). Typically, it requires specific simulations that reflect parameter uncertainty and process stochasticity. In some ways, such simulations are similar in spirit to the Bayesian projection (see e.g., Section 3.4.5), but accurately accounting for temporal correlation among demographic parameters can be very hard. In addition, strictly, the obtained quantities have a different meaning pertaining to some hypothetical replicate data sets and not directly to *your* data set and population (Wade, 2002).

Finally, a Bayesian PVA yields results that can be used directly in decision analysis (Possingham et al., 2002; Drechsler and Burgman, 2004; Green and Bailey, 2015) or adaptive management (McCarthy and Possingham, 2007). These approaches need as input exactly the type of probability statements *about the parameters* (and not about data or functions thereof) that Bayesian PVAs deliver by default. Probability statements about extinction risk are also used for classifications of species or populations according to the IUCN risk classification system (Taylor et al., 2002; Brooks et al., 2008). In summary, Bayesian PVAs offer several benefits over frequentist PVAs.

10.4 USE OF AN INTEGRATED POPULATION MODEL IN POPULATION VIABILITY ANALYSIS

The traditional approach to PVA consists of several steps. First, the available demographic or population count data are analyzed in a piecewise manner (and often in different software programs) to estimate demographic parameters or population sizes. Second, a stochastic population model is constructed that uses these estimates as inputs. This model is then used to predict the future population trajectory and derive relevant quantities such as extinction probabilities. A number of specialized software packages can be used to perform these stochastic projections and include VORTEX, RAMAS, and ALEX (Lindenmayer et al., 1995).

In contrast, when building your own Bayesian IPM, you can do it all at once, or in different words, when you have built an IPM, doing a PVA is almost trivially easy. The reason is that an IPM already contains all the elements that are needed for a PVA: an IPM represents a stochastic population model that links demography and population size and that fully accommodates parameter uncertainty. Thus, the IPM already summarizes in a probabilistic way all what we know about the population. The only part that is lacking is the population projection: in the usual IPM the time frame is given by the period over which we have some data. To perform the population projection as needed for the PVA we must make this time frame longer. Thus, the only change that is required when going from an IPM to a formal PVA is that we need to extend the time horizon such that predictions for the future are automatically obtained. This completely obviates the need for any stepwise and piecemeal approach, and we can do everything in one single software, BUGS. This is not only conceptually very elegant and satisfactory, but it has huge practical advantages, especially with respect to error propagation: running a single analysis ensures full and proper propagation into all estimated quantities of all uncertainty coming from process stochasticity and from parameter uncertainty. No information is lost, nor do any approximations need to be made, e.g., by summarizing the results from one model as input for another model further down the line in a stepwise analysis. And importantly for a user, you don't need to learn a different approach for every step in the PVA. If you can develop an IPM in the BUGS language, you will also master its use as a tool for a formal PVA—there is only a very small additional step, as we will soon show.

In BUGS, the use of an IPM as a PVA requires that the time loops for the models of the demographic rates and the population development are extended so that an additional number of years in the future are included and the desired quantities can be estimated for those years as well. This results in posterior distributions of demographic rates and population size in these future years. From these, other quantities can be derived that are the typical result from a PVA, such as extinction probability or the expected time to extinction. It is straightforward to compare different management scenarios in terms of their consequences for the demography of the projected population. Each scenario is included in the model by specifying the expected change in demographic rates after the implementation of an action. We can then predict future population sizes under each of these scenarios and compare them directly.

So far, a couple of studies have used IPMs for PVA, e.g., those of Hoyle and Maunder (2004), Maunder (2004), Tavecchia et al. (2009), Rhodes et al. (2011), Tenan et al. (2012), Oppel et al. (2014), Tenan et al. (2016), Saunders et al. (2018), Jansen et al. (2019), Plard et al. (2020a,b), and Schaub and Ullrich (2021). As a point of interest, Maunder (2004) presents several methods used in fisheries stock assessment that can of course be used for the PVA of a terrestrial species. He concluded that integrated analyses using Bayesian inference are flexible and powerful to merge different kinds of data (and in general, to synthesize quantitative knowledge) in one single analysis with proper uncertainty assessment and that

Bayesian IPMs are therefore particularly well suited for PVAs. Rhodes et al. (2011) studied the impact of different conservation strategies to increase the growth of an endangered population of koalas (*Phascolarctos cinereus*) in Australia. Oppel et al. (2014) conducted a classical PVA by estimating the extinction probability of a population of the endangered Montserrat oriole (*Icterus oberi*), and Saunders et al. (2018) assessed the viability of a population of piping plovers (*Charadrius melodus*) under different possible management actions. Schaub and Ullrich (2021) used a PVA to identify the most likely demographic process that has resulted in the extinction of a woodchat shrike (*Lanius senator*) population.

In the next section, we use simulated data sets to illustrate a PVA in practice and show how we can study the impact of different management actions. We do all of this with our wonderful and by now well-known model species, the woodchat shrike. After that, in Section 10.6 we show some challenges faced when we include a hidden parameter in the model, in this case immigration, and for this we use our real data set on the hoopoes.

10.5 A POPULATION VIABILITY ANALYSIS FOR SIMULATED WOODCHAT SHRIKE DATA

10.5.1 ESTIMATION OF EXTINCTION PROBABILITY AND RELATED QUANTITIES

We assume that we are interested in the viability of a small remnant population of woodchat shrikes (Fig. 10.1) that fluctuated between 4 and 20 breeding pairs over 20 years. We wish to assess population viability 15 years into the future. There are several useful metrics for such an assessment that can all be computed from the population projections, and here we demonstrate five examples:

- We predict population size and demographic rates into the future.
- We compute the extinction and quasi-extinction probabilities of the population.
- We calculate the probability that the population in 15 years is smaller than, equal to, or larger than the population in the final year of our study
- We calculate the conditional time to extinction.
- We quantify the increase in productivity required to achieve a population of 20 pairs in 15 years with a probability of 90%.

We assume that we have closely monitored our woodchat shrike population during the past 20 years, collected capture-recapture data, recorded the reproductive success of a number of surveyed broods, and counted the number of breeding pairs in each year. We load the (simulated) data set and produce an overview of what it contains.

```
# Load woodchat shrike data and produce data overview
library(IPMbook); library(jagsUI)
data(woodchat10)
str(woodchat10)

List of 5
 $ marr.a: num [1:19, 1:20] 8 0 0 0 0 0 0 0 0 0 ...
 $ marr.j: num [1:19, 1:20] 1 0 0 0 0 0 0 0 0 0 ...
 $ J     : num [1:20] 11 12 17 9 12 30 17 11 10 20 ...
 $ B     : num [1:20] 9 10 9 4 10 13 11 11 8 10 ...
 $ count : num [1:20] 10 12 9 4 11 14 18 11 8 13 ...
```

The capture-recapture data are summarized as age-dependent m-arrays (`marr.a`, `marr.j`). The annual number of fledglings (`J`) produced by a number (`B`) of surveyed broods and the number of breeding pairs recorded (`count`) are all stored in vectors.

The life history of the woodchat shrike was introduced in Section 5.2, and our IPM developed here reflects that knowledge with one simplification: we won't model any mother-age-dependence of productivity. Our female-based model assumes a prebreeding "census," i.e., males are sadly ignored in the model. In our process model, we include demographic and environmental stochasticity (see Chapter 3). For the sake of simplicity, we assume that the demographic rates vary independently from each other over time (i.e., their process correlation is zero) and that there is no density dependence in this small population.

We first develop a classical IPM as in Chapter 5. Then, we modify the model such that it produces forecasts of future population size, i.e., we use the IPM for a simple PVA. For this, we first extend the loop of the population model by the desired number of years for our forecast. Using the Markov property of the population model, the stage-specific population sizes are then computed for each year in the future. This projection depends on the values of the demographic rates for these future years, and hence, we also need to extend the loop for the annual demographic rates. We achieve the latter by specifying future demographic rates as being drawn from a normal distribution characterized by a mean and a temporal variance that are both estimated from the data as part of the fitting of the IPM to the observed data. In the code below, we

FIGURE 10.1

A beautiful woodchat shrike male (*Lanius senator*, photograph by Otto Samwald).

highlight in bold these minor changes required to move from an ordinary IPM to a simple PVA. Note that we call *K* the number of years for which we want to make predictions or produce the forecast. In the model code, we also add a derived quantity, `extinct`, which is equal to 1 when the total population size equals zero, and 0 otherwise. This quantity will produce an estimate of the population extinction rate.

```
# Bundle data
K <- 15                          # Number of years with predictions
jags.data <- list(marr.j=woodchat10$marr.j, marr.a=woodchat10$marr.a,
  n.occasions=ncol(woodchat10$marr.j), rel.j=rowSums(woodchat10$marr.j),
  rel.a=rowSums(woodchat10$marr.a), J=woodchat10$J, B=woodchat10$B, count=woodchat10$count,
  pNinit=dUnif(1, 50), K=K)
str(jags.data)

List of 10
 $ marr.j     : num [1:19, 1:20] 1 0 0 0 0 0 0 0 0 0 ...
 $ marr.a     : num [1:19, 1:20] 8 0 0 0 0 0 0 0 0 0 ...
 $ n.occasions: int 20
 $ rel.j      : num [1:19] 30 41 39 42 44 26 39 35 31 47 ...
 $ rel.a      : num [1:19] 18 16 14 20 21 21 20 17 19 18 ...
 $ J          : num [1:20] 11 12 17 9 12 30 17 11 10 20 ...
 $ B          : num [1:20] 9 10 9 4 10 13 11 11 8 10 ...
 $ count      : num [1:20] 10 12 9 4 11 14 18 11 8 13 ...
 $ pNinit     : num [1:50] 0.02 0.02 0.02 0.02 0.02 0.02 0.02 0.02 ...
 $ K          : num 15
```

```
# Write JAGS model file
cat(file="model1.txt", "
model {
# Priors and linear models
mean.logit.sj <- logit(mean.sj)
mean.sj ~ dunif(0, 1)
mean.logit.sa <- logit(mean.sa)
mean.sa ~ dunif(0, 1)
mean.p ~ dunif(0, 1)
mean.log.f <- log(mean.f)
mean.f ~ dunif(0, 10)

for (t in 1:(n.occasions-1)){
  p[t] <- mean.p
}

for (t in 1:(n.occasions-1+K)){  # Here we extend the loop to K more years
  logit.sj[t] <- mean.logit.sj + eps.sj[t]
  eps.sj[t] ~ dnorm(0, tau.sj)
  sj[t] <- ilogit(logit.sj[t])
  logit.sa[t] <- mean.logit.sa + eps.sa[t]
  eps.sa[t] ~ dnorm(0, tau.sa)
  sa[t] <- ilogit(logit.sa[t])
}

for (t in 1:(n.occasions+K)){    # Extended loop also here
  log.f[t] <- mean.log.f + eps.f[t]
  eps.f[t] ~ dnorm(0, tau.f)
  f[t] <- exp(log.f[t])
}

sigma.sj ~ dunif(0, 10)
tau.sj <- pow(sigma.sj, -2)
sigma.sa ~ dunif(0, 10)
tau.sa <- pow(sigma.sa, -2)
sigma.f ~ dunif(0, 10)
tau.f <- pow(sigma.f, -2)

sigma ~ dunif(0.5, 50)
tau <- pow(sigma, -2)

# Population count data (state-space model)
# Model for the initial population size: discrete uniform priors
N[1,1] ~ dcat(pNinit)
N[2,1] ~ dcat(pNinit)

# Process model over time: our model of population dynamics
for (t in 1:(n.occasions-1+K)){  # Note extended loop
  N[1,t+1] ~ dpois(sj[t] * f[t] * (N[1,t] + N[2,t]))
  N[2,t+1] ~ dbin(sa[t], (N[1,t] + N[2,t]))
}

# Observation model
for (t in 1:n.occasions){
  count[t] ~ dnorm(N[1,t] + N[2,t], tau)
}

# Productivity data (Poisson regression model)
for (t in 1:n.occasions){
  J[t] ~ dpois(f[t] * B[t])
}

# Capture-recapture data (CJS model with multinomial likelihood)
# Define the multinomial likelihood
for (t in 1:(n.occasions-1)){
  marr.j[t,1:n.occasions] ~ dmulti(pr.j[t,], rel.j[t])
  marr.a[t,1:n.occasions] ~ dmulti(pr.a[t,], rel.a[t])
}
# Define the cell probabilities of the m-arrays
for (t in 1:(n.occasions-1)){
  # Main diagonal
  q[t] <- 1-p[t]                # Probability of non-recapture
  pr.j[t,t] <- sj[t] * p[t]
  pr.a[t,t] <- sa[t] * p[t]
```

```
    # Above main diagonal
    for (j in (t+1):(n.occasions-1)){
      pr.j[t,j] <- sj[t] * prod(sa[(t+1):j]) * prod(q[t:(j-1)]) * p[j]
      pr.a[t,j] <- prod(sa[t:j]) * prod(q[t:(j-1)]) * p[j]
    } #j
    # Below main diagonal
    for (j in 1:(t-1)){
      pr.j[t,j] <- 0
      pr.a[t,j] <- 0
    } #j
  } #t
  # Last column: probability of non-recapture
  for (t in 1:(n.occasions-1)){
    pr.j[t,n.occasions] <- 1-sum(pr.j[t,1:(n.occasions-1)])
    pr.a[t,n.occasions] <- 1-sum(pr.a[t,1:(n.occasions-1)])
  }

  # Derived parameters
  # Total population size
  for (t in 1:(n.occasions+K)){
    Ntot[t] <- N[1,t] + N[2,t]
  }
  # Check whether the population is extinct in the future
  for (t in 1:K){
    extinct[t] <- equals(Ntot[n.occasions+t], 0)
  }
  }
  ")

  # Initial values
  inits <- function(){list(mean.sj=runif(1, 0, 0.5), mean.sa=runif(1, 0.4, 0.6),
    mean.f=runif(1, 1.3, 2))}

  # Parameters monitored
  parameters <- c("mean.sj", "sigma.sj", "mean.sa", "sigma.sa", "mean.p", "mean.f", "sigma.f", "sigma",
    "sj", "sa", "f", "N", "Ntot", "extinct")

  # MCMC settings
  ni <- 20000; nb <- 5000; nc <- 3; nt <- 3; na <- 1000

  # Call JAGS (ART 1 min), check convergence and summarize posteriors
  out1 <- jags(jags.data, inits, parameters, "model1.txt", n.iter=ni, n.burnin=nb, n.chains=nc,
    n.thin=nt, n.adapt=na, parallel=TRUE)
  traceplot(out1)
  print(out1, 3)

                  mean     sd    2.5%     50%   97.5% overlap0 f  Rhat n.eff
  mean.sj        0.284  0.029   0.230   0.284   0.343    FALSE 1 1.000 15000
  sigma.sj       0.413  0.160   0.122   0.402   0.764    FALSE 1 1.002  1046
  mean.sa        0.542  0.027   0.490   0.542   0.594    FALSE 1 1.000 11801
  sigma.sa       0.168  0.136   0.006   0.136   0.501    FALSE 1 1.006   566
  mean.p         0.633  0.033   0.568   0.633   0.696    FALSE 1 1.000 15000
  mean.f         1.529  0.093   1.347   1.529   1.714    FALSE 1 1.001  3187
  sigma.f        0.119  0.077   0.009   0.110   0.291    FALSE 1 1.005   407
  sigma          2.342  0.847   0.928   2.253   4.262    FALSE 1 1.001 14429
  sj[1]          0.231  0.060   0.114   0.230   0.349    FALSE 1 1.000  8420
  [ ...  output truncated... ]
  sj[34]         0.292  0.092   0.131   0.284   0.510    FALSE 1 1.000 15000
  sa[1]          0.555  0.052   0.464   0.550   0.678    FALSE 1 1.001 14315
  [ ...  output truncated... ]
  sa[34]         0.542  0.058   0.418   0.543   0.662    FALSE 1 1.001 15000
  f[1]           1.491  0.186   1.101   1.497   1.869    FALSE 1 1.001 12053
  [ ...  output truncated... ]
  f[35]          1.544  0.250   1.083   1.529   2.104    FALSE 1 1.005  3272
  N[1,1]         5.624  3.198   1.000   5.000  12.000    FALSE 1 1.000 15000
  N[2,1]         5.604  3.188   1.000   5.000  12.000    FALSE 1 1.000 15000
  [ ...  output truncated... ]
  N[1,35]        7.003 16.365   0.000   3.000  38.000     TRUE 1 1.012  9918
  N[2,35]        7.872 15.044   0.000   4.000  41.000     TRUE 1 1.003  8881
  Ntot[1]       11.229  2.046   8.000  11.000  16.000    FALSE 1 1.000 15000
  [ ...  output truncated... ]
  Ntot[35]      14.875 30.282   0.000   7.000  78.000     TRUE 1 1.006  8747
  extinct[1]     0.000  0.008   0.000   0.000   0.000    FALSE 1 1.291 15000
  [ ...  output truncated... ]
  extinct[15]    0.229  0.421   0.000   0.000   1.000     TRUE 1 1.000 15000
```

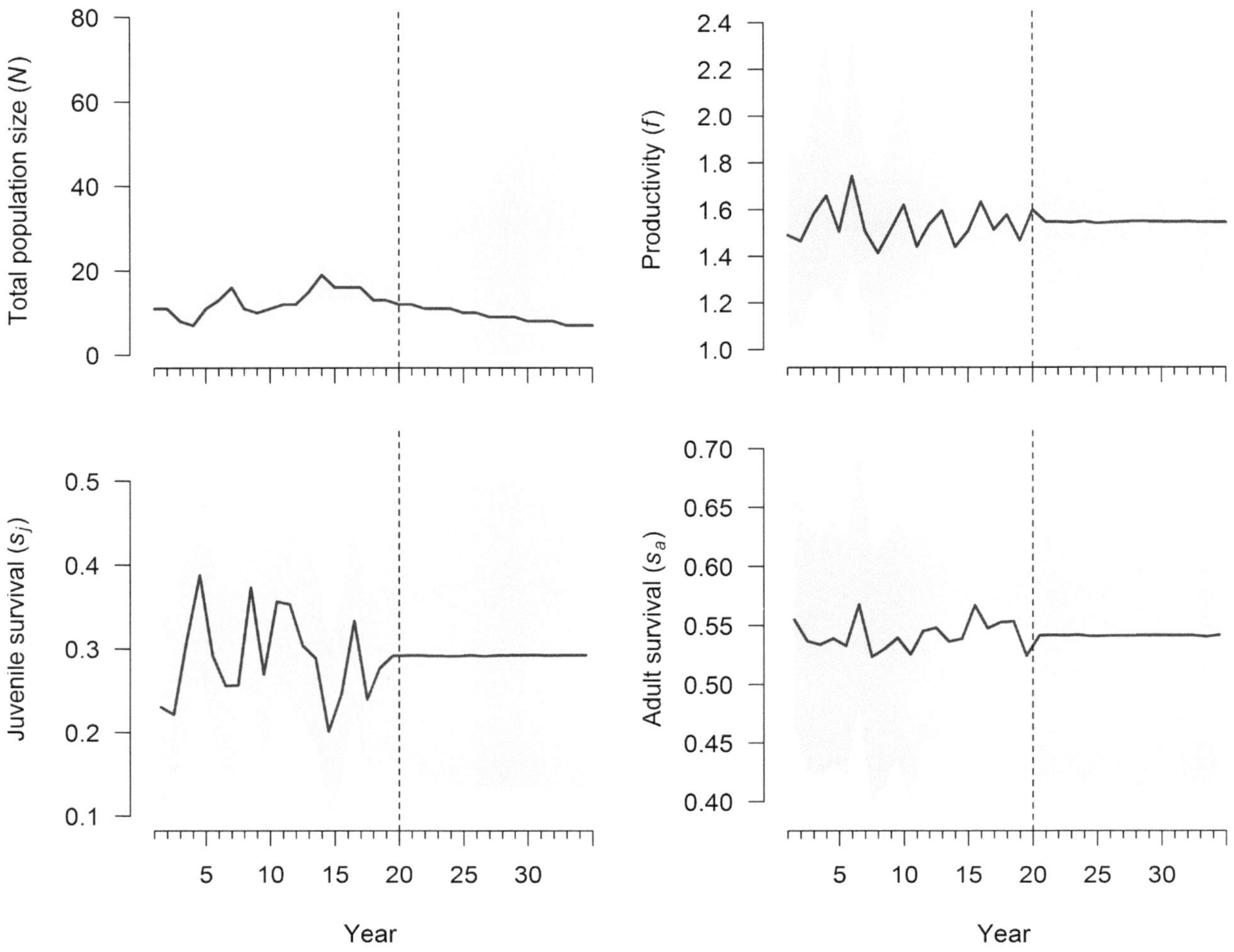

FIGURE 10.2

Example of the use of an integrated population model as population viability analysis: Estimated (years 1–20) and predicted, or forecast, population size (years 21–35) of a simulated woodchat shrike population (top left) and estimated and predicted demographic rates. Blue lines show the posterior median (for population size) and posterior means (for demographic rates), gray areas indicate 95% credible intervals, and the dashed vertical lines show the year when the forecast period begins.

We produce plots of the predicted population sizes and demographic rates. The uncertainty in the predicted population size increases strongly with increasing distance of the forecast into the future (Fig. 10.2). This corresponds to our expectations, since the uncertainty due to stochasticity is accumulating over time in this Markovian process. In Fig. 10.2, for population size we use the median instead of the mean as a measure of the central tendency of its posterior distribution because of the substantial skew in that distribution. Our population model does not include a density-dependent feedback, and hence, the predicted population sizes may become very large, as shown in Fig. 10.2, by a strong increase in the upper bound of the CRI. The temporal patterns of the uncertainty of the forecasts are fundamentally different for the demographic rates: uncertainty does not increase over time with increasing projection lengths. The reason for this discrepancy is that in our model, there is no Markovian dependence (or temporal autocorrelation) for the demographic rates, and hence there is no accumulation of the uncertainty over time. The width of the CRIs remains constant over time and depends on the temporal variability of the corresponding demographic rate and the estimation uncertainty. If temporal variability or parameter uncertainty were larger, the CRIs would become wider. And if we had specified temporal autocorrelations in the demographic rates (Chapter 12; Johnson and Hoeting, 2003), the uncertainty would also increase over time in the forecast period. Note that the uncertainty in the projected quantities adequately includes the contributions from both process and parameter uncertainty. This all occurs "automatically," and we don't have to do anything extra; neat, isn't it? The only neglected form of uncertainty is structural model uncertainty (an example of which could be the question of whether shrikes all start reproducing at age 1).

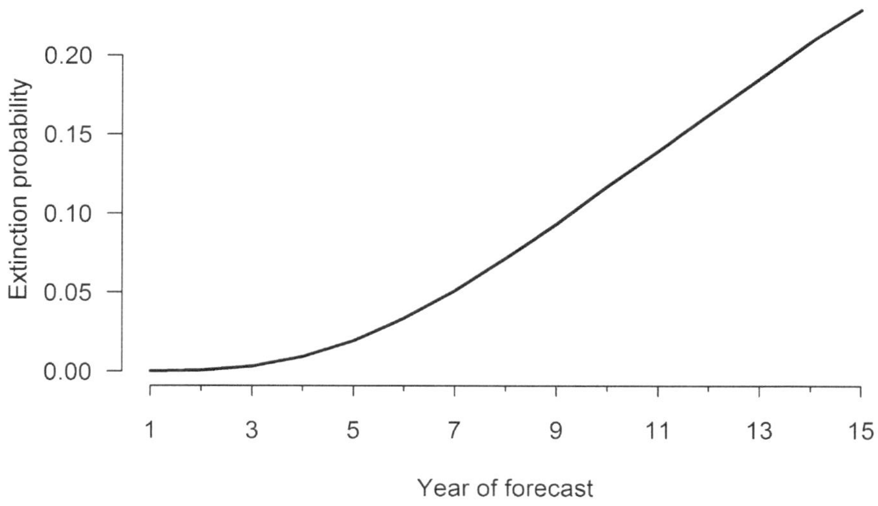

FIGURE 10.3

Cumulative extinction probability when the simulated woodchat shrike population is predicted into the future.

Next, we inspect the extinction probability, which is the traditional quantity reported in a PVA (Morris and Doak, 2002). In our PVA, this is a derived quantity that depends on predictions of population size, and we have calculated it as the quantity `extinct` using the BUGS function `equals`. For each year into the future and for every MCMC iteration, this function checks whether the total population is equal to zero, and if so, returns a 1, and otherwise a 0. Hence, taking the mean across all MCMC draws yields an estimate of the extinction probability for each year of the forecasting period. From the posterior summary, we find that the extinction probability 15 years into the future is 0.236. The summary also provides posterior standard deviations and CRIs of the extinction probability, although CRIs are fairly uninformative for a binary variable because they always range from 0 to 1. As a useful summary, we plot the cumulative extinction probability for the forecast period (Fig. 10.3).

So far, we have reported the absolute extinction probability, i.e., we called extinct a population of size zero. However, it may often be more meaningful to use an extinction threshold (D) that is larger than zero. A very small population is at risk in any case, and we may want to have reasonable faith in that the population is secure by choosing an extinction threshold greater than zero. We can define a *quasi-extinction* whenever the population drops to a size of D or lower. To compute a quasi-extinction probability, we could simply change one line in the BUGS code as follows:

```
extinct[t] <- step(D - Ntot[n.occasions+t])
```

Alternatively, we can calculate both absolute and quasi-extinction outside of BUGS in R based on the JAGS output. Here, we calculate the cumulative extinction probabilities for extinction thresholds of $D = \{1, 3, 5, 7\}$ and produce a plot to compare them (Fig. 10.4):

```
D <- c(1, 3, 5, 7)
T <- length(woodchat10$count)
library(scales)
color <- viridis_pal(option='E')(20)[c(18,13,7,1)]
plot(y=rep(0,K), x=1:K, type="n", ylim=c(0, 0.5), ylab="Quasi-extinction probability",
  xlab="Year of forecast", axes=FALSE)
axis(2, las=1)
axis(1, at=1:K, tck=-0.0125, labels=FALSE)
axis(1, at=c(1, 3, 5, 7, 9, 11, 13, 15), labels=c(1, 3, 5, 7, 9, 11, 13, 15), tck=-0.025)
for (i in 1:length(D)){
  qextinct <- out1$sims.list$Ntot[,(T+1):(T+K)] <= D[i]
  lines(apply(qextinct, 2, mean), lwd=2, col=color[i])
}
legend("topleft", legend=c("D = 7", "D = 5", "D = 3", "D = 1"), col=rev(color), lwd=2, bty="n")
```

In a Bayesian analysis, we represent our knowledge (or conversely, our uncertainty) about population size and other estimated quantities by posterior distributions. Hence, we can use them directly to make probability statements about these quantities. This is very useful and can provide valuable additional insights. Especially, such probability statements are very attractive when communicating with any kind of person outside of science. For example, we could assess the probability that our population at the end of the forecast horizon in 15 years is smaller than, equal to, or larger than it is now (i.e., at the end of the study period for which we have actual data). The challenge is that both population sizes compared are subject to

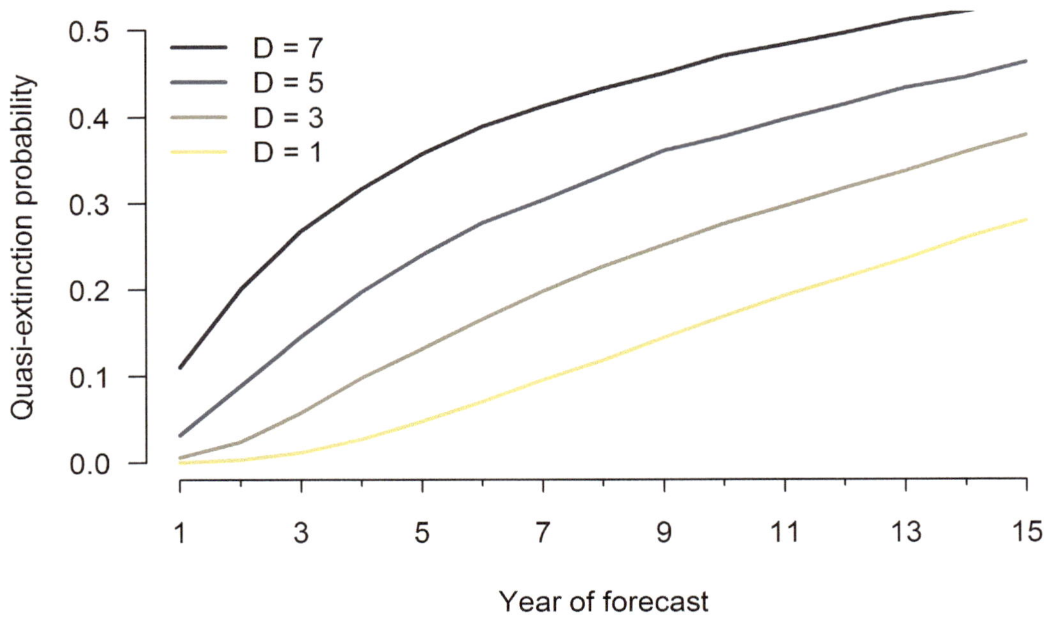

FIGURE 10.4

Cumulative quasi-extinction probabilities of the simulated woodchat shrike population for quasi-extinction thresholds *D* of 1–7.

uncertainty, but this is really simple in an MCMC-based analysis. We simply carry out a comparison in the form of a logical test of the population sizes in years 20 and 35 for each MCMC draw. This test returns a value of `FALSE` or `TRUE` in R. These results are interpreted as 0 and 1 and can therefore be used in a calculation. The mean across these binary values therefore returns the estimated probability that the logical test evaluates to true. The computation in R is simple as shown in a moment. Note that your result will be slightly different because of the stochastic nature of the MCMC solution and that we round to avoid meaningless precision.

```
# Probability of smaller future population size
round(mean(out1$sims.list$Ntot[,T] > out1$sims.list$Ntot[,T+K]), 2)
[1] 0.65

# Probability of the same population size in the future
round(mean(out1$sims.list$Ntot[,T] == out1$sims.list$Ntot[,T+K]), 2)
[1] 0.02

# Probability of a larger future population size
round(mean(out1$sims.list$Ntot[,T] < out1$sims.list$Ntot[,T+K]), 2)
[1] 0.33
```

Although our best guess is that the population size is declining somewhat in the future (Fig. 10.2), the probability of a smaller population size 15 years from now is not very far from 0.5, which is the expectation for a population that shows no clear trend, i.e., is stable.

A further useful approach is to restrict the calculation of derived quantities only to those simulated trajectories where the population went extinct somewhere between year 1 and 15 in the forecast period and compute the number of years it took to go extinct. The distribution of time to extinction is not normal or even symmetrical (Lande and Orzack, 1988; Ludwig, 1996a). Hence, we use the median rather than the mean to express its central tendency. However, we note that displaying the entire frequency distribution of the time to extinction is more informative still. Remember that this is the distribution of the time to extinction *conditional* on extinction: it depends on the extinction event as defined by the projection time *K* and the chosen extinction threshold *D*. For illustration, we here compute the median time to extinction and show the distribution of the conditional time to extinction (Fig. 10.5). We see a quick increase in the extinction with time up to year 7 and then a decline.

```
D <- 3
T <- length(woodchat10$count)
qextinct <- out1$sims.list$Ntot[,(T+1):(T+K)] <= D
ext <- qextinct[,K]          # extinct by year K, TRUE/FALSE
time.to.extinction <- K + 1 - apply(qextinct[ext,], 1, sum)
ns <- sum(ext)
```

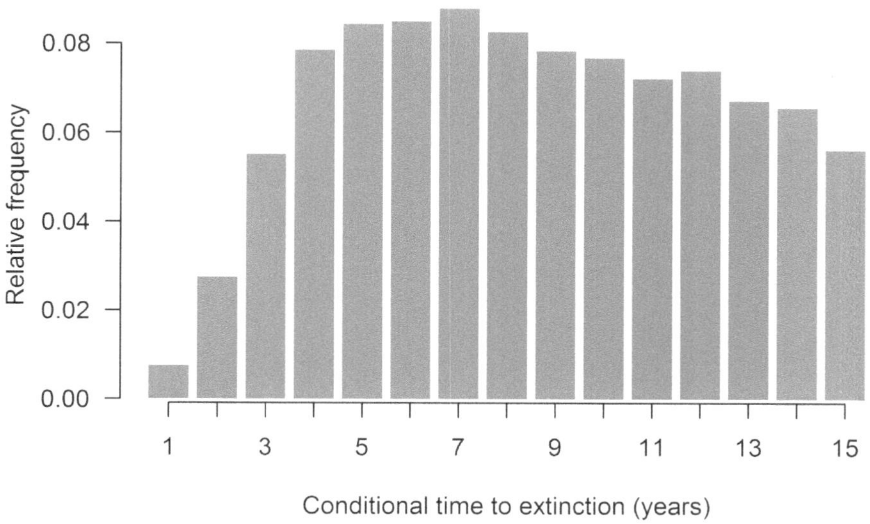

FIGURE 10.5

Frequency distribution of the conditional time to extinction in a simulated woodchat shrike population under an integrated population model. Conditional means that we include in this figure only the populations that went quasi-extinct (defined here by the quasi-extinction threshold of $D = 3$) within a $K = 15$-year forecast horizon.

```
a <- barplot(table(time.to.extinction) / ns, col="grey", ylab="Relative frequency",
  xlab="Conditional time to extinction (years)", axes=FALSE, names.arg=c(1, NA, 3, NA, 5, NA, 7,
  NA, 9, NA, 11, NA, 13, NA, 15), border=NA)
axis(1, at=a, labels=NA)
axis(2, las=1)
```

10.5.2 COMPARISON OF DIFFERENT MANAGEMENT OPTIONS

Quite often, computing absolute or quasi-extinction risks *per se* is not very interesting or insightful. Rather, a comparison of different management options in terms of the resulting extinction probability, time to extinction, or projected population size may be more meaningful to inform the management of a population. Moreover, the prediction of future population size is typically very imprecise, but a comparison of the predictions under different forecast options might result in more accurate inferences because some uncertainties may cancel out. As an illustration of what such a comparison of different management options in an IPM may look like, we use our simulated woodchat shrike example. We define a set of hypothetical conservation management options that are thought to be beneficial for the shrike population, and we want to compare them in terms of their efficiency by asking which is most efficient in reducing the extinction risk over a chosen forecast horizon (see also Saunders et al., 2018, for an application of this same idea in the context of an IPM). These management options are:

1. Reduction in nest predation by crows (*Corvidae*), which we assume results in a 20% increase in the number of fledglings;
2. Reduction of the temporal variability of adult survival by one half;
3. Release of three breeding pairs annually during 5 years, i.e., to perform translocations.
   For option 3, we assume that the breeding pairs are released in spring and that they immediately recruit into the population and reproduce. Of course, this means that the population cannot go extinct during the time when releases are conducted.
4. We also add the "do nothing" option as a control in our experiment.

We want to write a single model that yields forecasts under all four management options. Clearly, estimation of the demographic parameters and the population size during the study period is identical under all four options, and only the part of the model that does the prediction, or forecast, is different. Therefore, our code uses different loops over time when the past and future dynamics are computed (three management options and the control). We use the same data bundle packaged in Section 10.5.1.

```
# Write JAGS model file
cat(file="model2.txt", "
model {
# Priors and linear models
mean.logit.sj <- logit(mean.sj)
mean.sj ~ dunif(0, 1)
mean.logit.sa <- logit(mean.sa)
mean.sa ~ dunif(0, 1)
mean.p ~ dunif(0, 1)
mean.log.f <- log(mean.f)
mean.f ~ dunif(0, 10)

for (t in 1:(n.occasions-1)){
  p[t] <- mean.p
}

sigma.sj ~ dunif(0, 10)
tau.sj <- pow(sigma.sj, -2)
sigma.sa ~ dunif(0, 10)
tau.sa <- pow(sigma.sa, -2)
sigma.f ~ dunif(0, 10)
tau.f <- pow(sigma.f, -2)

sigma ~ dunif(0.5, 50)
tau <- pow(sigma, -2)

# Models for demographic rates
# Control, no change (option 4)
for (t in 1:(n.occasions-1+K)){
  logit.sj[t] <- mean.logit.sj + eps.sj[t]
  eps.sj[t] ~ dnorm(0, tau.sj)
  sj[t] <- ilogit(logit.sj[t])
  logit.sa[t,1] <- mean.logit.sa + eps.sa[t,1]
  eps.sa[t,1] ~ dnorm(0, tau.sa)
  sa[t,1] <- ilogit(logit.sa[t,1])
}
for (t in 1:(n.occasions+K)){
  log.f[t,1] <- mean.log.f + eps.f[t,1]
  eps.f[t,1] ~ dnorm(0, tau.f)
  f[t,1] <- exp(log.f[t,1])
}

# Option 1: increase of productivity
# Past: identical to control
for (t in 1:n.occasions){
  log.f[t,2] <- log.f[t,1]
  eps.f[t,2] <- eps.f[t,1]
  f[t,2] <- f[t,1]
}
# Future: increase mean productivity by 20%
for (t in (n.occasions+1):(n.occasions+K)){
  log.f[t,2] <- mean.log.f + log(1.2) + eps.f[t,2]
  eps.f[t,2] ~ dnorm(0, tau.f)
  f[t,2] <- exp(log.f[t,2])
}

# Option 2: reduction of temporal variability in adult survival
# Past: identical to control
for (t in 1:(n.occasions-1)){
  logit.sa[t,2] <- logit.sa[t,1]
  eps.sa[t,2] <- eps.sa[t,1]
  sa[t,2] <- sa[t,1]
}
# Future: reduction of temporal variability by half
for (t in n.occasions:(n.occasions-1+K)){
  logit.sa[t,2] <- mean.logit.sa + eps.sa[t,2]
  eps.sa[t,2] ~ dnorm(0, tau.sa*2)     # Temporal precision increased
  sa[t,2] <- ilogit(logit.sa[t,2])
}
```

```
# Population count data (state-space model)
# Model for the initial population size: discrete uniform priors
N[1,1,1] ~ dcat(pNinit)
N[2,1,1] ~ dcat(pNinit)

# Process model over time: our model of population dynamics
# Control, no change (option 4)
for (t in 1:(n.occasions-1+K)){
  N[1,t+1,1] ~ dpois(f[t,1] * sj[t] * (N[1,t,1] + N[2,t,1]))
  N[2,t+1,1] ~ dbin(sa[t,1], (N[1,t,1] + N[2,t,1]))
}

# Option 1: increase of productivity
# Past
for (t in 1:n.occasions){
  N[1,t,2] <- N[1,t,1]
  N[2,t,2] <- N[2,t,1]
}
# Future
for (t in n.occasions:(n.occasions-1+K)){
  N[1,t+1,2] ~ dpois(f[t,2] * sj[t] * (N[1,t,2] + N[2,t,2]))
  N[2,t+1,2] ~ dbin(sa[t,1], (N[1,t,2] + N[2,t,2]))
}

# Option 2: decrease of temporal variability in adult survival
# Past
for (t in 1:n.occasions){
  N[1,t,3] <- N[1,t,1]
  N[2,t,3] <- N[2,t,1]
}
# Future
for (t in n.occasions:(n.occasions-1+K)){
  N[1,t+1,3] ~ dpois(f[t,1] * sj[t] * (N[1,t,3] + N[2,t,3]))
  N[2,t+1,3] ~ dbin(sa[t,2], (N[1,t,3] + N[2,t,3]))
}

# Option 3: release of 3 females annually during 5 years
# Past
for (t in 1:n.occasions){
  N[1,t,4] <- N[1,t,1]
  N[2,t,4] <- N[2,t,1]
}
# Future, phase with releases
for (t in n.occasions:(n.occasions+5)){
  N[1,t+1,4] ~ dpois(f[t,1] * sj[t] * (N[1,t,4] + N[2,t,4] + 3))
  N[2,t+1,4] ~ dbin(sa[t,1], (N[1,t,4] + N[2,t,4] + 3))
}
# Future, after the phase with releases
for (t in (n.occasions+6):(n.occasions-1+K)){
  N[1,t+1,4] ~ dpois(f[t,1] * sj[t] * (N[1,t,4] + N[2,t,4]))
  N[2,t+1,4] ~ dbin(sa[t,1], (N[1,t,4] + N[2,t,4]))
}

# Observation model
for (t in 1:n.occasions){
  count[t] ~ dnorm(N[1,t,1] + N[2,t,1], tau)
}

# Productivity data (Poisson regression model)
for (t in 1:n.occasions){
  J[t] ~ dpois(f[t,1]*B[t])
}

# Capture-recapture data (CJS model with multinomial likelihood)
# Define the multinomial likelihood
for (t in 1:(n.occasions-1)){
  marr.j[t,1:n.occasions] ~ dmulti(pr.j[t,], rel.j[t])
  marr.a[t,1:n.occasions] ~ dmulti(pr.a[t,], rel.a[t])
}
```

```
# Define the cell probabilities of the m-arrays
for (t in 1:(n.occasions-1)){
  # Main diagonal
  q[t] <- 1-p[t]                        # Probability of non-recapture
  pr.j[t,t] <- sj[t] * p[t]
  pr.a[t,t] <- sa[t,1] * p[t]
  # Above main diagonal
  for (j in (t+1):(n.occasions-1)){
    pr.j[t,j] <- sj[t] * prod(sa[(t+1):j,1]) * prod(q[t:(j-1)]) * p[j]
    pr.a[t,j] <- prod(sa[t:j,1]) * prod(q[t:(j-1)]) * p[j]
  } #j
  # Below main diagonal
  for (j in 1:(t-1)){
    pr.j[t,j] <- 0
    pr.a[t,j] <- 0
  } #j
} #t
# Last column: probability of non-recapture
for (t in 1:(n.occasions-1)){
  pr.j[t,n.occasions] <- 1-sum(pr.j[t,1:(n.occasions-1)])
  pr.a[t,n.occasions] <- 1-sum(pr.a[t,1:(n.occasions-1)])
} #t

# Derived parameters
for (t in 1:(n.occasions+K)){
  Ntot[t,1] <- N[1,t,1] + N[2,t,1]      # Total population sizes control
  Ntot[t,2] <- N[1,t,2] + N[2,t,2]      # Total population sizes option 1
  Ntot[t,3] <- N[1,t,3] + N[2,t,3]      # Total population sizes option 2
  Ntot[t,4] <- N[1,t,4] + N[2,t,4]      # Total population sizes option 3
}
}
")

# Initial values
inits <- function(){list(mean.sj=runif(1, 0, 0.5), mean.sa=runif(1, 0.4, 0.6),
  mean.f=runif(1, 1.3, 2))}

# Parameters monitored
parameters <- c("mean.sj", "sigma.sj", "mean.sa", "sigma.sa", "mean.p", "mean.f", "sigma.f", "sigma",
  "sj", "sa", "f", "N", "Ntot")

# MCMC settings
ni <- 20000; nb <- 5000; nc <- 3; nt <- 3; na <- 1000

# Call JAGS (ART 1 min), check convergence and summarize posteriors
out2 <- jags(jags.data, inits, parameters, "model2.txt", n.iter=ni, n.burnin=nb, n.chains=nc,
  n.thin=nt, n.adapt=na, parallel=TRUE)
traceplot(out2)
print(out2, 3)
```

To compare the three management options and the control, we compute the extinction probabilities for each and plot them (Fig. 10.6).

```
D <- 0                                  # Define extinction threshold (here for absolute extinction)
T <- length(woodchat10$count)
extinct <- (out2$sims.list$Ntot[, (T+1):(T+K), ] <= D)
ext.prob <- apply(extinct, 2:3, mean)

color <- c("black", "skyblue", "seagreen", "orange")
par(las=1)
matplot(ext.prob, type="l", ylab="Extinction probability", lwd=2, lty=1, xlab="Year of forecast",
  axes=FALSE, col=color)
axis(1, at=1:K, tck=-0.0125, labels=FALSE)
axis(1, at=c(1, 3, 5, 7, 9, 11, 13, 15), labels=c(1, 3, 5, 7, 9, 11, 13, 15), tck=-0.025)
axis(2, las=1)
legend("topleft", lty=rep(1, 4), lwd=rep(2, 4), col=color, legend=c("Do nothing (control)",
  "Increase productivity", "Reduce variability", "Translocation"), bty="n")
```

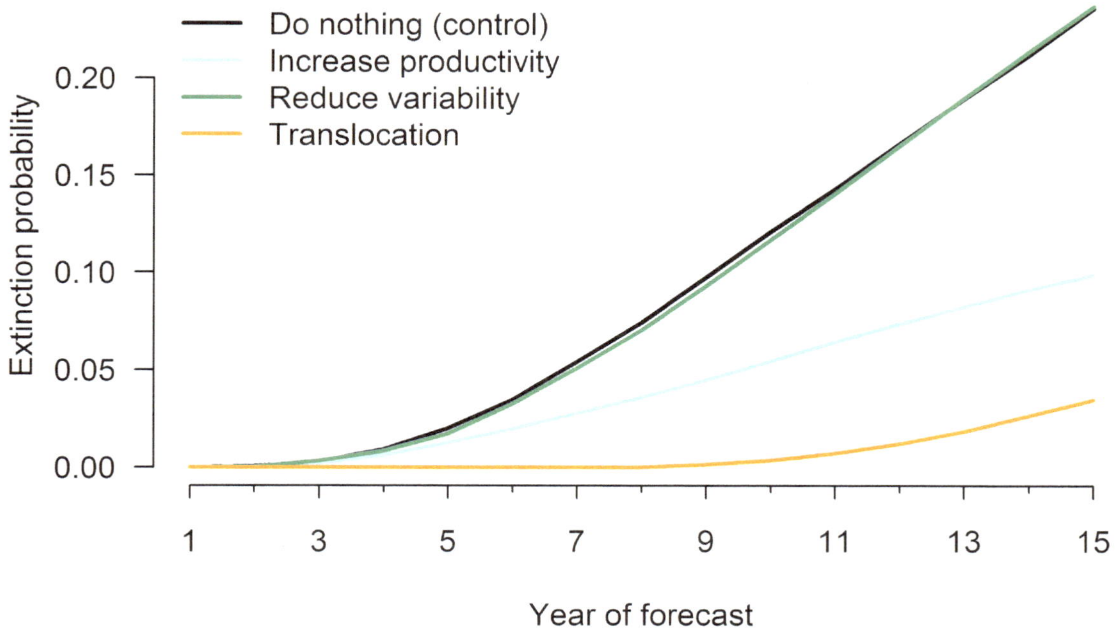

FIGURE 10.6

Illustration of an integrated population model in the context of a population viability analysis (PVA) to rank management options: comparison of the cumulative extinction probabilities in a simulated woodchat shrike population under four different management options (including "do nothing" as a control).

Translocations appear to be the most promising management option, i.e., the one that most reduces extinction probability (Fig. 10.6), followed by the option leading to increased productivity. In contrast, a reduction of the temporal variability of adult survival has almost no effect on extinction risk. We can also rank the different options in another way by computing the probability that a given option with a management action produces a larger population size in 15 years compared with the control where no action is taken.

```
last.year <- jags.data$n.occasions + jags.data$K
# Probability that option 1 (increased productivity) is better than control
round(mean(out2$sims.list$Ntot[,last.year,2] >
out2$sims.list$Ntot[,last.year,1]), 2)
[1] 0.77

# Probability that option 2 (reduced variability) is better than control
round(mean(out2$sims.list$Ntot[,last.year,3] >
out2$sims.list$Ntot[,last.year,1]), 2)
[1] 0.45

# Probability that option 3 (translocation) is better than control
round(mean(out2$sims.list$Ntot[,last.year,4] >
out2$sims.list$Ntot[,last.year,1]), 2)
[1] 0.80
```

When either productivity is increased or three individuals annually are translocated for five years, it is quite likely that the population size in 15 years will be larger than that of the control, where no management action is taken at all. Since the probabilities are similar for these two options, we can also compare the projected population size under the two options directly. We can compute the probability that translocation results in a larger population size in 15 years compared with predator control.

```
# Probability that translocation is better than increased productivity
round(mean(out2$sims.list$Ntot[,last.year,4] >
out2$sims.list$Ntot[,last.year,2]), 2)
[1] 0.43
```

Thus, although the extinction probability is lower with translocations than with predator control (Fig. 10.6), the projected population size in 15 years barely differs between these management options. This is because translocations only postpone possible extinction but do not affect the extinction risk in the long-term because the demographic rates governing population dynamics remain the same. You can see this better by extending the forecast horizon from 15 to, say, 30 years.

Thus, use of an IPM in a PVA to compare different possible management options for a population is straightforward, as just illustrated. The main practical difficulty is perhaps that we need to be careful with our bookkeeping for the multitude of different quantities that we calculate. Another, and perhaps more important, difficulty is that we need actual estimates or else reasonable (e.g., "educated") guesses about the magnitude of the change in a demographic rate associated with the implementation of each management action. Often we don't really have any empirical data on this and we also want to incorporate into our analysis the uncertainty in our guess about the magnitude of the effect. This form of uncertainty is an example of "partial controllability" (Williams, 1997; Nichols et al., 2011); see Section 10.2.

For an illustration, let's assume that in the predator control option we were uncertain about how much productivity can be increased. Instead of setting a fixed number of 20%, we could use a prior for the increase that reflects our uncertainty. For example, we may express our uncertainty by a standard deviation of 0.05, corresponding to 95% quantiles of an increase of about 10% and 30%. Consequently, for the prior of the expected increase in productivity we could use a normal distribution centered on the mean with a standard deviation of 0.05. The BUGS code would then be changed as follows:

```
# Option 1: increase of productivity
# Past: identical to control
for (t in 1:n.occasions){
  log.f[t,2] <- log.f[t,1]
  eps.f[t,2] <- eps.f[t,1]
  f[t,2] <- f[t,1]
}
# Future: increase mean productivity by 20%, but allow for uncertainty
incr.prod ~ dnorm(1.2, 1 / (0.05 * 0.05))    # Prior for expected increase
for (t in (n.occasions+1):(n.occasions+K)){
  log.f[t,2] <- mean.log.f + log(incr.prod) + eps.f[t,2]
  eps.f[t,2] ~ dnorm(0, tau.f)
  f[t,2] <- exp(log.f[t,2])
}
```

This will result in increased uncertainty of the total population size under this management option, and hence also in a slight increase in the estimated extinction probability. This type of uncertainty could of course also be accommodated in the other management options.

Environmental covariates that affect demographic rates or deterministic trends in demographic rates can be included in such an IPM as well: we simply specify linear models for the demographic rates. Generally, we would have to assume that the relationships remain the same in the future as what was observed in the past, i.e., that the estimated regression coefficients do not change over time. The future values of such covariates will typically be unknown and must therefore be estimated in some way and then have some associated uncertainty. If this estimation is simultaneously done in the same IPM, this uncertainty about the future values of the covariate is automatically accounted for in all estimated quantities of the model.

For example, let's assume that we have an environmental covariate that is spring temperature (`temp`) that we measured over our 20 year-population study, and now we want to use it in a population prediction over 10 years into the future. Spring temperature increases over time, and hence, we can assume that the rate of increase remains constant in the future as was observed in the past. We could then add the following BUGS code as a forecast model for spring temperature as a covariate in our demographic model:

```
for (t in 1:(20+10)){
  temp[t] ~ dnorm(beta[1] + beta[2] * t, tau.temp)
}
tau.temp <- pow(sigma.temp, -2)
sigma.temp ~ dunif(0, 5)
for (i in 1:2){
  beta[i] ~ dnorm(0, 0.001)
}
```

The first 20 elements of the vector `temp` will contain the measured temperatures, and the last 10 elements will be missing values that will be estimated using this part of the model. This estimation will have an associated uncertainty, and inclusion of this uncertainty into the linear model of a demographic rate `theta` (possibly on the scale of a link function) is straightforward.

```
for (t in 1:(20+10)){
  theta[t] ~ dnorm(alpha[1] + alpha[2] * temp[t], tau.theta)
}
tau.theta <- pow(sigma.theta, -2)
sigma. theta ~ dunif(0, 5)
for (i in 1:2){
  alpha[i] ~ dnorm(0, 0.001)
}
```

In general, the inclusion of relationships with environmental variables or of trends helps to reduce the forecast uncertainty in a PVA, which of course is always desirable.

You can read Saunders et al. (2018 and 2021) for examples of the inclusion of this type of uncertainty in an IPM-based PVA. In their first paper they used an IPM to compare the effects of the control of ground versus avian predators on the population viability of an endangered plover. They found that the control of the avian predator that affects adult survival of the plovers had a stronger impact than the control of the ground predators that affected productivity. In their second paper, they developed an IPM-based PVA that implements predicted trends in environmental variables which impact common loon (*Gavia immer*) demographic rates.

Sometimes, we want to know how much a demographic rate must be changed to reach a specific management goal that is specified in terms of some quantitative measure of a population (Rhodes et al., 2011; Plard et al., 2020a,b). This can be relevant to decide whether it is worth taking some action when we are uncertain by how much a given action would change the corresponding demographic rate. For illustration, let's assume that we want to know by how much productivity must be increased to obtain a population size of 20 breeding females (or pairs) in 15 years with a 90% probability. This exercise is easy conceptually but computationally intensive. We must increase productivity by very small amounts repeatedly, fit our IPM, and inspect the posterior distribution of the final population size. The answer to our question is the amount of increase that meets the objective.

```
# Write JAGS model file
cat(file="model3.txt", "
model {
# Priors and linear models
mean.logit.sj <- logit(mean.sj)
mean.sj ~ dunif(0, 1)
mean.logit.sa <- logit(mean.sa)
mean.sa ~ dunif(0, 1)
mean.p ~ dunif(0, 1)
mean.log.f <- log(mean.f)
mean.f ~ dunif(0, 10)

for (t in 1:(n.occasions-1)){
  p[t] <- mean.p
}

for (t in 1:(n.occasions-1+K)){   # Extend the loop to K more years
  logit.sj[t] <- mean.logit.sj + eps.sj[t]
  eps.sj[t] ~ dnorm(0, tau.sj)
  sj[t] <- ilogit(logit.sj[t])
  logit.sa[t] <- mean.logit.sa + eps.sa[t]
  eps.sa[t] ~ dnorm(0, tau.sa)
  sa[t] <- ilogit(logit.sa[t])
}

# Productivity in the past
for (t in 1:(n.occasions-1)){
  log.f[t] <- mean.log.f + eps.f[t]
  eps.f[t] ~ dnorm(0, tau.f)
  f[t] <- exp(log.f[t])
}

# Productivity in the future: increased by rep.inc
for (t in n.occasions:(n.occasions+K)){
  log.f[t] <- mean.log.f + eps.f[t] + log(rep.inc)
  eps.f[t] ~ dnorm(0, tau.f)
  f[t] <- exp(log.f[t])
}

sigma.sj ~ dunif(0, 10)
tau.sj <- pow(sigma.sj, -2)
sigma.sa ~ dunif(0, 10)
tau.sa <- pow(sigma.sa, -2)
sigma.f ~ dunif(0, 10)
tau.f <- pow(sigma.f, -2)

sigma ~ dunif(0.5, 50)
tau <- pow(sigma, -2)
```

```
# Population count data (state-space model)
# Model for the initial population size: discrete uniform priors
N[1,1] ~ dcat(pNinit)
N[2,1] ~ dcat(pNinit)

# Process model over time: our model of population dynamics
for (t in 1:(n.occasions-1+K)){  # Extend the loop to K more years
  N[1,t+1] ~ dpois(sj[t] * f[t] * (N[1,t] + N[2,t]))
  N[2,t+1] ~ dbin(sa[t], (N[1,t] + N[2,t]))
}

# Observation model
for (t in 1:n.occasions){
  count[t] ~ dnorm(N[1,t] + N[2,t], tau)
}

# Productivity data (Poisson regression model)
for (t in 1:n.occasions){
  J[t] ~ dpois(f[t] * B[t])
}

# Capture-recapture data (CJS model with multinomial likelihood)
# Define the multinomial likelihood
for (t in 1:(n.occasions-1)){
  marr.j[t,1:n.occasions] ~ dmulti(pr.j[t,], rel.j[t])
  marr.a[t,1:n.occasions] ~ dmulti(pr.a[t,], rel.a[t])
}
# Define the cell probabilities of the m-arrays
for (t in 1:(n.occasions-1)){
  # Main diagonal
  q[t] <- 1-p[t]                 # Probability of non-recapture
  pr.j[t,t] <- sj[t] * p[t]
  pr.a[t,t] <- sa[t] * p[t]
  # Above main diagonal
  for (j in (t+1):(n.occasions-1)){
    pr.j[t,j] <- sj[t] * prod(sa[(t+1):j]) * prod(q[t:(j-1)]) * p[j]
    pr.a[t,j] <- prod(sa[t:j]) * prod(q[t:(j-1)]) * p[j]
  } #j
  # Below main diagonal
  for (j in 1:(t-1)){
    pr.j[t,j] <- 0
    pr.a[t,j] <- 0
  } #j
} #t
# Last column: probability of non-recapture
for (t in 1:(n.occasions-1)){
  pr.j[t,n.occasions] <- 1-sum(pr.j[t,1:(n.occasions-1)])
  pr.a[t,n.occasions] <- 1-sum(pr.a[t,1:(n.occasions-1)])
} #t

# Derived parameters
# Total population size
for (t in 1:(n.occasions+K)){
  Ntot[t] <- N[1,t] + N[2,t]
}
}
")

# Parameters monitored
parameters <- c("mean.sj", "sigma.sj", "mean.sa", "sigma.sa", "mean.p", "mean.f", "sigma.f", "sigma",
  "sj", "sa", "f", "N", "Ntot")

# MCMC settings
ni <- 20000; nb <- 5000; nc <- 3; nt <- 3; na <- 1000
```

Basic calculations using a Leslie matrix lead us to think that the solution must be a little over a 50% increase in productivity. We therefore choose different values between a 40% and a 60% increase. We define a vector with these levels of increase and then fit the IPM for each of these levels.

```
# Define vector with the levels of increased productivity
rep.inc <- seq(1.4, 1.6, 0.025)

# Define matrix to store results
postN <- matrix(NA, nrow=(ni - nb) / nt * nc, ncol=length(rep.inc))

# Fit IPM for each level of increased productivity
for (i in 1:length(rep.inc)){   # Takes about 10 min total

  cat(paste('*** Running model for', (rep.inc[i]-1)*100, '% productivity increase ***\n'))

  # Bundle data
  K <- 15                      # Number of years with predictions
  jags.data <- list(marr.j=woodchat10$marr.j, marr.a=woodchat10$marr.a,
    n.occasions=ncol(woodchat10$marr.j), rel.j=rowSums(woodchat10$marr.j),
    rel.a=rowSums(woodchat10$marr.a), J=woodchat10$J, B=woodchat10$B, count=woodchat10$count,
    pNinit=dUnif(1, 50), K=K, rep.inc=rep.inc[i])

  # Initial values
  inits <- function(){list(mean.sj=runif(1, 0, 0.5), mean.sa=runif(1, 0.4, 0.6),
    mean.f=runif(1, 1.3, 2))}

  # Call JAGS (ART 1 min for each model)
  out3 <- jags(jags.data, inits, parameters, "model3.txt", n.iter=ni, n.burnin=nb, n.chains=nc,
    n.thin=nt, parallel=TRUE)

  # Save posterior distribution of pop. size in 15 years
  postN[,i] <- out3$sims.list$Ntot[,20+K]
}

# Calculate probability that population size is >20
prob <- colMeans(postN > 20)

# Produce figure 10.7
plot(x=rep.inc, y=prob, type="b", xlab="Increase in productivity (%)",
  ylab=expression(Prob (N[35] > 20)), axes=FALSE, pch=16)
axis(2, las=1)
axis(1, at=rep.inc[c(1,3,5,7,9)], labels=c(40, 45, 50, 55, 60))
axis(1, at=rep.inc, labels=NA, tcl=-0.25)
segments(0, 0.9, 10, 0.9, col="red")
```

Based on the posterior distributions of the population size in the final year of the forecast, we can easily obtain the probability that the population is greater than 20. We do this for all modeled levels of the increased productivity and then plot the relationship (Fig. 10.7). We find that productivity would need to be increased by about 53% to meet our defined

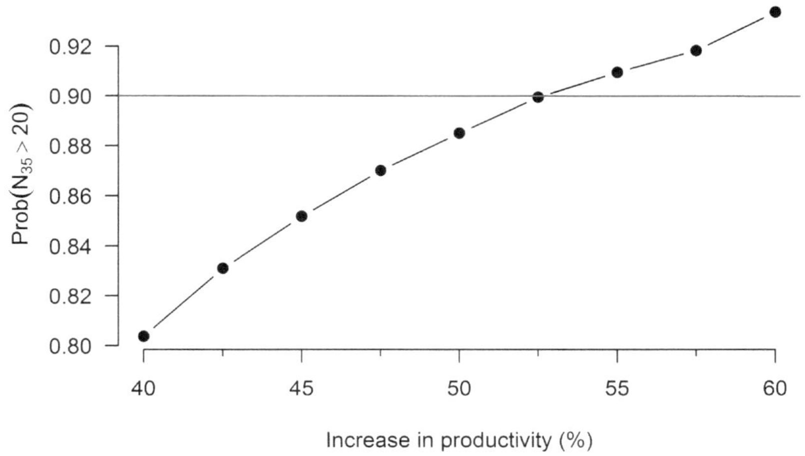

FIGURE 10.7

Relationship between the potential increase in productivity and probability to achieve a population size of at least 20 woodchat shrike pairs over a 15-year forecast horizon. The value on the x-axis that lies vertically below the place where the horizontal red line crosses the black curve is the solution to our exercise (see text).

objective. We could next ask whether we can conceive of any management action that is likely to produce such a high and persistent increase in productivity. Similar exercises could be done for other demographic rates and for the combination of multiple demographic rates. And as always in the Bayesian mode of inference, we obtain the full posterior distributions of all estimated quantities, including population size. Often, very rich inferences are possible that can simply be based on quantities derived from the basic estimated quantities (or parameters) in our IPM.

Before we look at one example of a PVA carried out with an empirical data set in the following section, we make a number of comments about the use of an IPM as a tool for conducting a PVA. First, we must not forget that in most IPMs, we will not estimate the true population size (see Section 7.4): when population counts are subject to false-negative errors only (i.e., with detection probability < 1) and we use a normal, log-normal, or Poisson distribution as an observation model for the population counts, true population size will be underestimated. As a result, we will then overestimate extinction probability to an unknown degree. To get a more reliable assessment of the extinction probability, we need to know (or be able to make an educated guess about) the size of the population in the final year of the period for which we have data. One solution would be to make the population forecast in a separate loop and start with that estimate of the true population size in the final year of study. It would also be possible to include uncertainty about the size of the population size by placing a prior on it. Exercise 10.1 provides an illustration and shows how the uncertainty associated with the estimated true population size at which we anchor the forecast, has a strong effect on the forecast uncertainty (McLoughlin and Messier, 2004). Clearly then, this means that we should try to reduce the uncertainty associated with that population size estimate as much as possible to achieve more accurate population projections. At the same time, it also emphasizes that the results of a PVA can be arbitrarily misleading if we ignore such uncertainty about the actual population sizes.

Second, a typical IPM combines population counts with demographic data, but sometimes we may have only population counts or demographic data but not both. In these cases, we cannot fit an IPM as defined in this book. However, it may still be possible to conduct a PVA by adopting a pure state-space or other suitable stochastic model for population counts (see Section 4.3.1 and Chapter 1 in Kéry and Royle, 2021) or by using simpler matrix population methods if some information about demographic parameters is available (see Chapter 3). One major benefit of a Bayesian approach, then, is that the propagation into all the estimated quantities of the errors due to process variability and parameter uncertainty is trivially easy in an MCMC-based analysis.

Third, we note that in our models, we haven't included one further type of uncertainty, structural uncertainty, which can arise when we are not sure about the required number of age classes in our model. To accommodate it, we could repeat the PVAs using different assumptions about population structure (or generally, model structure) and assess the degree to which the main results depend on the assumed model structure in a sort of model sensitivity analysis. A more formal way would be to perform some sort of model selection followed perhaps by an averaging of the main results (e.g., extinction probabilities) under the different models (Nichols et al., 2011).

10.6 POPULATION VIABILITY ANALYSIS OF A POPULATION WITH IMMIGRATION

The hoopoe (Fig. 9.1) was once a very widespread breeding bird in Switzerland but in recent times has had an extremely patchy distribution with a single stronghold in one major Alpine valley in southwestern Switzerland (Valais). This population has been the subject of an intensive demographic study since 2000, during which time it went through a severe initial bottleneck but then quickly increased sixfold due to the massive installation of nest boxes in areas where its preferred food, ground-dwelling large insects, is most common (Arlettaz et al., 2010). Unfortunately, the initial strong increase has been followed by a slow decline in more recent years. Our goal of the PVA in this section is to predict the likely hoopoe population trajectory over the next 10 years.

The hoopoe population in the Valais is part of a larger metapopulation. It receives immigrants every year but also produces emigrants that settle elsewhere. We need to include both processes in an IPM to provide adequate results. The capture-recapture data collected in this population provide estimates of apparent survival, and thus emigration is already accounted for in an IPM via this parameter even though we cannot estimate it as a parameter separate from mortality. We can accommodate immigrants in our model as a hidden parameter (see Chapter 8). We have some flexibility in terms of how we specify immigration in the model (Schaub and Fletcher, 2015), but basically, we can express it as either a rate or an absolute number. Both parameterizations typically yield similar results, and we can also calculate the rate from the number and vice-versa. However, once we want to forecast the population into the future, the manner in which we define immigration matters, and we will illustrate this here. Furthermore, in this example, we will explicitly include temporal correlations among the demographic rates because this may affect the predictions as well.

Both the data set and population model are the same as in Section 9.2. We load the data, produce the m-arrays from the capture-recapture data, and bundle all data needed into a list. We also include a constant K, which is our choice of the number of years over which we want to forecast the trajectory of the population.

```
# Load hoopoe data
library(IPMbook)
data(hoopoe)

# Produce m-array
marr <- marrayAge(hoopoe$ch, hoopoe$age)
marr.j <- marr[,,1]
marr.a <- marr[,,2]

# Bundle data
jags.data <- with(hoopoe$reproAgg, list(marr.j=marr.j, marr.a=marr.a, n.occasions=ncol(marr.j),
  rel.j=rowSums(marr.j), rel.a=rowSums(marr.a), J1=J1, B1=B1, J2=J2, B2=B2, count=hoopoe$count,
  pNinit=dUnif(1, 50), K=10))
str(jags.data)    # Remind ourselves of how the data look like
```

We first fit the IPM with immigration parameterized as a rate. The number of immigrants (N_3) in the population model is modeled with a Poisson distribution,

$$N_{3,t+1} \sim Poisson\left(\left(N_{1,t} + N_{2,t} + N_{3,t}\right)\omega_t\right),$$

where ω_t is the immigration rate, defined as the ratio of the number of immigrants in year $t+1$ and total population size in year t. Moreover, we model all demographic rates with correlated temporal random effect (see Section 6.4 for details).

```
# Write JAGS model file
cat(file="model4.txt", "
model {
# Priors and linear models
# For mean and variance parameters
mean.phij ~ dunif(0, 1)
mean.phia ~ dunif(0, 1)
mean.f1 ~ dunif(0, 10)
mean.f2 ~ dunif(0, 10)
mean.omega ~ dunif(0, 3)

mu[1] <- logit(mean.phij)
mu[2] <- logit(mean.phia)
mu[3] <- log(mean.omega)
mu[4] <- log(mean.f1)
mu[5] <- log(mean.f2)

mean.p ~ dunif(0, 1)
mean.pj ~ dunif(0, 1)

# Linear models for demographic parameters
for (t in 1:(n.occasions-1+K)){
  logit(phij[t]) <- eps[1,t]
  logit(phia[t]) <- eps[2,t]
  log(omega[t])  <- eps[3,t]
  log(f1[t]) <- eps[4,t]
  log(f2[t]) <- eps[5,t]
  eps[1:5,t] ~ dmnorm.vcov(mu[], sigma2[,])
}

for (t in 1:(n.occasions-1)){
  p[t] <- mean.p
  pj[t] <- mean.pj
}
```

```
# Reparameterize sigma2 in terms of SD and rho
for (i in 1:5){
  sigma2[i,i] <- sigma[i] * sigma[i]
}
for (i in 1:4){
  for (j in (i+1):5){
    sigma2[i,j] <- sigma[i] * sigma[j] * rho[i,j]
    sigma2[j,i] <- sigma2[i,j]
    rho[i,j] ~ dunif(-1, 1)              # prior for rho
  } #j
} #i

# Specify priors for SD
for (i in 1:5){
  sigma[i] ~ dunif(0, 5)
}

sigma.res ~ dunif(0.4, 20)
tau.res <- pow(sigma.res, -2)

# Population count data (state-space model)
# Model for initial stage-spec. population sizes: discrete uniform priors
N[1,1] ~ dcat(pNinit)
N[2,1] ~ dcat(pNinit)
N[3,1] ~ dcat(pNinit)

# Process model over time: our model of population dynamics
for (t in 1:(n.occasions-1+K)){
  N[1,t+1] ~ dpois(f1[t] / 2 * phij[t] * (N[1,t] + N[3,t]) + f2[t] / 2 * phij[t] * N[2,t])
  N[2,t+1] ~ dbin(phia[t], (N[1,t] + N[2,t] + N[3,t]))
  N[3,t+1] ~ dpois((N[1,t] + N[2,t] + N[3,t]) * omega[t])
}

# Observation model
for (t in 1:n.occasions){
  count[t] ~ dnorm(Ntot[t], tau.res)
}
for (t in 1:(n.occasions+K)){
  Ntot[t] <- N[1,t] + N[2,t] + N[3,t]     # total population size
}

# Productivity data (Poisson regression model)
for (t in 1:(n.occasions-1)){
  J1[t] ~ dpois(f1[t] * B1[t])
  J2[t] ~ dpois(f2[t] * B2[t])
}

# Capture-recapture data (CJS model with multinomial likelihood)
# Define the multinomial likelihood
for (t in 1:(n.occasions-1)){
  marr.j[t,1:n.occasions] ~ dmulti(pr.j[t,], rel.j[t])
  marr.a[t,1:n.occasions] ~ dmulti(pr.a[t,], rel.a[t])
}
# Define the cell probabilities of the m-arrays
for (t in 1:(n.occasions-1)){
  # Main diagonal
  qj[t] <- 1-pj[t]
  q[t] <- 1-p[t]
  pr.j[t,t] <- phij[t] * pj[t]
  pr.a[t,t] <- phia[t] * p[t]
  # Above main diagonal
  for (j in (t+1):(n.occasions-1)){
    pr.j[t,j] <- phij[t] * prod(phia[(t+1):j]) * qj[t] * prod(q[t:(j-1)]) * p[j] / q[t]
    pr.a[t,j] <- prod(phia[t:j]) * prod(q[t:(j-1)]) * p[j]
  } #j
  # Below main diagonal
  for (j in 1:(t-1)){
    pr.j[t,j] <- 0
    pr.a[t,j] <- 0
  } #j
} #t
```

```
# Last column: probability of non-recapture
for (t in 1:(n.occasions-1)){
  pr.j[t,n.occasions] <- 1-sum(pr.j[t,1:(n.occasions-1)])
  pr.a[t,n.occasions] <- 1-sum(pr.a[t,1:(n.occasions-1)])
} #t

# Derived parameters
# Annual population growth rate
for (t in 1:(n.occasions-1)){
  lambda[t] <- Ntot[t+1] / Ntot[t]
}
}
")

# Initial values
inits <- function(){list(mean.phij=runif(1, 0.05, 0.2), mean.phia=runif(1, 0.3, 0.5))}

# Parameters monitored
parameters <- c("mean.phij", "mean.phia", "mean.omega", "mean.f1", "mean.f2", "mean.p", "mean.pj",
  "sigma.res", "rho", "sigma2", "phij", "phia", "omega", "f1", "f2", "N", "Ntot", "lambda")

# MCMC settings
ni <- 110000; nb <- 10000; nc <- 3; nt <- 100; na <- 2000

# Call JAGS (ART 17 min), check convergence
out4 <- jags(jags.data, inits, parameters, "model4.txt", n.iter=ni, n.burnin=nb, n.chains=nc,
  n.thin=nt, n.adapt=na, parallel=TRUE)
traceplot(out4)
```

We don't look at the results yet but instead fit the next IPM immediately. In this IPM, we parameterize immigration in absolute terms, i.e., as a number, which we model as a Poisson random variable with intensity Ω corresponding to the expected number of immigrants.

$$N_{3,t} \sim Poisson(\Omega_t)$$

As a prior for the intensity, we use a uniform distribution between 0 and some reasonably large number, 100 in our case. If we model the number of immigrants with covariates, we use the log-link function for the intensity parameter. Thus, compared to the first IPM that we showed previously, we only need to change two lines of code. We replace

`N[3,t+1] ~ dpois((N[1,t] + N[2,t] + N[3,t]) * omega[t])` by `N[3,t+1] ~ dpois(omega[t])`
and
`mean.omega ~ dunif(0, 3)` by `mean.omega ~ dunif(0, 100)`.

We now also run this model (code shown on web appendix).

The two models converge fairly easily. In the posterior summaries (not shown here), we don't see much evidence for temporal cross-correlation between the demographic rates with the exception of the productivities between first-time and experienced breeders.

Fig. 10.8 shows estimates of total and stage-structured population size under the two IPMs that differ in the manner in which the immigration process is parameterized. We see hardly any difference between the two for the period with data, i.e., up to 2017. However, for the forecast period, there are very strong differences, especially in estimation uncertainty, which is much greater when immigration is specified as a rate rather than as a number. This discrepancy is most pronounced for the number of immigrants itself but also evident for the number of local recruits and surviving adults and hence also for total population size. The reason for this seemingly odd result is that in the specification of immigration as a rate, we include in the posterior distribution of the number of immigrants the uncertainty about the forecast population sizes, and their uncertainty increases over the forecast horizon (see, e.g., Figs. 10.2 and 10.8). This does *not* occur when we directly specify immigration as a number, rather, the estimation uncertainty then remains constant over the forecast horizon, as we can see clearly in Fig. 10.8.

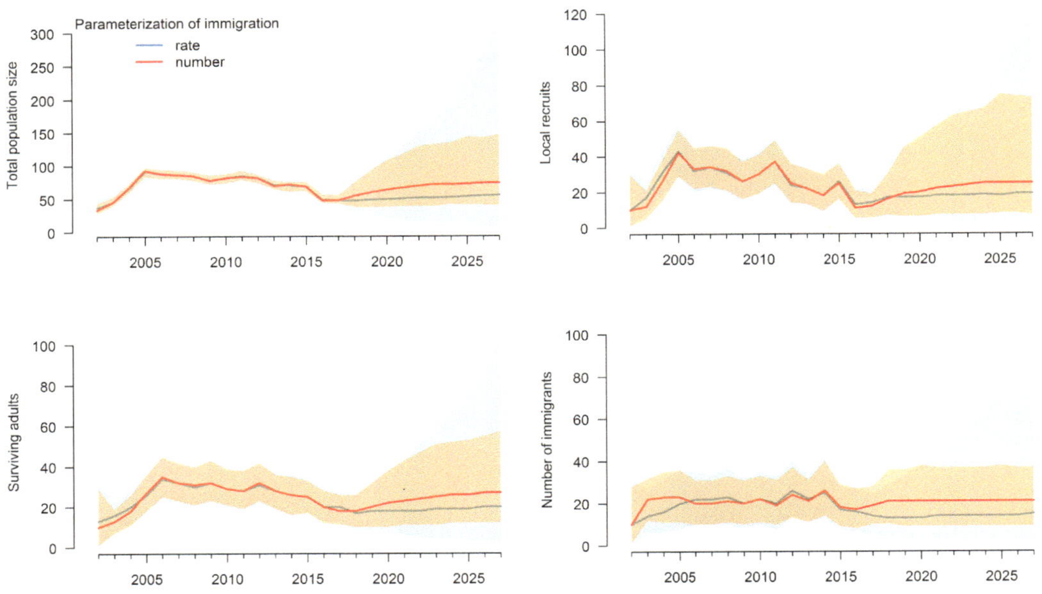

FIGURE 10.8

Estimates of the total and stage-structured population sizes of Swiss hoopoes under the two integrated population mode s with different parameterization of the immigration process: either as a rate (blue) or in absolute terms as a number (red). The parameterization does not matter for the study period when demographic data are available, but it makes a strong difference for the projection period starting after 2017. Lines show posterior medians, and polygons show 95% credible intervals.

It is perhaps not obvious to decide which option to choose. In terms of the uncertainty, we are tempted to favor the absolute specification. But is it plausible to assume that the *number* of immigrants remains constant in the future regardless of population size? Moreover, this specification amounts to that of a rescue effect, where the population can never go extinct because the number of immigrants does not change (i.e., does not go to zero along with population size). However, in reality, we think that it is quite plausible that the number of immigrants declines if the number of local recruits and surviving adults declines, perhaps because immigrants may be socially attracted (Dale and Steifetten, 2011; Szostek et al., 2014; Tenan et al. 2017). If immigration is defined as a rate, even if the rate is constant over time, the resulting number of immigrants will decline as population size declines. One drawback is that if all adults happen to die in some year and there are no recruits, under the model, immigration will be zero, and therefore the population must remain extinct forever. Presumably, this is not completely plausible either.

Clearly, immigration is a key process in the dynamics of most finite populations (Millon et al., 2019), and hence, we need a much better understanding of how it "works." That is, which cues are used by immigrants use for the settlement decision? Armed with this knowledge we may be better able to adopt an appropriate model for the immigration process. Currently, we do not yet have this knowledge for the Swiss hoopoes, but we wanted to emphasize the challenges that you are faced with when modeling immigration in a population forecasting exercise such as a PVA.

10.7 SUMMARY AND OUTLOOK

We have seen that conducting a PVA, i.e., the projection, or forecasting, of a population into the future within an IPM, is conceptually very easy. We simply need to increase the length of the time loops in the process models for both the population dynamics and the demographic rates. It is also straightforward in an IPM to include different management options and compare their efficiency in terms of predicted population sizes or the reduction in extinction probability and therefore to rank these options. Key advantages to performing a PVA within an IPM are the proper propagation of the errors and the ability to use the results for making direct probability statements. A further advantage is that the development of an IPM fosters a better understanding of the underlying population model. This may generally result in a better PVA (Chaudhary and Oli, 2020).

We have only shown some basic principles of formulating a PVA within an IPM, and numerous further extensions and refinements are possible. For example, we have only sketched the inclusion of driver variables that are observed with error or whose future values must be estimated (Saunders et al., 2018). In contrast, we have omitted showing inclusion of density dependence in a PVA (Gamelon et al., 2017). However, both are conceptually straightforward.

The uncertainty around estimated quantities in a PVA is typically substantial, and often so much so as to render the inferences virtually useless in practice. A future challenge, therefore, is to develop models such that this uncertainty can be reduced. One possible avenue is the explicit inclusion of individual heterogeneity. Current PVAs typically subsume such individual effects under "demographic stochasticity," but this may unduly increases uncertainty (Fox and Kendall, 2002). In Chapter 14, we show one possibility for including individual heterogeneity in an IPM, and such a model could also be the base for a PVA which accommodates individual effects.

Bayesian PVAs within an IPM can be used in a formal decision-making process, but so far, very few published examples exist of this. Fonnesbeck and Conroy (2004) used such an approach to study sustainable harvest quota for mallards (*Anas platyrhynchos*) and American black ducks (*A. rubripes*). Stochastic dynamic programming was used by Duarte et al. (2017) to evaluate the optimal translocation strategy among Oregon spotted frog (*Rana pretiosa*) populations as part of an IPM.

The types of models shown in this chapter can also be applied in the context of harvested populations. A challenge in harvested populations is the determination of the harvest quota such that the population size survives within a predefined size in the longer term, i.e., that harvesting is sustainable. IPMs with population predictions can be used for that purpose. For example, Hatter et al. (2017) developed an IPM to inform the harvest of mule deer (*Odocoileus hemionus*). They included harvest data and assessed whether the population projections remain within the specified bounds of their management objectives. The ability to forecast population size into the future while accounting for various types of uncertainty renders IPM a very useful tool for population management of rare, harvested, and pest species.

10.8 EXERCISES

1. We assume that we have collected demographic and population count data on just a part of a larger woodchat shrike population. In the last year of our study, we now assume that we surveyed the full area inhabited by the population and found that it consisted of between 25 and 30 breeding pairs, i.e., more than double the 12 pairs found in that year in the smaller area, which was surveyed in all 20 years. The exercise is to assess the quasi-extinction probability ($D = 5$) of the complete population within the next 15 years. The data are the same as those used in Section 10.5.
2. Let's assume that we want to create a self-sustainable hoopoe population, i.e., a population that can survive without immigration. How much does juvenile survival need to increase such that the extinction risk of the hoopoe population becomes less than 0.1 in 10 years?

PART

INTEGRATED POPULATION MODELS IN PRACTICE

2

CHAPTER

WOODCHAT SHRIKE

11

CHAPTER OUTLINE

IPM ingredients: population count data, capture-recapture data, productivity data
Location: Germany
Spatio-temporal scale: 120 km^2, 29 years

11.1 INTRODUCTION

The woodchat shrike (*Lanius senator*, Fig. 10.1) features prominently in this book, but so far we have only used simulated data that mimic a real population. Here now, we use an empirical data set collected by our colleague Bruno Ullrich over 34 years in southern Germany near Stuttgart. The 120 km^2 study area is described in detail in Ullrich (2017). Woodchat shrikes feed on large insects and nest mainly in the Mediterranean area. They are strongly migratory and return to their breeding grounds in late April. Southern Germany is at the northern border of their distributional range, and they were never abundant here (Lefranc and Worfalk, 1997). In the study area, the population showed strong temporal fluctuations, until in 1997, sadly, the last breeding pair was observed and from then on the woodchat shrike was locally extinct (Ullrich, 2017; Schaub and Ullrich, 2021). At the same time, the species has become a very rare and irregular breeder in the rest of Germany as well (Gedeon et al., 2015). There (Hölzinger, 1997), as in neighboring Eastern France (Lefranc, 1999) and in Switzerland (Schmid et al., 1998), extensive orchards were the typical breeding habitat of the species. The main reasons for the widespread long-term decline of woodchat shrikes in these regions are considered to be the sprawl of housing areas that destroy these habitats and the intensification of farming practices resulting in less heterogeneously managed meadows, increased application of pesticides, removal of orchards and reduced abundance of insect prey.

Each year during 1963–96, Bruno Ullrich searched all potential nesting territories in the study area, located the nests, counted the number of eggs, hatchlings and fledglings per nest, marked the nestlings with aluminum rings at the age of about 10 days, and captured the adults. Adults got additional individual color rings to facilitate identification in later years. Catching adults was time-consuming and therefore not conducted in every year. Moreover, the study area was smaller in the initial year, and from 1993 onwards, nests were no longer searched for. Therefore, here we restrict the analysis to 29 years from 1964 to 1992.

We analyze this data set to better understand the demography of this poorly studied species. We also emphasize estimation of immigration, since this demographic process is often important in small populations (Millon et al., 2019). Moreover we illustrate a somewhat more sophisticated model for the productivity data: a zero-inflated normal regression with censoring (see Section 4.4). This model allows separating the probability that a brood is successful (in the sense that

Integrated Population Models. https://doi.org/10.1016/B978-0-12-820564-8.00011-7

≥ 1 young fledge) from conditional brood size, given a brood was successful. Accounting for censoring is required because of a limitation in data sampling: sometimes when a fledged brood is observed, we are not sure about the exact number of fledglings, but can only say that there is at least some observed number. Finally, we also explore several more parsimonious variants of our first IPM in this chapter, which do not include temporal variability in survival and immigration, and/or exclude demographic stochasticity. We do this model sensitivity analysis to understand how inferences change when different models are adopted for the data sets, and therefore, to test how robust our conclusions are to our specific assumptions about these features of woodchat shrike demography.

11.2 DATA SETS

The data sets used in this IPM are fairly typical for an intense demographic study of a territorial species of bird. The first data set is the annual number of breeding pairs found in the study area, which ranged from 6 to 28. All suitable habitats (i.e., orchards) were visited multiple times during the breeding season (early May to late July) and the presence or absence of woodchat shrikes was recorded in each known territory. Hence, we assume that detection error was negligible.

Our second data set contains information about productivity. For each pair found, the nest was searched for, resulting in a total of 365 nests located. Woodchat shrikes build open nests, which in the study area were typically in fruit trees at relatively low height. Nests were inspected repeatedly, and their content (i.e., clutch size, number of nestlings) and outcome (i.e., number of fledglings produced) was recorded. In the study area woodchat shrikes were single brooders, but were able to produce a replacement clutch if the first brood failed. In our population model, we use as a measure of productivity the annual number of fledglings produced by a female, regardless of whether these were produced in the first or a replacement brood. For 48 out of the total of 365 surveyed broods, the number of fledglings was not known exactly, because the nest could not be inspected or the brood was only found after fledging. Counting the number of fledglings in these cases was difficult and typically it remained uncertain whether the recorded number was accurate or represented a minimum only.

Our third data set contains the capture-recapture data. A total of 1079 individuals was ringed, 967 as a nestling and 112 as an adult. Although adults can be sexed, we did not include sex in our model, since preliminary analyses found sex effects on survival or recapture to be negligible (Schaub and Ullrich, 2021). In 1973, 1978–81 and 1984–88 no resighting or capture of adults took place and we accounted for this in our model by setting some recapture parameters to zero, see below. A few dead recoveries (Ullrich, 1987) and some ringed nestlings that died before fledging were excluded from the data set.

11.3 POPULATION MODEL

Our first step towards an IPM is to develop a population model that links population size with the demographic rates. We assume a prebreeding "census" and adopt a female-based model where we distinguish three stage classes at the start of each month of May: local recruits, surviving adults and immigrants. Local recruits are locally-born, surviving individuals that return to the study area. Surviving adults are at least 2 years old and were present in the study area in the previous year already. Finally, immigrants are born outside of the study area, at least 1 year old and never reproduced in the study area before (but we assume nothing about any earlier reproductive events elsewhere). This third stage is important to include in our population model because of the relatively small size of the study area and the pronounced dispersal in most bird species. The life-cycle graph of our model for the woodchat shrike is shown in Fig. 11.1.

The number of local recruits in year t, R_t, is a function of the number of breeders in year $t-1$ (N_{t-1}), productivity (number of fledglings per female) in year $t-1$ (ρ_{t-1}), and the probability that a fledgling survives from $t-1$ to t and does not disperse outside of the study area ($\phi_{j,t-1}$, apparent juvenile survival). We include in our model demographic stochasticity and for the number of recruits therefore have

$$R_t \sim Poisson\left(\frac{\rho_{t-1}}{2}\phi_{j,t-1}N_{t-1}\right).$$

The number of surviving adults in year t, S_t, is a function of the number of breeders in year $t-1$ (N_{t-1}) and adult apparent survival probability for year $t-1$ to t ($\phi_{a,t-1}$). Again we include demographic stochasticity and get

$$S_t \sim Binomial\left(\phi_{a,t-1}, N_{t-1}\right).$$

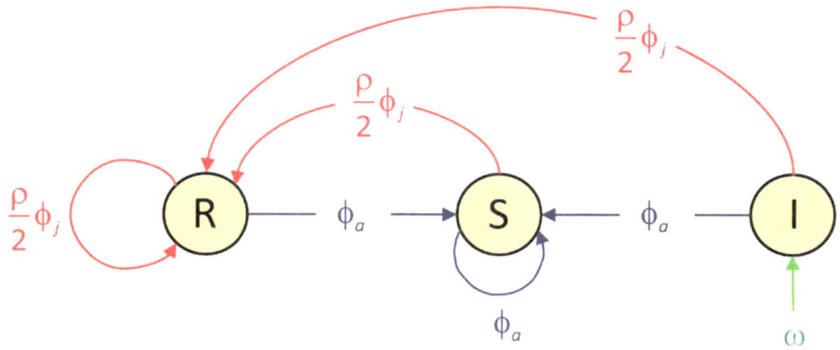

FIGURE 11.1

Life-cycle graph for the woodchat shrike model. The three population stages are local recruits (R), surviving adults (S) and immigrants (I), and the demographic processes are productivity (governed by productivity per female and year, ρ), apparent survival (governed by ϕ_j for juveniles and ϕ_a for adults), and immigration (governed by ω). Transitions involving local recruitment are shown as red arrows, pure survival transitions in blue, and immigration in green.

As we have seen at various places (e.g., Chapters 8 and 9, Section 10.6), we have two basic options for parameterizing the immigration process: as a rate or as an absolute number (Abadi et al., 2010; Schaub and Fletcher, 2015). The former has the advantage that the immigration rate can be used directly as a parameter in a matrix population model or that we could model it as a function of covariates to test hypotheses about drivers in the immigration process. Specification of the immigration process by the number of immigrants has the advantage that one can formally test whether immigration is occurring at all (Schaub and Fletcher, 2015). Moreover, the latter represents inclusion in the model of a possible rescue-effect (Brown and Kodric-Brown, 1977), such that an extinct population is allowed to be re-established by immigration. For more discussion about the parameterization of the immigration process in an IPM, see Schaub and Fletcher (2015).

In practice there is often not much of a difference in the inferences of two models that adopt one or the other parameterization (though see Section 10.6) and we can easily get one from the other as a derived quantity in the model. Here, we adopt the absolute parameterization and for the number of immigrants we write

$$I_t \sim Poisson(\omega_t),$$

where ω_t is the expected number of female immigrants in year t.

Naturally, the total number of female breeders, i.e., what we take to be (half of) the total population size, in year t is the sum of the three types of females, thus $N_t = R_t + S_t + I_t$.

11.4 COMPONENT DATA LIKELIHOODS

The three data sets for this analysis are stored in an R data file that we load and whose structure we inspect.

```
library(IPMbook); library(jagsUI)
data(woodchat11)
str(woodchat11)

List of 6
 $ ch   : int [1:1079, 1:29] 0 0 0 0 0 0 0 0 0 0 ...
 $ age  : num [1:1079] 1 1 1 1 1 1 1 1 1 1 ...
 $ count: num [1:29] 10 13 25 15 17 16 6 16 7 7 ...
 $ f    : int [1:365] 5 0 6 0 3 6 6 3 0 3 ...
 $ d    : int [1:365] 0 0 0 0 0 0 0 0 0 1 ...
 $ year : int [1:365] 1964 1964 1964 1964 1964 1964 1964 1964 1965 1965 ...
```

The data file contains a total of six objects that include the population count, capture-recapture and productivity data sets. Next we briefly describe each of them, along with the submodels that we use to analyze them.

11.4.1 POPULATION COUNT DATA

Annual population counts C, or more specifically the annual observed number of breeding pairs, are stored in a vector (count) of length 29, for the years 1964–92. We adopt for them a state-space model, with a state process model as defined in Section 11.3, and thus the only remaining thing to decide upon is the observation process. Due to the intense field work by Bruno Ullrich we are confident that the number of breeding pairs recorded is a fairly accurate count of population size in every year. Thus, we would not expect any systematic negative bias due to imperfect detection. Rather, we assume that the observation error in this part of the IPM would mostly reflect lack of fit and therefore that counts could deviate from the true (expected) latent states N in both directions. Moreover, in the field the stage of a female was not known, i.e., typically we do not know whether a female is a local recruit, a surviving adult, or an immigrant (except for some ringed individuals). Finally, the information about the recording of a ringed individual is already included in the capture-recapture data set, so to minimize the possible non-independence among the data sets, we won't use this information in this part of the model. We use an observation model with a Poisson error distribution and get

$$C_t \sim Poisson(N_t),$$

where C_t is the population count in year t.

11.4.2 PRODUCTIVITY DATA

The data collected on reproduction is the annual number of (male and female) fledglings produced by a female and is stored in vector f. The additional vectors year and d indicate for each brood the year in which it took place and whether the count of fledglings of brood i is accurate ($d_i = 0$) or represents a minimum ($d_i = 1$). In the latter case, we say that the fledgling count is *censored* and we have to deal with this feature in the observation process for the productivity data.

Before we decide upon a suitable model for the productivity data, we inspect the frequency distribution of the counts of fledglings, i.e., brood size (Fig. 11.2). Based on this distribution it is obvious that brood size cannot be approximated by a standard, unimodal statistical distribution. The main problem is an apparent excess of zeros, which occurs because some females failed to produce any fledgling. Thus, we can imagine that the observed data is actually the outcome of two sequential processes: the probability that a brood is successful, i.e., produces at least one fledgling (nest success probability), and the number of fledglings produced, given success. Here, we model these processes by a zero-inflated normal distribution (Kéry and Royle, 2016) which we specify hierarchically,

$$z_i \sim Bernoulli(\nu_t) \text{ and}$$

$$f_i \sim Normal\left(z_i \kappa_t, \sigma_f^2\right),$$

where f_i is the number of fledglings of brood i, z_i is a latent binary variable indicating the success of brood i ($z_i = 1$, if $f_i > 0$; $z_i = 0$, if $f_i = 0$), ν_t is the nest success probability, κ_t is the average number of fledglings produced per successful brood in

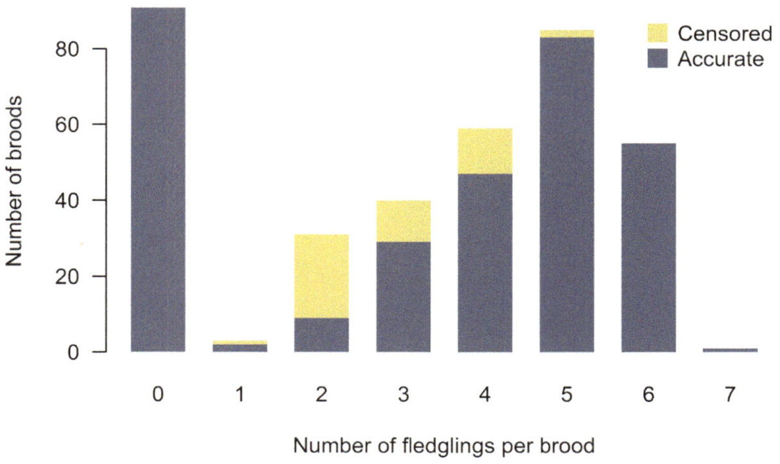

FIGURE 11.2

Frequency distribution of the observed brood size in a German study population of woodchat shrikes, distinguishing accurate and censored counts. In the latter, we only know the minimum number of fledglings produced by a female in a year.

year t, and σ_f^2 is the variance of brood size among these successful broods. Combined, the two processes result in an annual total productivity equal to $\rho_t = \nu_t \kappa_t$. Strictly, this interpretation is only correct if the mean and the variance of the conditional normal distribution are such that there is only a negligible mass near (and below) zero, but we will see below that this assumption is sensible for this data set.

We note that the annual number of offspring produced per successful female is a non-zero integer, and hence a more natural choice might appear to be the Poisson distribution. However, clutch size in birds is typically not well approximated by a Poisson distribution, because it is often less variable than stipulated under the Poisson. Consequently, an underdispersed Poisson distribution would typically be needed to model clutch size and there are several options for this (Ridout and Besbeas, 2004; Lynch et al., 2014; Brooks et al., 2019). Here, we used as a "cheaper" choice a zero-inflated normal, which we think is adequate in spite of being symmetrical around the mean, which does not quite fit the observed frequency distribution of the sizes of observed broods (Fig. 11.2). A sensitivity analysis varying the model for reproduction could tell us whether this choice has much effect on the overall inferences under the IPM, but our experience with many data sets like this suggests to us that we would find very little sensitivity.

There is also one additional challenge for the analysis of this data set: not all counts of fledglings are accurate, but some of them are in fact known to be minimum counts only (Fig. 11.2), i.e., to be right-censored. We could choose to toss out these more uncertain data, but here we prefer to keep them and instead *model* the censoring process for them using BUGS code due to Link and Barker (2010, p. 241). For this, we first replace the censored fledgling counts by a missing value. At the same time, we create an indicator variable called *censored*, which indicates for each brood the threshold value at which censoring occurred. Accurate counts are not censored, hence we set their value of *censored* at 1 plus the observed count, while for broods with censoring, we set *censored* equal to the observed count of fledglings. These calculations are done using the following R code.

```
censored <- woodchat11$f + 1
censored[woodchat11$d==1] <- woodchat11$f[woodchat11$d==1]
woodchat11$f[woodchat11$d==1] <- NA
```

Then, we extend the zero-inflated normal model by the following, where, perhaps a little surprisingly at first sight, the ultimate response variable is now the binary variable d, which says whether a fledgling count was censored or not as a function of the observed censoring threshold which is called *censored* and which we will provide to JAGS as data.

$$z_i \sim Bernoulli(\nu_t)$$

$$f_i \sim Normal\left(z_i \kappa_t, \sigma_f^2\right)$$

$$d_i \sim Bernoulli(z_i \delta_i)$$

$$\delta_i = step(f_i - censored_i)$$

The step function in JAGS is a logical test that returns a 0 if the argument is negative and a 1 if it evaluates to zero or a positive number. This function together with the Bernoulli trial d allows the estimation of the missing f values, i.e., of the censored counts, which will be identical or larger than the observed minimum count. Note that the Bernoulli success probability for response d contains a multiplication with the latent variable z, which ensures the inclusion of the zero-inflation part of the process.

The total annual reproductive output of a female is the product of the probability of successful reproduction (ν) and the number of fledglings given success (κ); thus $\rho = \nu\kappa$. Hence, we replace ρ in the state-process equation for the number of local recruits by this product to obtain

$$R_{t+1} \sim Poisson\left(\frac{\nu_t \kappa_t}{2} \phi_{j,t} N_t\right).$$

11.4.3 CAPTURE-RECAPTURE DATA

The capture histories of the 1079 marked individuals are stored in matrix `ch`. The vector `age` classifies the age of the shrikes at the time of marking by 1 (nestling) or 2 (adult, i.e., at least 1 year old). We transform the capture histories into the m-array format using function `marrayAge` and also calculate the number of released nestlings and adults.

```
marr <- marrayAge(woodchat11$ch, woodchat11$age)
rel.j <- rowSums(marr[,,1])
rel.a <- rowSums(marr[,,2])
```

In 10 years (1973, 1978–81, 1984–88) no adults were captured or resighted, so for these years we will fix recapture probability to zero. We create the vector `recap`, which has 28 elements (with 29 capture occasions we have 28 possible recapture occasions); these will be used as a covariate for modeling recapture. For each year with adult recaptures, we specify a different number (1–18), which allows full flexibility in the modeling of time effects, while for the years without captures/resightings we specify a value of 19 and will set the associated recapture probability to zero in the model (see below).

```
recap <- c(1, 2, 3, 4, 5, 6, 7, 8, 19, 9, 10, 11, 12, 19, 19, 19, 19, 13, 14, 19, 19, 19, 19, 19,
   15, 16, 17, 18)
```

We will use the m-array data to fit a CJS model with the multinomial likelihood (see Section 4.5.1.2).

11.5 THE INTEGRATED POPULATION MODEL

Now that we have seen all three data sets and know how to construct their likelihoods, we can put everything together in an IPM, where we make the usual independence assumption. Initial analyses (not shown) yielded evidence for constant recapture probability (unless of course when capture effort was zero), but for a difference between first-time and experienced breeders. Such an age effect may result from the need to physically catch the marked 1-year old individuals, while older individuals could be resighted due to their color rings, which resulted in a higher detection probability. For the demographic parameters we use temporal random effects. As usual, we specify vague priors for all parameters. For the stage-structured, initial population sizes we give discrete uniform priors. The total population count in year 1 was 10, hence, a uniform(1, 20) for each stage appears to be vague enough.

In Chapter 7, we stressed the importance of goodness-of-fit testing of all the main assumptions of an IPM. This is particularly important for IPMs that include a hidden parameter that is estimated without explicit data. A hidden parameter will compensate for any lack of fit in the other parts of the model and its estimates may therefore become biased (Chapter 7; Riecke et al., 2019). In our analysis, immigration is such a hidden parameter (Schaub and Fletcher, 2015). To assess the goodness-of-fit we use posterior predictive checks for all three component likelihoods, as explained in Chapter 7. We apply three commonly used discrepancy measures: the mean absolute percentage error for the state-space model as suggested by Besbeas and Morgan (2014), the deviance for the zero-inflated normal model and following Brooks et al. (2000) the Freeman-Tukey test statistics for the CJS model. Computation of these test statistics makes the BUGS code slightly more extensive.

Now, we bundle the data, write the model and launch it in JAGS.

```
# Bundle data and produce data overview
jags.data <- list(n.years=dim(marr)[2], marr.j=marr[,,1], marr.a=marr[,,2], rel.j=rel.j,
  rel.a=rel.a, recap=recap, n.recap=max(recap), f=woodchat11$f, d=woodchat11$d,
  censored=censored, year=woodchat11$year-1963, C=woodchat11$count, pNinit=dUnif(1, 20))
str(jags.data)

List of 13
 $ n.years : int 29
 $ marr.j  : num [1:28, 1:29] 1 0 0 0 0 0 0 0 0 0 ...
 $ marr.a  : num [1:28, 1:29] 3 0 0 0 0 0 0 0 0 0 ...
 $ rel.j   : num [1:28] 27 32 26 32 55 26 22 32 0 25 ...
 $ rel.a   : num [1:28] 9 14 18 16 15 13 8 17 2 0 ...
 $ recap   : num [1:28] 1 2 3 4 5 6 7 8 19 9 ...
 $ n.recap : num 19
 $ f       : int [1:365] 5 0 6 0 3 6 6 3 0 NA ...
 $ d       : int [1:365] 0 0 0 0 0 0 0 0 0 1 ...
 $ censored: num [1:365] 6 1 7 1 4 7 7 4 1 3 ...
 $ year    : num [1:365] 1 1 1 1 1 1 1 1 2 2 ...
 $ C       : num [1:29] 10 13 25 15 17 16 6 16 7 7 ...
 $ pNinit  : num [1:20] 0.05 0.05 0.05 0.05 0.05 0.05 0.05 0.05 0.05 ...
```

```
# Write JAGS model file
cat(file="model1.txt", "
model {
# Priors and linear models
for (t in 1:(n.years-1)){
  logit.phij[t] ~ dnorm(l.mean.phij, tau.phij)
  phij[t] <- ilogit(logit.phij[t])
  logit.phia[t] ~ dnorm(l.mean.phia, tau.phia)
  phia[t] <- ilogit(logit.phia[t])
  pj[t] <- p.j[recap[t]]
  pa[t] <- p.a[recap[t]]
}
for (t in 1:n.years){
  logit.nu[t] ~ dnorm(l.mean.nu, tau.nu)
  nu[t] <- ilogit(logit.nu[t])
  log.kappa[t] ~ dnorm(l.mean.kappa, tau.kappa)
  kappa[t] <- exp(log.kappa[t])
  log.omega[t] ~ dnorm(l.mean.omega, tau.omega)
  omega[t] <- exp(log.omega[t])
}
for (t in 1:(n.recap-1)){
  p.j[t] <- mean.p[1]
  p.a[t] <- mean.p[2]
}
# Fix at 0 recapture probability in years without capture/resighting
p.j[n.recap] <- 0
p.a[n.recap] <- 0

mean.phij ~ dunif(0, 1)
l.mean.phij <- logit(mean.phij)
mean.phia ~ dunif(0, 1)
l.mean.phia <- logit(mean.phia)
mean.nu ~ dunif(0, 1)
l.mean.nu <- logit(mean.nu)
mean.kappa ~ dunif(1, 10)
l.mean.kappa <- log(mean.kappa)
mean.omega ~ dunif(0.01, 30)
l.mean.omega <- log(mean.omega)
for (i in 1:2){
  mean.p[i] ~ dunif(0, 1)
}

sigma.phij ~ dunif(0, 5)
tau.phij <- pow(sigma.phij, -2)
sigma.phia ~ dunif(0, 5)
tau.phia <- pow(sigma.phia, -2)
sigma.nu ~ dunif(0, 5)
tau.nu <- pow(sigma.nu, -2)
sigma.kappa ~ dunif(0, 5)
tau.kappa <- pow(sigma.kappa, -2)
sigma.omega ~ dunif(0, 5)
tau.omega <- pow(sigma.omega, -2)
sigma.f ~ dunif(0, 5)
tau.f <- pow(sigma.f, -2)

# Population count data (state-space model)
# Model for the initial population size: uniform priors
R[1] ~ dcat(pNinit)                                          # Local recruits
S[1] ~ dcat(pNinit)                                          # Surviving adults
I[1] ~ dpois(omega[1])                                       # Immigrants

# Process model over time: our model of population dynamics
for (t in 2:n.years){
  R[t] ~ dpois(nu[t-1] * kappa[t-1]/2 * phij[t-1] * N[t-1])  # Local recruits
  S[t] ~ dbin(phia[t-1], N[t-1])                             # Surviving adults
  I[t] ~ dpois(omega[t])                                     # Immigrants
}

# Observation model
for (t in 1:n.years){
  N[t] <- S[t] + R[t] + I[t]                                 # Total number of breeding females
  C[t] ~ dpois(N[t])
```

```
  # GoF for population count data: mean absolute percentage error
  C.pred[t] ~ dpois(N[t])
  disc.C[t] <- pow(((C[t] - N[t]) / C[t]) * ((C[t] - N[t]) / (C[t] +
    0.001)), 0.5)    # Add a small number to avoid potential division by 0
  discN.C[t] <- pow(((C.pred[t] - N[t]) / (C.pred[t] + 0.001)) *
    ((C.pred[t] - N[t]) / (C.pred[t] + 0.001)), 0.5)
}
fit.C <- 100 / n.years * sum(disc.C)
fitN.C <- 100 / n.years * sum(discN.C)

# Productivity data (zero-inflated normal with right censoring)
for (i in 1:length(f)){
  z[i] ~ dbern(nu[year[i]])
  f[i] ~ dnorm(z[i] * kappa[year[i]], tau.f)
  delta[i] <- step(f[i] - censored[i])
  d[i] ~ dbern(z[i] * delta[i])

  # GoF for productivity data: deviance
  z.pred[i] ~ dbern(nu[year[i]])
  f.pred[i] ~ dnorm(z.pred[i] * kappa[year[i]], tau.f)
  f.exp[i] <- kappa[year[i]] * nu[year[i]]

  dev[i] <- (f[i] - f.exp[i]) * (f[i] - f.exp[i])
  devN[i] <- (f.pred[i] - f.exp[i]) * (f.pred[i] - f.exp[i])
}
fit.f <- 2 * sum(dev)
fitN.f <- 2 * sum(devN)

# Capture-recapture data (CJS model with multinomial likelihood)
# Define the multinomial likelihood
for (t in 1: (n.years-1)){
  marr.j[t,1:n.years] ~ dmulti(pr.j[t,], rel.j[t])
  marr.a[t,1:n.years] ~ dmulti(pr.a[t,], rel.a[t])
}
# Define the cell probabilities of the m-arrays
for (t in 1:(n.years-1)){
  # Main diagonal
  qj[t] <- 1-pj[t]
  qa[t] <- 1-pa[t]
  pr.j[t,t] <- phij[t] * pj[t]
  pr.a[t,t] <- phia[t] * pa[t]
  # Above main diagonal
  for (j in (t+1):(n.years-1)){
    pr.j[t,j] <- phij[t] * prod(phia[(t+1):j]) * qj[t] * prod(qa[t:(j-1)]) * pa[j] / qa[t]
    pr.a[t,j] <-prod(phia[t:j]) * prod(qa[t:(j-1)]) * pa[j]
  } #j
  # Below main diagonal
  for (j in 1:(t-1)){
    pr.j[t,j] <- 0
    pr.a[t,j] <- 0
  } #j
 } #t
 # Last column: probability of non-recapture
 for (t in 1:(n.years-1)){
   pr.j[t,n.years] <- 1-sum(pr.j[t,1:(n.years-1)])
   pr.a[t,n.years] <- 1-sum(pr.a[t,1:(n.years-1)])
 }

# GoF for cap.-recap. data: Freeman-Tukey test statistics
for (t in 1:(n.years-1)){
  # Simulated m-arrays
  marr.j.pred[t,1:n.years] ~ dmulti(pr.j[t,], rel.j[t])
  marr.a.pred[t,1:n.years] ~ dmulti(pr.a[t,], rel.a[t])

  # Expected values and test statistics
  for (j in 1:n.years){
    marr.jE[t,j] <- pr.j[t,j] * rel.j[t]
    E.org.j[t,j] <- pow((pow(marr.j[t,j], 0.5) - pow(marr.jE[t,j], 0.5)), 2)
    E.new.j[t,j] <- pow((pow(marr.j.pred[t,j], 0.5) - pow(marr.jE[t,j], 0.5)), 2)
    marr.aE[t,j] <- pr.a[t,j] * rel.a[t]
    E.org.a[t,j] <- pow((pow(marr.a[t,j], 0.5) - pow(marr.aE[t,j], 0.5)), 2)
    E.new.a[t,j] <- pow((pow(marr.a.pred[t,j], 0.5) - pow(marr.aE[t,j], 0.5)), 2)
  } #j
} #t
```

```
fit.CJS <- sum(E.org.j) + sum(E.org.a)
fitN.CJS <- sum(E.new.j) + sum(E.new.a)

# Compute derived parameters: total productivity and immigration rate
for (t in 1:n.years){
  rho[t] <- nu[t] * kappa[t]
}
mean.rho <- mean.nu * mean.kappa
for (t in 1:(n.years-1)){
  omega.rate[t] <- I[t+1] / N[t]
}
}
")
```

As we use a zero-inflated normal regression model with censoring for our productivity data, we must specify initial values for the latent variable z (the brood success indicator variable) and for the productivity of the censored broods, since otherwise, JAGS won't even start to update. For z we provide our observed data about brood success, and we initialize productivity with 1 plus the observed number of fledglings. Giving initial values for any additional parameters is not compulsory.

```
# Initial values
init.Juv <- censored+1
init.Juv[woodchat11$d==0] <- NA
s <- rep(1, length(woodchat11$f))
s[woodchat11$f==0] <- 0
inits <- function() {list(z=s, f=init.Juv)}

# Parameters monitored
parameters <- c("mean.phij", "mean.phia", "mean.nu", "mean.kappa", "mean.rho", "mean.omega",
  "sigma.phij", "sigma.phia", "sigma.nu", "sigma.kappa", "sigma.omega", "sigma.f", "mean.p",
  "fit.C", "fitN.C", "fit.f", "fitN.f", "fit.CJS", "fitN.CJS", "phij", "phia", "nu", "kappa",
  "omega", "rho", "omega.rate", "R", "S", "I", "N")

# MCMC settings
ni <- 30000; nb <- 10000; nc <- 3; nt <- 10; na <- 1000

# Call JAGS from R (ART 3 min) and check convergence
out1 <- jags(jags.data, inits, parameters, "model1.txt", n.iter=ni, n.burnin=nb, n.chains=nc,
  n.thin=nt, n.adapt=na, parallel=TRUE)
traceplot(out1)
```

11.6 RESULTS

The chains of all parameters show adequate convergence, as you can see when you inspect the trace-plots (not shown) and also by their values of R-hat that are all close to 1. Here is a partial (marginal) posterior summary for the estimated quantities in our model.

```
print(out1, 3)

                 mean       sd     2.5%      50%    97.5% overlap0 f  Rhat n.eff
mean.phij       0.052    0.024    0.016    0.048    0.107    FALSE 1 1.017   178
mean.phia       0.411    0.068    0.287    0.407    0.556    FALSE 1 1.005  2004
mean.nu         0.742    0.037    0.666    0.742    0.812    FALSE 1 1.000  6000
mean.kappa      4.757    0.083    4.593    4.757    4.923    FALSE 1 1.000  6000
mean.rho        3.528    0.184    3.151    3.534    3.875    FALSE 1 1.000  6000
mean.omega      6.817    1.424    4.370    6.683    9.600    FALSE 1 1.015  2040
sigma.phij      0.566    0.461    0.025    0.462    1.699    FALSE 1 1.010   287
sigma.phia      0.384    0.284    0.020    0.330    1.054    FALSE 1 1.002  1012
sigma.nu        0.705    0.220    0.293    0.696    1.156    FALSE 1 1.003   705
sigma.kappa     0.056    0.021    0.015    0.056    0.097    FALSE 1 1.001  3637
sigma.omega     0.276    0.198    0.019    0.242    0.731    FALSE 1 1.004   614
sigma.f         0.879    0.037    0.809    0.877    0.956    FALSE 1 1.000  4028
mean.p[1]       0.147    0.082    0.037    0.130    0.355    FALSE 1 1.002  1047
mean.p[2]       0.462    0.104    0.277    0.457    0.681    FALSE 1 1.000  6000
fit.C          22.401    3.506   16.117   22.290   29.613    FALSE 1 1.000  6000
fitN.C         74.001 1053.304   15.973   24.282   39.925    FALSE 1 1.005  6000
fit.f        3378.453   92.348 3221.118 3370.536 3586.062    FALSE 1 1.002  1319
```

```
fitN.f           3593.314  311.470 2992.793 3589.004 4213.025    FALSE 1 1.000  5265
fit.CJS            23.864    3.025   18.618   23.685   30.302    FALSE 1 1.002  1333
fitN.CJS           22.754    3.927   15.495   22.650   30.811    FALSE 1 1.001  1443
phij[1]             0.066    0.044    0.015    0.056    0.180    FALSE 1 1.007   619
[... output truncated ...]
phij[28]            0.056    0.041    0.006    0.048    0.159    FALSE 1 1.009   484
phia[1]             0.483    0.127    0.291    0.460    0.781    FALSE 1 1.001  5620
[... output truncated ...]
phia[28]            0.404    0.109    0.197    0.398    0.650    FALSE 1 1.001  1235
nu[1]               0.740    0.099    0.524    0.749    0.903    FALSE 1 1.000  6000
[... output truncated ...]
nu[29]              0.779    0.091    0.578    0.788    0.928    FALSE 1 1.000  3355
kappa[1]            4.792    0.221    4.363    4.786    5.251    FALSE 1 1.000  6000
[... output truncated ...]
kappa[29]           4.602    0.222    4.146    4.615    5.008    FALSE 1 1.000  6000
omega[1]            6.433    2.196    2.282    6.368   10.739    FALSE 1 1.005  2730
[... output truncated ...]
omega[29]           6.667    2.185    2.576    6.591   11.282    FALSE 1 1.003  6000
rho[1]              3.546    0.500    2.480    3.586    4.409    FALSE 1 1.000  6000
[... output truncated ...]
rho[29]             3.583    0.438    2.640    3.611    4.338    FALSE 1 1.001  2784
omega.rate[1]       0.567    0.314    0.115    0.500    1.333    FALSE 1 1.001  2628
[... output truncated ...]
omega.rate[28]      0.588    0.312    0.125    0.545    1.333    FALSE 1 1.000  6000
R[1]                4.192    2.921    1.000    4.000   11.000    FALSE 1 1.002  1537
[... output truncated ...]
R[29]               0.843    1.087    0.000    1.000    4.000     TRUE 1 1.002  1436
S[1]                4.149    2.912    1.000    3.000   11.000    FALSE 1 1.000  6000
[... output truncated ...]
S[29]               4.214    1.996    1.000    4.000    9.000    FALSE 1 1.000  5499
I[1]                5.138    2.590    1.000    5.000   11.000    FALSE 1 1.000  6000
[... output truncated ...]
I[29]               6.007    2.632    2.000    6.000   12.000    FALSE 1 1.001  6000
N[1]               13.480    3.209    8.000   13.000   20.000    FALSE 1 1.001  4489
[... output truncated ...]
N[29]              11.063    2.641    6.000   11.000   17.000    FALSE 1 1.000  6000
```

To assess the fit of the IPM we produce plots of the discrepancy measures computed for the actual data sets against those of the replicated data sets and calculate the resulting Bayesian p-values. All three test components show a good fit (Fig. 11.3), and the Bayesian p-values are far away from the values of 0 or 1, which both would indicate a model with suspect fit.

Once we are convinced of an adequate model fit and of Markov chain convergence, we inspect the estimates and see what we can learn about the terminal dynamics of this woodchat shrike population just before its extinction. The point estimates of the demographic rates confirm that the woodchat shrike has a life-history that is typical for a short-lived passerine. Adult survival is about 0.4 and mean total annual productivity about 3.5 fledglings per female. Apparent juvenile survival is only 0.05 and each year about seven females immigrated into the population; this shows that the

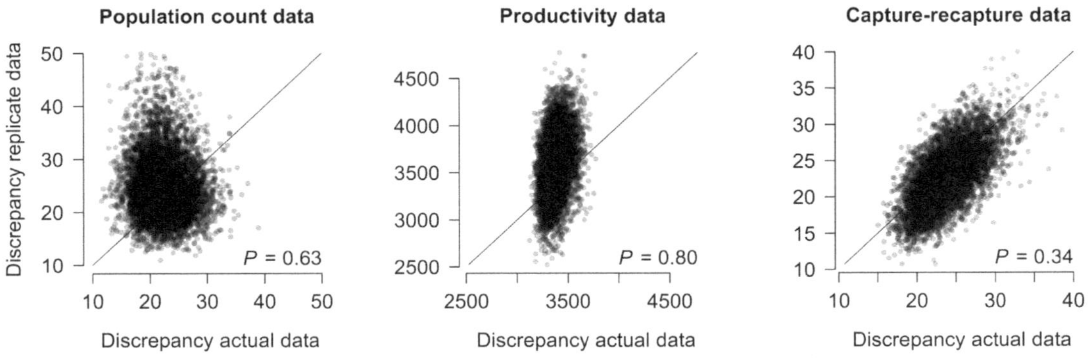

FIGURE 11.3

Scatterplots of discrepancy measures computed for the replicate (i.e., simulated) versus those for the actual (i.e., observed) data for the three model components of the IPM. The Bayesian p-values are the proportion of points above the 1:1 equality line.

study population was geographically open. The mean conditional productivity is estimated at 4.8 fledglings per successful brood, with a standard deviation (σ_f) of 0.9. A normal distribution with this mean and spread has virtually no mass near and below zero, hence our interpretation of parameter (ν) as giving the nest success probability is fully justified for our data set.

We can make some interesting calculations based on these results, which allow us to better understand this population. For instance, under the assumption that apparent adult survival is equal to true survival, i.e., that there was no breeding dispersal, we can make a rough calculation of the value of juvenile survival required to achieve a population that remains stable in size without the need of any immigration. Let's for a moment denote survival with s. The population growth rate then becomes $\lambda = 1 = \frac{\nu\kappa}{2}s_j + s_a$ (see Chapter 3), and hence $s_j = \frac{2(1-s_a)}{\nu\kappa} = 0.34$, with a CRI of 0.26–0.42. This is a liberal estimate, as we would expect some permanent emigration in the adults, i.e., that ϕ_a in our model is smaller than true adult survival s_a.

In addition, we may produce a rough estimate of the proportion of locally produced fledglings that settle outside of our study population (ξ, emigration) as follows: based on $\phi_j = (1-\xi)s_j$, we get $\xi = 1-\frac{\phi_j}{s_j} = 0.85$, with a CRI of 0.68–0.95. If the assumption (i.e., $\phi_a = s_s$, that there is no breeding, but only natal, dispersal beyond the study area) on which these simple calculations are based is true, this is a confirmation of strong natal dispersal in this species. If breeding dispersal was not zero, natal dispersal would be slightly less than the value calculated. Perhaps natal dispersal is so strong because the study population was at the edge of the species' distribution range, fairly isolated or perhaps because habitat quality had become insufficient.

```
# Compute CRI of juvenile survival and emigration
sj <- 2 * (1 - out1$sims.list$mean.phia) / out1$sims.list$mean.nu / out1$sims.list$mean.kappa
mean(sj); quantile(sj, c(0.025, 0.975))

xsi <- 1 - out1$sims.list$mean.phij / sj
mean(xsi); quantile(xsi, c(0.025, 0.975))
```

To better understand the temporal patterns in the demographic rates we produce some graphs. Juvenile apparent survival and conditional productivity (given nest success) varied only little over time (Fig. 11.4), while apparent adult survival, nest success probability and immigration rate fluctuated much more. Presumably, the negligible variability of juvenile survival is a consequence of the relatively small sample and our adoption of a random-effects specification for the time effects, resulting in strong shrinkage of the annual estimates toward the overall mean. If our goal is to get insights into the specific temporal patterns of such parameters, we might be tempted to specify a fixed-effects model then, but the resulting estimates may suffer from large uncertainty (due to sampling variability) and therefore not be very reliable in that sense. Thus, it may be challenging to obtain a reasonable description of the temporal variability when sample sizes are low.

We may also be interested in the composition of the woodchat shrike population and how it changes over time. This is one of the greatest assets of an IPM: that we can so easily estimate the stage composition of a population along with its temporal dynamics. Fig. 11.5 shows that the number of locally recruited birds was low in all years and that the numbers of surviving adults and of immigrants were about equal. We note also that the population composition in the first year was quite different—it appears as if the three types of females occurred in about equal proportion in that year. However, this is somewhat of an artifact: these estimates are different from those of the later years and in contrast to these latter should not be trusted too much. The difficulty is that in our Markovian population model, in the initial year there is no information about population structure from a previous year, and the IPM uses prior information of population structure in the first year. We used discrete uniform priors for the numbers of surviving adults and of local recruits, but a different prior for the number of immigrants. We modeled the number of immigrants and put a temporal random-effects structure on them, hence, we estimated the mean number of immigrants and we used that information in the prior of the first year. Hence, the number of immigrants in the first year is partially informed by the number of immigrants in later years and yielded a more precise estimate (SD = 2.50) compared to a variant of the analysis where we specified the analogous prior as for the number of individuals in the other two stage classes in the first year (SD = 3.13, code not shown). Although the difference in precision seems quite substantial, it is of little practical relevance. Generally, we advise to almost ignore estimates of population structure from the first year from a Markovian, stage-structured model such as ours. However, the estimates of the total population size are more reliable in such models, but often also less precise than those for later years.

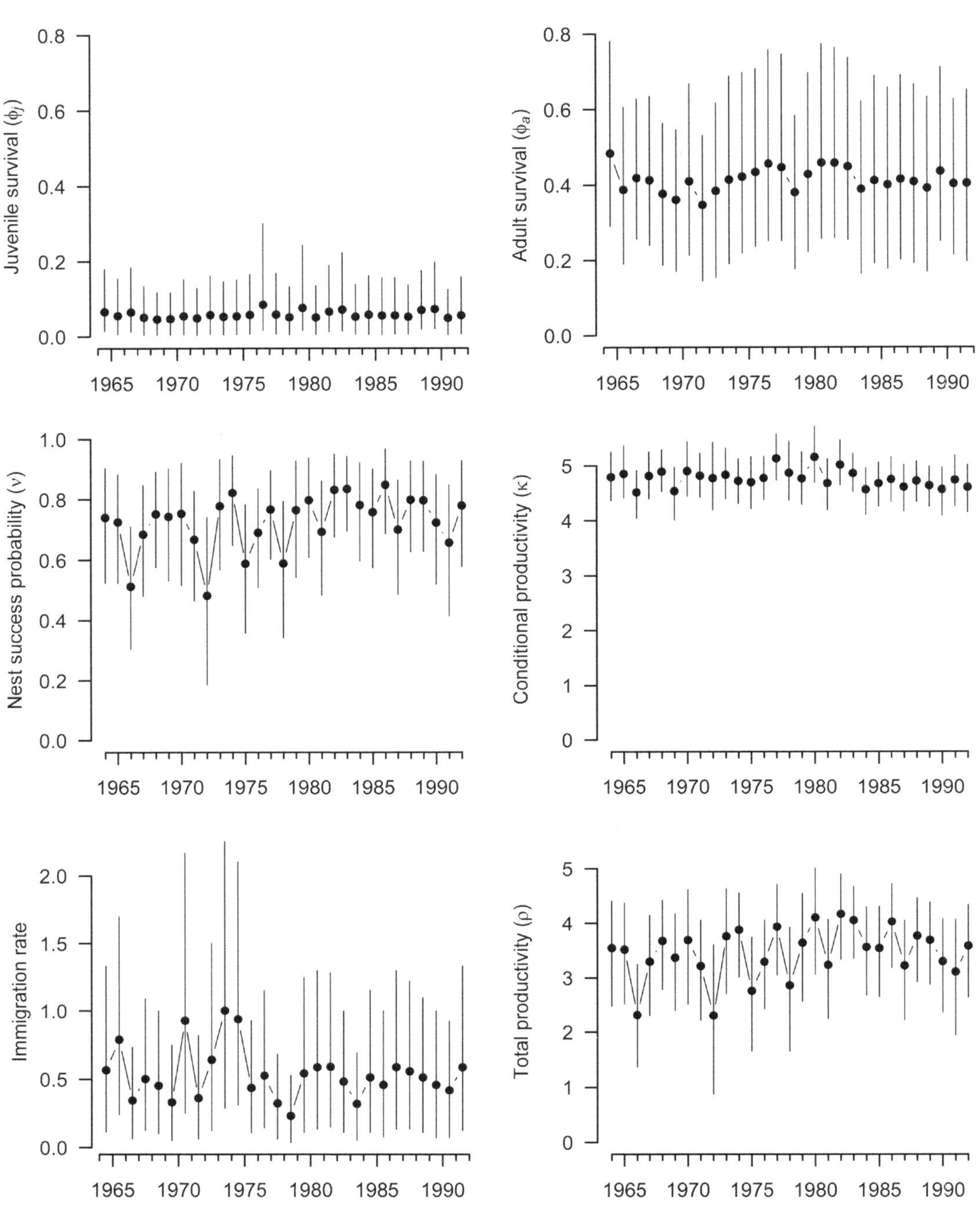

FIGURE 11.4

Annual estimates (posterior means with 95% CRIs) of demographic rates of the German population of woodchat shrike. Immigration rate and total productivity (ρ) are derived parameters.

11.7 MORE PARSIMONIOUS MODELS

Our data set covers a respectable 29 years, which seems like a long-enough time span for quantitative inferences about population dynamics. However, the sample sizes of the individual data sets are fairly modest. The number of breeding pairs varied between 6 and 28 and therefore the number of broods and the number of marked individuals were both relatively small. Moreover, in some years resighting effort was zero, which increased the uncertainty about survival. So far, we have fit a very flexible IPM that included temporal random effects for all demographic rates, and demographic stochasticity.

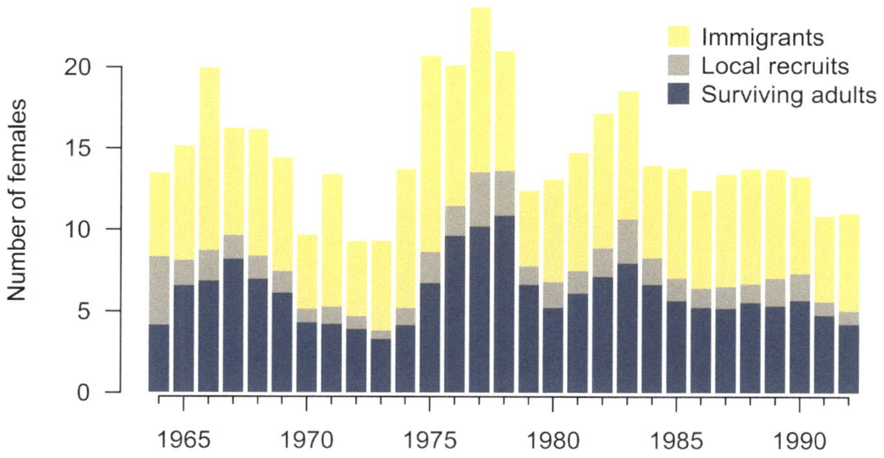

FIGURE 11.5

Estimates of the composition of the woodchat shrike population from 1964 to 1992 under the IPM.

This model may actually be too flexible relative to the amount of data and therefore we may wonder whether we shouldn't use a simpler, more parsimonious IPM for inference. Ideally, we could use some one-off model selection criterion to pick that model which is best according to some objective function such as success in prediction. However, although model selection is possible for IPMs (e.g., Besbeas et al., 2002; King et al., 2008; Davis et al., 2014), we think that uncertainty remains about how this is best done and no formal study so far has examined this question (e.g., which selection criteria should be applied or what strategy adopted). Hence, what we do next is an informal model sensitivity analysis, where we fit three less complex models and compare the inferences about key model output among all four of our models.

We focus on apparent survival and immigration, since these parameters are "weak" parameters, heuristically-speaking: the number of recaptures is relatively low and provides only limited information about survival, and immigration is estimated as a hidden parameter. We fit the three additional IPMs which are all variants of our first one in the last section, but where complexity is reduced in some way (in all three additional IPMs, temporal variability was retained for the two productivity parameters), see Table 11.1. We then compare the inferences under these three additional IPMs with the most complex IPM from the previous sections which we denote here as IPM_1.

It is illustrative to compare the estimates of adult survival and of immigration under the four models. Specifically we first compare the estimated annual survival probabilities (Fig. 11.6A). Obviously, they are temporally variable under IPM_1 and IPM_2, which both specify temporal random effects, and constant under IPM_3 and IPM_4, which both assume the absence of any temporal variability in this parameter. Both point estimates and width of CRIs remained about the same for IPM_1 and IPM_2, thus the inclusion of demographic stochasticity did not make much of a difference with respect to this parameter.

Secondly, we compare the ratio of the estimated number of survivors in year $t+1$ and population size in year t, which is the "realized" survival as a proportion. Realized survival under IPM_4 (without either temporal variability or demographic stochasticity) was constant over time and identical to the estimated survival probability under the same model (Fig. 11.6B). The temporal patterns of realized survival under the three other IPMs were almost identical. While this was to be expected for IPM_1 and IPM_2, which both allowed for annual variation in survival probability, we see the same pattern also for IPM_3 even though in that model we specified constant survival probability. This observed temporal pattern of realized survival under IPM_3 is the result of demographic stochasticity (see Section 3.3.4).

Table 11.1 Schematic of a model sensitivity analysis where we compare the most complex IPM from the previous section in this chapter (IPM_1) with three of its variants with reduced complexity.

	With Demographic Stochasticity	Without Demographic Stochasticity
Survival probability and expected number of immigrants variable over years	IPM_1	IPM_2
Survival probability and expected number of immigrants constant over years	IPM_3	IPM_4

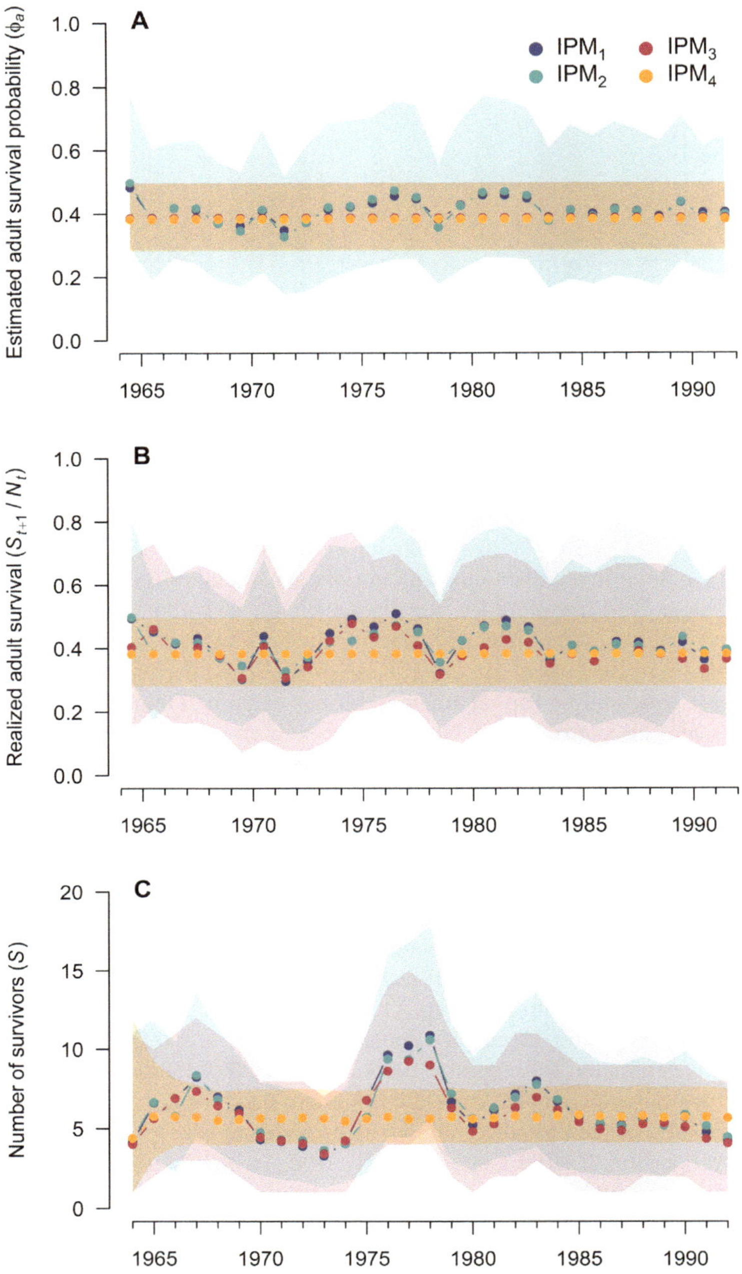

FIGURE 11.6

Example of a model sensitivity analysis showing estimates of three features of the apparent survival process for adult shrikes under four IPMs of different complexity: (A) annual adult survival probability (posterior means with 95% CRIs), (B) realized proportion of annual adult survival (i.e., ratio of the number of survivors at *t*+1 and population size at *t*), (C) realized number of survivors. See Table 11.1 for what each model means.

The estimates of the realized number of survivors yielded a similar picture (Fig. 11.6C): IPMs$_{1-3}$, with either temporal variability or with demographic stochasticity, resulted in very similar estimates that all varied over time, while IPM$_4$, without temporal variability and demographic stochasticity, estimated the number of survivors to be constant over time (except for year 1, because this estimate is not a function of survival). Thus, in terms of the realized adult survival we see that a simpler model with constant survival probability but with demographic stochasticity (IPM$_3$), produced essentially the same dynamics as the two more complex IPMs that did allow for temporal variability in survival probability, with

(IPM_1) or without demographic stochasticity (IPM_2). In contrast, the simplest IPM_4, without temporal variability and demographic stochasticity, stood out clearly from the rest and yielded completely different and arguably unrealistic estimates.

Next we compare estimates of the expected and the realized numbers of immigrants, and of total population size, under the four IPMs (Fig. 11.7). We see similar patterns in the immigration process as we just saw for the parameter estimates of three features of the survival process. The estimates of the expected number of immigrants varied over time under

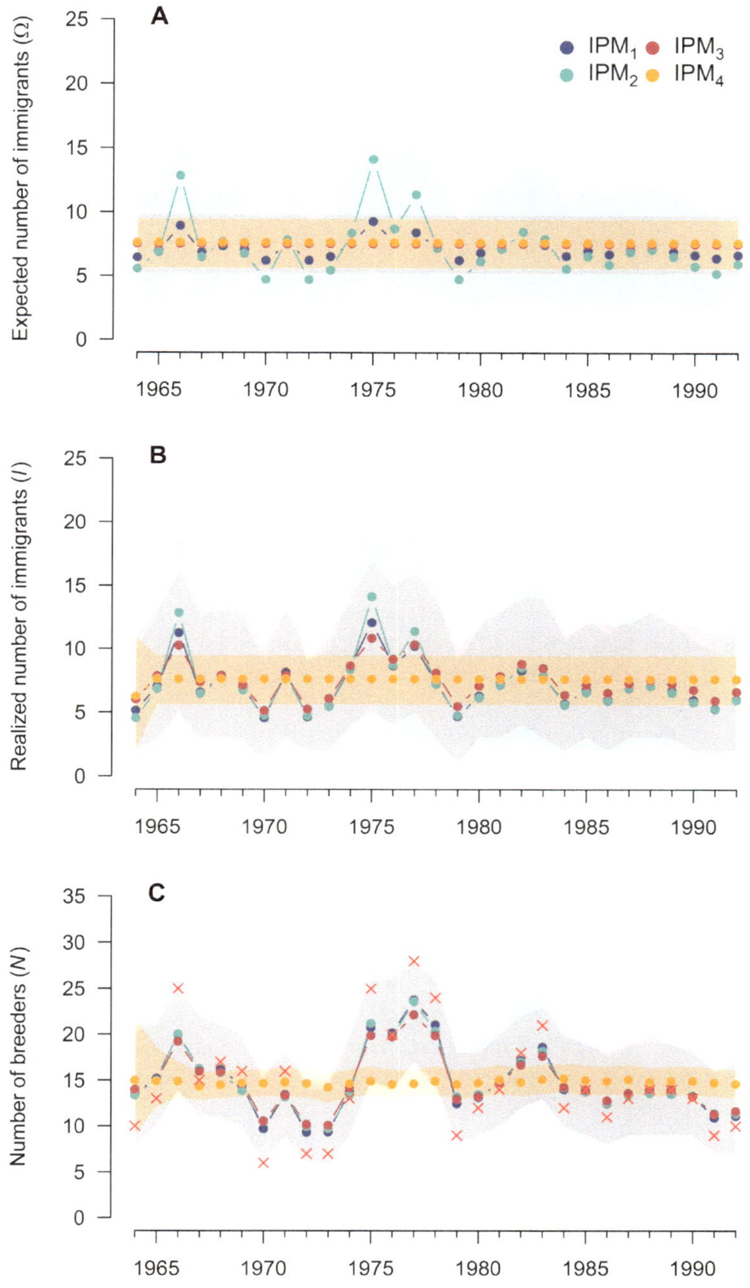

FIGURE 11.7

Example of a model sensitivity analysis showing estimates of the number of immigrants and of breeders under four IPMs of different complexity: (A) expected number of immigrants (posterior means and 95% CRIs), (B) realized number of immigrants, (C) number of breeders (total population size), together with the number of breeding pairs counted (red crosses). See Table 11.1 for what each model means.

IPM_1 and IPM_2, which both include temporal variability in immigration (Fig. 11.7A). The uncertainty about immigration was only slightly larger under IPM_1 which also included demographic stochasticity. Obviously, the estimated expected number of immigrants was constant over time under IPM_3 and IPM_4, since both assumed the absence of any temporal variability in this parameter. Again, the uncertainty around the estimates was only slightly increased when demographic stochasticity was included as opposed to when it was not. The estimates of the realized numbers of immigrants were exactly identical as those of the expected number of immigrants under IPM_2 and IPM_4, which both did not include demographic stochasticity (Fig. 11.7B). Under IPM_3, with demographic stochasticity and constant expected number of immigrants over time, there was a strong discrepancy between the estimates of the expected and of the realized numbers of immigrants. While the estimates of the expected number of immigrants remained constant over time, those of the realized number of immigrants showed a very similar temporal pattern as did the estimates under IPM_1 and IPM_2, which both included temporal variability. Again, this difference in the temporal pattern between the expected and the realized number of immigrants is the effect of demographic stochasticity (see Section 3.3.4). The uncertainty (95% CRI width) around the estimates of the realized number of immigrants was similar under $IPMs_{1-3}$, but slightly smaller under IPM_3 without temporal variability.

Finally, the estimates of the total population size were again similar under the three models with either demographic stochasticity or temporal variability ($IPMs_{1-3}$). Thus, as for survival and immigration, we can conclude that the simpler IPM_3, with demographic stochasticity but without temporal variability in survival or immigration, produces similar estimates of the structured population sizes as does the more complex model with temporal variability in these underlying parameters (see also Section 5.5).

This exercise shows that the simpler IPM_3 can produce similar estimates of structured population sizes (number of survivors, immigrants or total population size) as the more complex $IPMs_{1-2}$ and that the observed population counts agree well with the estimated population size under this simpler model. IPM_1 and IPM_2 appear to be too flexible (i.e., overfitted), but as we can see in the current example, estimates and precision were hardly affected by this overfitting. Besbeas and Morgan (2017) reached a similar and rather unexpected conclusion in their simulation study: that overfitting was not detrimental for the quality of parameter estimates in their IPMs.

In contrast, IPM_4 assumed neither temporal variability in survival and immigration nor demographic stochasticity and does not fit well the observed number of breeding pairs. In addition, its estimates of structured population size are very different from those under $IPMs_{1-3}$. We therefore would not recommend using IPM_4 for inference, but instead any one of the other three other IPMs. Possibly one could also model average with equal weights (Dormann et al., 2018), although in this case this would not produce much difference compared to inferences based on a single model, since their inferences were all so similar.

The decision question is whether and how our inferences would change if we based them on IPM_1, IPM_2 or IPM_3. The answer to this question depends on what we want to know. IPM_3 tells us that the observed dynamics of the population was possible even if the underlying survival probabilities and the expected number of immigrants were constant over time. Demographic stochasticity in this small population was strong enough to allow for the temporal variability that was required in the realized numbers of survivors and immigrants so that the resulting trajectory of the estimated total population agreed well with the observations. If we just want to describe the dynamics of the population and make inferences about temporal changes in its stage structure, we could therefore use IPM_3. Use of the more complex IPM_1 or IPM_2 for that purpose would not do any harm, because their parameter estimates are hardly more imprecise. In contrast, if we are interested in understanding how strongly the temporal variability of each demographic process has contributed to the variability of the population growth rate (for instance, in a retrospective analysis, see Chapter 10), then it does matter which IPM we use, because then we need temporally varying estimates of all demographic rates and would use IPM_1 or IPM_2. To avoid too strong temporal variation with contamination by sampling variability, specification of temporal random effects is recommended.

We stressed before that the reason why the estimated stage-structured population sizes are virtually identical under models IPM_1 and IPM_3 is because of demographic stochasticity, the effects of which are substantial in this small population. In a larger population, these effects would become relatively smaller and hence the results of IPM_1 and IPM_3 regarding stage-specific population sizes might differ more. Indeed, the estimates of stage-specific population sizes under IPM_3 would then approach those under IPM_4. This all means that if population size is low and, as a result, also the sample sizes of the capture-recapture and productivity data, then it may become hard to distinguish environmental (i.e., temporal) variability in the demographic rates from the effects of demographic stochasticity. Model choice may not matter for the estimation of the stage-specific population sizes, but it does if we want to identify drivers of past population dynamics or forecast future population trajectories. In these latter cases, it matters through which mechanisms temporal variability in the stage-specific population sizes is generated.

In addition, the observation model chosen for the population counts may also affect the results. Here, we used a Poisson model which is less flexible than the normal distribution with its extra dispersion parameter (Section 6.6.5). If similar IPMs are fitted to the woodchat shrike data set but using the normal instead of the Poisson observation model for the counts, the residual errors in IPM_1 and IPM_2 are estimated to be zero, indicating a perfect fit and therefore, quite certainly, a strongly overfitted model. In contrast, use of the normal distribution in IPM_3 works adequately since the residual error is not estimated to be zero. Thus, when small data sets are analyzed with IPMs, we advise care and the exploratory fitting of several variants of the target IPM to better understand the limits of each, and of the data sets used to inform their parameters. Although our example as well as the simulation study by Besbeas and Morgan (2017) shows that overfitting may not be detrimental for the inferences under an IPM in general, this issue is not yet well understood, and more research on the determination of the optimal degree of complexity in an IPM and, more generally, on model selection in IPMs, would certainly be extremely welcome.

11.8 DISCUSSION

We have fit four variants of an IPM to a fairly typical combination of three data sets that are produced by many bird population studies (population counts, capture-recapture and productivity data) and estimated immigration as a hidden parameter for the last three decades of a small population of woodchat shrikes before its extinction. Two special features of this IPM were the adoption of a zero-inflated model for the productivity data and the accommodation of right-censoring to accommodate some observations where only a minimum number of fledged young is known. Based on our biologist's intuition, we think that the IPM provided reasonable parameter estimates. We found that immigration was substantial in this small population, which is often typical in small populations of birds and also of mammals (reviewed by Millon et al., 2019).

In our model, we parameterized the immigration process in absolute terms and estimated the immigration rate as a derived parameter, but we could also have done the inverse (Schaub and Fletcher, 2015). Estimation of any hidden parameter in IPMs is akin to an additional residual term in the model for the population counts, and there is a risk of obtaining biased estimates for the hidden parameter if the model is seriously mis-specified (see Section 7.2.1.4; Schaub and Fletcher, 2015; Riecke et al., 2019). Therefore, assessment of the goodness-of-fit of the different model components is particularly important.

We modeled productivity as a zero-inflated normal random variable but, as mentioned earlier, other choices would have been possible. One particular case might have been a hurdle model (Potts and Elith, 2006; Dorazio et al., 2013) instead of a zero-inflated model. The difference between the two is that in the hurdle model we assume that there is only a single process producing the zeros, while in the zero-inflated model they may be generated by two processes. In our case, zero-inflation means that depending on its parameter values zeros could also have come from the normal distribution. If we wanted to make a clear distinction between the probability of breeding success and its outcome conditional on success, a hurdle model might generally be better. However, in our case a hurdle model would not have made any difference because the normal distribution estimated for the conditional brood size had virtually no mass near and below zero.

We might have been tempted to throw away the right-censored productivity data or, perhaps even worse, to include them in the model as if they were accurate data. Both approaches can lead to biased estimates of the components of productivity and should be avoided (especially the latter!). We have shown that the modeling of such censored data is fairly straightforward. However, it is important that the censoring limit is actually observed and not just a mere guess. That is, you must be able to say something like *"Here I saw exactly 3 fledglings, but I am not sure whether there are more than that"* rather than *"I saw several fledglings all over the place and think that they might have been about 3"*.

Estimates of stage-structured population sizes were nearly identical regardless of whether they were obtained from the most complex IPM with temporal random effects on all demographic rates, or under the more parsimonious IPM that assumed constant adult survival and immigration, but allowed for demographic stochasticity. The more complex model might therefore not be necessary, but its greater complexity did not compromise the precision of the parameter estimates. The more parsimonious IPM resulted in nearly the same estimates of stage-structure population size because the variability required in the realized number of surviving individuals and of immigrants was accommodated by the inclusion of demographic stochasticity. It appears that distinguishing between temporal variability and demographic stochasticity can be very difficult when such complex hierarchical models such as an IPM are fit to small data sets. We can hypothesize that this may not even be much alleviated with larger data sets and that there may be fundamental limits to distinguishing between different structures assumed in the deeper levels of a hierarchical model from the observed data (Conn et al., 2018). More research about effects of overfitting, separating temporal from demographic stochasticity, and generally on model selection in complex hierarchical models including IPMs will be valuable.

CHAPTER

PEREGRINE FALCON

12

CHAPTER OUTLINE

IPM ingredients: population count data, dead-recovery data, productivity data
Location: Switzerland and France
Spatio-temporal scale: 17,700 km^2, 43 years

12.1 INTRODUCTION

In this chapter, we develop an integrated population model (IPM) with the classical threefold combination of counts, productivity data, and survival data, specifically dead-recovery data. This combination is similar to that in the original IPM paper by Besbeas et al. (2002), the only difference being the lack of productivity data in the latter. As this is the only case study where we illustrate dead-recovery data, we will comment more on this type of data and the associated models than we will for the Cormack-Jolly-Seber (CJS) model that you see at many other places throughout this book.

Dead-recovery data, also known as ring-, band- or tag-recovery data, represent an important source of information for estimating survival and are often available in population studies of birds, including, among countless other examples, those of Thomson et al. (1999), Tavecchia et al. (2002), Robinson et al. (2007), Duriez et al. (2012), and Arnold et al. (2016), and sometimes also for mammals (Hall et al., 2001) or fish (Payton et al., 2019; Nater et al., 2020). Under this protocol, animals are marked and released once during their lives, and later, some of them are recovered again at the moment of their death or briefly afterward. This field protocol is different from live capture-recapture, where animals are marked once but can then be recaptured alive multiple times, and where no dead animal need ever be observed at all. Nevertheless, from an analytical point of view, statistical models for dead-recovery data can be viewed as a mere variation on a CJS model. Models can be fit to the individual encounter history or m-array data format using the state-space formulation or multinomial likelihood, respectively (see Section 4.5.3). Differences from the CJS model are the different expressions for the success parameters of the Bernoulli trials and the algebraic expectations for the multinomial cell probabilities in the m-array.

The groundwork for analyzing dead-recovery data was laid by Seber (1970, 1971) and Brownie et al. (1985). In principle, dead-recoveries can be found throughout the area used by the marked individuals; therefore, it is typically assumed that analysis of this data type yields estimates of true survival (i.e., where there is no confounding between survival and fidelity to the study area). This contrasts with the analysis of live capture-recapture data, where apparent survival is estimated because recaptures/resightings are typically restricted to a specific study area. Yet in reality, the probability of recovering a marked dead individual is spatially highly variable and may be effectively zero in many parts of the range of the individuals marked in a study. Clearly, the spatial (and also temporal; see below) variation of the recovery probability depends largely on the density and activity of the human population, but it has proven challenging to describe

Integrated Population Models. https://doi.org/10.1016/B978-0-12-820564-8.00005-1

and therefore account for such spatial variation (Korner-Nievergelt et al., 2010). Fortunately, unexplained spatial variation in recovery probabilities does not appear to result in appreciable bias in survival estimates (White et al., 2013). Besides survival, dead recoveries can be used to estimate the frequency of mortality causes (Schaub and Pradel, 2004; Nater et al., 2020), harvest rates (Nichols and Hines, 1983; Brownie et al., 1985; Calvert and Gauthier, 2005), dispersal (Paradis et al., 1998), and even productivity (Arnold, 2018). Thus, they can be a very informative source of information about population demography.

Often, dead-recoveries are analyzed jointly with capture-recapture data, and this permits estimation of site fidelity, i.e., the conditional probability for an individual to remain in a given study area given that it is alive (Burnham, 1993; Barker, 1997, 1999), which is the scaling factor between true survival and the apparent survival estimated in a CJS model from capture-recapture data. Dead-recoveries have frequently been used in IPMs either alone (e.g., Besbeas et al., 2002; Robinson et al., 2012; Koons et al., 2017; Zimmerman et al., 2017) or together with capture-recapture data (e.g., Reynolds et al., 2009; Tavecchia et al., 2009; Lieury et al., 2015; Bled et al., 2017).

In this chapter, we work with data from two long-term population studies of the wonderful peregrine falcon (*Falco peregrinus*, Fig. 12.1) in the Jura mountains on both sides of the border in France and Switzerland over 43 years (1965–2007). The peregrine is a medium-sized raptor (body mass 600 g for males and 1000 g for females; Ratcliffe, 1993; Monneret, 2017) that occurs worldwide and is celebrated as the fastest animal on Earth owing to the great terminal velocity reached during its stoop on mostly avian prey (in addition, in the humble opinion of one of the authors of this book, it is also by far the most fascinating and magnificent animal in the world). European peregrines usually nest on tall, vertical cliffs and more rarely on high buildings, which are typically occupied in an extremely traditional manner over many decades or even centuries (Kéry and Zuberogoitia, 2020). As top predators, peregrines naturally occur at low densities, and adjacent pairs are typically spaced several kilometers apart. Peregrines became a charismatic symbol of endangered species worldwide when their populations crashed and became regionally or even subcontinentally extinct during the 1960s as a result of the effects of organochlorine pesticides including DDT and Dieldrin (Newton, 1979; Ratcliffe, 1993).

FIGURE 12.1

Happy peregrine falcons (*Falco preregrinus*) in the Vosges mountains, France (photograph by Vincent Michel).

These pesticides caused females to lay thin-shelled eggs that broke during incubation and led to repeated reproductive failure. In addition, some of these pesticides may also have depressed the survival of adults, but this is somewhat debated (Ratcliffe, 1993).

We will focus on two topics in this case study: first, we investigate temporal patterns in demographic rates to see in particular whether survival, in addition to productivity, may also have played a role in population recovery after the DDT-crash, and by implication, in the crash itself, during which unfortunately no survival data seem to exist anywhere in the world. Reproductive failure of peregrine populations followed by a comeback of productivity as the population recovered has been very well documented (Ratcliffe, 1993), but not surprisingly, longer-term survival trajectories have hardly ever been published (McGrady et al., 2017; Robinson and Wilson, 2021). This is due in large part to the paucity of survival data time series that are long enough and extend back to the 1950s and 1960s.

Throughout this book, we frequently use unstructured random effects for annual time series of parameters such as survival or recapture. This typically results in strong shrinkage, such that with information-poor data sets, most annual estimates will oscillate fairly tightly around the hypermean. Here, we illustrate autoregressive smoothing of a time series of parameters where we describe *changes* in parameters rather than the parameters themselves as independent random variables (Link and Barker, 2010). Such a specification of temporal variation is more efficient for discerning finer-scale temporal patterns and we think should be used more often when there is an interest in these.

Second, we illustrate the ease with which really any age- or stage-structured population model, including an IPM, can yield estimates of the size of all population segments represented by the stage classes distinguished in the model, *even those for which no individual is ever observed* (see also Section 6.5). This is a considerable advantage, because it enables an accurate accounting of all classes of individuals in the population. This includes what in raptor studies are known as floaters, i.e., individuals that are not (yet) territorial but are, so to speak, "queuing" to obtain their own territory. Floaters will often be represented by younger individuals that have not yet recruited into the breeding population, but sometimes they may also contain older individuals that have bred earlier and then lost their territory, perhaps only temporarily. Floaters can form a substantial part of the total population of any long-lived animal and have been shown to be important for population dynamics and regulation (Newton, 1979; Hunt, 1988; Newton and Mearns, 1988; Rohner, 1996; Newton and Rothery, 2001; Bretagnolle et al., 2008; Penteriani et al., 2011). Often, they have an important role as a buffer against extreme fluctuations in the breeding population, and therefore, their assessment is important in judging the viability of a population and its resilience to negative impacts (Walters et al., 2002; Lopez-Sepulcre and Kokko, 2005; Kokko and Sutherland, 2007; Lee et al., 2017). Yet, the quantification of the number of floaters, or nonbreeders in general, is very difficult (Katzner et al., 2011; Lee et al., 2017).

12.2 DATA SETS

In this chapter, we combine part of the demographic data collected in two ongoing peregrine population studies. One is located in the Swiss Jura mountains and in an adjacent region of the Swiss Plateau and along the northern edge of the southwestern Swiss Alps. It was launched by the late Gaby Banderet and his colleagues in 1960. The other study is in the French part of the Jura mountains and was started by René-Jean Monneret, René Ruffinoni and their colleagues in 1965 (see Monneret, 2017; Monneret et al., 2018). Both studies must rank among the longest-term vertebrate population studies worldwide. The two study areas are immediate neighbors and share a border of about 90 km (see Monneret et al., 2018 and Kéry et al., 2018, 2021, for maps of both areas) and have a very similar environment, so we simply lump their data here. The convex hull defined by the locations of the territories in the French study yields a total occupied area of 12,714 km^2, while the Swiss study area measures 4993 km^2, which gives a combined area of 17,707 km^2.

During each breeding season in both parts of the study area, as many territories as logistically possible were surveyed, often multiple times, to assess territory occupancy by adult pairs. Breeding success was also recorded and expressed by the number of young per pair that were either newly fledged or still in the nest but in brown plumage (i.e., fewer than 10 days before fledging). Surveys were concentrated in February and March just before egg-laying to find pairs attending a nesting cliff and in May and June to count the young, typically with 8×–10× binoculars and 20×–60× telescopes from distances ranging from less than 100 m to just under 5 km. In addition, in the Swiss part of the study area, as many broods were ringed as possible, which means that Gaby Banderet or one of his colleagues abseiled down a cliff face and attached aluminum leg rings to young at an age between about 2 and 4 weeks.

In this chapter, we use three data sets spanning 43 years (1965–2007): the combined counts of territorial pairs and of fledged young (for a known sample of nests) and the dead-recovery data from 1810 birds ringed as nestling in the Swiss part of the study area. The annual counts ranged between 18 and 251 breeding pairs. For over 80% of these breeding pairs,

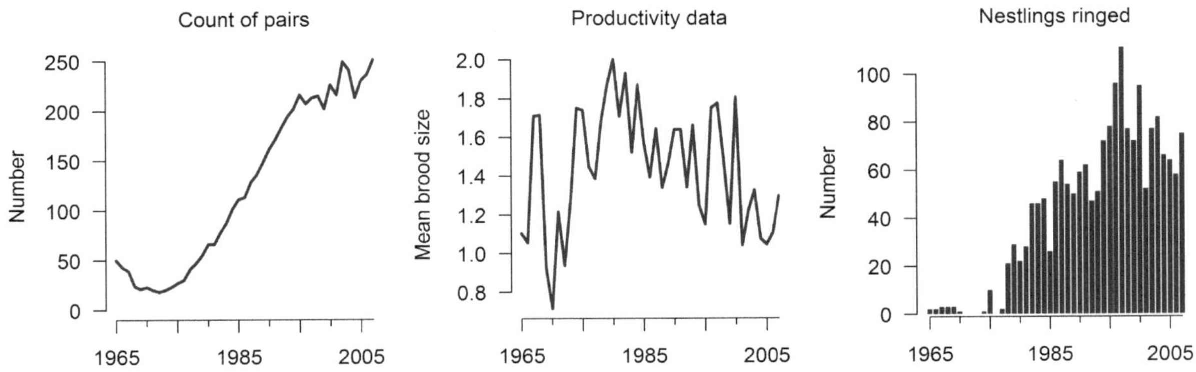

FIGURE 12.2

Raw data summaries in the peregrine IPM in this chapter: time series of the counts of territorial pairs (left), of mean brood size (middle), and of the number of ringed nestlings (right).

breeding success was recorded. The total number of broods for which breeding success was recorded amounts to 4692. Fig. 12.2 shows the trajectories of the counts of territorial pairs and of the number of ringed young per year. In this chapter, we use "pair," "breeding pair," and "territorial pair" more or less synonymously.

12.3 POPULATION MODEL

We use a prebreeding "census" female-based population model for the peregrines. The structure of the population model was inspired by other demographic studies on this species (Wootton and Bell, 1992; Kauffman et al., 2003; Altwegg et al., 2014). We assume that annual survival of fledglings (first-year survival, s_1) differs from the annual survival of birds that are ≥ 1 year old (adult survival, s_2). Clearly, there will be more age structure in the survival of a long-lived species such as the peregrine, but differences in age-dependent survival observed after the first year of life were only minor in a thorough study by Kauffman et al. (2003). Our dead-recovery data are fairly sparse, especially in the first half of the study, so we do not include additional age structure. Peregrines reproduce only very rarely when 1 year old (Wootton and Bell, 1992; Ratcliffe, 1993; Kauffman et al., 2003), and rather usually recruit into the breeding population from 2 years of age onward. In our model, we assume that a fraction α of the 2-year-old females reproduce but that all females reproduce when they are 3 years and older. Our model also assumes either that there is no dispersal into and out of the study population or that dispersal is at equilibrium, i.e., that the same number of birds enters and exits every year (see the Discussion for more on this assumption). In summary, our stage-structured population model assumes four demographic states (with their numbers in parentheses, and note that their subscripts do *not* indicate an age-class): 1-year-old birds (N_1), 2-year-old nonbreeders (N_2), 2-year-old breeders (N_3), and birds 3 years or older that are assumed to breed every year (N_4). The resulting life-cycle graph is shown in Fig. 12.3.

During the bottleneck phase at the nadir of the DDT crash in the early 1970s, the population size was very small, so we adopt a model with demographic stochasticity. The number of 1-year-old females in year $t+1$ is then

$$N_{1,t+1} \sim Poisson\left(\frac{\rho_t}{2} s_{1,t}\left(N_{3,t} + N_{4,t}\right)\right),$$

where ρ_t is the productivity in year t (i.e., the number of fledglings raised per breeding female) and $s_{1,t}$ is first-year survival from year t to $t+1$. The sum of the 2-year-old first-time breeders and the at-least 3-year-old breeders represents the total number of breeding females per year.

The number of 2-year-old nonbreeders is modeled using a binomial distribution as

$$N_{2,t+1} \sim Binomial\left(s_{2,t}(1 - \alpha_t), N_{1,t}\right),$$

where α_t is the probability that a 2-year-old female reproduces in year t, and $s_{2,t}$ is adult survival from year t to $t+1$.

The number of 2-year-old first-time breeder in year t is modeled using another binomial distribution.

$$N_{3,t+1} \sim Binomial\left(s_{2,t}\alpha_t, N_{1,t}\right)$$

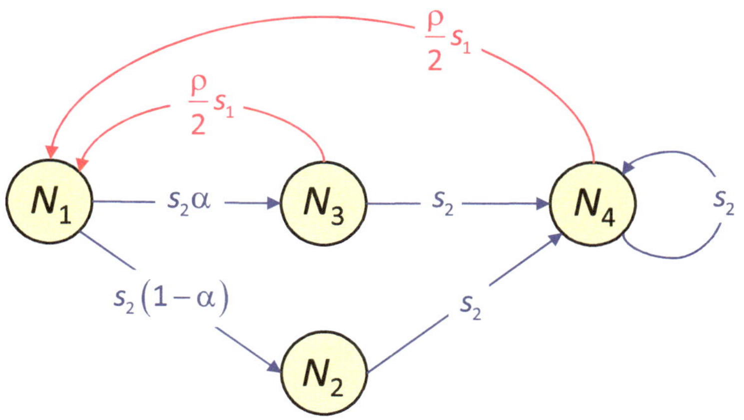

FIGURE 12.3

Life-cycle graph for the peregrine falcon population in the Jura mountains. The model is expressed for a postbreeding "census" and distinguishes four demographic states: 1-year-old nonbreeding birds (N_1), 2-year-old nonbreeders (N_2), 2-year-old first-time breeders (N_3), and birds that are 3 year or older and that all reproduce (N_4). The recruitment processes are the red arrows with first-year survival (s_1) and productivity (ρ), and the survival processes are the blue arrows with adult survival (s_2) and breeding probability at age 2 years (α).

Lastly, the number of 3-year-old females in year t is modeled using yet another binomial distribution, where now the success parameter is adult survival alone.

$$N_{4,t+1} \sim Binomial(s_{2,t}, N_{2,t} + N_{3,t} + N_{4,t})$$

12.4 COMPONENT DATA LIKELIHOODS

The three data sets for this analysis are stored in an R data file that we load and whose structure we inspect.

```
library(IPMbook); library(jagsUI)
data(peregrine)
str(peregrine)

List of 3
 $ count       : num [1:43, 1:2] 1965 1966 1967 1968 1969 ...
  ..- attr(*, "dimnames")=List of 2
  .. ..$ : NULL
  .. ..$ : chr [1:2] "Year" "Breeding_pairs"
 $ productivity: num [1:43, 1:3] 1965 1966 1967 1968 1969 ...
  ..- attr(*, "dimnames")=List of 2
  .. ..$ : NULL
  .. ..$ : chr [1:3] "Year" "No_surveyed_brood" "No_fledglings"
 $ recoveries  : int [1:1810, 1:43] 0 0 0 0 0 0 0 0 0 0 ...
  ..- attr(*, "dimnames")=List of 2
  .. ..$ : NULL
  .. ..$ : chr [1:43] "1965" "1966" "1967" "1968" ...
```

The data file contains three objects: time series of population counts, productivity, and a dead-recovery data set. Next, we briefly describe each along with the submodels we use to analyze them.

12.4.1 POPULATION COUNT DATA

Annual population counts C, or more specifically, the annual number of territorial (or breeding) pairs observed in the French and the Swiss parts of the study area combined, are stored in vector `count`, which has length 43 for the years 1965–2007. For the count data, we adopt a state-space model, as usual. The process model corresponds exactly to the population model that we defined in Section 12.3. For the observation model, we specify a Poisson distribution such that

$$C_t \sim Poisson(N_{3,t} + N_{4,t}).$$

Only the stage classes 3 and 4, corresponding to individuals that are recruited into the pool of breeders, are included in the observation model for the observed counts of territorial pairs. Nonbreeding birds in stage class 1 or 2 are not counted (in a sense, they are literally invisible during the population surveys), and therefore, they do not appear in the observation model for the population counts.

Remember that our use of a Poisson distribution implicitly assumes that our counts are correct on average, i.e., that there is no systematic error in the counting of breeding pairs. Given the fairly intensive surveys with multiple visits in a breeding season made to most known breeding cliffs, this seems a reasonable assumption. Moreover, the residual variance of the IPM is fixed by using the Poisson distribution, i.e., it is equal to the mean. This assumption would be hard to check, but we could try out other distributions such as normal, log-normal, or negative binomial to assess the robustness of the main inferences to this assumption (Section 6.6.5).

12.4.2 PRODUCTIVITY DATA

The productivity data are stored in matrix `productivity`, with 43 rows corresponding to years and with three columns indexing the year, the annual number of broods that were surveyed to record productivity in the French and the Swiss parts of the study area combined, and the annual total number of fledglings produced by the surveyed broods (failed broods are included here). We use a Poisson regression model to analyze these data,

$$J_t \sim Poisson(B_t \rho_t),$$

where J_t is the total number of fledglings counted in year t, B_t is the number of surveyed broods in t, and ρ_t is productivity in year t, i.e., number of fledglings per breeding female in the population.

12.4.3 DEAD-RECOVERY DATA

The dead-recovery data are stored in a matrix called `recoveries`, with rows that correspond to a total of 1810 nestlings ringed only in the Swiss part of the study area and columns to years (note that 75 nestlings were ringed in the last year and don't contribute any information to survival estimation—they are dropped when forming the m-array below). For all individuals, the year of marking is indicated by the first 1 in the row vector of its data in the matrix. A total of 145 individuals (8%) were recovered dead, and for them, there is an additional 1 in the data matrix in the corresponding year of death. In dead-recovery data, the correct assignment of a recovery date to a year requires some care. We cannot simply use the calendar year of the recovery date but must use the year as it is defined in the life-cycle graph and the model. For Swiss Jura peregrines, the mean date of ringing of a nestling was May 9, and hence the time length of the first occasion after marking would be only 8 months within a calendar year. We would therefore overestimate *annual* juvenile survival unless we accounted for this shorter duration. We would also need to change the population model that needed (annual) transitions from just before reproduction in 1 year to the next. To avoid these complications, we use May 9 as our annual divide between successive years, such that all occasions have a constant duration of 12 months, and the time span covered corresponds to that of the population model. We apply a simple rule for the coding of the dead-recovery data: any recovery that occurred between May 9 (our annual divide) of year t and May 8 of year $t+1$ is entered in the occasion/year $t+1$.

We illustrate this using the example of three imaginary birds that were all ringed on 2 May 1965. The first bird was recovered dead on 24 February 1966. Then its row vector in the recovery data matrix is:

```
ID     1965    1966    1967    1968 ...
1         1       1       0       0 ...
```

This is straightforward. But now look at bird 2, which was already recovered dead on 1 August 1965. By applying our rule, it ends up with the same history as bird 1:

```
ID     1965    1966    1967    1968 ...
2         1       1       0       0 ...
```

Finally, bird 3 was recovered dead on 6 September 1967. This falls into the period between 9 May 1967 and 8 May 1968, and hence belongs in occasion "1968":

```
ID     1965    1966    1967    1968 ...
3         1       0       0       1 ...
```

For computational reasons, we do not analyze the recovery histories directly using the state-space formulation (Section 4.5.3.1), but we use the multinomial likelihood for the usual m-array data aggregation (Brownie et al., 1985; Williams et al., 2002; Barry et al., 2003, Chapter 8 in Kéry and Schaub, 2012). The structure of the dead-recovery m-array was explained in Section 4.5.3.2, but here it is slightly different because we want to implement an age structure in the survival probabilities. The cell probabilities of the m-array are therefore:

	Recovery occasion					
Marking occasion	**2**	**3**	...	**T**	...	**Never**
1	$(1 - s_{1,1})r_1$	$s_{1,1}(1 - s_{2,2})r_2$	...	$s_{1,1}\prod_{j=2}^{t-1} s_{2,j}(1 - s_{2,t})r_t$	...	$1 - \Sigma_1$
2	0	$(1 - s_{1,2})r_2$	...	$s_{1,2}\prod_{j=3}^{t-1} s_{2,j}(1 - s_{2,t})r_t$	...	$1 - \Sigma_2$
...	...	...	...	...	...	

First-year survival in year t is $s_{1,t}$, adult survival in year t is $s_{2,t}$, and the recovery probability in year t is r_t. For the Jura peregrines we assume there is no age structure in the recovery probability. An age structure on recovery probabilities can be expected if the frequency of human-related mortality causes depends on the age of the birds (Schaub and Pradel, 2004). We do not really know that for the peregrine, but we think this would be very unlikely. One of the main reasons for why we have not included an age structure in the recovery probability is statistical convenience. When only young individuals are marked and no live resightings are made, dead-recovery models with an age structure in both the survival and the recovery parameters are not identifiable (Anderson et al., 1985; Cole et al., 2012). It is possible that such a model might be identifiable as part of an IPM, where we have additional information from counts and productivity data, but have not tested this here.

We use our function `marrayDead` to compute the m-array from the individual dead-recovery data. We also need to calculate the annual number of marked individuals, which is given by the row sums of the m-array.

```
# Create the dead-recovery m-array
m.dead <- marrayDead(peregrine$recoveries)

# Compute the annual number of marked individuals
rel <- rowSums(m.dead)
```

At the start of the study, only a few peregrine nestlings were marked, and in some years, none at all; these were the years when the Swiss part of the study population had in fact gone extinct (Fig. 12.2). The dead-recovery data therefore do not contribute much information about survival in the first part of the study, and we would *a priori* expect a lower precision of survival estimates in these years.

12.5 THE INTEGRATED POPULATION MODEL

Now that we have seen all three data sets and know their likelihoods, we can put everything together in an IPM while making the usual independence assumption (Abadi et al., 2010; Weegman et al., 2021). Our first goal in the analysis is to explore the temporal patterns in the demographic rates, in particular survival, to understand how they may have contributed to the recovery of the peregrine population in the Jura, and by implication to the population crash in the first place. We fit two variants of an IPM. In our first model, we specify unstructured temporal random effects for all demographic parameters with the only exception being the probability that 2-year-old females reproduce (α). This is a hidden parameter, and we would not expect to be able to gauge temporal variation in it given our sparse data, notably in the first part of the study period. Therefore, we keep α constant. We use normal distributions to model unstructured temporal random effects for a suitable transformation (or link function) of a parameter θ,

$$g(\theta_t) \sim Normal(\mu_\theta, \sigma_\theta^2),$$

where μ_θ is the mean of θ on the link scale, and σ_θ^2 is the temporal variance of $g(\theta)$. We use a logit-link for survival probabilities and a log-link for productivity.

Second, we fit a model that produces some temporal smoothing of the time series of these parameters. This can often improve our ability to recognize especially finer-scale temporal patterns in a parameter. Of course, we could also apply models with polynomial terms or piecewise regressions, but they may be too inflexible. Random walks are well suited for smoothing time series of parameters (see Chapter 13 in Link and Barker, 2010). In these models, we assume temporal dependence such that the closer two values are in time, the more similar they are. Other autoregressive models for survival probabilities have been described by Johnson and Hoeting (2003) and could be adopted for smoothing, as could splines (Crainiceanu et al., 2005; Gimenez et al., 2006; Hefley et al., 2017). To specify a random-walk smoother, we treat a parameter as a "free" value in its first year by specifying a suitable vague prior. For all later years, we specify its value in year t as a random draw from a normal distribution centered on the value in year $t-1$ with a variance that we estimate. In essence, we assume a normal prior for the first-differences of the time series of that parameter (see Link and Barker, 2010, for more details).

$$g(\theta_1) \sim \text{reasonable prior}$$

$$g(\theta_t) \sim Normal\left(g(\theta_{t-1}), \sigma_\theta^2\right)$$

Here, the noise parameter σ_θ^2 determines the strength of the temporal autocorrelation. If it is close to zero, autocorrelation is strong, and two neighboring values of θ will be more similar than when σ_θ^2 is large, in which case autocorrelation is weak, and two neighbors of θ will be more dissimilar.

Dead-recovery probabilities in Europe are typically found to be declining over time (Robinson et al., 2009). We therefore specify a linear trend for it.

$$\text{logit}(r_t) = \beta_1 + \beta_2 t$$

Here, β_1 is the intercept and β_2 the slope parameter.

For the population model, we must also define priors for the stage-specific population sizes in the first year. As we often do, we use a discrete uniform distribution for this purpose. The observed count in the first year was 50 pairs. To be fairly vague, we here use discrete uniform priors in the range between 1 and 70 for the initial abundance of all four stages in the model. Note that this corresponds to a discrete uniform prior between 2 and 140 for the sum of the stage classes 3 and 4, which are those that appear in our population count of territorial pairs.

At several places in this book, we have stressed the importance of goodness-of-fit (GoF) testing of all the main assumptions of an IPM. This is particularly important in IPMs that include a hidden parameter, as a hidden parameter will often soak up lack of fit from other parts of the model (Chapter 7; Riecke et al., 2019). In our peregrine model, the probability that a 2-year-old female starts to reproduce (α) is a hidden parameter. We use posterior predictive checks for all three component likelihoods to assess the goodness-of-fit, as explained in Chapter 7. We apply three commonly used discrepancy measures: the mean absolute percentage error for the state-space model as suggested by Besbeas and Morgan (2014), the deviance for the Poisson regression model, and the Freeman-Tukey test statistics for the dead-recovery model (Brooks et al., 2000). Computation of these test statistics makes the BUGS code slightly more extensive. We reiterate that a GoF test for the state-space model is likely to have little power, especially for an IPM with a hidden parameter that is estimated "by difference" (see Section 7.2.1.4). Therefore, we could also have chosen to conduct GoF tests only for the other responses in the model.

Next, we bundle the data into a list and briefly inspect the content of that list.

```
# Bundle data and produce data overview
jags.data <- with(peregrine, list(nyears=ncol(peregrine$recoveries), marr=m.dead, rel=rel,
  y=count[,2], B=productivity[,2], J=productivity[,3], pNinit=dUnif(1, 70)))
str(jags.data)

List of 7
 $ nyears: int 43
 $ marr  : num [1:42, 1:43] 0 0 0 0 0 0 0 0 0 0 ...
 $ rel   : num [1:42] 2 2 3 3 3 1 0 0 0 1 ...
 $ y     : num [1:43] 50 43 39 24 21 23 20 18 20 23 ...
 $ B     : num [1:43] 29 36 24 14 13 14 14 16 18 20 ...
 $ J     : num [1:43] 32 38 41 24 12 10 17 15 23 35 ...
 $ pNinit: num [1:70] 0.0143 0.0143 0.0143 0.0143 0.0143 ...
```

We then write code for IPM_1 with the temporal random effects on the demographic rates, define the initial value, and run it.

```
# Write JAGS model file
cat(file = "model1.txt", "
model {
# Priors and linear models
for (t in 1:(nyears-1)){
  logit.s[1,t] ~ dnorm(l.mean.s[1], tau.s[1])
  s[1,t] <- ilogit(logit.s[1,t])
  logit.s[2,t] ~ dnorm(l.mean.s[2], tau.s[2])
  s[2,t] <- ilogit(logit.s[2,t])
  alpha[t] <- mean.alpha
  logit(r[t]) <- beta[1] + beta[2] * t     # Linear trend in recovery
}
for (u in 1:2){
  mean.s[u] ~ dunif(0, 1)                  # Priors for mean age-dep. survival
  l.mean.s[u] <- logit(mean.s[u])
  sigma.s[u] ~ dunif(0, 5)
  tau.s[u] <- pow(sigma.s[u], -2)
}
mean.alpha ~ dunif(0, 1)                   # Prior for prob. start reproduction at age 2y
for (i in 1:2){                            # Priors for betas
  beta[i] ~ dnorm(0, 0.001)
}
for (t in 1:nyears){
  log.rho[t] ~ dnorm(l.mean.rho, tau.rho)
  rho[t] <- exp(log.rho[t])
}
mean.rho ~ dunif(0, 5)                     # Prior for mean productivity
l.mean.rho <- log(mean.rho)
sigma.rho ~ dunif(0, 5)
tau.rho <- pow(sigma.rho, -2)

# Population count data (state-space model)
# Model for the initial population size: discrete uniform priors
for (a in 1:4){
  N[a,1] ~ dcat(pNinit)
}

# Process model over time: our model of population dynamics
for (t in 1:(nyears-1)){
  N[1,t+1] ~ dpois(rho[t] / 2 * s[1,t] * (N[3,t] + N[4,t]))
  N[2,t+1] ~ dbin(s[2,t] * (1-alpha[t]), N[1,t])
  N[3,t+1] ~ dbin(s[2,t] * alpha[t], N[1,t])
  N[4,t+1] ~ dbin(s[2,t], (N[2,t] + N[3,t] + N[4,t]))
}

# Observation model
for (t in 1:nyears){
  NB[t] <- N[3,t] + N[4,t]
  y[t] ~ dpois(NB[t])

  # GOF for population count data: mean absolute percentage error
  y.pred[t] ~ dpois(NB[t])
  disc.y[t] <- pow(((y[t] - NB[t]) / y[t]) * ((y[t] - NB[t]) / (y[t] +
    0.001)), 0.5)                          # Add a small number to avoid potential division by 0
  discN.y[t] <- pow(((y.pred[t] - NB[t]) / (y.pred[t] + 0.001)) *
    ((y.pred[t] - NB[t]) / (y.pred[t] + 0.001)), 0.5)
}
fit.y <- 100 / nyears * sum(disc.y)
fitN.y <- 100 / nyears * sum(discN.y)

# Dead-recovery data (multinomial model)
# Define the multinomial likelihood
for (t in 1:(nyears-1)){
  marr[t,1:nyears] ~ dmulti(pr[t,], rel[t])
}
```

```
# Define the cell probabilities of the m-array
for (t in 1:(nyears-1)){
  # Main diagonal
  pr[t,t] <- (1-s[1,t]) * r[t]
  # Further than one above main diagonal
  for (j in (t+2):(nyears-1)){
    pr[t,j] <- s[1,t] * prod(s[2,(t+1):(j-1)]) * (1-s[2,j]) * r[j]
  } #j
  # Below main diagonal
  for (j in 1:(t-1)){
    pr[t,j] <- 0
  } #j
} #t
for (t in 1:(nyears-2)){
  # One above main diagonal
  pr[t,t+1] <- s[1,t] * (1-s[2,t+1]) * r[t+1]
} #t
# Last column: probability of non-recovery
for (t in 1:(nyears-1)){
  pr[t,nyears] <- 1-sum(pr[t,1:(nyears-1)])
} #t

# GOF for dead-recovery data: Freeman-Tukey test statistics
for (t in 1:(nyears-1)){
  # Simulated m-arrays
  marr.pred[t,1:nyears] ~ dmulti(pr[t,], rel[t])

  # Expected values and test statistics
  for (j in 1:nyears){
    marr.E[t,j] <- pr[t,j] * rel[t]
    E.org[t,j] <- pow((pow(marr[t,j], 0.5) - pow(marr.E[t,j], 0.5)), 2)
    E.new[t,j] <- pow((pow(marr.pred[t,j], 0.5) - pow(marr.E[t,j], 0.5)), 2)
  } #j
} #t

fit.DR <- sum(E.org)
fitN.DR <- sum(E.new)

# Productivity data (Poisson regression)
for (t in 1:nyears){
  J[t] ~ dpois(B[t] * rho[t])

  # GOF for productivity data: deviance
  J.pred[t] ~ dpois(B[t] * rho[t])
  J.exp[t] <- B[t] * rho[t]
  dev[t] <- J[t] * log(J[t] / J.exp[t]) - (J[t] - J.exp[t])
  devN[t] <- J.pred[t] * log(J.pred[t] / J.exp[t]) - (J.pred[t] - J.exp[t])
}
fit.J <- sum(dev)
fitN.J <- sum(devN)
}
")

# Initial values
inits <- function(){list(mean.s=runif(2, 0.6, 0.8))}

# Parameters monitored
parameters <- c("mean.s", "mean.alpha", "mean.rho", "beta", "sigma.rho", "sigma.s", "s", "rho", "r",
  "N", "NB", "fit.y", "fitN.y", "fit.DR", "fitN.DR", "fit.J", "fitN.J")

# MCMC settings
ni <- 100000; nb <- 20000; nc <- 3; nt <- 80; na <- 5000

# Call JAGS from R (ART 56 min) and check convergence
out1 <- jags(jags.data, inits, parameters, "model1.txt", n.iter=ni, n.burnin=nb, n.chains=nc,
  n.thin=nt, n.adapt=na, parallel=TRUE)
traceplot(out1)
```

Before we inspect the results, we also write code for IPM_2 that adopts random-walk smoothers for all demographic rates. Convergence is harder to obtain with this model, and therefore, we increase the length of the Markov chain Monte Carlo (MCMC) chains compared with those of IPM_1. We choose the MCMC settings in such a way that an identical nominal number of MCMC draws results for both models and that the effective sample sizes are also similar for the two models; see below for the reason why.

```
# Write JAGS model file
cat(file = "model2.txt", "
model {
# Priors and linear models
s[1,1] ~ dunif(0, 1)                # Prior of first-year survival in year 1
logit.s[1,1] <- logit(s[1,1])
s[2,1] ~ dunif(0, 1)                # Prior of adult survival in year 1
logit.s[2,1] <- logit(s[2,1])
rho[1] ~ dunif(0, 5)                # Prior of productivity in year 1
log.rho[1] <- log(rho[1])

for (t in 2:(nyears-1)){            # Autoregressive models for s
  logit.s[1,t] ~ dnorm(logit.s[1,t-1], tau.s[1])
  s[1,t] <- ilogit(logit.s[1,t])
  logit.s[2,t] ~ dnorm(logit.s[2,t-1], tau.s[2])
  s[2,t] <- ilogit(logit.s[2,t])
}
for (u in 1:2){
  sigma.s[u] ~ dunif(0, 5)
  tau.s[u] <- pow(sigma.s[u], -2)
}

for (t in 2:nyears){                # Autoregressive models for rho
  log.rho[t] ~ dnorm(log.rho[t-1], tau.rho)
  rho[t] <- exp(log.rho[t])
}
sigma.rho ~ dunif(0, 5)
tau.rho <- pow(sigma.rho, -2)

for (t in 1:(nyears-1)){            # Models for alpha and r
  alpha[t] <- mean.alpha
  logit(r[t]) <- beta[1] + beta[2] * t
}
mean.alpha ~ dunif(0, 1)            # Prior for prob. start reproduction at age 2y
for (i in 1:2){
  beta[i] ~ dnorm(0, 0.001)
}

# Population count data (state-space model)
# Model for the initial population size: discrete uniform priors
for (a in 1:4){
  N[a,1] ~ dcat(pNinit)
}

# Process model over time: our model of population dynamics
for (t in 1:(nyears-1)){
  N[1,t+1] ~ dpois(rho[t] / 2 * s[1,t] * (N[3,t] + N[4,t]))
  N[2,t+1] ~ dbin(s[2,t] * (1-alpha[t]), N[1,t])
  N[3,t+1] ~ dbin(s[2,t] * alpha[t], N[1,t])
  N[4,t+1] ~ dbin(s[2,t], (N[2,t] + N[3,t] + N[4,t]))
}

# Observation model
for (t in 1:nyears){
  NB[t] <- N[3,t] + N[4,t]
  y[t] ~ dpois(NB[t])

  # GOF for population count data: mean absolute percentage error
  y.pred[t] ~ dpois(NB[t])
  disc.y[t] <- pow(((y[t] - NB[t]) / y[t]) * ((y[t] - NB[t]) / (y[t] +
    0.001)), 0.5)                   # Add a small number to avoid potential division by 0
  discN.y[t] <- pow(((y.pred[t] - NB[t]) / (y.pred[t] + 0.001)) *
    ((y.pred[t] - NB[t]) / (y.pred[t] + 0.001)), 0.5)
}
fit.y <- 100 / nyears * sum(disc.y)
fitN.y <- 100 / nyears * sum(discN.y)

# Dead-recovery data (multinomial model)
# Define the multinomial likelihood
for (t in 1:(nyears-1)){
  marr[t,1:nyears] ~ dmulti(pr[t,], rel[t])
}
```

```
# Define the cell probabilities of the m-array
for (t in 1:(nyears-1)){
  # Main diagonal
  pr[t,t] <- (1-s[1,t]) * r[t]
  # Further than one above main diagonal
  for (j in (t+2):(nyears-1)){
    pr[t,j] <- s[1,t] * prod(s[2,(t+1):(j-1)]) * (1-s[2,j]) * r[j]
  } #j
  # Below main diagonal
  for (j in 1:(t-1)){
    pr[t,j] <- 0
  } #j
} #t
for (t in 1:(nyears-2)){
  # One above main diagonal
  pr[t,t+1] <- s[1,t] * (1-s[2,t+1]) * r[t+1]
} #t
# Last column: probability of non-recovery
for (t in 1:(nyears-1)){
  pr[t,nyears] <- 1-sum(pr[t,1:(nyears-1)])
} #t

# GOF for dead-recovery data: Freeman-Tukey test statistics
for (t in 1:(nyears-1)){
  # Simulated m-arrays
  marr.pred[t,1:nyears] ~ dmulti(pr[t,], rel[t])

  # Expected values and test statistics
  for (j in 1:nyears){
    marr.E[t,j] <- pr[t,j] * rel[t]
    E.org[t,j] <- pow((pow(marr[t,j], 0.5) - pow(marr.E[t,j], 0.5)), 2)
    E.new[t,j] <- pow((pow(marr.pred[t,j], 0.5) - pow(marr.E[t,j], 0.5)), 2)
  } #j
} #t

fit.DR <- sum(E.org)
fitN.DR <- sum(E.new)

# Productivity data (Poisson regression)

for (t in 1:nyears){
  J[t] ~ dpois(B[t] * rho[t])
  # GOF for productivity data: deviance
  J.pred[t] ~ dpois(B[t] * rho[t])
  J.exp[t] <- B[t] * rho[t]
  dev[t] <- J[t] * log(J[t] / J.exp[t]) - (J[t] - J.exp[t])
  devN[t] <- J.pred[t] * log(J.pred[t] / J.exp[t]) - (J.pred[t] - J.exp[t])
}
fit.J <- sum(dev)
fitN.J <- sum(devN)
}
")

# Initial values
s <- matrix(NA, nrow=2, ncol=jags.data$nyears-1)
s[1,1] <- runif(1, 0.5, 0.6)
s[2,1] <- runif(1, 0.7, 0.9)
inits <- function(){list(s=s)}

# Parameters monitored
parameters <-c("mean.alpha", "beta", "s", "rho", "r", "sigma.s", "sigma.rho", "N", "NB", "fit.y",
  "fitN.y", "fit.DR", "fitN.DR", "fit.J", "fitN.J")

# MCMC settings
ni <- 200000; nb <- 50000; nc <- 3; nt <- 150; na <- 5000

# Call JAGS from R (ART 103 min) and check convergence
out2 <- jags(jags.data, inits, parameters, "model2.txt", n.iter=ni, n.burnin=nb, n.chains=nc,
  n.thin=nt, n.adapt=na, parallel=TRUE)
traceplot(out2)
```

12.6 RESULTS

The chains of all parameters from IPM_1 show decent convergence, as you can see when you inspect the trace-plots (not shown) and by their R-hat values, which are all close to 1. Here is a partial posterior summary for the estimated quantities in our model.

```
print(out1, 3)

               mean     sd    2.5%     50%   97.5% overlap0     f  Rhat n.eff
mean.s[1]     0.595  0.049   0.501   0.594   0.692    FALSE 1.000 1.001  1801
mean.s[2]     0.780  0.016   0.748   0.780   0.810    FALSE 1.000 1.009   229
mean.alpha    0.476  0.184   0.144   0.468   0.861    FALSE 1.000 1.056    41
mean.rho      1.424  0.048   1.330   1.424   1.521    FALSE 1.000 1.000  3000
beta[1]      -1.713  0.348  -2.413  -1.707  -1.049    FALSE 1.000 1.001  1581
beta[2]      -0.019  0.011  -0.040  -0.019   0.004     TRUE 0.952 1.001  1690
sigma.rho     0.191  0.028   0.143   0.188   0.250    FALSE 1.000 1.000  3000
s[1,1]        0.479  0.173   0.124   0.496   0.788    FALSE 1.000 1.000  3000
s[2,1]        0.762  0.048   0.637   0.770   0.834    FALSE 1.000 1.004   763
[... output truncated ...]
s[1,42]       0.672  0.141   0.385   0.674   0.929    FALSE 1.000 1.000  3000
s[2,42]       0.774  0.031   0.705   0.775   0.832    FALSE 1.000 1.001  1505
rho[1]        1.200  0.151   0.928   1.199   1.514    FALSE 1.000 1.000  3000
rho[2]        1.157  0.141   0.899   1.151   1.447    FALSE 1.000 1.002   769
[... output truncated ...]
rho[42]       1.142  0.074   1.002   1.141   1.292    FALSE 1.000 1.001  2347
rho[43]       1.308  0.076   1.165   1.307   1.462    FALSE 1.000 1.000  3000
r[1]          0.155  0.044   0.082   0.151   0.252    FALSE 1.000 1.001  1842
r[2]          0.153  0.042   0.083   0.149   0.245    FALSE 1.000 1.001  1832
[... output truncated ...]
r[41]         0.078  0.011   0.058   0.077   0.101    FALSE 1.000 1.002  2152
r[42]         0.076  0.011   0.056   0.076   0.100    FALSE 1.000 1.002  2065
N[1,1]        8.257  7.172   1.000   6.000  26.025    FALSE 1.000 1.001  3000
N[2,1]        7.489  6.230   1.000   6.000  23.000    FALSE 1.000 1.004  1132
[... output truncated ...]
N[3,43]      24.739 12.885   4.000  23.000  53.000    FALSE 1.000 1.033    67
N[4,43]     216.995 16.523 184.000 217.000 249.000    FALSE 1.000 1.024    88
NB[1]        44.866  5.650  34.975  45.000  57.000    FALSE 1.000 1.000  3000
NB[2]        40.791  4.452  33.000  41.000  50.000    FALSE 1.000 1.001  1331
[... output truncated ...]
NB[42]      242.664 10.881 222.000 243.000 263.025    FALSE 1.000 1.000  3000
NB[43]      241.734 12.494 218.000 241.000 267.000    FALSE 1.000 1.000  3000
fit.y         7.623  1.272   5.411   7.542  10.459    FALSE 1.000 1.001  1935
fitN.y        9.571  1.475   7.031   9.419  12.804    FALSE 1.000 1.000  3000
fit.DR       83.500  4.303  75.718  83.238  92.400    FALSE 1.000 1.000  3000
fitN.DR      79.904  6.026  68.750  79.715  92.583    FALSE 1.000 1.002  1186
fit.J        21.533  4.343  13.864  21.347  30.728    FALSE 1.000 1.000  2079
fitN.J       21.365  4.645  13.245  21.051  31.076    FALSE 1.000 1.002  1062
```

The chains of IPM_2 also have converged with the higher number of MCMC draws that we specified. Here is a partial summary from IPM_2 with the autoregressive structures in the time series of the demographic rates.

```
print(out2, 3)

               mean     sd    2.5%     50%   97.5% overlap0     f  Rhat n.eff
mean.alpha    0.396  0.232   0.023   0.379   0.886    FALSE 1.000 1.003   568
beta[1]      -1.683  0.354  -2.412  -1.675  -0.995    FALSE 1.000 1.000  3000
beta[2]      -0.020  0.011  -0.041  -0.020   0.003     TRUE 0.956 1.000  3000
s[1,1]        0.579  0.136   0.253   0.592   0.826    FALSE 1.000 1.005   401
s[2,1]        0.615  0.091   0.436   0.616   0.784    FALSE 1.000 1.004   648
[... output truncated ...]
s[1,42]       0.587  0.087   0.416   0.585   0.783    FALSE 1.000 1.000  3000
s[2,42]       0.782  0.033   0.714   0.782   0.846    FALSE 1.000 1.000  3000
rho[1]        1.139  0.154   0.866   1.130   1.465    FALSE 1.000 1.000  3000
rho[2]        1.167  0.130   0.929   1.161   1.430    FALSE 1.000 1.000  3000
[... output truncated ...]
rho[42]       1.122  0.067   0.994   1.122   1.251    FALSE 1.000 1.000  3000
rho[43]       1.271  0.076   1.127   1.269   1.425    FALSE 1.000 1.000  2300
r[1]          0.159  0.045   0.083   0.155   0.261    FALSE 1.000 1.000  3000
r[2]          0.156  0.043   0.083   0.152   0.253    FALSE 1.000 1.000  3000
[... output truncated ...]
```

```
r[41]          0.077  0.011    0.058    0.077    0.099    FALSE 1.000 1.000  3000
r[42]          0.076  0.011    0.056    0.075    0.099    FALSE 1.000 1.000  3000
sigma.s[1]     0.164  0.154    0.006    0.117    0.581    FALSE 1.000 1.002  1579
sigma.s[2]     0.177  0.080    0.025    0.173    0.354    FALSE 1.000 1.001  3000
sigma.rho      0.174  0.031    0.120    0.172    0.238    FALSE 1.000 1.000  3000
N[1,1]        23.868 18.416    1.000   19.000   65.000    FALSE 1.000 1.001  1499
N[2,1]        17.664 14.066    1.000   14.000   53.000    FALSE 1.000 1.000  3000
[... output truncated ...]

N[3,43]       22.841 14.379    1.000   22.000   54.000    FALSE 1.000 1.002   727
N[4,43]      218.163 18.226  182.000  218.000  254.000    FALSE 1.000 1.003   703
NB[1]         49.149  6.600   37.000   49.000   63.000    FALSE 1.000 1.000  3000
NB[2]         43.923  5.349   34.000   44.000   55.000    FALSE 1.000 1.001  1598
[... output truncated ...]
NB[42]       240.518 10.260  221.000  240.000  261.000    FALSE 1.000 1.001  2142
NB[43]       241.003 12.953  216.000  241.000  267.000    FALSE 1.000 1.000  3000
fit.y          6.866  1.041    5.059    6.757    9.172    FALSE 1.000 1.000  3000
fitN.y         9.696  1.463    7.160    9.615   12.894    FALSE 1.000 1.000  3000
fit.DR        88.268  3.838   81.488   88.097   96.127    FALSE 1.000 1.001  2423
fitN.DR       81.585  6.074   70.321   81.355   93.988    FALSE 1.000 1.000  3000
fit.J         22.889  4.837   14.527   22.505   33.610    FALSE 1.000 1.001  3000
fitN.J        21.520  4.667   13.436   21.183   31.861    FALSE 1.000 1.002  1181
```

The three GoF components under IPM_1 reveal Bayesian *p*-values (*p*) that are neither too close to 0 nor too close to 1: state-space model, $p = 0.84$; dead-recovery model, $p = 0.25$; and Poisson regression, $p = 0.49$. This is similar also for IPM_2, except for the state-space (and perhaps also the dead-recovery) model, where there may be some room for improvement: state-space model, $p = 0.94$; dead-recovery model, $p = 0.10$; and Poisson regression, $p = 0.42$. We conclude that both models are an adequate representation of the main patterns in all three data sets.

Therefore, we now proceed to inspecting the estimates under IPM_1. At this stage, it may be interesting to inspect the long-term means of the demographic rates. As expected, first-year survival was substantially lower (0.59, SD: 0.049) than adult survival (0.78, SD: 0.016). Productivity per breeding pair was 1.42 (SD: 0.047) fledglings, and the probability to reproduce when 2 years old was estimated at 0.51 (SD: 0.191). We note that this parameter has a large uncertainty associated with it, which is not atypical for a hidden parameter. Finally, the dead-recovery probability was estimated to decline from 0.16 (SD: 0.044) in 1965 to 0.08 (SD: 0.011) in year 2007.

We are interested in the temporal pattern of the demographic rates, and we compare the patterns obtained by the two models (Fig. 12.4). The estimated trajectory of first-year survival under IPM_2 is much smoother than that under IPM_1. In addition, it is more precise. Neither shows very striking temporal patterns, although both seem to indicate an increase during the first 10 years of the study followed by a very slow decrease over the remaining years. For adult survival, there were more substantial differences between the two models in the estimated temporal pattern and the means in the first part of the study period, while later (with much more data), they were almost identical. Under both models, adult survival tended to be lower in the first half than in the second half of the study period and was more pronounced in IPM_2. Under that model, much more clearly than for first-year survival, there was an increase in adult survival during the first 10 years of study, from about 0.60 to more than 0.80.

Estimates of adult survival under IPM_2 were less precise at the beginning of the time series than under IPM_1, but not later. The autoregressive model requires an "independent" estimate of adult survival in the first year that is difficult to estimate with the very small sample sizes in our study during the 1960s. There is still little information in year 2, but this estimate also uses information from year 1, so there is a decrease in uncertainty over time. Once there are more data, the estimates and uncertainty converge to those under IPM_1. Under IPM_1, there is of course the same small amount of information in the first part of the study period, but in this model, the annual estimates are shrunk toward the overall mean, which is dominated by the data from the latter half of the study period—the smaller the dead-recovery sample size is, the stronger the shrinkage toward that mean. Thus, we see nicely in the adult survival parameter the different behavior of the two models when there are sparse data. Had the small sample size years occurred toward the end of the study period, we would expect smaller differences between the estimates of the two models due to the serial sharing of information in the Markovian model (i.e., IPM_2 with accommodation of serial autocorrelation).

The dead-recovery probabilities were estimated at nearly identical values under both models. They declined over time as expected and were naturally more imprecise at the beginning of the study, when the number of ringed nestlings was very low.

Finally, productivity is based on a large amount of data, and there is hardly any difference between the estimates of the two IPMs. Productivity increased until about 1980 and afterward declined again. We would interpret the increase as being due to the declining effects of the pesticides. The decline afterward may be the result of density dependence due to direct interference or to an increasing proportion of the population that inhabits less than optimal nesting cliffs. It may also be due

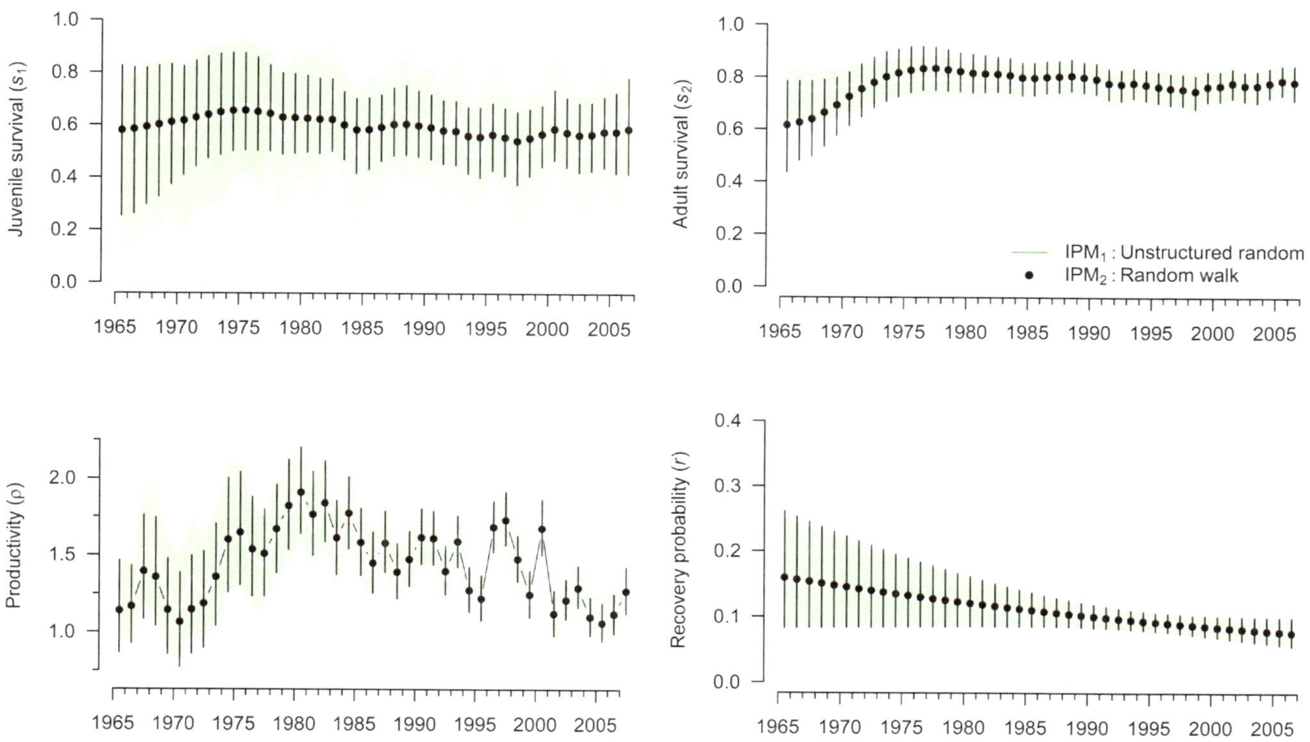

FIGURE 12.4

Estimates of survival, productivity, and recovery probability of peregrines under integrated population models (IPMs) with unstructured temporal random effects (IPM$_1$) and autoregressive structures for smoothing (random walk, IPM$_2$). The green areas show the 95% credible intervals and the white line the posterior means under IPM$_1$. The black dots and vertical lines show posterior means and the 95% credible intervals under IPM$_2$.

to increasing effects of the main natural predator of the peregrine, the eagle owl, *Bubo bubo* (Kéry et al., 2018). There were large annual fluctuations in productivity, presumably due in part to the negative effects of spring precipitation on the survival of small chicks in the nest.

The development of the peregrine population was spectacularly dynamic: the number of breeding pairs declined in the 1960s, reached an absolute minimum in year 1972, and increased thereafter (Fig. 12.5). The population counts agreed well with the estimated number of breeding pairs under both IPMs, and the estimates under both models are nearly identical. The increase was steeper in the two decades after 1972 and then leveled off somewhat after about 1995. Again, this might be due to density-dependent mechanisms, and the methods introduced in Chapter 8 could be used here to measure the evidence of density dependence in a quantitative way and try to identify the demographic mechanisms most affected by it.

Our second goal in this analysis was to assess the size of the population segment of the floaters, and in general, of all four stage classes that we distinguish in our population model. We don't do model selection between our two models, so rather than claiming that one was better than the other, we simply average their estimates in an unweighted manner (Dormann et al., 2018). We can do this by merging the MCMC draws from both models for each parameter and then summarizing them in the usual way by taking the mean and so forth. Note that we have chosen the MCMC settings above such that the same nominal sample sizes of 3000 resulted with every parameter in both models. In addition, we aimed to achieve broadly similar effective sample sizes to ensure that both models received about the same weight in this average.

```
# Estimate age/stage composition of population averaged over both models
str(out1$sims.list$N)                            # Draws of N from model 1
str(out2$sims.list$N)                            # Draws of N from model 2
Ndraws <- array(NA, dim=c(6000, 4, 43))          # Create empty array
Ndraws[1:3000,,] <- out1$sims.list$N             # Fill in draws from IPM1
Ndraws[3001:6000,,] <- out2$sims.list$N          #... and from IPM2
stage.compl <- apply(Ndraws, c(2,3), mean)       # Compute posterior mean
dimnames(stage.compl)<- list(c("N1 (Nonbr.)", "N2 (Nonbr.)", "N3 (Br.)", "N4 (Br)"), 1965:2007)
```

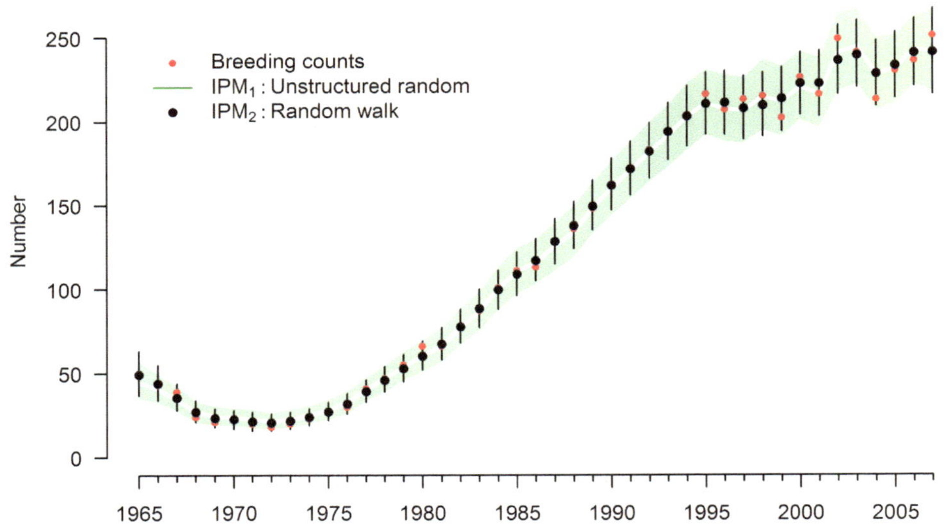

FIGURE 12.5

Estimates of the number of breeding female peregrines in the Jura mountains under two IPMs. The observed counts of breeders are also shown. The black dots and the white line are posterior means, and the vertical lines and the green polygon show 95% credible intervals.

Then, we combine two each of these four stage classes to obtain model-averaged estimates (posterior means) of the number of breeders and floaters in the Jura peregrines. We also add up the total population size and compute the proportion of floaters in the population.

```
floaters <- apply(apply(Ndraws[,1:2,], c(1,3), sum), 2, mean)
breeders <- apply(apply(Ndraws[,3:4,], c(1,3), sum), 2, mean)
stage.comp2 <- rbind(floaters, breeders)
dimnames(stage.comp2)<- list(c("Floaters", "Breeders"), 1965:2007)
# Total population size
all <- apply(apply(Ndraws, c(1,3), sum), 2, mean)
# Number of floaters
prop.floaters <- (Ndraws[,1,] + Ndraws[,2,]) / apply(Ndraws, c(1,3), sum)

round(range(all))
[1]  33 374
```

We see again the decline in the total population, which reached its lowest ebb in 1972 with an estimated 33 females in total, and then the recovery during the next 30 years, which reached its maximum in 2001 with an estimated 374 females (Fig. 12.6). Assuming an even sex ratio, this means that the total peregrine population amounted to 748 individuals in that year. Most females in the population are adult breeders (N_4), followed by first-year individuals (N_1), 2-year-old breeders (N_3), and 2-year-old nonbreeders (N_2, Fig. 12.6, top). The composition in terms of breeding and floating females is shown in the middle panel of Fig. 12.6. The proportion of floaters was estimated at around 30% in the mid-1960s and then increased to more than 40% 10–15 years later (around 1980) and then slowly declined again to about 30% after 2000 (Fig. 12.6, bottom).

The ability to also estimate the size of the unobserved parts of a population is a fascinating feature of IPMs and other structured population models—see Sections 6.5 and 11.6. However, if we were to inspect the uncertainty of the total population size, we would notice that it is much larger than that for the number of breeders only. This is to be expected, as the number of breeders is directly observed, while the nonbreeders are essentially invisible (and hence there are no data about them), and therefore the associated estimates will be more uncertain. This uncertainty is then propagated into estimates of the total number of individuals in the population.

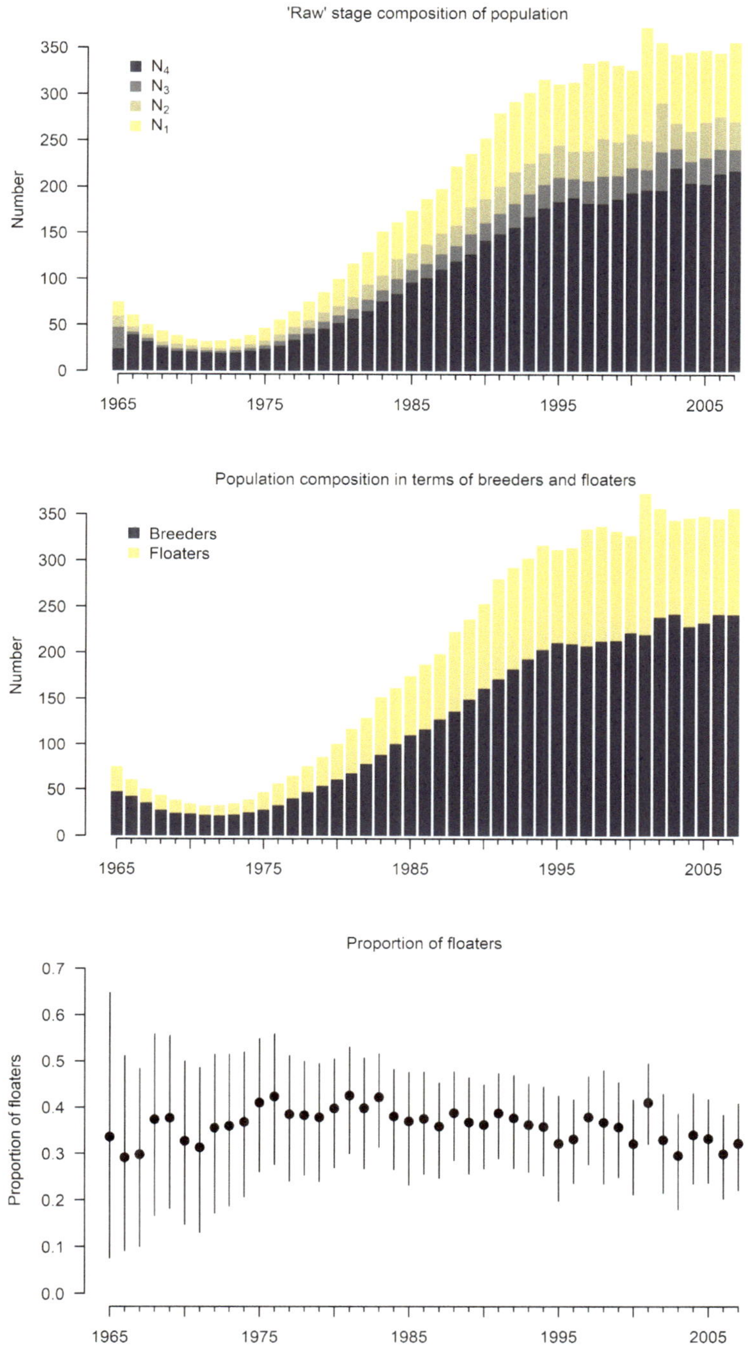

FIGURE 12.6

Estimated raw stage composition of the peregrine population in the Jura mountains (top; see text for stage definitions), population composition in terms of breeders and floaters (middle), and the proportion of floaters (bottom, dots are posterior means, vertical lines 95% credible intervals). Numbers in the first two plots refer to females only.

12.7 DISCUSSION

We have fit two IPMs with the classical three-way combination of counts, productivity, and survival data. For the latter, we had dead-recovery data, and since this is the only case study with this data type, we have spent a little more time commenting on some features of this useful source of information about the survival of wild animals. We have analyzed the population dynamics of a conservation flagship—the peregrine falcon. It became an icon of conservation in the 1960s when its populations crashed in many parts of the world due to unforeseen side-effects of pesticides such as the infamous DDT, but also others, that caused many bird- and fish-eating raptors to lay thin-shelled eggs that broke during incubation (Newton, 1979). The resulting reproductive failure caused their populations to crash and even to go regionally extinct, but after the use of these pesticides was banned, most populations bounced back to pre-DDT levels within the next 30–50 years. In this study, we were particularly interested in gauging a possible effect of changes in survival in this crash and in population composition in terms of breeders and "floaters" (i.e., nonbreeders). The latter typically cannot be surveyed in wild populations of most species, and yet they can be very important for the dynamics of long-lived bird populations. In addition, we have illustrated how time series of parameters can be smoothed in an autoregressive fashion and seen how this helps in detecting finer-scale temporal patterns.

We fit two IPMs to 43 years of data from two geographically adjacent peregrine population studies in the Jura mountains: one with unstructured annual random effects in most parameters (IPM_1) and another where we smoothed these trajectories with a random walk (IPM_2; as in Chapter 13 in Link and Barker, 2010). Both models fit the data adequately and were able to retrace the observed number of territorial pairs with the end of the crash until the early 1970s, recovery until about 2000, and approximate stability or only weak further increases thereafter (but see Kéry et al, 2021, for widespread postrecovery declines in this population). Interestingly, IPM_1, with unstructured temporal random effects, had strong shrinkage toward the overall means of all parameters and did not enable us to discern much temporal structure in the survival parameters. In contrast, IPM_2, with the random-walk smoother, more clearly showed some interesting temporal patterns in survival.

Although uncertainty around the point estimates was considerable, both juvenile and adult survival appeared to increase from the start of our study period (1965) until the middle of the 1970s with a very slow, drawn-out decline through all the remaining years of the study. Thus, it seems that we have some evidence for a contribution of survival in the recovery of a peregrine population from the DDT crash. Our data do not extend sufficiently far back in time, but clearly, this seems to imply that a reduction in survival, in addition to well-known decreased productivity, was responsible for the DDT crash in the first place.

One really basic, but exciting and practically very useful, asset of any stage-structured population model is that it enables us to obtain an accounting of the population in terms of the age (or other) classes distinguished in the model. In our IPMs with four classes in total, we found that birds 3 years and older were most numerous, followed by 1-year-old birds and finally followed in about equal numbers by 2-year-old breeders and nonbreeders (the proportion of 2-year-olds that started to reproduce was estimated at about 50%). The proportion of floaters, i.e., nonbreeding individuals 1 and 2 years old, was estimated to increase strongly from about 30% at the start (1965) to more than 40% 10 years later and thereafter to decline again. This suggests that a high proportion of floaters is a sign of a healthy population of long-lived raptors.

The estimates of the population structure are always conditional on the underlying population model. An important parameter in our peregrine population is the probability that the 2-year-old females start to reproduce (α) and the assumption that all females reproduce when they are 3 years old or older. We allowed no temporal variability in these parameters because of data limitations, but this might not have been true, and the estimated population structure might have been different had we been able to allow for yearly variation in α.

Our models assume that there is no immigration to our study population. If this assumption is violated, the estimate of the hidden parameter α would become biased. Or perhaps bias may not be the right term, as in fact the label that we give to the hidden parameter would not be correct in that case—it then would not measure the breeding propensity of 2-year-old individuals alone. In many IPMs that use capture-recapture data as a source of information about survival, an immigration parameter has to be included in the population model because the estimated apparent survival already accounts for emigrants. Here we use dead-recovery data that provide estimates of true survival. Hence, if there were emigrants from our study population, they would not automatically be accounted for in the population model. If we are unsure about whether emigration or immigration occurs, we can formulate models that estimate "net dispersal" (Fay et al., 2019). "Net dispersal" is the difference between the number of immigrants and emigrants and is a hidden parameter. This number might be negative or positive, and the prior must be chosen accordingly. For the peregrine population models, the estimation of "net dispersal" was not possible, because the IPMs already included one hidden parameter (α) and we generally cannot estimate two of them (Section 6.3). Yet again, if there was emigration and immigration in the peregrine population, this would be soaked up by α, which would then get a different interpretation. It is then better regarded as a "correction term" that accounts for multiple demographic processes that are not explicitly included in the population model (Robinson et al., 2014).

In addition, we reemphasize that demographic inferences about the numbers of floaters, or "(near-)invisible life stages" in general in populations, especially of long-lived species, are almost impossible to obtain without stage-structured population models, including IPMs. Importantly, we could have estimated the *entire age-structure of the population* by extending the number of age stages to, say, 25 (the maximum recorded life span of wild peregrines is about 21) but without necessarily estimating separate demographic parameters for each age class. Then we would have obtained estimates of how many birds there are in *every single age class*; see Margalida et al. (2020) for a neat example of such inferences on age-classified population sizes for a far more long-lived species, the bearded vulture (*Gypaetus barbatus*).

The analyses in this case study are just one example of a "demographic accountant's study" for one raptor population study (or actually, two). Worldwide, there must be hundreds of studies such as these peregrine population studies (albeit perhaps rarely *that* long), especially for, but of course not restricted to, raptor and owl species. We surmise that for the peregrine alone, there must be several dozen long-term population studies globally that have collected data similar to those of the Jura mountains. We believe that they all merit some form of demographic analysis such as the one in this chapter to achieve a much better quantitative understanding of their population dynamics, to discover scores of cool things, and hopefully to become better able to manage and conserve these populations.

DEDICATION

We dedicate this chapter to the dean of Swiss peregrinology, Gabriel ("Gaby") Banderet, who passed away on 23 August 2020.

CHAPTER

HORSESHOE BAT

13

CHAPTER OUTLINE

IPM ingredients: juvenile and adult population count data, capture-recapture data
Location: Switzerland
Spatio-temporal scale: a single colony, 29 years

13.1 INTRODUCTION

Bats are a speciose mammal family with a particular life history—relative to their body size, they are extremely long-lived mammals (Austad and Fischer, 1991). Studies on bat population dynamics used to be rare, but their prevalence has increased in recent years (e.g., Frick et al., 2007; Schorcht et al., 2009; Hayman et al., 2012; Papadatou et al., 2012; Giavi et al. 2014; Lentini et al. 2016; Arlettaz et al. 2017; Culina et al., 2019; Jan et al., 2019). The majority of demographic studies have focused on the estimation of survival from marked individuals, whereas other demographic parameters such as age at first reproduction or productivity have been investigated more rarely. In most bat species, reproducing females give birth to one offspring only, but in rare cases, twins are produced. The estimation of productivity can therefore often be simplified to estimation of the frequency with which females reproduce, i.e., to breeding propensity. For reproduction, most bats gather in nursery colonies, which contain reproducing females, often variable numbers of nonreproducing females, and depending on the species, also males. Therefore, the ratio of the young produced and the number of free-flying bats in a nursery colony is typically not a useful estimator of productivity. As a result, the estimation of productivity in bats is often a challenge.

Here, we analyze three demographic data sets on the endangered greater horseshoe bat (*Rhinolophus ferrumequinum*, Fig. 13.1) collected at a nursery colony located in a small church in the Swiss Alps. This species is one of the largest bats of Switzerland and mainly feeds on beetles and moths. Only two large nursery colonies are known in Switzerland. They are far apart from each other, and it is therefore fair to assume the absence of any dispersal between them. Greater horseshoe bats raise their young in nursery colonies that mostly consist of reproducing females, but some nonreproducing females and males may also be present. Some individuals may be temporarily absent from the nursery colony and use other roosts assumed to be located in caves. Hibernation also occurs in caves, but our knowledge about the behavior and ecology of the individuals when they are absent from the nursery colony is very limited. The data collected annually at the nursery colony consist of counts of bats emerging from the colony at dusk at the start of the nocturnal hunting period, counts of newborn juveniles, and capture-recapture data of adults. A shorter version of this data set was already analyzed with an integrated population model (IPM) by Schaub et al. (2007). The goal of this case study is estimation of the demographic rates and population structure as well as investigation of possible changes in both over time. Because sexing and aging of bats is impossible unless they are captured, the fraction of reproducing bats remains unknown, even if the total number of

Integrated Population Models. https://doi.org/10.1016/B978-0-12-820564-8.00015-4

FIGURE 13.1

Greater horseshoe bats (*Rhinolophus ferrumequinum*) in a wintering cave (photograph by Dietmar Nill).

emerging individuals can be counted quite accurately. Hence, any changes in these numbers need not necessarily reflect a change in the number of reproducing females. Application of an IPM permits insights into the population structure and hence estimation of the total population (i.e., the superpopulation size, which includes any individuals temporarily absent), not only the number of individuals present at the nursery colony.

From a more technical point of view, in this case study we show how to build a two-sex population model with accommodation in the analysis of temporally changing sampling protocols and a prior sensitivity analysis. Counts and recaptures were not conducted in all study years, and hence, missing data must be taken into account. Moreover, we include age-dependent productivity in our models. We carry out a model sensitivity analysis to gauge (1) whether age-dependence in productivity is informed by the data (i.e., is estimable) and (2) how different assumptions about productivity of 2-year-old females affect the estimates of other demographic rates and in general our inferences about the population dynamics of this bat species.

13.2 DATA SETS

Three data sets were collected by our colleagues Raphaël Arlettaz, François Biollaz, and Antoine Sierro for the 29 years from 1989 to 2017 in a small twelfth-century church in Vex, Valais, Switzerland (Schaub et al., 2007; Sierro et al., 2009). The first data set contains the counts of individuals leaving the nursery colony at dusk at the start of the nightly hunting period, which we call "population counts." These counts are carried out in early June before females give birth and thus refer to individuals that are at least 1 year old. These population counts are assumed to be quite accurate because individuals leave the church one after another. However, the bats cannot be sexed and aged, and thus, no insights into the colony structure are possible based on these counts alone. A few days after the population counts and about 1 week before birthing, the church was visited during the day and the colony entrance blocked. All adult bats present were captured with a handheld net. All individuals captured were sexed, age-classified (1 year old vs. >1 year old), and individually marked with a bat ring, while previously marked individuals were recorded. Adult capture was conducted for 11 years starting in 2004. In 2015, adult capture was carried out, but in contrast to all earlier years, it was incomplete because some bats moved to the clock tower where they could not be captured. The third data set, referred to as "juvenile count," was collected during a nocturnal visit to the nursery colony about 1–2 weeks after parturition when the adults had left the colony for hunting. The newborn juveniles were unattended and could be counted, sexed, and marked with a ring. In most years all juveniles were marked, but in a few years, some escaped capture. Combining data sets 2 and 3, a total of 574 individuals were marked: 279 juvenile females, 9 1-year-old females, 11 >1-year-old females, 269 juvenile males, 6 1-year-old males, and no >1-year-old males. A total of 190 individuals were recaptured at least once. The oldest known individual was a female marked as a juvenile in 1991 and last recaptured in 2016 when she was 25 years old, which we find quite astonishing for an animal with the body mass of a lean house sparrow!

13.3 POPULATION MODEL

From the literature and previous studies (Ransome, 1995; Schaub et al., 2007), we know that most females start to reproduce at 3 years of age and only rarely earlier. Therefore, we define a population model with three age classes and choose a prebreeding "census" model to match the timing of the population surveys. Moreover, we need to include males in the population model because the population counts contain both sexes. Only when males are included do we obtain an adequate estimate of the population structure and demographic rates. We allow survival to be different for all three age classes and the two sexes. We also assume no dispersal to and from the study population; this assumption appears fairly sensible due to the geographic isolation of the colony. The life-cycle graph represented by our population model is shown in Fig. 13.2. The three age classes are for individuals that are 1 year old (1y), 2 years old (2y), and more than 2 years old (>2y).

The modeling of recruitment and hence estimation of the number of 1y individuals is the trickiest part in this model. Because we want to include demographic stochasticity, we decompose the recruitment process, i.e., the red arrows in the life-cycle graph (Fig. 13.2), into three sequential subprocesses. First, we model the total number of juveniles produced. We then allocate them to a sex, and finally we model their survival until they become 1 year old. Reproducing females give birth to a single offspring only. Any difference between the number of mature females and that of juveniles must therefore be due to mature females that skip breeding and to early mortality before juveniles are counted. From the literature, we know that typically a smaller proportion of 2y than >2y females reproduce. Therefore, we model the recruitment process differentially by defining $\rho_{1,t}$ to be the productivity of 2y females in year t and $\rho_{2,t}$ as the productivity of >2y females. The total number of juveniles produced by 2y and >2y females ($G_{1,t}$ and $G_{2,t}$, respectively) in year t is modeled with binomial distributions as

$$G_{1,t} \sim Binomial(\rho_{1,t}, N_{2,f,t}) \text{ and}$$
$$G_{2,t} \sim Binomial(\rho_{2,t}, N_{3,f,t}),$$

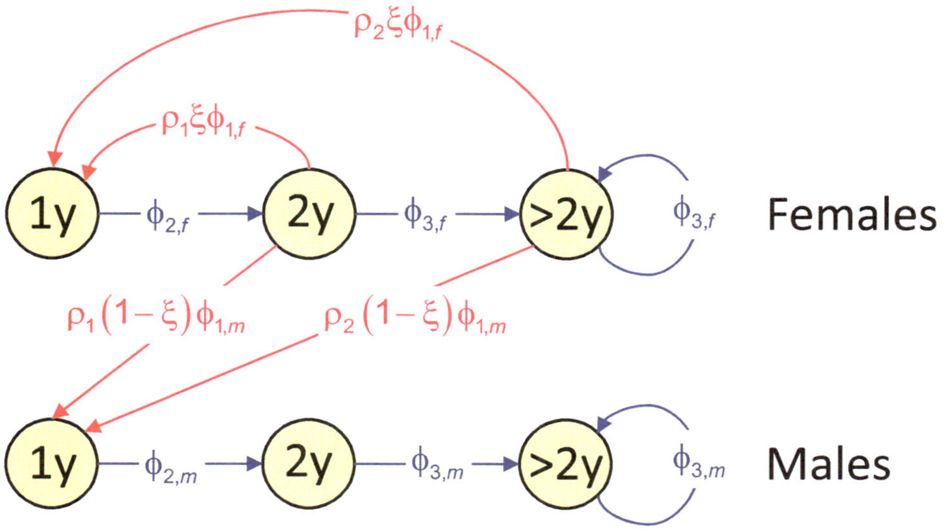

FIGURE 13.2

Life-cycle graph of greater horseshoe bats for a prebreeding "census" model that includes both sexes and three age classes for each; 1y, 2y, and >2y denote individuals that are 1 year old, 2 years old, and more than 2 years old, respectively. The demographic parameters are age- and sex-specific survival probabilities (ϕ), productivity (ρ), and the sex ratio of newborns (ξ). Red arrows indicate recruitment transitions, and blue arrows indicate pure survival transitions.

where $N_{2,f,t}$ is the number of 2y females in year t and $N_{3,f,t}$ the number of >2y females. The two productivity parameters are collected in ρ and mostly reflect the proportion of females of an age class that reproduce but only to a minor degree the juvenile mortality that takes place before the juveniles are counted.

The next subprocess for recruitment in the model is the allocation of juveniles to a sex. We define ξ to be the proportion of juvenile females and use a binomial distribution to allocate the number of juveniles to females in year t.

$$NB_{f,t} \sim Binomial(\xi, G_{1,t} + G_{2,t})$$

The number of juvenile males is obtained as a simple difference.

$$NB_{m,t} = G_{1,t} + G_{2,t} - NB_{f,t}$$

Finally, the juveniles of sex s in year t ($NB_{s,t}$) must survive for 1 year to become 1y individuals at $t+1$ ($N_{1,s,t+1}$). We model the third subprocess of recruitment using another binomial distribution with a success parameter, that is the sex- and time-specific juvenile survival probability ($\phi_{1,s,t}$).

$$N_{1,s,t+1} \sim Binomial(\phi_{1,s,t}, NB_{s,t})$$

The transitions of individuals within the other age classes over time are simply governed by survival, and hence we use even more binomial distributions. The number of 2y individuals of sex s in year $t+1$ ($N_{2,s,t+1}$) is

$$N_{2,s,t+1} \sim Binomial(\phi_{2,s,t}, N_{1,s,t}),$$

and that of >2y individuals ($N_{3,s,t+1}$) is

$$N_{3,s,t+1} \sim Binomial(\phi_{3,s,t}, N_{2,s,t} + N_{3,s,t}),$$

where $\phi_{2,s,t}$ is the annual survival probability of 1y individuals of sex s between t to $t+1$, and $\phi_{3,s,t}$ the annual survival probability of the 2y and >2y individuals of sex s between t and $t+1$ (adult survival).

13.4 SINGLE DATA LIKELIHOODS

We next describe all three data sets and define the likelihoods used for them. We load the data sets and produce an overview of their content.

```
# Load greater horseshoe bat data and produce data overview
library(IPMbook); library(jagsUI)
data(bats)
str(bats)

List of 7
 $ ch      : int [1:574, 1:29] 0 0 0 0 0 0 0 0 0 0 ...
 $ age     : num [1:574] 1 1 1 1 1 1 1 1 1 1 ...
 $ sex     : num [1:574] 2 2 2 1 2 1 1 2 2 1 ...
 $ J.count: num [1:29] 13 11 11 16 16 16 18 20 17 16 ...
 $ Jm      : num 293
 $ Jf      : num 289
 $ A.count: num [1:29] 27 33 NA 27 42 43 41 40 44 33 ...
```

The capture histories (`ch`) are collected in a 574 × 29 matrix, where the columns refer to the years 1989–2017. Two vectors, `sex` and `age`, indicate the sex (1: females; 2: males) and age (1: juvenile; 2: 1 year old; 3: at least 2 years old) at first capture of each of the 574 individuals within the matrix.

The second data set is the annual count of the juveniles (`J.count`). When the juveniles were counted and ringed, their sex was often, but not always, recorded. A total of 582 juveniles were sexed, and we can use their information to estimate the sex ratio: vectors `Jm` and `Jf` contain the numbers of sexed males and females among the counted juveniles.

Finally, the third data set is the population count of the adults. Vector `A.count` contains the annual number of adults counted when they were leaving the colony at dusk. In 1991 and 2001, no adult population count was conducted; therefore, this data set contains two missing values.

13.4.1 CAPTURE-RECAPTURE DATA

We want to use the capture-recapture data to estimate annual apparent, sex-specific survival for three age classes. Juvenile survival is defined as survival from the time when a juvenile is marked in year t until it is 1 year old in year $t+1$ ($\phi_{1,s,t}$), subadult survival as survival from 1 to 2 years of age ($\phi_{2,s,t}$), and adult survival as annual survival from 2 years of age onwards ($\phi_{3,s,t}$). We choose a state-space formulation of the Cormack-Jolly-Seber (CJS) model (Section 4.5.1.1). We must implement an age structure for the survival and recapture probabilities, so we define an age matrix $\boldsymbol{x}$ that indicates the age class at each recapture occasion for each individual. The following bit of code computes the age matrix $\boldsymbol{x}$ based on information about the occasion of first capture and age at first capture for each marked individual (note that this code is specific for our model with three age classes).

```
# Compute the age matrix x
f <- getFirst(bats$ch)
x <- matrix(NA, ncol=ncol(bats$ch)-1, nrow=nrow(bats$ch))
index <- cbind(1:nrow(bats$ch), f)
x[index] <- bats$age
for (t in 2:ncol(x)){
  x[is.na(x[,t]),t] <- x[is.na(x[,t]),t-1] + 1
}
x[x>3] <- 3
```

We assume time-dependent survival for each sex and age class. As survival of all individuals may be driven in part by the same environmental conditions, we adopt a model with temporally correlated random effects (Section 6.4) using a multivariate normal distribution.

$$\text{logit}(\boldsymbol{\phi}_{1...6,1...T}) \sim MVN(\boldsymbol{\mu}_{\phi}, \boldsymbol{\Sigma})$$

Here, $\boldsymbol{\phi}_{1...6,1...T}$ is a matrix with the survival probabilities for the $T = 28$ annual intervals and six sex-age class combinations, $\boldsymbol{\mu}_{\phi}$ is a vector with the mean survival of the six sex-age class combinations on the logit scale, and $\boldsymbol{\Sigma}$ is the corresponding variance-covariance matrix. The diagonal of $\boldsymbol{\Sigma}$ contains the temporal variance of the logit of survival, and the off-diagonals contain the covariances. There are several ways to implement this model in JAGS (see Section 6.4). Here, we use the JAGS function `dmnorm.vcov`, decompose $\boldsymbol{\Sigma}$ into temporal variances and correlations, and place vague priors on all of them.

While the ringing of young individuals was conducted in all years during 1989–2017 except 1990, adult recaptures are available for 11 years only. To accommodate this capture-recapture sampling design, we fix the recapture probability to zero in years without capture. We define vector $\boldsymbol{k}$ with length equal to the number of recapture occasions ($n = 28$) and use it as an explanatory variable in a linear model to fix to zero the recapture probability in some years. In this vector, we use the numbers 1 to 11 for the years with captures and label as 12 the years without recaptures.

```
k <- c(rep(12, 14), 1, 2, 3, 4, 5, 12, 12, 6, 7, 8, 12, 11, 9, 10)
```

In 2015, the entire population could not be marked because some smart individuals present in the church escaped capture. We therefore separately estimate recapture in 2015 from that in the other years, and to allow for that, label year 2015 as 11 in vector $\boldsymbol{k}$. The recapture probabilities of the 10 years with complete capture (labeled 1–10 in vector $\boldsymbol{k}$) were modeled with independent temporal random effects for each age and sex class.

$$\text{logit}(p_{s,a,k=\{1...10\}}) \sim Normal\left(\mu_{p_{s,a}}, \sigma^2_{p_{s,a}}\right)$$

$$p_{s,a,k=11} \sim Uniform(0, 1)$$

$$p_{s,a,k=12} = 0$$

In the model, we complete these with definitions of vague priors for the means ($\mu_{p_{s,a}}$) and the temporal variances ($\sigma^2_{p_{s,a}}$) of recapture probabilities.

13.4.2 JUVENILE AND POPULATION COUNT DATA

We have collected two count data sets, one for juveniles and one for adults. Therefore, we could say that we have age-structured counts. Alternatively, we could say we have one data set for productivity and another for population size as we have done previously.

We start with a probabilistic description of the juvenile counts. Newborn juveniles could not hide within the colony, and therefore, we assume their count was a census without any observation error. We thus treat the annual number of juveniles (J_t) as exactly the sum of the numbers of juveniles produced by 2y and >2y females (i.e., $J_t = G_{1,t} + G_{2,t}$) and for this use the function `dsum` in JAGS.

```
J[t] ~ dsum(G[1,t], G[2,t])
```

Only minor changes would be needed to accommodate observation errors for juveniles. For instance, we could write $J_t \sim Normal\left(G_{1,t} + G_{2,t}, \sigma_j^2\right)$, where σ_j^2 is the observation error associated the counts.

The observation model for the time series of the population counts requires some thought. There is the difficulty that the adult bats counted could not be sexed and aged. We know *a priori* that the proportions of the different age and sex classes present at the nursery colony differ and may vary over time. Hence, for each sex and age class and year, we need to know the colony attendance at the time when the count was conducted, and for this, we can use the estimated recapture probabilities. Generally, a recapture probability is the product of the probability to be available for capture (hence, in our case, the probability of colony attendance) and of being captured given availability (Kendall and Nichols, 2002; Schaub et al., 2004). For the latter in our case, we assume a value of 1 in most years; therefore, the estimated recapture probability will represent the probability of colony attendance. We only have information to estimate this parameter for 10 years but need it for all the study years. Therefore, we adopt a model with temporal random effects for the recapture probabilities so we can predict probabilities of presence for the remaining years.

Let us define $\kappa_{s,a,t}$ as the probability that an individual of sex s and age class a is present at the colony in year t. For years with a complete capture, κ equals the estimated recapture probability, and hence

$$\kappa_{s,a,k=\{1...10\}} = p_{s,a,k=\{1...10\}}.$$

For the other years, the probabilities of presence (on the logit scale) are assumed to be drawn from the following normal distribution.

$$\text{logit}\left(\kappa_{s,a,k=\{11,12\}}\right) \sim Normal\left(\mu_{p_{s,a}}, \sigma^2_{p_{s,a}}\right)$$

We can define the total number of bats present at the colony in year t (U_t) as

$$U_t = \sum_{s,a} N_{s,a,t}\kappa_{s,a,t}.$$

We adopt a normal distribution to model the population count in year t (C_t) as

$$C_t \sim Normal\left(U_t, \sigma_c^2\right),$$

where σ_c^2 is the "observation," or residual, error.

Finally, we can estimate the size of the total population, i.e., the superpopulation size, as $N_t^S = \sum_{s,a} N_{s,a,t}$.

13.5 THE INTEGRATED POPULATION MODELS

As always, the IPM contains the likelihoods of the different component data sets. We fit four IPMs that differ in their assumptions about the productivity of the 2y females, i.e., about parameter ρ_1. This parameter is difficult to estimate because we have no direct information about age-specific productivity. Hence, in all four models, we assume it is constant over time. IPM$_1$ assumes that 2y females do not reproduce ($\rho_1 = 0$), whereas IPM$_2$ assumes that they all reproduce successfully ($\rho_1 = 1$). Neither model is likely to reflect the truth, but importantly, they bracket all possible cases. The comparison of the other demographic rates and structured population sizes allows an assessment of how much our assumptions about the productivity of the 2y females affect our inferences about other quantities in the model. In IPM$_3$ and IPM$_4$, we attempt to estimate ρ_1. In IPM$_3$ we use a vague prior ($\rho_1 \sim Uniform(0, 1)$), whereas in IPM$_4$, we adopt a fairly strong, informative prior ($\rho_1 \sim Beta(3.87, 5.8)$) with a mean of 0.4 and variance of 0.15, which are educated guesses. Comparing the parameter estimates under IPM$_3$ and IPM$_4$ allows us to understand the effects of the prior, i.e., this is a prior sensitivity analysis. If the posteriors of ρ_1 are similar in IPM$_3$ and IPM$_4$ and different from their priors, we can conclude that the parameter is estimable from the available data.

The code of the four IPMs differs in just one line that defines the four assumptions about ρ_1. Therefore, we here show the code of just one of the models (IPM$_1$) but with the other three assumptions included and hashed out. In the data bundle, we add two further variables, the vectors `rpres` and `nrpres`, that indicate the years in which capture was complete and incomplete, respectively. They are needed for a proper matching of the recapture probabilities with `kappa`, the probability of colony attendance.

```
# Bundle data and produce a data overview
jags.data <- with(bats, list(ch=ch, f=getFirst(ch), nind=nrow(ch), n.occasions=ncol(ch), x=x,
  sex=sex, k=k, J=J.count, Jf=Jf, Jm=Jm, C=A.count, pinit1=dUnif(1, 50), pinit2=dUnif(1, 100),
  beta.priors=as.numeric(getBeta2Par(0.4, 0.15)), rpres=which(k<11), nrpres=which(k>10)))
```

```
str(jags.data)

List of 16
 $ ch         : int [1:574, 1:29] 0 0 0 0 0 0 0 0 0 0 ...
 $ f          : int [1:574] 4 4 4 4 4 4 4 4 4 4 ...
 $ nind       : int 574
 $ n.occasions: int 29
 $ x          : num [1:574, 1:28] NA NA NA NA NA NA NA NA NA NA ...
 $ sex        : num [1:574] 2 2 2 1 2 1 1 2 2 1 ...
 $ k          : num [1:28] 12 12 12 12 12 12 12 12 12 12 ...
 $ J          : num [1:29] 13 11 11 16 16 16 18 20 17 16 ...
 $ Jf         : num 289
 $ Jm         : num 293
 $ C          : num [1:29] 27 33 NA 27 42 43 41 40 44 33 ...
 $ pinit1     : num [1:50] 0.02 0.02 0.02 0.02 0.02 0.02 0.02 0.02 ...
 $ pinit2     : num [1:100] 0.01 0.01 0.01 0.01 0.01 0.01 0.01 0.01 ...
 $ beta.priors: num [1:2] 3.87 5.8
 $ rpres      : int [1:10] 15 16 17 18 19 22 23 24 27 28
 $ nrpres     : int [1:18] 1 2 3 4 5 6 7 8 9 10 ...
```

Here is the text file for the JAGS code of IPM_1.

```
# Write JAGS model file
cat(file="model1.txt", "
model {
# Priors and linear models
# Survival and recapture
for (i in 1:nind){
  for (t in f[i]:(n.occasions-1)){
    phi[i,t] <- sur[sex[i],x[i,t],t]
    p[i,t] <- pt[sex[i],x[i,t],k[t]]
  } #t
} #i

for (t in 1:(n.occasions-1)){
  logit(sur[1,1,t]) <- eps[1,t]
  logit(sur[1,2,t]) <- eps[2,t]
  logit(sur[1,3,t]) <- eps[3,t]
  logit(sur[2,1,t]) <- eps[4,t]
  logit(sur[2,2,t]) <- eps[5,t]
  logit(sur[2,3,t]) <- eps[6,t]
  eps[1:6,t] ~ dmnorm.vcov(mu.s[], Sigma[,])
}

# Recaptures
for (s in 1:2){
  for (a in 1:3){
    for (t in 1:10){
      logit.pt[s,a,t] ~ dnorm(mu.p[s,a], tau.p[s,a])
      pt[s,a,t] <- ilogit(logit.pt[s,a,t])
    } #t
    pt[s,a,11] ~ dunif(0, 1)              # rate in 2015, incomplete capture
    pt[s,a,12] <- 0                       # fix to 0 in years without capture
  } #a
} #s

mu.s[1] <- logit(mean.s[1,1])
mu.s[2] <- logit(mean.s[1,2])
mu.s[3] <- logit(mean.s[1,3])
mu.s[4] <- logit(mean.s[2,1])
mu.s[5] <- logit(mean.s[2,2])
mu.s[6] <- logit(mean.s[2,3])

for (j in 1:6){
  sd.s[j] ~ dunif(0.001, 10)              # Prior of the temporal SD
  Sigma[j,j] <- pow(sd.s[j], 2)
}

for (m in 1:5){
  for (j in (m+1):6){
    Sigma[m,j] <- zeta[m,j] * pow(Sigma[m,m] * Sigma[j,j], 0.5)
    Sigma[j,m] <- Sigma[m,j]
    zeta[m,j] ~ dunif(-1, 1)              # Priors for correlation coefficients
  } #j
} #m
```

```
for (s in 1:2){
  for (a in 1:3){
    mean.s[s,a] ~ dunif(0, 1)
    mean.p[s,a] ~ dunif(0, 1)
    mu.p[s,a] <- logit(mean.p[s,a])
    sigma.p[s,a] ~ dunif(0, 5)
    tau.p[s,a] <- pow(sigma.p[s,a],-2)
  } #a
} #s

# Productivity
for (t in 1:n.occasions){
  rho[1,t] <- mean.rho[1]                      # of 2y females
  logit(rho[2,t]) <- mu.rho + eps.rho[t]       # of >2y females
  eps.rho[t] ~ dnorm(0, tau.rho)
}
# Priors for productivity of 2y females
# To run IPMs 2-4, manually un-hash one line and hash out the others
mean.rho[1] <- 0
# mean.rho[1] <- 0.99                          # does not work if = 1 due to initial values
# mean.rho[1] ~ dunif(0, 1)
# mean.rho[1] ~ dbeta(beta.priors[1], beta.priors[2])

# Priors for productivity of >2y females
mean.rho[2] ~ dunif(0,1)
mu.rho <- logit(mean.rho[2])
sigma.rho ~ dunif(0, 10)
tau.rho <- pow(sigma.rho, -2)

# Variance of the observation error
sigma.y ~ dunif(0.5, 20)
tau.y <- pow(sigma.y, -2)

# Sex ratio of newborn individuals
xi ~ dbeta(Jf, Jm)

# Compute the proportion of bats present in the colony
for (s in 1:2){
  for (a in 1:3){
    lkappa[s,a,1] ~ dnorm(mu.p[s,a], tau.p[s,a])
    kappa[s,a,1] <- ilogit(lkappa[s,a,1])
    for (t in 1:length(nrpres)){
      lkappa[s,a,nrpres[t]+1] ~ dnorm(mu.p[s,a], tau.p[s,a])
      kappa[s,a,nrpres[t]+1] <- ilogit(lkappa[s,a,nrpres[t]+1])
    } #t
    for (t in 1:length(rpres)){
      kappa[s,a,rpres[t]+1] <- pt[s,a,t]
    } #t
  } #a
} #s

# Population count data (state-space model)
# Model for the initial population size: uniform priors
N[1,1,1] ~ dcat(pinit1)
N[1,2,1] ~ dcat(pinit1)
N[1,3,1] ~ dcat(pinit2)
N[2,1,1] ~ dcat(pinit1)
N[2,2,1] ~ dcat(pinit1)
N[2,3,1] ~ dcat(pinit2)

# Process model: population sizes of females and males
for(t in 2:n.occasions){
  N[1,1,t] ~ dbin(sur[1,1,t-1], NB[1,t-1])
  N[1,2,t] ~ dbin(sur[1,2,t-1], N[1,1,t-1])
  N[1,3,t] ~ dbin(sur[1,3,t-1], N[1,2,t-1] + N[1,3,t-1])
  N[2,1,t] ~ dbin(sur[2,1,t-1], NB[2,t-1])
  N[2,2,t] ~ dbin(sur[2,2,t-1], N[2,1,t-1])
  N[2,3,t] ~ dbin(sur[2,3,t-1], N[2,2,t-1] + N[2,3,t-1])
}

# Observation model
for (t in 1:n.occasions){
  Ns[t] <- sum(N[,,t])                         # total population size
  U[t] <- N[1,1,t] * kappa[1,1,t] + N[1,2,t] * kappa[1,2,t] + N[1,3,t] * kappa[1,3,t] + N[2,1,t] *
    kappa[2,1,t]  + N[2,2,t] * kappa[2,2,t] + N[2,3,t] * kappa[2,3,t]
  C[t] ~ dnorm(U[t], tau.y)
}
```

```
# Capture-recapture data (state-space model)
for (i in 1:nind){
  # Define latent state at first capture
  z[i,f[i]] <- 1
  for (t in (f[i]+1):n.occasions){
    # State process
    z[i,t] ~ dbern(phi[i,t-1] * z[i,t-1])
    # Observation process
    ch[i,t] ~ dbern(p[i,t-1] * z[i,t])
  } #t
} #i

# Newborn count data (binomial models)
for (t in 1:n.occasions){
  G[1,t] ~ dbin(rho[1,t], N[1,2,t])     # Newborn by 2y females
  G[2,t] ~ dbin(rho[2,t], N[1,3,t])     # Newborn by >2y females
  J[t] ~ dsum(G[1,t], G[2,t])           # Total newborn
  NB[1,t] ~ dbin(xi, J[t])              # Newborn females
  NB[2,t] <- J[t] - NB[1,t]             # Newborn males
}
}
")
```

Defining adequate initial values is important in this example, especially for all latent variables. First, we provide initial values for the latent states z in the capture-recapture model (see Section 4.5.1.1 for a discussion). Next, we provide initial values for the number of adult females. We specify 60 females in each year, a value that we determined by trial and error. Finally, we give initials for the number of juveniles produced by 2y and >2y females. We initialize the model such that 2y females do not reproduce and >2y females produce all juveniles counted.

In JAGS, providing initial values that work for the latent states, in particular for the structured population sizes, can sometimes be extremely difficult. If the initial values contradict the model, data, or initial values of other parameters, JAGS will throw an error such as the dreaded and far too common "*node inconsistent with parents*."

Here are a number of recommendations that might help you find working initial values:

- Use reasonable values for the initial stage-specific population sizes. These can be obtained by calculating the product of the counts and stable-state distribution, which you can obtain from a matrix population model parameterized with sensibly chosen demographic rates (see Section 3.3.1).
- Do not always provide initial values for the population size of all stages. Often, it is more important to give initial values for the older age classes.
- For latent population sizes, try out values larger than what you think they may be, and sometimes much larger. For instance, in our current example, we used 60 for the >2y females although we knew their true number must be less.
- Give high initial values for survival. This ensures that the majority of individuals survive at the initial stages of Markov chain Monte Carlo and can avoid a common problem when the number of survivors by chance exceeds the trial size in a binomial.

Sometimes, you may find it almost impossible to identify initial values that work for JAGS. For example, in IPM_2 we assume all 2y females reproduce successfully ($\rho_1 = 1$) and fail to find working initial values. The difficulty is that the number of juveniles produced by the 2y females becomes deterministic (fixed) if $\rho_1 = 1$. The observed number of juveniles, which is the sum of the contributions from 2y and >2y females, is also known without error. Consequently, the number of juveniles produced by older females is fixed and must be equal to or lower than the number of >2y females. Finding initial values that respect all these constraints is very hard. After much unsuccessful trial and error, we simply changed the assumption about the productivity of 2y females to $\rho_1 = 0.99$. This value is still very close to 1, but now the generation of G_1 becomes stochastic and thus has the flexibility we need for different combinations of initial values. The following initial values worked for all four IPMs.

```
# Initial values
Ninit <- array(NA, dim=c(2, 3, 29))
Ninit[1,3,] <- 60
G <- matrix(0, nrow=2, ncol=29)
G[2,] <- bats$J.count
inits <- function(){list(z=zInit(bats$ch), N=Ninit, G=G)}

# Parameters monitored
parameters <- c("mean.s", "Sigma", "mean.rho", "sigma.rho", "xi", "zeta", "mean.p", "sigma.p", "sigma.y",
  "sur", "pt", "rho", "kappa", "N", "U", "Ns", "NB", "G")
```

```
# MCMC settings
ni <- 500000; nb <- 200000; nc <- 3; nt <- 100; na <- 5000

# Call JAGS (ART 600 min) and check convergence
out1 <- jags(jags.data, inits, parameters, "model1.txt", n.iter=ni, n.burnin=nb, n.chains=nc,
  n.thin=nt, n.adapt=na, parallel=TRUE)
traceplot(out1)
```

13.6 RESULTS

All four models converged well (we don't show the output here). We first inspect parameter estimates from IPM_1 and then see how much the inferences differ among the four models. For both males and females, adult survival was clearly greater than survival of juveniles and 1y individuals (Fig. 13.3). Survival of adult males showed two lows, in 2005 and 2014. Females of all age classes were usually present at the nursery colony, while males, in particular the >2y males, were present less regularly. The number of females increased over time, whereas any such trend was less clear for the males due to high uncertainty. The estimates for the males were generally less precise than those for the females, as the recapture probabilities of males were lower than those of the females. Remember that in the majority of the years, no recaptures were conducted. As so often occurs (e.g., see Chapter 11), population size estimates for the first year of study were particularly imprecise because they cannot be informed by population size in earlier years in a Markov model.

The productivity of adult females was high and fairly constant over time, but the estimates were rather imprecise (Fig. 13.4). The estimated total population size of the nursery colony agreed very well with the population counts as we usually find in IPMs. The size of the nursery colony crashed in year 2013. The estimates of sex-specific survival

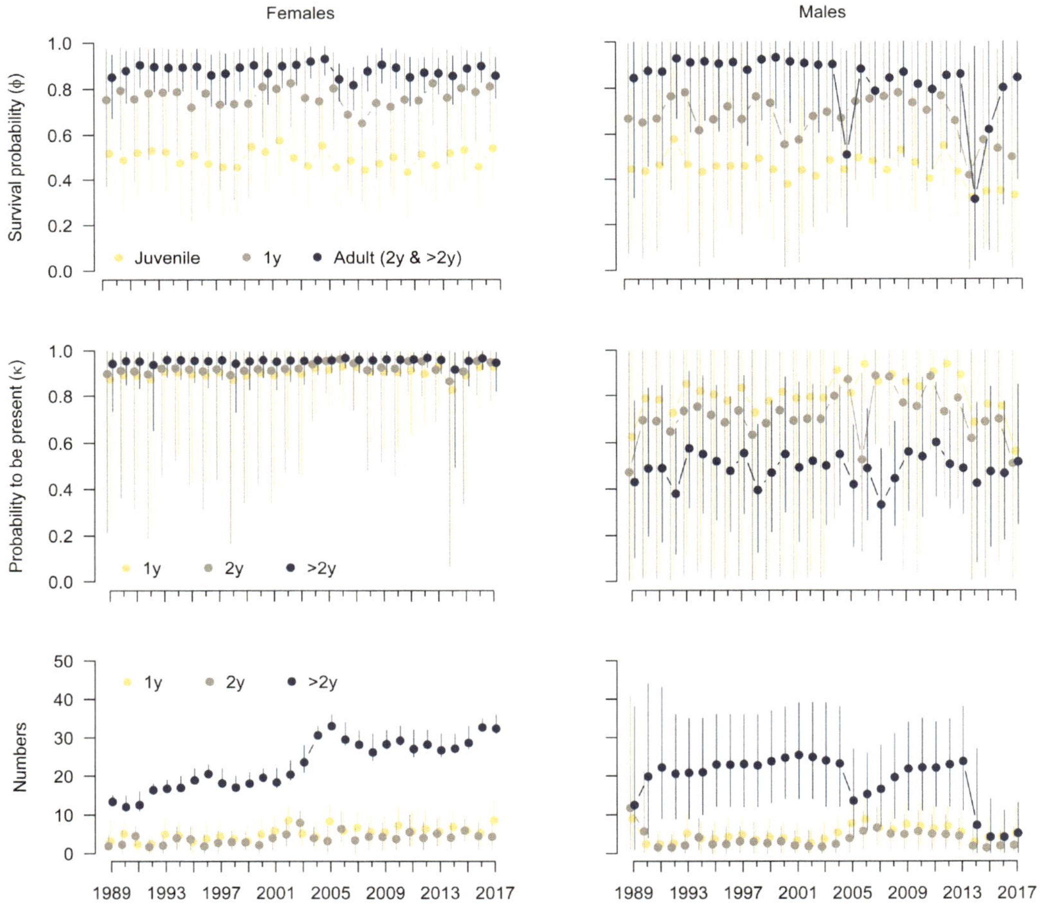

FIGURE 13.3

Estimates of age-specific probabilities of survival and presence at the nursery colony (top and middle) as well as the age-classified population sizes (bottom) of greater horseshoe bats in a colony in Vex/Valais under IPM_1. Left panels show females, the right panels show males, dots show posterior means and the vertical lines give the 95% credible intervals.

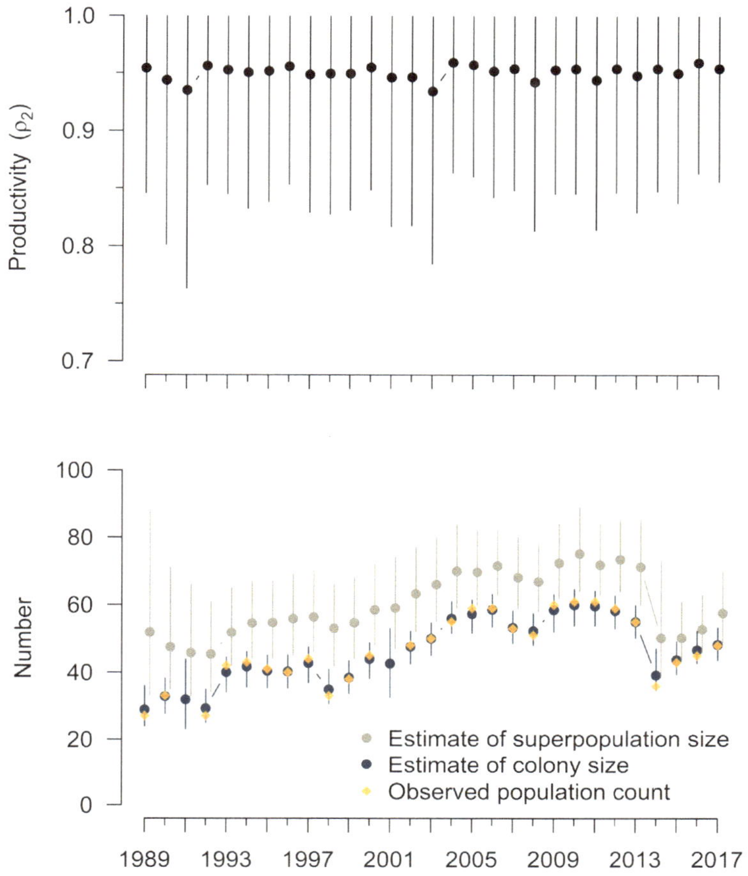

FIGURE 13.4

Estimates of the productivity of >2y female horseshoe bats (upper panel) and colony population and superpopulation sizes compared with observed population counts (lower panel). Estimates are obtained by IPM_1; dots show posteriors means and vertical lines the 95% credible intervals.

probabilities and population sizes clearly suggest that the crash was due to a severe drop in adult male survival. The reason for this is unknown, but it is certainly less alarming for the colony than if adult females had been affected. The size of the superpopulation, i.e., the number of individuals that have ever been present at the nursery colony during a year, showed the same temporal pattern as the nursery colony size but was larger by about 20 individuals on average. This means that about 20 adults on average roosted outside of this colony.

The temporal process correlations of sex- and age-specific survival were generally low (this is parameter `zeta`). The correlations among male survival were slightly larger than they were among female survival and across sexes. Thus, there was hardly any evidence for a common driver affecting the survival of all individuals.

In our model, we would expect the results to be particularly sensitive to the fit of the capture-recapture data, so we conducted the classical goodness-of-fit tests for a CJS model in the software U-Care (Choquet et al., 2009; Gimenez et al., 2018). The model with two age classes, sex, and time effects in survival and recapture probabilities fit the data well ($\chi^2_{30} = 30.47$, $P = 0.80$). Because we adopt a more complex version of a CJS model in our IPMs, our model is highly likely to fit the data at least as well. It is also possible to perform posterior predictive checks for the different data sets of the IPM (Section 7.2.1.1). However, this requires that the binary data (such as the detection histories) are aggregated in some way; if they are not, standard-fit statistics are just deterministic functions of sample size and do not contain any information about possible lack of fit; see section 4.4.5 in McCullagh and Nelder (1989) and section 8.4 in Royle et al. (2014).

13.7 PRIOR SENSITIVITY ANALYSIS

We fitted four IPMs with different assumptions about the productivity of 2y females (ρ_1). Therefore, one question is how much other estimates, and therefore our inferences, are affected by the different assumptions on ρ_1. A second question is whether parameter ρ_1 may be estimable at all.

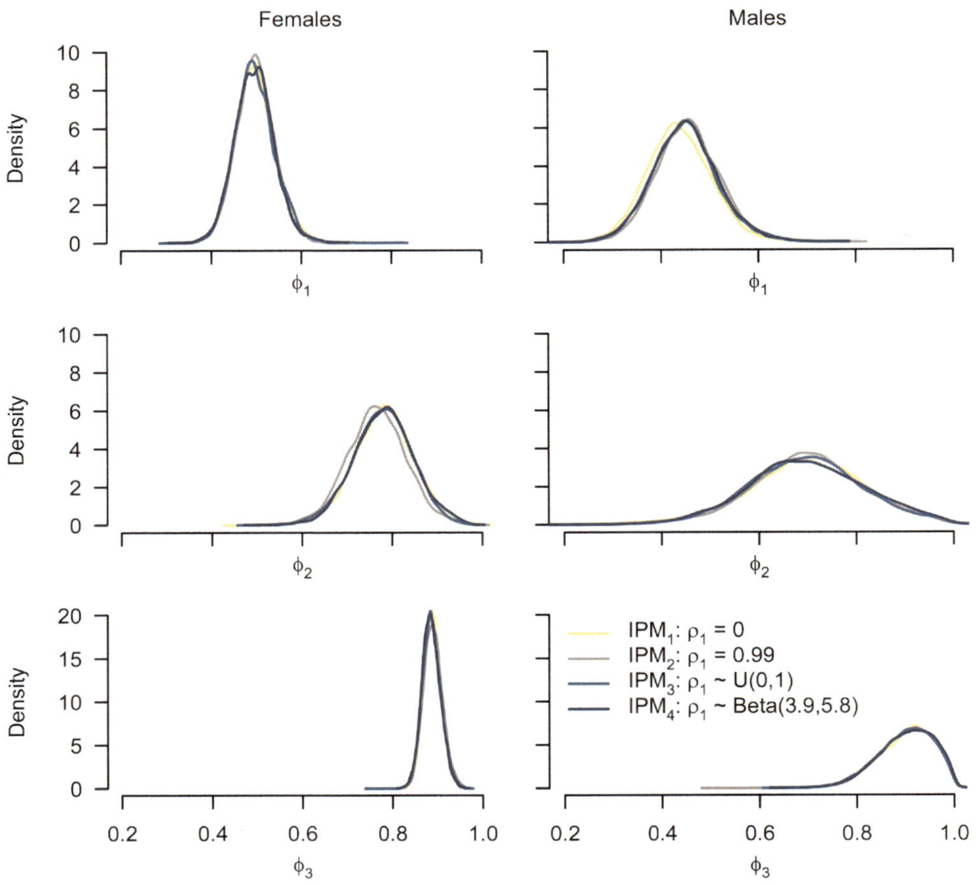

FIGURE 13.5

A model (or prior) sensitivity analysis—The posterior distributions of mean survival in each sex and age class estimated under the four IPMs that differ in their assumptions about age at first reproduction in greater horseshoe bats.

To address the first question, we inspect the posterior distributions of the demographic rates and of population size under the different models. A comparison of mean survival from the four IPMs shows them to be virtually identical—therefore, it seems that estimation of survival does not depend on our assumptions about the productivity of 2y females (Fig. 13.5). Likewise, the estimated colony size and observation error estimate are both insensitive to these model assumptions (Fig. 13.6). In contrast, the posterior distribution of the productivity of >2y females (ρ_2) is affected by our assumptions about ρ_1 (Fig. 13.6)—estimates of ρ_2 are greater when those of ρ_1 are smaller. Because we included the two extreme hypotheses about the productivity of 2y females ($\rho_1 = 0$ and $\rho_1 = 0.99$) in our comparison, we can infer that the true ρ_2 must lie between the estimates obtained under IPM$_1$ and IPM$_2$; i.e., between 0.88 and 0.96. The estimated number of adult females is also sensitive to the assumption about ρ_1. The lower bound of the posterior under IPM$_1$ is particularly striking, because the number of juveniles is known without error. Hence, if only >2y females reproduce, the number of >2y females must be greater than or equal to the number of juveniles counted because the number of juveniles is binary. In conclusion, we find that productivity-related parameters and population structure are only slightly sensitive to model assumptions about ρ_1, while survival and total colony size are essentially insensitive to them.

Second, we obviously want to know whether the productivity of 2y females (ρ_1) can be estimated at all. Parameter identifiability can be assessed with different approaches (Section 6.3), and here we compare the overlap of prior and posterior distributions (Gimenez et al., 2009). If priors and posteriors look the same when different priors are placed on a parameter, this indicates that the "information" in the posterior stems from the prior only and thus that estimates of a parameter are not informed by the data; therefore, the parameter should be considered unidentifiable. Remember that under models IPM$_3$ and IPM$_4$, we estimated ρ_1 using two different priors. Fig. 13.7 shows that with both, the posteriors are very similar to the corresponding priors, indicating that the posterior of ρ_1 was strongly informed by its prior but not by the data and hence suggesting that ρ_1 is not identifiable.

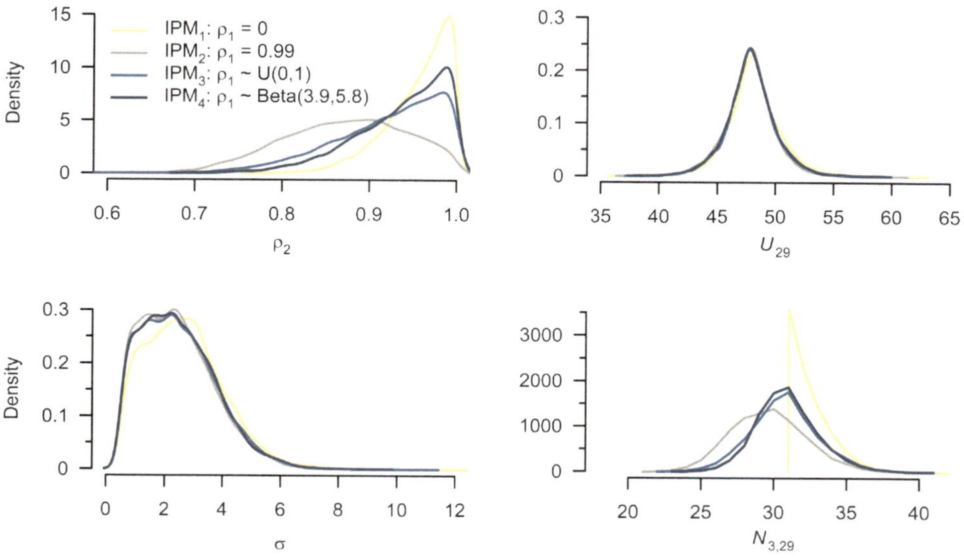

FIGURE 13.6

A model (or prior) sensitivity analysis—Posterior distributions of the productivity of >2y females (ρ_2), colony size of the last year (U_{29}), observation error (σ), and the number of >2y females of the last year of study ($N_{3,29}$) under four IPMs that differ in their assumptions about the productivity of 2y females (ρ_1); see legend inset.

This clearly suggests that we should not interpret the estimates of ρ_1 from our IPMs. However, the estimates of all other demographic parameters in the models allow us to learn about the biology of the bat, since they were hardly affected by the presumed nonidentifiability of ρ_1. The main reason that different assumptions about ρ_1 hardly had an impact on estimates of other demographic rates is the low growth rate sensitivity of 2y female productivity (you can check this out for yourself). Conversely, it also means that there was little information about this parameter in the count data of the adults (Fig. 13.7).

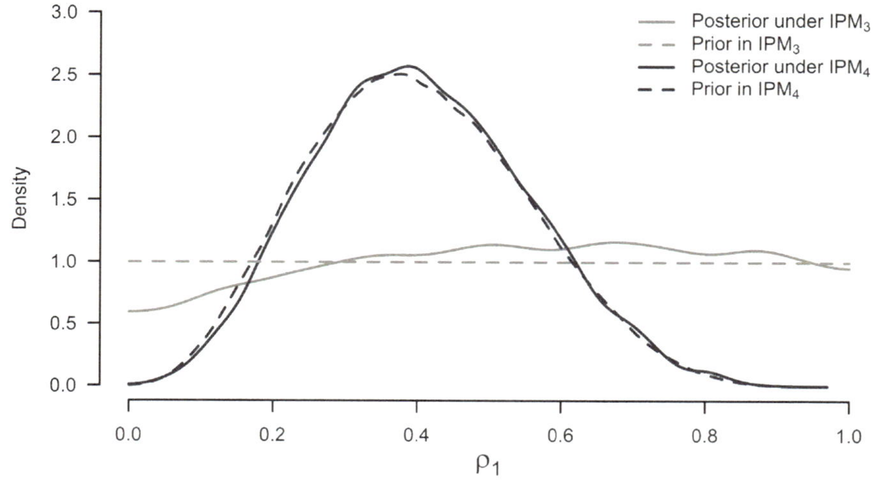

FIGURE 13.7

A visual check of the identifiability of the parameter for the productivity of 2y females (ρ_1) by comparison of the prior/posterior overlap in two of the fitted models. For both IPM$_3$ and IPM$_4$ with very different priors adopted for that parameter, there is extreme overlap between prior and posterior. This suggests no information in the data about ρ_1 and therefore that this parameter is unidentifiable under the model for our data set.

13.8 DISCUSSION

We applied an IPM to three data sets collected in a nursery colony of greater horseshoe bats and looked at estimates of survival, productivity, and structured population size. As we stress in almost every application of an IPM, such detailed results on the demography and the dynamics of a population would be nearly impossible without adoption of an integrated population model. In our study, we had to deal with several challenges.

First, estimation of productivity is difficult in bats in general because often mother-offspring pairs cannot be observed. Recall that the basic information about productivity needed in a population model is the number of offspring per female and year. One possibility, therefore, is to count the numbers of offspring and of females. While in our study it was straightforward to count the juveniles, the number of females producing them was unknown. A sex- and age-structured IPM allows estimation of the numbers of individuals in each sex and age class, but even then, the information about which female age class contributed how much to the total reproductive output of the colony would still be unknown. Therefore, we investigated several assumptions about the potential productivity of 2y females by specifying different models and evaluated how much they affected the inferences in the other parameters—we carried out a model (or prior) sensitivity analysis. Somewhat disappointingly, we found that the productivity of 2y and >2y females cannot be distinguished from our data sets but that inferences about the other parameters of interest were hardly affected by the non-identifiability of this parameter. This is an interesting illustration showing that in a complex hierarchical model, not all is necessarily lost when one (or possibly more) parameter is not identifiable from the data sets at hand; also, see a special "Forum" discussion in *Ecology* (Lavine, 2010, and following papers). What we can conclude for our specific bat study is this—if our aim is to investigate age-specific productivity, we need to collect additional data, i.e., we somehow need to determine which females reproduce in a given year. One possibility for this might be to catch females after they give birth and inspect for whether they are lactating. Another possibility might be the collection of genetic data from droppings and the determination of mother-offspring pairs. The latter approach was used in an IPM in a sister species of ours, the lesser horseshoe bat (*Rhinolophus hipposideros,* Jan et al., 2019).

A second challenge in our study was that the model for the counts of free-flying adults was complex because its composition in terms of demographic classes of individuals is unclear—both sexes and different age classes may be included, and their proportions may change over time. Hence, although we could obtain an accurate count of the adult bat numbers present at the colony, it was hard to interpret its meaning. This is very different from the population counts in the many bird examples in our book, where typically the number of territories, breeding pairs, or nests can be counted. Hence, in these examples we know that this number refers to the number of reproducing females. With the bats, we could solve this challenge because we had an additional data set, the capture-recapture data that contained information about the population composition, and we could leverage this information in our IPM. We were sure to detect all bats present in the church at the moment of capture, so we could interpret the estimated recapture probability as the probability of presence. This provided us with the necessary knowledge about the population composition. The use of the probability of presence for the population composition required the closure assumption, i.e., that population size and composition do not change between when it was counted and when it was captured. Therefore, the time interval between the collections of the two data sets had to be kept as short as possible.

Our ability to obtain a true population census, at least within the church, was rather unique in our study system and this might not commonly be possible in other bat populations. The challenge of estimating the population composition to make sense of the counts of free-flying adults remains. In such a case, bats could be captured repeatedly (at least twice) over a short time interval within a season. Assuming closure across this time interval, one could apply robust design models to obtain independent estimates of the probabilities of presence and capture (Kendall et al., 1997; Riecke et al., 2018).

The sampling protocol in the greater horseshoe bat study was not identical in every year of the 29-year study. Temporal changes in the sampling protocol are relatively common in long-term ecological studies, and it is good that IPMs are so flexible in this respect (Section 6.6.2). The adaptations to accommodate missing data due to changes in the sampling protocol needed in the bat IPM were easy to implement, but there was a small price to pay, which was lower precision of the parameter estimates for some years.

The bat example illustrates the principle of model (or prior) sensitivity analyses to assess the degree to which inferences are affected by specific assumptions and to determine whether a particular parameter is informed by the data at all, that is, whether a parameter is identifiable. Such model sensitivity analyses are important and can be particularly insightful for more complex models and when data are sparse. Here, we found that the productivity parameter of 2y females was not estimable from our data but that this did not compromise the inferences about the other parameters. It is important to know the limits of inference when some parameters cannot be estimated, and this can be studied by sensitivity analyses as demonstrated here.

CHAPTER

HOOPOE

14

CHAPTER OUTLINE

IPM ingredients: population count data, capture-recapture data, productivity data
Location: Switzerland
Spatio-temporal scale: 62 km^2, 16 years

14.1 INTRODUCTION

Individuals differ not only in morphological traits, coloration, or personality but also in their demographic performance (reviewed in Gimenez et al., 2018). If individual heterogeneity in demography is not adequately accounted for, erroneous or misleading inferences about demographic rates and population dynamics may be obtained. For one famous example, consider a cohort (year group) composed of two types of individuals, one with high and the other with low constant survival probability. At the cohort level, we will find an increasing trend in average survival over time simply because the proportion of high-survival individuals will increase in the population over time. The reason for this, of course, is that low-survival individuals disappear faster than high-survival individuals. Thus, individual demographic heterogeneity can prevent the detection of senescence unless properly accounted for in an analysis (Vaupel and Yashin, 1985). Demographic heterogeneity may also impact population dynamics (Łomnicki, 1978; Kendall et al., 2011). For example, the asymptotic growth rate of a population with members that differ in survival is higher than that of a homogenous population with the same average survival (Kendall et al., 2011). The reason for this result is the same as before: over time, the population proportion of "stronger" individuals will be increasing. Classical population models, and hence classical IPMs, stratify individuals by stage structure such as age class, sex, or reproductive state to account for demographic heterogeneity. However, within each class, all individuals are assumed to be identical from a demographic point of view. Demographic heterogeneity in these models is therefore captured only to a degree, and inferences may sometimes be biased by the lack of inclusion of "residual" variability within the groups created by the stratification chosen (Vaupel and Yashin, 1985; Kendall et al., 2011; Plard et al., 2019a). This can be avoided by constructing population models that account for additional demographic heterogeneity.

Heterogeneity within populations can be induced via a variety of mechanisms (e.g., Berger et al., 2015; Plard et al., 2018b; Acevedo et al., 2020), and the effects of such heterogeneity may change rapidly over time (Grant and Grant, 2002). To increase our understanding of evolutionary ecology and population dynamics, we need to identify mechanisms creating heterogeneity and be able to measure whether and how resulting effects change over time.

Plard et al. (2019a) reviewed the possibilities that IPMs offer for the study of demographic heterogeneity. The first is when demographic heterogeneity is related to a measurable, continuous individual trait such as size, mass, or laying date.

Integrated Population Models. https://doi.org/10.1016/B978-0-12-820564-8.00008-7

Models that describe individual demography as a function of one or several continuous individual traits are called integral projection models and are often also abbreviated IPM (Easterling et al., 2000; Ellner et al., 2016). They can be formulated as part of an IPM (as defined in this book, i.e., as an integrated population model), in which case they may be called an IPM^2: an integrated, integral projection model (Plard et al., 2019b).

Perhaps more often than not, we cannot relate demographic heterogeneity to a known and measurable continuous trait (Hamel et al., 2018). However, if individual heterogeneity is a fixed characteristic of an individual and does not change over time (or else we would have to describe that change as well), we may estimate it by specifying a latent variable for each individual in either finite- or continuous-mixture models. Here we give an example of a finite-mixture model as a simple and powerful method to accommodate individual heterogeneity (Pledger, 2000; Pledger and Schwarz, 2002; Pledger et al., 2003; Hamel et al., 2017; Gimenez et al., 2018). These models assume that a population can be split into a known number of groups (usually a small number such as 2 or 3), each of different quality, and that every member of a group has exactly the same demographic performance. At first sight, this may appear to be a shocking simplification of what in reality must surely be a continuous distribution. However, finite-mixture models have been shown to work quite well to explain residual individual variability (Pledger, 2000). In most cases (though not always), we should not think of the resulting groups as real entities but just as an additional individual stratification scheme to reduce unexplained heterogeneity at the level of the individual.

The purpose of this case study is the development of an IPM that includes a two-group, finite-mixture specification of the individual heterogeneity in reproductive performance. We want to estimate the mean reproductive performance and temporal variability of each group and the number of females belonging to each group. Thus, in our illustration we posit that the groups are "real" in a sense. We note that our model will have what might be called a "primary" and "secondary" stage structure, where the former are the usual stages that we would distinguish *a priori* in the life-cycle graph, while the latter are the two latent groups, the membership of which is *not* known *a priori* but will be estimated.

To illustrate the model, we use the same hoopoe (*Upupa epops*, Fig. 9.1) population data from the Swiss Alps that we analyzed already in Section 9.2 (where the general principle of data sampling has been described) and 10.6. Hoopoes are long-distance migrants (van Wijk et al., 2018), breed in cavities in trees and buildings, are short-lived, and have high productivity. The study population is geographically open; thus, it receives immigrants and produces emigrants (Schaub et al., 2012). About one third of the females raise two broods in a season (Hoffmann et al., 2015), which contributes to the substantial individual heterogeneity in the number of fledglings produced per female (Plard et al., 2018a,b). Part of this heterogeneity is due to the age of the mother, with 1-year-old mothers (first-time breeders) producing fewer young than older mothers (Section 9.2). But in addition, differences in reproductive performance might be due to fixed individual heterogeneity unrelated to age. Thus, instead of assuming that all females of a given age class have exactly the same productivity, we now assume the existence of two latent groups of females with different productivities. We estimate the fraction of females belonging to each group using a finite-mixture model for the productivity data. As a consequence, we must enlarge the population model to account for these two groups. This IPM requires a couple of further assumptions because of the additional structure, in particular about how local recruits and immigrants are allocated to the latent quality groups.

14.2 DATA SETS

Three data sets were collected in the upper Rhone valley in the Swiss Alps from 2002 to 2017 (see Schaub et al., 2012). First, the population count data refer to the number of breeding pairs in the nest boxes. Hoopoes may double-brood in the breeding season (Bötsch et al., 2012); hence, the total number of broods recorded in a year will normally overestimate population size. Owing to the frequent checks of all nest boxes, the dates of laying, failure, or fledging are known quite accurately for all broods. This allows the computation of the daily number of simultaneously active broods in every year. Here, we take the annual maximum number of simultaneously active broods as an index of hoopoe population size in the study area. Second, we collected capture-recapture data. All nestlings were ringed, and reproducing adults were captured at the nest box. Adult hoopoes, but not nestlings, were sexed. Our model will consider females only, and hence, we randomly allocated nestlings to a sex assuming an even sex ratio at ringing (Nichols et al., 2004). Third, we recorded the number of fledglings of each brood. Because we modeled the annual number of fledglings produced per female, we use only broods for which the identity of the mother was known. We then tally up the total annual number of fledglings raised by each mother, including any replacement and second broods.

14.3 POPULATION MODEL

The basic features of our hoopoe population model are identical to those in Chapters 9 and 10. We assume a prebreeding "census" for a short-lived bird species and a female-based population model with immigration. The key difference from these earlier chapters is that we now further subdivide the primary population stages into two latent groups defined by the reproductive quality of an individual. For each latent group, we define three primary stages: local recruits (R), surviving adults (S), and immigrants (I). This results in six stages in total, R_L, R_H, S_L, S_H, I_L, and I_H, where subscript L stands for low-quality and H for high-quality individuals; see Fig. 14.1 for a life-cycle graph. The resulting population equations for the local recruitment process are

$$R_t \sim Poisson\left(\frac{\phi_{j,t-1}}{2}\left(\rho_{H,t-1}N_{H,t-1} + \rho_{L,t-1}N_{L,t-1}\right)\right),$$
$$R_{H,t} \sim Binomial(\gamma, R_t), \text{ and}$$
$$R_{L,t} = R_t - R_{H,t},$$

where $\phi_{j,t}$ is juvenile apparent survival in year t, $\rho_{g,t}$ is productivity in year t of a female in group g, R_t is the total number of local recruits in year t, γ is the expected proportion of females in the higher-quality group (also called the group membership probability or mixture proportion, here expressed with respect to the high-quality group), $R_{g,t}$ is the number of local recruits in year t belonging to group g, and $N_{g,t}$ is the total number of females in year t in group g. Surviving adults (S) are modeled with binomial distributions.

$$S_{H,t} \sim Binomial(\phi_{a,t-1}, N_{H,t-1})$$
$$S_{L,t} \sim Binomial(\phi_{a,t-1}, N_{L,t-1})$$

Here, $\phi_{a,t}$ is the apparent adult survival in year t. Finally, we model the total number of immigrants using a Poisson distribution and allocate them to each of the latent groups using a binomial distribution with parameter γ. Thus,

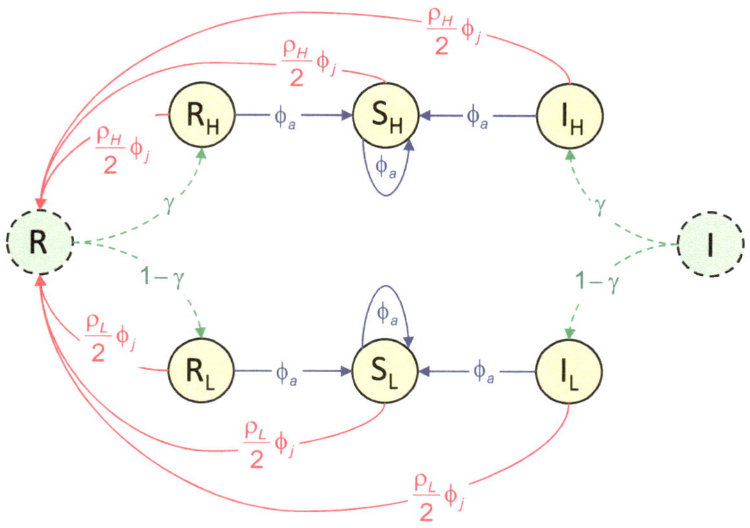

FIGURE 14.1

Life-cycle graph for a hoopoe population with three stages defined *a priori* as well as with two latent groups of individual quality. The nodes label the total number of recruits (R), local recruits of low (R_L) and high (R_H) quality, surviving adults of low (S_L) and high (S_H) quality, total number of immigrants (I), and immigrants of low (I_L) and high (I_H) quality. The dotted nodes refer to intermediate states, and the dotted arrows refer to the allocation of new individuals to the latent groups representing individual quality.

$$I_t \sim Poisson(\omega_t),$$
$$I_{H,t} \sim Binomial(\gamma, I_t), \text{ and}$$
$$I_{L,t} = I_t - I_{H,t},$$

where ω_t is the expected number of immigrants in year t. The group-specific total population sizes are then the sums of the three *a priori* stages.

$$N_{H,t} = R_{H,t} + S_{H,t} + I_{H,t}$$
$$N_{L,t} = R_{L,t} + S_{L,t} + I_{L,t}$$

In this model, we assume that local recruits are allocated randomly to the high-quality group with probability γ regardless of whether they were produced from a mother of low or high quality. Alternatively, we could formulate a model in which the quality of the mother is inherited by the offspring. In addition, our model assumes that individual quality in terms of group membership is estimated exclusively from the reproductive data and that we can allocate both immigrants and local recruits to these same groups according to mixture proportion γ estimated from the productivity data.

14.4 COMPONENT DATA LIKELIHOODS

We load the hoopoe data and have a quick look at their content to remind us of what they are.

```
# Load hoopoe data and produce data overview
library(IPMbook); library(jagsUI)
data(hoopoe)
str(hoopoe)

List of 5
 $ ch       : int [1:3844, 1:16] 0 0 0 0 0 0 0 0 0 0 ...
 $ age      : int [1:3844] 1 1 1 1 1 1 1 1 1 1 ...
 $ count    : num [1:16] 34 46 68 93 88 87 85 78 82 84 ...
 $ reproAgg:List of 4
  ..$ J1: num [1:16] 73 188 243 320 261 222 206 154 278 220 ...
  ..$ J2: num [1:16] 51 83 101 182 256 226 206 278 237 244 ...
  ..$ B1: num [1:16] 15 23 36 50 44 37 39 30 48 48 ...
  ..$ B2: num [1:16] 6 13 12 23 38 38 32 41 39 41 ...
 $ reproInd:List of 3
  ..$ f    : int [1:1092] 6 11 11 3 15 6 11 12 7 6 ...
  ..$ id   : num [1:1092] 483 486 530 531 532 533 534 535 536 538 ...
  ..$ year: int [1:1092] 2002 2002 2002 2002 2002 2002 2002 2002 2002 ...
```

In our model, we will use the capture-recapture data (`ch`) along with information about the age of the individuals at first capture (`age`), the population count (`count`), and the individual-level productivity data (`reproInd`), where `f`, `id`, and `year` contain, respectively, the total annual number of fledglings produced per female, female identity, and year.

14.4.1 POPULATION COUNT DATA

The population count data (C, denoted `count` in the data set) is a vector that contains the maximum number of simultaneously active broods for each year from 2002 to 2017. The quality of the individuals as well as their primary stage (i.e., R, S, or I) cannot be observed, and therefore, the counts are completely unstructured. As usual, we model population counts with a state-space model with the state-process equations shown in Section 14.3. For the observation model, we use a normal distribution.

$$C_t \sim Normal\left(N_{H,t} + N_{L,t}, \sigma^2\right)$$

In the state-process model, we must also estimate the states in the first year. Usually we specify discrete uniform priors for the initial stage-specific population sizes, and we also could do that for all six stages in our model. However, we can do better. Because the number of immigrants is modeled with a random year effect (see later), we have some information about the number of immigrants in the first year that we will use. Moreover, we estimate the mixture proportion γ and can use this information as well. Therefore, we place a discrete uniform prior on the total number of local recruits and of surviving adults and a Poisson prior on the total number of immigrants. Second, we allocate individuals to the latent quality classes with binomial distributions according to mixture proportion γ. Specifically, we make the following assumptions:

$$\begin{aligned}
\text{ini}_R &\sim DU(a,b)\\
\text{ini}_S &\sim DU(a,b)\\
\text{ini}_I &\sim Poisson(\omega_1)\\
R_{H,1} &\sim Binomial(\gamma, \text{ini}_R)\\
R_{L,1} &= \text{ini}_R - R_{H,1}\\
S_{H,1} &\sim Binomial(\gamma, \text{ini}_S)\\
S_{L,1} &= \text{ini}_S - S_{H,1}\\
I_{H,1} &\sim Binomial(\gamma, \text{ini}_I)\\
I_{L,1} &= \text{ini}_I - I_{H,1}
\end{aligned}$$

Here, a and b specify the range of the discrete uniform priors. Remember that we specify the discrete uniform prior in JAGS using a categorical distribution with cell probabilities π that we provide as data by calling function `dUnif` with arguments a and b.

By specifying the initial population sizes in this way, we use more informative priors than under an alternative formulation with discrete uniform priors for a suitable range (a, b) specified independently for all six stages. The prior chosen for the initial population sizes affects the estimates of stage-specific population sizes in the first two years but not those of demographic rates or later stage-specific population sizes. You can experience this model behavior by performing a prior sensitivity analysis.

14.4.2 CAPTURE-RECAPTURE DATA

We convert the capture-recapture data into an age-dependent m-array.

```
# Produce m-array
marr <- marrayAge(hoopoe$ch, hoopoe$age, 2)
marr.j <- marr[,,1]
marr.a <- marr[,,2]
```

We also need to tally the annual number of released individuals, which we do when we package the data list for JAGS.

We use a standard Cormack-Jolly-Seber (CJS) model with two age classes and temporal random effects for the apparent survival of both age classes. Based on previous results (Schaub et al., 2012; Plard et al., 2020), we specify an age structure for the recapture probabilities that were otherwise assumed to be constant. Individuals marked as nestlings are likely to have a lower recapture probability in the year they recruit into the population; they may fail to reproduce at all or at an earlier brood stage, both of which reduce their recapture probability under our sampling protocol. Moreover, based on the previous results (Schaub et al., 2012; Plard et al., 2020), we did not run additional goodness-of-fit tests.

14.4.3 PRODUCTIVITY DATA

Vector `f` contains the total annual number of fledglings produced by a known ID ($n = 1092$, do `plot(table(hoopoe$reproInd$f))` for a summary), with mother ID and year contained in vectors `year` and `id`. Note that `id` is coded as an integer that ranges from 1 to 767, which is the total number of identified females, because we use this as an index in the model. We use `year` as an index as well, and because the first year is 2002, we simply subtract 2001 from it before we package it into the data bundle.

For the productivity data, we use a finite-mixture model with two groups. The model assumes that each individual in a specific group has the same expected productivity but that expected productivity differs between the groups. Furthermore, individuals are not allowed to change groups; thus, group membership is fixed. Group membership is latent and needs to be estimated. We define h_i as a latent variable that takes on the value 1 if individual i belongs to the low-quality group and 2 if it belongs to the high-quality group. Further, we define γ as the mixture proportion with respect to the higher-quality group and use the Bernoulli distribution to model the individual group membership indicators h as follows:

$$k_i \sim Bernoulli(\gamma) \text{ and } h_i = k_i + 1.$$

The outcome of the Bernoulli distribution is a 0 or 1; hence, we need to add 1 to get the correct group membership indicators. We note that as an alternative, we could use a categorical distribution, i.e., $h_i \sim \text{Categorical}([1-\gamma, \quad \gamma])$, in which case the latent variable would correspond directly to the grouping labels. From a likelihood point of view, both formulations are identical.

The number of fledglings raised by individual i in year t is modeled with a normal distribution as $f_i \sim Normal(\rho_{h_i, year_i}, \sigma^2)$, where $\rho_{g,t}$ is the productivity of quality group g in year t and σ^2 is the residual variance. We prefer the normal over the possible alternative of a Poisson distribution here because it is more flexible. The Poisson is often overdispersed for bird productivity data, and the solution to an underdispersed Poisson distribution developed by Ridout and Besbeas (2004) is elegant, but it is more difficult (albeit not impossible, of course) to code in BUGS.

We assume that productivity of each quality group varies independently over time and therefore specify univariate temporal random effects.

$$\log(\rho_{g,t}) = \alpha_g + \varepsilon_{g,t}$$

$$\varepsilon_{g,t} \sim Normal\left(0, \sigma_g^2\right)$$

Here, α_g are the group-specific means on the log scale, and σ_g^2 are the group-specific temporal variances of productivity.

If we implement the finite-mixture model in the way written above and specify a vague uniform prior for the group membership probability γ in the 0−1 range, we are likely failing to obtain convergence because of the so-called label switching problem (Stephens, 2000). Inside the Markov chain Monte Carlo (MCMC) algorithm, the labels for the two groups are not fixed, that is, the group with the high-quality individuals might be labeled either 1 or 2, and *vice versa* for the other group. In our model, we defined γ as the probability of being in the high-quality group (which is label 2); hence, its estimate will differ if group 2 happens to represent the low-quality individuals compared with when it happens to denote the high-quality individuals. Indeed, the group membership probability for the high-quality individuals is the complement of that for the low-quality individuals. The likelihood of a two-group finite-mixture model is symmetric with respect to γ and the two α's, and has two optima: the first is $\alpha_1 < \alpha_2$ with γ being the proportion of high-quality individuals, and the second is $\alpha_1 > \alpha_2$, where the proportion of high-quality individuals is now $1-\gamma$. Typically, when we run multiple chains in an MCMC algorithm, we will see some convergence to one solution and some to the other with the result that overall, we diagnose dramatic nonconvergence (e.g., using the Books-Gelman-Rubin statistic). However, this behavior is easily spotted by inspection of the trace-plot for γ and the elements of α (Fig. 14.2).

Label switching is common in Bayesian MCMC applications of mixture models, and one way of preventing it is to impose identifiability constraints (Stephens, 2000). In our model, we defined γ to be the proportion of the high-quality individuals (Fig. 14.1). These have label 2, and hence, α_2 must be greater than α_1. We enforce this constraint by a reparameterization of α_2 as $\alpha_2 = \alpha_1 + \Delta\alpha$, where $\Delta\alpha \geq 0$. This can be achieved by specifying a prior with support only for nonnegative values. Moreover, from the trace-plots in Fig. 14.2 we see that if $\alpha_1 < \alpha_2$, γ is less than 0.5 (trace of the red chain). Therefore, for the mixture proportion, we choose a uniform prior between 0 and 0.5 to ensure that $\gamma \leq 0.5$.

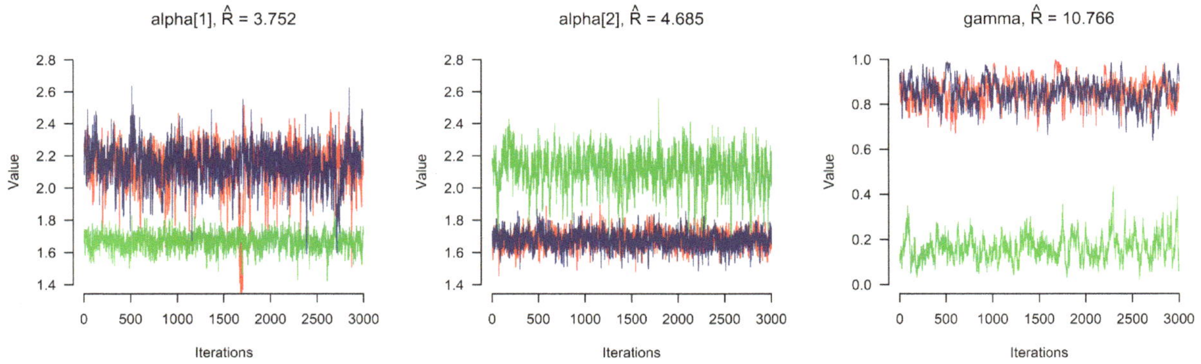

FIGURE 14.2

Illustration of label switching in a two-group mixture model with trace-plots of group mean values α and mixture proportion γ. For the blue and green traces, group 1 represents the high-quality individuals, while for the red trace these are represented by group 2, with 2 corresponding to solutions of γ. To achieve a unique solution and obtain convergence, identifiability constraints must be built into the model. (In some runs with three chains, you may not realize the two solutions, but with many more chains, say 10, you are almost certain to see them.)

14.5 INTEGRATED POPULATION MODEL

The JAGS code for the IPM contains the likelihoods for the three data sets, and the joint likelihood is specified through parameters that are shared among component data likelihoods. We specify vague priors for all parameters, but as discussed above, we do not claim to be completely ignorant about the initial structured population size.

We first create the list containing the data and then write code for the IPM.

```
# Bundle the data
jags.data <- with(hoopoe$reproInd, list(marr.j=marr.j, marr.a=marr.a, n.years=ncol(marr.j),
  rel.j=rowSums(marr.j), rel.a=rowSums(marr.a), nind=length(unique(id)), id=id, nrep=length(id),
  f=f, year=year-2001, C=hoopoe$count, pNinit=dUnif(1, 50)))

str(jags.data)                                          # Remind ourselves of how the data look like

List of 12
 $ marr.j : num [1:15, 1:16] 6 0 0 0 0 0 0 0 0 0 ...
 $ marr.a : num [1:15, 1:16] 6 0 0 0 0 0 0 0 0 0 ...
 $ n.years: int 16
 $ rel.j  : num [1:15] 104 165 256 285 265 264 237 222 260 242 ...
 $ rel.a  : num [1:15] 29 36 57 77 82 83 78 73 86 90 ...
 $ nind   : int 767
 $ id     : num [1:1092] 483 486 530 531 532 533 534 535 536 538 ...
 $ nrep   : int 1092
 $ f      : int [1:1092] 6 11 11 3 15 6 11 12 7 6 ...
 $ year   : num [1:1092] 1 1 1 1 1 1 1 1 1 1 ...
 $ C      : num [1:16] 34 46 68 93 88 87 85 78 82 84 ...
 $ pNinit : num [1:50] 0.02 0.02 0.02 0.02 0.02 0.02 0.02 0.02 0.02 ...


# Write JAGS model file
cat(file="model1.txt", "
model {
# Priors and linear models
# For mean and variance parameters
mean.logit.phij <- logit(mean.phij)
mean.phij ~ dunif(0, 1)
mean.logit.phia <- logit(mean.phia)
mean.phia ~ dunif(0, 1)
mean.log.omega <- log(mean.omega)
mean.omega ~ dunif(0, 50)
mean.pa ~ dunif(0, 1)
mean.pj ~ dunif(0, 1)

sigma.phij ~ dunif(0, 3)
tau.phij <- pow(sigma.phij, -2)
sigma.phia ~ dunif(0, 3)
tau.phia <- pow(sigma.phia, -2)
sigma.omega ~ dunif(0.001, 5)
tau.omega <- pow(sigma.omega, -2)

sigma.rep ~ dunif(0.001, 5)
tau.rep <- pow(sigma.rep, -2)

for (g in 1:2){
  sigma.rep.t[g] ~ dunif(0.001, 2)
  tau.rep.t[g] <- pow(sigma.rep.t[g], -2)
}

# Enforce order on the alphas: alpha[1] < alpha[2]
alpha[1] ~ dnorm(0, 0.01)                               # Not too low precision, for convergence
alpha[2] <- alpha[1] + diff.alpha
diff.alpha ~ dnorm(0, 0.01)T(0,)                        # Diff. must be positive; alpha[2] is
                                                        # defined to have higher values

# Mixture parameter
gamma ~ dunif(0, 0.5)

# Residual (observation) error
sigma ~ dunif(0.4, 20)                                  # Lower bound >0
tau <- pow(sigma, -2)
```

```
# Linear models for demographic parameters
# Survival and recapture
for (t in 1:(n.years-1)){
  logit.phij[t] ~ dnorm(mean.logit.phij, tau.phij)
  phij[t] <- ilogit(logit.phij[t])
  logit.phia[t] ~ dnorm(mean.logit.phia, tau.phia)
  phia[t] <- ilogit(logit.phia[t])
  pa[t] <- mean.pa
  pj[t] <- mean.pj
}

# Reproduction parameters
for (g in 1:2){                                        # Loop over two groups
  for (t in 1:n.years){
    log(rep[g,t]) <- alpha[g] + eps[g,t]
    eps[g,t] ~ dnorm(0, tau.rep.t[g])
  } #t
} #g

# Immigration parameters
for (t in 1:n.years){
  log.omega[t] ~ dnorm(mean.log.omega, tau.omega)
  omega[t] <- exp(log.omega[t])
}

# Population count data (state-space model)
# Model for initial stage-spec. population sizes
ini[1] ~ dcat(pNinit)                                  # Local recruits
ini[2] ~ dcat(pNinit)                                  # Surviving adults
ini[3] ~ dpois(omega[1])                               # Immigrants
R[1,1] ~ dbin(1-gamma, ini[1])
R[2,1] <- ini[1]-R[1,1]
S[1,1] ~ dbin(1-gamma, ini[2])
S[2,1] <- ini[2]-S[1,1]
I[1,1] ~ dbin(1-gamma, ini[3])
I[2,1] <- ini[3]-I[1,1]

# Process model over time: our model of population dynamics
for (t in 2:n.years){
  J[1,t] ~ dpois(rep[1,t-1] / 2 * phij[t-1] * N[1,t-1]) # Local recruits
  # produced by females of group 1 (L)
  J[2,t] ~ dpois(rep[2,t-1] / 2 * phij[t-1] * N[2,t-1]) # Local recruits
  # produced by females of group 2 (H)
  R[1,t] ~ dbin(1-gamma, J[1,t] + J[2,t])              # Allocate recruits to group 1 (L)
  R[2,t] <- J[1,t] + J[2,t] - R[1,t]                   # Allocate recruits to group 2 (H)
  S[1,t] ~ dbin(phia[t-1], N[1,t-1])                   # Surviving adults, group 1 (L)
  S[2,t] ~ dbin(phia[t-1], N[2,t-1])                   # Surviving adults, group 2 (H)
  Itot[t] ~ dpois(omega[t])                            # Total number of immigrants
  I[1,t] ~ dbin(1-gamma, Itot[t])                      # Immigrants of group 1 (L)
  I[2,t] <- Itot[t]-I[1,t]                             # Immigrants of group 2 (H)
}

# Observation model
for (t in 1:n.years){
  N[1,t] <- S[1,t] + R[1,t] + I[1,t]                   # Total no of females in group 1
  N[2,t] <- S[2,t] + R[2,t] + I[2,t]                   # Total no of females in group 2
  C[t] ~ dnorm(N[1,t] + N[2,t], tau)
}

# Capture-recapture data (CJS model with multinomial likelihood)
# Define the multinomial likelihood
for (t in 1:(n.years-1)){
  marr.j[t,1:n.years] ~ dmulti(pr.j[t,], rel.j[t])
  marr.a[t,1:n.years] ~ dmulti(pr.a[t,], rel.a[t])
}
# Define the cell probabilities of the m-arrays
for (t in 1:(n.years-1)){
  # Main diagonal
  qj[t] <- 1-pj[t]
  qa[t] <- 1-pa[t]
  pr.j[t,t] <- phij[t] * pj[t]
  pr.a[t,t] <- phia[t] * pa[t]
```

```
    # Above main diagonal
    for (j in (t+1):(n.years-1)){
      pr.j[t,j] <- phij[t] * prod(phia[(t+1):j]) * qj[t] * prod(qa[t:(j-1)]) * pa[j] / qa[t]
      pr.a[t,j] <- prod(phia[t:j]) * prod(qa[t:(j-1)]) * pa[j]
    } #j
    # Below main diagonal
    for (j in 1:(t-1)){
      pr.j[t,j] <- 0
      pr.a[t,j] <- 0
    } #j
  } #t
  # Last column: probability of non-recapture
  for (t in 1:(n.years-1)){
    pr.j[t,n.years] <- 1-sum(pr.j[t,1:(n.years-1)])
    pr.a[t,n.years] <- 1-sum(pr.a[t,1:(n.years-1)])
  }

  # Productivity data (finite-mixture model with Gaussian error)
  # Mixture
  for (i in 1:nind){              # Loop over all ID of mothers with known annual productivity
    k[i] ~ dbern(gamma)
    h[i] <- k[i] + 1
  }

  # Productivity
  for (i in 1:nrep){              # Loop over all known-ID annual productivity data
    f[i] ~ dnorm(rep[h[id[i]], year[i]], tau.rep)
  }
  }
  ")
```

We have found mixing to be improved when initial values for the number of immigrants given are "reasonable." We increase the number of MCMC chains to 10 here, which gives us more confidence that there are no alternative solutions. We prepare the rest of the analysis and launch the model.

```
# Initial values
inits <- function(){list(mean.omega=runif(1, 8, 12))}

# Parameters to be monitored
parameters <- c("mean.phij", "mean.phia", "alpha", "gamma", "diff.alpha", "mean.omega", "mean.pj",
  "mean.pa", "sigma.rep", "sigma.phij", "sigma.phia", "sigma.omega", "sigma.rep.t", "sigma",
  "phij", "phia", "rep", "R", "S", "I", "N")

# MCMC settings
ni <- 40000; nb <- 10000; nc <- 10; nt <- 10; na <- 5000

# Call JAGS (ART 15 min), check convergence and summarize posteriors
out1 <- jags(jags.data, inits, parameters, "model1.txt", n.iter=ni, n.burnin=nb, n.chains=nc,
  n.thin=nt, n.adapt=na, parallel=TRUE)
traceplot(out1)
```

14.6 RESULTS

The model takes a while to run until satisfactory convergence is achieved. The constraints imposed on α and γ were successful in preventing label switching. The trace-plots for these parameters now look completely satisfactory (Fig. 14.3).

Convergence for all the other parameters also looks good, which is formally confirmed by the R-hat values, which are all close to 1.

```
print(out1, 3)
                 mean    sd    2.5%     50%   97.5% overlap0 f  Rhat n.eff
mean.phij       0.122 0.012   0.099   0.122   0.147    FALSE 1 1.001  5725
mean.phia       0.363 0.020   0.324   0.363   0.403    FALSE 1 1.001  7546
alpha[1]        1.673 0.052   1.567   1.674   1.771    FALSE 1 1.001  5950
alpha[2]        2.132 0.117   1.878   2.139   2.346    FALSE 1 1.002  3125
gamma           0.154 0.055   0.051   0.152   0.269    FALSE 1 1.005  1315
diff.alpha      0.459 0.125   0.179   0.470   0.679    FALSE 1 1.001  5717
```

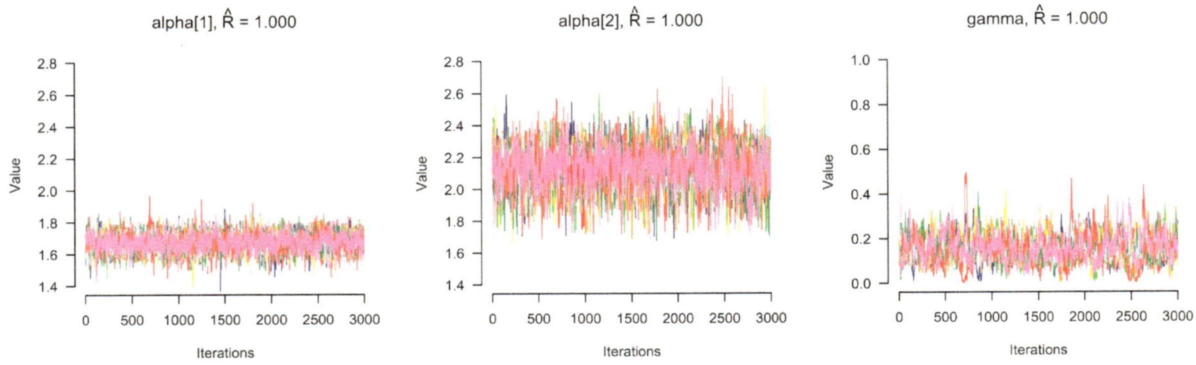

FIGURE 14.3

Trace-plots of α and γ based on an integrated population model with identifiability constraints imposed that prevent label switching. All 10 chains have converged to the same solution.

```
mean.omega        19.886  3.032   13.968   19.885   25.799    FALSE 1 1.002  2672
mean.pj            0.603  0.039    0.526    0.604    0.680    FALSE 1 1.001  6456
mean.pa            0.795  0.033    0.727    0.796    0.855    FALSE 1 1.000 14285
sigma.rep          2.931  0.092    2.759    2.927    3.122    FALSE 1 1.001  6303
sigma.phij         0.278  0.101    0.105    0.268    0.505    FALSE 1 1.000  9746
sigma.phia         0.164  0.103    0.008    0.154    0.395    FALSE 1 1.002  3011
sigma.omega        0.178  0.143    0.008    0.145    0.537    FALSE 1 1.003  1908
sigma.rep.t[1]     0.149  0.045    0.072    0.144    0.251    FALSE 1 1.001  4207
sigma.rep.t[2]     0.281  0.104    0.131    0.265    0.529    FALSE 1 1.002  7348
sigma              2.807  2.163    0.430    2.271    8.149    FALSE 1 1.002  3828
phij[1]            0.114  0.023    0.071    0.114    0.162    FALSE 1 1.001  6568
[... output truncated ...]
phij[15]           0.104  0.022    0.062    0.104    0.149    FALSE 1 1.001  7592
phia[1]            0.354  0.041    0.267    0.356    0.435    FALSE 1 1.001  7006
[... output truncated ...]
phia[15]           0.394  0.045    0.325    0.385    0.503    FALSE 1 1.001  7654
rep[1,1]           5.633  0.566    4.559    5.617    6.789    FALSE 1 1.001  4721
rep[2,1]          10.137  1.531    7.092   10.145   13.176    FALSE 1 1.000 15369
[... output truncated ...]
rep[1,16]          5.613  0.487    4.688    5.604    6.591    FALSE 1 1.000 16855
rep[2,16]          8.915  1.999    4.904    9.020   12.572    FALSE 1 1.001  6217
R[1,1]             7.361  5.144    1.000    6.000   19.000    FALSE 1 1.003  1731
R[2,1]             1.375  1.534    0.000    1.000    5.000     TRUE 1 1.001  5410
[... output truncated ...]
R[1,16]           10.498  3.704    4.000   10.000   18.000    FALSE 1 1.001  4311
R[2,16]            1.888  1.561    0.000    2.000    6.000     TRUE 1 1.001  7532
S[1,1]             7.452  5.259    1.000    6.000   20.000    FALSE 1 1.003  2333
S[2,1]             1.390  1.551    0.000    1.000    5.000     TRUE 1 1.002  3545
[... output truncated ...]
S[2,15]            3.085  2.061    0.000    3.000    8.000     TRUE 1 1.002  3401
S[1,16]           15.714  3.678    9.000   16.000   23.000    FALSE 1 1.001  4489
S[2,16]            2.835  1.950    0.000    3.000    7.000     TRUE 1 1.001  4372
I[1,1]            14.758  4.874    6.000   15.000   25.000    FALSE 1 1.001  4125
I[2,1]             2.770  2.011    0.000    2.000    7.000     TRUE 1 1.001  4911
[... output truncated ...]
I[1,16]           14.995  4.250    7.000   15.000   24.000    FALSE 1 1.001  4080
I[2,16]            2.738  1.948    0.000    2.000    7.000     TRUE 1 1.001  5716
N[1,1]            29.571  4.194   22.000   29.000   39.000    FALSE 1 1.003  3959
N[2,1]             5.535  2.950    1.000    5.000   12.000    FALSE 1 1.002  3206
[... output truncated ...]
N[1,16]           41.206  4.570   32.000   41.000   50.000    FALSE 1 1.002  3316
N[2,16]            7.462  3.700    1.000    7.000   16.000    FALSE 1 1.002  3310
```

The mixture proportion is estimated at $\gamma = 0.16$ (SD $= 0.06$); hence, high-quality individuals were much rarer in the population than were low-quality individuals. At the average population size, a female in the high-quality group annually produced on average 8.48 (SD $= 0.98$) fledglings (this is calculated as $\exp(\alpha_2)$) as opposed to 5.34 (SD $= 0.27$) for a female in the low-quality group. This is quite a remarkable difference that is likely to be biologically meaningful. Productivity of the high-quality females appeared to vary more over time than did that of the low-quality females (Fig. 14.4).

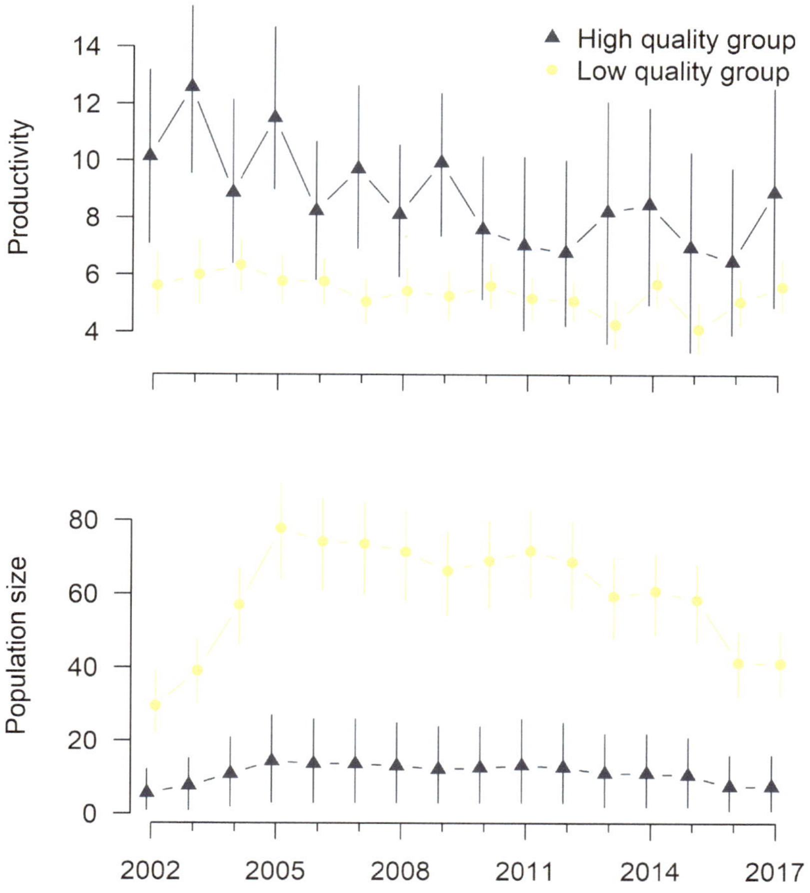

FIGURE 14.4

Estimates of the annual productivity in hoopoe females in the low- and high-quality groups and number of reproducing hoopoes in each (posterior means and 95% credible intervals).

The probability that the residual temporal variance was larger in the high-quality than in the low-quality group was 0.90 (do `mean(out1$sims.list$sigma.rep.t[,1] < out1 $sims.list$sigma.rep.t[,2])`).

14.7 DISCUSSION

The inclusion of the finite-mixture model in the IPM revealed that individual variation in productivity of the hoopoe can be described by two groups of individuals that differ in their mean annual output. Thus, the model allows insights into additional structure beyond that captured by stage and age structure. For the hoopoe example, it seems that incorporation of the additional structure in the submodel for the annual productivity of females was only of limited importance for the dynamics of the population because the high-quality individuals were a small minority and because local recruitment was generally low as a result of the strong natal dispersal, leading to permanent emigration from the study area. We could apply retrospective population analyses (Chapter 9) to more thoroughly study how much individuals of these stage classes contributed to the observed population dynamics. For conservation and management, it can be important to know that within a population, a small fraction of individuals exist that demographically outperform others, because in a next step we might perhaps want to identify such individuals and try to increase their frequency in the population.

Our mixture model relied on several assumptions; most of them are difficult to test statistically. Next, we discuss some of them and sketch out possible modeling alternatives.

First, we assumed *a priori* that there were two groups of individuals that differ in terms of their productivity but where all individuals in the same group were identical. This seems like a strong assumption, but in fact it is less so than the standard stage-structured models that assume identical reproductive output of *all* females within the same stage. The two-group mixture might not be flexible enough for existing individual differences, and perhaps we should assume more groups. We could use model selection to decide how many groups are necessary (Pledger, 2000; Cubaynes et al., 2012; McLachlan and Rathnayake, 2014). Generally, estimation becomes more difficult with more than two groups. We then must use the multinomial distribution for allocating individuals to their groups, but that does not work in BUGS because the multinomial total is a latent variable, and JAGS will report an error when we try to fit such a model. One workaround is the use of several conditional binomial distributions instead of the multinomial (Kéry and Royle, 2016). However, even if the finite-mixture model includes more than two groups, it may still be too restricting. Other mixture models that allow for the mixture of two distributions (instead of two fixed values) or models with individual random effects might be alternatives, i.e., continuous-mixture distributions (Coull and Agresti, 1999; Dorazio and Royle, 2003; Hamel et al., 2017).

A second assumption of our model is that inheritance of group membership from mother to offspring does not occur. As an alternative, we could assume that group membership is perfectly inherited such that local recruits are deterministically allocated to the same group as their mothers. It may also be possible to quantify the degree of heritability by parent-offspring regression or animal models (Villemereuil et al., 2013) and to define a more complex variant of our model that takes heritability into account.

A third assumption is that group membership of an individual does not change over time. It would be straightforward to relax this assumption, e.g., by using a temporal random effect as with the other demographic rates. We have opted for a simpler solution here for the sake of an easier presentation.

Fourth, the individual quality might not be well captured by productivity alone. Indeed, it appears quite likely that high-quality individuals not only have higher productivity, but also have higher survival (Aubry et al., 2011; Plard et al., 2018b). One way to model individual quality in this manner would be the use of a mixture model for the survival data as well, such that individuals in the high-quality group have higher survival and higher productivity. However, the capture-recapture data then need to be analyzed in the state-space formulation (Section 4.5.1.1), because the finite-mixture model is a model with individual effects. This significantly increases the computational burden and in addition, such models usually do not usually mix well at all. Plard et al. (2019a) used such a model to study individual effects in barn swallow (*Hirundo rustica*) males and found differences in productivity but not in adult survival.

Multievent capture-recapture models (Section 4.5.6) offer an alternative to model individual heterogeneity by specification of latent groups in terms of survival (Pradel, 2009; Péron et al., 2010a,b; Gimenez et al., 2018). These models are more flexible than a finite-mixture specification as part of a CJS model, because they can be applied when the latent group membership changes over the lifetime of an individual in a Markovian manner. In contrast, a finite mixture model does not model transitions explicitly, even though we might have specified it with temporally varying mixture proportions and group memberships, rather than keeping them constant over time as in our case study. Moreover, the multievent model can be combined with knowledge about the reproductive output of each individual by classifying the outcome into states. Interactions or consistency of individual survival and productivity can then be studied. A drawback is that productivity must be allocated to stages *a priori*, i.e., thresholds to define groups must be defined. All of this could in principle also be embedded into an IPM, but as far as we know, this has not been done before.

Finite-mixture or multievent models are both adopted when we assume that individual heterogeneity in demographic performance is not related to a measured individual trait. In contrast, if demographic heterogeneity can be predicted by a measured individual trait, this information should be used. If the trait is a categorical variable such as age, it is straightforward to account for it in an IPM; it is just a matter of defining states based on the levels of the categorical variable for the population model. If the trait is a continuous variable such as body mass (or an index for it), an integral projection model must be used instead of a stage-structured model. It is possible to combine an integral projection with an integrated population model (IPM^2), which offers interesting perspectives (Plard et al., 2019b).

Obtaining convergence in finite-mixture models can be challenging because of label switching, but nevertheless, this is easily diagnosed by inspecting the trace-plots, ideally for a model run with many more than our usual three chains (Fig. 14.2). We have shown how label switching can be avoided by imposing constraints. However, the challenges become more difficult when more than two groups are specified, and in general, label switching in Bayesian mixture models can be a challenging problem (Stephens, 2000).

Individual heterogeneity can have important effects on the dynamics of a population, and IPMs offer some interesting possibilities for further insights into population dynamics and this heterogeneity as well as the interaction between the two (Plard et al., 2019a). So far, only a few applications of IPMs have done much exploration of these topics, but we think we will see more such studies in the future.

CHAPTER

BLACK GROUSE

15

CHAPTER OUTLINE

IPM ingredients: spring male count data, sex- and age-classified summer population count data, radio tracking data
Location: Italy
Spatio-temporal scale: 16.2 km^2, 20 years

15.1 INTRODUCTION

The black grouse (*Lyrurus tetrix*) is a charismatic bird of Northern Eurasia that mainly inhabits the early stages of forest succession. In Southern and Western Europe, its distribution is extremely fragmented, as it occupies alpine habitats at the timberline and bogs. Black grouse are sexually dimorphic, with males being much larger and heavier than females. Males are almost completely blueish black apart from red eye combs, white wing bars, and a white under-tail fan (Fig. 15.1). Females are brown-greyish and much more camouflaged than males. In spring, males gather at communal display sites (called leks) in open areas and exhibit spectacular dances and fights. Females visit these leks for short times to mate. Not surprisingly, the strongest males mate with several females (Alatalo et al., 1991; Hovi et al., 1994; Höglund et al., 1997). Females then incubate and rear their young alone. Nests are placed on the ground, and chicks are precocious, i.e., they leave the nest immediately after hatching and follow their mother. In autumn, chicks are nearly fully grown and become independent. They can start to reproduce during the following spring at about 1 year of age. Black grouse are sedentary and show limited dispersal. In many areas of their vast distributional range, they are an important game species.

Black grouse population trends in the Alps are regionally variable, with declines in France and in Italy, but apparently more stable populations in Switzerland, Austria, and Germany (Knaus et al., 2018). Potential causes of threat are human disturbance, the intensification or abandonment of cattle grazing, and hunting (Arlettaz et al., 2007; Patthey et al., 2012; Zbinden et al., 2018), and possibly effects of climate change.

Here, we analyze a 20-year data set collected in the Italian Alps by our colleagues Luca Rotelli and Radames Bionda. They counted the birds in spring and late summer and also conducted a radio tracking study. As in most IPMs, our main interest lies in estimation of the demographic rates and structured population size.

In this case study, we develop a seasonal, two-sex integrated population model (IPM) that describes the dynamics of the black grouse population for each sex from spring to late summer in one calendar year and from late summer to the spring in the following calendar year. A seasonal IPM gives more insights into population dynamics and it becomes possible because the grouse were counted twice during a year and in addition seasonal survival could be estimated from the radio tracking data. In principle, developing a seasonal IPM is not more complicated than the more usual annual IPM, but care must be taken regarding the age class of an individual that may change seasonally. Moreover, we illustrate the adoption of different observation models for the count data sets of the two seasons to reflect different sampling protocols in each.

Integrated Population Models. https://doi.org/10.1016/B978-0-12-820564-8.00006-3

FIGURE 15.1

Fighting black grouse (*Lyrurus tetrix*) cocks at a lek in the Alps (Photograph by www.marcelburkhardt.ch).

15.2 DATA SETS

Luca Rotelli, Radames Bionda and their colleagues collected three different data sets over a total of 20 years (1997–2016) at the Alpe Devero located in the Piemont region of the Alps in Northern Italy (Rotelli et al., 2021). First, displaying males were counted every spring (around May 15) at all leks in the study area. We call this survey the spring survey. Black grouse leks are typically in open areas not far from the tree-line where displaying cocks will be visible and audible at a fairly long distance. Therefore, we assume that these counts are correct on average, i.e., that there is no systematic bias due to imperfect detection or incomplete coverage of the lekking areas.

The second data set was collected during another population survey in late summer (around August 22) when chicks are about 5 weeks old and still follow their mothers. This late summer survey was carried out with pointer dogs that searched and flushed the grouse in all areas that had suitable grouse habitat. Experienced observers determined the sex and age of all flushed birds and counted them; thus, a structured population count was obtained where adult (i.e., at least 1 year old) males and females, and chicks were distinguished. In some cases, the sex of chicks could be determined based on body size, and when this information was available for all chicks in a brood, we used it to estimate the chick sex ratio. The late summer survey did not cover the entire study area. Moreover, and in contrast to the counts of males at the open leks, we could not reasonably assume perfect detection. Rather, late summer counts presumably underestimated the true population size.

Third, during a subset of the study years (1998–2008), a total of 96 grouse of both sexes and age classes (chicks and adults) were captured and marked with VHF radio transmitters. Tagged grouse were located almost daily depending on the weather conditions. We aggregated these data in monthly intervals, i.e., for every month we determined whether an individual was alive or dead, or had lost its transmitter. The probability of detecting a live individual with a functioning radio transmitter is usually very high. In addition, the pooling of the data meant that there were typically many attempts to record an individual per occasion. Therefore, we assumed that detection was perfect in this data set.

15.3 POPULATION MODEL

Because the grouse population was surveyed twice per year in spring and late summer, we develop a seasonal population model that describes the dynamics from mid-May to late August in year *t* and from late August in year *t* to mid-May in year *t*+1. The duration from spring to late summer is 3.5 months and hence much shorter than that from late summer to the following spring (8.5 months). Different age classes occur in the two surveys: the youngest individuals during the spring survey are nearly 1 year old, whereas during the late summer survey, they are just a few weeks old. A potential difficulty in the formulation of a seasonal population model is the definition of the appropriate age classes for both survey periods. To keep our model as simple as possible and because we wanted to assume no age dependence in productivity and just two age classes for survival, we defined two age classes for each season. For the spring survey, we distinguish between 1-year-old individuals (1y) and those that are older (>1y), while for the late summer survey, we distinguish between juveniles that are just a few weeks old (j) and those that are 1 year old or older (≥1y). In our population model, we include both sexes explicitly because of the strong sex differences in behavior (e.g., only females raise broods) and body size, which make differences in demographic performance among the sexes highly likely. Moreover, males are hunted, whereas the killing of females is not allowed in the study area, and this may further add to differences in survival. Black grouse have a polygynous mating system. Therefore, we do not explicitly include a marriage function in our model. Thus, in the model, the number of females is assumed to evolve completely independently from that of males, while the number of males is a function of the number of females through reproduction. The corresponding life-cycle graph is shown in Fig. 15.2.

Seasonal population models are naturally specified using survival that is defined over monthly (or possibly daily) intervals, such that the survival over a season is obtained as a suitably scaled version of survival. In addition, we note that to indicate the seasonal, age, sex, and time attribution of the modeled quantities, we use both subscripts and superscripts, as shown here for the population size of females of age class 1 year during spring in year *t*: $N^{sp}_{1y,f,t}$, where *sp* means spring (and *ls* will be late summer; see below).

To develop the population model, we start by describing the transitions from spring in year *t* to late summer in the same calendar year corresponding to the dashed arrows in the life-cycle graph (Fig. 15.2) that involve survival and recruitment.

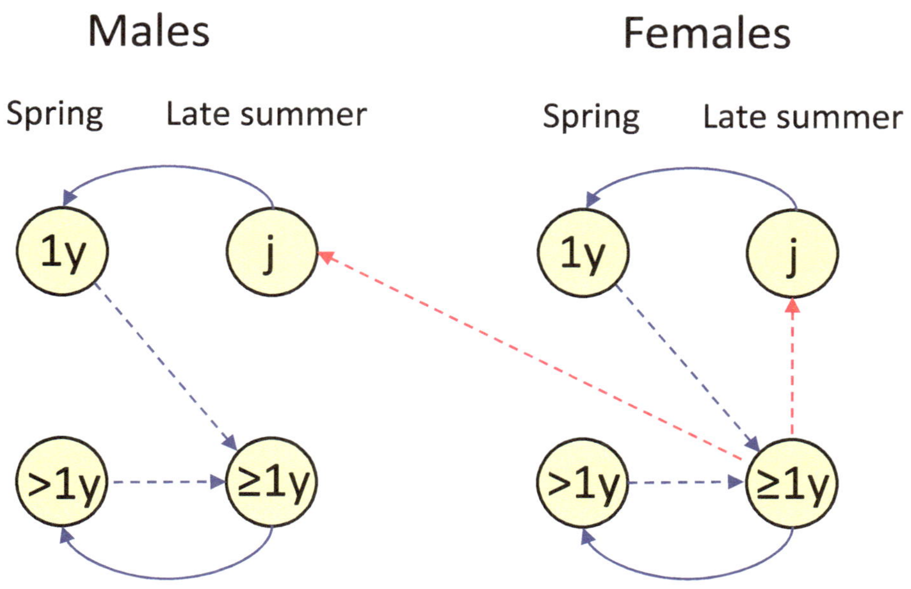

FIGURE 15.2

Life-cycle graph of a two-sex, seasonal population model with survey periods in spring and late summer for the Italian black grouse. The demographic processes are shown by arrows (red: recruitment; blue: survival); dashed blue arrows indicate the spring to late summer transition (3.5-month period), solid blue arrows indicate late summer to spring transitions (8.5-month period). The rates for each arrow are not shown because the expressions would become too long. There are a total of four age classes in the model: juveniles (j) that are just a few weeks old, 1-year-old individuals (1y), individuals that are at least 1 year old (≥1y), and those that are more than 1 year old (>1y).

The latter includes subprocesses for the production of the chicks and their allocation to a sex. We first model the total number of 5-week-old chicks in year t (F_t) using a Poisson distribution with an expected value given by the number of reproducing females in late summer and productivity. We define the productivity in year t (ρ_t) by the expected number of 5-week-old chicks per female. Hence, the number of chicks in year t is

$$F_t \sim Poisson\left(N^{ls}_{\geq 1y,f,t}, \rho_t\right).$$

The number of female chicks is modeled with a binomial distribution with a success rate that is the (female) sex ratio (γ).

$$N^{ls}_{j,f,t} \sim Binomial(\gamma, F_t)$$

The number of male chicks is then the difference between the total number of chicks and the number of females.

$$N^{ls}_{j,m,t} = F_t - N^{ls}_{j,f,t}$$

The second modeled process in the spring to late summer transition is the survival of adults, which we describe again with binomial distributions with success parameters that are the corresponding monthly survival probabilities raised to a power given by the length of the season in terms of the unit time (which we here take to be 1 month).

$$N^{ls}_{\geq 1y,f,t} \sim Binomial\left(s_{a,f,1}\left(s_{a,f,2}\right)^{2.5}, N^{sp}_{1y,f,t} + N^{sp}_{>1y,f,t}\right)$$

$$N^{ls}_{\geq 1y,m,t} \sim Binomial\left(s_{a,m,1}\left(s_{a,m,2}\right)^{2.5}, N^{sp}_{1y,m,t} + N^{sp}_{>1y,m,t}\right)$$

Next, we describe the transitions from the late summer in year t to the spring of year $t+1$. These correspond to the solid arrows in the life-cycle graph (Fig. 15.2), which involve only survival processes. The number of 1 year-old females in the spring of year $t+1$ ($N^{sp}_{1y,f,t+1}$) is modeled with a binomial distribution with a total that is the number of juvenile females in late summer of year t ($N^{ls}_{j,f,t}$) and with success probabilities that are again functions of appropriately scaled monthly survival probabilities (s).

$$N^{sp}_{1y,f,t+1} \sim Binomial\left(\left(s_{j,f,2}\right)^{0.5}\left(s_{j,f,3}\right)^{2}\left(s_{j,f,4}\right)^{5} s_{a,f,1}, N^{ls}_{j,f,t}\right)$$

So, again, in a seasonal population model the basic survival parameters refer to monthly survival. It can vary between age classes, sexes, and seasons, and therefore we have subscripts indicating the age class (j: juvenile; a: at least 1 year old), sex (f: female, m: male), and season (1: spring, defined as the period between April 15 and June 14 [2 months]; 2: summer, from June 15 to September 14 [3 months]; 3: autumn, from September 15 to November 14 [2 months]; 4: winter, from November 15 to April 14 [5 months]). The powers to which the monthly survival probabilities must be raised correspond to the length of the corresponding season in months.

The number of >1 year-old females in the spring in year $t+1$ ($N^{sp}_{>1y,f,t+1}$) is modeled similarly with a binomial distribution. The binomial total is now the number of $\geq$1y females in late summer of year t and the success parameters are adult female survival.

$$N^{sp}_{>1y,f,t+1} \sim Binomial\left(\left(s_{a,f,2}\right)^{0.5}\left(s_{a,f,3}\right)^{2}\left(s_{a,f,4}\right)^{5} s_{a,f,1}, N^{ls}_{\geq 1y,f,t}\right)$$

For the age-dependent number of males, we have the analogous equations.

$$N^{sp}_{1y,m,t+1} \sim Binomial\left(\left(s_{j,m,2}\right)^{0.5}\left(s_{j,m,3}\right)^{2}\left(s_{j,m,4}\right)^{5} s_{a,m,1}, N^{ls}_{j,m,t}\right)$$

$$N^{sp}_{>1y,m,t+1} \sim Binomial\left(\left(s_{a,m,2}\right)^{0.5}\left(s_{a,m,3}\right)^{2}\left(s_{a,m,4}\right)^{5} s_{a,m,1}, N^{ls}_{\geq 1y,m,t}\right)$$

15.4 COMPONENT DATA LIKELIHOODS

We next describe the basic models adopted for each of the three data sets and define their likelihoods. We load the data set and produce an overview of its content.

```
library(IPMbook); library(jagsUI)
data(grouse)
str(grouse)

List of 10
 $ ch        : int [1:96, 1:128] 1 1 NA NA NA NA NA NA NA NA ...
 $ age       : num [1:96, 1:127] 2 2 NA NA NA NA NA NA NA NA ...
 $ sex       : int [1:96] 2 2 2 2 2 2 2 2 2 2 ...
 $ season    : int [1:128] 1 1 2 2 2 3 3 4 4 4 ...
 $ count.sp  : int [1:20] 36 45 52 64 46 49 50 35 36 39 ...
 $ count.lsM: int [1:20] 15 11 14 14 14 5 16 19 19 16 ...
 $ count.lsF: int [1:20] 23 32 29 20 19 18 28 21 31 39 ...
 $ count.lsC: int [1:20] 44 86 76 28 18 24 99 53 101 82 ...
 $ v         : int [1:20] 30 39 21 10 11 6 67 32 65 55 ...
 $ u         : int [1:20] 18 22 11 5 4 2 25 15 29 26 ...
```

The "grouse" data set contains 10 objects including the counts during the two surveys and the radio tracking data.

15.4.1 POPULATION COUNT DATA

The annual counts of males during the spring survey (C^{sp}) are stored in vector `count.sp`. We use a Poisson distribution to model them because we assume there was no systematic error due to imperfect area coverage or detection.

$$C_t^{sp} \sim Poisson\left(N_{1y,m,t}^{sp} + N_{>1y,m,t}^{sp}\right)$$

The population counts obtained during the late summer surveys by flushing grouse with pointer dogs are structured by age and sex. The numbers of adult males (C_m^{ls}) and females (C_f^{ls}) are stored in vectors `count.lsM` and `count.lsF`, respectively. This survey did not cover the complete study area, and some individuals present in the surveyed area might have escaped detection. Therefore, we use binomial distributions to model the late summer counts with a success parameter (p_t) that is the product of incomplete survey area coverage and imperfect detection and that we here call observability. Therefore, for females and males, we write

$$C_{f,t}^{ls} \sim Binomial\left(p_t, N_{\geq 1y,f,t}^{ls}\right) \text{ and } C_{m,t}^{ls} \sim Binomial\left(p_t, N_{\geq 1y,m,t}^{ls}\right).$$

We use the same observability p_t for both sexes because we assume that males and females were present in the same areas (no sex-specific habitat selection) and because pointer dogs are similarly successful at detecting males and females (Jenkins et al., 1963). We allow observability to be variable over time using random time effects because we assume that survey coverage varied over time due to dog availability and perhaps also quality differences of different dogs, not all of which operated in all years.

15.4.2 RADIO TRACKING DATA

The radio tracking data `ch` of 96 individuals are summarized in capture histories with 128 monthly occasions between April 1998 and November 2008. Each occasion lasted from day 15 of a month to day 14 of the following month. If tracked individual i was found to be alive, $ch_{i,t} = 1$, and if it was found to be dead, $ch_{i,t} = 0$. A total of 63 individuals were eventually found dead, and hence, the data of the remaining individuals were right-censored at the occasion when they were seen alive for the last time. Reasons for censoring were mostly transmitter failure or loss. Vector $\boldsymbol{k}$ indicates the occasion of death or censoring for each individual, whereas vector $\boldsymbol{f}$ indicates its occasion of marking. Since most individuals were tracked several days per month, we assume perfect detection for each monthly occasion and adopt a "known-fate model" (Devineau et al., 2014; Murray and Bastille-Rousseau, 2020). This model is similar to the nest survival model described in Section 4.4.6 and consists of Bernoulli coin flips with the corresponding survival probability as a parameter.

$$ch_{i,t} \sim Bernoulli\left(s_{\text{age}(i,t-1),\text{sex}(i),\text{season}(t-1)}\right)$$

For each individual i, time t runs from 1 plus the marking occasion (f_i) until the occasion of censoring or death (k_i), matrix `age` indicates age class (1: juvenile, 2: adult) on each occasion, and the vectors `sex` (1: females, 2: males) and `season` (with four levels as defined above) denote the sex of an individual and the season of each capture occasion, respectively.

From the season-, sex-, and age-specific monthly survival probabilities, we obtain the sex- and age-specific annual survival probabilities as derived quantities. Annual juvenile survival was calculated from the age of 5 weeks until the spring survey in mid-May.

15.4.3 MODELING PRODUCTIVITY AND CHICK SEX RATIO

We estimate productivity from the late summer counts as a ratio of the number of flushed chicks $C^{ls}_{c,t}$ (in vector `count.lsC`) to that of flushed adult females ($C^{ls}_{f,t}$, as described above). Adopting a Poisson distribution, we model productivity (ρ) with annual variation according to random effects (not shown).

$$C^{ls}_{c,t} \sim Poisson\left(C^{ls}_{f,t}\rho_t\right)$$

The annual number of chicks in families where each member could be sexed is given in vector `v` and the number of the females among them in vector `u`. For the chicks we use a binomial distribution that depends on the proportion of females (γ), which we term "sex ratio" here and assume to be constant over time.

$$u_t \sim Binomial(\gamma, v_t)$$

15.5 INTEGRATED POPULATION MODEL

Using the component data likelihoods just described, we now develop the IPM by specification of a joint analysis where some parameters occur in more than one data set and have identical names there. We specify vague priors for all estimated quantities. For probabilities, we usually express lack of prior knowledge by adoption of a uniform distribution between 0 to 1. As an alternative, we here use a beta distribution with shape parameters 1 and 1, which looks exactly identical to such a uniform distribution. However, the beta distribution is the conjugate prior for the binomial distribution (Gelman et al., 2004). Using conjugate priors can have computational benefits, although in JAGS, it often appears to be irrelevant whether a conjugate or a nonconjugate prior is used. Because the first survey of the study was the spring lek count in 1998, the seasonal IPM starts with the population in spring. Consequently, we define priors for the initial sex- and age-structured population sizes in spring. To be vague, for each age and sex class, we use discrete uniforms between 1 and 100.

We also compute posterior predictive checks for the population counts, productivity, and sex ratio data. As discrepancy measures we use percentage error (Besbeas and Morgan, 2014), the deviance, and Freeman-Tukey statistics (Brooks et al., 2000), respectively. We don't compute any test on the radio tracking data.

Next, we define a function to compute the occasion of the last observation in a capture-history matrix, bundle the data (where we use that function), and write the BUGS code of the model.

```
# Bundle data
getLast <- function(x) max(which(!is.na(x)))                    # Last occasion function

jags.data <- with(grouse, list(ch=ch, age=age, sex=sex, season=season, f=getFirst(ch),
  k=apply(ch, 1, getLast), C.sp=count.sp, C.lsM=count.lsM, C.lsF=count.lsF, C.lsC=count.lsC,
  u=u, v=v, nind=nrow(ch), ny=length(count.sp), pinit=dUnif(1, 100)))

str(jags.data)

List of 15
 $ ch     : int [1:96, 1:128] 1 1 NA NA NA NA NA NA NA NA ...
 $ age    : num [1:96, 1:127] 2 2 NA NA NA NA NA NA NA NA ...
 $ sex    : int [1:96] 2 2 2 2 2 2 2 2 2 2 ...
 $ season: int [1:128] 1 1 2 2 2 3 3 4 4 4 ...
 $ f      : int [1:96] 1 1 2 2 2 2 14 14 14 14 ...
 $ k      : int [1:96] 8 15 9 4 6 27 25 18 23 25 ...
 $ C.sp   : int [1:20] 36 45 52 64 46 49 50 35 36 39 ...
 $ C.lsM : int [1:20] 15 11 14 14 14 5 16 19 19 16 ...
 $ C.lsF : int [1:20] 23 32 29 20 19 18 28 21 31 39 ...
 $ C.lsC : int [1:20] 44 86 76 28 18 24 99 53 101 82 ...
 $ u      : int [1:20] 18 22 11 5 4 2 25 15 29 26 ...
 $ v      : int [1:20] 30 39 21 10 11 6 67 32 65 55 ...
 $ nind   : int 96
```

```
 $ ny    : int 20
 $ pinit : num [1:100] 0.01 0.01 0.01 0.01 0.01 0.01 0.01 0.01 0.01 ...

# Write JAGS model file
cat(file="model1.txt", "
model {
# Priors and linear models
# Survival
for (a in 1:2){                                                         # age
  for (j in 1:2){                                                       # sex
    for (m in 1:4){                                                     # season
      s[a,j,m] ~ dbeta(1, 1)
    } #m
  } #j
} #a

# Productivity
for (t in 1:ny){
  log.rho[t] ~ dnorm(log.mean.rho, tau.rho)
  rho[t] <- exp(log.rho[t])
}
mean.rho ~ dunif(0, 5)
log.mean.rho <- log(mean.rho)
sigma.rho ~ dunif(0, 1)
tau.rho <- pow(sigma.rho, -2)

# Sex ratio of the 5 weeks old chicks
gamma ~ dbeta(1, 1)

# Resighting probabilities for the late summer counts
for (t in 1:ny){
  logit.p[t] ~ dnorm(lmean.p, tau.p)
  p[t] <- ilogit(logit.p[t])
}
mean.p ~ dbeta(1, 1)
lmean.p <- logit(mean.p)
sigma.p ~ dunif(0, 5)
tau.p <- pow(sigma.p, -2)

# Population count data (state-space model)
# Model for the initial population size in spring: uniform priors
Nsp[1,1,1] ~ dcat(pinit[])
Nsp[2,1,1] ~ dcat(pinit[])
Nsp[1,2,1] ~ dcat(pinit[])
Nsp[2,2,1] ~ dcat(pinit[])

# Process model: population sizes in spring
for (t in 1:(ny-1)){
  # Females
  Nsp[1,1,t+1] ~ dbin(s[1,1,2]^0.5 * s[1,1,3]^2 * s[1,1,4]^5 * s[2,1,1], Nls[1,1,t])  # 1y
  Nsp[2,1,t+1] ~ dbin(s[2,1,2]^0.5 * s[2,1,3]^2 * s[2,1,4]^5 * s[2,1,1], Nls[2,1,t])  # >1y

  # Males
  Nsp[1,2,t+1] ~ dbin(s[1,2,2]^0.5 * s[1,2,3]^2 * s[1,2,4]^5 * s[2,2,1], Nls[1,2,t])  # 1y
  Nsp[2,2,t+1] ~ dbin(s[2,2,2]^0.5 * s[2,2,3]^2 * s[2,2,4]^5 * s[2,2,1], Nls[2,2,t])  # >1y
}

# Process model: population sizes in late summer
for (t in 1:ny){
  # Total number of chicks
  F[t] ~ dpois(Nls[2,1,t] * rho[t])
  # Allocate chicks to a sex
  Nls[1,1,t] ~ dbin(gamma, F[t])                                       # Female chicks
  Nls[1,2,t] <- F[t] - Nls[1,1,t]                                      # Male chicks
  # Survival
  Nls[2,1,t] ~ dbin(s[2,1,1] * s[2,1,2]^2.5, (Nsp[1,1,t] + Nsp[2,1,t]))   # ≥1y females
  Nls[2,2,t] ~ dbin(s[2,2,1] * s[2,2,2]^2.5, (Nsp[1,2,t] + Nsp[2,2,t]))   # ≥1y males
}
```

```
# Observation models
for (t in 1:ny){
  # Spring counts of 1y and >1y males
  C.sp[t] ~ dpois((Nsp[1,2,t] + Nsp[2,2,t]))
  # Late summer counts of ≥1y females and of ≥1y males
  C.lsF[t] ~ dbin(p[t], Nls[2,1,t])
  C.lsM[t] ~ dbin(p[t], Nls[2,2,t])
}

# Radio tracking data (Bernoulli model)
for (i in 1:nind){
  for (t in (f[i]+1):k[i]){
    ch[i,t] ~ dbern(s[age[i,t-1], sex[i], season[t-1]])
  } #t
} #i

# Productivity data (Poisson model)
for (t in 1:ny){
  C.lsC[t] ~ dpois(C.lsF[t] * rho[t])
}

# Chick sex ratio data (binomial model)
for (t in 1:ny){
  u[t] ~ dbin(gamma, v[t])
}

# Derived quantities: annual survival
ann.s[1,1] <- s[1,1,2] * s[1,1,3]^2 * s[1,1,4]^5 * s[2,1,1]       # juv females
ann.s[2,1] <- s[2,1,1]^2 * s[2,1,2]^3 * s[2,1,3]^2 * s[2,1,4]^5  # ad females
ann.s[1,2] <- s[1,2,2] * s[1,2,3]^2 * s[1,2,4]^5 * s[2,2,1]       # juv males
ann.s[2,2] <- s[2,2,1]^2 * s[2,2,2]^3 * s[2,2,3]^2 * s[2,2,4]^5  # ad males
}
")
```

The choice of "good" initial values for the Markov chain Monte Carlo (MCMC) algorithm can become vital in more complex IPMs. If no initial values are supplied, JAGS draws them at random from the priors defined in the model, and this may then lead to conflicts between the initial values, data, and/or the model, in which case JAGS will crash. An example is when an observed number of survivors happens to be larger than the corresponding (latent) binomial total. We often had success using a small range of the population sizes in the first year and choosing hypermeans of the demographic rates close to the suspected values of the parameters. Here we choose such an approach. In choosing initial values that work for JAGS (or also in general, for other generic modeling software), some trial and error is always involved, and sometimes even extreme amounts of trial and error. (Remember that the initial values have nothing to do with specification of information by priors—at convergence of the Markov chains, the solutions are independent from the initial values, but they will always be affected by the priors as per Bayes rule).

```
# Initial values
inits <- function(){
  Nsp <- array(NA, dim=c(2, 2, 20))
  Nsp[,,1] <- round(runif(4, 30, 40))
  s <- array(runif(2*2*4, 0.94, 0.97), dim=c(2, 2, 4))
  list(Nsp=Nsp, s=s)
}
```

Finally, we list the parameters that we want to estimate, choose the MCMC settings, and launch the model in JAGS. Note that to avoid unwieldy BUGS code, we don't show the calculations required to compute the Bayesian *p*-values for the goodness-of-fit assessments; you can find that code on the book website.

```
# Parameters monitored
parameters <- c("s", "ann.s", "mean.rho", "gamma", "mean.p", "sigma.rho",  "sigma.p", "rho", "p",
  "Nsp", "Nls")

# MCMC settings
ni <- 120000; nb <- 20000; nc <- 3; nt <- 20; na <- 10000

# Call JAGS (ART 8 min), check convergence and summarize posteriors
set.seed(101)          # Reproducible results without crashing
out1 <- jags(jags.data, inits, parameters, "model1.txt", n.iter=ni, n.burnin=nb, n.chains=nc,
  n.thin=nt, n.adapt=na, parallel=TRUE)
traceplot(out1)
```

15.6 RESULTS

The model run converges reasonably quickly and yields the following results (with, as usual, not all rows of the posterior summary table shown).

```
print(out1, 3)
                  mean     sd     2.5%      50%      97.5% overlap0 f  Rhat n.eff
s[1,1,1]         0.501  0.290    0.024    0.503     0.976    FALSE 1 1.000 15000
s[2,1,1]         0.943  0.018    0.904    0.944     0.972    FALSE 1 1.000 15000
s[1,2,1]         0.501  0.288    0.025    0.506     0.974    FALSE 1 1.000 15000
s[2,2,1]         0.880  0.035    0.807    0.883     0.942    FALSE 1 1.000 15000
s[1,1,2]         0.930  0.065    0.759    0.948     0.998    FALSE 1 1.000 15000
s[2,1,2]         0.948  0.013    0.920    0.949     0.971    FALSE 1 1.000  8594
s[1,2,2]         0.843  0.097    0.612    0.861     0.979    FALSE 1 1.000 11434
s[2,2,2]         0.970  0.016    0.932    0.973     0.994    FALSE 1 1.000 15000
s[1,1,3]         0.878  0.055    0.759    0.884     0.969    FALSE 1 1.000 14762
s[2,1,3]         0.927  0.022    0.879    0.929     0.964    FALSE 1 1.000 15000
s[1,2,3]         0.863  0.063    0.728    0.868     0.966    FALSE 1 1.001  3878
s[2,2,3]         0.915  0.036    0.834    0.920     0.972    FALSE 1 1.000  5561
s[1,1,4]         0.937  0.027    0.881    0.939     0.983    FALSE 1 1.000  6933
s[2,1,4]         0.967  0.011    0.943    0.968     0.984    FALSE 1 1.000  8364
s[1,2,4]         0.927  0.032    0.858    0.930     0.980    FALSE 1 1.000 15000
s[2,2,4]         0.956  0.019    0.912    0.959     0.985    FALSE 1 1.000  5847
ann.s[1,1]       0.490  0.071    0.359    0.489     0.634    FALSE 1 1.000 10877
ann.s[2,1]       0.551  0.045    0.462    0.550     0.637    FALSE 1 1.000 15000
ann.s[1,2]       0.378  0.068    0.253    0.376     0.519    FALSE 1 1.000  7952
ann.s[2,2]       0.476  0.066    0.347    0.476     0.603    FALSE 1 1.000 14173
mean.rho         2.175  0.163    1.867    2.170     2.514    FALSE 1 1.000 15000
gamma            0.493  0.021    0.451    0.493     0.534    FALSE 1 1.000 15000
mean.p           0.430  0.038    0.361    0.428     0.510    FALSE 1 1.000 15000
sigma.rho        0.298  0.065    0.194    0.290     0.447    FALSE 1 1.000  8608
sigma.p          0.376  0.112    0.187    0.364     0.620    FALSE 1 1.001  2673
rho[1]           1.963  0.252    1.500    1.950     2.484    FALSE 1 1.000 14867
[... output truncated ...]
rho[20]          2.092  0.248    1.634    2.082     2.602    FALSE 1 1.000 10745
p[1]             0.408  0.059    0.298    0.405     0.531    FALSE 1 1.000  7604
[... output truncated ...]
p[20]            0.406  0.057    0.304    0.403     0.526    FALSE 1 1.000 15000
Nsp[1,1,1]      38.809 23.047    2.000   38.000    84.000    FALSE 1 1.000  3816
Nsp[2,1,1]      38.726 23.279    2.000   37.000    84.000    FALSE 1 1.000 14971
Nsp[1,2,1]      19.692 11.514    1.000   19.000    41.000    FALSE 1 1.001  3196
Nsp[2,2,1]      19.750 11.504    1.000   19.000    41.000    FALSE 1 1.001  2832
Nsp[1,1,2]      32.980  7.503   20.000   32.000    49.000    FALSE 1 1.000 12507
[... output truncated ...]
Nsp[1,1,20]     45.686 10.614   27.000   45.000    69.000    FALSE 1 1.000 15000
Nsp[2,1,20]     42.317  7.023   30.000   42.000    57.000    FALSE 1 1.000 15000
Nsp[1,2,20]     41.317  7.747   27.000   41.000    57.000    FALSE 1 1.000 11306
Nsp[2,2,20]     31.499  5.899   20.000   31.000    43.000    FALSE 1 1.000  7807
Nls[1,1,1]      63.130 12.022   42.000   62.000    89.000    FALSE 1 1.000 15000
Nls[2,1,1]      63.566  9.722   46.000   63.000    84.000    FALSE 1 1.000  5830
Nls[1,2,1]      60.773 12.075   40.000   60.000    87.000    FALSE 1 1.000 15000
Nls[2,2,1]      32.500  4.567   24.000   32.000    42.000    FALSE 1 1.000 15000
[... output truncated ...]
Nls[1,1,20]     74.401 17.089   45.000   73.000   111.000    FALSE 1 1.000 15000
Nls[2,1,20]     72.206 10.858   53.000   72.000    95.000    FALSE 1 1.000 15000
Nls[1,2,20]     76.579 17.259   47.000   75.000   114.000    FALSE 1 1.000 15000
Nls[2,2,20]     59.468  6.765   47.000   59.000    73.000    FALSE 1 1.000 11886
```

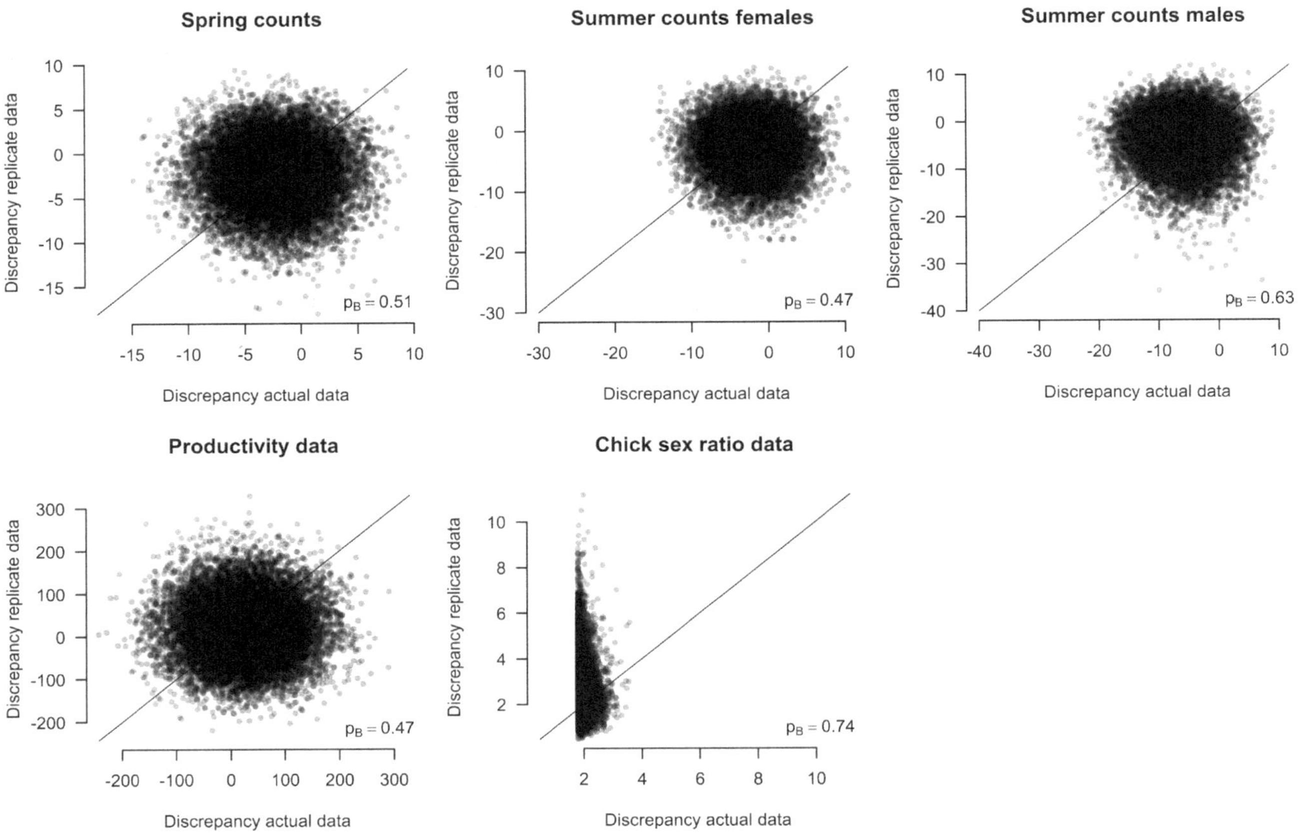

FIGURE 15.3

Scatter plots of the discrepancy measures of the actual data versus those for the replicated data for different model components. Insets give the values of the Bayesian *p*-values.

Before we look at individual estimates under the model, we look at the goodness-of-fit of the different model components. All Bayesian *p*-values and scatter plots look satisfactory for the chosen discrepancy measures (Fig. 15.3), suggesting an adequate fit of the model.

The estimated sex-specific population sizes in spring and late summer indicate different trajectories over time (Fig. 15.4). Females were more abundant than males in both seasons, but the difference tended to become smaller over time. As expected, population size was greater in late summer than in spring. The observed counts of cocks in spring agreed well with the estimated number of males at that time. We don't show the late summer counts in Fig. 15.4 because we knew *a priori* that they do not match the estimated numbers. With an average observability of about 0.4, the estimated population size is about 2.5 times greater than the counts.

In principle, our seasonal population model allows us to describe the trajectories of the age- and sex-specific population sizes for every month over 20 years, since we describe the population dynamics in terms of monthly survival. As an illustration, we produce a plot of the estimates of the annual female population size at the start of reproduction (during the spring survey) and after reproduction (during the late summer survey; Fig. 15.5). This yields the typical sawtooth pattern that we expect for a birth-pulse population (see Section 3.2.1). The increase from spring to summer mostly reflects reproduction, and we can recognize years with high reproduction as those with a larger difference between spring and summer.

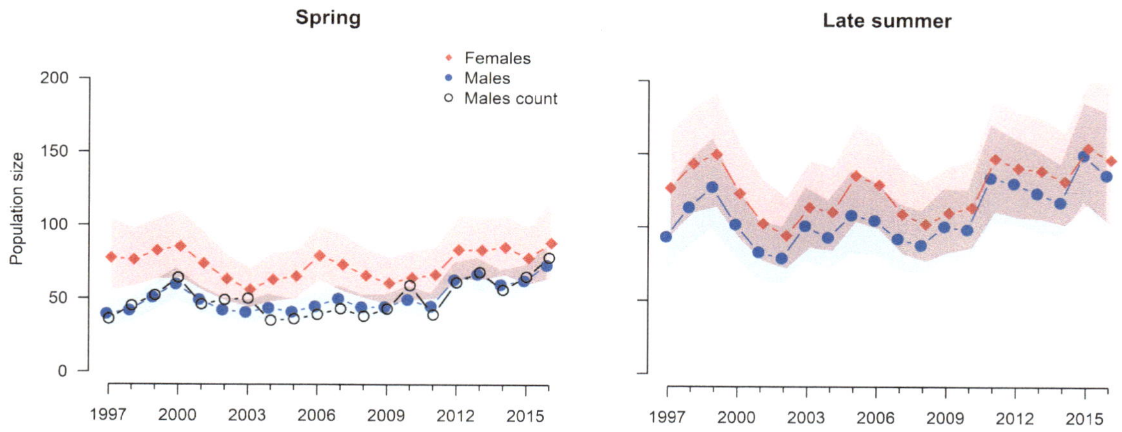

FIGURE 15.4

Estimates of the numbers of male and female black grouse in spring (left panel) and in late summer (right panel) on the Alpe Devero, Piemont, Italy. Diamonds and solid circles show the posterior means, with 95% credible intervals shown as shaded areas. Spring counts of males at the leks are also shown in the left panel.

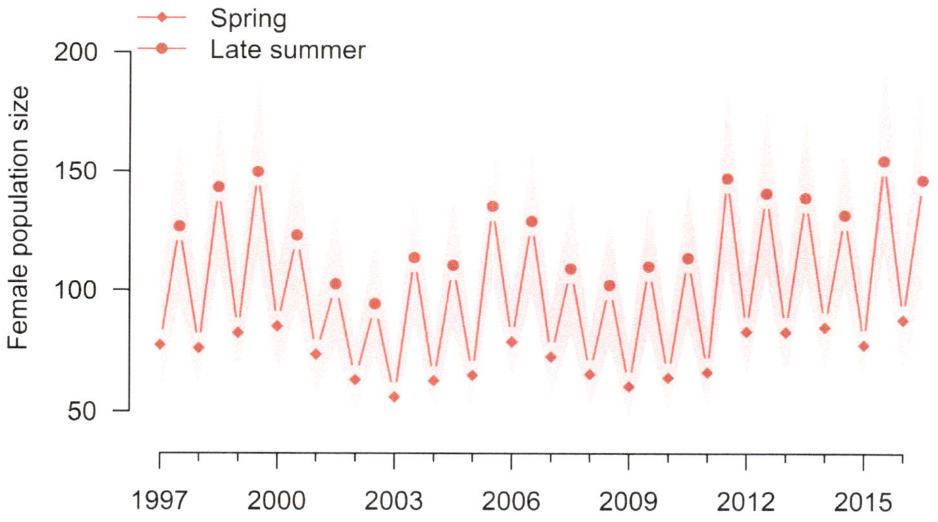

FIGURE 15.5

Estimates of the number of female black grouse in spring and late summer on the Alpe Devero over the 20-year study period. The sawtooth pattern is a classical feature of birth-pulse populations. We show posterior means and pointwise 95% credible intervals.

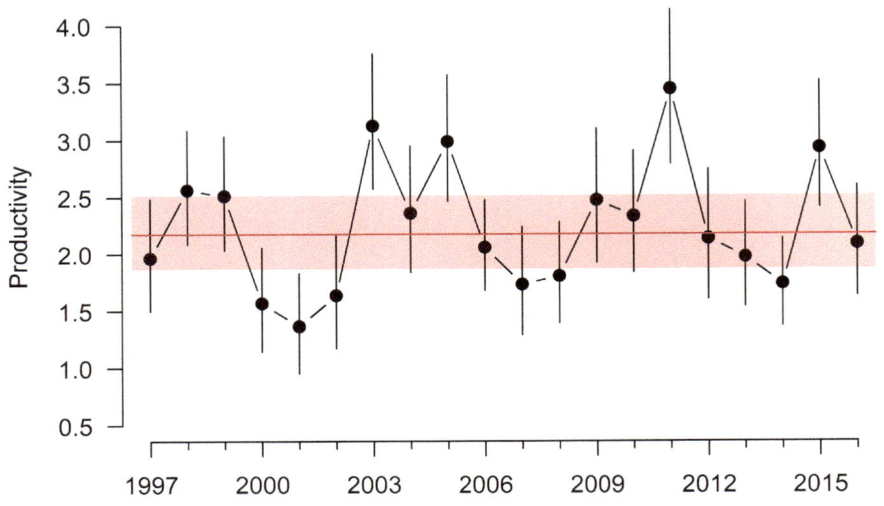

FIGURE 15.6

Estimates of the productivity of black grouse expressed as the number of 5-week-old chicks per female on the Alpe Devero. Solid circles are posterior means of the annual estimates along with their 95% credible intervals. The red horizontal line and the shaded area around it show the estimate of the hypermean of the random year effects with its 95% credible interval.

Productivity was on average 2.2 chicks per female, with strong annual fluctuations (Fig. 15.6). This is typical for black grouse and is often due to variable weather conditions. Newly hatched chicks are sensitive to cold and wet weather (Zbinden and Salvioni, 2004).

In our model, we estimate seasonal monthly survival, and from this, we compute annual survival probabilities. These tend to be higher in females than in males (Fig. 15.7), but the uncertainty associated with them is high. Consequently, the probability that annual survival of adult females is superior to that of adult males is only 0.83. Juvenile survival in both sexes is close to the adult survival of each. A possible reason for this slightly astonishing finding is that juvenile survival refers to the survival of chicks that are already 5 weeks old, thus excluding the most dangerous time in the life of a fledgling (Naef-Daenzer and Grüebler, 2016).

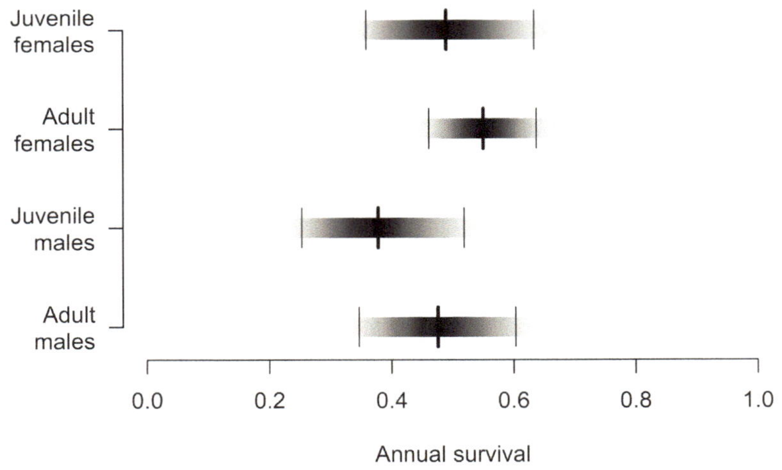

FIGURE 15.7

Estimates of the annual survival of black grouse according to age class and sex. The darkness of the strips is proportional to the posterior densities. The vertical lines show the 95% credible intervals and posterior means (thin lines).

15.7 DISCUSSION

We developed a seasonal IPM for a population of black grouse at the Alpe Devero (Piemont, Italy) to estimate demographic rates and structured population size and investigate population dynamics at the sub-annual scale. Such seasonal models are interesting and always boil down to describing the survival processes at a common, higher temporal resolution (here, 1 month) that can then be scaled up to the lengths of the seasons defined in the model. Seasonal IPMs are necessary when the birth-pulse does not occur at the same time of year as the population count (see Section 6.6.3).

Our estimates of annual survival are lower than those from another Alpine population in France (Caizergues and Ellison, 1997), but productivity is higher (Caizergues and Ellison, 2000). Females outnumbered males in the study population. As the chick sex ratio was even ($\gamma = 0.49$, SD: 0.02), the difference must be due to the slightly lower male survival probability.

The seasonal IPM included two annual surveys and hence two observation models for the state-space model. Because it is reasonable to assume that the spring count of males at the leks is unbiased on average, we used the Poisson distribution and estimated the true population of males in spring. This then enabled us to estimate the true population size in late summer, although that survey had incomplete observability. We used the binomial distribution for the observation model of the late summer survey and estimated an observability parameter. If the spring survey had not yielded true population size, this parameter would not have been estimable.

The IPM in this chapter is based on the assumption that observability during the late summer surveys is identical for both sexes. We think that this assumption is reasonable for our study. However, when in doubt, it would be straightforward in principle to relax the assumption by introducing a sex-specific observability parameter. This parameter may not be separately estimable for both sexes, and hence this should be checked, for instance, with a small simulation study or by a prior sensitivity analysis (see Chapter 13). In our case, we have no doubt that the parameter would be estimable for males but are less sure for the females.

Radio tracking surveys were very intense in our study. This led us to assume that detection was perfect and that we could always distinguish between radio failure and the death of an individual. In less intense studies, these assumptions may not be met, and the model would have to be adapted accordingly. Basically, there are three processes that would then have to be modeled: animal survival, the encounter of an individual with a working transmitter, and transmitter survival or failure. Imperfect detection of an individual that is alive and has a working transmitter is easily modeled by adopting a capture-recapture model (Section 4.5, Pollock et al., 1995). However, adequate modeling of transmitter failure can be more challenging (Tsai et al., 1999; Cowen and Schwarz, 2005; Murray, 2006).

The radio tracking data were used here to estimate survival only. However, they also contain information about productivity if additional information from tracked females can be collected, e.g., whether a female is reproducing, clutch size, hatching success, or renesting probability. Rotelli et al. (2021) developed an IPM that included this additional information, and therefore they could decompose overall productivity into different sub-processes. The recruitment processes then need to start from the number of females in spring (i.e., from the nodes 1y and >1y in Fig. 15.2) and must include the survival of the mothers from spring to the time when they initiate their clutches. Rotelli et al. (2021) performed a retrospective population analysis and found that chick survival during the first 5 weeks after hatching was the main driver of the population.

Seasonal or "full-annual-cycle" population models provide additional insights into population dynamics compared with annual models that typically focus on the population at one point in time, typically at the time of reproduction. For example, in migratory birds, density regulation may work completely differently on breeding than on nonbreeding grounds. This can only be studied using models with a seasonal resolution. Or if we want to use a model for setting a hunting quota in autumn but a population is surveyed in spring, a model with seasonal temporal resolution is again needed, since otherwise the number of individuals at risk of harvest in autumn will be unknown (Arnold et al., 2018). Hostetler et al. (2015) present a very insightful overview about population models with a seasonal resolution in general (i.e., not restricted to IPMs) and about what kind of additional insights they can offer.

Seasonal IPMs may need different and additional data over what we need in typical IPMs with an annual resolution in time. The most important requirement is that they need a finer temporal resolution of survival than 1 year. Ideally, data are collected that allow estimation of seasonal survival probabilities based on adequately sampled data, such as the radio tracking data here. Multiple population surveys within a season may also help in estimation of seasonal survival and moreover enhance estimates of population size. If only annual estimates of survival are available, we may still be able to develop a seasonal model, but we then must rely on the assumption that survival within a year is constant and not seasonally variable, or at least that we cannot estimate any seasonal differences.

The application of seasonal IPMs is still rare, most likely because the additional data they require are rarely available. Some notable exceptions are Hatter et al. (2017) and Zimmermann et al. (2017), who modeled harvested populations that required a seasonal decomposition of the demographic processes because the time of harvest was different from the time when the population surveys were conducted. Another interesting study estimated seasonal survival probabilities in a migratory bird whose population was surveyed during breeding and where capture-recapture data on an annual basis were collected both on the breeding and nonbreeding grounds (Rushing et al., 2017). This design allowed estimation of survival during the stationary phases as well as during migration (Rushing, 2019).

CHAPTER

BARN SWALLOW

16

CHAPTER OUTLINE

IPM ingredients: population count data, capture-recapture data, productivity data
Location: nine widely dispersed study areas in Switzerland
Spatio-temporal scale: approx. 20,000 km^2, 7 years

16.1 INTRODUCTION

The integrated population models (IPMs) developed in Chapters 11−15 jointly analyzed data from the individual and the population levels sampled in a single population. Dispersal movements of individuals within a study population as well as the potential impact of the dynamics of neighboring populations were assumed to be absent or at least negligible. The only spatial aspect in some of these IPMs was emigration from and immigration to the study population. We accounted for emigration in an indirect way without explicit estimation by using capture-recapture data that yield estimates of apparent survival. Immigration was estimated explicitly as a hidden parameter. Yet, the study of spatial aspects beyond emigration and immigration such as dispersal within the study area, metapopulation dynamics, or population synchrony may all enhance our understanding of the dynamics of a population of interest. Here we develop an IPM to investigate spatial population synchrony, i.e., coincident changes in abundance or demographic rates of geographically disjunct populations (Liebhold et al., 2004).

Population synchrony is common in many different taxa including insects, molluscs, and vertebrates (Liebhold et al., 2004) and can be important for conservation, as it increases the extinction risk of assemblages of populations (Harrison and Quinn, 1989). Three main mechanisms have been identified as causes of population synchrony (Liebhold et al., 2004). First, spatial autocorrelation of the environment, typically weather, can cause population synchrony over fairly large spatial scales (Koenig, 2002). Individuals from different populations are then exposed to similar environmental conditions over a given period, and therefore demographic rates show similar temporal fluctuations that lead to synchrony in the population growth rates. For this, at least one of the demographic rates must vary synchronously to induce population synchrony. This mechanism is also known as the Moran effect (Moran, 1953) and has been found repeatedly in different taxa (Grenfell et al., 1998; Tedesco et al., 2004; Sæther et al., 2007). Second, asymmetric dispersal among populations can cause population synchrony (Paradis et al., 1999; Ripa, 2000). Dispersal can partly even out differences in population densities over space, in particular when it is density-dependent. Under this mechanism, population synchrony is also possible in the absence of any demographic rates that covary among multiple populations. Finally, trophic interactions with a second species, whose populations vary in synchrony, may also result in population synchrony of a study species (Jones et al., 2003).

Usually, population synchrony is studied by the analyses of multivariate time-series of abundance data from multiple populations (Grenfell et al., 1998; Sæther et al., 2007). This allows the description of synchrony at the population level, but

Integrated Population Models. https://doi.org/10.1016/B978-0-12-820564-8.00007-5

FIGURE 16.1

Barn swallow (*Hirundo rustica*) feeding young in a nest in a goat stable (photograph by www.marcelburkhardt.ch).

the demographic mechanism(s) that induce population synchrony typically remain unknown. The study of the demographic mechanisms underlying population synchrony additionally requires sampling and analyzing of data from the individual level. If individual- and population-level data are available from several populations for time periods that are at least partly overlapping, we can adopt an IPM to study synchrony of demographic rates and of the population growth rate. The development of such an IPM is the goal of this case study.

We analyze data sampled in nine Swiss populations of barn swallows (*Hirundo rustica*, Fig. 16.1) using an IPM originally developed by Schaub et al. (2015). Barn swallows are small, short-lived, long-distance migratory passerine birds. They live in agricultural areas and often occur in association with cattle or dairy farming. About 75% of pairs in Switzerland are double-brooded (Grüebler et al., 2010). Barn swallows exhibit strong natal, but limited breeding dispersal (Turner, 2006; Schaub and Hirschheydt, 2009). The IPM developed here to study population synchrony jointly analyzes the data from all nine populations. The population model for all populations is identical, but the estimates of the demographic rates vary both temporally and spatially (among populations). We study synchrony by decomposing the total spatio-temporal variance into a synchronous and an asynchronous part. As a measure of synchrony, we compute the intraclass correlation coefficient (ICC) from this variance decomposition (Grosbois et al., 2009) for each demographic rate that varies over time.

16.2 DATA SETS

The data were collected primarily by volunteer bird ringers in a collaborative project coordinated by the Swiss Ornithological Institute. The bird ringers followed a standardized protocol to collect the data, which we briefly describe. A total of nine study sites were defined based on logistics and practical considerations (Fig. 16.2). The sizes of these study sites ranged between 1 and 20 km^2, and the pairwise distances between them varied between 19 and 224 km. In five study sites,

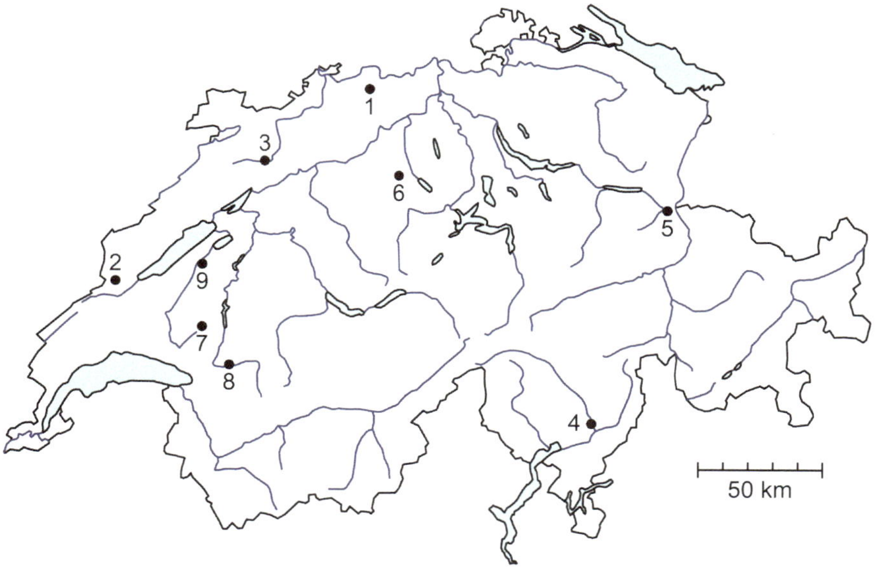

FIGURE 16.2

Map of Switzerland showing the locations of the nine study sites: (1) Buus; (2) Baulmes; (3) Tavannes; (4) Riviera; (5) Sargans; (6) Wauwil; (7) Vaulruz; (8) Enhaut; (9) Dompierre.

data were collected over seven years (1997–2003): three study sites entered the project in 1999, resulting in five study years, and one entered in 2000, resulting in only four study years. In all study sites and years, all potential barn swallow breeding sites, mostly farm buildings, were visited once a week during the entire breeding season (April to September). Occupied nests were counted, and the number of eggs and nestlings in occupied nests was recorded. A total of 17,691 nestlings were ringed 6–15 days after hatching. Most of the time, nests were rechecked after the nestlings fledged to obtain an accurate count of the fledglings in each brood. Adults were trapped during the night with a special handheld net when they were roosting on or near their nests and were ringed and sexed. A total of 2639 adults were ringed, but here we use only the data from 1607 females. Subsequent broods in the same building during the same season were assigned to breeding pairs of the first brood, if adults were trapped in both broods, or if the same nest was used again and the onset of the second brood was within 3 weeks from the fledging of the first. In this way, data on 2740 first and 1738 second broods were collected, and a total of 459 females were recaptured in a subsequent year after marking.

16.3 POPULATION MODEL

We defined the same prebreeding "census" female-based population model for all nine study sites. The barn swallow is a typical short-lived passerine species that starts to reproduce at age 1 and has fairly low survival. While adult barn swallows are generally philopatric to their breeding sites (Schaub and Hirschheydt, 2009), they exhibit extensive natal dispersal (Turner, 2006). Given the relatively small size of each study site, our population model ought to formally accommodate immigrants. We assume that demographic rates do not depend on age, with the exception of apparent survival, which we allow to differ between fledglings (i.e., survival from fledging to age 1 year) and adults (annual survival of females that are 1 year or older). We defined a life-cycle graph with three stages: local recruits, surviving adults, and immigrants (Fig. 16.3). A local recruit is a female born at the study site in the previous year and present at the study site during the current year. A surviving adult is a female that is 2 years or older in the current year and was already present at the study site during the previous year. Finally, an immigrant is a female born elsewhere during a previous year and currently present at the study site for the first time.

The life-cycle graph in Fig. 16.3 refers to a single study site only, because we assume that there is no dispersal among the studied sites (which is supported by our data; Schaub et al., 2015) and because the population model at all sites is identical. However, the population dynamics equations are developed for all sites, meaning that an index s (for site) is added to all parameters. The number of local recruits at site s in year t, $R_{s,t}$, is a function of the number of breeders at site s in year $t-1$, $N_{s,t-1}$, productivity, and the probability of a fledgling surviving and returning to the same study site (apparent

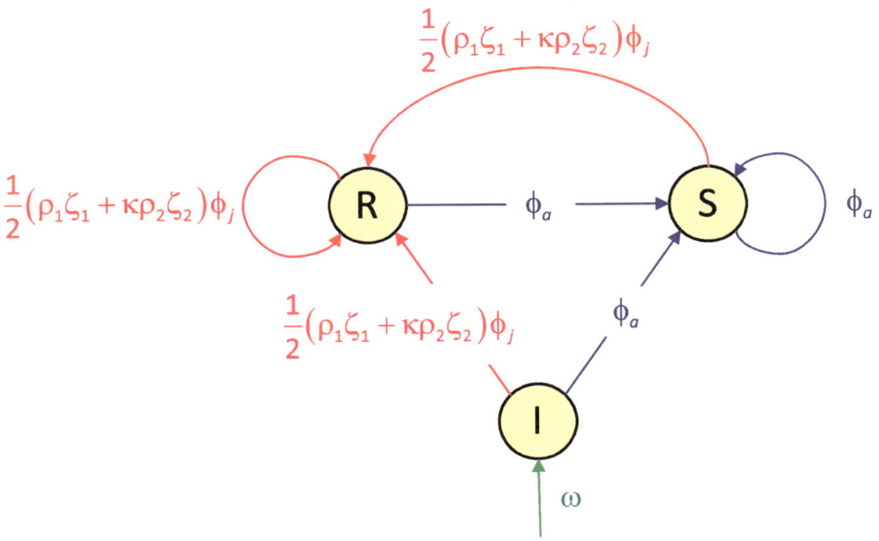

FIGURE 16.3

Prebreeding "census" life-cycle graph for female barn swallows in one population. The three stages are local recruits (*R*), surviving adults (*S*), and immigrants (*I*), and the demographic processes are productivity (governed by the clutch size and fledging success of the first brood ($\rho_1\zeta_1$), probability of double brooding (κ), and by clutch size and fledging success of the second brood ($\rho_2\zeta_2$)), apparent survival (governed by ϕ_j for juveniles and ϕ_a for adults), and immigration (governed by ω). Recruitment transitions are shown as red arrows, pure survival transitions in blue, and immigration in green.

juvenile survival: $\phi_{j,s,t-1}$). Total productivity is composed of the outcomes of the first and possibly a second brood in year $t-1$. The number of fledglings from the first brood depends on clutch size of the first brood (clutch size first brood: $\rho_{1,s,t-1}$) and on the probability that an egg produces a fledgling (fledging success first brood: $\zeta_{1,s,t-1}$; where the funny Greek letter is pronounced "zeta"). Likewise, the number of fledglings from a second brood depends on the clutch size of the second brood (clutch size second brood: $\rho_{2,s,t-1}$) and on the fledging success of the second brood (fledging success second brood: $\zeta_{2,s,t-1}$). Not all females are able to produce a second brood, and hence a further component of productivity is the probability that a second brood occurs (double brooding probability: $\kappa_{s,t-1}$). We assume an even nestling sex ratio (evidence for this is provided by Saino et al., 2002) and account for demographic stochasticity with a Poisson distribution.

$$R_{s,t} \sim Poisson\left(\frac{\rho_{1,s,t-1}\zeta_{1,s,t-1} + \kappa_{s,t-1}\rho_{2,s,t-1}\zeta_{2,s,t-1}}{2}\phi_{j,s,t-1}N_{s,t-1}\right)$$

The number of adults at site s in year t is given by the number of local recruits, adults, and immigrants in year $t-1$ (i.e., population size in year $t-1$) and the probability that individuals in these stages survive and do not disperse to a location outside the study site (apparent adult survival: $\phi_{a,s,t-1}$). Demographic stochasticity is accounted for by adopting a binomial distribution, and thus

$$S_{s,t} \sim Binomial(\phi_{a,s,t-1}, N_{s,t-1}).$$

The number of immigrants at site s in year t is modeled using a Poisson distribution as

$$I_{s,t} \sim Poisson(\omega_{s,t}),$$

where $\omega_{s,t}$ is the expected number of female immigrants in population s in year t.

Finally, the total number of females in site s during year t is the sum of the numbers of local recruits, adults, and immigrants: $N_{s,t} = R_{s,t} + S_{s,t} + I_{s,t}$. Note that this population model is similar to the population model for the woodchat shrike (Chapter 11). The main difference is that here we have multiple populations and a different decomposition of the annual productivity.

16.4 COMPONENT DATA LIKELIHOODS

For each of the nine sites, five different data sets were sampled: capture-recapture data of females, population count data, data about clutch size, the number of fledglings for first and second broods, and data about the frequency of double brooding. All data sets are stored in arrays, such that an array contains data from all years and sites, and unequal study duration is accommodated by filling in NAs where necessary. These data are taken from Schaub et al. (2015), but here are organized in a slightly different format and restricted to females. We load these data and then have a look at each part of them:

```
library(IPMbook); library(jagsUI)
data(swallow)
str(swallow)

List of 5
 $ marr.j      : num [1:9, 1:6, 1:7] 0 0 0 0 1 NA NA NA NA 0 ...
  ..- attr(*, "dimnames")=List of 3
  .. ..$ : chr [1:9] "Buus" "Baulmes" "Tavannes" "Riviera" ...
  .. ..$ : chr [1:6] "1997" "1998" "1999" "2000" ...
  .. ..$ : chr [1:7] "1998" "1999" "2000" "2001" ...
 $ marr.a      : num [1:9, 1:6, 1:7] 19 7 9 8 9 NA NA NA NA 0 ...
  ..- attr(*, "dimnames")=List of 3
  .. ..$ : chr [1:9] "Buus" "Baulmes" "Tavannes" "Riviera" ...
  .. ..$ : chr [1:6] "1997" "1998" "1999" "2000" ...
  .. ..$ : chr [1:7] "1998" "1999" "2000" "2001" ...
 $ counts      : num [1:9, 1:7] 75 NA 43 NA 39 NA NA NA NA 77 ...
  ..- attr(*, "dimnames")=List of 2
  .. ..$ : chr [1:9] "Buus" "Baulmes" "Tavannes" "Riviera" ...
  .. ..$ : chr [1:7] "1997" "1998" "1999" "2000" ...
 $ productivity: num [1:2, 1:9, 1:7, 1:3] 359 208 233 79 183 93 92 48 ...
  ..- attr(*, "dimnames")=List of 4
  .. ..$ : chr [1:2] "First" "Second"
  .. ..$ : chr [1:9] "Buus" "Baulmes" "Tavannes" "Riviera" ...
  .. ..$ : chr [1:7] "1997" "1998" "1999" "2000" ...
  .. ..$ : chr [1:3] "Eggs" "Fledglings" "Broods"
 $ second      : num [1:9, 1:7, 1:2] 77 52 39 20 17 0 0 0 0 80 ...
  ..- attr(*, "dimnames")=List of 3
  .. ..$ : chr [1:9] "Buus" "Baulmes" "Tavannes" "Riviera" ...
  .. ..$ : chr [1:7] "1997" "1998" "1999" "2000" ...
  .. ..$ : chr [1:2] "First broods" "Second broods"
```

16.4.1 POPULATION COUNT DATA

The site-specific annual population counts y, or, more specifically, the annual number of breeding pairs observed, are stored in a matrix (`counts`) with dimension of nine sites × seven years (1997–2003). The matrix contains a missing value ("NA") for site and year combinations when no population count was conducted. Usually, this means that the study at a given site and year had not started yet, but for Baulmes in 1997 and Wauwil in 1999, it meant that the number of breeding pairs had not been recorded, though all other data were. To model these counts, we adopt a state-space model as usual with a process model that we already defined in Section 16.3, but we also need an observation model. Owing to the intense field work, we are fairly confident that counts of breeding pairs are correct on average, i.e., that there is no detection bias and that the observation errors only reflect lack of fit for the most part. The state of a female during the time when occupied nests were counted was unknown, and hence we have an unstructured count. To account for the possibility that absolute errors might increase with population size, we use a log-normal distribution with a different variance for each site, since model fit may differ among sites. Thus,

$$y_{s,t} \sim logNormal\left(N_{s,t}, \sigma_s^2\right),$$

where $y_{s,t}$ is the count at site s in year t and σ_s^2 is the magnitude of the site-specific residual error.

16.4.2 PRODUCTIVITY DATA

The two objects `productivity` and `second` contain the data to model the different components of productivity. Object `productivity` is a four-dimensional array. The first dimension stands for brood type (first vs. second brood), the second dimension indexes the nine study sites, and the third dimension is the seven study years. The fourth dimension has three levels, which are the total number of eggs, fledglings, and of broods. These totals contain the sums across all nests for each study site, year, and brood type. The object `second` is a three-dimensional array that contains information about the frequency of second broods. The first and second dimensions again indicate study sites and years, respectively, while in the third dimension, we make a distinction between the number of "first broods" and "second broods" observed.

We model the clutch size of first broods at site s in year t using a Poisson distribution as

$$c_{1,s,t} \sim Poisson\left(\rho_{1,s,t} b_{1,s,t}\right),$$

where $\rho_{1,s,t}$ is expected clutch size of a first brood at site s in year t, $c_{1,s,t}$ is the total number of observed eggs in all first broods at site s in year t, and $b_{1,s,t}$ is the total number of first broods observed at site s in year t. Similarly, the expected clutch size of second broods is modeled as

$$c_{2,s,t} \sim Poisson\left(\rho_{2,s,t} b_{2,s,t}\right).$$

Fledging success is expressed as the probability that an egg produces a fledgling and is modeled using a binomial distribution. For first broods, we have

$$f_{1,s,t} \sim Binomial\left(\zeta_{1,s,t}, c_{1,s,t}\right),$$

where $\zeta_{1,s,t}$ is the fledging success of a first brood at site s in year t, and $f_{1,s,t}$ is the total number of fledglings in all first broods observed at site s in year t. Similarly, fledging success of second broods is modeled as

$$f_{2,s,t} \sim Binomial\left(\zeta_{2,s,t}, c_{2,s,t}\right).$$

The probability of double brooding is modeled using a binomial distribution as

$$w_{s,t} \sim Binomial\left(\kappa_{s,t}, v_{s,t}\right),$$

where $\kappa_{s,t}$ is the probability that a female barn swallow at site s in year t produces two clutches, $v_{s,t}$ is the number of first broods observed at site s in year t, and $w_{s,t}$ is the number of second broods observed at site s in year t.

16.4.3 CAPTURE-RECAPTURE DATA

The capture-recapture data already come summarized in the m-array format (see Section 4.5.1.2), and we have one m-array for females marked as nestlings (`marr.j`) and another for females released as adults (i.e., at age 1 or older; `marr.a`). These arrays contain the data from all nine populations, and hence they have an additional dimension. The first dimension indexes study sites, and the second and third index the release and recapture years, respectively, as usual.

The capture-recapture data include females only. While adult barn swallows can be sexed morphologically, sexing of nestlings is not possible based on morphology. Therefore, the sex of all ringed nestlings was unknown except for those that were later recaptured as adults. Under the assumption of an even nestling sex ratio, it is possible to assign a sex to the unsexed nestlings and to estimate sex-specific survival (Nichols et al., 2004). Here, we use m-arrays in which such an allocation of each individual to a sex has already been done.

For the m-arrays, we specify multinomial likelihoods (see Section 4.5.1.2) and fit a model with site- and year-specific apparent survival for juvenile and adult females. Recapture probabilities are assumed to be site- and year-specific and also include an age effect, as a previous study showed lower recapture probabilities for first-year than for older females (Schaub and Hirschheydt, 2009).

16.5 THE INTEGRATED POPULATION MODEL

We are now ready to put everything together in an IPM, making the usual independence assumption. But first, we define the linear models for the demographic rates. Remember that our primary goal is the study of synchrony in the demographic rates.

As just written, all demographic rates can vary by site and year, i.e., they are completely independent from one another. If we fit such a model, we will obtain the same results as when we analyze the data sets for each site separately. However, we want to decompose the spatio-temporal variability in the demographic rates into a main temporal component that is shared by all sites (a synchronous part) and a local component that is different at each site (an asynchronous part). In linear modeling terms, you can also think of these as a main effect of time and as the time-site interaction, respectively. This requires that the data from all sites are analyzed in the same model with some parameters shared. For a parameter θ, we then define

$$g(\theta_{s,t}) = \mu_{\theta_s} + \varepsilon_{\theta_t} + \eta_{\theta_{s,t}},$$

where g is a suitable link function, μ_{θ_s} is the link-scale mean of θ at site s, ε_{θ_t} is the temporal effect of year t on θ that is shared across sites (i.e., the main effects), and $\eta_{\theta_{s,t}}$ are the site-specific effects of year t on θ (i.e., the interaction effects). Both of these time effects are modeled with zero-centered normal distributions as

$$\varepsilon_{\theta_t} \sim Normal(0, \tau_\theta^2) \text{ and } \eta_{\theta_{s,t}} \sim Normal(0, \upsilon_\theta^2),$$

where τ_θ^2 is the shared temporal variability and υ_θ^2 is the site-specific (i.e., local) temporal variability of θ, with the latter capturing what is not explained by the former.

As a measure of synchrony, we compute the ICC that expresses the ratio of the part of temporal variability that is shared among sites (synchronous) to the total spatio-temporal variability (Grosbois et al., 2009). For a parameter θ, this is

$$ICC_\theta = \frac{\tau_\theta^2}{\tau_\theta^2 + \upsilon_\theta^2}.$$

An ICC close to 1 indicates strong synchrony, and an ICC close to 0 indicates asynchrony.

To avoid even longer code for our IPM, we do not add goodness-of-fit tests here. Schaub et al. (2015) used posterior predictive checks with standard discrepancy measures (Section 7.2.1) for that purpose and identified some underdispersion of the clutch size data, but otherwise found no lack of fit in the model.

Besides the data sets described above, the data bundle also contains a variable that defines the starting year of each study, variables containing the total number of individuals released, and a variable that defines the ranges of the discrete uniform priors that we will adopt for the initial population sizes. To count as vague, we define them by the range of 1 to the maximum number of breeding pairs counted at a site. Since the number of breeding pairs is the sum of local recruits, surviving adults, and immigrants, and we specify these priors for each of them, these upper limits should definitely be wide enough to count as vague. We bundle the data, write BUGS code for the IPM, and launch the model.

```
# Define the starting year of each site
start <- c(1, 1, 1, 1, 1, 3, 3, 3, 4)

# Compute the total number of released individuals
rel.j <- apply(swallow$marr.j, c(1,2), sum, na.rm=TRUE)
rel.a <- apply(swallow$marr.a, c(1,2), sum, na.rm=TRUE)

# Define ranges of discrete uniform priors for initial stage-structured population sizes per site
pinit <- dUnif(lower=rep(1,9), upper=apply(swallow$counts, 1, max, na.rm=TRUE))

# Bundle data and produce data overview
jags.data <- with(swallow, list(marr.j=marr.j, rel.j=rel.j, marr.a=marr.a, rel.a=rel.a, y=counts,
  c=productivity[,,,'Eggs'], b=productivity[,,,'Broods'], f=productivity[,,,'Fledglings'],
  v=second[,,'First broods'], w=second[,,'Second broods'], pinit=pinit, nsites=9,
  nyears=7, start=start))
str(jags.data)

List of 14
 $ marr.j: num [1:9, 1:6, 1:7] 0 0 0 0 1 NA NA NA NA 0 ...
 $ rel.j : num [1:9, 1:6] 229 161 128 66 92 0 0 0 0 269 ...
 $ marr.a: num [1:9, 1:6, 1:7] 19 7 9 8 9 NA NA NA NA 0 ...
 $ rel.a : num [1:9, 1:6] 57 44 31 19 29 0 0 0 0 62 ...
 $ y     : num [1:9, 1:7] 75 NA 43 NA 39 NA NA NA NA 77 ...
 $ c     : num [1:2, 1:9, 1:7] 359 208 233 79 183 93 92 48 71 4 ...
 $ b     : num [1:2, 1:9, 1:7] 77 47 52 18 39 21 20 12 17 1 ...
 $ f     : num [1:2, 1:9, 1:7] 281 179 178 61 148 78 66 37 60 4 ...
 $ v     : num [1:9, 1:7] 77 52 39 20 17 0 0 0 0 80 ...
 $ w     : num [1:9, 1:7] 47 18 21 12 1 0 0 0 0 55 ...
 $ pinit : num [1:9, 1:115] 0.0116 0.013 0.0227 0.0161 0.0256 ...
 $ nsites: num 9
 $ nyears: num 7
 $ start : num [1:9] 1 1 1 1 1 3 3 3 4
```

```
# Write JAGS model file
cat(file = "model1.txt", "
model {
# Priors and linear models
# Linear models for the eight demographic parameters
for (s in 1:nsites){
  for (t in start[s]:(nyears-1)){
    logit(phij[s,t]) <- mu.phij[s] + eps.phij[t] + eta.phij[s,t]
    logit(phia[s,t]) <- mu.phia[s] + eps.phia[t] + eta.phia[s,t]
    log(omega[s,t]) <- mu.omega[s] + eps.omega[t] + eta.omega[s,t]
    pj[s,t] ~ dunif(0, 1)
    pa[s,t] ~ dunif(0, 1)
  } # t
  for (t in start[s]:nyears){
    log(rho[1,s,t]) <- mu.rho[1,s] + eps.rho[1,t] + eta.rho[1,s,t]
    log(rho[2,s,t]) <- mu.rho[2,s] + eps.rho[2,t] + eta.rho[2,s,t]
    logit(zeta[1,s,t]) <- mu.zeta[1,s] + eps.zeta[1,t] + eta.zeta[1,s,t]
    logit(zeta[2,s,t]) <- mu.zeta[2,s] + eps.zeta[2,t] + eta.zeta[2,s,t]
    logit(kappa[s,t]) <- mu.kappa[s] + eps.kappa[t] + eta.kappa[s,t]
  } #t
} #s

# Priors for the means of the eight demographic parameters
for (s in 1:nsites){
  mu.phij[s] ~ dnorm(0, 0.001)
  overall.mean[1,s] <- ilogit(mu.phij[s])
  mu.phia[s] ~ dnorm(0, 0.001)
  overall.mean[2,s] <- ilogit(mu.phia[s])
  mu.omega[s] ~ dnorm(0, 0.001)
  overall.mean[3,s] <- exp(mu.omega[s])
  mu.rho[1,s] ~ dnorm(0, 0.001)
  overall.mean[4,s] <- exp(mu.rho[1,s])
  mu.rho[2,s] ~ dnorm(0, 0.001)
  overall.mean[5,s] <- exp(mu.rho[2,s])
  mu.zeta[1,s] ~ dnorm(0, 0.001)
  overall.mean[6,s] <- ilogit(mu.zeta[1,s])
  mu.zeta[2,s] ~ dnorm(0, 0.001)
  overall.mean[7,s] <- ilogit(mu.zeta[2,s])
  mu.kappa[s] ~ dnorm(0, 0.001)
  overall.mean[8,s] <- ilogit(mu.kappa[s])
}

# Priors for temporal random effects (common across sites)
for (t in 1:(nyears-1)){
  eps.phij[t] ~ dnorm(0, tau.t[1])
  eps.phia[t] ~ dnorm(0, tau.t[2])
  eps.omega[t] ~ dnorm(0, tau.t[3])
}
for (t in 1:nyears){
  eps.rho[1,t] ~ dnorm(0, tau.t[4])
  eps.rho[2,t] ~ dnorm(0, tau.t[5])
  eps.zeta[1,t] ~ dnorm(0, tau.t[6])
  eps.zeta[2,t] ~ dnorm(0, tau.t[7])
  eps.kappa[t] ~ dnorm(0, tau.t[8])
}

# Priors for within-site temporal random effects
for (s in 1:nsites){
  for (t in start[s]:(nyears-1)){
    eta.phij[s,t] ~ dnorm(0, tau.st[1])
    eta.phia[s,t] ~ dnorm(0, tau.st[2])
    eta.omega[s,t] ~ dnorm(0, tau.st[3])
  } #t
  for (t in start[s]:nyears){
    eta.rho[1,s,t] ~ dnorm(0, tau.st[4])
    eta.rho[2,s,t] ~ dnorm(0, tau.st[5])
    eta.zeta[1,s,t] ~ dnorm(0, tau.st[6])
    eta.zeta[2,s,t] ~ dnorm(0, tau.st[7])
    eta.kappa[s,t] ~ dnorm(0, tau.st[8])
  } #t
} #s
```

```
# Priors for precisions of the eight demographic parameters
for (i in 1:8){
  tau.t[i] ~ dgamma(0.001, 0.001)
  tau2[i] <- 1 / tau.t[i]
  tau.st[i] ~ dgamma(0.001, 0.001)
  ypsilon2[i] <- 1 / tau.st[i]

  # Intraclass correlation coefficients (ICC)
  ICC[i] <- tau2[i] / (tau2[i] + ypsilon2[i])
}

# Observation error for state-space model
for (s in 1:nsites){
  tau.y[s] ~ dgamma(0.001, 0.001)
  sigma2[s] <- 1 / tau.y[s]
}

# Population count data (state-space model)
# Model for the initial population size: uniform priors
for (s in 1:nsites){
  R[s,start[s]] ~ dcat(pinit[s,])   # Local recruits
  S[s,start[s]] ~ dcat(pinit[s,])   # Surviving adults
  I[s,start[s]] ~ dcat(pinit[s,])   # Immigrants
}
# Process model over time: our model of population dynamics
for (s in 1:nsites){
  for (t in (start[s]+1):nyears){
    R[s,t] ~ dpois((rho[1,s,t-1] * zeta[1,s,t-1] + kappa[s,t-1] * rho[2,s,t-1] * zeta[2,s,t-1]) *
      0.5 * phij[s,t-1] * N[s,t-1])
    S[s,t] ~ dbin(phia[s,t-1], N[s,t-1])
    I[s,t] ~ dpois(N[s,t-1] * omega[s,t-1])
  } #t
} #s

# Observation model
for (s in 1:nsites){
  for(t in start[s]:nyears){
    N[s,t] <- R[s,t] + S[s,t] + I[s,t]
    logN[s,t] <- log(N[s,t])
    y[s,t] ~ dlnorm(logN[s,t], tau.y[s])
  } #t
} #s

# Productivity data (Poisson and binomial models)
for (s in 1:nsites){
  for (t in start[s]:nyears){
    # Clutch size and fledging success of first brood
    c[1,s,t] ~ dpois(rho[1,s,t] * b[1,s,t])
    f[1,s,t] ~ dbin(zeta[1,s,t], c[1,s,t])

    # Clutch size and fledging success of second brood
    c[2,s,t] ~ dpois(rho[2,s,t] * b[2,s,t])
    f[2,s,t] ~ dbin(zeta[2,s,t], c[2,s,t])

    # Probability to conduct a second brood
    w[s,t] ~ dbin(kappa[s,t], v[s,t])
  } #t
} #s

# Capture-recapture data (CJS model with multinomial likelihood)
for (s in 1:nsites){
  # Define the multinomial likelihood
  for (t in start[s]:(nyears-1)){
    marr.j[s,t,start[s]:nyears] ~ dmulti(pr.j[s,t,start[s]:nyears], rel.j[s,t])
    marr.a[s,t,start[s]:nyears] ~ dmulti(pr.a[s,t,start[s]:nyears], rel.a[s,t])
  } #t
  # Define the cell probabilities of the m-arrays
  for (t in start[s]:(nyears-1)){
    # Main diagonal
```

```
      qj[s,t] <- 1-pj[s,t]
      qa[s,t] <- 1-pa[s,t]
      pr.j[s,t,t] <- phij[s,t] * pj[s,t]
      pr.a[s,t,t] <- phia[s,t] * pa[s,t]
      # Above main diagonal
      for (j in (t+1):(nyears-1)){
        pr.j[s,t,j] <- phij[s,t] * prod(phia[s,(t+1):j]) * qj[s,t] *  prod(qa[s,t:(j-1)]) *
          pa[s,j] / qa[s,t]
        pr.a[s,t,j] <- prod(phia[s,t:j]) * prod(qa[s,t:(j-1)]) * pa[s,j]
      } #j
      # Below main diagonal
      for (j in 1:(t-1)){
        pr.j[s,t,j] <- 0
        pr.a[s,t,j] <- 0
      } #j
      # Last column: probability of non-recapture
      pr.j[s,t,nyears] <- 1-sum(pr.j[s,t,start[s]:(nyears-1)])
      pr.a[s,t,nyears] <- 1-sum(pr.a[s,t,start[s]:(nyears-1)])
    } #t
} #s
}
")

# Initial values
inits <- function(){list(mu.phia=runif(9, -1, 0))}

# Parameters monitored
parameters <- c("phij", "phia", "omega", "rho", "zeta", "kappa", "pj", "pa", "R", "S", "I", "N",
  "overall.mean", "eps.phij", "eps.phia", "eps.omega", "eps.rho", "eps.zeta", "eps.kappa", "eta.phij",
  "eta.phia", "eta.omega", "eta.rho", "eta.zeta", "eta.kappa", "tau2", "ypsilon2", "ICC", "sigma2")

# MCMC settings
ni <- 50000; nb <- 10000; nc <- 3; nt <- 40; na <- 5000

# Call JAGS from R (ART 12 min) and check convergence
out1 <- jags(jags.data, inits, parameters, "model1.txt", n.iter=ni, n.burnin=nb, n.chains=nc,
  n.thin=nt, n.adapt=na, parallel=TRUE)
traceplot(out1)
```

The IPM yields estimates of the demographic rates, their ICCs, and of structured population sizes directly. It does not, however, provide an estimate of the ICC for the population growth rate. We need an extra step to compute this, where we calculate the population growth rate and quantify its ICC based on the results from the fitted IPM. We do this next.

First, we use the posterior samples of population size to calculate the posterior distribution of the site- and year-specific population growth rates on the log-scale as a derived quantity,

$$r_{i,s,t} = \log(N_{i,s,t+1}) - \log(N_{i,s,t}).$$

where i indexes the Markov chain Monte Carlo (MCMC) draws, and s and t index site and time as usual. We prefer to work on the log-scale to model population growth rates to avoid the bounds imposed by the nonnegativity of the growth rate on the natural scale. We then model the posterior samples of the growth rate r with a linear mixed-effects model with the same structure as assumed for the demographic rates,

$$r_{i,s,t} = \mu_s + \varepsilon_t + \eta_{s,t},$$

where μ_s is the mean growth rate at site s on the log-scale, ε_t represents the effect of the year in r that is shared among sites, and $\eta_{s,t}$ expresses the effect of the year that is specific to each site. The latter two are random effects for which we assume zero-mean normal distributions,

$$\varepsilon_t \sim Normal(0, \tau^2) \text{ and } \eta_{s,t} \sim Normal(0, \upsilon^2).$$

The ICC of the population growth rate is then calculated again as the ratio of the shared temporal variance (τ^2) to the total temporal variance ($\tau^2 + \upsilon^2$).

We bundle the data, write BUGS code for this model, and launch the analysis.

```
# Bundle data
# Calculate population growth rate
r <- log(out1$sims.list$N[,,2:7]) - log(out1$sims.list$N[,,1:6])
jags.data <- list(r=r, start=start, nsites=9, nyears=6, nsamp=dim(r)[1])

cat(file = "model2.txt", "
model {
# Priors and linear models
# Fixed site effects
for (s in 1:nsites){
  mu[s] ~ dnorm(0, 0.001)
}

# Temporal random effects
for (t in 1:nyears){
  epsilon[t] ~ dnorm(0, tau.t)
}

# Priors for precisions
tau.t ~ dgamma(0.001, 0.001)
tau.ts ~ dgamma(0.001, 0.001)  # Residual, site-time variation

# Likelihood
for (i in 1:nsamp){
  for (s in 1:nsites){
    for (t in start[s]:nyears){
      r[i,s,t] ~ dnorm(mu[s] + epsilon[t], tau.ts)
    } #t
  } #s
} #i

# Derived quantities
# Variances
tau2 <- 1 / tau.t
ypsilon2 <- 1 / tau.ts

# Intraclass correlation coefficient (ICC)
ICC <- tau2 / (tau2 + ypsilon2)
}
")

# Initial values
inits <- function(){list(tau.t=0.01)}

# Parameters monitored
parameters <- c("ICC", "tau2", "ypsilon2", "mu")

# MCMC settings
ni <- 10000; nb <- 5000; nc <- 3; nt <- 1; na <- 2000

# Call JAGS from R (ART 54 min) and check convergence
out2 <- jags(jags.data, inits, parameters, "model2.txt", n.iter=ni, n.burnin=nb, n.chains=nc,
  n.thin=nt, n.adapt=na, parallel=TRUE)
traceplot(out2)
```

16.6 RESULTS

The chains for the IPM show good convergence, as you can see when you inspect the trace-plots (not shown) of the parameters and the values of R-hat. Here is just a smallish part of the very large posterior summary of the IPM.

```
print(out1, 3)

              mean    sd   2.5%    50%  97.5% overlap0     f  Rhat n.eff
phij[1,1]    0.004 0.003  0.000  0.003  0.011    FALSE 1.000 1.003   759
phij[2,1]    0.005 0.004  0.000  0.004  0.014    FALSE 1.000 1.003  3000
phij[3,1]    0.000 0.000  0.000  0.000  0.001    FALSE 1.000 1.089  1624
phij[4,1]    0.006 0.005  0.000  0.005  0.018    FALSE 1.000 1.001  2262
phij[5,1]    0.011 0.008  0.002  0.009  0.031    FALSE 1.000 1.002  3000
phij[1,2]    0.005 0.003  0.001  0.004  0.013    FALSE 1.000 1.004   913
phij[2,2]    0.007 0.006  0.001  0.006  0.023    FALSE 1.000 1.001  3000
phij[3,2]    0.000 0.000  0.000  0.000  0.001    FALSE 1.000 1.029  2816
phij[4,2]    0.009 0.007  0.001  0.007  0.026    FALSE 1.000 1.002  1658
phij[5,2]    0.012 0.007  0.003  0.010  0.031    FALSE 1.000 1.000  3000
phij[1,3]    0.005 0.003  0.001  0.004  0.012    FALSE 1.000 1.002  1109
phij[2,3]    0.005 0.004  0.000  0.004  0.013    FALSE 1.000 1.001  1383
phij[3,3]    0.000 0.000  0.000  0.000  0.001    FALSE 1.000 1.026  3000
phij[4,3]    0.006 0.005  0.001  0.005  0.017    FALSE 1.000 1.002  3000
[... output truncated ...]
```

Our second model, which was used to estimate the ICC of the population growth rate, converged nicely and yields the following summary output.

```
print(out2, 3)

           mean    sd   2.5%    50%  97.5% overlap0     f  Rhat n.eff
ICC       0.320 0.147  0.120  0.288  0.685    FALSE 1.000 1.000 15000
tau2      0.011 0.017  0.003  0.008  0.041    FALSE 1.000 1.056 13423
ypsilon2  0.019 0.000  0.019  0.019  0.019    FALSE 1.000 1.000     1
mu[1]    -0.099 0.044 -0.184 -0.099 -0.014    FALSE 0.984 1.001 15000
mu[2]    -0.138 0.044 -0.224 -0.138 -0.053    FALSE 0.996 1.001 15000
mu[3]    -0.045 0.044 -0.130 -0.045  0.041     TRUE 0.880 1.001 15000
mu[4]    -0.173 0.044 -0.258 -0.173 -0.088    FALSE 0.998 1.001 15000
mu[5]    -0.128 0.044 -0.214 -0.128 -0.043    FALSE 0.995 1.001 15000
mu[6]    -0.247 0.044 -0.332 -0.247 -0.162    FALSE 1.000 1.001 15000
mu[7]    -0.120 0.044 -0.205 -0.120 -0.034    FALSE 0.992 1.001 15000
mu[8]    -0.078 0.044 -0.163 -0.078  0.007     TRUE 0.968 1.001 15000
mu[9]    -0.217 0.044 -0.302 -0.217 -0.132    FALSE 0.999 1.001 15000
```

The total population size at all nine sites was declining over the study period, even "significantly" so in seven out of the nine sites, judging from the column in the posterior summary named "overlap0." In addition, the population growth rates at most sites exhibited substantial temporal variation (Fig. 16.4).

Most of the estimated site- and year-specific demographic rates showed considerable variation over time (Fig. 16.5), but the sizes of first and second clutches were remarkably constant. Juvenile apparent survival was very low at all sites, suggesting that natal dispersal of female barn swallows was strong. This means that there was almost no local recruitment, where, as always, "local" is defined implicitly by the scale of the chosen study area. Furthermore, temporal variability in juvenile apparent survival was low. Immigration was very pronounced. Presumably, immigrants were predominantly first-time breeders, i.e., natal dispersers from locations outside the study sites. The size of first clutches was generally larger than that of second clutches, but fledging success of first and second broods was similar. The probability of double brooding was high at around 0.6.

Visually synchrony of population growth rates and of demographic rates is manifest in the parallelism of the lines for the nine populations (Figs. 16.4 and 16.5). For example, the graphs suggest synchrony in the probability of double brooding, but more asynchrony in the fledging success of second broods. Numerically, the degree of synchrony is quantified by the estimated ICC shown in Fig. 16.6. We notice that the posteriors of ICC of all parameters are wide and hence the estimates fairly uncertain. This can be explained by the short study duration of no more than seven years, which naturally renders difficult the estimation of temporal random effects. Riecke et al. (2019) suggested that more than ten annual estimates should be available for reliable estimation of the temporal variance. The posterior distribution of the ICC for juvenile apparent survival is even bimodal, with peaks close to the boundaries at 0 and 1. This shows that the estimation of the temporal variance components for juvenile survival was particularly challenging. Presumably, we should not try to come up with very elaborate biological stories around the ICC estimate of that parameter.

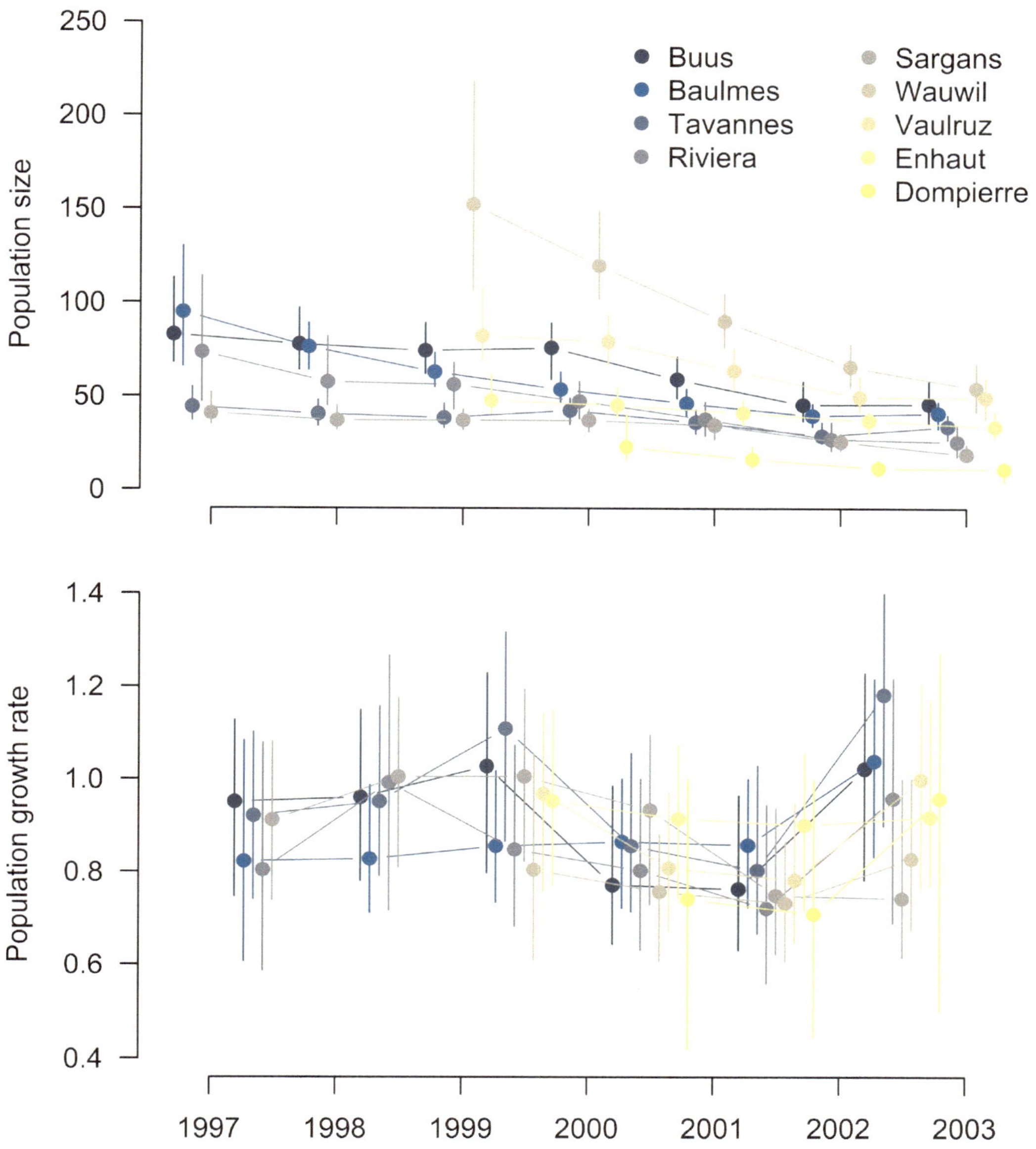

FIGURE 16.4

Estimates of barn swallow population sizes and growth rates from nine populations in Switzerland during 1997–2003. Dots show posterior means, and vertical lines 95% credible intervals.

Nevertheless, the median ICCs suggest that the demographic rates differ in terms of their synchrony across the nine study sites. Fledging success of both first and second broods showed the weakest synchrony, indicating that their temporal variability was driven mostly by local factors such as local weather (Turner, 2006; Grüebler et al., 2010). The probability of double brooding showed the strongest synchrony. Whether or not a female is able to start a second brood depends primarily on the timing of the first brood; if it fledges early, a second brood is more likely (Grüebler and Naef-Daenzer, 2008). The timing of the first brood in turn depends on the arrival time at the breeding grounds (Turner, 2006), which itself is affected by the weather at the nonbreeding grounds (Saino et al., 2004; Gordo et al., 2005). All of this indicates that the causes of the synchrony in the probability of double brooding may well be environmental effects at the nonbreeding grounds. Interestingly, despite considerable synchrony in some demographic rates, the synchrony in the population growth rates was not particularly strong (Fig. 16.6).

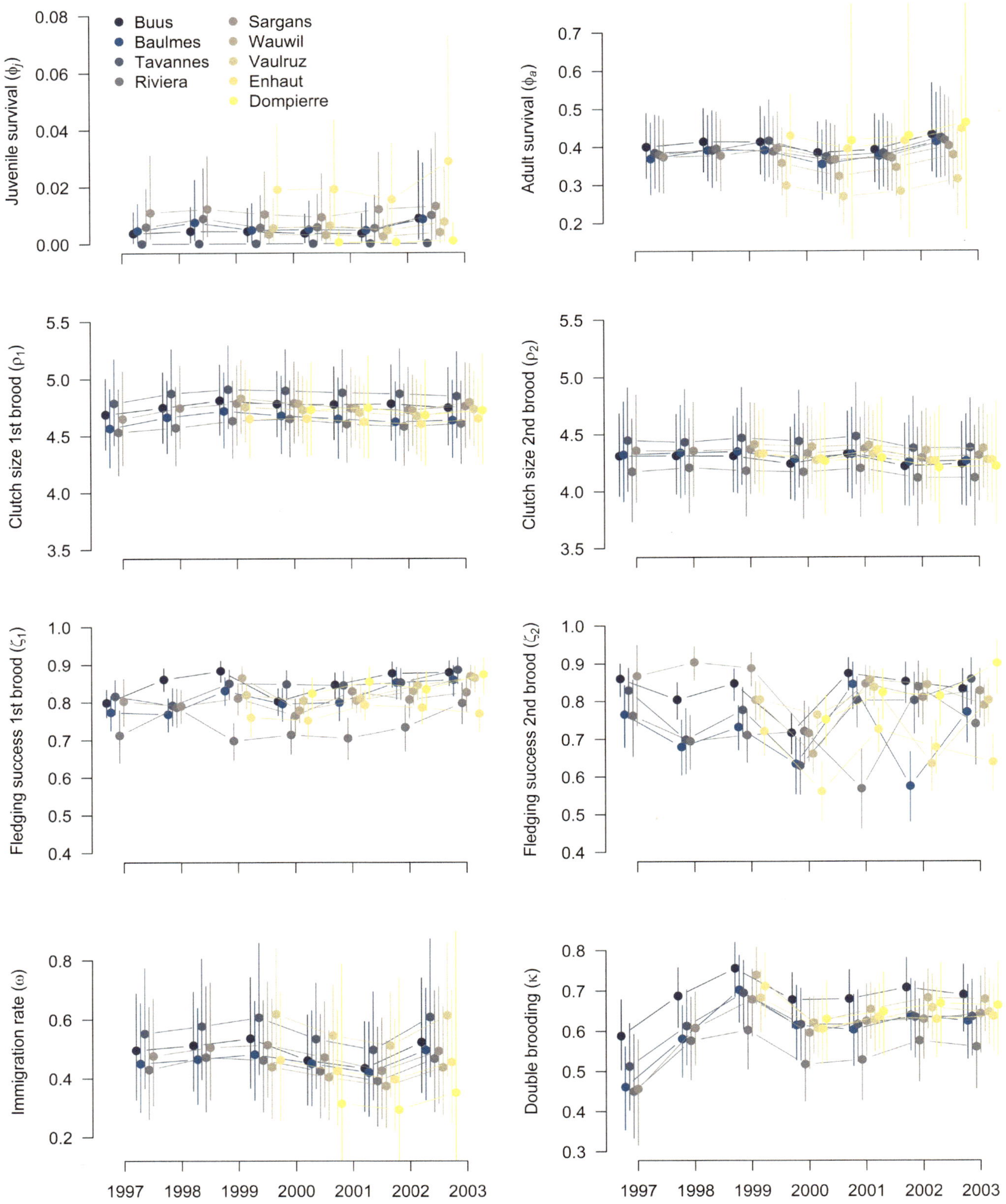

FIGURE 16.5

Estimates of demographic rates of barn swallows from nine populations in Switzerland. Dots show posterior means, and vertical lines 95% credible intervals.

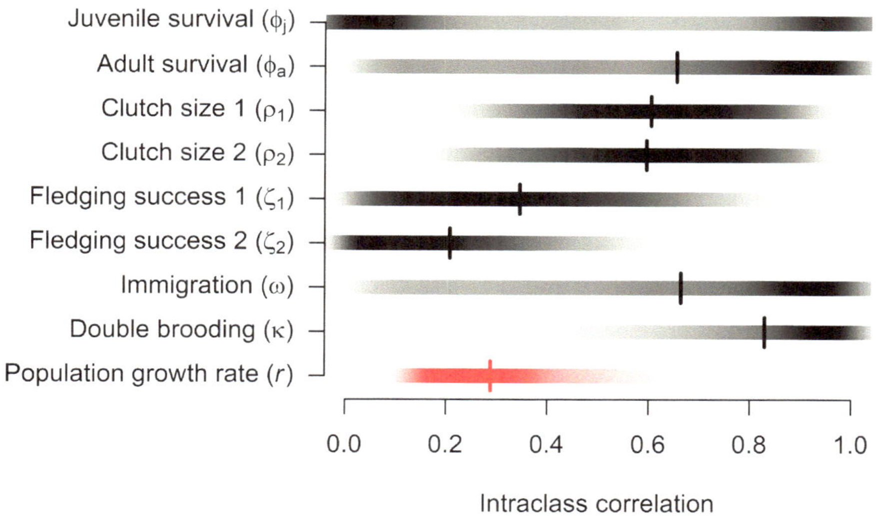

FIGURE 16.6

Posterior densities of the intraclass correlation coefficient (ICC) in eight demographic parameters and in the annual population growth rates (red) of female barn swallows at nine study sites across Switzerland (darker means a higher density). Vertical lines show the posterior medians. The posterior of the ICC for juvenile survival is bimodal, and hence no median is shown. Note that the ICC of the first eight parameters were computed in the integrated population model (IPM), while that of the population growth rate (red) was obtained from a linear mixed model fit to a function of the posterior samples produced in the IPM.

16.7 DISCUSSION

We jointly analyzed different types of data sets from nine populations of barn swallows widely dispersed over Switzerland to investigate a particular aspect of spatial population dynamics—the synchrony of demographic rates among multiple populations. We analyzed all data sets from all nine sites in a single IPM and assessed synchrony by a decomposition of the total temporal variance in demographic rates and population growth rates into a synchronous component that is shared among the nine populations and another that is asynchronous and specific to each population, or, in other words, into a main effect of time and an interaction effect between time and study site. The degree of synchrony can then be expressed by the intraclass correlation coefficient (ICC), which is simply the proportion of the shared variance (Grosbois et al., 2009). To assess the synchrony in the population growth rate, we fit a second model (a standard linear mixed model) to the MCMC output from the first by first computing the annual growth rates for each posterior draw of the estimated population sizes. This illustrates a very useful feature in a Bayesian analysis: that in a second step, we can conduct valuable and insightful statistical analysis on the output from a first-step statistical analysis, and all that of it will have proper error propagation.

Use of an IPM offers some interesting opportunities for the study of population synchrony. First of all, we can estimate the numerical strength of the synchrony in all demographic rates or other population quantities (such as population size) that vary in space and time. Therefore, it becomes possible to identify the demographic mechanisms underpinning the synchrony of populations. Second, the usual benefits of IPMs extend to this analysis. The uncertainty of parameter estimates is reduced due to inclusion of the information about demography contained in the counts. Hierarchical modeling of the demographic rates leads to sharing of information across populations which again reduces estimation uncertainty. Data collection to study population synchrony using an IPM may be costly, because we need to collect demographic data sets at multiple sites for time periods with at least some overlap (where of course, complete temporal overlap is best). This requirement holds not only for the study periods of the different sites, but also within a site when the different component data sets are collected. We have demonstrated in Section 6.6.3 how to deal with less than complete temporal overlap of component data sets within one site when fitting an IPM. An IPM offers the flexibility needed to include different data sets sampled during partially different time periods at different sites. This may help in analyzing existing data sets to study population synchrony.

The IPM developed here yields estimates of synchrony for each demographic trait in the form of ICCs. We found rather strong synchrony in the probability of double brooding and weak synchrony in fledging success. Because of the short study duration of no more than seven years, the estimates of these ICCs were rather imprecise. Clearly, we would have benefitted from much longer time series.

As a next step, we might now investigate the environmental factors that may be the synchronizing agents for the demographic rates studied. There are several ways to do this; and one is the inclusion of environmental covariates in the linear model that decomposes the spatio-temporal variance of a parameter. This then allows one to gauge how much the synchronous and the asynchronous variation is reduced by including this covariate, i.e., how much variation that covariate can explain (Lahoz-Monfort et al., 2011, 2013). Another obvious next step would be to understand which demographic rate has contributed how much to the observed population synchrony. This could be done via a retrospective population analysis (Chapter 9), the principles of which also apply in the context of population synchrony. The contribution to synchrony of a demographic rate depends on (1) the growth rate sensitivity of that demographic rate, (2) on the temporal variability of the rate, and (3) on its synchrony. Schaub et al. (2015) used simulation to assess the contribution of the synchrony in each demographic rate to overall population synchrony. There, we found that the synchrony in apparent adult survival and in immigration contributed most to the overall synchrony of Swiss barn swallow populations.

Other measures of synchrony such as cross-correlations (Björnstad et al., 1999; Paradis et al., 2000) or the correlations between distances and pairwise correlation coefficients can also be computed from the output of the IPM (see Schaub et al., 2015). The former has an interpretation similar to that of the ICC as a general measure of synchrony, whereas the latter quantifies the spatial scale of synchrony by expressing how quickly synchrony decays with increasing distance between two populations. In the future, it would be valuable to directly model spatial correlations of demographic rates among multiple sites as a function of the pairwise distances among these sites using the multivariate normal distribution (Riecke et al., 2019) in exactly the same way as spatial autocorrelation is commonly modeled in spatial statistics (Cressie and Wikle, 2011).

For the assessment of the synchrony in the population growth rate, an extra step was required. We could not compute that as part of the same IPM, since the population growth rate in our Bayesian analysis is a derived (and in that sense fixed) quantity and therefore cannot also be declared to be stochastic, as is necessary when we want to model its variation in JAGS. Computing the ICC of the population growth rate then becomes an exercise in "doing statistics on statistics" (Link, 1999), in which it can be challenging to properly propagate the uncertainty. However, in a Bayesian analysis based on MCMC, this is actually quite trivial, since we merely need to repeat all the required calculations or statistical analyses for every single draw from the posterior distribution of the parameter in question. In our case, we applied a linear mixed model to all MCMC draws of the log-scale population growth rates, which were themselves derived quantities based on the MCMC output for population size in the original IPM. Thus, we were able to fully account for the whole uncertainty in the estimation of the ICC for the population growth rate. Schaub et al. (2015) did not account for the full uncertainty when calculating the ICC of population growth rate. Rather, they only used the posterior means of the population growth rates and then fit a linear mixed model to them, thus ignoring part of the total variability in the population growth rates. Their approach resulted in a higher estimate of the ICC (0.43) compared to the one we found here (0.32). This shows that results from the analysis on estimated quantities can differ depending on whether all the uncertainty is properly incorporated or not. Remember that we found a similar pattern when we computed the correlation coefficients between demographic rates and the population growth rate (Section 9.2).

The ability of Bayesian MCMC-based analyses, to properly "do statistics on statistics" (Link, 1999) with full incorporation of all the uncertainties involved, is underutilized in our opinion (see, e.g., Chapter 7 in Hooten et al., 2017, for some nice further examples). There must be many cases in which very interesting analyses can be carried out as a second step based on the estimates of a first-step statistical analysis. Carrying out the necessary calculations for the second-step analysis for all (or even a random sample) of the MCMC draws produced by the first-step analysis guarantees proper propagation of all the estimation uncertainty.

The same principles applied here to study the synchrony among populations of one species could also be applied to multiple species studied at a single site, i.e., to study the synchrony among the species in a community. In a neat series of studies, José Lahoz-Monfort and colleagues used this approach to study synchrony in survival (Lahoz-Monfort et al., 2011), components of productivity (Lahoz-Monfort et al., 2013), and in population dynamics (Lahoz-Monfort et al., 2017) of a seabird community in the Scottish Sea. Obviously, one could then go a step further and study the dynamics of a community at multiple sites.

CHAPTER

ELK

17

CHAPTER OUTLINE

IPM ingredients: age-at-death data, radio tracking data, hunter survey data
Location: USA (Idaho)
Spatio-temporal scale: 5877 km^2, 6 years

17.1 INTRODUCTION

Intriguingly, the exploitation of animal populations through hunting and fishing may produce data that are informative about population size and demographic processes. Typical data sets collected by hunters or fishermen are the numbers and/or biomass of the individuals harvested, often structured according to age, sex, or size. Such data are known as *age-at-harvest* data when harvested individuals are structured by their age. The collection of harvesting data is often organized by agencies such as the US Fish and Wildlife Service, which uses them to monitor and manage harvested populations.

A number of statistical methods have been developed for the analysis of age-at-harvest data; see Skalski et al. (2005) for a review. One analysis aims to estimate the size and trend of a harvested population and is known as statistical population reconstruction, virtual population analysis, or cohort analysis (Fry, 1949; Deriso et al., 1985; Fryxell et al., 1988; Maunder and Punt, 2013). The principle of a population reconstruction is simple: if the age at death of all individuals in a population is known, the size of a cohort can be deduced from the total number of dead individuals belonging to that cohort. Consequently, the number of individuals alive in each age class can be reconstructed, and the year-specific population sizes can be derived as well. In reality, however, far from all dead individuals in a population are ever observed; age-at-death data typically include only killed individuals, i.e., those that died as a result of hunting or fishing. All individuals that died from other causes, especially from natural mortality, must typically be ignored in such analyses. Early applications of population reconstruction simply assumed that natural mortality was absent or at least negligible and constant across age classes and over time. But even the constancy assumption is hardly ever met in practice, and hence, population reconstruction typically produces biased estimates of population size and often also of population trends (Skalski et al., 2005).

To relax the assumption that all dead individuals are included in the age-at-harvest data, age-specific survival and harvest rates must be known to estimate age- and time-specific population size (Skalski et al., 2005). Therefore, models have been developed that use additional ("auxiliary") data that are informative about these parameters (Gove et al., 2002; Skalski et al., 2007; Broms et al., 2010; Clawson et al., 2013; Berg et al., 2017); i.e., in a spirit very similar to that of an integrated population model (IPM). Such auxiliary data can be radio tracking data complemented with information about

Integrated Population Models. https://doi.org/10.1016/B978-0-12-820564-8.00002-6

the source of mortality (Gove et al., 2002; Broms et al., 2010; Berg et al., 2017), catch effort data (Gast et al., 2013), or population index data (Skalski et al., 2007; Broms et al., 2010). Sometimes not all harvested individuals are reported, and additional data are needed that are informative about the fraction of reported individuals among those harvested (Gove et al., 2002). Skalski et al. (2005) provides a valuable review of the different types of population reconstruction methods for data types typically available for terrestrial species and contains a discussion of the critical assumptions needed for each. Program PopRecon 2.0 (Clawson et al., 2017) is a user-friendly interface to the AD Model Builder software (Fournier et al., 2012) to analyze age-at-harvest data jointly with auxiliary data.

Here we develop an IPM for the analysis of age-at-harvest data. The key idea is that age-at-harvest data can be regarded as biased count observations made on an age-structured population. The observations are biased because they only include harvested individuals and not those dying from any other causes. Nevertheless, age-at-harvest data contain some information about the size and dynamics of a population. We use the age-at-harvest data as the population count data in a state-space model, the state process of which is exactly an age- or stage-stratified population model. The required integration with other data sets that are informative about individual demographic processes or the observation process is straightforward. The explicit adoption of a population model is a main difference from earlier population reconstruction methods that did not formally accommodate such structure due to population dynamics. Conn et al. (2008) and Fieberg et al. (2010) realized that age-at-harvest data can be analyzed jointly with an additional data set as an IPM. In this chapter, we illustrate the application of an IPM for age-at-harvest data. We perform a prior sensitivity analysis and check parameter estimability by data cloning. Finally, we present a reparameterization of the model in terms of mortality hazard rates and discuss advantages of the IPM over conventional population reconstruction. We conclude that this is an interesting model for the data commonly available for harvested populations, but a good understanding of the proportion of population dynamics information that comes from the data as opposed to the priors is required in any application of these models, and case-specific simulations as well as prior sensitivity analyses may have to be carried out before you can believe the inferences from your own analysis.

17.2 ELK IN IDAHO

As an example, we use data from an elk (*Cervus canadensis*) population in northern Idaho, USA (Fig. 17.1). Besides the age-at-harvest data, radio tracking data and data on hunter reporting rates are available. These data were published and analyzed using population reconstruction by Gove et al. (2002). For each data set, the likelihood was constructed, and they were jointly analyzed in a frequentist mode.

All three data sets were sampled over 6 years from 1988 to 1993 and include only females, i.e., elk cows. The age-at-harvest data contain the number of elk shot and reported in each year stratified by age. Most elk harvest occurs in October, and hunters are requested to register their quarry at a checking station. All animals registered are sexed and aged based on tooth inspection. Some rare data on elk older than 18 years were ignored because of their paucity.

Although hunters are requested to report their quarry at a checking station, some fail to comply. Ignoring such non-reporting would negatively bias the estimates of harvest rates and population sizes (Gove et al., 2002). To estimate the reporting rate, a telephone survey was conducted each year. A sample of registered hunters was interviewed, and it was noted whether a specific hunter killed an elk cow and whether he/she then indeed reported it.

Finally, a telemetry study of radio-collared elk was conducted during the same period. The annual number of tagged individuals was recorded for those that were killed, died from a cause other than hunting, or survived. Unfortunately, the age of the tagged elk cows was unknown, preventing the estimation of age-specific survival and harvest rates.

17.3 POPULATION MODEL

We first decide upon a sensible population model. The observed population data, namely the age-at-harvest data, have a resolution of 17 1-year age classes, and we develop a population model with the same 17 age classes, i.e., we do not lump some of them into multiyear age classes.

We define a preharvest "census" model for females only. The youngest age therefore contains 1-year-old cows; actually, they are slightly older than 1 year because elk typically give birth in June, and the preharvest "census" is assumed to take place in early October. The oldest age class contains 17-year-old females. Since harvested cows older than 17 years were excluded from the data, our population model assumes that elk cannot become older than 17 years. This assumption could easily be relaxed, but for our example, results not shown here suggest that this would have no impact on the main inferences. The life-cycle graph for our population model is shown in Fig. 17.2.

The number of 1-year-old elk cows in year t ($N_{1,t}$) is governed by the recruitment rate (f), which is assumed to be identical for all elk older than 1 year, while 1-year-old elk are assumed not to have reproduced yet. The recruitment rate itself is composed of the annual number of female calves produced by a cow as well as annual calf survival. Here we only

FIGURE 17.1

Impressive elk (*Cervus canadensis*) bull from the Canadian Rockies (photograph by John E. Marriott).

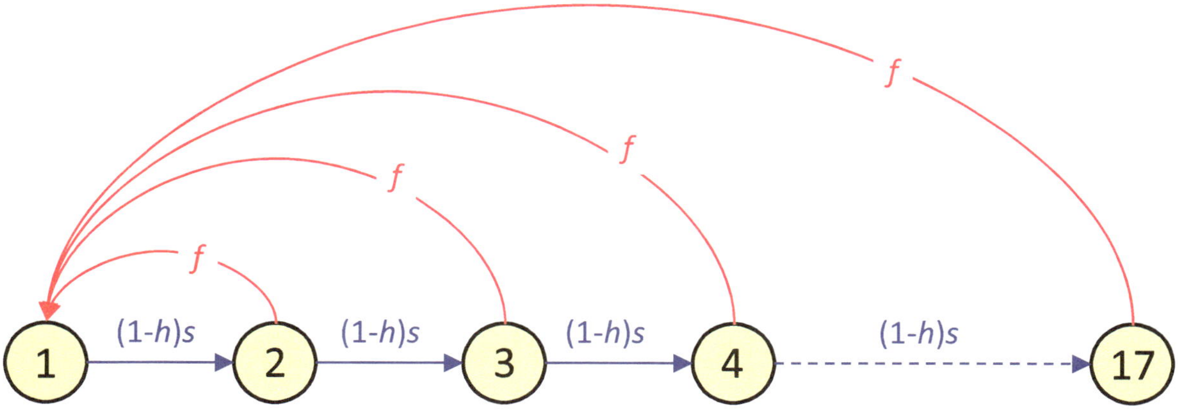

FIGURE 17.2

Preharvest "census" life-cycle graph of female elk in Idaho. The graph contains 17 stages corresponding to elk in different annual age classes, the labels being the age in years (note that ages 5–16 are omitted). The demographic processes are recruitment (f), hunting mortality (h), and natural survival (s); the last of these is the complement of natural mortality. Note that total annual survival is $s^* = (1 - h)s$, i.e., an individual must fail to be killed and then survive. Recruitment transitions are shown as red arrows and pure survival transitions in blue.

estimate a single recruitment parameter because the two components are not separately estimable with our data. To accommodate demographic stochasticity in our model, we use a Poisson distribution.

$$N_{1,t} \sim Poisson\left(f_t \sum_{a=2}^{17} N_{a,t-1}\right)$$

The number of elk cows that are a years old in year t ($N_{a,t}$) is modeled with a binomial distribution with a total that is the number of cows in age $a-1$ in year $t-1$ and a success parameter that is the product of surviving the fall hunting season and the rest of the year (i.e., total annual survival s_t^*).

$$N_{a,t} \sim Binomial\left((1 - h_{t-1})s_{t-1}, N_{a-1,t-1}\right)$$

Hunting mortality (h_t) is defined as the probability that an individual alive at t is killed during the hunting season in year t, and natural survival probability (s_t) is the probability that an individual alive in year t just after the hunting season survives until the start of the subsequent hunting season in year $t+1$. This formulation thus assumes that during the comparatively short hunting period, only hunting-related mortality occurs, while during the rest of the year, the only mortality is from causes other than hunting. Overall annual survival is then given by $s_t^* = (1 - h_t)s_t$.

This population model is extremely simple and perhaps unrealistic to some degree, because it lacks any age structure in natural survival and hunting mortality and has only a simple age structure in recruitment. Large ungulates including elk usually exhibit strong age-related patterns in demographic rates (Eberhardt et al., 1996; Gaillard et al., 1998; Loison et al., 1999; Larkin et al., 2003). For some applications of such models, age structure should ideally be included. Our choice of a very simple model is motivated by considerations of parameter identifiability: demographic rates cannot be estimated from the age-at-harvest data alone; instead, additional information either in the form of auxiliary data and/or strong priors is needed (Skalski et al., 2005). In our integrated analysis, we use additional radio tracking data that provide information about hunting mortality and natural survival. However, since the ages of the radio-tracked individuals are unknown, no age structure in these parameters can be included. Of course, this is specific to this example and is not generally a problem in this type of IPM when the right data are available.

17.4 COMPONENT DATA LIKELIHOODS

The three 6-year data sets for this analysis are stored in an R data file that we load and whose structure we inspect. The data are taken from Gove et al. (2002), who published them in tables 1–3 of their paper.

```
library(IPMbook); library(jagsUI)
data(elk)
str(elk)

List of 3
 $ C: num [1:17, 1:6] 21 36 15 8 7 8 6 5 3 3 ...
  ..- attr(*, "dimnames")=List of 2
  .. ..$ : chr [1:17] "1" "2" "3" "4" ...
  .. ..$ : chr [1:6] "1988" "1989" "1990" "1991" ...
 $ H: num [1:2, 1:6] 275 143 290 154 211 211 360 272 201 201 ...
  ..- attr(*, "dimnames")=List of 2
  .. ..$ : chr [1:2] "Hunters surveyed" "Registered harvest"
  .. ..$ : chr [1:6] "1988" "1989" "1990" "1991" ...
 $ R: num [1:3, 1:6] 1 2 10 4 1 20 0 1 24 0 ...
  ..- attr(*, "dimnames")=List of 2
  .. ..$ : chr [1:3] "hunted" "died naturally" "survived"
  .. ..$ : chr [1:6] "1988" "1989" "1990" "1991" ...
```

17.4.1 AGE-AT-HARVEST DATA

When developing an IPM, we usually have some population counts or population indices and thus statistics about the number of individuals *alive* in a population. This is quite different here, as we use a count data set of *dead* individuals. The age-at-harvest data (`C`) are a matrix where columns refer to the 6 years 1988–93, and rows refer to age. The entries are the number of elk cows killed and reported per year and annual age classes up to the maximum of 17.

For these data, we adopt a state-space model with a state model corresponding to the population model described in Section 17.3. For an observation model, we use the binomial distribution with a total that is the age-specific population size just prior to the hunting period and a success parameter that is assumed to consist exclusively of hunting mortality. Yet since not all hunters report their quarry, the age-at-harvest data do not include all harvested individuals. Therefore, the reporting rate (r) must be included in the model as an additional nuisance parameter; it is defined as the probability that a hunter reports its quarry in year t (r_t). Hence, the number of reported, harvested elk of age class a in year t is

$$C_{a,t} \sim Binomial\left(h_t r_t, N_{a,t}\right).$$

Here, the reporting rate is assumed to vary by year only and not by age. If the likelihood of a hunter reporting a killed elk depends on elk age, one would need to include an age structure in the reporting rate as well. Again, mostly due to our sparse data, we don't do this in our example.

As always, the state-space model contains a likelihood contribution of the initial population size. In typical IPMs, we usually place vague priors on these quantities, e.g., a uniform distribution with a wide range. Our elk data are relatively sparse, and there is less information about population sizes in age-at-harvest data than in the typical count data of live individuals. In this example, therefore, we specify informative priors to facilitate estimation. We choose the priors for the initial $\boldsymbol{N}$ vector to be quite narrow in the range of values where the likelihood is expected to have some support for that parameter. In addition, we conduct a prior sensitivity analysis, i.e., explore the changes in our key inferences when using different priors and use data cloning to gauge parameter identifiability.

In our main analysis, we use Poisson distributions with rate parameters that correspond to the age-specific population sizes in the first year. To gauge age-specific population sizes in the first year, we build a quick matrix population model parameterized with educated guesses of the demographic rates and evaluate the stable age distribution ($\boldsymbol{d}$). Using another set of educated guesses at the hunting mortality and reporting rate, we calculate expected population size of the first age class in year 1 (n_1) from the number of 1-year-old individuals harvested in that year ($C_{1,1}$). Under the assumption of a stable age distribution, we can then calculate the expected number of individuals in the other age classes as $n_a = n_1 \frac{d_a}{d_1}$ and then choose a prior as $N_{a,1} \sim Poisson(n_a)$. We can derive different priors by using another set of *a priori*, educated guesses about survival, recruitment, or hunting mortality to calculate $\boldsymbol{n}$ or by adoption of uniform or negative binomial distributions instead of the Poisson. We note that how we chose priors for the population size in the first year is not innocuous because to some degree, we make use of the data, here of $C_{1,1}$, twice. We think this is not particularly problematic here because it is just one datum, and moreover our prior does not fix the associated parameter but has some dispersion. Moreover, to assess the impact of the prior, we will perform a prior sensitivity analysis (see Section 17.7).

17.4.2 HUNTER SURVEY DATA

To obtain information about the probability that successful hunters report their quarry, a telephone survey was conducted each year. The resulting data appear in the form of the matrix `H` with six columns for the years 1988–93. The first row contains the number of successful hunters included in the survey ($\boldsymbol{a}$), and the second row is the number of those hunters who dutifully reported their quarry ($\boldsymbol{b}$). We adopt a binomial distribution for these data,

$$b_t \sim Binomial(r_t, a_t),$$

where r_t is the reporting rate in year t.

17.4.3 RADIO TRACKING DATA

Each year, between 13 and 25 elk cows were equipped with a radio transmitter for an assessment of the number that died from hunting or other causes of mortality. For the latter, we use the terms "background mortality" or "natural mortality": this is the complement of natural survival. Individuals whose radio tag failed in the course of a year were excluded in order to simplify the model. These data are arranged in matrix `R` with six columns for the years 1988–93 and with rows that distinguish individuals as having "died from hunting," "died naturally," and "survived." For these data, we specify a multinomial distribution with cell probabilities that are functions of h and s and with a total given by the total number of tracked cows in year t.

$$R_{1...3,t} \sim Multinomial\left([h_t, (1-h_t)(1-s_t), (1-h_t)s_t], \sum_i R_{i,t}\right)$$

17.5 THE INTEGRATED POPULATION MODEL

Now that we understand all three data sets and know how to construct the likelihoods of the associated component models individually, we can put everything together in an IPM by making the usual independence assumption. Before we bundle the data, we calculate the age-specific population sizes in the first year, which we will use for specification of the prior for the initial age-specific population size. For this, we define a prebreeding "census" matrix population model with 17 age classes, assume that annual survival is 0.8 and annual recruitment is 0.25, and calculate the stable age distribution.

```
s <- 0.8        # Guess of annual survival
f <- 0.25       # Guess of recruitment
```

```
# Create matrix population model (transition matrix A)
A <- matrix(0, ncol=17, nrow=17)
A[1,2:17] <- f
for (a in 2:17){
  A[a,a-1] <- s
}

# Compute stable age distribution (right eigenvector of A)
z <- which.max(Re(eigen(A)$values))
revec <- Re(eigen(A)$vectors[,z])
```

Based on reasonable guesses of the reporting rate and hunting mortality, we compute the expected population size of the first age class in year 1 and then the age-specific vector with population sizes in the first year.

```
# Population size in first age class of first year
r <- 0.5      # Guess of reporting rate
h <- 0.1      # Guess of hunting mortality
N1 <- elk$C[1,1] / (h * r)

# Compute age-specific population sizes in first year
n <- N1 * revec / revec[1]
```

We have now defined the elements for prior specifications of the age-specific population sizes in the first year; later, when we perform a prior sensitivity analysis, we will alter this computation. For now we bundle the data, write the model, and launch it in JAGS.

```
# Bundle data and produce data overview
jags.data <- with(elk, list(C=C, a=H[1,], b=H[2,], R=R, total=colSums(R), n.years=ncol(C),
  n.age=nrow(C), n=n))
str(jags.data)

List of 8
 $ C       : num [1:17, 1:6] 21 36 15 8 7 8 6 5 3 3 ...
  ..- attr(*, "dimnames")=List of 2
  .. ..$ : chr [1:17] "1" "2" "3" "4" ...
  .. ..$ : chr [1:6] "1988" "1989" "1990" "1991" ...
 $ a       : Named num [1:6] 275 290 211 360 201 325
  ..- attr(*, "names")= chr [1:6] "1988" "1989" "1990" "1991" ...
 $ b       : Named num [1:6] 143 154 211 272 201 155
  ..- attr(*, "names")= chr [1:6] "1988" "1989" "1990" "1991" ...
 $ R       : num [1:3, 1:6] 1 2 10 4 1 20 0 1 24 0 ...
  ..- attr(*, "dimnames")=List of 2
  .. ..$ : chr [1:3] "hunted" "died naturally" "survived"
  .. ..$ : chr [1:6] "1988" "1989" "1990" "1991" ...
 $ total   : Named num [1:6] 13 25 25 16 25 23
  ..- attr(*, "names")= chr [1:6] "1988" "1989" "1990" "1991" ...
 $ n.years: int 6
 $ n.age   : int 17
 $ n       : num [1:17] 420 338 272 218 176 ...


# Write JAGS model file
cat(file = "model1.txt", "
model {
# Priors and linear models
for (t in 1:(n.years-1)){
  f[t] ~ dunif(0, 1)
}
for (t in 1:n.years){
  s[t] ~ dunif(0, 1)
  h[t] ~ dunif(0, 1)
  r[t] ~ dunif(0, 1)
}

# Age-at-harvest data (state-space model)
# Model for the initial population size: Poisson priors
for (a in 1:n.age){
  N[a,1] ~ dpois(n[a])
}

# Process model over time: our model of population dynamics
for (t in 1:(n.years-1)){
  N[1,t+1] ~ dpois((Ntot[t] - N[1,t]) * f[t])
  for (a in 2:n.age){
    N[a,t+1] ~ dbin((1-h[t]) * s[t], N[a-1,t])
  } #a
} #t
```

```
# Derived quantity: total year-specific population size
for (t in 1:n.years){
  Ntot[t] <- sum(N[,t])
}

# Observation model
for (t in 1:n.years){
  for (a in 1:n.age){
    C[a,t] ~ dbin(h[t] * r[t], N[a,t])
  } #a
} #t

# Hunter survey data (logistic regression model)
for (t in 1:n.years){
  b[t] ~ dbin(r[t], a[t])
}

# Radio tracking data (multinomial model)
for (t in 1:n.years){
  R[,t] ~ dmulti(prt[,t], total[t])
  prt[1,t] <- h[t]
  prt[2,t] <- (1-h[t]) * (1-s[t])
  prt[3,t] <- (1-h[t]) * s[t]
}
}
")


# Initial values
inits <- function() {list(s=runif(jags.data$n.years, 0.8, 1), h=runif(jags.data$n.years,
  0.05, 0.15))}

# Parameters monitored
parameters <- c("h", "s", "f", "r", "Ntot", "N")

# MCMC settings
ni <- 250000; nb <- 50000; nc <- 3; nt <- 20; na <- 5000

# Call JAGS from R (ART 4 min) and check convergence
out1 <- jags(jags.data, inits, parameters, "model1.txt", n.iter=ni, n.burnin=nb, n.chains=nc,
  n.thin=nt, n.adapt=na, parallel=TRUE)
traceplot(out1)
```

17.6 RESULTS ON ELK POPULATION DYNAMICS

The chains of all parameters show good convergence after many iterations, as you can see when you inspect the traceplots (not shown) as well as by the values of R-hat. Here is a partial posterior summary for the estimated quantities in our model.

```
print(out1, 3)

            mean    sd    2.5%     50%   97.5% overlap0 f  Rhat n.eff
h[1]       0.116 0.012   0.094   0.115   0.141    FALSE 1 1.000 30000
h[2]       0.145 0.018   0.114   0.144   0.184    FALSE 1 1.001  6631
h[3]       0.086 0.011   0.067   0.085   0.111    FALSE 1 1.001  3978
h[4]       0.085 0.013   0.063   0.084   0.115    FALSE 1 1.002  1590
h[5]       0.080 0.013   0.058   0.079   0.110    FALSE 1 1.001  1802
h[6]       0.182 0.036   0.124   0.178   0.265    FALSE 1 1.003  1062
s[1]       0.851 0.068   0.701   0.857   0.963    FALSE 1 1.001  2988
s[2]       0.929 0.046   0.817   0.938   0.991    FALSE 1 1.000 10767
s[3]       0.934 0.044   0.824   0.943   0.991    FALSE 1 1.001  7929
s[4]       0.951 0.046   0.828   0.965   0.999    FALSE 1 1.001  4161
s[5]       0.817 0.071   0.662   0.823   0.935    FALSE 1 1.001  1967
s[6]       0.792 0.082   0.611   0.800   0.926    FALSE 1 1.000 30000
f[1]       0.259 0.035   0.193   0.258   0.331    FALSE 1 1.000  4754
f[2]       0.300 0.038   0.231   0.298   0.378    FALSE 1 1.000 30000
f[3]       0.316 0.043   0.240   0.314   0.407    FALSE 1 1.000  4837
f[4]       0.207 0.034   0.145   0.205   0.280    FALSE 1 1.000 29810
f[5]       0.140 0.034   0.083   0.137   0.215    FALSE 1 1.001  2521
r[1]       0.520 0.030   0.462   0.520   0.579    FALSE 1 1.000 30000
r[2]       0.530 0.029   0.473   0.530   0.587    FALSE 1 1.000 14354
r[3]       0.995 0.005   0.983   0.997   1.000    FALSE 1 1.000 15095
r[4]       0.755 0.023   0.709   0.755   0.797    FALSE 1 1.000 30000
r[5]       0.995 0.005   0.982   0.997   1.000    FALSE 1 1.001 10259
r[6]       0.482 0.027   0.429   0.482   0.536    FALSE 1 1.000 22795
```

```
Ntot[1]  2096.276  45.462 2007.000 2096.000 2187.000    FALSE 1 1.000 23146
Ntot[2]  1997.991 170.065 1641.975 2007.000 2302.025    FALSE 1 1.001  2137
Ntot[3]  2048.117 222.540 1608.000 2051.000 2478.000    FALSE 1 1.001  1819
Ntot[4]  2239.809 285.800 1693.000 2236.000 2815.000    FALSE 1 1.002  1255
Ntot[5]  2299.903 336.905 1675.000 2291.000 2995.025    FALSE 1 1.001  1903
Ntot[6]  1991.210 355.567 1352.000 1973.000 2742.000    FALSE 1 1.002   964
N[1,1]    428.762  18.718  393.000  429.000  466.000    FALSE 1 1.000  9726
N[2,1]    355.155  16.736  323.000  355.000  388.000    FALSE 1 1.000  8812
[...output truncated...]
N[16,6]    15.846   4.218    8.000   16.000   25.000    FALSE 1 1.001  2284
N[17,6]    12.726   3.854    6.000   12.000   21.000    FALSE 1 1.001  3377
```

The estimates are similar to those reported from the original analyses using the population reconstruction method (Gove et al., 2002), although we here use a model with more parameters. We allow temporal variation in all model parameters, whereas Gove et al. (2002) assumed constant survival probabilities. We produce plots to show estimates of cause-specific mortality, total annual survival, and recruitment from our IPM (Fig. 17.3). Hunting and background (natural) mortality are broadly similar and vary only moderately over time even though we estimated them as fixed effects (which

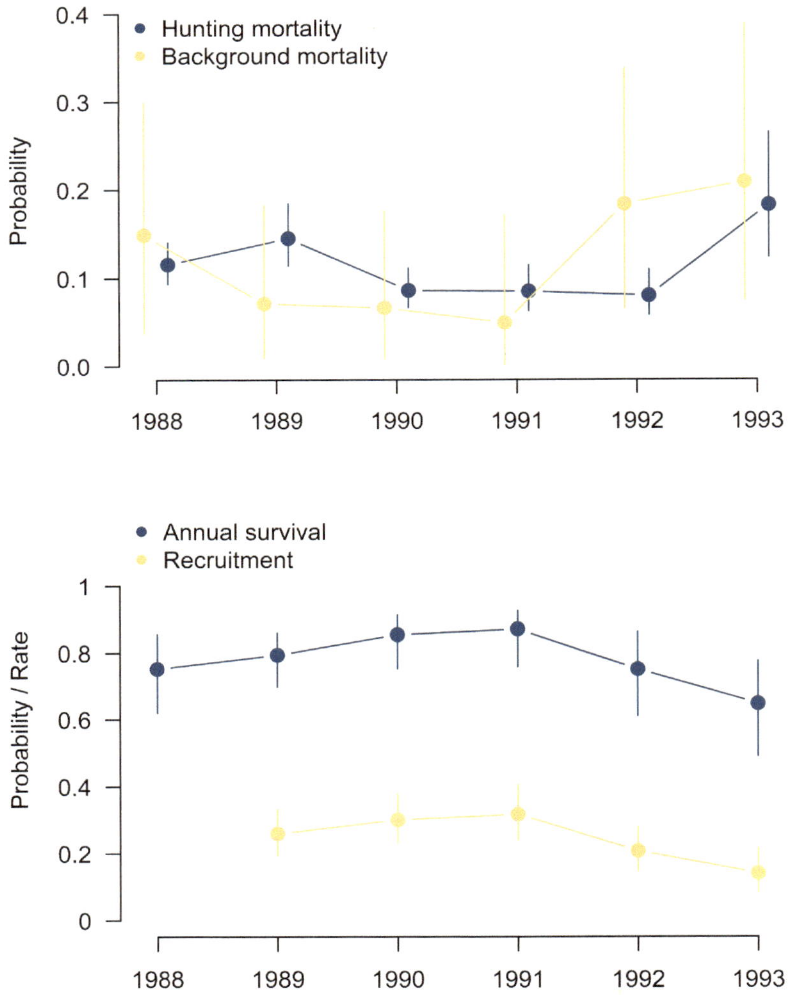

FIGURE 17.3

Estimates of cause-specific mortality probabilities (top panel) and annual survival probabilities and recruitment rates (bottom panel) of elk cows in Idaho. Dots show posterior means, and vertical lines show 95% credible intervals.

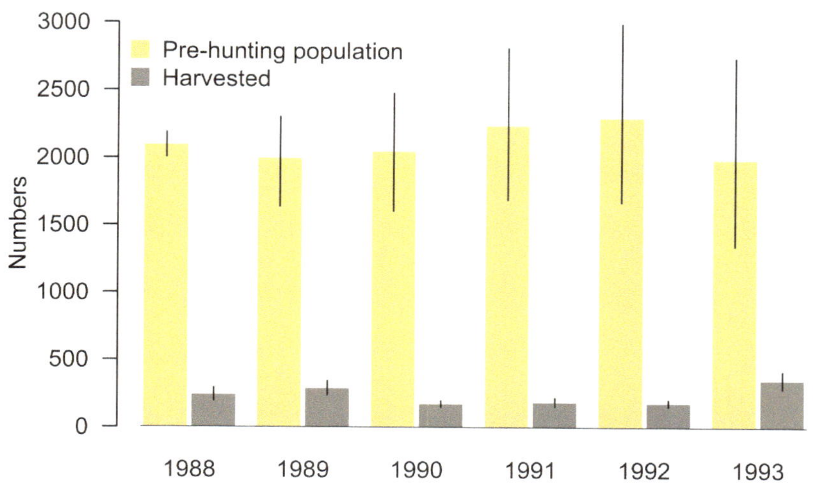

FIGURE 17.4

Estimates of the total preharvest population size of elk in northern Idaho and the number of harvested individuals. Bars show posterior means, and vertical lines show 95% credible intervals.

will make the point estimates vary more than under a random-effects model). Estimates of annual survival are calculated from hunting mortality and natural survival as a derived quantity. Annual survival tends to first increase and then decrease. Estimates of recruitment are available for 5 years only and indicate a temporal pattern similar to that of annual survival—first a slight increase, and then a decline.

```
# Calculation of annual survival (s*)
s.star <- (1-out1$sims.list$h) * out1$sims.list$s
```

The model provides estimates of age-specific and total annual population sizes. Under the assumption that the demographic rates and reporting rate are accurate, we obtain estimates of true rather than relative population size. This is important because this allows the setting of a proper harvest quota in a management context. Based on the estimates of total population size and hunting mortality, we can estimate the total number of elk killed. These latter estimates are larger than the observed age-at-harvest data because the latter include only the killed elk that were actually reported (Fig. 17.4).

17.7 PRIOR SENSITIVITY ANALYSIS

Above, we briefly mentioned that the priors adopted for the age-specific initial population size may be critical for inferences under the model for most real-life data sets that are not "huge." Here, we first carry out a formal prior sensitivity analysis to understand the impact of these priors on the estimates of the model. Hence, we reanalyze the data with three different sets of priors for the initial population sizes. Second, we investigate how prior sensitivity changes when sample size increases in order to gauge whether parameters are at all identifiable.

Of course, there are myriad possible priors for these parameters, and here we restrict ourselves to just two types of priors. First, we use the same idea as the one developed for the priors in our first model (Section 17.4.1) but now assume that our educated guess about the harvest rate was not 0.1 as before but either 0.075 or 0.125. In this way, the age-specific *distribution* of the population sizes in the first year will not change compared with our original formulation, but the actual population sizes will be different. For the priors, we then adopt Poisson distributions centered on these expected population sizes. Second, we use discrete uniform distributions centered on the same expected population sizes as in the first model. We use a fairly wide range (±50% of the mean) for the bounds of these uniform distributions. Compared with the Poisson priors from the first IPM, the mean is still the same, but we allow for more uncertainty when using the discrete uniform distributions. All the priors that we specified for the initial population size are truly informative.

Comparison of the posteriors for hunting mortality and total population size under the four different sets of priors for the initial population size clearly show considerable prior sensitivity in these parameters (Fig. 17.5). Although specification of the different priors was based on only a moderate range of guesses about hunting mortality (between 0.075 and

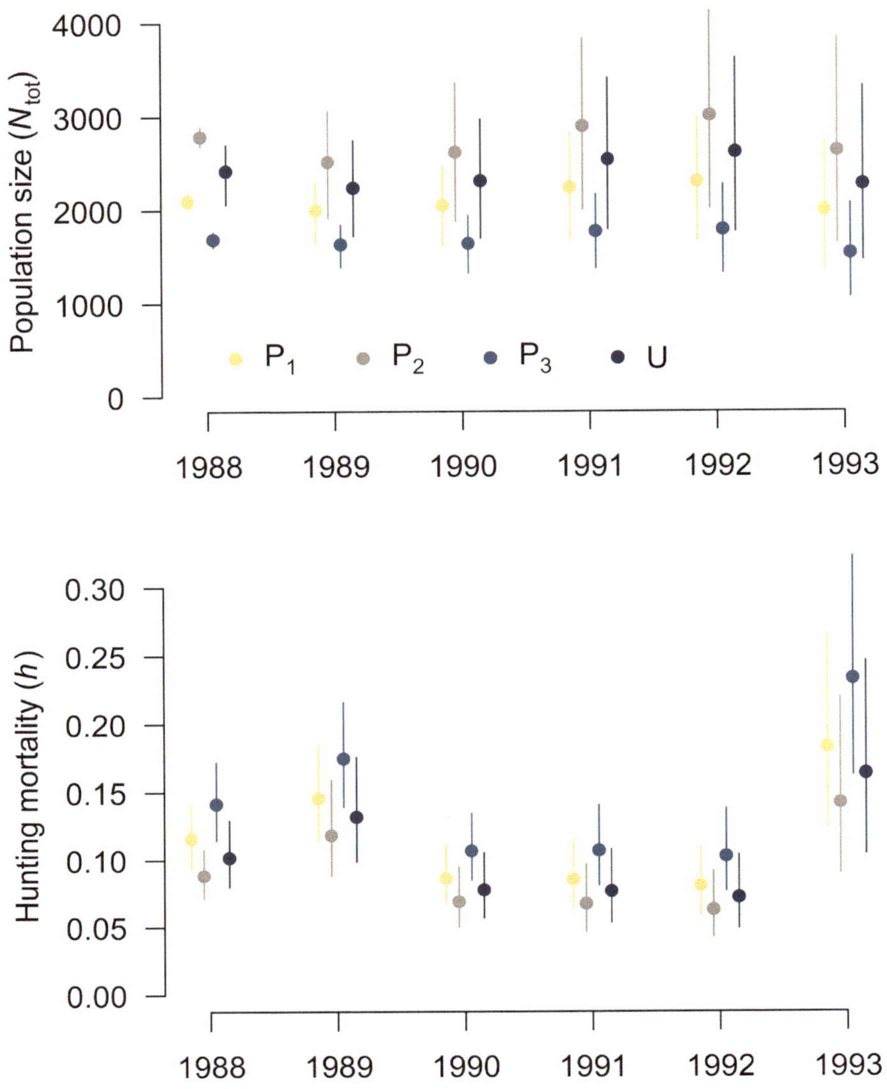

FIGURE 17.5

A prior sensitivity analysis – Estimates of the total preharvest population size of northern Idaho elk and hunting mortality under four different sets of priors for the age-specific population sizes in the first year. Dots show posterior means, and vertical lines show 95% credible intervals. The three different Poisson priors are labeled P_{1-3}, and U indicates the analysis with a discrete uniform prior. P_1 is derived from an educated guess of hunting mortality as $h = 0.1$, while P_2 assumes $h = 0.075$, and P_3 assumes $h = 0.125$. Results appear in the order P_1, P_2, P_3 and U.

0.125), the resulting posteriors of hunting mortality were rather different from one another. The uniform priors centered on expected population sizes based on a hunting mortality of 0.1 resulted in posteriors of hunting mortality similar to those under P_1 in most years (Fig. 17.5), but the posteriors of population size were more diffuse, i.e., had more uncertainty. Estimates of total population size and hunting mortality appear to be negatively correlated: when estimates of hunting mortality are high, those of population size are low.

This prior sensitivity may worry some, in particular if we have no *a priori* ideas about sensible values for such informative priors. However, not all parameters are equally sensitive to the choice of priors for initial population size. For instance, the population growth rates in our analysis turn out to be much less sensitive to prior choice (Fig. 17.6). Hence, population trends can be assessed much more reliably with the data at hand.

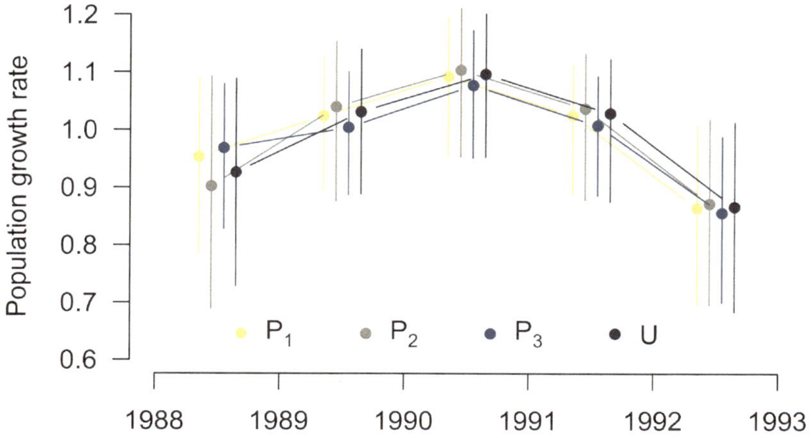

FIGURE 17.6

A prior sensitivity analysis – Estimates of annual population growth rates of northern Idaho elk under four different sets of priors for age-specific, initial population sizes. Dots show posterior means, and vertical lines show 95% credible intervals. The three different Poisson priors are labeled P_{1-3}, and U indicates the analysis with a discrete uniform prior. P_1 is derived from an educated guess of hunting mortality as $h = 0.1$, while P_2 assumes $h = 0.075$, and P_3 assumes $h = 0.125$.

When a data set contains no information at all about a parameter in a model, then the posterior will be identical to the prior—the parameter is not identifiable (Gimenez et al., 2009). Hence, when a posterior is very sensitive to the choice of its prior, this typically suggests little information in the data about that particular parameter. We can check whether prior sensitivity is reduced by increasing the sample size. If this is the case, we know there is at least *some* information in the data about the parameter; this is the principle behind the use of data cloning for the assessment of parameter estimability (Lele et al., 2007, 2010). We did this by supposing that we have 20 times as much radio tracking data as in our actual application (we can easily simulate this by multiplying `elk$R` by 20) and then fit two IPMs with priors for the age-specific population sizes in the first year that are supposed to be noninformative as opposed to the strongly informative priors used before. That is, we now use uniform priors in the range between 1 and 1000 for the initial population size of each of the 17 age classes, and hence we do not use any information about the presumed stable age distribution in the analysis. We then use priors with an even larger range (i.e., that are even more vague) by specifying uniform distributions between 1 and 3000 to gauge prior sensitivity.

We need reasonable starting values for the age-specific population size in the first year for these models, which we pick based on the presumed stable age distribution of the elk population because otherwise the Markov chain Monte Carlo (MCMC) sampling will not start (remember that at convergence, use of stable age computations for selecting where to start the MC chains will *not* affect the solutions, unlike the priors). The models run fairly fast, but when the prior for the population size in the first year is most diffuse, many MCMC draws are required to obtain convergence. Code for these analyses is provided on the book web page.

We focus our attention on the posterior distributions of the population in the first year for the age classes 1 to 4 under the four scenarios; it is within these age classes that the major differences appear (Fig. 17.7). With the original data, the posterior of the population size in the first year of the first age class (parameter $N_{1,1}$) hits the upper bound of the uniform prior regardless of whether it is 1000 or 3000 (Fig. 17.7). A hard upper bound is visible also for the second age class but not for the older age classes (note that age classes 5–17 are not shown in Fig. 17.7). The posterior means for the population sizes in the first year differ strongly for the two priors, and hence the results are very sensitive to the prior of the initial population size. This is true not only for the population sizes in the first year but also for the age-specific population sizes in years 2–6 and for hunting mortality.

The prior sensitivity disappears when we assume we had 20 times more radio tracking data (Fig. 17.7). Now the posteriors of the population size in the first year no longer hit the upper limit of the prior, and the results are the same regardless of whether the upper limit of the prior is 1000 or 3000. These results indicate that the parameters of the IPM are identifiable with sufficient data and that the strong prior sensitivity in the analysis of the original data is a consequence of not enough information in our original analysis.

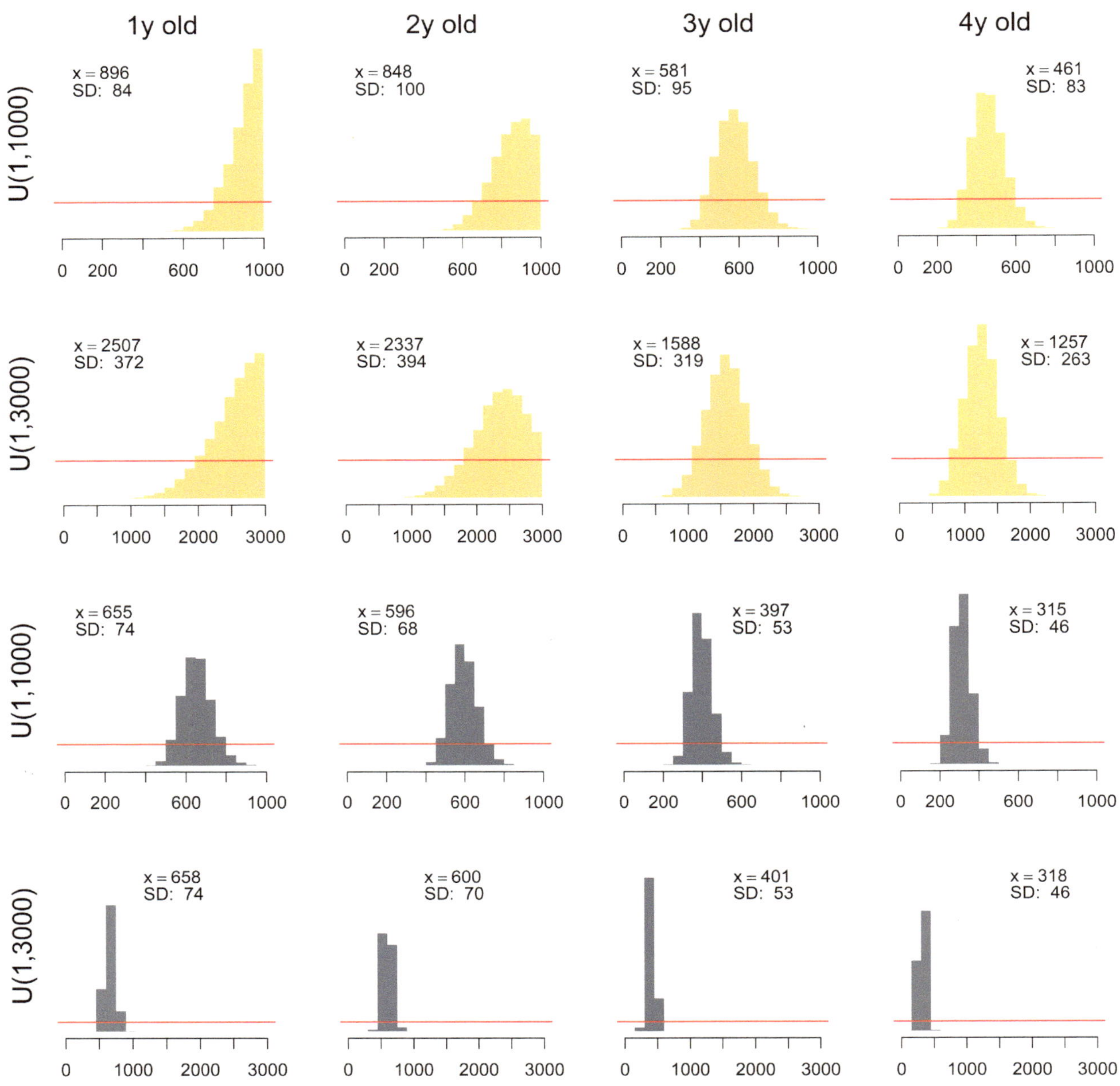

FIGURE 17.7

A prior sensitivity analysis combined with data cloning for the assessment of parameter identifiability – Posterior distributions of population sizes in the first year of the first four age classes (columns) for different priors (rows) when the original data (yellow histograms) and 20 times larger sample of radio tracking data (dark gray histograms) are used. Also shown are the posterior means ($\overline{x}$) and posterior standard deviations (SDs). The prior distribution is shown by the red line, and the *x*-axes of all graphs cover the full range of priors.

17.8 SPECIFICATION OF THE SURVIVAL PROCESS WITH HAZARD RATES

The survival, or mortality, process operates in continuous time, but in most statistical models, mortality is defined for discrete time intervals. In this book, we usually work with the discrete-time formulation of survival processes, in which case mortality probability is defined as the probability that an individual dies sometime within a defined time interval. In many cases, it may not be relevant whether mortality is defined in continuous or in discrete time, but sometimes it does matter. One important example is the study of different sources of mortality and their temporal interaction, e.g., the assessment of whether different sources of mortality act in an additive or compensatory manner (Péron, 2013). Knowledge about whether, and if so, how strongly, hunting mortality is compensated by other sources of mortality is extremely relevant for the management of harvested populations (Burnham and Anderson, 1984; Nichols et al., 1984; Boyce et al., 1999; Lebreton, 2005; Sandercock et al., 2011). Differences between continuous- and discrete-time formulations can also occur when covariates are modeled and when mortality is used in elasticity analyses of the population growth rate. Ergon et al. (2018) have nicely summarized the differences between continuous- and discrete-time survival models.

Continuous-time survival models are specified in terms of the mortality hazard rate (η_t), which is defined as the instantaneous (i.e., the limit at time t) intensity of deadly events that an individual is exposed to (Skalski et al., 2005; Ergon et al., 2018; Murray and Bastille-Rousseau, 2020). The probability of surviving any period between t_1 and t_2 can then be calculated from the hazard rate as

$$s = \exp\left(- \int_{t_1}^{t_2} \eta_t dt \right).$$

If we assume that the mortality hazard within a time interval is approximately constant, the integral in the formula can be replaced by the average hazard rate ($\overline{\eta}$) multiplied with the width of the interval (i.e., its duration). Furthermore, if the time interval corresponds to the desired unit of time in an analysis (e.g., a year in most demographic studies), survival probability over this period is given as $s = \exp(-\overline{\eta})$. Hence, survival probability is then a simple transformation of the time-averaged mortality hazard rate.

In our elk case study, we estimate two mortality rates, one due to hunting and the other form all other causes of death. Originally, we applied the discrete time scale in the IPM for formulating hunting mortality probability (h) and natural survival (s). Background mortality is therefore given by $(1-h)\,(1-s)$. The two mortality probabilities are "linked" statistically by the way in which we express them. Hence, even if s remains constant, background mortality will decline when hunting mortality increases and vice versa. This undesirable feature can be avoided if mortality is expressed as a function of the two mortality hazard rates. In that case, if the hunting hazard rate increases, for instance because of a longer hunting season, the background mortality hazard will remain unaffected.

The formulation of an IPM using mortality hazard rates instead of hunting mortality and natural survival is really not complicated (Skalski et al., 2005; Ergon et al., 2018; Nater et al., 2020). We define the two mortality hazard rates κ and ν for hunting and background mortality, respectively. The overall survival is then defined as $s_t^* = \exp(-(\kappa_t + \nu_t))$, the mortality probability due to hunting as $h_t = \left(1 - s_t^*\right)\frac{\kappa_t}{\kappa_t+\nu_t}$, and the probability of dying from any other cause is $o_t = \left(1 - s_t^*\right)\frac{\nu_t}{\kappa_t+\nu_t}$. Note that here we estimate overall survival (s^*), which is different from natural survival (s).

For illustration, we now write BUGS code for our IPM using mortality hazards instead of mortality probabilities. Compared with the original IPM, only a few changes are needed. Overall survival and cause-specific mortality probabilities are expressed in terms of the mortality hazards using the above formulae. The primary parameters to describe the mortality process are now the two mortality hazard rates, for which we must specify priors. Mortality hazard rates are nonnegative; hence, a gamma distribution is a natural possible choice. The population model is directly parameterized with overall survival. Moreover, the cell probabilities of the multinomial distribution for the radio tracking data are parameterized in terms of the two cause-specific mortality probabilities and overall survival. We use the same priors for the age-specific initial population sizes as we did in our IPM earlier in this chapter.

```
# Bundle data and produce data overview
jags.data <- with(elk, list(C=C, a=H[1,], b=H[2,], R=R, total=colSums(R), n.years=ncol(C),
  n.age=nrow(C), n=n))

# Write JAGS model file
cat(file = "model3.txt", "
model {
# Priors and linear models
for (t in 1:(n.years-1)){
  f[t] ~ dunif(0, 1)
}
for (t in 1:n.years){
  # Priors for hazard rates
  mh[t] ~ dgamma(0.1, 0.1)                          # Hunting mortality hazard rate
  mo[t] ~ dgamma(0.1, 0.1)                          # Background mortality hazard rate

  # Calculate probabilities from hazard rates
  s[t] <- exp(-(mh[t] + mo[t]))                     # Overall survival
  h[t] <- (1 - s[t]) * (mh[t] / (mh[t] + mo[t]))    # Hunting mortality
  o[t] <- (1 - s[t]) * (mo[t] / (mh[t] + mo[t]))    # Background mortality

  # Prior for reporting rate
  r[t] ~ dunif(0, 1)
}

# Age-at-harvest data (state-space model)
# Model for the initial population size: Poisson priors
for (a in 1:n.age){
  N[a,1] ~ dpois(n[a])
}

# Process model over time: our model of population dynamics
for (t in 1:(n.years-1)){
  N[1,t+1] ~ dpois((Ntot[t] - N[1,t]) * f[t])
  for (a in 2:n.age){
    N[a,t+1] ~ dbin(s[t], N[a-1,t])
  } #a
} #t

# Derived quantity: total year-specific population size
for (t in 1:n.years){
  Ntot[t] <- sum(N[,t])
}

# Observation model
for (t in 1:n.years){
  for (a in 1:n.age){
    C[a,t] ~ dbin(h[t] * r[t], N[a,t])
  } #a
} #t

# Hunter survey data (logistic regression model)
for (t in 1:n.years){
  b[t] ~ dbin(r[t], a[t])
}

# Radio tracking data (multinomial model)
for (t in 1:n.years){
  R[,t] ~ dmulti(prt[,t], total[t])
  prt[1,t] <- h[t]
  prt[2,t] <- o[t]
  prt[3,t] <- s[t]
}
}
")
```

```
# Initial values
inits <- function() {list(mh=runif(jags.data$n.years, 0.01, 0.1), mo=runif(jags.data$n.years,
  0.01, 0.1))}

# Parameters monitored
parameters <- c("mh", "mo", "h", "s", "o", "f", "r", "Ntot", "N")

# MCMC settings
ni <- 450000; nb <- 50000; nc <- 3; nt <- 40; na <- 5000

# Call JAGS from R (ART 9 min) and check convergence
out9 <- jags(jags.data, inits, parameters, "model3.txt", n.iter=ni, n.burnin=nb, n.chains=nc,
  n.thin=nt, n.adapt=na, parallel=TRUE)
traceplot(out9)

print(out9, 3)
```

```
             mean      sd     2.5%      50%    97.5% overlap0 f  Rhat n.eff
mh[1]       0.129   0.015    0.102    0.128    0.161    FALSE 1 1.001  8101
mh[2]       0.152   0.020    0.118    0.150    0.197    FALSE 1 1.003  1669
mh[3]       0.083   0.011    0.065    0.082    0.110    FALSE 1 1.004  1791
mh[4]       0.078   0.012    0.058    0.076    0.106    FALSE 1 1.004  1279
mh[5]       0.075   0.012    0.056    0.073    0.104    FALSE 1 1.005  1123
mh[6]       0.174   0.038    0.118    0.169    0.265    FALSE 1 1.004  1287
mo[1]       0.112   0.071    0.016    0.098    0.283    FALSE 1 1.004  1132
mo[2]       0.039   0.036    0.001    0.029    0.133    FALSE 1 1.000 30000
mo[3]       0.038   0.036    0.001    0.028    0.134    FALSE 1 1.001 30000
mo[4]       0.004   0.010    0.000    0.000    0.036    FALSE 1 1.192    64
mo[5]       0.173   0.083    0.049    0.160    0.371    FALSE 1 1.001  6370
mo[6]       0.189   0.094    0.052    0.175    0.412    FALSE 1 1.000 15474
h[1]        0.114   0.012    0.092    0.114    0.139    FALSE 1 1.000 30000
h[2]        0.138   0.017    0.109    0.137    0.176    FALSE 1 1.002  1774
h[3]        0.079   0.010    0.062    0.077    0.102    FALSE 1 1.004  1949
h[4]        0.075   0.011    0.056    0.073    0.101    FALSE 1 1.004  1174
h[5]        0.066   0.010    0.050    0.065    0.090    FALSE 1 1.005  1314
h[6]        0.146   0.028    0.103    0.142    0.212    FALSE 1 1.004  1159
s[1]        0.788   0.058    0.655    0.797    0.874    FALSE 1 1.003  1236
s[2]        0.827   0.035    0.742    0.834    0.877    FALSE 1 1.001  5186
s[3]        0.886   0.034    0.801    0.894    0.928    FALSE 1 1.001  7629
s[4]        0.922   0.015    0.885    0.924    0.943    FALSE 1 1.041   166
s[5]        0.783   0.066    0.637    0.790    0.890    FALSE 1 1.001  4348
s[6]        0.699   0.072    0.540    0.705    0.819    FALSE 1 1.000 27082
o[1]        0.097   0.057    0.015    0.088    0.231    FALSE 1 1.003  1230
o[2]        0.035   0.031    0.001    0.026    0.116    FALSE 1 1.000 30000
o[3]        0.035   0.032    0.001    0.026    0.120    FALSE 1 1.001 30000
o[4]        0.003   0.010    0.000    0.000    0.034    FALSE 1 1.188    64
o[5]        0.150   0.066    0.046    0.143    0.299    FALSE 1 1.000  7247
o[6]        0.155   0.068    0.047    0.147    0.309    FALSE 1 1.000 11278
f[1]        0.271   0.036    0.203    0.270    0.343    FALSE 1 1.002  1744
f[2]        0.312   0.038    0.242    0.311    0.391    FALSE 1 1.001  2894
f[3]        0.327   0.043    0.250    0.325    0.418    FALSE 1 1.001  1956
f[4]        0.219   0.034    0.157    0.217    0.292    FALSE 1 1.000  5287
f[5]        0.146   0.035    0.087    0.143    0.223    FALSE 1 1.000 29143
r[1]        0.522   0.030    0.464    0.522    0.581    FALSE 1 1.000  6106
r[2]        0.531   0.029    0.473    0.530    0.587    FALSE 1 1.000 30000
r[3]        0.995   0.005    0.983    0.997    1.000    FALSE 1 1.000 30000
r[4]        0.755   0.023    0.710    0.756    0.798    FALSE 1 1.000 30000
r[5]        0.995   0.005    0.981    0.997    1.000    FALSE 1 1.000 22544
r[6]        0.481   0.027    0.428    0.481    0.535    FALSE 1 1.000 30000
Ntot[1]  2097.226  44.984 2010.000 2097.000 2186.000    FALSE 1 1.000 30000
Ntot[2]  2093.870 165.500 1727.000 2111.000 2371.000    FALSE 1 1.003  1318
Ntot[3]  2236.494 225.383 1756.000 2252.000 2637.025    FALSE 1 1.004  1308
Ntot[4]  2537.690 295.944 1917.000 2553.000 3080.000    FALSE 1 1.004  1048
Ntot[5]  2761.377 352.628 2031.975 2776.000 3406.025    FALSE 1 1.003  1260
Ntot[6]  2491.857 398.151 1714.000 2491.000 3268.000    FALSE 1 1.002  1298
N[1,1]    428.932  18.692  392.000  429.000  466.000    FALSE 1 1.000 30000
N[2,1]    355.061  16.618  323.000  355.000  388.000    FALSE 1 1.000 30000
[...output truncated...]
N[16,6]    19.665   4.792   11.000   19.000   30.000    FALSE 1 1.000  5106
N[17,6]    16.137   4.328    8.000   16.000   25.000    FALSE 1 1.000  5597
```

The estimates of hunting and background mortality from the IPM parameterized with mortality hazard rates are fairly similar to those from the original IPM. This was expected because we did not include any covariates or complex model structures with additive effects (Ergon et al., 2018). The main benefit of the IPM formulated in terms of mortality hazard rates is that we can directly assess whether hunting mortality is compensated by reduced background mortality: we simply check whether their estimates have a negative cross-correlation (Ergon et al., 2018). If the model is not parameterized with mortality hazard rates, complicated corrections must be made to assess such a correlation because of inbuilt statistical dependence (Schaub and Lebreton, 2004; Servanty et al., 2010). However, we do not inspect this correlation, because its proper evaluation would require the specification of temporal random effects to eliminate sampling variation (Ergon et al., 2018). Unfortunately, the elk time series are too short to expect reasonable estimates of temporal random effects.

17.9 DISCUSSION

Age-at-harvest data are frequently collected by agencies to obtain the information necessary for evidence-based population management of harvested species. Often these data are analyzed with so-called population reconstruction methods that can produce reliable population estimates *if* additional information about hunting mortality, survival, and reporting rates is included (Williams et al., 2002; Conroy and Carroll, 2009). Formal joint analyses (but not IPMs) of age-at-harvest and different kinds of auxiliary data have been developed in the past (Gove et al., 2002; Broms et al., 2010; Berg et al., 2017). Here we showed how to formulate a formal IPM for such data. IPMs that use age-at-harvest data are rare so far, and we only know of two such studies with empirical data (Conn et al., 2008; Fieberg et al., 2010).

The IPM approach uses the age-at-harvest data analogous to the population counts in a standard IPM except that now dead instead of live individuals are counted, and the observation model in the state-space model must be adapted accordingly. The formulation as an IPM requires that a population model is formulated including all the demographic rates needed in its description. This is not necessary in the classical population reconstruction approaches, but we see the explicit inclusion of a population model as an advantage: the population model confers the ability to include auxiliary data that are informative about demographic processes and allows us to use informative priors for some demographic rates if this is deemed useful. Moreover, we can perform further analyses such as retrospective population analyses (Chapter 9) or population viability analyses (Chapter 10) directly from the IPM.

The IPM for age-at-harvest data relies on the same main assumptions as those of other IPMs (Chapter 7) but includes other assumptions as well. Conn et al. (2009) used simulation to assess the performance of an IPM combining age-at-harvest and dead recovery data. They found that the model produced reasonable estimates when all assumptions were met. Yet they noticed that the precision of the parameter estimates was too optimistic if the independence assumption (Section 7.2.2) was violated. One assumption specific to the age-at-harvest data is that the age of all individuals must be identified correctly. However, Conn et al. (2009) used simulation to study the effects of aging errors in the age-at-harvest data and found that they had little effect on the coverage of the credible intervals of recruitment. Determination of age often becomes less reliable with increasing age of an individual, and if there are doubts in age determination, we might pool all individuals with an age greater than some maximum "ageable" age into one single age class, "oldies." The observation model of the IPM then needs to be adapted accordingly.

For the elk data set, we noticed that estimates of population size and hunting mortality were sensitive to the chosen prior for the age-specific population size in the first year. Indeed, strongly informative priors for the vector of initial population sizes appear to be necessary for Bayesian formulation of the IPM applied to the elk data. In most IPMs, priors for the initial population size are vague and do not include information about age distribution. However, when we specified vague priors for initial population size in the elk example, the resulting parameter estimates were unreasonable (Fig. 17.7). This must be kept in mind when using such models and may somewhat limit the scope of the inferences that can be drawn from them for many data sets.

As expected, prior sensitivity declined with the increasing sample size of the radio tracking data, indicating that the model does not suffer from parameter nonidentifiability but rather that our specific analysis suffered from a relative paucity of data. The prior sensitivity suggests that the auxiliary data (or component data sets other than the population counts in our usual IPM terminology) play a key role for the analysis of age-at-harvest data. It seems that rather strong additional information is needed to exploit age-at-harvest data for the estimation of demographic population parameters (Williams et al., 2002; Skalski et al., 2005). It is encouraging, though, that the population trend, which is an important parameter in many applications, was much less sensitive to the prior and thus less dependent on additional information. More research on the impact of the prior for age-specific population sizes in the first year and on strategies about how such priors are best specified in IPM using age-at-harvest data is clearly needed.

Here, we made use of radio tracking data that are informative about survival and mortality causes as well as hunter survey data that are informative about the reporting rate. In other studies, different types of auxiliary data were used. Conn et al. (2008) combined age-at-harvest data with dead recovery data of black bears (*Ursus americanus*) in Pennsylvania. Fieberg et al. (2010) combined age-at-harvest data from Minnesota black bears with data on hunter effort and natural food supply as well as with state-wide capture-recapture estimates of abundance. Broms et al. (2010) studied the dynamics of sage grouse (*Centrocercus urophasianus*) based on a combination of age-at-harvest data, a population index derived from lek counts, data on hunter effort, and radio tracking data. Finally, Berg et al. (2017) used age-at-harvest data combined with radio tracking data to study the demography of the American marten (*Martes americana*).

We concur with Conn et al. (2008), Fieberg et al. (2010) and Berg et al. (2017) that at present, poor use may be made of age-at-harvest data that are frequently and routinely collected. We think one reason for this is that often the required additional data that are informative about demographic rates or reporting rates are lacking. Another reason for the rarity of use of such data may be that proper analysis using population reconstruction methods may appear complicated. At least to us, the conceptualization of such a model as an IPM seems more accessible and may result in a more transparent model in which it is easier to see where to integrate the necessary additional information through component data or informative priors. Moreover, all the usual benefits of IPMs come into play here. For instance, it would be possible to compare different management options via a population viability analysis within the same framework (Chapter 10), which should be particularly appealing in the typical hunting applications of an IPM for such data. However, IPMs based on age-at-harvest data are not well studied apart from the simulation studies in Conn et al. (2009) and Fieberg et al. (2010), and there is certainly considerable scope for additional studies of the performance of these models in different circumstances. For example, it would be relevant for management to know how well such models are suited for the estimation of density-dependence. Another fruitful area of research might be the prior sensitivity of the population estimates: How should meaningful priors be formulated, and how much or which auxiliary data have to be collected to minimize prior sensitivity? We hope that the ability to use an IPM for the analysis of age-at-harvest data may encourage increased exploration of the theoretical properties of this type of model and more routine use of this valuable type of data in actual applications.

CHAPTER

CORMORANT

18

CHAPTER OUTLINE

IPM ingredients: population count data, capture-recapture data
Location: Denmark
Spatio-temporal scale: three colonies, 14 years

18.1 INTRODUCTION

Dispersal strongly affects the spatial configuration of populations as well as their temporal dynamics, and to a large degree also determines gene flow among populations. At the scale of a single site (or population), dispersal behavior is manifest as emigration and immigration, both of which can be estimated in an integrated population model (IPM) provided that the right kind of data are available (see, e.g., Chapter 11). If multiple sites are studied simultaneously among which individuals may move, dispersal can be studied more explicitly. Such a setup is the focus of this case study.

We study the dynamics of three Danish breeding colonies of great cormorants (*Phalacrocorax carbo*) that exchange dispersing individuals (Fig. 18.1). We develop an IPM that jointly analyzes multistate capture-recapture data and population counts sampled at the three colonies. Because cormorants only recruit into the breeding population at age 2 or older, we estimate not only natal and breeding dispersal among the three colonies but also recruitment. The underlying population model therefore becomes slightly complex. Data on productivity are not available, but colony-specific estimates can be obtained from the IPM as hidden parameters. The goal of the study is the estimation of the demographic rates, in particular the movement probabilities among colonies. These data have already been analyzed using an IPM within the frequentist framework by Borysiewicz et al. (2009) and McCrea et al. (2010). Here, we follow the analysis of McCrea et al. (2010) but do not replicate their model exactly, basically to keep things simple.

18.2 DATA SETS

The data were collected during the 14 years from 1991 to 2004 at the three Danish cormorant breeding colonies Vorsø (labeled *V*), Mågeøerne (*M*), and Stavns Fjord (*S*), which are all located in the southwestern Kattegat in the inner Danish waters. In all colonies, a large number of nestlings was marked with a standard metal ring on one leg and an alphanumeric colored plastic ring on the other; in total, 12,659 individuals were marked. The first data set was provided by resightings of marked individuals at the breeding colonies by trained field workers from towers, hides, or the ground using a telescope. Resightings were restricted to breeding individuals. Marking of nestlings had already started before 1991, and hence, some individuals in the data set looked as if they had been ringed as breeders. The second data set contains the colony counts, the sizes of which were assessed each year by counting the number of active nests in early May. These

Integrated Population Models. https://doi.org/10.1016/B978-0-12-820564-8.00018-X

FIGURE 18.1

Great cormorant (*Phalacrocorax carbo*) taking off (photograph by www.marcelburkhardt.ch).

population counts ranged between about 1900 and 5000 at Vorsø, 1800 and 2600 at Mågeøerne, and 800 and 3300 at Stavns Fjord. Although dead-recovery data are also available in this study, we did not use them so we could keep the model simpler (see Borysiewicz et al., 2009 for an IPM that uses all three sources of demographic information).

18.3 POPULATION MODEL

The population model for the three cormorant colonies is quite complex because individuals can move among colonies and because of delayed recruitment. We must also distinguish between natal and breeding dispersal (i.e., age-specific movement rates), as natal is typically more pronounced in birds than breeding dispersal is (Greenwood and Harvey, 1982). We reduce the complexity of the model by making some restrictive but (we hope) judicious assumptions about survival and movement. For both parameters, we distinguish between two age classes only; thus, we estimate juvenile survival (between fledging and 1-year-old) and adult survival (after 1 year old), and natal and breeding dispersal. Making these assumptions is sufficient for us to define only two stages per colony, nonbreeders (N) and breeders (B). Nonbreeders are defined here as individuals at least 1 year old that so far have not reproduced. Breeders are defined as individuals 2 years or older that have either just started to reproduce or have reproduced already in earlier years. We further assume that individuals that recruit into the breeding stage remain there even if they may actually skip breeding in some years. If we had wanted to model recruitment with parameters that change with age, we would have had to define more stages; see Szostek et al. (2014) and Margalida et al. (2020) for examples of population models with a total of 19 and 37 stages, respectively. A further simplifying assumption in our current IPM is that natal dispersal only occurs in the first year after birth. Recruitment (i.e., the transition from the nonbreeding to the breeding stage) starts at age 2 and occurs in that colony to which an individual has dispersed by then. Breeding dispersal only occurs when individuals are in the breeding stage. Finally, the model is female-based, and thus the estimated stage-specific population sizes refer to females only.

In summary, our model assumes that the life of a female after birth can be described as follows: in the first year she may survive with first-year survival probability and may disperse to another colony (this is natal dispersal). In the second year the female may survive with adult survival probability and may recruit into the breeding pool of that colony to which she dispersed previously. If she does not recruit, then her life history over the next years remains the same; i.e., she may survive and then will have another chance to recruit, and so forth. There is no age at which all females are assumed to have recruited, and hence, some females may remain in the nonbreeding stage until their death. Once a female has recruited, she may produce offspring, survive with adult survival probability, and move among colonies (this is breeding dispersal), and she keeps this life-history pattern until her eventual death. The life-cycle graph on which our model is based is shown in Fig. 18.2.

In order to translate this verbal description into mathematics, we start by defining a fair number of parameters:

- Natal dispersal ($\eta_{XY,t}$, 'eta') is the probability that a nestling from colony X in year t moves to colony Y by year $t+1$ given that it survives from t to $t+1$.
- Breeding dispersal ($\nu_{XY,t}$, 'nu') is the probability that a breeder from colony X in year t moves to colony Y by year $t+1$ given that it survives from t to $t+1$.
- Colony-specific recruitment ($\kappa_{X,t}$, 'kappa') is the probability that an individual in the nonbreeding stage in year t recruits into the breeding pool of colony X in year $t+1$ given that it survives from t to $t+1$.

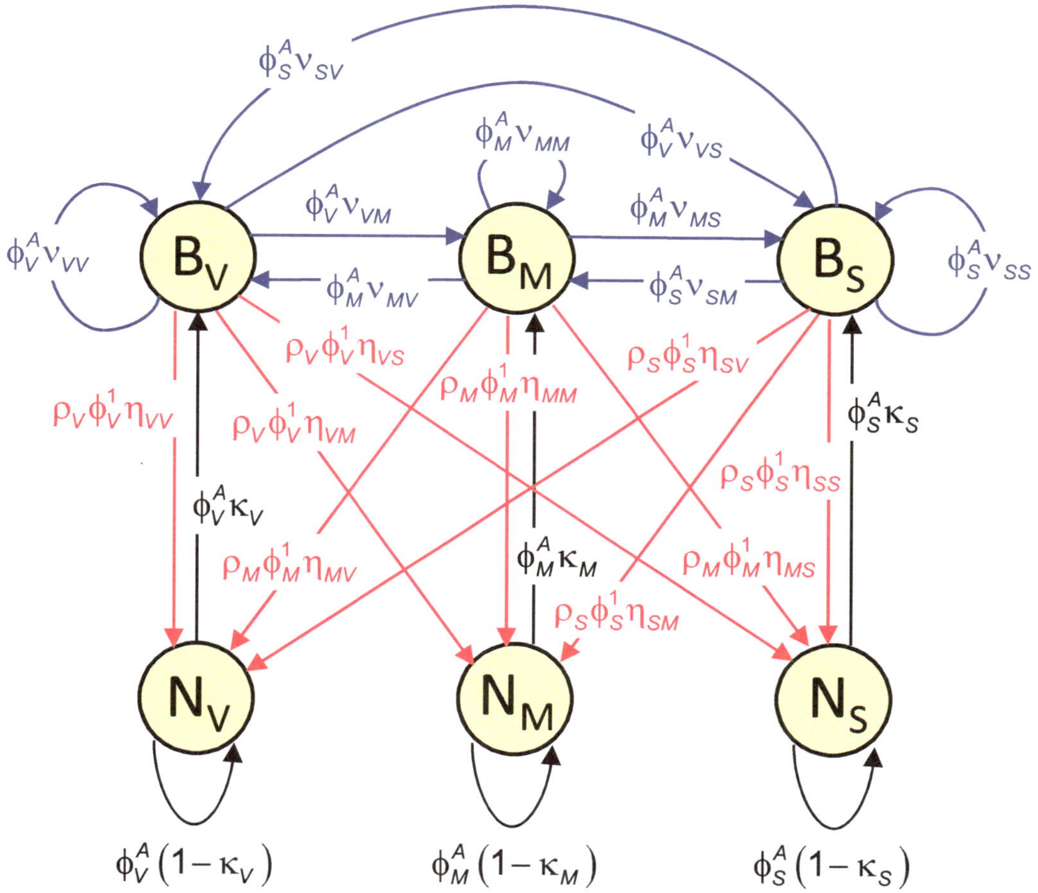

FIGURE 18.2

Life-cycle graph for the multicolony cormorant model. The nodes refer to breeders (B) and nonbreeders (N) in the three Danish colonies at Vorsø (*V*), Mågeøerne (*M*) and Stavns Fjord (*S*). Parameters are defined in the main text. Colors index the following three demographic subprocesses—red: productivity, juvenile survival, and natal dispersal; black: adult survival and recruitment; blue: adult survival and breeding dispersal.

- Colony-specific productivity ($\rho_{X,t}$, 'rho') is the expected number of female nestlings raised by a breeding pair in colony X in year t.
- Colony-specific juvenile survival ($\phi^1_{X,t}$, 'phi1') is the probability that a nestling born in year t in colony X survives until year $t+1$.
- Colony-specific adult survival ($\phi^A_{X,t}$, 'phiA') is the probability that an individual of at least 1 year of age present in colony X in year t survives until year $t+1$.
- Colony-specific number of nonbreeders ($N_{X,t}$): the number of female cormorants in colony X in year t that have so far never reproduced.
- Colony-specific number of breeders ($B_{X,t}$): the number of female cormorants in colony X in year t that reproduce in year t or have already reproduced earlier.

Note that the time index for the transition probabilities (η, ν, κ) could also be defined in a different manner such that, e.g., $\eta_{XY,t}$ is the probability that a nestling from colony X in year $t-1$ is in colony Y in year t, given that it has survived from $t-1$ to t. The reason for our choice above is that it matches with the time indices of the survival probabilities. Generally, a correct indexing of time can become challenging in complex IPMs, and more so if a postbreeding "census" model is adopted (Section 3.2.4).

Our population model is prebreeding "census"-based. Following the life-cycle graph shown in Fig. 18.2, it can be represented by the following equations:

$$N_{V,t+1} = B_{V,t}\rho_{V,t}\phi^1_{V,t}\eta_{VV,t} + B_{M,t}\rho_{M,t}\phi^1_{M,t}\eta_{MV,t} + B_{S,t}\rho_{S,t}\phi^1_{S,t}\eta_{SV,t} + N_{V,t}\phi^A_{V,t}(1-\kappa_{V,t})$$

$$N_{M,t+1} = B_{V,t}\rho_{V,t}\phi^1_{V,t}\eta_{VM,t} + B_{M,t}\rho_{M,t}\phi^1_{M,t}\eta_{MM,t} + B_{S,t}\rho_{S,t}\phi^1_{S,t}\eta_{SM,t} + N_{M,t}\phi^A_{M,t}(1-\kappa_{M,t})$$

$$N_{S,t+1} = B_{V,t}\rho_{V,t}\phi^1_{V,t}\eta_{VS,t} + B_{M,t}\rho_{M,t}\phi^1_{M,t}\eta_{MS,t} + B_{S,t}\rho_{S,t}\phi^1_{S,t}\eta_{SS,t} + N_{S,t}\phi^A_{S,t}(1-\kappa_{S,t})$$

$$B_{V,t+1} = B_{V,t}\phi^A_{V,t}\nu_{VV,t} + B_{M,t}\phi^A_{M,t}\nu_{MV,t} + B_{S,t}\phi^A_{S,t}\nu_{SV,t} + N_{V,t}\phi^A_{V,t}\kappa_{V,t}$$

$$B_{M,t+1} = B_{V,t}\phi^A_{V,t}\nu_{VM,t} + B_{M,t}\phi^A_{M,t}\nu_{MM,t} + B_{S,t}\phi^A_{S,t}\nu_{SM,t} + N_{M,t}\phi^A_{M,t}\kappa_{M,t}$$

$$B_{S,t+1} = B_{V,t}\phi^A_{V,t}\nu_{VS,t} + B_{M,t}\phi^A_{M,t}\nu_{MS,t} + B_{S,t}\phi^A_{S,t}\nu_{SS,t} + N_{S,t}\phi^A_{S,t}\kappa_{S,t}$$

Because the colonies are fairly large, we do not account for demographic stochasticity in the model (see the Discussion for more on this). A model with demographic stochasticity becomes substantially more complex and is likely to be much harder to get to convergence in an Markov chain Monte Carlo (MCMC)-based analysis. In contrast to the paper by McCrea et al. (2010), we here impose an age structure on survival and distinguish juvenile from adult survival but assume that survival is independent of breeding status (i.e., nonbreeder vs. breeder).

18.4 COMPONENT DATA LIKELIHOODS

The two data sets for this analysis are stored in an R data file that we load from `IPMbook` and whose structure we inspect.

```
library(IPMbook); library(jagsUI)
data(cormorant)
str(cormorant)

List of 2
$ count: int [1:3, 1:14] 5048 1982 804 4321 1860 1350 4634 2170 1848 4318 ...
  ..- attr(*, "dimnames")=List of 2
  .. ..$ : chr [1:3] "V" "M" "S"
  .. ..$ : chr [1:14] "1991" "1992" "1993" "1994" ...
$ ms.ch: int [1:12659, 1:14] 1 1 1 0 0 0 0 0 0 0 ...
  ..- attr(*, "dimnames")=List of 2
  .. ..$ : NULL
  .. ..$ : chr [1:14] "1991" "1992" "1993" "1994" ...
```

The first object in the list is a matrix containing the population counts. The three rows refer to the three colonies and the 14 columns to the years 1991–2004. The second object is an individual x year matrix with the multistate capture histories. The state codes 1 to 3 index nestlings in colonies V, M, and S, respectively, and the state codes 4 to 6 do the same for the breeders.

We start by transforming the individual capture histories into the m-array format. As we will explain below, the multistate model includes three unobserved states, and this information must be provided for computation of the m-array.

```
marr <- marray(cormorant$ms.ch, unobs=3)
```

Next, we describe the construction of the component data likelihoods.

18.4.1 POPULATION COUNT DATA

For the population count data, we use a state-space model as usual, where the process model is already defined by the population model (Section 18.3). For the observation model, we use a normal distribution such that we write for the counts of breeders in colony X,

$$C_{X,t} \sim Normal\left(B_{X,t}, \sigma_X^2\right).$$

Remember that only breeders are assumed to appear in the counts, while nonbreeders are treated as unobserved. We allow the observation error (σ_X^2) to be colony specific.

As always, we require a model for the stage-specific population sizes in the initial year of the study. We intend to use uniform priors for that purpose with ranges that differ by colony and stage. As only the number of breeding pairs is observed but not the number of nonbreeders, it is challenging to find appropriate ranges for the stage-specific population sizes. A useful approach is the evaluation of the stable stage distribution from a matrix population model parameterized with the approximate demographic rates of a species. To keep the model simple, here we use a matrix population model from just a single site and ignore dispersal. We pick reasonable values for the age-specific survival probabilities, recruitment probability, and productivity. Then we write the matrix population model and calculate the stable stage distribution:

```
phi <- c(0.4, 0.8)
kappa <- 0.25
rho <- 1.5

A <- matrix(c(
  phi[2] * (1-kappa), phi[1] * rho,
  phi[2] * kappa, phi[2]), byrow=TRUE, ncol=2)

z <- which.max(Re(eigen(A)$values))
revec <- Re(eigen(A)$vectors[,z])
matrix(revec / sum(revec))      # Standardized right eigenvector

          [,1]
[1,] 0.5657415
[2,] 0.4342585
```

This suggests that about 55% of the individuals in a cormorant population described with a prebreeding "census" will be nonbreeders. In population Vorsø, about 5000 breeding pairs were counted in the first study year, and therefore, the expected number of nonbreeding females is about $5000 \times 0.55/0.45 \cong 6100$. We use $\pm 30\%$ of the population counts and of the calculated number of nonbreeders to determine suitable ranges for the uniform distributions intended to be vague for both breeders and nonbreeders. Clearly, there is some degree of arbitrariness in this choice, and it is advisable to conduct sensitivity analyses, i.e., use different ranges of the uniform priors and see how by much the inferences about key parameters in the model change. Nevertheless, in our experience, estimates from the first year may be affected by that choice, but hardly any estimates from later years will be.

18.4.2 MULTISTATE CAPTURE-RECAPTURE DATA

We use a multistate capture-recapture model formulated with the multinomial likelihood to analyze the capture-recapture data. The multinomial likelihood has substantial computational benefits compared with the state-space formulation (Section 4.5.2.2) because the dimension of the data set is greatly reduced, and convergence is typically achieved much faster. The multistate capture-recapture data contain six observed states (1: nestling at V, 2: nestling at M, 3: nestling at S, 4: breeder at V, 5: breeder at M, 6: breeder at S). There are no resightings in states 1 to 3 because these states refer to nestlings that cannot be resighted: if an individual marked as a nestling is resighted in a following year, it has become a breeder and is therefore in one of the states 4, 5, or 6. Since recruitment starts at age 2 and all transitions refer to 1-year time steps, we also need to include states for individuals at least 1 year old but not yet recruited into the breeding pool. These are unobservable states that we label 7 to 9 for the colonies V, M, and S, respectively.

To describe the multistate capture-recapture model, we define a matrix with the state transitions and a vector with the state-specific recapture probabilities. The state-transition matrix $\mathbf{\Psi}$ is a 9 × 9 matrix defining the transition probability of an individual in a given state (row) at time t to a state (column) at time $t+1$.

$$\mathbf{\Psi}_t = \begin{bmatrix} 0 & 0 & 0 & 0 & 0 & 0 & \phi_V^1\eta_{VV} & \phi_V^1\eta_{VM} & \phi_V^1\eta_{VS} \\ 0 & 0 & 0 & 0 & 0 & 0 & \phi_M^1\eta_{MV} & \phi_M^1\eta_{MM} & \phi_M^1\eta_{MS} \\ 0 & 0 & 0 & 0 & 0 & 0 & \phi_S^1\eta_{SV} & \phi_S^1\eta_{SM} & \phi_S^1\eta_{SS} \\ 0 & 0 & 0 & \phi_V^A\nu_{VV} & \phi_V^A\nu_{VM} & \phi_V^A\nu_{VS} & 0 & 0 & 0 \\ 0 & 0 & 0 & \phi_M^A\nu_{MV} & \phi_M^A\nu_{MM} & \phi_M^A\nu_{MS} & 0 & 0 & 0 \\ 0 & 0 & 0 & \phi_S^A\nu_{SV} & \phi_S^A\nu_{SM} & \phi_S^A\nu_{SS} & 0 & 0 & 0 \\ 0 & 0 & 0 & \phi_V^A\kappa_V & 0 & 0 & \phi_V^A(1-\kappa_V) & 0 & 0 \\ 0 & 0 & 0 & 0 & \phi_M^A\kappa_M & 0 & 0 & \phi_M^A(1-\kappa_M) & 0 \\ 0 & 0 & 0 & 0 & 0 & \phi_S^A\kappa_S & 0 & 0 & \phi_S^A(1-\kappa_S) \end{bmatrix}_t$$

The parameters of the transition matrix are defined in Section 18.3. The vector with the state-specific recapture probabilities is

$$\boldsymbol{p}_t = [0 \quad 0 \quad 0 \quad p_V \quad p_M \quad p_S \quad 0 \quad 0 \quad 0]_t,$$

where p_V, p_M and p_S are the colony-specific resighting probabilities.

So far, we have written all parameters as time-dependent, but we follow McCrea et al. (2010) and implement a simpler model. Specifically, we assume that natal dispersal is constant over time and only depends on the colony to which an individual disperses, i.e., $\eta_{MV} = \eta_{SV}$, $\eta_{VM} = \eta_{SM}$ and $\eta_{VS} = \eta_{MS}$. The probability of being philopatric (i.e., of not dispersing) in colony V is $\eta_{VV} = 1 - \eta_{VM} - \eta_{VS}$, in colony M is $\eta_{MM} = 1 - \eta_{MV} - \eta_{MS}$, and in colony S it is $\eta_{SS} = 1 - \eta_{SV} - \eta_{SM}$. The sum of the dispersal probabilities from a site must be ≤ 1 because site fidelity would otherwise become negative. We use a multinomial logit link function to impose this constraint (Kéry and Schaub, 2012).

We further assume the breeding dispersal probabilities to be constant over time and to depend only on the departure colony such that $\nu_{VM} = \nu_{VS}$, $\nu_{MV} = \nu_{MS}$ and $\nu_{SV} = \nu_{SM}$. The probability of site fidelity is $\nu_{VV} = 1 - \nu_{VM} - \nu_{VS}$ at colony V, $\nu_{MM} = 1 - \nu_{MV} - \nu_{MS}$ at colony M and $\nu_{SS} = 1 - \nu_{SV} - \nu_{SM}$ at colony V, and we again use the multinomial logit link function to ensure that the sum of the movement probabilities from a site ≤ 1.

The recruitment probabilities are assumed to change linearly over time on the logit link scale but for each colony at a different level. Thus, for colony X, we adopt the following logit-linear model with additive effects of colony and time,

$$\text{logit}(\kappa_{X,t}) = \alpha_X + \beta t,$$

where α_X is the colony-specific intercept and β the change in recruitment per year.

Survival is assumed to vary over time as a fixed effect but in an additive fashion for bird age and colony. We implement this by first providing a uniform prior for juvenile survival of each year at colony Vorsø. Survival of the other age classes and/or colonies is then expressed as the sum of the values at Vorsø and an age- and colony-specific contribution (on the logit scale, and there are five such latter values). Finally, the recapture probabilities are modeled for each colony with independent, fixed time effects.

18.5 THE INTEGRATED POPULATION MODEL

We are now ready to combine the component data likelihoods in the IPM. No direct data about productivity is available, and hence, we estimate it as a hidden parameter for which we also need to specify a prior. We assume that productivity is constant over time but may differ among colonies.

$$\rho_{X,t} = \mu_X$$
$$\mu_X \sim Uniform(0, 5)$$

We store the data in object `jags.data`. In addition to the population counts and the multistate m-array, we provide the number of years, the number of released individuals, and some matrices needed to calculate the cell probabilities in the m-array (see Section 4.5.2.2 for details).

```
# Bundle data and produce data overview
jags.data <- list(marr=marr, n.years=ncol(cormorant$ms.ch), rel=rowSums(marr), ns=9,
  zero=matrix(0, ncol=9, nrow=9), ones=diag(9), C=cormorant$count)
str(jags.data)

List of 7
 $ marr    : num [1:117, 1:118] 0 0 0 0 0 0 0 0 0 0 ...
 $ n.years: int 14
 $ rel     : num [1:117] 379 0 0 532 39 0 0 0 0 375 ...
 $ ns      : num 9
 $ zero    : num [1:9, 1:9] 0 0 0 0 0 0 0 0 0 0 ...
 $ ones    : num [1:9, 1:9] 1 0 0 0 0 0 0 0 0 0 ...
 $ C       : int [1:3, 1:14] 5048 1982 804 4321 1860 1350 4634 2170 ...
```

Then we are ready to write the model code into a text file.

```
# Write JAGS model file
cat(file = "model1.txt", "
model {
# -------------------------------------------------
# Stages:
# N: not-yet recruited individuals
# B: breeders
# Parameters:
# phi[age, site, time]: survival probability
# eta[departure site, arrival site, time]: natal dispersal
# nu[departure site, arrival site, time]: breeding dispersal
# kappa[site, time]: recruitment probability
# rho[site, time]: productivity
# p[site, time]: recapture probability
# -------------------------------------------------

# Priors and linear models
# Productivity
for (t in 1:n.years){
  for (s in 1:3){
    rho[s,t] <- mean.rho[s]
  } #s
} #t
mean.rho[1] ~ dunif(0, 4)
mean.rho[2] ~ dunif(0, 4)
mean.rho[3] ~ dunif(0, 4)

# Parameters of multistate model
for (t in 1:(n.years-1)){
  phi[1,1,t] ~ dunif(0, 1)
  logit.phi[1,1,t] <- logit(phi[1,1,t])
  logit.phi[2,1,t] <- logit.phi[1,1,t] + beta.phi[2,1]
  phi[2,1,t] <- ilogit(logit.phi[2,1,t])
  logit.kappa[1,t] <- mu.kappa[1] + mu.kappa[2] * t
  kappa[1,t] <- ilogit(logit.kappa[1,t])
  for (s in 2:3){
    logit.phi[1,s,t] <- logit.phi[1,1,t] + beta.phi[1,s]
    phi[1,s,t] <- ilogit(logit.phi[1,s,t])
    logit.phi[2,s,t] <- logit.phi[1,1,t] + beta.phi[2,s]
    phi[2,s,t] <- ilogit(logit.phi[2,s,t])
    logit.kappa[s,t] <- logit.kappa[1,t] + mu.kappa[s+1]
    kappa[s,t] <- ilogit(logit.kappa[s,t])
  } #s
  for (s in 1:3){
    p[s,t] ~ dunif(0, 1)
    eta[1,s,t] <- mean.eta[1,s]
    eta[2,s,t] <- mean.eta[2,s]
    eta[3,s,t] <- mean.eta[3,s]
    nu[1,s,t] <- mean.nu[1,s]
    nu[2,s,t] <- mean.nu[2,s]
    nu[3,s,t] <- mean.nu[3,s]
  } #s
} #t
```

```
# Multinomial logit link to impose the constraint that natal dispersal does
# only depend on the site of departure
logit.mean.eta[1,2] ~ dnorm(0, 0.01)
logit.mean.eta[1,3] <- logit.mean.eta[1,2]
mean.eta[1,2] <- exp(logit.mean.eta[1,2]) / (1 + exp(logit.mean.eta[1,2]) +
  exp(logit.mean.eta[1,3]))
mean.eta[1,3] <- exp(logit.mean.eta[1,3]) / (1 + exp(logit.mean.eta[1,2]) +
  exp(logit.mean.eta[1,3]))
mean.eta[1,1] <- 1 - mean.eta[1,2] - mean.eta[1,3]
logit.mean.eta[2,1] ~ dnorm(0, 0.01)
logit.mean.eta[2,3] <- logit.mean.eta[2,1]
mean.eta[2,1] <- exp(logit.mean.eta[2,1]) / (1 + exp(logit.mean.eta[2,1]) +
  exp(logit.mean.eta[2,3]))
mean.eta[2,3] <- exp(logit.mean.eta[2,3]) / (1 + exp(logit.mean.eta[2,1]) +
  exp(logit.mean.eta[2,3]))
mean.eta[2,2] <- 1 - mean.eta[2,1] - mean.eta[2,3]
logit.mean.eta[3,1] ~ dnorm(0, 0.01)
logit.mean.eta[3,2] <- logit.mean.eta[3,1]
mean.eta[3,1] <- exp(logit.mean.eta[3,1]) / (1 + exp(logit.mean.eta[3,1]) +
  exp(logit.mean.eta[3,2]))
mean.eta[3,2] <- exp(logit.mean.eta[3,2]) / (1 + exp(logit.mean.eta[3,1]) +
  exp(logit.mean.eta[3,2]))
mean.eta[3,3] <- 1 - mean.eta[3,1] - mean.eta[3,2]

# Multinomial logit link to impose the constraint that breeding dispersal does only depend on the
# site of arrival
logit.mean.nu[1,2] ~ dnorm(0, 0.01)
logit.mean.nu[1,3] ~ dnorm(0, 0.01)
logit.mean.nu[2,1] ~ dnorm(0, 0.01)
mean.nu[1,2] <- exp(logit.mean.nu[1,2]) / (1 + exp(logit.mean.nu[1,2]) +
  exp(logit.mean.nu[1,3]) + exp(logit.mean.nu[2,1]))
mean.nu[1,3] <- exp(logit.mean.nu[1,3]) / (1 + exp(logit.mean.nu[1,2]) +
  exp(logit.mean.nu[1,3]) + exp(logit.mean.nu[2,1]))
mean.nu[2,1] <- exp(logit.mean.nu[2,1]) / (1 + exp(logit.mean.nu[1,2]) +
  exp(logit.mean.nu[1,3]) + exp(logit.mean.nu[2,1]))
mean.nu[1,1] <- 1 - mean.nu[1,2] - mean.nu[1,3]
mean.nu[2,2] <- 1 - mean.nu[2,1] - mean.nu[1,3]
mean.nu[2,3] <- mean.nu[1,3]
mean.nu[3,1] <- mean.nu[2,1]
mean.nu[3,2] <- mean.nu[1,2]
mean.nu[3,3] <- 1 - mean.nu[3,1] - mean.nu[3,2]

for (i in 1:4){
  mu.kappa[i] ~ dnorm(0, 0.01)
}

beta.phi[2,1] ~ dnorm(0, 0.01)
beta.phi[1,2] ~ dnorm(0, 0.01)
beta.phi[2,2] ~ dnorm(0, 0.01)
beta.phi[1,3] ~ dnorm(0, 0.01)
beta.phi[2,3] ~ dnorm(0, 0.01)

# Residual (observation) error
for (s in 1:3){
  sigma[s] ~ dunif(0.05, 1000)
  tau[s] <- pow(sigma[s], -2)
}

# Population count data (state-space model)
# Models for the initial population size: uniform priors
N[1,1] ~ dunif(4270, 7940)
B[1,1] ~ dunif(3500, 6500)
N[2,1] ~ dunif(1710, 3180)
B[2,1] ~ dunif(1400, 2600)
N[3,1] ~ dunif(680, 1270)
B[3,1] ~ dunif(560, 1040)
```

```
# Process model over time: our model of population dynamics
for (t in 1:(n.years-1)){
  N[1,t+1] <- B[1,t] * rho[1,t] * phi[1,1,t] * eta[1,1,t] + B[2,t] * rho[2,t] * phi[1,2,t] *
    eta[2,1,t] + B[3,t] * rho[3,t] * phi[1,3,t] * eta[3,1,t] + N[1,t] * phi[2,1,t] *
    (1 - kappa[1,t])
  N[2,t+1] <- B[1,t] * rho[1,t] * phi[1,1,t] * eta[1,2,t] + B[2,t] * rho[2,t] * phi[1,2,t] *
    eta[2,2,t] + B[3,t] * rho[3,t] * phi[1,3,t] * eta[3,2,t] + N[2,t] * phi[2,2,t] *
    (1 - kappa[2,t])
  N[3,t+1] <- B[1,t] * rho[1,t] * phi[1,1,t] * eta[1,3,t] + B[2,t] * rho[2,t] * phi[1,2,t] *
    eta[2,3,t] + B[3,t] * rho[3,t] * phi[1,3,t] * eta[3,3,t] + N[3,t] * phi[2,3,t] *
    (1 - kappa[3,t])
  B[1,t+1] <- B[1,t] * phi[2,1,t] * nu[1,1,t] + B[2,t] * phi[2,2,t] * nu[2,1,t] + B[3,t] *
    phi[2,3,t] * nu[3,1,t] + N[1,t] * phi[2,1,t] * kappa[1,t]
  B[2,t+1] <- B[1,t] * phi[2,1,t] * nu[1,2,t] + B[2,t] * phi[2,2,t] * nu[2,2,t] + B[3,t] *
    phi[2,3,t] * nu[3,2,t] + N[2,t] * phi[2,2,t] * kappa[2,t]
  B[3,t+1] <- B[1,t] * phi[2,1,t] * nu[1,3,t] + B[2,t] * phi[2,2,t] * nu[2,3,t] + B[3,t] *
    phi[2,3,t] * nu[3,3,t] + N[3,t] * phi[2,3,t] * kappa[3,t]
}

# Observation model
for (t in 1:n.years){
  for (s in 1:3){
    C[s,t] ~ dnorm(B[s,t], tau[s])
  } #s
} #t

# Multistate capture-recapture data (with multinomial likelihood)
# Define state-transition and re-encounter probabilities
for (t in 1:(n.years-1)){
  psi[1,t,1] <- 0
  psi[1,t,2] <- 0
  psi[1,t,3] <- 0
  psi[1,t,4] <- 0
  psi[1,t,5] <- 0
  psi[1,t,6] <- 0
  psi[1,t,7] <- phi[1,1,t] * eta[1,1,t]
  psi[1,t,8] <- phi[1,1,t] * eta[1,2,t]
  psi[1,t,9] <- phi[1,1,t] * eta[1,3,t]
  psi[2,t,1] <- 0
  psi[2,t,2] <- 0
  psi[2,t,3] <- 0
  psi[2,t,4] <- 0
  psi[2,t,5] <- 0
  psi[2,t,6] <- 0
  psi[2,t,7] <- phi[1,2,t] * eta[2,1,t]
  psi[2,t,8] <- phi[1,2,t] * eta[2,2,t]
  psi[2,t,9] <- phi[1,2,t] * eta[2,3,t]
  psi[3,t,1] <- 0
  psi[3,t,2] <- 0
  psi[3,t,3] <- 0
  psi[3,t,4] <- 0
  psi[3,t,5] <- 0
  psi[3,t,6] <- 0
  psi[3,t,7] <- phi[1,3,t] * eta[3,1,t]
  psi[3,t,8] <- phi[1,3,t] * eta[3,2,t]
  psi[3,t,9] <- phi[1,3,t] * eta[3,3,t]
  psi[4,t,1] <- 0
  psi[4,t,2] <- 0
  psi[4,t,3] <- 0
  psi[4,t,4] <- phi[2,1,t] * nu[1,1,t]
  psi[4,t,5] <- phi[2,1,t] * nu[1,2,t]
  psi[4,t,6] <- phi[2,1,t] * nu[1,3,t]
  psi[4,t,7] <- 0
  psi[4,t,8] <- 0
  psi[4,t,9] <- 0
  psi[5,t,1] <- 0
  psi[5,t,2] <- 0
  psi[5,t,3] <- 0
  psi[5,t,4] <- phi[2,2,t] * nu[2,1,t]
  psi[5,t,5] <- phi[2,2,t] * nu[2,2,t]
  psi[5,t,6] <- phi[2,2,t] * nu[2,3,t]
```

```
    psi[5,t,7] <- 0
    psi[5,t,8] <- 0
    psi[5,t,9] <- 0
    psi[6,t,1] <- 0
    psi[6,t,2] <- 0
    psi[6,t,3] <- 0
    psi[6,t,4] <- phi[2,3,t] * nu[3,1,t]
    psi[6,t,5] <- phi[2,3,t] * nu[3,2,t]
    psi[6,t,6] <- phi[2,3,t] * nu[3,3,t]
    psi[6,t,7] <- 0
    psi[6,t,8] <- 0
    psi[6,t,9] <- 0
    psi[7,t,1] <- 0
    psi[7,t,2] <- 0
    psi[7,t,3] <- 0
    psi[7,t,4] <- phi[2,1,t] * kappa[1,t]
    psi[7,t,5] <- 0
    psi[7,t,6] <- 0
    psi[7,t,7] <- phi[2,1,t] * (1-kappa[1,t])
    psi[7,t,8] <- 0
    psi[7,t,9] <- 0
    psi[8,t,1] <- 0
    psi[8,t,2] <- 0
    psi[8,t,3] <- 0
    psi[8,t,4] <- 0
    psi[8,t,5] <- phi[2,2,t] * kappa[2,t]
    psi[8,t,6] <- 0
    psi[8,t,7] <- 0
    psi[8,t,8] <- phi[2,2,t] * (1-kappa[2,t])
    psi[8,t,9] <- 0
    psi[9,t,1] <- 0
    psi[9,t,2] <- 0
    psi[9,t,3] <- 0
    psi[9,t,4] <- 0
    psi[9,t,5] <- 0
    psi[9,t,6] <- phi[2,3,t] * kappa[3,t]
    psi[9,t,7] <- 0
    psi[9,t,8] <- 0
    psi[9,t,9] <- phi[2,3,t] * (1-kappa[3,t])

    po[1,t] <- 0
    po[2,t] <- 0
    po[3,t] <- 0
    po[4,t] <- p[1,t]
    po[5,t] <- p[2,t]
    po[6,t] <- p[3,t]
    po[7,t] <- 0
    po[8,t] <- 0
    po[9,t] <- 0

    # Calculate probability of non-encounter (dq) and reshape the array for the encounter
    # probabilities
    for (s in 1:ns){
      dp[s,t,s] <- po[s,t]
      dq[s,t,s] <- 1-po[s,t]
    } #s
    for (s in 1:(ns-1)){
      for (m in (s+1):ns){
        dp[s,t,m] <- 0
        dq[s,t,m] <- 0
      } #m
    } #s
    for (s in 2:ns){
      for (m in 1:(s-1)){
        dp[s,t,m] <- 0
        dq[s,t,m] <- 0
      } #m
    } #s
  } #t
```

```
# Define the multinomial likelihood
for (t in 1:((n.years-1)*ns)){
  marr[t,1:(n.years*ns-(ns-1))] ~ dmulti(pr[t,], rel[t])
}

# Define the cell probabilities of the multistate m-array
for (t in 1:(n.years-2)){
  U[(t-1)*ns+(1:ns), (t-1)*ns+(1:ns)] <- ones
  for (j in (t+1):(n.years-1)){
    U[(t-1)*ns+(1:ns), (j-1)*ns+(1:ns)] <- U[(t-1)*ns+(1:ns), (j-2)*ns+(1:ns)] %*%
      psi[,t,] %*% dq[,t,]
  } #j
} #t
U[(n.years-2)*ns+(1:ns), (n.years-2)*ns+(1:ns)] <- ones

# Diagonal
for (t in 1:(n.years-2)){
  pr[(t-1)*ns+(1:ns),(t-1)*ns+(1:ns)] <- U[(t-1)*ns+(1:ns),(t-1)*ns+(1:ns)]
    %*% psi[,t,] %*% dp[,t,]
  # Above main diagonal
  for (j in (t+1):(n.years-1)){
    pr[(t-1)*ns+(1:ns), (j-1)*ns+(1:ns)] <- U[(t-1)*ns+(1:ns), (j-1)*ns+(1:ns)] %*%
      psi[,j,] %*% dp[,j,]
  } #j
} #t
pr[(n.years-2)*ns+(1:ns), (n.years-2)*ns+(1:ns)] <- psi[,n.years-1,] %*% dp[,n.years-1,]

# Below main diagonal
for (t in 2:(n.years-1)){
  for (j in 1:(t-1)){
    pr[(t-1)*ns+(1:ns),(j-1)*ns+(1:ns)] <- zero
  } #j
} #t

# Last column: probability of non-recapture
for (t in 1:((n.years-1)*ns)){
  pr[t,(n.years*ns-(ns-1))] <- 1-sum(pr[t,1:((n.years-1)*ns)])
} #t
}
")
```

Choosing initial values can be difficult in IPMs with multistate models, but here we found that it is sufficient to just provide values for the number of breeders and nonbreeders for each colony in the first year. To ensure some dispersion of the initial values among MCMC chains, we use Poisson random values with expected values corresponding to the observed counts. The numbers of nonbreeders are then scaled up proportionally according to the stable stage distribution (see above).

```
# Initial values
inits <- function(cc=cormorant$count){
  B <- array(NA, dim=c(3, 14))
  B[1,1] <- rpois(1, cc[1,1])
  B[2,1] <- rpois(1, cc[2,1])
  B[3,1] <- rpois(1, cc[3,1])
  N <- B * 0.55/0.45
  return(list(B=B, N=N))
}
```

Now we only need to define the list with the monitored parameters and the MCMC specifications before we launch the model.

```
# Parameters monitored
parameters <- c("beta.phi", "phi", "mu.kappa", "kappa", "p", "mean.eta", "mean.nu", "mean.rho", "sigma",
  "N", "B")

# MCMC settings
ni <- 150000; nb <- 50000; nc <- 3; nt <- 100; na <- 3000

# Call JAGS from R (ART 143 min) and check convergence
out1 <- jags(jags.data, inits, parameters, "model1.txt", n.iter=ni, n.burnin=nb, n.chains=nc,
  n.thin=nt, n.adapt=na, parallel=TRUE)
traceplot(out1)
```

18.6 RESULTS

The model takes a while to run but converges relatively easily. Here is a partial posterior summary for the estimated quantities in our model.

```
print(out1, 3)
```

```
                    mean      sd     2.5%       50%     97.5% overlap0     f  Rhat n.eff
mu.phi[2,1]        0.201   0.298   -0.410     0.212     0.753     TRUE 0.765 1.016   167
mu.phi[1,2]       -1.654   0.333   -2.320    -1.652    -0.999    FALSE 1.000 1.011   223
mu.phi[2,2]        0.244   0.291   -0.347     0.258     0.779     TRUE 0.802 1.016   171
mu.phi[1,3]       -0.648   0.432   -1.470    -0.658     0.267     TRUE 0.936 1.002   834
mu.phi[2,3]        0.452   0.300   -0.167     0.456     1.017     TRUE 0.929 1.019   138
phi[1,1,1]         0.686   0.063    0.562     0.687     0.807    FALSE 1.000 1.017   140
phi[2,1,1]         0.731   0.016    0.698     0.731     0.761    FALSE 1.000 1.001  3000
[...output truncated...]
phi[1,3,13]        0.582   0.094    0.416     0.577     0.778    FALSE 1.000 1.000  2855
phi[2,3,13]        0.806   0.040    0.737     0.802     0.895    FALSE 1.000 1.002  1077
mu.kappa[1]       -0.774   0.189   -1.118    -0.778    -0.381    FALSE 1.000 1.006   325
mu.kappa[2]       -0.160   0.019   -0.198    -0.160    -0.123    FALSE 1.000 1.000  3000
mu.kappa[3]        2.135   0.301    1.525     2.141     2.706    FALSE 1.000 1.003   713
mu.kappa[4]        1.336   0.249    0.846     1.330     1.834    FALSE 1.000 1.001  1486
kappa[1,1]         0.283   0.036    0.221     0.281     0.363    FALSE 1.000 1.007   286
kappa[2,1]         0.764   0.054    0.645     0.769     0.855    FALSE 1.000 1.001  3000
[...output truncated...]
kappa[2,13]        0.329   0.056    0.228     0.328     0.443    FALSE 1.000 1.000  3000
kappa[3,13]        0.181   0.033    0.123     0.178     0.252    FALSE 1.000 1.000  3000
p[1,1]             0.648   0.021    0.607     0.648     0.688    FALSE 1.000 1.000  3000
p[2,1]             0.036   0.028    0.001     0.029     0.105    FALSE 1.000 1.001  1883
[...output truncated...]
p[2,13]            0.947   0.042    0.841     0.956     0.998    FALSE 1.000 1.001  2164
p[3,13]            0.650   0.051    0.552     0.651     0.751    FALSE 1.000 1.003   637
mean.eta[1,1]      0.978   0.005    0.966     0.979     0.987    FALSE 1.000 1.000  3000
mean.eta[2,1]      0.086   0.018    0.054     0.085     0.124    FALSE 1.000 1.002   932
mean.eta[3,1]      0.005   0.002    0.001     0.004     0.011    FALSE 1.000 1.001  2490
mean.eta[1,2]      0.011   0.003    0.006     0.011     0.017    FALSE 1.000 1.000  3000
mean.eta[2,2]      0.827   0.035    0.752     0.829     0.892    FALSE 1.000 1.002   932
mean.eta[3,2]      0.005   0.002    0.001     0.004     0.011    FALSE 1.000 1.001  2490
mean.eta[1,3]      0.011   0.003    0.006     0.011     0.017    FALSE 1.000 1.000  3000
mean.eta[2,3]      0.086   0.018    0.054     0.085     0.124    FALSE 1.000 1.002   932
mean.eta[3,3]      0.991   0.005    0.979     0.992     0.998    FALSE 1.000 1.001  2490
mean.nu[1,1]       0.989   0.002    0.984     0.989     0.993    FALSE 1.000 1.001  1674
mean.nu[2,1]       0.006   0.002    0.003     0.006     0.011    FALSE 1.000 1.002  1362
mean.nu[3,1]       0.006   0.002    0.003     0.006     0.011    FALSE 1.000 1.002  1362
mean.nu[1,2]       0.005   0.001    0.003     0.005     0.008    FALSE 1.000 1.001  3000
mean.nu[2,2]       0.988   0.003    0.982     0.988     0.993    FALSE 1.000 1.002  1139
mean.nu[3,2]       0.005   0.001    0.003     0.005     0.008    FALSE 1.000 1.001  3000
mean.nu[1,3]       0.006   0.002    0.003     0.006     0.010    FALSE 1.000 1.001  1898
mean.nu[2,3]       0.006   0.002    0.003     0.006     0.010    FALSE 1.000 1.001  1898
mean.nu[3,3]       0.989   0.002    0.983     0.989     0.993    FALSE 1.000 1.001  1398
mean.rho[1]        1.650   0.133    1.390     1.647     1.908    FALSE 1.000 1.001  2853
mean.rho[2]        2.220   0.204    1.840     2.212     2.630    FALSE 1.000 1.001  1994
mean.rho[3]        1.641   0.159    1.347     1.631     1.966    FALSE 1.000 1.001  1301
sigma[1]         303.682  91.371  171.488   289.210   526.688    FALSE 1.000 1.002   839
sigma[2]         265.069  71.951  163.375   251.908   441.373    FALSE 1.000 1.000  3000
sigma[3]         320.671  82.020  202.394   306.157   524.126    FALSE 1.000 1.000  3000
N[1,1]          4902.564 595.877 4288.845  4723.980  6474.122    FALSE 1.000 1.000  3000
N[2,1]          1867.905 150.467 1713.960  1821.940  2269.922    FALSE 1.000 1.001  3000
[...output truncated...]
N[2,14]         2237.194  516.186 1416.561 2178.520  3424.011    FALSE 1.000 1.001  3000
N[3,14]         7498.992 1512.295 4967.528 7374.228 10861.533    FALSE 1.000 1.000  3000
B[1,1]          4678.304  270.860 4089.630 4705.478  5148.747    FALSE 1.000 1.002  1119
B[2,1]          1604.252  132.981 1409.873 1586.594  1901.241    FALSE 1.000 1.000  3000
[...output truncated...]
B[2,14]         1931.980  153.149 1640.733 1928.051  2237.172    FALSE 1.000 1.000  3000
B[3,14]         3564.089  211.356 3187.816 3552.110  4016.581    FALSE 1.000 1.000  2507
```

The estimated number of breeders closely follows the observed colony counts (Fig. 18.3), which is not surprising because in the model we estimate productivity as a hidden parameter. The number of nonbreeders associated with each colony can be estimated even though nonbreeders are not included in the colony counts. Indeed, many nonbreeders are

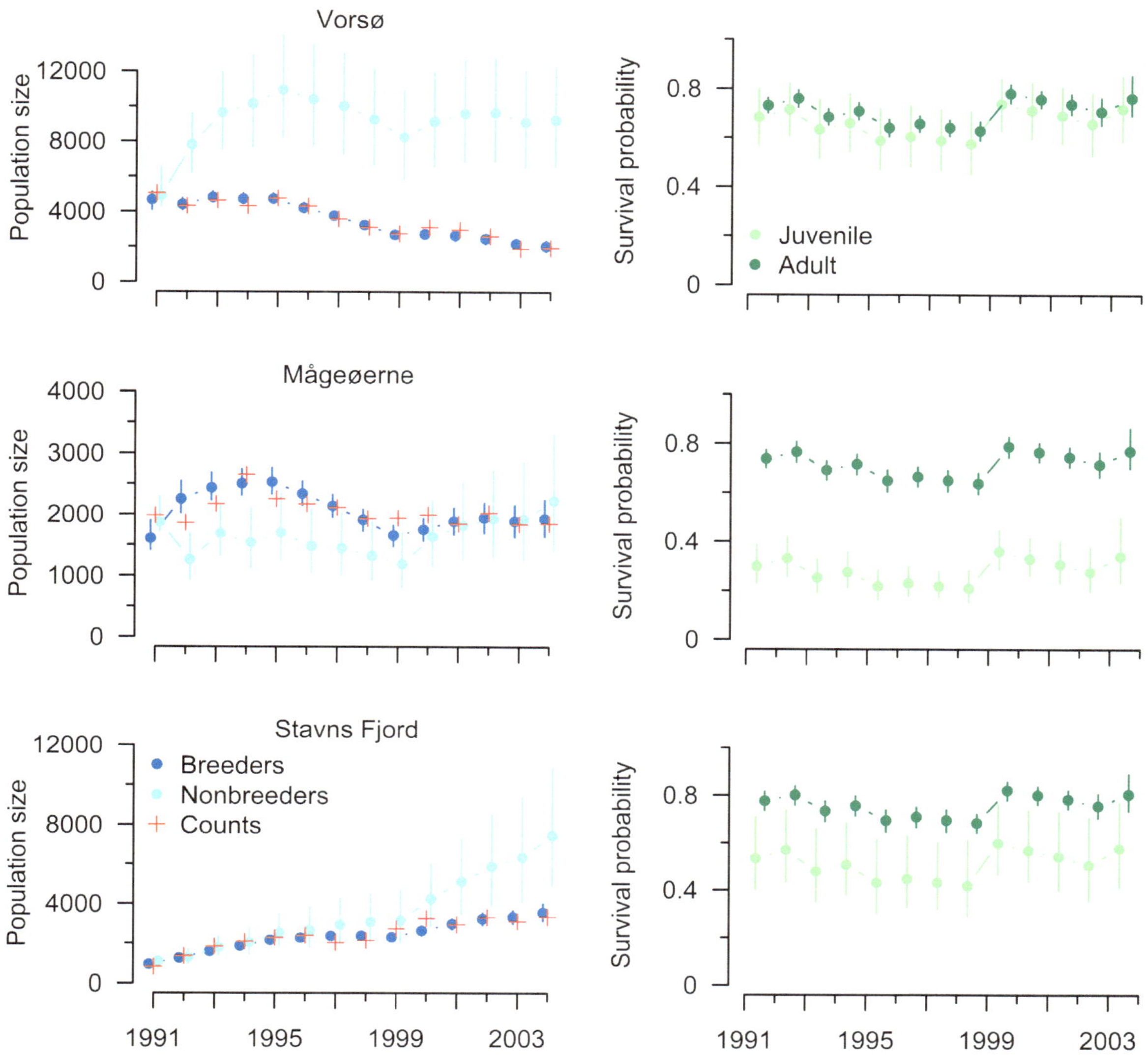

FIGURE 18.3

Estimates of population size and annual apparent survival probabilities of cormorants in three Danish breeding colonies, 1991–2004. Dots show posterior means, and vertical lines show the 95% credible intervals.

absent from the breeding colonies, which prevents their inclusion in a count at the colony. The estimated number of nonbreeders is a function of the size of the population segment of the breeders and the demographic rates. The uncertainty in these estimates is therefore much larger than in those of the breeders, which are informed directly by the observed colony counts. Moreover, the numbers in the first year seem particularly unreliable, and the chosen priors have a strong impact on them. From the second year onwards, the number of nonbreeders becomes more precise, as it is now a function of the number of fledglings the year before and of juvenile survival. Over time, the precision should further increase as it tallies up the individuals that have not yet recruited. Interestingly, though, this is not reflected in the credible intervals because they are even smaller in the first year than in later years. The estimated number of breeders is about the same as the number of nonbreeders at the colonies Mågeøerne and Stavns Fjord during the first half of the study period, while nonbreeders clearly outnumbered breeders at Vorsø and at Stavns Fjord during the second half.

The annual adult survival probabilities are very similar at all three colonies (Fig. 18.3). Juvenile survival is clearly lower than adult survival at Mågeøerne and Stavns Fjord but similar to adult survival at Vorsø. The high juvenile survival at Vorsø was probably the reason for why the number of nonbreeders in this colony was much larger than the number of breeders.

Dispersal among the three colonies is weak; thus, estimated natal and breeding site fidelities are very high (Fig. 18.4A and B). The only exception is natal site fidelity at Mågeøerne: while still relatively high, it is clearly lower than in the other colonies. Although productivities are hidden parameters, their precision is quite good (Fig. 18.4C). The estimates from all

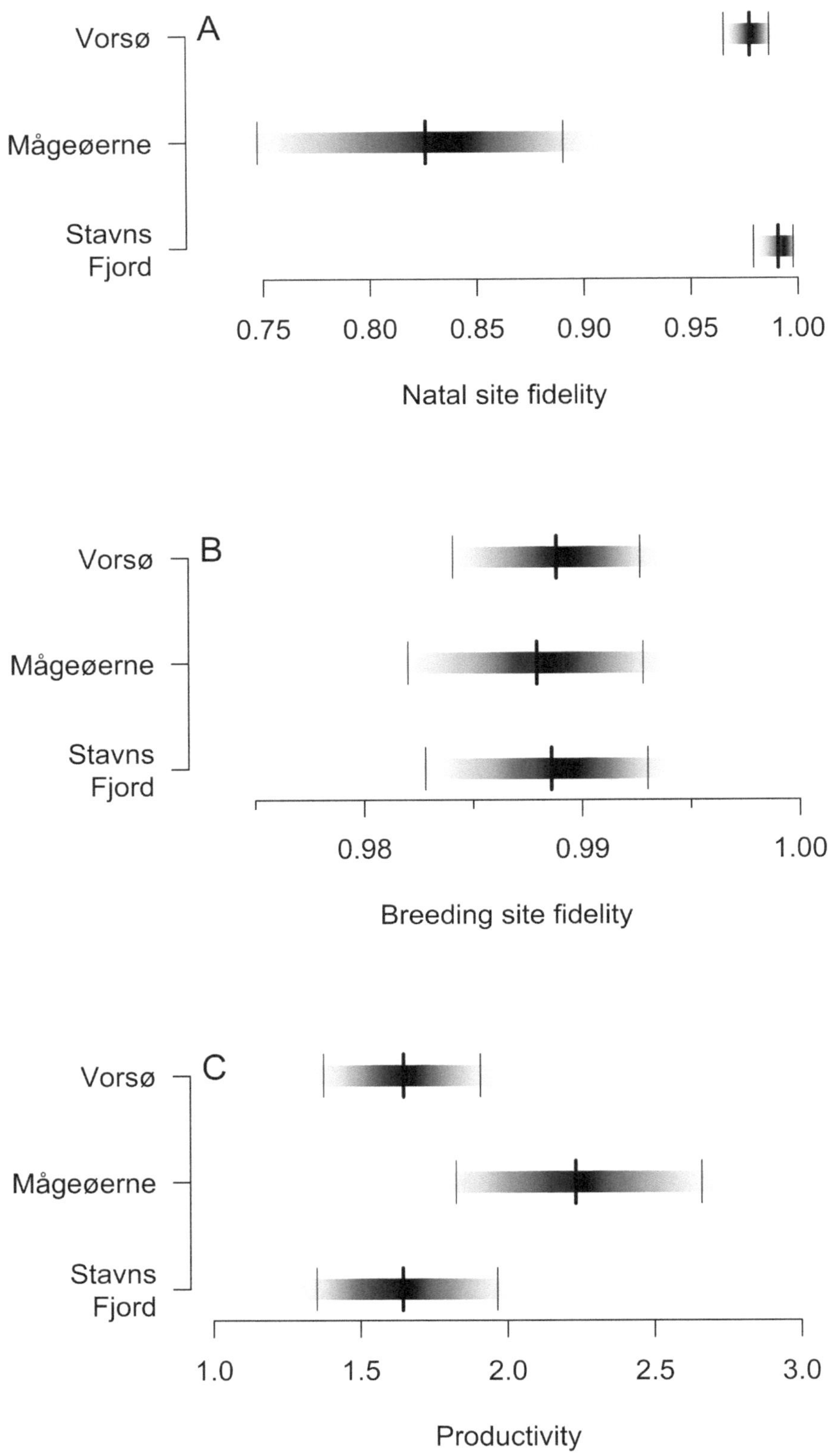

FIGURE 18.4

Natal and breeding site fidelity, and productivity of cormorants at three Danish breeding colonies, 1991–2004 (note widely different scales of x-axes). The darkness of the strips is proportional to the posterior densities. Vertical lines show the 95% credible intervals and posterior means (thick lines).

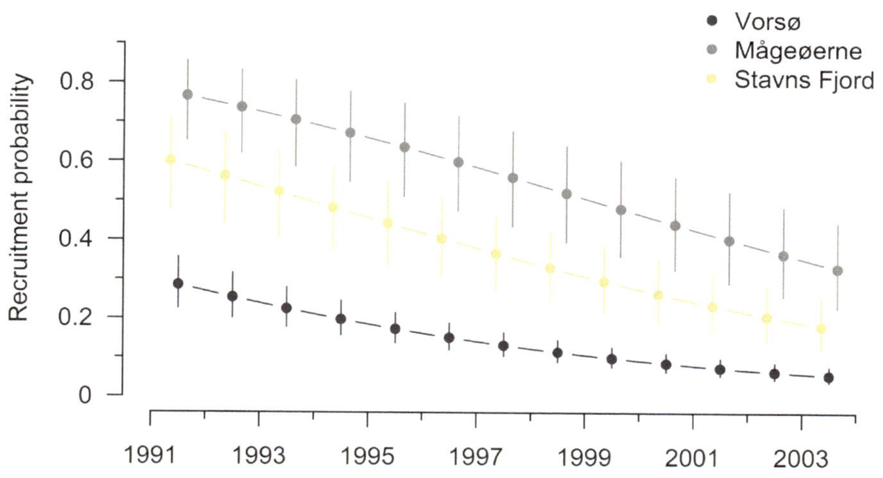

FIGURE 18.5

Recruitment probabilities of cormorants at three Danish colonies, 1991–2004. Dots are posterior means, and vertical lines show the 95% credible intervals.

colonies are similar, with slightly higher values at Mågeøerne. Note that our estimates of productivity refer only to the number of females produced.

Recruitment probabilities declined over time (Fig. 18.5), and thus cormorants were recruited into the breeding populations at a progressively later age as time went on. Recruitment was highest at Mågeøerne and lowest at Vorsø. Newly recruited individuals were therefore older on average at Vorsø than at the two other colonies.

18.7 DISCUSSION

We analyzed colony counts and capture-recapture data of great cormorants from three Danish breeding colonies over 14 years (1991–2004). The population model is relatively complex because of dispersal among the colonies and delayed recruitment at ages older than 1 year. No data on productivity were available, and therefore we estimated colony-specific productivity as a hidden parameter. Our estimates were similar to those of McCrea et al. (2010), who analyzed the same data set with a similar IPM using maximum likelihood.

Our multisite IPM makes some fairly strong assumptions about movements underlying dispersal: that they are restricted to the three studied colonies and that there is no immigration from outside. Although multisite capture-recapture data allow the modeling of dispersal among sites, the estimated survival is still apparent rather than true, exactly as for a single-site capture-recapture model without dead recoveries. But "apparent" now refers to the set of three colonies. Thus, if some birds emigrate permanently away from this system, apparent survival will be reduced relative to true survival. The difference between the two depends on the amount of permanent emigration (Cilimburg et al., 2002; Zimmerman et al., 2007; Schaub and Royle, 2014). In other analyzes combining capture-recapture and dead recoveries, estimates of site fidelity in the largest colony, Vorsø, declined over time, with values of 0.85 and 0.9 in first-year and adult birds, respectively (Frederiksen and Bregnballe, 2000a,b). Since cormorants hardly moved from Vorsø to any of the other studied colonies, these results suggest that there was indeed some emigration from the studied system and that therefore true survival was greater than the estimated apparent survival.

Presumably the system also received some immigrants, i.e., recruits born outside the three studied colonies (Hénaux et al., 2007). In principle, the estimation of immigration in the multisite IPM is possible in a similar way as in a single population (Chapter 11). Moreover, with multiple study sites, we can estimate immigration into each. A difference from the single-site case is that estimated immigration is not the total immigration received by a site. Rather, it refers to arriving individuals born outside the three study populations. A focal site will also receive immigrants that originate from one of the studied sites, but here this is estimated by the dispersal parameters. Total immigration to a single site is then the sum of the two. Such an IPM with dispersal among several sites including immigration from "the rest of the world" has been developed to study multiple colonies of roseate terns (*Sterna dougallii*, Seward et al., 2019). In our cormorant model, estimation of immigration was not possible because the model already included one hidden parameter (productivity), and we cannot normally estimate two hidden parameters in an IPM (Section 6.3). If the study system receives immigrants that were not properly modeled—i.e., the model assumes that there were none—we expect that the estimate of the hidden parameter will be biased (Section 7.3.3). Indeed, our estimates of productivity appear too high for cormorants, with the typical range being between 0.8 and 1.5 fledglings per breeding pair (Bregnballe and Gregersen, 2003). Thus, this may be an artifact caused by immigration that was not properly accommodated in our model.

We have not included demographic stochasticity in our model to keep it simple. To include demographic stochasticity, we would specify Poisson and binomial random variables with suitable expectations for the elements of the population model. For example, the number of nonbreeders is composed of the sum of 1-year-old individuals from different origins and the number of nonbreeders that have not yet recruited. We would then use a Poisson distribution for the former and a binomial distribution for the latter. Hence, as an example, the number of nonbreeders at Vorsø would be expressed as follows:

$$N_{V,t+1} \sim Poisson\left(B_{V,t}\rho_{V,t}\phi^{1}_{V,t}\eta_{VV,t} + B_{M,t}\rho_{M,t}\phi^{1}_{M,t}\eta_{MV,t} + B_{S,t}\rho_{S,t}\phi^{1}_{S,t}\eta_{SV,t}\right) + Binomial\left(\phi^{A}_{V,t}(1-\kappa_{V,t}), N_{V,t}\right).$$

In the model fit in JAGS, we would first evaluate the outcomes of the Poisson and binomial distributions separately and then add the two realizations to obtain $N_{V,t+1}$. The number of breeders would then be expressed similarly by a series of binomial distributions. We have experimented with an IPM with demographic stochasticity for this data set and found results similar to those shown here. This is not surprising, since the study colonies were quite large. Computationally, the IPM with demographic stochasticity was much more demanding—it took more than 10 times longer to converge.

A simpler variant to implement variability due to demographic stochasticity with substantial computational benefits is a normal approximation. Instead of using Poisson and binomial distributions, we use normal distributions that approximate them. The variance of the Poisson distribution is the same as its mean, so we can replace $\sim Poisson(\lambda)$ with $\sim Normal(\lambda, \lambda)$. Likewise, the variance of a binomial distribution is a function of the trial size and the success parameter, and we replace $\sim Binomial(p, N)$ by $\sim Normal(Np, Np(1-p))$. We also analyzed the cormorant data set with this variant of the model. We found that we needed to increase the number of posterior draws to 400,000 to obtain satisfactory convergence, but the run time was still only about one-fifth of what it was when demographic stochasticity was modeled with binomial and Poisson distributions. A normal approximation as shown here is standard in IPMs fitted with maximum likelihood methods (Besbeas et al., 2002; Tavecchia et al., 2009) because the Kalman filter for the state-space model does not work with distributions other than the normal. The normal approximation has also been used fruitfully for the computationally demanding demographic state-space model of Dail and Madsen (2011) by Brintz et al. (2018) and Zhao et al. (2019).

Multisite IPMs of a species with delayed recruitment typically become complex and require many parameters. The risk that the model then includes nonidentifiable parameters increases, and we need to check by simulation or using more formal approaches (Cole and McCrea, 2016) whether and which parameters are informed by the data. For this case study, we know from McCrea et al. (2010) that the parameters are all identifiable.

The use of multistate capture-recapture models within IPMs has great potential because of the great flexibility in multistate models to define states in all sorts of ways to address the particular features required in a population model. In the cormorant case study, states were used to define geographical locations and breeding stages. Other states that can be defined and are often relevant in population studies include for example age and size classes, habitat types, disease status, mortality causes, or states related to temporary emigration as well as combinations thereof (Lebreton et al., 2009). The ability to estimate transition probabilities among these states allows a more complex and potentially more relevant structure of the population model in the IPM. Even greater flexibility is offered by multievent capture-recapture models (Pradel, 2005), which are multistate models that allow for state uncertainty. So far, they have only rarely been used as parts of an IPM (Tenan et al., 2017; Jan et al., 2019), and their potential has not been fully explored. We expect that both multistate and multievent models will be used increasingly as component models in IPMs.

So far, most published IPMs have focused on a single "site" or single population and have not included space explicitly. Individuals emigrating from or immigrating to a population often need to be included in such a model, as we have seen in the woodchat shrike case study, for example (Chapter 11). Space can be incorporated in an IPM in many different ways. One is to distinguish multiple sites or populations and allow them to exchange individuals. The typical model for this is a multistate capture-recapture model as in the cormorant IPM in this chapter. Space is treated in a discrete manner, the number of studied populations is typically low, and inference is restricted to the spatial scale of the studied sites. For making inferences about multiple and notably larger spatial scales, space must be represented either in a continuous way or by gridding space into many discrete spatial units. IPMs that accommodate such a representation of space are the focus of the following case studies (Chapters 19–21).

CHAPTER

GRAY CATBIRD

19

CHAPTER OUTLINE

IPM ingredients: population count data, capture-recapture data
Location: USA
Spatio-temporal scale: 75,634 km^2, 17 years

19.1 INTRODUCTION

For an improved understanding of population dynamics, it is often valuable to study population processes at multiple spatial scales (Baillie et al., 2000; Melbourne and Chesson, 2005; Chave, 2013; Chandler and Hepinstall-Cymerman, 2016). Processes that are important at small spatial scales such as immigration (Millon et al., 2019) become less important at larger spatial scales. Conversely, factors that are important at large spatial scales such as local recruitment may be of more minor importance at small spatial scales (Schaub et al., 2015). Most case studies in Part 2 of this book focus on a single population or at best on a handful of populations—naturally, inferences from such studies are restricted to the spatial scale of the studied population(s). Typically, these "local populations" extend over a relatively small area, and therefore, at least geographically, the inferences from these studies may be somewhat limited. In addition, they do not normally allow us to say anything about how the landscape affects demography because these models are aspatial and do not accommodate any spatial variation in demography that could be explained using measured or unmeasured covariates associated with space. Here we present a study that includes multiple spatial scales with a particular focus on a large spatial scale.

Specifically, we would like to estimate abundance trends over larger areas such as an entire country and be able to say how the underlying demographic processes (gains and losses) contribute to changes in abundance over time. In principle, we could simply enlarge a study area until it covers the entire desired area of interest and then apply the same type of integrated population model (IPM) we have now seen so many times in this book. However, it is usually impossible to sample population data over very large areas, e.g., it is usually not possible to survey all breeding pairs of a target species in, say, a whole country. Moreover, even if such a survey was possible, we still would not be able to move between different spatial scales, and therefore, our inferences would be restricted to one scale represented by the entire study area. Therefore, if we require inferences at multiple scales that cover a large spatial extent, we must use another strategy. A natural possibility is to define a number of smaller sampling areas within the target area and collect population and demographic data in each. These are spatially replicated samples, or geographic strata, and we have to treat them as such to make inferences about the larger spatial scale. Ideally, these sampling units represent a random sample of space, i.e., are chosen randomly from a larger statistical "population" of such units. Inferences are then possible at the scale of the studied sampling units and at larger spatial scales by extrapolation or summing up. These sampling principles are applied at least to some degree in well-designed monitoring programs, such as the national breeding bird surveys of North America (BBS; Sauer et al., 2013), Great Britain (Baillie, 1990), and Switzerland (Schmid et al., 2004). In contrast, citizen science data

Integrated Population Models. https://doi.org/10.1016/B978-0-12-820564-8.00017-8

such as eBird suffer from various biases, one of which is the nonrandom spatial sample, which needs to be accommodated by statistical adjustments in the proper analyses of such data (Fink et al., 2020).

The same principles hold for the collection of data for the estimation of demographic rates, which again should ideally be carried out at multiple locations spanning the entire extent of an area of interest. Examples of programs that cover a large spatial extent are the "Monitoring Avian Productivity and Survivorship" (MAPS) program in North America (DeSante et al., 1993) and the Constant Effort Sites (CES) scheme in the UK (Baillie, 1990; Baillie and Schaub, 2009).

When trying to understand population dynamics at a large spatial scale in an IPM, we are faced with two challenges for model specification: (1) how to represent space and (2) how to extrapolate from the sampled sites to the unsampled areas within the entire region for which inferences are desired. We deal with each in turn.

With extremely few exceptions, published IPMs are thus far nonspatial in the sense that the individuals in the modeled populations have no spatial attribution nor does the model accommodate spatial variation in the parameters governing population dynamics. Some IPMs for multiple discrete sites have been published (e.g., McCrea et al., 2010; Péron et al., 2010; Seward et al., 2019, and see also our cormorant case study in Chapter 18), and usually account for differences between these specific sites along with dispersal among them. In this chapter, however, we are dealing with models for population dynamics in large *continuous* rather than discrete areas, where any spatial discretization scheme is not immediately obvious. The simplest approach to space would again be to ignore it and assume there is no spatial variability in population dynamics. We may argue that a model with parameters that are constant over space will allow inferences about some average of these parameters in the likely case of spatial variation. However, the latter means that the common demography assumption will be violated to at least some degree, and we have seen how this can be a problem for IPMs (see Section 7.2.3).

Hence, a more natural and perhaps better approach is to subdivide the entire region into discrete strata and then make an common demography assumption for each of these strata separately. This is relatively straightforward and results in a model with spatial stratification. However, the choice of how to stratify is mostly arbitrary. Moreover, different stratification schemes may well lead to different inferences. Hence, a second variant to represent space in a model that is much more refined is to avoid any spatial discretization (at least conceptually) and instead represent space in a continuous manner. In such a continuous field, smooth changes in density and demography can be specified either via models with environmental covariates as drivers or via a spatial random field (see Chapters 3 and 9 in Kéry and Royle, 2021). In practice, the latter means modeling the spatial autocorrelation in a quantity such as density, survival, or recruitment.

To date, most published large-scale IPMs have ignored space (e.g., Besbeas et al., 2002; Robinson et al., 2014; Plard et al., 2020b), and only a very few have defined spatial strata as the simplest representation of spatial variation in demography (Morrison et al., 2016; Ahrestani et al., 2017; Plard et al., 2020a; Zhao, 2020). Chandler and Clark (2014), Chandler et al. (2018) and Sun et al. (2019) are the only researchers who have chosen a fully continuous representation of space in an IPM (see also the black bear case study in Chapter 21).

However, there is a myriad of examples of continuous representations of space in simpler models for individual components in an IPM, such as density, survival, and recruitment. Indeed, the modeling of spatial autocorrelation is a dominant interest in spatial statistics (Cressie and Wikle, 2011; Banerjee and Gelfand, 2014; Wikle et al., 2019), and presumably any of the models and techniques in these and many other textbooks could be incorporated in any parameter in an IPM, including density, survival, or recruitment. Examples of the smooth spatial modeling via autocorrelation (with or without effects of additional spatially indexed covariates) can be found, e.g., in the works of Royle and Dubovsky (2001), Saracco et al. (2010) or in Section 3.4 in Kéry and Royle (2021).

A second, and partly related, challenge in an IPM fit to data at large spatial scales is extrapolation. Since data are virtually always collected in only a subset of the entire area for which inferences are desired (for instance, about total abundance or abundance trends), some decision has to be made about how to extrapolate from the surveyed to unsurveyed parts of the study area (or perhaps the strata defined; see above). One of the simplest manners of doing this extrapolation is simple area expansion, which implies the random sampling of space in a study. In models with a continuous representation of space, we can use our model with covariates and/or spatial autocorrelation to estimate quantities for the entire area. In practice, space in such models is still represented by very small, discrete spatial pixels. Quantities are predicted for each pixel, and we add them up as desired either for regional totals or for the grand total to make inferences at the desired spatial scale.

To demonstrate the potential of this approach, we analyze North American Breeding Bird Survey (BBS) data together with MAPS data of the gray catbird (*Dumetella carolinensis*, Fig. 19.1) over a 17-year period (1992–2008) from the New England/Mid-Atlantic region of the United States. The catbird is a widely distributed territorial passerine that breeds in dense shrubs or thickets of young trees. The goal of the study is the estimation of the spatially stratified abundance trends of the species across this large area and of the underlying demographic rates, specifically survival and recruitment.

FIGURE 19.1

Gray catbird (*Dumetella carolinensis*) perched on a bush (photograph by Robert Royse).

Recruitment is a hidden parameter and cannot be estimated from either data set alone. These data were analyzed with an IPM by Ahrestani et al. (2017), and we basically reproduce their analysis.

19.2 DATA SETS

We analyze two types of data originating from the BBS and MAPS programs. The BBS is an avian monitoring program designed to provide relative abundance and trend estimates for all North American breeding birds at the subcontinental scale. Once each spring, in late May or June, all breeding birds are counted at 50 approximately equidistant points along more than 3000 transects of 24.5 miles ($\sim$39 km) in length on secondary roads, with routes being established at haphazard locations. At each point, all birds heard or seen are counted for 3 min as far away as 400 m (Sauer et al., 2013). The total number of individuals of a given species detected over all 50 points per route and year is the index of relative annual abundance on that route.

The annual number of surveyed BBS routes in the New England/Mid-Atlantic region (75,634 km^2) varied between 64 and 90 during 1992–2008, with a total of 115. This region is number 30 in the Bird Conservation Region (BCR) scheme, which distinguishes large physiographic areas that define major habitats for birds across North America (Sauer et al., 2013). A total of 1298 counts were conducted by 172 different observers in this region (Fig. 19.2); the median count of gray catbirds was 18.3 (range 0–224). In order to account for the substantial spatial heterogeneity in abundance and its change over time in this large area, the routes were assigned to nine strata defined by the intersection of the US states within that region.

The second data set originates from the MAPS program. This is a network of more than 1200 banding stations that use a constant effort mist netting protocol to ring and monitor birds across all of North America (DeSante and Kaschube, 2009). Typically, a MAPS station consists of 10 12-m long mist nets distributed over an area of about 8 ha. Stations are operated during 6–10 days of a breeding season (DeSante et al., 2004).

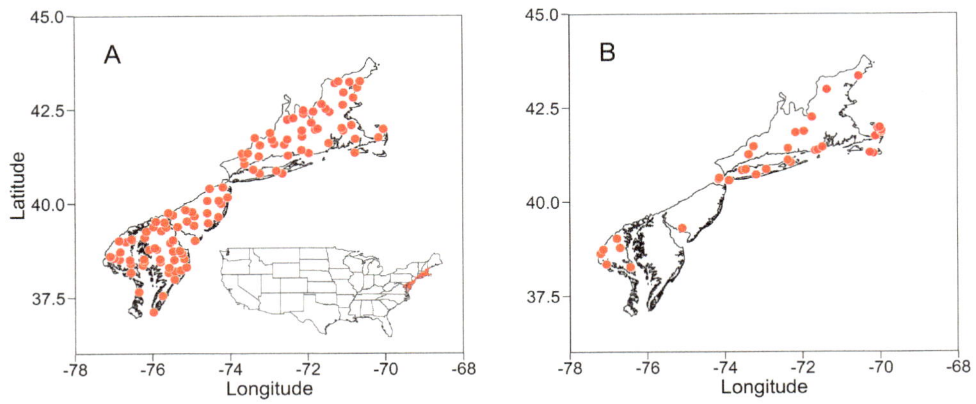

FIGURE 19.2

Spatial configuration of the integrated population model and the utilized data sets with outline of the New England/Mid-Atlantic region (75,634 km^2) and location of 115 North American Breeding Bird Survey transects (A) and 38 Monitoring Avian Productivity and Survivorship banding stations (B). *(This is a modified version of fig. 2 in Ahrestani et al., 2017.)*

The gray catbird capture-recapture data originate from 38 stations in the New England/Mid-Atlantic region that captured between 1 and 381 individuals (Fig. 19.2). In total, 4276 individuals were marked (banded) during the 17 study years, of which 686 were recaptured in at least one later year. Banding stations were active for an average of 10.5 years, and five stations operated in all 17 years. The capture-recapture data are provided at an annual base, i.e., a 1 in the detection history matrix denotes an individual that was captured at least once during the respective breeding season.

A captured individual either belongs to the local breeding pool (i.e., is a "resident") or is a "transient" that merely uses the banding station for stopover during migration or *en route* during the settlement process. Apparent survival of a transient is zero by definition (Pradel et al., 1997). Our goal in the analysis is estimation of the survival of residents only, and therefore, transients had to be filtered out in some way. Some individuals were captured multiple times during the breeding season. We defined individuals as confirmed residents if they produced multiple within-year recaptures that occurred at least 10 days apart. For all other individuals, the residency state was unknown and estimated as a fixed (time invariant) latent individual covariate in the model (see below).

19.3 POPULATION MODEL

Our population model for the gray catbird is fairly simple because it does not include an age structure. In contrast, the model contains spatial structure associated with the spatial units, which are the nine defined strata. This spatial stratification is included to allow for spatially different dynamics such that the common demography assumption is considerably relaxed. The within-stratum dynamics are assumed to be independent from the dynamics in other strata, as there is no explicit link between them, i.e., the model contains neither dispersal nor spatial autocorrelation. For each stratum, we assume a prebreeding "census" model and distinguish between two types of individuals, "survivors" and "recruits," which is the state process in the Dail and Madsen (2011) model. Survivors at time t were in the pool of individuals already in year $t-1$; i.e., they survived and returned to the same stratum. Recruits are the new individuals in a stratum. They could be immigrants or local recruits, i.e., individuals that were born in a stratum that survived and remained in the same stratum.

The gray catbird life-cycle graph is shown in Fig. 19.3. The numbers of recruits ($R_{s,t+1}$) and survivors ($S_{s,t+1}$) in year $t+1$ in stratum s as a function of the total number of individuals present in stratum s in year t are given by

$$R_{s,t+1} \sim Poisson(\rho_t N_{s,t}),$$
$$S_{s,t+1} \sim Binomial(\phi_t, N_{s,t}), \text{ and}$$
$$N_{s,t} = R_{s,t} + S_{s,t},$$

where ϕ_t is apparent adult survival from year t to $t+1$, and ρ_t is the fraction of new individuals in year $t+1$ with respect to the population size in t (recruitment). The definition of apparent survival corresponds to the classical definition of this parameter in capture-recapture studies (Lebreton et al., 1992); i.e., it is the joint probability to survive 1 year and be

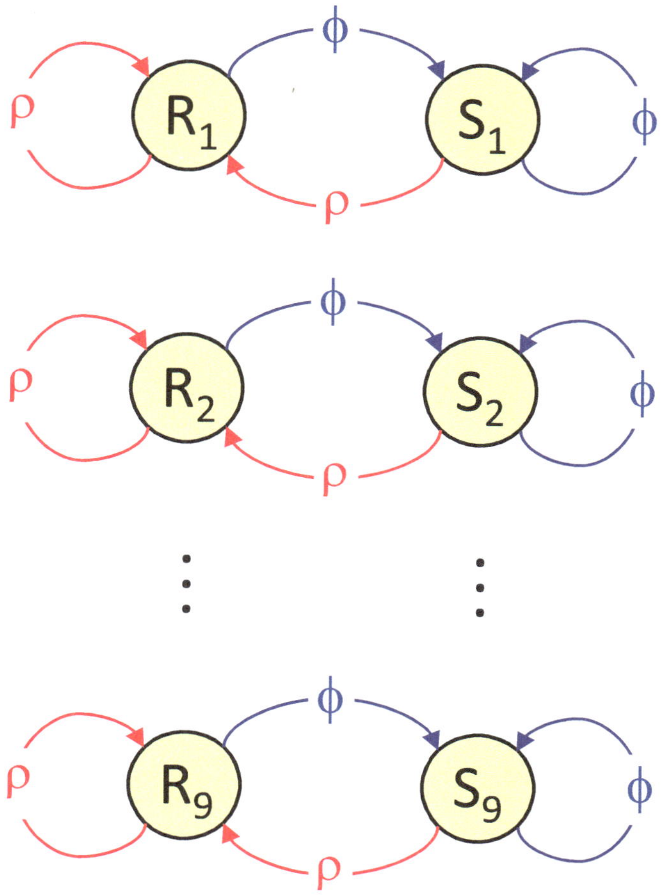

FIGURE 19.3

Life-cycle graph for the gray catbird population model. In each of the nine strata, the population is composed of two classes of individuals, recruits (*R*) and survivors (*S*). The demographic processes are recruitment (ρ, red) and apparent survival (ϕ, blue). In our model, these demographic processes are assumed to be identical in all strata, but this assumption could also be relaxed (and this might be a good exercise for you to do).

philopatric, i.e., remain in an area. It is not completely clear to which area philopatry refers, and we offer some thoughts about that in the discussion. Recruitment includes several components of "gains" to the studied populations/sites, being a function of productivity, juvenile survival, site fidelity, and immigration. In our model, we assume both demographic rates are identical across all nine strata, but this is our choice, not a requirement of the model. Our use of the Poisson and binomial distributions means that effects of demographic stochasticity are accommodated in the model. The abundance index in stratum s of year t ($N_{s,t}$) is the sum of the numbers of recruits and survivors. Note that we use the terms "number" and "index" somewhat interchangeably here. Strictly speaking, we estimate the expected number of individuals *detected* on a stratum-specific route, which is clearly not the true population size in that stratum but only an index for it (i.e., a measure of relative abundance; see Chapter 1 in Kéry and Royle, 2021).

19.4 COMPONENT DATA LIKELIHOODS

We next describe the basic models adopted for each data set and define their likelihoods. We load the data set and produce an overview of its content.

```
library(IPMbook); library(jagsUI)
data(catbird)
str(catbird)

List of 9
$ y        : num [1:4276, 1:17] 0 1 0 0 1 1 0 0 0 0 ...
  ..- attr(*, "dimnames")=List of 2
  .. ..$ : chr [1:4276] "427696" "427697" "427699" "427703" ...
  .. ..$ : chr [1:17] "1992" "1993" "1994" "1995" ...
 $ r        : num [1:4276] 0 0 0 0 0 0 0 0 1 0 ...
 $ station  : num [1:4276] 1 1 1 1 1 1 1 1 1 1 ...
 $ count    : num [1:1298] 23 10 14 39 22 26 28 30 23 27 ...
 $ stratum  : num [1:1298] 1 1 1 1 1 1 1 1 1 1 ...
 $ year     : num [1:1298] 1992 1992 1992 1992 1992 ...
 $ observer: num [1:1298] 1 2 3 4 5 6 7 8 9 8 ...
 $ firstyr  : num [1:1298] 0 1 0 0 0 0 0 0 1 0 ...
 $ area     : num [1:9] 10982 14868 14329 2132 4192 ...
```

The `catbird` data contain a total of nine objects that include the population counts and associated variables (for the BBS data) and the capture-recapture data with associated variables (for the MAPS data). We briefly describe each along with the submodels we use for their analyses.

19.4.1 POPULATION COUNT DATA (BBS DATA)

The BBS count data (C) are stored in the vector `count` with length 1298. Each element of this vector is the annual number of individuals detected along a route. The vectors `stratum` and `year` indicate the stratum and year in which each count was made. Two further vectors provide the observer identity (`observer`) and a binary indicator for whether the observer was counting on a given route for the first time (`firstyr`). The former represents unique combinations of observers and routes, and thus the same observer on different routes is given a different identity. In principle, the count data could be arranged in a 3-D array with dimensions stratum x year x route (within a stratum). One major advantage of this is that the array dimensions then carry the information about these "covariates." However, since not every route was surveyed in every year, this array would then have to be filled up with a large number of missing values (NAs). Dealing with them would be very hard in a Bayesian analysis from either a coding or a computational point of view. Hence, it is much more efficient to get rid of the implied NAs and string out such data in simple vectors containing each individual count. When the data come in this "long format," we need additional vectors that carry the information about stratum, year and route associated with each element in the count vector.

As always, we model the counts with a state-space model with a process model defined by our population model (Section 19.3). The observation model is an overdispersed Poisson model with observer effects corresponding to the standard analyses of the BBS data alone (Sauer and Link, 2011). Hence, for each route and year combination j, we have

$$C_j \sim Poisson(\lambda_j), \text{ and}$$
$$\log(\lambda_j) = \log(N_{s(j),t(j)}) + \omega_{o(j)} + \eta F_j + \varepsilon_j,$$

where s and t are vectors providing the stratum and year information for each j, o is the observer identity, and F is the first-year indicator variable indicating whether an observer is new on a route. We estimate two effects associated with an observer, ω and η, where ω is a Gaussian observer random effect, which for observer i is

$$\omega_i \sim Normal(0, \sigma^2_\omega)$$

and η is the first-year observer effect, since typically observers count more birds on a route once they know it better, especially after their first year of service (Kendall et al., 1996). Experienced observers should miss fewer birds, and hence we expect η to be negative. Finally, to account for overdispersion at the level of each individual count, we add a Gaussian residual.

$$\varepsilon_j \sim Normal(0, \sigma^2_\varepsilon)$$

19.4.2 CAPTURE-RECAPTURE DATA (MAPS DATA)

The capture-recapture data (y) from all 38 stations are stored in a single matrix (y). We adopt a Cormack-Jolly-Seber (CJS) model for them in the state-space likelihood (Section 4.5.1.1) since we want to model survival of the resident individuals. This requires estimation of the residency state of each individual that cannot be done in a reasonable manner with the multinomial likelihood. We also have a vector with the identifier of the banding station (`station`) and a vector containing the known residency information (`r`), which contains a 1 for individuals captured more than once at least 10 days apart in the same year and a 0 otherwise.

To model the capture-recapture data, we start by defining the latent state variable z, which is 1 if individual i is alive at occasion t and 0 otherwise. We define a further latent state variable u, which is 1 if i is a resident and 0 if it is a transient; note that u cannot change over time, in contrast to z. As always in the CJS model, state z_{i,f_i} of individual i at its marking occasion f_i is deterministic.

$$z_{i,f_i} = 1$$

Over time, z is modeled as a sequence of Bernoulli trials.

$$z_{i,t} \sim Bernoulli\left(u_i z_{i,t-1} \phi_{t-1}\right)$$

The Bernoulli success probability is a product of three quantities: apparent survival probability (ϕ) and the latent individual states z and u, where the multiplication with u ensures that we estimate the survival of residents only.

The residency state u_i of individual i is modeled as another Bernoulli trial.

$$u_i \sim Bernoulli(\pi_t)$$

Here, π_t is the probability that individual i marked at occasion t is a resident. The probability that a newly marked individual is a resident can vary over time, but the residency state of an individual is fixed and does not change over its life.

The residency state (r) is known for some individuals because they are captured twice within the same year and at least 10 days apart. To estimate the residency state for all individuals (u), we use an observation model with the Bernoulli distribution,

$$r_i \sim Bernoulli(u_i \kappa_t),$$

where κ_t is the probability of being able to predetermine the residency state of individual i that was marked at occasion t. Finally, our CJS model needs its usual observation model in the form of yet another Bernoulli trial.

$$y_{i,t} \sim Bernoulli\left(z_{i,t} p_{i,t}\right)$$

We model recapture probability p with a fixed year effect and an additive effect of the banding station.

$$\text{logit}\left(p_{i,t}\right) = \mu_t + \alpha_{sta_i}$$

Here, μ_t is the fixed year effect on the logit scale, and α_{sta} represents the random banding station effects, and thus $\alpha_{sta} \sim Normal\left(0, \sigma_\alpha^2\right)$.

19.5 THE INTEGRATED POPULATION MODEL

Now that we have inspected all the different data sets and know how to construct their component likelihoods, we are ready to assemble the full IPM. We make the conventional independence assumption for the two data sets and specify the joint likelihood, i.e., the IPM, as the product of the component data likelihoods.

Besides estimation of the primary parameters of survival and recruitment and underlying the stratum-specific abundance index, we also want to estimate an overall abundance index for the New England/Mid-Atlantic region. We have to make a choice about how to combine the stratum-specific abundance indices. One common choice is to view the stratum-specific indices as representative for the entire stratum and then give them a weight according to their area. This is the information stored in the vector `area`. Moreover, we compute stratum-specific and overall abundance trends calculated as the geometric means of annual changes in abundance indices. We do these calculations inside the model in JAGS, but we also could have done them in R afterward.

We bundle the data, write the model, and launch it in JAGS.

```
# Bundle data and produce data overview
jags.data <- with(catbird, list(y=y, f=getFirst(y), r=r, sta=station,
  n.sta=length(unique(station)), n.years=ncol(y), n.ind=nrow(y), C=count, n.counts=length(count),
  str=stratum, yr=year-1991, n.str=length(unique(stratum)), I=firstyr, obs=observer,
  n.obs=length(unique(observer)), area=area, pNinit=dUnif(1, 50)))

str(jags.data)

List of 17
 $ y         : num [1:4276, 1:17] 0 1 0 0 1 1 0 0 0 0 ...
 $ f        : Named int [1:4276] 4 1 4 4 1 1 4 4 4 4 ...
 $ r         : num [1:4276] 0 0 0 0 0 0 0 0 1 0 ...
 $ sta       : num [1:4276] 1 1 1 1 1 1 1 1 1 1 ...
 $ n.sta     : int 38
 $ n.years   : int 17
 $ n.ind     : int 4276
 $ C         : num [1:1298] 23 10 14 39 22 26 28 30 23 27 ...
 $ n.counts  : int 1298
 $ str       : num [1:1298] 1 1 1 1 1 1 1 1 1 1 ...
 $ yr        : num [1:1298] 1 1 1 1 1 1 1 1 1 1 ...
 $ n.str     : int 9
 $ I         : num [1:1298] 0 1 0 0 0 0 0 0 1 0 ...
 $ obs       : num [1:1298] 1 2 3 4 5 6 7 8 9 8 ...
 $ n.obs     : int 172
 $ area      : num [1:9] 10982 14868 14329 2132 4192 ...
 $ pNinit   : num [1:50] 0.02 0.02 0.02 0.02 0.02 0.02 0.02 0.02 ...


# Write JAGS model file
cat(file="model1.txt", "
model {
# Priors and linear models
for (i in 1:n.ind){
  for (t in f[i]:(n.years-1)){
    logit(p[i,t]) <- lp0[t] + alpha[sta[i]]
  } #t
} #i

for (t in 1:(n.years-1)){
  phi[t] ~ dunif(0, 1)          # Survival probability
  lp0[t] <- logit(p0[t])        # Recapture probability
  p0[t] ~ dunif(0, 1)
  kappa[t] ~ dunif(0, 1)        # Predetermined residency probability
  pi[t] ~ dunif(0, 1)           # Residency probability
  rho[t] ~ dunif(0, 5)          # Recruitment
}

# Random station effect for recapture probability
for (j in 1:n.sta){
  alpha[j] ~ dnorm(0, tau.p)
}
sigma.p ~ dunif(0, 10)
tau.p <- pow(sigma.p, -2)

# Random observer effect for survey observation model
for (i in 1:n.obs){
  omega[i] ~ dnorm(0, tau.om)
}
tau.om ~ dgamma(0.001,0.001)
sd.om <- pow(tau.om, -0.5)

# Overdispersion
for (k in 1:n.counts){
  eps[k] ~ dnorm(0, tau.eps)
}
tau.eps ~ dgamma(0.001,0.001)
sd.eps <- pow(tau.eps, -0.5)

# Start-up (novice) effect for survey
eta ~ dnorm(0, 0.001)
```

```
# BBS data (state-space model)
# Model for initial abundance in each stratum: uniform priors
for (s in 1:n.str){
  N[s,1] ~ dcat(pNinit)
}
# Process model over space and time: our model of population dynamics
for (s in 1:n.str){
  for (t in 2:n.years){
    S[s,t] ~ dbin(phi[t-1], N[s,t-1])
    R[s,t] ~ dpois(rho[t-1] * N[s,t-1])
    N[s,t] <- S[s,t] + R[s,t]
  } #t
} #s
# Observation model: overdispersed Poisson
for (k in 1:n.counts){
  log(lambda[k]) <- log(N[str[k],yr[k]]) + omega[obs[k]] + eta * I[k] + eps[k]
  C[k] ~ dpois(lambda[k])
}

# MAPS data (CJS model with state-space likelihood)
for (i in 1:n.ind){
  # Model for initial state
  for (t in 1:f[i]){
    z[i,t] <- 1
  } #t
  # Model for residency state
  u[i] ~ dbern(pi[f[i]])
  r[i] ~ dbern(u[i] * kappa[f[i]])                    # Observation model for residency
  # State process
  for (t in (f[i]+1):n.years){
    z[i,t] ~ dbern(z[i,t-1] * phi[t-1] * u[i])
    # Observation process
    y[i,t] ~ dbern(p[i,t-1] * z[i,t])
  } #t
} #i

# Derived parameters
for (s in 1:n.str){
  for (t in 1:n.years){
    wN[s,t] <- area[s] * N[s,t] / sum(area)
  } #t
  trend[s] <- pow(N[s,n.years] / N[s,1], 1/(n.years-1))       # Strata-spe.trend
} #s
for (t in 1:n.years){
  wIndex[t] <- sum(wN[1:n.str,t])                             # Weighted abundance index
}
trend.com <- pow(wIndex[n.years] / wIndex[1], 1/(n.years-1)) # Combined trend
}
")
```

We need to give suitable initial values for the latent state variables z and u. For the former, we use the function `zInit` (Section 4.5.1.1), and for the latter, we provide a 1 for all individuals. Further, we specify reasonable values for the initial abundance in each stratum and for recruitment.

```
# Initial values
ny <- jags.data$n.years
ns <- jags.data$n.str
uini <- rep(1, nrow(catbird$y))
inits <- function(){
  Nini <- matrix(NA, nrow=ns, ncol=ny)
  Nini[,1] <- rpois(ns, 17)
  list(z=zInit(catbird$y), u=uini, N=Nini, rho=runif(ny-1, 0.3, 1.3))
}


# Parameters monitored
parameters <- c("phi", "rho", "pi", "p0", "kappa", "eta", "alpha", "sd.om", "sd.eps", "wIndex", "trend",
  "trend.com", "N")

# MCMC settings
ni <- 70000; nb <- 20000; nc <- 3; nt <- 25; na <- 2000

# Call JAGS from R (ART 191 min) and check convergence
out1 <- jags(jags.data, inits, parameters, "model1.txt", n.iter=ni, n.burnin=nb, n.chains=nc,
  n.thin=nt, n.adapt=na, parallel=TRUE)
traceplot(out1)
```

19.6 RESULTS

The model runs quite slowly, but after about 3 h, the chains have converged.

```
print(out1)
                 mean     sd      2.5%      50%    97.5% overlap0     f  Rhat n.eff
phi[1]          0.804  0.137     0.484    0.828    0.991    FALSE 1.000 1.001  2576
phi[2]          0.719  0.157     0.409    0.727    0.978    FALSE 1.000 1.002   987
[... output truncated ...]
phi[15]         0.561  0.151     0.323    0.539    0.916    FALSE 1.000 1.004   511
phi[16]         0.707  0.211     0.269    0.759    0.985    FALSE 1.000 1.017   153
rho[1]          0.266  0.161     0.029    0.245    0.629    FALSE 1.000 1.001  2296
rho[2]          0.280  0.177     0.017    0.265    0.659    FALSE 1.000 1.003   916
[... output truncated ...]
rho[15]         0.484  0.187     0.098    0.493    0.830    FALSE 1.000 1.004   594
rho[16]         0.284  0.223     0.010    0.224    0.777    FALSE 1.000 1.016   151
pi[1]           0.322  0.075     0.203    0.312    0.496    FALSE 1.000 1.000  6000
pi[2]           0.519  0.106     0.354    0.507    0.770    FALSE 1.000 1.004   569
[... output truncated ...]
pi[15]          0.405  0.088     0.262    0.396    0.604    FALSE 1.000 1.001  5991
pi[16]          0.538  0.126     0.335    0.524    0.825    FALSE 1.000 1.001  2668
p0[1]           0.494  0.118     0.269    0.492    0.728    FALSE 1.000 1.002  1002
p0[2]           0.301  0.083     0.163    0.294    0.483    FALSE 1.000 1.000  6000
[... output truncated ...]
p0[15]          0.309  0.090     0.158    0.300    0.505    FALSE 1.000 1.002  2213
p0[16]          0.296  0.163     0.117    0.245    0.764    FALSE 1.000 1.018   171
kappa[1]        0.372  0.098     0.198    0.366    0.573    FALSE 1.000 1.000  6000
kappa[2]        0.304  0.071     0.179    0.299    0.450    FALSE 1.000 1.003   755
[... output truncated ...]
kappa[15]       0.236  0.061     0.133    0.230    0.371    FALSE 1.000 1.000  6000
kappa[16]       0.283  0.075     0.158    0.276    0.453    FALSE 1.000 1.002  1087
eta             0.008  0.054    -0.096    0.008    0.112     TRUE 0.561 1.001  2773
alpha[1]       -0.063  0.245    -0.529   -0.070    0.417     TRUE 0.607 1.000  3298
alpha[2]       -0.111  0.731    -1.567   -0.103    1.325     TRUE 0.561 1.000  6000
[... output truncated ...]
alpha[37]      -0.199  0.713    -1.687   -0.177    1.192     TRUE 0.607 1.001  2609
alpha[38]      -0.098  0.718    -1.552   -0.075    1.314     TRUE 0.550 1.001  6000
sd.om           0.672  0.046     0.589    0.669    0.768    FALSE 1.000 1.000  6000
sd.eps          0.294  0.014     0.268    0.294    0.321    FALSE 1.000 1.000  3926
wIndex[1]      19.198  1.498    16.393   19.158   22.287    FALSE 1.000 1.003   850
wIndex[2]      20.443  1.586    17.508   20.357   23.706    FALSE 1.000 1.003   747
[... output truncated ...]
wIndex[16]     20.775  1.694    17.676   20.712   24.341    FALSE 1.000 1.002  1261
wIndex[17]     20.448  1.775    17.177   20.363   24.195    FALSE 1.000 1.003   795
trend[1]        1.020  0.011     0.998    1.020    1.042    FALSE 1.000 1.000  6000
trend[2]        0.998  0.009     0.981    0.998    1.016    FALSE 1.000 1.000  1846
[... output truncated ...]
trend[8]          1.010  0.023   0.965    1.010    1.055    FALSE 1.000 1.000  4797
trend[9]          0.961  0.028   0.917    0.958    1.000    FALSE 1.000 1.000  1245
trend.com         1.004  0.006   0.9993   1.004    1.015    FALSE 1.000 1.000  6000
N[1,1]         25.999  4.257    19.000   26.000   35.000    FALSE 1.000 1.003   717
N[2,1]         23.695  3.433    18.000   23.000   31.000    FALSE 1.000 1.009   280
[... output truncated ...]
N[8,17]        36.925 14.075    16.000   35.000   70.000    FALSE 1.000 1.001  4363
N[9,17]         1.082  0.291     1.000    1.000    2.000    FALSE 1.000 1.000  6000
```

The estimates of adult apparent survival of the gray catbird varied strongly over time (Fig. 19.4) and averaged 0.59 (arithmetic mean of the annual estimates, credible interval (CRI): 0.54−0.63). Recruitment was also variable over time (Fig. 19.4) and averaged 0.42 (CRI: 0.37−0.48). The resulting area-weighted abundance index for the New England/Mid-Atlantic region was fairly stable over the study period, though it increased slightly in the first half and then declined again somewhat (Fig. 19.4).

The mean residency probability was 0.59 (CRI: 0.54−0.64), indicating that a small majority of the marked individuals were residents. This estimate and its uncertainty are numerically nearly identical to those of apparent survival, but this is of course a mere coincidence.

The stratum-specific and overall abundance trends are shown in Fig. 19.5. Estimated trends in most strata are not different from 1, indicating that gray catbird populations in most strata were stable. The density strips of the abundance trends in strata 3 and 9 look patchy, indicating that the corresponding posteriors have several peaks. Multiple peaks of a

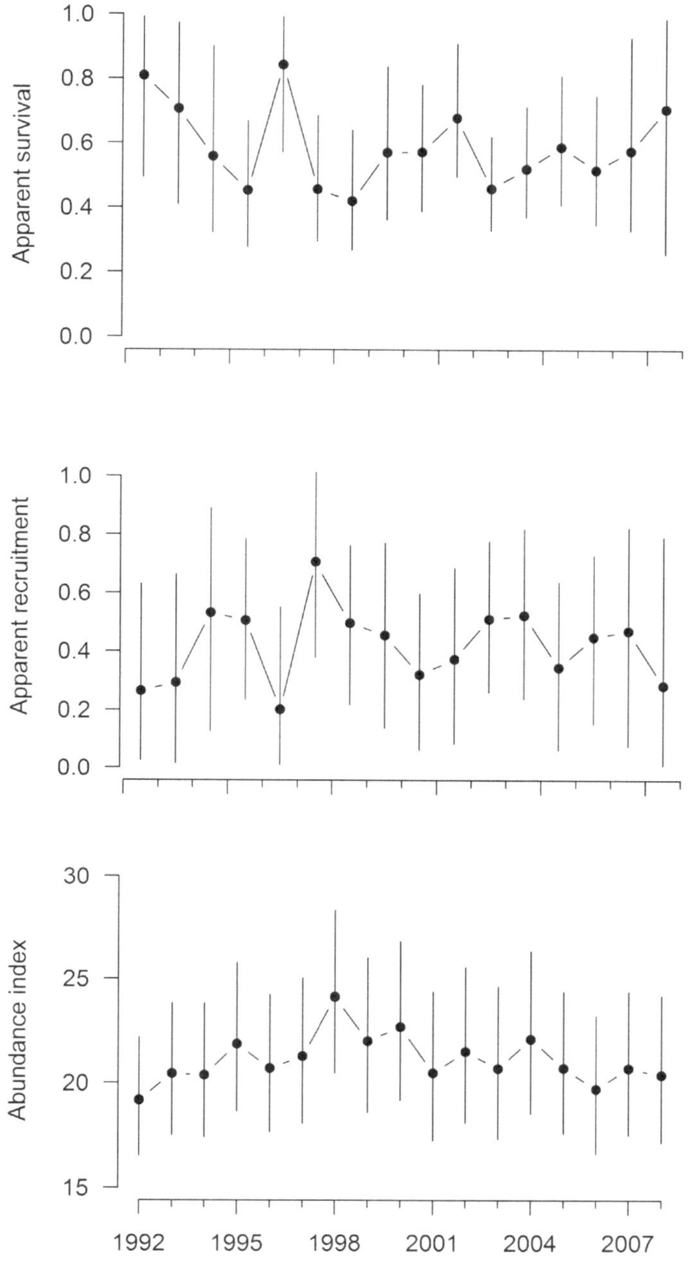

FIGURE 19.4

Posterior means and 95% credible intervals of adult apparent survival (top), recruitment (middle), and the area-weighted abundance index (bottom) of gray catbirds in the New England/Mid-Atlantic region.

posterior are usually an indication that the chains have not converged, but here the reason is different. In both stratum 3 and 9 the estimated abundance indices are very small (see Fig. 19.6). As the indices are discrete and the abundance trend calculated as the ratio of successive indices, there are only a few possible values for that ratio, and hence the posterior is essentially discrete. The area-weighted abundance trend across the complete New England/Mid-Atlantic region also suggests a stable gray catbird population.

Besides the area-weighted annual index of abundance (Fig. 19.4), we can compute the stratum-specific annual indices of abundance. This gives us the possibility to compare these estimates with the average (across routes) counts (Fig. 19.6). In most strata, the mean counts and estimated population indices correspond quite well, but the estimates in some strata are strikingly larger than the counts. If there is such a difference, the number of surveyed routes per year is typically low

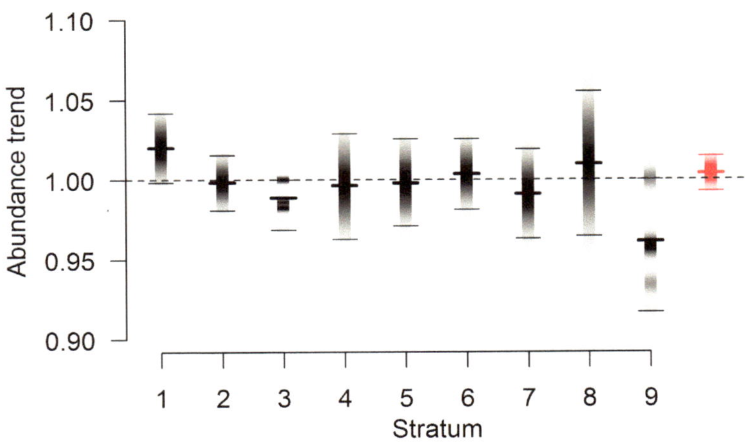

FIGURE 19.5

Stratum-specific values (black) and regional total (red) of abundance trends of gray catbirds in the New England/Mid-Atlantic region between 1992 and 2008. The darkness of the strips is proportional to the posterior densities. The short horizontal lines show the 95% credible intervals and the posterior means (think lines). The horizontal dashed line represents a stable population. (See text for why densities in stratum 3 and 9 appear multimodal.)

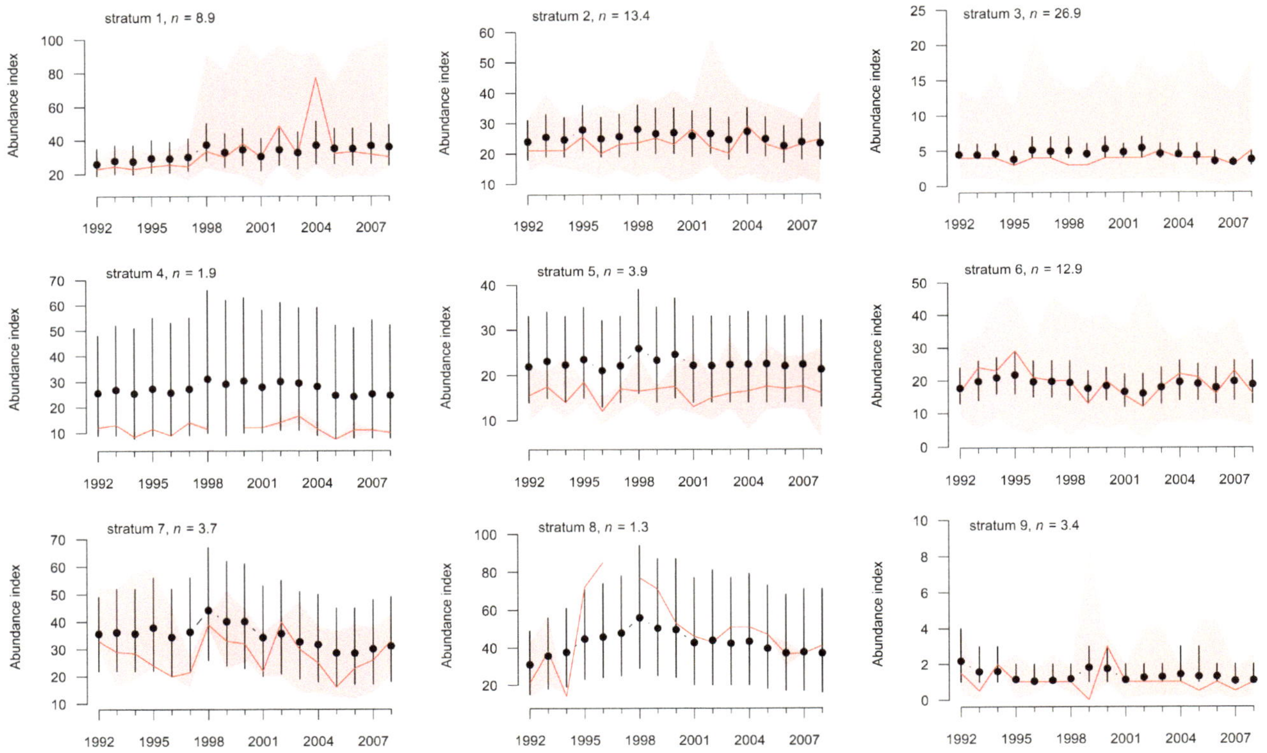

FIGURE 19.6

Posterior means and 95% credible intervals of gray catbird abundance in the nine strata of the New England/Mid-Atlantic region (black dots and vertical lines). Red lines show the mean counts averaged over all routes of a stratum, and the pink area gives the 10% and 90% percentiles of these counts. For each stratum, the average number of surveyed routes per year is given (*n*).

(Fig. 19.6, e.g., stratum 4, 5, 7, 8). We notice that for a large number of routes in a stratum (i.e., a large sample size; e.g., stratum 1 to 3, 6), there is more variability in the counts (large reddish area in Fig. 19.6) but more precision in the abundance estimates and vice versa.

19.7 DISCUSSION

The focus of this case study is the population dynamics of gray catbirds across a large spatial scale. To this end, we combined data collected at many smaller-scale, local study areas (BBS routes and MAPS banding stations), which enabled us to infer abundance dynamics at the desired large spatial extent and at multiple spatial scales. The ability to understand population dynamics at multiple spatial scales is valuable both scientifically and from a management perspective, but so far, most IPMs are developed for a single spatial scale only, and this prevents inferences at multiple scales.

The ability to make inferences about multiple spatial scales depends on how space is represented in a model. Here we assumed that the underlying demographic processes do not vary spatially and just included spatial random effects for the recapture probabilities. Differences in the dynamics among strata at the smallest spatial scale are then due to demographic stochasticity that imposes temporal variability in demographic processes, the magnitude of which depends on population size. The dynamics at the scale of the entire region may differ from those at the scale of the stratum because of differential weighing of the different strata based on their area.

Generally, spatial scaling can be done in at least two ways. The first is to regard the spatial units from where the data originate as representative of a larger area. The scaling up is then a matter of weighting by the area of interest. This is the simplest approach and the one we have adopted here. A second possibility is spatial prediction of the abundance and demographic rates across the entire area of interest. The spatial scaling of variables of interest is then a matter of summation. Spatial prediction is possible by using models that accommodate spatial autocorrelation, via the modeling of spatial covariates, or both (Guélat and Kéry, 2018, Chapters 3 and 9 in Kéry and Royle, 2021). This second approach appears to be a more elegant solution and one that avoids having to define any artificial spatial boundaries, such as in the definition of the strata in our case study. So far, we have never seen an IPM that implemented this second possibility, but the necessary elements such as smooth mapping of survival probabilities based on spatial autocorrelation and or covariates are available (Royle and Dubovsky, 2001; Saracco et al., 2010, 2012, Chapter 3 in Kéry and Royle, 2021). A limitation of both approaches is that scaling down is only possible down to the spatial scale of the smallest study area, but not to, say, the scale of an individual territory.

An exciting new spatial IPM that allows smooth spatial scaling at arbitrary grains has been developed by Chandler et al. (2018). In order to scale down to single individuals, the population model is not a matrix population model as in classical IPMs that aggregates individuals over some larger space ("the population") but rather an individual-based model that describes, among other things, the location of each individual. Individuals appear, move, and disappear in space according to latent spatial point process models. Observations of these latent processes are formulated for spatial capture-recapture data and distance sampling data. Chandler et al. (2018) applied their model to data on Canada warblers (*Cardellina canadensis*) and modeled density and demographic rates as a function of elevation. Survival increased with elevation, while recruitment was highly variable in space but did not respond to elevation. A similar spatial IPM is the focus of Chapter 21.

Generally, spatial IPMs can describe space in either a discrete or a continuous manner. In the discrete-space formulation, the entire area of inference is stratified in spatial units. These can be regularly or irregularly shaped, and there may be many or just a few (Ahrestani et al., 2017; Plard et al., 2020a; Zhao, 2020). The underlying population model is written in terms of the *number* of individuals per spatial unit, and these units may (Plard et al., 2020a; Zhao, 2020) or may not (Ahrestani et al., 2017) be demographically connected by dispersing individuals. The spatial IPMs that describe space in a continuous manner do not need an "arbitrary" gridding of space (at least conceptually), and the underlying population model is written in terms of the *locations* of each individual in an individual-based model (Chandler and Clark, 2014; Chandler et al., 2018; Sun et al., 2019). These models are computationally very demanding such that they can hardly be used for inferences about very large spatial scales such as whole countries with currently available computational power. The discrete space formulation converges to the continuous space formulation when the spatial units become quite small. Indeed, computationally even a continuous-space model is implemented in practice for a discrete grid of discrete spatial units, i.e., for a fine pixellation of space.

One challenge for the type of IPM adopted here is that the spatial scale of the population count data may not match that associated with the capture-recapture data. This is sometimes called spatial misalignment; see Pacifici et al. (2019) for an excellent overview for ecologists. Adult survival in the gray catbird model is apparent, but it is not entirely clear to which area it refers. Obviously, and perhaps primarily, it refers to some implied "zone of influence" around the MAPS banding stations, i.e., an area of a few hectares. On the other hand, the population model is defined for the BBS data, and there the spatial unit is a BBS route, which surveys an area of about 39 km $\times$ 2 $\times$ 0.4 km $=$ 31 km^2. The resulting estimate from the IPM is a weighted average (Section 6.2.2), hence we are uncertain as to which spatial scale the estimate of adult survival refers. Generally, if capture-recapture data and count data are combined in an IPM and have very different spatial references, the common demography assumption will be violated because the underlying apparent survivals are different (Section 7.2.3). For the gray catbird study, we may argue that apparent survival closely approaches true survival because breeding dispersal is weak as in many bird species (Greenwood and Harvey, 1982). As a consequence, spatial upscaling is easily possible, as true survival is invariant to spatial scales, and there is no problem with the violation of the common demography assumption.

The explicit modeling of different spatial scales to which the sampled data refer to is possible when they are spatially referenced, e.g., when spatial capture-recapture and distance-sampling data are collected. Spatial capture-recapture models estimate dispersal kernels, i.e., some sort of average dispersal distances, in addition to survival (Ergon and Gardner, 2014; Royle et al., 2014; Schaub and Royle, 2014; Chandler et al., 2018; Efford and Schofield, 2020). Spatial scaling of survival, and emigration and immigration, then becomes possible. Furthermore, the location of the individuals (or of their home ranges) is estimated and hence population density for any desired spatial scale can be computed. This then allows accurate spatial alignment with, and integration of, the information on population density coming from distance-sampling data (Chandler et al., 2018).

A different type of IPM for inference about large spatial scales has been developed by Robinson et al. (2014) for 17 British breeding bird species. The data were also collected at multiple sites in the course of different monitoring programs. The breeding bird surveys are based on a territory mapping protocol and on distance sampling along transects. From the number of breeding individuals recorded per site a large-scale population index was estimated using a Poisson GLM with year and site as fixed effects (see Chapter 1 in Kéry and Royle, 2021), which was then used as the population counts in the state-space model part of the IPM. Components of productivity were based on ringing data (number of ringed nestlings in a brood) and on data from the nest record scheme (Crick et al., 2003). Observers repeatedly visit nests and record their state and content, allowing for the estimation of nest survival (Mayfield, 1961; Rotella et al., 2000; Dinsmore et al., 2002). Survival was estimated from nationwide dead recoveries of ringed birds. Because dead recoveries are received from anywhere and not just in a specific study area as are life re-encounters, true survival is usually assumed to be estimated from such data (Burnham, 1993). None of the estimated demographic rates have any spatial restriction, i.e., they are associated with the very large spatial scale of Britain of about 230,000 km^2. Because of the lack of spatial structure, no spatial scaling is possible in this IPM. Hence, this large-scale IPM is a nonspatial IPM in contrast to the spatial IPM of Ahrestani et al. (2017) showcased in this chapter.

Recruitment as defined in the gray catbird IPM incorporates different demographic processes such as reproduction, first-year survival, breeding propensity, and immigration. If we want to disentangle these processes, additional data such as nest records, age ratios, or capture-recapture of nestlings must be available. Nevertheless, recruitment is a useful quantity to estimate because it still allows understanding of whether population size at large spatial scales is primarily driven by losses (the complement of apparent survival) or by gains. Retrospective analyses such as those shown in Chapter 9 could also be carried out for this purpose. Moreover, the estimated recruitment is analogous to the "gain" quantity estimated in capture-recapture (Jolly-Seber model, reverse time models (Williams et al., 2002)) or population dynamics models (Dail and Madsen, 2011; Hostetler and Chandler, 2015; Bellier et al., 2016).

The case study in this chapter uses a process model formulated for the aggregation of individuals within a "population," and spatial scaling is possible because observations originate from several locations. Technically, this IPM is similar to classical IPMs seen many times in this book and is computationally not particularly demanding. Yet the spatial scaling has limits (scaling down is not possible) and remains an approximation. For a full and exact spatial scaling, the process model would have to be written for (unaggregated) individuals, i.e., must be an individual-based point process model (see Chapter 21 and Chandler and Clark, 2014; Chandler et al., 2018; Sun et al., 2019). These models are computationally much more demanding but have the benefit that unrestricted spatial scaling is possible and component data sampled at different spatial scales can be included without violating the common demography assumption. Such spatial IPMs are still very rare (Chandler and Clark, 2014; Chandler et al., 2018; Sun et al., 2019), but we expect that the inclusion of space is one of the major future developmental avenues for IPMs. There is vast literature on spatial statistical models, and plenty of its ideas and concepts await incorporation into IPMs (e.g., Cressie and Wikle, 2011; Baddeley et al., 2016; Wikle et al., 2019; see also Chapter 9 in Kéry and Royle, 2021, for a short and gentle introduction to the topic of spatial modeling).

CHAPTER

KESTREL

20

CHAPTER OUTLINE

IPM ingredients: time series of counts (monitoring data), replicated point counts (atlas data), dead-recovery data
Location: Switzerland
Spatio-temporal scale: 16,000 km^2, 19 years

20.1 INTRODUCTION

Space can safely be predicted to become a big topic for integrated population models (IPMs) as it has in many other branches of ecological statistics over the last 20 years or so. In the case of IPMs, one reason for space becoming more important, of course, is that its inclusion increases our understanding of population dynamics—populations are typically spatially structured, and hence, their dynamics can be better understood when space is included. However, the main reasons for the rise of space that can be predicted for IPMs are arguably the increasing availability of spatially replicated data and great progress in computational power and understanding in the treatment of space in statistics. Hence, we now see increasing numbers of locations from which data on the demography of a species become available. While most published IPMs have dealt with a single population only and a very few with a small handful of populations in the context of a multistate model as in Chapter 18, we now increasingly see moderate or even massive spatial replication, where population counts or even capture-recapture and productivity data are available from more than a single site or location (Morrison et al., 2016; Ahrestani et al., 2017; Zhao et al., 2019a, 2019b; Zhao, 2020). The treatment of space in the statistical sciences has made great progress over the past 30 years (Illian et al., 2008; Cressie and Wikle, 2011; Banerjee and Gelfand, 2014; Cressie, 2015; Baddeley et al., 2016; Wikle et al., 2019), and this trickles into the modeling of distribution, abundance, and population dynamics in population ecology (e.g., Royle and Dubovsky, 2001; Saracco et al., 2010; Péron et al., 2011, Chapter 3 in Kéry and Royle, 2021), though as yet it has hardly done so in the field of IPMs (but see, e.g., Chandler and Clark, 2014; Zhao, 2020).

Integrated Population Models. https://doi.org/10.1016/B978-0-12-820564-8.24001-3

When faced with the task of building a demographic model with data from multiple locations, an important question is how to combine these data. We argue that this should be done in the same way as spatial statisticians have been dealing with space for some time now (at least since the first edition of Cressie's (2015) classic book in 1993): either by invoking a latent spatial field, i.e., a set of correlated random variables in two dimensions, or alternatively by adopting a more complex, process-based model where spatial autocorrelation arises as a function of demographic processes. Here, we introduce the former type of spatial modeling in the context of an IPM that we develop on a large spatial domain. This allows us to make an important step toward that grand goal in the population ecology of *landscape demography*: viz., the study of demography at a large or landscape scale (Gurevitch et al., 2016). We focus on large-scale spatial modeling, which contrasts with the more local-scale point-pattern approach to abundance models, including IPMs, adopted by Chandler and Clark (2014), which is the focus of Chapter 21.

One of the first attempts toward this grand goal was made by Ahrestani et al. (2017), who modeled the demography of gray catbirds (*Dumetella carolinensis*) and wood thrushes (*Hylocichla mustelina*) in a large study area in the US mid-Atlantic region (see Chapter 19) using an as yet highly underutilized source of information for the fitting of IPMs—data from large-scale biodiversity monitoring programs. They chose a simple representation of space by defining nine large strata, assuming that all sites within a stratum had identical demographic rates, and they estimated stratum-specific parameters as fixed effects, i.e., completely unrelated constants. Zhao et al. (2019a, 2019b) chose a similar approach in their spatial, space-stratified IPMs, although they estimated stratum-level effects as random and added in spatio-temporal covariates. All three are important papers conceptually because they point the way toward what we believe will be a critical part of the future of IPMs—large areas and many "local" populations—because we will increasingly be using data from large-scale monitoring programs. Biodiversity monitoring has become a huge enterprise worldwide, and it produces Big Data about the occurrence, abundance, and population dynamics of many species. Most of this is as yet completely unused for demographic modeling, though there are notable exceptions including the works of Wikle (2003), Hooten et al. (2007), Hooten and Wikle (2008), Pagel and Schurr (2012), Pagel et al. (2014), Ahrestani et al. (2017), and Zhao et al. (2017, 2019a, 2019b).

The representation of space in the spatially stratified IPMs of Ahrestani et al. (2017) and Zhao et al. (2019a, 2019b) is rather crude, and the constancy assumption for parameters is unlikely to hold within the entirety of a stratum. In addition, the choice of fixed effects for stratum-level parameters by Ahrestani et al. (2017) requires at least moderate sample sizes in terms of locations within each stratum and thus runs counter to a natural desire to have small strata to achieve increased spatial resolution in an analysis. Finally, in their approach, any spatial autocorrelation is ignored, i.e., that almost universal observation in nature that things closer in space tend to be more alike than things farther apart (Tobler, 1970). Spatial autocorrelation in population density and demographic parameters may arise broadly by two different mechanisms: either by the impact of environmental covariates that are not in the model and are themselves spatially structured or by biotic mechanisms such as dispersal or conspecific attraction (Guélat and Kéry, 2018). In parametric statistical models, spatial autocorrelation is usually addressed either by phenomenological models that simply describe how the similarity between two sites decreases with increasing intersite distance or else by process-based models where the similarity between two sites may arise as a function of the movement of individuals among sites, i.e., dispersal. Thus, we can adopt either a phenomenological or a process-based representation of spatial autocorrelation in a spatial demographic model (Cressie and Wikle, 2011, Chapter 9 in Kéry and Royle, 2021).

An improved approach over that involving the crude spatial strata in studies by Ahrestani et al. (2017) and Zhao et al. (2019a, 2019b) might be to make spatial strata much smaller such that we obtain a near-continuous representation of space and then describe the spatial autocorrelation among these spatial units either phenomenologically or by a population model involving dispersal. Here, we choose the former, make strata very small, treat stratum-level parameters as random effects, and allow for spatial autocorrelation among the set of stratum-level parameters. We do this by choosing a (quasi-) continuous representation of space and by specification in the IPM of a spatial random field for each of the main demographic parameters, i.e., a set of random effects in space that may be correlated along their two dimensions. A spatial random field is a customary representation of space in spatial statistics (Diggle et al., 1998; Cressie and Wikle, 2011; Cressie, 2015). In addition, we use a common dimension-reduction technique known as *basis functions* or (two-dimensional) *splines* (Crainiceanu et al., 2005; Gimenez et al., 2006; Collier et al., 2012; Hefley et al., 2017a; Wood, 2017). This considerably alleviates the substantial computational burden typical of parametric models of spatial autocorrelation (Johnson et al., 2013; Heaton et al., 2019, Chapter 9 in Kéry and Royle, 2021).

Here, we develop a spatial IPM with a continuous, spline-based representation of space by fusing three data sets of the Eurasian kestrel (*Falco tinnunculus*, Fig. 20.1) in Switzerland. This study can be viewed as a spatial generalization of the type of large-scale IPM developed by Fay et al. (2019) for partially the same data. We integrate three sources of information about the demography of Swiss kestrels in the northwestern quadrant of Switzerland—longitudinal population count data from 102 1 km^2 survey quadrats in the Swiss breeding bird survey "Monitoring Häufige Brutvögel" (MHB)

FIGURE 20.1

Three nestling Eurasian kestrel (*Falco tinnunculus*) at the entry of a nest box for a barn owl (*Tyto alba*) that had been installed in a barn in Biel-Benken, Switzerland, by the second author in the 1990s (photograph by Andi Meier).

during 1999–2017, cross-sectional, or snapshot, population counts from 574 1 km^2 quadrats in the latest Swiss breeding bird atlas during 2013–15 (Knaus et al., 2018), and dead-recovery data from seven combined ringing sites throughout the studied region. The first and third data source, but not the second, were also used by Fay et al. (2019).

The kestrel is a small falcon that is the most widespread and common member of its genus over large parts of Europe, including Switzerland. It feeds mostly on rodents and small birds and occurs in a wide variety of open landscapes. In Switzerland, it occurs in all nonforest habitats below 3000 m (Knaus et al., 2018). Most kestrels start reproducing at age 1 or 2. They nest in the tree nests of other species, on ledges in cliffs, and often on buildings, especially when nest boxes are provided (Fay et al., 2019).

Our continuous representation of space neatly solves one ubiquitous challenge in spatial analyses—spatial or temporal mismatch among multiple data sets used in a single model, or what is known as *change of support* (see Pacifici et al., 2019 for an excellent introduction to this topic for ecologists). This denotes the situation where a spatio-temporal process is not observed at exactly the same locations and times for all the data sets from which we infer it. Spatial mismatch among combined data sets is arguably the rule for any spatial IPM, since we will very rarely be able to collect the different types of data at exactly the same set of locations. Adopting a continuous representation of space and specifying in the model an underlying, continuous random field for one or more parameters of interest enables us to let each data set inform the field estimates at exactly the right location. Pacifici et al. (2019) call this approach the "shared model."

In addition to the reconciliation of spatially mismatched data sets in an IPM, this study provides two illustrations of the reconciliation of a temporal mismatch (see also Section 6.6.3). The first instance of such a mismatch is that the years during which the population count data were collected overlap only partially. While the monitoring data were collected annually during 1999–2017, the atlas data were collected in only 1 year during 2013–15 (we ignored the final atlas field season of 2016; Knaus et al., 2018). Thus, in our model, we will show how we can "plug in" the atlas data at the "right place in time" so as to reconcile the temporal scale of the two data sets in terms of the underlying demographic process.

The second example of reconciliation of a temporal mismatch between the monitoring and atlas data sets concerns their observation processes. The actual field protocol was identical in both schemes, i.e., two or three surveys were conducted during a breeding season in every 1 km^2 survey quadrat. However, here we used the same version of the monitoring data that were used in the IPM by Fay et al. (2019). In this variant of the MHB, data results from replicate surveys within a breeding season were aggregated to a single total count per quadrat and year. In principle, we also could have obtained the raw, disaggregated data for the MHB but chose not to do so because this type of temporal mismatch in the observation model of a species distribution model (and our spatial IPM *is* a kind of demographic species distribution model; see also Pagel and Schurr, 2012) is quite common, and its reconciliation is in fact quite trivial. Specifically, we will resolve it by defining the aggregate detection probability in the monitoring data as a function of the latent survey-level detection probability (p) and the known number of surveys (n) per quadrat assuming independent surveys, where, in the atlas data with observations available at the level of each individual survey, p is estimable directly. See also Section 10.3 in Kéry and Royle (2021) for another example of this type of an integrated species distribution model.

In this chapter, we greatly extend the spatial IPMs of Ahrestani et al. (2017) and Zhao et al. (2019a, 2019b) as well as the nonspatial, large-scale IPM on Swiss kestrels by Fay et al. (2019). We integrate three types of data collected at a total of 683 study locations by specification of latent random fields for all three key model parameters via two-dimensional splines as in a generalized additive model (GAM; Hastie and Tibshirani, 1990; Hefley et al., 2017a; Wood, 2017). This allows us to fully reconcile the spatial mismatches that exist among these data sets, deal with spatial autocorrelation in a computationally efficient manner, and interpolate the demographic parameters in between all study sites (i.e., the 683 locations) in an exciting instance of *landscape demography* (Gurevitch et al., 2016). We believe that our study, along with so far only a handful others that include Morrison et al. (2016), Ahrestani et al., (2017), Zhao et al. (2019a, 2019b), Pacifici et al. (2019), Sun et al. (2019) and Zhao (2020), will provide a blueprint for many future spatial IPMs that exploit that hitherto virtually untapped data resource for the fitting of an IPM: data from large-scale monitoring programs.

20.2 DATA SETS

We combine three different data sets to model the spatial population dynamics of Swiss kestrels in a northwestern quadrant of Switzerland spanning nearly 16,000 km^2 (Fig. 20.2). The first data set stems from the Swiss breeding bird survey "Monitoring Häufige Brutvögel" (MHB; Schmid et al., 2004), which is described in Kéry et al. (2005), Royle and Kéry (2007) and Royle et al. (2007) and in many chapters in Kéry and Royle (2016, 2021). In this scheme, 267 1 km^2 quadrats have been surveyed 2–3 times in every breeding season (mid-April to early July) since 1999 using territory (or spot) mapping along a quadrat-specific irregular route measuring, on average, 4–6 km in length. High-elevation quadrats are

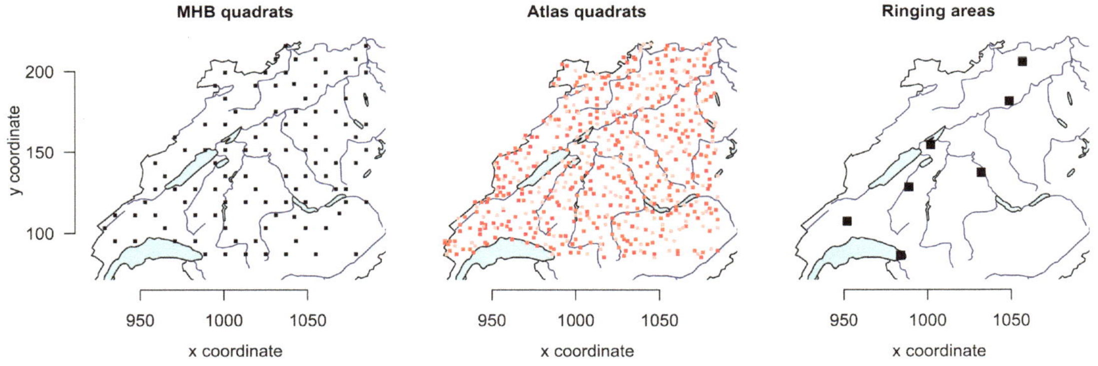

FIGURE 20.2

Spatial configurations of the spatial integrated population model and the utilized data sets. Shown is the northwestern quadrant of Switzerland (15,734 km^2). Survey quadrats in the longitudinal Monitoring Häufige Brutvögel breeding bird survey data ($n = 102$, 1999–2017) and in the cross-sectional atlas project ($n = 574$, 2013–15) are quadrats with nominal 1 km^2 area, with atlas quadrats surveyed in different years depicted in different colors. Ringing areas ($n = 7$, 2002–16) are represented by central coordinates of their nest boxes (and for this reason, one of them appears to be in Lake Geneva). Coordinates are in kilometric units of an undisclosed projection.

only surveyed twice due to logistical challenges with snow cover and avalanche hazards. Successive surveys within the same quadrat and year are typically spaced in time by only 2–3 weeks, so in our analysis, we make the closure assumption for each quadrat and breeding season. After completion of the annual surveys, all bird detections are overlaid, and clusters of detections or detections sufficiently far from other detections (given the typical territory sizes of a species) are declared to belong to the same territory. This procedure yields an observed count of territories for each quadrat and year, which is what we used as our longitudinal population count data from the 102 quadrats that lay within the chosen region in northwestern Switzerland.

The second data set stems from the third Swiss breeding bird atlas (Knaus et al., 2018). Field work was conducted in a total of about 2300 1 km^2 quadrats in a single year from 2013 to 2016 using a field protocol identical to that used in the MHB. However, we used the disaggregated, i.e., survey-specific, counts of kestrels in each quadrat rather than the total count as in the MHB data, for a total of 574 quadrats that lay within the chosen geographic region and that were surveyed during the first 3 years of the atlas field work (2013–15).

One important feature shared by both the MHB and the atlas data is that the species detections we utilize here are not spatially explicit in the sense of spatial capture-recapture (SCR; Efford, 2004; Borchers and Efford, 2008; Royle and Young, 2008; Royle et al., 2014); see also Chandler and Clark (2014) and Chapter 21 for a different example of SCR data being utilized in an IPM. That means that the effective sample area is unknown in both the MHB and the atlas data. Moreover, it will be different for each quadrat depending on the length and spatial configuration of the survey route, the hearing and visual detection abilities of the surveyor, and importantly, the home-range size of kestrels and typical distance over which an individual kestrel can be detected by eye or ear. Hence, counts in the MHB and the atlas data will certainly include at least *some* detections of kestrels that in reality reside mostly outside of the 1 km^2 nominal sample area, thus inflating the observed and estimated abundance with respect to that 1 km^2 area. On the other hand, the survey transects don't always cover the entire 1 km^2 sample area. Hence, the effective sample area will typically extend beyond the limits of the quadrat, looking something like a sunflower, and possibly include some "holes" that are effectively unsurveyed inside of the 1 km^2 survey quadrat; see also the discussion in Section 6.10 in Kéry and Royle (2016).

This should not affect the estimates of the vital rates obtained in our model (see below), but it is relevant for estimates of population size. For instance, when we want to add up the total population size over the entire region, we want to correct the quadrat-specific abundance estimates such that they are scaled to an area of exactly 1 km^2—otherwise, we will overestimate the total abundance when the effective sample area is greater than the nominal 1 km^2 sample area and underestimate it when it is smaller than the nominal sample area (see Section 6.10 in Kéry and Royle, 2016). It would be possible to estimate the effective sampling area by using the most highly resolved data, spatially speaking, i.e., the actual locations of every single detection of a kestrel, using a variant of an SCR model (see Royle et al., 2011, and Chapter 11 in Kéry and Royle, 2021) for just such a model applicable to MHB and atlas data. However, this is not the format of the data we use here.

So we must scale our abundance by the size of the effective sampling area (A) associated with the 1 km^2 sampling unit in the MHB and the atlas design, and we want to fit the following model for the abundance at site i (N_i).

$$N_i \sim Poisson(A\lambda)$$

Here, A is the effective sample area in 1 km^2 units and λ is the expected abundance per 1 km^2 unit (i.e., density); thus, we want to use A as an offset (Chapter 14 in Kéry, 2010). But the problem is that we don't know the value of A. Most likely, A will be > 1. For instance, if we detect inside of the survey quadrat kestrels that reside up to 250 m outside most of the times, we can imagine that the nominal unit is buffered by a strip of 250 m around it. This will yield $A = 1.5^2 = 2.25$ km^2. If we easily detect inside of the survey quadrat kestrels that have their home-range center up to 500 m outside of the survey quadrat, that imaginary buffer will be 500 m, and the resulting effective sample area will be $A = 2^2 = 4$ km^2 (always assuming there are no 'holes').

In our analysis below, we place an informative prior on A that covers all values that appear reasonable to us based on our personal knowledge of both the kestrel and the survey protocol. Hence, we choose a uniform prior on A from 0.8 to 4, where the former allows for more "holes" than "sunflower petals" and the latter accommodates extensive overlap of external kestrel territories within the survey quadrat. Our analysis will then average over the uncertainty in the value of A in a sort of automatic sensitivity analysis for a parameter that is not estimated but instead fixed or, in this case, constrained to lay within some reasonable range of values.

The third and final data set is the dead-recovery data for 24,561 individuals ringed in seven aggregate ringing areas during 2002–16. These data were collected by volunteer ringers that maintained nest boxes for kestrels and ringed nestlings and some adults. Some of the ringed kestrels were found dead and reported to the ringing office. Fig. 20.2 shows the sampling locations of all three data sets that were combined in our IPM.

20.3 POPULATION MODEL

Our population model for the kestrel has a spatial structure referring to the 1 km^2 quadrats in our study area. The model describes the dynamics of the number of breeding females in each quadrat independently from each other as a function of quadrat-specific probabilities of survival and absolute recruitment for which we will allow spatial dependence; see below. We assume a prebreeding "census" model and distinguish between two classes of females, "survivors" and "recruits." Survivors in a quadrat at time t are females that were already in the pool of individuals in that quadrat at time $t-1$. Recruits at time t are new breeding females in a quadrat: they are either local recruits (i.e., that were born in that quadrat and have survived) or immigrants (i.e., born outside a given quadrat).

The life-cycle graph is shown in Fig. 20.3. The number of survivors in year $t+1$ in quadrat i is a function of the total number of females present in quadrat i in year t, while the number of recruits at time $t+1$ is specified as independent of the population size,

$$S_{i,t+1} \sim Binomial(\phi_i, N_{i,t}),$$

$$R_{i,t+1} \sim Poisson(\gamma_i), \text{ and}$$

$$N_{i,t} = R_{i,t} + S_{i,t},$$

where ϕ_i is annual adult survival and γ_i is absolute recruitment (i.e., the annual number of new breeding females) in quadrat i.

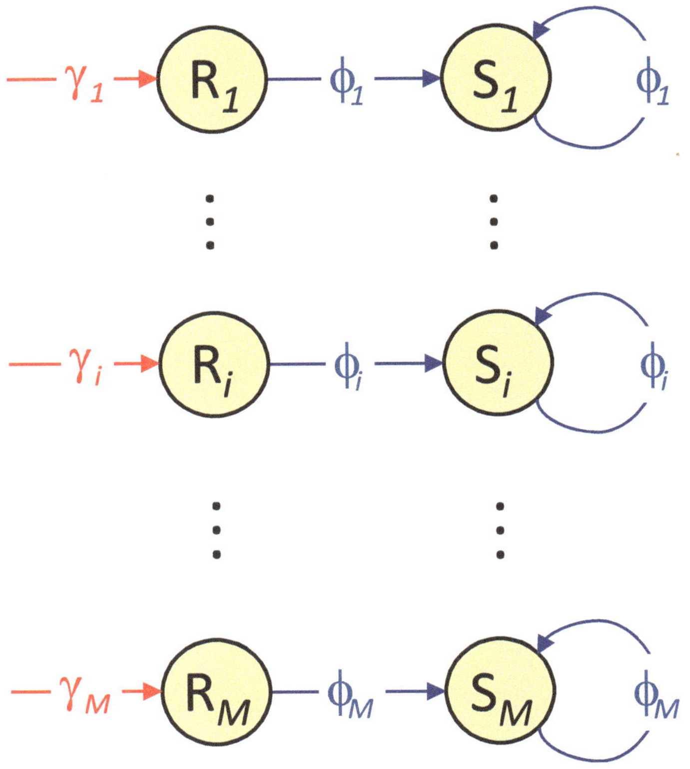

FIGURE 20.3

Life-cycle graph for the kestrel population model. In each of the M quadrats, the population is composed of two classes of individuals: recruits (R) and survivors (S). The demographic processes are absolute recruitment (γ, red) and adult survival (ϕ, blue); both are allowed to vary spatially and be spatially autocorrelated.

The population model is similar to that for the gray catbird (Section 19.3). Key differences are that for the kestrel adult survival is assumed to be true rather than apparent and that recruitment is specified in absolute terms and not per capita. Kestrels show short breeding dispersal (Vasko et al., 2011); hence, the population count data are informative about true adult survival, as are the dead-recovery data (Section 4.5.3). We have chosen absolute instead of per-capita recruitment because it allows unoccupied quadrats to be recolonized. Specification of absolute recruitment has the consequence that the population growth at a quadrat is density-dependent with an equilibrium density given by $N_i^* = \gamma_i/1 - \phi_i$.

20.4 COMPONENT DATA LIKELIHOODS

We load the data sets and produce an overview of its contents. We also conduct some data preparations that are necessary for the analyses. All areal data refer to units of 1 km^2, and coordinates are likewise in units of 1 km but for an undisclosed projection to protect the locational information of the survey quadrats.

```
library(IPMbook); library(jagsUI)
data(kestrel)
str(kestrel)

List of 4
 $ landData :'data.frame':      15734 obs. of  4 variables:
  ..$ x        : num [1:15734] 920 920 921 921 921 ...
  ..$ y        : num [1:15734] 86.3 87.3 85.3 86.3 87.3 ...
  ..$ elevation: int [1:15734] 1270 1283 1371 1342 1384 1292 1265 1219 ...
  ..$ lakes    : num [1:15734] 0 0 0 0 0 0 0 0 0 0 ...
 $ mhbData   :List of 3
  ..$ x     : num [1:102] 929 935 935 947 947 ...
  ..$ y     : num [1:102] 103.3 95.3 111.3 95.3 111.3 ...
  ..$ tcount: int [1:102, 1:19] NA 0 0 NA NA 0 0 0 0 0 ...
  .. ..- attr(*, "dimnames")=List of 2
  .. .. ..$ : NULL
  .. .. ..$ : chr [1:19] "1999" "2000" "2001" "2002" ...
 $ atlasData:List of 4
  ..$ x    : num [1:574] 923 929 947 953 953 ...
  ..$ y    : num [1:574] 87.3 87.3 119.3 103.3 127.3 ...
  ..$ year : num [1:574] 2015 2013 2013 2013 2015 ...
  ..$ count: num [1:574, 1:3] 0 0 0 1 1 0 0 0 0 2 ...
  .. ..- attr(*, "dimnames")=List of 2
  .. .. ..$ : NULL
  .. .. ..$ : chr [1:3] "count1" "count2" "count3"
 $ drData    :List of 3
  ..$ site         : num [1:24561, 1:3] 1 1 1 1 1 1 1 1 1 1 ...
  .. ..- attr(*, "dimnames")=List of 2
  .. .. ..$ : NULL
  .. .. ..$ : chr [1:3] "site" "x" "y"
  ..$ deadrecovery: num [1:24561, 1:15] 0 0 0 0 0 0 0 0 0 0 ...
  .. ..- attr(*, "dimnames")=List of 2
  .. .. ..$ : NULL
  .. .. ..$ : chr [1:15] "2002" "2003" "2004" "2005" ...
  ..$ age          : chr [1:24561] "juv" "juv" "juv" "juv" ...
```

The data file consists of a data frame and three lists. The data frame contains the landscape data (`landData`), which contain coordinates, mean elevation (in meters), and lake cover as proportions for each of 15,734 1 km^2 quadrats that form the domain of our spatial model. The three lists refer to the MHB, atlas, and dead-recovery data that we describe in more detail in the following sections.

We next describe the basic models adopted for each data set and define their likelihoods. In summary, the first submodel for the population counts, and hence the core of the IPM, is a demographic state-space model (Section 4.3.6), i.e., a traditional Dail-Madsen (DM) model (Dail and Madsen, 2011), as in Ahrestani et al. (2017), Zhao et al. (2019a, 2019b) and in Chapter 19; for an IPM with the spatial variant of that model as developed by Zhao et al. (2017), see Zhao (2020). The second submodel, for the atlas data, is a static binomial N-mixture model (Royle, 2004). The third submodel, for the dead-recovery data, is a dead-recovery model (see Section 4.5.3) similar to but simpler than the model for what partly are the same data in Fay et al. (2019). Here, these data are spatially stratified, so we can link the survival data with the appropriate location in the spatial field for survival.

20.4.1 MONITORING HÄUFIGE BRUTVÖGEL POPULATION COUNT DATA

The longitudinal population count data from the MHB (`mhbData`) contain the annual total count of kestrels in 102 1 km^2 MHB survey quadrats during 1999–2017 (`tcount`) and the coordinates of these quadrats (`x`, `y`). We isolate the counts in a matrix and then align these counts within a larger matrix that contains one row for each quadrat in the landscape data and has NAs for counts at all quadrats that do not belong to the sample of the 102 MHB quadrats.

```
# Grab MHB total count data as a matrix
C <- as.matrix(kestrel$mhbData$tcount)
dimnames(C) <- NULL

# Produce 'stretched counts', filled up with NAs for unsurveyed quadrats and where the MHB counts
  # are filled in at the right place
mhb.coord <- paste(kestrel$mhbData$x, kestrel$mhbData$y, sep=".")
lscape.coord <- paste(kestrel$landData$x, kestrel$landData$y, sep=".")
C.mhb <- array(NA, dim=c(length(kestrel$landData$x), ncol(C)))
mhb.positions <- pmatch(mhb.coord, lscape.coord)
for (i in 1:nrow(C)){
  C.mhb[mhb.positions[i],] <- C[i,]
}
```

Thus, `C.mhb` is a matrix of dimensions 15,734 × 19, which is the number of 1 km^2 quadrats in the modeled landscape and the number of years in the MHB population counts. It would be computationally more efficient to only model the 102 quadrats with MHB counts, but "stretching" these data to comprise the entire modeled domain provides an easy way of obtaining spatial predictions in the model. This response variable contains the total counts of kestrels in each quadrat of the entire modeled landscape, where all counts are NAs in the vast majority of quadrats, and only counts in the 102 surveyed quadrats are non-NA. We adopted the likelihood of a DM model for this submodel whose state process model is defined in Section 20.3 and where the abundance $N_{i,1}^{MHB}$ at site i in year 1 (initial abundance) is assumed to be a Poisson random variable with an offset A that is the effective sample area of the quadrat and initial density λ_i^{MHB} (see above and Section 20.4.5).

$$N_{i,1}^{MHB} \sim Poisson\left(A\lambda_i^{MHB}\right)$$

For the observation model, we expressed detection probability for the three surveys combined as a function of the per-survey detection probability under an assumption of independent surveys and constant detection probability (p) (note that in this sample of MHB quadrats, we have three rather than two surveys throughout).

$$C_{i,t}^{MHB} \sim Binomial\left(1-(1-p)^3, N_{i,t}^{MHB}\right)$$

For simplicity, we chose detection probability as a constant, but we note that all three parameters in the state model (initial density λ, survival ϕ, and recruitment γ) are indexed by quadrat i. For each of the three, we specify a spatial random field through penalized splines constructed from radial basis functions associated with 50 knots. That is, we specify the following three spatial random fields on a link-transformed scale:

$$\log(\lambda_i) = \alpha^{\lambda} + s_i^{\lambda}$$

$$\log(\phi_i) = \alpha^{\phi} + s_i^{\phi}$$

$$\log(\gamma_i) = \alpha^{\phi} + s_i^{\gamma}$$

We describe these random fields s in Section 20.4.4.

We obtained an estimate of the spatially explicit annual population growth rate based on the quadrat-level expected population size (basing it on the realized abundance would lead to problems whenever $N_{i,t} = 0$). Thus, the growth rate (gr) at site i is given by $\sqrt[18]{E(N_{i,19})/E(N_{i,1})}$, where $E(N_{i,1}) = A\lambda_i$ and $E(N_{i,19})$ is expressed as a function of the parameters for the population dynamics model, i.e., $E(N_{i,19}) = \left(\phi_i^{18}\lambda_i + \gamma_i \sum_{u=0}^{17} \phi_i^u\right)A$, analogously to the expressions derived in the next section. The formula to express expected abundance in a given year is developed from the population model in Section 20.3.

20.4.2 ATLAS POPULATION COUNT DATA

The second data set to be integrated is the cross-sectional population counts from the Swiss atlas survey (atlasData) from those 1 km^2 quadrats located inside of the modeled region (Knaus et al., 2018). The data set contains coordinates (x, y), year of the survey (year), and the number of kestrels counted during the two to three surveys in a quadrat (count, high-elevation quadrats will have an NA in the last column). The rows in the data set are ordered by survey years, with rows 1–182 containing quadrats surveyed in 2013, 183–387 those surveyed in 2014, and 388–574 quadrats surveyed in 2015. We will need this information in the submodel specification below. We also want to align the data spatially, which we do next.

```
# Grab atlas count data as a matrix
C.atlas <- kestrel$atlasData$count
dimnames(C.atlas) <- NULL

# Align the atlas data with the modeled domain by determining their position within the landscape
atlas.coord <- paste(kestrel$atlasData$x, kestrel$atlasData$y, sep=".")
atlas.positions <- pmatch(atlas.coord, lscape.coord)
```

The response variable (C.atlas) is a site-by-visit matrix of the kestrel counts for a total of 574 1 km^2 quadrats that were surveyed in the modeled part of Switzerland during the first 3 out of 4 years during which the field work for the last atlas was conducted. Note that this number excluded the MHB quadrats in this region (which were also utilized in the modeling for Knaus et al., 2018) as well as all quadrats outside of Switzerland in Germany and France. The survey years for the atlas data were 2013, 2014, and 2015, representing years 15, 16, and 17 in our basic population model above.

We chose to share all the parameters between the MHB and the atlas data, i.e., both the constant detection probability as well as the population dynamics parameters (see the Discussion for a simpler model where p is the only parameter shared between the MHB and the atlas data). Hence, the atlas counts needed to be "inserted at the right" places in the model. We achieved this by projecting the expected abundance forwards for each atlas quadrat using the quadrat-specific estimates of the initial density, survival and recruitment, as obtained from the estimates of the spatial fields in the model. Thus, the abundance in atlas quadrat i (where i is now a different indexing scheme) depends on the same effective survey area A discussed above and on the expected density, λ_i^{Atlas}, given by the expressions below:

$$N_i^{Atlas} \sim Poisson\left(A\lambda_i^{Atlas}\right).$$

The expected density in a given year is calculated using the same formula as in Section 20.4.1. The atlas data used here were collected in three different years. The density for site i in year t is then (where 1999 is $t = 1$, and hence, 2013 is $t = 15$ and so forth)

$$\lambda_i^{atlas} = \phi_i^{t-1}\lambda_i + \gamma_i \sum_{u=0}^{t-2} \phi_i^u.$$

For the observation model of the atlas data, we have the standard formulation for a static N-mixture model where j indexes the two to three visits made at each atlas quadrat.

$$C_{i,j}^{Atlas} \sim Binomial\left(p, N_i^{Atlas}\right)$$

Thus, we see that both the abundance parameters and the detection probability are shared between the MHB and atlas data sets.

20.4.3 DEAD-RECOVERY DATA

The third data set we use in our IPM is the dead-recovery data (drData) from a total of 24,561 individual kestrels that were ringed in a total of seven combined ringing areas in the modeled part of Switzerland, during 2002–16 using data from kestrels breeding in nest boxes. Some of the ringed kestrels were later recovered dead. We used a subset of the same data that were utilized by Fay et al. (2019), but for the spatial attribution of the dead-recovery data lumped all individual ringing sites (i.e., nest boxes) within 50 × 50 km squares. We assigned the average coordinate of all the ringing sites in that area to be "the" location of that subset of the ringing data. In six areas, only nestlings were ringed, while in one of them, adult kestrels were ringed in addition to nestlings.

The kestrel dead-recovery data (deadrecovery) is a matrix with the 24,561 ringed individuals in the rows and the 15 years in the columns. The years in which an individual was ringed and recovered dead is indicated by a 1, all the other cells of the matrix contain a 0. The two objects `age` and `site` indicate for each individual the age at marking (with two levels, "juv" and "ad," the latter meaning ≥ 1 year old) and the location of marking (7 sites). Juveniles were marked at all sites, but marking of adults was performed at site 3 only, which we can see by a tabulation of the number of marked individuals for each site and age class.

```
# Check which birds (age) were marked at which site
table(kestrel$drData$site[,1], kestrel$drData$age)      # Not shown
```

We transform the individual dead-recovery data in the m-array format (Section 4.5.3.2), such that they can be analyzed using the computationally efficient multinomial likelihood. For each age class and site, a different m-array is constructed using the `marrayDead` function of the `IPMbook` package.

```
# Compute the m-arrays, one for each site and age at ringing
# 1. For individuals ringed as adults (from site 3 only)
a <- with(kestrel$drData, which(age=="ad" & site==3))
ma <- with(kestrel$drData, marrayDead(deadrecovery[a,]))

# 2. For individuals ringed as juveniles (all 7 sites)
mj <- array(0, dim=c(7, 14, 15))
for (i in 1:7){
  a <- with(kestrel$drData, which(age=="juv" & site[,1]==i))
  mj[i,,] <- with(kestrel$drData, marrayDead(deadrecovery[a,]))
}
```

We calculate the number of released individuals in each year and obtain the position of each ringing site.

```
# Compute the annual number of marked individuals
rel.a <- rowSums(ma)
rel.j <- apply(mj, 2:1, sum)

# Get position of each ringing area within modeled landscape
dr.sites <- unique(kestrel$drData$site)
dr.site.coord <- paste(dr.sites[,"x"], dr.sites[,"y"], sep=".")
dr.positions <- pmatch(dr.site.coord, lscape.coord)
nsites.dr <- length(dr.positions)
```

The m-arrays are analyzed with the multinomial likelihood (see Section 4.5.3.2 for details).

$$\boldsymbol{m}_{i,t} \sim Multinomial(\pi_{i,t}, R_{i,t})$$

Here, $\boldsymbol{m}_{i,t}$ is the m-array count at site i for individuals marked in year t, $R_{i,t}$ is the total number of marked individuals in year t at site i, and $\pi_{i,t}$ are the cell probabilities in the m-array. The cell probabilities are functions of survival and recovery parameters and the structure differs between the m-arrays for juveniles and adults.

For birds ringed and released as juveniles, the cell probabilities of the juvenile m-array are:

Release occasion	Recovery occasion			
	2	3	4	Never recovered
1	$\left(1-\phi_i^j\right)r$	$\phi_i^j(1-\phi_i)r$	$\phi_i^j\phi_i(1-\phi_i)r$	$1-\sum$ rel. occ. 1
2	0	$\left(1-\phi_i^j\right)r$	$\phi_i^j(1-\phi_i)r$	$1-\sum$ rel. occ. 2
3	0	0	$\left(1-\phi_i^j\right)r$	$1-\left(1-\phi_i^j\right)r$

For birds ringed and released as adults the cell probabilities of the adult m-array are:

Release occasion	Recovery occasion			
	2	3	4	Never recovered
1	$(1-\phi_i)r$	$\phi_i(1-\phi_i)r$	$\phi_i\phi_i(1-\phi_i)r$	$1-\sum$ rel. occ. 1
2	0	$(1-\phi_i)r$	$\phi_i(1-\phi_i)r$	$1-\sum$ rel. occ. 2
3	0	0	$(1-\phi_i)r$	$1-(1-\phi_i)r$

Here, ϕ_i is annual adult survival at site i, ϕ_i^j is annual juvenile survival at site i, and r is the recovery probability. In our model, we assume that the survival probabilities vary spatially and by age but not temporally. By contrast, we assume the recovery probability to be independent of age, location, and year. Note that we estimate juvenile survival and recovery probability as mere nuisance parameters in the model, but they do not appear in the population model of our IPM.

20.4.4 BASIS FUNCTION APPROACH TO THE MODELING OF SPATIAL AUTOCORRELATION

We adopt a basis function approach (Hefley et al., 2017a) and use penalized splines to model spatial autocorrelation in all three main parameters of the population model to define a spatial random field for each of initial density (λ), survival (ϕ), and recruitment (γ). In other words, we add into the linear predictors of the three demographic parameters λ, ϕ, and γ a two-dimensional GAM using radial basis functions of the quadrat coordinates (Hastie and Tibshirani, 1990; Wood, 2017). We follow Collier et al. (2012; see also Chapter 9 in Kéry and Royle, 2021) to construct two-dimensional, or radial, splines centered at selected spatial locations, referred to as "knots." These are locations distributed in the modeled landscape and associated with each of which we compute a spatially indexed covariate that is defined in the entire landscape and depends on the distance from the knot location (see Fig. 20.5 for a sample). In the model, we estimate a parameter b for each basis function. For instance, the model for initial abundance is

$$\log(\lambda_i) = \alpha^\lambda + s_i^\lambda,$$

where s_i^λ is the random field, which is simply a sum of the products of the value of the basis function $Z_{i,k}$ for knot k and basis function coefficient b_k, i.e.,

$$s_i^\lambda = \sum_{k=1}^{K} Z_{i,k} b_k^\lambda.$$

The parameter b_k^λ is the coefficient of the k-th basis function in the spatial model for λ and is estimated in the model subject to a constraint: $b_k^\lambda \sim Normal(0, \sigma_b^\lambda)$. Thus, the estimated basis function coefficients are considered random effects, and we estimate the standard deviation or the variance of their hyperdistribution, which governs the wiggliness of the random field. Note that we used the same basis expansion, i.e., collection of basis functions $Z_{i,k}$, for all three parameters, but in the model estimated different sets of parameters for b_k^λ, b_k^ϕ, and b_k^γ. Thus, the resulting spatial fields were different.

We start by constructing the splines and for this, first need to decide on the number and location of the knots. We have a sample size of $n = 102$ in the smaller MHB data. Ruppert's (2002) rule suggests that in a spatial application of such thin-plate splines with n sites the number of knots is best taken as $\max(20, \min(n/4, 150))$. This would suggest 25 knots, but to be sure we have high enough spatial resolution, we go for 50. We use the space-filling algorithm of Royle and Nychka (1998), which is implemented in function `cover.design` in library `fields`, to place these knots in the landscape and then look at their locations in an elevation map of the study region (Fig. 20.4).

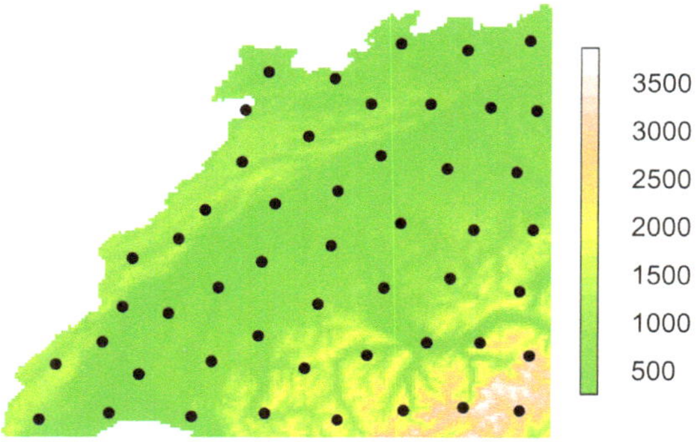

FIGURE 20.4

The chosen knot design shown against a background of the elevation of the modeled landscape in the northwestern quadrant of Switzerland of almost 16,000 km^2.

```
library(AHMbook)
library(fields)
coordgrid <- cbind(kestrel$landData$x, kestrel$landData$y)

# Run a knot design
nknots <- 50                              # This is K in the algebra above
set.seed(50)
knots <- cover.design(R=coordgrid, nd=nknots, nruns=10, num.nn=200,
  max.loop=20)                            # Takes about 25 min

# Plot the knot design (Fig. 20.4)
library(raster)
r <- rasterFromXYZ(data.frame(x=coordgrid[,1], y=coordgrid[,2], z=kestrel$landData$elevation))
plot(r, col=terrain.colors(20), axes=FALSE, box=FALSE, main="", asp=1)
points(knots$design, col="black", pch=16, cex=1.5)
```

Next, we produce the basis expansion using a radial basis function for the 50 chosen knots and collect the 50 basis vectors in a matrix **Z**, using code by Crainiceanu et al. (2005), Gimenez et al. (2006) and Collier et al. (2012).

```
# Define the Z matrix for the random effects/knot coefficients
knotlocs <- knots$design
omega <- (e2dist(knotlocs, knotlocs)/100)^3    # Scale distances
svd.omega <- svd(omega)
sqrt.omega <- t(svd.omega$v %*% (t(svd.omega$u) * sqrt(svd.omega$d)))
Zk <- (e2dist(coordgrid, knotlocs) / 100)^3    # Scale distances
Zmat <- t(solve(sqrt.omega, t(Zk)))
```

To better understand the basis function approach, we plot the basis vectors and produce really fancy maps such as those in Fig. 20.5.

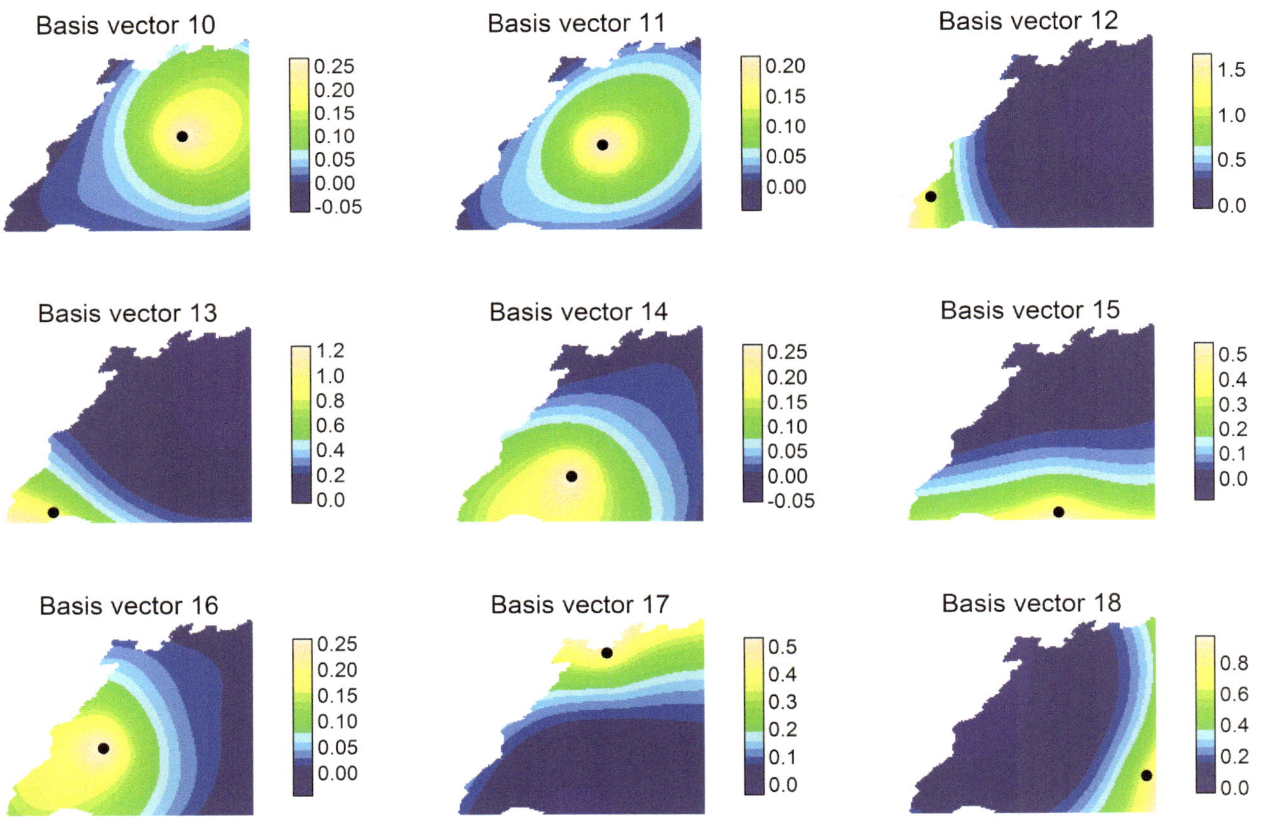

FIGURE 20.5

Maps of the basis vectors for a selection of nine among the 50 knots in our analysis. You can think of each of these maps as simply another spatial covariate we use in a big multiple regression and that collectively define the spatial random field for the three demographic parameters in our spatial IPM.

```
# Visualize basis vectors
head(Zmat)                                          # Look at first couple of values
par(mfrow=c(3, 3), mar=c(2, 2, 3, 8), cex.main=1.5, ask=TRUE)
for (i in 1:nknots){
  r <- rasterFromXYZ(data.frame(x=coordgrid[,1], y=coordgrid[,2], z=Zmat[,i]))
  plot(r, col=topo.colors(20), axes=FALSE, box=FALSE, main=paste("Basis vector", i),
    legend=TRUE,axis.args=list(cex.axis=1.5), legend.width=1.5)
  points(knots$design[i,1], knots$design[i,2], col='black', pch=16, cex=1.5)
}
```

To fit the model, we simply package these 50 additional covariates into a matrix, add them to the data bundle, and specify them as covariates with a common zero-mean normal distribution as a prior for the associated 50 basis vector coefficients. We are now ready to use these basis vectors as merely another spatially explicit covariate in the linear models for the parameters in our IPM.

To allow the atlas and the dead-recovery data to inform the estimation of the three random fields, we create a separate set of basis function vectors for each. Hence, from the full basis expansion just defined, we pick the values at those locations where the atlas data and the dead-recovery data were collected.

```
# Get values of the Zmat for the atlas data (n = 574 sites)
str(Zmat.at <- Zmat[atlas.positions,])

# Get values of the Zmat for the dead-recovery data (n = 7 sites)
str(Zmat.dr <- Zmat[dr.positions,])
```

20.4.5 SCALING THE MODELED POPULATION SIZE TO THE NOMINAL 1 KM^2 AREA

As described above, we must scale the quadrat-specific population size estimates by the effective survey area (A) if we want to produce an estimate of the total population size in the study area. This means that we define A to be an offset in the submodel for the initial abundance in 1999. For all later years, we divide the estimate of $N_{i,t}$ by A before adding up the estimates over all "land quadrats," i.e., minus those that have 50% or more lake coverage.

```
land.quad <- which(kestrel$landData$lakes < 0.5)
```

20.5 THE INTEGRATED POPULATION MODEL

We bundle the data to a list and inspect its structure.

```
# Bundle data and produce data overview
jags.data <- list(nsites.region=nrow(C.mhb), nyears.mhb=ncol(C.mhb), C.mhb=C.mhb, nknots=nknots,
  Zmat=Zmat, land.quad=land.quad, nsites.atlas=nrow(C.atlas), nsurveys.atlas=ncol(C.atlas),
  C.atlas=C.atlas, year.atlas=kestrel$atlasData$year, Zmat.at=Zmat.at,nsites.dr=nsites.dr,
  nyears.dr=ncol(kestrel$drData$deadrecovery), mj=mj, rel.j=rel.j, ma=ma,
  rel.a=rel.a, Zmat.dr=Zmat.dr)
str(jags.data)

List of 18
 $ nsites.region : int 15734
 $ nyears.mhb    : int 19
 $ C.mhb         : int [1:15734, 1:19] NA NA NA NA NA NA NA NA NA NA ...
 $ nknots        : num 50
 $ Zmat          : num [1:15734, 1:50] -0.0666 -0.0647 -0.0654 -0.0635 ...
 $ land.quad     : int [1:15055] 1 2 3 4 5 6 7 8 9 10 ...
 $ nsites.atlas  : int 574
 $ nsurveys.atlas: int 3
 $ C.atlas       : num [1:574, 1:3] 0 0 0 1 1 0 0 0 0 2 ...
 $ year.atlas    : num [1:574] 2015 2013 2013 2013 2015 ...
 $ Zmat.at       : num [1:574, 1:50] -0.05587 -0.04094 0.00386 0.0013 ...
 $ nsites.dr     : int 7
 $ nyears.dr     : int 15
```

```
$ mj          : num [1:7, 1:14, 1:15] 0 0 4 1 4 0 0 0 0 0 ...
$ rel.j       : num [1:14, 1:7] 22 37 50 152 148 235 117 163 289 318 ...
$ ma          : num [1:14, 1:15] 0 0 0 0 0 0 0 0 0 0 ...
$ rel.a       : num [1:14] 69 74 86 64 81 89 64 86 77 45 ...
$ Zmat.dr     : num [1:7, 1:50] 0.37579 0.42542 0.01412 -0.00927...
```

In initial model variants not shown here, we had explored what we can do with fairly vague priors (especially, with uniform(0,1) for the intercept of ϕ and p) and found that convergence of such a model was very hard to obtain. Hence, we choose a moderately informative prior for the intercept of *phi* that gives less mass to very small and very large values of survival. In addition, we give weakly informative priors for the intercepts of λ and γ (Fig. 20.6).

```
# Explore shape and location of some distributions for priors
par(mfrow=c(2, 2), mar=c(6, 6, 5, 2), cex.axis=1.5, cex.lab=1.5, cex.main=1.5, las=1)
curve(dnorm(x, mean=0, sd=sqrt(1 / 0.1)), 0, 10, col="blue", lwd=3, xlab="", ylab="Density",
  frame=FALSE)
mtext(expression(paste("Prior for intercept of ", lambda)), side=3, line=0)
curve(dbeta(x, shape1=6, shape2=4), 0, 1, col="blue", lwd=3, xlab="", ylab="Density", frame=FALSE)
mtext(expression(paste("Prior for intercept of ", phi)), side=3, line=0)
curve(dnorm(x, mean=0, sd=sqrt(1 / 1)), 0, 3, col="blue", lwd=3, xlab="", ylab="Density", frame=FALSE)
mtext(expression(paste("Prior for intercept of ", gamma)), side=3, line=0)
curve(dbeta(x, shape1=1, shape2=1), 0, 1, col="blue", lwd=3, xlab="", ylab="Density", frame=FALSE)
mtext(expression(paste("Prior for constant ", italic(p))), side=3, line=0)
```

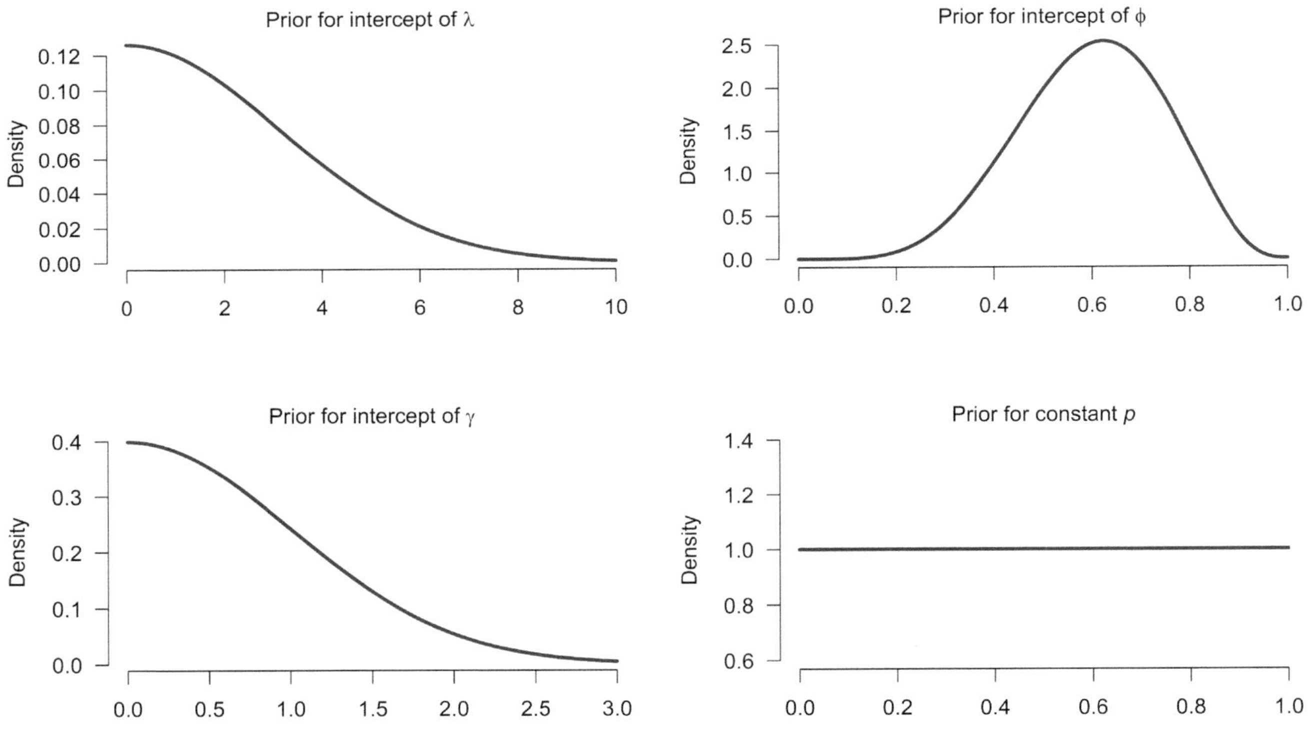

FIGURE 20.6

Priors chosen for the intercepts of the four parameters in the spatial integrated population model. We consider them to be moderately informative for ϕ, weakly informative for λ and γ, and completely vague for p.

This model has pretty extensive code.

```
# Write JAGS model file
cat(file = "model1.txt","
model {
# Priors and linear models
lambda.int ~ dnorm(0, 0.1)I(0,)              # Intercept of initial site-specific abundance
alpha.lam <- log(lambda.int)
phi.int ~ dbeta(6, 4)                         # Intercept of survival
alpha.phi <- logit(phi.int)
gamma.int ~ dnorm(0, 1)I(0,)                  # Intercept of recruitment
alpha.gam <- log(gamma.int)
p ~ dunif(0, 1)                               # Constant detection prob. (per survey)

# Fixed parameter for effective survey quadrat size (in km2)
# with uniform prior for automatic sensitivity analysis/error propagation
A ~ dunif(0.8, 4)                             # Effective sampling area of survey

# Priors on random effects parameters representing the splines
for (k in 1:nknots){
  b.lam[k] ~ dnorm(0, tau.b.lam)
  b.phi[k] ~ dnorm(0, tau.b.phi)
  b.gam[k] ~ dnorm(0, tau.b.gam)
}

# Priors on random effects dispersion (variance among values at knots)
tau.b.lam <- pow(sd.b.lam, -2)
tau.b.phi <- pow(sd.b.phi, -2)
tau.b.gam <- pow(sd.b.gam, -2)
sd.b.lam ~ dnorm(0, 0.01)I(0,)
sd.b.phi ~ dnorm(0, 0.01)I(0,)
sd.b.gam ~ dnorm(0, 0.1)I(0,)

# Linear models for lambda, phi and gamma for MHB sites
for (i in 1:nsites.region){                   # Loop over all sites in the modeled domain,
                                              # including the 102 sites with MHB counts
  lambda[i] <- exp(alpha.lam + smooth.lam[i])
  phi[i] <- ilogit(alpha.phi + smooth.phi[i])
  gamma[i] <- exp(alpha.gam + smooth.gam[i])
  smooth.lam[i] <- inprod(Zmat[i,], b.lam[])  # Random field in lambda
  smooth.phi[i] <- inprod(Zmat[i,], b.phi[])  # Random field in phi
  smooth.gam[i] <- inprod(Zmat[i,], b.gam[])  # Random field in gamma
}

# Calculate lambda, phi and gamma for atlas sites
for (i in 1:nsites.atlas){                    # Loop over 574 sites in the atlas data set
  lamA[i] <- exp(alpha.lam + inprod(Zmat.at[i,], b.lam[]))
  phiA[i] <- ilogit(alpha.phi + inprod(Zmat.at[i,], b.phi[]))
  gamA[i] <- exp(alpha.gam + inprod(Zmat.at[i,], b.gam[]))
}

# Linear models for parameters of the dead-recovery model
for (t in 1:(nyears.dr-1)){
  for (i in 1:nsites.dr){
    logit(sj[i,t]) <- mu.sj + eta.sj[i]
    sa[i,t] <- mean.sa[i]
  } #i
  r[t] <- mean.r
} #t
for (i in 1:nsites.dr){
  eta.sj[i] ~ dnorm(0, tau.sj)
  site.sj[i] <- ilogit(mu.sj + eta.sj[i])
  mean.sa[i] <- ilogit(alpha.phi + smooth.lsa[i]) # alpha.phi shared
  smooth.lsa[i] <- inprod(Zmat.dr[i,], b.phi[])   # b.phi also shared
}

# Priors for juvenile survival and dead-recovery probability
mean.sj ~ dunif(0, 1)
mu.sj <- logit(mean.sj)
sigma.sj ~ dunif(0, 5)
tau.sj <- pow(sigma.sj, -2)
mean.r ~ dunif(0, 1)
```

```
# Longitudinal population count data (MHB, demogr. state-space model)
for (i in 1:nsites.region){                          # Loop over entire region
  # Model for initial condition for MHB data
  N[i,1] ~ dpois(A * lambda[i])

  # Process model over time: our model for population dynamics
  for (t in 1:(nyears.mhb-1)){
    S[i,t+1] ~ dbin(phi[i], N[i,t])                  # Survival process
    R[i,t+1] ~ dpois(gamma[i])                       # 'absolute' recruitment
    N[i,t+1] <- S[i,t+1] + R[i,t+1]
  } #t
} #i

# Observation model for the MHB data
for (i in 1:nsites.region){
  for (t in 1:nyears.mhb){
    C.mhb[i,t] ~ dbin(1 - (1-p)^3, N[i,t])           # Reconcile temporal mismatch
  } #t
} #i

# Cross-sectional population count data (atlas, static N-mixture model)
# Define expected abundance for each year
for (i in 1:nsites.atlas){                           # Loop over 574 sites in the atlas data set
  # Project expected abundance from 1999 to 2013/2014/2015 (part of calcs)
  for (t in 1:16){
    su[i,t] <- pow(phiA[i], t-1)
  } #t
} #i

# Reconcile temporal mismatch between atlas and MHB data
for (i in 1:182){                                    # Quadrats surveyed in 2013
  lambdaAtlas[i] <- lamA[i] * phiA[i]^14 + gamA[i] * sum(su[i,1:14])
  N.atlas[i] ~ dpois(A * lambdaAtlas[i])
}
for (i in 183:387){                                  # Quadrats surveyed in 2014
  lambdaAtlas[i] <- lamA[i] * phiA[i]^15 + gamA[i] * sum(su[i,1:15])
  N.atlas[i] ~ dpois(A * lambdaAtlas[i])
}
for (i in 388:574){                                  # Quadrats surveyed in 2015
  lambdaAtlas[i] <- lamA[i] * phiA[i]^16 + gamA[i] * sum(su[i,1:16])
  N.atlas[i] ~ dpois(A * lambdaAtlas[i])
}

# Observation model for the atlas data
for (i in 1:nsites.atlas){                           # Loop over all 574 sites in the atlas data set
  for (j in 1:nsurveys.atlas){                       # Loop over 2-3 occasions
    C.atlas[i,j] ~ dbin(p, N.atlas[i])               # Assume constant p
  } #j
} #i

# Dead-recovery data (dead-recovery model with multinomial likelihood)
# m-array for adults
for (t in 1:(nyears.dr-1)){
  ma[t,1:nyears.dr] ~ dmulti(pr.a[t,], rel.a[t])
}
for (t in 1:(nyears.dr-1)){
  pr.a[t,t] <- (1-sa[3,t])*r[t]
  for (j in (t+1):(nyears.dr-1)){
    pr.a[t,j] <- prod(sa[3,t:(j-1)]) * (1-sa[3,j]) * r[j]
  } #j
  for (j in 1:(t-1)){
    pr.a[t,j] <- 0
  } #j
} #t
for (t in 1:(nyears.dr-1)){
  pr.a[t,nyears.dr] <- 1-sum(pr.a[t,1:(nyears.dr-1)])
} #t
```

```
# m-array for juveniles (7 sites)
for (k in 1:nsites.dr){
  for (t in 1:(nyears.dr-1)){
    mj[k,t,1:nyears.dr] ~ dmulti(pr.j[k,t,], rel.j[t,k])
  } #t
  for (t in 1:(nyears.dr-1)){
    pr.j[k,t,t] <- (1-sj[k,t])*r[t]
    for (j in (t+2):(nyears.dr -1)){
      pr.j[k,t,j] <- sj[k,t] * prod(sa[k,(t+1):(j-1)]) * (1-sa[k,j]) * r[j]
    } #j
    for (j in 1:(t-1)){
      pr.j[k,t,j] <- 0
    } #j
  } #t
  for (t in 1:(nyears.dr-2)){
    pr.j[k,t,t+1] <- sj[k,t] * (1-sa[k,t+1]) * r[t+1]
  } #t
  for (t in 1:(nyears.dr-1)){
    pr.j[k,t,nyears.dr] <- 1-sum(pr.j[k,t,1:(nyears.dr-1)])
  } #t
} #k

# Derived quantities
# Estimate population growth rate for each quadrat based on
# expected abundances rather than on N (to avoid division by zero)
for (i in 1:nsites.region){
  for (t in 1:(nyears.mhb-1)){
    g[i,t] <- pow(phi[i], t-1)
  } #t
  gr[i] <- pow((lambda[i] * phi[i]^(nyears.mhb-1) + gamma[i] * sum(g[i,]) /
    lambda[i]), 1 / (nyears.mhb-1))
}

# Estimate annual abundance in the entire region (excluding lakes)
Ntot[1] <- sum(lambda[land.quad])                   # Total population size in NW quarter
for (t in 2:nyears.mhb){
  Ntot[t] <- sum(N[land.quad,t] / A)
}
}
")
```

The models need care for the initial values, especially for the latent states, which may include S and R.

```
# Initial values
Nst.atlas <- as.numeric(apply(C.atlas, 1, max, na.rm=TRUE) + 1)
Nst.mhb <- C.mhb + 5
Nst.mhb[, 2:19] <- NA # cols 2:19 of N are deterministic, N <- S + R.
R1 <- C.mhb
R1[,1] <- NA
inits <- function(){list(lambda.int=runif(1, 0.4, 0.6), phi.int=runif(1, 0.8, 0.9),
  gamma.int=runif(1, 0.5, 0.9), p=runif(1, 0.6, 0.8), N=Nst.mhb, R=R1 + 1,
  b.lam=runif(jags.data$nknots, -0.5, 0.5), b.phi=runif(jags.data$nknots, -0.5, 0.5),
  b.gam=runif(jags.data$nknots, -0.5, 0.5), mean.r=runif(1, 0, 0.2), N.atlas=Nst.atlas)}

# Parameters monitored
parameters <-c("lambda.int", "alpha.lam", "phi.int", "alpha.phi", "gamma.int", "alpha.gam", "p",
  "sd.b.lam", "sd.b.phi", "sd.b.gam", "mean.sj", "site.sj", "sigma.sj", "mean.sa", "mean.r", "Ntot",
  "b.lam", "b.phi", "b.gam", "lambda", "phi", "gamma", "gr")

# MCMC settings
ni <- 150000; nb <- 50000; nc <- 3; nt <- 200; na <- 10000

# Call JAGS from R (ART 88 h!) and check convergence
out1 <- jags(jags.data, inits, parameters, "model1.txt", n.iter=ni, n.burnin=nb, n.chains=nc,
  n.thin=nt, n.adapt=na, parallel=TRUE)
traceplot(out1)
```

This model takes almost 4 days to run, but then we find that the convergence is acceptable (see Fig. 20.7 for a sample of trace-plots for key parameters). Hence, we go on and produce a posterior summary for some key parameters; see next section.

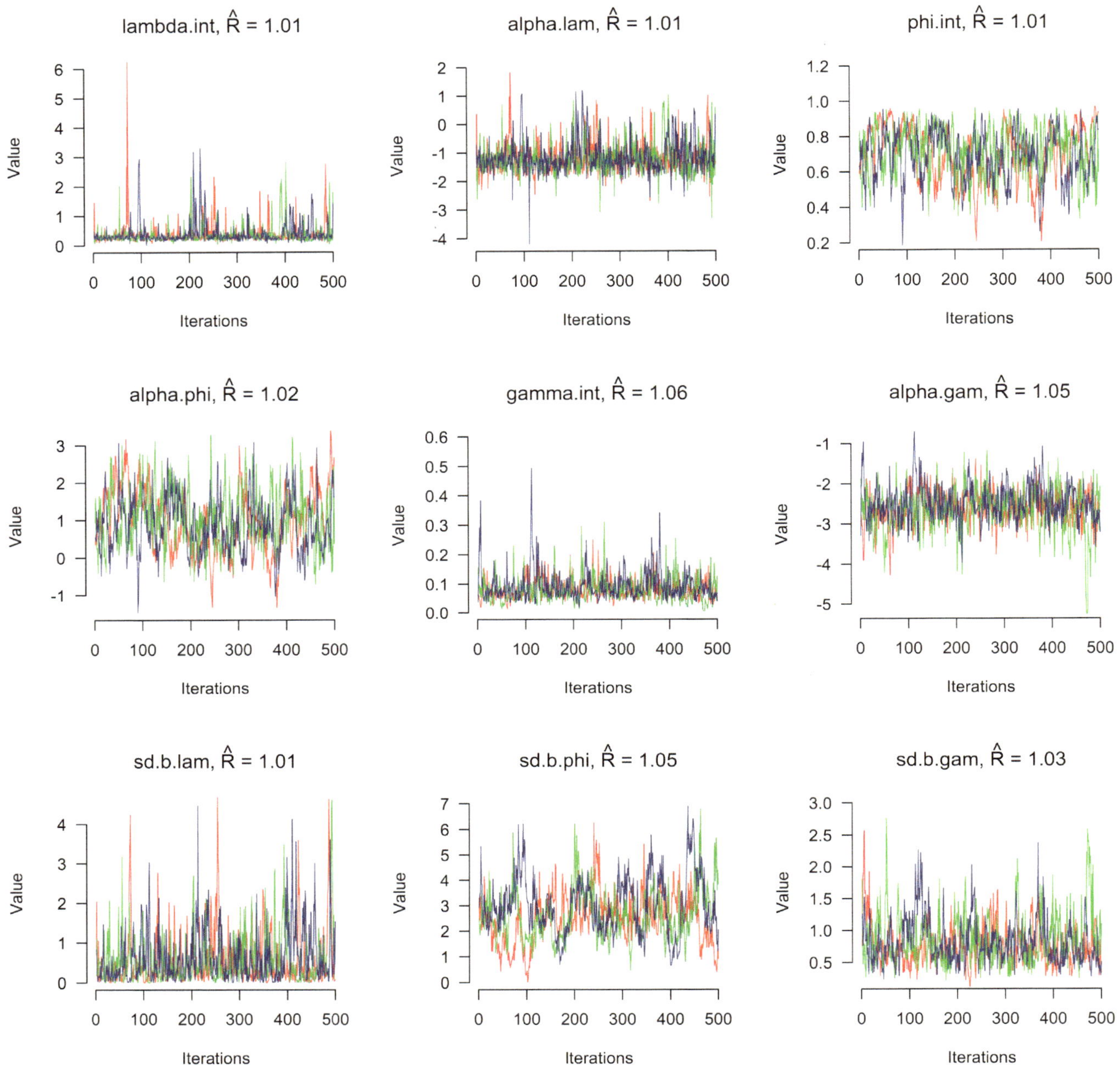

FIGURE 20.7

Trace-plots for some key parameters in the model.

20.6 RESULTS

The next table shows the posterior summary for some of the key parameters and derived quantities.

```
print(out1$summary[1:46,-c(4,6)], 3)

                 mean          sd         2.5%         50%        97.5% Rhat n.eff overlap0     f
lambda.int     0.3771     0.37939     0.1384      0.2763       1.4014 1.01   664        0 1.000
alpha.lam     -1.1866     0.56794    -1.9779     -1.2862       0.3375 1.01   414        1 0.949
phi.int        0.7015     0.14452     0.4001      0.7124       0.9297 1.01   172        0 1.000
alpha.phi      0.9692     0.77920    -0.4051      0.9072       2.5815 1.02   162        1 0.900
gamma.int      0.0856     0.04414     0.0278      0.0766       0.1967 1.06    82        0 1.000
alpha.gam     -2.5742     0.49159    -3.5831     -2.5697      -1.6260 1.05    75        0 1.000
p              0.2777     0.01882     0.2427      0.2766       0.3178 1.01   185        0 1.000
sd.b.lam       0.5732     0.66389     0.0143      0.3371       2.3572 1.01   450        0 1.000
sd.b.phi       2.7273     1.05800     0.8968      2.6225       5.1486 1.05    48        0 1.000
sd.b.gam       0.8178     0.36308     0.3109      0.7451       1.7188 1.03   127        0 1.000
mean.sj        0.6231     0.08182     0.4486      0.6261       0.7795 1.00  1500        0 1.000
site.sj[1]     0.6702     0.06186     0.5354      0.6725       0.7809 1.00   674        0 1.000
site.sj[2]     0.5112     0.04943     0.4132      0.5114       0.6081 1.00  1389        0 1.000
site.sj[3]     0.8078     0.03296     0.7341      0.8104       0.8615 1.01   388        0 1.000
site.sj[4]     0.5677     0.07773     0.4115      0.5697       0.7155 1.00  1500        0 1.000
site.sj[5]     0.5694     0.04736     0.4733      0.5698       0.6597 1.00  1280        0 1.000
site.sj[6]     0.6458     0.16850     0.2660      0.6616       0.9439 1.00   441        0 1.000
site.sj[7]     0.5693     0.08679     0.3872      0.5756       0.7224 1.00   680        0 1.000
sigma.sj       0.7901     0.42492     0.3249      0.6845       1.8401 1.01   602        0 1.000
mean.sa[1]     0.8952     0.03187     0.8203      0.9000       0.9424 1.00   510        0 1.000
mean.sa[2]     0.8146     0.04639     0.7120      0.8195       0.8913 1.01   235        0 1.000
mean.sa[3]     0.9474     0.01518     0.9137      0.9495       0.9666 1.05   207        0 1.000
mean.sa[4]     0.8309     0.05409     0.7049      0.8374       0.9168 1.00   832        0 1.000
mean.sa[5]     0.7689     0.04217     0.6840      0.7688       0.8487 1.00   514        0 1.000
mean.sa[6]     0.9167     0.03380     0.8361      0.9228       0.9648 1.00   363        0 1.000
mean.sa[7]     0.8736     0.05797     0.7252      0.8861       0.9504 1.05    67        0 1.000
mean.r         0.0396     0.00383     0.0326      0.0394       0.0477 1.01   289        0 1.000
Ntot[1]     3757.6250   879.61423 2288.8791 3667.5283  5674.0263 1.00  1500        0 1.000
Ntot[2]     4586.0053   856.73316 3195.7486 4496.2487  6475.2987 1.00  1500        0 1.000
Ntot[3]     5236.5288   892.89131 3750.7405 5160.4451  7234.0252 1.00  1500        0 1.000
Ntot[4]     5773.9687   949.04202 4164.5854 5683.4273  7935.3141 1.00  1500        0 1.000
Ntot[5]     6229.0835  1014.34675 4483.0932 6124.4378  8545.4490 1.00  1500        0 1.000
Ntot[6]     6628.2767  1083.24952 4752.5860 6551.7435  9174.1480 1.00  1500        0 1.000
Ntot[7]     6971.9769  1148.07832 4954.9725 6877.1834  9607.1014 1.00  1500        0 1.000
Ntot[8]     7274.4552  1211.63479 5162.8650 7165.3406  9999.6642 1.00  1500        0 1.000
Ntot[9]     7543.5614  1270.96517 5295.5986 7432.9123 10429.5929 1.00  1500        0 1.000
Ntot[10]    7783.4523  1326.65793 5443.9203 7664.0278 10809.4471 1.00  1500        0 1.000
Ntot[11]    7997.1026  1378.66310 5574.9320 7883.5742 11130.7045 1.00  1500        0 1.000
Ntot[12]    8192.6759  1427.22263 5685.0568 8086.6733 11396.9888 1.00  1500        0 1.000
Ntot[13]    8367.8689  1474.01509 5786.9385 8245.9030 11610.4482 1.00  1500        0 1.000
Ntot[14]    8532.3876  1517.62460 5887.2966 8416.9704 11879.4728 1.00  1500        0 1.000
Ntot[15]    8675.6203  1559.33544 5959.8361 8558.4323 12121.7904 1.00  1500        0 1.000
Ntot[16]    8804.8099  1596.13958 6045.7682 8678.9731 12362.7632 1.00  1500        0 1.000
Ntot[17]    8923.5676  1630.29313 6109.6724 8808.8073 12619.2903 1.00  1500        0 1.000
Ntot[18]    9035.4767  1664.85550 6191.8468 8912.2445 12822.9241 1.00  1500        0 1.000
Ntot[19]    9135.8852  1696.37655 6220.5593 8989.3095 13014.6297 1.00  1500        0 1.000
```

When inspecting the model intercepts, we see that we estimate about one kestrel per 3 km^2 in 1999, annual survival of 0.70, and a recruitment rate of 0.09 kestrels per year and quadrat. Detection probability is estimated at 0.28 per survey, which translates to $1-(1-p)^3 = 0.62$ annual detection in our version of the MHB data with one aggregate count per site and year. The parameters `mean.sj` and `sigma.sj`, and parameter vector `site.sj`, are needed to quantify juvenile survival in our population model but are not parameters of direct interest to us, so we don't discuss them in detail. Parameter vector `mean.sa` contains the mean adult survival at the central locations of the seven ringing areas defined in our study; we see that they range from 0.76 to 0.94. The recovery probability (`mean.r`) is estimated at about 0.04. Finally, the next 19 parameters in the posterior summary contain the estimates of the total population size in the modeled landscape; we will discuss and plot them at the end of this section.

The three parameters `sd.b.lam`, `sd.b.phi`, and `sd.b.gam` represent the standard deviations of the three normal distributions that we place on the 50 basis function coefficients in the splines employed to specify the three random fields for initial abundance, survival, and recruitment. We see much smaller values for abundance and recruitment than for survival, but the meaning of these values is hard to gauge and will depend in part on the link scale at which the splines are entered in

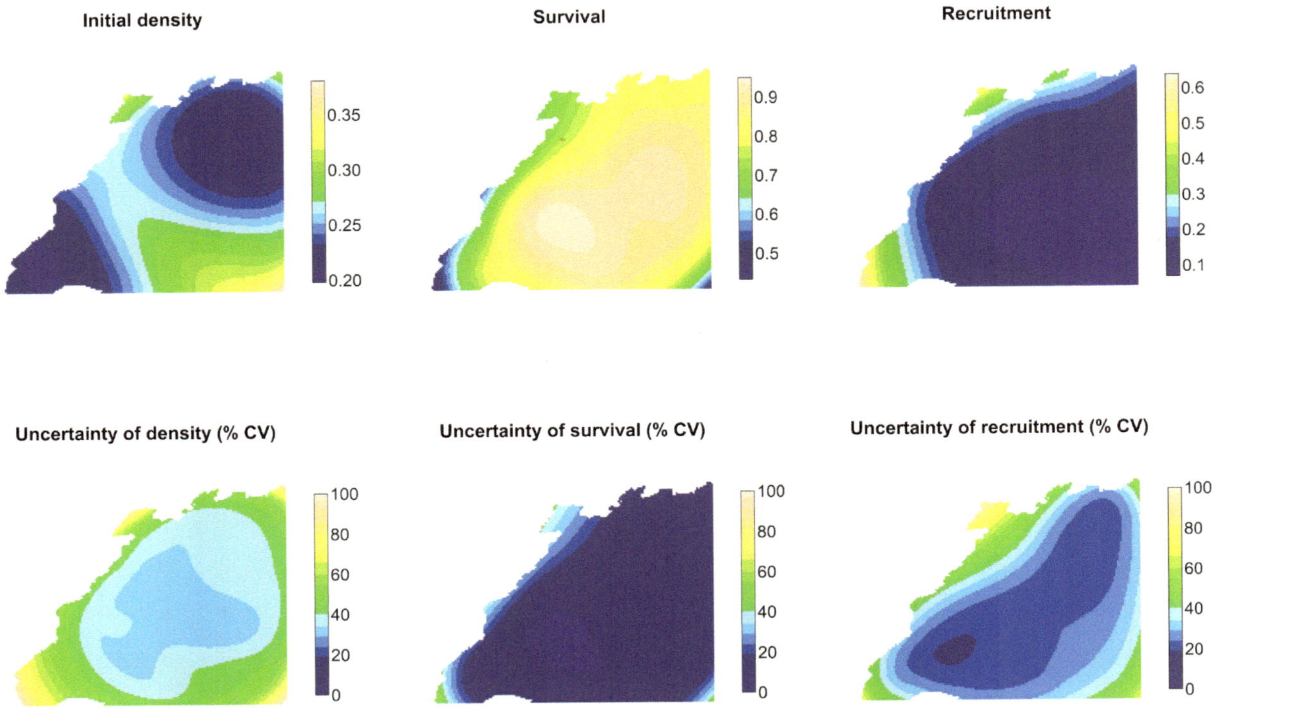

FIGURE 20.8

Maps of the estimated random fields for the three parameters in the population model for Swiss kestrels (top row) and associated uncertainties (bottom row). *What we see here is landscape demography*!

the linear predictor as well as at the magnitude of the intercept of the model. It is best to inspect plots of the spatial fields that they code for along with a map of their estimation uncertainty (expressed here as the % CV); see Fig. 20.8.

```
# Produce Fig. 20.8
par(mfrow=c(2, 3), mar=c(4, 4, 6, 9), cex.main=1.6)
# Random field of initial abundance
r <- rasterFromXYZ(data.frame(x=coordgrid[,1], y=coordgrid[,2], z=out1$mean$lambda))
plot(r, col=topo.colors(20), axes=FALSE, box=FALSE, asp=1, main="Initial density",
  axis.args=list(cex.axis=2), legend.width=3)

# Random field of apparent survival
r <- rasterFromXYZ(data.frame(x=coordgrid[,1], y=coordgrid[,2], z=out1$mean$phi))
plot(r, col=topo.colors(20), axes=FALSE, box=FALSE, asp=1, main="Survival",
  axis.args=list(cex.axis=2), legend.width=3)

# Random field of (absolute) recruitment
r <- rasterFromXYZ(data.frame(x=coordgrid[,1], y=coordgrid[,2], z=out1$mean$gamma))
plot(r, col=topo.colors(20), axes=FALSE, box=FALSE, asp=1, main="Recruitment",
  axis.args=list(cex.axis=2), legend.width=3)

# Uncertainty in the random field of initial abundance (% CV)
r <- rasterFromXYZ(data.frame(x=coordgrid[,1], y=coordgrid[,2], z=100 *
  out1$sd$lambda / out1$mean$lambda))
plot(r, col=topo.colors(20), axes=FALSE, box=FALSE, asp=1, main="Uncertainty of density (% CV)",
  axis.args=list(cex.axis=2), legend.width=3)

# Uncertainty in the random field of survival (% CV)
r <- rasterFromXYZ(data.frame(x=coordgrid[,1], y=coordgrid[,2], z=100 *
  out1$sd$phi /out1$mean$phi))
plot(r, col=topo.colors(20), axes=FALSE, box=FALSE, asp=1, main="Uncertainty of survival (% CV)",
  axis.args=list(cex.axis=2), legend.width=3)

# Uncertainty in the random field of recruitment (% CV)
r <- rasterFromXYZ(data.frame(x=coordgrid[,1], y=coordgrid[,2], z=100 *
  out1$sd$gamma / out1$mean$gamma))
plot(r, col=topo.colors(20), axes=FALSE, box=FALSE, asp=1, main="Uncertainty of recruitment (% CV)",
  axis.args=list(cex.axis=2), legend.width=3)
```

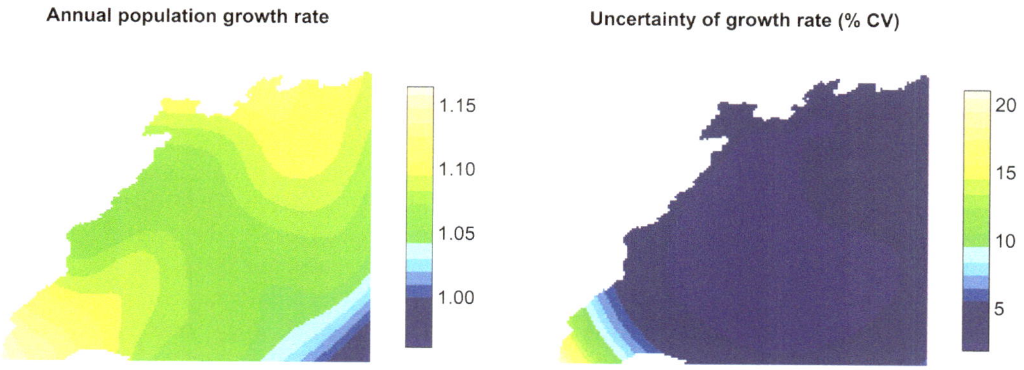

FIGURE 20.9

Map of the estimated average annual population growth rate of the kestrel population in northwestern Switzerland (left) with a map of the associated uncertainty shown on the right.

We emphasize that beyond their considerable artistic appeal, we consider the maps in Fig. 20.8 quite revolutionary in principle, since they are an example of that holy grail in population ecology—landscape demography (Gurevitch et al., 2016). That is, the representation of spatial variation in demographic variables at a large spatial scale. Despite this, we have to be careful when trying to interpret these maps and should presumably only take some of the large patterns at face value. In addition, several of the extreme values at the borders of the modeled domain are most likely mere artifacts due to our choice of a spline spatial modeling framework (Trevor Hefley, pers. comm.). Accordingly, we see that the estimation uncertainty for all three fields blows up at these same edges. We don't want to interpret the random fields much here but merely draw attention to the survival map, which suggests values of annual survival that are greater than 0.9 in the majority of the study area. This is suspiciously high for kestrels, and we will say more about this in the Discussion.

We also produce a map of the average annual population growth between 1999 and 2017 (Fig. 20.9). This suggests considerable spatial variation of kestrel population dynamics but overall a population increase almost throughout the study area.

Finally, we plot an estimate of the trajectory of the total population size in the study area (Fig. 20.10 left) and of the total magnitude of the multiplication rate of the population between 1999 and 2017 (Fig. 20.10 right). Things look good for Swiss kestrels currently; we see what we had already seen in the posterior summary above: the population is estimated to increase from about 3800 to 9100 females between 1999 and 2017. That is, we estimate a 2.5-fold population increase over the 19 years of study.

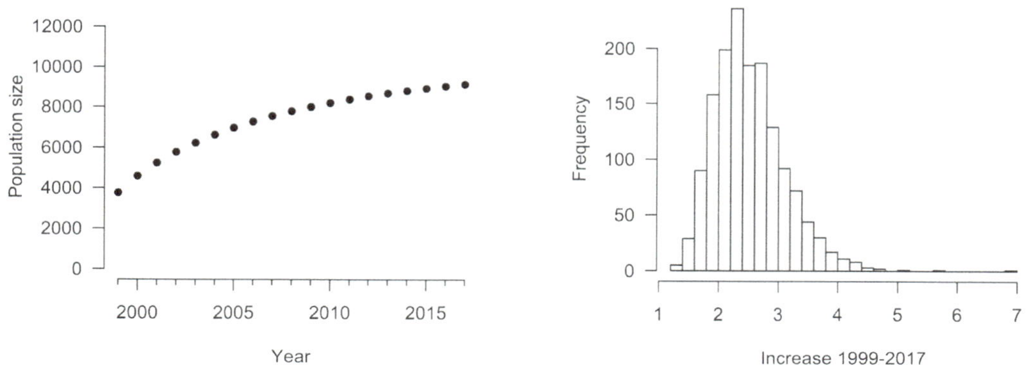

FIGURE 20.10

Estimates of the trajectory of the total population size in the northwestern quarter of Switzerland (left) and total population increase as a multiplication factor between 1999 and 2017 (right).

20.7 DISCUSSION

In this chapter, we have developed a spatial IPM where we have for the first time introduced into a population-based IPM the notion of a spatial random field. This is one conventional way in which "space" is treated in all branches of spatial statistics (Diggle et al., 1998; Illian et al., 2008; Cressie and Wikle, 2011; Diggle, 2013; Banerjee et al., 2014; Cressie, 2015; Baddeley et al., 2016; Wikle et al., 2019). An important feature of our IPM and most other broad- and narrow-sense spatial IPMs (e.g., Péron et al., 2010; Ahrestani et al., 2017; Zhao et al., 2019a, 2019b; Zhao, 2020, Chapters 18 and 19) is that they incorporate "space" by retaining the usual population-level perspective, i.e., the description of an aggregate of individuals at a site. Spatial dependence of population dynamics in an IPM can be treated by defining discrete spatial units with interactions (McCrea et al., 2010; Péron et al., 2010) or with mere spatial stratification of parameters either without spatially indexed environmental covariates (Morrison et al., 2016; Ahrestani et al., 2017; Zhao et al., 2019b) or including them (Zhao et al., 2019a; Zhao, 2020). In a stricter sense, these different manners of dealing with space in the model do not all result in what a statistician would call a "spatial model," since coordinates do not appear in the model. In contrast, one type of IPM that is a spatial model in the narrow sense of the term is represented by Zhao (2020), who used in an IPM a spatial extension of the DM model (Zhao et al., 2017) that expresses the immigration to a cell as the sum of the emigrants expected to come from all neighboring cells. In such a spatio-temporal Markov model, the actual spatial configuration of each "local" population matters. The random field approach to space in an IPM that we introduce here results in another type of truly spatial IPM. Both the spatial IPM in Zhao (2020) and that in this chapter retain the population-level perspective of all typical IPMs. In contrast, in the spatial IPMs of Chandler and Clark (2014), Chandler et al. (2018) and Sun et al. (2019), an individual-based perspective is adopted; see Chapter 21.

Despite our great enthusiasm about our model, we see this study as somewhat of a "concept IPM." We believe it illustrates several important features of an IPM that integrates data from a large number of sites and shows the direction to a kind of large-scale, spatial IPMs that we predict we will see increasingly in the near future. But at the same time, there are some serious deficiencies in our model that need to be mended, and we think that a host of improvements and extensions is possible. All of that is discussed in the following.

20.7.1 THE PATH TO LANDSCAPE DEMOGRAPHY

Novel in our approach is the large sample size of sites, or local populations, integrated into a single model, and especially our use of a latent spatial random field as the vehicle for that integration. This allows for a seamless reconciliation of the spatial mismatches between our data sets; these are almost guaranteed to occur when such a large number of study sites are integrated in a single model. Adoption of the random field allows each site to inform the parameter estimates at the "right" place. At the same time, it enables us to explore spatial variation in the demography of our species and to relax the "common demography assumption" (Section 7.2.3) to a large extent. Finally, our modeling framework leads the way toward that fascinating vision of *landscape demography*, i.e., the study of population dynamics and its spatial variation at large spatial scales (Gurevitch et al., 2016).

Relatively new in our spatial IPM is the use of that common source of information about the demography of a species contained in the data produced by large-scale biodiversity monitoring schemes (see also Pagel and Schurr, 2012; Zhao et al., 2019a, 2019b, 2020). In our model, we even integrate two different kinds of data from such large-scale monitoring programs, one longitudinal (i.e., the MHB data, a typical sort of breeding bird survey) and the other cross-sectional (i.e., a typical breeding bird atlas project). To do so, we had to solve two further challenges in data mismatch, both temporal in this case, and in both models, for the underlying states and the observation in these two data sets. First, to be able to specify shared abundance parameters between the MHB and the atlas data sets, we needed to propagate forwards the abundance model to express the expected abundance for the atlas data as a function of the initial abundance parameter, which refers to a year more than 1 decade before the first year of the atlas surveys, and the population dynamics parameters for survival and recruitment. Second, in our IPM we used a variant of the MHB data that had been used by the recent nonspatial IPM on Swiss kestrels by Fay et al. (2019) and that contained only the aggregate count of territories per year and site, although these are in fact based on three surveys in all the 102 MHB sites used in our model. In contrast, for the atlas data we have used the full temporal resolution for these data. We resolved the resulting temporal mismatch by expressing the combined detection probability in the aggregated data as a function of the per-survey detection probability parameter assuming independent surveys and a basic binomial argument.

20.7.2 THE SPATIAL FIELD APPROACH FOR AN INTEGRATED POPULATION MODEL USING LARGE-SCALE MONITORING DATA

Ours is certainly not the first example of the use of large-scale monitoring programs in an IPM. Even in the foundational paper by Besbeas et al. (2002), a population index derived from the national British breeding bird survey was used as the population count data. In their nonspatial IPM for Swiss kestrels, Fay et al. (2019) have used the same type of MHB data as we have in our model. However, in both of these, a very crude geographical approach was adopted to modeling these data wherein the spatial replicate counts were distilled into just a single time series for the entire modeled domain (Britain in Besbeas et al., 2002; and Switzerland in Fay et al., 2019). We believe that our model here demonstrates a more interesting approach to the modeling of such large-scale monitoring data.

In addition, since the work of Besbeas et al. (2002), a majority of published IPMs have taken a more local view and have described the dynamics of populations at local scales. If they did have a larger spatial extent (such as Koons et al., 2017 and others), space was not treated in the same flexible way that our conventional spatially statistical device of a latent random field does or else the more process-based approach of Zhao et al. (2020). In both approaches, one or more parameters are allowed to change smoothly in two dimensions over a possibly large region, but they are informed locally at the specific places where each data set has been collected.

Thus, we invoke a set of spatially correlated random variables that are defined for the entire modeled domain but are informed only at a small subset of all sites in the domain, which is the dominant approach to the modeling of spatial structure in spatial statistics. In addition, we could employ environmental covariates to explain part of the spatial variation in density or any other parameter in the model. These could of course be used alongside a latent spatial field, which would in that case become a "residual" spatial field, thus "smearing out" spatially structured unexplained variation after we have explained all spatial variation that can be explained by the covariates. See the maps in the most recent Swiss breeding bird atlas (Knaus et al., 2018) for a large-scale application of spatial modeling, using JAGS, of abundance and occupancy involving both known and measured covariates and a spatial field that is defined by two-dimensional splines, exactly as in this study.

We stress that it will be very useful in the future to see what spatial statistics has to offer for population modelers interested in the adoption of IPMs at large spatial scales. Of course, this will include the issues known as *dimension reduction*: fitting spatial models is notoriously very computationally expensive. Thus, much effort is spent in this field to find methods that reduce this computational burden by reducing the computational dimension of a modeling problem. One such approach is the spline (or GAM) approach that we have adopted here, where instead of modeling the "spatial relations" at each individual spatial location for which we have data, it is defined for a smaller number of spatial locations known as knots plus the estimable effects of a distance-dependent covariate that links each knot location to the locations where our data are (for estimation) or, in general, to *every* location in the domain of interest, for inter- and extrapolation and to produce nice maps.

20.7.3 COMPARISON OF DIFFERENT SPATIAL INTEGRATED POPULATION MODELS

The IPMs in this chapter and those of Chandler and Clark (2014; Chapter 21) and Zhao (2020), are the most spatially explicit in the narrow sense of the term as used in statistics, as we have explained above. So what is the distinction between these three models? At a certain level, our model and that of Chandler and Clark (2014) are very similar in that a key ingredient is a latent spatial field, which in Chandler and Clark (2014) is the intensity of the point process model describing the number and locations of the activity ranges of each individual in the modeled population. In the model in this chapter, a latent spatial field is defined both for the initial expected density per 1 km^2 spatial unit, as well as for survival probability and recruitment. The main distinction between these two types of models is that the model of Chandler and Clark (2014) is individual-based, while in that in this chapter, the "points" are the study locations, or spatial replicates, of which there is a large number (nearly 700). These are used to infer estimates of three fields that represent the aggregate over all individuals present at each study site location. Thus, the model in this chapter still retains the population-level perspective that is adopted in virtually all IPMs to date. The model in Zhao (2020) is also a population-level model, but rather than simply adopting the descriptive treatment of space by a random field, it describes some of the processes that create spatial structure in demographic parameter. Specifically, it models dispersal in a spatially Markovian manner within the neighborhood of each cell.

Demographic data with individual resolution are much harder to obtain, and hence, we will usually see individual-based, spatial IPMs developed only at comparatively small spatial extents. In contrast, data wherein individuals are aggregated over to counts of unmarked individuals, such as produced by countless biodiversity monitoring programs the world over, are much more available than individual-level demographic data and are much "cheaper" in this sense. This means that they are typically available over much larger spatial extents than are the individual-level demographic data. Therefore, the type of population-level spatial IPMs developed in this chapter will likely have a fairly larger scope of application.

One further distinction between the individual-based spatial IPMs of Chandler and Clark (2014; Chapter 21) and the population-based spatial IPMs as developed here is that the computational burden of the individual-based models will typically be even larger and increase much faster as the extent of the modeled population grows. This may thus limit somewhat the scope of application of individual-based spatial IPMs. In contrast, for a population-based spatial IPM data are much more widely available, they are available at often much larger spatial extents and although still very computationally expensive to fit, they are less so than the individual-based models. Therefore, we envision many more population-level spatial IPMs in the future.

It would be possible to merge the two approaches. That is, we might try to model surfaces of density and of the demographic rates in an individual-based manner at a set of study sites where individual recognition of the animals allows such spatial capture-recapture- (SCR-) type of modeling. Then we use standard methods of spatial statistics, i.e., the invoking of spatial random fields, to link these local studies across the landscape and to interpolate between them. In the context of a "simple" SCR model, i.e., not technically in an IPM as defined in this book, this has been achieved by Bischof et al. (2020), though not with a continuous latent random field spanning all local studies, but by the use of spatially indexed covariates for the interpolation between different study sites.

20.7.4 PROCESS-BASED SPATIAL INTEGRATED POPULATION MODELS

In future spatial IPMs, we can envision more process-based representations of space, where the spatial and spatio-temporal autocorrelation in the demographic parameters will be described by dispersal processes, as in reaction-diffusion models of population dynamics (Wikle, 2003; Hooten et al., 2007; Pagel and Schurr, 2012; Williams et al., 2017; Hefley et al., 2017b, 2017c; Louvrier et al., 2020) and related models including Zhao et al. (2017) or Howell et al. (2020). A process-based representation of the spatial autocorrelation in a demographic landscape has multiple advantages such as an improved ability to transfer results to new areas and predict a population into the future (e.g., in a PVA context), but most of all, a deeper understanding of the population processes.

20.7.5 ALTERNATIVE MODELS FOR THE KESTRELS

During the development of the model in this chapter, we experimented with many model variants and would like to briefly comment on two of them. Both were simpler models motivated by the difficulty to achieve convergence in the model that we now present and which we intuited was due to the complex forward-propagation of the expected abundance for the atlas data as a function of the three parameters in the population model between 1999 and the period during which the atlas data utilized here were collected (2013–15).

In the first simplification, we tossed out the atlas data set altogether and only fitted the model to the MHB and the dead-recovery data. This model ran more quickly, but we failed to reach convergence for a number of parameters, including initial abundance (λ) and detection probability (p). This was not so surprising to us, since although under ideal conditions it is possible to estimate abundance and detection in the DM model from data that are not in the robust-design format (i.e., with a single count per site and year as in our format of the MHB data used here), estimability must certainly be tenuous in this situation. Quite likely, it will also depend on the magnitude of abundance. Our kestrel counts are binary for the most part and thus contain less information than would, say, counts of a small passerine that might easily have counts that are 10 times greater. Thus, we decided that we needed those atlas data, which not only add a large sample size but in addition have the short-term temporal replication in the counts that allow us to separate abundance and detection according to the standard static binomial N-mixture model (Royle, 2004).

Thus, the second simplified model that we experimented with did include the atlas data, but in it we specified a simpler integration where only the detection parameter was shared between the MHB and the atlas data. In this simpler variant of the three-way combination of our data sets, we estimated the abundance parameters in the atlas data set to be free parameters independent from the population dynamics estimated from the MHB data, but there is a shared detection parameter.

$$N_i^{Atlas} \sim Poisson\left(A\lambda_i^{Atlas}\right)$$

$$C_{i,j}^{Atlas} \sim Binomial\left(p, N_i^{Atlas}\right)$$

Thus, in this variant of the IPM, *p* was the only parameter shared between the MHB and the atlas data. This model converged, and much more rapidly than the full model presented in this chapter (i.e., in about 24 instead of 88 h). Convergence was fine for all parameters. The main patterns in the parameters were similar, although total population sizes were estimated about 50% higher.

We like to mention these two model variants since you will sometimes have to make a choice of whether to include in an IPM a particular data set. Even if you choose to include a certain data set, there may be more than one parameter that you may choose as a link with one or more of the other data sets used to estimate the parameters in the IPM.

20.7.6 COMMENTS ON USE OF THE DAIL-MADSEN MODEL AS THE CORE OF A SPATIAL IPM

The core of our model, representing the demographic population model, is a simple DM model (Dail and Madsen, 2011) as in the models of Ahrestani et al. (2017; Chapter 19) and Zhao et al. (2019a, 2019b) or similar as in the more complex, spatial DM model of Zhao (2020). This is a very natural model for the multivariate, spatially indexed population count data in these studies and to serve for estimation of population size based on these spatially replicated counts of unmarked individuals in a sort of binomial *N*-mixture model (Royle, 2004). There exists an increasing number of variants of the DM model now, based on generalizations both of the state and the observational part of the model (Chapter 2 in Kéry and Royle, 2021). For instance, it is straightforward to extend the basic model to data collected in other protocols that produce the required extra information to estimate detection probability including distance sampling (Sollmann et al., 2015), removal and other types of protocols that lead to a multinomial, capture-recapture-type of likelihood (Ken Kellner, cited in Chapter 2 in Kéry and Royle, 2021). In addition, the original single-state model has been extended to a multistate model by Zipkin et al. (2014a, 2014b, Chapter 2 in Kéry and Royle, 2021), thus opening up this modeling framework to yield the whole tremendous power of multistate models in general, which in turn can be viewed as a sort of extension of a matrix population model where the stage-classified population model is embedded in a likelihood to estimate its parameters. The main distinction between the multistate DM model and the classical multistate capture-recapture models is again that between a population-level and an individual-level formulation. In the population-level variant of the model, the multistate DM model, replicate data for multiple sites are required, while for the individual-based model, replicate data for multiple, individually identifiable animals (or plants) at multiple sites are required.

We have argued already in Chapter 4 that the DM model offers one alternative to the usual Gaussian state-space model so commonly employed in contemporary IPMs that has a more natural representation of the observation process by a discrete-valued distribution, but which is otherwise exactly the same type of state-space model (see also Section 1.7 in Kéry and Royle, 2021). We think that the DM is a great model in principle, but unfortunately, it may suffer from severe problems in practice: while the estimation of abundance and detection (and consequently of the annual population growth rates) appears fairly unproblematic[1], the parsing out of population growth into a contribution from survival and one from recruitment appears to be extremely complicated with empirical data. We have fitted DM models to hundreds of data sets (mostly for small to medium-sized birds) and in virtually all cases, we found that the estimate of survival was far too high. To provide an illustration, fitting a DM model to MHB data for the winter wren (*Troglodytes troglodytes*), the third-smallest bird species in Switzerland, produces a survival estimate of about 0.94, of which an albatross would be fairly proud. That is, in this and many other cases, we *know* either from capture-recapture models fit to other data from these species or else from simple allometric considerations that such values are too high for their annual survival. Consequently, of course, estimates of recruitment will be drastically too low, although we wouldn't be able to say so from just their magnitude, since recruitment is so hard to estimate in natural populations. Hence, we usually don't have any benchmark with which to compare any recruitment estimate.

Some progress has been made to improve the basic DM model by adding in more demographic realism such as density dependence or temporal variation in the demographic rates (Bellier et al., 2016, 2018), but we are not entirely sure that this solves the major problem that the estimates of the demographic processes are often biased. We suggest that this aberrant behavior of the model for many real data sets ought to be studied in more detail, for instance by using simulation to identify mechanisms that can lead to this pattern when data sets are analyzed with the DM model. This might provide pointers to how the model would have to be changed to avoid such bias or even if this is possible in principle.

[1]We dare say this at a time when this model has not yet come under intense scrutiny by Bill Link and Richard Barker.

20.7.7 FURTHER POSSIBLE WORK WITH THE KESTREL IPM

Our case study here provides a nice illustration of this fairly typical DM behavior of producing too high survival estimates, even when we included data from 24,000 ringed kestrels and when using what one might think of as moderately strongly informative priors—although see Section 2.4 and specifically Table 2.1 and Fig. 2.5, which may suggest that the prior for survival adopted in our model is not that strong after all. Anyway, we obtain what we are fairly certain will be estimates of survival that are much too high: the random field for survival in Fig. 20.8 shows values > 0.9 for most parts of the modeled domain, which again is in the albatross region and disagrees strongly with the recent (nonspatial) survival estimate from partly the same data of 0.72 (Fay et al., 2019).

We estimated a 2.5-fold increase in the population over 19 years, resulting in a total population of about 9100 pairs throughout the almost 16,000 km^2 study area in 2017. A simple area expansion would therefore suggest about $9100/16 \times 41.3$ and thus a little over 23,000 kestrel pairs in all of Switzerland (in practice, we would expect fewer, since our study area has higher-than-average habitat quality for the species). It is interesting to compare this estimate with the estimate of the national population total in the latest Swiss breeding bird atlas (Knaus et al., 2018): 5000–7000 pairs! So either we greatly underestimated the national population size in the atlas or else here we overestimate density greatly, or there are major errors in both. In contrast, there is no disagreement between the estimate of the population increase in our analysis with that reported in Knaus et al. (2018).

Thus, our analysis is clearly unsatisfactory in several respect and its actual results should be viewed with caution and as being somewhat provisional. We would hope that we simply overestimate the level of survival by some amount but that the spatial patterns revealed remain unaffected by this bias, and analogously for the levels and the patterns in the other parameters, especially recruitment which is most directly affected. However, at this stage this is no more than a hopeful conjecture which should ideally be confirmed or rejected by specialized simulations.

In the future, we plan to explore the following topics in this IPM:

- Explore ways to modify the process model so that more sensible estimates of the mean level of survival result, i.e., that overestimates of survival can be avoided. In this, we think about the DM model both alone (i.e., outside of the context of an IPM) and when specified as part of a population-level, spatial IPM. As suggested by the work of Bellier et al. (2016, 2018), one line of attack might be to introduce more demographic realism in the form of density dependence and temporal variation. In our case study, we tried the former for survival (in addition to recruitment, which is naturally density-dependent in our formulation of the recruitment process; see above) but failed with that most dreaded of all JAGS error messages ("invalid parent value for node X") and then gave up.
- Add environmental covariates to explain part of the spatial (and possibly of the temporal) variation in the parameters, such as elevation, forest cover, proportion of arable land, or some measure of winter hardness for the temporal variation in population.
- Relax the strict constancy assumption of p (which at face value appears ridiculous anyway), perhaps by representing p as the sum of the effects of some covariates, a spatial (residual) random field, and especially by adding year effects. Zhao and Royle (2019) suggest that one always allows for temporal variability in detection probability in a DM model.
- We could try to add age structure into the model by using a variant of the multistate DM models (Zipkin et al., 2014a, 2014b). The counts used in our IPM are for adults only, but the dead-recovery data produce estimates of juvenile survival anyway.
- The dead-recovery data enter the analysis at a fairly crude spatial grain in our analysis, namely aggregated per 50×50 km^2 square. We could try to increase their information content about spatial variation by working with a more highly resolved aggregation scheme, for instance by 10×10 km^2 squares. In addition, our current model doesn't allow for temporal variation in survival, and it would appear like a natural candidate to extend the model to allow for this by adding random year effects for instance.
- Finally, we could try to move away from a mere phenomenological representation of the spatial structure in the demographic landscape to a more process-based one. For the population model this would mean that we include the exchange of individuals among neighboring sites, i.e., dispersal, as for instance, in the models by Wikle (2003), Pagel and Schurr (2012), Hefley et al. (2017b; 2017c), Zhao et al. (2017) or Howell et al. (2020).

In summary, despite some misgivings having to do with that well-known behavior of the DM model to produce estimates of survival that often have a (strong) positive bias and therefore a negative bias for recruitment rates, we think the model developed in this chapter is a blueprint for many future population-level spatial IPMs, which really are nothing more than dynamic species distribution models, that have also been called dynamic range models (Pagel and Schurr, 2012). Increasing the spatial extent of the study area in an IPM in combination with the framework developed here or elsewhere (e.g., Ahrestani et al., 2017; Zhao et al., 2019a, 2019b; Zhao, 2020) produces a host of advantages in population modeling and invites us to use that hitherto almost untapped source of demographic information for an IPM: data from large-scale, longitudinal and cross-sectional biodiversity monitoring programs.

CHAPTER

BLACK BEAR

21

CHAPTER OUTLINE

IPM ingredients: spatial capture-recapture (SCR) data, occupancy data
Location: USA
Spatio-temporal scale: 1200 km^2, 6 years

21.1 INTRODUCTION

All the integrated population models (IPMs) you have seen so far in the second part of our book have specified the change in population size at the scale of some largish spatial areal unit. In other words, individuals are spatially aggregated into what we typically called a population or a colony. The definition of the borders of such a population may be based on ecological considerations. For example, a species may only occur in a specific type of habitat that is patchily distributed, or a study is conducted on an island. Most of the time, however, a population is delimited in a purely arbitrary manner such as "the area containing all the individuals I can survey in a breeding season." Importantly, our inferences about the spatial scale and movements into and away from the population (i.e., immigration and emigration) are conditional on the spatial definition of our population. Consequently, generalizations from other spatial scales or comparisons with other populations of the same species are notoriously difficult and may to some degree remain speculative.

One possibility to address these difficulties is to work in a larger area and divide it up into smaller spatial units for which distinct process models are developed but where these process models are still formulated for individuals that are aggregated into these areal units. Inferences are then possible about these smaller units as well as for the larger unit comprising the sets of smaller units, and therefore questions of spatial scaling can be addressed. You have seen such a variant of a spatial IPM in Chapters 19 and 20.

A completely different class of IPMs specifies process models that are individual-based. That is, they are no longer specified for summaries of individuals over some areal aggregation unit, i.e., for some "population." In an individual-based process model, the fate of every individual is described, e.g., when and where it is born, where it moves to, and when it dies. Typically, for each of these random processes, a point process model is adopted (Chandler and Clark, 2014; Chandler et al., 2018b). Individual-based process models are completely general and allow for inferences about demographic processes and their consequences (e.g., population size) at any desired spatial scale. A key requirement for the adoption of such models is that data are available about the *locations* of all individuals observed. Such an IPM, which we call an *individual-based spatial IPM*, is the focus of this last case study in our book.

We showcase the individual-based spatial IPM for black bears (*Ursus americanus luteolus*, Fig. 21.1) in Louisiana, USA, developed by Chandler and Clark (2014). We jointly analyze spatial capture-recapture (SCR) and occupancy data with the goal of estimating survival, recruitment, and population density. We also want to produce density maps of bears

Integrated Population Models. https://doi.org/10.1016/B978-0-12-820564-8.00012-9

FIGURE 21.1

Black bear (*Ursus americanus*; photograph by John E. Marriott).

over the 6 years of study and gauge whether the average change in population density was spatially uniform or not, i.e., to produce a map of population trend at a high spatial resolution. Our analysis is a variant of the one in Chandler and Clark (2014).

21.2 DATA SETS

Data on the Louisiana black bear were collected during 6 years (2007–2012) in the Upper Atchafalaya River Basin in south-central Louisiana, USA (Chandler and Clark, 2014). The study area was largely covered by bottomland hardwood forest. It was gridded into cells measuring 1.6×1.6 km^2 (i.e., one square mile), and within each grid cell, one suitable location for placing traps for snaring bear hair ("hair snares") was identified. The size of the study area defined as the sum of all grid cells equipped with a trap amounted to about 270 km^2. Sampling was conducted annually at 115 hair snares, which were strands of barbed wire around a bait (two donuts—American bears *love* fast food ...) and a scent lure. Some traps later had to be moved because the originals were destroyed, so a total of 128 locations were eventually included in the 6-year study. We often simply call these trap locations "locations," and sometimes just "traps." Hair samples were collected on eight consecutive weeks in each year according to a robust design (Pollock, 1982) where the 6 years represent the primary sampling occasions, and the nested eight weekly periods are the secondary occasions. In the model, population closure is assumed among secondary but not among primary occasions.

This design produced a large number of collected hair samples. Many of these samples were likely redundant in the sense that they were from the same individual at the same location during an occasion. Genotyping all samples would have resulted in high costs, which were avoided by thinning in two steps: first, only one sample per location and occasion (week) was retained, and second, from the samples retained each week, a random subsample of at most 38 samples were genotyped for individual identity and sex, producing non-invasive genetic sampling data (see Chandler and Clark, 2014, for details).

Genetic capture-recapture data are rich in information, but they are expensive to obtain, as genotyping is very costly. In order to explore a more cost-effective scenario, we assume here (as did Chandler and Clark, 2014) that hair samples collected were only genotyped every other year. Thus, in 2007, 2009, and 2011, the hair samples were genotyped, resulting in SCR data, while in 2008, 2010, and 2012, they were not genotyped, resulting in simple "occupancy" data only, i.e., location- and occasion-specific detection or nondetection of black bears. Chandler and Clark (2014) developed an IPM with which the two data sets could be jointly analyzed. This model qualifies as an individual-based spatial IPM because the occupancy data represent a sort of longitudinal population count, and they are augmented with SCR data that directly inform the two demographic processes in the model: survival and recruitment.

21.3 POPULATION MODEL

The population model of an individual-based spatial IPM differs from that of traditional, nonspatial IPMs, which specify a population model that aggregates individuals into age or stage classes but is ignorant about individual locations and describes numerical changes as a function of demographic rates that are assumed to be identical for all individuals within the same class. In contrast, individual-based spatial IPMs have an individual-based formulation at their core; thus, there is no aggregation of individuals into stage classes. The process model is a spatial point process model (Baddeley et al., 2016; González et al., 2016) that describes the arrival, location, movement, and eventual disappearance of each individual over time as a function of demographic processes that are likewise specified at the level of the individual (Chandler et al., 2018a). One obvious consequence is that we can no longer show a life-cycle graph, as the individuals are no longer aggregated into stages. But other than that, the development of such a model consists of the same conceptual steps as for any conventional stage-structured Markovian population model—it describes the population in its first year of study and its evolution over time as a function of demographic processes and individual states during the previous time step. For an individual-based model, we focus on the location of an individual, and hence, we first model the initial location of all individuals and then how these change over time as a function of survival, recruitment, and movement.

The state of the population in the first year (i.e., the initial state) is modeled using a spatial point process model that describes as random quantities both the number of individuals (N_1) and their locations in a two-dimensional spatial region called the *state-space* (S). The state-space is a precisely defined area that includes the possible location of all observed individuals. Note that its definition need not be identical to that of the study area, and indeed, it usually is not. Rather, the state-space is frequently chosen to be larger than the study area, and in some common forms of Bayesian analysis essentially amounts to a vague prior on the initial population size. Thus, we usually define it to be just large enough that it does not affect the posteriors any more (and thus can be called "vague") but not so large as to unduly increase the computational burden (Royle et al., 2014). We then describe the coordinates ($s_{i,1}$) of the average location of individual i in year $t = 1$. This average location is often called the home-range center or the activity center. The intensity function $\lambda(s)$ is the expected density of activity centers exactly at location s in state-space S and is a key element in a spatial point process model. If the density is constant in space, the point process is said to be homogeneous, and the individual activity centers are uniformly distributed, and thus $s_{i,1} \sim Uniform(S)$. The expected population size in year $t = 1$ is conceptualized as the number of activity centers within S and is given by $N_1 = \Lambda = \int_S \lambda(s)ds$, which you can think of as the sum over all pixels of a continuous "intensity landscape" represented by λ, when that landscape is approximated by small discrete spatial units (pixels).

Direct estimation of N_1 poses problems when using Markov chain Monte Carlo (MCMC) because the dimension of the parameter space will typically vary with every iteration of the algorithm. Therefore, Chandler and Clark (2014) use data augmentation, also known as parameter-expanded data augmentation (Royle et al., 2007; Royle and Dorazio, 2012), where the observed n individuals are augmented with an arbitrary but large number $M-n$ of "potential" individuals with all-zero detection histories and missing locational information. The model is then expanded to include an additional hierarchical level, which is a binary indicator variable that describes whether an augmented individual is a "real" individual and is governed by an estimable probability ψ. Under the above assumption of a homogenous point process, initial population size is given by $N_1 \sim Binomial(\psi, M)$, where M is a large integer number and $\psi = \Lambda/M$ (Royle et al., 2007; Royle and Dorazio, 2012). Data augmentation can be thought of as inducing a discrete uniform prior between 0 and M for the number of individuals ever alive in the state-space during the study period (i.e., the superpopulation size). Hence, M must be chosen such that the posterior of the superpopulation size is not truncated. This can be easily diagnosed by fitting a model, inspecting a summary of the posterior samples of the superpopulation size, and if it "hits" the boundary of M, refitting the model with a larger value of M. Remember that too-large values of M don't hurt the estimates, but they result in longer run times for the model.

Conceptually, in the model of Chandler and Clark (2014), we assume that there are M individuals that could potentially be recruited into the population over the course of the entire study. We define $z_{i,t}$ to be the latent state of individual i at time t indicating whether i is alive at t ($z_{i,t} = 1$) or whether i is either dead or not yet recruited into the population at t ($z_{i,t} = 0$). We use a Bernoulli trial to decide whether a member of M is already recruited into the population at $t = 1$, i.e.,

$$z_{i,1} \sim Bernoulli(\psi).$$

The population size in the first year is then given by $N_1 = \sum_{i=1}^{M} z_{i,1}$. Models of inhomogeneous point processes, i.e., where the intensity function is spatially variable, can also be specified with data augmentation, e.g., when the intensity function is expressed as a function of spatially indexed covariates or random effects (Chandler et al., 2018b).

The state process model describes changes in abundance and distribution over time as a function of probabilities of survival (ϕ) and recruitment (δ). In principle, in such a spatio-temporal point process model the individuals could also move between time steps (representing dispersal, Gardner et al., 2010; Ergon and Gardner, 2014; Schaub and Royle, 2014; Chandler et al., 2018b), but in our case, we assume no dispersal over the years. We model the arrival (recruitment) and disappearance (death) of each member in the set of M with a sequence of Bernoulli distributions as

$$z_{i,t} \sim Bernoulli\left(z_{i,t-1}\phi_t + \left(1 - a_{i,t-1}\right)\delta_t\right).$$

The argument of this Bernoulli trial is the sum of two components and is best thought of as a sort of logical "switch" to turn the survival and recruitment processes on and off. The first component applies when individual i was alive and already recruited during the previous occasion $t-1$, i.e., when $z_{i,t-1} = 1$. That individual will then still be alive at time t with probability ϕ_t. The second component models the arrival of new individuals not yet recruited into the population. The variable a is a function of z, indicates whether individual i is available to be recruited in year $t-1$, and is given by $a_{i,t-1} = \max\left(z_{i,1}, z_{i,2}, \ldots, z_{i,t-1}\right)$. If an individual is still available for recruitment, i.e., when $1 - a_{i,t-1} = 1$, it may be recruited with probability δ_t. However, as soon as it has been recruited, it gets $z = 1$ and therefore will no longer be available for recruitment, and δ_t will then become zeroed out. In addition, once the individual has been recruited and dies in occasion $t-1$, both $1 - a_{i,t-1}$ and $z_{i,t-1}$ will become zero, and all future states of the individual will be drawn from a Bernoulli with probability zero.

Recruitment is governed by probability δ_t and is expressed at the individual level. This parameterization of recruitment is not commonly seen in population models, but we can express per-capita recruitment (γ_t) as $\gamma_t = A_{t-1}\delta_t / N_{t-1}$, where $A_{t-1} = M - \sum_{i=1}^{M} a_{i,t-1}$ is the number of individuals still available to be recruited at time $t-1$. Per-capita recruitment can also be estimated directly in individual-based spatial IPMs, which then also allows modeling of spatial variation in this parameter (Chandler et al., 2018b). Per-capita recruitment is a function of productivity and juvenile survival and is a more familiar parameter in population models than individual-level recruitment.

At the population rather than individual level, the model of Chandler and Clark (2014) corresponds to the following simple population model that lacks an age structure:

$$S_{t+1} \sim Binomial(\phi_t, N_t)$$
$$R_{t+1} \sim Poisson(\gamma_t N_t)$$
$$N_t = S_t + R_t.$$

21.4 COMPONENT DATA LIKELIHOODS

We load the R data file, which contains a total of three objects. They are the SCR data, the occupancy data, and the locational information for the entire trap array. To better understand spatial data, mapping them is highly recommended, so we will produce a number of maps as well.

```
library(IPMbook); library(jagsUI)
data(bear)
str(bear)
```

```
List of 3
 $ scr       : int [1:78, 1:128, 1:8, 1:3] 0 0 0 0 0 0 0 0 0 0 ...
 $ occ       : num [1:128, 1:8, 1:3] 0 0 0 0 0 NA 1 1 0 0 ...
 $ trap.coord: num [1:128, 1:2] 618 619 618 620 621 ...
```

21.4.1 SPATIAL CAPTURE-RECAPTURE DATA

Object `scr` is a four-dimensional array that contains the spatial capture-recapture data. Its first dimension refers to the total of 78 bears identified by genotyping, its second dimension to the 128 trap locations, its third dimension to the eight (weekly) secondary occasions, and its fourth dimension is the three primary occasions, i.e., the 3 years 2007, 2009, and 2011 (note that this last dimension is not 6 because in this study, we assume we had the SCR data from only 3 out of the 6 years). If bear *i* was captured at trap *j* during secondary occasion *k* in year *t*, $scr_{i,j,k,t} = 1$. Otherwise, the array entry is 0. A trap *j* may have been closed or was not functioning during secondary occasion *k* in year *t*, in which case `scr` contains a missing value ("NA") at the corresponding place for all 78 individuals.

The object `trap.coord` contains the coordinates of each of the 128 trap locations. We produce a picture of the configuration of the 128 traps with coordinate units in kilometers (Fig. 21.2). We define the state-space as a rectangle around all trap locations and add it to the plot. Our state-space measures 1200 km^2.

```
library(scales)
plot(bear$trap.coord, xlab='x coordinate', ylab='y coordinate', frame=FALSE, las=1, pch=16,
  col='#002A64FF', asp=1)
# Define limits of the state-space and add it to the plot
xlims <- c(615, 645)
ylims <- c(3380, 3420)
rect(xlims[1], ylims[1], xlims[2], ylims[2], col=alpha('grey', 0.3), border=NA)
```

The capture-recapture data are spatially indexed, i.e., we keep track of the identity of the trap at which each individual bear was detected for each occasion and year. A total of 78 bears were detected in the 128 traps over 8 occasions in each of the 3 years when genotyping took place for a total of 626 detections.

```
table(bear$scr)              # Total number of detections
     0      1
174640    626
```

It is informative (and fun!) to check out the observed "distribution map" of individual bears, which is given by the locations at which an individual was detected. We produce such a map for all occasions and years combined, but you could

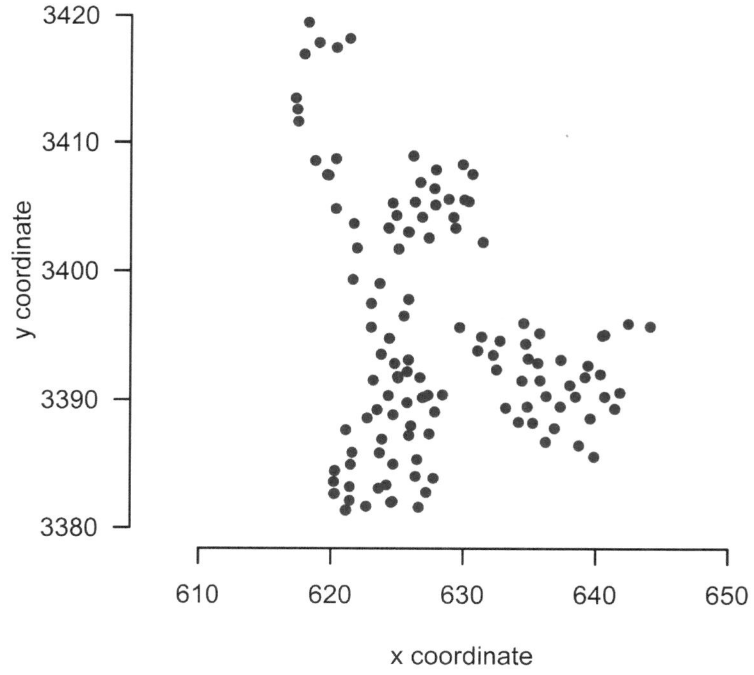

FIGURE 21.2

A map of the black bear study area of Chandler and Clark (2014) showing the locations of all 128 hair snare-traps (dots) and the 1200 km^2 state-space (*S*; gray rectangle) specified in the model. Axes are in units of kilometers in the Universal Transverse Mercator (UTM) coordinate system.

also generate it for each year separately. We need to first create a table (`bearXtrap`) indicating, for each individual bear and each trap, the number of times a bear was detected.

```
# Replace the missing values by zero
bear.sum <- bear$scr
bear.sum[is.na(bear.sum)] <- 0
# Produce table
bearXtrap <- apply(bear.sum, c(1, 2), sum)
```

We can then compute the total number of detections per bear and trap, and we discover the very skewed distribution of detection frequencies that is typical for SCR data.

```
table(bearXtrap)
   0     1    2    3    4    5    6    7    8    9   10
9638   209   82   23   10    6    8    3    2    1    2
```

We may also tally up the total number of times each bear was detected.

```
table(apply(bearXtrap, 1, sum))
 1  2  3  4  5  6  7  8  9 11 13 14 18 20 21 23 25 27 38 91
10  7 11  7  9  5  7  3  6  1  1  1  2  2  1  1  1  1  1  1
```

So 10 bears were detected on only a single occasion, but one bear was detected on a total of 91 occasions (all years and traps combined). We can now browse through all individuals and see where they were detected (again, here we combine all 3 years, which really cover a 5-year duration).

```
# Loop over 78 individuals and map their locations (individually)
# combined over all 3 years (which represent a 5-year duration)
op <- par(ask=TRUE)      # Press Esc to interrupt
for (i in 1:nrow(bearXtrap)){
  cat(paste("\n\n*** Plot bear number", i, "***\n\n"))
  lucky.traps <- which(bearXtrap[i,] > 0)
  plot(bear$trap.coord, xlab='x coordinate', ylab='y coordinate', frame=FALSE, pch=1,
    col=rgb(0,0,0,1), cex=1.5, asp=1, main=paste('Bear number', i))
  points((bear$trap.coord)[lucky.traps,1], (bear$trap.coord)[lucky.traps,2], pch=16, col='red',
    cex=1.5)
}
par(op)
```

When you execute this code, you will see extreme variation in the apparent ranging behavior of these bears over the 5 years.

Because the population process model is defined at the scale of the individual, there is no need for an additional model for the survival process. This is different from a classical IPM that models the population dynamics separately from the capture-recapture model, in which case both the population and the survival model require a state process model. In our spatial IPM, it is sufficient to just define an observation model for the SCR data.

The key idea of SCR models is that an individual i that is alive in year t, having activity center with coordinates s_i, may be captured during secondary occasion k in year t at trap j with coordinates x_j with a probability $p_{i,j,k,t}$. This capture probability is then defined to be a function of the distance between the activity center and the trap location—the further the trap is from the individual's activity center, the lower the capture probability becomes. There are countless possible functions that could describe such a distance-dependent decay in p, but the usual choice in SCR models is a half-normal detection function (Royle et al., 2014),

$$p_{i,j,k,t} = p_0 \exp\left(-\frac{\left\|x_j - s_i\right\|^2}{2\sigma^2}\right),$$

where $\left\|x_j - s_i\right\|$ is the Euclidean distance between the location of trap j and the activity center of individual i, p_0 is the baseline capture probability at distance zero, and σ is the scale parameter determining how fast the capture probability declines with increasing distance between the trap and the activity center. The observed SCR data (now denoted as y) are then modeled with a Bernoulli distribution as

$$y_{i,j,k,t} \sim Bernoulli\left(p_{i,j} z_{i,t}\right).$$

Here, we assume constant capture probability for all study years and secondary occasions and thus drop the indices k and t, but in principle, p could vary as a function of individual, trap, year, and secondary occasion in the design of this study.

21.4.2 OCCUPANCY DATA

The occupancy data (`occ`) are formatted as a three-dimensional array. The first dimension refers to the 128 trap locations, the second to the eight secondary occasions, and the third to the three primary occasions 2008, 2010, and 2012. If a bear is detected at trap j during secondary occasion k in year t, $occ_{j,k,t} = 1$, and otherwise it is 0. As for the SCR data, the occupancy data only need their own observation model, but their process model is exactly identical to that of the SCR data. Hence, in this IPM, the two integrated data sets share not just one or two but all parameters of their process model. In this way, both observation models are conditionally specified on the whole population process model defined in Section 21.3.

We use a Bernoulli distribution as an observation model for the occupancy data with a parameter that is the probability of observing at least one individual bear at a given trap j in secondary occasion k and year t. From the process model, the state $z_{i,t}$ in year t of each individual i and its location s_i are both known (or rather, can be estimated), and hence we can also estimate the capture probability $p_{i,j}$. The probability that at least one bear is captured is then the complement to 1 of the probability that none of the bears present and alive is captured. In summary, the observed occupancy data (here denoted as o) are modeled as

$$o_{j,k,t} \sim Bernoulli\left(1 - \prod_{i=1}^{M}\left(1 - p_{i,j}z_{i,t}\right)\right).$$

21.5 THE INTEGRATED POPULATION MODEL

The above description completes our individual-based spatial IPM; there is no extra integration step necessary as in traditional, nonspatial IPMs. This is because the observation models of both data sets are conditional on the same underlying process model and not on just a single parameter as in most conventional IPMs. The process model yields estimates of the individual state variables z and s, which also occur in the observation models for both data sets.

As explained in Section 21.3, we need to augment the SCR data with "imaginary" individuals that *could* have been present in the study area. During the 3 years when we assume that SCR data were available, a total of 78 individual bears were detected, so we think that augmenting the data set by 122 such "potential" individuals, for a complete data set of 200, will be enough. We will monitor the size of the superpopulation in the model below to let us decide whether this is sufficient to represent a vague prior.

```
# Do data augmentation of the SCR data
M <- 200
y <- array(0, dim=c(M, dim(bear$scr)[2:4]))
y[1:dim(bear$scr)[1],,,] <- bear$scr
```

We then bundle the data for JAGS. We also provide information about state-space S (Fig. 21.2) in the data bundle.

```
# Bundle data and produce data overview
jags.data <- list(y=y, o=bear$occ, T=6, K=dim(y)[3], J=dim(y)[2], M=M, x=bear$trap.coord, xlims=xlims,
  ylims=ylims)
str(jags.data)

List of 9
 $ y     : num [1:200, 1:128, 1:8, 1:3] 0 0 0 0 0 0 0 0 0 0 ...
 $ o     : num [1:128, 1:8, 1:3] 0 0 0 0 0 NA 1 1 0 0 ...
 $ T     : num 6
 $ K     : int 8
 $ J     : int 128
 $ M     : num 200
 $ x     : num [1:128, 1:2] 618 619 618 620 621 ...
 $ xlims: num [1:2] 615 645
 $ ylims: num [1:2] 3380 3420
```

We then write the model in the BUGS language. The main section headings in the model are now slightly different from those we had in the other IPMs in this book. We now have one individual-based population model to describe the latent process and one observation model for each data set.

```
# Write JAGS model file
cat(file="model1.txt", "
model {
# Priors
psi ~ dunif(0, 1)                          # M*psi = E[N(1)], for data augmentation
phi ~ dunif(0, 1)                          # Survival
gamma ~ dunif(0, 3)                        # Per-capita recruitment
p0 ~ dunif(0, 1)                           # Baseline capture probability
sigma ~ dunif(0, 50)                       # Scale parameter of detection function

# Derived parameters
for (t in 1:T){
  N[t] <- sum(z[,t])                       # Population size at time t
  EB[t] <- N[t] * gamma                    # Expected number of recruits
  A[t] <- max(M - sum(a[,t]), 0.001)       # Bears available to be recruited
  b[t] <- min(EB[t] / A[t], 0.999)         # Probability of being recruited
}

# Population model (individual-based model)
# Initial state
for (i in 1:M){
  z[i,1] ~ dbern(psi)                      # Is a member of M alive?
  a[i,1] <- z[i,1]                         # Recruited yet?
  s[i,1] ~ dunif(xlims[1], xlims[2])       # Activity center, homogeneous
  s[i,2] ~ dunif(ylims[1], ylims[2])

# Dynamics over time
  for (t in 2:T){
    z[i,t] ~ dbern(z[i,t-1] * phi + (1 - a[i,t-1]) * b[t-1])
    a[i,t] <- max(z[i,1:t])
  } #t
} #i

# Observation models
# Spatial capture-recapture data
for (i in 1:M){
  for (j in 1:J){
    d2[i,j] <- (s[i,1] - x[j,1])^2 + (s[i,2] - x[j,2])^2  # Distance squared
    p[i,j] <- p0 * exp(-d2[i,j] / (2 * sigma^2))
  } #j
  for (j in 1:J){                          # Trap
    for (k in 1:K){                        # Secondary occasions
      # The years with SCR data
      y[i,j,k,1] ~ dbern(p[i,j] * z[i,1])
      y[i,j,k,2] ~ dbern(p[i,j] * z[i,3])
      y[i,j,k,3] ~ dbern(p[i,j] * z[i,5])
    } #k
  } #j
  zi[i] <- (sum(z[i,]) > 0)                # Was this bear ever alive?
} #i

# Occupancy data
for (j in 1:J){
  for (k in 1:K){
    o[j,k,1] ~ dbern(1-prod(1-p[,j] * z[,2]))
    o[j,k,2] ~ dbern(1-prod(1-p[,j] * z[,4]))
    o[j,k,3] ~ dbern(1-prod(1-p[,j] * z[,6]))
  } #k
} #j
N.ever <- sum(zi[])                        # Bears ever alive - superpopulation size
}
")
```

We need to give initial values for the latent state variable z; otherwise, the model will not run. We define $z = 1$ for all of the known and a fraction of the "potential" individuals and $z = 0$ for the remaining "potential" individuals. In this model, it is really important to not provide $z = 1$ for all individuals, because otherwise there would be no individual that is ready to be recruited, and the MCMC will not work properly. This differs from data augmentation in simpler models, for instance, for closed-capture population size estimation, where initialization for z with a vector of 1s is fine (Chapter 6 in Kéry and Schaub, 2012).

```
# Initial values
zin <- array(0, dim=c(dim(y)[1],6))
zin[1:150,] <- 1
inits <- function() {list(z=zin, phi=runif(1, 0.7, 0.9), p0=runif(1, 0.01, 0.1),
  psi=runif(1, 0.2, 0.4), sigma=runif(1, 3, 4), gamma=runif(1, 0.2, 0.3))}

# Parameters monitored
parameters <- c("psi", "phi", "gamma", "p0", "sigma", "N", "EB", "b", "A", "s", "zi", "z", "N.ever")

# MCMC settings
ni <- 5000; nb <- 2000; nc <- 3; nt <- 3; na <- 2000

# Call JAGS from R (ART 1400 min) and check convergence
out1 <- jags(jags.data, inits, parameters, "model1.txt", n.iter=ni, n.burnin=nb, n.chains=nc,
  n.thin=nt, n.adapt=na, parallel=TRUE)
traceplot(out1)
```

21.6 RESULTS

The model converges after only 5000 iterations, but the updating in JAGS is extremely slow and takes almost a day on our computer. A partial summary output is shown here.

```
print(out1, 3)

                mean     sd      2.5%       50%     97.5% overlap0 f  Rhat n.eff
psi            0.243  0.037     0.175     0.241     0.320    FALSE 1 1.001  3000
phi            0.807  0.035     0.736     0.808     0.871    FALSE 1 1.001  1485
gamma          0.235  0.037     0.166     0.233     0.312    FALSE 1 1.001  3000
p0             0.115  0.007     0.101     0.115     0.129    FALSE 1 1.004   491
sigma          3.205  0.093     3.042     3.198     3.404    FALSE 1 1.018   129
N[1]          48.017  4.074    41.000    48.000    57.000    FALSE 1 1.003   665
N[2]          59.722  4.278    52.000    60.000    69.000    FALSE 1 1.005   357
N[3]          71.046  3.386    65.000    71.000    78.000    FALSE 1 1.008   248
N[4]          64.342  4.704    56.000    64.000    74.000    FALSE 1 1.007   287
N[5]          58.501  5.098    49.000    58.000    69.000    FALSE 1 1.011   210
N[6]          61.190  6.112    50.000    61.000    74.000    FALSE 1 1.008   312
EB[1]         11.218  1.688     7.999    11.163    14.642    FALSE 1 1.000  3000
EB[2]         13.976  2.161     9.932    13.858    18.382    FALSE 1 1.000  3000
EB[3]         16.654  2.606    11.743    16.560    21.847    FALSE 1 1.000  3000
EB[4]         15.085  2.516    10.371    14.963    20.281    FALSE 1 1.000  3000
EB[5]         13.708  2.339     9.453    13.563    18.695    FALSE 1 1.001  2231
EB[6]         14.377  2.763     9.430    14.140    20.296    FALSE 1 1.001  2915
b[1]           0.074  0.012     0.053     0.073     0.098    FALSE 1 1.000  3000
b[2]           0.104  0.017     0.073     0.104     0.141    FALSE 1 1.000  3000
b[3]           0.146  0.025     0.101     0.145     0.199    FALSE 1 1.000  3000
b[4]           0.147  0.029     0.096     0.144     0.212    FALSE 1 1.001  2379
b[5]           0.145  0.031     0.093     0.143     0.214    FALSE 1 1.002  1177
b[6]           0.181  0.050     0.103     0.174     0.298    FALSE 1 1.002  1710
A[1]         151.983  4.074   143.000   152.000   159.000    FALSE 1 1.003   665
[... output truncated ...]
A[6]          81.080  8.215    64.000    81.000    96.000    FALSE 1 1.002  1000
s[1,1]       624.062  0.791   622.593   624.058   625.670    FALSE 1 1.001  3000
s[2,1]       629.740  1.956   625.371   629.806   633.406    FALSE 1 1.003   718
[... output truncated ...]
s[199,2]    3401.807 11.931  3381.122  3403.032  3419.268    FALSE 1 1.000  3000
s[200,2]    3401.362 11.748  3380.839  3402.505  3419.104    FALSE 1 1.000  3000
zi[1]          1.000  0.000     1.000     1.000     1.000    FALSE 1    NA     1
[... output truncated ...]
zi[200]        0.313  0.464     0.000     0.000     1.000     TRUE 1 1.000  3000
N.ever       118.920  8.215   104.000   119.000   136.000    FALSE 1 1.002  1000
z[1,1]         1.000  0.000     1.000     1.000     1.000    FALSE 1    NA     1
z[2,1]         1.000  0.000     1.000     1.000     1.000    FALSE 1    NA     1
[... output truncated ...]
z[199,6]       0.205  0.404     0.000     0.000     1.000     TRUE 1 1.001  2279
z[200,6]       0.182  0.386     0.000     0.000     1.000     TRUE 1 1.000  3000
```

The first thing we check is whether our degree of data augmentation was sufficient by inspecting the posterior summary for the size of the superpopulation (`N.ever`). The number of bears ever alive and with an activity center within the state-space is 118.9 (SD: 8.2), and the upper bound of its 95% credible interval is far below 200, and hence, we conclude that the

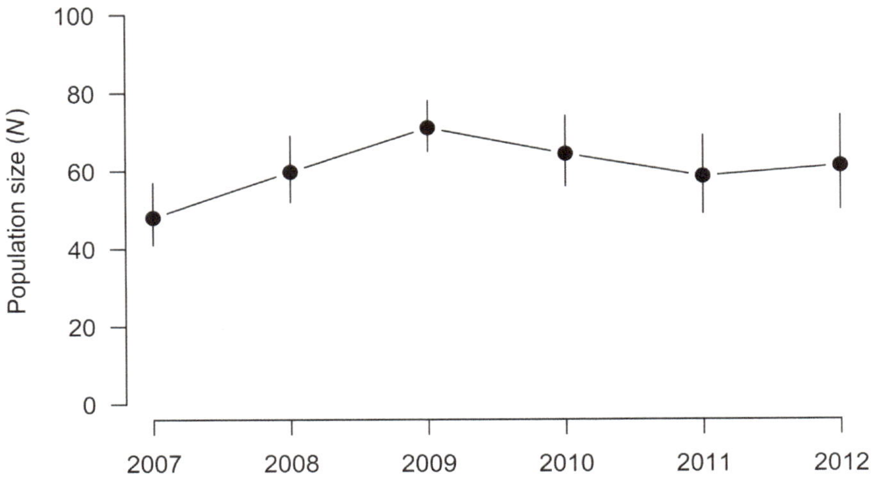

FIGURE 21.3

Estimated number of black bears within the state-space of our analysis during the 6 study years. Dots show posterior means, and vertical lines the 95% credible intervals.

answer is yes. Next, we note that estimated annual population sizes refer to the defined state-space (S). The population first increased, reached a peak in year 2009, and then declined again somewhat (Fig. 21.3).

The posterior distributions of annual adult survival and per-capita recruitment are nicely peaked, indicating that plenty of information was available to estimate these demographic parameters (Fig. 21.4). In our model, we assumed that adult bears do not disperse; that is why the locations of their activity centers do not change over time. The estimated adult survival therefore corresponds to true survival, although it is based on capture-recapture data that are usually known to yield only estimates of apparent survival. Per-capita recruitment refers to adults and thus expresses how many adult bears recruit into the state-space for every adult bear already established. Recruitment may be local, in which case it is productivity per bear multiplied by the survival between birth and the recruitment age, and by site fidelity. Alternatively, recruitment may be due to immigration from outside the state-space. With the data at hand, we cannot separate these two possibilities.

The individual-based spatial IPM is formulated with an individual-based model at its core. We therefore know *where* each individual lives, which allows us to compute interesting summaries of the estimates of the locations of these activity centers using output from the MCMC simulations. These include maps of density (i.e., the number of activity centers per unit area) or estimates of population size for any defined polygon of interest (Chapter 5 in Royle et al., 2014). We just need

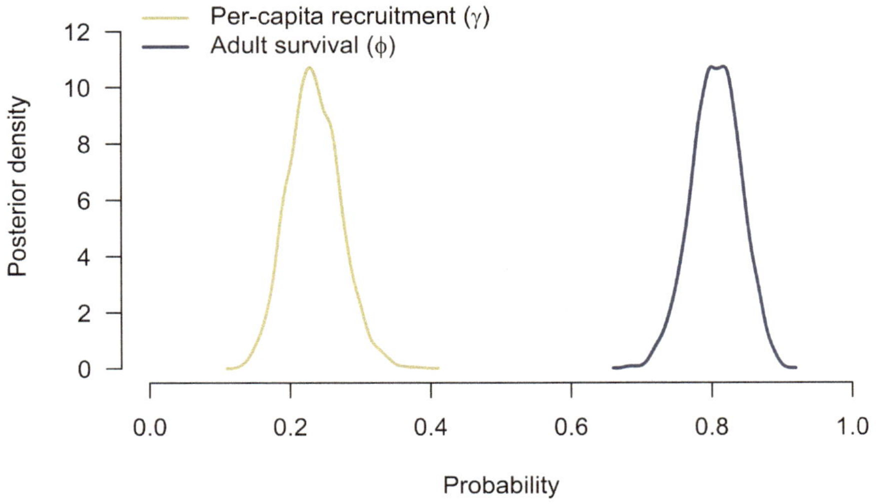

FIGURE 21.4

Posterior distributions of per-capita recruitment and annual adult survival of Louisiana black bears.

to grid the state-space or define some polygon; then we simply tally up the number of activity centers in each grid while doing this for each MCMC draw separately.

For illustration, we here produce density maps of each year based on pixel areas of 500 × 500 m (our grid cells). Our state-space is 645−615 = 30 km in an east−west direction. Hence, we need 61 bins to define the limits for a regular 500 m increment. Similarly, in the north−south dimension, we find that we need 81 bins. To allocate the locations of all bears to the grid cells, we use the R function `cut`. We do this allocation for each MCMC draw but include only individuals that are alive in a given year, and hence we make a restriction with $z_{i,t} = 1$.

The resulting two objects `sx` and `sy` are vectors with entries that are the bins thus defined in which there was at least one MCMC draw of an activity center. The number of times a bin occurs corresponds to the bear density in that bin. To summarize, we produce a cross-tabulation from these two vectors. This yields the number of times a combination of bins, and hence a grid cell, occurs. Then, we scale that number by dividing it by the total number of MCMC draws. Finally, we use the `image` function to create density maps for each year. These calculations are included in a loop over years to produce one map for each year.

```
# Make Fig. 21.5
# Define the grid
xgrid <- seq(xlims[1], xlims[2], length.out=61)
ygrid <- seq(ylims[1], ylims[2], length.out=81)

# Grab the MCMC draws for the x and y coordinates of all activity centres (ac) and of z
ac.x <- out1$sims.list$s[,,1]
ac.y <- out1$sims.list$s[,,2]
z <- out1$sims.list$z

library(scales)
library(AHMbook)
cl <- terrain.colors(100, rev=TRUE)[c(1, 1:5*10, 51:100)]
par(mar=c(4, 4, 2, 4), mfrow=c(3, 2))
year <- 2007:2012
# Define density classes
h <- seq(0, 0.3, length.out=100)
for (t in 1:6){
  # Allocate the locations to the defined grid cells
  sx <- cut(ac.x[z[,,t]==1], breaks=xgrid, include.lowest=TRUE)
  sy <- cut(ac.y[z[,,t]==1], breaks=ygrid, include.lowest=TRUE)

  # Cross-tabulation and down-scaling by the number of MCMC draws
  dens <- table(sx, sy) / out1$mcmc.info$n.samples

  # Draw map
  image(x=xgrid, y=ygrid, z=dens, col=cl, asp=1, axes=FALSE, xlim=xlims, ylim=ylims,
    xlab='x coordinate', ylab='y coordinate', main=year[t], zlim=c(0, 0.3))
  axis(1); axis(2, las=1)
}
image_scale(c(0, 0.3), col=cl[seq(1, length(cl), length=13)], digits=2, labels ="breaks",
  cex.legend=0.8)
```

Fig. 21.5 shows substantial spatial variation in black bear density, or more precisely, in the density of the activity centers of the bears. In areas of the state-space where no traps were placed, e.g., in the upper right corner (Fig. 21.2), bear density is spatially extrapolated (Fig. 21.5). In the absence of spatial covariates, this reduces to basically the mean density without much spatial structure. The spatial structure of population density appears to be less clear, and the density is more spread out, in the last year, 2012. This is because in that year, only occupancy data were available, and these contain less information than SCR data. This spreading out is reduced in other years with only occupancy data available, but when a preceding and a following year had SCR data, i.e., in years 2008 and 2010 (Fig. 21.5).

From the information of these bear locations, we can also produce estimates of the average density in the study area or parts thereof. The number of bears alive in a given year refers to the defined state-space and hence to 1200 km^2 in our case. The average density of black bears in year 2009 (the year of the population peak, Fig. 21.2) per 100 km^2 along with the 95% credible interval can therefore be calculated as follows:

```
out1$mean$N[3] / 1200 * 100                # Posterior mean
> 5.92

out1$q2.5$N[3] / 1200 * 100                # Lower credible limit
> 5.42

out1$q97.5$N[3] / 1200 * 100               # Upper credible limit
> 6.50
```

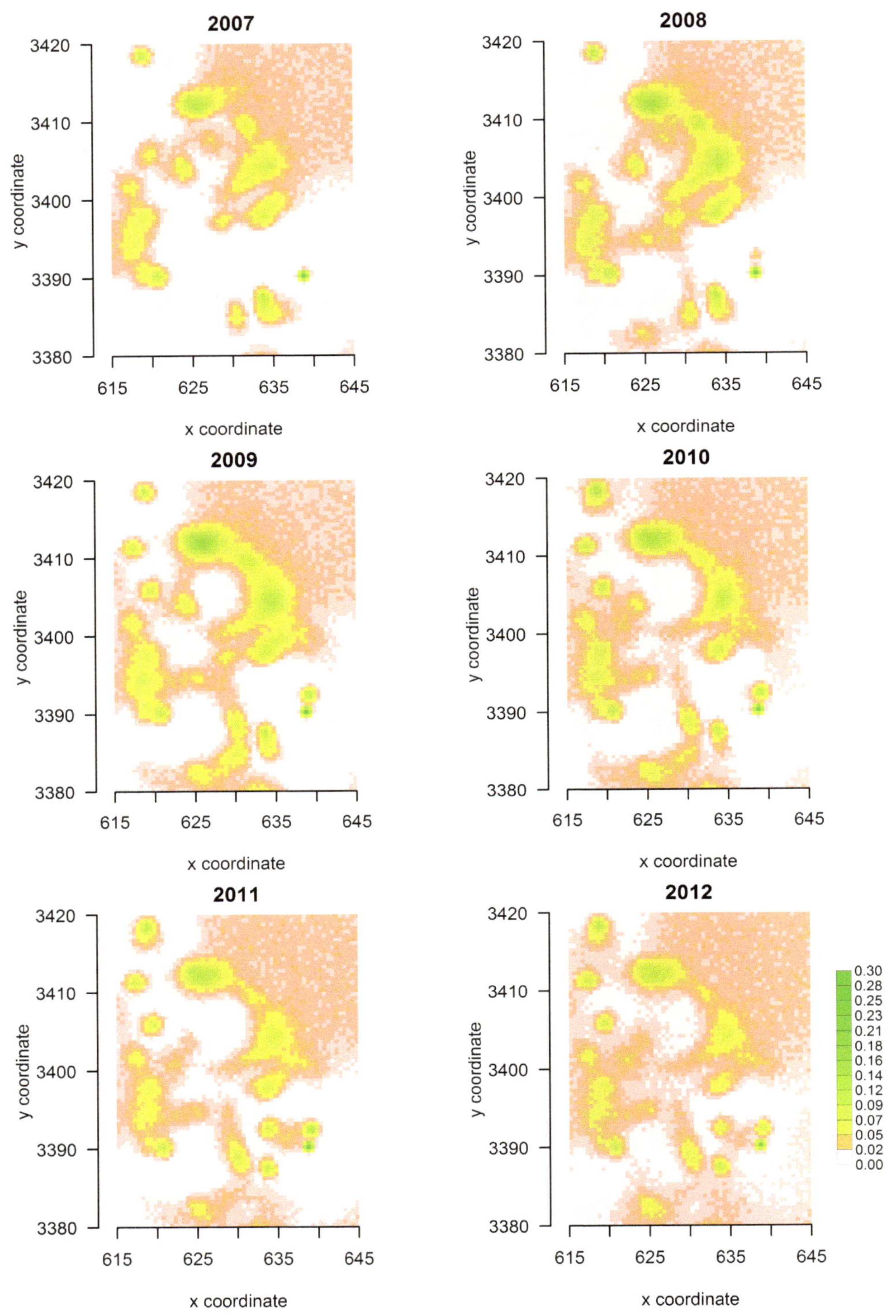

FIGURE 21.5

Density maps of Louisiana black bears for a 500 × 500 m resolution. Density is the expected number of activity centres per pixel. Note that in areas of the state-space without traps (e.g., in the upper right corner, Fig. 21.2), the estimated density is a strong spatial extrapolation and thus should probably not be interpreted too deeply.

Besides the annual density maps, it may be valuable to produce a map with density trends to highlight areas of population increases and declines. We want to show grid cell-wise mean trends over all 6 years of study and we calculate the geometric mean for each cell and produce a map of these means (though we note that trend could also be calculated based on a variety of regression approaches (Dennis et al., 1991; Thomas and Martin, 1996)). We are not interested in fine spatial detail about the trends, so we choose a scale that bins trends in 0.05 units. We see that in most parts of the

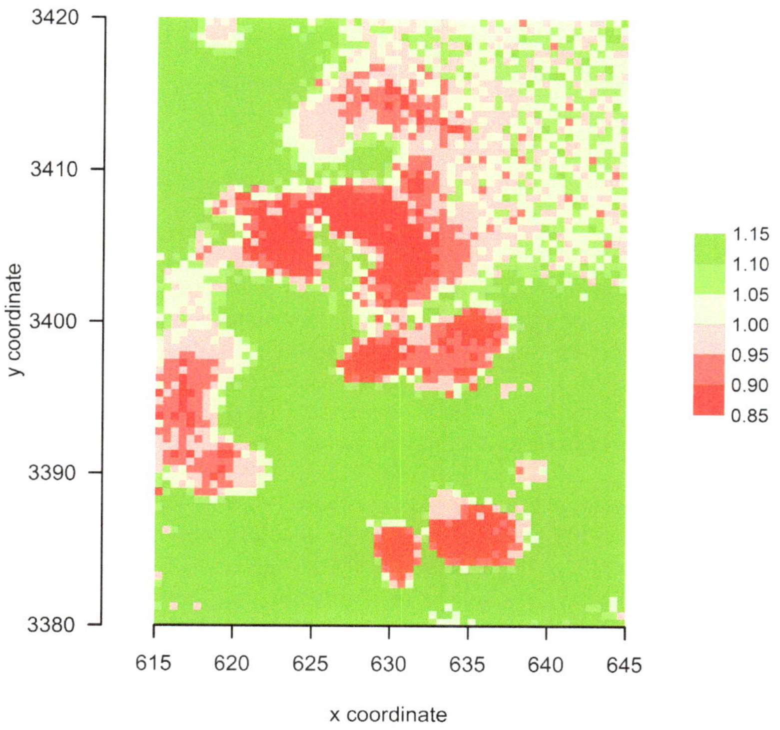

FIGURE 21.6

Spatial variation in geometric average population trends from 2007 to 2012 of black bear density. Green cells indicate an increase in bear density, while red cells show a local decline in bear density (expressed as the number of activity centers per cell). For interpretation of this map, don't forget to look back at Fig. 21.2, which shows the configuration of the trapping array.

state-space, the density trend was positive (Fig. 21.6) between 2007 and 2012. Interestingly, the trends seem to be positive at places where population density is low and negative at places where population density is high (compare Figs. 21.5 and 21.6). However, it may be that this is partially an artifact because the density in the last study year is somewhat smeared out; see above.

```
# Make Fig. 21.6
par(mar=c(4, 4, 2, 4), las=1)
# To compute the geometric mean, we only need the density in the first and the last year
# Allocate the locations to the defined grid cells in year 2007
sx <- cut(ac.x[z[,,1]==1], breaks=xgrid, include.lowest=TRUE)
sy <- cut(ac.y[z[,,1]==1], breaks=ygrid, include.lowest=TRUE)
# Cross-tabulation
dens07 <- table(sx, sy) + 0.001

# Allocate the locations to the defined grid cells in year 2012
sx <- cut(ac.x[z[,,6]==1], breaks=xgrid, include.lowest=TRUE)
sy <- cut(ac.y[z[,,6]==1], breaks=ygrid, include.lowest=TRUE)
# Cross-tabulation
dens12 <- table(sx, sy) + 0.001

# Compute geometric mean for each grid
lam <- matrix(NA, nrow=60, ncol=80)
for (i in 1:60){
  for (j in 1:80){
    lam[i,j] <- exp((log(dens12[i,j]) - log(dens07[i,j])) / 5)
  } #j
} #i

# Define the limits of the scale
h <- c(0, 0.9, 0.95, 1, 1.05, 1.10, 20)

# Define colors for that scale
cl <- c(alpha('red', c(0.9, 0.6, 0.3)), alpha('green', c(0.3, 0.6, 0.9)))

# Allocate the actual geom. means into categories
glam <- matrix(.bincode(lam, h), nrow=60, ncol=80)

# Draw map
image(x=xgrid, y=ygrid, z=glam, col=cl, asp=1, axes=FALSE, xlim=xlims, ylim=ylims, xlab='x coordinate',
  ylab='y coordinate', zlim=c(1, 6))
axis(1); axis(2)
```

21.7 DISCUSSION

We have fit an individual-based spatial IPM jointly to spatial capture-recapture (SCR) and occupancy data of black bears in Louisiana. We estimated survival and per-capita recruitment of adult bears and have demonstrated how we can produce maps of annual bear densities and of population trends. These trend maps identified areas where density declined and others where density increased over the study period. The IPM itself is relatively simple, as we assumed no spatial or temporal variation in any demographic process and the absence of any dispersal. In principle, and for the right kind and quantity of data, these assumptions could be relaxed. It would also be possible to model demographic processes as a function of spatial environmental covariates such as habitat in such an IPM, which at its core is simply an open SCR model (see e.g., Chandler et al., 2018b).

Traditional IPMs, including those that address features of "space" such as those in Chapters 16–20, all have process models that only describe the numbers of individuals in some spatially aggregated fashion. That is, individuals are summarized into a "population" inhabiting some defined "areal unit" such as a colony, a region, or a pixel of some chosen resolution. All these IPMs need some sort of abundance or occupancy data that are informative about either absolute or relative population size in these spatial units, and these are then combined with other data that are directly informative about some demographic process(es). Individual-based spatial IPMs differ from these IPMs by adopting a process model that describes each individual; no aggregation of individuals into any spatial units is required. Besides the demographic processes of survival and recruitment, we estimate the location of each individual, unlike in traditional IPMs. These individual-based spatial IPMs require spatial capture-recapture data that can be combined with other spatially indexed data such as occupancy data here, spatial distance-sampling data (Chandler et al., 2018b), or opportunistically sampled presence-absence data (Sun et al., 2019; Section 10.8 in Kéry and Royle, 2021).

The individual-based formulation has a number of advantages. First, these models enable us to relax the independence assumption made for the different data sets in traditional IPMs. Because the observation models of the component data sets are all conditional on a single process model, any nonindependence between them is fully accounted for (see Section 7.2.2).

Second, any desired spatial scaling is possible, hence, we can make inferences about any spatial scale or for any spatial unit (e.g., polygon) we are interested in. In our bear example, we have assumed the absence of dispersal, which then manifests itself by activity centers with locations that do not change over time. However, provided that enough data are available, movements of activity centers (i.e., dispersal) can also be included in the SCR process model (Royle et al., 2016; Gardner et al., 2018).

Third, covariates describing the habitat or other spatial features can be included in a model. This allows the estimation of habitat-specific densities in a model-based manner, to test for effects of the environment on features of the demography, and to model habitat-specific demography. It also opens the door to assess spatial variation in source-sink dynamics (Furrer and Pasinelli, 2016).

Fourth, because of the individual-based process model, individual heterogeneity in survival or reproduction can readily be accounted for. In the bear example we used the location as our sole individual covariate, but others such as sex, age or size could be included. The inclusion of individual heterogeneity in traditional IPMs may require mixture models that allocate individuals into groups (see Chapter 14), a continuous mixture model such as the logit-normal or beta-binomial (Dorazio and Royle, 2003), or the combination of an IPM with "the other IPM", an integral projection model (Plard et al., 2019).

Fifth, the individual-based spatial IPM provides estimates of true population size and density that are correct in the sense that the model has a known spatial attribution, as is always the case in SCR models. Estimates of population size in traditional IPMs are often relative (Section 7.4), and because strictly the sampling area is typically unknown, accurate computation of density is not possible, as in any case of nonspatial capture-recapture (Royle et al., 2014, Chapter 6 in Kéry and Royle, 2016); see also the discussion about the unknown effective sampling area in Chapter 20.

Naturally, there may be drawbacks of individual-based spatial IPMs compared with traditional IPMs. The most important is perhaps the increased computational burden, which may make fitting such a model prohibitive for a large number of individuals or over large spatial and/or temporal scales (Linden and McKinney, 2016). Yet with the increasing speed of computers, this caveat is likely to become less important in the future. Obviously, individual-based spatial IPMs need data with spatial information—in the first place, SCR data—and these will typically be more costly to sample than nonspatial capture-recapture data. Spatial information is often not available in historical data sets, or no clear sampling protocol with respect to the spatial aspect of the data was respected. Thus, for many existing data sets, it may be hard or impossible to apply individual-based spatial IPMs, while traditional IPMs can often be adopted for them.

The process model of the individual-based spatial IPM corresponds to an individual-based, spatial Jolly-Seber model with the "restricted occupancy formulation" (Royle and Dorazio, 2008; Kéry and Schaub, 2012; Royle et al., 2014). Using the SCR data alone, it is possible to estimate the same parameters by fitting a spatial Jolly-Seber model as with the individual-based spatial IPM that in addition uses the occupancy data. Hence, the addition of the occupancy data set does not allow the estimation of any additional parameters as we typically see in traditional IPMs. We may therefore wonder why we should adopt an individual-based spatial IPM at all? One added value of an individual-based spatial IPM over a simpler spatial Jolly-Seber model is certainly the increased scope of inference. Obtaining SCR data is costly, while obtaining spatial data of unmarked individuals such as occupancy data or opportunistically sampled presence-absence data is much cheaper and can typically be conducted over larger areas or for longer periods. The individual-based spatial IPM then allows us to jointly analyze all these data. Another advantage is the increased precision of parameter estimates (Sun et al., 2019), a feature shared with nonspatial IPMs (Section 6.2).

As far as we know, at the time of writing, only three individual-based spatial IPMs have been published: Chandler and Clark (2014), Chandler et al. (2018b) and Sun et al. (2019). Therefore, these models are still in their infancy, and we will certainly see many more developments in the near future. We think this will include models that incorporate other types of spatial data, in particular productivity data. This would enable one to decompose per-capita recruitment into components due to productivity, juvenile survival, and natal dispersal. The underlying individual-based model then needs to be written in a different way, and the sex of each individual must be accounted for. It might also be interesting to include population count data for which the location of each individual is unknown and just the grid cell in which a count was performed is known. A prototype of such a model is the model of Chandler and Royle (2013) for spatial counts of unmarked individuals, and the variant of such a model for occupancy data developed by Ramsey et al. (2015). Such data are commonly collected in national bird monitoring programs (Baillie, 1990; Sauer et al., 2013) and might offer the potential for large-scale spatial inference on animal demography.

We clearly think that the inclusion of "space" is one of the new big topics in the further development of IPMs in general. It remains to be seen whether this will be in models that use an individual-based process model or in the more traditional summary process model that includes spatial aspects by mere spatial stratification with or without explicit modeling of spatial autocorrelation, as in Chapters 19 and 20.

CHAPTER

CONCLUSIONS

22

CHAPTER OUTLINE

22.1 FITTING INTEGRATED POPULATION MODELS: A STEEP MOUNTAIN ... BUT ONE THAT'S REALLY WORTH THE CLIMB!

In the two parts of our book, we have provided an introduction and synthesis of integrated population models, or IPMs, within the Bayesian framework of statistical inference. An IPM is a complex hierarchical statistical model that is a combination of other statistical models that may already be fairly complex themselves. Thus, the successful development and fitting of an IPM to your own data set requires a fairly sound understanding of a whole array of topics, including statistical inference, computation, population modeling in general, and the typical models used to estimate demographic quantities from actual data and that are integrated in the joint likelihood of an IPM. Learning IPMs from scratch is no doubt a tall and daunting mountain to climb, but we hope that by now, you have been able to obtain the necessary knowledge by walking with us through our book. We also hope that this hike has been pleasant, enjoyable, and most of all, inspiring. We are convinced that if you master the fitting of an IPM to your data, you have become a mature quantitative ecologist with fairly advanced applied skills in statistical modeling. Of course, these skills will hopefully allow you to do the integrated population modeling you are interested in. In addition, they will be useful when you analyze other data sets with statistical models that may be quite different from an IPM. Hence, we believe there are many beneficial side effects of learning at least some of the material in our book.

At this point, it may be valuable to briefly look back and see where we have come from. After a short introduction in Chapter 1, the book itself really began in Chapter 2 with an introduction to Bayesian inference and Bayesian computing, since in our treatment of IPMs we have restricted ourselves to this mode of inference. IPMs can also be fitted with likelihood methods (e.g., Besbeas et al., 2002; Gauthier et al., 2007; Véran and Lebreton, 2008; Borysiewicz et al., 2009; Tavecchia et al., 2009; Fieberg et al., 2010; McCrea et al., 2010; Besbeas and Morgan, 2012; de Valpine, 2012; Newman et al., 2014; Dennis et al., 2021), but we don't present them here. This is not because we claim that the Bayesian approach to fitting IPMs is intrinsically better than the use of maximum likelihood. The main reason for our choice of the Bayesian over a classical (frequentist) statistical treatment of IPMs is a practical one: we are convinced that the use of statistical modeling software that adopts the BUGS model description language is currently the only way in which a reasonably large number of nonstatisticians can learn how to fit fully flexible IPMs to their data sets and answer their questions without enrolling a professional statistician on their team. The BUGS language greatly simplifies the implementation of hierarchical models, and IPMs represent a prime example of big and "ugly" hierarchical models in ecology. Any of the BUGS

Integrated Population Models. https://doi.org/10.1016/B978-0-12-820564-8.24002-5

software including WinBUGS and OpenBUGS (Lunn et al., 2000; Lunn et al., 2009), JAGS (Plummer, 2017), multiBUGS (Goudie et al., 2020), and NIMBLE (NIMBLE Development Team, 2018) could be used. We are particularly excited about the new NIMBLE project, which offers substantially more flexibility than any of the other programs in the BUGS family for tuning MCMC algorithms to a particular problem.

In Chapter 2, we also briefly introduced linear models, since they are so very important for fitting more interesting and relevant models by modeling patterns in their parameters. Indeed, the basic structure of different models, i.e., their likelihoods, can be used/copied from elsewhere, e.g., from some template that you find in the appendix of a published paper or in this book. However, the specific form of the model, i.e., the linear models with covariates as well as random or fixed effects for the spatial or spatio-temporal patterns in the parameters is *certain* to be different in your model owing to your study system and your research questions. In the courses we teach, we always find that a large proportion of the practical problems encountered by course participants in defining the model they want is simply due to the challenge of fitting linear models that corresponds to their questions of interest. Nevertheless, this is a book about IPMs and not about linear models, so we had only space to scratch the surface a bit in this respect. If you need to understand more about the topics of linear and generalized linear and simple mixed models and their implementation in the BUGS language, we refer you to the works of Ntzoufras (2009), Kéry (2010), or Kéry and Schaub (2012). Chapter 6 in Cooch and White (2020) is an excellent introduction to linear modeling for ecologists as is Chapter 1 in Wood (2017), albeit both outside of BUGS.

At the end of Chapter 2, we showed the fitting of an integrated model for estimating abundance based on a count and a detection-nondetection data set in both the frequentist and the Bayesian modes of inference. For the former, no canned function is available in R, and consequently, we wrote the joint likelihood explicitly and then used the R function `optim` to maximize it ourselves. An advantage of that exercise was that the construction of the joint likelihood as a product of the likelihoods of the two separate data sets became fully transparent. The Bayesian solution of specifying the joint likelihood in a hierarchical manner in BUGS is much simpler to code because we just have to write down the likelihood definitions for each of the single-data models. The integration then "happens" implicitly by giving shared parameters the same name. In this process, the nature of the joint likelihood is hidden, but of course this approach is fantastic for actually fitting an IPM to your data.

In Chapter 3, and indeed throughout the book, we have stressed the importance of a sound understanding of stage-structured, or matrix, population models (Caswell, 2001). These models are the backbone of almost all current IPMs. Although the connection of IPMs with stage-structured population models has been emphasized previously (Buckland et al., 2004, 2007; Thomas et al., 2005), we feel that the consequences of this realization and the opportunities this recognition offers have perhaps not been understood well enough within the IPM community: *the fact that most IPMs contain at their core a stage-structured population model means that IPMs can be used in exactly the same way as a stage-structured population model to analyze the dynamics of a population*. For instance, we can estimate life-history summaries such as generation time or net reproductive rate or conduct prospective and retrospective population analyses. When we carry out such matrix population model analyses as part of an IPM, we obtain measures of uncertainty associated with all calculated quantities for free. Moreover, such a one-step model is statistically more efficient than a solution with multiple steps, as we have stressed many times.

We have demonstrated the huge benefits that arise from the combination of matrix population modeling with the MCMC algorithms implemented in JAGS. We have identified IPMs as a natural generalization of the powerful matrix population models and MCMC algorithms as our generic simulation tool for the following, all of which is achieved by analysis in JAGS:

- analysis of the matrix population model (e.g., sensitivity/elasticity)
- sensitivity analysis for parameters that are unknown and cannot be estimated and so must be fixed
- simulation-based projection/forecasting
- computation of derived quantities such as population growth rates or extinction probabilities including full error propagation
- parameter estimation

So instead of using MARK or E-SURGE for parameter estimation and R or Matlab for the matrix population model analysis, we can do everything in JAGS or any other member of the BUGS software family. We find this truly wonderful.

In Chapter 4, we provided a large, but of course not comprehensive, catalog of the main statistical models that form the "ingredients" of an IPM. A sound applied knowledge of such models is important, since they are the building blocks of all IPMs. Of course, there are many more models that could also be part of an IPM, and here we focused on those models that are most frequently used or perhaps that we think *should* be used more often. For the most part, we have shown only the basic structure of such component models. That is, we have not shown any modeling of the parameters, e.g., by introducing covariates. The modeling of parameters is merely a matter of fitting linear models, and we have covered this briefly in Chapter 2.

In Chapters 5–7, we defined an IPM as a demographic population model embedded in several submodels to simultaneously estimate parameters and analyze the population model itself. An IPM uses data from both the individual and the population levels, and the stage-structured population model connects the two. We showed how to fit IPMs and highlighted their advantages over piecewise analyses of the component data. Key advantages are typically the increased precision of individual parameter estimates and often the ability to estimate additional parameters that may otherwise not be identifiable. The latter appears to be a main advantage of an IPM, but inferences from such parameters must be made with care, as they can easily be biased and thus misleading when model assumptions are violated (Schaub and Fletcher, 2015; Riecke et al., 2019). In these chapters, we also covered goodness-of-fit testing. Much of that material is novel, since this is an understudied and challenging aspect of IPMs.

In Chapters 9 and 10, we showed how IPMs can be used for both retrospective and prospective population analyses. Both are of great practical importance, especially in conservation and management. Retrospective analyses come with associated errors, and often the uncertainties are so large that making inferences becomes challenging. In the past, uncertainty in retrospective population analyses has often simply been ignored, and this is made particularly easy to do in non-Bayesian analyses, where obtaining proper measures of uncertainty in these analyses is more challenging. But in the era of IPMs fit with BUGS, we are now essentially getting these uncertainties for free. Of course, one price to pay will be that the inferences may not be as unambiguous as we might have hoped—often, it is sobering to see the magnitude of the uncertainty around such retrospective inferences. But of course, this is not a mistake of the model or of Bayesian inference. Rather, the model, along with the MCMC fitting techniques, is merely a messenger that tells you honestly that there is a great deal of uncertainty in such inferences, especially for small populations and with limited data. We have also shown how easily population viability analyses can be carried out as part of an IPM. Once again, you get the necessary uncertainty assessment for the population forecasts for free in a Bayesian MCMC-based analysis, and this is a huge practical advantage. The ability to make probability statements about future population sizes or dynamics is very useful for applied purposes, especially when communicating with a nonscientific audience (Wade, 2002).

In Part 2 of the book, we presented 11 case studies that show applications of IPMs to actual data sets, often based on published studies. In their selection, we strived for a diversity of data sets, model types, and study systems, and in each case study, we illustrate every step in an analysis going from data management in R, to the specification and fitting of a model, and finally to output processing. These analyses are meant to provide an illustration of the wide range of problems and model types that can be tackled with an IPM. In many cases, they also further illustrate some topics that have been dealt with in Part 1 of the book.

Overall, we think that the advantages of a joint likelihood approach to the demographic modeling of animal (and plant) populations are so compelling that the phrase "IPM" may well eventually disappear from population ecology and management. Namely, we predict that in the future it will become so commonplace to simply use all the available data in an estimation problem in the field of population dynamics that a special term for this highly sensible approach will no longer be needed. Then, everybody in applied population modeling will be fitting IPMs, and the latter will simply be called "population models."

22.2 SHOULD WE ALWAYS INTEGRATE?

On the other hand, it is important to moderate our excitement somewhat and clearly see some dangers and limitations of IPMs. In addition to the detailed discussion in Chapter 7, we would like to mention two challenges, both of which have to do with growth in the complexity of our population model when specifying an IPM. The first is about parameter non-identifiability, and the second is about increasing rather than decreasing the uncertainty in the estimates of an IPM.

IPMs typically are statistical models for highly parameterized, i.e., complex, stochastic systems. We have seen that BUGS software is a very powerful tool to specify, fit, and analyze such complex statistical models. But there is a downside to this power: it is quite possible with BUGS to fit nonsensical models, i.e., models that have one or more parameters that are not identifiable with your data. This means there is no information in your data to estimate them, and hence any Bayesian learning (i.e., the change from a relatively flat prior for a parameter to a somewhat peaked posterior; see Fig. 2.6) will be due to the imposed model structure and perhaps complicated ways in which the posterior of such a parameter may be constrained by the priors for other parameters in the model, even if these were meant to be vague.

Assessing the identifiability of parameters is always a challenge in complex statistical models, and these challenges increase with the complexity of a model. For an interesting discussion in this context, see the "Big fancy models" forum in *Ecology* in December 2010 with the introduction, "Living dangerously with big fancy models," by Lavine (2010). Understanding parameter identifiability is perhaps even more challenging in a Bayesian analysis where you will always get "some numbers out"—if there is no information about a parameter in the data, there still will be information coming from the prior and model structure, including the priors for the other parameters. It is not wrong to use statistical models where one or more parameters are unidentifiable, and we may still be able to use the estimates of the other parameters. The Cormack-Jolly-Seber (CJS) model with fully time-dependent parameters provides a famous example of this, where the last survival and recapture parameters are only estimable as a product and not individually (Lebreton et al., 1992). But if we have nonidentifiable parameters in our model, we would like to *know* about them because otherwise we risk inventing biological stories about them, when in a sense they are simply a reflection of our priors and misunderstanding and do not tell us anything about our data.

The risks of having nonidentifiable parameters increase with the complexity of a model, and since IPMs are always more complex than their component models (e.g., a CJS model), we run a higher risk in them, and this is also a potential cost of fitting an IPM. We must be aware of this risk and try to assess parameter identifiability. There are several formal possibilities for doing that (Catchpole and Morgan, 1997; Gimenez et al., 2009; Cole and McCrea, 2016; Cole, 2020; Ogle and Barber, 2020). In practice, however, the only simple, applied way of assessing parameter identifiability is by analysis of simulated data sets within a range of values for the parameter in question (Section 6.3).

The second risk is that sometimes estimates of the parameters of interest may become more, rather than less, uncertain in an integrated analysis. To warrant data integration, the increased complexity of the model that accompanies the inclusion of an additional data set must be more than compensated for by the increased amount of information in the added data set. We think it is possible that introduction into a model of an information-poor data set may actually make standard errors larger rather than smaller—e.g., when the data set added is small and does not contain much information or contains much noise, but its inclusion requires a substantial increase in the complexity of the model. Some data sets may just not be good enough to warrant their integration into a model.

The decision of whether to integrate should also depend on the objective of an analysis. For instance, if you are primarily interested in the assessment of a population trend, the best approach may simply be the analysis of your population count data in some time series model or Poisson generalized linear model, for instance, rather than also adding a small capture-recapture data set that you have lying around somewhere at the cost of greatly increasing the model complexity of the resulting IPM. Similarly, when your main interest is in survival estimates, you may be better off with just a plain old CJS model fit to a capture-recapture data set alone than with an IPM where you also include some noisy population count data that typically contain only very scant information about survival.

Hence, as always, the excitement about a new technique should not gain the upper hand over common sense, and more specifically, over considerations of the fundamental bias-variance tradeoff in statistics. We do think it very often makes sense to add all available information to a model of population dynamics. An IPM is a way we can do this. But sometimes, it may also make sense not to integrate, and we have to do some thinking about this decision before embarking on a population modeling project.

22.3 THE GREAT IMPORTANCE OF LONG-TERM ECOLOGICAL RESEARCH

Only very few IPMs could ever be fitted without the data from specialized surveys conducted in targeted studies, usually for single species over many years and using a more or less standardized protocol. That is, the field of IPMs would be fairly barren without intensive, long-term population studies. This is a very important type of long-term ecological research, and its value for population ecology cannot be overemphasized (Magurran et al., 2010; Lindenmayer et al., 2012). Astonishingly, a large number of such population studies are conducted by what nowadays are called citizen scientists, i.e., people outside of academia who just love to observe some species and are relentless enough to continue a study with some sort of protocol over many years without being paid (Greenwood, 2007), and sometimes for more than half a century (for an example see our peregrine case study in Chapter 12). We like to think of their contributions as *high-value citizen science*. It is very hard to envision many other cases of citizen science activities that require more dedication and greater time investments. But it is also hard to see any citizen science contributions from a single person or a small group of individuals that is more valuable in terms of adding information to both scientific research and wildlife management. Such population studies have a huge value. Consequently, we should strive to continue as many as we can and launch new ones whenever there is a possibility to do so.

22.4 FUTURE DIRECTIONS IN INTEGRATED POPULATION MODELING

IPMs are increasingly used for population analysis because multiple data sets are often available in a demographic estimation problem in population ecology or management. We have seen how the joint analysis of the component data sets offers a number of important advantages over a piecemeal, i.e., separate, analysis of the component data sets. Since the foundational paper by Besbeas et al. (2002), we have now witnessed nearly 20 years of vigorous development of IPMs, showing a transition from mostly maximum likelihood to many more Bayesian implementations and a vast diversification in terms of the fused data types and taxonomy of the study species. We have also seen the development of individual-based IPMs (Chandler and Clark, 2014), which replace the matrix population model at the core of most current IPMs with a spatio-temporal point process model. Goodness-of-fit methods have been introduced for IPMs (Besbeas and Morgan, 2014), and alternative fitting algorithms for Hidden Markov models (HMMs) have been explored (Besbeas and Morgan, 2019). Data from large-scale biodiversity monitoring programs are increasingly being used in IPMs, where the model by Dail and Madsen (2011) serves as the motherboard for fusing different data sets in the face of sometimes massive spatial replication (Ahrestani et al., 2017; Zhao et al., 2019a, 2019b, Chapter 20). Finally, it is important to realize that IPMs are a special case of general integrated models (Dorazio et al., 2017; Pacifici et al., 2017; Fletcher et al., 2019; Miller et al., 2019; Zipkin et al., 2019; see Chapter 10 in Kéry and Royle (2021); see also Section 2.8 in this book). We believe that a broader vision of IPMs in the context of data integration in statistical models in general may be fruitful and may lead to new developments in IPMs.

These achievements will certainly not be the end of the development of IPMs, and next we sketch out possible future research avenues that we foresee for IPMs. We envision three main directions of such developments. The first focuses on the population model, the second on the observation model, and the third on computational and statistical advances in general. For the population models, most of the developments that we expect are along the individual, temporal, and spatial axes that can be made more highly resolved, i.e., less aggregated. In a sense, we expect an increased move from discrete to more continuous scales along these axes and increasingly a replacement of state-structured population models by individual-based models. In addition, we may start to see models specified in continuous rather than discrete time. Finally, we expect that more IPMs will be developed that simultaneously include multiple species for the study of species interactions at a very mechanistic level.

For observation models, we expect to see more IPMs being developed with mechanistically more realistic descriptions of the false-negative and possibly the false-positive measurement error processes that produce all ecological field data, including (of course) population counts. In a sense, the Gaussian state-space model typically adopted for this part of an IPM is a misspecification, and a discrete observation model would be a much more natural candidate. Binomial and Poisson distributions are by far the most common probability models for the observation model in capture-recapture, distance sampling, and occupancy modeling, and we think that over time, these will be more commonly adopted as part of an IPM.

Finally, we imagine much progress on the computational and general statistical side of things. We now look at some of these possible developments in more detail.

22.4.1 INCREASED SPATIALIZATION OF INTEGRATED POPULATION MODELS

Dispersal is a common behavioral process in animals with profound consequences for the spatial distribution of individuals (Hanski, 1998; Clobert et al., 2001; Bowler and Benton, 2005; Hooten et al., 2017). At the population level, i.e., when we impose a boundary to define the spatial extent of a population that we intend to study, individual dispersal is usually reduced to emigration and immigration. Population models therefore need to account for these processes in some way. We have done that in all of our case studies but in several different ways.

The simplest approach is arguably to simply ignore emigration and immigration in the population model, but of course, this must be justifiable. In our case studies on peregrines, horseshoe bats, black grouse, and elk (Chapters 12, 13, 15, and 17), we justified the assumption about an absence of emigration and immigration by (1) a study area with a spatial extent so large that the number of individuals arriving and departing from the study population can be assumed as small relative to the number of individuals residing within the study area (Chapters 12 and 17), (2) a study population so isolated that no or only very scant immigration and emigration is likely to occur (Chapters 13 and 15) and (3) the claim that emigration and immigration cancel out (Chapter 12).

Another, and perhaps better, approach is the explicit inclusion of emigration and immigration in the model. We did this in our case studies on the woodchat shrike and hoopoe (Chapters 11 and 14). Emigration is not explicitly estimated but accounted for by the use of capture-recapture data and resulting estimates of apparent survival such that immigration can be estimated explicitly as a hidden parameter. If additional data are available that allow the estimation of true survival (e.g., dead-recoveries, radio tracking data), emigration could also be estimated explicitly (Reynolds et al., 2009).

In all of these cases, the emigration and immigration processes depend on how a population is spatially defined. Indeed, both dispersal processes are scale dependent: at small spatial scales, immigration and emigration can both be very important for the dynamics of a population, but with increasing spatial scales, they will eventually vanish (Millon et al., 2019). It could be said that immigration and emigration are mere scale effects and the consequence of the limited spatial extent of the defined population (Rushing et al., 2016). As the spatial definition of a population is often based on practical considerations and therefore is typically arbitrary to some degree, such a scale dependence of demographic population analysis results will not be entirely satisfactory. There are two possible approaches to improve on this. The first is the use of spatially stratified grids (or other areal subunits) for the population-level data. In that case, the spatial definition of populations may be even more arbitrary (e.g., being the cells in a grid), but since there are many cells, we can better understand spatial dynamics at the chosen scale. This approach has been used in Chapters 19 and 20 and in a small number of published IPMs (Morrison et al., 2016; Ahrestani et al., 2017; Zhao et al., 2019a, 2019b; Zhao, 2020).

The second possibility is the adoption of an individual-based spatial IPM (Chandler and Clark, 2014; Chandler et al., 2018; Sun et al., 2019; Section 10.8 in Kéry and Royle, 2021, and Chapter 21)—actually, this class of models offers a complete solution to the problem of spatial scale dependence by doing away with the definition of a study area in the foregoing sense. Individual-based spatial IPMs describe demographic processes for each individual in continuous space, and hence, they remain at the individual level and avoid a rigid spatial definition of a population. Nevertheless, these models allow inference at any chosen population level by simply summarizing the point pattern for any chosen discretization scheme or region while the underlying demographic processes are not affected/changed by this choice of scale. This is merely a consequence of the point process view taken in these models (Dorazio, 2014; Renner et al., 2014; Baddeley et al., 2016; Chandler et al., 2018; Section 10.6 in Kéry and Royle, 2021).

We are convinced that more explicit inclusion of space will be one of the major topics where we will see much progress in future IPMs as well as demographic models in general. But at the same time, we think that both individual-based and spatially stratified summary-process models will be important as quantitative ecology progresses. While spatially stratified grids may be most useful for data that were collected previously but without continuous point process models in mind, the continuing and spectacular progress of wildlife tracking devices, computational resources, and quantitative methods will undoubtedly lead to the improvement and increased importance of individual-based models.

22.4.2 BETTER REPRESENTATION OF INDIVIDUAL HETEROGENEITY

Individuals differ in almost any trait with which we may choose to describe them, including personality, quality, condition, size, age, and genetics. This has the consequence that their demographic performance and individual fitness differs accordingly (Cam et al., 2013; Senner et al., 2015; Gimenez et al., 2018). Individual heterogeneity in demographic rates affects population dynamics (Łomnicki, 1978; Björnstad and Hansen, 1994; Kendall et al., 2011), and consequently, it may be misleading to ignore this heterogeneity. Stage-structured population models capture some of this heterogeneity by a mere stratification, but they still assume that all individuals within one stratum, i.e., stage class, have identical demographic rates. Models that adopt demographic stochasticity also account for the consequences of individual heterogeneity to some degree, although the underlying mechanisms are different here. Hence, if we want to understand the effects of individual heterogeneity on population dynamics, individuality must be accounted for properly.

IPMs offer interesting opportunities to understand the effects of individual heterogeneity on population dynamics (reviewed by Plard et al., 2019a). First, we could capture extra heterogeneity within the defined stage classes in a matrix population model by using finite mixture models for the demographic rates (Hamel et al., 2017; Gimenez et al., 2018; Hamel et al., 2018) and enlarge the population model by these latent groups. The case study on hoopoes in Chapter 14 illustrates this for an IPM. Second, part of the heterogeneity in individual demographic performance can often be ascribed to continuous individuals traits that we can measure (Ellner and Rees, 2006; Coulson et al., 2010; Plard et al., 2015). Integral projection models (or "the other IPMs," as we like to think of them) can be used to model population dynamics when demographic rates depend on continuous individual traits (Easterling et al., 2000; Ellner et al., 2016). Plard et al. (2019b) developed an *integrated integral population model* (IPM^2), which is a merger of the two IPMs: the population model is an integral projection model, population counts are included in the model as well, and of course, the latter is the hallmark of "real" IPMs. This results in the usual benefits of more precision of the parameter estimates and the ability to estimate additional parameters. We also see as a third possibility the adoption of an individual-based process model similar to individual-based spatial IPMs (Chapter 21). It offers the possibility to include individual heterogeneity in demographic performance, knowledge about which may originate from models with individual random effects on demographic rates (Gimenez et al., 2018; Hamel et al., 2018). To our knowledge, such an IPM has so far not been developed.

22.4.3 FINER TEMPORAL SCALES

Most classical IPMs adopt a projection interval of 1 year corresponding to the annual birth pulse so typical of most organisms in temperate and cold latitudes. This allows many insights into population dynamics, but clearly, even more detailed insights can be obtained by models with a finer temporal resolution than 1 year, as important processes may operate seasonally only, i.e., at a sub-annual temporal scale. In particular, identifying the drivers of population dynamics and modeling density dependence may require finer temporal resolutions (Webster and Marra, 2005; Calvert et al., 2009; Hostetler et al., 2015). For example, survival may differ seasonally or may be subject to density dependence operating seasonally (e.g., only during breeding, only during nonbreeding, or just differentially between the two). In migratory species, individuals from one breeding population may spend the nonbreeding period at different locations and be exposed to very different environments or schedules and strengths of density dependence, resulting in individual differences in survival (Newton, 2004). If several populations congregate at the same nonbreeding location, individuals from one population may compete with individuals from another population. This may result in complex patterns of density-dependent survival. Finally, environmental conditions that affect an individual during parts of the year may affect subsequent reproduction or survival (carry-over effects; Norris and Taylor, 2006; Inger et al., 2010; Souchay et al., 2018; Robinson et al., 2020).

The modeling of time scales finer than 1 year requires data that allow the estimation of seasonal survival or population count data from several points in time over the course of a year. The former can be achieved when data on continuous time scales such as radio or satellite tracking data are available. Our case study on black grouse (Chapter 15) uses radio tracking data sampled throughout the year that allows the estimation of seasonal survival. With the technological progress to track animals, we expect to see more seasonal IPMs in the future. Another possibility to estimate seasonal survival is the combination of capture-recapture data from the breeding and nonbreeding periods, which is particularly useful for migratory birds (Rushing et al., 2017; Rushing, 2019).

All current sub-annual modeling of processes in IPMs is still based on a discrete representation of time. But time can also be represented more realistically as continuous. Such a continuous time representation has increasingly appeared in models of movement (Johnson et al., 2008; Michelot and Blackwell, 2019), individual survival (Barbour et al., 2013; Fouchet et al., 2016; Ergon et al., 2018; Murray and Bastille-Rousseau, 2020), and survival and observation processes in spatial capture-recapture models (Borchers et al., 2014; Dorazio and Karanth, 2017). If we have data that are very highly resolved along the temporal dimension, we may in the future see IPMs that are formulated not in discrete but in continuous time. This is likely to lead to better inferences in some ways but will also come at the cost of greatly increased computational time and much more complex models, which may also limit somewhat their use for answering ecological questions.

22.4.4 MULTIPLE SPECIES

Individuals of a species may not only affect the behavior and food intake of conspecifics, eventually resulting in density-dependent population regulation, but may also affect individuals from other species through interactions that include predation, parasitism, or interspecific competition. If we want to better understand one study species or are interested in the demography of multiple species or even an entire community, IPMs for several species may be developed. We may distinguish between two types of multispecies IPMs. The first aims to study temporal synchrony of demographic rates and population growth rates of multiple species (Lahoz-Monfort et al., 2017). This approach models correlations among species but not any directed interactions between them (see also Chapter 8 in Kéry and Royle, 2021, for a conceptual overview of the types of models for species interactions). Therefore, these models only allow for the identification of potential common or divergent environmental drivers for the demography of the studied species, including interactions that may go in one or both directions, without being able to disentangle the latter.

In contrast, a second type of IPM includes a direct link of the abundance of one species on a demographic rate on the other species, i.e., uses the abundance or occurrence of one species as a covariate for one or more parameters in the model of the other species. Conceptually, this is similar to the modeling of density dependence (Chapter 8). The first published IPM of this second type is due to Péron and Koons (2012), who developed an IPM for two interacting duck species. They included specific links between reproduction of one species and the density of the other via a linear model and found a negative relationship possibly driven by nest parasitism. Gamelon et al. (2019) simultaneously estimated inter- and intraspecific competition between two tit species, and Quéroué et al. (2021) studied predator-prey interactions in two seabird species. Barraquand and Gimenez (2019) summarized the possible structure of such multispecies IPMs with direct links and investigated what kinds of data are required for parameter estimation. So clearly, we expect to see an increasing number of multispecies IPMs. Not only can the results they offer be fascinating, but also, they may offer the potential to study species demography with little data. The link with other species and the perceived similarity may help with parameter estimation through shared information. Such sharing may be weighted by phylogenetic similarities (Abadi et al., 2014) or be mediated by estimated allometric relationships (McCarthy and Masters, 2005).

22.4.5 BETTER OBSERVATION MODELS FOR POPULATION COUNT DATA

The observation model for the population count data in IPMs is often only implicit in the sense that the observation error model is not a truly mechanistic representation of the observation process as it typically is in other common models in population ecology such as capture-recapture, distance sampling, or occupancy modeling (Williams et al., 2002; Royle and Dorazio, 2008; Buckland et al., 2015; MacKenzie et al., 2018). In current IPMs, a normal, log-normal, or Poisson distribution is typically used, resulting in smoothed estimates of "relative" population size. Whether this quantity corresponds to the true population size depends on the type of error in the collection of the count data (see Sections 4.3.2 and 7.4). If the effects of counting errors are completely symmetric around the true population size, these distributions will successfully recover the true population size on average. Such counting errors may be realistic in some situations, e.g., when large numbers of individuals are counted in a habitat that allows good visibility on average (e.g., a flock of ducks on a large lake), where double counts may approximately cancel out the effects of individuals that are missed (i.e., nondetection). Yet in most cases, imperfect detection will be the main process producing errors in the population count data. The use of the above-mentioned statistical distributions can only "shave away" some temporal variability due to the stochasticity represented by binomial sampling error, but they don't allow the proper estimation of true population size (Sections 4.3.2 and 7.4). If the sampling error lacks any deterministic trends, the resulting "relative" population size will have the same temporal pattern as the true population size, and thus it will be useful for inferences about population change. However, we typically do not know how large the detection error is or whether it shows any temporal or spatial trends. Indeed, this can only be known if the detection error is properly estimated. Unfortunately, this requires additional information—from a single time series of counts, we cannot estimate detection probability.

The required additional information can come in many forms, including from distance information (Buckland et al., 2015), replicated counts (Royle, 2004) or time-to-detection data (Garrard et al., 2008; Halstead et al., 2018). This has two implications. First, at the sampling design stage of a study, we need to think hard about which kind of population count data we ought to collect. In most IPMs in the past, people simply used the data that were available, but we may change that in the future and adopt a better sampling design for what we call the "population counts." Second, IPMs must be developed that allow the proper inclusion of data that carry the required extra information for the estimation of detection error rates. The problem is that population size is typically estimated from such data, and in an IPM, population size is a function of demographic processes. When implementing IPMs in the way we do in this book, we do not know how to combine them in a single model, but we have shown a two-step approach as a workaround (Sections 4.3.3 and 6.6.4). An interesting option is the demographic state-space models that use spatially replicated counts (Dail and Madsen, 2011; Zipkin et al., 2014). They can fully accommodate imperfect detection and have a demographic population model such that other data that provide information about demographic rates can be easily linked (see, e.g., Zhao, 2020). But these models require the existence of multivariate population counts, i.e., replication across a set of sites. We expect that such models will be increasingly used for modeling population dynamics at larger spatial scales. As an alternative, it should be possible to explicitly develop a joint likelihood for an IPM where a proper estimation model, say, of the capture-recapture or distance sampling type, is adopted for the population counts and then maximize that joint likelihood for the IPM in a single step. But we surmise that this will not result in code that is easily understandable for nonstatisticians as is the BUGS code that we offer throughout this book.

22.4.6 IMPROVING SAMPLING DESIGNS FOR INTEGRATED POPULATION MODELS

IPMs have mostly been applied to already existing data sets, and owing to their great flexibility, it has been possible to make use of such data even if they might have been collected in a less than ideal way. For example, there might be spatial or temporal mismatches between the different data sets, or data about an important parameter may be lacking. In the future, we can try to do better and start thinking about optimal sampling designs for IPMs and the data they require.

The development of a sampling design must start from the objectives of a study or monitoring program. When more data are sampled, the precision of the parameter estimates will tend to increase, so we can generally say "more is better." However, there are also costs involved, and the goal is typically the development of a cost-effective sampling design that best allows the stated goals to be achieved at the lowest cost. This seems particularly important for long-term studies such as national programs for demographic monitoring or data collection to manage harvested populations. Given that IPMs are so flexible in combining different data sets, we may want to find out the optimal temporal and spatial combination of data collection of different data sets in terms of costs and objectives. Do we need all data types from each year and from all sites? What is the optimal allocation of data collection effort in terms of which data are sampled when and where? As far as we know, only studies by Zhao (2020) and Johnson et al. (2020) have so far investigated sampling designs for

IPMs. Ross et al. (2018) performed a power analysis to identify the required sample sizes to estimate an effect size with desired precision. We think that in the future we will see more such studies. Another interesting question to address is the degree to which high-quality data sets are necessary for a joint analysis of low-quality citizen science data (Zipkin and Saunders, 2018). That is, what are the optimal "mixing proportions" of the different kinds of data combined in an integrated model? See the work of Dorazio (2014) for a neat example of this in the context of an integrated species distribution model.

22.4.7 STATISTICAL AND COMPUTATIONAL ADVANCES

We imagine that there will also be much progress in IPMs in terms of the general statistical modeling and computation involved. First, more research on goodness-of-fit testing is certainly needed, as we have argued in Chapter 7, and we hope to see progress in this important field in the future. Second, most IPMs assume independence among the likelihoods of the component data, and this assumption is often not completely met. Also, we think it is fair to say that the very concept of dependence among the component data sets being used in an IPM is not entirely clear thus far. The consequences of violating this assumption often seem to be fairly irrelevant (Abadi et al., 2010; Weegman et al., 2021), but there are examples where the lack of independence did make a difference (Besbeas et al., 2009). At any rate, it seems cleaner from a statistical point of view to develop models that properly account for dependence if this is needed in a study. A very small number of such models have already been developed (Chandler and Clark, 2014; Lee et al., 2015; Béliveau, 2016). Not surprisingly, they are more complex than models assuming independence. We expect to see more such models but welcome an evaluation of whether it is worth going in that direction simply because of possible dependence. Third, the fitting of IPMs by representing them as HMMs has just started to be explored (Besbeas and Morgan, 2019). HMMs are very well studied in engineering and other fields (McClintock et al., 2020), and perhaps the formulation of an IPM as an HMM will prove to be a way forward to general fitting algorithms that employ maximum likelihood with the major attendant benefit being increased computational speed.

Fitting complex IPMs or fitting any IPM when sample size is large can be computationally very expensive, and you must be prepared to sit and wait for a long time. Of course, computer power is steadily increasing, and the ability to parallelize computing is helpful. NIMBLE (NIMBLE Development Team, 2018) offers the possibility to expert-tune the algorithms, which can speed up a model fitting problem by orders of magnitude (Bischof et al., 2020). However, it must be said that to exploit these benefits of NIMBLE, you will again need a greater understanding of statistics and computation than what the average population ecologist has, even the statistically savvy one, and for this, you may again have to run to your statistician friend for help. The same goes for any implementation of an IPM in software outside the BUGS family such as TMB (Kristensen et al., 2016) or Stan (Carpenter et al., 2017).

In the future, it may be possible to fit IPMs using maximum likelihood in NIMBLE while still being able to specify the model in that *lingua franca* of hierarchical models—the BUGS language. In principle, NIMBLE has shown some ability to fit hierarchical models using maximum likelihood, and it would be great if this could be used to fit general IPMs in a flexible way. Alternatively, the development of other non-Bayesian software for complex statistical models such as IPMs would be welcome. At the risk of sounding repetitive, from an applied ecologist user perspective, we might add that such software would be especially valuable if it had a simple "model description interface" as the BUGS language has proven to be.

22.5 CONCLUDING REMARKS

IPMs are complex hierarchical models, but they are very powerful, and their development can be a great deal of fun and give you deep satisfaction when things appear to be working as they should. When we first encountered BUGS code to fit an IPM (in Brooks et al., 2004), we felt that we were staring at a piece of fairly long computer code that was difficult for us to understand. But after awhile, we realized that this complex model is in fact defined by relatively simple component likelihoods, each of which is associated with one of the data sets combined in that model. The joint likelihood is defined implicitly by the shared parameters. This opened our eyes to the feasibility of fitting IPMs in practice and allowed us to start thinking about our own data sets in a creative way, about their likelihoods, and about how to connect them within a single population model. We see the development of an IPM in the BUGS language as akin to a LEGO game—you have several pieces that represent different data sets and their associated likelihoods and combine them in a creative way to build the IPM. This allows the utilization of all available information in a statistically rigorous and transparent way. We truly hope that our book provides you with the same happy and empowering experience we have had—and that it leads to more exciting discoveries and better conclusions in your science and to improved decision-making in your conservation and wildlife management.

References

Abadi, F., Gimenez, O., Ullrich, B., Arlettaz, R., Schaub, M., 2010a. Estimation of immigration rate using integrated population models. J. Appl. Ecol. 47, 393–400.

Abadi, F., Gimenez, O., Arlettaz, R., Schaub, M., 2010b. An assessment of integrated population models: bias, accuracy, and violation of the assumption of independence. Ecology 91, 7–14.

Abadi, F., Gimenez, O., Jakober, H., Stauber, W., Arlettaz, R., Schaub, M., 2012. Estimating the strength of density dependence in the presence of observation errors using integrated population models. Ecol. Model. 242, 1–9.

Abadi, F., Barbraud, C., Besson, D., Bried, J., Crochet, P.-A., Delord, K., Forcada, J., Grosbois, V., Phillips, R.A., Sagar, P., Thompson, P., Waugh, S., Weimerskirch, H., Wood, A.G., Gimenez, O., 2014. Importance of accounting for phylogenetic dependence in multi-species mark-recapture studies. Ecol. Model. 273, 236–241.

Acevedo, C.R., Riecke, T.V., Leach, A.G., Lohman, M.G., Williams, P.J., Sedinger, J.S., Koons, D.N., 2020. Long-term research and hierarchical models reveal consistent fitness costs of being the last egg in a clutch. J. Anim. Ecol. 89, 1978–1987.

Aebischer, N.J., 1999. Multi-way comparisons and generalized linear models of nest success: extensions of the Mayfield method. Bird Study 46 (suppl. 1), 22–31.

Ahrestani, F.S., Saracco, J.F., Sauer, J.R., Pardieck, K.L., Royle, J.A., 2017. An integrated population model for bird monitoring in North America. Ecol. Appl. 27, 916–924.

Alatalo, R.V., Höglund, J., Lundberg, A., 1991. Lekking in the black grouse - a test of male viability. Nature 352, 155–156.

Allendorf, F.W., Ryman, N., 2002. The role of genetics in population viability analysis. In: Beissinger, S.R., McCullough, D. (Eds.), Population Viability Analysis. The University of Chicago Press, Chicago, pp. 50–85.

Altwegg, R., Roulin, A., Kestenholz, M., Jenni, L., 2003. Variation and covariation in survival, dispersal, and population size in barn owls *Tyto alba*. J. Anim. Ecol. 72, 391–399.

Altwegg, R., Roulin, A., Kestenholz, M., Jenni, L., 2006. Demographic effects of extreme winter weather in the barn owl. Oecologia 149, 44–51.

Altwegg, R., Schaub, M., Roulin, A., 2007. Age-specific fitness components and their temporal variation in the barn owl. Am. Nat. 169, 47–61.

Altwegg, R., Jenkins, A., Abadi, F., 2014. Nestboxes and immigration drive the growth of an urban Peregrine Falcon *Falco peregrinus* population. Ibis 156, 107–115.

Alvarez, I., Niemi, J., Simpson, M., 2016. Bayesian Inference for a Covariance Matrix. arXiv:1408.4050.

Anderson, D.R., Burnham, K.P., White, G.C., 1985. Problems in estimating age-specific survival rates from recovery data of birds ringed as young. J. Anim. Ecol. 54, 89–98.

Anderson, D.L., Bente, P.J., Booms, T.L., Dunn, L., McClure, C.J., 2019. Nonrandom territory occupancy by nesting Gyrfalcons (*Falco rusticolus*). Arctic Sci. 5, 148–160.

Andreassen, H.P., Ims, R.A., 2001. Dispersal in patchy vole populations: role and patch configuration, density dependence, and demography. Ecology 82, 2911–2926.

Arlettaz, R., Patthey, P., Baltic, M., Leu, T., Schaub, M., Palme, R., Jenni-Eiermann, S., 2007. Spreading free-riding snow sports represent a novel serious threat for wildlife. Proc. R. Soc. Lond. B 274, 1219–1224.

Arlettaz, R., Schaub, M., Fournier, J., Reichlin, T.S., Sierro, A., Watson, J.E.M., Braunisch, V., 2010. From publications to public actions: when conservation biologists Bridge the gap between research and implementation. BioScience 60, 835–842.

Arlettaz, R., Christe, P., Schaub, M., 2017. Food availability as a major driver in the evolution of life-history strategies of sibling species. Ecol. Evol. 7, 4163–4172.

Arnason, A.N., 1972. Parameter estimates from mark-recapture experiments on two populations subject to migration and death. Res. Pop. Ecol. 13, 97–113.

Arnason, A.N., 1973. The estimation of population size, migration rates and survival in a stratified population. Res. Pop. Ecol. 15, 1–8.

Arnold, T.W., 2018. Using ring-recovery and within-season recapture data to estimate fecundity and population growth. Ecol. Evol. 8, 10298–10305.

Arnold, T.W., Devries, J.H., Howerter, D.W., 2010. Factors that affect renesting in Mallards (*Anas platyrhynchos*). Auk 127, 212–221.

Arnold, T.W., Afton, A.D., Anteau, M.J., Koons, D.N., Nicolai, C.A., 2016. Temporal variation in survival and recovery rates of lesser scaup. J. Wildl. Manag. 80, 850–861.

Arnold, T.W., Clark, R.G., Koons, D.N., Schaub, M., 2018. Integrated population models facilitate ecological understanding and improved management decisions. J. Wildl. Manag. 82, 266–274.

Aubry, L.M., Koons, D.N., Monnat, J.Y., Cam, E., 2009. Consequences of recruitment decisions and heterogeneity on age-specfic breeding success in a long-lived seabird. Ecology 90, 2491–2502.

Aubry, L.M., Cam, E., Koons, D.N., Monnat, J.Y., Pavard, S., 2011. Drivers of age-specific survival in a long-lived seabird: contributions of observed and hidden sources of heterogeneity. J. Anim. Ecol. 80, 375–383.

Auger-Méthé, M., Field, C., Albertsen, C.M., Derocher, A.E., Lewis, M.A., Jonsen, I.D., Mills Flemming, J., 2016. State-space models' dirty little secrets: even simple linear Gaussian models can have estimation problems. Sci. Rep. 6, 26677.

Auger-Méthé, M., Newman, K., Cole, D., Empacher, F., Gryba, R., King, A.A., Leos-Barajas, V., Flemming, J.M., Nielsen, A., Petris, G., Thomas, L., 2021. An Introduction to State-Space Modeling of Ecological Time Series. Ecol. Monogr. https://doi.org/10.1002/ecm.1470.

Augustine, B.C., Royle, J.A., Kelly, M.J., Satter, C.B., Alonso, R.S., Boydston, E.E., Crooks, K.R., 2018. Spatial capture-recapture with partial identity: an application to camera traps. Ann. Appl. Stat. 12, 67–95.

Augustine, B.C., Kéry, M., Olano Marin, J., Mollet, P., Pasinelli, G., Sutherland, C., 2020a. Sex-specific population dynamics and demography of capercaillie (*Tetrao urogallus* L.) in a patchy environment. Popul. Ecol. 62, 80–90.

Augustine, B.C., Royle, J.A., Linden, D.W., Fuller, A.K., 2020b. Spatial proximity moderates genotype uncertainty in genetic tagging studies. Proc. Nat. Acad. Sci. U.S.A. 117, 17903–17912.

Austad, S.N., Fischer, K.E., 1991. Mammalian aging, metabolism, and ecology: evidence from the bats and marsupials. J. Gerontol. 46, 47–53.

Baddeley, A., Rubak, E., Turner, R., 2016. Spatial Point Patterns. Methodology and Applications With R. CRC Press Taylor & Francis Group, Boca Raton, London, New York.

Baillie, S.R., 1990. Integrated population monitoring of breeding birds in Britain and Ireland. Ibis 132, 151–166.

Baillie, S.R., Schaub, M., 2009. Understanding changes in bird populations - the role of bird marking. Ringing Migr. 24, 189–198.

Baillie, S.R., Sutherland, B.J., Freeman, S.N., Gregory, R.D., Paradis, E., 2000. Consequences of large-scale processes for the conservation of bird populations. J. Appl. Ecol. 37, 88–102.

Baillie, S.R., Brooks, S.P., King, R., Thomas, L., 2009. Using a state-space model of the British song thrush *Turdus pilomenos* population to diagnose the causes of a population decline. In: Thomson, D., Cooch, E., Conroy, M. (Eds.), Modeling Demographic Processes in Marked Populations. Springer, New York, pp. 541–561.

Banerjee, S., Carlin, B.P., Gelfand, A.E., 2014. Hierarchical Modeling and Analysis for Spatial Data. CRC Press, Boca Raton.

Barbour, A.B., Ponciano, J.M., Lorenzen, K., 2013. Apparent survival estimation from continuous mark-recapture/resighting data. Methods Ecol. Evol. 4, 846–853.

Barbraud, C., Weimerskirch, H., 2003. Climate change and density shape population dynamics of a marine top predator. Proc. R. Soc. Lond. B 270, 2111–2116.

Barker, R.J., 1997. Joint modeling of live-recapture, tag-resight, and tag-recovery data. Biometrics 53, 666–677.

Barker, R.J., 1999. Joint analysis of mark-recapture, resighting and ring-recovery data with age-dependence and marking effect. Bird Study 46 (suppl. 1), 82–91.

Barker, R.J., Schofield, M.R., Link, W.A., Sauer, J.R., 2018. On the reliability of N-mixture models for count data. Biometrics 74 (suppl. 1), 369–377.

Barnard, J., McCulloch, R., Meng, X.-L., 2000. Modeling covariance matrices in terms of standard deviations and correlations, with application to shrinkage. Stat. Sin. 10, 1281–1312.

Barraquand, F., Gimenez, O., 2019. Integrating multiple data sources to fit matrix population models for interacting species. Ecol. Model. 411, 108713.

Barry, S.C., Brooks, S.P., Catchpole, E.A., Morgan, B.J.T., 2003. The analysis of ring-recovery data using random effects. Biometrics 59, 54–65.

Beissinger, S.R., 2002. Population viability analysis: past, present, future. In: Beissinger, S.R., McCullough, D. (Eds.), Population Viability Analysis. The University of Chicago Press, Chicago, pp. 5–17.

Beissinger, S.R., Westphal, M.I., 1998. On the use of demographic models of population viability in endangered species management. J. Wildl. Manag. 62, 821–841.

Béliveau, A., 2016. Data Integration Methods for Studying Animal Population Dynamics. PhD thesis, Vancouver.

Bellier, E., Kéry, M., Schaub, M., 2016. Simulation-based assessment of dynamic *N*-mixture models in the presence of density dependence and environmental stochasticity. Methods Ecol. Evol. 7, 1029–1040.

Bellier, E., Kéry, M., Schaub, M., 2018. Relationships between vital rates and ecological traits in an avian community. J. Anim. Ecol. 87, 1172–1181.

Berg, S.S., Erb, J.D., Fieberg, J.R., Forester, J.D., 2017. Utility of radio-telemetry data for improving statistical population reconstruction. J. Wildl. Manag. 81, 535–544.

Berger, V., Lemaître, J.-F., Allainé, D., Gaillard, J.-M., Cohas, A., 2015. Early and adult social environments have independent effects on individual fitness in a social vertebrate. Proc. R. Soc. Lond. B 282, 20151167.

Berliner, L.M., 1996. Hierarchical Bayesian time series models. In: Hanson, K.M., Silver, R.N. (Eds.), Maximum Entropy and Bayesian Methods. Santa Fe, New Mexico, U.S.A., 1995 Proceedings of the Fifteenth International Workshop on Maximum Entropy and Bayesian Methods. Springer Netherlands, Dordrecht, pp. 15–22.

Besbeas, P., Freeman, S.N., 2006. Methods for joint inference from panel survey and demographic data. Ecology 87, 1138–1145.

Besbeas, P., Morgan, B.J.T., 2012a. A threshold model for heron productivity. JABES 17, 128–141.

Besbeas, P., Morgan, B.J.T., 2012b. Kalman filter initialisation for integrated population modelling. App. Stat. 61, 151–162.

Besbeas, P., Morgan, B.J.T., 2014. Goodness of fit of integrated population models using calibrated simulation. Methods Ecol. Evol. 5, 1373–1382.

Besbeas, P., Morgan, B.J.T., 2017. Variance estimation for integrated population models. Adv. Stat. Anal. 101, 439–460.

Besbeas, P., Morgan, B.J.T., 2019. Exact inference for integrated population modelling. Biometrics 75, 475–484.

Besbeas, P., Freeman, S.N., Morgan, B.J.T., Catchpole, E.A., 2002. Integrating mark-recapture-recovery and census data to estimate animal abundance and demographic parameters. Biometrics 58, 540–547.

Besbeas, P., Lebreton, J.-D., Morgan, B.J.T., 2003. The efficient integration of abundance and demographic data. App. Stat. 52, 95–102.

Besbeas, P., Freeman, S.N., Morgan, B.J.T., 2005. The potential of integrated population modelling. Aust. N. Z. J. Stat. 47, 35–48.

Besbeas, P., Borysiewicz, R.S., Morgan, B.J.T., 2009. Completing the ecological jigsaw. In: Thomson, D., Cooch, E., Conroy, M. (Eds.), Modeling Demographic Processes in Marked Populations. Springer, New York, pp. 513–539.

Birch, L.C., 1953. Experimental background to the study of the distribution and abundance of insects. I. The influence of temperature, moisture and food on the innate capacity for increase of three grain beetles. Ecology 34, 698–711.

Bischof, R., Milleret, C., Dupont, P., Chipperfield, J., Tourani, M., Ordiz, A., Valpine de, P., Turek, D., Royle, J.A., Gimenez, O., Flagstad, Ø., Åkesson, M., Svensson, L., Brøseth, H., Kindberg, J., 2020. Estimating and forecasting spatial population dynamics of apex predators using transnational genetic monitoring. Proc. Natl. Acad. Sci. U.S.A. 117, 30531–30538.

Björnstad, O.N., Hansen, T.F., 1994. Individual variation and population dynamics. Oikos 69, 167–171.

Björnstad, O.N., Ims, R.A., Lambin, X., 1999. Spatial population dynamics: analysing patterns and processes of population synchrony. Trends Ecol. Evol. 14, 427–432.

Bled, F., Royle, J.A., Cam, E., 2011. Assessing hypotheses about nesting site occupancy dynamics. Ecology 92, 938–951.

Bled, F., Belant, J.L., van Daele, L.J., Svoboda, N., Gustine, D., Hilderbrand, G., Barnes, V.G., 2017. Using multiple data types and integrated population models to improve our knowledge of apex predator population dynamics. Ecol. Evol. 91, 9531–9543.

Bolker, B.M., 2008. Ecological Models and Data in R. Princeton University Press, Princeton, New Jersey.

Bolker, B.M., Brooks, M.E., Clark, C.J., Geange, S.W., Poulsen, J.R., Stevens, M.H.H., White, J.-S.S., 2009. Generalized linear mixed models: a practical guide for ecology and evolution. Trends Ecol. Evol. 24, 127–135.

Bonenfant, C., Gaillard, J.M., Coulson, T., Festa-Bianchet, M., Loison, A., Garel, M., Loe, L.E., Blanchard, P., Pettorelli, N., Owen-Smith, N., Du Toit, J., Duncan, P., 2009. Empirical evidence of density-dependence in populations of large herbivores. Adv. Ecol. Res. 41, 313–357.

Bonnet, X., Naulleau, G., 1996. Catchability in snakes: consequences for estimates of breeding frequency. Can. J. Zool. 74, 233–239.

Borchers, D.L., Efford, M.G., 2008. Spatially explicit maximum likelihood methods for capture-recapture studies. Biometrics 64, 377–385.

Borchers, D.L., Marques, T.A., 2017. From distance sampling to spatial capture-recapture. Adv. Stat. Anal. 101, 475–494.

Borchers, D.L., Buckland, S.T., Zucchini, W., 2002. Estimating Animal Abundance. Springer, London.

Borchers, D., Distiller, G., Foster, R., Harmsen, B., Milazzo, L., 2014. Continuous-time spatially explicit capture-recapture models, with an application to a jaguar camera-trap survey. Methods Ecol. Evol. 5, 656–665.

Borchers, D.L., Stevenson, B.C., Kidney, D., Thomas, L., Marques, T.A., 2015. A unifying model for capture-recapture and distance sampling surveys of wildlife populations. J. Am. Stat. Assoc. 110, 195–204.

Borysiewicz, R.S., Morgan, B.J.T., Hénaux, V., Bregnballe, T., Lebreton, J.-D., Gimenez, O., 2009. An integrated analysis of multisite recruitment, mark-recapture-recovery and multisite census data. In: Thomson, D., Cooch, E., Conroy, M. (Eds.), Modeling Demographic Processes in Marked Populations. Springer, New York, pp. 579–591.

Bötsch, Y., Arlettaz, R., Schaub, M., 2012. Breeding dispersal of Eurasian Hoopoes (*Upupa epops*) within and between years in relation to reproductive success, sex, and age. Auk 129, 283–295.

Bouwhuis, S., Choquet, R., Sheldon, B.C., Verhulst, S., 2012. The forms and fitness cost of senescence: age-specific recapture, survival, reproduction, and reproductive value in a wild bird population. Am. Nat. 179, E15–E27.

Bowler, D.E., Benton, T.G., 2005. Causes and consequences of animal dispersal strategies: relating individual behaviour to spatial dynamics. Biol. Rev. 80, 205–225.

Boyce, M.S., 1992. Population viability analysis. Annu. Rev. Ecol. Syst. 23, 481–506.

Boyce, M.S., Sinclair, A.R.E., White, G.C., 1999. Seasonal compensation of predation and harvesting. Oikos 87, 419–426.

Boyce, M.S., Haridas, C.V., Lee, C.T., The NCEAS Stochastic Demography Working Group, 2006. Demography in an increasingly variable world. Trend. Ecol. Evol. 21, 141–148.

Brault, S., Caswell, H., 1993. Pod-specific demography of resident killer whales *Orcinus orca* in British Columbia and Washington State. Ecology 74, 1444–1454.

Bregnballe, T., Gregersen, J., 2003. Within-colony variation in breeding success in a Great Cormorant colony in Denmark. Vogelwelt 124, 115–122.

Breiman, L., 2001. Statistical modeling: the two cultures (with comments and a rejoinder by the author). Stat. Sci. 16, 199–231.

Bretagnolle, V., Mougeot, F., Thibault, J.-C., 2008. Density dependence in a recovering osprey population: demographic and behavioural processes. J. Anim. Ecol. 77, 998–1007.

Brintz, B., Fuentes, C., Madsen, L., 2018. An asymptotic approximation to the N-mixture model for the estimation of disease prevalence. Biometrics 74, 1512–1518.

Brommer, J.E., Wistbacka, R., Selonen, V., 2017. Immigration ensures population survival in the Siberian flying squirrel. Ecol. Evol. 7, 1858–1868.

Broms, K., Skalski, J.R., Millspaugh, J.J., Hagen, C.A., Schulz, J.H., 2010. Using statistical population reconstruction to estimate demographic trends in small game populations. J. Wildl. Manag. 74, 310–317.

Brook, B.W., Bradshaw, C.J.A., 2006. Strength of evidence for density dependence in abundance time series of 1198 species. Ecology 87, 1445–1451.

Brooks, S.P., 2003. Bayesian computation: a statistical revolution. Philos. Trans. R. Soc. A 361, 2681–2697.

Brooks, S.P., Gelman, A., 1998. General methods for monitoring convergence of iterative simulations. J. Comput. Graph. Stat. 7, 434–455.

Brooks, S.P., Catchpole, E.A., Morgan, B.J.T., 2000a. Bayesian animal survival estimation. Stat. Sci. 15, 357–376.

Brooks, S.P., Catchpole, E.A., Morgan, B.J.T., Barry, S.C., 2000b. On the Bayesian analysis of ring-recovery data. Biometrics 56, 951–956.

Brooks, S.P., Catchpole, E.A., Morgan, B.J.T., Harris, M.P., 2002. Bayesian methods for analysing ringing data. J. Appl. Stat. 29, 187–206.

Brooks, S.P., King, R., Morgan, B.J.T., 2004. A Bayesian approach to combining animal abundance and demographic data. Anim. Biodivers. Conserv. 27.1, 515–529.

Brooks, S.P., Freeman, S.N., Greenwood, J.J., King, R., Mazzetta, C., 2008. Quantifying conservation concern – Bayesian statistics, birds and the red lists. Biol. Conserv. 141, 1436–1441.

Brooks, S., Gelman, A., Jones, G.L., Meng, X.-L. (Eds.), 2011. Handbook of Markov Chain Monte Carlo. Chapman & Hall/CRC, Hoboken.

Brooks, M.E., Kristensen, K., Darrigo, M.R., Rubim, P., Uriarte, M., Bruna, E., Bolker, B.M., 2019. Statistical modeling of patterns in annual reproductive rates. Ecology 100, e02706.

Brown, J.H., Kodric-Brown, A., 1977. Turnover rates in insular biogeography: effect of immigration on extinction. Ecology 58, 445–449.

Brownie, C., Anderson, D.R., Burnham, K.P., Robson, D.S., 1985. Statistical Inference from Band Recovery Data - a Handbook. US Fish and Wildlife Service, vol. 156. Duxbury Press,, Washington DC.

Brownie, C., Hines, J.E., Nichols, J.D., Pollock, K.H., Hestbeck, J.B., 1993. Capture-recapture studies for multiple strata including non-Markovian transitions. Biometrics 49, 1173–1187.

Bruggeman, J.E., Swem, T., Andersen, D.E., Kennedy, P.L., Nigro, D., 2015. Dynamics of a recovering arctic bird population: the importance of climate, density dependence, and site quality. Ecol. Appl. 25, 1932–1943.

Bruggeman, J.E., Swem, T., Andersen, D.E., Kennedy, P.L., Nigro, D., Sanchez-Zapata, J.A., Smart, J., 2016. Multi-season occupancy models identify biotic and abiotic factors influencing a recovering arctic peregrine falcon *Falco peregrinus tundrius* population. Ibis 158, 61–74.

Buckland, S.T., Anderson, D.R., Burnham, K.P., Laake, J.L., Borchers, D.L., Thomas, L., 2001. Introduction to Distance Sampling. Oxford University Press, Oxford.

Buckland, S.T., Newman, K.B., Thomas, L., Koesters, N.B., 2004. State-space models for the dynamics of wild animal populations. Ecol. Model. 171, 157–175.

Buckland, S.T., Newman, K.B., Fernández, C., Thomas, L., Harwood, J., 2007. Embedding population dynamics models in inference. Stat. Sci. 22, 44–58.

Buckland, S.T., Rexstad, E.A., Oedekoven, C.S., Marques, T.A., 2015. Distance Sampling: Methods and Applications. Springer, Cham.

Budden, M., Hadavas, P., Hoffman, L., Pretz, C., 2007. Generating valid 4×4 correlation matrices. Appl. Math. E-Notes 7, 53–59.

Burnham, K.P., 1993. A theory for combined analysis of ring recovery and recapture data. In: Lebreton, J.-D., North, P.M. (Eds.), Marked Individuals in the Study of Bird Populations. Birkhäuser, Basel, pp. 199–213.

Burnham, K.P., Anderson, D.R., 1984. Tests of compensatory vs. additive hypotheses of mortality in Mallards. Ecology 65, 105–112.

Burnham, K.P., White, G.C., 2002. Evaluation of some random effects methodology applicable to bird ringing data. J. Appl. Stat. 29, 245–264.

Burnham, K.P., Anderson, D.R., White, G.C., Brownie, C., Pollock, K.H., 1987. Design and analysis methods for fish survival experiments based on release-recapture. Am. Fish. Soc. Monogr. 5.

Caizergues, A., Ellison, L.N., 1997. Survival of black grouse *Tetrao tetrix* in the French Alps. Wildl. Biol. 3, 177–186.

Caizergues, A., Ellison, L.N., 2000. Age-specific reproductive performance of Black Grouse *Tetrao tetrix* females. Bird Study 47, 344–351.

Calvert, A.M., Gauthier, G., 2005. Effects of exceptional conservation measures on survival and seasonal hunting mortality in greater snow geese. J. Appl. Ecol. 42, 442–452.

Calvert, A.M., Walde, S.J., Taylor, P.D., 2009. Nonbreeding-season drivers of population dynamics in seasonal migrants: conservation parallels across taxa. Anim. Conserv. Ecol. 4, 5.

Cam, E., Gimenez, O., Alpizar-Jara, R., Aubry, L.M., Authier, M., Cooch, E.G., Koons, D.N., Link, W.A., Monnat, J.Y., Nichols, J.D., Rotella, J.J., Royle, J.A., Pradel, R., 2013. Looking for a needle in a haystack: inference about individual fitness components in a heterogeneous population. Oikos 122, 739–753.

Cao, J., He, C.Z., McCoy, T.D., 2008. Bayesian estimation of age-specific bird nest survival rates with categorical covariates. Environ. Ecol. Stat. 15, 49–58.

Cao, J., He, C.Z., Suedkamp Wells, K.M., Millspaugh, J.J., Ryan, M.R., 2009. Modeling age and nest-specific survival using a hierarchical Bayesian approach. Biometrics 65, 1052–1062.

Carlin, B.P., Louis, T.A., 2009. Bayesian Methods for Data Analysis. CRC Press/Taylor & Francis Group, Boca Raton.

Carpenter, B., Gelman, A., Hoffman, M.D., Lee, D., Goodrich, B., Betancourt, M., Brubaker, M., Guo, J., Li, P., Riddell, A., 2017. Stan: a probabilistic programming language. J. Stat. Software 76.

Carvalho, F., Punt, A.E., Chang, Y.-J., Maunder, M.N., Piner, K.R., 2017. Can diagnostic tests help identify model misspecification in integrated stock assessments? Fish. Res. 192, 28–40.

Casella, G., Berger, R.L., 2002. Statistical Inference. Duxbury Press, Pacific Grove, CA, USA.

Caswell, H., 1989. Analysis of life table response experiments I. Decomposition of effects on population growth rate. Ecol. Model. 46, 221–237.

Caswell, H., 1996. Analysis of life table response experiments. II. Alternative parameterizations for size- and stage-structured models. Ecol. Model. 88, 73–82.

Caswell, H., 2000. Prospective and retrospective perturbation analyses: their roles in conservation biology. Ecology 81, 619–627.

Caswell, H., 2001. Matrix Population Models. Construction, Analysis, and Interpretation. Sinauer Associates, Sunderland, Massachusetts.

Caswell, H., 2007. Sensitivity analysis of transient population dynamics. Ecol. Lett. 10, 1–15.

Caswell, H., 2010. Life table response experiment analysis of the stochastic growth rate. J. Ecol. 98, 324–333.

Caswell, H., Brault, S., Read, A.J., Smith, T.D., 1998. Harbor porpoise and fisheries: an uncertainty analysis of incidental mortality. Ecol. Appl. 8, 1226–1238.

Catchpole, E.A., Morgan, B.J.T., 1997. Detecting parameter redundancy. Biometrika 84, 187–196.

Catlin, D.H., Gibson, D., Hunt, K.L., Friedrich, M.J., Weithman, C.E., Karpanty, S.M., Fraser, J.D., 2019. Direct and indirect effects of nesting density on survival and breeding propensity of an endangered shorebird. Ecosphere 10, e02740.

Caughley, G., 1994. Directions in conservation biology. J. Anim. Ecol. 63, 215–244.

Cave, V.M., King, R., Freeman, S.N., 2010. An integrated population model from constant effort bird-ringing data. JABES 15, 119–137.

Celeux, G., Forbes, F., Robert, C.P., Titterington, D.M., 2006. Deviance information criteria for missing data models. Bayes. Anal. 1, 651–674.

Chambert, T., Miller, D.A.W., Nichols, J.D., 2015. Modeling false positive detections in species occurrence data under different study designs. Ecology 96, 332–339.

Chandler, R.B., Clark, J.D., 2014. Spatially explicit integrated population models. Methods Ecol. Evol. 5, 1351–1360.

Chandler, R.B., Hepinstall-Cymerman, J., 2016. Estimating the spatial scales of landscape effects on abundance. Landsc. Ecol. 31, 1383–1394.

Chandler, R.B., Royle, J.A., 2013. Spatially explicit models for inference about density in unmarked or partially marked populations. Ann. Appl. Stat. 7, 936–954.

Chandler, R.B., Engebretsen, K., Cherry, M.J., Garrison, E.P., Miller, K.V., Fisher, D., 2018a. Estimating recruitment from capture-recapture data by modelling spatio-temporal variation in birth and age-specific survival rates. Methods Ecol. Evol. 9, 2115–2130.

Chandler, R.B., Hepinstall-Cymerman, J., Merker, S., Abernathy-Conners, H., Cooper, R.J., 2018b. Characterizing spatio-temporal variation in survival and recruitment with integrated population models. Auk 135, 409–426.

Charlesworth, B., 1972. Selection in populations with overlapping generations. III. Conditions for genetic equilibrium. Theor. Popul. Biol. 3, 377–395.

Chaudhary, V., Oli, M.K., 2020. A critical appraisal of population viability analysis. Conserv. Biol. 34, 26–40.

Chave, J., 2013. The problem of pattern and scale in ecology: what have we learned in 20 years? Ecol. Lett. 16, 4–16.

Choquet, R., Lebreton, J.-D., Gimenez, O., Reboulet, A.M., Pradel, R., 2009a. U-CARE: Utilities for performing goodness of fit tests and manipulating CApture-REcapture data. Ecography 32, 1071–1074.

Choquet, R., Rouan, L., Pradel, R., 2009b. Program E-SURGE: a software application for fitting multievent models. In: Thomson, D., Cooch, E., Conroy, M. (Eds.), Modeling Demographic Processes in Marked Populations. Springer, New York, pp. 845–865.

Christians, J.K., 1999. Controlling for body mass effects: is part-whole correlation important? Physiol. Biochem. Zool. 72, 250–253.

Cilimburg, A.B., Lindberg, M.S., Tewksbury, J.J., Hejl, S.J., 2002. Effects of dispersal on survival probability of adult yellow warblers (*Dendroica petechia*). Auk 119, 778–789.

Clawson, M.V., Skalski, J.R., Millspaugh, J.J., 2013. The utility of auxiliary data in statistical population reconstruction. Wildl. Biol. 19, 147–155.

Clawson, M.V., Skalski, J.R., Lady, J.M., Hagen, C.A., Millspaugh, J.J., Budeau, D., Severson, J.P., 2017. Performing statistical population reconstruction using Program PopRecon 2.0. Wildl. Soc. Bull. 41, 581–589.

Clobert, J., Danchin, E., Dhondt, A.A., Nichols, J.D. (Eds.), 2001. Dispersal. Oxford University Press, Oxford.

Clutton-Brock, T.H., Coulson, T., 2002. Comparative ungulate dynamics: the devil is in the detail. Philos. Trans. R. Soc. Lond. B 357, 1285–1298.

Coates, P.S., Prochazka, B.G., Ricca, M.A., Halstead, B.J., Casazza, M.L., Blomberg, E.J., Brussee, B.E., Wiechman, L., Tebbenkamp, J., Gardner, S.C., Reese, K.P., 2018. The relative importance of intrinsic and extrinsic drivers to population growth vary among local populations of greater Sage-Grouse: an integrated population modeling approach. Auk 135, 240−261.

Cohen, J.E., 1979. Ergodic theorems in demography. Bull. Am. Math. Soc. 1, 275−295.

Colchero, F., Clark, J.S., 2012. Bayesian inference on age-specific survival for censored and truncated data. J. Anim. Ecol. 81, 139−149.

Cole, D., 2020. Parameter Redundancy and Identifiability. Chapman and Hall/CRC, Boca Raton.

Cole, D.J., McCrea, R.S., 2016. Parameter redundancy in discrete state-space and integrated models. Biom. J. 5, 1071−1090.

Cole, D.J., Morgan, B.J.T., Catchpole, E.A., Hubbard, B.A., 2012. Parameter redundancy in mark-recovery models. Biom. J. 54, 507−523.

Collier, B.A., Groce, J.E., Morrison, M.L., Newnam, J.C., Campomizzi, A.J., Farrell, S.L., Mathewson, H.A., Snelgrove, R.T., Carroll, R.J., Wilkins, R.N., 2012. Hierarchical modeling of species occupancy data using semiparametric regression. Ecol. Appl. 18, 158−167.

Conn, P.B., Cooch, E.G., 2009. Multistate capture-recapture analysis under imperfect state observation: an application to disease models. J. Appl. Ecol. 46, 486−492.

Conn, P.B., Diefenbach, D.R., Laake, J.L., Ternent, M.A., White, G.C., 2008. Bayesian analysis of wildlife age-at-harvest data. Biometrics 64, 1170−1177.

Conn, P.B., White, G.C., Laake, J.L., 2009. Simulation performance of Bayesian estimators of abundance employing age-at-harvest and mark-recovery data. In: Thomson, D., Cooch, E., Conroy, M. (Eds.), Modeling Demographic Processes in Marked Populations. Springer, New York, pp. 965−986.

Conn, P.B., Thorson, J.T., Johnson, D.S., 2017. Confronting preferential sampling when analysing population distributions: diagnosis and model-based triage. Methods Ecol. Evol. 8, 1535−1546.

Conn, P.B., Johnson, D.S., Williams, P.J., Melin, S.R., Hooten, M.B., 2018. A guide to Bayesian model checking for ecologists. Ecol. Monogr. 88, 526−542.

Conroy, M.J., Carroll, J.P., 2009. Quantitative Conservation of Vertebrates. Wiley-Blackwell, Oxford.

Converse, S.J., Kendall, W.L., Doherty, P.F., Ryan, P.G., 2009. Multistate models for estimation of survival and reproduction in the grey-headed albatross (*Thalassarche chrysostoma*). Auk 126, 77−88.

Converse, S.J., Royle, J.A., Adler, P.H., Urbanek, R.P., Barzen, J.A., 2013. A hierarchical nest survival model integrating incomplete temporally varying covariates. Ecol. Evol. 3, 4439−4447.

Cooch, E.G., White, G.C., 2020. Program MARK - a Gentle Introduction. http://www.phidot.org/software/mark/docs/book/.

Cooch, E., Rockwell, R.F., Brault, S., 2001. Retrospective analysis of demographic responses to environmental change: a lesser snow goose example. Ecol. Monogr. 71, 377−400.

Cormack, R.M., 1964. Estimates of survival from the sighting of marked animals. Biometrika 51, 429−438.

Cornulier, T., Elston, D.A., Arcese, P., Benton, T.G., Douglas, D.J.T., Lambin, X., Reid, J., Robinson, R.A., Sutherland, B.J., 2009. Estimating the annual number of breeding attempts from breeding dates using mixture models. Ecol. Lett. 12, 1184−1193.

Couet, P., Gally, F., Canonne, C., Besnard, A., 2019. Joint estimation of survival and breeding probability in female dolphins and calves with uncertainty in state assignment. Ecol. Evol. 9, 13043−13055.

Coull, B.A., Agresti, A., 1999. The use of mixed logit models to reflect heterogeneity in capture-recapture studies. Biometrics 55, 294−301.

Coulson, T., 2012. Integral projection models, their construction and use in posing hypotheses in ecology. Oikos 121, 1337−1350.

Coulson, T., Tuljapurkar, S., Childs, D.Z., 2010. Using evolutionary demography to link life history theory, quantitative genetics and population ecology. J. Anim. Ecol. 79, 1226−1240.

Courchamp, F., Clutton-Brock, T., Grenfell, B.T., 1999. Inverse density dependence and the Allee effect. Trends Ecol. Evol. 14, 405−410.

Cowen, L., Schwarz, C.J., 2005. Capture-recapture studies using radio telemetry with premature radio-tag failure. Biometrics 61, 657−664.

Crainiceanu, C.M., Ruppert, D., Wand, M.P., 2005. Bayesian analysis for penalized spline regression using WinBUGS. J. Stat. Software 14, 1−24.

Cressie, N.A.C., 2015. Statistics for Spatial Data. John Wiley & Sons Inc., New York, Chichester, Toronto, Brisbane, Singapore.

Cressie, N.A.C., Wikle, C.K., 2011. Statistics for Spatio-Temporal Data. Wiley, Hoboken, N.J.

Cressie, N., Calder, C.A., Clark, J.S., ver Hoef, J.M., Wikle, C.K., 2009. Accounting for uncertainty in ecological analysis: the strengths and limitations of hierarchical statistical modeling. Ecol. Appl. 19, 553−570.

Crick, H.Q.P., Baillie, S.R., Leech, D.I., 2003. The UK nest record scheme: its value for science and conservation. Bird Study 50, 254−270.

Crouse, D.T., Crowder, L.B., Caswell, H., 1987. A stage-based population model for loggerhead sea turtles and implications for conservation. Ecology 68, 1412−1423.

Cubaynes, S., Lavergne, C., Marboutin, E., Gimenez, O., 2012. Assessing individual heterogeneity using model selection criteria: how many mixture components in capture-recapture models? Methods Ecol. Evol. 3, 564−573.

Culina, A., Lachish, S., Pradel, R., Choquet, R., Sheldon, B.C., 2013. A multievent approach to estimating pair fidelity and heterogeneity in state transitions. Ecol. Evol. 3, 4326–4338.

Culina, A., Linton, D.M., Pradel, R., Bouwhuis, S., MacDonald, D.W., 2019. Live fast, don't die young: survival-reproduction trade-offs in long-lived income breeders. J. Anim. Ecol. 88, 746–756.

Dail, D., Madsen, L., 2011. Models for estimating abundance from repeated counts of an open metapopulation. Biometrics 67, 577–587.

Dale, S., Steifetten, Ø., 2011. The rise and fall of local populations of ortolan buntings *Emberiza hortulana:* importance of movements of adult males. J. Avian Biol. 42, 114–122.

Daniels, R.E., Allan, J.D., 1981. Life table evaluation of chronic exposure to a pesticide. Can. J. Fish. Aquat. Sci. 38, 485–494.

Darrah, A.J., Cohen, J.B., Castelli, P.M., 2018. A Bayesian multinomial logistic exposure model for estimating probabilities of competing sources of nest failure. Ibis 160, 23–35.

Davis, A.J., Hooten, M.B., Phillips, M.L., Doherty, P.F., 2014. An integrated modeling approach to estimating Gunnision sage-grouse population dynamics: combining index and demographic data. Ecol. Evol. 4, 4247–4257.

de Valpine, P., 2003. Better inferences from population-dynamics experiments using Monte Carlo state-space likelihood methods. Ecology 84, 3064–3077.

de Valpine, P., 2012. Frequentist analysis of hierarchical models for population dynamics and demographic data. J. Ornithol. 152, 393–408.

de Valpine, P., Hastings, A., 2002. Fitting population models incorporating process noise and observation error. Ecol. Monogr. 72, 57–76.

de Valpine, P., Turek, D., Paciorek, C.J., Anderson-Bergman, C., Lang, D.T., Bodik, R., 2017. Programming with models: writing statistical algorithms for general model structures with NIMBLE. J. Comput. Graph Stat. 26, 403–413.

de Valpine, P., Paciorek, C.J., Turek, D., Michaud, N., Anderson-Bergman, C., Obermeyer, F., Wehrhahn Cortes, C., Rodriguez, A., Temple Lang, D., Paganin, S., 2020a. NIMBLE User Manual. R Package Manual Version 0.10.1.

de Valpine, P., Paciorek, C.J., Turek, D., Michaud, N., Anderson-Bergman, C., Obermeyer, F., Wehrhahn Cortes, C., Rodriguez, A., Temple Lang, D., Paganin, S., 2020b. NIMBLE: MCMC, Particle Filtering, and Programmable Hierarchical Modeling. R Package Version 0.10.1.

Dennis, B., Taper, M.L., 1994. Density dependence in time series observations of natural populations: estimation and testing. Ecol. Monogr. 64, 205–224.

Dennis, B., Munholland, P.L., Scott, J.M., 1991. Estimation of growth and extinction parameters for endangered species. Ecol. Monogr. 61, 115–143.

Dennis, B., Ponciano, J.M., Lele, S.R., Taper, M.L., Staples, D.F., 2006. Estimating density dependence, process noise, and observation error. Ecol. Monogr. 76, 323–341.

Dennis, E.B., Kéry, M., Morgan, B.J.T., Coray, A., Schaub, M., Baur, B., 2021. Integrated modelling of insect population dynamics at two temporal scales. Ecol. Model. 441, 109408.

Denwood, M., 2019. Runjags: Interface Utilities, Model Templates, Parallel Computing Methods and Additional Distributions for MCMC Models in JAGS. R Package Version 2.0.4-6.

Deriso, R.B., Quinn II, T.J., Neal, P.T., 1985. Catch-age analysis with auxiliary information. Can. J. Fish. Aquat. Sci. 42, 815–824.

DeSante, D.F., Kaschube, D.R., 2009. The monitoring avian productivity and survivorship (MAPS) program 2004, 2005, and 2006 report. Bird. Popul. 9, 86–169.

DeSante, D.F., Williams, O.E., Burton, K.M., 1993. The monitoring avian productivity and survivorship (MAPS) program: overview and progress. In: Finch, D.M., Stangel, P.W. (Eds.), Status and Management of Neotropical Migratory Birds. U.S. Dept. of Agriculture, Forest Service, pp. 208–222.

DeSante, D.F., Saracco, J.F., O'Grady, D.R., Burton, K.M., 2004. Methodological considerations of the monitoring avian productivity and survivorship (MAPS) program. Stud. Avian Biol. 29, 28–45.

Desprez, M., Pradel, R., Cam, E., Monnat, J.-Y., Gimenez, O., 2011. Now you see him, now you don't: experience, not age, is related to reproduction in kittiwakes. Proc. R. Soc. Lond. B 278, 3060–3066.

Devineau, O., Kendall, W.L., Doherty, P.F., Shenk, T.M., White, G.C., Lukacs, P.M., Burnham, K.P., 2014. Increased flexibility for modeling telemetry and nest-survival data using the multistate framework. J. Wildl. Manag. 78, 224–230.

Dhondt, A.A., Kempenaers, B., Adriaensen, F., 1992. Density-dependent clutch size caused by habitat heterogeneity. J. Anim. Ecol. 61, 643–648.

Dice, L.R., 1938. Some census methods for mammals. J. Wildl. Manag. 2, 119–130.

Diggle, P.J., 2013. Statistical Analysis of Spatial and Spatio-Temporal Point Patterns. CRC Press, Boca Raton.

Diggle, P.J., Tawn, J.A., Moyeed, R.A., 1998. Model-based geostatistics. J. R. Stat. Soc. 47, 299–350.

Dinsmore, S.J., White, G.C., Knopf, F.L., 2002. Advanced techniques for modelling avian nest survival. Ecology 83, 3476–3488.

Dixon, P.M., 2006. Bootstrap resampling. Encycl. Environmetrics. https://doi.org/10.1002/9780470057339.vab028.

Dobson, A., Barnett, A., 2008. An Introduction to Generalized Linear Models. CRC/Chapmann & Hall, Boca Raton.

Dorazio, R.M., 2014. Accounting for imperfect detection and survey bias in statistical analysis of presence-only data. Global Ecol. Biogeogr. 23, 1472–1484.

Dorazio, R.M., Karanth, K.U., 2017. A hierarchical model for estimating the spatial distribution and abundance of animals detected by continuous-time recorders. PLoS ONE 12, e0176966.

Dorazio, R.M., Royle, J.A., 2003. Mixture models for estimating the size of a closed population when capture rates vary among individuals. Biometrics 59, 351–364.

Dorazio, R.M., Jelks, H.L., Jordan, F., 2005. Improving removal-based estimates of abundance by sampling a population of spatially distinct subpopulations. Biometrics 61, 1093–1101.

Dorazio, R.M., Martin, J., Edwards, H.H., 2013. Estimating abundance while accounting for rarity, correlated behavior, and other sources of variation in counts. Ecology 94, 1472–1478.

Dorazio, R.M., Delampady, M., Dey, S., Gopalaswamy, A.M., 2017. Concepts: integrating population survey data from different spatial scales, sampling methods, and species. In: Karanth, K.U., Nichols, J.D. (Eds.), Methods for Monitoring Tiger and Prey Populations. Springer Singapore, Singapore, pp. 247–254.

Dormann, C.F., Calabrese, J.M., Guillera-Arroita, G., Matechou, E., Bahn, V., Bartoń, K., Beale, C.M., Ciuti, S., Elith, J., Gerstner, K., Guélat, J., Keil, P., Lahoz-Monfort, J.J., Pollock, L.J., Reineking, B., Roberts, D.R., Schröder, B., Thuiller, W., Warton, D.I., Wintle, B.A., Wood, S.N., Wüest, R.O., Hartig, F., 2018. Model averaging in ecology: a review of Bayesian, information-theoretic, and tactical approaches for predictive inference. Ecol. Monogr. 88, 485–504.

Drechsler, M., Burgman, M.A., 2004. Combining population viability analysis with decision analysis. Biodivers. Conserv. 13, 115–139.

Du Feu, C.R., Clark, J.A., Schaub, M., Fiedler, W., Baillie, S.R., 2016. The EURING Data Bank – a critical tool for continental-scale studies of marked birds. Ringing Migr. 31, 1–18.

Duarte, A., Weckerly, F.W., Schaub, M., Hatfield, J.S., 2016. Estimating golden-cheeked warbler immigration: implications for the spatial scale of conservation. Anim. Conserv. 19, 65–74.

Duarte, A., Pearl, C.A., Adams, M.J., Peterson, J.T., 2017. A new parameterization for integrated population models to document amphibian reintroductions. Ecol. Appl. 27, 1761–1775.

Duriez, O., Sæther, S.A., Ens, B.J., Choquet, R., Pradel, R., Lambeck, R.H.D., Klaassen, M., 2009. Estimating survival and movements using both live and dead recoveries: a case study of oystercatchers confronted with habitat change. J. Appl. Ecol. 46, 144–153.

Duriez, O., Ens, B.J., Choquet, R., Pradel, R., Klaassen, M., 2012. Comparing the seasonal survival of resident and migratory oystercatchers:carry-over effects of habitat quality and weather conditions. Oikos 121, 862–873.

Eacker, D.R., Lukacs, P.M., Proffitt, K.M., Hebblewhite, M., 2017. Assessing the importance of demographic parameters for population dynamics using Bayesian integrated population modeling. Ecol. Appl. 27, 1280–1293.

Easterling, M.R., Ellner, S.P., Dixon, P.M., 2000. Size-specific sensitivity: applying a new structured population model. Ecology 81, 694–708.

Eberhardt, L.E., Eberhardt, L.L., Tiller, B.L., Cadwell, L.L., 1996. Growth of an isolated elk population. J. Wildl. Manag. 60, 369–373.

Edwards, A.W.F., 1992. Likelihood, Expanded edition. The Johns Hopkins Univ. Press, Baltimore.

Efford, M., 2004. Density estimation in live-trapping studies. Oikos 106, 598–610.

Efford, M.G., 2019. Non-circular home ranges and the estimation of population density. Ecology 100, e02580.

Efford, M.G., Mowat, G., 2014. Compensatory heterogeneity in spatially explicit capture-recapture data. Ecology 95, 1341–1348.

Efford, M.G., Schofield, M.R., 2020. A spatial open-population capture-recapture model. Biometrics 76, 392–402.

Efron, B., 1986. Why isn't everyone a Bayesian? Am. Statistician 40, 1–5.

Ehrlen, J., van Groenendael, J., 1998. Direct perturbation analysis for better conservation. Conserv. Biol. 12, 470–474.

Ellis, M.M., Crone, E.E., 2013. The role of transient dynamics in stochastic population growth of nine perennial plants. Ecology 94, 1681–1686.

Ellner, S.P., Rees, M., 2006. Integral projection models for species with complex demography. Am. Nat. 167, 410–428.

Ellner, S.P., Childs, D.Z., Rees, M., 2016. Data-driven Modelling of Structured Populations: A Practical Guide to the Integral Projection Model. Springer, Switzerland.

Ergon, T., Gardner, B., 2014. Separating mortality and emigration:modelling space use, dispersal and survival with robust-design spatial capture-recapture data. Methods Ecol. Evol. 5, 1327–1336.

Ergon, T., Borgan, Ø., Nater, C.R., Vindenes, Y., Yoccoz, N., 2018. The utility of mortality hazard rates in population analyses. Methods Ecol. Evol. 9, 2046–2056.

Etterson, M.A., Ellis-Felege, S.N., Evers, D., Gauthier, G., Grzybowski, J.A., Mattsson, B.J., Nagy, L.R., Olsen, B.J., Pease, C.M., van der Burg, M.P., Potvien, A., 2011. Modeling fecundity in birds: conceptual overview, current models, and considerations for future developments. Ecol. Model. 222, 2178–2190.

Ezard, T.H.G., Bullock, J.M., Dalgleish, H.J., Millon, A., Pelletier, F., Ozgul, A., Koons, D.N., 2010. Matrix models for a changeable world: the importance of transient dynamics in population management. J. Appl. Ecol. 47, 515–523.

Fahse, L., Wissel, C., Grimm, V., 1998. Reconciling classical and individual-based approaches in theoretical population ecology: a protocol for extracting population parameters from individual-based models. Am. Nat. 152, 838–852.

Fay, R., Michler, S., Laesser, J., Schaub, M., 2019. Integrated population model reveals that kestrels breeding in nest boxes operate as a source population. Ecography 42, 2122–2131.

Fieberg, J., Ellner, S.P., 2000. When is it meaningful to estimate an extinction probability? Ecology 81, 2040–2047.

Fieberg, J., Ellner, S.P., 2001. Stochastic matrix models for conservation and management: a comparative review of methods. Ecol. Lett. 4, 244–266.

Fieberg, J.R., Shertzer, K.W., Conn, P.B., Noyce, K.V., Garshelis, D.L., 2010. Integrated population modeling of black bears in Minnesota: implications for monitoring and management. PLoS ONE 5, e12114.

Fink, D., Auer, T., Johnston, A., Ruiz-Gutierrez, V., Hochachka, W.M., Kelling, S., 2020. Modeling avian full annual cycle distribution and population trends with citizen science data. Ecol. Appl. 30, e02056.

Finke, A., King, R., Beskos, A., Dellaportas, P., 2019. Efficient sequential Monte Carlo algorithms for integrated population models. JABES 24, 204–224.

Fiske, I.J., Chandler, R., 2011. unmarked: an R package for the analysis of wildlife occurrence and abundance data. J. Stat. Software 43.

Flanders-Wanner, B.L., White, G.C., McDaniel, L.L., 2004. Validity of prairie grouse harvest-age ratios as production indices. J. Wildl. Manag. 68, 1088–1094.

Fletcher, R.J., Hefley, T.J., Robertson, E.P., Zuckerberg, B., McCleery, R.A., Dorazio, R.M., 2019. A practical guide for combining data to model species distributions. Ecology 100, e02710.

Fonnesbeck, C.J., Conroy, M.J., 2004. Application of integrated Bayesian modeling and Markov chain Monte Carlo methods to the conservation of a harvested species. Anim. Biodivers. Conserv. 27.1, 267–281.

Forslund, P., Pärt, T., 1995. Age and reproduction in birds - hypotheses and tests. Trends Ecol. Evol. 10, 374–378.

Fouchet, D., Santin-Janin, H., Sauvage, F., Yoccoz, N.G., Pontier, D., 2016. An R package for analysing survival using continuous time open capture-recapture models. Methods Ecol. Evol. 7, 518–528.

Fournier, D., Archibald, C.P., 1982. A general theory for analyzing catch at age data. Can. J. Fish. Aquat. Sci. 39, 1195–1207.

Fournier, D.A., Skaug, H.J., Ancheta, J., Ianelli, J., Magnusson, A., Maunder, M.N., Nielsen, A., Sibert, J., 2012. AD Model Builder: using automatic differentiation for statistical inference of highly parameterized complex nonlinear models. Optim. Methods Software 27, 233–249.

Fox, G.A., Kendall, B.E., 2002. Demographic stochasticity and the variance reduction effect. Ecology 83, 1928–1934.

Francis, R.C., 2011. Data weighting in statistical fisheries stock assessment models. Can. J. Fish. Aquat. Sci. 68, 1124–1138.

Francis, R., 2017. Revisiting data weighting in fisheries stock assessment models. Fish. Res. 192, 5–15.

Freckleton, R.P., Watkinson, A.R., Green, R.E., Sutherland, B.J., 2006. Census error and the detection of density dependence. J. Anim. Ecol. 75, 837–851.

Frederiksen, M., Bregnballe, T., 2000. Evidence for density-dependent survival in adult cormorants from a combined analysis of recoveries and resightings. J. Anim. Ecol. 69, 737–752.

Frederiksen, M., Bregnballe, T., 2000a. Diagnosing a decline in return rate of 1-year-old cormorants: mortality, emigration or delayed return? J. Anim. Ecol. 69.

Freeman, S.N., Robinson, R.A., Clark, J.A., Griffin, B.M., Adams, S.Y., 2007. Changing demography and population decline in the common Starling *Sturnus vulgaris*: a multisite approach to integrated population monitoring. Ibis 149, 587–596.

Fretwell, D.S., Lucas Jr., H.L., 1969. On territorial behavior and other factors influencing habitat distribution in birds. I. Theoretical development. Acta Biotheor. 19, 16–36.

Frick, W.F., Rainey, W.E., Pierson, E.D., 2007. Potential effects of environmental contamination on Yuma myotis demography and population growth. Ecol. Appl. 17, 1213–1222.

Fry, F., 1949. Statistics of a lake trout fishery. Biometrics 5, 27–67.

Fryxell, J.M., Mercer, W.E., Gellately, R.B., 1988. Population dynamics of Newfoundland moose using cohort analysis. J. Wildl. Manag. 52, 14–21.

Furrer, R.D., Pasinelli, G., 2016. Empirical evidence for source-sink populations: a review on occurrence, assessments and implications. Biol. Rev. 91, 782–795.

Furrer, R., Schaub, M., Bossert, A., Isler, R., Jenny, H., Jonas, T., Marti, C., Jenni, L., 2016. Variable decline of Alpine Rock Ptarmigan (*Lagopus muta helvetica*) in Switzerland between regions and sites. J. Ornithol. 157, 787–796.

Gaillard, J.M., Yoccoz, N.G., 2003. Temporal variation in survival of mammals: a case of environmental canalization? Ecology 84, 3294–3306.

Gaillard, J.M., Allaine, D., Pontier, D., Yoccoz, N.G., Promislow, D.E.L., 1994. Senescence in natural populations of mammals: a reanalysis. Evolution 48, 509–516.

Gaillard, J.M., Festa-Bianchet, M., Yoccoz, N.G., 1998. Population dynamics of large herbivores: variable recruitment with constant adult survival. Trends Ecol. Evol. 13, 58–63.

Gaillard, J.M., Festa-Bianchet, M., Yoccoz, N.G., Loison, A., Toigo, C., 2000. Temporal variation in fitness components and population dynamics of large herbivores. Annu. Rev. Ecol. Syst. 31, 367–393.

Gaillard, J.M., Yoccoz, N.G., Lebreton, J.-D., Bonenfant, C., Devillard, S., Loison, A., Pontier, D., Allaine, D., 2005. Generation time: a reliable metric to measure life-history variation among mammalian populations. Am. Nat. 166, 119–123.

Gamelon, M., Gimenez, O., Baubet, E., Coulson, T., Tuljapurkar, S., Gaillard, J.M., 2014. Influence of life-history tactics on transient dynamics: a comparative analysis across mammalian populations. Am. Nat. 184, 673–683.

Gamelon, M., Grøtan, V., Engen, S., Björkvoll, E., Visser, M.E., Sæther, B.-E., 2016. Density dependence in an age-structure population of great tits: identifying the critical age classes. Ecology 97, 2479–2490.

Gamelon, M., Grøtan, V., Nilsson, A.L.K., Engen, S., Hurrell, J.W., Jerstad, K., Phillips, A.S., Røstad, O.W., Slagsvold, T., Walseng, B., Stenseth, N.C., Sæther, B.-E., 2017. Interactions between demography and environmental effects are important determinants of population dynamics. Sci. Adv. 3, e1602298.

Gamelon, M., Vriend, S.J.G., Engen, S., Adriaensen, F., Dhondt, A.A., Evans, S.R., Matthysen, E., Sheldon, B.C., Sæther, B.-E., 2019. Accounting for interspecific competition and age structure in demographic analyses of density dependence improves predictions of fluctuations in population size. Ecol. Lett. 22, 797–806.

Gardner, B., Reppucci, J., Lucherini, M., Royle, J.A., 2010. Spatially explicit inference for open populations: estimating demographic parameters from camera-trap studies. Ecology 91, 3376–3383.

Gardner, B., Sollmann, R., Kumar, N.S., Jathanna, D., Karanth, K.U., 2018. State space and movement specification in open population spatial capture-recapture models. Ecol. Evol. 8, 10336–10344.

Garrard, G.E., Bekessy, S.A., McCarthy, M.A., Wintle, B.A., 2008. When have we looked hard enough? A novel method for setting minimum survey effort protocols for flora surveys. Austral Ecol. 33, 986–998.

Gast, C., Skalski, J.R., Beyer, D.E., 2013. Evaluation of fixed- and random-effects models and multistage estimation procedures in statistical population reconstruction. J. Wildl. Manag. 77, 1258–1270.

Gauthier, G., Pradel, R., Menu, S., Lebreton, J.-D., 2001. Seasonal survival of greater snow geese and effect of hunting under dependence in sighting probability. Ecology 82, 3105–3119.

Gauthier, G., Besbeas, P., Lebreton, J.-D., Morgan, B.J.T., 2007. Population growth in snow geese: a modeling approach integrating demographic and survey information. Ecology 88, 1420–1429.

Gauthier, G., Milot, E., Weimerskirch, H., 2012. Estimating dispersal, recruitment and survival in a biennially breeding species, the Wandering Albatross. J. Ornithol. 152, 457–467.

Gedeon, K., Sudfeldt, C., Dougalis, P. (Eds.), 2015. Atlas Deutscher Brutvogelarten. Atlas of German Breeding Birds. Dachverband Deutscher Avifaunisten, Münster, Westf.

Geiser, S., Arlettaz, R., Schaub, M., 2008. Impact of weather variation on feeding behaviour, nestling growth and brood survival in Wrynecks *Jynx torquilla*. J. Ornithol. 149, 597–606.

Gelfand, A.E., Smith, A.F.M., 1990. Sampling-based approaches to calculating marginal densities. J. Am. Stat. Assoc. 85, 398–409.

Gelman, A., Hill, J., 2007. Data Analysis Using Regression and Multilevel/Hierarchical Models. Cambridge University Press, Cambridge.

Gelman, A., Rubin, D.B., 1992. Inference from iterative simulation using multiple sequences. Stat. Sci. 7, 457–511.

Gelman, A., Shalizi, C.R., 2013. Philosophy and the practice of Bayesian statistics. Br. J. Math. Stat. Psychol. 66, 8–38.

Gelman, A., Meng, X.-L., Stern, H.S., 1996. Posterior predictive assessment of model fitness via realized discrepancies (with discussion). Stat. Sin. 6, 733–807.

Gelman, A., Carlin, J.B., Stern, H.S., Rubin, D.B., 2004. Bayesian Data Analysis. CRC/Chapman & Hall, Boca Raton.

Gelman, A., Carlin, J.B., Stern, H.S., Dunson, D.B., Vehtari, A., Rubin, D.B., 2013. Bayesian Data Analysis, 3rd edition. CRC Press Taylor and Francis Group, Boca Raton, London, New York.

Geman, S., Geman, D., 1984. Stochastic relaxation, Gibbs distributions, and the Bayesian restoration of images. IEEE Trans. Pattern Anal. Mach. Intell. 6, 721–741.

Genovart, M., Oro, D., Tenan, S., 2018. Immature survival, fertility, and density dependence drive global population dynamics in a long-lived species. Ecology 99, 2823–2832.

Giavi, S., Moretti, M., Bontadina, F., Zambelli, N., Schaub, M., 2014. Seasonal survival probabilities suggest low migration mortality in migrating bats. PLoS ONE 9, e85628.

Gilks, W.R., Thomas, A., Spiegelhalter, D.J., 1994. A language and program for complex Bayesian modelling. Statistician 43, 169–177.

Gill, J.A., Norris, K., Potts, P.M., Gunnarsson, T.G., Atkinson, P.W., Sutherland, W.J., 2001. The buffer effect and large-scale population regulation in migratory birds. Nature 412, 436–438.

Gilpin, M.E., Ayala, F.J., 1973. Global models of growth and competition. Proc. Nat. Acad. Sci. U.S.A. 70, 3590–3593.

Gilpin, M.E., Soulé, M.E., 1986. Minimum viable populations: processes of species extinction. In: Soulé, M. (Ed.), Conservation Biology: The Science of Scarcity and Diversity. Sinauer, Sunderland, pp. 19–34.

Gimenez, O., Viallefont, A., Catchpole, E.A., Choquet, R., Morgan, B.J.T., 2004. Methods for investigating parameter redundancy. Anim. Biodivers. Conserv. 27.1, 561–572.

Gimenez, O., Crainiceanu, C., Barbraud, C., Jenouvrier, S., Morgan, B.J.T., 2006. Semiparametric regression in capture-recapture modeling. Biometrics 62, 691–698.

Gimenez, O., Rossi, V., Choquet, R., Dehais, C., Doris, B., Varella, H., Vila, J.-P., Pradel, R., 2007. State-space modelling of data on marked individuals. Ecol. Model. 206, 431–438.

Gimenez, O., Bonner, S.J., King, R., Parker, R.A., Brooks, S.P., Jamieson, L.E., Grosbois, V., Morgan, B.J.T., Thomas, L., 2009a. WinBUGS for population ecologists: Bayesian modeling using Markov Chain Monte Carlo methods. In: Thomson, D., Cooch, E., Conroy, M. (Eds.), Modeling Demographic Processes in Marked Populations. Springer, New York, pp. 883–915.

Gimenez, O., Morgan, B.J.T., Brooks, S.P., 2009b. Weak identifiability in models for mark-recapture-recovery data. In: Thomson, D., Cooch, E., Conroy, M. (Eds.), Modeling Demographic Processes in Marked Populations. Springer, New York, pp. 1055–1067.

Gimenez, O., Lebreton, J.-D., Gaillard, J.M., Choquet, R., Pradel, R., 2012. Estimating demographic parameters using hidden process dynamic models. Theor. Popul. Biol. 82, 307–316.

Gimenez, O., Lebreton, J.-D., Choquet, R., Pradel, R., Poisot, T., 2018a. R2ucare: an R package to perform goodness-of-fit tests for capture-recapture models. Methods Ecol. Evol. 9, 1749–1754.

Gimenez, O., Cam, E., Gaillard, J.-M., 2018b. Individual heterogeneity and capture-recapture models: what, why and how? Oikos 127, 664–686.

Ginzburg, L.R., Ferson, S., Akçakaya, H.R., 1990. Reconstructibility of density dependence and the conservative assessment of extinction risks. Conserv. Biol. 4, 63–70.

Golding, N., 2019. greta: simple and scalable statistical modelling in R. JOSS 4, 1601.

González, J.A., Rodríguez-Cortés, F.J., Cronie, O., Mateu, J., 2016. Spatio-temporal point process statistics: a review. Spat. Stat. 18, 505–544.

Gordo, O., Brotons, L., Ferrer, X., Comas, P., 2005. Do changes in climate patterns in wintering areas affect the timing of the spring arrival of trans-Saharan migrant birds? Global Change Biol. 11, 12–21.

Goudie, R.J.B., Turner, R.M., Angelis de, D., Thomas, A., 2020. MultiBUGS: a parallel implementation of the BUGS modelling framework for faster Bayesian inference. J. Stat. Software 95. https://doi.org/10.18637/jss.v095.i07.

Gould, W.R., Nichols, J.D., 1998. Estimation of temporal variability of survival in animal populations. Ecology 79, 2531–2538.

Gove, N.E., Skalski, J.R., Zager, P., Townsend, R.L., 2002. Statistical models for population reconstruction using age-at-harvest data. J. Wildl. Manag. 66, 310–320.

Grant, P.R., Grant, B.R., 2002. Unpredictable evolution in a 30-year study of Darwin's finches. Science 296, 707–711.

Green, A.W., Bailey, L.L., 2015. Using Bayesian population viability analysis to define relevant conservation objectives. PLoS ONE 10, e0144786.

Greenwood, J.J.D., 2007. Citizens, science and bird conservation. J. Ornithol. 148, 77–124.

Greenwood, P.J., Harvey, P.H., 1982. The natal and breeding dispersal of birds. Annu. Rev. Ecol. Syst. 13, 1–21.

Grenfell, B.T., Wilson, K., Finckenstädt, B.F., Coulson, T.N., Murray, S., Albon, S.D., Pemberton, J.M., Clutton-Brock, T.H., Crawley, M.J., 1998. Noise and determinism in synchronized sheep dynamics. Nature 394, 674–677.

Grosbois, V., Harris, M.P., Anker-Nilssen, T., McCleery, R.H., Shaw, D.N., Morgan, B.J.T., Gimenez, O., 2009. Modeling survival at multi-population scales using mark-recapture data. Ecology 90, 2922–2932.

Gross, K., Craig, B.A., Hutchison, W.D., 2002. Bayesian estimation of a demographic matrix model from stage-frequency data. Ecology 83, 3285–3298.

Grüebler, M.U., Naef-Daenzer, B., 2008. Fitness consequences of pre- and post-fledging timing decisions in a double-brooded passerine. Ecology 89, 2736–2745.

Grüebler, M.U., Korner-Nievergelt, F., Hirschheydt von, J., 2010. The reproductive benefits of livestock farming in barn swallows *Hirundo rustica*: quality of nest site or foraging habitat? J. Appl. Ecol. 47, 1340–1347.

Guélat, J., Kéry, M., 2018. Effects of spatial autocorrelation and imperfect detection on species distribution models. Methods Ecol. Evol. 9, 1614–1625.

Gunnarsson, T.G., Gill, J.A., Petersen, A., Appleton, G.F., Sutherland, B.J., 2005. A double buffer effect in a migratory shorebird population. J. Anim. Ecol. 74, 965–971.

Gurevitch, J., Fox, G.A., Fowler, N.L., Graham, C.H., 2016. Landscape demography: population change and its drivers across spatial scales. Q. Rev. Biol. 91, 459–485.

Hadley, G.L., Rotella, J.J., Garrott, R.A., Nichols, J.D., 2006. Variation in probability of first reproduction of Weddell seals. J. Anim. Ecol. 75, 1058–1070.

Hall, A.J., McConnell, B.J., Barker, R.J., 2001. Factors affecting first-year survival in grey seals and their implications for life history strategy. J. Anim. Ecol. 70, 138–149.

Halstead, B.J., Kleeman, P.M., Rose, J.P., 2018. Time-to-detection occupancy modeling: an efficient method for analyzing the occurrence of amphibians and reptiles. J. Herpetol. 52, 415–424.

Hamel, S., Yoccoz, N.G., Gaillard, J.-M., 2017. Assessing variation in life-history tactics within a population using mixture regression models: a practical guide for evolutionary ecologists. Biol. Rev. 92, 754–775.

Hamel, S., Gaillard, J.-M., Douhard, M., Festa-Bianchet, M., Pelletier, F., Yoccoz, N.G., 2018. Quantifying individual heterogeneity and its influence on life-history trajectories: different methods for different questions and contexts. Oikos 127, 687–704.

Hanks, E.M., Hooten, M.B., Baker, F.A., 2011. Reconciling multiple data sources to improve accuracy of large-scale prediction of forest disease incidence. Ecol. Appl. 21, 1173–1188.

Hanski, I., 1998. Metapopulation dynamics. Nature 396, 41–49.

Hanson, W.R., 1963. Calculation of productivity, survival, and abundance of selected vertebrates from sex and age ratios. Wildl. Monogr. 9, 1–60.

Harrison, S., Quinn, J.F., 1989. Correlated environments and the persistence of metapopulations. Oikos 56, 293–298.

Harvey, A.C., 2014. Forecasting, Structural Time Series Models and the Kalman Filter. Cambridge University Press, Cambridge.

Harvey, P.H., Zammuto, R.M., 1985. Patterns of mortality and age at first reproduction in natural populations of mammals. Nature 315, 319–320.

Hastie, T.J., Tibshirani, R.J., 1990. Generalized Additive Models. Chapman & Hall/CRC, Boca Raton.
Hastings, W.K., 1970. Monte Carlo sampling methods using Markov chains and their applications. Biometrika 57, 97–109.
Hastings, A., 1997. Population Biology. Springer, New York.
Hastings, A., 2001. Transient dynamics and persistence of ecological systems. Ecol. Lett. 4, 215–220.
Hastings, A., 2004. Transients: the key to long-term ecological understanding? Trends Ecol. Evol. 19, 39–45.
Hatter, I.W., Dielman, P., Kuzyk, G.W., 2017. An integrated modeling approach for assessing management objectives for mule deer in central British Columbia. Wildl. Soc. Bull. 41, 508–515.
Hayman, D.T.S., McCrea, R.S., Restif, O., Suu-Ire, R., Fooks, A.R., Wood, J.L.N., Cunningham, A.A., Rowcliffe, J.M., 2012. Demography of straw-colored fruit bats in Ghana. J. Mammal. 93, 1393–1404.
He, C.Z., 2003. Bayesian modeling of age-specific survival in bird nesting studies under irregular visits. Biometrics 59, 962–973.
Heard, G.W., McCarthy, M.A., Scroggie, M.P., Baumgartner, J.B., Parris, K.M., 2013. A Bayesian model of metapopulation viability, with application to an endangered amphibian. Divers. Distrib. 19, 555–566.
Heaton, M.J., Datta, A., Finley, A.O., Furrer, R., Guinness, J., Guhaniyogi, R., Gerber, F., Gramacy, R.B., Hammerling, D., Katzfuss, M., Lindgren, F., Nychka, D.W., Sun, F., Zammit-Mangion, A., 2019. A case study competition among methods for analyzing large spatial data. JABES 24, 398–425.
Hefley, T.J., Broms, K.M., Brost, B.M., Buderman, F.E., Kay, S.L., Scharf, H.R., Tipton, J.R., Williams, P.J., Hooten, M.B., 2017a. The basis function approach for modeling autocorrelation in ecological data. Ecology 98, 632–646.
Hefley, T.J., Hooten, M.B., Hanks, E.M., Russell, R.E., Walsh, D.P., 2017b. Dynamic spatio-temporal models for spatial data. Spat. Stat. 20, 206–220.
Hefley, T.J., Hooten, M.B., Russell, R.E., Walsh, D.P., Powell, J.A., 2017c. When mechanism matters: Bayesian forecasting using models of ecological diffusion. Ecol. Lett. 20, 640–650.
Hegg, D., MacKenzie, D.I., Jamieson, I.G., 2013. Use of Bayesian population viability analysis to assess multiple management decisions in the recovery programme for the Endangered takahe *Porphyrio hochstetteri*. Oryx 47, 144–152.
Hénaux, V., Bregnballe, T., Lebreton, J.-D., 2007. Dispersal and recruitment during population growth in a colonial bird, the great cormorant *Phalacrocorax carbo sinensis*. J. Avian Biol. 38, 44–57.
Hensler, G.L., Nichols, J.D., 1981. The Mayfield method of estimating nesting success: a model, estimators and simulation results. Wilson Bull. 93, 42–53.
Heppell, S.S., Crouse, D.T., Crowder, L.B., 2000. Using matrix models to focus research and management efforts in conservation. In: Ferson, S. (Ed.), Quantitative Methods for Conservation Biology. Springer, New York, pp. 148–168.
Hestbeck, J.B., Nichols, J.D., Malecki, R.A., 1991. Estimates of movement and site fidelity using mark-resight data of wintering Canada Geese. Ecology 72, 523–533.
Hines, J.E., 1994. MSSURVIV User's Manual. Laurel, ML.
Hines, J.E., 2006. PRESENCE2- Software to Estimate Patch Occupancy and Related Parameters. USGS-PWRC.
Hixon, M.A., Pacala, S.W., Sandin, S.A., 2002. Population regulation: historical context and contemporary challenges of open vs. closed systems. Ecology 83, 1490–1508.
Hobbs, N.T., Hooten, M.B., 2015. Bayesian Models: A Statistical Primer for Ecologists. Princeton University Press.
Hobbs, N.T., Geremia, C., Treanor, J., Wallen, R., White, P.J., Hooten, M.B., Rhyan, J.C., 2015. State-space modeling to support management of brucellosis in the Yellowstone bison population. Ecol. Monogr. 85, 525–556.
Hoffmann, J., Postma, E., Schaub, M., 2015. Factors influencing double brooding in Eurasian Hoopoes *Upupa epops*. Ibis 157, 17–30.
Höglund, J., Johansson, T., Pelabon, C., 1997. Behaviourally mediated sexual selection: characteristics of successful male black grouse. Anim. Behav. 54, 255–264.
Holbrook, S.J., Schmitt, R.J., 2002. Competition for shelter space causes density-dependent predation mortality in damselfishes. Ecology 83, 2855–2868.
Hölzinger, J. (Ed.), 1997. Die Vögel Baden-Württembergs. Band 3.2: Singvögel 2. Ulmer-Verlag, Stuttgart.
Hooten, M.B., Hefley, T.J., 2019. Bringing Bayesian Models to Life. CRC Press, Boca Raton.
Hooten, M.B., Hobbs, N.T., 2015. A guide to Bayesian model selection for ecologists. Ecol. Monogr. 85, 3–28.
Hooten, M.B., Wikle, C.K., 2008. A hierarchical Bayesian non-linear spatio-temporal model for the spread of invasive species with application to the Eurasian Collored Dove. Environ. Ecol. Stat. 15, 59–70.
Hooten, M.B., Wikle, C.K., Dorazio, R.M., Royle, J.A., 2007. Hierarchical spatiotemporal matrix models for characterizing invasions. Biometrics 63, 558–567.
Hooten, M.B., Johnson, D.S., McClintock, B.T., Morales, J.M., 2017. Animal Movement. Statistical Models for Telemetry Data. CRC Press, Boca Raton.
Horvitz, C., Schemske, D.W., Caswell, H., 1997. The relative "importance" of life-history stages to population growth: prospective and retrospective analyses. In: Tuljapurkar, S. (Ed.), Structured Population Models in Marine, Terrestrial, and Freshwater Systems. Chapman & Hall, New York, pp. 247–271.
Hostetler, J.A., Chandler, R.B., 2015. Improved state-space models for inference about spatial and temporal variation in abundance from count data. Ecology 96, 1713–1723.
Hostetler, J.A., Sillett, T.S., Marra, P.P., 2015. Full-annual-cycle population models for migratory birds. Auk 132, 433–449.

Hovi, M., Alatalo, R.V., Höglund, J., Lundberg, A., Rintamäki, P.T., 1994. Lek center attracts black grouse females. Proc. R. Soc. Lond. B 258, 303–305.
Howell, P.E., Hossack, B.R., Muths, E., Sigafus, B.H., Chandler, R.B., 2020. Informing amphibian conservation efforts with abundance-based metapopulation models. Herpetologica 76, 240–250.
Hoyle, S.D., Maunder, M.N., 2004. A Bayesian integrated population dynamics model to analyze data for protected species. Anim. Biodivers. Conserv. 27.1, 247–266.
Humbert, J.-Y., Mills, L.S., Horne, J.S., Dennis, B., 2009. A better way to estimate population trends. Oikos 118, 1940–1946.
Hunt, G., 1988. The natural regulation of peregrine falcon populations. In: Cade, T.J., Enderson, J.H., Thelander, C.G., White, C.M. (Eds.), Peregrine Falcon Populations: Their Management and Recovery. The Peregrine Fund, Boise, Idaho, pp. 667–676.
Husby, A., Kruuk, L.E.B., Visser, M.E., 2009. Decline in the frequency and benefits of multiple brooding in great tits as a consequence of a changing environment. Proc. R. Soc. Lond. B 276, 1845–1854.
Huston, M., DeAngelis, D., Post, W., 1988. New computer models unify ecological theory. BioScience 38, 682–691.
Illian, J., Penttinen, A., Stoyan, H., Stoyan, D., 2008. Statistical Analysis and Modelling of Spatial Point Patterns. Wiley, Chichester.
Inger, R., Harrison, X.A., Ruxton, G.D., Newton, J., Colhoun, K., Gudmundsson, G.A., McElwaine, G., Pickford, M., Hodgson, D., Bearhop, S., 2010. Carry-over effects reveal reproductive costs in a long-distance migrant. J. Anim. Ecol. 79, 974–982.
Jacobson, E.K., Boyd, C., McGuire, T.L., Shelden, K.E.W., Himes Boor, G.K., Punt, A.E., 2020. Assessing cetacean populations using integrated population models: an example with Cook Inlet beluga whales. Ecol. Appl. 30, e02114.
Jamieson, L.E., Brooks, S.P., 2004. Density dependence in North American ducks. Anim. Biodivers. Conserv. 27.1, 113–128.
Jan, P.-L., Lehnen, L., Besnard, A., Kerth, G., Biedermann, M., Schorcht, W., Petit, E.J., Le Gouar, P., Puechmaille, S.J., 2019. Range expansion is associated with increased survival and fecundity in a long-lived bat species. Proc. R. Soc. Lond. B 286, 20190384.
Jansen, D.Y.M., Pradel, R., Mares, R., Doutrelant, C., Spottiswoode, C.N., Covas, R., Altwegg, R., 2019. An integrated population model sheds light on the complex population dynamics of a unique colonial breeder. Popul. Ecol. 61, 406–420.
Jenkins, D., Watson, A., Miller, G.R., 1963. Population studies on red grouse, *Lagopus lagopus scoticus* (Lath.) in north-east Scotland. J. Anim. Ecol. 32, 317–367.
Johnson, D.H., 1979. Estimating nest success: the Mayfield method and an alternative. Auk 96, 651–661.
Johnson, D.H., 2008. In defense of indices. the case of bird surveys. J. Wildl. Manag. 72, 857–868.
Johnson, D.S., Hoeting, J.A., 2003. Autoregressive models for capture-recapture data: a Bayesian approach. Biometrics 59, 341–350.
Johnson, D.S., London, J.M., Lea, M.-A., Durban, J.W., 2008. Continuous-time correlated random walk model for animal telemetry data. Ecology 89, 1208–1215.
Johnson, D.S., Conn, P.B., Hooten, M.B., Ray, J.C., Pond, B.A., 2013. Spatial occupancy models for large data sets. Ecology 94, 801–808.
Johnson, F.A., Zimmerman, G.S., Jensen, G.H., Clausen, K.K., Frederiksen, M., Madsen, J., 2020. Using integrated population models for insights into monitoring programs: an application using pink-footed geese. Ecol. Model. 415, 108869.
Johnson, H.E., Mills, L.S., Wehausen, J.D., Stephenson, T.R., 2010. Combining ground count, telemetry, and mark-resight data to infer population dynamics in an endangered species. J. Appl. Ecol. 47, 1083–1093.
Jolly, G.M., 1965. Explicit estimates from capture-recapture data with both death and immigration-stochastic model. Biometrika 52, 225–247.
Jones, J.H., 2007. demogR: a package for the construction and analysis of age-structured demographic models in R. J. Stat. Software 22, 1–28.
Jones, J., Doran, P.J., Holmes, R.T., 2003. Climate and food synchronize regional forest bird abundances. Ecology 84, 3024–3032.
Karanth, K.U., Nichols, J.D., 1998. Estimation of tiger densities in India using photographic captures and recaptures. Ecology 79, 2852–2862.
Katzner, T.E., Ivy, J.A., Bragin, E.A., Milner-Gulland, E.J., DeWoody, J.A., 2011. Conservation implications of inaccurate estimation of cryptic population size. Anim. Conserv. 14, 328–332.
Kauffman, M.J., Frick, W.F., Linthicum, J., 2003. Estimation of habitat-specific demography and population growth for peregrine falcons in California. Ecol. Appl. 13, 1802–1816.
Keller, L.F., 1998. Inbreeding and its fitness effects in an insular population of song sparrows. Evolution 52, 240–250.
Kellner, K.F., 2019. Package 'jagsUI': A Wrapper Around 'rjags' to Streamline 'JAGS' Analyses.
Kellner, K.F., Swihart, R.K., 2014. Accounting for imperfect detection in ecology: a quantitative review. PLoS ONE 9, e111436.
Kendall, B.E., Fox, G.A., Fujiwara, M., Nogeire, T.M., 2011. Demographic heterogeneity, cohort selection, and population growth. Ecology 92, 1985–1993.
Kendall, B.E., Fujiwara, M., Diaz-Lopez, J., Schneider, S., Voigt, J., Wiesner, S., 2019. Persistent problems in the construction of matrix population models. Ecol. Model. 406, 33–43.
Kendall, B.E., Wittmann, M.E., 2010. A stochastic model for annual reproductive success. Am. Nat. 175, 461–468.
Kendall, W.L., Bjorkland, R., 2001. Using open robust design models to estimate temporary emigration from capture-recapture data. Biometrics 57, 1113–1122.
Kendall, W.L., Nichols, J.D., 2002. Estimating state-transition probabilities for unobservable states using capture-recapture/resighting data. Ecology 83, 3276–3284.

Kendall, W.L., Pollock, K.H., Brownie, C., 1995. A likelihood-based approach to capture-recapture estimation of demographic parameters under the robust design. Biometrics 51, 293–308.

Kendall, W.L., Peterjohn, B.G., Sauer, J.R., 1996. First-time observer effects in the north American breeding bird survey. Auk 113, 823–829.

Kendall, W.L., Nichols, J.D., Hines, J.E., 1997. Estimating temporary emigration using capture-recapture data with Pollock's robust design. Ecology 78, 563–578.

Kendall, W.L., Stapleton, S., White, G.C., Richardson, J.I., Pearson, K.N., Mason, P., 2019. A multistate open robust design: population dynamics, reproductive effort, and phenology of sea turtles from tagging data. Ecol. Monogr. 89, e01329.

Kenward, R.E., Marcström, V., Karlbom, M., 1999. Demographic estimates from radio-tagging: models of age-specific survival and breeding in the goshawk. J. Anim. Ecol. 68, 1020–1033.

Kenward, R.E., Walls, S.S., Hodder, K.H., Pahkala, M., Freeman, S.N., Simpson, V.R., 2000. The prevalence of non-breeders in raptor populations: evidence from rings, radio-tags and transect surveys. Oikos 91, 271–279.

Kéry, M., 2010. Introduction to WinBUGS for Ecologists. - A Bayesian Approach to Regression, ANOVA, Mixed Models and Related Analyses. Academic Press, Burlington.

Kéry, M., 2011. Towards the modeling of true species distributions. J. Biogeogr. 38, 617–618.

Kéry, M., Royle, J.A., 2016. Applied hierarchical modeling in ecology - Analysis of distribution, abundance and species richness in R and BUGS. In: Prelude and Static Models, vol. 1. Elsevier/Academic Press, London.

Kéry, M., Royle, J.A., 2021. Applied hierarchical modeling in ecology - Analysis of distribution, abundance and species richness in R and BUGS. In: Dynamic and Advanced Models, vol. 2. Elsevier/Academic Press, London.

Kéry, M., Schaub, M., 2012. Bayesian Population Analysis Using WinBUGS - A Hierarchical Perspective. Academic Press, Boston.

Kéry, M., Zuberogoitia, I., 2020. Peregrine falcon. In: Keller, V., Herrando, S., Voříšek, P., Franch, M., Kipson, M., Milanesi, P., Martí, D., Anton, M., Klvaňová, A., Kalyakin, M.V., Bauer, H.-G., Foppen, R. (Eds.), European Breeding Bird Atlas 2: Distribution, Abundance and Change. European Bird Census Council & Lynx Edicions, Barcelona, pp. 526–527.

Kéry, M., Matthies, D., Spillmann, H.H., 2000. Reduced fecundity and offspring performance in small populations of the declining grassland plants *Primula veris* and *Gentiana lutea*. J. Ecol. 88, 17–30.

Kéry, M., Matthies, D., Fischer, M., 2001. The effect of plant population size on the interactions between the rare plant *Gentiana cruciata* and its specialized herbivore *Maculinea rebeli*. J. Ecol. 89, 418–427.

Kéry, M., Royle, J.A., Schmid, H., 2005. Modeling avian abundance from replicated counts using binomial mixture models. Ecol. Appl. 15, 1450–1461.

Kéry, M., Royle, J.A., Plattner, M., Dorazio, R.M., 2009. Species richness and occupancy estimation in communities subject to temporary emigration. Ecology 90, 1279–1290.

Kéry, M., Banderet, G., Neuhaus, M., Weggler, M., Schmid, H., Sattler, T., Parish, D., 2018. Population trends of the Peregrine Falcon in Switzerland with special reference to the period 2005–2016. Ornis Hung. 26, 91–103.

Kéry, M., Banderet, G., Müller, C., Pinaud, D., Savioz, J., Schmid, H., Werner, S., Monneret, R.-J., 2021. Spatio-temporal variation in post-recovery dynamics in a large Peregrine Falcon (*Falco peregrinus*) population in the Jura mountains 2000–2020. Ibis. (in press). https://doi.org/10.1111/ibi.12999.

Keyfitz, N., Caswell, H., 2005. Applied Mathematical Demography. Springer Science+Business Media Inc., New York, NY.

King, R., 2012. A review of Bayesian state-space modelling of capture-recapture-recovery data. Interface Focus 2, 190–204.

King, R., 2014. Statistical ecology. Annu. Rev. Stat. Appl. 1, 401–426.

King, R., Brooks, S.P., Mazzetta, C., Freeman, S.N., Morgan, B.J.T., 2008. Identifying and diagnosing population declines: a Bayesian assessment of lapwings in the UK. App. Stat. 57, 609–632.

King, R., Morgan, B.J.T., Gimenez, O., Brooks, S.P., 2010. Bayesian Analysis for Population Ecology. Chapmann & Hall, Boca Raton.

Knape, J., 2008. Estimability of density dependence in models of time series data. Ecology 89, 2994–3000.

Knape, J., de Valpine, P., 2012. Are patterns of density dependence in the global population dynamics database driven by uncertainty about population abundance? Ecol. Lett. 15, 17–23.

Knape, J., Jonzen, N., Sköld, M., 2011. On observation distributions for state space models of population survey data. J. Anim. Ecol. 80, 1269–1277.

Knape, J., Besbeas, P., de Valpine, P., 2013. Using uncertainty estimates in analyses of population time series. Ecology 94, 2097–2107.

Knaus, P., Antoniazza, S., Wechsler, S., Guelat, J., Kéry, M., Strebel, N., Sattler, T., 2018. Schweizer Brutvogelatlas 2013–2016. Verbreitung und Bestandsentwicklung der Vögel in der Schweiz und im Fürstentum Liechtenstein. Sempach.

Koenig, W.D., 2002. Global patterns of environmental synchrony and the Moran effect. Ecography 25, 283–288.

Kokko, H., Lindström, J., 1998. Seasonal density dependence, timing of mortality, and sustainable harvesting. Ecol. Model. 110, 293–304.

Kokko, H., Sutherland, W.J., 2007. Optimal floating and queuing strategies: consequences for density dependence and habitat loss. Am. Nat. 152, 354–366.

Koons, D.N., Grand, J.B., Zinner, B., Rockwell, R.F., 2005. Transient population dynamics: Relations to life history and initial population state. Ecol. Model. 185, 283–297.

Koons, D.N., Colchero, F., Hersey, K., Gimenez, O., 2015. Disentangling the effects of climate, density dependence, and harvest on an iconic large herbivore's population dynamics. Ecol. Appl. 25, 956–967.

Koons, D.N., Iles, D.T., Schaub, M., Caswell, H., 2016. A life-history perspective on the demographic drivers of structured population dynamics in changing environments. Ecol. Lett. 19, 1023–1031.

Koons, D.N., Arnold, T.W., Schaub, M., 2017. Understanding the demographic drivers of realized population growth rates. Ecol. Appl. 27, 2102–2115.

Korner-Nievergelt, F., Sauter, A., Atkinson, P.W., Guélat, J., Kania, W., Kéry, M., Köppen, U., Robinson, R.A., Schaub, M., Thorup, K., van der Jeugd, H., van Noordwijk, A.J., 2010. Improving the analysis of movement data from marked individuals through explicit estimation of observer heterogeneity. J. Avian Biol. 41, 8–17.

Korner-Nievergelt, F., Felten von, S., Roth, T., Almasi, B., Guélat, J., Korner-Nievergelt, P., 2015. Bayesian Data Analysis in Ecology Using Linear Models with R, BUGS, and Stan. Academic Press, Amsterdam, Boston.

Kramer, A.M., Dennis, B., Liebhold, A.M., Drake, J.M., 2009. The evidence for Allee effects. Popul. Ecol. 51, 341–354.

Kristensen, K., Nielsen, A., Berg, C.W., Skaug, H., Bell, B.M., 2016. TMB: automatic differentiation and Laplace approximation. J. Stat. Software 70. https://doi.org/10.18637/jss.v070.i05.

Kuhnert, P.M., Martin, T.G., Griffiths, S.P., 2010. A guide to eliciting and using expert knowledge in Bayesian ecological models. Ecol. Lett. 13, 900–914.

Lack, D., 1954. The Natural Regulation of Animal Numbers. Oxford University Press, Oxford.

Lahoz-Monfort, J.J., Morgan, B.J.T., Harris, M.P., Wanless, S., Freeman, S.N., 2011. A capture-recapture model for exploring multi-species synchrony in survival. Methods Ecol. Evol. 2, 116–124.

Lahoz-Monfort, J.J., Morgan, B.J.T., Harris, M.P., Daunt, F., Wanless, S., Freeman, S.N., 2013. Breeding together: modeling synchrony in productivity in a seabird community. Ecology 94, 3–10.

Lahoz-Monfort, J.J., Harris, M.P., Morgan, B.J.T., Freeman, S.N., Wanless, S., 2014. Exploring the consequences of reducing survey effort for detecting individual and temporal variability in survival. J. Appl. Ecol. 51, 534–543.

Lahoz-Monfort, J.J., Harris, M.P., Wanless, S., Freeman, S.N., Morgan, B.J.T., 2017. Bringing it all together: multi-species integrated population modelling of a breeding community. JABES 22, 140–160.

Lande, R., 1982. A quantitative genetic theory of life history evolution. Ecology 63, 607–615.

Lande, R., 1988. Genetics and demography in biological conservation. Science 241, 1455–1460.

Lande, R., 1993. Risks of population extinction from demographic and environmental stochasticity and random catastrophes. Am. Nat. 142, 911–927.

Lande, R., Orzack, S.H., 1988. Extinction dynamics of age-structured populations in a fluctuating environment. Proc. Nat. Acad. Sci. U.S.A. 85, 7418–7421.

Lande, R., Engen, S., Sæther, B.-E., 2003. Stochastic Population Dynamics in Ecology and Conservation. Oxford University Press, Oxford.

Landes, J., Engelhardt, S.C., Pelletier, F., 2020. An introduction to event history analyses for ecologists. Ecosphere 11, 457.

Larkin, J.L., Maehr, D.S., Cox, J.J., Bolin, D.C., Wichrowski, M.W., 2003. Demographic characteristics of a reintroduced elk population in Kentucky. J. Wildl. Manag. 67, 467–476.

Lavine, M., 2010. Living dangerously with big fancy models. Ecology 91, 3487.

Lebreton, J.-D., 2005. Dynamical and statistical models for exploited populations. Aust. N. Z. J. Stat. 47, 49–63.

Lebreton, J.-D., 2009. Assessing density-dependence: where are we left? In: Thomson, D., Cooch, E., Conroy, M. (Eds.), Modeling Demographic Processes in Marked Populations. Springer, New York, pp. 19–42.

Lebreton, J.-D., Gaillard, J.-M., 2016. Wildlife demography: population processes, analytical tools and management applications. In: Mateo, R., Arroyo, B., Garcia, J.T. (Eds.), Current Trends in Wildlife Research. Springer, Cham, pp. 29–54.

Lebreton, J.-D., Gimenez, O., 2013. Detecting and estimating density dependence in wildlife populations. J. Wildl. Manag. 77, 12–23.

Lebreton, J.-D., Burnham, K.P., Clobert, J., Anderson, D.R., 1992. Modeling survival and testing biological hypothesis using marked animals: a unified approach with case studies. Ecol. Monogr. 62, 67–118.

Lebreton, J.-D., Almeras, T., Pradel, R., 1999. Competing events, mixtures of information and multistratum recapture models. Bird Study 46 (suppl. 1), 39–46.

Lebreton, J.-D., Nichols, J.D., Barker, R.J., Pradel, R., Spendelow, J.A., 2009. Modeling individual animal histories with multistate capture-recapture models. Adv. Ecol. Res. 41, 87–173.

Lee, A.M., Björkvoll, E., Hansen, B.B., Albon, S.D., Stien, A., Sæther, B.-E., Engen, S., Veiberg, V., Loe, L.E., Grøtan, V., 2015. An integrated population model for a long-lived ungulate: more efficient data use with Bayesian methods. Oikos 124, 806–816.

Lee, A.M., Reid, J.M., Beissinger, S.R., 2017. Modeling effects of nonbreeders on population growth estimates. J. Anim. Ecol. 86, 75–87.

Lefkovitch, L.P., 1965. The study of population growth in organisms grouped by stages. Biometrics 21, 1–18.

Lefranc, N., 1999. Les pies-grièches *Lanius sp.* en France: répartition et statut actuels, histoire récente, habitats. Ornithos 6, 58–82.

Lefranc, N., Worfalk, T., 1997. Shrikes. A Guide to the Shrikes of the World. Pica Press, East Sussex.

Lele, S.R., Dennis, B., Lutscher, F., 2007. Data cloning: easy maximum likelihood estimation for complex ecological models using Bayesian Markov chain Monte Carlo methods. Ecol. Lett. 10, 551–563.

Lele, S.R., Nadeem, K., Schmuland, B., 2010. Estimability and likelihood inference for generalized linear mixed models using data cloning. J. Am. Stat. Assoc. 105, 1617–1625.

Lemoine, N.P., 2019. Moving beyond noninformative priors: why and how to choose weakly informative priors in Bayesian analyses. Oikos 128, 912–928.

Lentini, P.E., Bird, T.J., Griffith, S.R., Godinho, L.N., Wintle, B.A., 2016. A global synthesis of survival estimates for microbats. Biol. Lett. 11, 1–5.

León-Ortega, M., Jiménez-Franco, M.V., Martínez, J.E., Calvo, J.F., 2017. Factors influencing territorial occupancy and reproductive success in a Eurasian eagle-owl (*Bubo bubo*) population. PLoS ONE 12, e0175597.

Leslie, P.H., 1945. On the use of matrices in certain population mathematics. Biometrika 33, 183–212.

Levin, L.A., Caswell, H., DePatra, K.D., Creed, E.L., 1987. Demographic consequences of larval development mode: planktotrophy vs. lecithotrophy in *Streblospio benedicti*. Ecology 68, 1877–1886.

Liebhold, A., Koenig, W.D., Björnstad, O.N., 2004. Spatial synchrony in population dynamics. Annu. Rev. Ecol. Syst. 35, 467–490.

Lieury, N., Gallardo, M., Ponchon, C., Besnard, A., Millon, A., 2015. Relative contribution of local demography and immigration in the recovery of a geographically-isolated population of the endangered Egyptian vulture. Biol. Conserv. 191, 349–356.

Lieury, N., Besnard, A., Ponchon, C., Ravayrol, A., Millon, A., 2016. Geographically isolated but demographically connected: immigration supports efficient conservation actions in the recovery of a range-margin population of the Bonelli's eagle in France. Biol. Conserv. 195, 272–278.

Linden, D.W., McKinney, S.T., 2016. Evaluation of the Maine Department of Inland Fisheries and Wildlife black bear management program: Integrated population modeling of intensive monitoring data. Final Report (v.2016.06.08). University of Maine. https://doi.org/10.13140/RG.2.1.4479.1921.

Lindenmayer, D.B., Burgman, M.A., Akçakaya, H.R., Lacy, R.C., Possingham, H.P., 1995. A review of the generic computer programs ALEX, RAMAS/space and VORTEX for modelling the viability of wildlife metapopulations. Ecol. Model. 82, 161–174.

Lindenmayer, D.B., Likens, G.E., Andersen, A., Bowman, D., Bull, C.M., Burns, E., Dickman, C.R., Hoffmann, A.A., Keith, D.A., Liddell, M.J., Lowe, A.J., Metcalfe, D.J., Phinn, S.R., Russell-Smith, J., Thurgate, N., Wardle, G.M., 2012. Value of long-term ecological studies. Austral Ecol. 37, 745–757.

Lindley, D.V., 1983. Theory and practice of Bayesian statistics. Statistician 32, 1–11.

Lindley, D.V., 2006. Understanding Uncertainty. Wiley, Hoboken, New Jersey.

Lindley, S.T., 2003. Estimation of population growth and extinction parameters from noisy data. Ecol. Appl. 13, 806–813.

Link, W.A., 1999. Modeling pattern in collections of parameters. J. Wildl. Manag. 63, 1017–1027.

Link, W.A., 2013. A cautionary note on the discrete uniform prior for the Binomial N. Ecology 94, 2173–2179.

Link, W.A., Barker, R.J., 2005. Modeling association among demographic parameters in analysis of open population capture-recapture data. Biometrics 61, 46–54.

Link, W.A., Barker, R.J., 2010. Bayesian Inference with Ecological Applications. Academic Press, London.

Link, W.A., Cam, E., Nichols, J.D., Cooch, E.G., 2002. Of *BUGS* and birds: Markov chain Monte Carlo for hierarchical modeling in wildlife research. J. Wildl. Manag. 66, 277–291.

Link, W.A., Royle, J.A., Hatfield, J.S., 2003. Demographic analysis from summaries of an age-structured population. Biometrics 59, 778–785.

Loison, A., Festa-Bianchet, M., Gaillard, J.M., Jorgenson, J.T., Jullien, J.-M., 1999. Age-specific survival in five populations of ungulates: evidence of senescence. Ecology 80, 2539–2554.

Łomnicki, A., 1978. Individual differences between animals and the natural regulation of their numbers. J. Anim. Ecol. 47, 461–475.

Lopez-Sepulcre, A., Kokko, H., 2005. Territorial defense, territory size, and population regulation. Am. Nat. 166, 317–329.

Lotka, A.J., 1907. Studies on the mode of growth of material aggregates. Am. J. Sci. 24, 199–216.

Louvrier, J., Papaïx, J., Duchamp, C., Gimenez, O., 2020. A mechanistic-statistical species distribution model to explain and forecast wolf (*Canis lupus*) colonization in South-Eastern France. Spat. Stat. 36, 100428.

Low Choy, S.J., O'Leary, R., Mengersen, K., 2009. Elicitation by design in ecology: using expert opinion to inform priors for Bayesian statistical models. Ecology 90, 265–277.

Ludwig, D., 1996a. The distribution of population survival times. Am. Nat. 147, 506–525.

Ludwig, D., 1996b. Uncertainty and assessment of extinction probabilities. Ecol. Appl. 6, 1067–1076.

Ludwig, D., 1999. Is it meaningful to estimate a probability of extinction? Ecology 80, 298–310.

Luijten, S., Kéry, M., Oostermeijer, G., den Nils, H., 2002. Demographic consequences of inbreeding and outbreeding in *Arnica montanus*: a field experiment. J. Ecol. 90, 593–603.

Lunn, D.J., Thomas, A., Best, N., Spiegelhalter, D., 2000. WinBUGS - a Bayesian modelling framework: concepts, structure, and extensibility. Stat. Comput. 10, 325–337.

Lunn, D.J., Spiegelhalter, D., Thomas, A., Best, N., 2009. The BUGS project: evaluation, critique and future directions. Stat. Med. 28, 3049–3067.

Lunn, D., Jackson, C., Best, N., Thomas, A., Spiegelhalter, D., 2013. The BUGS Book. A Practical Introduction to Bayesian Analysis. CRC Press, Boca Raton.

Lynch, H.J., Thorson, J.T., Shelton, A.O., 2014. Dealing with under- and over-dispersed count data in life history, spatial, and community ecology. Ecology 95, 3173–3180.

Mace, G.M., Collar, N.J., Gaston, K.J., Hilton-Taylor, C., Akçakaya, H.R., Leader-Williams, N., Milner-Gulland, E.J., Stuart, S.N., 2008. Quantification of extinction risk: IUCN's system for classifying threatened species. Conserv. Biol. 22, 1424–1442.

MacKenzie, D.I., Nichols, J.D., Lachman, G.B., Droege, S., Royle, J.A., Langtimm, C.A., 2002. Estimating site occupancy rates when detection probabilities are less than one. Ecology 83, 2248–2255.

MacKenzie, D.I., Nichols, J.D., Hines, J.E., Knutson, M.G., Franklin, A.B., 2003. Estimating site occupancy, colonization, and local extinction when a species is detected imperfectly. Ecology 84, 2200–2207.

MacKenzie, D.I., Nichols, J.D., Seamans, M.E., Gutierrez, R.J., 2009. Modeling species occurrence dynamics with multiple states and imperfect detection. Ecology 90, 823–835.

MacKenzie, D.I., Nichols, J.D., Royle, J.A., Pollock, K.H., Bailey, L.L., Hines, J.E., 2018. Occupancy Estimation and Modeling: Inferring Patterns and Dynamics of Species Occurrence, Second Edition. Academic Press, San Diego.

Magurran, A.E., Baillie, S.R., Buckland, S.T., Dick, J.M., Elston, D.A., Scott, E.M., Smith, R.I., Somerfield, P.J., Watt, A.D., 2010. Long-term datasets in biodiversity research and monitoring: assessing change in ecological communities through time. Trends Ecol. Evol. 25, 574–582.

Margalida, A., Jiménez, J., Martínez, J.M., Sesé, J.A., García-Ferré, D., Llamas, A., Razin, M., Colomer, M., Arroyo, B., 2020. An assessment of population size and demographic drivers of the Bearded Vulture using integrated population models. Ecol. Monogr. 90, e01414.

Maronde, L., McClintock, B.T., Breitenmoser, U., Zimmermann, F., 2020. Spatial capture-recapture with multiple noninvasive marks: An application to camera-trapping data of the European wildcat (*Felis silvestris*) using R package multimark. Ecol. Evol. 10, 13968–13979.

Marolla, F., Aarvak, T., Hamel, S., Ims, R.A., Kéry, M., Mellard, J.P., Nater, C.R., Schaub, M., Øien, I.J., Vougioukalou, M., Yoccoz, N.G. Life-cycle analysis of an endangered migratory bird shows no evidence that predator control drove population recovery. (in prep).

Martin, T.G., Kuhnert, P.M., Mengersen, K., Possingham, H., 2005. The power of expert opinion in ecological models using Bayesian methods: impact of grazing on birds. Ecol. Appl. 15, 266–280.

Martin, J., McIntyre, C.L., Hines, J.E., Nichols, J.D., Schmutz, J.A., MacCluskie, M.C., 2009a. Dynamic multistate site occupancy models to evaluate hypotheses relevant to conservation of Golden Eagles in Denali National Park, Alaska. Biol. Conserv. 142, 2726–2731.

Martin, J., Nichols, J.D., McIntyre, C.L., Ferraz, G., Hines, J.E., 2009b. Perturbation analysis for patch occupancy dynamics. Ecology 90, 10–16.

Matthiopoulos, J., Cordes, L., Mackey, B., Thompson, D., Duck, C., Smout, S., Caillat, M., Thompson, P., 2014. State-space modelling reveals proximate causes of harbour seal population declines. Oecologia 174, 151–162.

Mattsson, B.J., Cooper, R.J., 2007. Which life-history components determine breeding productivity for individual songbirds? A case study of the Louisiana Waterthrush (*Seiurus motacilla*). Auk 124, 1186–1200.

Maunder, M.N., 2004. Population viability analysis based on combining Bayesian, integrated, and hierarchical analyses. Acta Oecol. 26, 85–94.

Maunder, M.N., Piner, K.R., 2015. Contemporary fisheries stock assessment: many issues still remain. ICES J. Mar. Sci. 72, 7–18.

Maunder, M.N., Piner, K.R., 2017. Dealing with data conflicts in statistical inference of population assessment models that integrate information from multiple diverse data sets. Fish. Res. 192, 16–27.

Maunder, M.N., Punt, A.E., 2004. Standardizing catch and effort data: a review of recent approaches. Fish. Res. 70, 141–159.

Maunder, M.N., Punt, A.E., 2013. A review of integrated analysis in fisheries stock assessment. Fish. Res. 142, 61–74.

Maunder, M.N., Starr, P.J., 2001. Bayesian assessment of the SNA1 snapper (*Pagrus auratus*) stock on the north-east coast of New Zealand. N. Z. J. Mar. Freshw. Res. 35, 87–110.

Mayfield, H.F., 1961. Nesting success calculated from exposure. Wilson Bull. 73, 255–261.

Mayfield, H.F., 1975. Suggestions for calculating nest success. Wilson Bull. 87, 456–466.

McCaffery, R., Lukacs, P.M., 2016. A generalized integrated population model to estimate greater sage-grouse population dynamics. Ecosphere 7, e01585.

McCaffrey, R., Solonen, A., Crone, E., 2012. Frog population viability under present and future climate conditions: a Bayesian state-space approach. J. Anim. Ecol. 81, 978–985.

McCarthy, M.A., 2007. Bayesian Methods for Ecology. Cambridge University Press, Cambridge.

McCarthy, M.A., Masters, P., 2005. Profiting from prior information in Bayesian analyses of ecological data. J. Appl. Ecol. 42, 1012–1019.

McCarthy, M.A., Possingham, H.P., 2007. Active adaptive management for conservation. Conserv. Biol. 21, 956–963.

McCarthy, M.A., Andelman, S.J., Possingham, H.P., 2003. Reliability of relative predictions in population viability analysis. Conserv. Biol. 17, 982–989.

McClintock, B.T., 2015. multimark: an R package for analysis of capture-recapture data consisting of multiple "noninvasive" marks. Ecol. Evol. 5, 4920–4931.

McClintock, B.T., White, G.C., 2009. A less field-intensive robust design for estimating demographic parameters with mark-resight data. Ecology 90, 313–320.
McClintock, B.T., Conn, P.B., Alonso, R.S., Crooks, K.R., 2013. Integrated modeling of bilateral photo-identification data in mark-recapture analyses. Ecology 94, 1464–1471.
McClintock, B.T., White, G.C., Pryde, M.A., 2019. Improved methods for estimating abundance and related demographic parameters from mark-resight data. Biometrics 75, 799–809.
McClintock, B.T., Langrock, R., Gimenez, O., Cam, E., Borchers, D.L., Glennie, R., Patterson, T.A., 2020. Uncovering ecological state dynamics with hidden Markov models. Ecol. Lett. 23, 1878–1903.
McCrea, R.S., Morgan, B.J.T., 2015. Analysis of Capture-Recapture Data. Chapman & Hall/CRC, Boca Raton.
McCrea, R.S., Morgan, B.J.T., Gimenez, O., Besbeas, P., Bregnballe, T., Hénaux, V., Lebreton, J.-D., 2010. Multi-site integrated population modelling. JABES 15, 539–561.
McCrea, R.S., Morgan, B.J.T., Pradel, R., 2014. Diagnostic goodness-of-fit tests for joint recapture and recovery models. JABES 19, 338–356.
McCullagh, P., Nelder, J.A., 1989. Generalized Linear Models. Chapman & Hall, London.
McCulloch, C.E., Searle, S.R., Neuhaus, J.M., 2008. Generalized, Linear, and Mixed Models. John Wiley & Sons, Hoboken.
McGowan, C.P., Runge, M.C., Larson, M.A., 2011. Incorporating parameteric uncertainty into population viability analysis models. Biol. Conserv. 144, 1400–1408.
McGrady, M.J., Hines, J.E., Rollie, C.J., Smith, G.D., Morton, E.R., Moore, J.F., Mearns, R.M., Newton, I., Murillo-García, O.E., Oli, M.K., 2017. Territory occupancy and breeding success of peregrine falcons *Falco peregrinus* at various stages of population recovery. Ibis 159, 285–296.
McLachlan, G.J., Rathnayake, S., 2014. On the number of components in a Gaussian mixture model. WIREs Data Mining Knowl. Discov. 4, 341–355.
McLoughlin, P.D., Messier, F., 2004. Relative contributions of sampling error in initial population size and vital rates to outcomes of population viability analysis. Conserv. Biol. 18, 1665–1669.
Melbourne, B.A., Chesson, P., 2005. Scaling up population dynamics: integrating theory and data. Oecologia 145, 179–187.
Metropolis, N., Ulam, S., 1949. The Monte Carlo method. J. Am. Stat. Assoc. 44, 335–341.
Metropolis, N., Rosenbluth, A.W., Rosenbluth, M.N., Teller, A.H., Teller, E., 1953. Equation of state calculations by fast computing machines. J. Chem. Phys. 21, 1087–1092.
Michelot, T., Blackwell, P.G., 2019. State-switching continuous-time correlated random walks. Methods Ecol. Evol. 10, 637–649.
Middleton, D.A.J., Nisbet, R.M., 1997. Population persistence time: estimates, models, and mechanisms. Ecol. Appl. 7, 107–117.
Millar, R.B., Meyer, R., 2000. Bayesian state-space modeling of age-structured data: fitting a model is just the beginning. Can. J. Fish. Aquat. Sci. 57, 43–50.
Miller, D.A., Nichols, J.D., McClintock, B.T., Grant, E.H.C., Bailey, L.L., Weir, L., 2011. Improving occupancy estimation when two types of observational errors occur: non-detection and species misidentification. Ecology 92, 1422–1428.
Miller, M.W., Leech, D.I., Pearce-Higgins, J.W., Robinson, R.A., 2017. Multi-state, multi-stage modeling of nest-success suggests interaction between weather and land-use. Ecology 98, 175–186.
Miller, D.A.W., Pacifici, K., Sanderlin, J.S., Reich, B.J., Gardner, B., 2019. The recent past and promising future for data integration methods to estimate species' distributions. Methods Ecol. Evol. 10, 22–37.
Millon, A., Lambin, X., Devillard, S., Schaub, M., 2019. Quantifying the contribution of immigration to population dynamics: a review of methods, evidence and perspectives in birds and mammals. Biol. Rev. 94, 2049–2067.
Mills, L.S., 2013. Conservation of Wildlife Populations. Demography, Genetics and Management. Wiley-Blackwell, West Sussex.
Mills, L.S., Hayes, S.G., Baldwin, C., Wisdom, M.J., Citta, J., Mattson, D.J., Murphy, K., 1996. Factors leading to different viability predictions for a grizzly bear data set. Conserv. Biol. 10, 863–873.
Mizel, J.D., McIntyre, C.L., Lewis, S.B., Lindberg, M.S., Schmidt, J.H., 2018. A multi-state, time-removal model for population dynamics of cliff-nesting raptors. J. Wildl. Manag. 82, 1701–1710.
Mollet, P., Kéry, M., Gardner, B., Pasinelli, G., Royle, J.A., 2015. Estimating population size for capercaillie (*Tetrao urogallus* L.) with spatial capture-recapture models based on genotypes from one field sample. PLoS ONE 10, e0129020.
Monneret, R.-J., 2017. Le Faucon Pèlerin: Description, Moeurs, Observation, Protection, Mythologie. Delachaux et Niestlé, Paris.
Monneret, R.-J., Ruffinoni, R., Parish, D., Pinaud, D., Kéry, M., 2018. The Peregrine population study in the French Jura mountains 1964–2016: use of occupancy modeling to estimate population size and analyze site persistence and colonization rates. Ornis Hung. 26, 69–90.
Moran, P.A.P., 1953. The statistical analysis of the Canadian lynx cycle. II. Synchronization and meteorology. Aust. J. Zool. 1, 291–298.
Moreno, M., Lele, S.R., 2010. Improved estimation of site occupancy using penalized likelihood. Ecology 91, 341–346.
Morris, W.F., Doak, D.F., 2002. Quantitative Conservation Biology. Sinauer, Sunderland, Massachusetts, USA.
Morrison, C.A., Robinson, R.A., Butler, S.J., Clark, J.A., Gill, J.A., 2016. Demographic drivers of decline and recovery in an Afro-Palaearctic migratory bird population. Proc. R. Soc. Lond. B 283, 20161387.

Mosnier, A., Doniol-Valcroze, T., Gosselin, J.-F., Lesage, V., Measures, L.N., Hammill, M.O., 2015. Insights into processes of population decline using an integrated population model: the case of the St. Lawrence Estuary beluga (*Delphinapterus leucas*). Ecol. Model. 314, 15–31.

Murray, D.L., 2006. On improving telemetry-based survival estimation. J. Wildl. Manag. 70, 1530–1543.

Murray, D.L., Bastille-Rousseau, G., 2020. Estimating survival and cause-specific mortality from continuous time observations. In: Murray, D., Sandercock, B.K. (Eds.), Population Ecology in Practice. Wiley-Blackwell, Hoboken, NJ, pp. 123–155.

Naef-Daenzer, B., Grüebler, M.U., 2016. Post-fledging survival of altricial birds: ecological determinants and adaptation. J. Field Ornithol. 87, 227–250.

Nater, C.R., Vindenes, Y., Aass, P., Cole, D., Langangen, Ø., Moe, S.J., Rustadbakken, A., Turek, D., Vøllestad, L.A., Ergon, T., 2020. Size- and stage-dependence in cause-specific mortality of migratory brown trout. J. Anim. Ecol. 89, 2122–2133.

Nelder, J.A., Wedderburn, R.W.M., 1972. Generalized linear models. J. R. Stat. Soc. A 135, 370–384.

Nevoux, M., Gimenez, O., Arlt, D., Nicoll, M., Jones, C., Norris, K., 2011. Population regulation of territorial species: both site dependence and interference mechanisms matter. Proc. R. Soc. Lond. B 178, 2173–2181.

Newman, K.B., Buckland, S.T., Morgan, B.J.T., King, R., Borchers, D.L., Cole, D.J., Besbeas, P., Gimenez, O., Thomas, L., 2014. Modelling Population Dynamics. Model Formulation, Fitting and Assessment Using State-Space Methods. Springer, New York.

Newton, I., 1979. Population Ecology of Raptors. Berkhamsted, Poyser.

Newton, I., 1998. Population Limitation in Birds. Academic Press, London.

Newton, I., 2004. Population limitation in migrants. Ibis 146, 197–226.

Newton, I., 2010. The Sparrowhawk. T & A D Poyser, Calton , UK.

Newton, I., Mearns, R., 1988. Population ecology of peregrines in Scotland. In: Cade, T.J., Enderson, J.H., Thelander, C.G., White, C.M. (Eds.), Peregrine Falcon Populations: Their Management and Recovery. The Peregrine Fund, Boise, Idaho, pp. 651–666.

Newton, I., Rothery, P., 2001. Estimation and limitation of numbers of floaters in a Eurasian Sparrowhawk population. Ibis 143, 442–449.

Nichols, J.D., 1992. Capture-recapture models. Using marked individuals to study population dynamics. Bioscience 42, 94–102.

Nichols, J.D., Hines, J.E., 1983. The relationship between harvest and survival rates of mallards: a straightforward approach with partitioned data sets. J. Wildl. Manag. 47, 334–348.

Nichols, J.D., Conroy, M.J., Anderson, D.R., Burnham, K.P., 1984. Compensatory mortality in waterfowl populations: a review of the evidence and implications for research and management. Trans. N. Am. Wildl. Nat. Resour. Conf. 49, 535–554.

Nichols, J.D., Sauer, J.R., Pollock, K.H., Hestbeck, J.B., 1992. Estimating transition probabilities for stage-based population projection matrices using capture-recapture data. Ecology 73, 306–312.

Nichols, J.D., Hines, J.E., Lebreton, J.-D., Pradel, R., 2000a. Estimation of contributions to population growth: a reverse-time capture-recapture approach. Ecology 81, 3362–3376.

Nichols, J.D., Hines, J.E., Sauer, J.R., Fallon, F.W., Fallon, J.E., Heglund, P.J., 2000b. A double-observer approach for estimating detection probability and abundance from point counts. Auk 117, 393–408.

Nichols, J.D., Kendall, W.L., Hines, J.E., Spendelow, J.A., 2004. Estimation of sex-specific survival from capture-recapture data when sex is not always known. Ecology 85, 3192–3201.

Nichols, J.D., Hines, J.E., MacKenzie, D.I., Seamans, M.E., Gutierrez, R.J., 2007. Occupancy estimation and modeling with multiple states and state uncertainty. Ecology 88, 1395–1400.

Nichols, J.D., Koneff, M.D., Heglund, P.J., Knutson, M.G., Seamans, M.E., Lyons, J.E., Morton, J.M., Jones, M.T., Boomer, G.S., Williams, B.K., 2011. Climate change, uncertainty, and natural resource management. J. Wildl. Manag. 75, 6–18.

Nilsson, S.G., 1987. Limitation and regulation of population density in the Nuthatch *Sitta europea* (Aves) breeding in natural cavities. J. Anim. Ecol. 56, 921–937.

NIMBLE Development Team, 2018. NIMBLE User Manual. Version 0.6-12.

Norris, D.R., Taylor, C.M., 2006. Predicting the consequences of carry-over effects for migratory populations. Biol. Lett. 2, 148–151.

Norris, K., 2004. Managing threatened species: the ecological toolbox, evolutionary theory and declining-population paradigm. J. Appl. Ecol. 41, 413–426.

Northrup, J.M., Gerber, B.D., 2018. A comment on priors for Bayesian occupancy models. PLoS ONE 13, e0192819.

Ntzoufras, I., 2009. Bayesian Modeling Using WinBUGS. Wiley, Hoboken, New Jersey.

Ogle, K., Barber, J.J., 2020. Ensuring identifiability in hierarchical mixed effects Bayesian models. Ecol. Appl. 30, e02159.

Oli, M.K., Dobson, F.S., 2003. The relative importance of life-history variables to population growth rate in mammals: Cole's prediction revisited. Am. Nat. 161, 422–440.

Oosthuizen, W.C., Pradel, R., Bester, M.N., Bruyn, P.J.N., 2019. Making use of multiple surveys: estimating breeding probability using a multievent-robust design capture-recapture model. Ecol. Evol. 9, 836–848.

Oppel, S., Powell, A.N., 2010. Age-specific survival estimates of King Eiders derived from satellite telemetry. Condor 112, 323–330.

Oppel, S., Hilton, G.M., Allcorn, R., Fenton, C., Matthews, A.J., Gibbons, D.W., Stewart, I., 2013. The effects of rainfall on different components of seasonal fecundity in a tropical forest passerine. Ibis 155, 464–475.

Oppel, S., Hilton, G., Ratcliffe, N., Fenton, C., Daley, J., Gray, G., Vickery, J., Gibbons, D., 2014. Assessing population viability while accounting for demographic and environmental uncertainty. Ecology 95, 1809−1818.

Otis, D.L., Burnham, K.P., White, G.C., Anderson, D.R., 1978. Statistical inference from capture data on closed animal populations. Wildl. Monogr. 62, 1−135.

Ozgul, A., Oli, M.K., Armitage, K.B., Blumstein, D.T., van Vuren, D.H., 2009. Influence of local demography on asymptotic and transient dynamics of a yellow-bellied marmot metapopulation. Am. Nat. 173, 517−530.

O'Hagan, A., Buck, C.E., Daneshkhah, A., Eiser, J.R., Garthwaite, P.H., Jenkinson, D.J., Oakley, J.E., Radkow, T., 2006. Uncertain Judgements. Eliciting Experts' Probabilities. Wiley, Chichester.

Pacifici, K., Reich, B.J., Miller, D.A.W., Gardner, B., Stauffer, G., Singh, S., McKerrow, A., Collazo, J.A., 2017. Integrating multiple data sources in species distribution modeling: a framework for data fusion. Ecology 98, 840−850.

Pacifici, K., Reich, B.J., Miller, D.A.W., Pease, B.S., 2019. Resolving misaligned spatial data with integrated species distribution models. Ecology 100, e02709.

Pagel, J., Schurr, F.M., 2012. Forecasting species ranges by statistical estimation of ecological niches and spatial population dynamics. Global Ecol. Biogeogr. 21, 293−304.

Pagel, J., Anderson, B.J., O'Hara, R.B., Cramer, W., Fox, R., Jeltsch, F., Roy, D.B., Thomas, C.D., Schurr, F.M., McMahon, S., 2014. Quantifying range-wide variation in population trends from local abundance surveys and widespread opportunistic occurrence records. Methods Ecol. Evol. 5, 751−760.

Papadatou, E., Pradel, R., Schaub, M., Dolch, D., Geiger, H., Ibañez, C., Kerth, G., Popa-Lisseanu, A., Schorcht, W., Teubner, J., Gimenez, O., 2012. Comparing survival among species with imperfect detection using multilevel analysis of mark-recapture data: a case study on bats. Ecography 35, 153−161.

Paquet, M., Arlt, D., Knape, J., Low, M., Forslund, P., Pärt, T., 2019. Quantifying the links between land use and population growth rate in a declining farmland bird. Ecol. Evol. 9, 868−879.

Paradis, E., Baillie, S.R., Sutherland, W.J., Gregory, R.D., 1998. Patterns of natal and breeding dispersal in birds. J. Anim. Ecol. 67, 518−536.

Paradis, E., Baillie, S.R., Sutherland, W.J., Gregory, R.D., 1999. Dispersal and spatial scale affect synchrony in spatial population dynamics. Ecol. Lett. 2, 114−120.

Paradis, E., Baillie, S.R., Sutherland, W.J., Gregory, R.D., 2000. Spatial synchrony in populations of birds: effects of habitat, population trend, and spatial scale. Ecology 81, 2112−2125.

Parmenter, R.R., Yates, T.L., Anderson, D.R., Burnham, K.P., Dunnum, J.L., Franklin, A.B., Friggens, M.T., Lubow, B.C., Miller, M., Olson, G.S., Permenter, C.A., Pollard, J., Rexstad, E., Shenk, T.M., Stanley, T.R., White, G.C., 2003. Small-mammal density estimation: a field comparison of grid-based vs. web-based density estimators. Ecol. Monogr. 73, 1−26.

Pasinelli, G., Schaub, M., Häfliger, G., Frey, M., Jakober, H., Muller, M., Stauber, W., Tryjanowski, P., Zollinger, J.-L., Jenni, L., 2011. Impact of density and environmental factors on population fluctuations in a migratory passerine. J. Anim. Ecol. 80, 225−234.

Patthey, P., Signorell, N., Rotelli, L., Arlettaz, R., 2012. Vegetation structural and compositional heterogeneity as a key feature in Alpine black grouse microhabitat selection: conservation management implications. Eur. J. Wildl. Res. 58, 59−70.

Payton, Q., Hostetter, N.J., Evans, A.F., 2019. Jointly estimating survival and mortality: integrating recapture and recovery data from complex multiple predator systems. Environ. Ecol. Stat. 26, 107−125.

Penteriani, V., Ferrer, M., Delgado, M.M., 2011. Floater strategies and dynamics in birds, and their importance in conservation biology: towards an understanding of nonbreeders in avian populations. Anim. Conserv. 14, 233−241.

Péron, G., 2013. Compensation and additivity of anthropogenic mortality: life-history effects and review of methods. J. Anim. Ecol. 82, 408−417.

Péron, G., Koons, D.N., 2012. Integrated modeling of communities: parasitism, competition, and demographic synchrony in sympatric species. Ecology 93, 2456−2464.

Péron, G., Crochet, P.-A., Choquet, R., Pradel, R., Lebreton, J.-D., Gimenez, O., 2010a. Capture-recapture models with heterogeneity to study survival senescence in the wild. Oikos 119, 524−532.

Péron, G., Crochet, P.-A., Doherty, P.F., Lebreton, J.-D., 2010b. Studying dispersal at the landscape scale: efficient combination of population surveys and capture-recapture data. Ecology 91, 3365−3375.

Péron, G., Ferrand, Y., Gossmann, F., Bastat, C., Guenezan, M., Gimenez, O., 2011. Nonparametric spatial regression of survival probability: visualization of population sinks in Eurasian woodcock. Ecology 92, 1672−1679.

Péron, G., Nicolai, C.A., Koons, D.N., 2012. Demographic response to perturbations: the role of compensatory density dependence in a North American duck under variable harvest regulations and changing habitat. J. Anim. Ecol. 81, 960−969.

Petchey, O.L., Pontarp, M., Massie, T.M., Kéfi, S., Ozgul, A., Weilenmann, M., Palamara, G.M., Altermatt, F., Matthews, B., Levine, J.M., Childs, D.Z., McGill, B.J., Schaepman, M.E., Schmid, B., Spaak, P., Beckerman, A.P., Pennekamp, F., Pearse, I.S., 2015. The ecological forecast horizon, and examples of its uses and determinants. Ecol. Lett. 18, 597−611.

Pettorelli, N., Gaillard, J.M., Duncan, P., Maillard, D., van Laere, G., Delorme, D., 2003. Age and density modify the effects of habitat quality on survival and movements of roe deer. Ecology 84, 3307−3316.

Pilastro, A., Tavecchia, G., Marin, G., 2003. Long living and reproduction skipping in the fat dormouse. Ecology 84, 1784−1792.

Piner, K.R., Lee, H.-H., Maunder, M.N., Methot, R.D., 2011. A simulation-based method to determine model misspecification: examples using natural mortality and population dynamics models. Mar. Coast. Fish. 3, 336–343.

Plard, F., Gaillard, J.M., Coulson, T., Hewison, A.J.M., Douhard, M., Klein, F., Delorme, D., Warnant, C., Bonenfant, C., 2015. The influence of birth date via body mass on individual fitness in a long-lived mammal. Ecology 96, 1516–1528.

Plard, F., Arlettaz, R., Schaub, M., 2018a. Hoopoe males experience intra-seasonal while females experience inter-seasonal reproductive costs. Oecologia 186, 665–675.

Plard, F., Schindler, S., Arlettaz, R., Schaub, M., 2018b. Sex-specific heterogeneity in fixed morphological traits influences individual fitness in a monogamous bird population. Am. Nat. 191, 106–119.

Plard, F., Turek, D., Grüebler, M.U., Schaub, M., 2019. IPM^2: toward better understanding and forecasting of population dynamics. Ecol. Monogr. 89, e01364.

Plard, F., Fay, R., Kéry, M., Cohas, A., Schaub, M., 2019a. Integrated population models: powerful methods to embed individual processes in population dynamics models. Ecology 100, e02715.

Plard, F., Arlettaz, R., Jacot, A., Schaub, M., 2020a. Disentangling the spatial and temporal causes of decline in a bird population. Ecol. Evol. 10, 6906–6918.

Plard, F., Bruns, H.A., Cimiotti, D.V., Helmecke, A., Hötker, H., Jeromin, H., Roodbergen, M., Schekkerman, H., Teunissen, W., van der Jeugd, H., Schaub, M., 2020b. Low productivity and unsuitable management drive the decline of central European lapwing populations. Anim. Conserv. 23, 286–296.

Plard, F., Turek, D., Schaub, M., 2021. Consequences of violating assumptions of integrated population models on parameter estimates. Environ. Ecol. Stat. 28, 667–695.

Pledger, S., 2000. Unified maximum likelihood estimates for closed capture-recapture models using mixtures. Biometrics 56, 434–442.

Pledger, S., Schwarz, C.J., 2002. Modelling heterogeneity of survival in band-recovery data using mixtures. J. Appl. Stat. 29, 315–327.

Pledger, S., Pollock, K.H., Norris, J.L., 2003. Open capture-recapture models with heterogeneity: I. Cormack-Jolly-Seber model. Biometrics 59, 786–794.

Plummer, M., 2003. JAGS: A Program for Analysis of Bayesian Graphical Models Using Gibbs Sampling. Technische Universität, Wien, Vienna, Austria.

Plummer, M., 2017. JAGS Version 4.3.0 User Manual.

Plummer, M., 2019. Rjags: Bayesian Graphical Models Using MCMC. R Package Version 4-10.

Polansky, L., de Valpine, P., Lloyd-Smith, J.O., Getz, W.M., 2009. Likelihood ridges and multimodality in population growth rate models. Ecology 90, 2313–2320.

Pollock, K.H., 1982. A capture-recapture design robust to unequal probability of capture. J. Wildl. Manag. 46, 752–757.

Pollock, K.H., Hines, J.E., Nichols, J.D., 1985. Goodness-of-fit tests for open capture-recapture models. Biometrics 41, 399–410.

Pollock, K.H., Bunck, C.M., Winterstein, S.R., Chen, C.L., 1995. A capture-recapture survival analysis model for radio-tagged animals. J. Appl. Stat. 22, 661–672.

Ponisio, L.C., Valpine de, P., Michaud, N., Turek, D., 2020. One size does not fit all: customizing MCMC methods for hierarchical models using NIMBLE. Ecol. Evol. 10, 2385–2416.

Possingham, H.P., Lindenmayer, D.B., Tuck, G.N., 2002. Decision theory for population viability analysis. In: Beissinger, S.R., McCullough, D. (Eds.), Population Viability Analysis. The University of Chicago Press, Chicago, pp. 470–489.

Potts, J.M., Elith, J., 2006. Comparing species abundance models. Ecol. Model. 199, 153–163.

Powell, L.A., 2007. Approximating variance of demographic parameters using the delta method: a reference for avian biologists. Condor 109, 949–954.

Powell, L.A., Conroy, M.J., Krementz, D.G., Lang, J.D., 1999. A model to predict breeding-season productivity for multibrooded songbirds. Auk 116, 1001–1008.

Pradel, R., 1993. Flexibility in survival analysis from recapture data: handling trap-dependence. In: Lebreton, J.-D., North, P.M. (Eds.), Marked Individuals in the Study of Bird Population. Birkhäuser-Verlag, Basel, pp. 29–37.

Pradel, R., 2005. Multievent: an extension of multistate capture-recapture models to uncertain states. Biometrics 61, 442–447.

Pradel, R., 2009. The stakes of capture-recapture models with state uncertainty. In: Thomson, D., Cooch, E., Conroy, M. (Eds.), Modeling Demographic Processes in Marked Populations. Springer, New York, pp. 781–795.

Pradel, R., Lebreton, J.-D., 1999. Comparison of different approaches to the study of local recruitment of breeders. Bird Study 46 (suppl. 1), 74–81.

Pradel, R., Hines, J.E., Lebreton, J.-D., Nichols, J.D., 1997a. Capture-recapture survival models taking account of transients. Biometrics 53, 60–72.

Pradel, R., Johnson, A.R., Viallefont, A., Nager, R.G., Cézilly, F., 1997b. Local recruitment in the Greater Flamingo: a new approach using capture-recapture data. Ecology 78, 1431–1445.

Pradel, R., Wintrebert, C.M.A., Gimenez, O., 2003. A proposal for a goodness-of-fit test to the Arnason-Schwarz multistate capture-recapture model. Biometrics 59, 43–53.

Pradel, R., Gimenez, O., Lebreton, J.-D., 2005. Principles and interest of GOF tests for multistate capture-recapture models. Anim. Biodivers. Conserv. 28.2, 189–204.

Pradel, R., Maurin-Bernier, L., Gimenez, O., Genovart, M., Choquet, R., Oro, D., 2008. Estimation of sex-specific survival with uncertainty in sex assessment. Can. J. Stat. 36, 29–42.

Quéroué, M., Barbraud, C., Barraquand, F., Turek, D., Delord, K., Pacoureau, N., Gimenez, O., 2021. Multispecies integrated population model reveals bottom-up dynamics in a seabird predator-prey system. Ecol. Monogr. 91, e01459.

Ralls, K., Beissinger, S.R., Cochrane, J.F., 2002. Guidelines for using population viability analysis in endangered-species management. In: Beissinger, S.R., McCullough, D. (Eds.), Population Viability Analysis. The University of Chicago Press, Chicago, pp. 521–550.

Ramsey, D.S.L., Caley, P.A., Robley, A., 2015. Estimating population density from presence-absence data using a spatially explicit model. J. Wildl. Manag. 79, 491–499.

Rankin, R.W., Nicholson, K.E., Allen, S.J., Krützen, M., Bejder, L., Pollock, K.H., 2016. A full-capture hierarchical Bayesian model of Pollock's closed robust design and application to dolphins. Front. Mar. Sci. 3, 1–18.

Ransome, R.D., 1995. Earlier breeding shortens life in female greater horseshoe bats. Phil. Trans. R. Soc. B 350, 153–161.

Ratcliffe, D.A., 1993. The Peregrine Falcon. T & AD Poyser, London.

Reed, J.M., Mills, L.S., Dunning, J.B., Menges, E.S., McKelvey, K.S., Frye, R., Beissinger, S.R., Anstett, M.C., Miller, P., 2002. Emerging issues in population viability analysis. Conserv. Biol. 16, 7–19.

Rees, M., Childs, D.Z., Ellner, S.P., 2014. Building integral projection models: a user's guide. J. Anim. Ecol. 83, 528–545.

Regehr, E.V., Hostetter, N.J., Wilson, R.R., Rode, K.D., Martin, M.S., Converse, S.J., 2018. Integrated population modeling provides the first empirical estimates of vital rates and abundance for polar bears in the Chukchi Sea. Sci. Rep. 8, 16780.

Renner, I.W., Elith, J., Baddeley, A., Fithian, W., Hastie, T., Phillips, S.J., Popovic, G., Warton, D.I., O'Hara, R.B., 2014. Point process models for presence-only analysis. Methods Ecol. Evol. 6, 366–379.

Reynolds, T.J., King, R., Harwood, J., Frederiksen, M., Harris, M.P., Wanless, S., 2009. Integrated data analysis in the presence of emigration and mark loss. JABES 14, 411–431.

Reznick, D., Nunney, L., Tessier, A., 2000. Big houses, big cars, superfleas and the costs of reproduction. Trends Ecol. Evol. 15, 421–425.

Rhodes, J.R., Ng, C.F., Villiers, D. L. de, Preece, H.J., McAlpine, C.A., Possingham, H.P., 2011. Using integrated population modelling to quantify the implications of multiple threatening processes for a rapidly declining population. Biol. Conserv. 144, 1081–1088.

Ricklefs, R.E., Bloom, G., 1977. Components of avian breeding productivity. Auk 94, 86–96.

Ridout, M.S., Besbeas, P., 2004. An empirical model for underdispersed count data. Stat. Model. 4, 77–89.

Riecke, T.V., Leach, A.G., Gibson, D., Sedinger, J.S., Freckleton, R., 2018. Parameterizing the robust design in the BUGS language: Lifetime carry-over effects of environmental conditions during growth on a long-lived bird. Methods Ecol. Evol. 9, 2294–2305.

Riecke, T.V., Sedinger, B.S., Williams, P.J., Leach, A.G., Sedinger, J.S., 2019a. Estimating correlations among demographic parameters in population models. Ecol. Evol. 10, 13521–13531.

Riecke, T.V., Williams, P.J., Behnke, T.L., Gibson, D., Leach, A.G., Sedinger, B.S., Street, P.A., Sedinger, J.S., 2019b. Integrated population models: model assumptions and inference. Methods Ecol. Evol. 10, 1072–1082.

Ripa, J., 2000. Analysing the Moran effect and dispersal: their significance and interaction in synchronous dynamics. Oikos 90, 175–187.

Rivalan, P., Prévot-Julliard, A.C., Choquet, R., Pradel, R., Jacquemin, B., Girondot, M., 2005. Trade-off between current reproductive effort and delay to next reproduction in the leatherback sea turtle. Oecologia 145, 564–574.

Rivalan, P., Pradel, R., Choquet, R., Girondot, M., Prévot-Julliard, A.C., 2006. Estimating clutch frequency in the sea turtle *Dermochelys coriacea* using stopover duration. Mar. Ecol. Prog. Ser. 317, 285–295.

Rivalan, P., Barbraud, C., Inchausti, P., Weimerskirch, H., 2010. Combined impacts of longline fisheries and climate on the persistence of the Amsterdam Albatross *Diomedia amsterdamensis*. Ibis 152, 6–18.

Robinson, R.A., Green, R.E., Baillie, S.R., Peach, W.J., Thomson, D.L., 2004. Demographic mechanisms of the population decline of the song thrush *Turdus philomelos* in Britain. J. Anim. Ecol. 73, 670–682.

Robinson, R.A., Baillie, S.R., Crick, H.Q.P., 2007. Weather-dependent survival: implications of climate change for passerine population processes. Ibis 149, 357–364.

Robinson, R.A., Grantham, M.J., Clark, J.A., 2009. Declining rates of ring recovery in British birds. Ringing Migr. 24, 266–272.

Robinson, R.A., Baillie, S.R., King, R., 2012. Population processes in European blackbirds *Turdus merula*: a state-space approach. J. Ornithol. 152, 419–433.

Robinson, R.A., Morrison, C.A., Baillie, S.R., 2014. Integrating demographic data: towards a framework for monitoring wildlife populations at large spatial scales. Methods Ecol. Evol. 5, 1361–1372.

Robinson, R.A., Meier, C.M., Witvliet, W., Kéry, M., Schaub, M., 2020. Survival varies seasonally in a migratory bird: linkages between breeding and non-breeding periods. J. Anim. Ecol. 89, 2111–2121.

Robinson, R.A., Wilson, M., 2021. Contrasting long-term trends in age-specific survival of Peregrine Falcons (*Falco peregrinus*) in Britain using smoothed estimates of recovery probabilities. Ibis 163, 890–898.

Robinson, O.J., Ruiz-Gutierrez, V., Fink, D., Meese, R.J., Holyoak, M., Cooch, E.G., 2018. Using citizen science data in integrated population models to inform conservation. Biol. Conserv. 227, 361–368.

Rockwood, L.L., 2006. Introduction to Population Ecology. Blackwell, Malden, Mass.

Rodenhouse, N.L., Sherry, T.W., Holmes, R.T., 1997. Site-dependent regulation of population size: a new synthesis. Ecology 78, 2025–2042.

Rodenhouse, N.L., Sillett, T.S., Doran, P.J., Holmes, R.T., 2003. Multiple density-dependence mechanisms regulate a migratory bird population during the breeding season. Proc. R. Soc. Lond. B 270, 2105–2110.

Rohner, C., 1996. The numerical response of great horned owls to the snowshoe hare cycle: consequences of non-territorial 'floaters' on demography. J. Anim. Ecol. 65, 359–370.

Ross, B.E., Haukos, D.A., Hagen, C.A., Pitman, J., 2018. Combining multiple sources of data to inform conservation of Lesser Prairie-Chicken populations. Auk 135, 228–239.

Rotella, J.J., Taper, M.L., Hansen, A.J., 2000. Correcting nesting-success estimates for observer effects: maximum-likelihood estimates of daily survival rates with reduced bias. Auk 117, 92–109.

Rotella, J.J., Dinsmore, S.J., Shaffer, T.L., 2004. Modeling nest-survival data: a comparison of recently developed methods that can be implemented in MARK and SAS. Anim. Biodivers. Conserv. 27.1, 187–205.

Rotelli, L., Bionda, R., Zbinden, N., Schaub, M., 2021. Chick survival and hunting are important drivers for the dynamics of two Alpine black grouse *Lyrurus tetrix* populations. Wildl. Biol. 2021, wlb.00874.

Roth, T., Amrhein, V., 2009. Estimating individual survival using territory occupancy data on unmarked animals. J. Appl. Ecol. 47, 386–392.

Rouan, L., Choquet, R., Pradel, R., 2009. A general framework for modeling memory in capture-recapture data. JABES 14, 338–355.

Royle, J.A., 2004a. N-mixture models for estimating population size from spatially replicated counts. Biometrics 60, 108–115.

Royle, J.A., 2004b. Generalized estimators of avian abundance from count survey data. Anim. Biodivers. Conserv. 27.1, 375–386.

Royle, J.A., 2004c. Modeling abundance index data from anuran calling surveys. Conserv. Biol. 18, 1378–1385.

Royle, J.A., 2008. Modeling individual effects in the Cormack-Jolly-Seber model: a state-space formulation. Biometrics 64, 364–370.

Royle, J.A., Dorazio, R.M., 2008. Hierarchical Modeling and Inference in Ecology. Academic Press, Amsterdam.

Royle, J.A., Dorazio, R.M., 2012. Parameter-expanded data augmentation for Bayesian analysis of capture-recapture models. J. Ornithol. 152, 521–537.

Royle, J.A., Dubovsky, J.A., 2001. Modeling spatial variation in waterfowl band-recovery data. J. Wildl. Manag. 65, 726–737.

Royle, J.A., Kéry, M., 2007. A Bayesian state-space formulation of dynamics occupancy models. Ecology 88, 1813–1823.

Royle, J.A., Link, W.A., 2005. A general class of multinomial mixture models for anuran calling survey data. Ecology 86, 2505–2512.

Royle, J.A., Link, W.A., 2006. Generalized site occupancy models allowing for false positive and false negative errors. Ecology 87, 835–841.

Royle, J., Nychka, D., 1998. An algorithm for the construction of spatial coverage designs with implementation in SPLUS. Comput. Geosci. 24, 479–488.

Royle, J.A., Young, K.G., 2008. A hierarchical model for spatial capture-recapture data. Ecology 89, 2281–2289.

Royle, J.A., Dorazio, R.M., Link, W.A., 2007a. Analysis of multinomial models with unknown index using data augmentation. J. Comput. Graph Stat. 16, 67–85.

Royle, J.A., Kéry, M., Gauthier, R., Schmid, H., 2007b. Hierarchical spatial models of abundance and occurrence from imperfect survey data. Ecol. Monogr. 77, 465–481.

Royle, J.A., Kéry, M., Guélat, J., 2011. Spatial capture-recapture models for search-encounter data. Methods Ecol. Evol. 2, 602–611.

Royle, J.A., Chandler, R.B., Gazenski, K.D., Graves, T.A., 2013. Spatial capture-recapture models for jointly estimating population density and landscape connectivity. Ecology 94, 287–294.

Royle, J.A., Chandler, R.B., Sollmann, R., Gardner, B., 2014. Spatial Capture-Recapture. Academic Press/Elsevier, Waltham.

Royle, J.A., Fuller, A.K., Sutherland, C., 2016. Spatial capture-recapture models allowing Markovian transience or dispersal. Popul. Ecol. 58, 53–62.

Rubin, D.B., 1976. Inference and missing data. Biometrika 63, 581–592.

Rue, H., Martino, S., Chopin, N., 2009. Approximate Bayesian inference for latent Gaussian models by using integrated nested Laplace approximations (with discussion). J. R. Stat. Soc. B 71, 319–392.

Ruppert, D., 2002. Selecting the number of knots for penalized splines. J. Comput. Graph Stat. 11, 735–757.

Rushing, C.S., 2019. Estimability of migration survival rates from integrated breeding and winter capture–recapture data. Ecol. Evol. 9, 849–858.

Rushing, C.S., Ryder, T.B., Scarpignato, A.L., Saracco, J.F., Marra, P.P., Siriwardena, G., 2016. Using demographic attributes from long-term monitoring data to delineate natural population structure. J. Appl. Ecol. 53, 491–500.

Rushing, C.S., Hostetler, J.A., Sillett, T.S., Marra, P.P., Rotenberg, J.A., Ryder, T.B., 2017. Spatial and temporal drivers of avian population dynamics across the annual cycle. Ecology 98, 2837–2850.

Saino, N., Ambrosini, R., Martinelli, R., Calza, S., Möller, A.P., Pilastro, A., 2002. Offspring sexual dimorphism and sex-allocation in relation to parental age and paternal ornamentation in the barn swallow. Mol. Ecol. 11, 1533–1544.

Saino, N., Szép, T., Romano, M., Rubolini, D., Spina, F., Möller, A.P., 2004. Ecological conditions during winter predict arrival date at the breeding quarters in a trans-Saharan migratory bird. Ecol. Lett. 7, 21–25.

Sandercock, B.K., 2020. Mark-recapture models for estimation of demographic parameters. In: Murray, D., Sandercock, B.K. (Eds.), Population Ecology in Practice. Wiley-Blackwell, Hoboken, NJ, pp. 157–190.

Sandercock, B.K., Nilsen, E.B., Brøseth, H., Pedersen, H.C., 2011. Is hunting mortality additive or compensatory to natural mortality? Effects of experimental harvest on the survival and cause-specific mortality of willow ptarmigan. J. Anim. Ecol. 80, 244–258.

Sanderlin, J.S., Block, W.M., Strohmeyer, B.E., Saab, V.A., Ganey, J.L., 2019. Precision gain versus effort with joint models using detection/non-detection and banding data. Ecol. Evol. 9, 804–817.

Sanderson, F.J., Donald, P.F., Pain, D.J., Burfield, I.J., van Bommel, F.P.J., 2006. Long-term population declines in Afro-Palaearctic migrant birds. Biol. Conserv. 131, 93–105.

Saracco, J.F., Rubenstein, M., 2020. Integrating broad-scale data to assess demographic and climatic contributions to population change in a declining songbird. Ecol. Evol. 10, 1804–1816.

Saracco, J.F., Royle, J.A., DeSante, D.F., Gardner, B., 2010. Modeling spatial variation in avian survival and residency probabilities. Ecology 91, 1885–1891.

Saracco, J.F., Royle, J.A., DeSante, D.F., Gardner, B., 2012. Spatial modeling of survival and residency and application to the monitoring avian productivity and survivorship program. J. Ornithol. 152, 469–476.

Sæther, B.-E., Engen, S., Lande, R., 1996. Density-dependence and optimal harvesting of fluctuating populations. Oikos 76, 40–46.

Sæther, B.-E., Engen, S., Islam, A., McCleery, R.H., Perrins, C., 1998. Environmental stochasticity and extinction risk in a population of a small songbird, the great tit. Am. Nat. 151, 441–450.

Sæther, B.-E., Engen, S., Lande, R., Möller, A.P., Bensch, S., Hasselquist, D., Beier, J., Leisler, B., 2004. Time to extinction in relation to mating system and type of density regulation in populations with two sexes. J. Anim. Ecol. 73, 925–934.

Sæther, B.-E., Engen, S., Solberg, E.J., Heim, M., 2007a. Estimating the growth of a newly established moose population using reproductive value. Ecography 30, 417–421.

Sæther, B.-E., Engen, S., Grøtan, V., Fiedler, W., Matthysen, E., Visser, M.E., Wright, J., Moller, A.P., Adriaensen, F., van Balen, J.H., Balmer, D., Mainwaring, M.C., McCleery, R.H., Pampus, M., Winkel, W., 2007b. The extended Moran effect and large-scale synchronous fluctuations in the size of great tit and blue tit populations. J. Anim. Ecol. 76, 315–325.

Sæther, B.-E., Engen, S., Grøtan, V., Bregnballe, T., Both, C., Tryjanowski, P., Agu, L., Wright, J., Möller, A.P., Visser, M.E., Winkel, W., 2008. Forms of density regulation and (quasi-) stationary distributions of population sizes in birds. Oikos 117, 1197–1208.

Sauer, J.R., Link, W.A., 2002. Hierarchical modeling of population stability and species group attributes from survey data. Ecology 86, 1743–1751.

Sauer, J.R., Link, W.A., 2011. Analysis of the North American breeding bird survey using hierarchical models. Auk 128, 87–98.

Sauer, J.R., Link, W.A., Fallon, J.E., Pardieck, K.L., Ziolkowski, D.J., 2013. The North American breeding bird survey 1966–2011: summary analysis and species accounts. N. Am. Fauna 79, 1–32.

Saunders, S.P., Cuthbert, F.J., Zipkin, E.F., 2018. Evaluating population viability and efficacy of conservation management using integrated population models. J. Appl. Ecol. 55, 1380–1392.

Saunders, S.P., Farr, M.T., Wright, A.D., Bahlai, C.A., Ribeiro, J.W., Rossman, S., Sussman, A.L., Arnold, T.W., Zipkin, E.F., 2019. Disentangling data discrepancies with integrated population models. Ecology 100, e02714.

Saunders, S.P., Piper, W., Farr, M.T., Bateman, B.L., Michel, N.L., Westerkam, H., Wilsey, C.B., 2021. Interrelated impacts of climate and land-use change on a widespread waterbird. J. Anim. Ecol. 90, 1165–1176.

Scharf, H.R., Lu, X., Williams, P.J., Hooten, M.B., 2021. Hierarchical approaches for flexible and interpretable binary regression models. https://arxiv.org/abs/1905.05242.

Schaub, M., 2009. Evaluation of bias, precision and accuracy of mortality cause proportion estimators from ring recovery data. In: Thomson, D., Cooch, E., Conroy, M. (Eds.), Modeling Demographic Processes in Marked Populations. Springer, New York, pp. 1081–1097.

Schaub, M., 2020. Combining counts of unmarked individuals and demographic data using integrated population models. In: Murray, D., Sandercock, B.K. (Eds.), Population Ecology in Practice. Wiley-Blackwell, Hoboken, NJ, pp. 215–236.

Schaub, M., Abadi, F., 2011. Integrated population models: a novel analysis framework for deeper insights into population dynamics. J. Ornithol. 152, 227–237.

Schaub, M., Fletcher, D., 2015. Estimating immigration using a Bayesian integrated population model: choice of parametrization and priors. Environ. Ecol. Stat. 22, 535–549.

Schaub, M., Hirschheydt von, J., 2009. Effect of current reproduction on apparent survival, breeding dispersal, and future reproduction in barn swallows assessed by multistate capture-recapture models. J. Anim. Ecol. 78, 625–635.

Schaub, M., Lebreton, J.-D., 2004. Testing the additive versus the compensatory hypothesis of mortality from ring recovery data using a random effects model. Anim. Biodivers. Conserv. 27.1, 73–85.

Schaub, M., Pradel, R., 2004. Assessing the relative importance of different sources of mortality from recoveries of marked animals. Ecology 85, 930–938.

Schaub, M., Royle, J.A., 2014. Estimating true instead of apparent survival using spatial Cormack-Jolly-Seber models. Methods Ecol. Evol. 5, 1316–1326.

Schaub, M., Ullrich, B., 2021. A drop in immigration results in the extinction of a local woodchat shrike population. Anim. Conserv. 24, 335–345.

Schaub, M., Gimenez, O., Schmidt, B.R., Pradel, R., 2004. Estimating survival and temporary emigration in the multistate capture-recapture framework. Ecology 85, 2107–2113.

Schaub, M., Kania, W., Köppen, U., 2005. Variation of primary production during winter induces synchrony in survival rates in migratory white storks *Ciconia ciconia*. J. Anim. Ecol. 74, 656–666.

Schaub, M., Gimenez, O., Sierro, A., Arlettaz, R., 2007. Use of integrated modeling to enhance estimates of population dynamics obtained from limited data. Conserv. Biol. 21, 945–955.

Schaub, M., Reichlin, T.S., Abadi, F., Kéry, M., Jenni, L., Arlettaz, R., 2012. The demographic drivers of local population dynamics in two rare migratory birds. Oecologia 168, 97–108.

Schaub, M., Jakober, H., Stauber, W., 2013. Strong contribution of immigration to local population regulation: evidence from a migratory passerine. Ecology 94, 1828–1838.

Schaub, M., Hirschheydt von, J., Grüebler, M.U., 2015. Differential contribution of demographic rate synchrony to population synchrony in barn swallows. J. Anim. Ecol. 84, 1530–1541.

Schmid, H., Luder, R., Naef-Daenzer, B., Graf, R., Zbinden, N., 1998. Schweizer Brutvogelatlas. Schweizerische Vogelwarte, Sempach.

Schmid, H., Zbinden, N., Keller, V., 2004. Überwachung der Bestandsentwicklung häufiger Brutvögel in der Schweiz. Schweizerische Vogelwarte, Sempach.

Schmidt, J.H., Robison, H.L., 2020. Using distance sampling-based integrated population models to identify key demographic parameters. J. Wildl. Manag. 84, 372–381.

Schmidt, J.H., Walker, J.A., Lindberg, M.S., Johnson, D.S., Stephens, S.E., 2010. A general Bayesian hierarchical model for estimating survival of nests and young. Auk 127, 379–386.

Schorcht, W., Bontadina, F., Schaub, M., 2009. Variation of adult survival drives population dynamics in a migrating forest bat. J. Anim. Ecol. 78, 1182–1190.

Schwarz, C.J., Schweigert, J.F., Arnason, A.N., 1993. Estimating migration rates using tag recovery data. Biometrics 49, 177–193.

Seber, G.A.F., 1965. A note on the multiple recapture census. Biometrika 52, 249–259.

Seber, G.A.F., 1970. Estimating time-specific survival and reporting rates for adult birds from band returns. Biometrika 57, 313–318.

Seber, G.A.F., 1971. Estimating age-specific survival rates for birds from bird-band returns when the reporting rate is constant. Biometrika 58, 491–497.

Seber, G.A.F., 1982. The Estimation of Animal Abundance and Related Parameters. Charles Griffin & Company Ltd., London.

Seber, G.A.F., Schofield, M.R., 2019. Capture-recapture: Parameter Estimation for Open Animal Populations. Springer, Cham, Switzerland.

Senner, N.R., Conklin, J.R., Piersma, T., 2015. An ontogenetic perspective on individual differences. Proc. R. Soc. Lond. B 282, 20151050.

Servanty, S., Choquet, R., Baubet, E., Brandt, S., Gaillard, J.M., Schaub, M., Toigo, C., Lebreton, J.-D., Buoro, M., Gimenez, O., 2010. Assessing whether mortality is additive using marked animals: a Bayesian state-space modeling approach. Ecology 91, 1916–1923.

Seward, A., Ratcliffe, N., Newton, S., Caldow, R., Piec, D., Morrison, P., Cadwallender, T., Davies, W., Bolton, M., 2019. Metapopulation dynamics of roseate terns: sources, sinks and implications for conservation management decisions. J. Anim. Ecol. 88, 138–153.

Shaffer, M.L., 1981. Minimum population sizes for species conservation. BioScience 31, 131–134.

Shenk, T.M., White, G.C., Burnham, K.P., 1998. Sampling-variance effects on detecting density-dependence from temporal trends in natural populations. Ecol. Monogr. 68, 445–463.

Shmueli, G., 2010. To explain or to predict? Stat. Sci. 25, 289–310.

Sibly, R.M., Barker, D., Denham, M.C., Hone, J., Pagel, M., 2005. On the regulation of populations of mammals, birds, fish, and insects. Science 309, 607–610.

Sierro, A., Lugon, A., Arlettaz, R., 2009. La colonie de grands rhinolophes *Rhinolophus ferrumequinum* de l'église St-Sylvie à Vex (Valais, Suisse): évolution sur deux décennies (1968–2006). Le Rhinolophe 18, 75–82.

Sillett, T.S., Rodenhouse, N.L., Holmes, R.T., 2004. Experimentally reducing neighbor density affects reproduction and behavior of a migratory songbird. Ecology 85, 2467–2477.

Sillett, T.S., Chandler, R.B., Royle, J.A., Kéry, M., Morrison, S.A., 2012. Hierarchical distance-sampling models to estimate population size and habitat-specific abundance of an island endemic. Ecol. Appl. 22, 1997–2006.

Skalski, J.R., Ryding, K.E., Millspaugh, J.J., 2005. Wildlife Demography. Analysis of Sex, Age, and Count Data. Elsevier Academic Press, Amsterdam, Boston.

Skalski, J.R., Townsend, R.L., Gilbert, B.A., 2007. Calibrating statistical population reconstruction models using catch-effort and index data. J. Wildl. Manag. 71, 1309–1316.

Sollmann, R., Gardner, B., Chandler, R.B., Shindle, D.B., Onorato, D.P., Royle, J.A., O'Connell, A.F., 2013. Using multiple data sources provides density estimates for endangered Florida panther. J. Appl. Ecol. 50, 961–968.

Sollmann, R., Gardner, B., Chandler, R.B., Royle, J.A., Sillett, T.S., 2015. An open-population hierarchical distance sampling model. Ecology 96, 325–331.

Solymos, P., 2019. dclone: Data Cloning and MCMC Tools for Maximum Likelihood Methods. R Package Version 2.3-0.

Souchay, G., Gauthier, G., Pradel, R., 2014. To breed or not: a novel approach to estimate breeding propensity and potential trade-offs in an Arctic-nesting species. Ecology 95, 2745–2756.

Souchay, G., van Wijk, R.E., Schaub, M., Bauer, S., 2018. Identifying drivers of breeding success in a long-distance migrant using structural equation modelling. Oikos 127, 125–133.

Specht, H.M., Arnold, T.W., 2018. Banding age ratios reveal prairie waterfowl fecundity is affected by climate, density dependence and predator-prey dynamics. J. Appl. Ecol. 55, 2854–2864.

Spiegelhalter, D.J., Best, N.G., Carlin, B.P., van der Linde, A., 2002. Bayesian measure of model complexity and fit. J. R. Stat. Soc. B 64, 583–639.

Stan Development Team, 2020. Stan User's Guide. Version 2.25.

Stearns, S.C., 1992. The Evolution of Life Histories. Oxford University Press, Oxford.

Stephens, M., 2000. Dealing with label switching in mixture models. J. R. Stat. Soc. B 62, 795–809.

Steward, J., 1976. Positive definite functions and generalizations, an historical overview. Rocky Mt. J. Math. 6, 409–434.

Stott, I., Townley, S., Hodgson, D.J., 2011. A framework for studying transient dynamics of population projection matrix models. Ecol. Lett. 14, 959–970.

Streby, H.M., Refsnider, J.M., Andersen, D.E., 2014. Redefining reproductive success in songbirds: moving beyond the nest success paradigm. Auk 131, 718–726.

Stubben, C., Milligan, B., 2007. Estimating and analyzing demographic models using the popbio package in R. J. Stat. Software 22, 1–23.

Sturtz, S., Ligges, U., Gelman, A., 2005. R2WinBUGS: a package for running WinBUGS from R. J. Stat. Software 12, 1–16.

Su, Y.-S., Yajima, M., 2015. R2jags: Using R to Run 'JAGS'. R Package Version 0.5-7.

Sullivan, B.L., Aycrigg, J.L., Barry, J.H., Bonney, R.E., Bruns, N., Cooper, C.B., Damoulas, T., Dhondt, A.A., Dietterich, T., Farnsworth, A., Fink, D., Fitzpatrick, J.W., Fredericks, T., Gerbracht, J., Gomes, C., Hochachka, W.M., Iliff, M.J., Lagoze, C., La Sorte, F.A., Merrifield, M., Morris, W., Phillips, T.B., Reynolds, M., Rodewald, A.D., Rosenberg, K.V., Trautmann, N.M., Wiggins, A., Winkler, D.W., Wong, W.-K., Wood, C.L., Yu, J., Kelling, S., 2014. The eBird enterprise: an integrated approach to development and application of citizen science. Biol. Conserv. 169, 31–40.

Sun, C.C., Fuller, A.K., Royle, J.A., 2019. Incorporating citizen science data in spatially explicit integrated population models. Ecology 100, e02777.

Sutherland, W.J., 1996. From Individual Behaviour to Population Ecology. Oxford University Press, Oxford.

Sutherland, W.J., Norris, K., 2002. Behavioural models of population growth rates: implications for conservation and prediction. Philos. Trans. R. Soc. B 357, 1237–1284.

Sutherland, C., Fuller, A.K., Royle, J.A., 2015. Modelling non-Euclidean movement and landscape connectivity in highly structured ecological networks. Methods Ecol. Evol. 6, 169–177.

Szostek, K.L., Schaub, M., Becker, P.H., 2014. Immigrants are attracted by local pre-breeders and recruits in a seabird colony. J. Anim. Ecol. 83, 1015–1024.

Tavecchia, G., Pradel, R., Boy, V., Johnson, A.R., Cézilly, F., 2001. Sex- and age-related variation in survival and cost of first reproduction in greater Flamingos. Ecology 82, 165–174.

Tavecchia, G., Pradel, R., Gossmann, F., Bastat, C., Ferrand, Y., Lebreton, J.-D., 2002. Temporal variation in annual survival probability of the Eurasian woodcock *Scolopax rusticola* wintering in France. Wildl. Biol. 8, 21–30.

Tavecchia, G., Besbeas, P., Coulson, T., Morgan, B.J.T., Clutton-Brock, T.H., 2009. Estimating population size and hidden demographic parameters with state-space modeling. Am. Nat. 173, 722–733.

Taylor, R.L., Udevitz, M.S., 2015. Demography of the Pacific walrus (*Odobenus rosmarus divergens*): 1974–2006. Mar. Mamm. Sci. 31, 231–254.

Taylor, B.L., Wade, P.R., Ramakrishnan, U., Gilpin, M., Akçakaya, H.R., 2002. Incorporating uncertainity in population viability analyses for the purpose of classifying species by risk. In: Beissinger, S.R., McCullough, D. (Eds.), Population Viability Analysis. The University of Chicago Press, Chicago, pp. 239–252.

Tedesco, P.A., Hugueny, B., Paugy, D., Fermon, Y., 2004. Spatial synchrony in population dynamics of West African fishes: a demonstration of an interspecific and interspecific Moran effect. J. Anim. Ecol. 73, 693–705.

Tempel, D.J., Peery, M.Z., Gutierrez, R.J., 2014. Using integrated population models to improve conservation monitoring: California spotted owls as a case study. Ecol. Model. 289, 86–95.

Tenan, S., Adrover, J., Navarro, A.M., Sergio, F., Tavecchia, G., 2012. Demographic consequences of poison-related mortality in a threatened bird of prey. PLoS ONE 7, e49187.

Tenan, S., Iemma, A., Bragalanti, N., Pedrini, P., Barba de, M., Randi, E., Groff, C., Genovart, M., 2016. Evaluating mortality rates with a novel integrated framework for nonmonogamous species. Conserv. Biol. 30, 1307–1319.

Tenan, S., Fasola, M., Volponi, S., Tavecchia, G., 2017. Conspecific and not performance-based attraction on immigrants drives colony growth in a waterbird. J. Anim. Ecol. 86, 1074–1081.

Tenan, S., Becker, D., Tolkmitt, D., Schaub, M., 2021. Decomposing fecundity and evaluating demographic influence of multiple broods in a migratory bird. J. Anim. Ecol. 90, 1071–1084.

Thomas, L., Martin, K., 1996. The importance of analysis method for breeding bird survey population trend estimates. Conserv. Biol. 10, 479–490.

Thomas, L., Buckland, S.T., Newman, K.B., Harwood, J., 2005. A unified framework for modelling wildlife population dynamics. Aust. N. Z. J. Stat. 47, 19–34.

Thomas, A., O'Hara, R.B., Ligges, U., Sturtz, S., 2006. Making BUGS open. R. News 6, 12–17.

Thomson, D.L., Baillie, S.R., Peach, W.J., 1999. A method for studying post-fledged survival rates using data from ringing recoveries. Bird Study 46, 104–111.

Tierney, L., 1994. Markov chains for exploring posterior distributions. Ann. Stat. 22, 1701–1728.

Tobler, W.R., 1970. A computer movie simulating urban growth in the Detroit region. Econ. Geogr. 46, 234–240.

Tobler, M.W., Kéry, M., Hui, F.K.C., Guillera-Arroita, G., Knaus, P., Sattler, T., 2019. Joint species distribution models with species correlations and imperfect detection. Ecology 100, e02754.

Tredennick, A.T., Hooker, G., Ellner, S.P., Adler, P.B., 2021. A practical guide to selecting models for exploration, inference, and prediction in ecology. Ecology 102, e03336.

Trolle, M., Kéry, M., 2003. Estimation of ocelot density in the Pantanal using capture-recapture analysis of camera-trapping data. J. Mammal. 84, 607–614.

Tsai, K., Pollock, K.H., Brownie, C., 1999. Effects of violation of assumptions for survival analysis methods in radiotelemetry studies. J. Wildl. Manag. 63, 1369–1375.

Tuljapurkar, S.D., Orzack, S.H., 1980. Population dynamics in variable environments. 1. Long-run growth rates and extinction. Theor. Popul. Biol. 18, 314–342.

Tuljapurkar, S., Horvitz, C.C., Pascarella, J.B., 2003. The many growth rates and elasticities of populations in random environments. Am. Nat. 162, 489–502.

Turek, D., de Valpine, P., Paciorek, C., 2016. Efficient Markov chain Monte Carlo sampling for hierarchical hidden Markov models. Environ. Ecol. Stat. 23, 549–564.

Turner, A.K., 2006. The Barn Swallow. Poyser Monographs, London.

Tyre, A.J., Tenhumberg, B., Field, S.A., Niejalke, D., Parris, K., Possingham, H.P., 2003. Improving precision and reducing bias in biological surveys: estimating false-negative error rates. Ecol. Appl. 13, 1790–1801.

Udevitz, M.S., Gogan, P.J.P., 2012. Estimating survival rates with time series of standing age-structure data. Ecology 93, 726–732.

Ullrich, B., 1987. Beringungsergebnisse aus einer Brut-Population des Rotkopfwürgers (*Lanius senator*) im Mittleren Albvorland, Kreis Göppingen und Esslingen. Ornithol. Jh. Baden-Württ. 3, 107–112.

Ullrich, B., 2017. Entdeckung und Aussterben einer Population des Rotkopfwürgers *Lanius senator* in den Streuobstwiesen des Mittleren Albvorlandes, Landkreise Göppingen und Esslingen am Neckar. Ökologie der Vögel 35, 273–335.

van Wijk, R.E., Schaub, M., Hahn, S., Juárez-García-Pelayo, N., Schäfer, B., Viktora, L., Martín-Vivaldi, M., Zischewski, M., Bauer, S., 2018. Diverse migration strategies in hoopoes (*Upupa epops*) lead to weak spatial but strong temporal connectivity. Sci. Nat. 105, 42.

Vasko, V., Laaksonen, T., Valkama, J., Korpimäki, E., 2011. Breeding dispersal of Eurasian kestrels *Falco tinnunculus* under temporally fluctuating food abundance. J. Avian Biol. 42, 552–563.

Vaupel, J.W., Yashin, A.I., 1985. Heterogeneity's ruses: some surprising effects of selection on population dynamics. Am. Stat. 39, 176–185.

Véran, S., Lebreton, J.-D., 2008. The potential of integrated modelling in conservation biology: a case study of the black-footed albatross (*Phoebastria nigripes*). Can. J. Stat. 36, 85–98.

Villemereuil de, P., Gimenez, O., Doligez, B., 2013. Comparing parent-offspring regression with frequentist and Bayesian animal models to estimate heritability in wild populations: a simulation study for Gaussian and binary traits. Methods Ecol. Evol. 4, 260–275.

Wade, P.R., 2000. Bayesian methods in conservation biology. Conserv. Biol. 14, 1308–1316.

Wade, P.R., 2002. Bayesian population viability analysis. In: Beissinger, S.R., McCullough, D. (Eds.), Population Viability Analysis. The University of Chicago Press, Chicago, pp. 213–238.

Walters, J.R., Crowder, L.B., Priddy, J.A., 2002. Population viability analysis for red-cockaded woodpeckers using an individual-based model. Ecol. Appl. 12, 249–260.

Wang, G., Hobbs, N.T., Boone, R.B., Illius, A.W., Gordon, I.J., Gross, J.E., Hamlin, K.L., 2006. Spatial and temporal variability modify density dependence in populations of large herbivores. Ecology 87, 95–102.

Wang, S.-P., Maunder, M.N., Piner, K.R., Aires-da-Silva, A., Lee, H.-H., 2014. Evaluation of virgin recruitment profiling as a diagnostic for selectivity curve structure in integrated stock assessment models. Fish. Res. 158, 158–164.

Webster, M.S., Marra, P.P., 2005. The importance of understanding migratory connectivity and seasonal interactions. In: Greenberg, R., Marra, P. (Eds.), Birds of Two Worlds: The Ecology and Evolution of Temperate-Tropical Migration Systems. Johns Hopkins University Press, Baltimore, MD, pp. 199–209.

Weegman, M.D., Bearhop, S., Fox, A.D., Hilton, G.M., Walsh, A.J., McDonald, J.L., Hodgson, D.J., 2016. Integrated population modelling reveals a perceived source to be a cryptic sink. J. Anim. Ecol. 85, 467–475.

Weegman, M.D., Arnold, T.W., Clark, R.G., Schaub, M., 2021. Partial and complete dependency among data sets has minimal consequence on estimates from integrated population models. Ecol. Appl. 31, e2258.

White, G.C., 1983. Numerical estimation of survival rates from band-recovery and biotelemetry data. J. Wildl. Manag. 47, 716–728.

White, G.C., Burnham, K.P., 1999. Program MARK: survival estimation from populations of marked animals. Bird Study 46 (suppl. 1), 120–139.
White, G.C., Cordes, L.S., Arnold, T.W., 2013. Band reporting rates of waterfowl: does individual heterogeneity bias estimated survival rates? Ecol. Evol. 3, 4215–4220.
Whitfield, D.P., 2003. Predation by Eurasian sparrowhawks produces density-dependent mortality of wintering redshanks. J. Anim. Ecol. 72, 27–35.
Wielgus, J., Gonzalez-Suarez, M., Aurioles-Gamboa, D., Gerber, L.R., 2008. A noninvasive demographic assessment of sea lions based on stage-specific abundances. Ecol. Appl. 18, 1287–1296.
Wikle, C.K., 2003. Hierarchical Bayesian models for predicting the spread of ecological processes. Ecology 84, 1382–1394.
Wikle, C.K., Zammit Mangion, A., Cressie, N.A.C., 2019. Spatio-temporal Statistics with R. CRC Press, Boca Raton, Florida.
Williams, B.K., 1997. Approaches to the management of waterfowl under uncertainty. J. Wildl. Manag. 25, 714–720.
Williams, B.K., Nichols, J.D., Conroy, M.J., 2002. Analysis and Management of Animal Populations. Modeling, Estimation, and Decision Making. Academic Press, San Diego.
Williams, C.K., Ives, A.R., Applegate, R.D., 2003. Population dynamics across geographical ranges: time-series analyses of three small game species. Ecology 84, 2654–2667.
Williams, P.J., Hooten, M.B., Womble, J.N., Esslinger, G.G., Bower, M.R., Hefley, T.J., 2017. An integrated data model to estimate spatiotemporal occupancy, abundance, and colonization dynamics. Ecology 98, 328–336.
Wilson, S., Gil-Weir, K.C., Clark, R.G., Robertson, G.J., Bidwell, M.T., 2016. Integrated population modeling to assess demographic variation and contributions to population growth for endangered whooping cranes. Biol. Conserv. 197, 1–7.
Wilson, S., McKellar, A.E., Reudink, M.W., Marra, P.P., Ratcliffe, L.M., 2017. Density-dependent immigration promotes population stability in a long-distance migratory bird. Popul. Ecol. 59, 169–178.
Wintle, B.A., Walshe, T.V., Parris, K.M., McCarthy, M.A., 2012. Designing occupancy surveys and interpreting non-detection when observations are imperfect. Divers. Distrib. 18, 417–424.
Wisdom, M.J., Mills, L.S., Doak, D.F., 2000. Life stage simulation analysis: estimating vital-rate effects on population growth for conservation. Ecology 81, 628–641.
Wonnacott, T.H., Wonnacott, R.J., 1990. Introductory Statistics. Wiley, New York, NY.
Wood, S.N., 2017. Generalized Additive Models. An Introduction with R, second ed. CRC Press, Portland.
Woodworth, B.K., Wheelwright, N.T., Newman, A.E., Schaub, M., Norris, D.R., 2017. Winter temperatures limit population growth rate of a migratory songbird. Nat. Commun. 8, 14812.
Wootton, J.T., Bell, D.A., 1992. A metapopulation model of the peregrine falcon in California: viability and management strategies. Ecol. Appl. 2, 307–321.
Wyatt, R.J., 2002. Estimating riverine fish population size from single- and multiple-pass removal sampling using a hierarchical model. Can. J. Fish. Aquat. Sci. 59, 695–706.
Yackulic, C.B., Dodrill, M., Dzul, M., Sanderlin, J.S., Reid, J.A., 2020. A need for speed in Bayesian population models: a practical guide to marginalizing and recovering discrete latent states. Ecol. Appl. 30, 1–19.
Zbinden, N., Salvioni, M., 2004. Bedeutung der Temperatur in der frühen Aufzuchtzeit für den Fortpflanzungserfolg des Birkhuhns *Tetrao tetrix* auf verschiedenen Höhenstufen im Tessin, Südschweiz. Ornithol.Beob. 101, 307–318.
Zbinden, N., Salvioni, M., Korner-Nievergelt, F., Keller, V., 2018. Evidence for an additive effect of hunting mortality in an alpine black grouse *Lyrurus tetrix* population. Wildl. Biol. 2018, wlb.00418.
Zhao, Q., 2020. On the sampling design of spatially explicit integrated population models. Methods Ecol. Evol. 27, 515.
Zhao, Q., Royle, J.A., 2019. Dynamic N-mixture models with temporal variability in detection probability. Ecol. Model. 393, 20–24.
Zhao, Q., Royle, J.A., Boomer, G.S., 2017. Spatially explicit dynamic N-mixture models. Popul. Ecol. 59, 293–300.
Zhao, Q., Arnold, T.W., Devries, J.H., Howerter, D.W., Clark, R.G., Weegman, M.D., 2019a. Land use change increases climatic vulnerability of migratory birds: insights from integrated population modelling. J. Anim. Ecol. 88, 1625–1637.
Zhao, Q., Boomer, G.S., Royle, J.A., 2019b. Integrated modeling predicts shifts in waterbird population dynamics under climate change. Ecography 42, 1470–1481.
Zimmerman, G.S., Gutierrez, R.J., LaHaye, W.S., 2007. Finite study areas and vital rates: sampling effects on estimates of spotted owl survival and population trends. J. Appl. Ecol. 44, 963–971.
Zimmerman, G.S., Link, W.A., Conroy, M.J., Sauer, J.R., Richkus, K.D., Boomer, G.S., 2010. Estimating migratory game-bird productivity by integrating age ratio and banding data. Wildl. Res. 37, 612–622.
Zimmerman, G.S., Sauer, J.R., Boomer, G.S., Devers, P.K., Garrettson, P.R., 2017. Integrating breeding bird survey and demographic data to estimate Wood Duck population size in the Atlantic Flyway. Condor 119, 616–628.
Zipkin, E.F., Saunders, S.P., 2018. Synthesizing multiple data types for biological conservation using integrated population models. Biol. Conserv. 217, 240–250.
Zipkin, E.F., Sillett, T.S., Grant, E.H.C., Chandler, R.B., Royle, J.A., 2014a. Inferences about population dynamics from count data using multistate models: a comparison to capture-recapture approaches. Ecol. Evol. 4, 417–426.
Zipkin, E.F., Thorson, J.T., See, K., Lynch, H.J., Grant, E.H.C., Kanno, Y., Chandler, R.B., Letcher, B.H., Royle, J.A., 2014b. Modeling structured population dynamics using data from unmarked individuals. Ecology 95, 22–29.
Zipkin, E.F., Inouye, B.D., Beissinger, S.R., 2019. Innovations in data integration for modeling populations. Ecology 100, e02713.

Author Index

'*Note:* Page numbers followed by "f" indicate figures and "t" indicate tables.'

C

D

E

F

G

H

I

J

K

L

M

N

O

R

S

Subject Index

'*Note:* Page numbers followed by "f" indicate figures and "t" indicate tables.'

D

E

I

J

K

L

M

N

Q

R

S

T

U

V

W